PLANT SYSTEMATICS

Plant Systematics

Third Edition

Michael G. Simpson

AMSTERDAM • BOSTON • HEIDELBERG • LONDON
NEW YORK • OXFORD • PARIS • SAN DIEGO
SAN FRANCISCO • SINGAPORE • SYDNEY • TOKYO
Academic Press is an imprint of Elsevier

Cover Images are all photographs by the author.

Academic Press is an imprint of Elsevier
30 Corporate Drive, Suite 400, Burlington, MA 01803, USA
525 B Street, Suite 1900, San Diego, California 92101-4495, USA
The Boulevard, Langford Lane, Kidlington, Oxford, OX5 1GB, UK

Library of Congress Cataloging-in-Publication Data
Simpson, Michael G. (Michael George), 1953-
Plant systematics / Michael G. Simpson. – 3rd ed.
p. cm.
ISBN 978-0-12-812628-8
1. Plants–Classification. I. Title.
QK95.S566 2019
580.1'2–dc22

2010009204

British Library Cataloguing-in-Publication Data
A catalogue record for this book is available from the British Library.

For information on all Academic Press publications visit our website at https://www.elsevier.com/books-and-journals

Printed in the United States of America
12 11 10 9 8 7 6 5 4

The first edition of this book was dedicated to three mentors I have been very fortunate to know: Albert Radford, who taught critical thinking; P. Barry Tomlinson, who taught the fine art of careful observation; and Rolf Dahlgren, whose magnetic personality was inspirational. I also wish to thank my many students who have provided useful suggestions over the years, plus three writers who captured my interest in science and the wonder of it all: Isaac Asimov, Richard Feynman, and Carl Sagan.

I wish to dedicate the second edition of this book to my wonderful family: Anna, Bonnie, Claire, Lee, and Lori.

The third edition is dedicated to Stella, Layla, and Quinn, who have brought so much joy into our lives.

Contents

UNIT IV RESOURCES IN PLANT SYSTEMATICS

UNIT V SPECIES CONCEPTS AND CONSERVATION BIOLOGY

Preface

Plant Systematics is an introduction to the morphology, evolution, and classification of land plants. My objective is to present a foundation of the approach, methods, research goals, evidence, and terminology of plant systematics and to summarize information on the most recent knowledge of evolutionary relationships of plants as well as practical information vital to the field. I have tried to present the material in a condensed, clear manner, such that the beginning student can better digest the more important parts of the voluminous information in the field and acquire more detailed information from the literature.

The book is meant to serve students at the college upper undergraduate and graduate levels in plant systematics or taxonomy courses, although portions of the book may be used in flora courses and much of the book could be used in courses in plant morphology, diversity, or general botany.

Each chapter has an expanded Table of Contents on the first page, a feature my students recommended as very useful. Numerous line drawings and color photographs are used throughout. A key feature is that illustrated plant material is often dissected and labeled to show important diagnostic features. At the end of each chapter are (1) Review Questions, which go over the chapter material; (2) Exercises, whereby a student may apply the material; and (3) References for Further Study, listing some of the basic and recent references. Literature cited in the references is not exhaustive, so students are encouraged to do literature searches on their own (see Appendix 3). Websites are listed for some chapters.

The book is classified into units, which consist of one or more chapters. Of course, a given instructor may choose to vary the sequence of these units or the chapters within, depending on personal preference and the availability of plant material. There is a slight amount of repetition between chapters of different units, but this was done so that chapters could be used independently of one another.

Unit 1, Systematics, gives a general overview of the concepts and methods of the field. Chapter 1 serves as an introduction to the definition, relationships, classification, and importance of plants and summarizes the basic concepts and principles of systematics, taxonomy, evolution, and phylogeny. Chapter 2 introduces taxonomy as continuing, active field of research and covers the details of phylogenetic systematics, and the theory and methodology for inferring phylogenetic trees or cladograms.

Unit 2, Evolution and Diversity of Plants, describes in detail the characteristics and classification of plants. The six chapters of this unit are intended to give the beginning student a basic understanding of the evolution of Green and Land Plants (Chapter 3), Vascular Plants (Chapter 4), Woody and Seed Plants (Chapter 5), and Flowering Plants (Chapters 6–8). Chapters 3–5 are formatted into two major sections. The first section presents cladograms (phylogenetic trees), which portray the evolutionary history of the group. Each of the major derived evolutionary features (apomorphies) from that cladogram is described and illustrated, with emphasis on its possible adaptive significance. This evolutionary approach to plant systematics makes learning the major plant groups and their features conceptually easier than simply memorizing a static list of characteristics. Treating these features as the products of unique evolutionary events brings them "to life," especially when their possible functional significance is pondered. The second section of Chapters 3–5 presents a survey of the diversity of the group in question. Etymologies of the type genus for families are included. Rare conditions and synonyms are enclosed by square brackets. Exemplars within major groups are described and illustrated, such that the student may learn to recognize and know the basic features of the major lineages of plants.

Because they constitute the great majority of plants, the flowering plants, or angiosperms, are covered in three chapters. Chapter 6 deals with the evolution of flowering plants, describing the apomorphies for that group and presenting a synopsis of their origin. Chapters 7 and 8 describe specific groups of flowering plants. In Chapter 7 the non-eudicot groups are treated, including the earliest diverging angiosperm lineages and the monocotyledons. Chapter 8 covers the eudicots, which make up the great majority of angiosperms. In these two chapters numerous flowering plant families (166 of the 472 recognized here) are described in detail, accompanied by photographs and illustrations; these are mostly families that are commonly encountered or for which material is usually available to the beginning student. Additional families are not described, but are illustrated with one or more exemplars. I have tried to emphasize diagnostic features a student might use to recognize a plant family, and have included some economically important uses of family members. Reference to Chapter 9 and occasionally to Chapters 10–14 (or use of the comprehensive Glossary) may be needed with regard to the technical terms. The Angiosperm Phylogeny Group IV system of classification is used throughout, with some exceptions. This system uses orders as the major taxonomic rank in grouping families of close relationship and has proven extremely useful in dealing with the tremendous diversity of the flowering plants.

Unit 3, Systematic Evidence and Descriptive Terminology, begins with a chapter on plant morphology (Chapter 9). Explanatory text, numerous diagrammatic illustrations, and photographs are used to train beginning students to precisely and thoroughly describe a plant morphologically. Appendices 1 and 2 (see below) are designed to be used along with Chapter 9. The other chapters in this unit cover the basic descriptive terminology of plant anatomy (Chapter 10), plant embryology (Chapter 11), palynology (Chapter 12), plant reproductive biology (Chapter 13), and plant molecular systematics (Chapter 14). The rationale for including these in a textbook on plant systematics is that features from these various fields are described in systematic research and are commonly utilized in phylogenetic reconstruction and taxonomic delimitation. In particular, in the last chapter on plant molecular systematics, I have attempted to update techniques and methodologies acquired in what has become in recent years the most fruitful of endeavors in phylogenetic reconstruction.

Unit 4, Resources in Plant Systematics, discusses some basics that are essential in everyday systematic research. Plant identification (Chapter 15) contains a summary of both standard dichotomous keys and computerized polythetic keys and reviews practical identification methods. The chapter on nomenclature (Chapter 16) summarizes the basic rules of the most recent International Code of Nomenclature for Algae, Fungi, and Plants, including the steps needed in the valid publication of a new species plus a review of botanical names. A chapter on plant collecting and documentation (Chapter 17) emphasizes both correct techniques for collecting plants and thorough data acquisition, the latter of which has become increasingly important today in biodiversity studies and conservation biology. Finally, the chapter on herbaria and data information systems (Chapter 18) reviews the basics of herbarium management, emphasizing the role of computerized database systems in plant collections for analyzing and synthesizing morphological, ecological, and biogeographic data.

Unit 5, Species Concepts and Conservation Biology, contains a chapter (Chapter 19) that reviews basic plant reproduction and the criteria and concepts of species and infraspecies definitions. In addition, a section on conservation biology reviews the basic concepts of this field, how it relates to taxonomy and systematics, and its importance to biologists and society.

Lastly, four Appendices and a Glossary are included. I have personally found each of these addenda to be of value in my own plant systematics courses. Appendix 1 is a list of characters used for detailed plant descriptions (available on the Plant Systematics Resources website). This list is useful in training students to write descriptions suitable for publication. Appendix 2 is a brief discussion of botanical illustration. I have found that drawing what one sees helps to develop observational skills. Appendix 3 is a listing of scientific journals in plant systematics, with literature exercises. Appendix 4 gives a brief overview of statistical and morphometric methods and how those may be applied in addressing questions in taxonomy and phylogenetic systematics. The Glossary defines all terms used in the book and indicates synonyms, adjectival forms, plurals, abbreviations, and terms to compare.

Three web sites will be available to be used in conjunction with the textbook: (1) a Plant Systematics Resources site (*http://www.sci.sdsu.edu/plants/plantsystematics*), with web links and materials that are universally available; (2) a companion website (*http://www.elsevierdirect.com/companions/9780123743800*) that includes the chapter figures, appendix material from the textbook, and links to the author's website; and (3) an Instructor Resources site (*http://textbooks.elsevier.com/web/Login.aspx*), with material that is password protected. Please contact your sales representative at textbooks@elsevier.com for access to the Instructor Resources site.

Throughout the book, I have attempted to adhere to W–H–Y, What–How–Why, in organizing and clarifying chapter topics: (1) What is it? What is the topic, the basic definition? (Many scientific arguments could have been resolved at the start by a clear statement or definition of terms.) (2) How is it done? What are the materials and methods, the techniques of data acquisition, the types of data analysis? (3) Why is it done? What is the purpose, objective, or goal? What is the overriding paradigm involved? How does the current study or topic relate to others? This simple W–H–Y method, first presented to me by one of my mentors, A. E. Radford, is useful to follow in any intellectual endeavor. It is a good lesson to teach one's students, and helps both in developing good writing skills and in critically evaluating a topic.

Finally, I would like to propose that each of us, instructors and students, pause occasionally to evaluate why it is that we do what we do. Over the years I have refined my ideas and offer these suggestions as possible goals: (1) to realize and explore the beauty, grandeur, and intricacy of nature; (2) to engage in the excitement of scientific discovery; (3) to experience and share the joy of learning. It is in this spirit that I sincerely hope the book may be of use to others.

Acknowledgments

I sincerely thank Andy Bohonak, Bruce Baldwin, Lisa Campbell, Travis Columbus, Gary Emberger, Lluvia Flores-Rentería, Chrissen Gemmill, Matt Guilliams, Robert Hattaway, Bruce Kirchoff, Eric Knox, Kristen Hasenstab-Lehman, Makenzie Mabry, Lucinda McDade, Steve O'Kane, Kathleen Pryer (and her lab group), Jon Rebman, Tanya Renner, P. van Rijckevorsel, Paula Rudall, Dennis Stevenson, Livia Wanntorp, Annette Winner, and several anonymous reviewers for their comments on various chapters or appendices of the first, second, or third editions of this book. I thank Anna C. Simpson and Lee M. Simpson for technical help. I am grateful to Peter Stevens for up-to-date information on higher level classification of angiosperms from his excellent Angiosperm Phylogeny Website. As always, none of these bear responsibility for any mistakes, omissions, incongruities, misinterpretations, or general stupidities.

Almost all of the illustrations and photographs are the product of the author. I thank the following for additions to these (in order of appearance in text):

The "tree" of Unit 1 opening page (at left) is from Augier, A. 1801. Essai d'une nouvelle classification des végétaux. Lyon, Bruyset Ainé. (See Stevens, P. F. 1983. Taxon 32: 203–211.) The cladogram (at right) was an unpublished tree from the study by Simpson et al. 2017. Taxon 66: 1406–1420..

The Jepson Herbarium (University of California Press) gave special permission to reproduce the key to the Oleaceae (Thomas J. Rosatti, author) in Figure 1.7.

Madroño (California Botanical Society) gave permission to reproduce Figure 4C of Simpson, et al. 2016. Madroño 63: 39-54, and Figure 2 of Dodero and Simpson. 2012. Madroño 59:223-229 in Figure 2.1 of Chapter 2.

Rick Bizzoco contributed the images of *Chlamydomonas reinhardtii* in Figures 3.2C and 3.3A.

Linda Graham contributed the image of *Coleochaete* in Figure 3.5B.

Figure 4.12A was reproduced from Kidston, R. and W. H. Lang. 1921. Transactions of the Royal Society of Edinburgh 52(4): 831–902.

Figure 4.20C was redrawn from Wakasugi, T., M. Sugita, T. Tsudzuki, and M. Sugiura. 1998. Plant Molecular Biology Reporter 16: 231–241, by permission.

Figure 4.20D was reproduced from Banks et al. 1975. Palaeontographica Americana 8: 77–126, with permission from Paleontological Research Institution, Ithaca, New York.

John Braggins contributed the images of Figure 4.26H–J.

Figure 4.31 was redrawn from Smith, G. M. 1955. Cryptogamic Botany, McGraw-Hill Book Company, Inc., New York.

Lawrence Jensen contributed the images of Figure 4.34A,B,E, 4.35D–E,H–J, 4.36C–D, and 4.37C–F.

Vera Svobodova contributed the images of 4.35A–C.

Gerald Carr contributed the image of 4.36F.

Figure 5.9 was reproduced and modified from Swamy, B. G. L. 1948. American Journal of Botany 35: 77–88, by permission.

Figure 5.13A,B was reproduced from: Beck, C. B. 1962. American Journal of Botany 49: 373–382, by permission.

Figure 5.13C was reproduced from Stewart, W. N., and T. Delevoryas. 1956. Botanical Review 22: 45–80, by permission.

Figure 5.19 was redrawn from Florin, R. 1951. Evolution in Cordaites and Conifers. Acta Horti Bergiani 15: 285–388.

John Braggins contributed the images of Figure 5.22D and 5.24D–F,L.

Figure 5.26B was reproduced from Esau, K. 1965. Plant Anatomy. J. Wiley and Sons, New York, by permission.

Mark Olson contributed the images of *Welwitschia mirabilis* in Figure 5.27B–E.

Figure 6.5 was based upon Jack, T. 2001. Trends in Plant Science 6: 310–316.

Figure 6.18A–C was redrawn from Thomas, H. H. 1925. Philosophical Transactions of the Royal Society of London 213: 299–363.

Figure 6.18D–F was redrawn from Gould, R. E. and T. Delevoryas. 1977. Alcheringa 1: 387–399, by permission.

Figure 6.19A was contributed by K. Simons and David Dilcher (©); Figure 6.19B was contributed by David Dilcher (©) and Ge Sun.

Stephen McCabe contributed the images of *Amborella* in Figures 7.3A,C.

The Arboretum at the University of California-Santa Cruz contributed the image of *Amborella* in Figure 7.3B.

Sandra Floyd provided the image of *Amborella* in Figure 7.3D.

Jeffrey M. Osborn and Mackenzie L. Taylor contributed the images of the Cabombaceae in Figure 7.4.

Jack Scheper contributed the image of *Illicium floridanum* in Figure 7.6A.

Figure 7.18A was reproduced from Behnke, H.-D. 1972. Botanical Review 38: 155–197, by permission.

Constance Gramlich contributed the image of *Amorphophallus* in Figure 7.25C.

Wayne Armstrong contributed the image of a flowering *Wolffia* in Figure 7.25G.

Benjamin Lowe contributed the image of *Scoliopus bigelovii* (Liliaceae) in Figure 7.28H.

John Kress contributed the Zingiberales drawing in Figure 7.54.

Will Cook contributed the images of *Hamamelis virginiana* in Figure 8.13A–C and of *Ulmus alata* in 8.39G.

Gerald Carr contributed the images of *Hillebrandia sandwicensis* in Figure 8.42A,B and of *Juglans hindsii* in Figure 8.47D–H.

The Rampant Gardener contributed the image of *Juglans regia* in Figure 8.47J.

Jerry Green contributed the image of *Crossosoma bigelovii* in Figure 8.48A.

Reid Moran contributed the image of *Crossosoma californicum* in Figure 8.48B–D, of *Koeberlinia spinosa* in Figure 8.57J–L, and of *Eucnide urens* in Figure 8.88B.

Mark Olson contributed the images of *Moringa* spp. in Figure 8.57M–Q.

Figure 8.72B was reproduced from Behnke, H.-D. 1972. Botanical Review 38: 155–197, by permission.

Gerald Carr contributed the images of *Ardisia crenata* in Figure 8.93A,B.

Steven Swartz contributed the images of *Heliamphora* sp. in Figure 8.97F,G.

Matt Guilliams contributed the image of *Pholisma sonorae* in Figure 8.100F,G.

David G. Smith contributed the images of *Phryma leptostachya* in Figure 8.116I,J.

Michael Silveira contributed the image of *Linnaea borealis* in Figure 8.127C,D.

Michael Mayer contributed material of *Scabiosa*, photographed in Figure 8.128E–L.

On the Unit III opening page the brass microscope image is courtesy of Allan Wissner (*www.antique-microscopes.com*), and the image of an Illumina sequencer is a general use image (*https://www.illumina.com/company/news-center/multimedia-images.html*)

Figure 9.6 was reproduced and condensed from Hallé, F. et al. 1978. *Tropical Trees and Forests: An Architectural Analysis*. Springer, Berlin., by permission.

Figure 9.13 was redrawn from Hickey, L. J. 1973. American Journal of Botany 60: 17–33, by permission.

Darren Burton prepared several illustrations in Chapter 9.

Figure 13.4A was redrawn from Weberling. 1989. *Morphology of Flowers and Inflorescences*. Cambridge University Press, Cambridge, New York, by permission.

Figure 13.4B was redrawn from Kohn et al. 1996. Evolution 50: 1454–1469, by permission.

Jon Rebman contributed the images in Figure 13.7D,E.

Figure 14.4 was redrawn from Wakasugi, T., M. Sugita, T. Tsudzuki, and M. Sugiura. 1998. Plant Molecular Biology Reporter 16: 231–241, by permission.

The Herbarium at the San Diego Natural History Museum contributed the images in Figure 17.2.

Jon Rebman contributed the image of the herbarium sheet in Figure 18.2.

Figure 19.6 was redrawn from Huang et al. 2005. Journal of Plant Research 118: 1–11, by permission.

Figure 19.9 was redrawn from Baldwin. 2000. Madroño 47: 219–229, by permission.

Dinna Estrella contributed the stippled line drawing in Appendix 2.

I

SYSTEMATICS

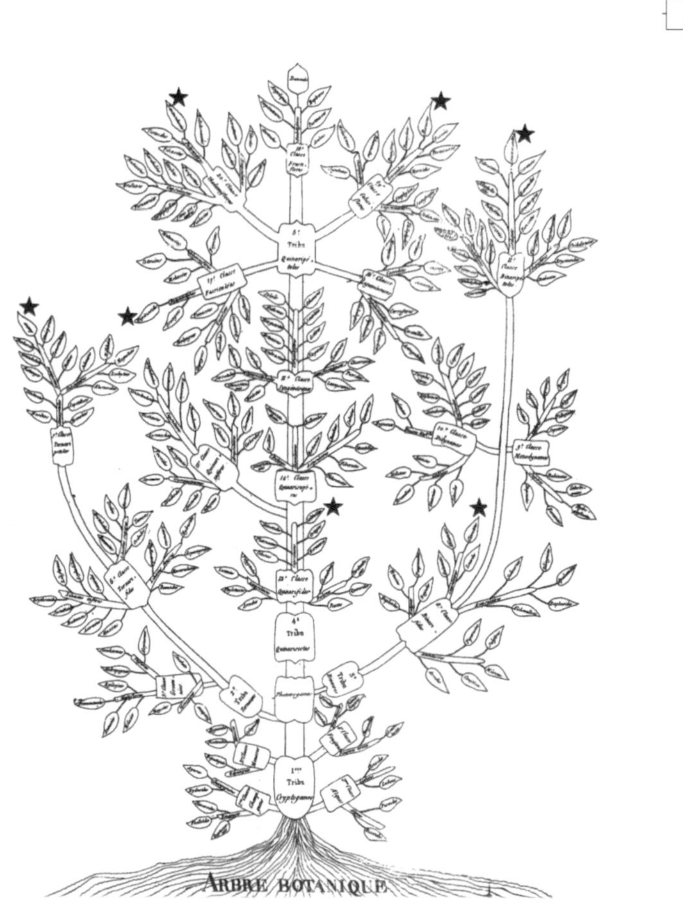

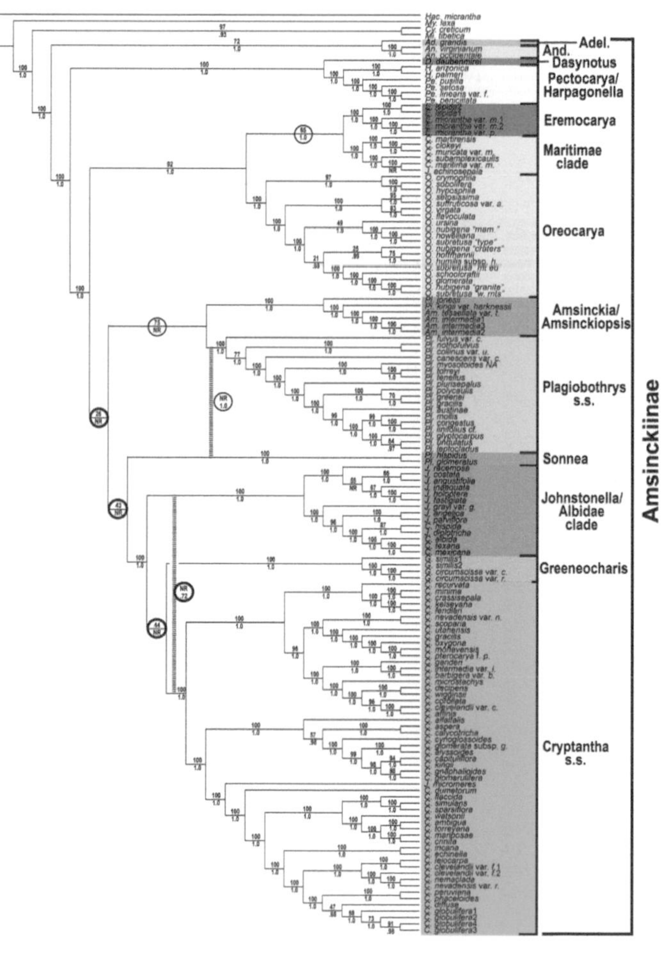

1

PLANT SYSTEMATICS: AN OVERVIEW

This book is about a fascinating field of biology called plant systematics. The purpose of this chapter is to introduce the basics: what a plant is, what systematics is, and the reasons for studying plant systematics.

PLANTS

WHAT IS A PLANT?

This question can be answered in either of two conceptual ways. One way, the traditional way, is to define groups of organisms such as plants by the characteristics they possess. Thus, historically, "plants" included those organisms that possess photosynthesis, cell walls, spores, and a more or less sedentary behavior. This traditional grouping of plants contained a variety of microscopic organisms, all of the "algae," and the more familiar plants that live on land. A second way to answer the question "What is a plant?" is to evaluate the evolutionary history of life and to use that history to delimit the groups of life. We now know from repeated research studies that some of the photosynthetic organisms evolved independently of one another and are not closely related.

Thus, the meaning or definition of the word *plant* can be ambiguous and can vary from person to person. Some still like to treat plants as a "polyphyletic" assemblage (see later discussion), defined by the common (but independently evolved) characteristic of photosynthesis. However, delimiting organismal groups based on evolutionary history has gained almost universal acceptance. This latter type of classification directly reflects the patterns of that evolutionary history and can be used to explicitly test evolutionary hypotheses (discussed later; see Chapter 2).

An understanding of what plants are requires an explanation of the evolution of life in general.

PLANTS AND THE EVOLUTION OF LIFE

Life is currently classified as three major groups (sometimes called domains) of organisms: **Archaea** (also called **Archaebacteria**), **Bacteria** (also called **Eubacteria**), and **Eukarya** or **eukaryotes** (also spelled eucaryotes). The evolutionary relationships of these groups are summarized in the simplified evolutionary tree or cladogram of Figure 1.1. The Archaea and Bacteria are small, mostly unicellular organisms that possess circular DNA, replicate by fission, and lack membrane-bound organelles. The two groups differ from one another in the chemical structure of certain cellular components. Eukaryotes are unicellular or multicellular organisms that possess linear DNA (organized as histone-bound chromosomes), replicate by mitotic and often meiotic division, and possess membrane-bound organelles such as nuclei, cytoskeletal structures, and (in almost all) mitochondria (Figure 1.1).

https://doi.org/10.1016/B978-0-12-812628-8.50001-8,

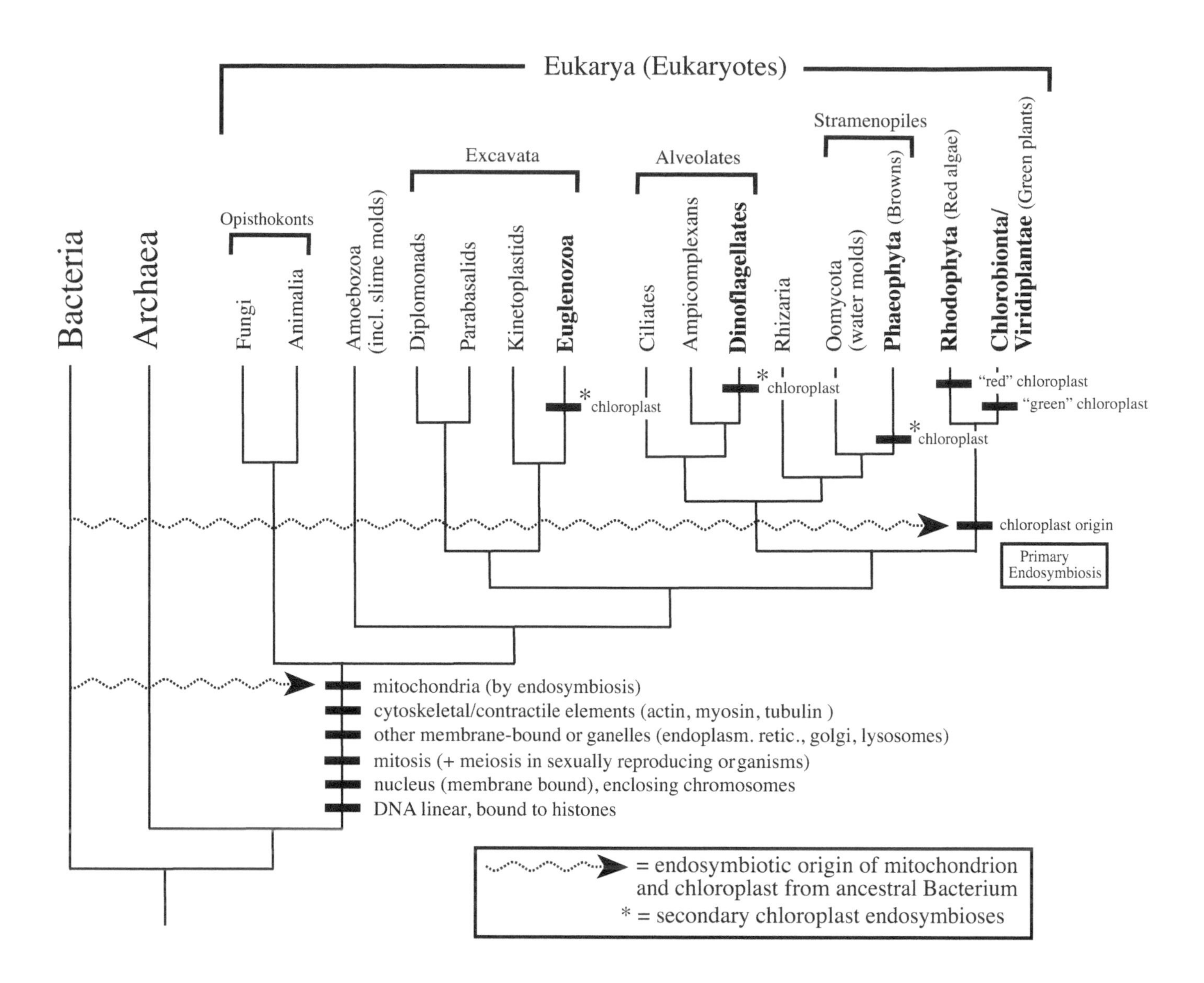

Figure 1.1 Simplified cladogram (evolutionary tree) of life (modified from Parfrey et al. 2010; see also Williams et al. 2018), illustrating eukaryotic apomorphies (the relative order of which is unknown) and the hypothesis of a single origin of mitochondria and chloroplasts via endosymbiosis (arrows). Note modification of chloroplast structure in the red and green plants, and subsequent secondary endosymbiosis in numerous other lineages (indicated by *). Eukaryotic groups with photosynthetic members in bold.

Some of the unicellular bacteria (e.g., the Cyanobacteria, or blue-greens) carry on photosynthesis, a biochemical system in which light energy is used to synthesize high-energy compounds from simpler starting compounds, carbon dioxide and water. These photosynthetic bacteria have a system of internal membranes called thylakoids, within which are embedded photosynthetic pigments, compounds that convert light energy to chemical energy. Of the several groups of eukaryotes that are photosynthetic, all have specialized photosynthetic organelles called **chloroplasts**, which resemble photosynthetic bacteria in having pigment-containing thylakoid membranes.

How did chloroplasts evolve? It is now largely accepted that chloroplasts of eukaryotes originated by the engulfment of an ancestral photosynthetic bacterium (probably a cyanobacterium) by an ancestral eukaryotic cell such that the photosynthetic bacterium continued to live and ultimately multiply *inside* the eukaryotic cell (Figures 1.1, 1.2). (Mitochondria also evolved by this process, from an ancestral, non-photosynthetic bacterium; see Figure 1.1.) The evidence for this is the fact that chloroplasts, like bacteria today, (a) have their own single-stranded, circular DNA; (b) have a smaller sized, 70S ribosome; and (c) replicate by fission. These engulfed photosynthetic bacteria

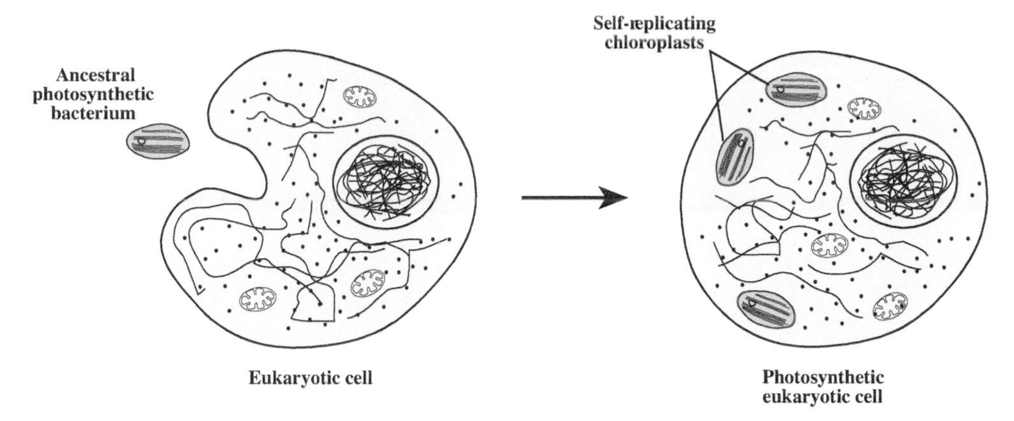

Figure 1.2 Diagrammatic illustration of the origin of chloroplasts by endosymbiosis of ancestral photosynthetic bacterium within ancestral eukaryotic cell.

provided high-energy products to the eukaryotic cell; the "host" eukaryotic cell provided a beneficial environment for the photosynthetic bacteria. The condition of two species living together in close contact is termed symbiosis, and the process in which symbiosis results by the engulfment and retention of one cell by another is termed **endosymbiosis**. Over time, these endosymbiotic, photosynthetic bacteria became transformed structurally and functionally, retaining their own DNA and the ability to replicate, but losing the ability to live independently of the host cell. In fact, over time there has been a transfer of some genes from the DNA of the chloroplast to the nuclear DNA of the eukaryotic host cell, making the two biochemically interdependent.

Although knowledge of eukaryotic relationships is still in flux, the most recent data from molecular systematic studies indicates that this so-called "primary" endosymbiosis of the chloroplast probably occurred one time, a shared evolutionary novelty of the red algae (Rhodophyta) and green plants (Viridiplantae or Chlorobionta; Figure 1.1). This early chloroplast became modified with regard to photosynthetic pigments, thylakoid structure, and storage products into forms characteristic of the red algae and green plants (see Figure 1.1). In addition, several lineages of photosynthetic organisms – including the euglenoids (Euglenozoa), dinoflagellates, and brown algae (Phaeophyta), and a few other lineages – may have acquired chloroplasts via "secondary" endosymbiosis, which occurred by the engulfment of an ancestral chloroplast-containing *eukaryote* by another eukaryotic cell (Figure 1.1).

LAND PLANTS

Of the major groups of photosynthetic eukaryotes, the green plants (Viridiplantae or Chlorobionta) are united primarily by distinctive characteristics of the green plant chloroplast with respect to photosynthetic pigments, thylakoid structure, and storage compounds (see Chapter 3 for details). Green plants include both the predominately aquatic "green algae" and a group known as embryophytes (formally, the Embryophyta), usually referred to as the land plants (Figure 1.3). The land plants are united by several evolutionary novelties that were adaptations to the transition from an aquatic environment to living on land. These include (1) an outer cuticle, which aids in protecting tissues from desiccation; (2) specialized gametangia (egg and sperm producing organs) that have an outer, protective layer of sterile cells; and (3) an intercalated diploid phase (sporophyte) in the life cycle, the early, immature component of which is termed the embryo (hence, "embryophytes"; see Chapter 3 for details).

Just as the green plants include the land plants, the land plants are inclusive of the vascular plants (Figure 1.3), the latter being united by the evolution of an independent sporophyte and xylem and phloem vascular conductive tissue (see Chapter 4). The vascular plants are inclusive of the seed plants (Figure 1.3), which are united by the evolution of wood and seeds (see Chapter 5). Finally, seed plants include the angiosperms (Figure 1.3), united by the evolution of the flower, including carpels and stamens, and by a number of other specialized features (see Chapters 6–8).

For the remainder of this book, the term *plant* is treated as equivalent to the embryophytes, the land plants. The rationale for this is partly that land plants make up a so-called monophyletic group, whereas the photosynthetic eukaryotes as a whole are not monophyletic, and do not accurately reflect evolutionary history (see later discussion, Chapter 2). And, practically, it is land plants that most people are talking about when they refer to "plants," including those in the field of plant systematics. However, as noted before, the word *plant* can be used by some to refer to other groupings; when in doubt, get a precise clarification.

WHY STUDY PLANTS?

The tremendous importance of plants cannot be overstated. Without them, we and most other species of animals (and of many other groups of organisms) would not be here. Photosyn-

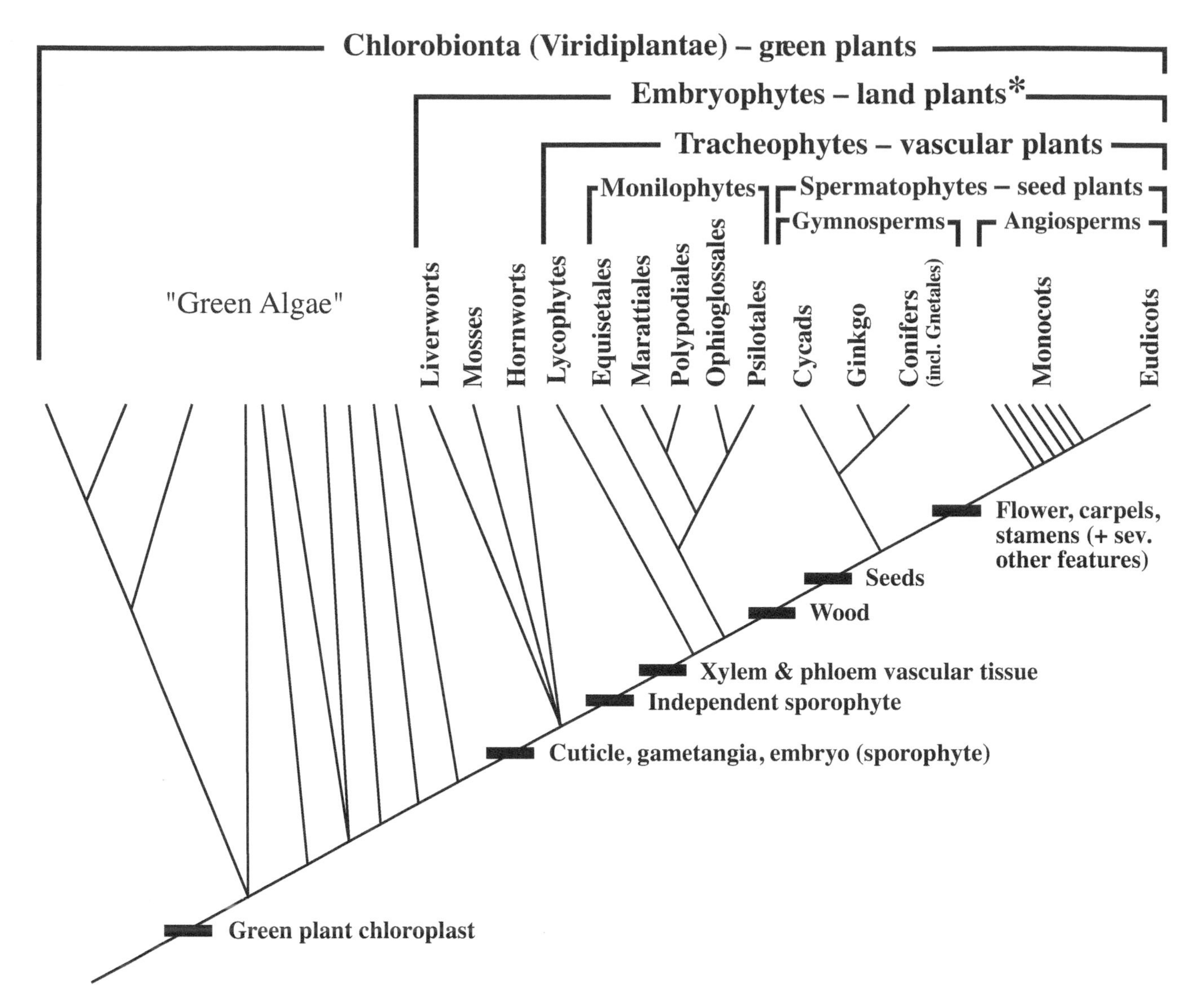

Figure 1.3 Simplified cladogram (evolutionary tree) of the green plants, illustrating major extant groups and evolutionary events (or "apomorphies," hash marks). *Embryophytes are treated as "plants" in this book.

thesis in plants and the other photosynthetic organisms changed the earth in two major ways. First, the fixation of carbon dioxide and the release of molecular oxygen in photosynthesis directly altered the earth's atmosphere over billions of years. What used to be an atmosphere deficient in oxygen underwent a gradual change. As a critical mass of oxygen accumulated in the atmosphere, selection for oxygen-dependent respiration occurred (via oxidative phosphorylation in mitochondria), which may have been a necessary precursor to the evolution of many multicellular organisms, including all animals. In addition, an oxygen-rich atmosphere permitted the establishment of an upper atmosphere ozone layer, which shielded life from excess UV radiation. This allowed organisms to inhabit more exposed niches that were previously inaccessible.

Second, the compounds that photosynthetic species produce are utilized, directly or indirectly, by nonphotosynthetic (heterotrophic) organisms. For virtually all land creatures and many aquatic ones as well, land plants make up the so-called primary producers in the food chain, the source of high-energy compounds such as carbohydrates, structural compounds such as certain amino acids, and other compounds essential to metabolism in some heterotrophs. Thus, most species on land today, including the millions of species of animals, are absolutely dependent on plants for their survival. As primary producers, plants are the major components, and definers, of many communities and ecosystems. The survival of plants is essential to maintaining the health of those ecosystems, the severe disruption of which could bring about rampant species extirpation or extinction and disastrous changes in erosion, water flow, and ultimately climate.

To humans, plants are also monumentally important in numerous, direct ways (Figures 1.4, 1.5). Agricultural plants, most of which are flowering plants, are our major source of food. We utilize all plant parts as food products: roots (e.g., sweet potatoes and carrots; Figure 1.4A,B); stems (e.g., yams, cassava/manioc, potatoes; Figure 1.4C); leaves (e.g., cabbage, celery, lettuce; Figure 1.4D); flowers (e.g., cauliflower and broccoli; Figure 1.4E); and fruits and seeds, including grains such as rice (Figure 1.4F), wheat (Figure 1.4G), corn (Figure 1.4H), rye, barley, and oats, legumes such as beans and peas (Figure 1.4I), and a plethora of fruits such as bananas (Figure 1.4J), tomatoes, peppers, pineapples (Figure 1.4K), apples (Figure 1.4L), cherries, peaches, melons, kiwis, citrus, olives (Figure 1.4M), and others too numerous to mention. Other plants are used as flavoring agents, such as herbs (Figure 1.5A–D) and spices (Figure 1.5E), as stimulating beverages, such as chocolate, coffee, tea, and cola (Figure 1.5F), or as alcoholic drinks, such as beer, wine, distilled liquors, and sweet liqueurs. Woody trees of both conifers and flowering plants are used structurally for lumber and for pulp products such as paper (Figure 1.5G). Non-woody plants, such as bamboos, palms, and a variety of other species, serve as construction materials for a great variety of purposes. Plant fibers are used to make thread for cordage (such as sisal), for sacks (such as jute for burlap), and for textiles (most notably cotton, Figure 1.5H, but also linen and hemp, Figure 1.5I). Extracts from plants, which include essential oils, latex (for rubber or balata), vegetable oils, pectins, starches, and waxes, have a plethora of uses in industry, food, perfume, and cosmetics. In many cultures, plants or plant products are used as euphorics or hallucinogenics (whether legally or illegally), such as marijuana (Figure 1.5I), opium, cocaine, and a great variety of other species that have been used by indigenous peoples for centuries. Plants are important for their aesthetic beauty, and the cultivation of plants as ornamentals is an important industry. Finally, plants have great medicinal significance, to treat a variety of illnesses or to maintain good health. Plant products are very important in the pharmaceutical industry; their compounds are extracted, semisynthesized, or used as templates to synthesize new drugs. Many "modern" drugs, from aspirin (originally derived from the bark of willow trees) to vincristine and vinblastine (obtained from the Madagascar periwinkle, used to treat childhood leukemia; Figure 1.5J), are ultimately derived from plants. In addition, various plant parts of a great number of species are used whole or are processed as so-called herbal supplements, which have become tremendously popular of late.

The people, methods, and rationale concerned with the **plant sciences** (defined here as the study of land plants) are as diverse as are the uses and importance of plants. Some of the fields in the plant sciences are very practically oriented. Agriculture and horticulture deal with improving the yield or disease resistance of food crops or cultivated ornamental plants, e.g., through breeding studies and identifying new cultivars. Forestry is concerned with the cultivation and harvesting of trees used for lumber and pulp. Pharmacognosy deals with crude natural drugs, often of plant origin. In contrast to these more practical fields of the plant sciences, the "pure" sciences have as their goal the advancement of scientific knowledge (understanding how nature works) through research, regardless of the practical implications. But many aspects of the pure sciences also have important practical applications, either directly by applicable discovery or indirectly by providing the foundation of knowledge used in the more practical sciences. Among these are plant anatomy, dealing with cell and tissue structure and development; plant chemistry and physiology, dealing with biochemical and biophysical processes and products; plant molecular biology, dealing with the structure and function of genetic material; plant ecology, dealing with interactions of plants with their environment; and, of course, plant systematics.

Note that a distinction should be made between "botany" and "plant sciences." **Plant sciences** is the study of plants, treated as equivalent to land plants here. **Botany** is the study of most organisms traditionally treated as plants, including virtually all eukaryotic photosynthetic organisms (land plants and the several groups of "algae") plus other eukaryotic organisms with cell walls and spores (true fungi and groups formerly treated as fungi, such as the water molds, Oomycota, and the slime molds, included within the Amoebozoa; Figure 1.1). Thus, in this sense, botany is inclusive of but broader than the plant sciences. Recognition of both botany and plant sciences as fields of study can be useful, although how these fields are defined can vary and may require clarification.

SYSTEMATICS

WHAT IS SYSTEMATICS?

Systematics is defined in this book as a science that includes and encompasses traditional **taxonomy**, the description, identification, nomenclature, and classification of organisms, and that has as its primary goal the reconstruction of **phylogeny**, or evolutionary history, of life. This definition of *systematics* is not novel, but neither is it universal. Others in the field would treat taxonomy and systematics as separate but overlapping areas; still others argue that historical usage necessitates what is in essence a reversal of the definitions used here. But words, like organisms, evolve. The use of *systematics* to describe an all-encompassing field of endeavor is both most useful and

Figure 1.4 Examples of economically important plants. **A–E.** Vegetables. **A.** *Ipomoea batatas*, sweet potato (root). **B.** *Daucus carota*, carrot (root). **C.** *Solanum tuberosum*, potato (stem). **D.** *Lactuca sativa*, lettuce (leaves). **E.** *Brassica oleracea*, broccoli (flower buds). **F–H.** Fruits, dry (grains). **F.** *Oryza sativa*, rice. **G.** *Triticum aestivum*, bread wheat. **H.** *Zea mays*, corn. **I.** Seeds (pulse legumes), from top, clockwise to center: *Glycine max*, soybean; *Lens culinaris*, lentil; *Phaseolus aureus*, mung bean; *Phaseolus vulgaris*, pinto bean; *Phaseolus vulgaris*, black bean; *Cicer areitinum*, chick-pea/garbanzo bean; *Vigna unguiculata*, black-eyed pea; *Phaseolus lunatus*, lima bean. **J–M.** Fruits, fleshy. **J.** *Musa paradisiaca*, banana. **K.** *Ananas comosus*, pineapple. **L.** *Malus pumila*, apple. **M.** *Olea europaea*, olive.

Figure 1.5 Further examples of economically important plants. **A–D.** Herbs. **A.** *Petroselinum crispum*, parsley. **B.** *Salvia officinalis*, sage. **C.** *Salvia rosmarinus*, rosemary. **D.** *Thymus vulgaris*, thyme. **E.** Spices and herbs, from upper left: *Cinnamomum cassia/zeylanicum*, cinnamon (bark); *Vanilla planifolia*; vanilla (fruit); *Laurus nobilis*, laurel (leaf); *Syzygium aromaticum*, cloves (flower buds); *Myristica fragrans*, nutmeg (seed); *Carum carvi*, caraway (fruit); *Anethum graveolens*, dill (fruit); *Pimenta dioica*, allspice (seed); *Piper nigrum*, pepper (seed). **F.** Flavoring plants, from upper left, clockwise. *Theobroma cacao*, chocolate (seeds); *Coffea arabica*, coffee (seeds); *Camellia sinensis*, tea (leaves). **G.** Wood products: lumber (*Sequoia sempervirens*, redwood), and paper derived from wood pulp. **H.** Fiber plant. *Gossypium* sp., cotton (seed trichomes), one of the most important natural fibers. **I.** Euphoric, medicinal, and fiber plant. *Cannabis sativa*, marijuana, hemp; stem fibers used in twine, rope, and cloth; resins contain the euphoric and medicinal compound tetrahydrocannabinol. **J.** Medicinal plant. *Catharanthus roseus*, Madagascar periwinkle, from which is derived vincristine and vinblastine, used to treat childhood leukemia.

represents the consensus of how most specialists in the field use the term, an example being the journal *Systematic Botany*, which contains articles both in traditional taxonomy and phylogenetic reconstruction. Plant systematics is studied by acquiring, analyzing, and synthesizing information about plants and plant parts, the content and methodology of which is the topic for the remainder of this book.

Systematics is founded in the principles of **evolution**, its major premise being that there is one phylogeny of life. The goal of systematists is, in part, to discover that phylogeny.

EVOLUTION

Evolution, in the broadest sense, means "change" and can be viewed as the cumulative changes occurring since the origin of the universe some 14 billion years ago. Biological evolution, the evolution of life, may be defined (as it was by Charles Darwin) as "descent with modification." **Descent** is the transfer of genetic material (enclosed within a cell, the unit of life) from parent(s) to offspring over time. This is a simple concept, but one that is important to grasp and ponder thoroughly. Since the time that life first originated some 3.8 billion years ago, *all life has been derived from preexisting life*. Organisms come to exist by the transfer of genetic material, within a surrounding cell, from one or more parents. Descent may occur by simple clonal reproduction, such as a single bacterial cell "parent" dividing by fission to form two "offspring" cells or a land plant giving rise to a vegetative propagule. It may also occur by complex sexual reproduction (Figure 1.6A), in which each of two parents produces specialized gametes (e.g., sperm and egg cells), which contain half the complement of genetic material, the result of meiosis. Two of the gametes fuse together to form a new cell, the zygote, which may develop into a new individual (as occurs in plants; see Chapter 3) or may itself divide by meiosis to form gametes. Descent through time results in the formation of a **lineage** (Figure 1.6B), a set of organisms interconnected through time and space by the transfer of genetic material from parents to offspring. So, in a very literal sense, we and all other forms of life on earth are connected by descent, the transfer of DNA (actually the pattern of DNA) from parent to offspring (ancestor to descendant), generation after generation.

The **modification** component of evolution refers to a change in the genetic material that is transferred from parent(s) to offspring, such that the genetic material of the offspring is different from that of the parent(s). This modification may occur either by mutation, which is a direct alteration of DNA, or by genetic recombination, whereby existing genes are reshuffled in different combinations (during meiosis in eukaryotes, by the genetic processes of crossing over and independent assortment). Systematics is concerned with the identification of the unique modifications of evolution.

It should also be asked, *what* evolves? Although genetic modification may occur in offspring relative to their parents, individual organisms do not generally evolve. This is because a new individual begins when it receives its complement of DNA from the parent(s); that individual's DNA does not change during its/his/her lifetime (with the exception of relatively rare, nonreproductive "somatic" mutations). The general units of evolution are populations and species. A **population** is a group of individuals of the same species that is usually geographically delimited and that typically have a significant amount of gene exchange. **Species** are groups of populations that are related to one another by various criteria and that have evolutionarily diverged from other such groups. There are a number of different species concepts of definitions, dependent on the biological system and on the criteria used to recognize them (see Chapter 19). With changes in the genetic makeup of offspring (relative to parents), the genetic makeup of populations and species changes over time.

In summary, evolution is descent with modification occurring by a change in the genetic makeup (DNA) of populations or species over time. How does evolution occur? Evolutionary change may come about by two major mechanisms: (1) **genetic drift**, in which genetic modification is random; or (2) **natural selection**, in which genetic change is directed and nonrandom. Natural selection is the differential contribution of genetic material from one generation to the next, differential in the sense that genetic components of the population or species are contributed in different amounts to the next generation; those genetic combinations resulting in *increased survival or reproduction* are contributed to a greater degree. (A quantitative measure of this differential contribution is known as **fitness**.) Natural selection results in an **adaptation**, a structure or feature that performs a particular function and which itself brings about increased survival or reproduction. In a consideration of the evolution of any feature in systematics, the possible adaptive significance of that feature should be explored.

Finally, an ultimate result of evolution is **speciation**, the formation of new species from preexisting species. Speciation can follow lineage divergence, the splitting of one lineage into two, separate lineages (Figure 1.6D). Lineage divergence is itself a means of increasing evolutionary diversity. If two, divergent lineages remain relatively distinct, they may change independently of one another, into what may be designated as separate species (see Chapter 19).

TAXONOMY

Taxonomy is a major part of systematics that includes four components: **D**escription, **I**dentification, **N**omenclature, and **C**lassification. (Remember the mnemonic device: **DINC**.) The general subjects of study are **taxa** (singular, **taxon**), which are

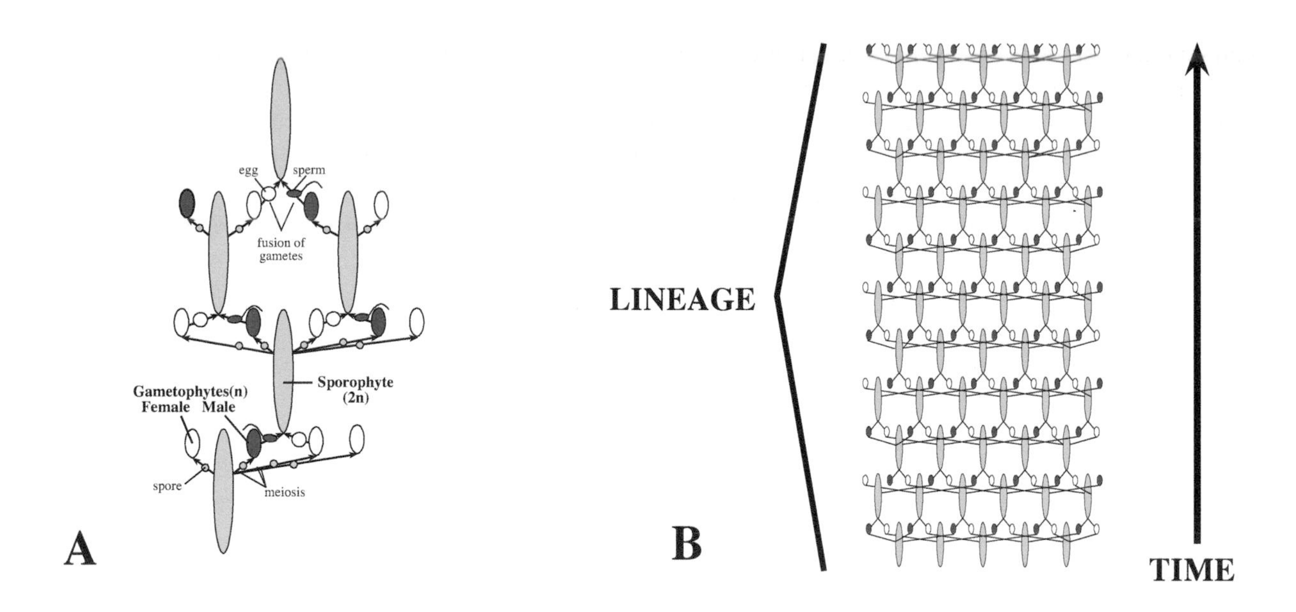

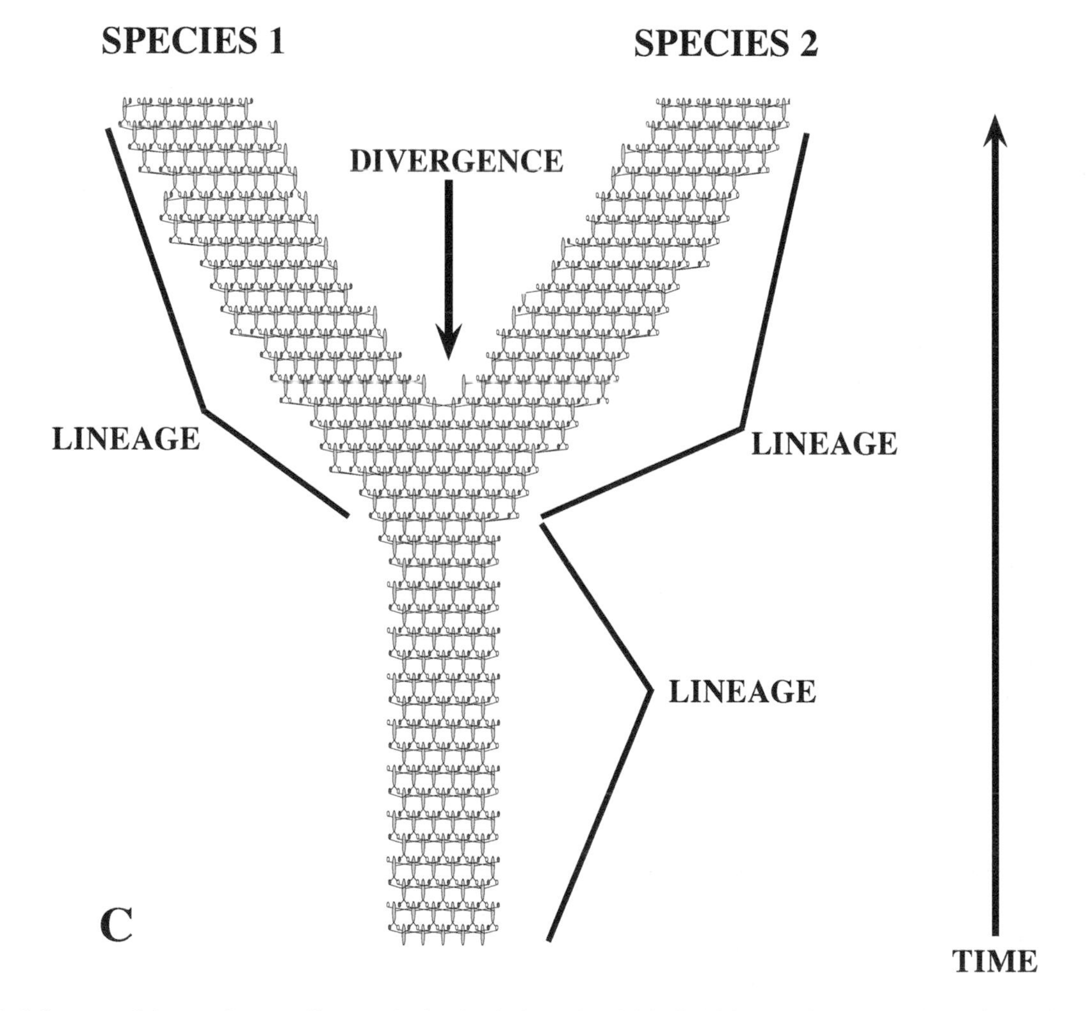

Figure 1.6 **A.** Simplified diagram of descent in sexually reproducing land plants, in which diploid sporophytes give rise to haploid spores (through meiosis), which develop into haploid gametophytes; the latter produce egg and sperm, fusing to form a diploid zygote, which develops into a diploid sporophyte. **B.** A lineage, the result of transfer of genetic material over time and space. **C.** Divergence of one lineage into two, which may result in speciation (illustrated here).

defined or delimited groups of organisms. Ideally, taxa should have a property known as **monophyly** (discussed later; Chapter 2) and are traditionally treated at a particular rank (see later discussion). It should be pointed out that the four components of taxonomy are not limited to formal systematic studies but are the foundation of virtually all intellectual endeavors of all fields, in which conceptual entities are described, identified, named, and classified. In fact, the ability to describe, identify, name, and classify things undoubtedly has been selected for in humans and, in part, in other organisms as well.

Description is the assignment of features or attributes to a taxon. The features are called **characters**. Two or more forms of a character are **character states**. One example of a character is "petal color," for which two character states are "yellow" and "blue." Another character is "leaf shape," for which possible character states are "elliptic," "lanceolate," and "ovate." Numerous character and character state terms are used in plant systematics, both for general plant morphology (see Chapter 9) and for specialized types of data (Chapters 10–14). The purpose of these descriptive character and character state terms is to use them as tools of communication, for concisely categorizing and for delimiting the attributes of a taxon, an organism, or some part of the organism. An accurate and complete listing of these features is one of the major objectives and contributions of taxonomy.

Identification is the process of associating an unknown taxon with a known one, or recognizing that the unknown is new to science and warrants formal description and naming. One generally identifies an unknown by first noting its characteristics, that is, by describing it. Then, these features are compared with those of other taxa to see if they conform. Plant taxa can be identified in many ways (see Chapter 15). A taxonomic key is perhaps the most utilized of identification devices. Of the different types of taxonomic keys, the most common, used in virtually all floras, is a dichotomous key. A **dichotomous key** consists of a series of two contrasting statements. Each statement is a **lead**; the pair of leads constitutes a **couplet** (Figure 1.7). That lead which best fits the specimen to be identified is selected; then all couplets hierarchically beneath that lead (by indentation and/or numbering) are sequentially checked for agreement until an identification is reached (Figure 1.7).

Nomenclature is the formal naming of taxa according to some standardized system. For algae, fungi, and plants, the rules and regulations for the naming of taxa are provided by the *International Code of Nomenclature for algae, fungi, and plants*. These formal names are known as **scientific names**, which by convention are translated into the Latin language. The fundamental principle of nomenclature is that all taxa may bear *only one "correct" scientific name* (Chapter 16). Although they may seem difficult to learn at first, scientific names are much preferable to common (vernacular) names. (Note that another system of nomenclature, the *Phylocode*, is also available and has been used especially for names of higher rank; see Chapter 16.)

The scientific name of a species traditionally consists of two parts (typically underlined or italicized): the genus name, which is always capitalized, e.g., *Quercus*, plus the specific epithet, which by general consensus is not capitalized, e.g., *agrifolia*. Thus, the species name for what is commonly called California live oak is *Quercus agrifolia*. Species names are known as **binomials** (literally meaning "two names") and this type of nomenclature is called binomial nomenclature, first formalized in the mid-18th century by Carolus Linnaeus.

Classification is the arrangement of entities (in this case, taxa) into some type of order. The purpose of classification is to provide a system for cataloguing and expressing relationships between these entities. Taxonomists have traditionally agreed upon a method for classifying organisms that utilizes categories called **ranks**. These taxonomic ranks are hierarchical, meaning that each rank is inclusive of all other ranks beneath it (Figure 1.8).

As defined earlier, a **taxon** is a group of organisms typically treated at a given rank. Thus, in the example of Figure 1.8, Magnoliophyta is a taxon placed at the rank of phylum; Liliopsida is a taxon placed at the rank of class; Arecaceae is a taxon placed at the rank of family; etc. Note that taxa of a particular rank

Couplet:
Lead: 1. Leaf simple of generally pinnately compound, leaflets 0 or (1)3–9; fruit winged achene .... *Fraxinus*
Lead: 1' Leaf simple; fruit capsule or drupe
  2. Leaves generally alternate; fruit a deeply 2-lobed capsule .... *Menodora*
  2' Leaves opposite or clustered; fruit a drupe
    3. Flowers unisexual; corolla 0; stamens 0 or 4–5 .... *Forestiera*
    3' Flowers bisexual; corolla funnel-shaped, salverform, or rotate; stamens 2
      4. Corolla funnel-shaped or salverform; fruit 1–4-seeded .... *Ligustrum*
      4' Corolla rotate; fruit 1-seeded .... *Olea*

Figure 1.7 Dichotomous key to the genera of the Oleaceae of California, by Thomas J. Rosatti, Oleaceae, in Jepson Flora Project (eds.) Jepson eFlora, *http://ucjeps.berkeley.edu/cgi-bin/get_IJM.pl?key=207*, accessed on May 31, 2019, reprinted by special permission.

Major Taxonomic Ranks	Taxa
Kingdom	Plantae
Phylum ("**Division**" also acceptable)	Magnoliophyta
Class	Liliopsida (Monocots)
Order	Arecales
Family	Arecaceae
Genus (plural: genera)	*Cocos*
Species (plural: species)	*Cocos nucifera*

Figure 1.8 The primary taxonomic ranks accepted by the *International Code of Nomenclature for algae, fungi, and plants*, with some examples of taxa.

generally end in a particular suffix (Chapter 16). There is a trend among systematic biologists to eliminate the rank system of classification (see Chapter 16). In this book, ranks are generally used for naming groups but not emphasized as ranks.

There are two major means of arriving at a classification of life: phenetic and phylogenetic. **Phenetic** classification is that based on overall similarities. Most of our everyday classifications are phenetic. For efficiency of organization (e.g., storing and retrieving objects, like nuts and bolts in a hardware store) we group similar objects together and dissimilar objects apart. Many traditional classifications in plant systematics are phenetic, based on noted similarities between and among taxa. **Phylogenetic** classification is that which is based on evolutionary history, or pattern of descent, which may or may not correspond to overall similarity (see later discussion, Chapter 2).

PHYLOGENY

Phylogeny, the primary goal of systematics, refers to the evolutionary history of a group of organisms. Phylogeny is commonly represented in the form of a **cladogram** (or phylogenetic tree), a branching diagram that conceptually represents the evolutionary pattern of descent (see Figure 1.9). The lines of a cladogram represent **lineages**, which (as discussed earlier) denote descent, the sequence of ancestral-descendant populations through time (Figure 1.9A). Thus, cladograms have an implied (relative) time scale. Any branching of the cladogram represents lineage **divergence**, the diversification of lineages from one **common ancestor**.

Changes in the genetic makeup of populations, i.e., evolution, may occur in lineages over time. Evolution is recognized as a change from a preexisting, or **ancestral**, character state to a new, **derived** character state. The derived character state is an evolutionary novelty, also called an **apomorphy** (Figure 1.9A). **Phylogenetic systematics**, or **cladistics**, is a methodology for inferring the pattern of evolutionary history of a group of organisms, utilizing these apomorphies (Chapter 2).

As cited earlier, cladograms serve as the basis for phylogenetic classification. A key component in this classification system is the recognition of what are termed monophyletic groups of taxa. A **monophyletic group**, or **clade**, is a group consisting of a common ancestor plus all (and only all) descendants of that common ancestor. For example, the monophyletic groups of the cladogram in Figure 1.9B are circled. A phylogenetic classification recognizes only monophyletic groups. Note that some monophyletic groups are included within others (e.g., in Figure 1.9B the group containing only taxa *E* and *F* is included within the group containing only taxa *D*, *E*, and *F*, which is included within the group containing only taxa *B*, *C*, *D*, *E*, and *F*, etc.). The sequential listing of clades can serve as a phylogenetic classification scheme (see Chapter 2).

In contrast to a monophyletic group, a **paraphyletic group** is one consisting of a common ancestor but *not all* descendants of that common ancestor; a **polyphyletic group** is one in which there are two or more separate groups, each with a separate common ancestor. Paraphyletic and polyphyletic groups distort the accurate portrayal of evolutionary history and should be abandoned (see Chapter 2).

Knowing the phylogeny of a group, in the form of a cladogram, can be viewed as an important end in itself. As discussed earlier, the cladogram may be used to devise a system of classification, one of the primary goals of taxonomy. The cladogram also can be used as a tool for addressing several interesting biological questions, including biogeographic or ecological history, processes of speciation, and adaptive character evolution. A thorough discussion of the principles and methodology of phylogenetic systematics is discussed in Chapter 2.

WHY STUDY SYSTEMATICS?

The rationale and motives for engaging the field of systematics are worth examining. For one, systematics is important in providing a foundation of information about the tremendous diversity of life. Virtually all fields of biology are dependent on the correct taxonomic determination of a given study organism, which relies on formal description, identification, naming, and classification. Systematic research is the basis for acquiring, cataloguing, and retrieving information about life's diversity. Essential to this research is documentation, through collection (Chapter 17) and storage of reference specimens, e.g., for plants in an accredited herbarium (Chapter 18). Computerized data entry of this collection information is now vital to cataloguing

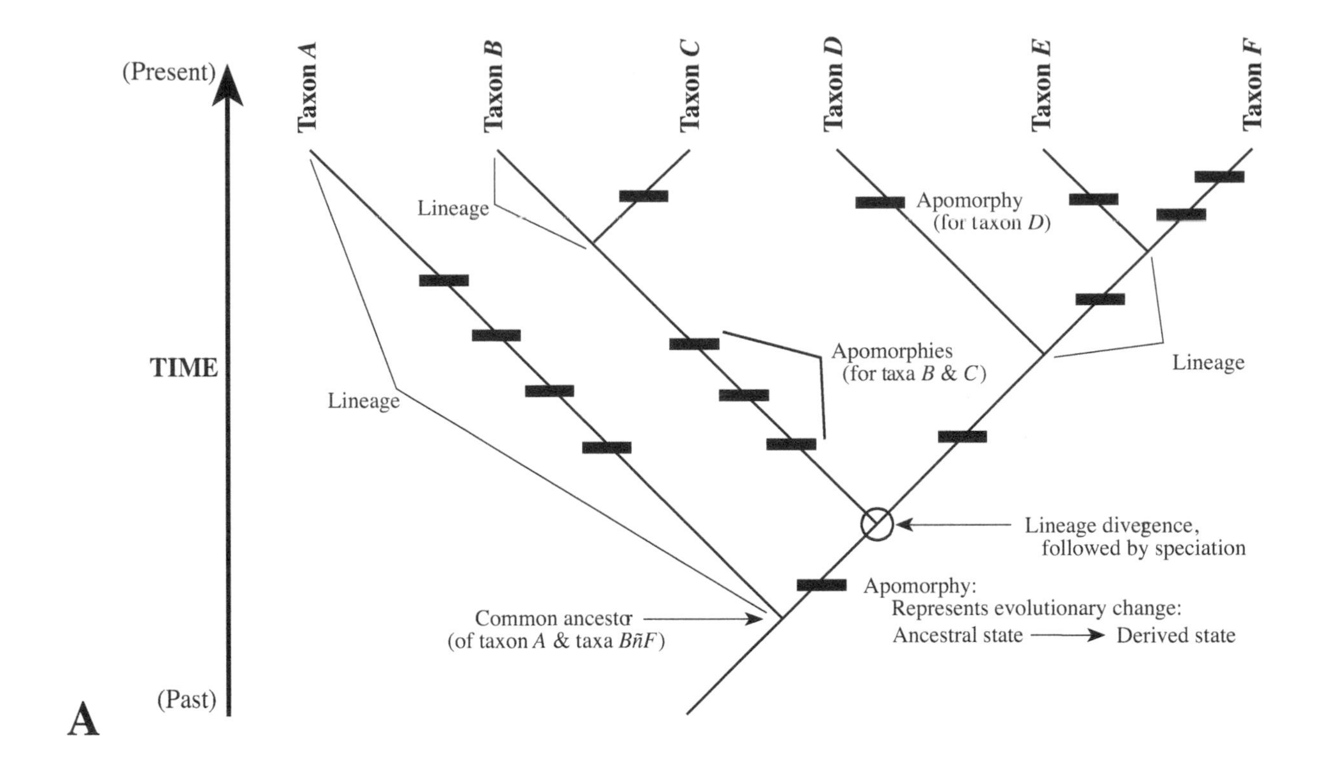

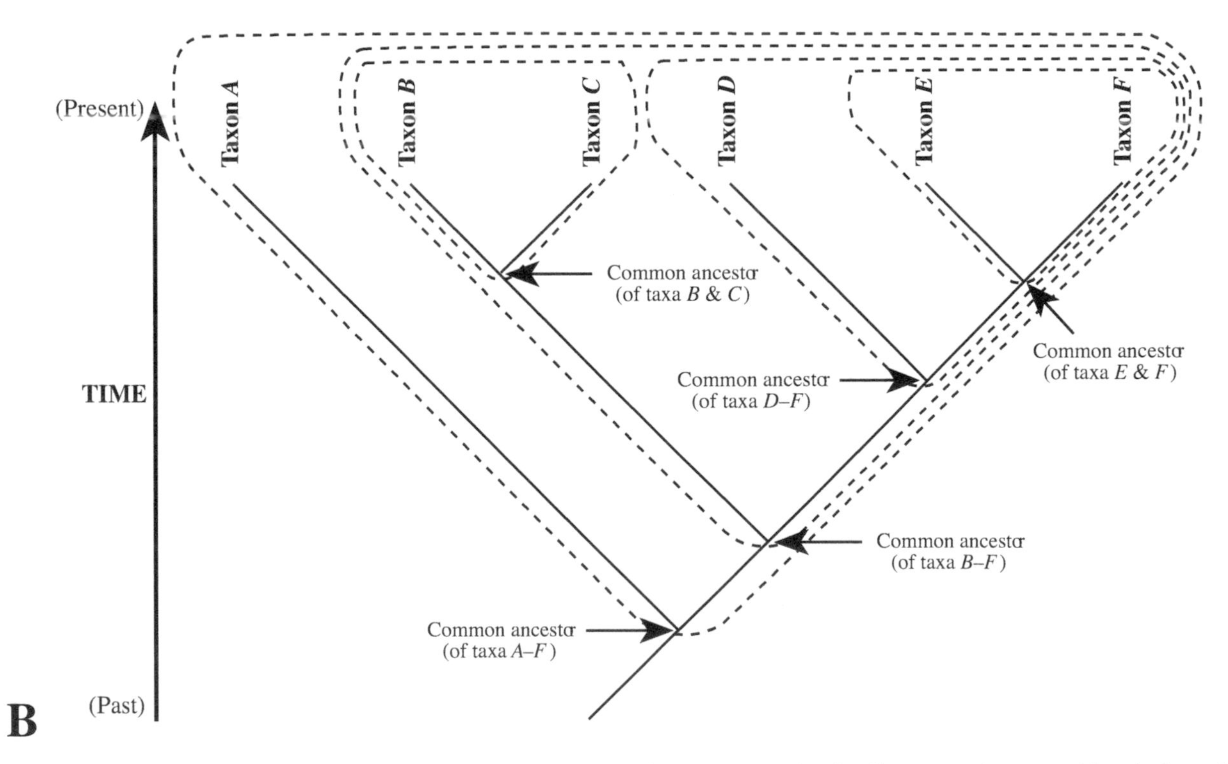

Figure 1.9 Example of a cladogram or phylogenetic tree for taxa *A–F*. **A.** Cladogram showing lineages and apomorphies, the latter indicated by thick hash marks. **B.** Cladogram with common ancestors shown and monophyletic groups (clades) circled.

and retrieving the vast amount of information dealing with biodiversity (Chapter 18).

Systematics is also an integrative and unifying science. One of the "fun" aspects of systematics is that it may utilize data from all fields of biology: morphology, anatomy, embryology/development, ultrastructure, paleontology, ecology, geography, chemistry, physiology, genetics, karyology, and cell/molecular biology. The systematist has an opportunity to understand all aspects of his/her group of interest in an overall synthesis of what is known from all biological specialties, with the goal being to understand the evolutionary history and relationships of the group.

Knowing the phylogeny of life can give insight into other fields and have significant practical value. For example, when a species of *Dioscorea*, wild yam, was discovered to possess steroid compounds (used first in birth control pills), examination of other closely related species revealed species that contained even greater quantities and diversity of these compounds. Other examples corroborate the practical importance of knowing phylogenetic relationships among plant species. The methodology of phylogenetics is now an important part of comparative biology, used by, for example, evolutionary ecologists, functional biologists, and parasitologists, all of whom need to take history into account in formulating and testing hypotheses.

The study of systematics provides the scientific basis for defining or delimiting species and infraspecific taxa (subspecies or varieties) and for establishing that these are distinct from other, closely related and similar taxa. Such studies are especially important today in conservation biology (Chapter 19). In order to determine whether a species or infraspecific taxon of plant is rare or endangered and warrants protection, one must first know the limits of that species or infraspecific taxon. In addition, understanding the history of evolution and geography may aid in conservation and management decisions, where priorities must be set as to which regions to preserve.

Finally, perhaps the primary motivation for many, if not most, in the field of systematics has been the joy of exploring the intricate complexity and incredible diversity of life. This sense of wonder and amazement about the natural world is worth cultivating (or occasionally rekindling). Systematics also can be a challenging intellectual activity, generally requiring acute and patient skills of observation. Reconstruction of phylogenetic relationships and ascertaining the significance of those relationships can be especially challenging and rewarding. But today we also face a moral issue: the tragic and irrevocable loss of species, particularly accelerated by rampant destruction of habitat, such as deforestation in the tropics. We can all try to help, both on a personal and a professional level. Systematics, which has been called simply "the study of biodiversity," is *the* major tool for documenting that biodiversity and can be a major tool for helping to save it. Perhaps we can all consider reassessing our own personal priorities in order to help conserve the life that we study.

REVIEW QUESTIONS

PLANTS

1. What is a "plant"? In what two conceptual ways can the answer to this question be approached?
2. What are the three major groups of life currently accepted?
3. Name and define the mechanism for the evolution of chloroplasts.
4. Name some chlorophyllous organismal groups that have traditionally been called "plants" but that evolved or acquired chloroplasts independently.
5. Draw a simplified cladogram showing the relative relationships among the green plants (Chlorobionta/Viridiplantae), land plants (embryophytes), vascular plants (tracheophytes), seed plants (spermatophytes), gymnosperms, and angiosperms (flowering plants).
6. Why are land plants treated as equivalent to "plants" in this book?
7. List the many ways that plants are important, both in evolution of life on earth and in terms of direct benefits to humans.

SYSTEMATICS

8. What is systematics and what is its primary emphasis?
9. Define biological evolution, describing what is meant both by descent and by modification.
10. What is a lineage?
11. Name and define the units that undergo evolutionary change.
12. What are the two major mechanisms for evolutionary change?
13. What is a functional feature that results in increased survival or reproduction called?
14. Name and define the four components of taxonomy.
15. Define character and character state.

16. Give one example of a character and character state from morphology or from some type of specialized data.
17. What is a dichotomous key? A couplet? A lead?
18. What is a scientific name?
19. Define binomial and indicate what each part of the binomial is called.
20. What is the difference between rank and taxon?
21. What is the plural of taxon?
22. Name the two main ways to classify organisms and describe how they differ.
23. Define phylogeny and give the name of the branching diagram that represents phylogeny.
24. What does a split, from one lineage to two, represent?
25. Name the term for both a preexisting feature and a new feature.
26. What is phylogenetic systematics (cladistics)?
27. What is a monophyletic group or clade? A paraphyletic group? A polyphyletic group?
28. For what can phylogenetic methods be used?
29. How is systematics the foundation of the biological sciences?
30. How can systematics be viewed as unifying the biological sciences?
31. How is systematics of value in conservation biology?
32. Of what benefit is plant systematics to you?

EXERCISES

1. Obtain definitions of the word *plant* by asking various people (lay persons or biologists) or looking in reference sources, such as dictionaries or textbooks. Tabulate the various definitions into classes. What are the advantages and disadvantages of each?
2. Take a day to note and list the uses and importance of plants in your everyday life.
3. Pick a subject, such as history or astronomy, and cite how the principles of taxonomy are used in its study.
4. Do a Web search for a particular plant species (try common and scientific name) and note what aspect of plant biology each site covers.
5. Peruse five articles in a systematics journal and tabulate the different types of research questions that are addressed.

REFERENCES FOR FURTHER STUDY

GENERAL

Daly, M., P. S. Herendeen, R. P. Guralnick, M. W. Westneat, and L. McDade. 2012. Systematics agenda 2020: The mission evolves. Systematic Biology 61:549-552.

Parfrey, L. W., J. Grant, Y. I. Tekle, E. Lasek-Nesselquist, H. G. Morrison, M. L. Sogin, D. J. Patterson, and L. A. Katz. 2010. Broadly sampled multigene analyses yield a well-resolved eukaryotic tree of life. Systematic Biology 59:518–533.

Reaka-Kudla, M. L., D. E. Wilson, and E. O. Wilson (eds.). 1997. Biodiversity II: Understanding and Protecting Our Biological Resources. Joseph Henry Press, Washington, DC.

Sauquet, H. and S. W. Graham. 2016. Planning the future of plant systematics: Report on a special colloquium at the Royal Netherlands Academy of Arts and Sciences. American Journal of Botany 103:2022-2027.

Simpson, B. B., and M. C. Ogorzaly. 2001. Economic Botany: Plants in Our World. 3rd ed. McGraw-Hill, New York.

Stevens, P. F. 1994. The Development of Biological Systematics: Antoine-Laurent de Jussieu, Nature, and the Natural System. Columbia University Press, New York.

Systematics Agenda 2000: Charting the Biosphere. 1994. Produced by Systematics Agenda 2000. Wilson, E. O. (ed.), and F. M. Peter (assoc. ed.). 1988. Biodiversity. National Academy Press, Washington, DC.

Williams, T. A., P. G. Foster, C. J. Cox, and T. M. Embley. 2018. An archaeal origin of eukaryotes supports only two primary domains of life. Nature 504:231-236.

OTHER PLANT SYSTEMATICS TEXTBOOKS

Judd, W. S., C. S. Campbell, E. A. Kellogg, P. F. Stevens, and M. J. Donoghue. 2015. Plant Systematics: A Phylogenetic Approach, 4th edition. Sinauer Associates, Sunderland, Mass.

Singh, G. 2004. Plant Systematics: An integrated approach. Science Publishers, Enfield, N.H.

Woodland, D. W. 2009. Contemporary Plant Systematics, 4th edition. Andrews University Press, Berrien Springs, Mich.

Walters, D. R., and D. J. Keil. 1996. Vascular Plant Taxonomy, 4th edition. Kendall/Hunt Publishing Co., Dubuque, Iowa.

2

TAXONOMY AND PHYLOGENETIC SYSTEMATICS

TAXONOMY

As introduced in Chapter 1, **taxonomy** is defined as the field of science dealing with description, identification, nomenclature, and classification. By itself, plant taxonomy is not concerned with the evolutionary relationships of a group of species, which is in the realm of systematics. Yet, taxonomic research is still a valid and vibrant field of endeavor. Almost all of our named species have been and continue to be defined based on taxonomic studies, generally with no direct knowledge of evolutionary relationships, although closeness of relationship may certainly be inferred from similarities noted from those studies.

Taxonomic research generally entails determining if a particular set of populations, generally studied from specimens, is discrete enough from others to warrant its own taxonomic status, e.g., as a separate species, infraspecies (subspecies, variety, or form), or higher ranked taxon. A taxonomic study may include a clarification of already named taxa, or may likely entail the formal naming of a new taxon. Comparisons are made between and among the taxa of study with respect to overall similarity, generally using morphological features (see Chapter 9), but other types of features may be assessed as well (Chapters 10–13). These comparisons may be qualitative and/or quantitative; the latter often include measurements that are analyzed by various statistical methods (e.g., Figure 2.1A; see Appendix 4). Illustrations or photographs are often used to document the distinctiveness of a new taxon (Figure 2.1B). Any new taxon name must be validly published according to specified rules (see Chapter 16, Nomenclature and Botanical Names). Proposing a new taxonomic entity includes all of the components of the field:

https://doi.org/10.1016/B978-0-12-812628-8.50002-X,

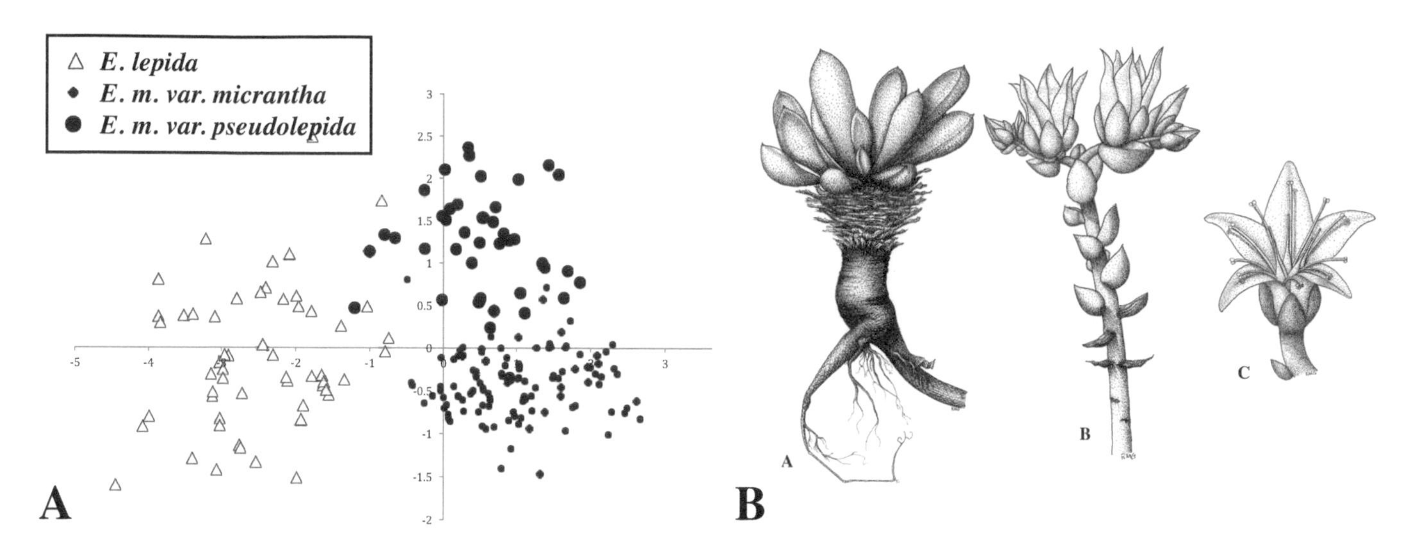

Figure 2.1 Examples of results of taxonomic studies. **A.** Principal components analysis, a statistical procedure demonstrating the distinctiveness of a new plant variety (from Simpson et al. 2016. Madroño 63: 39–54). **B.** Line drawings of vegetative and floral parts, illustrating distinctiveness from closely related taxa (from Dodero and Simpson 2012. Madroño 59:223-229). Courtesy of Madroño.

describing the plant, suggesting a means of identification (usually providing a taxonomic key), formally naming it, and designating, as part of that name, its classification within higher ranked taxa.

Taxonomic research will continue to provide the foundation of our knowledge of the biodiversity of life. It can stand alone, or it may be viewed as a starting point for future phylogenetic studies that aim to assess more precise species delimitations or to elucidate the evolutionary relationships of the taxa of study.

PHYLOGENETIC SYSTEMATICS

Also as introduced in the previous chapter, **systematics** is the field that is inclusive of taxonomy but having the primary objective of elucidating **phylogeny**, the evolutionary history or pattern of descent of a group of organisms. **Phylogenetic systematics**, or **cladistics**, is that specific branch of systematics concerned with inferring phylogeny. Ever since Darwin laid down the fundamental principles of evolutionary theory, a major goal of the biological sciences has been the elucidation of the pattern of life's history of descent. This phylogeny of life, visualized as a branching pattern, can be inferred by an analysis of characters from living or fossil organisms, utilizing phylogenetic principles and methodology.

Recall that a phylogeny is commonly represented in the form of a **cladogram**, or **phylogenetic tree**, a branching diagram that conceptually represents an estimate of phylogeny (Figure 2.2). The lines of a cladogram are known as **lineages**, often referred to simply as "branches." Lineages represent the sequence of ancestral-descendant populations through time, ultimately denoting descent. (The term "lineage" is treated here as a single branch; "clade" is defined as a given common ancestor plus all descendants, including two or more lineages, essentially equivalent to a monophyletic group.)

Evolution may occur within lineages over time and is recognized as a change from a preexisting **ancestral** (also called **plesiomorphic**) condition to a new, **derived** (also called **apomorphic**) condition. The derived condition, or **apomorphy**, represents an evolutionary novelty. As seen in Figure 2.2A, an apomorphy that unites two or more lineages is known as a **synapomorphy** (*syn*, together); one that occurs within a single lineage for a single terminal taxon is called an **autapomorphy** (*aut*, self). Either may be referred to simply as an **apomorphy**, a convention used throughout this book. (Note that many people use **synapomorphy** in this general way.)

Any branching of the cladogram represents lineage **divergence** or **diversification**, the formation of two separate lineages from one **common ancestor**. The two lineages could diverge into what would be designated separate species, the process of forming two species from one termed **speciation**. The *point* of divergence of one clade into two (where the most recent common ancestor of the two divergent clades is located) is termed a **node**; the region between two nodes is an **internode** (Figure 2.2A). Cladograms may be represented in different ways. Figure 2.2B shows the same cladogram as in Figure 2.2A, but shifted 90° clockwise and with the lineages drawn perpendicular to one another and of a length reflective of the number of apomorphic changes. These two representa-

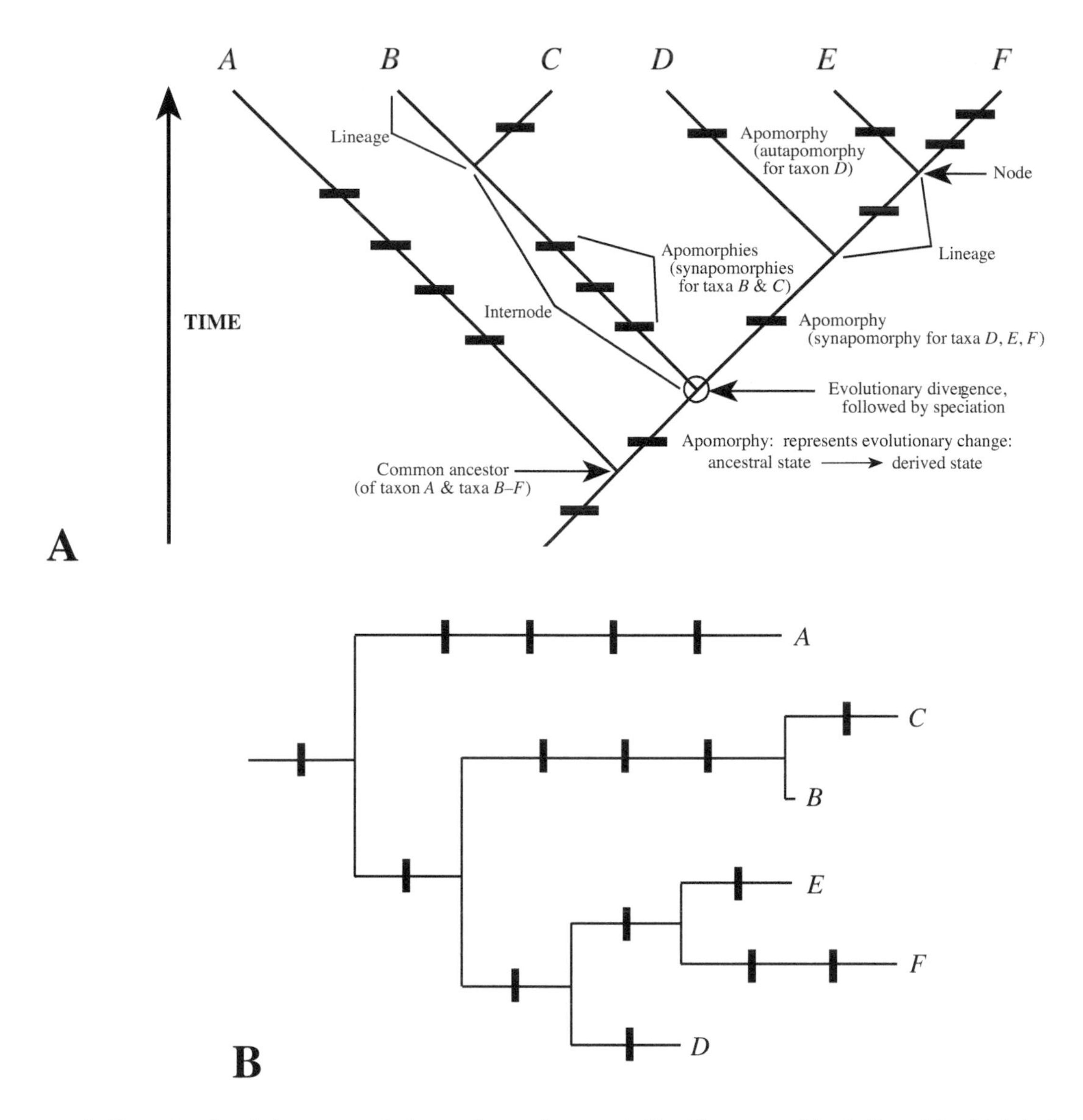

Figure 2.2 **A.** Example of a cladogram or phylogenetic tree for taxa *A–F*, with apomorphies indicated by thick hash marks. See text for explanation of terms. **B.** Same cladogram topology but drawn horizontally, with branch lengths scaled to number of apomorphic changes.

tions of a cladogram have the same **topology**, which is the structure of the branching diagram, i.e., how lineages, including those terminating in taxa, are connected together.

Cladograms have an implied, but *relative*, time scale. For example, in Figure 2.2A, the common ancestor giving rise to taxa E and F occurred later in time than that giving rise to taxa D, E, and F, but we do not know when the lineage split at these nodes occurred or how long the lineages are in terms of real time. The term **phylogram** is often used for a cladogram that has an estimated *absolute* time scale, such that nodes and branch lengths are calibrated and correspond more closely to real elapsed time. (See later discussion.)

Why study phylogeny? Elucidating the pattern of descent, in the form of a cladogram, can be viewed as an important end in itself. The branching pattern derived from a phylogenetic analysis may be used to infer the collective evolutionary changes that have occurred in ancestral/descendant populations through time. Thus, a knowledge of phylogenetic relationships may be invaluable in understanding structural or DNA sequence evolution as well as in gaining insight into the possible functional, adaptive significance of hypothesized evolutionary changes. The cladogram can also be used to classify life in a way that directly reflects evolutionary history. Cladistic analysis may also serve as a tool for inferring biogeographic and ecological history, assessing evolutionary processes such as species diversification rates, and making decisions in the conservation of threatened or endangered species (see Chapter 19).

The principles, methodology, and applications of phylogenetic analyses are described in the remainder of this chapter.

TAXA AND CHARACTERS

TAXON SELECTION

The study of phylogeny begins with the selection of **taxa** (taxonomic groups) to be analyzed, which may include living and/or fossil organisms. Taxon selection includes both the group as a whole, called the study group or **ingroup**, and the individual unit taxa, often termed **O**perational **T**axonomic **U**nits, or **OTUs**. The rationale as to *which* taxa are selected from among many rests by necessity on previous classifications or phylogenetic hypotheses. The ingroup is often a traditionally defined taxonomic group for which there are competing or uncertain classification schemes, the objective being to test the bases of those different classification systems or to provide a new classification system derived from the phylogenetic analysis. The OTUs are previously classified members of the study group and may be species, infraspecies, or taxa consisting of groups of species (e.g., named genera or families or clades within). Sometimes populations, if distinctive and presumed to be on their own evolutionary track, can be used as OTUs in a phylogenetic analysis.

In addition, outgroup OTUs are selected. An **outgroup** is a taxon that is closely related to but not a member of the ingroup. Outgroups are used to "root" a tree (see later discussion).

Some caution should be taken in choosing which taxa to study. First, the OTUs must be *correctly identified* and well circumscribed and delimited from one another. Second, the study group itself should be large enough so that all probable closely related OTUs are included in the analysis. Stated strictly, both OTUs and the ingroup as a whole must be assessed for **monophyly** before the analysis (see later discussion). In summary, initial selection of taxa in a cladistic analysis, both study group and OTUs, should be questioned beforehand to avoid the bias of blindly following past classification systems.

DESCRIPTION

Fundamental in any systematic study is description, the characterization of the attributes or features of taxa using any number of types of evidence (see Chapters 9–14). A systematist may make original descriptions (for example, acquisition of DNA sequence data) or rely partly or entirely on previously published or catalogued data (such as repositories like GenBank; see Chapter 14). In any case, it cannot be overemphasized that the ultimate validity of a phylogenetic study depends on the descriptive accuracy and completeness of the primary source of descriptive data. Thorough research and a comprehensive familiarity with the literature on the taxa and characters of concern are prerequisites to a phylogenetic study.

CHARACTER SELECTION AND DEFINITION

After taxa are selected and the basic research and literature surveys are completed, the next step in a phylogenetic study is the actual selection and definition of **characters** and **character states** from the descriptive data. (Recall that a character is an attribute or feature; character states are two or more forms of a character.) Generally, those features that (1) are genetically determined and heritable (termed "intrinsic"), (2) are relatively invariable within an OTU, and (3) denote clear discontinuities from other similar characters and character states should be utilized. However, the selection of a finite number of characters from the virtually infinite number that could be used adds an element of subjectivity to the study. Thus, it is important to realize that any analysis is inherently biased simply by *which* characters are selected and *how* the characters and character states are defined.

Characters used in phylogenetic analyses are usually conceptually divided into two classes: "**morphological**," essentially equivalent to non-molecular features, such as morphology (Chapter 9), anatomy (Chapter 10), embryology (Chapter 11), palynology (Chapter 12), and some aspects of reproductive biology (Chapter 13); and "**molecular**," derived from genetic data, such as DNA sequences (Chapter 14).

Morphological features are generally the manifestation of numerous intercoordinated genes, and because evolution occurs by a change in one or more of those genes, the precise definition of a morphological feature in terms of characters and character states may be problematic. A structure may be defined broadly as a whole entity with several components. Alternatively, discrete features of a structure may be defined individually as separate characters and character states. For example, in comparing the evolution of fruit morphology within some study group, the character "fruit type" might be designated as two character states: berry versus capsule, *or* the characteristics of the fruit may be subdivided into a host of characters with their corresponding states, such as "fruit shape," "fruit wall texture," "fruit dehiscence," and "seed number." (These characters may be correlated, however; see later discussion.) In practice, characters are divided only enough to communicate differences between two or more taxa. However, this type of terminological atomization may be misleading with reference to the effect of specific genetic changes in evolution, as genes do not normally correspond one for one with taxonomic characters. The morphology of a structure is the end product of development, involving a host of complex interactions of the genome.

Molecular characters may be less "subjective" than morphological ones, but they are not full proof. Polymorphisms or uncertainties in base determination may occur for DNA sequence data. Sequence alignment, in particular, may not be clear-cut if sequences between taxa are very different (e.g., some taxa having significant deletions or insertions), necessitating often "black-box" sequence alignment programs. And the possibility of paralogy due to ancestral gene duplication or hybridization may confound comparison of sequences that are homologous. (See Chapter 14.)

In real practice today, molecular characters are almost exclusively used to infer phylogenies. Morphological characters are typically "traced" on cladograms derived from molecular analyses (see Character Evolution). However, the points above still apply in that application.

CHARACTER STATE DISCRETENESS

Because phylogenetic systematics entails the recognition of an evolutionary transformation from one state to another, an important requirement of character analysis is that character states be discrete or discontinuous from one another. Molecular characters and their states are usually discrete (see Chapter 14), although polymorphism of nucleotide base sites can occur. For some morphological, qualitative characters such as corolla color, the discontinuity of states is clear; e.g., the corolla is yellow in some taxa and blue in others. But for other features, character states may not actually be clearly distinguishable from one another. This lack of discontinuity often limits the number of available characters and is often the result of variation of a feature either within a taxon or between taxa. Statistical tests, such as ANOVAS, *t*-tests, or multivariate statistics, may be used as other criteria for evaluating morphological character state discontinuity. (See Appendix 4 for details.)

CHARACTER CORRELATION

Another point to consider in character selection and definition is whether there is possible correlation of characters. Character correlation refers to an association between what are defined as separate characters, but which are actually components of a common structure, the manifestation of a single evolutionary novelty. Two or more characters are correlated if a change in one always accompanies a corresponding change in the other, bringing up the possibility that they are genetically linked. When characters defined in a cladistic analysis are correlated, including them in the analysis (as two or more separate characters) may inadvertently add weight to what could otherwise be listed as a single character.

HOMOLOGY ASSESSMENT

One concept critical to cladistics is that of **homology**, which can be defined as similarity resulting from common ancestry. Characters or character states of two or more taxa are homologous if those same features were present in the common ancestor of the taxa. For example, the flower of a daisy and the flower of an orchid are homologous as flowers because their common ancestor had flowers, which the two taxa share by continuity of descent. Taxa with homologous morphological features are presumed to share, by common ancestry, the same or similar DNA sequences or gene assemblages are involved in the development of a common structure such as a flower. (Unfortunately, molecular biologists may use the term *homology* to denote similarity in DNA sequence, even though the common ancestry of these sequences may not have been tested; using the term *sequence similarity* in this case is preferred.)

Homology may also be defined with reference to similar structures within the same individual; two or more structures are homologous if the genetic mechanisms that determine their similarity share a common evolutionary history. For example, conduplicate carpels of flowering plants may be hypothesized as homologous with leaves because of a basic similarity between the two in form, anatomy, and development. Their similarity may be hypothesized to be the result of a "sharing" of common genes (or of duplicated genes) that direct their development. The duplication and subsequent divergence of genes is a type of intra-individual or intra-species homology; the genes are similar because of origin from a common ancestor, in this case the gene prior to duplication.

Molecular data must also be assessed for homology. Generally, a particular nucleotide base site for a given gene or intergenic region is assumed to be homologous among the OTUs of a study. However, gene duplication or past hybridization (e.g., resulting in polyploidy; see Chapter 13) may confound homology of DNA sequence data in that non-homologous, paralogous genes or sequences are unknowingly being compared (Chapter 14).

Similarity between taxa can arise not only by common ancestry, but also by independent evolutionary origin. Similarity that is *not* the result of homology is termed **homoplasy** (also sometimes termed *analogy*). Homoplasy may arise in two ways: **convergence** (equivalent to "parallelism," here) or **reversal**. **Convergence** is the independent evolution of a similar feature in two or more lineages. Thus, liverwort gametophytic leaves and lycophyte sporophytic leaves evolved independently as dorsiventral, photosynthetic appendages; their similarity is homoplasious by convergent evolution. (However, although "leaves" in the two groups evolved independently, they could possibly be homologous in the sense of utilizing at least some gene complexes of common origin that function in the development of bifacial organs. This is unknown at present.)

Reversal is the loss of a derived feature with the re-establishment of a feature that resembles an ancestral condition. For example, the reduced flowers of many angiosperm taxa, such as *Lemna* (Araceae; see Chapter 7), lack a perianth; comparative and phylogenetic studies have shown that flowers of these taxa lack the perianth by secondary loss, i.e., via a reversal, reverting to a condition prior to the evolution of a reproductive shoot having a perianth-like structure. However, it should be noted that even if a reversed feature is identical to that of the ancestor, they do not represent the same historical product. As based on a given phylogenetic analysis, a reversal constitutes a unique evolutionary event in itself.

The determination of homology is one of the most challenging aspects of a phylogenetic study and may involve a variety of criteria. Generally, homology is hypothesized based on some evidence of similarity, either direct similarity (e.g., of structure, position, or development) or similarity via a gradation series (e.g., intermediate forms between character states). In tracing morphological characters, homology should be assessed for each character of all taxa in a study, particularly of those taxa having *similarly termed* character states. For example, both the cacti and stem-succulent euphorbs have spines (Figure 2.3). Thus, for the character "spine presence/absence," the character state "spines present" may be assigned to both of these two taxa in a broad cladistic analysis. Whether intended or not, this designation of the same character state for two or more taxa presupposes that these features are homologous in those taxa and arose by common evolutionary origin. Thus, a careful distinction should be made between terminological similarity and similarity by homology. In the above example, more detailed study demonstrates that the spines of cacti and euphorbs are quite different in origin, cacti having leaf spines arising from an areole (a type of short shoot), euphorbs having spines derived from modified stipules. Despite the similarity between spines of cacti and stem-succulent euphorbs, their structural and developmental *dissimilarity* indicates that they are homoplasious and had independent evolutionary origins (with similar selective pressures, i.e., protection from herbivores). This hypothesis necessitates a redefinition of the characters and character states, such that the two taxa are not a priori coded the same.

Hypotheses of homology are tested by means of the cladistic analysis. The *totality* of characters are used to infer the most likely evolutionary tree, and the original assessment of homology is checked by determining if convergences or reversals must be invoked to explain the distribution of character states on the final cladogram (see later discussion).

TRANSFORMATION SERIES AND POLARITY

After the characters and character states have been selected and defined and homology assessed, the character states for each character are arranged in a sequence, known as a **transformation series**. Transformation series represent the hypothesized sequence of evolutionary change, from one character state to another, in terms of direction and probability. For a character with only two character states, known as a **binary character**, obviously only one transformation series exists. For example, for the character "presence or absence of the chloroplast inverted repeat," only two states are possible: presence and absence.

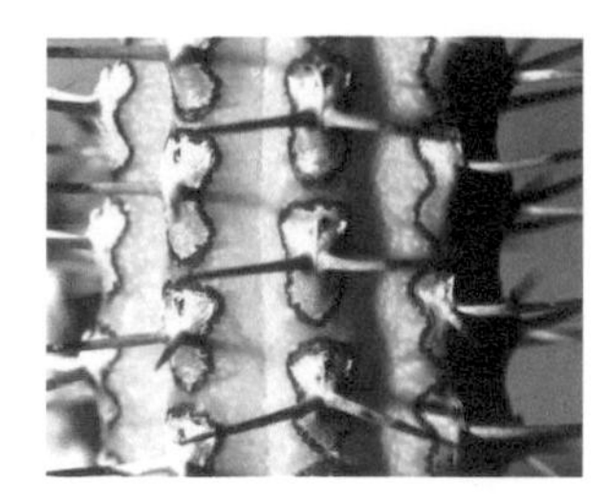

Figure 2.3 Comparison of cactus (left) and euphorb (right) spines, which are not homologous as spines.

Characters having three or more character states, known as **multistate characters**, can be arranged in transformation series that are either ordered or unordered. An **unordered** transformation series allows for each character state to evolve into every other character state with equal probability, i.e., in a single evolutionary step, as exemplified for morphological data (Figure 2.4A) or for the four DNA nucleotides (Figure 2.4B). In contrast, an **ordered** transformation series limits character state transformations to a predetermined sequence, of specific character state changes. For example, in Figure 2.4C, the evolution of unifoliolate leaves from simple leaves takes two evolutionary steps and necessitates passing through the intermediate condition, compound leaves.

The rationale for an ordered series is the assumption or hypothesis that evolutionary change proceeds gradually, such that going from one extreme to another most likely entails passing through some recognizable intermediate condition. Ordered transformation series or morphological features are generally postulated vis-à-vis some obvious intergradation of character states or stages in, say, the ontogeny of a character. A general suggestion in cladistic analyses is to code all characters as unordered unless there is compelling evidence for an ordered transformation, such as the presence of a vestigial feature in a derived structure.

A final aspect of character state transformations is the assignment of polarity. **Polarity** is the designation of relative ancestry to the character states of a morphocline. As summarized earlier, a change in character state represents a heritable evolutionary modification from a preexisting structure or feature (**ancestral** or **plesiomorphic**) to a new structure or feature (**derived** or **apomorphic**). For example, for the character "ovary position," with character states "superior" and "inferior," if a superior ovary is hypothesized as ancestral, the resultant "polarized" character state transformation would be "superior → inferior." For a multistate character (e.g., "leaf type" of Figure 2.4C), an example of a polarized, ordered

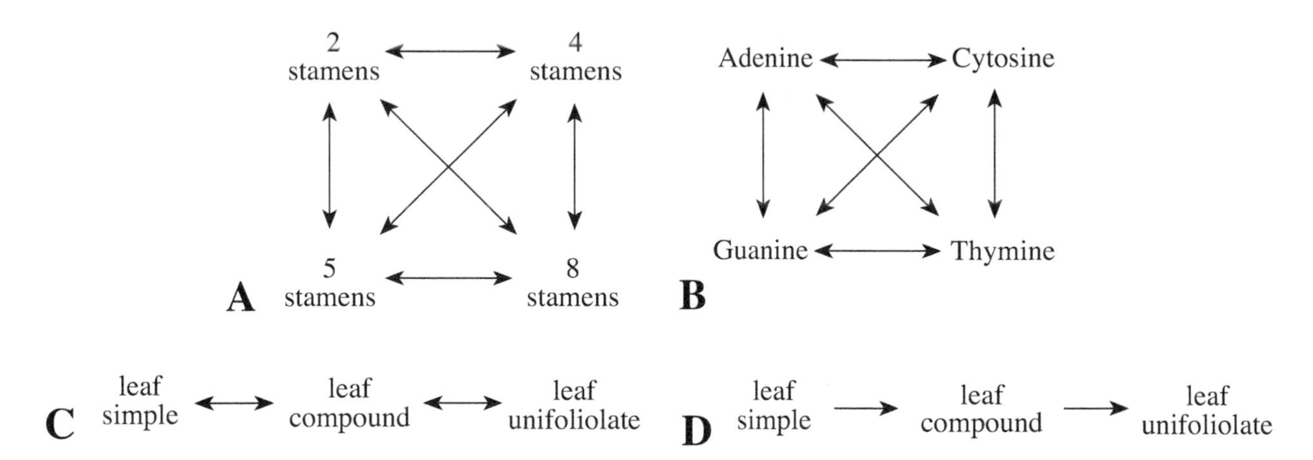

Figure 2.4 Examples of character state transformations. **A.** Unordered, four-state character, stamen number. **B.** Unordered, four-state character, nucleotide type. **C.** Ordered, four-state character, leaf type. **D.** Ordered and polarized character of "C."

transformation series is seen in Figure 2.4D. In the latter example, the unifoliolate leaf possesses a vestige of an ancestrally compound condition, evidence that it should terminate in the ordered character state transformation. In practice polarity determination of characters is inferred by assigning one or more outgroups in a phylogenetic analysis (see **Outgroup Comparison**).

CHARACTER WEIGHTING

As part of a phylogenetic analysis, one or more characters or character states may be assigned a weight, the effect being greater or lesser taxonomic importance relative to other characters in determining phylogenetic relationships. Assigning a character greater weight has the effect of listing it more than once in the character x taxon matrix (see later section) in order to possibly "override" competing changes in unweighted characters. (Note that fractional weights can also be assigned using computer algorithms.)

In practice, character weighting of morphological data is rarely done, in part because of the arbitrariness of determining the amount of weight a character or state should have. A frequent exception, however, is molecular data, for which empirical studies may justify the rationale for and degree of weighting. *Evolutionary models* utilize a sophisticated type of character weighting (see later discussion).

CHARACTER STEP MATRIX

As reviewed earlier assigning a character state transformation determines the number of steps that may occur when going from one character state to another. Computerized phylogeny reconstruction algorithms available today permit a more precise tabulation of the number of steps occurring between each pair of character states through a **character step matrix**. The matrix consists of a listing of character states in a top row and left column; intersecting numbers within the matrix indicate the number of steps required, going from states in the left column to states in the top row. For example, the character step matrix of Figure 2.5A illustrates an ordered character state transformation series such that a single step is required when going from state 0 to state 1 (or state 1 to state 0), two steps are required when going from state 0 to state 2, etc. The character step matrix of Figure 2.5B shows an unordered transformation series, in which a single step is required when going from one state to any other (nonidentical) state. Character step matrices need not be symmetrical; that of Figure 2.5C illustrates an ordered transformation series but one that is irreversible, disallowing a change from a higher state number to a lower state number (e.g., from state 2 to state 1) by requiring a large number of step changes (symbolized by "∞"). Character step matrices are most useful with specialized types of data. For example, the matrix of Figure 2.5D could represent DNA sequence data, where 0 and 1 are the states for the two purines (adenine and guanine) and 2 and 3 are the states for the two pyrimidines (cytosine and thymine; see Chapter 14). Note that in this matrix the change from one purine to another purine or one pyrimidine to another pyrimidine (each of these known as a "transition") requires only one step, hypothesized to be biochemically more probable to occur, whereas a change from a purine to a pyrimidine or from a pyrimidine to a purine (termed a "transversion") is given five steps, being more biochemically less likely. Thus, in a cladistic analysis, the latter change will be given substantially more weight.

As mentioned earlier, DNA sequence data may be transformed in a more complicated *evolutionary model*, based on a number of parameters, such as branch length, codon position, base frequency, or transition/transversion ratio. Such models of evolution are an integral component of maximum likelihood and Bayesian analyses (see later discussion).

A

	0	1	2	3
0	0	1	2	3
1	1	0	1	2
2	2	1	0	1
3	3	2	1	0

B

	0	1	2	3
0	0	1	1	1
1	1	0	1	1
2	1	1	0	1
3	1	1	1	0

C

	0	1	2	3
0	0	1	2	3
1	∞	0	1	2
2	∞	∞	0	1
3	∞	∞	∞	0

D

	0	1	2	3
0	0	1	5	5
1	1	0	5	5
2	5	5	0	1
3	5	5	1	0

Figure 2.5 Character step matrices for: **A.** Ordered character. **B.** Unordered character. **C.** Irreversible character. **D.** Differentially weighted character.

CHARACTER X TAXON MATRIX

Prerequisite to a phylogenetic analysis, characters and character states for each taxon are tabulated in a **character x taxon matrix**, as illustrated in Figure 2.6A for morphological features. In order to analyze the data using computer algorithms, the characters and character states must be assigned a numerical value. In doing so, character states are assigned nonnegative integer values, typically beginning with 0. Figure 2.6B shows the numerical coding of the matrix of Figure 2.6A, the states numerically coded in sequence to correspond with the hypothesized transformation series for that character. For example, character 5 represents an ordered transformation series for stamen number, proceeding from "five" to "four" to "two" stamens (Figure 2.6A), enumerated as 0, 1, and 2.

Figure 2.6G shows a character x taxon matrix for a molecular data set, illustrating variable nucleotide bases as given sites for a gene.

In both character x taxon matrices, polarity is established by including one or more outgroup taxa as part of the character x taxon matrix (as in Figure 2.6A,B,G) and by subsequently "rooting" the tree by placing the outgroups as sister to the ingroup of the final cladogram (see later discussion).

FEATURES OF CLADOGRAMS

APOMORPHIES AS EVOLUTIONARY NOVELTIES

A primary tenet of phylogenetic systematics is that derived character states, or **apomorphies**, that are *shared* between two or more taxa (OTUs) constitute evidence that these taxa share them because of common ancestry. These shared derived character states are hypothesized to be the products of unique evolutionary events that may be used to link the taxa in a common evolutionary history. Thus, by sequentially linking taxa together based on their common possession of shared apomorphies, the evolutionary history of the study group can be inferred.

The character x taxon matrix supplies the data for constructing a phylogenetic tree or cladogram. For example, Figure 2.6C–E illustrates construction of the cladogram for the five species of the hypothetical genus *Xida* from the character x taxon matrix at Figure 2.6A,B. The OTUs may be visualized as terminating lineages arising from a single common ancestor above the point of attachment of the outgroup (Figure 2.6C). This unresolved complex of lineages is known as a **polytomy** (see later discussion). Next, *derived* character states are identified and used to sequentially link sets of taxa (Figure 2.6D,E). In this example, **apomorphies** include (1) the derived states of characters 1 and 3 that group together *X. nigra*, *X. purpurea*, and *X. rubens*; (2) the derived state of character 4 that groups together *X. alba* and *X. lutea*; (3) the derived state "four stamens" of character 5, which is found in all ingroup OTUs and constitutes a **synapomorphy** for the entire study group; and the derived state "two stamens" of character 5 (an ordered character) that groups *X. nigra* and *X. purpurea*. The derived state of character 2 is restricted to *X. lutea* and is therefore an **autapomorphy**. Autapomorphies occur for a single OTU and are not informative in cladogram construction. Finally, the derived state of character 6 evolved twice, in the lineages leading to both species *X. alba* and *X. purpurea*; these independent evolutionary changes constitute homoplasies due to convergence.

One important principle is illustrated in Figure 2.6E for character 5, in which the derived state "four stamens" is an apomorphy for *all* species of the study group, including *X. nigra* and *X. purpurea*. Although the latter two species lack the state "four stamens" for that character, they still *share the evolutionary event* (five → four stamens) in common with the other three species. The lineage terminating in *X. nigra* and *X. purpurea* has simply undergone additional evolutionary change in this character, transforming from four to two stamens (Figure 2.6E).

RECENCY OF COMMON ANCESTRY

Phylogenetic analysis allows for a precise definition of biological relationship. *Relationship* in phylogenetic systematics is a measure of **recency of common ancestry**. Two taxa are more closely related to one another if they share a common ancestor that is more recent in time than the common ancestor they share with other taxa. For example, in Figure 2.7A taxon *C* is more closely related to taxon *D* than it is to taxon *E* or *F*. This is true because the common ancestor of taxa *C* and *D* is more recent in time (closer to the present) than is the common ancestor of taxa *C, D, E,* and *F* (Figure 2.7A). In the earlier example of Figure 2.6E, it is evident that

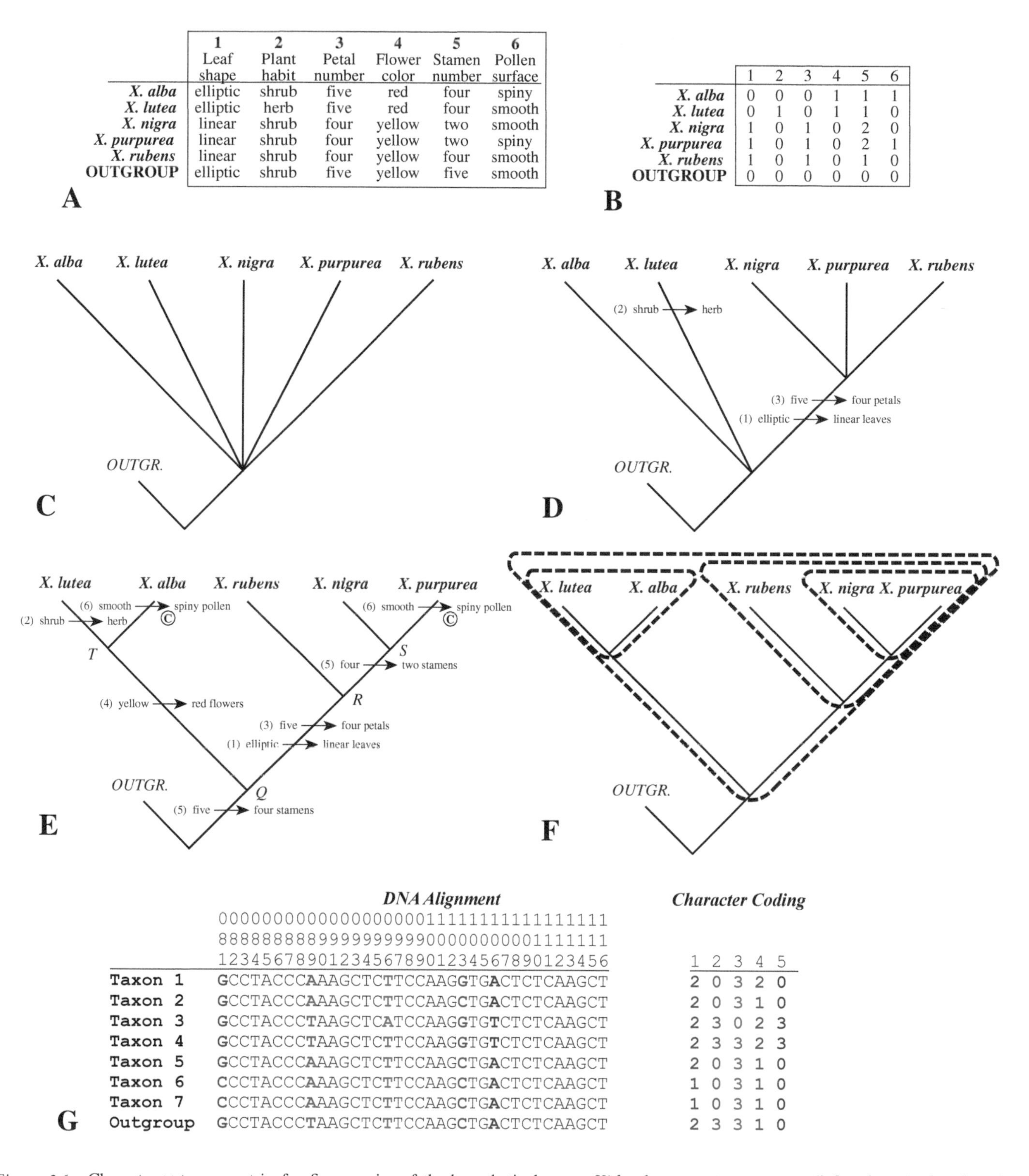

	1 Leaf shape	2 Plant habit	3 Petal number	4 Flower color	5 Stamen number	6 Pollen surface
X. alba	elliptic	shrub	five	red	four	spiny
X. lutea	elliptic	herb	five	red	four	smooth
X. nigra	linear	shrub	four	yellow	two	smooth
X. purpurea	linear	shrub	four	yellow	two	spiny
X. rubens	linear	shrub	four	yellow	four	smooth
OUTGROUP	elliptic	shrub	five	yellow	five	smooth

	1	2	3	4	5	6
X. alba	0	0	0	1	1	1
X. lutea	0	1	0	1	1	0
X. nigra	1	0	1	0	2	0
X. purpurea	1	0	1	0	2	1
X. rubens	1	0	1	0	1	0
OUTGROUP	0	0	0	0	0	0

	DNA Alignment	Character Coding
	000000000000000000111111111111111111	
	888888888999999999900000000001111111	
	123456789012345678901234567890123456	1 2 3 4 5
Taxon 1	GCCTACCCAAAGCTCTTCCAAGGTGACTCTCAAGCT	2 0 3 2 0
Taxon 2	GCCTACCCAAAGCTCTTCCAAGCTGACTCTCAAGCT	2 0 3 1 0
Taxon 3	GCCTACCCTAAGCTCATCCAAGGTGTCTCTCAAGCT	2 3 0 2 3
Taxon 4	GCCTACCCTAAGCTCTTCCAAGGTGTCTCTCAAGCT	2 3 3 2 3
Taxon 5	GCCTACCCAAAGCTCTTCCAAGCTGACTCTCAAGCT	2 0 3 1 0
Taxon 6	CCCTACCCAAAGCTCTTCCAAGCTGACTCTCAAGCT	1 0 3 1 0
Taxon 7	CCCTACCCAAAGCTCTTCCAAGCTGACTCTCAAGCT	1 0 3 1 0
Outgroup	GCCTACCCTAAGCTCTTCCAAGCTGACTCTCAAGCT	2 3 3 1 0

Figure 2.6 Character × taxon matrix for five species of the hypothetical genus *Xida* plus an outgroup taxon (left column), showing six characters (top row) and their character states (inner columns). **A.** Character state names listed, character 5 an ordered character: five → four → two stamens. **B.** Characters and character states converted to numerical values. **C.** Unresolved cladogram. **D.** Addition of characters 1–3. **E.** Most parsimonious cladogram, with addition of other characters. Note common ancestors *Q*, *R*, *S*, *T*, shown for illustrative purposes. **F.** Cladogram at E, with all monophyletic groups circled. **G.** Character × taxon matrix for seven hypothetical taxa plus an outgroup taxon, illustrating DNA sequence data. Nucleotide bases at a given site showing variation among taxa are colored, these alone used in numerical

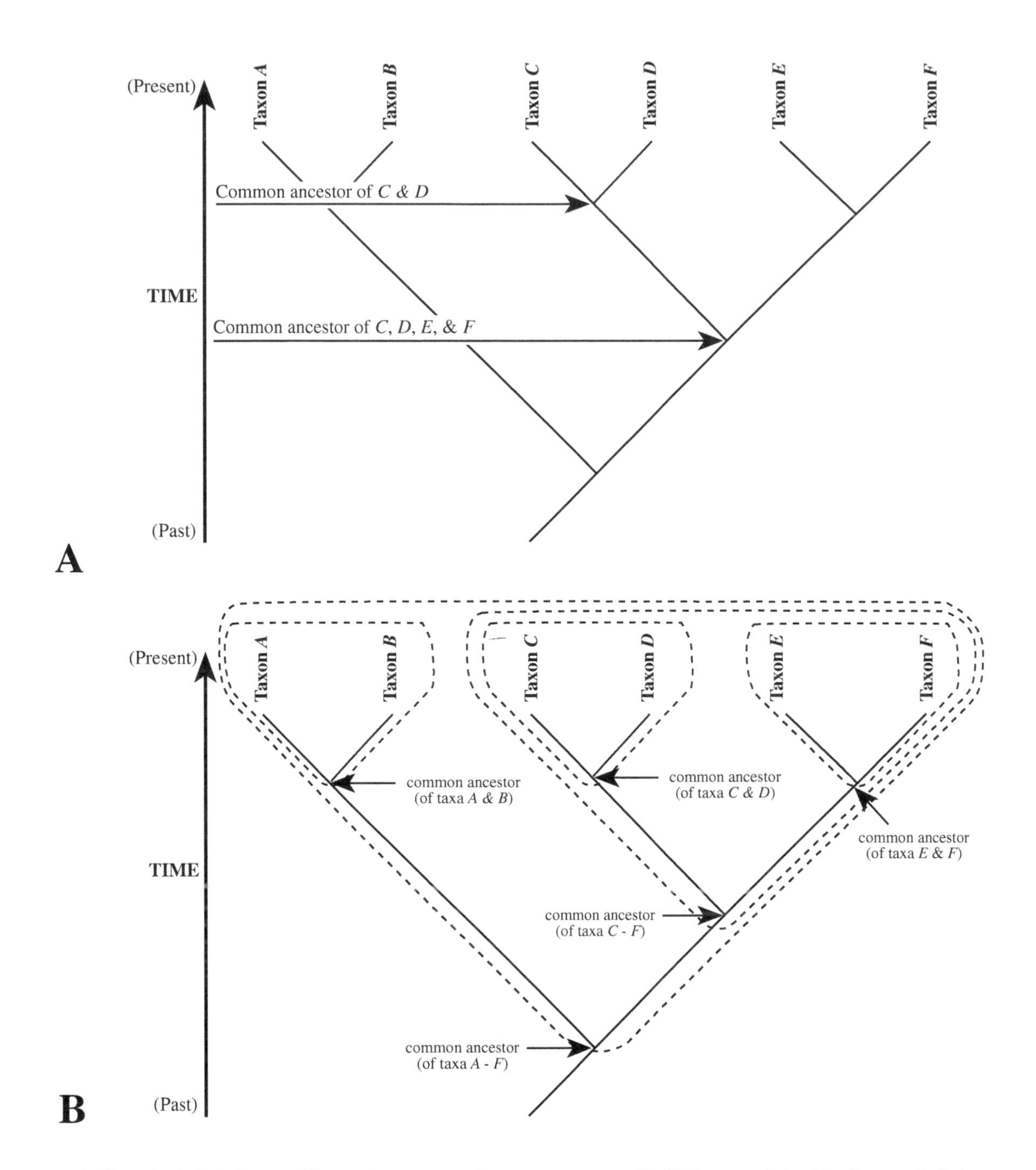

Figure 2.7 **A.** Hypothetical cladogram, illustrating recency of common ancestry. **B.** Cladogram of A with all monophyletic groups circled.

X. nigra and *X. purpurea* are more closely related to one another than either is to *X. rubens*. This is because the former two species together share a common ancestor (*S*) that is more recent in time than the common ancestor (*R*) that they share with *X. rubens*. Similarly *X. rubens* is more closely related to *X. nigra* and *X. purpurea* than it is to either *X. lutea* or *X. alba* because the former three taxa share a common ancestor (*R*) that is more recent in time than *Q*, the common ancestor shared by all five species.

The fact that descent is assessed by means of recency of common ancestry gives the rationale that the branches of a given cladogram may be visually rotated around their junction point or "node" (at the common ancestor) with no change in phylogenetic relationships. For example, the cladograms portrayed in Figure 2.8A–C are all the same as that in Figure 2.6E, differing only in that the lineages have been rotated about their common ancestors. As discussed earlier, the **topology** of all these cladograms is exactly the same; only the visual structure of branches varies. (Again note that cladograms can be portrayed in different manners, with taxa at the top, bottom, or sides and with lineages drawn as vertical, horizontal, or angled lines, as in Figure 2.8A–C.)

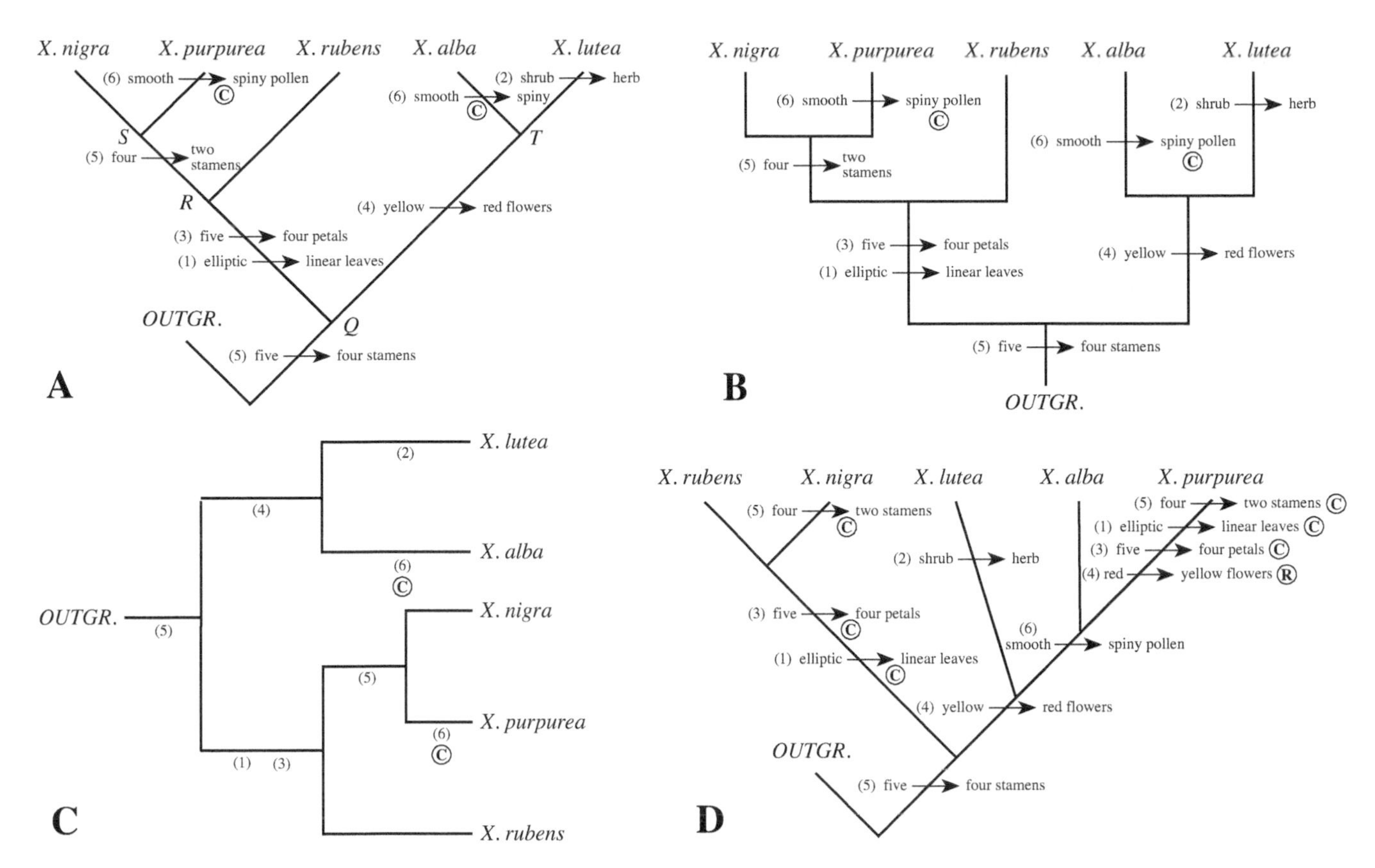

Figure 2.8 **A–C.** Most parsimonious cladogram of Fig. 2.6E. **A.** Cladogram with diagonal lines, but lineages rotated about common ancestor. **B.** Cladogram portrayed with perpendicular lines. **C.** As in B but rotated 90°. **D.** Alternative cladogram for the data set of Figure 2.6A, showing a different relationship among the five species of genus *Xida*, requiring 11 character state changes, three more than the most parsimonious cladogram at Figure 2.6E. Convergent and reversal homoplasies are denoted by circled **C** and **R,** respectively.

MONOPHYLY

A very important concept in phylogenetic systematics is that of monophyly, or monophyletic groups. As introduced earlier, a **monophyletic group** or **clade** is a group that consists of a common ancestor plus *all* descendants of that ancestor. The rationale for monophyly is based on the concept of recency of common ancestry. Members of a monophyletic group share one or more unique evolutionary events; otherwise, the group could not generally be identified as monophyletic. For example, four monophyletic groups can be delimited from the cladogram of Figure 2.6E; these are circled in Figure 2.6F. In another example, the monophyletic groups of the cladogram of Figure 2.7A are shown in Figure 2.7B. Note that all monophyletic groups include the common ancestor plus all lineages derived from the common ancestor, with each lineage terminating in an OTU.

The two descendant lineages or clades from a single common ancestor are known as **sister groups** or **sister taxa**. For example, in Figure 2.6E and F, sister group pairs are: (1) *X. lutea* and *X. alba*; (2) *X. nigra* and *X. purpurea*; (3) *X. nigra* + *X. purpurea* and *X. rubens*; and (4) *X. lutea* + *X. alba* and *X. nigra* + *X. purpurea* + *X. rubens*.

The converse of monophyly is paraphyly. A **paraphyletic group** is one that includes a common ancestor and some, *but not all*, known descendants of that ancestor. For example, in Figure 2.6E, a group including ancestor *Q* plus descendants *X. lutea*, *X. alba*, and *X. rubens* alone is paraphyletic because this group has omitted two taxa (*X. nigra* and *X. purpurea*), which are also descendants of common ancestor *Q*. Paraphyletic groups have often been historically designated by the absence of an apomorphy; in the example of Figure 2.6E, *X. lutea*, *X. alba*, and *X. rubens* may have originally been grouped by their lack of the derived condition (two stamens, character 5) that links *X. nigra* and *X. purpurea*.

Similarly, a **polyphyletic group** is one containing two or more common ancestors. For example, in Figure 2.6E, a group containing *X. alba* and *X. purpurea* alone could be interpreted as polyphyletic because these two taxa do not share a common ancestor apart from that shared by the other taxa, *X. lutea*, *X. nigra*, and *X. rubens*. Polyphyletic groups have typically been defined based on convergences; in the example of Figure 2.6E, *X. alba* and *X. purpurea* share a feature (spiny pollen of character 6), which has been determined from the analysis to be convergent in the two taxa.

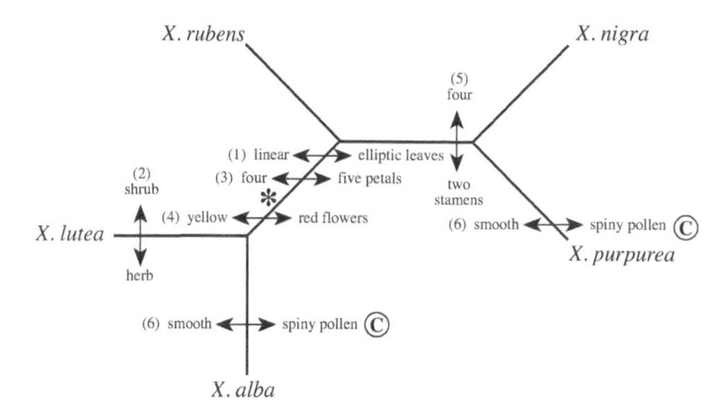

Figure 2.9 Unrooted tree for the data set of Figure 2.7A (minus the *Outgroup* taxon). Direction of evolutionary change is not indicated and monophyletic groups cannot be defined. The "*" indicates the point of rooting that yields the tree of Figure 2.7E.

"Paraphyletic" and "polyphyletic" as designates for a group may intergrade, dependent on the designation of common ancestor; the term **non-monophyletic** may be used to refer to either.

Non-monophyletic groups do not accurately portray evolutionary history and should be abandoned in formal classification systems (see **Phylogenetic Classification**). Their use in comparative studies of character evolution, evolutionary processes, ecology, or biogeography will likely bias the results. A good example of a paraphyletic group is the traditionally defined "dicots" (Dicotyledonae). Because all recent analyses show that some members of the dicots are more closely related to, e.g., monocots (Monocotyledonae) than they are to other dicots, the dicots are paraphyletic (see Chapters 7, 8) and should no longer be recognized as a taxonomic group.

UNROOTED TREES

In contrast to a cladogram, a method for the representation of relative character state changes between taxa is the unrooted tree, sometimes called a "network." Unrooted trees are constructed by grouping taxa from a matrix in which polarity is not indicated (in which no hypothetical ancestor or outgroup is designated), perhaps because the polarity of one or more characters cannot be ascertained or because outgroups were not designated. Because no assumptions of polarity are made, no evolutionary hypotheses are implicit in an unrooted tree. Figure 2.9 illustrates the unrooted tree for the data set of Figure 2.6A,B. Note that monophyletic groups cannot be recognized in unrooted trees because relative ancestry is not indicated. The character state changes noted on the unrooted tree simply denote evolutionary changes when going from one group of taxa to another, without reference to direction of change. After an unrooted tree is constructed, it may be rooted and portrayed as a cladogram. If the relative ancestry of one or more characters can be established, a point on the unrooted tree may be designated as most ancestral, forming the root of the cladogram. For example, if the unrooted tree of Figure 2.9 is rooted (at * in the figure), the result is the tree of Figure 2.6E. However, rooting is effectively done by simply including one or more outgroup(s) in the analysis and placing these outgroups to denote the base (the "root") of the tree.

POLYTOMY

Occasionally, the relationships among taxa cannot be resolved. A **polytomy** is a branching diagram in which the lineages of three or more taxa arise from a single hypothetical ancestor.

A polytomy may arise because of missing data, whereby there are no derived character states identifying the monophyly of two or more taxa among the ingroup. For example, from the **character x taxon matrix** of Figure 2.10A, the relationships among taxa *W*, *X*, and *Y* cannot be resolved; synapomorphies link none of the taxon pairs. Thus, *W*, *X*, and *Y* are grouped as a polytomy in the most parsimonious cladogram (Figure 2.10B).

Polytomies may also occur because of conflicting data. Two or more cladograms with conflicting branching patterns may be portrayed in a *consensus tree* (see later discussion), which may illustrate the conflicting patterns in a polytomy.

Lastly, another possible reason for the occurrence of a polytomy is that three or more taxa actually diverged from a single ancestral species, or diverged dichotomously but within such a short time period that no apomorphic evolutionary event can be detected to link any two of the taxa as a monophyletic group. (See Chapter 19.) The occurrence of polytomies in phylogenetic analysis may be indicative of very rapid evolutionary divergence, a significant finding in itself.

RETICULATION

The methodology of phylogenetic systematics generally presumes the dichotomous or polytomous splitting of taxa, representing putative ancestral speciation events. However, another possibility in the evolution of plants is **reticulation**, which could involve hybridization or horizontal gene transfer (e.g., as occurred in the evolution of chloroplasts and mitochondria; Chapter 1). A reticulation event involving hybridization between two ancestral taxa (*E* and *F*) is exemplified in Figure 2.10D, resulting in the hybrid ancestral taxon *G*, which is the immediate ancestor of extant taxon *X*. Many standard phylogenetic analyses do not consider reticulation as an option and would yield an incorrect cladogram if such a process had occurred. For example, the character x taxon matrix of Figure 2.10C is perfectly compatible with the reticulate cladogram of Figure 2.10D. However, most phylogenetic algorithms would construct the most parsimonious *dichotomously branching* cladogram of Figure 2.10E or

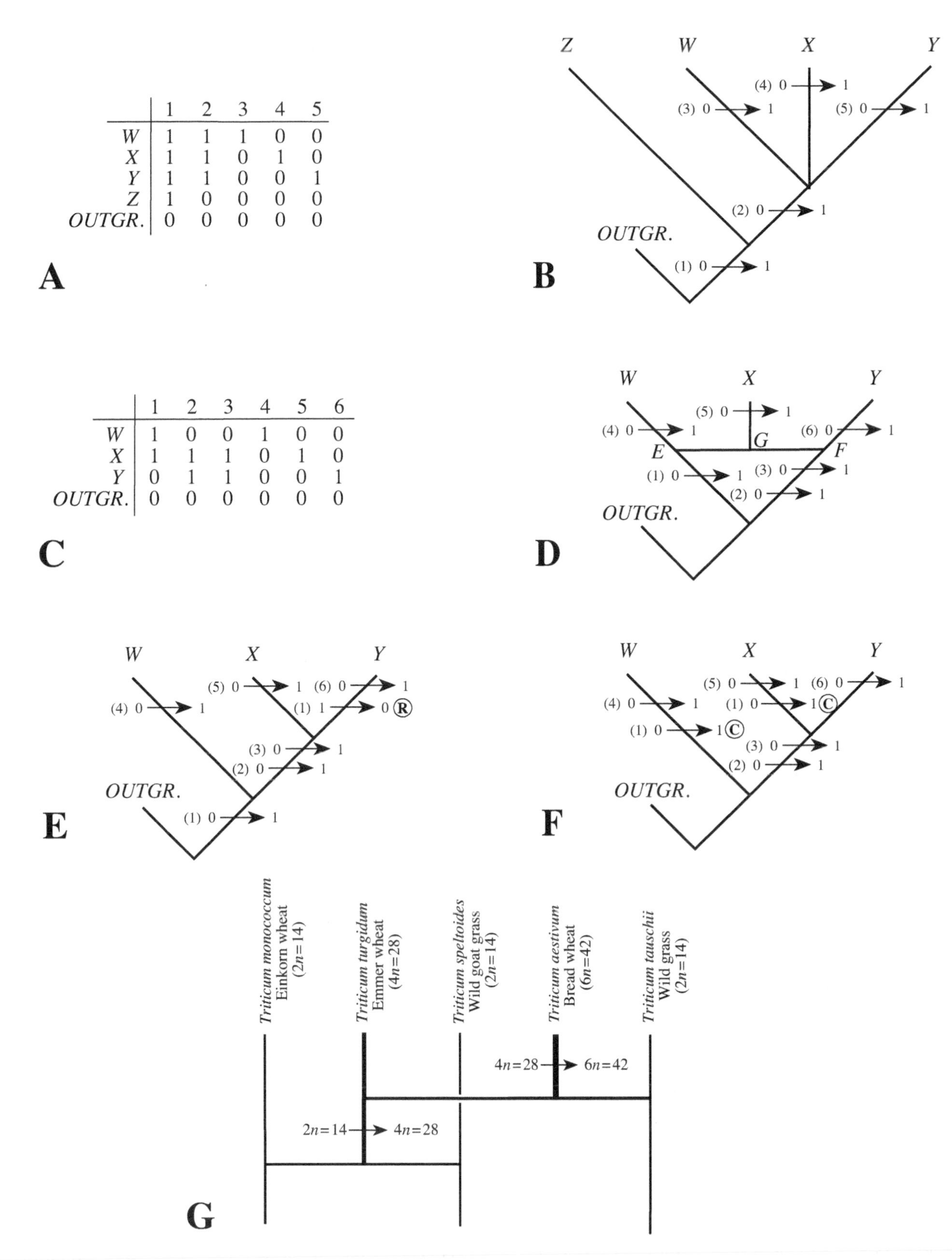

	1	2	3	4	5
W	1	1	1	0	0
X	1	1	0	1	0
Y	1	1	0	0	1
Z	1	0	0	0	0
OUTGR.	0	0	0	0	0

	1	2	3	4	5	6
W	1	0	0	1	0	0
X	1	1	1	0	1	0
Y	0	1	1	0	0	1
OUTGR.	0	0	0	0	0	0

Figure 2.10 **A.** Hypothetical data set. **B.** Resultant tree from data set at A. Note polytomy of lineages to *W*, *X*, and *Y*. **C.** Hypothetical data set. **D.** Cladogram exhibiting reticulation that is compatible with data set at C. **E,F.** Dichotomously branching cladograms arising from data set at C, showing two alternative distributions of character state changes. **G.** Evolution of wheat via ancestral hybridization and polyploidy.

2.10F, which show homoplasy and require one additional character state change than Figure 2.10D.

Data, such as chromosome analysis or certain molecular markers, may provide compelling evidence for past hybridization. A good example of this is the evolution of durum and bread wheat (*Triticum* spp.) via past hybridization and polyploidy (Figure 2.10G). Algorithms that create **networks**, as in nested clade analysis (Chapter 19), are used to study reticulation events.

TAXON SELECTION AND POLYMORPHIC CHARACTERS

As alluded to earlier, the initial selection of taxa to be studied may introduce bias in a phylogenetic analysis. Prior to a phylogenetic analysis, each of the smallest unit taxa under study (OTUs) *and* the group as a whole must be assessed for monophyly prior to the analysis. Monophyly is ascertained by the recognition of one or more unique, shared derived character states that argue for most recent common ancestry of all and only all members of the taxon in question. If such an apomorphy cannot be identified, any relationships denoted from the phylogenetic analysis might be in doubt. For example, in a cladistic analysis of several angiosperm genera (Figure 2.11A), only if each of the unit taxa (genera in this case) is monophyletic will the resultant cladogram be unbiased. If, however, genus *A* is not monophyletic, then it may be possible for some species of genus *A* to be more closely related to (i.e., have more recent common ancestry with) a species of another genus than to the other species of genus *A* (e.g., Figure 2.11B). Therefore, if any doubt exists as to the monophyly of component taxa to be analyzed, the taxa in question should be subdivided until the monophyly of these subtaxa is reasonably certain. If this is not possible, an exemplar species (selected as representative of a higher taxon and assumed to be monophyletic) may be chosen for a first approximation of relationships.

Related to the requirement of OTU monophyly is the problem of **polymorphic** characters, i.e., those that have variable character state values *within* an OTU. If an OTU for which monophyly has been established is polymorphic for a given character, then it may be subdivided into smaller taxonomic groups until each of these groups is monomorphic (i.e., invariable) for the character. If an OTU at the level of species is polymorphic, it is generally listed as such in computer algorithms.

If the ingroup as a whole is not monophyletic, the effect is identical to excluding taxa from the analysis, which could give erroneous results under certain conditions. For example, the most parsimonious cladogram constructed from the data matrix of Figure 2.11C is that of Figure 2.11D. However, if taxon *W* is inadvertently omitted from the ingroup (which is now not monophyletic; Figure 2.11E), then a different, most parsimonious cladogram topology may result for taxa *X*, *Y*, and *Z* (Figure 2.11F). The question of monophyly may be a serious problem for traditionally recognized taxa that were generally not defined by demonstrable apomorphies.

CLADOGRAM CONSTRUCTION

MAXIMUM PARSIMONY

In constructing a cladogram, a single branching pattern may be selected from among many, many possibilities. The number of possible "rooted" dichotomously branching cladograms increases dramatically with a corresponding increase in the number of taxa. For two taxa, there is only one rooted cladogram (Figure 2.12A); for three taxa, there are three (Figure 2.12B); and for four taxa there are 15 rooted, dichotomously branched cladograms (Figure 2.12C). (See later discussion for explanation of rooted versus unrooted trees.) The formula for the number of rooted trees is $\prod (2i-1)$, with $\prod$ being the product of all the factors $(2i-1)$ from $i = 1$ to $i = n-1$, where n is the number of OTUs. For a cladistic analysis involving 54 OTUs, the number of possible dichotomously branching trees is 3×10^{84} (which is greater than the estimated number of atoms in the universe!). The number of trees is even greater when the additional possibilities of reticulation or polytomies are taken into account (see later discussion).

Because there are generally many possible trees for any given data set, one of the major methods of reconstructing phylogenetic relationships is known as the **maximum parsimony** (often abbreviated as **MP**). The **principle of parsimony** states that of the numerous possible cladograms for a given group of OTUs, the one (or more) exhibiting the fewest number of evolutionary steps is accepted as being the best estimate of phylogeny. (Note that there may be two or more cladograms that are equally most parsimonious.) The principle of parsimony is actually a specific example of a general tenet of science known as Ockham's Razor (*Entia non sunt multiplicanda praeter necessitatem*), which states that given two or more competing hypotheses, each of which can explain the facts, the simplest one is accepted. The rationale for parsimony analysis is that the simplest explanation minimizes the number of ad hoc hypotheses, i.e., hypotheses for which there is no direct evidence. In other words, of all possible cladograms for a given group of taxa, the one (or more) implying the fewest number of character state changes is accepted. A consequence of minimizing the total number of character state changes is also to minimize the number of homoplasious reversals or convergences. The principle of

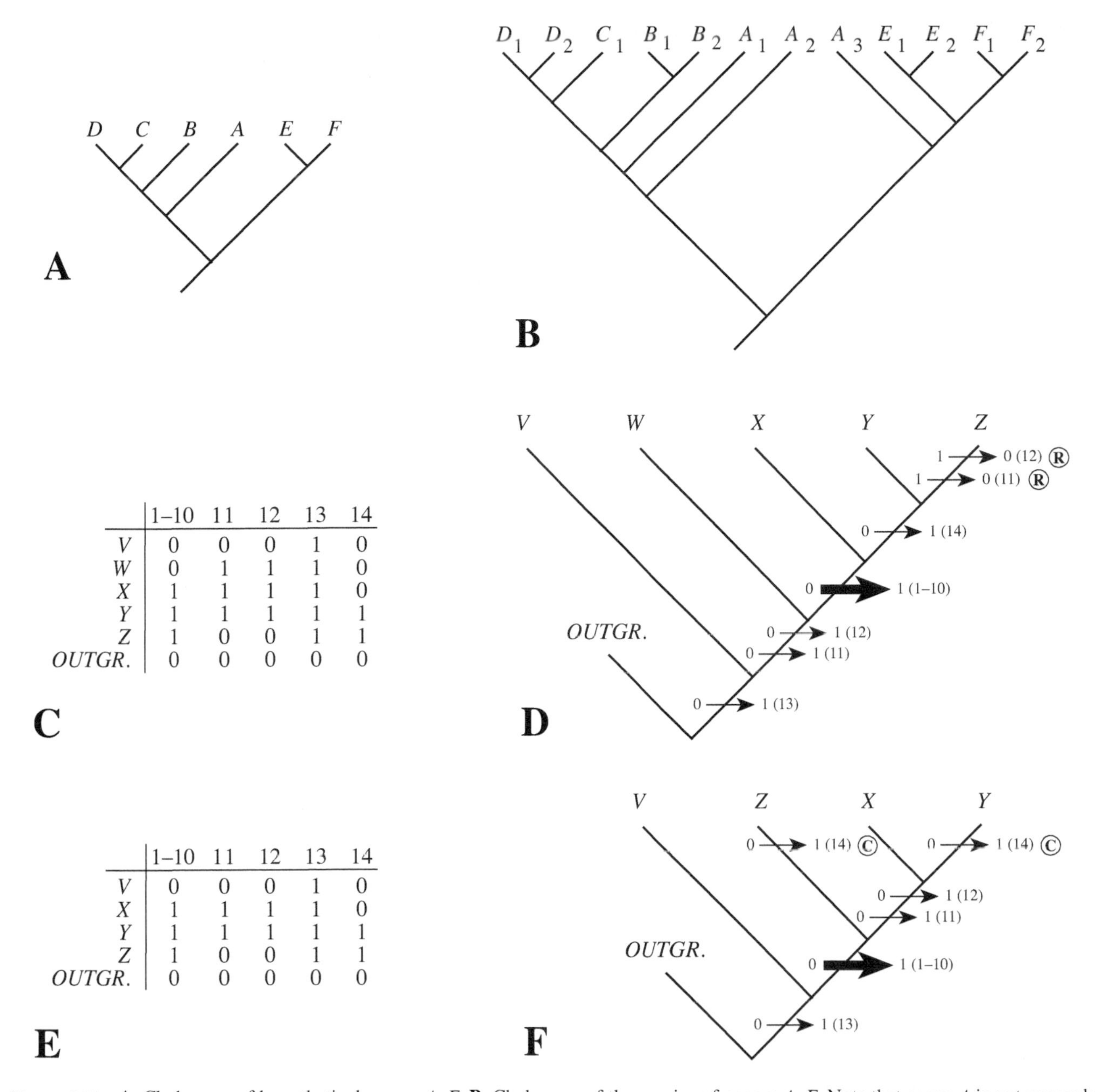

	1–10	11	12	13	14
V	0	0	0	1	0
W	0	1	1	1	0
X	1	1	1	1	0
Y	1	1	1	1	1
Z	1	0	0	1	1
OUTGR.	0	0	0	0	0

	1–10	11	12	13	14
V	0	0	0	1	0
X	1	1	1	1	0
Y	1	1	1	1	1
Z	1	0	0	1	1
OUTGR.	0	0	0	0	0

Figure 2.11 **A.** Cladogram of hypothetical genera *A–E*. **B.** Cladogram of the species of genera *A–E*. Note that genus *A* is not monophyletic. **C.** Character × taxon matrix for taxa *V–Z* plus *OUTGROUP*. **D.** Most parsimonious cladogram for taxa *V*, *W*, *X*, *Y*, and *Z*. **E.** Character × taxon matrix for same taxa, minus taxon *W*, omitted because it is not considered as part of the ingroup. **F.** Most parsimonious cladogram for taxa *V*, *X*, *Y*, and *Z*. Note different branching pattern for taxa *X*, *Y*, and *Z*.

parsimony is a valid working hypothesis because it minimizes uncorroborated hypotheses, thus assuming no additional evolutionary events for which there is no evidence.

Parsimony analysis can be illustrated as follows: For the example data set of Figure 2.6A, which includes five taxa (plus an outgroup), there are actually 105 *possible* dichotomously branching, rooted cladograms; the cladogram at Figures 2.6E and 2.8A (having a total of eight character state changes) is only one of these. One of the other 104 alternative cladistic hypotheses is illustrated in Figure 2.8D. Note, however, that for this cladogram, there are a minimum of 11 character state changes (including three pairs of convergent evolutionary events and one reversal). Thus, of all the possible cladograms for the data set of Figure 2.6A, the *one* shown in Figures 2.6E and 2.8A is the shortest, containing the fewest number of evolutionary steps, and would be accepted as the best estimate of phylogeny by maximum parsimony.

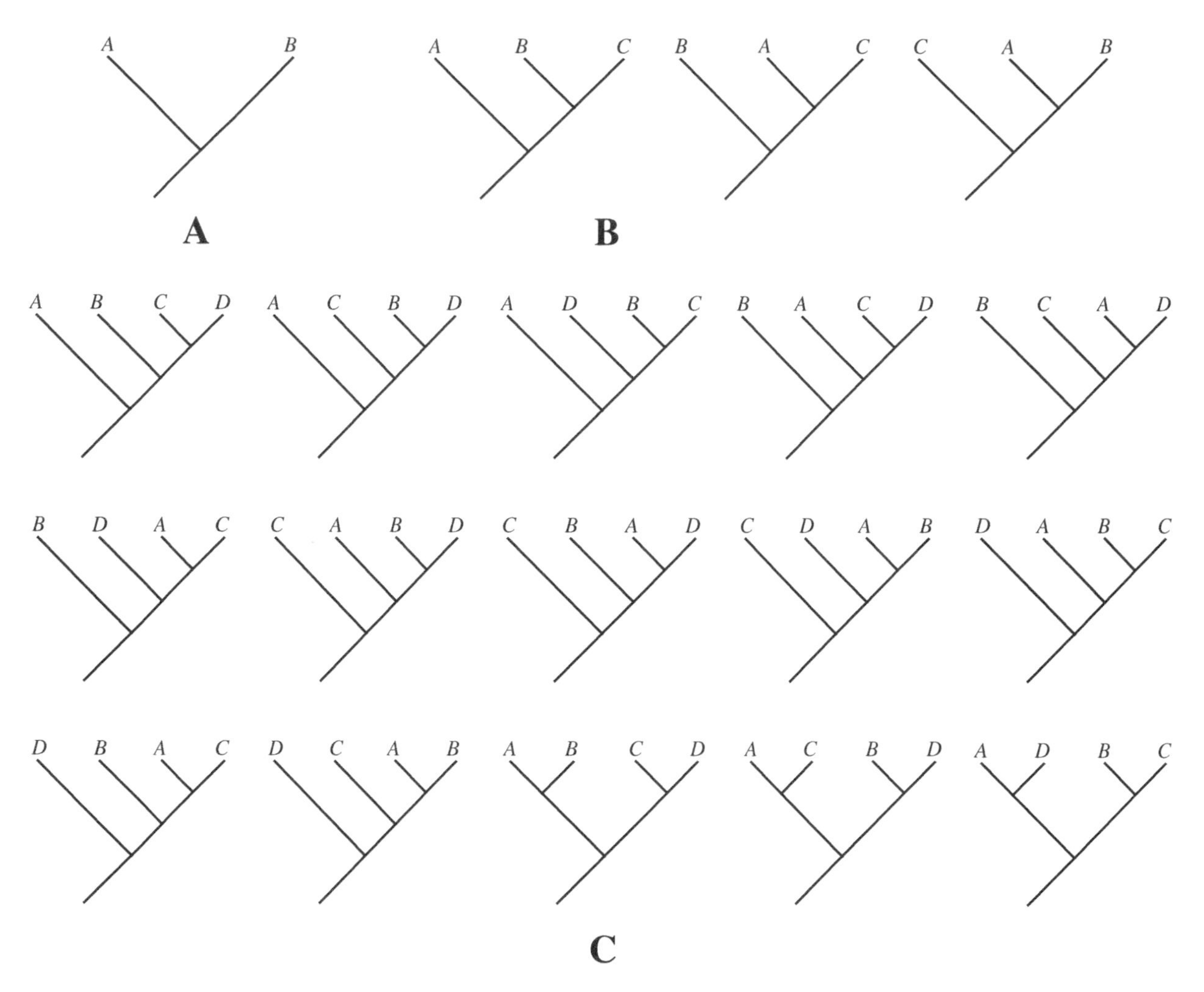

Figure 2.12 All possible dichotomously branched cladograms for a group consisting of the following. **A.** Two taxa (*A* and *B*). **B.** Three taxa (*A*, *B*, and *C*). **C.** Four taxa (*A*, *B*, *C*, and *D*).

Various computer programs (algorithms) are used to determine the most parsimonious cladogram from a given character x taxon matrix. (See **Phylogeny Computer Programs** at the end of this chapter.) These use algorithms that obviate the need to calculate the length of every possible tree, as this would quickly become prohibitively time intensive.

OUTGROUP COMPARISON

As mentioned earlier in the discussion on character analysis, knowledge of character polarity is necessary to recognize shared derived character states that define monophyletic taxa. The primary criterion for ascertaining polarity is outgroup comparison. As discussed earlier, an **outgroup** is a taxon that is not a member of the study group under investigation (the **ingroup**). Outgroup comparison entails character assessment of the *closest* outgroups to the ingroup. Those character states possessed by the closest outgroups are considered to be ancestral; character states present in the ingroup, but not occurring in the closest outgroups, are derived.

The rationale for outgroup comparison is founded in the principle of parsimony (see below). For example, given some monophyletic ingroup *X* (Figure 2.13A), members of which possess either state 0 or 1 of a character, and given that taxon *Y* (nearest outgroup to *X*) possesses only character state 1, then the most parsimonious solution (requiring a single character change: $1 \rightarrow 0$) is that state 1 is ancestral and present in the common ancestor *M* (the "outgroup node"); character state 0 is derived within taxon *X* (Figure 2.13A). The alternative, that state 0 is ancestral, requires at least two character state changes (Figure 2.13B). Verification is made by considering an additional outgroup (e.g., taxon *Z* in Figure 2.13C). If this next outgroup possesses only character state 1, then the ancestral status of state 1 for taxon *Y* is substantiated (Figure 2.13C). If, however, outgroup *Z* contains only character state 0, then it is equally parsimonious to assume that state 1 is ancestral to ingroup *X* (Figure 2.13D) versus derived within ingroup *X* (Figure 2.13E). In this case, consideration of additional outgroups may resolve polarity. The major problem with out-

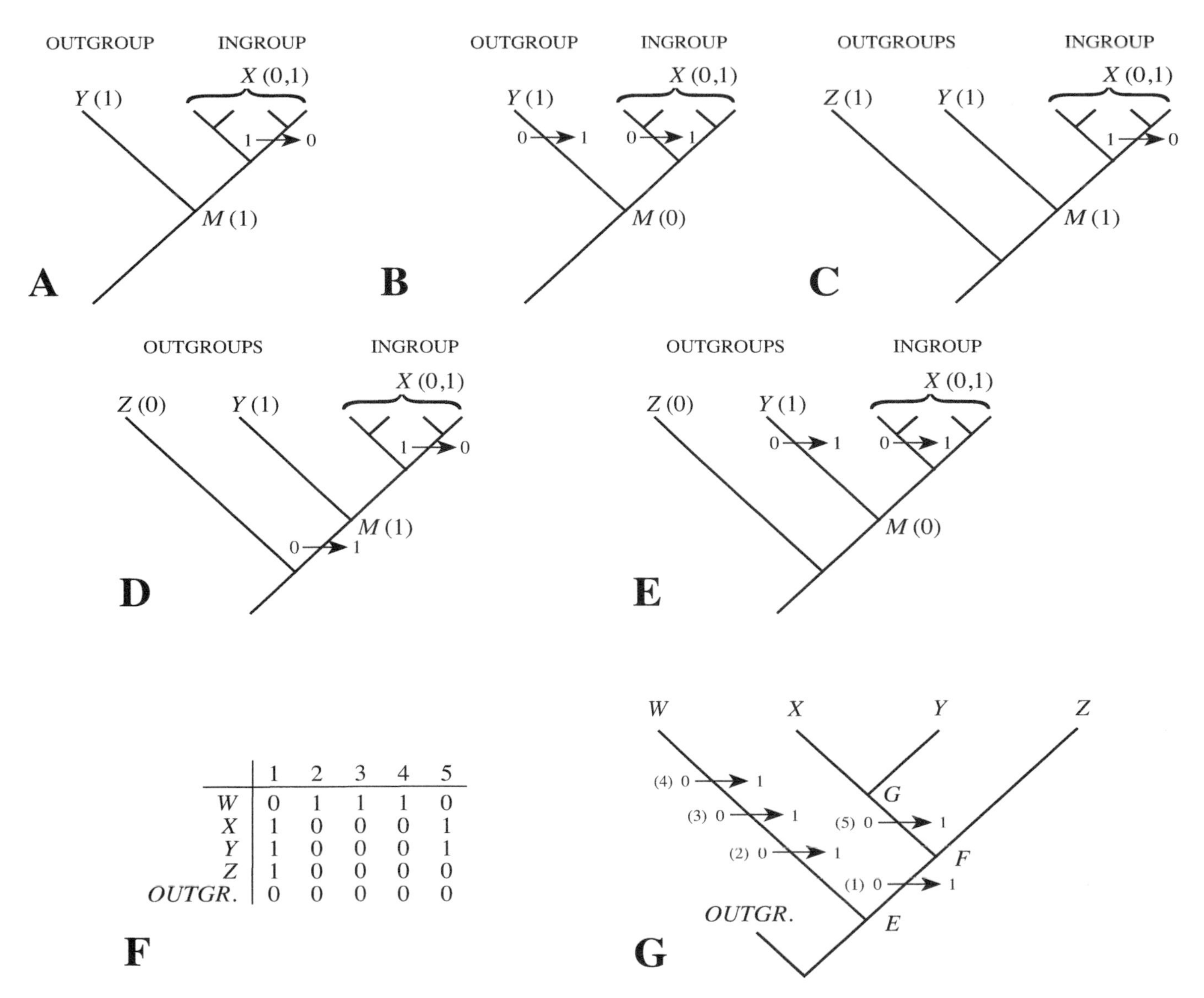

	1	2	3	4	5
W	0	1	1	1	0
X	1	0	0	0	1
Y	1	0	0	0	1
Z	1	0	0	0	0
OUTGR.	0	0	0	0	0

Figure 2.13 Determination of character state polarity using outgroup comparison. **A.** Most parsimonious assumption in which character state 1 is ancestral and present in ancestor *M*. **B.** Alternative, less parsimonious cladogram, in which state 0 is assumed to be ancestral. **C.** Verification of cladogram at A by addition of next outgroup *Z*, which also has state 1. **D,E.** Cladograms in which additional outgroup *Z* has state 0, showing that assumption of polarity is equivocal; ancestor *M* is equally likely to possess state 0 as opposed to state 1. **F.** Character × taxon matrix for taxa *W–Z* plus *OUTGROUP*. **G.** Most parsimonious cladogram of taxa *W*, *X*, *Y*, and *Z* and ancestors *E*, *F*, and *G*.

group comparison is that the cladistic relationships of outgroup taxa may be unknown; in such a case, all reasonably close outgroups (in all possible combinations) may be tested. In practice, prior studies at a higher taxonomic level are often used to establish near outgroups for a phylogenetic analysis.

ANCESTRAL VERSUS DERIVED CHARACTERS

A point of confusion may occur in the use of the terms *ancestral* (plesiomorphic or primitive) and *derived* (apomorphic or advanced). It is advisable that these terms be limited to the description of characters (not taxa) and then only relative to monophyletic groups. For example, in the cladogram of Figure 2.13G (constructed from the matrix of Figure 2.13F), state 1 of character 1 is derived *within* the group including *W*, *X*, *Y*, and *Z* (i.e., state 1 is absent in common ancestor *E*), but it is ancestral with regard to the monophyletic group *X*, *Y*, *Z* (i.e., state 1 is present in *F*, the common ancestor of *X*, *Y*, and *Z*). The use of the terms ancestral and derived to describe *taxa* should be avoided to prevent ambiguity. For example, from Figure 2.13G, it might be asked which taxon is most "primitive"? Confusion is avoided by describing, e.g., taxon *W* as sister to the *X–Y–Z* clade and, e.g., taxon *Z* as possessing the fewest number of observed apomorphic states, relative to a common ancestor (ancestral taxon *E*).

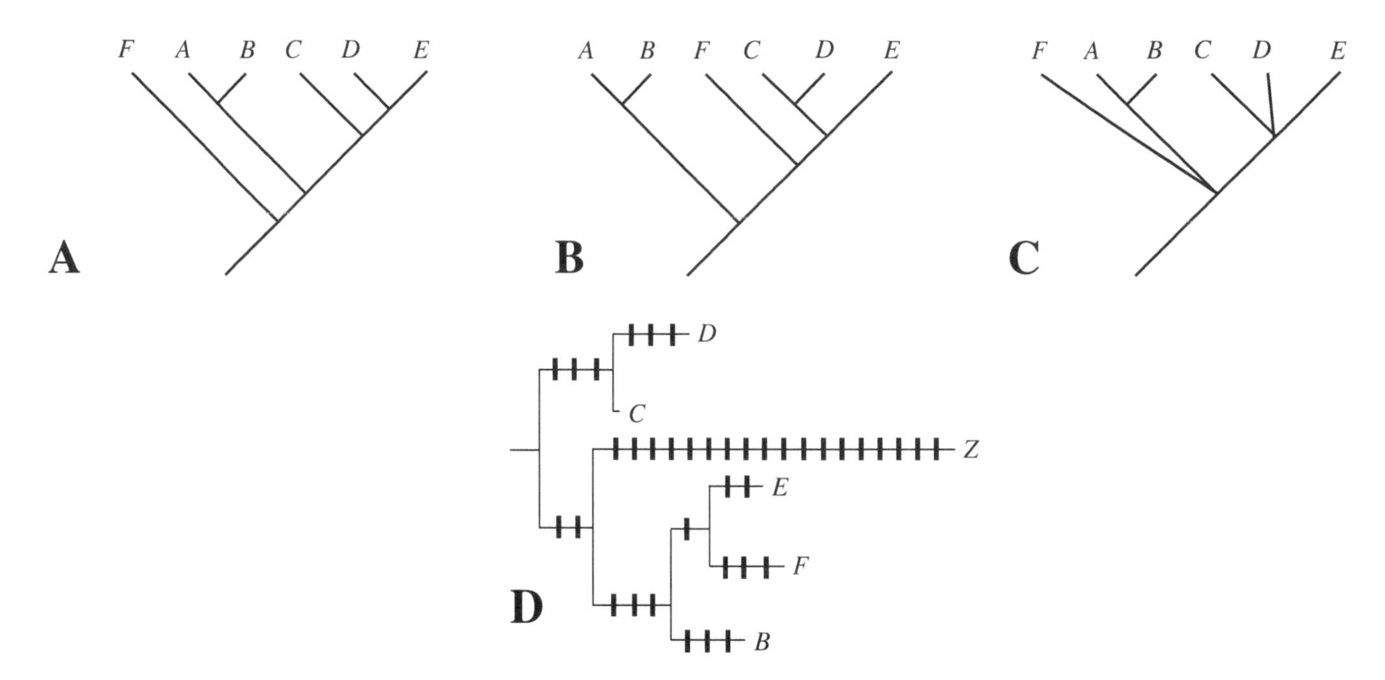

Figure 2.14 **A,B.** Two equally most parsimonious cladograms resulting from cladistic analysis. **C.** Strict consensus tree of cladograms at A and B. **D.** Cladogram illustrating taxon *Z* with very long branch.

CONSENSUS TREES

In practice, most parsimony analyses yield numerous cladograms that are equally most parsimonious. Rather than view and discuss each of these cladograms, it is usually convenient to visualize one tree that is compatible with all equally most parsimonious trees. A **consensus tree** is a cladogram derived by combining the features in common between two or more cladograms. There are several types of consensus trees. One of the most commonly portrayed is the **strict consensus tree**, which collapses differences in branching pattern between two or more cladograms to a polytomy. Thus, the two equally parsimonious cladograms of Figure 2.14A,B are collapsible to the strict consensus tree of Figure 2.14C. Another type of consensus tree is the **50% majority consensus tree**, in which only those clades that occur in 50% or more of a given set of trees are retained. Consensus trees may be valuable for assessing those clades that are robust, i.e., have strong support (see **Cladogram Robustness**). Greater confidence may be given to such clades in terms of recognition of accepted and named monophyletic groupings.

LONG BRANCH ATTRACTION

Sometimes, e.g., with molecular sequence data, one or more taxa will have a very long branch, meaning that these taxa have a large number of autapomorphies relative to other taxa in the analysis (e.g., taxon *Z* of Figure 2.14D). This can be caused by unequal rates of evolution among the taxa examined or can be the by-product of the particular data used. Such a situation can result in "long branch attraction," in which taxa with relatively long branches tend to come out as close relatives of one another (or, if only one taxon has a long branch, its phylogenetic placement may easily shift from one analysis to another). Long branch attraction occurs because when relatively numerous state changes occur along lineages, random changes can begin to outweigh nonrandom, phylogenetically informative ones. The phylogenetic placement of a taxon with a long branch can be uncertain and can unduly influence the placement of other taxa.

Maximum parsimony methods are particularly susceptible to long branch attraction. Alternatively, taxa with long branches may be analyzed using a different data set, or even omitted from an analysis to see what the effect is on cladogram robustness (see later discussion).

MAXIMUM LIKELIHOOD

The principle of parsimony can be viewed as evaluating all alternative trees (or as many subsets as feasible), calculating the length of those trees, and selecting those trees that are shortest, i.e., require the minimum number of character state changes under the set of conditions (character coding) specified. Another method of phylogenetic inference is termed **maximum likelihood** (abbreviated as **ML**). Maximum likelihood, like parsimony methods, also evaluates alternative trees (hypotheses of relationship), but considers the *probability*, based on some selected *model of evolution*, that each tree explains the data. That tree which has the highest probability of explaining the data is preferred over trees having a lower probability. The appropriate model of evolution used is typi-

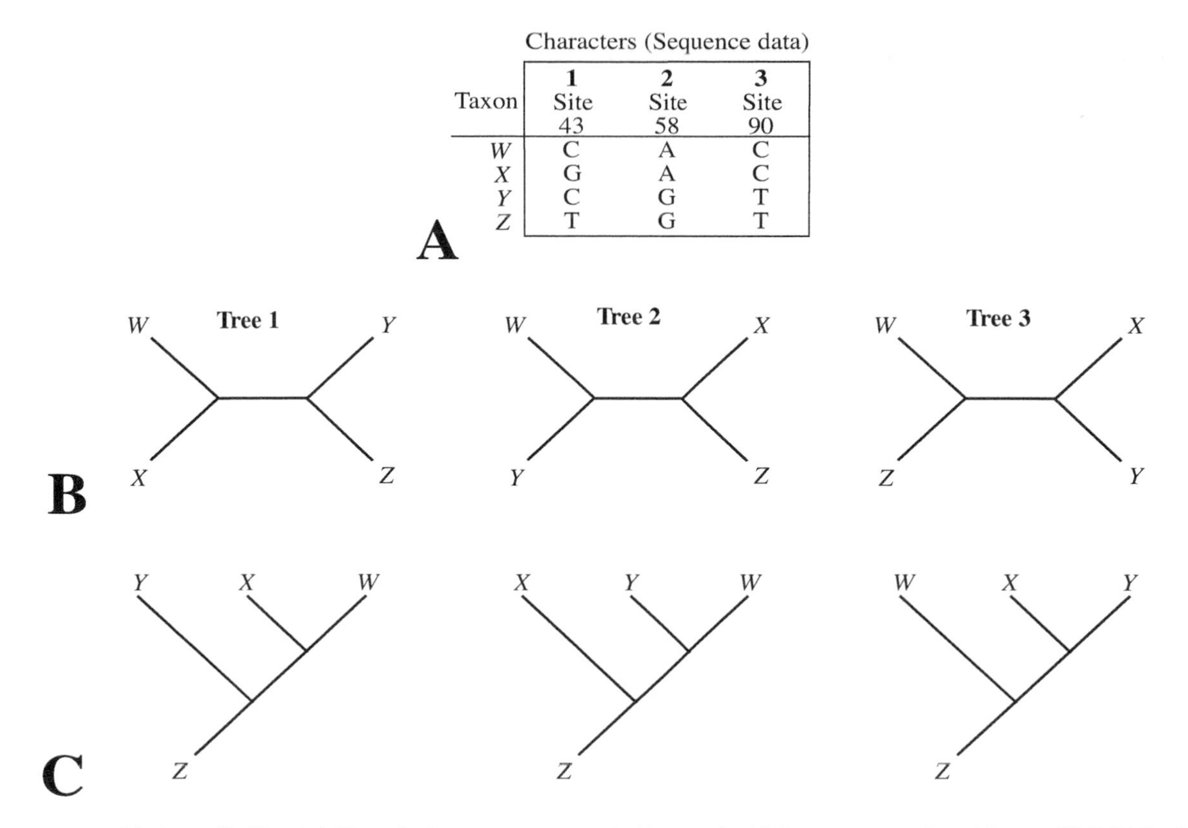

Figure 2.15 Maximum likelihood. **A.** Example character × taxon matrix (three nucleotide base sequences shown) for taxa *W–Z*. **B.** The three possible unrooted trees for taxa *W–Z*. **C.** Three possible rooted trees for taxa, with *Z* set as the root.

cally based on the data of the current analysis, but may be based on other data sets.

Maximum likelihood is used in practice for molecular sequence data, although morphological data or a combination of the two can be used. Figure 2.15A shows a simple molecular data set of three characters (three nucleotide sites of some gene or gene region). In this example, there are three possible trees, shown as unrooted in Figure 2.15B and rooted at taxon *Z* (assuming this is the outgroup) in Figure 2.15C. Maximum likelihood evaluates each tree and calculates, for each character, the total *probability* that each node of the tree possesses a given nucleotide. (See Chapter 14 for information on molecular data.)

In this same example, Figure 2.16A shows the actual probability of a change from one base to another, i.e., via nucleotide substitutions. Thus, a change from **A** (adenine) to **C** (cytosine) has a probability of 0.1 (10%), that from an **A** to **G** has a probability of 0.2 (20%), etc. Consider the first unrooted tree of Figure 2.13B for the first character (base 43) of Figure 2.15A. Nucleotide bases (A, C, G, or T) are substituted for taxon names (W–Z), and each of the two internal nodes (ultimately corresponding to hypothetical common ancestors) is arbitrarily assumed to possess **A** (adenine), shown in Figure 2.16B. The overall probability for this nucleotide combination on this particular tree is the starting probability of any particular nucleotide (0.25 in this example, with the assumption that nucleotide bases are in equal frequency, each being 25% given there are 4 bases) x the probability of going from an **A** to a **C** (=0.1) x the probability of going from an **A** to a **G** (=0.2) x the probability of going from an **A** to an **A** (0.60), and so on; the total probability for this tree topology and base combination is P_1=0.00003 (Figure 2.16B). Now, the total probabilities for all 16 possible combinations of nucleotide bases at the internal nodes are seen in Figure 2.16C (P_1 ... P_{16}). The *likelihood* score for this tree and site (character 1) is calculated by adding all of these individual probabilities ($P_1 + P_{2+} \dots P_{16}$) = 0.0026 in this example (Figure 2.16C).

Likelihood scores for each of the other two characters (sites 58 and 90 in Figure 2.17A) of this first tree topology are then calculated (Figure 2.16D,E). The total likelihood for this tree topology is obtained by multiplying the likelihood scores of each of the three characters. (Normally, the negative natural log of each probability is calculated and these added together, due to the often small numbers obtained for L.) In the example of Figure 2.16, L= 0.0026 x 0.01132 x .01132 = 3.332 x 10^{-7}, or −LnL = −(Ln 0.0026 + Ln 0.1132 + Ln 0.1132) = 14.915. It turns out that the first tree represented [(W,X)(Y,Z)] has the highest likelihood (L) of the three possible unrooted trees and would be accepted over the other two.

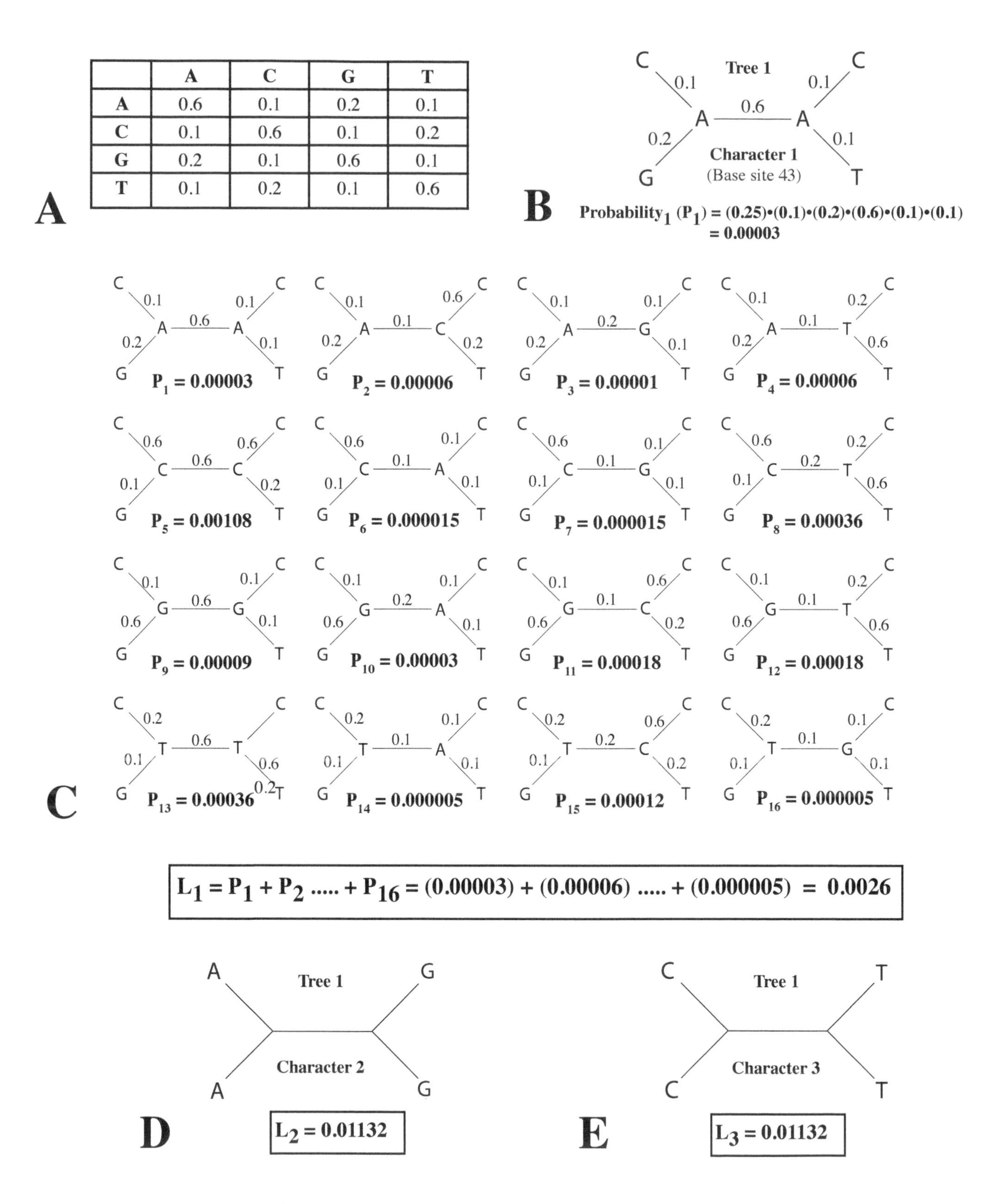

	A	C	G	T
A	0.6	0.1	0.2	0.1
C	0.1	0.6	0.1	0.2
G	0.2	0.1	0.6	0.1
T	0.1	0.2	0.1	0.6

Figure 2.16 Maximum likelihood. **A.** Matrix showing the probabilities of a change from one base to another. **B.** Tree 1 (one of three possible unrooted trees from Figure 2.17B), with nucleotide bases of character 1 substituted for terminal taxa and with internal nodes arbitrarily set to A (adenine). Note calculation of probability of site 1 (P_1). **C.** The sixteen possibilities for internal node bases for tree 1, with probabilities calculated. **D,E.** Representation of nucleotides at terminal taxa in tree 1 for characters 2 (D) and 3 (E), with likelihood calculations below.

A

	A	C	G	T
A	$-\mu(a\pi_C + b\pi_G + c\pi_T)$	$\mu a\pi_C$	$\mu b\pi_G$	$\mu c\pi_T$
C	$\mu a\pi_A$	$-\mu(a\pi_A + d\pi_G + e\pi_T)$	$\mu d\pi_G$	$\mu e\pi_T$
G	$\mu b\pi_A$	$\mu d\pi_C$	$-\mu(b\pi_A + d\pi_C + f\pi_T)$	$\mu f\pi_T$
T	$\mu c\pi_A$	$\mu e\pi_C$	$\mu f\pi_G$	$-\mu(c\pi_A + d\pi_C + e\pi_G)$

B

	A	C	G	T
A	$-3/4\mu$	$1/4\mu$	$1/4\mu$	$1/4\mu$
C	$1/4\mu$	$-3/4\mu$	$1/4\mu$	$1/4\mu$
G	$1/4\mu$	$1/4\mu$	$-3/4\mu$	$1/4\mu$
T	$1/4\mu$	$1/4\mu$	$1/4\mu$	$-3/4\mu$

C

	A	C	G	T
A	$-1/4\mu(\kappa+2)$	$1/4\mu$	$1/4\mu\kappa$	$1/4\mu$
C	$1/4\mu$	$-1/4\mu(\kappa+2)$	$1/4\mu$	$1/4\mu\kappa$
G	$1/4\mu\kappa$	$1/4\mu$	$-1/4\mu(\kappa+2)$	$1/4\mu$
T	$1/4\mu$	$1/4\mu\kappa$	$1/4\mu$	$-1/4\mu(\kappa+2)$

Figure 2.17 Models of base substitution. **A.** General time reversible model, in which probabilities of change from one base to another are a function of mean instantaneous base substitution rate (μ), relative rate parameters (a, b, c, d, e, f), and base frequencies (π_A, π_C, π_G, π_T). **B.** Jukes-Cantor (**JC**) model, in which substitution rates are the same. **C.** Kimura's two-parameter model (**K2P**), in which base frequencies are the same but transitions (in red) and transversions (in blue) occur at different rates.

Maximum likelihood uses a DNA substitution **model** to determine the probabilities of going from one nucleotide to another. These models and algorithms are complicated, but the very basics are important to grasp. One that is commonly used and serves as the basis for other specific models is the **general time-reversible** model (**GTR**), in which a change from one base to another (e.g., A to C) is equivalent to the reverse (e.g., C to A). As seen in Figure 2.17, the GTR model is based on substitution probabilities that are influenced by the *rate parameter* (e.g., μa), which is the product of the mean instantaneous substitution rate (μ) and the relative rate parameters (*a*, *b*, ...*f*), those for each substitution type (e.g., A to C) and the *frequency parameters* (π_A, π_C, π_G, and π_T), which are the frequencies of the nucleotide bases A, C, G, and T. (Other assumptions are made in this model; see Hillis et al., 1996.) Specific models may be derived from the GTR model. For example, if the frequency parameters are equivalent ($\pi_A = \pi_C = \pi_G = \pi_T = 0.25$, given there are 4 bases), and if all substitutions occur at the same rate (a = b = c = d = e = f = 1), then the Jukes-Cantor (JC) model is obtained (Figure 2.17B). If the frequency parameters are equal, but substitutions occur at different rates, such that all transition rates (A→G and C→T) are equivalent but potentially different from all transversions (A→C, A→T, G→C, and G→T), then Kimura's two parameter model (K2P) is obtained (Figure 2.17C). Other models might take into account, e.g., the codon position of a base. The model that is used in a maximum likelihood analysis is calculated from the actual sequence data using a computer algorithm. For example, if significantly more transitions occur than transversions, the K2P model might be selected. (See WEBSITES, Phylogeny Programs for listing of phylogeny computer programs, including those determining the model from a data set.)

Maximum likelihood methods have an advantage over parsimony in that the estimation of the pattern of evolutionary history can take into account probabilities of character state changes from a precise evolutionary model, one that is based and evaluated from the data at hand. Maximum likelihood methods also help eliminate the problem of long branch attraction (discussed later), as the probabilities of base change from one node to another are influenced by the length of that branch. (Generally, as the length of a branch increases, the probabilities of state changes along that branch decrease.) A disadvantage of Maximum Likelihood has been that both the analyses and calculation of confidence measures (usually bootstrap calculations; see **Cladogram Robustness**, below) have been very "computer-intensive" and generally not feasible for large data sets; however, new computer programs have dramatically increased the calculation speeds. (See **Phylogeny Computer Programs**.)

BAYESIAN INFERENCE

Another widely used method for phylogenetic analysis is **Bayesian inference** (often abbreviated **BI**; see the references at the end of this chapter for a detailed understanding). Bayesian inference is similar to maximum likelihood (ML) in using a model of evolution based on nucleotide substitution, discussed earlier. However, Bayesian inference differs in making calculations of **posterior probability**, these based upon prior knowledge, utilizing a probability formula devised by T. Bayes in 1763. This is done by incorporating a prior probability distribution of trees in calculations.

Bayesian inference calculates the posterior probability of the phylogeny, branch lengths, and various parameters of the data. As with maximum parsimony, it is computationally infeasible to evaluate the posterior probability of all possible trees in a Bayesian analysis. In practice, the posterior probabilities of phylogenies are approximated by sampling trees from the distribution, using algorithms such as Markov chain Monte Carlo (MCMC). These algorithms have made it possible to perform Bayesian inference of phylogeny.

The results of a Bayesian analysis yield the posterior probabilities for each of the branches of a given tree, which is typically summarized on a 50% majority consensus tree of sampled trees. Bayesian algorithms are relatively rapid, and the posterior probabilities that are generated for each clade are direct measures of robustness. Generally, a Bayesian posterior probability of 95% or greater is considered robust for a particular clade. However, studies have claimed that Bayesian posterior probabilities may be unrealistically inflated (see **Cladogram Robustness**).

COALESCENT METHODS

A third, general class of algorithms for reconstruction phylogenies, are broadly known as **coalescent methods**. These methods make a distinction between gene trees, trees of one or more gene sequences of a group of taxa, and species trees, inference of the evolutionary history of species and populations. Three phenomena are commonly cited as the cause of discordance between gene trees and species trees: gene duplication, horizontal gene transfer, and deep coalescence (see Maddison 1997; Chapter 19), the latter of most concern in coalescent methods.

The aforementioned methods of Maximum Parsimony, Maximum Likelihood, and Bayesian analyses utilize genetic sequence data, from one or typically numerous genes, but are basically creating single gene trees. If sequence data are **concatenated**, whereby sequences from two or more genes (e.g., chloroplast, mitochondrial, cistron, or nuclear coding genes; see Chapter 14) are combined into one data set, these methods will then yield a **super-gene tree**. Although each gene sequence may use a different model of evolution, gene trees, whether concatenated or not, may or may not resemble the actual species tree, the latter the objective of phylogenetic systematics.

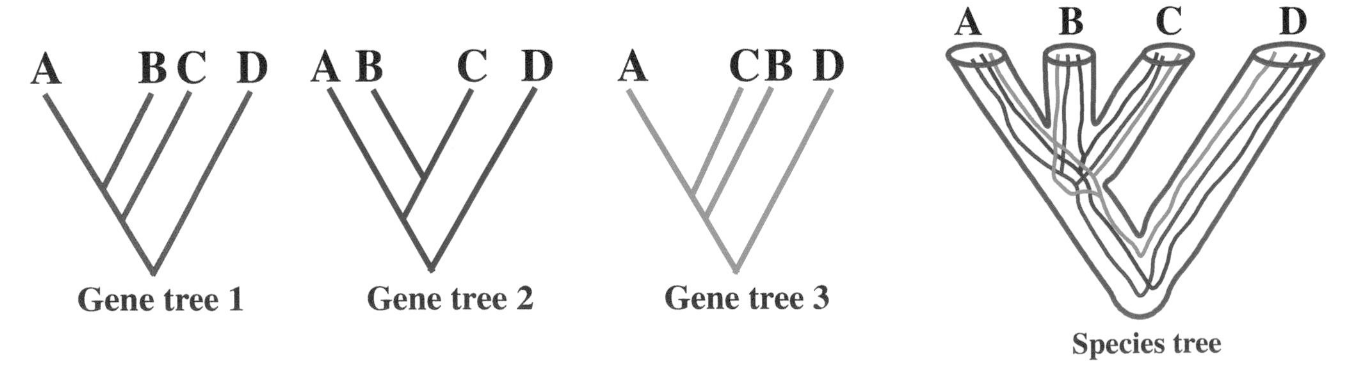

Figure 2.18 Simple illustration of three separate gene trees (left) used to derive the species tree (right). Coalescent methods attempt to derive the most likely species tree from separate gene tree data. Redrawn from Edwards (2009).

Coalescence can be defined as the merging of lineages of a single gene locus into a common ancestor, as viewed backwards in time. Deep coalescence occurs when two gene lineages do not coalesce before a given speciation event. The larger the population size, and the shorter the phylogenetic branches, the less is the chance the two gene lineages will coalesce before a speciation event. This may confound phylogenetic inference, as the common occurrence of what is interpreted as a derived genetic feature may not represent an apomorphy for that clade.

Coalescent methods attempt to infer the most likely species tree, given separate gene trees (conceptually illustrated in Figure 2.18). A given hypothetical species tree is used to generate gene trees, using the available sequence data and a model of evolution. Some algorithms (e.g., *BEAST) incorporate uncertainty in the gene tree estimates but are computationally intensive. Others (e.g., ASTRAL) use a summary method and are computationally feasible but cannot incorporate uncertainty in the gene tree estimates, making species tree inference less certain. Network analyses (see Chapter 19) also use coalescent methods.

MEASURES OF HOMOPLASY

If significant homoplasy occurs in a cladistic analysis, the data might be viewed as less than reliable for reconstructing phylogeny. One measure of the relative amount of homoplasy in the cladogram is the consistency index. **Consistency index (CI)** is equal to the ratio m/s, where m is the *minimum* number of character state changes that must occur and s is the *actual* number of changes that occur. The minimum number of changes is that needed to account for a single transformation between all character states of all characters. For example, a three-state character transformation, $0 \rightarrow 1 \rightarrow 2$, requires a minimum of two steps; e.g., one possibility (of several) is the change $0 \rightarrow 1$ (first step) and then $1 \rightarrow 2$ (second step).

A consistency index close to 1 indicates little to no homoplasy; a CI close to 0 is indicative of considerable homoplasy. As an example, the character x taxon matrix of Figure 2.6A,B necessitates a minimum of seven changes; i.e., there must be at least seven character state transformations to explain the distribution of states in the taxa. The actual number of changes in the most parsimonious cladogram is eight because of homoplasy (Figure 2.6E). Thus, the CI for this cladogram is 7/8 = 0.875. The consistency index may be viewed as a gauge of confidence in the data in reconstructing phylogenetic relationships.

A consistency index may be calculated for individual characters as well. For example, relative to the most parsimonious cladogram of Figure 2.6E, the CI of all characters is equal to 1, except for character 6, which has a CI of 0.5 (because of two convergent character state changes).

Two other measures of homoplasy may be calculated: the **retention index (RI)** and the **rescaled consistency index (RC)**. The retention index is calculated as the ratio $(g - s)/(g - m)$, where g is the maximum possible number of state changes that could occur on any conceivable tree. Thus, the retention index is influenced by the number of taxa in the study. The rescaled consistency index (RC) is equal to the product of the CI and RI.

These various measures of homoplasy are still used in analyses today but do not have the importance they used to.

CLADOGRAM ROBUSTNESS

It is very important to calculcate metrics of **robustness**, the confidence for which a tree or particular clade actually denotes true phylogenetic relationships. A common way to evaluate cladogram robustness is the **bootstrap**, which can be used in both parsimony and maximum likelihood inference methods. **Bootstrapping** is a method that reanalyzes the data of the original character x taxon matrix by selecting (resampling) characters *at random* such that a given character can be selected more than once. The effect of this resampling is that some characters are given greater weight than others, but the total number of characters used is the same as that of the original matrix. (See example at Figure 2.19A,B.) This resampled data is then used to construct new trees. Many sequential bootstrapping analyses are generated (often 100–1,000 runs), and all most parsimonious or greatest likelihood trees are determined. From all of these trees, a 50% majority consensus tree is constructed; the values placed along each internode of the cladogram represent the percentage, from the bootstrap runs, that a particular clade is maintained from all resampled runs (e.g., Figure 2.19D). A bootstrap value of 70%–90% or more is often considered a robustly supported node. The rationale for bootstrapping is that differential weighting by resampling of the original data will tend to produce the same clades if the data are "good," i.e., reflect the actual phylogeny and exhibit little homoplasy. One problem with the bootstrapping method is that it technically requires a random distribution of the data, with no character correlation. These criteria are almost never verified in a cladistic analysis. However, bootstrapping is still the most used method to evaluate tree robustness.

Another method of measuring cladogram robustness occasionally used is the so-called **jackknife** (or **jackknifing**), which is similar to the bootstrap but differs in that each randomly selected character may be resampled only once (not multiple times), and the resultant resampled data matrix is smaller than the original. (See example at Figure 2.19A,C.) The resampled matrix is used to generate a tree or trees. This

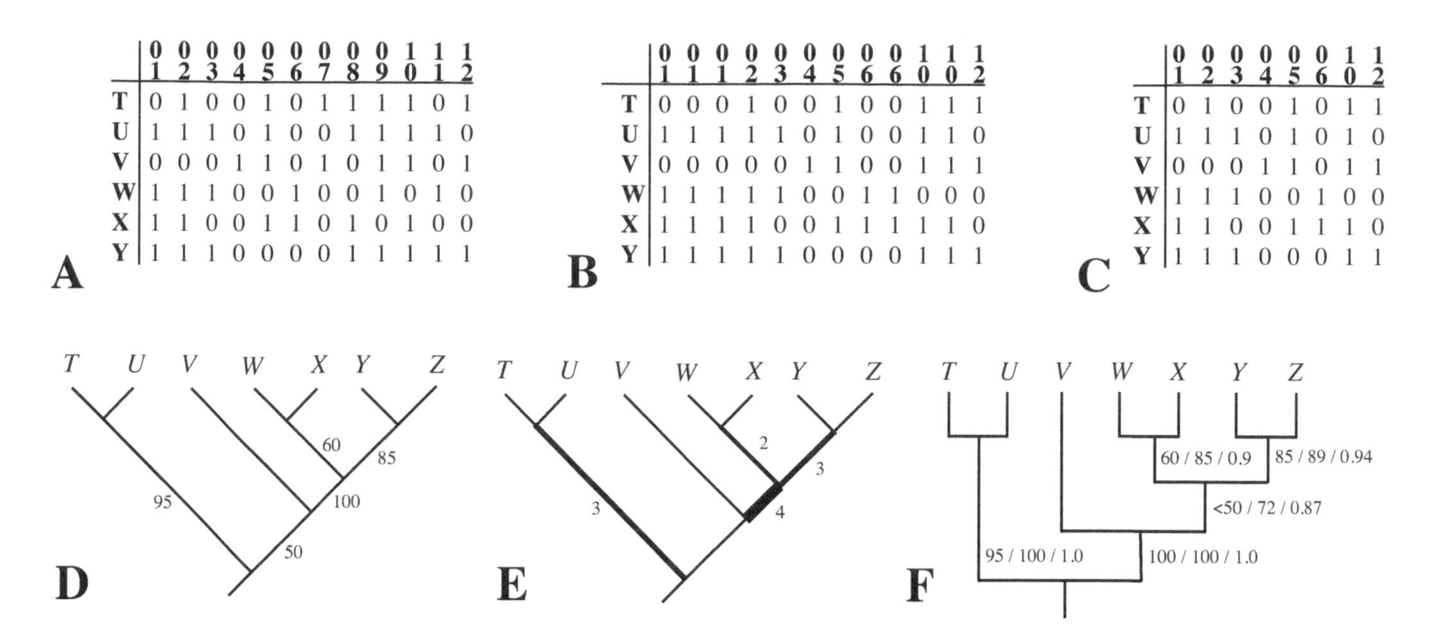

A

	01	02	03	04	05	06	07	08	09	10	11	12
T	0	1	0	0	1	0	1	1	1	1	0	1
U	1	1	1	0	1	0	0	1	1	1	1	0
V	0	0	0	1	1	0	1	0	1	1	0	1
W	1	1	1	0	0	1	0	0	1	0	1	0
X	1	1	0	0	1	1	0	1	0	1	0	0
Y	1	1	1	0	0	0	0	1	1	1	1	1

B

	01	01	01	02	03	04	05	06	06	10	10	12
T	0	0	0	1	0	0	1	0	0	1	1	1
U	1	1	1	1	1	0	1	0	0	1	1	0
V	0	0	0	0	0	1	1	0	0	1	1	1
W	1	1	1	1	1	0	0	1	1	0	0	0
X	1	1	1	1	0	0	1	1	1	1	1	0
Y	1	1	1	1	1	0	0	0	0	1	1	1

C

	01	02	03	04	05	06	10	12
T	0	1	0	0	1	0	1	1
U	1	1	1	0	1	0	1	0
V	0	0	0	1	1	0	1	1
W	1	1	1	0	0	1	0	0
X	1	1	0	0	1	1	1	0
Y	1	1	1	0	0	0	1	1

Figure 2.19 CLADOGRAM ROBUSTNESS. **A.** Data matrix of six taxa (T–Y) and 12 characters. **B.** Resampling of matrix, to be used in a bootstrap analysis. Note that the number of characters is the same and that some characters are repeated, some deleted. **C.** Resampling of matrix, to be used in a jackknife analysis. Note that no characters are repeated and the number of characters has been reduced. **D.** Cladogram showing parsimony bootstrap values at internodes with values ≥ 50. **E.** Same cladogram as in D showing decay index values. (Increasing numbers correspond to increasing line thickness; internodes not numbered have a decay index of 1.) **F.** Same cladogram showing (left to right) parsimony bootstrap values, maximum likelihood bootstrap values, and Bayesian posterior probabilities.

is repeated multiple times and, like the bootstrap, a 50% majority tree is created to generate jackknife values. Jackknifes are considered more rigorous in evaluating cladogram robustness, but are not often used.

Yet another way to evaluate clade confidence is by measuring clade "decay." A **decay index** (also called "Bremer support") is a measure of how many extra steps are needed (beyond the number in the most parsimonious cladograms) before the original clade is no longer retained. Thus, if a given cladogram internode has a decay index of 4, then the monophyletic group arising from it is maintained even in cladograms that are four steps longer than the most parsimonious (e.g., Figure 2.19E). The greater the decay index value, the greater the "confidence" in a given clade. Calculation of decay indices is rarely done today, however.

Bayesian analysis provides a measure of robustness in calculating posterior probabilities for each of the clades generated. Any branch with a posterior probability of 95% or greater is considered statistically well supported. However, this method has come under some scrutiny because it may generates artificially high values of support.

In some analyses, parsimony bootstraps, maximum likelihood bootstraps, and/or Bayesian posterior probabilities may be indicated on the same cladogram, (e.g., a consensus tree), illustrating clade support from the different analyses (see Figure 2.19F).

CLADOGRAM ANALYSIS

Cladograms represent an estimate of the pattern of evolutionary descent, both in terms of recency of common ancestry and in the distribution of derived (apomorphic) character states, which represent unique evolutionary events. Once a robust cladogram is derived, the pattern of relationships and evolutionary change may be used for a variety of purposes.

From one or more cladogram(s) derived from cladistic analyses, it is informative to trace character state changes. In addition, all monophyletic groupings should be evaluated in terms of their overall robustness (e.g., bootstrap support) and the specific apomorphies that link them together. Homoplasies (convergences or reversals) should also be assessed. A homoplasy may represent an error in the initial analysis of that character that may warrant reconsideration of character state definition, intergradation, homology, or polarity. Thus, cladogram construction can be viewed as an interative process, but as a means of pointing out those areas where additional research is needed to resolve satisfactorily the phylogeny of a group of organisms, or indicating biologically interesting aspects of evolution.

PHYLOGENETIC CLASSIFICATION

One of the most important uses of cladograms is to serve as a basis for classification. The pattern of evolutionary history

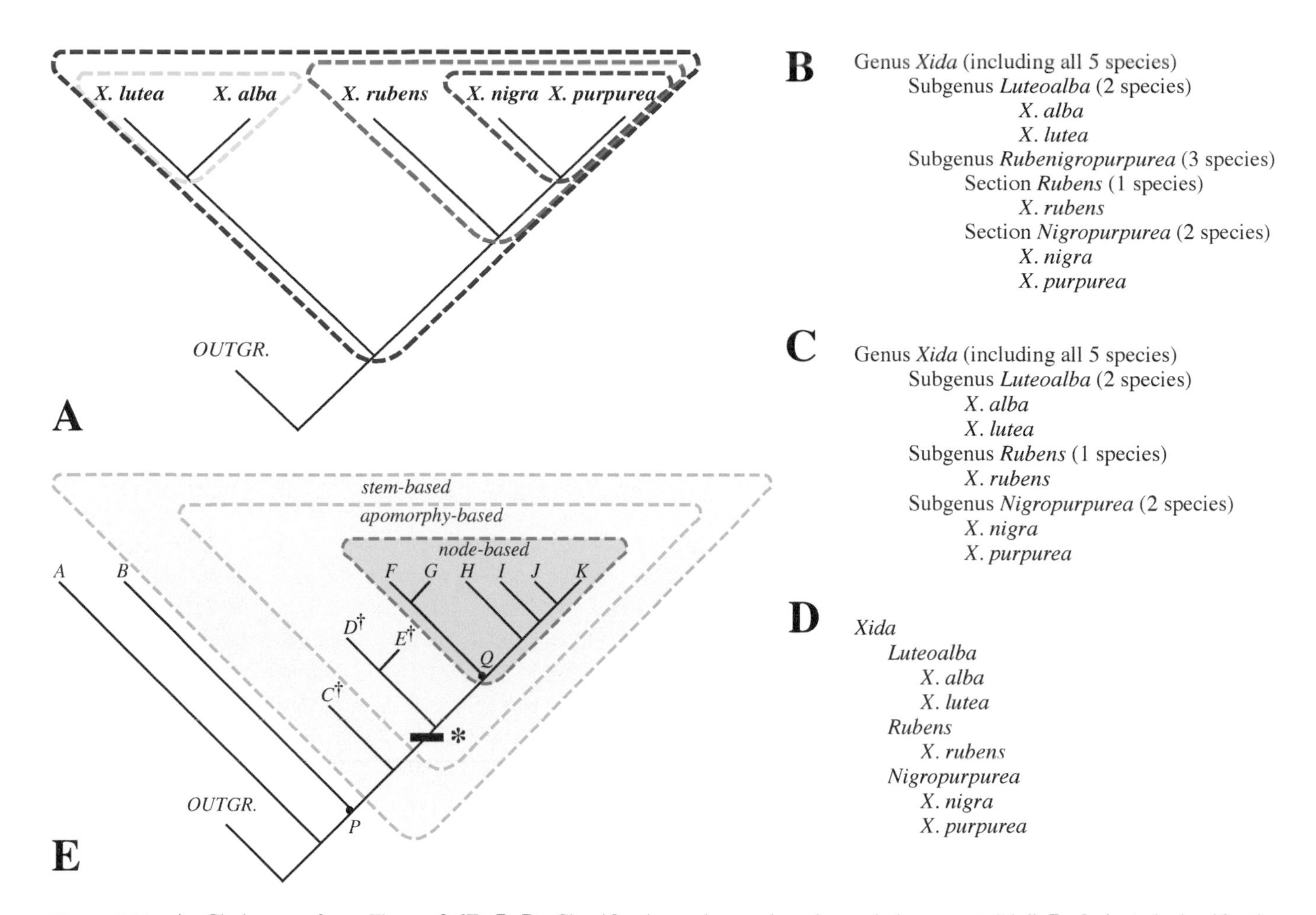

Figure 2.20 **A.** Cladogram from Figure 2.6E. **B–D.** Classification schemes based on cladogram at "A." **B.** Indented classification. **C.** Annotated classification. **D.** Indented, but rankless, classification. **E.** Cladogram illustrating node-based, apomorphy-based, and stem-based classification. †Extinct taxon. *Major evolutionary change, used as the basis for an apomorphy-based group.

portrayed in a cladogram may be used to classify taxa phylogenetically. A phylogenetic classification may be devised by naming and ordering monophyletic groups in a sequential, hierarchical classification, sometimes termed an **indented** method. The hierarchically arranged monophyletic groups may be assigned standard taxonomic ranks. For example, for the most parsimonious cladogram of Figure 2.20A, one possible classification of hypothetical genus *Xida* is seen in Figure 2.20B. Note that in this example, each named taxon corresponds to a monophyletic group (Figure 2.20A) and that these groups are sequentially nested such that the original cladogram may be directly reconstructed from this classification system. Two taxa of the same rank (e.g., sections *Rubens* and *Nigropurpurea*) are automatically sister groups. Each higher taxon above (e.g., subgenus *Luteoalba*) would also include automatically created lower taxa (e.g., species *Xida alba* and *Xida lutea* in this case).

An alternative, and often more practical, means of deriving a classification scheme from a cladogram is by annotation. **Annotation** is the sequential listing of derivative lineages from the base to the apex of the cladogram, each derivative lineage receiving the same hierarchical rank. The sequence of listing of taxa may be used to reconstruct their evolutionary relationships. For example, an annotated classification of the taxa from Figure 2.20A is seen in Figure 2.20C. In this case all named taxa are monophyletic, but taxa at the same rank are not necessarily sister groups. This is common, for example, in the naming of angiosperm orders (see Chapters 7 and 8).

The particular rank at which any given monophyletic group is given is arbitrary and is often done to conserve a past, traditional classification. A recent trend in systematics is to eliminate ranks altogether or, alternatively, to permit unranked names between the major rank names (see Chapter 16). In either case, the taxon names, minus ranks, would still retain their hierarchical, evolutionary relationship (e.g., as in Figure 2.20D).

Perhaps the most common type of phylogenetic classification is sometimes termed **node-based**, because it recognizes a node (common ancestor) of the cladogram and all descendants of that common ancestor as the basis for grouping (Figure 2.19E). A node-based classification may specify a **crown**

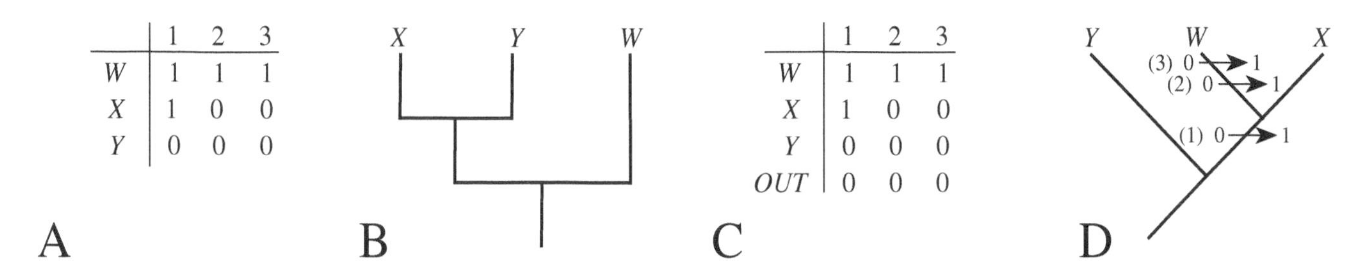

Figure 2.21 **A.** Character × taxon matrix for taxa *W–Y*. **B.** Phenogram of taxa *W*, *X*, and *Y*. **C.** Character × taxon matrix for taxa *W–Y* plus *OUTGROUP*. **D.** Most parsimonious cladogram of taxa *W*, *X*, and *Y*. Note different branching pattern.

clade, one in which both or all branches from the common ancestor contain extant members. In some cases, it may be valuable to recognize a group that is **stem-based**, i.e., one that includes the "stem" (internode) region just above a common ancestor plus all descendants of that stem (Figure 2.20E). A stem-based group may be equivalent to a **total clade**, one that includes a crown clade plus all other taxa that share a recent common ancestor with the crown clade but not with other crown clades. A stem-based classification might be useful, for example, in including both a well-defined and corroborated node-based monophyletic group (crown clade) plus one or more fossil lineages that arise along the stem, the lineage below the crown clade. The paraphyletic stem may contain some, but not all, of the apomorphies possessed by the node-based crown clade. Yet a third general type of phylogenetic classification is **apomorphy-based**, in which all members of a monophyletic group that share a given, unique evolutionary event (illustrated by an "*" in Figure 2.19E) are grouped together and named. (See Cantino et al. 2007 for an explanation of these types of phylogenetic classification.)

Last, it should be mentioned that a monophyletic group can be recognized with a phylogenetic "definition" in a *Phylocode* (see Cantino et al. 2007; Cantino and de Queroz 2010). For example, in Figure 2.20A, the monophyletic *Xida* might be "defined" as the "least inclusive monophyletic group containing the common ancestor of *X. lutea* and *X. nigra*." The rationale is that this presents a more explicit and stable means of classification of taxa. However, any given phylogenetic definition is based on some cladistic analysis. If future cladistic analyses portray a somewhat different relationship of taxa, then the phylogenetically defined groups may contain taxa that were unintended, making them less useful and less stable than more standard classifications.

As mentioned in Chapter 1, a second major type of analysis and classification is **phenetic**, in which taxa are grouped by overall similarity. This phenetic grouping may be represented in the form of a branching diagram known as a **phenogram**. For example, for the data matrix of Figure 2.21A, the resultant phenogram is seen in Figure 2.21B. In this case taxa *X* and *Y* share more similar features (state 0 of characters 2 and 3) than either does with taxon *W*; thus, *X* and *Y* are more similar and are grouped together. (Note that no outgroup is included in the matrix.) Phenetic classifications will often be quite different from phylogenetic ones because in a phenetic analysis, taxa may be grouped together by shared *ancestral* features (known as **symplesiomorphies**) as well as by shared derived character states (synapomorphies). For example, the data matrix of Figure 2.21C (identical to that of 2.21A except for the addition of an outgroup) yields the most parsimonious cladogram at Figure 2.21D, which has a different branching pattern from the phenogram of Figure 2.21B. Note that in the cladogram, taxa *W* and *X* are grouped as sister taxa because they share the derived state of character 1, which is a synapomorphy for *W* and *X*. In contrast, the phenogram of Figure 2.21B groups together taxa *X* and *Y* because they are more similar, having in common state "0" of characters 2 and 3; however, these are shared ancestral states (symplesiomorphies) and cannot be used to recognize monophyletic groups. Because many past classification systems have been based on overall phenetic similarity, great caution should be taken in evaluating "relationship." Taxa that are most similar to one another may not, in fact, be particularly close relatives in a phylogenetic sense (i.e., by recency of common ancestry).

In summary, phylogenetic classification of taxa has the tremendous advantage of being based upon and of reflecting the evolutionary history of the group in question. The *International Code of Nomenclature for Algae, Fungi, and Plants* (Chapter 16) has been used very successfully to assign taxonomic names based on the criterion of monophyly. The *Phylocode* (Cantino and de Queroz 2010) has been useful as a supplement to the former, particularly in naming certain higher-ranked clades. Phylogenetic classifications have resulted in many name changes in certain groups, but these are gradually beginning to stabilize, particularly as additional, robust molecular studies are implemented. In practice, assigning a name to every monophyletic group, whether ranked or not, is unwieldy, impractical, and unnecessary. Generally, only monophyletic groups that are well supported (and ideally that have a well-recognized apomorphy) should be formally named, and every effort should be made to retain (or modify) former classification systems, where possible.

CHARACTER EVOLUTION

Cladograms can be used as an analytical device to evaluate the ancestral conditions at the cladogram nodes and the evolutionary change (apomorphies) occurring from one node to another. This may be done using the character x taxon matrix and a pre-existing tree, one inferred, e.g., by parsimony, maximum likelihood, or Bayesian methods. The character(s) evaluated may or may not have been included in the original tree reconstruction.

A standard way to evaluate character evolution is by **parsimony optimization**. Optimization of characters refers to their representation (or "plotting") on a cladogram in the most parsimonious way such that the minimal number of character state changes occurs between nodes. This method assigns those character states at ancestral nodes that minimize the number of state changes between nodes, i.e., that minimize the tree length. (In the optimization procedure, a given type of character coding is selected, such as ordered, unordered, step matrices, dollo coding.) For example, Figures 2.22A,B show a cladogram in which the evolution of a character is explained in two different ways, but neither of which is the most parsimonious explanation. In Figures 2.22C,D the character is optimized, showing the fewest possible number of state changes. In these last two examples, character state evolution can be optimized in either of the two equally parsimonious ways (with ancestral nodes assigned a different set of states). **Acctran** (accelerated transformation) optimization hypothesizes an earlier initial state change with a later *reversal* of the same character (Figure 2.22C). **Deltran** (delayed transformation) optimization hypothesizes two later, *convergent* state changes (Figure 2.22D). Note that when alternative character optimization exists, there are nodes in the cladogram that are *equivocal*, i.e., for which the character state cannot be definitively determined. Optimization is automatically performed by computer algorithms that trace characters and character states. (See end of chapter.)

Another way of assessing character evolution is using maximum likelihood in **ancestral state reconstruction**, called this because it emphasizes determining the character condition at each ancestral node rather than changes between nodes. For a given tree and character distribution of the terminal taxa, this method calculates the maximum probability of a state at each node, using a selected model of evolution, generally the one used to construct the tree. An example of maximum likelihood ancestral state reconstruction is seen in Figure 2.22E. Note that ancestral nodes often do not have discrete states, but a probability of a given state, between 0 and 100%. As with maximum likelihood tree construction, branch lengths influence the probability of state changes in ancestral state reconstruction. A relatively long branch may introduce a higher or lower probability of a state, which would not be evident in parsimony optimization.

Assessment of character evolution often yields insight into the possible adaptive significance of a feature. For example, Figure 2.22F shows parsimony optimization of chromosome number shifts for a given tree and state distribution. Studies of character evolution may allow detection of the correlation of character shifts, indicative of, say, a genetic or adaptive linkage. It may also give insight into past classification, e.g., as to whether a particular taxon was historically grouped by an apomorphy or plesiomorphy.

BIOGEOGRAPHY, ECOLOGY, DATING METHODS, AND SPECIATION RATES

A phylogenetic analysis can be used to evaluate past changes in biogeographic distribution, assess ecological changes in habitat, evaluate changes in rates of speciation, and estimate dates of clade divergence.

Analysis of biogeographic data can give insight into the direction of change in biogeographic distribution. A change from one distribution to another can occur by either of two general ways: dispersal or vicariance. **Dispersal** is the movement of an organism or propagule from one region to another, such as the transport of a seed or fruit (by wind, water, or bird) from a continent to an island (Figure 2.23A). **Vicariance**, in contrast, is the splitting of one ancestral population into two (or more) populations, e.g., by continental drift or the formation of a new waterway or mountain range, resulting in a barrier between the split populations; this barrier prevents gene flow between these populations, allowing them to diverge independently (Figure 2.23B).

Determining vicariance versus dispersal as an explanation for biogeographic change is often not obvioius, and requires additional evidence, including knowledge of past geologic history. For example, Figure 2.23C illustrates a cladogram of taxa endemic to the Hawaiian archipelago, in which the ranges (by island) are superposed. A simple optimization shows the changes in geographic ranges that would be needed to explain the data. In this case, a shift from the island of Kauai to Maui and one from Maui to the island of Hawaii constitutes the simplest explanation needed to account for the current distribution of taxa. Because geologic data firmly suggests that the Hawaiian islands arose from sequential "hot-spot" volcanic activity and that the major islands were never connected, vicariance as an explanation is ruled out, leaving dispersal as the mechanism for biogeographic change. The hypothetical example of Figure 2.23D shows another cladogram in which both biogeographic distributions are superposed. A likely explanation for change in biogeographic distributions in this example is the splitting of the three continents from an ancestral Gondwana (Figure 2.23D). Although

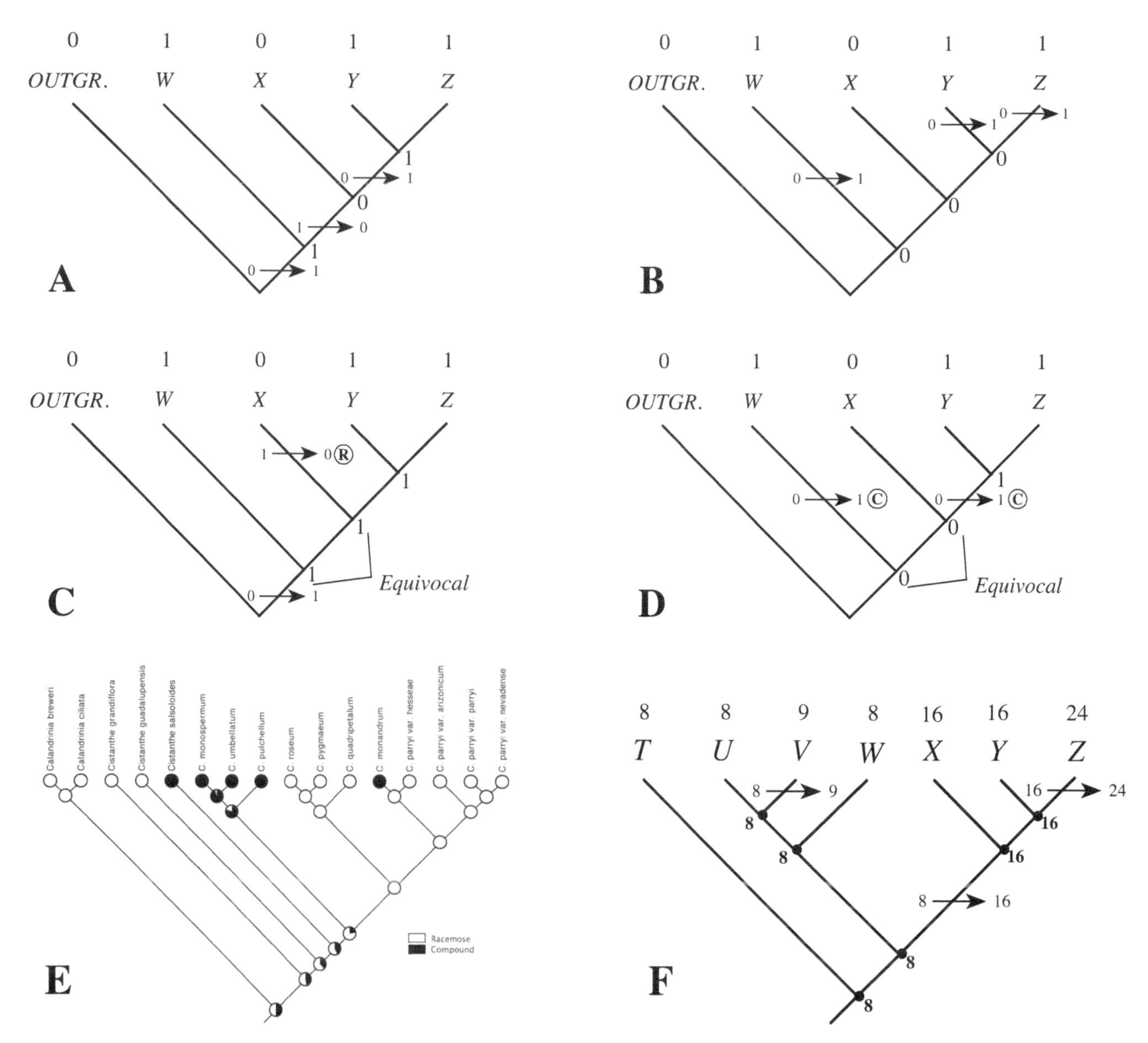

Figure 2.22 **A,B.** Cladograms for taxa *W–Z* and Outgroup, in which character states of a character (superposed above taxa) are accounted for by hypothesizing three state changes, not optimized. **C.** Parsimony optimization ("Acctran") of character, hypothesizing two state changes, including one reversal. **D.** Parsimony optimization ("Deltran") of character, hypothesizing two convergent state changes. **E.** Character evolution assessed by likelihood ancestral state reconstruction. Note that ancestral nodes show a probability between ca. 25% and 80% for a given state. (Example courtesy of M. Guilliams.) **F.** Character evolution using parsimony optimization, illustrated for haploid chromosome number. States at ancestral nodes in bold. Note that parsimony optimization minimizes tree length, requiring a total of three state changes. Note two polyploid events in the clade containing *X*, *Y*, and *Z*, and an aneuploid event for taxon *V*.

dispersal across oceans cannot be ruled out, vicariance might be more likely because the changes in distribution correspond to a hypothesis of continental drift. (Note that the continentally delimited groups need not be monophyletic.)

More sophisticated methods of tracing biogeographic history can estimate the ancestral ranges of taxa, e.g., defined bioregions (Figure 2.24A). These may utilize likelihood estimations for the bioregions of a common ancestor, e.g., following a dispersal event (Figure 2.24B).

Ecological habitat data may be traced from a phylogenetic analysis. A simple example is seen in Figure 2.23E, in which habitat types are superposed on the taxa from a cladistic analysis. Note here the shift from a terrestrial to an aquatic habitat. Analyses such as this may yield insight into the adaptive significance of evolutionary changes in anatomy, morphology, physiology, or genetics relative to differing habitat requirements. Note that ecological data in the simple sense of the habitat a taxon occupies, such as "desert" or "salt marsh,"

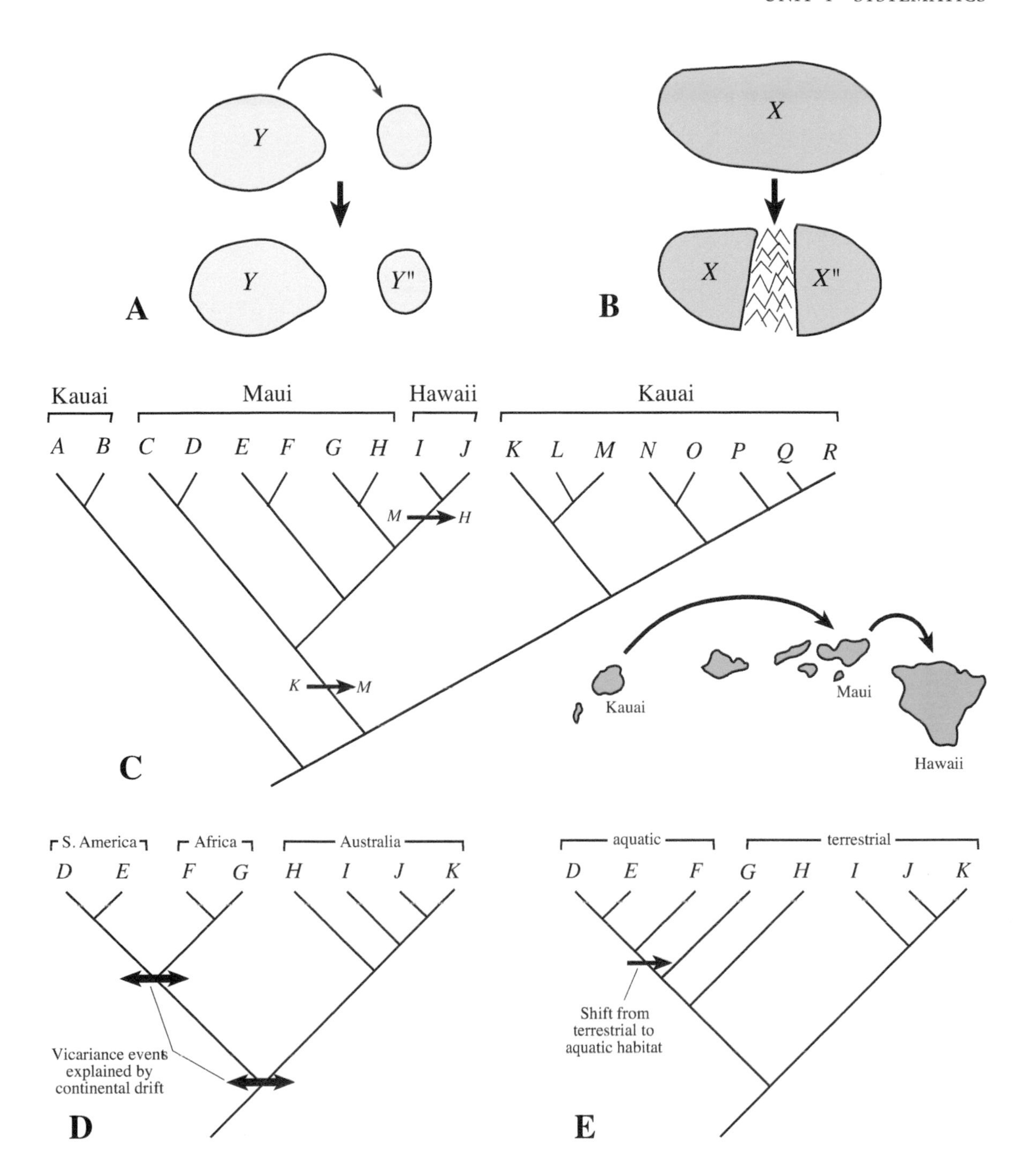

Figure 2.23 Cladistic analysis of biogeographic data. **A.** Hypothesis of dispersal, in which a propagule of species *Y* lands on another island. The isolated population subsequently diverges into *Y"*. **B.** Hypothesis of vicariance, in which ancestral population *X* is divided into two populations by a mountain range. The two populations, now isolated, can subsequently diverge, one becoming *X"*. **C.** Cladogram of taxa *A-R* in which geographic distributions are superposed atop lineages, illustrating dispersal in Hawaiian archipelago. Optimized explanation is dispersal of ancestral taxa from Kauai to Maui and then from Maui to Hawaii. **D.** Cladogram in which geographic distributions are superposed atop lineages, illustrating vicariance. Continental drift explains current distribution of taxa *D–K*. **E.** Superposition of ecological habitat data, illustrating use of cladogram to deduce history of ecological change.

is extrinsic. However, the propensity and capability to survive in a particular habitat, e.g., physiological or morphological adaptations that allow survival in these extreme environments, are intrinsic and may be used directly as characters in an analysis. Much more sophisticated algorithms have been developed for including habitat and climatic data in a phylogenetic analysis. These methods can infer **niche shifts**, changes in the range of a species, e.g., via adaptation following a dispersal event. Estimation of niche shifts in plant groups could be particularly pertinent as climate change alters their habitat and living conditions.

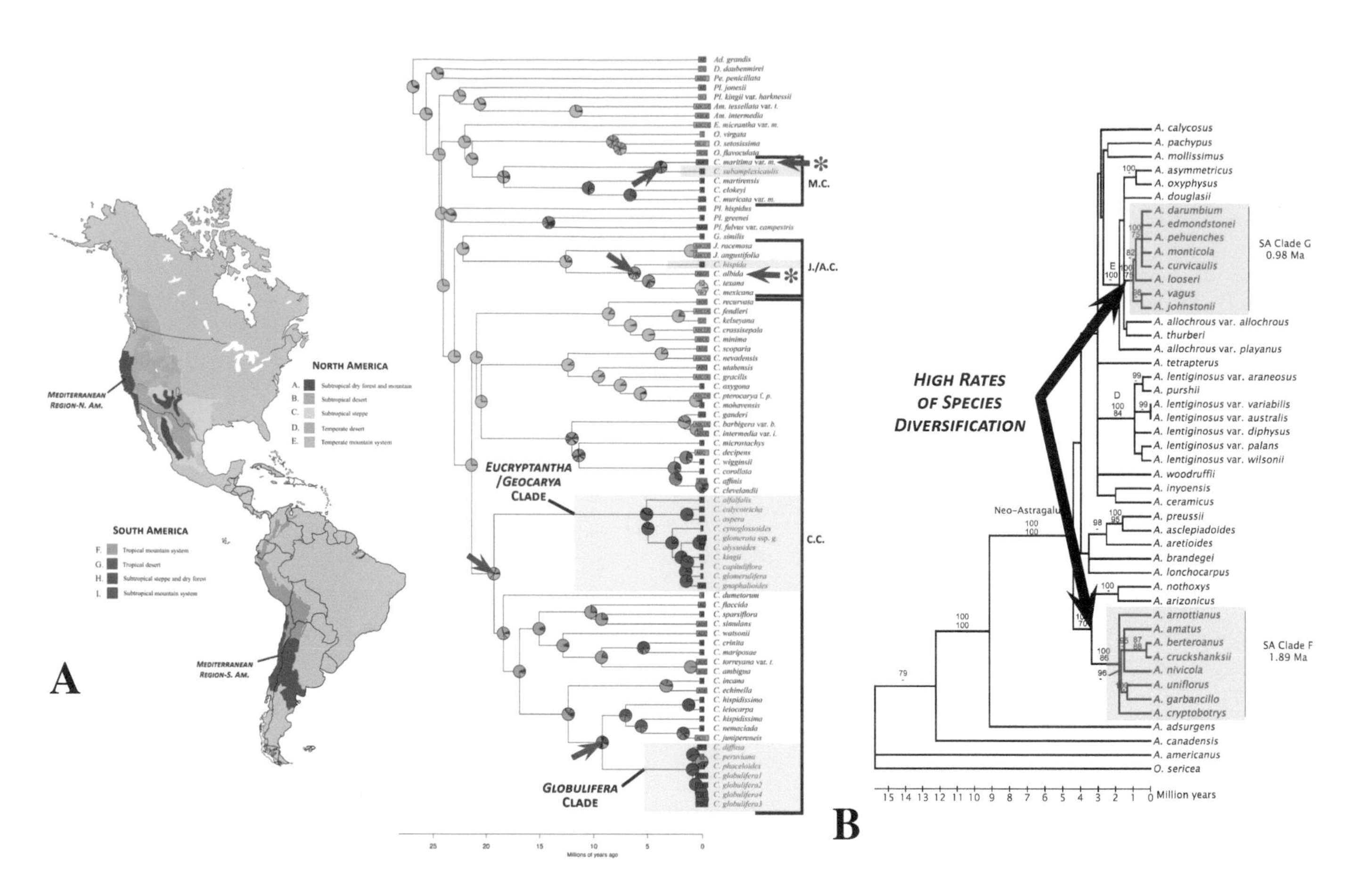

Figure 2.24 **A.** Cladistic analysis of biogeographic data (subtribe Amsinckiinae, Boraginaceae). Left: Global Ecological Zones of North and South America, used for determining species boundaries for BioGeoBEARS (Matzke 2012, 2013). Right: Cladogram, showing most likely ancestral ranges (colored circles) and mean dates of divergence of clades (lower bar). Arrows=inferred long-distance dispersals of clades or individual species (the latter indicated with "*"). From Figure 10 of Mabry and Simpson 2018. Systematic Botany 43: 53-76. Figure courtesy of the American Society of Plant Taxonomists. **B.** Cladogram of *Astragalus* (Fabaceae), showing shift of species diversifi cation rate correlated with two independent dispersals from North to South America (arrows). Modified from Figure 2 of Scherson et al. 2008. American Journal of Botany 95:1030–1039, courtesy of the Botanical Society of America.

The **timing** of clade divergence can be estimated from phylogenetic analyses. This can be done in theory by estimating the average rate of change of nucleotides of one or more gene sequences (e.g., ITS or intergenic chloroplast markers; see Chapter 14) and using this **molecular clock** to determine the time required to go from one node to the next. However, because these molecular clock estimates are not constant from taxon to taxon or even between two lineages derived from the same common ancestor, relaxed clock methods have been developed to take into account this rate variation. In addition, dating studies are considered more accurate if they utilize **fossil calibration**, using one or more fossil taxa to estimate the age of a tree node. Given that the age of the fossil is known from isotopic analysis, one can use that fossil age to date a particular node on the tree, e.g., that considered closest to the extant taxa derived from it (see Figure 2.24A).

Knowing the absolute ages of divergence allows one to test hypothesis, such as: (1) co-evolution and co-diversification of species, e.g., between two symbionts (parasite and host or two mutualists); (2) biogeographic history as correlated with past geologic or climatic changes; (3) changes in morphology correlated with geologic or climatic events; and (4) changes in **species diversification rate**. An increase in species diversification rate can be evaluated to see if it is correlated with past geologic changes or climatic changes, with an adaptive evolutionary novelty, or with species or ecological interactions. Such a rate increase could also be related to, e.g., dispersal into a new habitat, with subsubsequent adaptive radiation (see Figure 2.24B).

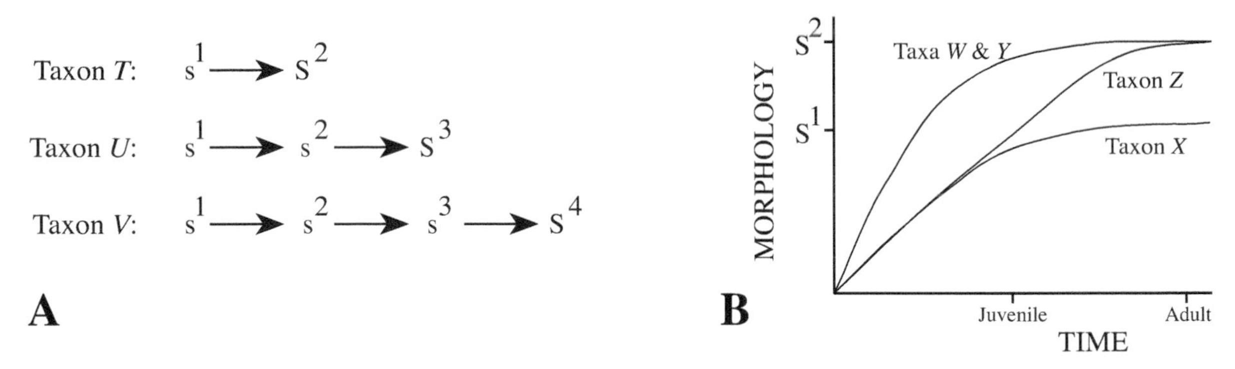

Figure 2.25 **A.** Representation of an ontogenetic sequence, a change from one discrete stage to another in various taxa. **B.** Ontogenetic trajectories of various taxa. Note juvenile and adult stages.

ONTOGENY AND HETEROCHRONY

Phylogeny and character evolution are normally studied only with regard to the mature features of adult individuals. However, a mature structure, whether organ, tissue, or cell, is the end product of **ontogeny**, the developmental sequence under the control of a number of genes. Ontogeny may be visualized in either of two ways. First, a study of the developmental pattern may reveal a series of discrete structural stages or entities, one transforming into the next until the end point (the mature adult structure) is obtained. These discrete stages are identified and named and the transformation in **ontogenetic sequence**, from one stage to the next, is compared in different taxa (Figure 2.25A). Second, some feature of the developmental change of a structure may be measured quantitatively as a function of real time. This plot of morphology as a function of time is called an **ontogenetic trajectory** (Figure 2.25B). Ontogenetic trajectories may be compared between different taxa. Note, e.g., in Figure 2.23B that taxon *Z* and taxa *W* and *Y* have the same adult structures but differing ontogenetic trajectories.

Ontogenetic data may be used in a cladistic analysis like any other character. Thus, two or more discrete ontogenetic sequences (Figure 2.25A) or ontogenetic trajectories (Figure 2.25B) may be defined as separate character states of a developmental character. (See Appendix 4.) The polarity of ontogenetic character states may be assessed by outgroup comparison as can be done for any other character.

Evolution may often be manifested by a change in ontogeny. An evolutionary change in the rate or timing of development is known as **heterochrony**. Heterochrony has apparently been an important evolutionary mechanism in many groups, in which the relatively simple evolutionary alteration of a regulatory gene results in often profound changes in the morphology of a descendant. Heterochrony can be assessed by performing a cladistic analysis and determining from this the ancestral versus the derived condition of an ontogenetic sequence or trajectory. The two major categories of heterochrony are peramorphosis and paedomorphosis. **Peramorphosis** is a derived type of heterochrony in which ontogeny passes through *and goes beyond* the stages or trajectory of the ancestral condition, resulting in the addition of a new stage or ontogenetic trajectory.

Paedomorphosis is a type of heterochrony in which the mature or adult stage of the derived ontogenetic sequence resembles a juvenile ontogenetic stage of the ancestral condition. **Neoteny** is one type of paedomorphosis that is caused by a *decrease* in the rate of development of a structure. Tracing developmental features on a robust cladogram may give insight into the evolutionary mechanism resulting in a novel, adaptive feature. Research that elucidates the molecular basis for those developmental shifts is an exciting, new field in plant systematics.

REVIEW QUESTIONS

TAXONOMY

1. What is the focus and importance of taxonomic research and how does taxonomy differ from phylogenetic systematics?

PHYLOGENETIC SYSTEMATICS

2. What is phylogenetic systematics and what are its goals?
3. What is the name of the branching diagram that represents the pattern of descent of organisms?
4. What are the lines of a cladogram called and what do they represent?
5. What does a split, from one lineage to two, represent?
6. Name the term for both a preexisting feature and a new feature with respect to an evolutionary change.
7. What is the difference between an autapomorphy and a synapomorphy?
8. What does topology refer to and what is its significance in displaying cladograms?
9. What is a phylogram and how does it differ from a typical cladogram?

TAXA AND CHARACTERS

10. What names are given to both the group as a whole and the individual component taxa in a cladistic analysis?
11. What must be taken into account with respect to taxon selection?
12. What precautions must be taken with regard to character state discreteness and character correlation?
13. What is homology and how may it be assessed?
14. What is homoplasy? Name and define the two types of homoplasy and give an example of each.
15. What is a transformation series? Give an example.
16. What is character state polarity and what is the commonly accepted method for establishing polarity?
17. What is character weighting? How is character weighting done with molecular data?
18. What is a character step matrix? Give an example of one for an unordered character.
19. What is a character x taxon matrix? Give a simple example for molecular sequence data.

FEATURES OF CLADOGRAMS

20. What is a primary tenet of phylogenetic systematics with respect to apomorphies?
21. What is meant by recency of common ancestry?
22. What is a monophyletic group? What is the rationale for their recognition?
23. What are sister groups?
24. What is a paraphyletic group? A polyphyletic group?
25. Name a traditionally named taxonomic plant group that is not monophyletic. (Refer to Chapters 3–6.)
26. What is an unrooted tree and what can it not represent?
27. What is a polytomy and how may polytomies arise in cladistic analyses?
28. What is reticulation? How might it be detected?
29. Why do the OTUs and the ingroup itself need to be verified for monophyly?

CLADOGRAM CONSTRUCTION METHODS

30. What is meant by maximum parsimony, and what is the rationale of this principle?
31. From the data set of Figure 2.6, construct five trees that are different from the one in Figure 2.6E, draw in all character state changes, and calculate the total length of these trees. Are these trees of a different length than that of Figure 2.6E?
32. What is outgroup comparison and what is the rationale for using it to determine character state polarity?
33. Why should the terms ancestral/plesiomorphic and derived/apomorphic not be applied to taxa?
34. What is a consensus tree? Name two types of consensus trees.
35. What is long branch attraction and why is it a problem in phylogenetic analysis?
36. Briefly describe the rationale and methodology of maximum likelihood. How are likelihood values calculated?
37. What are the advantages of maximum likelihood over parsimony?
38. What is the general time reversible (GTR) model?
39. Briefly describe the basic rationale of Bayesian analysis in cladogram construction.
40. What are coalescent methods of cladogram construction and how might a species tree differ from gene trees?
41. What is a consistency index and what does it measure?
42. What are bootstrap, jackknife, decay index, and posterior probability? What do these assess?

CLADOGRAM ANALYSIS

43. Describe ways in which a phylogenetic classification system may be derived from a cladistic analysis.
44. What are the differences between a node-based, apomorphy-based, and stem-based classification system?
45. What is parsimony optimization and how is it used to assess character evolution?
46. How does ancestral state reconstruction differ in assessing character evolution?
47. Name the two primary explanations for changes in biogeographic distribution and indicate how they differ.
48. Give an example as to how a cladistic analysis can be used to assess biogeographic history.
49. How might a phylogenetic analysis be used to assess shifts in habitat of the OTUs of a study?
50. What is a molecular clock and how can it be used to assess divergence times in a phylogenetic analysis?
51. How is fossil calibration used in refining divergence times?
52. What might be the significance of an increase in species diversification rate in a phylogenetic analysis?
53. What is ontogeny and how may ontogeny be assessed in a phylogenetic analysis?
54. Define heterochrony, peramorphosis, paedomorphosis, and neoteny.

EXERCISES

55. For the following data sets: (a) draw the three possible (dichotomously branching) cladograms; (b) for *each* of the three cladograms indicate (with arrows and corresponding characters and states) the minimum character state changes that are needed to explain the data; (c) indicate which of the three trees would be accepted by a cladist as the best estimate of phylogeny and why.

	1	2	3	4	5
A	1	1	1	1	1
B	1	0	0	0	0
C	0	0	1	1	1
OUTGROUP	0	0	0	0	0

	1	2	3	4	5
A	0	1	1	1	0
B	0	0	1	0	0
C	1	0	1	1	1
OUTGROUP	0	0	0	0	0

56. For each of the following data sets: (a) draw the most parsimonious cladogram; (b) indicate all character state changes; (c) circle all monophyletic groups; (d) derive a hypothetical classification scheme.

GENERA:	**1** Flower symmetry	**2** Perianth tube	**3** Perianth aestivation	**4** Stamen number	**5** Anther shape	**6** Pollen exine
Aahh	bilateral	present	valvate	6	oblong	homogeneous
Batahr	bilateral	present	valvate	6	oblong	homogeneous
Conarus	radial	present	valvate	6	oblong	homogeneous
Phlebus	radial	absent	imbricate	6	oblong	tectate
Tribus	radial	present	imbricate	6	fringed	homogeneous
OUTGROUP	radial	absent	imbricate	3	oblong	tectate

GENERA:	**1** Nucleotide site	**2** Nucleotide site	**3** Nucleotide site	**4** Nucleotide site	**5** Nucleotide site	**6** Nucleotide site	**7** Nucleotide site
Queesus	T	G	A	T	G	C	A
Racamupa	T	G	A	T	G	C	G
Shoota	T	C	G	T	G	T	A
Tumblus	T	C	G	T	G	T	A
Uvulus	G	C	A	T	A	T	A
Vertex	G	C	A	T	A	T	G
OUTGROUP	A	C	A	C	A	T	A

57. Given the following data matrix and model of evolution, calculate the maximum likelihood values for at least one of the three possible unrooted trees.

Taxon	Gene Site
W	A
X	C
Y	A
Z	G

	A	C	G	T
A	0.9	0.1	0.4	0.1
C	0.1	0.9	0.1	0.4
G	0.4	0.1	0.9	0.1
T	0.1	0.4	0.1	0.9

58. Explore one of the commonly used phylogeny software applications, such as Mesquite (see Phylogeny Computer Programs). These programs allow the user to input data, including taxa names and their characters and character states, and enable both the phylogenetic relationships of taxa and specific character state changes to be visualized. You may use the data matrix below for the families of the Zingiberales. (Note: Root the tree at **Musaceae**.)

Examine the optimal (e.g., most parsimonious) tree. Engage the function that displays characters and visualize several, noting the distribution of their states. You may also "swap branches" on the cladogram, exploring alternative evolutionary hypotheses and noting the change in tree length.

Review as a class the following terms: cladogram, lineage/clade, common ancestor, lineage divergence/diversification, apomorphy, synapomorphy, autapomorphy, monophyletic, paraphyletic.

Example data set of the families of the Zingiberales.

	LEAF ARRANGEMENT	SEED ARIL	POLYARC ROOT	INNER MED. STAMEN	RAPHIDES	SILICA CRYSTALS
Cannaceae	distichous	present	present	present	absent	present
Costaceae	monistichous	present	present	present	absent	present
Heliconiaceae	distichous	present	present	present	present	absent
Lowiaceae	distichous	present	absent	absent	present	absent
Marantaceae	distichous	present	present	present	absent	present
Musaceae	spiral	absent	absent	absent	present	absent
Strelitziaceae	distichous	present	present	absent	present	absent
Zingiberaceae	distichous	present	present	present	absent	present

	STAMEN NUMBER	STAMINODE PETALOID	PERISPERM	OUT. TEPALS FUSED	ANTHER TYPE
Cannaceae	1	present	present	absent	monothecal
Costaceae	1	present	present	present	bithecal
Heliconiaceae	5	absent	absent	absent	bithecal
Lowiaceae	5	absent	absent	absent	bithecal
Marantaceae	1	present	present	absent	monothecal
Musaceae	5	absent	absent	absent	bithecal
Strelitziaceae	5	absent	absent	absent	bithecal
Zingiberaceae	1	present	present	present	bithecal

59. **Website trees.**
Log onto The Tree of Life (*http://tolweb.org*), TreeBASE (*http://www.treebase.org*), Angiosperm Phylogeny Website (*http://www.mobot.org/MOBOT/research/Apweb*), or a similar Web page. These Web pages contain up-to-date information on the relationships of organismal groups and plants, respectively. Browse through the trees illustrated on the sites and note the source of the data. Examine the apomorphies denoted at the nodes for these trees.

REFERENCES FOR FURTHER STUDY

Akaike, H. 1974. A new look at statistical model identification. IEEE transactions on Automatic Control 19.

Archibald, J. K., M. E. Mort, and D. J. Crawford. 2003. Bayesian inference of phylogeny: a non-technical primer. Taxon 52:187–191.

Bell, C. D. 2015. Between a rock and a hard place: Applications of the "molecular clock" in systematic biology. Systematic Botany 40:6-13.

Bouckaert, R., T. G. Vaughan, J. Barido-Sottani, S. Duchêne, M. Fourment, A. Gavryushkina, J. Heled, G. Jones, D. Kühnert, N. D. Maio, M. Matschiner, F. K. Mendes, N. F. Müller, H. A. Ogilvie, L. d. Plessis, A. Popinga, A. Rambaut, D. Rasmussen, I. Siveroni, M. A. Suchard, C.-H. Wu, D. Xie, C. Zhang, T. Stadler, and A. J. Drummond. 2019. BEAST 2.5: An advanced software platform for Bayesian evolutionary analysis. PLoS computational biology 15:e1006650.

Brooks, D. R., and D. A. McLennan. 1991. Phylogeny, Ecology, and Behavior: A Research Program in Comparative Biology. Univ. Chicago Press, Chicago.

Cantino, P. D. and K. de Queiroz. 2010. International Code of Phylogenetic Nomenclature (Phylocode). Version 4. https://www.ohio.edu/phylocode/index.html. Accessed on

Cantino, P. D., J. A. Doyle, S. W. Graham, W. S. Judd, R. G. Olmstead, D. E. Soltis, P. S. Soltis, and M. J. Donoghue. 2007. Towards a phylogenetic nomenclature of Tracheophyta. Taxon 56: 822-846.

Drummond, A. J., M. A. Suchard, D. Xie, and A. Rambaut. 2012. Bayesian phylogenetics with BEAUti and the BEAST 1.7. Molecular Biology And Evolution 29:1969-1973.

Edwards, S. V. 2009. Is a new and general theory of molecular systematics emerging? Evolution 63:1–19.

Edwards, S. V., Z. Xi, A. Janke, B. C. Faircloth, J. E. McCormack, T. C. Glenn, B. Zhong, S. Wug, E. M. Lemmon, A. R. Lemmon, A. D. Leaché, L. Liu, and C. C. Davis. 2016. Implementing and testing the multispecies coalescent model: A valuable paradigm for phylogenomics. Molecular Phylogenetics and Evolution 94:447-462.

Felsenstein, J. 2003. Inferring Phylogenies. Sinauer Associates. Sunderland, Massachusetts.

Gould, S. J. 1977. Ontogeny and Phylogeny. Belknap Press of Harvard University, Cambridge, Massachusetts.

Heled, J. and A. J. Drummond. 2010. Bayesian inference of species trees from multilocus data. Molecular Biology and Evolution 27:570-580.

Hennig, W. 1966. Phylogenetic Systematics. University of Illinois Press, Urbana.

Hillis, D. M., C. Moritz, and B. Mable (eds.). 1996. Molecular Systematics. Second edition. Sinauer, Sunderland, Massachusetts.

Huelsenbeck, J. P., and J. P. Bollback. 2001a. Empirical and hierarchical Bayesian estimation of ancestral states. Systematic Biology 50: 351–366.

Huelsenbeck, J. P., F. Ronquist, R. Nielsen, and J. P. Bollback. 2001b. Bayesian influence of phylogeny and its impact on evolutionary biology. Science 294:2310-2314.

Huelsenbeck, J. P., B. Larget, R. E. Miller, and F. Ronquist. 2002. Potential application and pitfalls of Bayesian inference of phylogeny. Systematic Biology 51:673-688.

Kitching, I. J. 1998. Cladistics: The Theory and Practice of Parsimony Analysis, 2nd ed. Oxford University Press, Oxford.

Lanfear, R., P. B. Frandsen, A. M. Wright, T. Senfeld, and B. Calcott. 2016. PartitionFinder 2: new methods for selecting partitioned models of evolution for molecular and morphological phylogenetic analyses. Molecular Biology and Evolution 34:772-773.

Li, W. 1997. Molecular Evolution. Sinauer Associates. Sunderland, Massachusetts.

Li, D., L. Trotta, H. E. Marx, J. M. Allen, M. Sun, D. E. Soltis, P. S. Soltis, R. P. Guralnick, and B. Baiser. 2019. For common community phylogenetic analyses, go ahead and use synthesis phylogenies. Ecology 0:e02788.

Liu, L., L. Yu, D. K. Pearl, and S. V. Edwards. 2009. Estimating species phylogenies using coalescence times among sequences. Systematic Biology 58:468-477.

Liu, L. and L. Yu. 2010. Phybase, an R package for species tree analysis. Bioinformatics 26:962–963.

Maddison, W. P. 1997. Gene trees in species trees. Systematic Biology 46:523-536.

Maddison, W. P. and D. R. Maddison. 2017. Mesquite: a modular system for evolutionary analysis. Version 3.2. http://mesquiteproject.org. http://mesquiteproject.org. Accessed on

Matzke, N. J. 2012. Founder-event speciation in BioGeoBEARS package dramatically improves likelihoods and alters parameter inference in dispersal–extinction–cladogenesis (DEC) analyses. Frontiers of Biogeography 4:210.

Matzke, N. J. 2014. Model selection in historical biogeography reveals that founder-event speciation is a crucial process in island clades. Systematic Biology 63:951-970.

Mirarab, S. and T. Warnow. 2015. ASTRAL-II: coalescent-based species tree estimation with many hundreds of taxa and thousands of genes. Bioinformatics 31:i44-52.

Mirarab, S., R. Reaz, M. S. Bayzid, T. Zimmerman, M. S. Swenson, and T. Warnow. 2014. ASTRAL: genome-scale coalescent-based species tree estimation. Bioinformatics 30:541-548.

Nei, M. and S. Kumar. 2000. Molecular Evolution and Phylogenetics. Oxford University Press, New York.

Page, R. D., and E. C. Holmes. 1998. Molecular Evolution: A Phylogenetic Approach. Blackwell Science, Oxford.

Rabosky, D. L., M. Grundler, C. Anderson, P. Title, J. J. Shi, J. W. Brown, H. Huang, and J. G. Larson. 2014. BAMMtools: an R package for the analysis of evolutionary dynamics on phylogenetic trees. Methods in Ecology and Evolution 5:701-707.

Rambaut, A. 2014. FigTree. 2014. v.1.4.2: tree drawing tool. Available at: http://tree.bio.ed.ac.uk/software/figtree/. Accessed on

Sanderson, M. 2002. Estimating absolute rates of molecular evolution and divergence times: A penalized likelihood approach. Molecular Biology and Evolution 14:1218-1231.

Semple, C., and M. A. Steel. 2003. Phylogenetics. Oxford University Press, Oxford.
Seo, T. K. 2008. Calculating bootstrap probabilities of phylogeny using multilocus sequence data. Molecular biology and evolution 25:960-971.
Slowinski, J. and R. D. M. Page. 1999. How should species phylogenies be inferred from sequence data? Systematic Biology 48:814-825.
Smith, S. A. and B. C. O'Meara. 2012. TreePL: Divergence time estimation using penalized likelihood for large phylogenies. Bioinformatics 28:2689-2690.
Stamatakis, A. 2014. RAxML Version 8: A tool for phylogenetic analysis and post-analysis of large phylogenies. Bioinformatics 30:1312-1313.
Wiley, E. O., D. Siegel-Causey, D. R. Brooks, and V. A. Funk. 1991. The Compleat Cladist: A Primer of Phylogenetic Procedures. Univ. Kansas Museum Nat. History Sp. Publ. no. 19.
Xi, Z., J. S. Rest, and C. C. Davis. 2013. Phylogenomics and coalescent analyses resolve extant seed plant relationships. PLoS ONE 8:e80870.
Zhang, C., M. Rabiee, E. Sayyari, and S. Mirarab. 2018. ASTRAL-III: Polynomial Time Species Tree Reconstruction from Partially Resolved Gene Trees. BMC Bioinformatics 19 (S6): 153 19:153.

PHYLOGENY COMPUTER PROGRAMS

ASTRAL. Estimating species tree given a set of gene trees, using a multi-species coalescent model. *https://github.com/smirarab/ASTRAL*
BAMM. Bayesian Analysis of Macroevolutionary Mixtures. Modeling speciation rates, extinction, and trait evolution on phylogenetic trees. *http://bamm-project.org*
BEAST: Bayesian Evolutionary Analysis Sampling Trees. Bayesian analysis of molecular sequences using MCMC, rooted, time-measured phylogenies inferred using strict or relaxed molecular clock models. *http://beast.community*
BEAST 2 (including *BEAST): An advanced software platform for Bayesian evolutionary analysis. *https://www.beast2.org*
BioGeoBEARS: BioGeography with Bayesian (and Likelihood) Evolutionary Analysis in R Scripts. Inference of biogeographic history on phylogenies, model testing, and biogeography model choice (dispersal, vicariance, founder-event speciation. *http://phylo.wikidot.com/biogeobears*
CIPRES Science Gateway. Resource for inference of large phylogenetic trees, providing large computational resources, with parallel versions of RAxML, MrBayes, and GARLI. *http://www.phylo.org/sub_sections/portal*
FigTree. A graphical viewer of phylogenetic trees and for producing publication-ready figures. *http://tree.bio.ed.ac.uk/software/figtree*
MrBayes: Bayesian inference of phylogeny. *www.mrbayes.net*
ParitionFinder. Software to select partitioning schemes and models of molecular evolution. *http://www.robertlanfear.com/partitionfinder*
Mesquite. Display and manipulation of trees, reconstruction and display of ancestral states, character evolution, comparative methods, parametric bootstrapping. morphometrics, coalescence, tree comparisons and simulations, cluster analysis. *http://mesquiteproject.org*
PHYLIP: Phylogeny Inference Package. Parsimony, distance matrix, and likelihood methods, including bootstrapping and consensus trees *http://evolution.gs.washington.edu/phylip.html*
Phylogeny Programs (J. Felsenstein). *http://evolution.genetics.washington.edu/phylip/software.html*
An exhaustive list and links to just about every phylogeny computer program available, a very few of which are listed here.
RAxML: Randomized Axelerated Maximum Likelihood. Maximum likelihood, plus parsimony, bootstrapping, and consensus tree methods *https://cme.h-its.org/exelixis/web/software/raxml/index.html*
Tree PL. Divergence time estimation using penalized likelihood for large phylogenies. *https://github.com/blackrim/treePL*

II

EVOLUTION AND DIVERSITY OF PLANTS

3

EVOLUTION AND DIVERSITY OF GREEN AND LAND PLANTS

THE GREEN PLANTS

The green plants, formally called the **Viridiplantae** or **Chlorobionta**, are a monophyletic group of eukaryotic organisms that includes what have traditionally been called "green algae" plus the land plants or embryophytes (Figure 3.1). Like all eukaryotes, the Viridiplantae have cells with membrane-bound organelles, including a nucleus (containing chromosomes composed of linear chains of DNA bound to proteins, that are sorted during cell division by mitosis), microtubules, mitochondria, an endoplasmic reticulum, vesicles, and golgi bodies. Although the interrelationships of the non-land plant Viridiplantae will not be covered in detail here, it is important to realize that some of the evolutionary innovations, or apomorphies, that we normally associate with land plants actually arose before plants colonized the land.

Several apomorphies unite the Viridiplantae (Figure 3.1). One possible novelty for this group is a **cellulosic cell wall** (Figure 3.2A). Cellulose, like starch, is a polysaccharide, but one in which the glucose sugar units are bonded in the beta-1,4 position (=β-1,4-glucopyranoside). This slight change in chemical bond position results in a very different molecule. Cellulose is secreted outside the plasma membrane as microscopic fiber-like units called **microfibrils** that are further intertwined into larger fibril units, forming a supportive meshwork. The function of cellulose is to impart rigidity to the cells, acting as a sort of cellular exoskeleton. The evolution of a cellulosic cell wall was a preamble to the further evolution of more complex types of growth, particularly of self-supporting shoot systems. It is not clear whether a cellulosic cell wall constitutes an apomorphy for the Viridiplantae alone, as it may have evolved much earlier, constituting an apomorphy for the Viridiplantae plus one or more other groups; in any case, its adaptive significance seems clear.

Perhaps the primary apomorphy for the Viridiplantae is a specialized type of **chloroplast** (Figure 3.2). As discussed in Chapter 1, chloroplasts are one of the major defining characteristics of traditionally defined "plants"; their adaptive significance as organelles functioning in photosynthesis, the conversion of light energy to chemical energy, is unquestioned. Chloroplasts in the Viridiplantae, the green plants, differ from those of most other organisms, such as the red and brown "algae," in (1) containing **chlorophyll b** in addition to chlorophyll a, the former of which acts as an accessory pigment in light capture; (2) having **thylakoids**, the chlorophyll-containing membranes that are **stacked into grana**, which are pancake-like aggregations (see Figure 3.2B); and (3) manufacturing as a storage product true **starch**, a polymer of glucose sugar units (= *polysaccharide*) in which the glucose molecules are chemically bonded in the alpha-1,4 position (α-1,4-glucopyranoside). Thus, all green plants, from filamentous green "algae" in a pond or tide pool to giant sequoia or

https://doi.org/10.1016/B978-0-12-812628-8.50003-1,

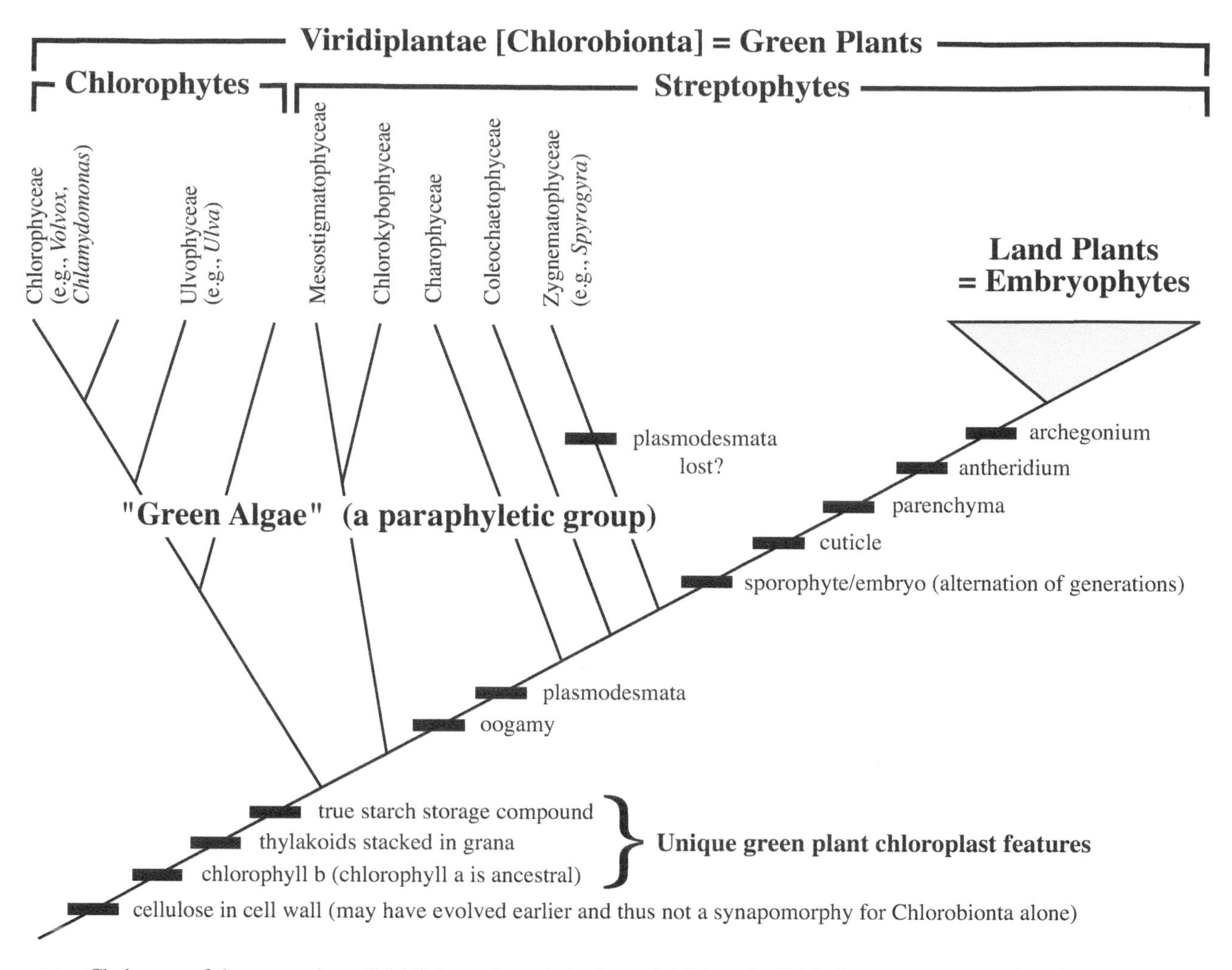

Figure 3.1 Cladogram of the green plants (Viridiplantae), modified from Ruhfel et al. (2014). Important apomorphies discussed in the text are listed beside thick hash marks.

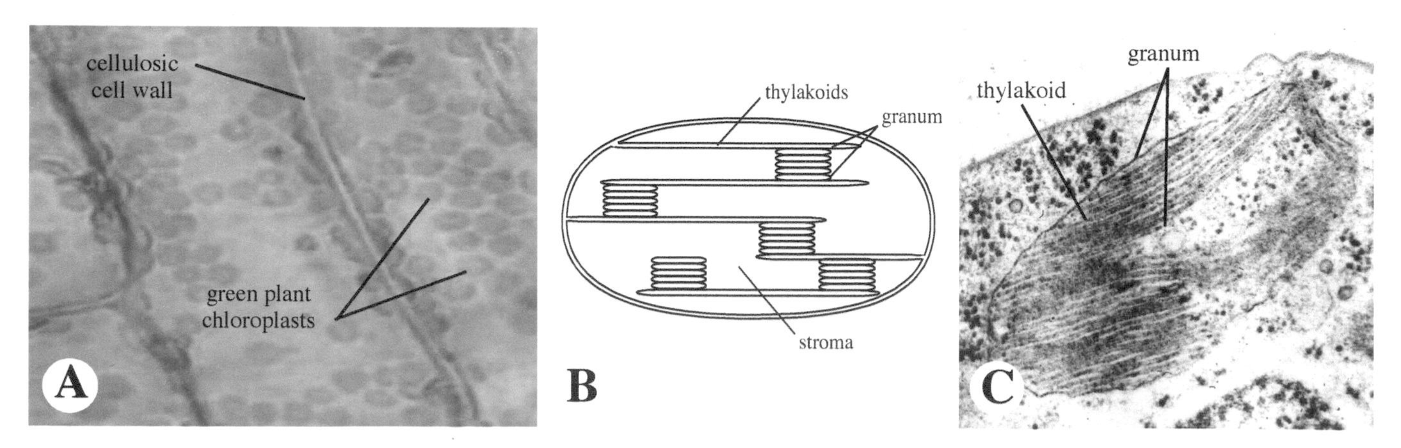

Figure 3.2 **A.** *Elodea*, whole leaf in face view, showing apomorphies of the Viridiplantae: a cellulosic cell wall and green plant chloroplasts. **B.** Diagram of chloroplast structure of green plants, showing thylakoids and grana. **C.** Electron micrograph of *Chlamydomonas reinhardtii*, a unicellular "green alga," showing granum of chloroplast. (Photo courtesy of Rick Bizzoco.)

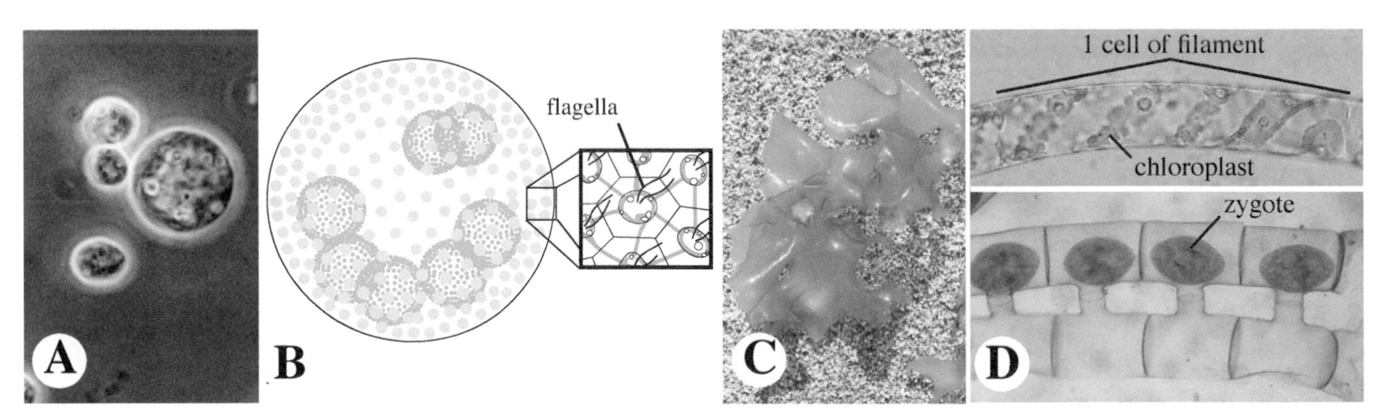

Figure 3.3 Examples of non-land plant Viridiplantae. **A,B.** Chlorophyceae. **A.** *Chlamydomonas reinhardtii*, a unicellular form. (Photo courtesy of Rick Bizzoco.) **B.** *Volvox*, a colonial form. **C.** Ulvophyceae. *Ulva*, a thalloid form. **D.** Zygnematophyceae. *Spirogyra*, a filamentous form. *Above:* vegetative form, with large, spiral chloroplasts. *Below:* reproductive conjugation stage, showing + and – mating strains and nonmotile zygotes.

Eucalyptus trees have this same type of chloroplast. Recent data imply that chloroplasts found in the green plants today were modified from those that evolved via **endosymbiosis**, the intracellular cohabitation of an independently living, unicellular prokaryote inside a eukaryotic cell (see Chapter 1).

The Viridiplantae as a whole are classified as two sister groups: Chlorophytes, or Chlorophyta, and Streptophytes, or Streptophyta (Figure 3.1). The traditional "green algae" are a paraphyletic group (which is why the name is placed in quotation marks) and are defined as the primarily aquatic Viridiplantae, consisting of all chlorophytes and the non-land plant streptophytes. "Green algae" occur in a tremendous variety of morphological forms. These include flagellated unicells (Figure 3.3A) with or without flagella, thalloid forms (Figure 3.3B), motile and nonmotile colonies (Figure 3.3C), and nonmotile filaments (Figure 3.3D). Many have flagellated motile cells in at least one phase of their life history. "Green algae" mostly inhabit fresh and marine waters, but some live in or on soil (or even on snow!) or in other terrestrial habitats, some able to withstand extreme desiccation.

The primitive type of green plant sexual reproduction seems to have been the production of flagellate, haploid (n) **gametes** (reproductive cells) that are "isomorphic," that is, that look identical. **Fertilization** is the union of two of these gametes, resulting in a diploid (2n) **zygote** (Figure 3.4A). The zygote, which is free-living, then divides by meiosis to form four haploid **spores**, each of which may germinate and develop into a new haploid individual, which produces more gametes, completing what is termed a **haplontic** (or "haplobiontic") life cycle (Figure 3.4A).

Within the streptophyte lineage that gave rise to the land plants, a few innovations evolved that may have been

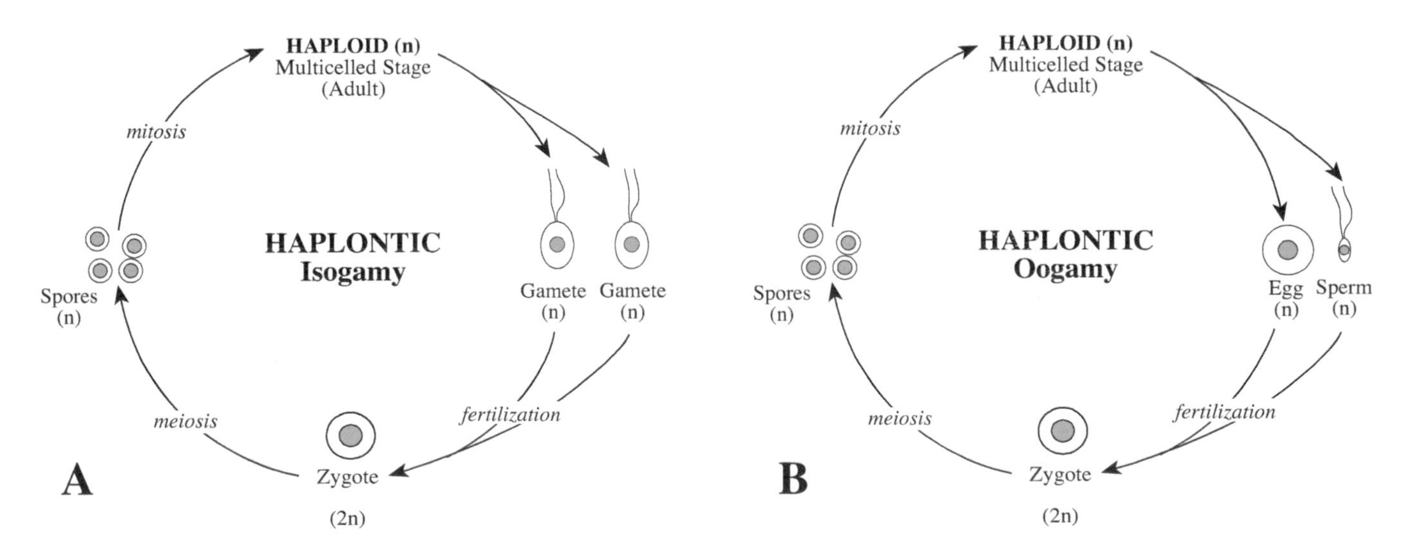

Figure 3.4 Haplontic life cycles in some of the green plants. **A.** Isogamy. **B.** Oogamy.

"preadaptations" to survival on land. First of these was the evolution of **oogamy**, a type of sexual reproduction in which one gamete, the **egg**, becomes larger and nonflagellate; the other gamete is, by default, called a **sperm** cell (Figure 3.4B). Oogamy is found in all land plants but independently evolved in many other groups, including many other "algae" and in the animals.

Several other apomorphies of and within the Viridiplantae include ultrastructural specializations of flagella and some features of biochemistry. Although these have been valuable in elucidating phylogenetic relationships, their adaptive significance is unclear, and they will not be considered further here.

An apomorphy within the Charophytes that apparently was transferred to the land plants (Figure 3.1) are **plasmodesmata**. Plasmodesmata are essentially pores in the primary (1°) cell wall through which membranes traverse between cells, allowing for transfer of compounds between cells (Figure 3.5A). Plasmodesmata may function in more efficient or rapid transport of solutes, including regulatory and growth-mediating compounds, such as hormones.

The closest Streptophyte relative to the land plants is unclear, but three non-land plant lineages appear to share oogamy and plasmodesmata (with the possible loss of the latter in one; see Figure 3.1, showing the results of one recent study): Charophyceae (Figure 3.5C–E), Coleochaetophyceae (Figure 3.5B), and Zygnematophyceae (Figure 3.3D). The Charophyceae can serve as a model for potential adaptations that may have been precursors to the land plants. Charophyceae, including the genera *Chara* and *Nitella*, are fresh water, aquatic organisms with a haplontic life cycle, their bodies consisting of a central axis bearing whorls of lateral branches (Figure 3.5D) or (if small) "leaves" on the haploid body. Some Charophyceae are capable of precipitating calcium carbonate as an outer layer of the plant body (accounting for the common names "brittleworts" or "stoneworts"). Members of the Charophyceae grow by means of a single apical cell, similar to that of some land plants and representing a possible synapomorphy with them. However, the Charophyceae differ from land plants in lacking true parenchyma (see later discussion). The Charophyceae have specialized male and female gametangia, termed antheridia and

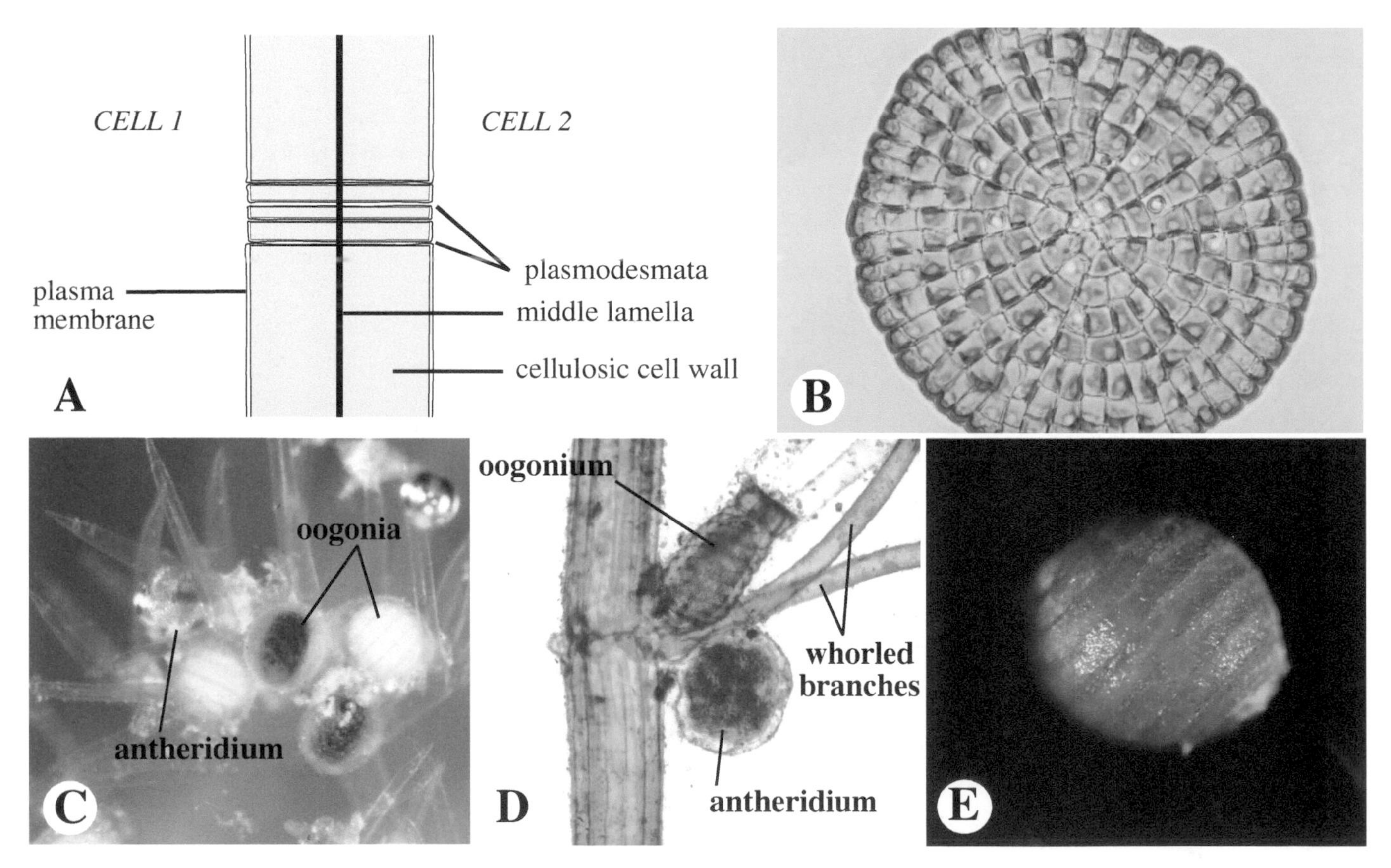

Figure 3.5 **A.** Diagram of plasmodesmata in cellulosic cell wall, an apomorphy of some green plants, including the land plants. **B.** *Coleochaete* sp., a close relative to the embryophytes. (Photo courtesy of Linda Graham.) **B–D.** Charales. **B,C.** *Nitella* sp. **B.** Whole plant. **C.** Oogonia and antheridia. **D.** *Chara* sp., oogonium and antheridium. Note spiral tube cells of oogonia. **E.** *Tectochara helicteres*, a fossil oogonium from the Eocene, showing remnants of spiral tube cells.

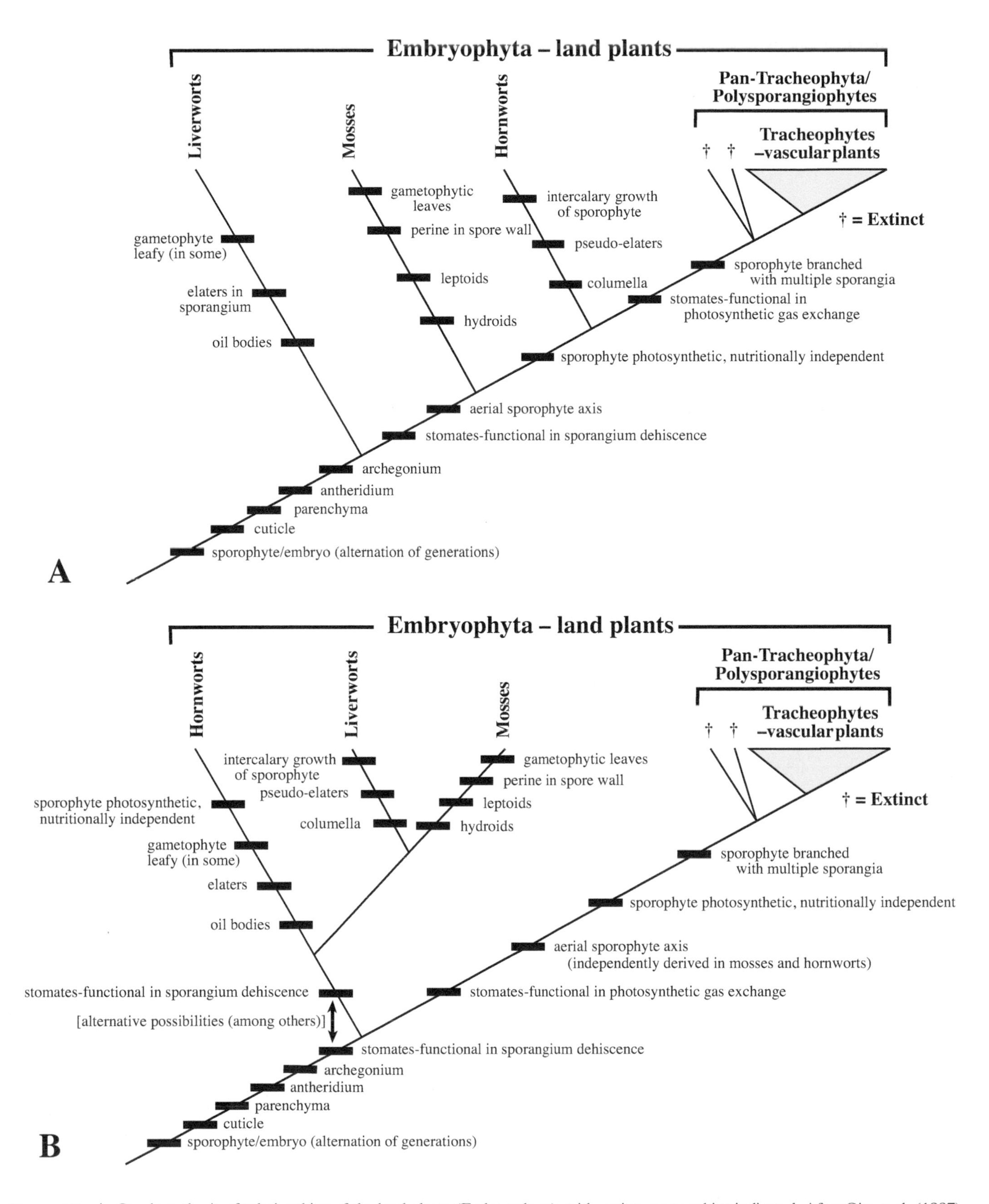

Figure 3.6 **A.** One hypothesis of relationships of the land plants (Embryophyta), with major apomorphies indicated. After Qiu et al. (1997), some apomorphies after Bremer (1985), Mishler and Churchill (1985), and Mishler et al. (1994). **B.** One of many alternative hypothesis, this showing a monophyletic bryophytes (after Puttick et al. 2018), with different scenarios of the evolution of stomates and some other charac-

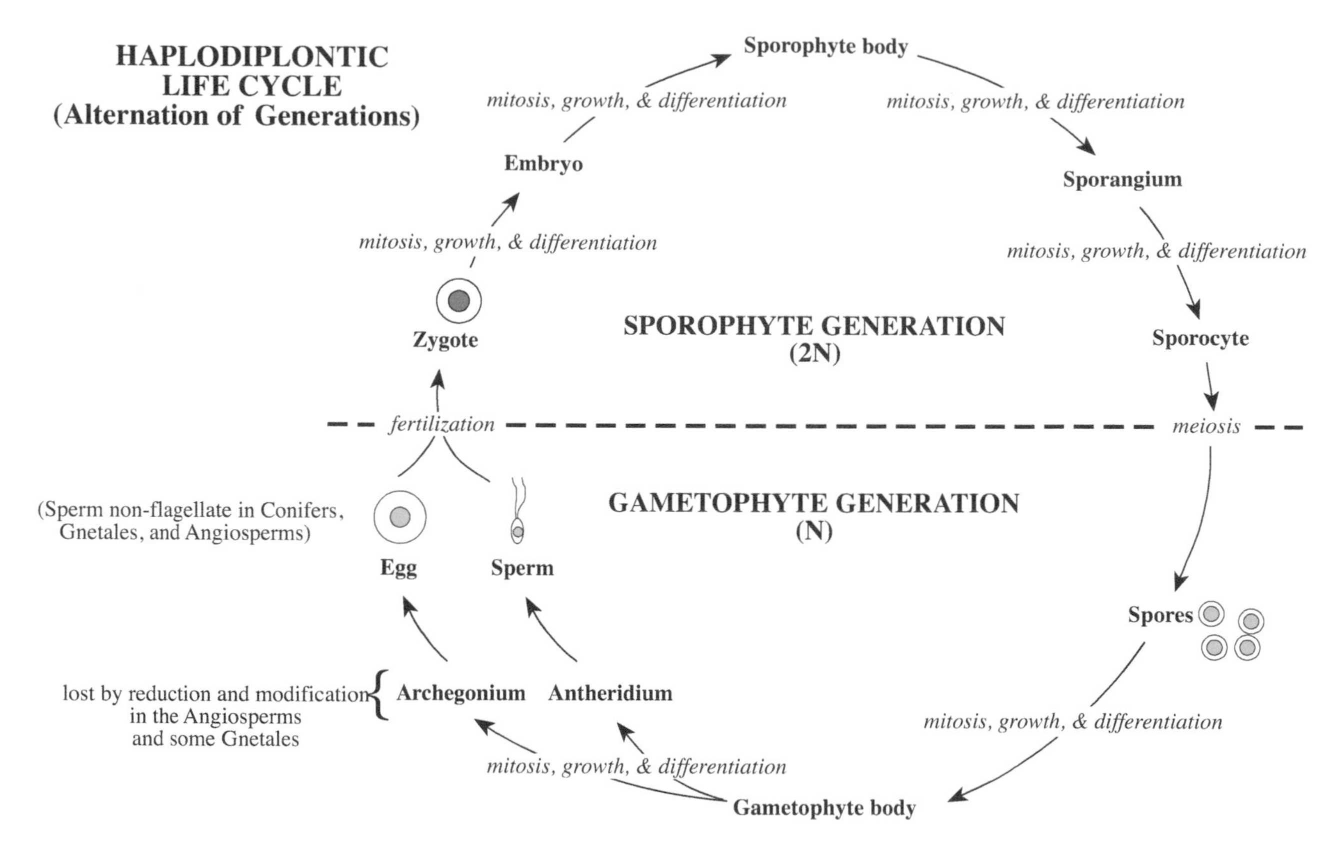

Figure 3.7 Haplodiplontic "alternation of generations" in the land plants (Embryophytes).

oogonia (Figure 3.5C,D). The oogonia are distinctive in having a spirally arranged group of outer "tube" cells (Figure 3.5D); fossilized casts of oogonia retain the outline of these tube cells (Figure 3.5E). Oogonia and antheridia of the Charophyceae resemble the archegonia and antheridia of land plants (see later discussion) in having an outer layer of sterile cells, but the gametangia of the two groups are generally thought not to be directly homologous because of major differences in structure and development. However, members of the Charales retain the egg and zygote (although the latter only briefly) on the plant body. This retention of egg and zygote on the haploid body may represent a transition to their permanent retention on the gametophyte of land plants (see later discussion).

EMBRYOPHYTA - LAND PLANTS

The Embryophyta, or embryophytes (commonly known as land plants), are a monophyletic assemblage within the green plants (Figures 3.1, 3.6). The first colonization of plants on land during the Silurian period, ca. 400 million years ago, was concomitant with the evolution of several important features. These shared, evolutionary novelties (Figure 3.6) constituted major adaptations that enabled formerly aquatic green plants to survive and reproduce in the absence of a surrounding water medium.

One major innovation of land plants was the evolution of the **embryo** and **sporophyte** (Figures 3.6, 3.7). The sporophyte is a separate diploid (2n) phase in the life cycle of all land plants. The corresponding haploid, gamete-producing part of the life cycle is the **gametophyte**. The life cycle of land plants, having both a haploid gametophyte and a diploid sporophyte, is an example of a **haplodiplontic** (also called "diplobiontic") life cycle, commonly called **alternation of**

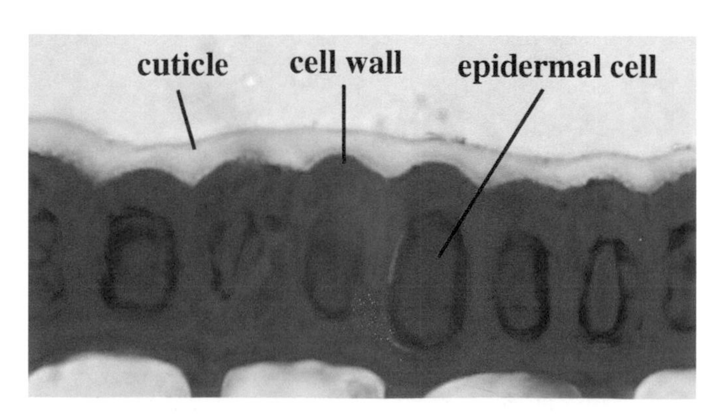

Figure 3.8 The cuticle, an apomorphy for the land plants.

generations (Figure 3.7). Note that alternation of generations does not necessarily mean that the two phases occur at different points in time; at any given time, *both* phases may occur in a population.

The sporophyte can be viewed as forming from the zygote by the delay of meiosis and spore production. Instead of meiosis, the zygote undergoes numerous *mitotic* divisions, which result in the development of a separate entity. The **embryo** is defined as an immature sporophyte that is attached to or surrounded by the gametophyte. In many land plants, such as the seed plants, the embryo will remain dormant for a period of time and will begin growth only after the proper environmental conditions are met. As the embryo grows into a mature sporophyte, a portion of the sporophyte differentiates as the spore-producing region. This spore-producing region of the sporophyte is called the **sporangium**. A sporangium contains **sporogenous tissue**, which matures into **sporocytes**, the cells that undergo meiosis. Each sporocyte produces, by meiosis, four haploid **spores**. The sporangium is enveloped by a **sporangial wall**, which consists of one or more layers of sterile, non-spore-producing cells.

One adaptive advantage of a sporophyte generation as a separate phase of the life cycle is the large increase in spore production. In the absence of a sporophyte, a single zygote (the result of fertilization of egg and sperm) will produce four spores. The elaboration of the zygote into a sporophyte and sporangium can result in the production of literally millions of spores, a potentially tremendous advantage in reproductive output and increased genetic variation.

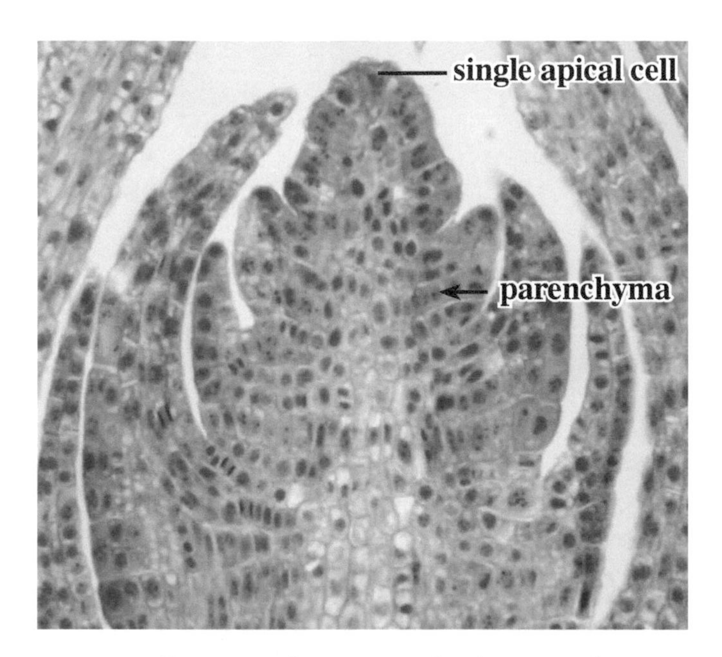

Figure 3.9 *Equisetum* shoot apex, showing parenchymatous growth form, from an apical meristem.

Another possible adaptive value of the sporophyte is associated with its diploid ploidy level. The fact that a sporophyte has two copies of each gene may give this diploid phase an increased fitness in either of two ways: (1) by potentially preventing the expression of recessive, deleterious alleles (which, in the sporophyte, may be "shielded" by dominant alleles, but which, in the gametophyte, would always be expressed); and (2) by permitting increased genetic variability in the sporophyte generation (via genetic recombination from two "parents") upon which natural selection acts, thus increasing the potential for evolutionary change.

A second innovation in land plant was the evolution of **cutin** and the **cuticle** (Figure 3.8). A **cuticle** is a protective layer that is secreted to the outside of the cells of the **epidermis** (Gr. *epi*, "upon" + *derma*, "skin"), the outermost layer of land plant organs. The epidermis functions to provide mechanical protection of inner tissue and to inhibit water loss. The cuticle consists of a thin, homogeneous, transparent layer of **cutin**, a polymer of fatty acids, and functions as a sealant, preventing excess water loss. Cutin also impregnates the outer cellulosic cell walls of epidermal cells; these are known as "cutinized" cell walls. The adaptive advantage of cutin and the cuticle is obvious: prevention of desiccation outside the ancestral water medium. In fact, plants that are adapted to very dry environments will often have a particularly thick cuticle (as in Figure 3.8) to inhibit water loss.

A third apomorphy for the land plants was the evolution of parenchyma tissue (Figure 3.9). All land plants grow by means of rapid cell divisions at the apex of the stem, shoot, and thallus or (in most vascular plants) of the root. This region of actively dividing cells is the **apical meristem**. The apical meristem of liverworts, hornworts, and mosses (discussed later), and of the monilophytes (see Chapter 4) have a single apical cell (Figure 3.9), probably the ancestral condition for the land plants. In all land plants the cells derived from the apical meristem region form a solid mass of tissue known as **parenchyma** (Gr. *para*, "beside" + *enchyma*, "an infusion"; in reference to a concept that parenchyma infuses or fills up space beside and between the other cells). Parenchyma tissue consists of cells that most resemble the unspecialized, undifferentiated cells of actively dividing meristematic tissue. Structurally, parenchyma cells are (1) elongate to isodiametric; (2) have a primary (1°) cell wall only (rarely a secondary wall); and (3) are living at maturity and potentially capable of continued cell divisions. Parenchyma cells function in metabolic activities such as respiration, photosynthesis, lateral transport, storage, and regeneration/wound healing. Parenchyma cells may further differentiate into other specialized cell types. It is not clear whether the evolution of both apical

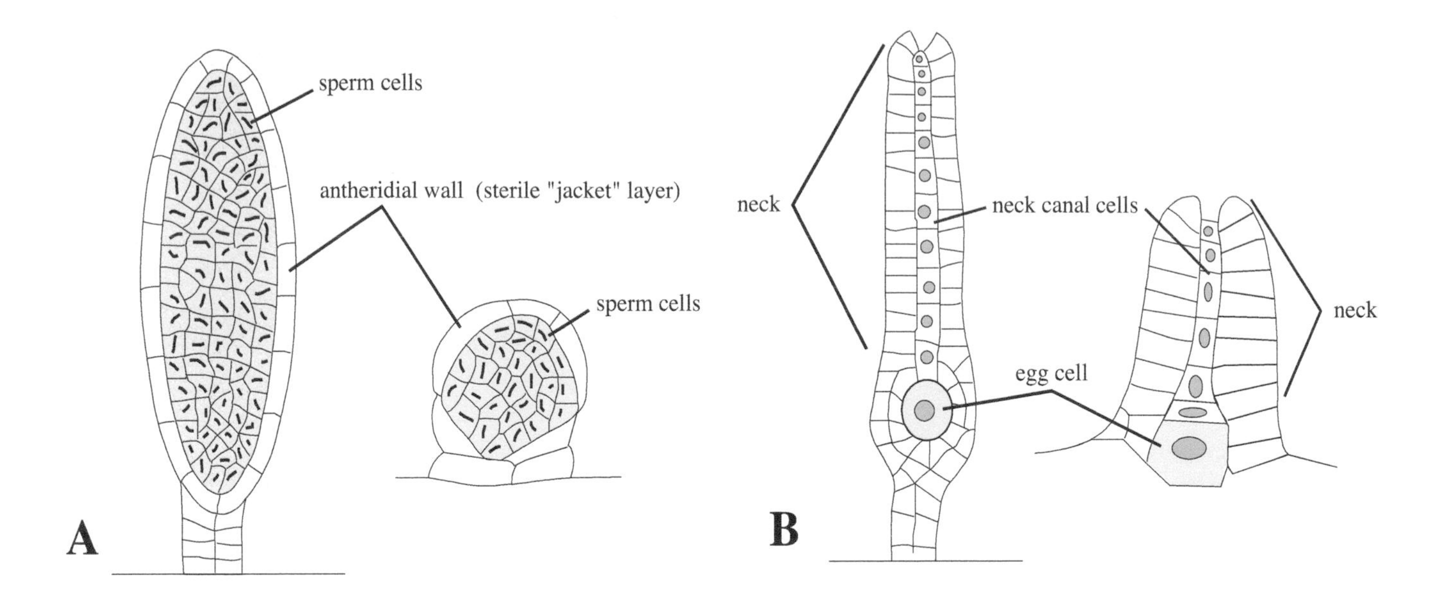

Figure 3.10 **A.** Antheridia. **B.** Achegonia. Both are apomorphies of land plants.

growth and true parenchyma is an apomorphy for the land plants alone, as shown here (Figure 3.6). Both may be interpreted to occur in certain closely related green plants, including the Charales.

Correlated with the evolution of parenchyma may have been the evolution of a **middle lamella** in land plants. The middle lamella is a pectic-rich layer that develops between the primary cell walls of adjacent cells (Figure 3.5A). Its function is to bind adjacent cells together, perhaps a prerequisite to the evolution of solid masses of parenchyma tissue.

Another evolutionary innovation for the land plants was the **antheridium** (Figure 3.10A). The antheridium is a type of specialized gametangium of the haploid (n) gametophyte, one that contains the sperm-producing cells. It is distinguished from similar structures in the Viridiplantae in being surrounded by a layer of sterile cells, the antheridial wall. The evolution of the surrounding layer of sterile wall cells, which is often called a sterile "jacket" layer, was probably adaptive in protecting the developing sperm cells from desiccation. In all of the nonseed land plants, the sperm cells are released from the antheridium into the external environment and must swim to the egg in a thin film of water. Thus, a wet environment is needed for fertilization to be effected in the nonseed plants, a vestige of their aquatic ancestry. Members of the Charales also have a structure termed an antheridium, which has an outer layer of sterile cells (Figure 3.5C,D). However, because of its differing anatomy, the Charales antheridium may not be homologous with that of the land plants, and thus may have evolved independently.

Another land plant innovation was the evolution of the **archegonium**, a specialized female gametangium (Figure 3.10B). The archegonium consists of an outer layer of sterile cells, termed the **venter**, that immediately surround the **egg** plus others that extend outward as a tube-like **neck**. The archegonium is stalked in some taxa; in others the egg is rather deeply embedded in the parent gametophyte. The egg cell is located inside and at the base of the archegonium. Immediately above the egg is a second cell, called the **ventral canal cell**, and above this and within the neck region, there may be several **neck canal cells**. The archegonium may have several adaptive functions. It may serve to protect the developing egg. It may also function in fertilization. Before fertilization occurs, the neck canal cells and ventral canal cell break down and are secreted from the terminal pore of the neck itself; the chemical compounds released function as an attractant, acting as a homing device for the swimming sperm. Sperm cells enter the neck of the archegonium and fertilize the egg cell to form a diploid (2n) zygote. In addition to effecting fertilization, the archegonium serves as a site for embryo/sporophyte development and the establishment of a nutritional dependence of the sporophyte upon gametophytic tissue.

The land plants share other possible apomorphies: the presence of various ultrastructural modifications of the sperm cells, flavonoid chemical compounds, and a proliferation of heat shock proteins. These are not discussed here.

DIVERSITY OF NONVASCULAR LAND PLANTS

During the early evolution of land plants, three major, monophyletic lineages diverged before or concurrent with the vascular plants (discussed in Chapter 4). These lineages may collectively be called the nonvascular land plants or "bryophytes" and include the liverworts, mosses, and hornworts. Although the monophyly of each of these groups is well supported, their interrelationships and relationships to the land plants are not, with almost every possible iteration having been proposed. The relationships portrayed in Figure 3.6A should be viewed as suspect, and to be treated as one of several possible phylogenies, including the option that the bryophytes are monophyletic (as in Figure 3.6B), necessitating a reinterpretation of homology and possible convergent evolution of some structures.

Liverworts, mosses, and hornworts differ from the vascular plants in lacking true vascular tissue and in having the gametophyte as the dominant, photosynthetic, persistent, and free-living phase of the life cycle; it is likely that the ancestral gametophyte of the land plants was thalloid in nature, similar to that of the hornworts and many liverworts. The sporophyte of the liverworts, hornworts, and mosses is relatively small, ephemeral, and attached to and nutritionally dependent upon the gametophyte (see later discussion).

LIVERWORTS

Liverworts, also traditionally called the **Hepaticae**, are one of the monophyletic groups that are descendants of some of the first land plants. Today, liverworts are relatively minor components of the land plant flora, growing mostly in moist, shaded areas (although some are adapted to periodically dry, hot habitats). Among the apomorphies of liverworts are (1) distinctive **oil bodies** and (2) specialized structures called **elaters**, elongate, nonsporogenous cells with spiral wall thickenings, found inside the sporangium. Elaters are hygroscopic, meaning that they change shape and move in response to changes in moisture content. Elaters function in spore dispersal; as the sporangium dries out, the elaters twist out of the capsule, carrying spores with them (Figures 3.11, 3.12K).

There are two basic morphological types of liverwort gametophytes: thalloid and leafy (Figures 3.11–3.13). **Thalloid** liverworts consist of a **thallus,** a flattened mass of tissue; this is likely the ancestral form, based on cladistic studies. As in hornworts and mosses, the gametophyte bears **rhizoids,** uniseriate, filamentous processes that function in anchorage and absorption. **Pores** in the upper surface of the thallus function in gas exchange (Figure 3.12B,L). These pores are not true stomata (discussed later), as they have no regulating guard cells. Some liverworts, as in the hornworts (discussed next), have a symbiotic relationship with Cyanobacteria. On the upper surface of the gametophytes of some thalloid liverworts, such as *Marchantia*, are specialized structures called **gemma cups**, which contain propagules called **gemmae**. These structures function in vegetative (asexual) reproduction; when a droplet of water falls into the gemma cup, the gemmae themselves may be dispersed some distance away, growing into a haploid genetic clone of the parent.

Leafy liverworts have gametophytes consisting of a stem axis bearing three rows of thin leaves. In most leafy liverworts, the stem is prostrate and the leaves are modified such that the upper two rows of leaves are larger and the lowermost row (on the stem underside) are reduced (Figures 3.11, 3.13). Other leafy liverworts are more erect, with the three rows of leaves similar. The leaves of leafy liverworts evolved independently from those of mosses (discussed later) or vascular plants (Chapter 4).

As in all of the early diverging land plant lineages, liverworts have antheridia and archegonia that develop on the gametophyte. In some liverwort taxa (e.g., *Marchantia*), the gametangia form as part of stalked, peltate structures: **antheridiophores** bearing antheridia and **archegoniophores** bearing archegonia (Figures 3.11, 3.12). Sperm released from an antheridium of the antheridiophore swims in a film of water to the archegonia of the archegoniophore, effecting fertilization.

After fertilization the zygote divides mitotically and eventually differentiates into a diploid (2n) embryo, which matures into the diploid (2n) sporophyte. This sporophyte is relatively small, nonphotosynthetic, and short lived. It consists almost entirely of a **sporangium**, also called a **capsule** (Figure 3.12F,J). At a certain stage, the internal cells of the capsule divide meiotically, forming haploid (n) spores (see Figure 3.7). In liverworts the spores are released by a splitting of the capsule into four valves. The spores may land on a substrate, germinate (under the right conditions), and grow into a new gametophyte, completing the life cycle.

MOSSES

The mosses, or **Musci**, are by far the most speciose and diverse of the three major groups of nonvascular land plants and inhabit a number of ecological niches. Mosses may share one apomorphy with the hornworts (discussed later) and vascular plants: possession of stomates (Figure 3.6A). However, it should be noted that the evolution of stomates in these bryophytes may have differed from that of the land plants (see, e.g., Duckett and Pressel 2017 for details). In this scenario, stomates either evolved independently in the two groups, or, alternatively, evolved once, functioning in sporagium dehis-

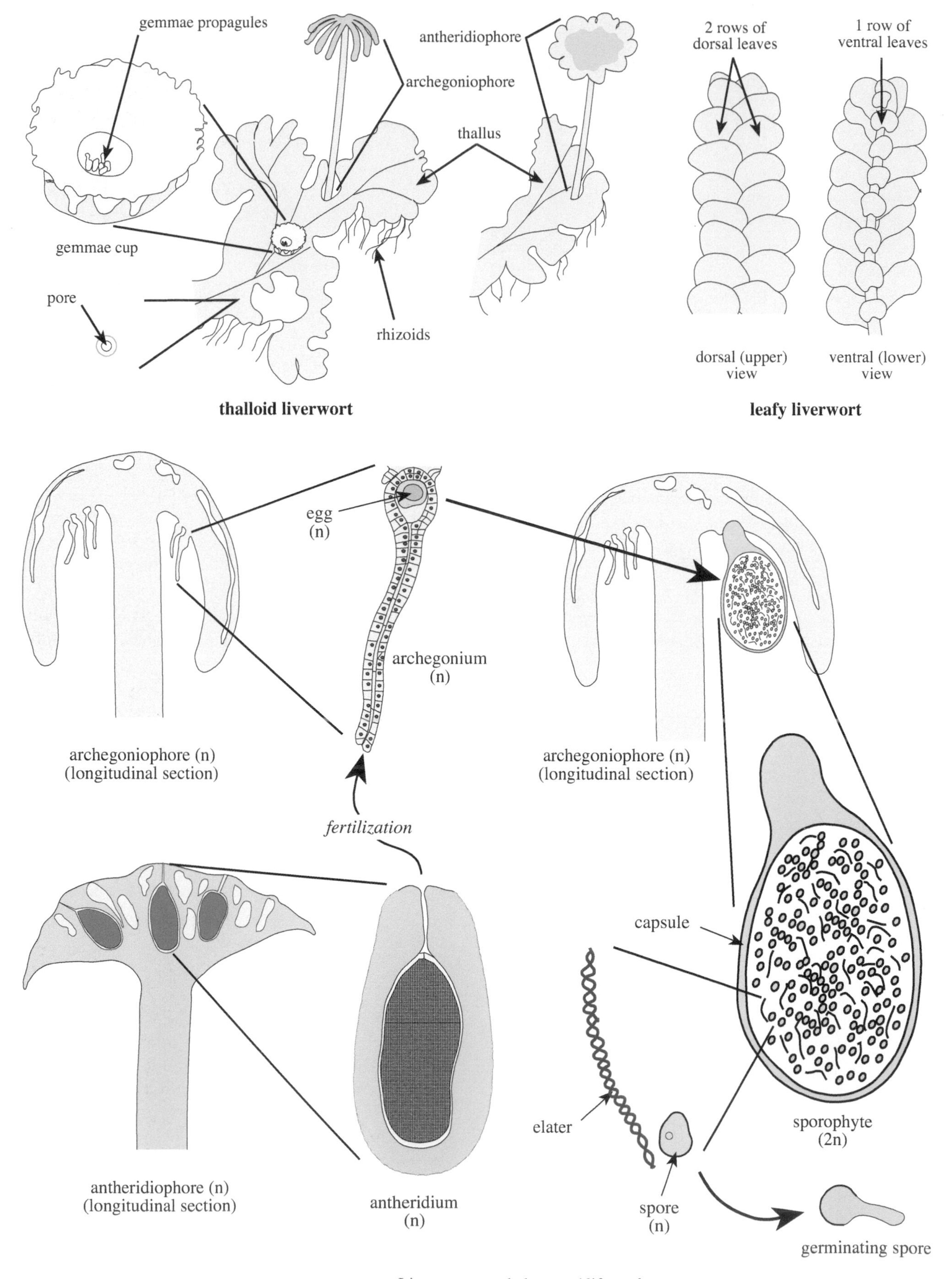

Figure 3.11 Liverwort morphology and life cycle.

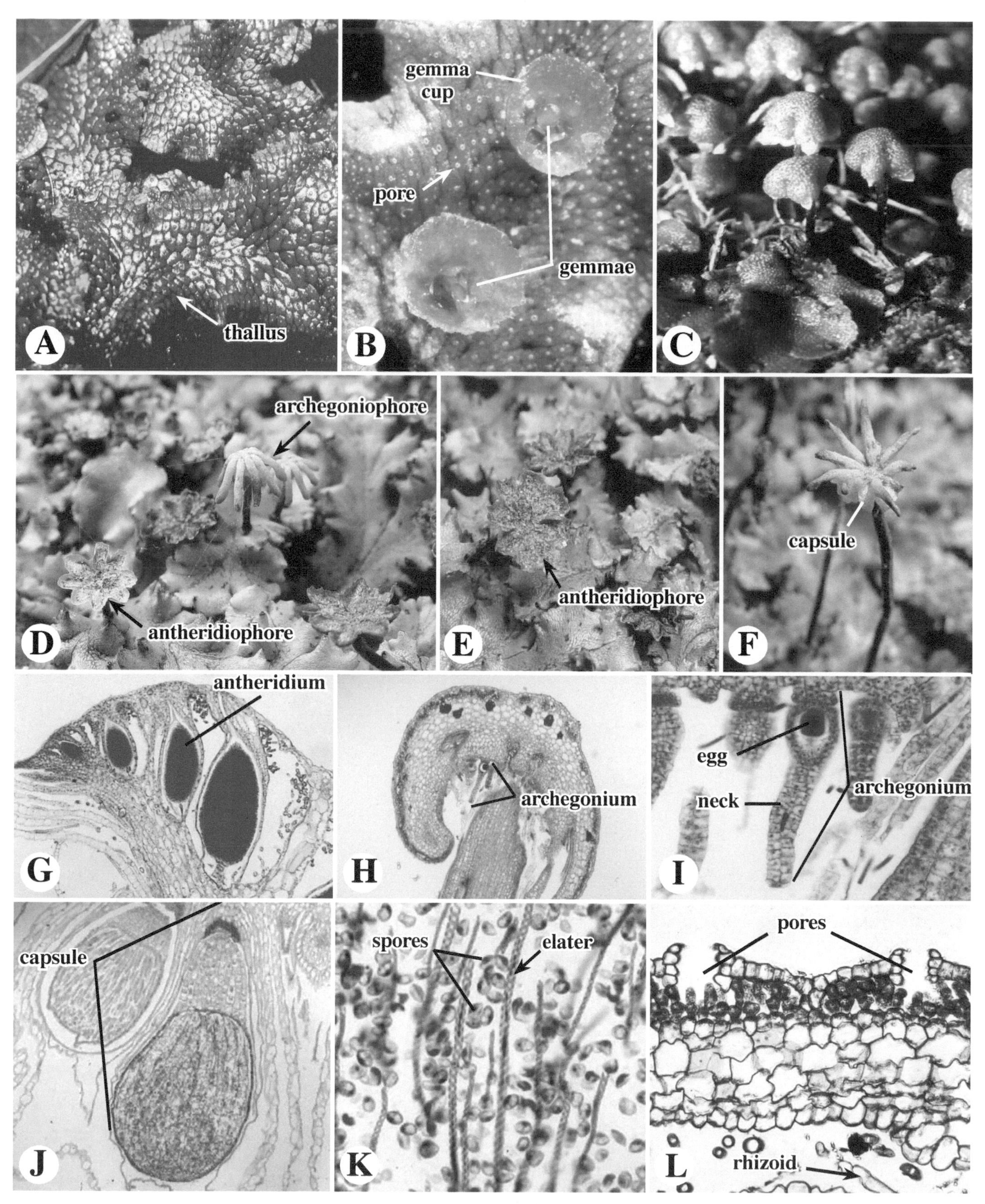

Figure 3.12 Hepaticae – Liverworts. **A.** *Conocephalum* sp., a thalloid liverwort. **B.** *Marchantia*, thallus with gemma cups and gemmae. Note whitish pores. **C.** *Asterella*, a thalloid liverwort with archegoniophores. **D–L.** *Marchantia*. **D.** Thallus with antheridiophores and archegoniophores. **E.** Antheridiophore, close-up. **F.** Archegoniophore, showing capsules beneath lobes. **G.** Antheridiophore, longitudinal section. **H.** Archegoniophore, longitudinal section. **I.** Archegonium. **J.** Capsule, longitudinal section, showing sporogenous tissue. **K.** Close-up, sporogenous tissue, showing spores and elaters. **L.** Cross-section of thallus, showing rhizoids and upper pores.

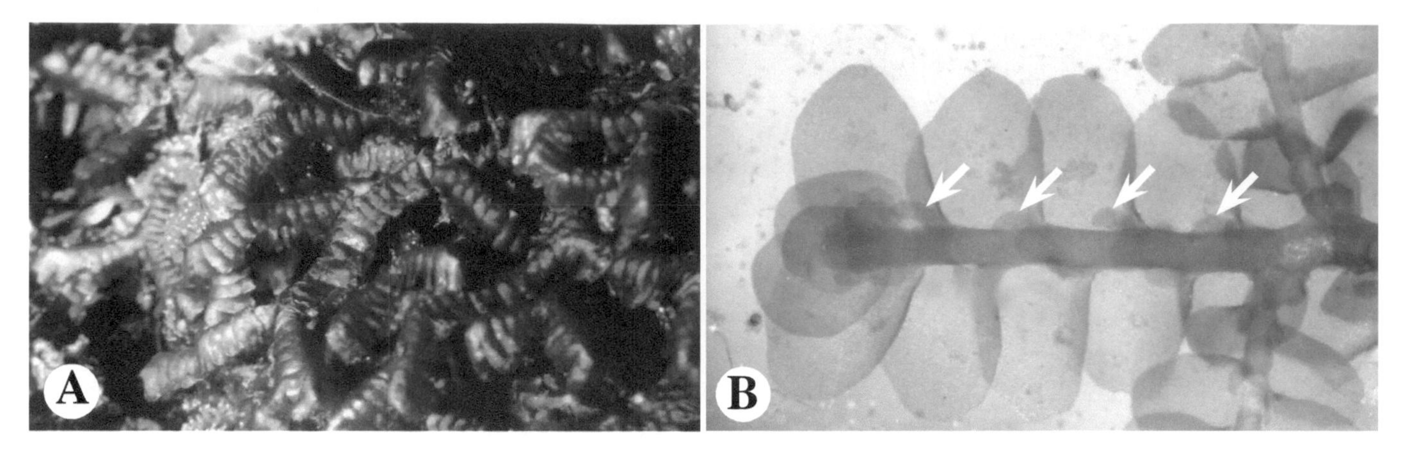

Figure 3.13 Hepaticae – Leafy liverworts. **A.** *Bazania trilobata*, a leafy liverwort. **B.** *Porella*, a leafy liverwort, showing third row of reduced leaves at arrows (lower side facing).

cence in bryophytes and later acquiring function in photosynthetic gas exchange in the vascular plants (see Figure 3.6B).

Stomates (also termed **stomata**) are specialized epidermal cells generally found on leaves, but sometimes on stems. Stomata consist of two chloroplast-containing cells, the **guard cells**, which, by changes in turgor pressure, can increase or decrease the size of the opening between them, the **stoma** (Figure 3.14). Each guard cell has one or more ridge-like deposits on the side facing the stoma (Figure 3.14). This material, which is rich in **suberin**, a waxy, water-resistant substance, functions to better seal the stoma. Stomata generally function in regulation of gas exchange, in terms of both photosynthesis and water uptake. Carbon dioxide passing through the stoma diffuses to the chloroplasts of photosynthetic cells within and is used in the dark reactions of photosynthesis. Oxygen, a by-product of photosynthesis, exits via the stoma. Stomata also allow water vapor to escape from the leaf. In most plants stomata open during the day when photosynthesis takes place; thus, heat from the sun may cause considerable water loss through stomata. In some plants, loss of water via stomata is simply a by-product, a price to be paid for entry of carbon dioxide, which is essential for photosynthesis. However, in other plants, such as tall trees, stomatal water loss may actually be adaptive and functional, as a large quantity of water must flow through the leaves in order to supply sufficient quantities of mineral nutrients absorbed via the roots.

A second apomorphy, possibly shared among mosses, hornworts, and vascular plants (in the scenario of Figure 3.6A), is an **elongate, aerial sporophyte axis**. The elongate, aerial sporophyte seen in mosses and hornworts may be a possible precursor to the evolution of the dominant, aerial sporophytic stem in vascular plants (see Chapter 4).

Mosses have a number of autapomorphies. First, some mosses have specialized conductive cells called **hydroids**, which function in water conduction, and **leptoids**, which function in sugar conduction. These cells resemble typical xylem tracheary elements and phloem sieve elements (Chapter 4), but lack the specializations of the latter cell types. They likely evolved independently of vascular tissue (Figure 3.6), although alternative hypotheses of "bryophyte" relationships argue that hydroids and leptoids may represent intermediate structures in the evolution of true vascular tissue. Second, the spores of mosses have a thick outer layer called a **perine layer** (Figure 3.15), which may be apomorphic for the mosses alone (Figure 3.7) or possibly for the

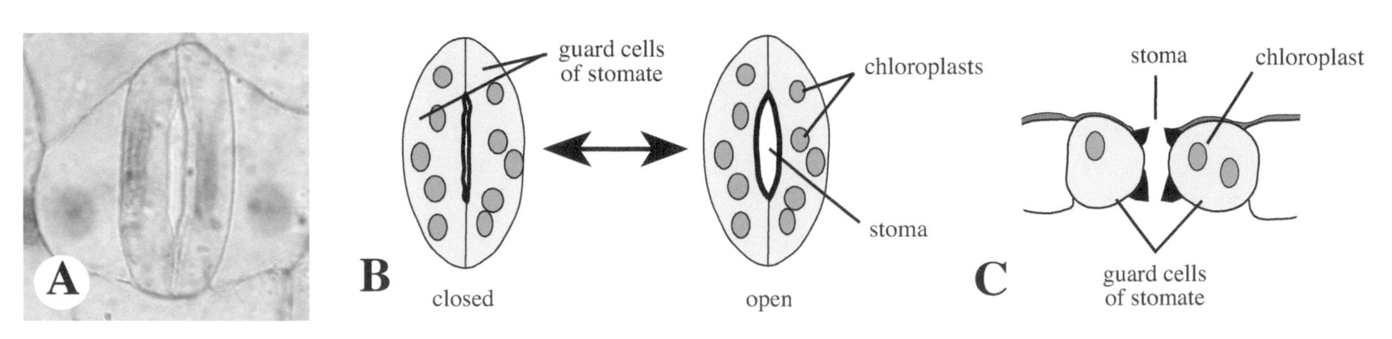

Figure 3.14 The stomate, an innovation for mosses, hornworts, and vascular plants. **A.** Face view, slightly open. **B.** Diagram, face view, open and closed. **C.** Diagram, cross section.

mosses and vascular plants combined. The perine layer may function in preventing excess desiccation and provide additional mechanical protection of the spore cytoplasm. As with liverworts and hornworts, a three-lined structure, called a **trilete mark**, develops on the spore wall; the trilete mark is the scar of attachment of the adjacent three spores of the four spores produced at meiosis (Figure 3.16; see also Chapter 4). Third, moss gametophytes are always leafy, with a variable number of leaf ranks or rows (Figures 3.16, 3.17B). The leaves of mosses are thought to have evolved independently from those in liverworts and, thus, constitute an apomorphy for the mosses alone. Moss leaves are mostly quite small and thin, but may have a central **costa** (Figure 3.17C), composed of conductive cells, that resembles a true vein.

Antheridia and archegonia in mosses are usually produced at the apex of gametophytic stems (Figures 3.16, 3.17D–F). After fertilization, the sporophyte grows upward (Figures 3.16, 3.17G) and often carries the apical portion of the original archegonium, which continues to grow. This apical archegonial tissue, known as a **calyptra** (Figures 3.16, 3.17H), may function in protecting the young sporophyte apex. The sporophyte generally develops a long stalk, known as a **stipe**, at the apex of which is born the sporangium or capsule (Figures 3.16, 3.17G,H). The capsule of most mosses has a specialized mechanism of dehiscence. At the time of spore release, a lid known as an **operculum** falls off the capsule apex, revealing a whorl of **peristome teeth**. The peristome teeth, like the elaters of liverworts, are hygroscopic. As the capsule dries up, the peristome teeth retract, effecting release of the spores (Figures 3.16, 3.17H,I).

Under the right environmental conditions, moss spores will germinate and begin to grow into a new gametophyte. The initial development of the gametophyte results in the formation of filamentous structure, known as a **protonema** (Figures 3.16, 3.17A). The protonema probably represents an ancestral vestige, resembling a filamentous green "alga." After a period of growth, the protonema grows into a parenchymatous gametophyte.

One economically important moss worth mentioning is the genus *Sphagnum*, or peat moss, containing numerous species. *Sphagnum* grow in wet bogs and chemically modifies its environment by making the surrounding water acidic. The leaves of *Sphagnum* are unusual in having two cell types: **chlorophyllous cells**, which form a network, and large, clear **hyaline cells**, having characteristic pores and helical thickenings (Figure 3.18). The pores of the hyaline cells give *Sphagnum* remarkable properties of water absorption and retention, making it quite valuable horticulturally in potting mixtures. **Peat** is fossilized and partially decomposed *Sphagnum* and is mined for use in potting mixtures and as an important fuel source in parts of the world. See Liu et al. (2019) for details on the classification of mosses.

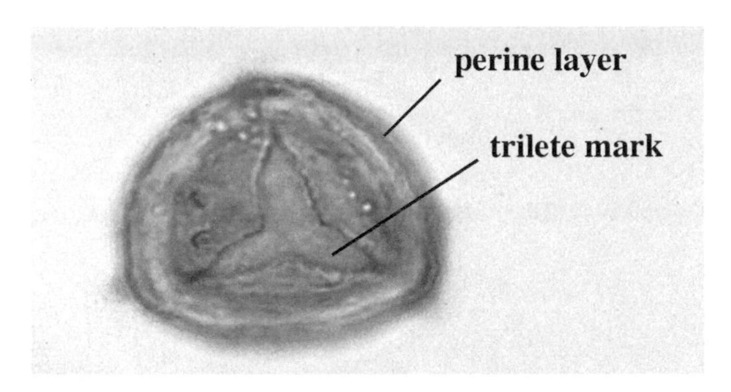

Figure 3.15 Moss spore. Note protective perine layer and trilete mark.

HORNWORTS

The hornworts, formally known as **Anthocerotae**, are a monophyletic group comprising a third extant lineage of non-vascular land plants. Hornworts are similar to the thalloid liverworts in gametophyte morphology and are found in similar habitats. Hornworts differ from liverworts, however, in lacking pores, with some species having stomates, a presumed apomorphy of all land plants except liverworts. All hornworts have a symbiotic relationship with cyanobacteria (blue-greens), which live inside cavities of the thallus. This relationship is found in a few thalloid liverworts as well (probably evolving independently), but not in mosses. Interestingly, hornworts and liverworts may also have a symbiotic association between the gametophytes and a fungus, similar to the mycorrhizal association with the roots of vascular plants.

The basic life cycle of hornworts is similar to that of liverworts and mosses. The sporophyte of hornworts is similar to that of mosses in being aerial and elongate, but unique in being cylindrical, and photosynthetic (Figure 3.19A,B). This cylindrical sporophyte has indeterminate (potentially continuous) growth, via a basal, **intercalary meristem** (Figure 3.19E). The intercalary meristem is a region of actively dividing cells near the base of the sporophyte (just above the point of attachment to the gametophyte), constituting an apomorphy for the hornworts. This region is surrounded by a protective collar of gametophytic tissue (Figure 3.19C). The proximal end of the sporophyte is embedded within gametophytic tissue, its surface somewhat lobed (Figure 3.19D,E). Other apomorphies include a unique central column of sterile (non-spore-producing) tissue called a **columella** and the production of specialized structures in the sporangium called **pseudo-elaters** (Figure 3.19F), groups of cohering, nonsporogenous, elongate, generally hygroscopic cells, which are

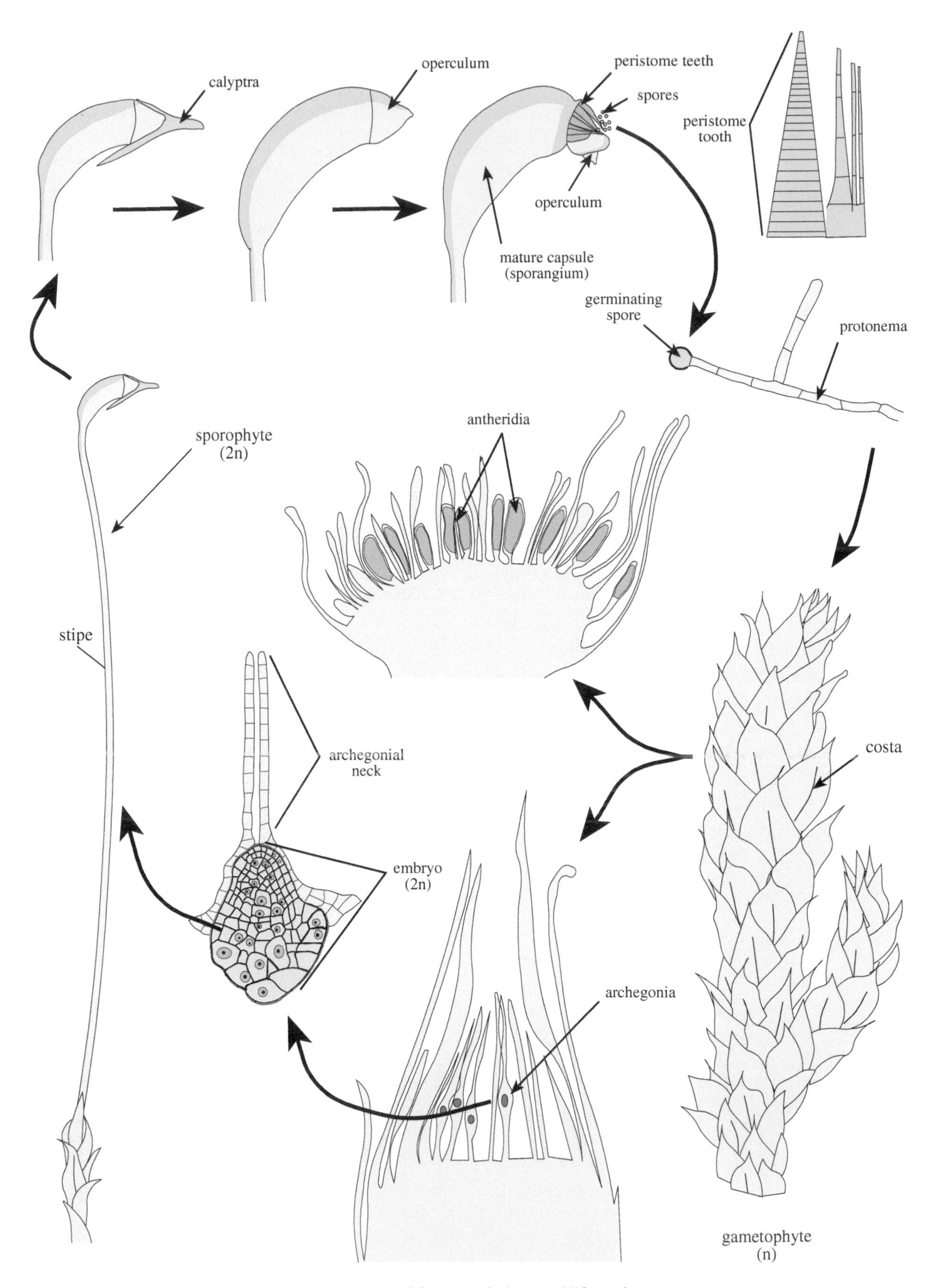

Figure 3.16 Moss morphology and life cycle.

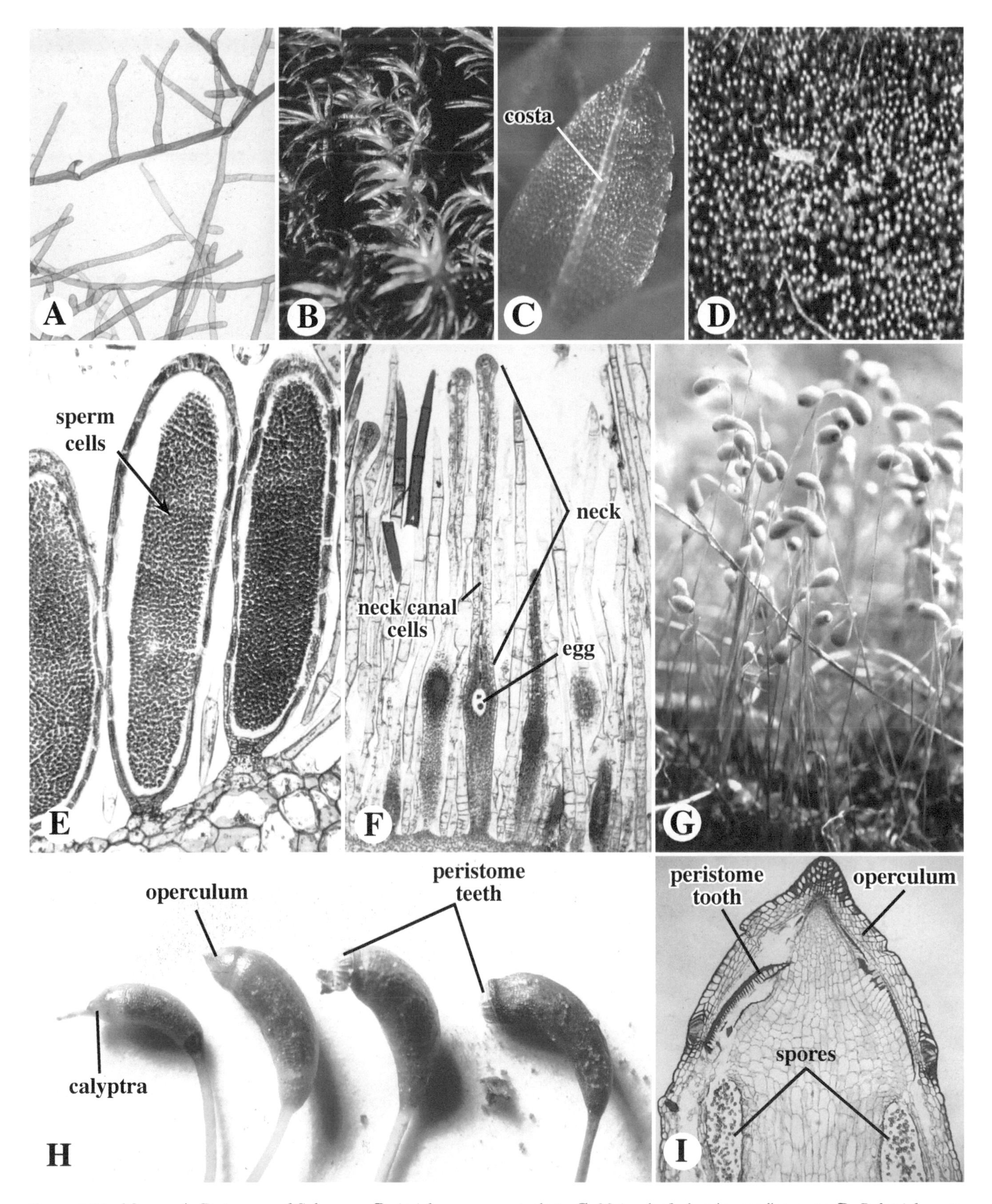

Figure 3.17 Mosses. **A.** Protonema of *Sphagnum*. **B.** *Atrichum* sp. gametophyte. **C.** *Mnium* leaf, showing median costa. **D.** *Polytrichum* sp. gametophyte, face view, showing antheridia at tips of branches. **E,F.** *Mnium* sp. **E.** Antheridia, longitudinal section, showing external capsule wall (sterile layer) and internal sporogenous tissue. **F.** Archegonia, showing stalk, egg cell, neck, and neck canal cells. **G.** Sporophytes of moss, showing capsules. **H.** Moss sporophyte close-up, showing developmental series (left to right). **I.** *Mnium*, capsule (sporangium) longitudinal section, showing operculum, one of several peristome teeth, and spores within sporangium.

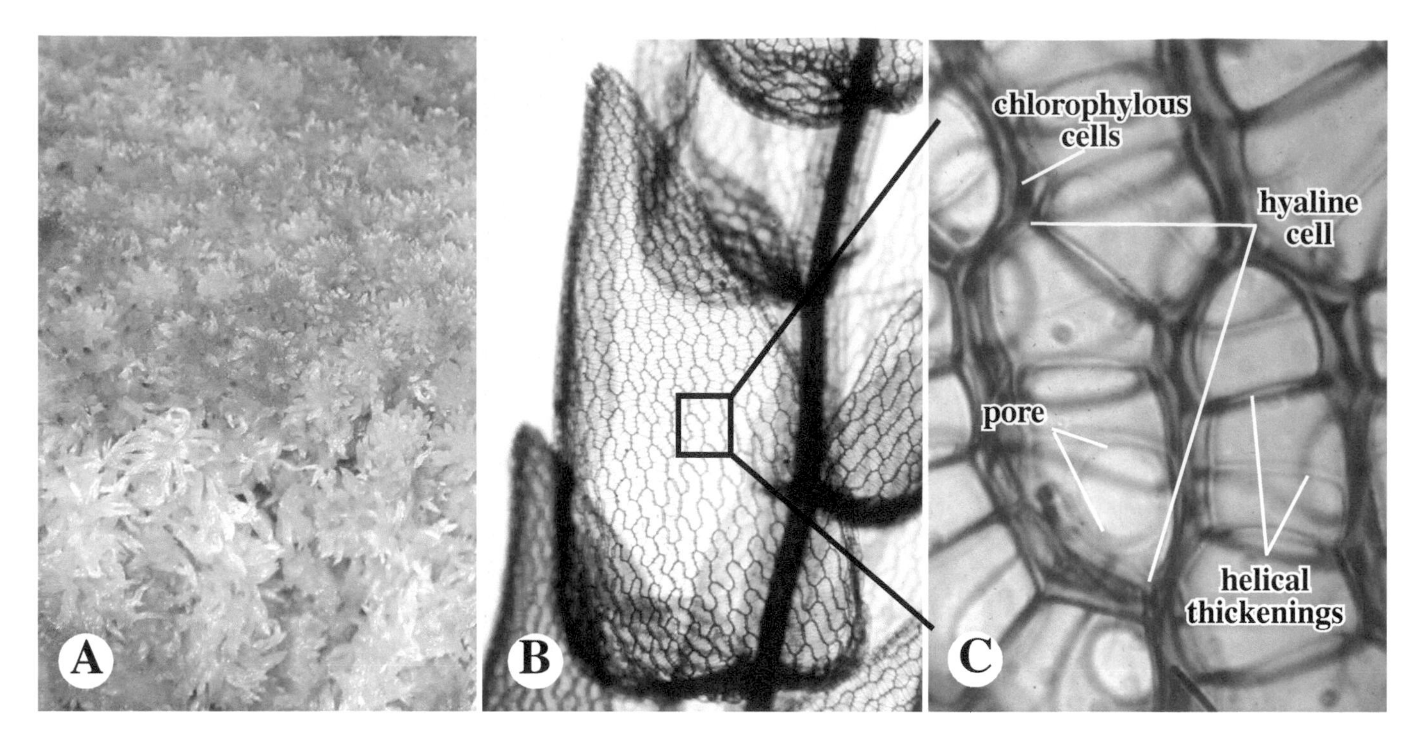

Figure 3.18 *Sphagnum*, or peat moss. **A.** Clonal population. **B.** Individual leaf at center, showing the specialized chlorophyllous and hyaline cells. **C.** Leaf close-up, showing chlorophyllous cells, hyaline cells, pores, and spiral wall thickenings of hyaline cells.

nonhomologous but have a similar function to the elaters of liverworts.

Some molecular analyses place the hornworts as sister to the vascular plants (Figure 3.6A). If true, one possible apomorphy shared between hornworts and vascular plants is related to the sporophyte. The sporophyte of hornworts is photosynthetic and relatively long-lived. In fact, the sporophyte of some hornworts is capable of persisting independent of the gametophyte for long periods. In addition, the foot of hornworts is somewhat lobed and the surface is compared to incipient rhizoids. Thus, by this hypothesis, hornwort sporophytes may represent a transition to the very dominant, long-lived sporophytes of vascular plants (Chapter 4).

POLYSPORANGIOPHYTES/PAN-TRACHEOPHYTA

This group is inclusive of a few, basal fossil taxa plus all of the true vascular plants, or tracheophytes (Chapter 4). The basal (first-evolving) polysporangiophytes (formally called Polysporangiomorpha or Pan-Tracheophyta; see Chapter 4), such as the genus *Horneophyton* (not illustrated), were similar to hornworts, liverworts, and mosses in lacking vascular tissue. However, they are different from "bryophytes," and linked to the vascular plants, in having branched stems with multiple sporangia. Thus, the polysporangiophytes include taxa that were transitional to the evolution of tracheophytes.

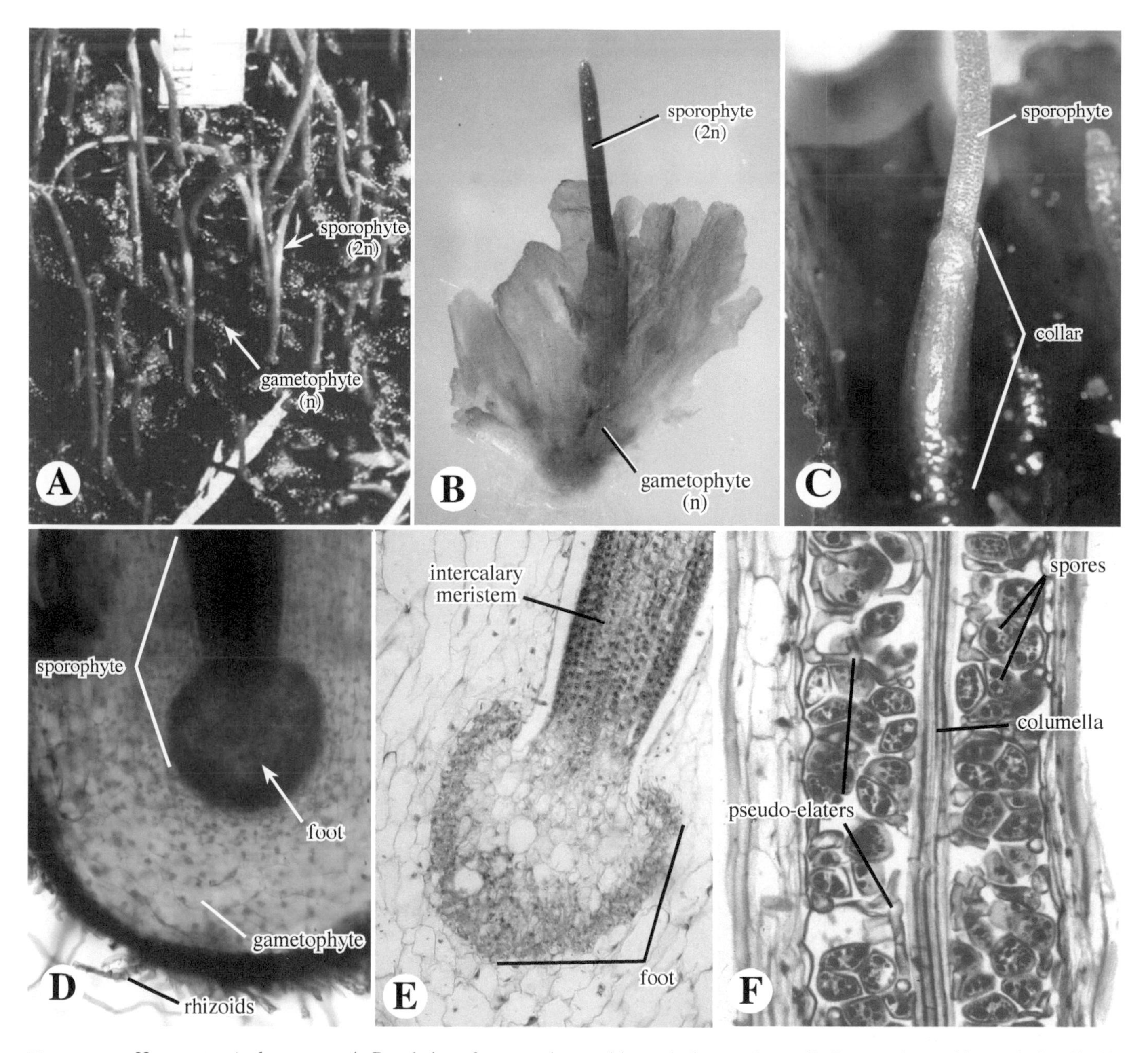

Figure 3.19 Hornworts, *Anthoceros* sp. **A.** Population of gametophytes with attached sporophytes. **B.** Gametophyte with attached, cylindrical sporophyte. **C.** Close-up of sporophyte base, showing ensheathing collar of gametophytic tissue surrounding intercalary meristem of sporophyte. **D.** Whole mount of sporophyte base, showing foot embedding within gametophyte. **E.** Longituidnal section of sporophyte base. Note basal foot and actively dividing cells of intercalary meristem at sporophyte base. **F.** Sporophyte longitudinal section, showing columella, spores, and pseudo-elaters.

REVIEW QUESTIONS

GREEN PLANTS

1. What are two formal names for the green plants?
2. What are apomorphies for the green plants?
3. The bulk of the primary cell wall of green plants is composed of what substance? (Give the common name and chemical name.)
4. Is the cell wall synthesized inside or outside the plasma membrane?
5. What are the unique features of green plant chloroplasts?
6. How are chloroplasts thought to have originated (i.e., by what evolutionary process)?
7. What is a haplontic life cycle? Draw and label.
8. Name the three groups of "green algae" that are most closely related to the land plants.
9. What is oogamy?
10. Describe and give the function of plasmodesmata.

LAND PLANTS

11. What is the formal name for the land plants?
12. Name the major apomorphies of the land plants.
13. Draw and label the basic haplodiplontic life cycle (alternation of generations) of all land plants, illustrating all structures, processes, and ploidy levels.
14. What is an embryo?
15. What is a sporangium?
16. Name the possible adaptive features of the sporophyte.
17. What are cutin and cuticle and what are their adaptive significance?
18. Define apical growth and parenchyma.
19. In land plants what is the name of the pectic-rich layer between adjacent cell walls that functions to bind them together?
20. What is an antheridium? Draw and label the parts.
21. What is an archegonium? Draw and label the parts.

LIVERWORTS, MOSSES, AND HORNWORTS

22. What is the formal name of the liverworts?
23. Name two apomorphies of the liverworts.
24. What is the function of elaters?
25. What are the two major morphological forms of liverworts? Which is likely ancestral?
26. What are gemmae and gemma cups?
27. Describe the morphology of the leaves of leafy liverworts.
28. What is an antheridiophore? An archegoniophore?
29. Describe the structural makeup and function of a stomate.
30. What land plant groups possess stomates?
31. What possible apomorphies may be shared by the mosses, hornworts, and vascular plants?
32. What is the formal name of the mosses?
33. Name major apomorphies shared by the mosses alone.
34. What is a calyptra, stipe, operculum, peristome tooth?
35. What is the scientific name of peat moss?
36. What feature of the leaf anatomy of peat moss enables the leaves to absorb and retain water?
37. How is peat moss of economic importance?
38. What is the formal name of hornworts?
39. Describe the major features of hornworts, citing how they differ from the liverworts and mosses.
40. What is the function of pseudoelaters, and how do they differ structurally from the elaters of liverworts?
41. What feature of the sporophyte might unite the hornworts with the vascular plants?
42. What apomorphy links the Pan-Tracheophyta / Polysporangiophytes with the vascular plants?

EXERCISES

1. Peruse the recent literature on phylogenetic relationships of the "green algae" relative to the land plants. Are there any differences relative to Figure 3.1?
2. Peruse the recent literature on phylogenetic relationships of the hornworts, liverworts, and mosses (see References below). Review the relationships and the evidence thereof of the three bryophyte lineages; which of them differ from that portrayed in Figure 3.6?
3. Peruse botanical journals and find a systematic article on a moss, liverwort, or hornwort. What is the objective of the article and what techniques were used to address it?
4. Collect and identify local liverworts, hornworts, and mosses. What features are used to distinguish among families, genera, and species?

REFERENCES FOR FURTHER STUDY

Bremer, Kåre. 1985. Summary of green plant phylogeny and classification. Cladistics 1(4): 369–385.

Burleigh, J. G., M. S. Bansal, O. Eulenstein, S. Hartmann, A. Wehe, and T. J. Vision. 2011. Genome-scale phylogenetics: Inferring the plant tree of life from 18,896 gene trees. Systematic Biology 60:117-125.

Cox, C. J., B. Goffinet, A. J. Shaw, and S. B. Boles 2004 Phylogenetic relationships among the mosses based on heterogeneous Bayesian analysis of multiple genes from multiple genomic compartments. Systematic Botany 29: 234–250.

Cox, C. J. 2018. Land plant molecular phylogenetics: A review with comments on evaluating incongruence among phylogenies. Critical Reviews in Plant Sciences 37:113-127.

Crandall-Stotler B., and R. E. Stotler. 2000. Morphology and classification of the Marchantiophyta. In: AJ Shaw, B Goffinet, eds. Bryophyte biology. Pp. 21–70. Cambridge University Press, Cambridge.

Crum, H. 2001. Structural diversity of bryophytes. University of Michigan Herbarium, Ann Arbor.

de Sousa, F., P. G. Foster, P. C. J. Donoghue, H. Schneider, and C. J. Cox. 2019. Nuclear protein phylogenies support the monophyly of the three bryophyte groups (Bryophyta Schimp.). New Phytologist 222:565-575.

Duckett, J. G. and S. Pressel. 2018. The evolution of the stomatal apparatus: Intercellular spaces and sporophyte water relations in bryophytes—two ignored dimensions. Philosophical Transactions of the Royal Society B: Biological Sciences 373:20160498.

Duff, R. J., D. C. Cargill, J. C. C. Villarreal, and K. S. Renzaglia. 2004. Phylogenetic relationships of the hornworts based on rbcL sequence data: novel relationships and new insights. In: B. Goffinet, V. Hollowell, and R. Magill (eds.). Molecular systematics of bryophytes. Pp. 41–58. Missouri Botanical Garden, St. Louis.

Forrest, L.L., E. C. Davis, D. G. Long, B. J. Crandall-Stotler, A. Clark, and M. L. Hollingsworth. 2006. Unraveling the evolutionary history of the liverworts (Marchantiophyta): multiple taxa, genomes and analyses. Bryologist 109: 303–334.

Garbary, D. J., K. S. Renzaglia, and J. G. Duckett. 1993. The phylogeny of land plants-a cladistic analysis based on male gametogenesis. Plant Systematics and Evolution 188:237-269.

Gensel, P. G., and D. Edwards (eds.). 2001. Plants invade the land: evolutionary and environmental perspectives. Columbia University Press, New York.

Goffinet, B., and W. R. Buck. 2004. Systematics of the bryophyta (mosses): from molecules to a revised classification. In: B. Goffinet, V Hollowell, and R Magill (eds.) Molecular systematics of bryophytes. Missouri Botanical Garden, St. Louis.

Goremykin, V. V., and F. H. Hellwig. 2005. Evidence for the most basal split in land plants dividing bryophyte and tracheophyte lineages. Plant Systematics and Evolution 254:93–103.

Graham, L. E. 1985. The origin of the life cycle of land plants. American Scientist 73: 178–186.

Graham, L. E. 1993 Origin of land plants. Wiley, New York.

Groth-Malonek, M., D. Pruchner, F. Grewe, and V. Knoop. 2005. Ancestors of trans-splicing mitochondrial introns support serial sister group relationships of hornworts and mosses with vascular plants. Molecular Biology and Evolution 22:117-125.

He-Nygren, X., A. Juslen, I. Ahonen, D. Glenny, and S. Piippo. 2006. Illuminating the evolutionary history of liverworts (Marchantiophyta): towards a natural classification. Cladistics 22:1–31.

Heinrichs, J., S. R. Gradstein, R. Wilson, and H. Schneider. 2005. Towards a natural classification of liverworts (Marchantiophyta) based on the chloroplast gene rbcL. Cryptogamie Bryologie 26:131–150.

Karol, K. G., R. M. McCourt, M. T. Cimino, and C. F. Delwiche. 2001. The closest living relatives of land plants. Science 294: 2351–2353.

Karol, K. G., K. Arumuganathan, J. L. Boore, A. M. Duffy, K. D. E. Everett, J. D. Hall, S. K. Hansen, J. V. Kuehl, D. F. Mandoli, B. D. Mishler, R. G. Olmstead, K. S. Renzaglia, and P. G. Wolf. 2010. Complete plastome sequences of Equisetum arvense and Isoetes flaccida: implications for phylogeny and plastid genome evolution of early land plant lineages. BMC Evolutionary Biology 10:321.

Kelch, D. G., A. Driskell, and B. D. Mishler 2004 Inferring phylogeny using genomic characters: a case study using land plant plastomes. In: B. Goffinet, V. Hollowell, and R. Magill (eds.) Molecular systematics of bryophytes. Pp. 3–11. Missouri Botanical Garden Press, St. Louis.

Kenrick, P., and P. R. Crane. 1997. The origin and early diversification of land plants: a cladistic study. Smithsonian Institution Press, Washington, DC.

Liu, Y., M. G. Johnson, C. J. Cox, R. Medina, N. Devos, A. Vanderpoorten, L. Hedenäs, N. E. Bell, J. R. Shevock, B. Aguero, D. Quandt, N. J. Wickett, A. J. Shaw, and B. Goffinet. 2019. Resolution of the ordinal phylogeny of mosses using targeted exons from organellar and nuclear genomes. Nature Communications 10: 1485.

Mishler, B. D., and S. P. Churchill. 1984. A cladistic approach to the phylogeny of the "Bryophytes." Brittonia 36(4): 406–424.

Mishler, B. D., and S. P. Churchill. 1985. Transition to a land flora: phylogenetic relationships of the green algae and bryophytes. Cladistics 1(4): 305–328.

Mishler, B. D., L. A. Lewis, M. A. Buchheim, K. S. Renzaglia, D. J. Garbary, C. F. Delwiche, F. W. Zechman, T. S. Kantz, and R. L. Chapman. 1994. Phylogenetic relationships of the "Green Algae" and "Bryophytes." Annals of the Missouri Botanical Garden 81: 451–483.

Nickrent, D. L., C. L. Parkinson, J. D. Palmer, and R. J. Duff. 2000. Multigene phylogeny of land plants with special reference to bryophytes and the earliest land plants. Molecular Biology and Evolution 17: 1885–1895.

Nishiyama, T., Paul G. Wolf, M. Kugita, R. B. Sinclair, M. Sugita, C. Sugiura, T. Wakasugi, K. Yamada, K. Yoshinaga, K. Yamaguchi, K. Ueda, and M. Hasebe. 2004. Chloroplast phylogeny indicates that bryophytes are monophyletic. Molecular Biology and Evolution 21:1813-1819.

Pickett-Heaps, J. D. 1975. Green algae: structure, reproduction and evolution in selected genera. Sinauer, Sunderland, MA.

Qiu, Y. L., Y. Cho, J. C. Cox, and J. D. Palmer. 1998. The gain of three mitochondrial introns identifies liverworts as the earliest land plants. Nature 394:671-674.

Qiu, Y.-L., L. Li, B. Wang, Z. Chen, V. Knoop, M. Groth-Malonek, O. Dombrovska, J. Lee, L. Kent, J. Rest, G. F. Estabrook, T. A. Hendry, D. W. Taylor, C. M. Testa, M. Ambros, B. Crandall-Stotler, R. J. Duff, M. Stech, W. Frey, D. Quandt, and C. C. Davis. 2006. The deepest divergences in land plants inferred from phylogenomic evidence. Proceedings of the National Academy of Sciences 103:15511.

Qiu, Y-L., L. Li, B. Wang, Z. Chen, O. Dombrovska, J. Lee, L. Kent, R. Li, R. W. Jobson, T. A. Hendry, D. W. Taylor, C. M. Testa, and M. Ambros. 2007. A nonflowering land plant phylogeny inferred from nucleotide sequences of seven chloroplast, mitochondrial, and nuclear genes. International Journal of Plant Sciences 168: 691-708.

Qiu, Y. L. 2008. Phylogeny and evolution of charophytic algae and land plants. Journal of Systematics and Evolution 46:287-306.

Rensing, S. A. 2018. Plant evolution: Phylogenetic relationships between the earliest land plants. Current Biology 28:R210-R213.

Renzaglia, K. S., R. J. Duff, D. L. Nickrent, and D. J. Garbary. 2000. Vegetative and reproductive innovations of early land plants: implications for a unified phylogeny. Philosophical Transactions: Biological Sciences 355: 769–793.

Renzaglia, K. S. and D. J. Garbary. 2001. Motile gametes of land plants: Diversity, development, and evolution. Critical Reviews in Plant Sciences 20:107-213.

Renzaglia, K. S., S. Schuette, R. J. Duff, R. Ligrone, A. J. Shaw, B. D. Mishler, and J. G. Duckett. 2007. Bryophyte phylogeny: Advancing the molecular and morphological frontiers. The Bryologist 100:179-213.

Ruhfel, B. R., M. A. Gitzendanner, P. S. Soltis, D. E. Soltis, and J. G. Burleigh. 2014. From algae to angiosperms - inferring the phylogeny of green plants (Viridiplantae) from 360 plastid genomes. BMC Evolutionary Biology 14:23.

Shaw, J. and K. Renzaglia. 2004. Phylogeny and diversification of bryophytes. American Journal of Botany 91:1557-1581.

Wickett, N. J., S. Mirarab, N. Nguyen, T. Warnow, E. Carpenter, N. Matasci, S. Ayyampalayam, M. S. Barker, J. G. Burleigh, M. A. Gitzendanner, B. R. Ruhfel, E. Wafula, J. P. Der, S. W. Graham, S. Mathews, M. Melkonian, D. E. Soltis, P. S. Soltis, N. W. Miles, C. J. Rothfels, L. Pokorny, A. J. Shaw, L. DeGironimo, D. W. Stevenson, B. Surek, J. C. Villarreal, B. Roure, H. Philippe, C. W. dePamphilis, T. Chen, M. K. Deyholos, R. S. Baucom, T. M. Kutchan, M. M. Augustin, J. Wang, Y. Zhang, Z. Tian, Z. Yan, X. Wu, X. Sun, G. K.-S. Wong, and J. Leebens-Mack. 2014. Phylotranscriptomic analysis of the origin and early diversification of land plants. Proceedings of the National Academy of Sciences 111:E4859.

4

EVOLUTION AND DIVERSITY OF VASCULAR PLANTS

VASCULAR PLANT APOMORPHIES

The vascular plants, or Tracheophyta (also called tracheophytes), are a monophyletic subgroup of the land plants. The major lineages of tracheophytes (excluding many fossil groups) are seen in Figure 4.1 (after Pryer et al. 2001, 2004a,b; Qiu et al. 2006, 2007; and Rothfels et al. 2015). Vascular plants together share a number of apomorphies, including (1) an **independent, long-lived sporophyte**; (2) a **branched sporophyte**; (3) **lignified secondary walls**, with pits, in certain specialized cells; (4) **sclerenchyma**, specialized cells that function in structural support; (5) **tracheary elements**, cells of **xylem** tissue, involved in water transport; (6) **sieve elements**, cells of **phloem** tissue, involved in sugar transport (the xylem and phloem comprising the **vascular tissue**); (7) an **endodermis**, involved in selective transfer of compounds; and (8) **roots**, functioning in anchorage and absorption of water and nutrients. See Kenrick and Crane (1997) and Pryer et al. (2004b) for detailed information.

https://doi.org/10.1016/B978-0-12-812628-8.50004-3,

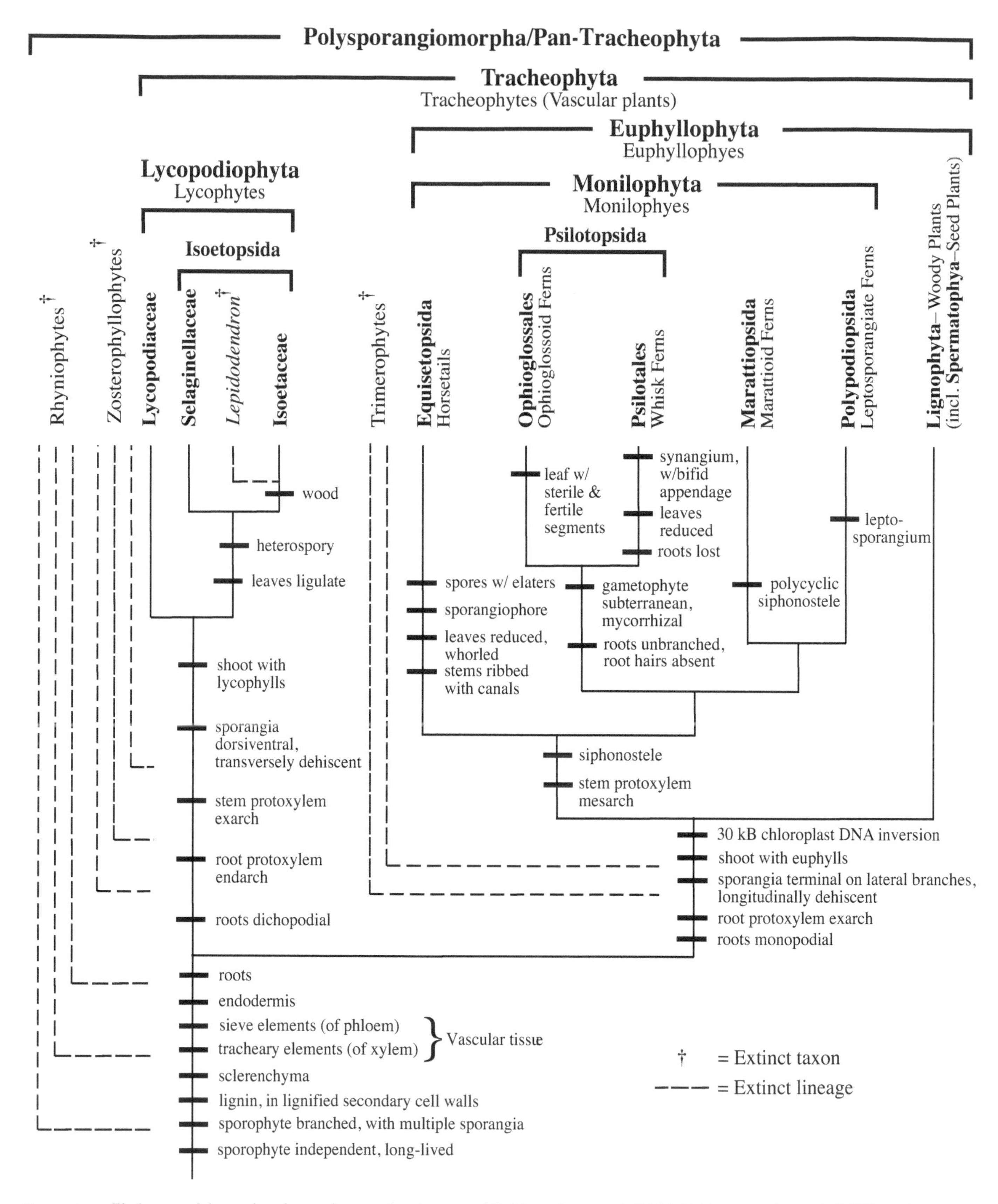

Figure 4.1 Phylogeny of the tracheophytes, the vascular plants, modified from Pryer et al. (2001, 2004a,b) and Qiu et al. (2007), and Rothfels et al. (2015), with selected apomorphies.

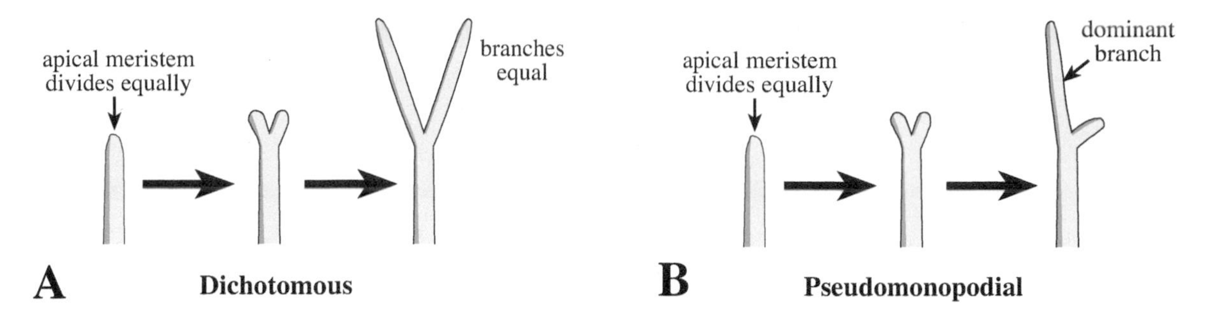

Figure 4.2 Dichotomous (**A**) and pseudomonopodial (**B**) branching patterns in vascular plants.

INDEPENDENT, LONG-LIVED SPOROPHYTE

Like all land plants, the vascular plants have a haplodiplontic "alternation of generations," with a haploid gametophyte and a diploid sporophyte. Unlike the liverworts, hornworts, and mosses, however, vascular plants have a dominant, free-living, photosynthetic, relatively persistent *sporophyte* generation, although, as discussed in Chapter 3, the hornworts have a sporophyte that is photosynthetic and relatively long-persistent. In the vascular plants, the gametophyte generation is also (ancestrally) free-living and may be photosynthetic, but it is smaller (often much more so) and much shorter lived than the sporophyte generation, although the gametophyte may be relatively long lived. In all land plants, the sporophyte is initially attached to and nutritionally dependent upon the gametophyte. However, in the vascular plants, the sporophyte soon grows larger and becomes nutritionally independent, usually with the subsequent death of the gametophyte. (In seed plants the female gametophyte is attached to and nutritionally dependent upon the sporophyte; see Chapter 5.)

BRANCHED SPOROPHYTE

The sporophytic axes, or **stems**, of vascular plants are different from those of liverworts, hornworts, and mosses in that they are branched and bear multiple (not just one) sporangia. Extant vascular plants share this apomorphy with some fossil plants that are transitional between the "bryophytes" and the tracheophytes. This more inclusive group, including fossil and extant taxa having branched sporophytic stems and multiple sporangia, has been called the **Polysporangiomorpha** (Kenrick and Crane 1997) or "polysporangiophytes." The even more inclusive **Pan-Tracheophyta** (Cantino et al. 2007) encompasses all descendants exclusive of the liverworts, mosses, and hornworts.

The earliest vascular plant stems had branching that was **dichotomous**, in which the apical meristem splits into two, equal meristems, each of which grows independently more or less equally (Figure 4.2A). Later lineages evolved a modified growth pattern, called **pseudomonopodial**, which starts out dichotomous, but then one branch becomes dominant and overtops the other, the latter appearing lateral (Figure 4.2B). Subsequent vascular plant lineages evolved **monopodial** growth. (See Euphyllophytes.)

The sporophytic stems of vascular plants function as supportive organs, bearing and usually elevating reproductive organs and leaves (see below). They also function as conductive organs, via vascular tissue, of water, minerals, and sugars between roots, leaves, and reproductive organs. Structurally, stems can be distinguished from roots by several anatomical features (to be discussed).

LIGNIFIED SECONDARY CELL WALLS

Vascular plants have evolved a chemical known as **lignin**, which is a complex polymer of phenolic compounds. Lignin is incorporated into an additional cell wall layer, known as the **secondary (2°) wall** (Figure 4.3), which is found in certain, specialized cells of vascular plants. Secondary walls are secreted to the outside of the plasma membrane (between the plasma membrane and the primary cell wall) *after* the primary wall has been secreted, which is also after the cell ceases to elongate. Secondary cell walls are usually much thicker than primary walls and, like primary walls, contain cellulose. However, in secondary walls, lignin is secreted into the space between the cellulose microfibrils, forming a sort of interbinding cement. Thus, lignin imparts significant strength and rigidity to the cell wall.

In virtually all plant cells with secondary, lignified cell walls, there are holes in the secondary wall called **pits** (Figure 4.3). Pits commonly occur in pairs opposite the sites of numerous plasmodesmata in the primary cell wall. This group of plasmodesmata is called a **primary pit field**. Pits function in allowing chemical "communication" between cells, via the plasmodesmata of the primary pit field, during their development and differentiation. They may also have specialized functions in water conducting cells (discussed later). Plant cells with secondary walls include **sclerenchyma** and **tracheary elements** (see later discussion).

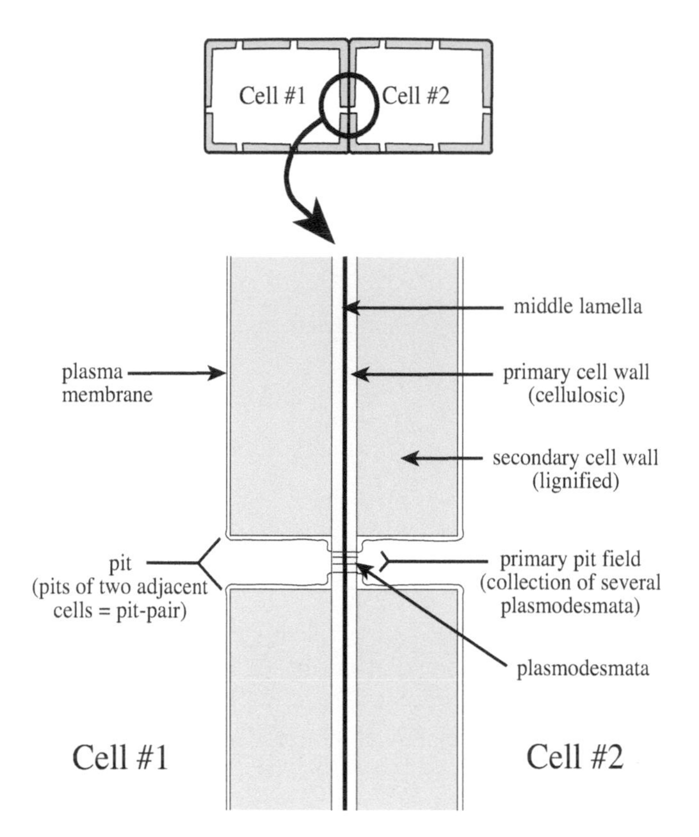

Figure 4.3 Lignified secondary cell wall of specialized cells of vascular plants. Note pit pair adjacent to primary pit field.

SCLERENCHYMA

Sclerenchyma (Gr. *scleros*, hard + *enchyma*, infusion, in reference to the infusion of lignin in the secondary cell walls) are nonconductive cells that have a thick, lignified secondary cell wall, typically with pits, and that are dead at maturity. There are two types of sclerenchyma (Figure 4.4): (1) **fibers**, which are long, very narrow cells with sharply tapering end walls; and (2) **sclereids**, which are isodiametric to irregular or branched in shape. Fibers function in mechanical support of various organs and tissues, sometimes making up the bulk of the tissue. Fibers often occur in groups or bundles. They may be components of the xylem and/or phloem or may occur independently of vascular tissue. Sclereids may also function in structural support, but their role in some plant organs is unclear; they may possibly help to deter herbivory in some plants. The evolution of sclerenchyma, especially fibers, with lignified secondary cell walls, constitutes a major plant adaptation permitting the structural support needed to attain greater stem height.

Another tissue type that functions in structural support is **collenchyma**, consisting of live cells with unevenly thickened, pectin-rich, primary cell walls (see Chapter 10). Collenchyma is found in many vascular plants, but is probably not an apomorphy for the group.

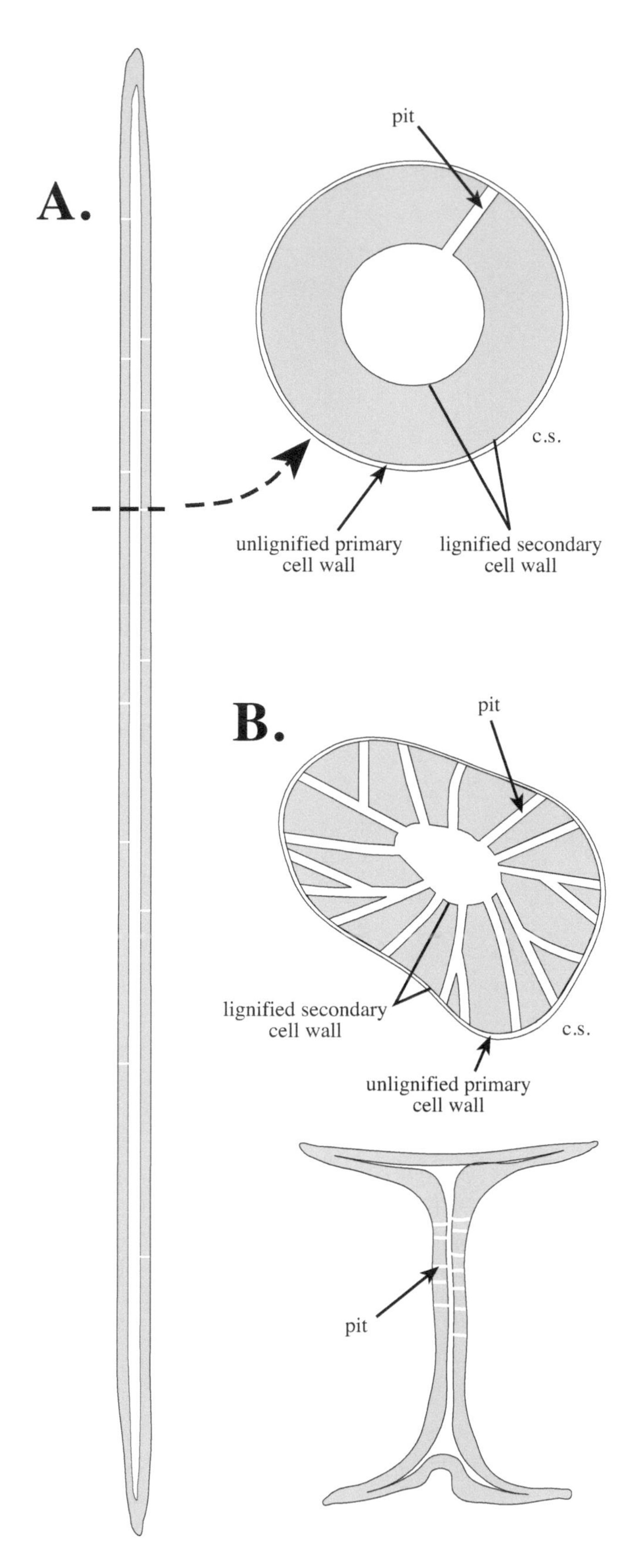

Figure 4.4 Sclerenchyma. **A.** Fiber cell. **B.** Sclereid cells. c.s = cross section

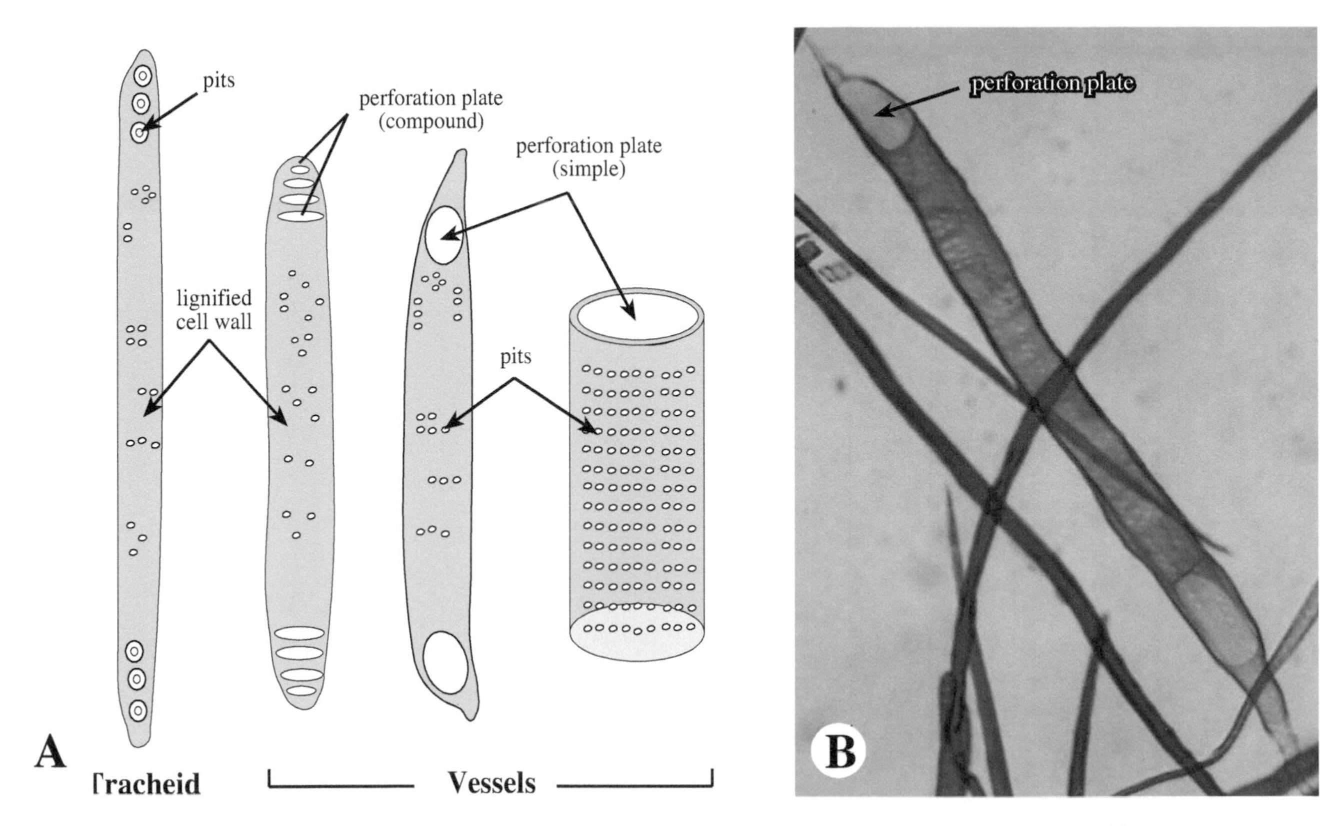

Figure 4.5 Conductive cells of vascular plants: tracheary elements. **A.** Types of tracheary elements. **B.** Vessel.

TRACHEARY ELEMENTS (OF XYLEM)

The vascular plants, as the name states, have true vascular tissue, consisting of cells that have become highly specialized for conduction of fluids. (A **tissue** consists of two or more cell types that have a common function and often a common developmental history; see Chapter 10.) Vascular tissue was a major adaptive breakthrough in plant evolution; more efficient conductivity allowed for the evolution of much greater plant height and diversity of form.

Tracheary elements are specialized cells that function in water and mineral conduction. Tracheary elements are generally elongate cells, are dead at maturity, and have lignified 2° cell walls (Figure 4.5A,B). They are joined end-to-end, forming a tube-like continuum. Tracheary elements are typically associated with parenchyma and often some sclerenchyma in a common tissue known as **xylem** (Gr. *xylo*, wood, after the fact that wood is composed of secondary xylem). The function of tracheary elements is to conduct water and dissolved essential mineral nutrients, generally from the roots to other parts of the plant.

There are two types of tracheary elements: **tracheids** and **vessel members** (Figure 4.5A). These differ with regard to the junction between adjacent end-to-end cells, whether *imperforate* or *perforate*. Tracheids are imperforate, meaning that water and mineral nutrients flow between adjacent cells through the primary cell walls at pit pairs, which are adjacent holes in the lignified 2° cell wall. Vessel members are perforate, meaning that there are one or more continuous holes or perforations, with no intervening 1° or 2° wall between adjacent cells through which water and minerals may pass. The contact area of two adjacent vessel members is called the **perforation plate**. The perforation plate may be **compound** if composed of several perforations, or **simple** if composed of a single opening (see Chapter 10). Vessels may differ considerably in length, width, angle of the end walls, and degree of perforation.

Tracheids are the primitive type of tracheary element. Vessels are thought to have evolved from preexisting tracheids *independently* in several different groups, including a few species of *Equisetum*, a few leptosporangiate ferns, all Gnetales (Chapter 5), and almost all angiosperms (Chapter 6).

SIEVE ELEMENTS (OF PHLOEM)

Sieve elements are specialized cells that function in the conduction of sugars. They are typically associated with parenchyma and often some sclerenchyma in a common tissue known as **phloem** (Gr. *phloe*, bark, after the location of secondary phloem in the inner bark). Sieve elements are elongate cells having only a primary (1°) wall with no lignified 2° cell wall. This primary wall has specialized pores

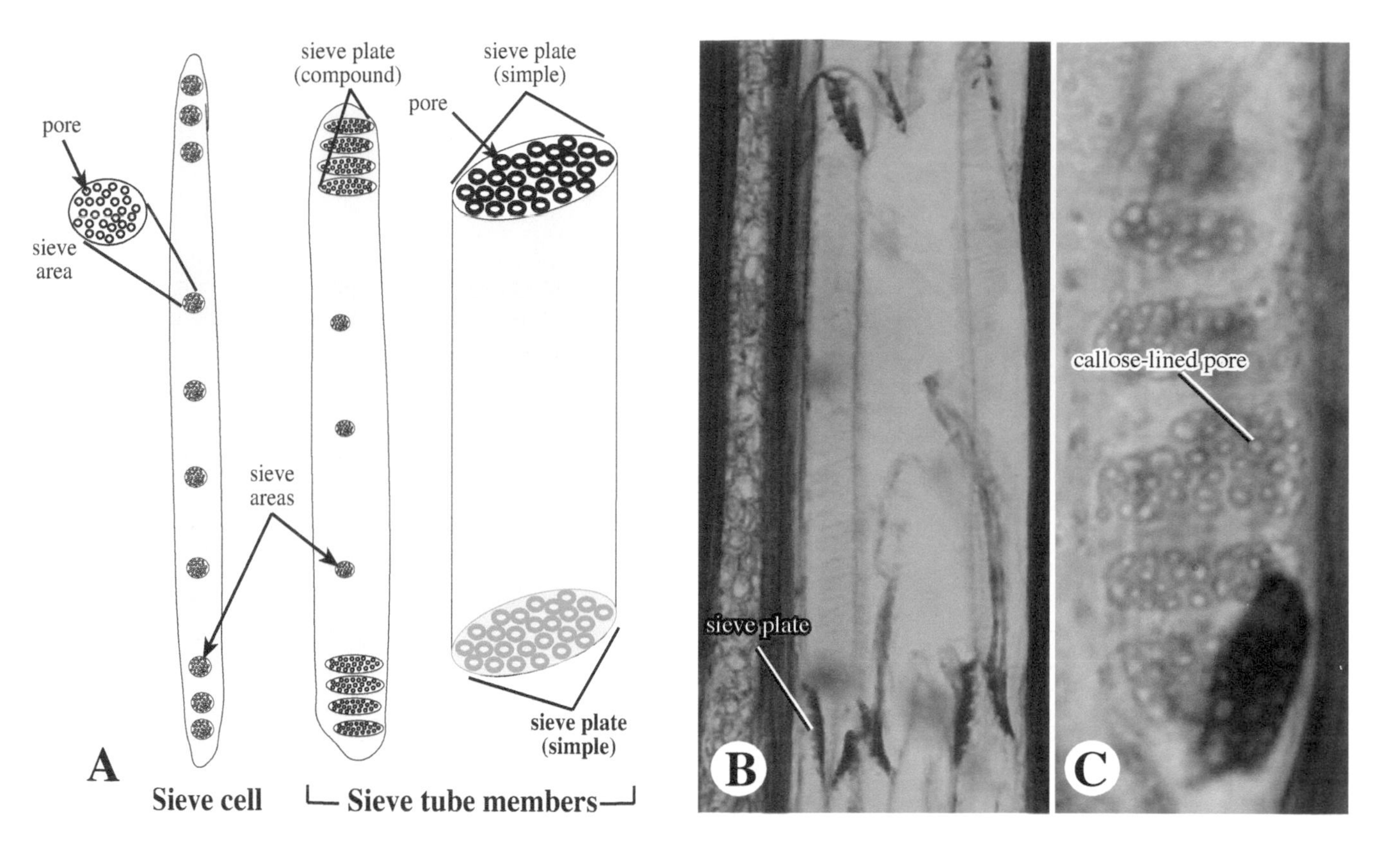

Figure 4.6 Conductive cells of vascular plants: sieve elements. **A.** Types of sieve elements. **B,C.** Sieve tube members.

(Figure 4.6C), which are aggregated together into **sieve areas** (Figure 4.6A). Each pore of the sieve area is a continuous hole in the 1° cell wall that is lined with a substance called **callose**, a polysaccharide composed of β-1,3-glucose units. (Note the difference in chemical linkage from cellulose, which is a polymer of β-1,4-glucose.) Sieve elements are "semi-alive" at maturity. They lose their nucleus and other organelles but retain the endoplasmic reticulum, mitochondria, and plastids. Like tracheary elements, sieve elements are oriented end-to-end, forming a tubelike continuum. Sieve elements function by conducting dissolved sugars from a sugar-rich "source" to a sugar-poor "sink" region of the plant. Source regions include the leaves, where sugars are synthesized during photosynthesis, or mature storage organs, where sugars may be released by the hydrolysis of starch. Sinks can include actively dividing cells, developing storage organs, or reproductive organs such as flowers or fruits.

There are two types of sieve elements: **sieve cells** and **sieve tube members** (Figure 4.6A). **Sieve cells** have only sieve areas on both end and side walls. **Sieve tube members** have both sieve areas and sieve plates (Figure 4.5B). **Sieve plates** consist of one or more sieve areas at the end-wall junction of two sieve tube members; the pores of a sieve plate, however, are significantly larger than are those of sieve areas located on the side wall (Figure 4.6C). Both sieve cells and sieve tube members have parenchyma cells associated with them. Parenchyma cells associated with sieve cells are called **albuminous cells**; those associated with sieve tube members are called **companion cells**. The two differ in that companion cells are derived from the same parent cell as are sieve tube members, whereas albuminous cells and sieve cells are usually derived from different parent cells. Both albuminous cells and companion cells function to load and unload sugars into the cavity of the sieve cells or sieve tube members. Sieve cells (and associated albuminous cells) are the ancestral sugar-conducting cells and are found in all nonflowering vascular plants. Sieve tube members were derived from sieve cells and are found only in flowering plants, the angiosperms (see Chapter 6).

Stems of the vascular plants typically have a consistent and characteristic spatial arrangement of xylem and phloem. This organization of xylem and phloem in the stem is known as a **stele**. In several groups of early vascular plant lineages, the stelar type is a **protostele**, with a central solid cylinder of xylem and phloem (Figure 4.7). The largely parenchymatous tissue between the epidermis and vascular tissue defines the **cortex**. Protosteles, thought to be the most ancestral type of stem vasculature, are found, e.g., in the rhyniophytes (see later discussion).

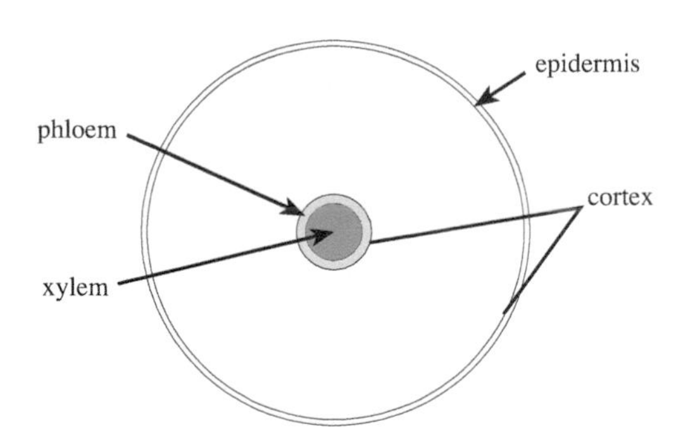

Figure 4.7 Example of a protostele, an ancestral vasculature of vascular plants.

ENDODERMIS

Another apparent apomorphy for the vascular plants is the occurrence, in some (especially underground) stems and all roots, of a special cylinder of cells known as the **endodermis** (Figure 4.8). Each cell of the endodermis possesses a **Casparian strip**, which is a band or ring of lignin and **suberin** (chemically similar to lignin) that infiltrates the cell wall, oriented tangentially (along the two transverse walls) and axially (vertically, along the two radial walls; Figure 4.6C). The Casparian strip acts as a water-impermeable material that binds to the plasma membrane of the endodermal cells. Because of the presence of the Casparian strip, absorbed water and minerals that flow from the outside environment to the central vascular tissue must flow through the plasma membrane of the endodermal cells (as opposed to flowing through the intercellular spaces, i.e., between the cells or through the cell wall). Because the plasma membrane may differentially control solute transfer, the endodermis (with Casparian strips) selectively controls *which* mineral nutrients are or are not absorbed by the plant; thus, toxic or unneeded minerals may be differentially excluded.

ROOT

A major novelty in the evolution of vascular plants was the differentiation between stems and roots. **Roots** are specialized plant organs that function in anchorage and absorption of water and minerals. Roots are found in all vascular plants except for the Psilotales, Salviniales, and a few other specialized groups, all of which lost roots secondarily (see later discussion). Other fossil groups of vascular plants may have lacked roots; plants lacking roots generally have uniseriate (one cell thick), filamentous **rhizoids** (similar to those of "bryophytes"), which assume a similar absorptive function. Roots constituted a major adaptive advance in enabling much more efficient water and mineral acquisition and conduction, permitting the evolution of plants in more extreme habitats.

Roots, like stems, develop by the formation of new cells within the actively growing **apical meristem** of the root tip, a region of continuous mitotic divisions (Figure 4.9B). At a later growth stage and further up the root, these cell derivatives elongate significantly. This cell growth, which occurs by considerable expansion both horizontally and vertically, pushes the apical meristem tissue downward. At an even later stage and further up the root, the fully grown cells differentiate into specialized cells. The ancestral apical meristem of roots most likely consisted of a single, apical cell, a feature

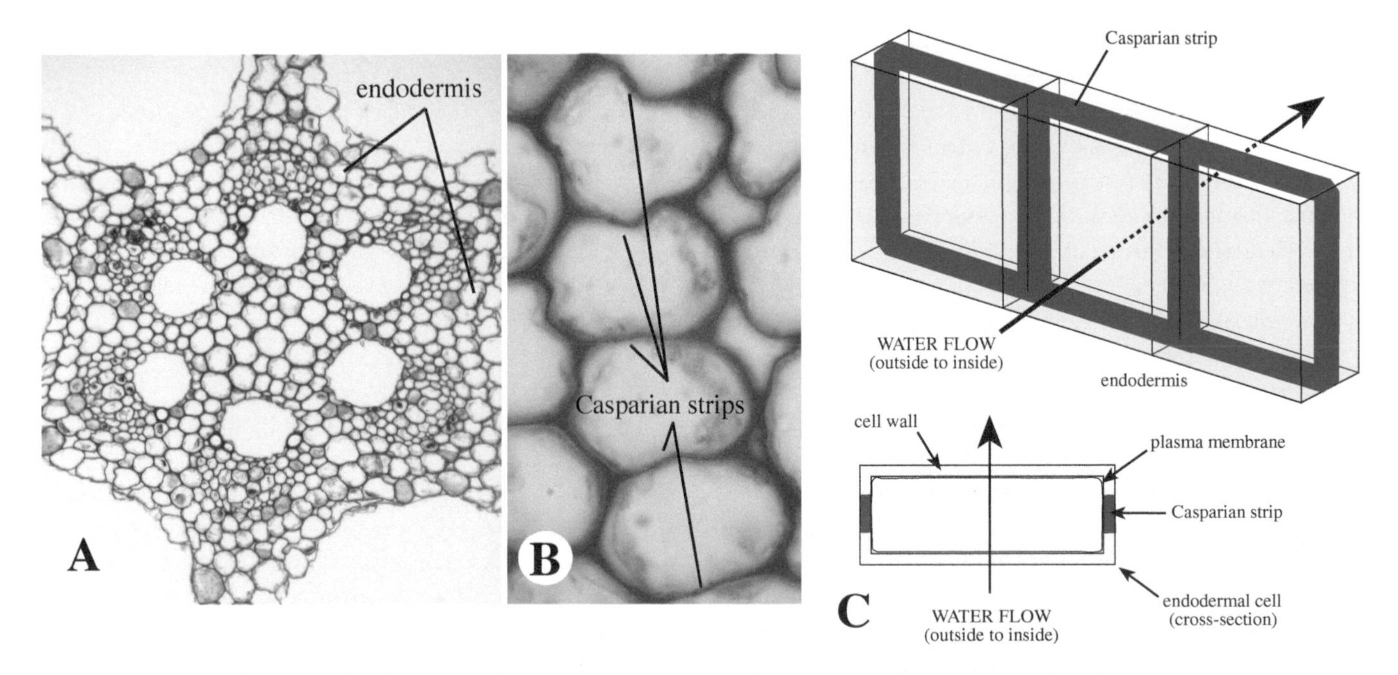

Figure 4.8 Endodermis of vascular plants. **A,B.** *Equisetum* rhizome. **A.** Rhizome cross section, showing single layer of endodermal cells. **B.** Close-up of endodermal cells (in cross section), showing Casparian strip thickenings. **C.** Diagram of Casparian strip, indicating function.

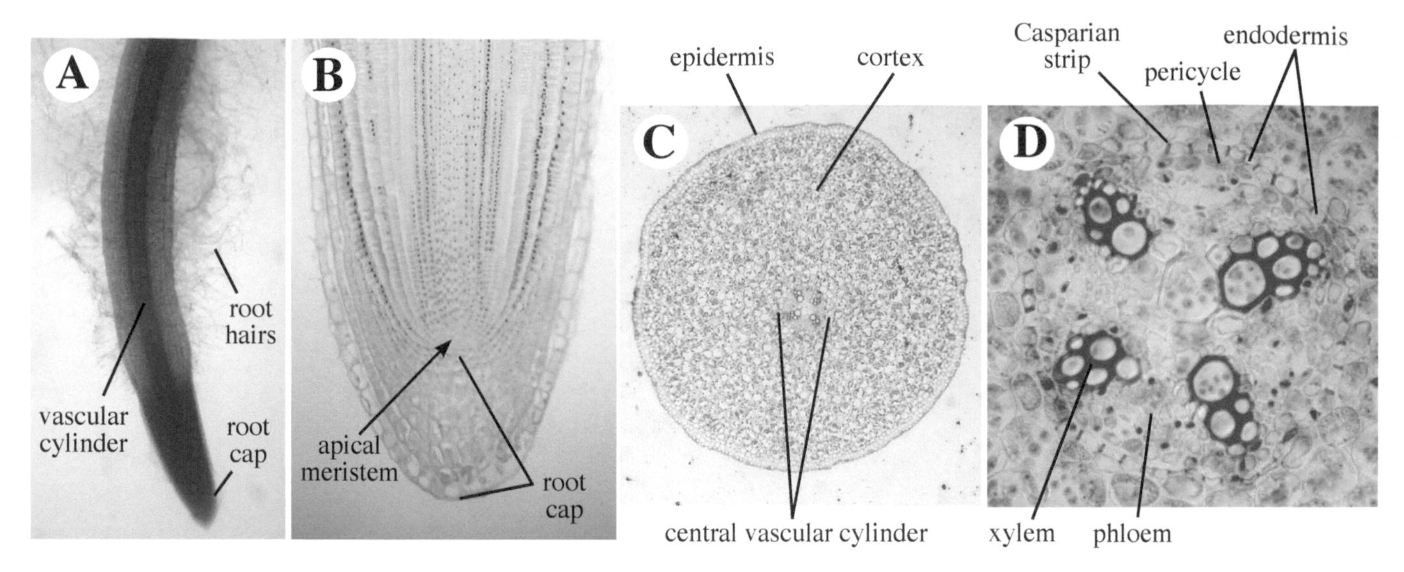

Figure 4.9 Anatomy of the root, an apomorphy of the vascular plants. **A.** Root whole mount. **B.** Root longitudinal section. **C.** Whole root cross section. **D.** Close-up of central vascular cylinder, showing tissues.

found today in the Selaginellaceae of the Lycophytes and all Monilophytes (discussed later). In the Lycopodiaceae, Isoetaceae, and seed plants (see Chapter 5), the apical meristem is complex, consisting of a group of continuously dividing cells.

Roots are characterized by several anatomical features. First, the apical meristem is covered on the outside by a **root cap** (also called a **calyptra**; Figure 4.9B); stems lack such a cell layer. The root cap functions both to protect the root apical meristem from mechanical damage as the root grows into the soil and to provide lubrication as the outer cells slough off. Second, with the exception of the Psilotopsida (Psilotales and Ophioglossales), the epidermal cells away from the root tip develop hairlike extensions called **root hairs** (Figure 4.9A); these are absent from stems (although underground stems of the Psilotales bear **rhizoids**, which resemble root hairs). Root hairs function to *greatly* increase the surface area available for water and mineral absorption. Third, roots always have a **central vascular cylinder** (Figure 4.9C,D). As in stems, the mostly parenchymatous region between the vasculature and epidermis is called the **cortex** (Figure 4.9C); the center of the vascular cylinder, if vascular tissue is lacking, is called a **pith**. Fourth, the vascular cylinder of roots is surrounded by an **endodermis** with **Casparian strips** (Figure 4.9D). As with some stems, the endodermis in roots selectively controls which chemicals are and are not absorbed by the plant, functioning in selective absorption. (An undifferentiated layer internal to the endodermis, called the **pericycle**, is also typically present.) Fifth, roots generally have **endogenous lateral roots** (Figure 4.10), in which new lateral roots originate by means of actively growing meristems, arising at the pericycle or endodermis. Lateral roots penetrate the tissues of the cortex before exiting to the outside.

Numerous modifications of roots have evolved, most of these restricted to the flowering plants (see Chapter 9). Roots of many, if not most, vascular plants have an interesting symbiotic interaction with various species of fungi; this association between the two is known as **mycorrhizae**. The fungal component of mycorrhizae appears to aid the plant in both increasing overall surface area for water and mineral absorption and increasing the efficiency of selective mineral absorption, such as of phosphorus. The fungus benefits in obtaining photosynthates (sugars and other nutrients) from the plant.

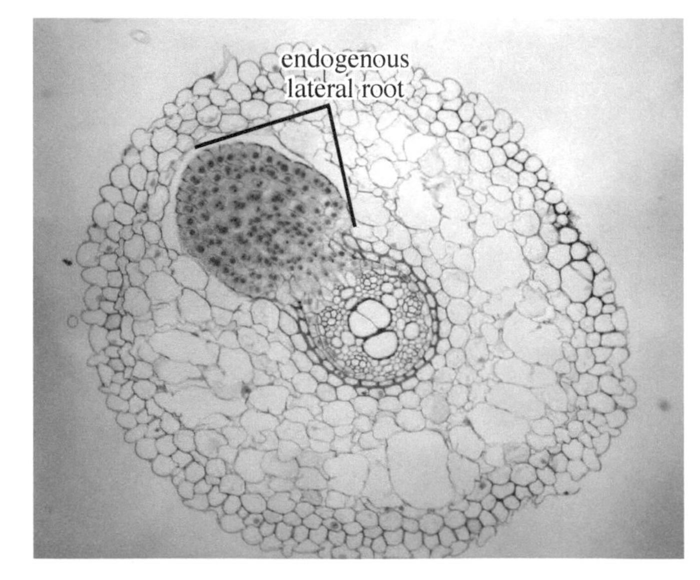

Figure 4.10 Root cross section (*Lilium* sp.), showing endogenous lateral root, a characteristic of vascular plant roots.

- **LYCOPODIOPHYTA/LYCOPODIOPSIDA**
  - **LYCOPODIALES**
    - **Lycopodiaceae** (16/388)
  - **ISOETALES**
    - **Isoetaceae** (1/250)
    - **Selaginellaceae** (1/700)
- **EUPHYLLOPHYTA**
  - **MONILOPHYTA**
    - **EQUISETOPSIDA/EQUISETIDAE**
      - **Equisetaceae** (1/15)
    - **PSILOTOPSIDA/OPHIOGLOSSIDAE**
      - **Ophioglossaceae** (10/112)
      - **Psilotaceae** (2/17)
    - **MARATTIOPSIDA/MARATTIIDAE**
      - **Marattiaceae** (6/111)
    - **POLYPODIOPSIDA/POLYPODIIDAE**
      - **Osmundales**
        - **Osmundaceae** (6/18)
      - **Hymenophyllales**
        - **Hymenophyllaceae** (9/434)
      - **Gleicheniales**
        - Dipteridaceae (2/11)
        - **Gleicheniaceae** (6/157)
        - Matoniaceae (2/4)
      - **Schizaeales**
        - Anemiaceae (1/115)
        - **Lygodiaceae** (1/40)
        - Schizaeaceae (2/35)
      - **Salviniales**
        - **Marsileaceae** (3/61)
        - **Salviniaceae** (2/21)
      - **Cyatheales**
        - Cibotiaceae (1/9)
        - Culcitaceae (1/2)
        - **Cyatheaceae** (3/643)
        - Dicksoniaceae (3/35)
        - Loxosomataceae (2/2)
        - Metaxyaceae (1/6)
        - Plagiogyriaceae (1/15)
        - Thyrsopteridaceae (1/1)
      - **Polypodiales**
        - **Saccolomatineae**
          - Saccolomataceae (1/18)
        - **Lindsaeineae**
          - Cystodiaceae (1/1)
          - Lindsaeaceae (7/234)
          - Lonchitidaceae (1/2)
        - **Pteridineae**
          - **Pteridaceae** (53/1211)
        - **Dennstaedtiineae**
          - Dennstaedtiaceae (10/265)
        - **Aspleniineae**
          - **Aspleniaceae** (2/730)
          - Athyriaceae (3/650)
          - Blechnaceae (24/265)
          - Cystopteridaceae (3/37)
          - Desmophlebiaceae (1/2)
          - Diplaziopsidaceae (2/4)
          - Hemidictyaceae (1/1)
          - Onocleaceae (4/5)
          - Rhachidosoraceae (1/8)
          - Thelypteridaceae (30/1034)
          - Woodsiaceae (1/39)
        - **Polypodiineae**
          - Davalliaceae (1/65)
          - Didymochlaenaceae (1/1)
          - **Dryopteridaceae** (26/2115)
          - Hypodematiaceae (2/22)
          - Lomariopsidaceae (4/69)
          - Nephrolepidaceae (1/19)
          - Oleandraceae (1/15)
          - **Polypodiaceae** (65/1652)
          - Tectariaceae (7/250)
  - **SPERMATOPHYTA** (See Chapter 5)

Table 4.1 Taxonomic groups of Tracheophyta, vascular plants (minus those of Spermatophyta, seed plants). Classes, orders, and family names after PPGI (2016). Higher groups (traditionally treated as phyla) after Cantino et al. (2007). Families in bold are described in detail. Number of genera and species (often approximate), respectively, are indicated in parentheses, separated by slash mark.

VASCULAR PLANT DIVERSITY

A classification scheme of vascular plants, after Smith et al. (2006) and Cantino et al. (2007), is seen in Table 4.1. Of the tremendous diversity of vascular plants that has arisen since their first appearance some 400 million years ago, only the major lineages will be described here. These include the rhyniophytes, known only from fossils, plus clades that have modern-day descendants: the Lycopodiophyta (lycophytes) and Euphyllophyta (euphyllophytes; Figure 4.1, Table 4.1). See Bierhorst (1971) and Foster and Gifford (1974) for general information on vascular plant morphology.

Features that have been used to classify vascular plants include sporophyte vegetative morphology (branching pattern, leaf type/shape/arrangement/venation, stem and leaf anatomy), life cycle and reproductive morphology (homospory/heterospory, sporophyll morphology, sporangium shape/dehiscence/attachment, spore morphology), and gametophyte morphology (whether green and photosynthetic or non-green and saprophytic or mycorrhizal). Spore morphology in particular has been useful in the classification of vascular plant groups. (See Chapter 12.) Features include spore size, shape (e.g., reniform, tetrahedral, globose), sculpturing patterns, and whether green (photosynthetic) or not. One major spore feature is related to the **laesura** (plural **laesurae**), the differentially thickened wall region corresponding to the tetrad attachment scar on each of the four immature spores following meiosis. Three basic spore types are recognized: (1) **trilete** spores, with a 3-branched laesura (Figure 4.11A); (2) **monolete** spores, with an unbranched, linear laesura that is linear and unbranched (Figure 4.11B); and (3) **alete**, lacking any evidence of a laesura.

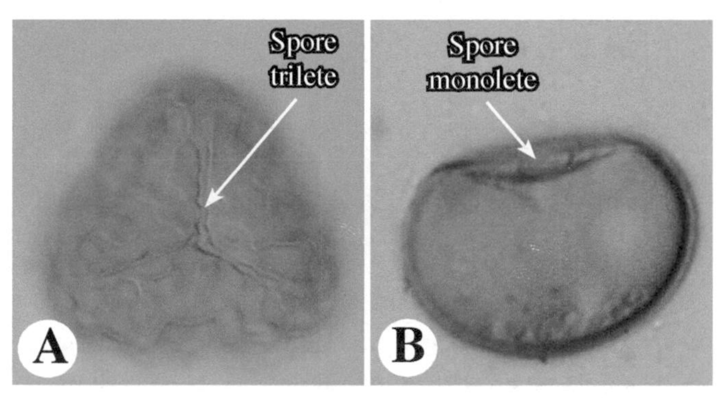

Figure 4.11 MONILOPHYTA. Spore morphology. **A.** Spore with trilete scar (*Pentagramma triangularis*, Pteridaceae). **B.** Spore with monolete scar (*Asplenium nidus*, Aspleniaceae).

Figure 4.12 **A,B.** Rhyniophytes. **A.** Reconstruction of *Rhynia major*, an early, extinct vascular plant. Note erect, branched stem (without leaves) bearing terminal sporangia. [Reproduced from Kidston, R. and W. H. Lang. 1921. Transactions of the Royal Society of Edinburgh. vol. 52(4):831–902.] **B.** *Rhynia* stem axes embedded in "Rhynie" chert. **C–E.** Lycophytes. **C,D.** *Sigillaria*, an extinct, woody lycophyte. **C.** Stem cross section showing outer wood. **D.** Fossil impression of lycophyll leaf showing single vein. **E.** Fossil cast of *Lepidodendron*, an extinct, woody, tree-sized lycophyte. Note lycophyll scars.

RHYNIOPHYTES

Rhyniophytes are a paraphyletic assemblage that included the first land plants with branched sporophytic axes, some of which (but not all) also had vascular tissue. Rhyniophytes include the genus *Rhynia* (Figure 4.12A,B), a well-known vascular plant from the early Devonian, ca. 416–369 million years ago. Rhyniophyte sporophytes consisted of dichotomously branching axes bearing terminal sporangia that dehisced longitudinally.

Rhyniophytes ancestrally lacked both roots and a leaf-bearing shoot system; these two features evolved later, prior to or within the Lycophyte and Euphyllophyte lineages (Figure 4.1). The stems of rhyniophytes were protostelic (Figure 4.7) in which the first-formed xylem (known as protoxylem) was "centrarch" (positioned at the center).

LYCOPODIOPHYTA — LYCOPHYTES

The Lycopodiophyta, or lycophytes (also commonly called lycopods), are a lineage of plants that diverged after the rhyniophytes. An extinct, probably paraphyletic, fossil group, known as the zosterophyllophytes [Zosterophyllophytina], diverged along the immediate lineage leading to lycophytes (Figure 4.1). Zosterophyllophytes had no leaves, but possessed lateral sporangia, similar to those of the lycophytes (see later discussion). Within the lycophytes, the now extinct *Lepidodendron*, *Sigillaria*, and relatives (Figure 4.12C–E) were woody trees that comprised a large portion of the primary biomass of forests during the Carboniferous, approximately 300 million years ago. Fossil remains of these plants today make up much of the Earth's coal deposits.

A number of apomorphies characterize the lycophytes (Figure 4.1). First, the roots of lycophytes are **dichopodial**, meaning that the root apical meristem may branch into two roots (Figure 4.13); no lateral roots develop, as they do in euphyllophytes (see later discussion). Second, lycophyte

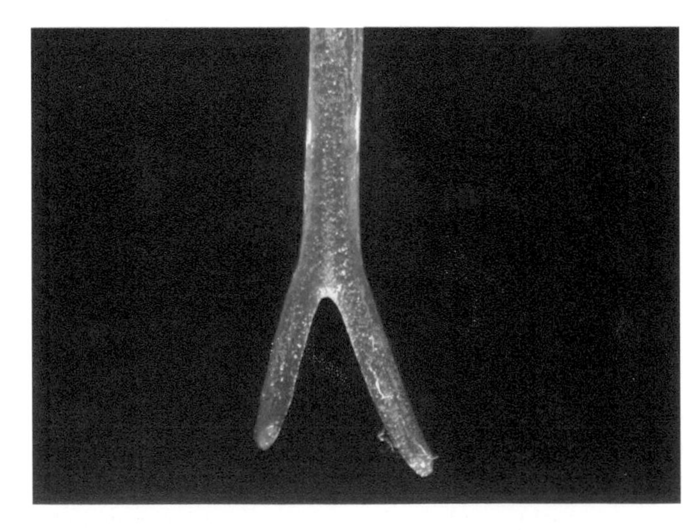

Figure 4.13 LYCOPODIOPHYTA. Tip of dichopodial root (*Selaginella kraussiana*), apomorphic for the lycophytes.

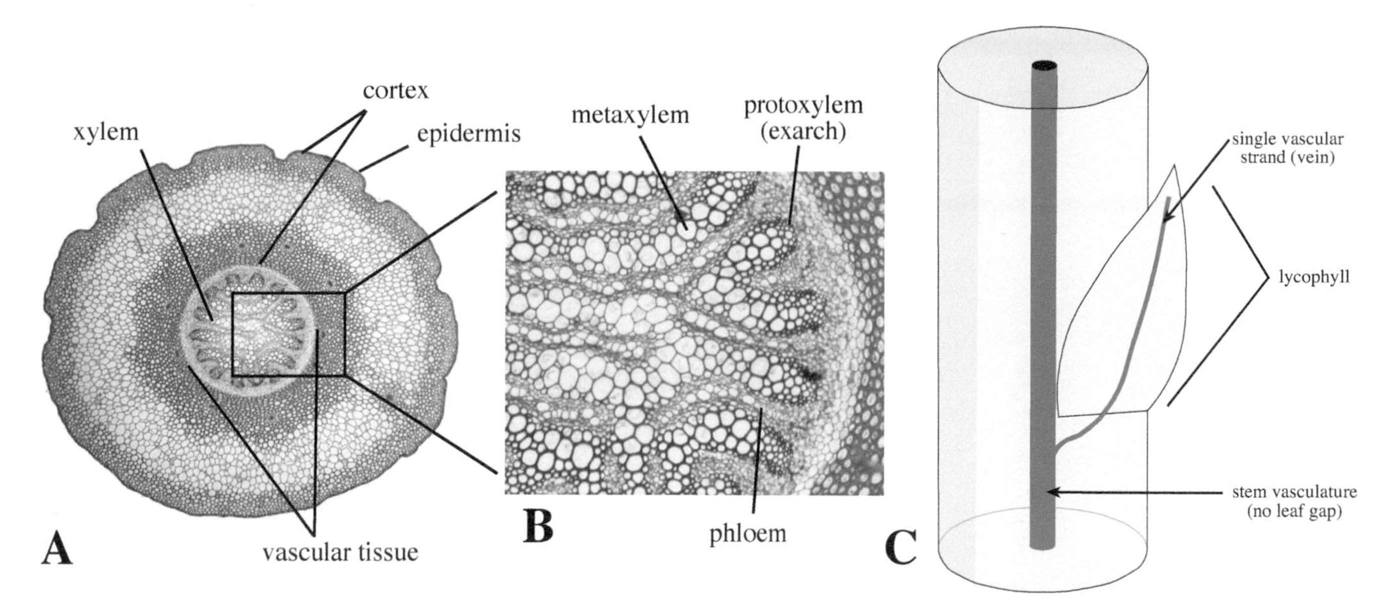

Figure 4.14 **A,B.** *Lycopodium* stem cross section showing protoxylem that is exarch (to periphery of stem). **C.** Lycophyll structure.

roots usually have an **endarch protoxylem**. **Protoxylem** refers to the first tracheary cells that develop within a patch of xylem and that are typically smaller and have thinner cell walls than the later formed **metaxylem**. In the roots of lycophytes, the protoxylem forms in a position *interior* to the metaxylem (i.e., toward the stem center), relative to the phloem tissue. Third, the stems of lycophytes have an **exarch protoxylem** (just the reverse of the roots). In the stems of lycophytes, the protoxylem forms in a position *exterior* to the metaxylem (i.e., away from the stem center; Figure 4.14A,B). Fourth, lycophytes, at least ancestrally, have sporangia that are **dorsiventral** (i.e., flattened and having a dorsal, upper, and ventral, lower, surface) and **dehisce transversely** relative to the axis of the stem or subtending leaf (see *Lycopodium*, Figure 4.15E). Fifth, lycophytes have sporophytic leaves, usually just called "leaves." (Although some liverworts and all mosses have "leaves," these occur on gametophytes only and are not strictly homologous with the sporophytic leaves of vascular plants.) Leaves are typically dorsiventral organs that are the primary site for photosynthesis. The evolution of sporophytic leaves constituted a major adaptive innovation by greatly increasing the area available for light capture in photosynthesis. This paved the way for the evolution of various ecological adaptive strategies, enabling some vascular plants to survive in previously inaccessible habitats. In addition, leaves or leaf-like homologs have become evolutionarily modified for numerous other functions in plants (see later discussion).

The sporophytic leaves of lycophytes are called **lycophylls** (essentially synonymous with "microphyll"). **Lycophylls** are distinctive in having a single, unbranched (very rarely branched) vein, lacking a gap in the vasculature of the stem, and developing by an intercalary meristem, i.e., at the base of the leaf (Figure 4.14C). Lycophylls, like all sporophytic leaves, develop from a shoot apex. A **shoot** is defined as a stem plus associated leaves. Sporophytic leaves originate from actively dividing cells very near the stem/shoot apical meristem. (For a more detailed description of shoots and leaves, see Euphyllophyta.)

Although all vascular plants have shoots, fossil evidence suggests that shoot systems evolved independently in the lycophytes and euphyllophytes (see later discussion), because their associated leaves evolved independently. Lycophylls possibly originated from the transformation of small appendages called **enations** (found, e.g., in fossil zosterophyllophytes and relatives), which are external, peg-like appendages that lack vascular tissue. Lycophylls may have evolved via the development of vasculature tissue leading from the stem into the enation, allowing for more efficient transfer of water and solutes; this was associated with flattening ("**planation**") of the enation into a dorsiventral, planar posture. Such a gradation, from enation to lycophyll, may be seen in some fossil plants. (Alternatively, lycophylls may have evolved by the sterilization and planation of sporangia; see Kenrick and Crane, 1997.)

The only lycophytes that survived to the present are small, nonwoody, herbaceous plants, grouped into three extant families: Lycopodiaceae of the Lycopodiales, and Sellaginellaceae and Isoetaceae of the Isoetales (Figure 4.1). These are discussed below.

Figure 4.15 LYCOPODIOPHYTA. Lycopodiales. Lycopodiaceae. **A.** *Huperzia lucidula*, a species with unspecialized reproductive organs. **B.** *Lycopodium clavatum*, a species with strobili. **C.** *Huperzia lucidula*, showing sporangia in leaf axils with no specialized cones. **D.** *Lycopodium annotinum*, strobilus close-up showing sporophylls. **E.** *Lycopodium clavatum*, sporophylls removed from strobilus showing sporangia having lateral dehiscence, adaxial view (left) and abaxial view (right). **F,G.** *Huperzia squarrosa*, a large, epiphytic lycopod with pendant branches and specialized strobili (**G**). **H.** *Phylloglossum drummondii*, a small, cormose lycopod.

LYCOPODIALES

The Lycopodiales consist solely of the family Lycopodiaceae, described in detail below. These plants are often commonly called club-mosses, and are distinguished in having one type of spore, a condition known as **homospory**. Some family members may in fact resemble a large moss (e.g., Figure 14.15A), but they are true vascular plants, the persistent, long-lived phase being sporophytic. Sporangia of the Lycopodiaceae, like those of all lycophytes, develop laterally (relative to the stem) in the axils of specialized leaves termed **sporophylls** (Figure 4.15E). In some members of the family, the sporophylls are similar to the vegetative leaves (Figure 4.15C) and co-occur with them on shoots that are indeterminate, i.e., with continuous growth. In other family members, the sporophylls differ in size or shape from vegetative leaves and are aggregated into a terminal shoot system that is determinate, meaning that it terminates growth after formation. This determinate reproductive shoot, consisting of a terminal aggregate of sporophylls with associated sporangia, is known as a **strobilus** or **cone** (Figure 4.15B,D,G,H).

Lycopodiaceae–Club-Moss family (type *Lycopodium,* after Greek *lykos*, wolf, + *podion*, foot, from resemblance to a wolf's foot). 16 genera/ca. 388 species (Figure 4.15).

The Lycopodiaceae are terrestrial or epiphytic and pendulous, perennial herbs. The **roots** are adventitious and endogenous, dichotomously branched, and grow from underground portions of the stem. The **stems** are dichotomously branched rhizomes or corms, with a protostelic vasculature (sometimes plectostelic); in some taxa the stems are pseudomonopodial, in which one branch of a dichotomy is dominant, the other appearing lateral, forming a flattened branch system in some species (e.g., *L. complanatum*); in some species specialized short shoots (known as "gemmae" or "bulbils") may detach, functioning as vegetative propagules. The **leaves** are simple, sessile, spiral, or whorled, the blades scale-like to acicular, heteromorphic in some species, all with a single midrib (microphyllous), ligule absent (eligulate). **Sporangia** are homosporous, generally reniform, occurring on short stalks in axils of leaves (sporophylls); sporophylls are photosynthetic, resembling and dispersed among vegetative leaves, or scale-like and non-photosynthetic, organized in terminal strobili; dehiscence occurs along distal margin of the sporangium, transverse relative to sporophyll axis. Spores are globose or tetrahedral, with a trilete laesura. **Gametophytes** are mycorrhizal, either epiterranean and photosynthetic or subterranean and saprophytic.

The Lycopodiaceae are classified by PPGI (2016) into 16 genera, the largest being *Huperzia* (ca. 25 spp., mostly worldwide), *Lycopodiella* (ca. 15 spp., tropical and temperate regions), *Lycopodium* (15 spp., tropical and temperate regions), *Phlegmariurus* (ca. 250 spp.), *Pseudolycopodiella* (10 spp., worldwide), and *Phylloglossum* (1 sp., Australia, New Zealand). The family has a worldwide distribution. Economic importance includes cultivated ornamentals, local medicinal plants (e.g., *Lycopodiella* spp.; *Huperzia selago* used as an emetic; *H. serrata* experimentally used to treat Alzheimer's), fiber plants (used as stuffing material, baskets, nets), dyes, and mordants; dried spores, which have been used as a lubricant (in condoms, rubber gloves) and, because they are very flammable, in fireworks and lights, including early flash photography. See Øllgaard (1990) for general information and Wikström and Kenrick (2000a,b; 2001) for phylogenetic studies of the family.

The Lycopodiaceae are distinctive in being ***homosporous, dichotomously branched***, erect, prostrate, or pendulous, perennial, ***microphyllous*** herbs, the leaves ***eligulate***, the sporangia ***reniform and transversely dehiscing***, born on sporophylls that are ***photosynthetic*** and resemble ***vegetative leaves*** or that are ***non-photosynthetic and scale-like in terminal strobili***, the gametophytes mycorrhizal, photosynthetic, or saprophytic.

ISOETALES

The two other extant lycophyte families, classified in the Isoetales, are the Selaginellaceae and Isoetaceae (described in detail below). Members of the Isoetales differ from those of the Lycopodiales in having leaf **ligules** and in being **heterosporous**, both of which are apomorphies within the lycophytes (Figure 4.1). Ligules are tiny appendages on the upper (adaxial) side of the leaf (both vegetative and reproductive), near the leaf base (Figures 4.16, 4.18D,H, 4.19D). The function of ligules is not clear; one proposal is that they act as glands, providing hydration for young, developing lycophylls. **Heterospory** (Figure 4.17) refers to

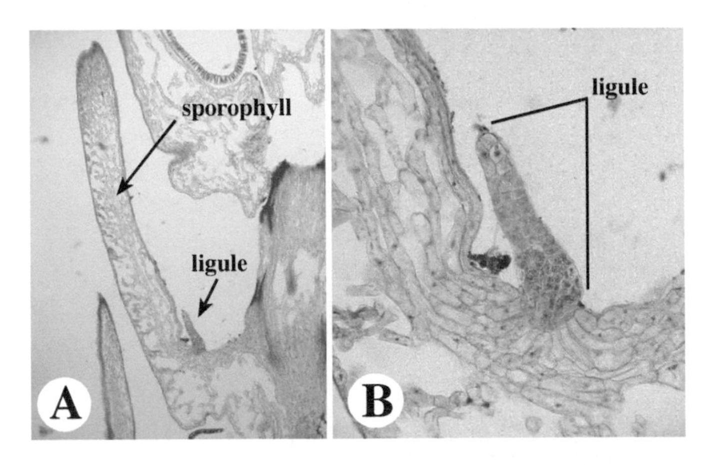

Figure 4.16 LYCOPODIOPHYTA. Isoetales. **A.** Longitudinal section of *Selaginella* strobilus showing sporophyll and ligule. **B.** Ligule, close-up.

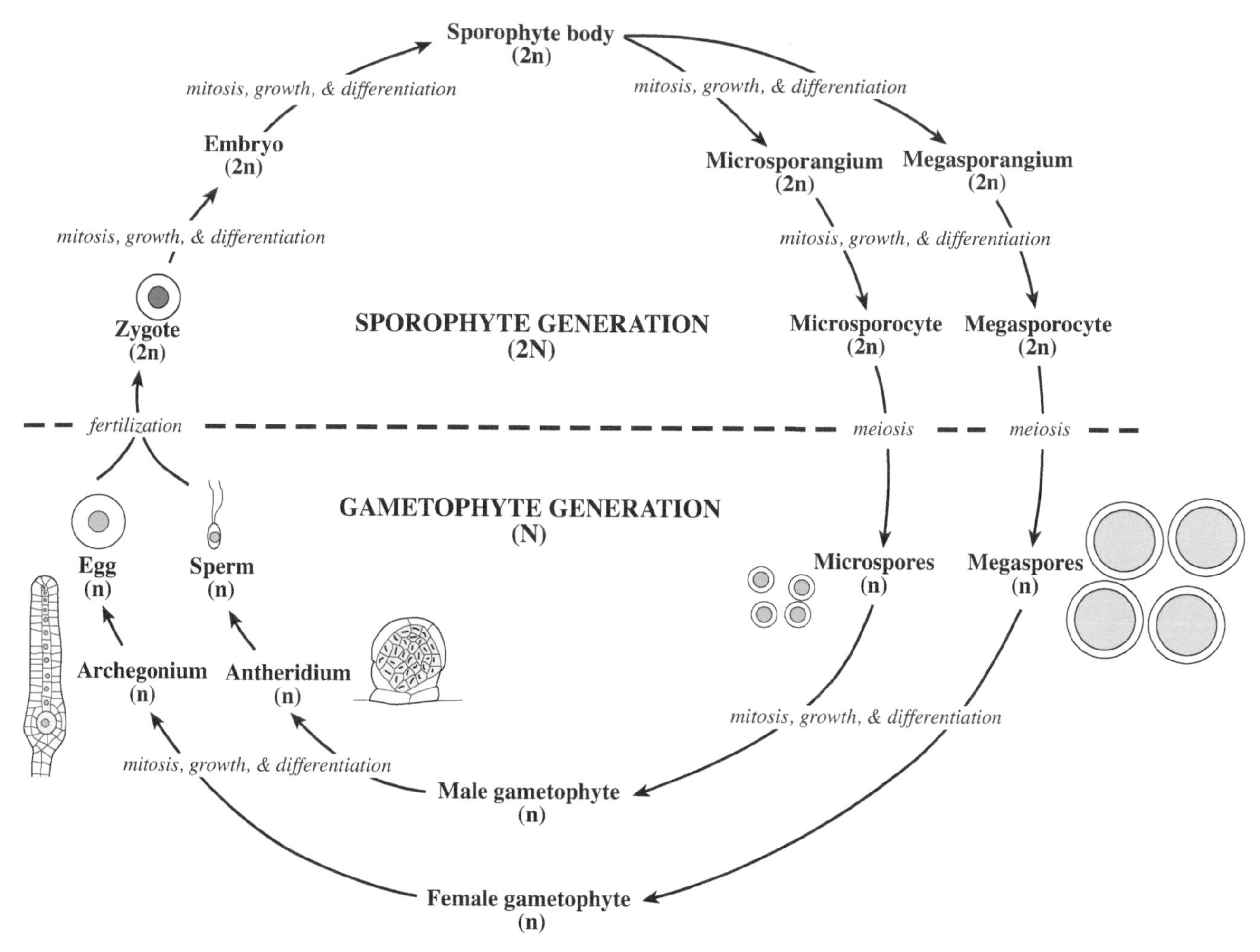

Figure 4.17 Heterospory, a characteristic of the Isoetales (Isoetaceae and Selaginellaceae) of the lycophytes.

the production of *two* types of spores, **microspores** and **megaspores**, which develop within specialized sporangia, **microsporangia** and **megasporangia**, via meiois of specialized cells, **microsporocytes** and **megasporocytes** (Figures 4.18, 4.19). Microspores are relatively small (Figures 4.18I, 4.19I,K) and are produced in large numbers. Megaspores (Figures 4.18E, 4.19J,K) are much larger in size and are produced in fewer numbers (typically four per megasporangium in *Selaginella*, more in *Isoetes*). Megasporangia and microsporangia may be produced in the same shoot or on different shoots. Some species of *Selaginella* have **strobili**, with specialized sporophylls subtending the sporangia on a determinate shoot (Figure 4.19E). In *Isoetes*, the sporophylls bear enlarged microsporangia or megasporangia on the upper (adaxial) side of the leaf base (Figure 4.18B,C). In both *Selaginella* and *Isoetes*, the megaspore develops into a **female gametophyte**, which contains only archegonia, housing the egg cell. Each microspore germinates to form a **male gametophyte**, which produces only antheridia, the sperm-manufacturing organs (Figure 4.17). The gametophytes of *Selaginella* and *Isoetes* are **endosporic**, meaning that the gametophytes develop entirely *within* the original spore wall. Heterospory and endospory also evolved independently in the seed plants (see Chapter 5).

Interestingly, the fossil tree *Lepidodendron* and relatives (Figure 4.12C–E) belongs to the "ligulate" lycophytes, being most closely related to *Isoetes* among the extant lycophytes (Figure 4.1). *Lepidodendron* possessed leaf ligules and was heterosporous. It and other close relatives were woody, via the development of a vascular cambium (but one that was "unifacial," producing only cells in one direction; see Chapter 5). Wood was likely an apomorphy for the group including *Lepidodendron* and *Isoetes* (Figure 4.1) but was lost in the extant members of the latter.

Isoetaceae–Quillwort family (type *Isoetes*, from Greek *isos*, ever, + *etas*, green, in reference to apparent evergreen duration of some). 1 genus (*Isoetes* [incl. *Stylites*])/ca. 250 species (Figure 4.18).

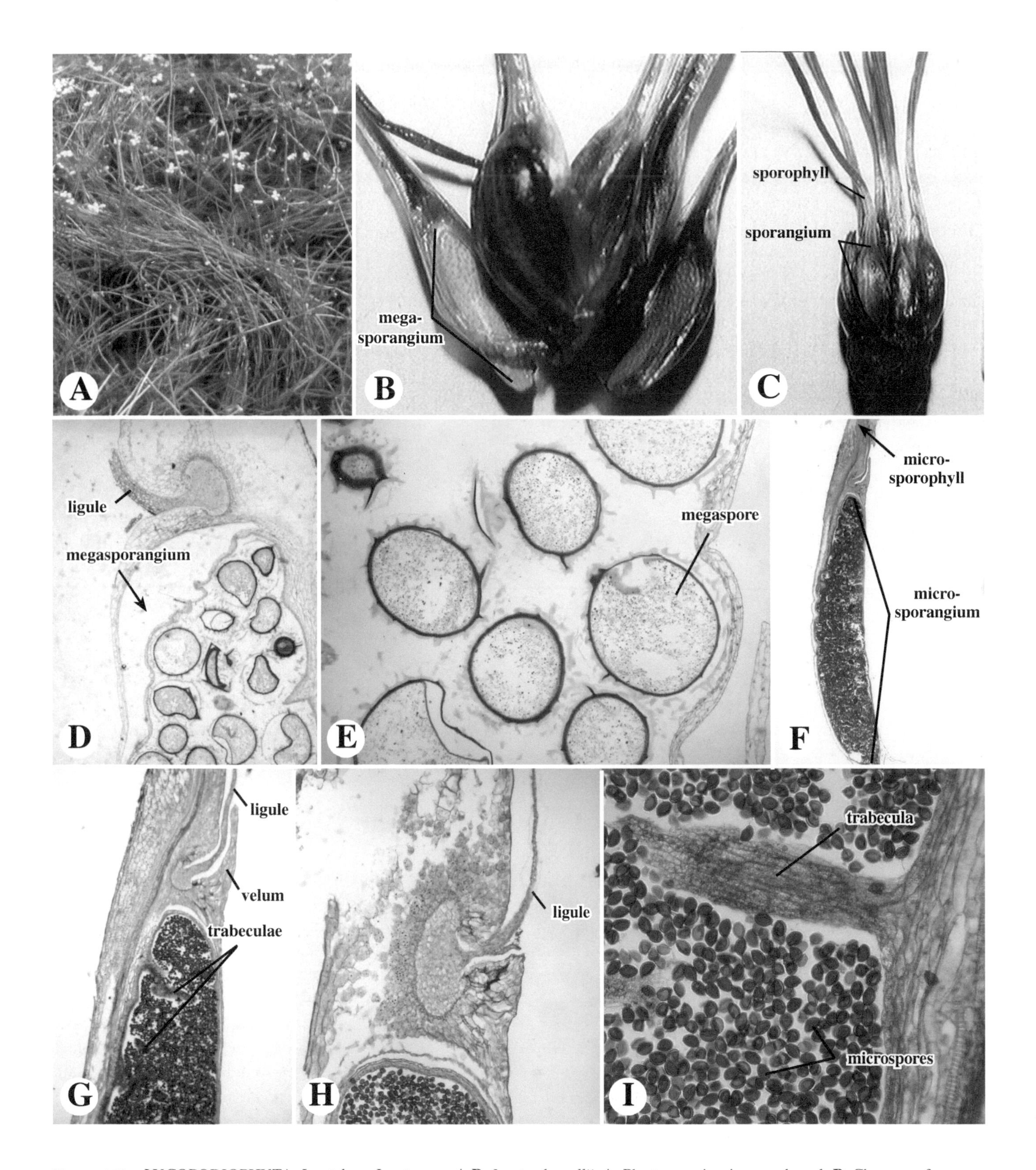

Figure 4.18 LYCOPODIOPHYTA. Isoetales—Isoetaceae. **A,B.** *Isoetes howellii*. **A.** Plants growing in vernal pool. **B.** Close-up of sporangia, megasporangium containing megaspores visible. **C.** *Isoetes orcuttii*, showing sporophylls with basal microsporangia or megasporangia. **D–I.** *Isoetes* sp., reproductive. **D,E.** Megasporangium longitudinal section with ligule and spinose megaspores. **F,I.** Microsporangium longitudinal-section showing trabeculae, ligule, and numerous, smooth microspores.

The Isoetaceae consist of usually aquatic or terrestrial, perennial herbs. The **roots** are adventitious and endogenous from lower grooves in the stem and branch dichotomously. The **stems** are protostelic, vertically oriented and corm-like, rarely rhizomatous, with an apical and lateral meristem, becoming lobed at the base, the lobes sometimes elongate. The **leaves** are simple, spiral, in a basal rosette, the blades basally widened, sheathing, apically linear to acicular, flat to terete (generally short, but > 50 cm in some species), with a single midrib (microphyllous), ligulate at the apex of the expanded base, most sporophyllous. **Sporangia** are heterosporous, and are located on the adaxial side of leaf (sporophyll) bases; **megasporangia** occur on outer leaves of a flush of growth, the **megaspores** large (50–300 per sporangium), trilete, spore sculpturing used in species identification; **microsporangia** occur on inner leaves (or in alternating cycle with megasporangia), the **microspores** small, monolete, very numerous (up to 1 million per sporangium); both sporangia are marginally covered by a membrane, the "velum," and are internally traversed by sterile strands (trabeculae); sporangia lack a precise dehiscence mechanism and open by tissue degradation. **Gametophytes** are endosporic. Plants have CAM photosynthesis. Air chambers occur in roots and leaves.

The Isoetaceae have a worldwide distribution. Economic importance is limited to some cultivated ornamentals. See Jermy (1990a) for general information and Rydin and Wikstrom (2002) and Hoot et al. (2006) for phylogenetic and biogeographic studies of the family.

The Isoetaceae are distinctive in being ***cormose to rhizomatous*** plants with a ***basal rosette of microphyllous, ligulate*** leaves, the leaves ***basally sheathing***, ***apically linear to acicular***, ***heterosporous***, bearing adaxial megasporangia or microsporangia within sheathing leaf base.

[Note that *Isoetes* and Isoetaceae can be spelled *Isoëtes* and Isoëtaceae, respectively, the umlaut indicating that the "e" is a separate vowel and should be pronounced, not part of the diphthong "oe." See Botanical Names, in Chapter 16.]

Selaginellaceae–Spike-Moss family (type *Selaginella*, from Latin *Selago*, a moss-like plant of the Scrophulariaceae + *ella*, diminutive). 1 genus /ca. 700 species (Figure 4.19).

The Selaginellaceae consist of perennial herbs, rarely tree-like, some species xeric-adapted "resurrection plants" (e.g., *S. lepidophylla*). The **roots** are adventitious and dichotomously branching, in some taxa arising from branch junctions and growing downward (formerly interpreted as leafless stems, termed "rhizophores"). The **stems** are generally dichotomously branching, with erect, cespitose, prostrate/repand, or climbing habit; the stems may be pseudomonopodial or sympodial, forming a very flattened, "fern-like" branch system in some species; some with aerial tubers; the stem vasculature is a protostele (exarch or mesarch). The **leaves** are simple, sessile, spiral, with a single midrib (microphyllous), adaxially ligulate, blades generally small, either homomorphic ("isophyllous") or, in some prostrate taxa, dimorphic ("anisophyllous") and in four rows, leaves of two upper (dorsal) rows smaller, those of the other two lower (ventral or lateral) rows larger. **Sporangia** are heterosporous; **microsporangia** (bearing numerous, small, trilete **microspores**) and **megasporangia** (bearing usually 4 [numerous], large, trilete, gen. ornamented **megaspores**) occur on short stalks in the axils of ligulate sporophylls (termed **microsporophylls** and **megasporophylls**, respectively), grouped together in terminal strobili, the sporophylls in 4 rows, not much differentiated from vegetative leaves. **Gametophytes** are endosporic.

The Selaginellaceae are mostly distributed in tropical and warm regions, worldwide. Economic importance includes cultivated ornamentals and local medicinal plants. See Jermy (1990b) for general information and Korall and Kenrick (2002, 2004) for phylogenetic analyses of the family.

The Selaginellaceae are distinctive in being ***erect to prostrate*** herbs, with ***dichotomously branched*** stems, sometimes forming ***planar branch systems***, the leaves ***microphyllous***, spiral, either ***homomorphic*** or ***dimorphic and 4-rowed*** (with 2 upper rows of leaves smaller than the 2 lower rows), sporangia ***heterosporous***, microsporangia and megasporangia borne in axils of ligulate sporophylls of ***terminal strobili***; gametophytes ***endosporic***.

EUPHYLLOPHYTA—EUPHYLLOPHYTES

The sister group of the lycophytes are the euphyllophytes, including all the other extant vascular plants (Figure 4.1). Several major apomorphies that unite the euphyllophytes are mentioned here. First, in contrast to the lycophytes, the roots of Euphyllophytes are **monopodial**, meaning that they do not dichotomously branch at the apical meristem. Lateral roots arise endogenously from either the **endodermis** (in Monilophytes) or the **pericycle** (in spermatophytes, Chapter 5). Second, the roots of euphyllophytes have an **exarch protoxylem**, in which the protoxylem is positioned outer to the metaxylem (Figure 4.20A,B); lycophytes have an endarch protoxylem. Third, the ancestral sporangia in Euphyllophytes were terminal with longitudinal dehiscence (although these features have undergone considerable modification in some groups). Fourth, extant euphyllophytes have a molecular apomorphy, a 30-kilobase **inversion** located in the large single-copy region of **chloroplast DNA** (Figure 4.20C; see Figure 14.4 of Chapter 14). Fifth, the leaves of euphyllophytes, termed **euphylls**, are distinctive. (Note that *euphyll* is essentially synonymous with *megaphyll*, a more traditional term.) **Euphylls**, like lycophylls, are generally dorsiventral organs, functioning as the primary organ of photosynthesis. Euphylls

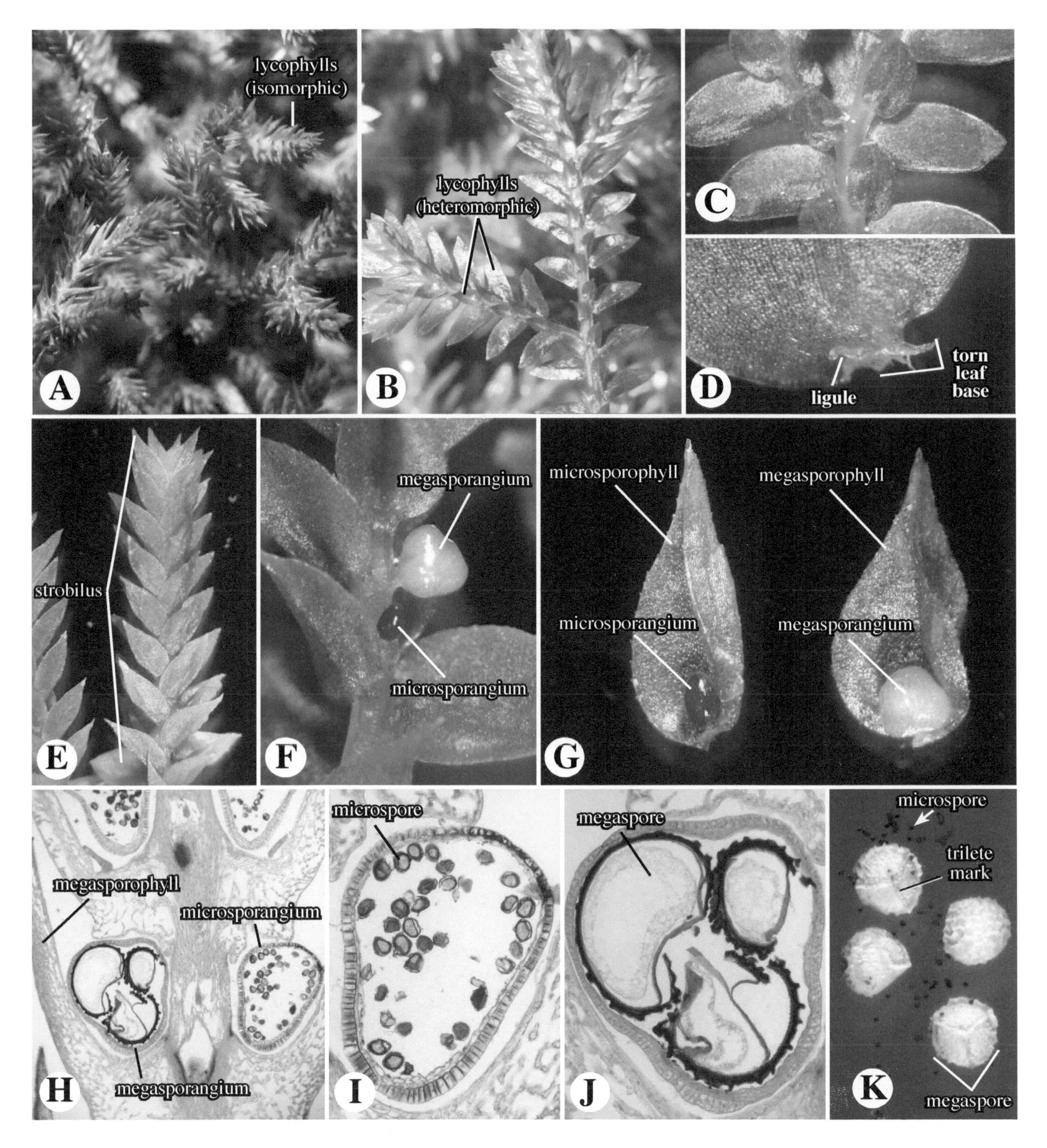

Figure 4.19 LYCOPODIOPHYTA. Isoetales—Selaginellaceae. **A.** *Selaginella bigelovii*, with isomorphic leaves. **B.** *Selaginella apoda*, with dimorphic leaves. **C.** Close-up of vegetative shoot, showing 2 rows of large and 2 rows of small leaves. **D.** Close-up of ligule, adaxial side of leaf base. **E.** Cone (strobilus), an axis bearing microsporophylls and megasporophylls. **F.** Close-up of microsporangium and megasporangium (sporophylls removed). **G.** Adaxial view of microsporophyll and megasporophyll with axillary microsporangium and megasporangium, respectively. **H.** Strobilus longitudinal section, showing sporophylls, megasporangia, and microsporangia. **I.** Close-up of microsporangium, containing numerous microspores. **J.** Close-up of megasporangium, containing 4 megaspores. **K.** Dispersed microspores and megaspores, the latter showing trilete mark. Note great size difference.

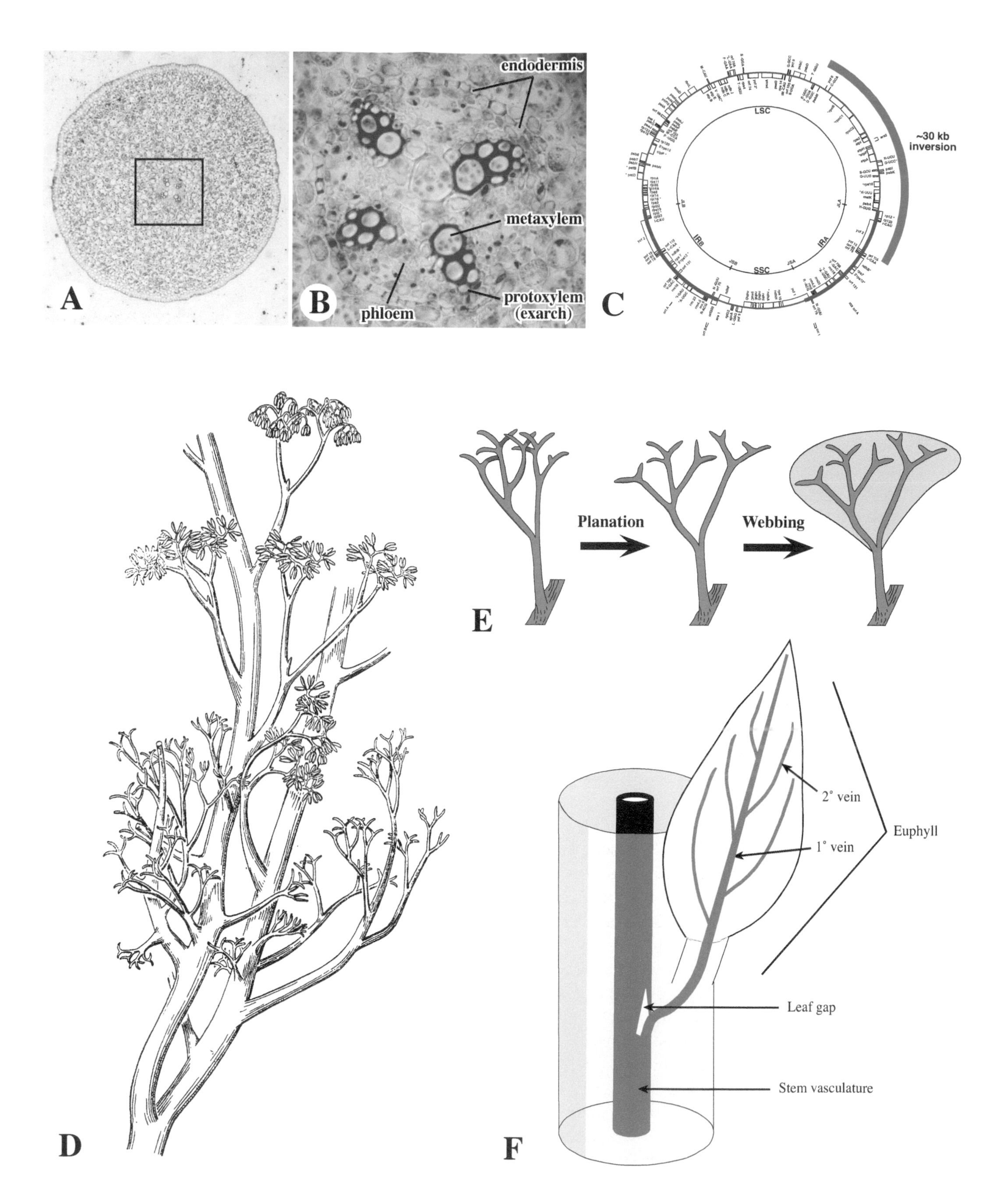

Figure 4.20 **A–C.** Apomorphies of Euphyllophyta. **A,B.** Root with exarch protoxylem. **A.** Root cross section with central vascular cylinder; insert showing area at "B." **B.** Close-up of cylinder. Note protoxylem external to metaxylem. **C.** 30kb inversion of chloroplast DNA (genes *ycf2-psb*M). **D.** *Psilophyton dawsonii* reconstruction, a trimerophyte with fertile and sterile branch systems. Reproduced from Banks et al. (1975), with permission from Paleontological Research Institution, Ithaca, New York. **E.** Hypothetical evolutionary transformation of 3-dimensional branch system into leaf by planation and webbing. **F.** Euphyll structure and anatomy, showing multiple veins and leaf gap.

are different in being associated with a **leaf gap**, a region of nonvascular, parenchyma tissue interrupting the vasculature of the stem, and in (usually) having more than one vein per leaf (Figure 4.20F). Euphylls generally have a highly branched system of veins, between which is the **mesophyll**, the chloroplast-containing tissue. (Note that in a few euphyllous taxa, the veins have become secondarily reduced to a single mid-vein, an evolutionary reversal.) In addition, euphylls, in contrast to lycophylls, grow by means of either marginal or apical meristems.

Euphylls are thought to have *evolved independently* of lycophylls. Among the many fossil plants thought to have diverged from the lineage leading to the common ancestor of the euphyllophytes were the trimerophytes, an extinct, paraphyletic group (Figure 4.1). Trimerophytes (illustrated by *Psilophyton dawsonii* in Figure 4.20D) had sporophytes with no leaves; the stems were photosynthetic. The stems of Trimerophytes had **pseudomonopodial growth**, in which one axis from a bifurcating stem apex overtops the other (see Figure 4.2). The functionally lateral axes became highly branched, and were either sterile or bore longitudinally dehiscent sporangia (Figure 4.20D). One theory is that these sterile, highly branched lateral axes were the precursors of euphyllous leaves. By this hypothesis, euphylls evolved via the transformation of 3-dimensional (non-planar) lateral branch system (as seen in Trimerophytes) into a leaf, by the processes of **planation**, flattening of the axes into a 2-dimensional plane, and **webbing**, the development of thin tissue between the axes of the branches (Figure 4.20E). The original axes become transformed into veins, and the "web" functioned as a photosynthetic mesophyll. (Note that this hypothesis of euphyll evolution is not universally accepted, but it remains perhaps the most viable notion.)

Given that lycophylls and euphylls evolved independently, then shoots evolved twice in the vascular plants, in both the lycophyte and euphyllophyte lineages. Despite their presumed separate origin, vascular plant shoots develop very similarly. As mentioned earlier, a **shoot** is defined as a stem plus associated leaves, both organ types ultimately developing from a region of actively dividing cells at the tip of the shoot known as the **apical meristem** (Figure 4.21A,D,E,G). One or more ultimate cells of the apical meristem undergo sequential mitotic cell divisions, resulting in a proliferation of cell derivatives. [The apical meristem may contain one, dominant apical cell (Figure 4.21C), found in most of the Selaginellaceae and the monilophytes, or a complex of several, actively dividing cells (Figure 4.21E,G) , found in the Lycopodiaceae, Isoetaceae, and the euphyllophytes.] One or more of these derivatives maintains the position and function of the apical meristem; the others continue to divide, and their derivatives may continue to divide. Vertically down from the apical meristem, the cell derivatives undergo considerable elongation, literally pushing the cells of the apical meristem upward or forward. Even further down from the shoot tip, the fully grown cells differentiate into their mature, specialized form. To the sides of the apical meristem region, certain localized regions of the outermost cell layers of a shoot undergo cell division and elongation. Further growth and differentiation in these regions result in the formation of a **leaf primordium** (Figure 4.21A,B,D,E), which matures into a **leaf**. The point of attachment of a leaf to the stem is known as the **node**; the region between two nodes is called an **internode** (Figure 4.21D). As the shoot matures, the leaves fully differentiate into an amazing variety of forms (see Chapter 9), and the stem differentiates a vascular system. Vascular strands run between stem and leaf providing a connection for fluid transport. (See Chapter 10 for more details.)

Leaves have a characteristic anatomy (Figure 4.21H). Because they are usually dorsiventral organs (with some exceptions), both upper and lower epidermis can be defined. As with all land plants, a cuticle covers the outer cell wall of the epidermal cells. One or more **vascular bundles**, or **veins**, contain xylem and phloem tissue and conduct water and sugars to and from the chloroplast-containing **mesophyll** cells. The mesophyll of some leaves is specialized into upper, columnar palisade mesophyll cells and lower, irregularly shaped spongy mesophyll cells, the latter with large intercellular spaces (Figure 4.21H). Stomata, which function in gas exchange (see Chapters 3, 10), are found typically only in the lower epidermis of leaves (Figure 4.21H).

Later in shoot development, the tissue at or above the region of the junction of stem and upper leaf, termed the **axil**, may begin to divide and differentiate into a **bud** (Figure 4.21F), defined as an immature shoot system. Buds have an architectural form identical to that of the original shoot. They may develop into a lateral branch or may terminate by developing into a reproductive structure. This pattern of growth, in which lateral branches develop from axillary buds, is known as **monopodial**. Monopodial growth is responsible for lateral branching in most of the euphyllophytes. (See Chapter 9 for modifications of the monopodial growth pattern.)

Euphyllophytes are composed of two major groups, which are sister groups to one another: monilophytes (ferns, in the broad sense) and spermatophytes (seed plants, Chapter 5).

MONILOPHYTA—MONILOPHYTES, FERNS

Morphological and molecular phylogenetic studies (e.g., Kenrick and Crane 1997; Pryer et al. 2001; Grewe et al. 2013; Ruhfel et al. 2014; Rothfels et al. 2015; PPGI 2016) support the recognition of a monophyletic group of vascular plants that are inclusive of four major lineages, classified here as: Equisetopsida (horsetails), Marattiopsida (marattioid ferns),

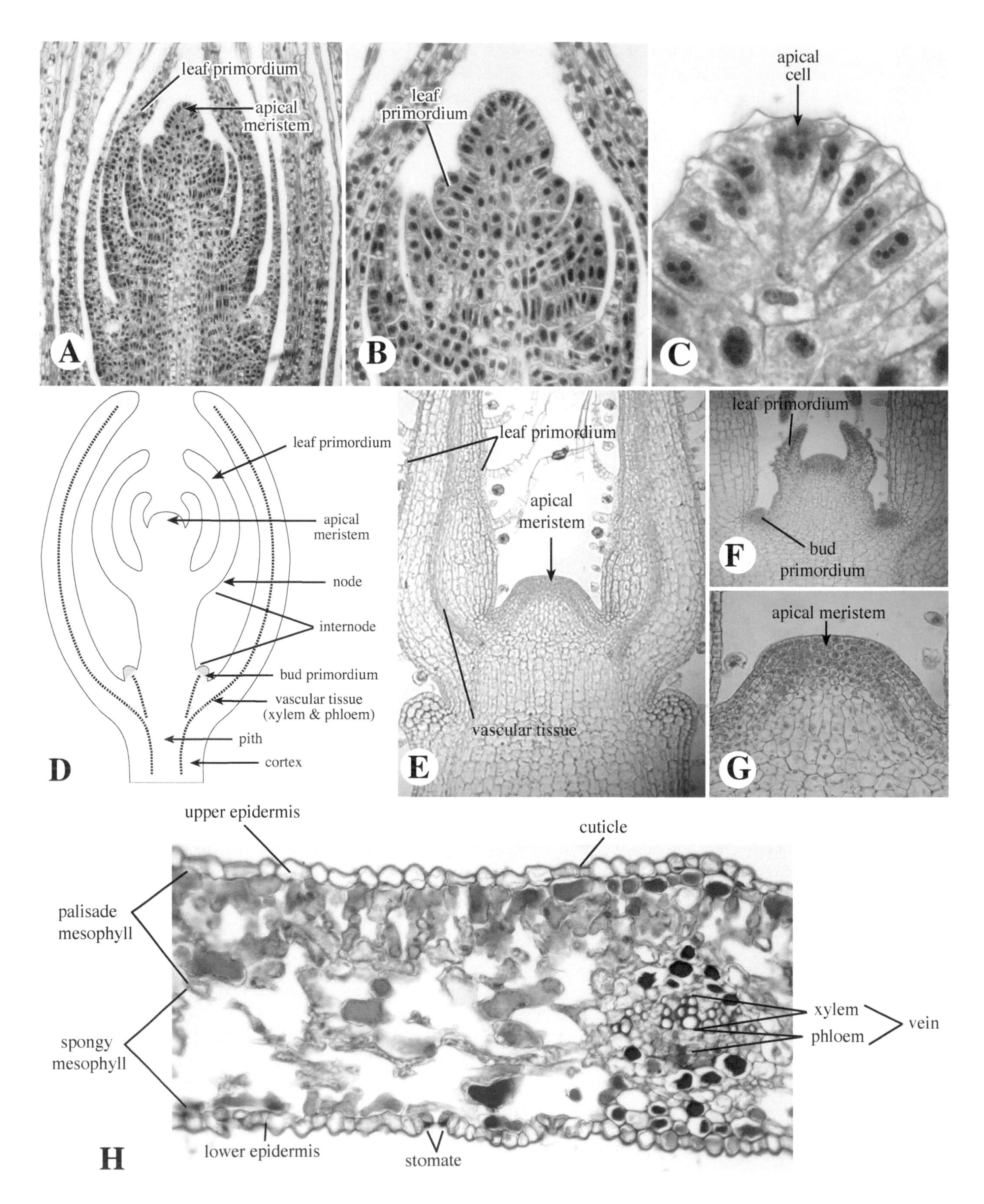

Figure 4.21 **A–G.** Shoot longitudinal section. **A–C.** *Equisetum*, showing single, apical meristem cell. **D.** Diagram of shoot longitudinal section. **E–G.** *Plectranthus* [*Coleus*] shoot. Note complex apical meristem, leaf and bud primordia, and vasculature. **H.** Cross section of a vascular plant leaf (*Ligustrum*).

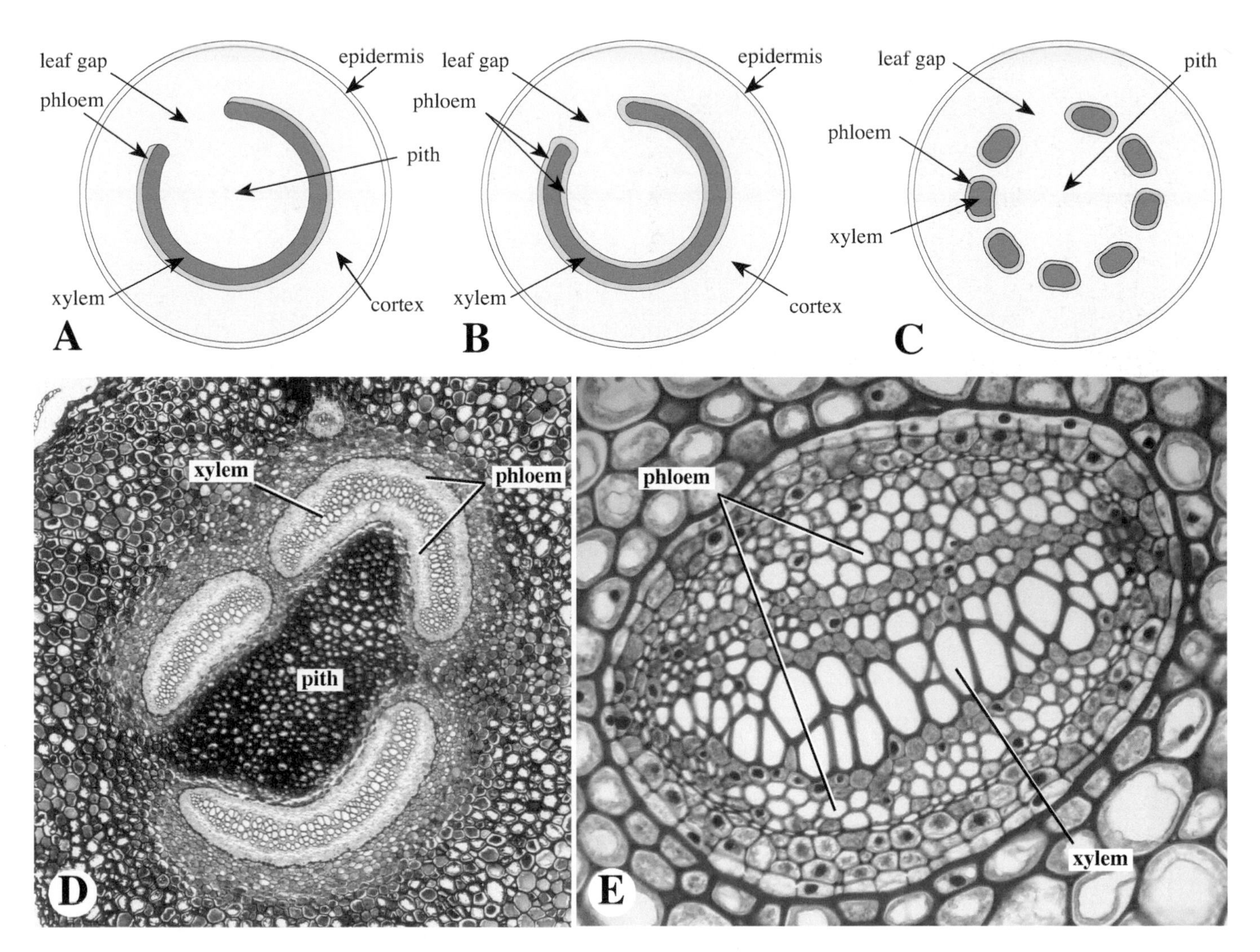

Figure 4.22 **A–C.** Siphonostele types. **A.** Ectophloic siphonostele, with phloem to outside of xylem. **B.** Amphiphloic siphonostele, with phloem to outside and inside. **C.** Dictyostele, a dissected amphiphloic siphonostele. **D.** *Adiantum* rhizome, an amphiphloic siphonostele. **E.** *Polypodium* rhizome, close-up of vasculature showing mesarch protoxylem, an apomorphy for the monilophytes (ferns).

Psilotopsida (whisk ferns and ophioglossoid ferns), and Polypodiopsida or Leptosporangiatae (leptosporangiate ferns). The entire monophyletic group has been termed the Monilophyta (or moniliformopses); the common name is often now termed monilophytes or just "ferns," in the broad sense of the word. Classification and family circumscription here largely follows the PPG1 (2016).

One apomorphy of the monilophytes, found in most (but not all) extant members, is the **siphonostele**. A **siphonostele** (Figure 4.22A–D) is a type of stem vasculature in which a ring of xylem is surrounded by an outer layer of phloem ("ectophloic siphonostele," Figure 4.22A) or by outer and inner layers of phloem ("amphiphloic siphonostele," Figure 4.22B; if dissected, called a "dictyostele," Figure 4.22C). Siphonosteles have a central, parenchymatous pith (Figure 4.22A–D). Siphonosteles have evidently become secondarily modified in some monilophytes. A second anatomical apomorphy of the monilophytes is that the **stem protoxylem is mesarch** in position (Figure 4.22E), meaning that tracheary elements first mature near the middle of a patch of xylem; this protoxylem (unlike that of some related fossil taxa) is restricted to the lobes of the xylem. The derivation of monilophyte (*L. monilo*, necklace or string of beads + Gr. *phyt*, plant) is in reference to this anatomy in fossil plants.

EQUISETOPSIDA—HORSETAILS/SCOURING-RUSHES

The Equisetopsida [Equisetidae, Equisetales], also called the equisetophytes, sphenophytes, or sphenopsids, are a monophyletic group that diverged early in the evolution of vascular plants. As with the lycophytes, some equisetophytes of the Carboniferous period, approximately 300 million years ago, were large woody trees. Among these was *Calamites* (Figure 4.23), another contributor to coal deposits. Most current molecular systematic studies place the equisetophytes firmly

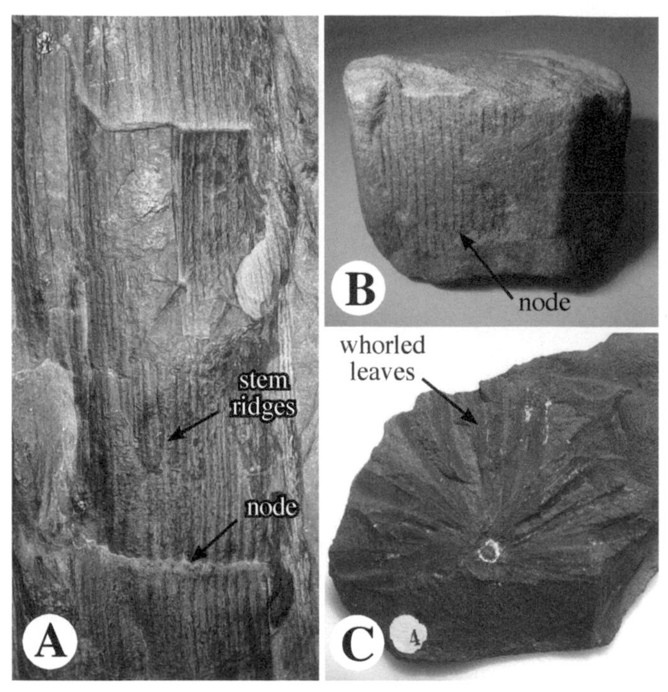

Figure 4.23 *Calamites*, an extinct, tree-sized equisetophyte. **A.** Fossil impression, showing nodes and stem ridges. **B.** Fossil cast of stem. **C.** Fossil impression showing whorled leaves of branch.

in the monilophytes but in various positions, one shown in Figure 4.1, others as sister to the Psilotopsida (e.g., Ruhfel et al. 2014). However, incorporation of morphological and, in particular, fossil data may yield different results (see Rothwell and Nixon 2006).

The equisetophytes are united by several apomorphies, four of which are cited here (Figure 4.1): (1) **ribbed stems** (Figure 4.24A,J), these often associated with internal **hollow canals** (Figure 4.24C); (2) **reduced, whorled leaves** that are usually marginally fused (Figure 4.24A,J); (3) **sporangiophores**, each of which consists of a peltate axis bearing pendant longitudinally dehiscent sporangia, arranged in a **cone/strobilus** (Figure 4.24F,L); and (4) photosynthetic spores with **elaters** (Figure 4.24G,H; see later discussion).

The only extant equisetophytes are species of the genus *Equisetum*, in the single family Equisetaceae, described here.

Equisetaceae–Horsetail/Scouring-Rush family (type *Equisetum*, from Latin *equus*, horse, + *seta*, bristle). 1 genus/15 species (+ some hybrids) (Figure 4.24).

The Equisetaceae consist of perennial herbs, often growing in or near wet [sometimes xeric] habitats. The underground **stems** are rhizomes (often forming extensive colonies), with tubers produced in some taxa. The aerial stems are ridged and photosynthetic, with an epidermis containing silica inclusions. Stems are distinctive in having a hollow pith and two rings of canals (lacunae): an inner ring of "carinal" canals (opposite the stem ridges) and an outer ring of "vallecular" canals (between stem ridges). Some taxa have lateral branches, either in dense whorls, or irregularly, sometimes due to injury of the apical meristem. The lateral branches develop from buds forming just above the nodal region; they erupt through the leaf sheath between adjacent leaves. A few taxa have dimorphic aerial shoots: photosynthetic vegetative shoots and non-photosynthetic reproductive shoots. The **leaves** are small, simple, whorled (several per node), non-photosynthetic at maturity, and laterally connate into a sheath, with distinct, tooth-like apices. **Sporangia** are homosporous, born in terminal **strobili** (subtended by a whorl of sheathing leaves), consisting of an axis bearing numerous, peltate sporangiophores, each (at maturity) bearing 5–10 sporangia beneath the distal, hexagonal outer portion; dehiscence longitudinal (parallel to axis of the elongate sporangium). **Spores**, which lack an attachment scar, are spherical, green (with chloroplasts), each bearing four, spatulate, hygroscopic elaters (derived from the spore wall), which coil and uncoil with changes in humidity, functioning in spore dispersal. **Gametophytes** are photosynthetic and generally cushion-like.

The Equisetaceae have a mostly worldwide distribution (absent in Australasia). The family is classified into two monophyletic groups: subgenus *Equisetum*, the horsetails, having numerous, whorled, lateral branches and stomates flush with the epidermis, and subgenus *Hippochaete*, the scouring-rushes, which mostly lack lateral branches (occasionally forming, particularly after injury) and have sunken stomates. Economic importance includes local medicinal uses, dye and fiber plants, weeds (some toxic to livestock), and edible plant parts (cones and young shoots); aerial shoots have been used for scouring/polishing. *Equisetum giganteum* has sprawling stems > 10 m long. See Hauke (1990) for general information and Des Marais et al. (2003) for a phylogenetic analysis of the family.

The Equisetaceae are readily distinguished in being ***rhizomatous***, perennial herbs, the aerial shoots ***hollow***, ***ridged***, with ***siliceous epidermal cells*** and internal ***canals***, the leaves small, non-photosynthetic, simple, microphyllous, ***whorled,*** and ***laterally connate (sheathed)***, the apices ***teeth-like***; sporangia in ***terminal strobili***, ***homosporous***, several born underneath ***peltate sporangiophores***, the spores ***green, with 4, spatulate, hygroscopic elaters***.

PSILOTOPSIDA

This group consists of two orders, Psilotales and Ophioglossales. In addition to DNA sequence data, two features may constitute apomorphies linking the two orders. First, the roots of Ophioglossales are unusual in lacking both root branches and root hairs. This may represent a transitional stage to the total loss of roots in the Psilotales. Second, the gametophytes of both orders are nonphotosynthetic (heterotrophic), contain mycorrhizal fungi, and are often subterranean (Figure 4.1).

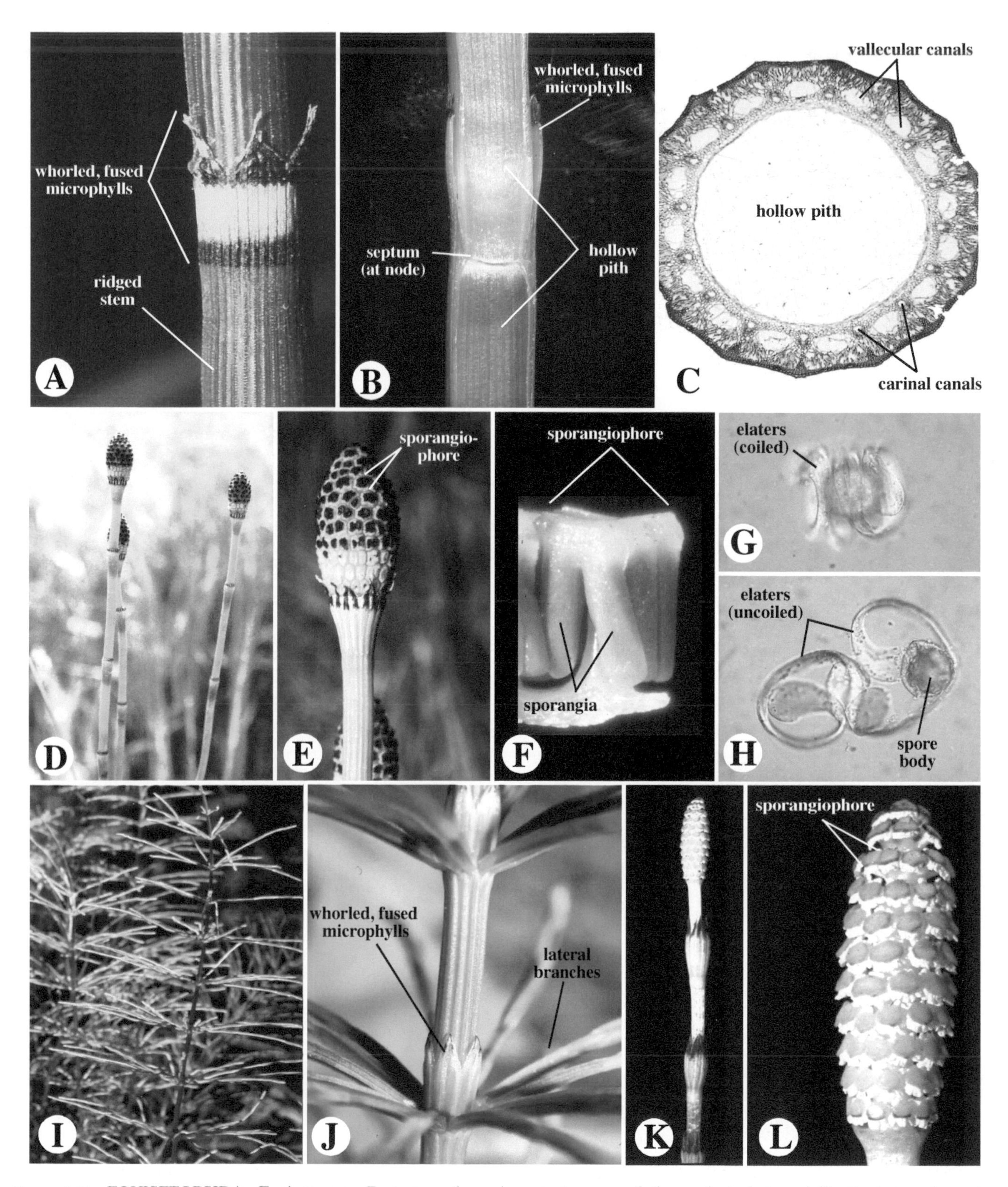

Figure 4.24 EQUISETOPSIDA. Equisetaceae. *Equisetum*, the only extant genus of the equisetophytes. **A,B.** *Equisetum hyemale*. **A.** Vegetative stem. Note ridged stem and whorled microphylls. **B.** Stem longitudinal section showing central hollow pith and septum at nodes. **C.** Stem cross section of *Equisetum* sp., showing central, hollow pith and peripheral, vallecular canals. **D,E.** *Equisetum laevigatum*, a "scouring rush," having photosynthetic, generally unbranched aerial stems. **F.** Sporangiophore, with several pendant sporangia. **G,H.** Spores, each with four elaters. **G.** Elaters coiled. **H.** Elaters uncoiled. **I–L.** *Equisetum arvense*, a "horsetail," with dimorphic aerial stems. **I,J.** Sterile, photosynthetic stems with whorls of lateral branches. **K.** Reproductive, nonphotosynthetic aerial stem, lacking whorls of branches and terminating in a strobilus. **L.** Strobilus close-up, showing sporangiophores.

Figure 4.25 OPHIOGLOSSALES. Ophioglossaceae. **A,B.** *Botrychium* species, showing vegetative lamina and fertile segment. **A.** *B. multifidum*. **B.** *B. lunaria*. **C.** *Ophioglossum californicum*, adder's tongue. Note elongate fertile segment bearing eusporangia.

OPHIOGLOSSALES — OPHIOGLOSSOID FERNS

The Ophioglossales [=Ophioglossidae], or ophioglossoid ferns, consist of a single family of fernlike plants. The ophioglossoid ferns are unique in that each leaf (or "frond") consists of a **sterile segment**, which contains the photosynthetic blade or lamina, and a **fertile segment**, bearing the sporangia. The underground rhizome gives rise to unbranched roots that lack root hairs. The sporangia of the Ophioglossales, and of all land plants except for the leptosporangiate ferns, are often termed **eusporangia** (or "eusporangiate sporangia"), in contrast to leptosporangia (see later discussion). A **eusporangium** is relatively large, is derived from several epidermal cells, has a sporangial wall comprised of more than one cell layer, and produces very many (usually thousands of) spores (Figures 4.25B,C, 4.26E). Eusporangia are the ancestral condition of the land plants.

Ophioglossaceae–Adder's Tongue family (type *Ophioglossum*, from Greek *ophis*, snake + *glossa*, tongue, in reference to the shape of fertile segments). 10 genera/112 species (Figure 4.25).

The Ophioglossaceae consist of terrestrial, perennial herbs. The **roots** are fleshy, mycorrhizal, lacking root hairs, sometimes bearing adventitious buds (that may grow into a new plantlet). The **stems** are subterranean, erect; the vasculature is a protostele or ectophloic siphonostele. The **leaf** is often solitary, lacking circinate vernation, the blade simple and unlobed or compound to divided (1-2-pinnatifid); ventation is open-dichotomous or reticulate. **Sori** are lacking. **Sporangia** are homosporous and eusporangiate, born on a stalked, spike-like (with two vertical rows of sporangia, e.g., *Ophioglossum*) or branched and panicle-like fertile segment; sporangial dehiscence is transverse (relative to fertile segment axis). **Spores** are tetrahedral and trilete. **Gametophytes** are non-photosynthetic and mycorrhizal or mycotrophic.

The Ophioglossaceae have a worldwide distribution. The genera include *Botrychium* (ca. 35 spp.), *Helminthostachys* (1 sp., *H. zeylanica*, Indo-Australasia), *Mankyua* (1 sp., *M. chejuense*, Jeju Island, S. Korea), and *Ophioglossum* (ca. 41 spp.). Economic importance includes local uses as food or medicine. *Ophioglossum reticulatum* has the highest recorded chromosome count of any organism (cited in different studies as 2n=1,260 or, recently, as n=760, the latter equivalent to 2n=1,520!). See Wagner (1990) for general information and Hauk et al. (2003) for a phylogenetic analysis of the Ophioglossaceae.

The Ophioglossaceae are distinctive in being ***homosporous***, ***eusporangiate***, perennial herbs; leaves simple to compound, bearing a stalked ***fertile segment*** containing eusporangia, the gametophytes ***non-photosynthetic***.

PSILOTALES — WHISK FERNS

The Psilotales, or psilotophytes (commonly called "whisk ferns"), consist of the single family Psilotaceae, with two genera (*Psilotum* and *Tmesipteris*). Like all vascular plants, the whisk ferns have an independent, dominant, free-living sporophyte; the haploid gametophyte is small, obscure, and free-living in or on the soil. The sporophyte consists of a horizontal rhizome that gives rise to aerial, photosynthetic, generally dichotomously branching stems (Figure 4.26A,B).

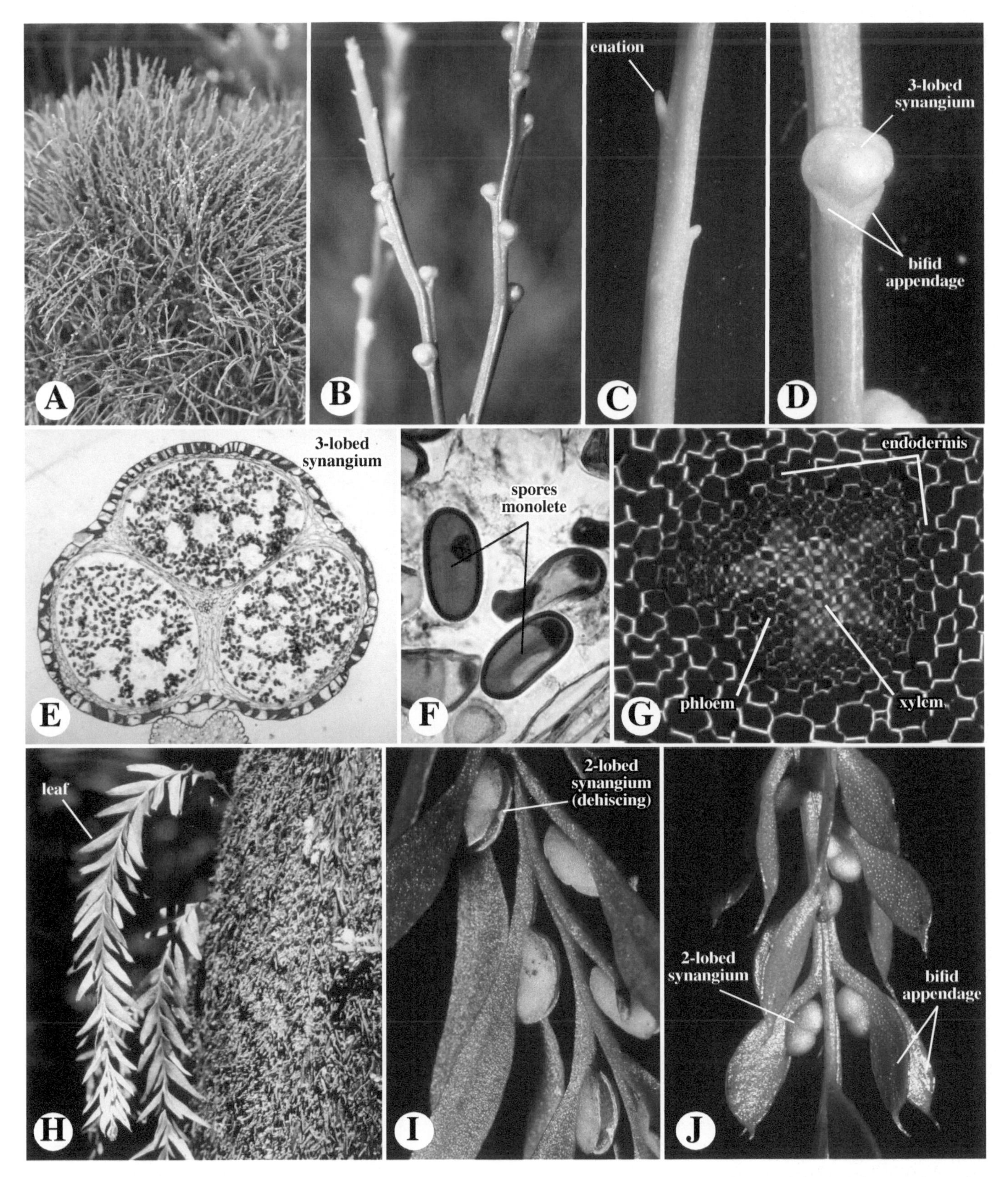

Figure 4.26 PSILOTALES. Psilotaceae. **A–G.** *Psilotum nudum*. **A.** Whole plant showing dichotomous branching. **B.** Close-up of dichotomous aerial shoots. **C.** Vegetative stem close-up, showing reduced leaves or enations. **D.** Close-up of synangium, subtended by bifid appendage. **E.** Synangium cross section. **F.** Spores, showing monolete aperture. **G.** Cross section of stem, a protostele with central xylem and peripheral phloem (surrounded by endodermis). **H,I.** *Tmesipteris elongata*. **H.** Epiphytic plant. **I.** Close-up of dehiscing, 2-lobed synangia. **J.** *Tmesipteris lanceolata*. Two-lobed synangia, each subtended by bifid appendage. (Images H–J courtesy of John Braggins.)

Plants are often epiphytic, with rhizomes having symbiotic mycorrhizal associations. All psilotophytes **lack true roots**, an apomorphy for the group; only absorptive rhizoids arise from the rhizome. The absence of roots in the psilotophytes has often been considered to be a primitive retention, the psilotophytes having being viewed as direct descendants of the rhyniophytes. However, molecular studies clearly indicate that psilotophytes are sister to the Ophioglossales (Figure 4.1) and likely lost roots secondarily.

The leaves of psilotophytes are very reduced and peglike (Figure 4.26C) and may lack a vascular strand, in which case they are termed **enations**. The sporangia (which, like the Ophioglossales, could be termed eusporangia) are two- or three-lobed, which is interpreted as a **synangium**, a fusion product of two or three sporangia (Figure 4.26D,J). The synangia are yellowish at maturity and are subtended by a forked appendage, an apomorphy for the group. As in the Ophioglossales, the gametophytes of the Psilotales are nonphotosynthetic (subterranean or surface-dwelling) and may contain mycorrhizal fungi.

Psilotum nudum, the "whisk broom," is the most widespread species of the psilotophytes, and commonly serves as an exemplar for the group (Figure 4.26A–G). *Psilotum nudum*, native to tropical regions, is cultivated in greenhouses and naturalized in warm climates worldwide.

Psilotaceae–Whisk Fern family (type *Psilotum*, from Greek *psilos*, naked, alluding to leafless stems). 2 genera/ca. 17 species (Figure 4.26).

The Psilotaceae consist of terrestrial or epiphytic, perennial herbs. **Roots** are absent. The underground **stems** are rhizoid-bearing rhizomes with endophytic mycorrhizae; aerial stems are photosynthetic, terete, ridged, or flattened, and are dichotomously branched or unbranched, the vasculature a protostele or solenostele. The **leaves** are simple, spiral, or distichous; blades small and subulate to scale like, or (in *Tmesipteris*) some larger (to 2 cm long) and vertically oriented, either with a single midrib (microphyllous) or lacking vasculature, although sometimes vasculature runs to base (these termed "enations" or "prophylls"). **Sporangia** are eusporangiate, homosporous, arising from short, lateral branches, subtended by a bifid appendage, synangiate, the product of fusion of three (*Psilotum*) or two (*Tmesipteris*) sporangia. **Spores** are reniform and monolete. **Gametophytes** are non-photosynthetic, cylindrical, saprophytic, and mycorrhizal; sperm are multiflagellate.

The Psilotaceae are distributed worldwide in tropical or warm temperate regions. The two genera are: *Psilotum* (2–3 spp., tropical and warm regions worldwide) and *Tmesipteris* (ca. 15 spp., s.e. Asia to Australasia). Economic importance includes cultivated ornamentals (esp. *Psilotum nudum*, the whisk broom). See Kramer (1990e) for general information.

The Psilotaceae are distinctive in being ***rootless***, rhizomatous, perennial herbs, with unbranched or ***dichotomously branched*** stems, aerial stems ***photosynthetic***, the leaves simple, spiral or distichous, ***lacking vasculature (enations)*** or with a single vein (***microphyllous***), sporangia 2- or 3-***synangiate***, born on a short lateral axis subtended by a ***bifid appendage***, the gametophytes ***non-photosynthetic, mycorrhizal***.

MARATTIOPSIDA—MARATTIOID FERNS

The Marattiopsida contain the single order Marattiales and family Marattiaceae, consisting of six genera: *Angiopteris*, *Christensenia*, *Danaea*, *Eupodium*, *Marattia*, and *Ptisana*. Marattioid ferns are very similar to the Polypodiopsida or leptosporangiate ferns (discussed later) in general form, having large pinnate or bipinnate leaves (Figure 4.27A,F) with circinate vernation, sporangia located on the abaxial surface of leaflet blades, and a photosynthetic gametophyte (see later discussion). However, the sporangia of the Marattiales are **eusporangiate**, like those of all vascular plants except for the leptosporangiate ferns. In some taxa of the Marattiales, the sporangia are fused into a common structure, a **synangium** (Figures 4.27G,J). A distinctive apomorphy of the Marattiales is the occurrence of a "polycyclic siphonostele" (Figures 4.1, 4.27C), which appears as concentric rings of siphonosteles in cross section (the vasculature is, however, connected at a lower level).

Marattiaceae–Marattia family (type *Marattia*, after G. F. Maratti, Italian botanist (1723–1777). 6 genera/ca. 111 species (Figure 4.27).

The Marattiaceae are large, terrestrial ferns, tissues with mucilage canals. The **stems** are rhizomatous or erect, stout, and trunk-like, the anatomy a polycyclic dictyostele. The **leaves** develop by circinate vernation, mature leaves large, stipulate (stipules large, persistent on stems), simple or 1-4-pinnate, the petioles and rachillae with swollen pulvini and prominent pneumatodes (lenticels). **Sori** are abaxial, intramarginal, exindusiate. **Sporangia** are eusporangiate, homosporous, distinct, vertically dehiscent (plane of dehiscence perpendicular to blade surface), and arranged in a narrow ring or fused into a raised or sunken synangium, which dehisces tangentially into two valves. **Spores** are trilete, monolete, or alete. **Gametophytes** are large, thalloid, and photosynthetic.

The Marattiaceae have a worldwide distribution in tropical and warm regions. Economic importance includes cultivated ornamentals and food (from edible stems), perfume oil, and an alcoholic drink (from stem starch). The six genera of the family are: *Angiopteris*, *Christenssenia* (2 spp.), *Danaea* (30 spp.), *Eupodium* [*Marattia*] (neotropics), *Marattia* s.s., and *Ptisana* [*Marattia*] (paleotropics). See Camus (1990) for general information and Murdock (2008a,b) for a phylogenetic

Figure 4.27 **A–D.** MARATTIALES. Marattiaceae. **A–E.** *Angiopteris* sp. **A.** Plant from stout, erect trunk with bipinnately compound leaves. **B.** Base of leaves showing swollen (pulvinal) petiole bases and persistent stipules. **C.** Rhizome cross section with polycyclic siphonostele. **D.** Sori, each consisting of an ellipse of eusporangia. **E.** Sori cross section, showing unfused eusporangia. **F,G.** *Danaea* sp. **F.** Pinnately compound leaf with lower, fertile pinnae having numerous synangia on abaxial surface. **G.** Synangium containing several eusporangia. **H–K.** *Marattia* spp. **H.** Whole plant with large, bipinnately compound leaves. **I.** Close-up of leaflets. **J.** Leaflet, abaxial view, showing intramarginal synangia. **K.** Synangia dehiscing in a bivalvate manner, each valve containing several eusporangia.

analysis and taxonomic revision of the family.

The Marattiaceae are distinctive in being large, terrestrial ferns with ***mucilage canals***, the stems with a ***polycyclic dictyostele***, the leaves generally large, simple to several-pinnate, with ***abaxial, intramarginal eusporangia***, sometimes fused into ***synangia***.

POLYPODIOPSIDA–LEPTOSPORANGIATE FERNS

The Polypodiopsida or Leptosporangiatae (also known as Filiopsida or Filicales) correspond to what are commonly known as the **leptosporangiate ferns**. Of the major monilophyte groups, the leptosporangiate ferns contain by far the greatest diversity, estimated from at least 8,800 to over 12,000 species.

Leptosporangiate ferns have a specialized terminology (see Chapter 9), which may be slightly different from that of other vascular plants. (See Lellinger 2002 for an excellent compendium of fern terminology, in four languages.) The sporophytes of almost all leptosporangiate ferns are perennial herbs or trees, an exception being some of the aquatic ferns (Salviniales; see later discussion), which may be annuals. Most leptosporangiate ferns have a horizontally oriented stem, the **rhizome** (Figure 4.28A,B), which may grow under or upon the ground (**terrestrial**), on or in cracks of rocks (**epipetric**), on or in water (**aquatic**), or upon another plant (**epiphytic**). Some leptosporangiate ferns are **arborescent**, with a tall, erect, aerial stem, which in the tree ferns (Figure 4.40) can attain heights of up to 20 meters (66 feet). A few ferns are **vines** (Figure 4.37), with weak stems or with elongate, vine-like leaves that sprawl on the ground or upon another plant. The stem anatomy of leptosporangiate ferns can be diagnostic, either an ectophloic or amphiphloic **siphonostele** (Figure 4.22A,B), **dictyostele** (Figure 4.22C), or **protostele** (Figure 4.7).

The immature leaves of leptosporangiate ferns, like those of the Marattiales, have a type of development known as **circinate vernation**, in which both major and minor axes or leaf divisions are coiled early in development and uncoiled at maturity. The young, coiled leaves are known as **fiddleheads** or **croziers** (Figure 4.28E). Circinate vernation with its associated crozier formation may constitute an apomorphy for the Polypodiopsida and Marattiopsida together; however, this feature is also shared with the cycads of the seed plants (see Chapter 5).

The leaves of ferns come in a great variety of forms. The leaf itself is often called a **frond**; the petiole is often called a **stipe**. (However, note that "leaf" and "petiole" are used throughout the descriptions here.) The first discrete leaflets or blade divisions of a fern leaf are called **pinnae** (singular **pinna**; Figure 4.28C). If there is more than one division, the terms **1° pinna**, **2° pinna**, and so forth may be used. The ultimate leaflets or blade divisions are called **pinnules** or **segments** (Figure 4.28C; see also Chapter 9). Variations in leaf size, type (e.g., simple, pinnate, bipinnate, or pedate), and division (e.g., pinnatifid, bipinnatifid; Figure 4.28C; see Chapter 9) constitute the primary features in fern identification. The leaves of most ferns are similar to one another (**monomorphic**), but some ferns have leaves that are **dimorphic**, in which fertile leaves differ (other than just the presence of reproductive organs) from sterile, vegetative leaves. Other ferns may have dimorphic leaf segments, in which fertile segments (usually modified pinnae or pinnules) differ significantly in blade morphology from sterile segments.

Leaf venation can be valuable in fern classification and identification. In most ferns the overall venation is usually **pinnate**, with a central vein giving rise to veinlets on either side, or less commonly **palmate**, with more than one main vein arising from the base (see Chapter 9, leaf venation types). However, the most important venation feature is the ultimate vein pattern (smallest veins) of the pinnules or smallest segments of the leaf. The two general types referring to this ultimate venation are **open** (**free**) and **reticulate** (**anastomosing**). **Open** or **free** venation is that in which the veins arising from the midvein or base of a pinnule do not join back together. These free veins may be **simple**, in which they do not branch (Figure 4.29A), or (more commonly) **forked** or **bifurcate**, in which each vein gives rise to pairs of veins toward the margin (Figure 4.29B). (If the two veins of a fork are equal throughout, the venation may be termed **dichotomous**.) **Reticulate** or **anastomosing** venation is that in which the veins appear to join back together, forming a net-like "reticulum," enclosing an area sometimes known as an areole (Figure 4.29C). Reticulate patterns can be complicated (e.g., some having areoles within areoles) but diagnostic for a given taxon. See Kramer and Green (1990) for detailed fern venation terminology.

Many leptosporangiate ferns have **trichomes** (hair-like structures) or **scales** (flattened, minute, leaf-like structures on the stem, shoot apex, petiole, or blade; e.g., Figure 4.29D), the cellular anatomy of which can be a valuable taxonomic character. For example, scales with the cell walls of adjacent cells ("anticlinal" walls) thick are termed **clathrate** (Figure 4.29E); those with thin anticlinal cell walls are **non-clathrate** (Figure 4.29F), which may be further classified into sub-types (e.g., fibrillose, denticulate, or marginate; see Lellinger, 2002).

The primary apomorphy of the leptosporangiate ferns is, of course, the **leptosporangium** (Figures 4.1, 4.29G–I). Leptosporangia are unique among vascular plants in (1) developing from a single cell, and (2) having a single layer of cells making up the sporangium wall. Leptosporangia also tend to have a much smaller spore number than eusporangia. (Note that at least some members of the family Osmundaceae, discussed below, have sporangia that are intermediate in that they

Figure 4.28 Leptosporangiate Ferns—Characters. **A,B.** Rhizomes, the most common type of stem in the group. **A.** Underground rhizome (*Nephrolepis cordifolia*, Lomariopsidaceae). **B.** Above-ground rhizome (*Davallia trichomanoides*, Davalliaceae). **C.** Leaf morphology of ferns, illustrating specialized terminology. **D.** Fern petiole (stipe) covered with scales (*Nephrolepis cordifolia*, Lomariopsidaceae). **E.** Croziers or fiddleheads (*Polypodium aureum*, Polypodiaceae), the result of circinnate vernation, in early (left) and later (right) stages.

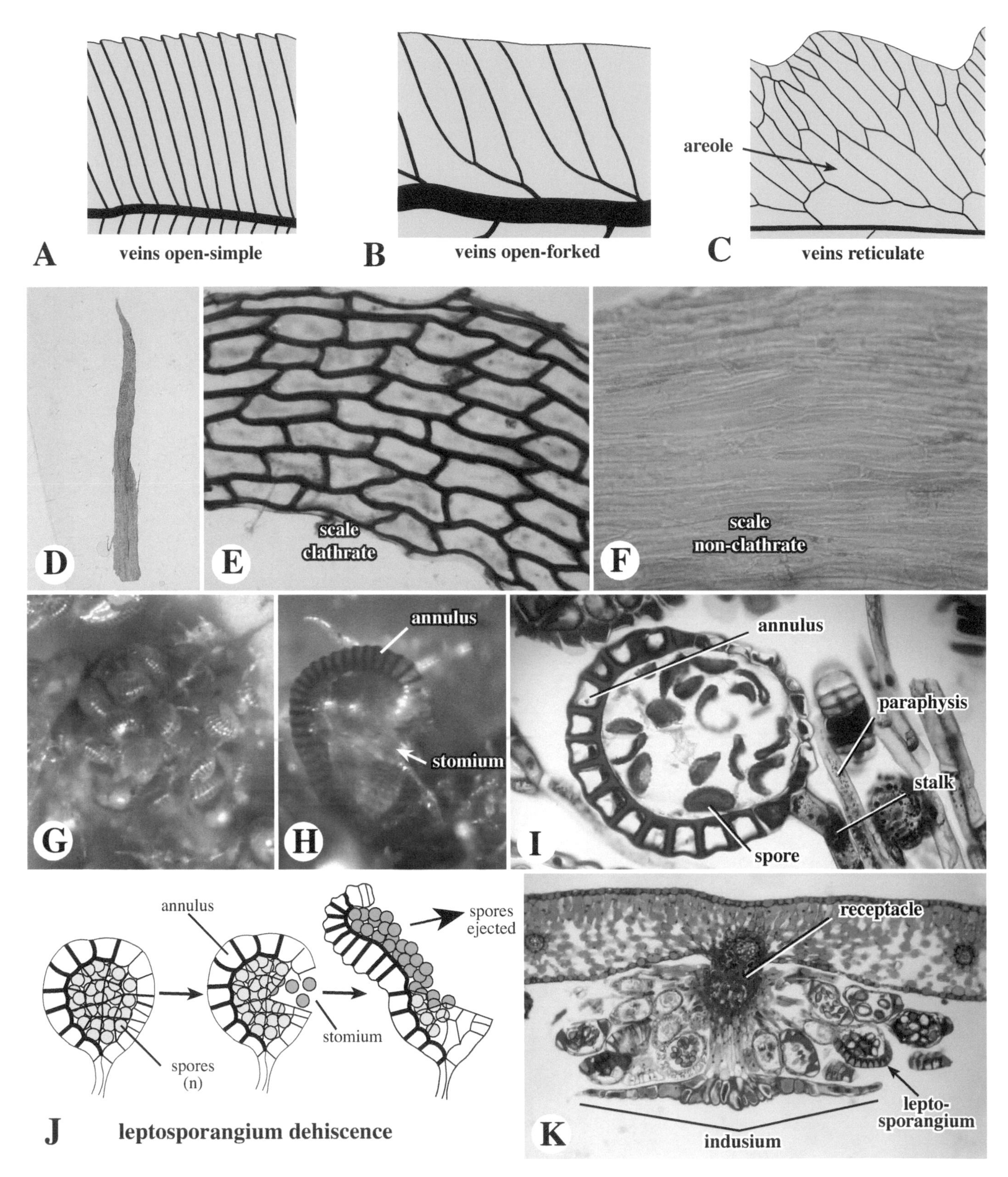

Figure 4.29 Leptosporangiate Ferns—Characters. **A.** Open, simple venation (*Blechnum* [*Lomarea*] *procera*, Blechnaceae). **B.** Open, forked venation (*Dicranopteris linearis*, Gleicheniaceae). **C.** Reticulate venation (*Onoclea sensibilis*, Onocleaceae). **D.** Example of fern scale. **E.** Clathrate scale (from surface rhizome) with thick adjacent (anticlinal) walls (*Asplenium nidus*, Aspleniaceae). **F.** Non-clathrate scale without thickened anticlinal walls (*Dryopteris arguta*, Dryopteridaceae). **G–I.** Leptosporangia. **G.** Sorus, a cluster of leptosporangia. **H.** Side view of dehiscing leptosporangium showing annulus and stomium. **I.** Leptosporangium showing stalk, annulus, and paraphysis. **J.** Leptosporangium dehiscence. (See text.) **K.** Sorus, longitudinal section, showing receptacle, leptosporangia, and indusium.

may develop from more than one initial cell and often contain a larger number of spores.) Leptosporangia typically have a proximal stalk and distal sporangial body. The leptosporangium may have been an important adaptation in the ferns because of a unique mechanism of spore dispersal, at least in most taxa. Part of the wall of the sporangial body develops into a single row of specialized cells, collectively known as an **annulus**, in which the cell walls are differentially thickened on the inner cell face and on the faces between adjacent annular cells (Figure 4.29J). As the leptosporangium matures and begins to dry, water evaporates from the cells of the annulus. The force of capillarity, in which water molecules strongly attract to one another and to the inner cell wall surfaces, causes the cells to buckle on the *outer* faces, as these are regions in which the cell wall is not thickened and thus structurally weakest. This buckling provides a force that causes splitting, or dehiscence, of the leptosporangium (typically occurring at a region of thin-walled cells called the **stomium**), followed by a backward retraction of the annulus (Figure 4.29J). A short time after the annular cells fully retract, total evaporation of water within the cells releases the tensile force of capillarity, resulting in the annulus catapulting forward, ejecting the spores in the process (Figure 4.29J). (Note that leptosporangia of the Salviniales, the water ferns, are highly modified and lack an annulus.)

Leptosporangia are often aggregated into discrete clusters, known as **sori** (singular **sorus**; Figure 4.30A–E). Leptosporangia are attached within the sorus at a common region, the **receptacle** (Figure 4.29K), which, in some ferns, can become quite elongate (e.g., *Trichomanes*, Hymenophyllaceae; see Figure 4.35H–J). Sterile, hair-like structures, known as **paraphyses**, may arise from the receptacle and be intermixed with the leptosporangia. Sori, when present, are often covered by a flap of tissue arising from the blade surface known as an **indusium** (Figures 4.29K, 4.30C–E), which may function to protect the young leptosporangia or to control the dispersal of spores. The presence (and morphology; see later discussion) of an indusium is an important character in fern taxonomy. For example, the Aspleniaceae are **indusiate**, having an indusium, whereas the Polypodiaceae are **exindusiate**, lacking an indusium (although some fern families may have both conditions). If leptosporangia are not aggregated into definable sori, and appear scattered on the leaf surface, they may be termed **acrostichoid** (after the genus *Acrostichum* having this condition; Figure 4.30F). Some taxa lack an indusium, but have a reflexed extension of the blade margin called a **false indusium**, which overlaps the sorus (Figure 4.30G,H).

Several features of the indusia, sori, and leptosporangia may be important taxonomic characters. The position and shape of the sorus and indusium, especially relative to the blade margin or veins, are useful delimiting features. For example, the indusium may be **reniform** (kidney-shaped; Figure 4.30C), **orbicular** (circular; Figure 4.30D), or **linear** (narrowly elongate; Figure 4.30E). (See Chapter 9, Shapes.) The attachment of the indusium (if present) may be **peltate**, with a central stalk (Figure 4.30D), or **lateral**, attached at the side (Figure 4.30E). Features of the sorus itself include (1) shape (in outline); (2) variation in the shape and size of the **receptacle**; and (3) presence and morphology of **paraphyses**. The development of the leptosporangia within the sorus can be: (1) **gradate** (**sequential**), in which the sporangia of a sorus mature in succession from the base (periphery) toward the apex (**acropetalous**) or from the apex toward the base (**basipetalous**); (2) **simultaneous** (**simple**), in which sprorangia or a sorus mature at the same time; (3) **mixed**, a combination of gradate and simultaneous; or (4) **intermingled**, with no consistent developmental pattern. There may also be variation in the leptosporangia in length and number of stalk cells, body size and shape, and morphology and position of the annulus (e.g., whether lateral, transverse, apical, oblique), stomium, and dehiscence line (slit; see Figure 4.31). Finally, the spores of leptosporangiate ferns can vary in many features, including size, shape, sculpturing of the outer wall layer (known as **perine**; see Chapter 12); whether green (chlorophyllous) or not; number per sporangium; and spore type, whether **trilete**, **monolete**, or **alete** (see Figure 4.11).

Leptosporangiate ferns, like most non-seed tracheophytes, have a haploid gametophyte phase that is separate from the "dominant" sporophyte phase. Fern gametophytes are quite small and generally consist of a thin flat sheet of green, photosynthetic cells, the shape of which is often cordate, but can vary and be of taxonomic importance. Gametophytes typically are **surficial** (grow upon the ground) and bear **rhizoids**, filamentous cells that function in attachment and absorption (resembling and likely homologous with those of liverworts, mosses, and hornworts). (The gametophytes of the heterosporous Salviniales are quite different from other leptosporangiate ferns; see later discussion.) Fern gametophytes bear sperm-producing **antheridia** and egg-producing **archegonia** (Figure 4.32). Sperm cells are coiled and multi-flagellate, sometimes attached to a cytoplasmic vesicle. As in virtually all non-flowering land plants, a sperm cell fertilizes an egg cell of the archegonium. The resultant zygote divides and differentiates into a new embryo (immature sporophyte), which initially remains attached to the gametophyte (Figure 4.32). Soon, however, the sporophyte attains independence of the gametophyte (which subsequently dies), and the sporophyte becomes the persistent, "dominant" phase of the life cycle, a characteristic of all vascular plants (Figure 4.1). (Note that fern sporophytes can develop from gametophytes asexually, termed "apogamy/apogamous.")

Finally, chromosome numbers, whether diploid (**2n**),

Figure 4.30 Leptosporangiate Ferns—Characters. Sori, indusia, and sporangia. **A,B.** Exindusiate sorus (*Polypodium* spp., Polypodiaceae). **C.** Indusiate sorus, indusium reniform (*Nephrolepis cordifolia*, Lomariopsidaceae). **D.** Indusiate sorus, indusium orbicular, peltately attached (*Cyrtomium falcatum*, Dryopteridaceae). **E.** Indusiate sorus, indusium linear (*Asplenium bulbiferum*, Aspleniaceae). **F.** Acrostichoid sporangia (*Platycerium* sp., Polypodiaceae). **G,H.** False indusium (*Adiantum* spp., Pteridaceae).

haploid (**n**), or the base number for a group (**x**), have traditionally been important in fern taxonomy, both at higher and lower levels of classification.

The economic importance of leptosporangiate ferns is primarily as cultivated ornamentals in the horticultural trade. However, some ferns (usually the croziers) are edible, and many have local uses as medicines, fibers and matting, or flavoring. (See family descriptions for economic uses.)

The classification of the families of leptosporangiate ferns has varied considerably in the past and is still an active area of research. The following descriptions are based on the treatment of Smith et al. (2006), which recognizes 7 orders and 31 families of leptosporangiate ferns (Table 4.1), derived from phylogenetic analyses using molecular data (including Pryer et al. 2001a, 2004, Schuettpelz et al. 2006; see Smith et al. 2006 for additional references and Kramer 1990d for general

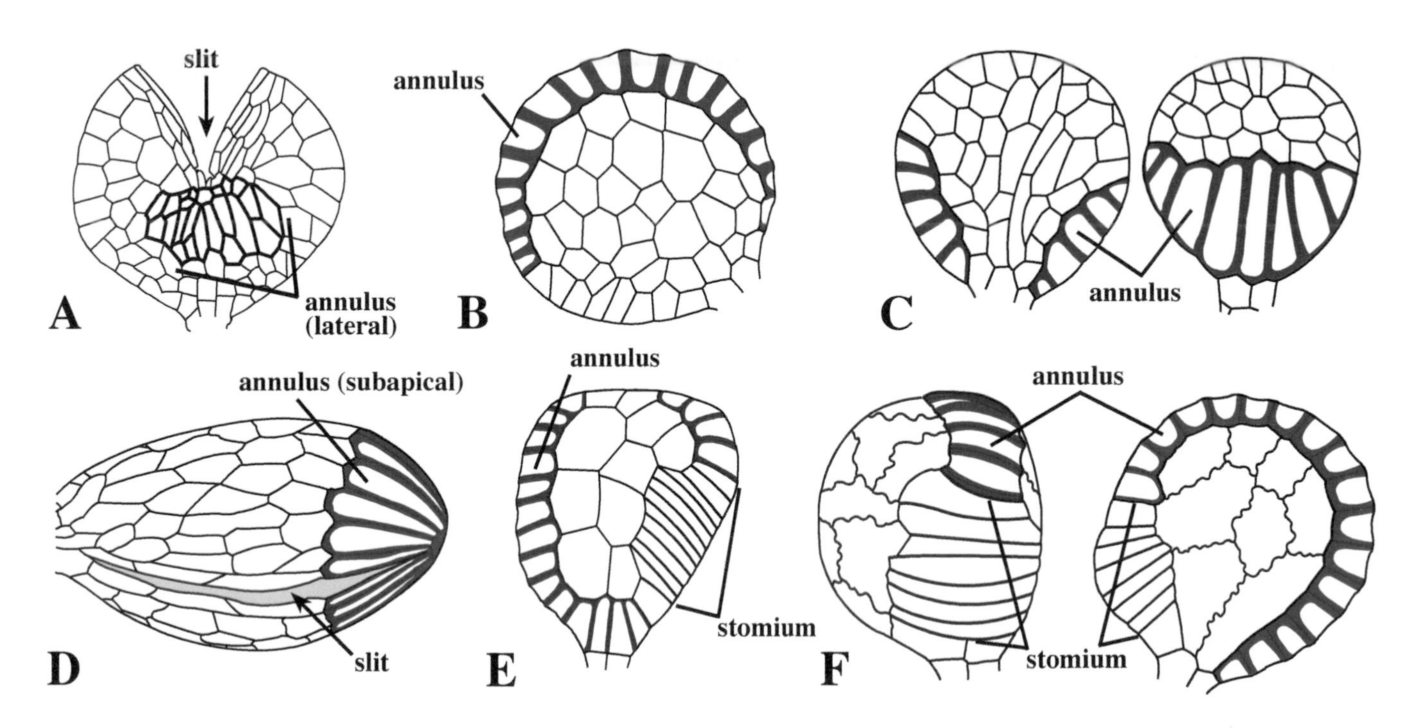

Figure 4.31 Leptosporangiate Ferns—Leptosproangium morphology. **A.** *Osmunda claytoniana* (Osmundaceae), with lateral annulus and apical slit. **B.** *Hymenophyllum australe* (Hymenophyllaceae), with an oblique annulus. **C.** *Gleichenia pectinata* (Gleicheniaceae), with an obliquely transverse annulus. **D.** *Schizaea bifida* (Schizaeaceae), with a transverse, subapical, continuous annulus. **E.** *Cyathea medullaris* (Cyatheaceae), with oblique annulus. **F.** *Pteridium aquilinum* (Dennstaedtiaceae), with oblique annulus. Redrawn from Smith (1955).

information; see cladogram at Figure 4.33). Although these orders and families appear to be monophyletic, some of them have few diagnostic morphological features and may be hard to characterize, e.g., certain families of the Polypodiales (see later discussion). See Tryon and Tryon (1982) and Kramer and Green (1990) for major compendia on ferns. (See Chapter 9 and Glossary for questions on terminology.)

OSMUNDALES — OSMUNDACEOUS FERNS

The Osmundales consists of the single family Osmundaceae.

Osmundaceae–Cinnamon Fern family (type *Osmunda*, possibly named for the Scandinavian writer *Asmund*, ca. 1025 A.D.). 6 genera, including *Leptopteris, Osmunda,* and *Todea*)/ ca. 18 species (Figures 4.31A, 4.34).

The Osmundaceae are terrestrial plants. The **stems** are erect (some taxa being arborescent "tree ferns"), with an ectophloic siphonostele having a ring of discrete xylem strands, these often conduplicate or twice conduplicate in cross section. The **leaves** are 1-2-pinnate or pinnatifid, stipulate, with stipules at the base of petioles, and dimorphic with either fertile and sterile leaves or with fertile and sterile leaf segments. **Sori** and **indusia** are absent, the sporangia occuring on the abaxial surface of leaves or leaf segments. **Sporangia** have large bodies and short stalks, dehiscing by an apical slit, the **annulus** lateral. **Spores** are green, subglobose, and trilete, 128–512 per sporangium. **Gametophytes** are relatively large, green, cordate, and surficial. Chromosome number: x=22.

Members of the Osmundaceae are found is tropical and temperate regions. Economic importance includes some cultivated ornamentals (*Osmunda* spp., *Todea barbara*) and local uses for fiber and food. *Osmunda regalis* is used in brewing a Celtic ale. See Kramer and Viane (1990b) for general information and Yatabe et al. (1999) for a phylogenetic study of the family.

The Osmundaceae are distinctive in having ***erect stems*** (sometimes arborescent), an ectophloic siphonostele with separate xylem strands, ***dimorphic leaves or leaf segments*** with ***sori and indusia absent***, and short-stalked, large-bodied sporangia with a ***lateral annulus***.

HYMENOPHYLLALES — FILMY FERNS

The Hymenophyllales consist solely of the Hymenophyllaceae.

Hymenophyllaceae [incl. Trichomanaceae]–Filmy Fern family (type *Hymenophyllum*, from Greek *hymen*, "membrane" + *phyllum*, "leaf," after the very thin leaves of family members). 9 genera/ca. 434 species (Figure 4.35).

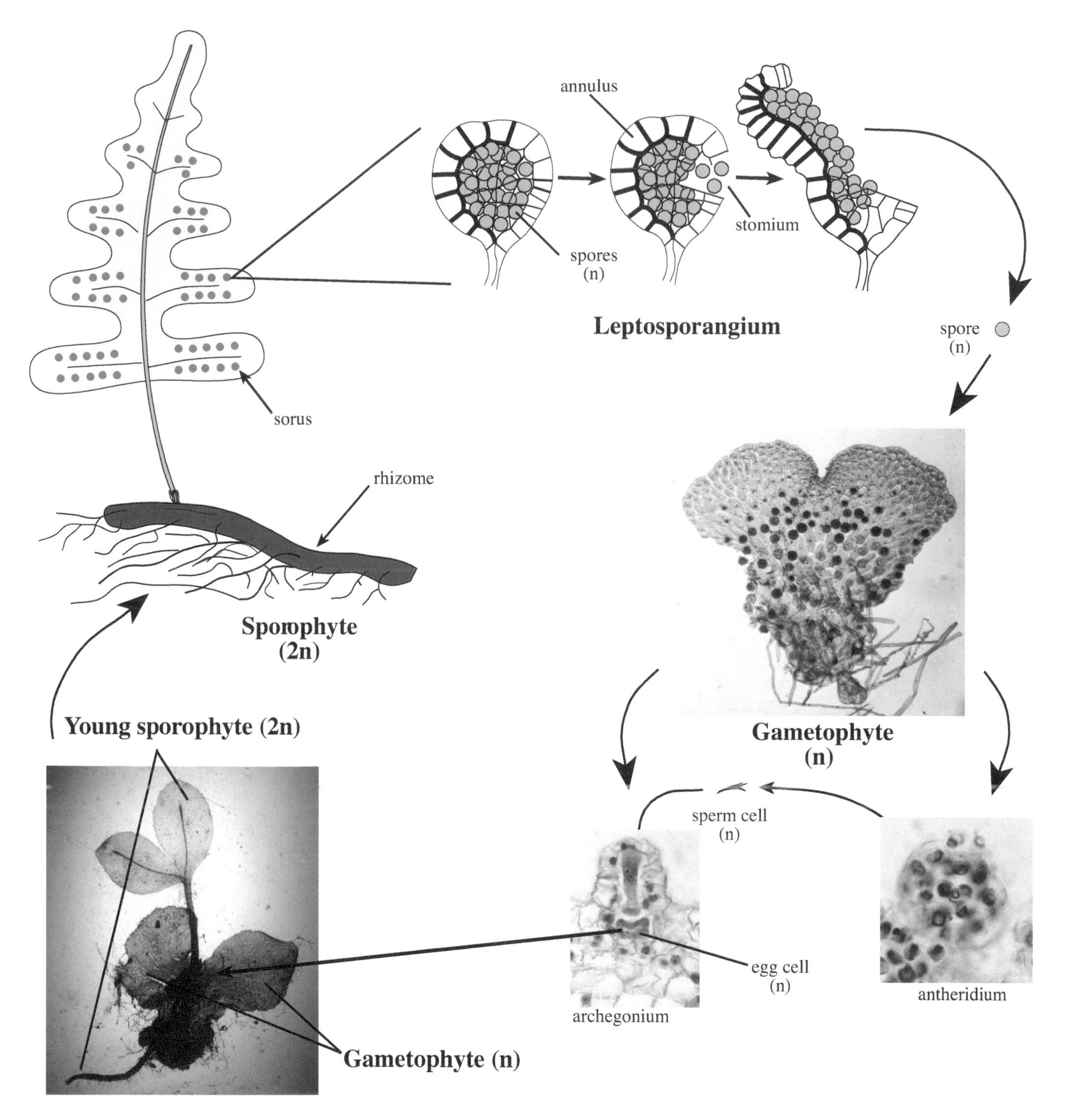

Figure 4.32 Life cycle of leptosporangiate ferns. Note mechanism of spore dispersal, gametophyte development, fertilization, and sporophyte development.

The Hymenophyllaceae [incl. Trichomanaceae] contain both epiphytic (usually) and terrestrial plants. The **stems** are rhizomatous, rhizomes without scales, slender and creeping or stout and erect, protostelic. The **leaves** are usually 1-cell thick, with stomata absent, cuticle absent or reduced, blade scales usually absent, trichomes sometimes present, venation open. **Sori** are marginal, the **receptacle** elongate and continuous with vein tips. **Indusia** are present and conical, tubular, or 2-lobed (bivalvate). **Sporangia** are basipetalous, dehiscence is uninterrupted, the **annulus** oblique. **Spores** are green, globose, and trilete. **Gametophytes** are filamentous or ribbon-like, often reproducing by fragmentation or gemmae. Chromosome numbers: x = 11, 12, 18, 28, 32, 33, 34, 36.

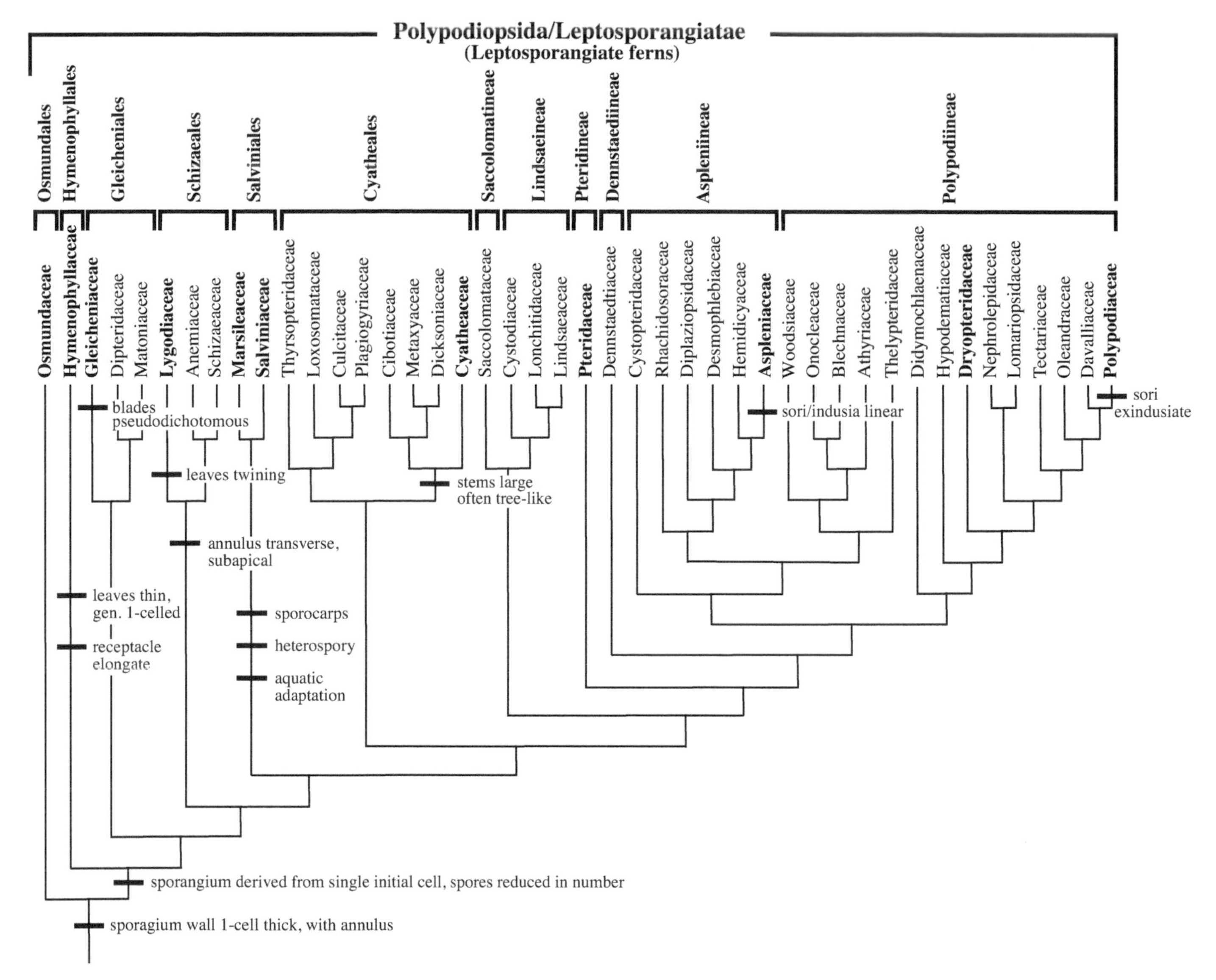

Figure 4.33 Cladogram of families and orders of leptosporangiate ferns (Polypodiopsida/Leptosporangiatae), after Smith et al. (2006), with selected apomorphies added. Families in bold are described in detail.

The Hymenophyllaceae contain two, sister groups: the "trichomanoid" clade (incl. *Trichomanes*) and the "hymenophylloid" clade (incl. *Hymenophyllum*). Family members are found in humid regions (e.g., cloud forests, streamsides) of pantropical and south-temperate regions (gametophytes surviving in north-temperate areas). Economic importance is limited to a few cultivated ornamental, e.g., *Trichomanes* spp. (bristle, kidney ferns). See Iwatsuki (1990) for general information and Pryer et al. (2001b), Hennequin et al. (2003), and Ebihara et al. (2006) for recent phylogenetic studies.

The Hymenophyllaceae [incl. Trichomanaceae] are distinctive in having ***scale-less***, usually slender, creeping, ***protostelic rhizomes***, ***thin leaves usually 1-cell thick, lacking stomata***, and marginal sori with ***conical, tubular, or 2-lobed indusia*** and ***elongate receptacles***.

GLEICHENIALES — GLEICHENIOID FERNS

The Gleicheniales contain over 170 species in three families, one of which is described here. Morphological features which may unite the order include root steles with 3–5 protoxylem poles and antheridia with walls containing 6–12 narrow, twisted, or curved cells.

Gleicheniaceae [incl. Dicranopteridaceae & Stromatopteridaceae]–Forking Fern family (type *Gleichenia*, after German botanist Friedrich Wilhelm von Gleichen, 1717–1783). 6 genera (including *Dicranopteris*, *Gleichenia*, *Stromatopteris*)/ca. 157 species (Figure 4.36).

The Gleicheniaceae are terrestrial, often growing in dense, open thickets. The **stems** are rhizomatous, the rhizomes creeping, dichotomously branched, with a "vitalized" pro-

Figure 4.34 Leptosporangiate Ferns–OSMUNDALES. Osmundaceae. **A,B.** *Leptopteris superba*, with 2-3-pinnatifid leaves. Note relatively large, globose sporangia. **C.** *Osmunda cinnamomaea*, cinnamon fern. **D.** *Osmunda claytoniana*, interrupted fern. **E.** *Todea barbara*, with bipinnately compound leaves. (Images A, B, & E courtesy of Lawrence Jensen.)

Figure 4.35 Leptosporangiate Ferns–HYMENOPHYLLALES. Hymenophyllaceae. **A–C.** *Hymenophyllum tunbrigense*. **A.** Plants growing on log. **B.** Leaf, adaxial surface. **C.** Leaf, abaxial surface, showing sori with bivalvate indusia. **D,E.** *Hymenophyllum dilatatum*, leaf and bivalvate indusium. **F.** *Hymenophyllum flabelliforme*, leaf. **G,H.** *Trichomanes reniforme*. **G.** Immature leaf; note open, dichotomous venation. **H.** Eruptive, elongate receptacles bearing leptosporangia. **I,J.** *Trichomanes endlicherianum*, showing elongate receptacles of sori. (Images A–C, courtesy of Vera Svobodova; D,E, H–J courtesy of Lawrence Jensen.)

tostele or rarely a solenostele. The **leaves** are often indeterminate in growth (sometimes reaching great lengths), with the rachises pseudodichotomously branched, 1-2-pinnate, veins free. **Sori** are round, abaxial, not marginal, exindusiate. **Sporangia** are round to pear-shaped, up to 5–15 sporangia per sorus, developing simultaneously, the **annulus** transverse-oblique. **Spores** are globose-tetrahedral or bilateral, trilete or monolete, 128–800 per sporangium. **Gametophytes** are large, green, surficial, with club-shaped hairs, developing endotrophic mycorrhizae. Chromosome numbers: x = 22, 34, 39, 43, 56.

The Gleicheniaceae have a largely pantropical distribution. Economic importance includes the use of *Dicranopteris linearis* stems as a fiber plant (for cordage, mats, fish traps, etc.), and use of some taxa as cultivated ornamentals (e.g., *Sticherus* spp.). See Kramer (1990b) for general information and Pryer et al. (2004) and Schuettpelz (2006) for recent phylogenetic studies of the family and order.

The Gleicheniaceae are distinctive in ***often forming dense, open thickets***, the leaves usually ***long, indeterminate, pseudodichotomously branched***, with ***round, exindusiate sori***, leptosporangia with a ***transverse-oblique annulus***.

SCHIZAEALES — SCHIZAEOID FERNS

The Schizaeales contain about 190 species in three families–Anemiaceae, Lygodiaceae, and Schizaeaceae–the second described here. Common morphological features of these taxa include dimorphic leaves, lack of well-defined sori, and sporangia having a transverse, subapical, continuous annulus (Figures 4.31D, 4.37). See Skog et al. (2002) for information about relationships in the order.

Lygodiaceae–Climbing Fern family (type *Lygodium*, from Greek *lygodes*, "flexible," after the flexuous rachis). 1 genus (*Lygodium*)/ca. 40 species (Figure 4.37).

The Lygodiaceae are terrestrial and climbing plants. The **stems** are rhizomatous, the rhizomes slender, creeping, bearing hairs, and protostelic. The **leaves** are mostly indeterminate with an elongate, twining and climbing rachis that bears pinnae alternately, the pinnae pseudodichotomously branching, veins free or anastomosing. **Sori** are abaxial, on lobes of ultimate leaf segments, an indusium-like flange covering the sporangium. **Sporangia** are 1 per sorus, the **annulus** transverse, subapical, and continuous. **Spores** are tetrahedral, trilete, 128–256 per sporangium. **Gametophytes** are green, cordate, and surficial. Chromosome numbers: x = 29, 30.

The Lygodiaceae are pantropical in distribution. Economic importance includes use of the twining leaf rachis as a fiber/mat material; some species are invasive weeds. See Kramer (1990f) for general information on the family (in this reference included within the Schizaeaceae).

The Lygodiaceae are distinctive in having ***indeterminate leaves****, with* ***twining/climbing rachises****, alternately bearing* ***pseudodichotomously-branching pinnae***, leaf segments ***dimorphic***, the sori at the ***tips of ultimate segments***, each with only ***1 sporangium covered by indusium-like flap***, sporangia with a ***transverse, subapical continuous*** annulus.

SALVINIALES — AQUATIC/HETEROSPOROUS FERNS

The Salviniales, classified as two families–Marsileaceae and Salviniaceae–is unusual among leptosporangiate ferns in being aquatic (either floating or rooted and emergent), with members sometimes cultivated for small ponds or aquaria. As in *Selaginella* and *Isoetes* of the Lycophytes, all members of Salviniales are virtually unique among the leptosporangiate ferns in being **heterosporous**, producing two types of spores and sporangia: **megaspores**, produced within **megasporangia**, and **microspores**, forming within **microsporangia**. Both sporangia types are leptosporangiate in development and form (together or separately) within a **sporocarp**, a rounded, seed-like structure with a hard outer layer. The sporocarp functions in both protection (being resistant to desiccation) and dormancy. Sporocarps produced at the end of the season will remain dormant until conditions are right for growth; this is generally the start of the next growing season, but some sporocarps have been observed to remain dormant for decades, at which point they "germinate" (open up) and release the spores, which are typically embedded in a gelatinous exudate (known as a sorophore or massula; see later discussion). As in the Lycophytes, megaspores are large and produced in few numbers (typically only 1 per megasporangium). The single, haploid nucleus of the megaspore gives rise (via mitotic cell divisions and differentiation) to a female gametophyte, which may at least partially develop within the megaspore wall (as in the Lycophytes, this process known as **endospory**). The female gametophyte bears one or more archegonia. Microspores, in contrast, are small and produced in large numbers. Each microspore develops into a male gametophyte, which bears one or more sperm-producing antheridia. Interestingly, these reproductive features–heterospory, reduction of megaspore number per megasporangium, and endospory–also occurred in the evolution of seeds (see Chapter 5).

Marsileaceae–Clover Fern family (type *Marsilea*, after Italian Count Luigi Ferdinando Marsigli (1658–1730), Latinized as Marsilius). 3 genera (*Marsilea*, leaves with 4 pinnae, *Pilularia*, leaves filiform or thread-like, and *Regnellidium*, leaves with 2 pinnae)/ca. 61 species (Figure 4.38).

The Marsileaceae consist of rooted, aquatic herbs with emergent leaves, the blade (if present) sometimes floating. The **stems** are elongate, slender, creeping rhizomes, often bearing hairs, with aerenchyma and a solenostelic anatomy. The **leaves**

Figure 4.36 Leptosporangiate Ferns–GLEICHENIALES. Gleicheniaceae. **A,B.** *Gleichenia microphylla*. **A.** Leaf showing pseudodichotomous branching of segments. **B.** Close-up of leaf, abaxial surface with dense trichomes. **C,D.** *Sticherus cunninghamii*. **C.** Leaf with pseudodichotomous branching of segments. **D.** Close up of sori, each with 3-5 sporangia. **E,F.** *Dicranopteris linearis*. **E.** Long, sprawling, pseudodichotomously branching leaf rachis. **F.** Pinnule close-up, abaxial surface, showing sori, each with numerous sporangia. (Images C,D courtesy of Lawrence Jensen; F courtesy of Gerald Carr.)

Figure 4.37 Leptosporangiate Ferns–SCHIZAEALES. Lygodiaceae. **A,B.** *Lygodium japonicum*, climbing fern. **A.** Plant climbing by means of elongate, indeterminate leaf rachises. **B.** Close-up of fertile leaves with sori at tips of ultimate leaf lobes. **C–F.** *Lygodium articulatum*. **C.** Single leaf, the rachis twining around another plant and giving rise to alternate pinnae. **D.** Fertile leaf segments (at right), bearing abaxial sporangia clusters. **E.** Ultimate fertile leaf segments, abaxial view, bearing rows of sori. **F.** Close-up of sori, each one consisting of an indusium-like flange subtending a single leptosporangium. (Images C–F, courtesy of Lawrence Jensen.)

are circinate, simple, or palmate with 2 or 4 sessile leaflets, veins dichotomous, often fusing apically. **Sporocarps** (interpreted as modified pinnate leaves or pinnae) are reniform with a stalk arising from the petiole base or leaf axil, each sporocarp bearing two halves, each with several rows of internal sori. **Sori** consist of a column of megasporangia and microsporangia that lack an annulus (therefore indehiscent) and that are enveloped by a hood-like **indusium**. At germination (in water) the sporocarp releases an elongate, gelatinous structure (*sorophore*), with several pairs of sori attached. Each **megasporangium** bears a single, trilete **megaspore**. After imbibing water, the megaspore releases a gelatinous mass, called *acrolamellae*. The acrolamellae, with apical longitudinal folds and basal horizontal folds, contains a central, liquid-filled region called the *sperm lake*, into which the sperm migrate. The megasporangial wall breaks away and the endosporic, female gametophyte forms a single archegonium at the megaspore apex, rupturing that region of the spore wall and protruding into the sperm lake. **Microsporangia** produce several trilete **microspores**, each microspore forming an endosporic, male gametophyte that bears and releases (by breakdown of sporangial and spore walls) numerous coiled, multi-flagellate sperm cells, some of which enter the opening of the acrolamellae into the sperm lake region, which leads to archegonium. Chromosome numbers: x = 10 (*Pilularia*) or 20 (*Marsilea*).

Distribution of the Marsileaceae is subcosmopolitan. Economic importance of family members includes use of *Marsilea* species as food (sporocarps or leaves) and cultivated ornamentals (especially *Marsilea* spp. & *Regnellidium diphyllum*). See Kramer (1990c) for general information, Pryer (1999) for a study of phylogenetic relationships, and Schneider and Pryer (2002) for studies of spore morphology in the family.

The Marsileaceae are distinctive in being rhizomatous, ***aquatic*** ferns, the ***leaves lacking blade tissue or palmate with 2 or 4, sessile leaflets***, sori developing within seed-like, dessication-resistant ***sporocarps***, which, upon imbibing water, each release an ***elongate, gelatinous structure*** bearing the sori and sporangia, the spores ***heterosporous***.

Salviniaceae [incl. Azollaceae]–Floating Fern family (type *Salvinia*, after Italian Antonio Maria Salvini, 1633–1729). 2 genera (*Azolla*, mosquito ferns, & *Salvinia*, water spangles)/ ca. 21 species (Figure 4.39).

The Salviniaceae consist of floating, aquatic herbs. **Roots** are absent in *Salvinia*, present in *Azolla*, the latter aerenchymatous, without root cap, and growing in water medium. The **stems** are dichotomously branched rhizomes, protostelic, aerenchymatous. The **leaves** are simple, dimorphic (fertile different from sterile), aerenchymatous, at maturity distichous (2-ranked) in *Azolla*, in whorls of 3 in *Salvinia* (2 floating, 1-submerged and root-like), blades round to oblong, entire, with water-repellant trichomes on upper surface in *Salvinia*, leaves of *Azolla* 2-lobed, the lower lobe submersed and largely achlorophyllous, the upper lobe aerial and chlorophyllous with a large, mucilage-filled cavity containing colonies of the nitrogen-fixing cyanobacterium *Anabaena azollae*, veins free or anastomosing. **Sporocarps** (each interpreted as a modified sorus with the indusium functioning as protective wall) globose, heterosporous, each bearing (at maturity) either one megasporangium or several microsporangia. **Megasporangium** with one functional megaspore (surrounded by "massulae," gelatinous masses of tissue from multinucleate plasmodium), each megaspore forming an endosporic, female gametophyte with several, protruding, apical archegonia. **Microsporangia** each bearing several microspores (each microspore developing into a male gametophyte with antheridia forming sperm cells), surrounded by and becoming embedded within gelatinous massulae, which may form hook-like glochidia. Chromosome numbers: x = 9 (*Salvinia*), 22 (*Azolla*).

Distribution of the Salviniaceae is subcosmopolitan. Economic importance includes *Salvinia* species used as cultivated ornamentals (some species weeds of water bodies) and use of *Azolla* to control mosquitoes (by covering the water surface), as animal fodder, and as a "seeded" addition to rice paddies, enhancing rice growth by release of nitrogens from symbiotic cyanobacteria. See Schneller (1990a,b) for general information and Reid et al. (2006) for a phylogenetic study of the family.

The Salviniaceae are distinctive in being ***floating, aquatic herbs***, the leaves simple, either in ***whorls of 3*** (2 floating, 1 root-like) bearing ***water-repellant trichomes*** (*Salvinia*) or ***2-ranked and 2-lobed***, the upper lobes housing cyanobacteria (*Azolla*), ***sori modified as sporocarps***, each bearing either one megaspore or several microspores, surrounded by ***gelatinous massulae***.

CYATHEALES — TREE FERNS

The Cyatheales are a group of over 700 species in eight families (Table 4.1). Stems may be arborescent or rhizomatous, bearing hairs or prominent scales. Sori are marginal or abaxial, indusiate or exindusiate. Spores are trilete, and gametophytes are green and cordate. The great majority of species occur in the Cyatheaceae, described below. Of the families not described here, the Cibotiaceae and Dicksoniaceae also include many arborescent "tree ferns," some of horticultural value. See Wolf et al. (1999) for a study of phylogenetic relationships in the order.

Figure 4.38 Leptosporangiate Ferns–SALVINIALES. Marsileaceae. **A–I.** *Marsilea* sp., clover fern. **A.** Plant showing rhizome bearing roots and leaves. **B,C.** Close-ups showing leaves with 4, distal pinnae. **D.** Young leaf showing coiled circinate vernation. **E.** Sporocarp, sagittal section, with thick wall and internal microsporangia and megasporangia. **F.** Sporocarp close-up, showing microsporangia (containing numerous microspores) and megasporangia (each containing one megaspore). **G.** Female gametophyte, with distal acrolamellae and apical sperm lake. **H.** Sperm cells, close-up, embedded within acrolamellae. **I.** Herbarium specimen of *Marsilea quadrifolia*, bearing sporocarps from rhizome. **J,K.** *Pilularia americana*, pillwort. **J.** Plant in habitat. **K.** Soil removed, showing subterranean sporocarps.

Figure 4.39 Leptosporangiate Ferns–SALVINIALES. Salviniaceae. **A–E.** *Azolla*, mosquito fern. **A.** Vegetative, prostrate shoots. **B.** Close-up of 2-ranked (distichous) leaves. **C.** Underside of shoot, showing lower achlorophyllous and upper chlorophyllous of leaf lobes. **D.** View illustrating upper leaf lobe cavity housing the cyanobacterium *Anabaena azollae*. **E.** *Anabaena azollae* removed from upper leaf cavity. Note nitrogen-fixing heterocyst (inset) of filaments. **F–H.** *Salvinia*. **F,G.** Close-up of leaves, showing water-repellent trichomes. **H.** Shoot showing two floating leaves and one submersed and root-like, at each rhizome node.

Cyatheaceae–Scaly Tree Fern family (type *Cyathea*, from Greek *kyatheion*, a little cup, after the spore cups in this group). 3 genera (*Alsophila, Cyathea*, and *Sphaeropteris*)/643 species (Figure 4.40).

The Cyatheaceae are mostly terrestrial, some epiphytic. The **stems** are mostly arborescent (tree ferns), the trunks often with marcescent (persisting) leaves or leaf bases; shoot apices and petiole bases are covered with large scales or hairs, the stem anatomy a polycyclic dictyostele. The **leaves** are usu. large (up to 5 m long), blades 1-3-pinnate (rarely simple), petioles with obvious, usually discontinuous pneumathodes (tissue with air spaces) in two lines; blade veins free, simple to forked, rarely anastomosing. **Sori** are abaxial, round, superficial or terminal on veins and marginal or submarginal, the receptacle raised, paraphyses present, exindusiate or insudiate. **Indusium**, when present, saucer-like, cup-like, bivalvate, or globose and completely surrounding sporangia. **Sporangia** maturing gradately, the **annulus** oblique. **Spores** are tetrahedral, trilete, variously ornamented. **Gametophytes** are green, cordate. Chromosome number: x = 69.

Distribution of the Cyatheaceae is pantropical, especially in montane forests. Economic importance includes use of the trunk for construction material or bee-hives, the starchy pith for animal food, and various plant parts for medicinal purposes; some species are invasive weeds; several *Cyathea* species are important ornamental cultivars. See Kramer (1990a) for general information and Korall et al. (2007) for a recent phylogenetic analysis of the family.

The Cyatheaceae are distinctive in being mostly ***arborescent***, the shoots generally covered with ***trichomes or scales***, ***leaves very large***, usually 1-3-pinnate leaves, the sori ***exindusiate or indusiate***; indusia, when present, ***saucer-like, cup-like, bivalvate, or globose*** and completely surrounding sporangia, the sporangium ***annulus oblique***.

POLYPODIALES — POLYPOD FERNS

The Polypodiales are a large group of over 8,700 species in 26 families, grouped in six suborders (Table 4.1), four of which are described here. Members of the order are indusiate (laterally or centrally attached) or exindusiate. The sporangia are distinctive in having a ***thin*** (1-3-celled), generally ***long stalk***, a ***lateral stomium***, and an annulus that is ***vertically oriented and interrupted by the stalk and stomium***. Gametophytes are ***green, usually cordate, and surficial***. Several of the families not described here contain ornamental cultivars, such as Nephrolepidaceae (*Nephrolepis*, sword fern, boston fern) and Blechnaceae (*Blechnum* and *Woodwardia*). *Matteuccia struthiopteris* (ostrich fern) of the Onocleaceae, has edible croziers.

Aspleniaceae–Spleenwort family (type *Asplenium*, from Greek *a*, "without" + *splen*, "spleen," in reference to its use to treat ailments of that organ). 2 genera (*Asplenium, Hymenasplenium*)/730 species (most in *Asplenium*) (Figure 4.41).

The Aspleniaceae consist of terrestrial, epipetric, or epiphytic perennials. The **stems** are rhizomatous, the rhizomes creeping, climbing, ascending, or suberect, bearing clathrate scales at shoot apices and petiole bases. The **leaves** are monomorphic, simple to multi-pinnate, often with small clavate hairs, two back-to-back C-shaped vascular strands at petiole base fusing distally into an X-shape, venation pinnate or forking, usually free, less often reticulate and without included veinlets. **Sori** and **indusia** are elongated (linear) along veins. **Sporangia** are mixed, the sporangial stalks in 1-row, long. **Spores** are reniform and monolete, with a winged perine. Chromosome numbers: x = 36, rarely 38, 39.

Distribution of the Aspleniaceae is subcosmopolitan, being most numerous in the tropics. Economic importance includes some local medicinals and many important cultivars, such as *Asplenium bulbiferum*, mother fern, with marginal plantlets on leaves; *A. nidus*, bird's nest fern; *A. rhizophyllum*, walking fern, with leaf tips rooting, forming a new plantlet; and *A. trichomanes*, maidenhair spleenwort. See Kramer and Viane (1990a) for general information and Schneider et al. (2004a) and Perrie et al. (2005) for recent phylogenetic analyses of the family.

The Aspleniaceae are distinctive within the Polypodiales in having shoot apices with ***clathrate scales*** and leaves with ***elongate, linear sori and indusia***.

Dryopteridaceae–Dryopteroid family (type *Dryopteris*, from Greek *drys*, "oak" + *pteris*, "fern," presumably in reference to habitat of taxa in genus). 26 genera (including *Ctenitis, Dryopteris, Elaphoglossum*, and *Polystichum*) / 2,115 species. (Figure 4.42).

The Dryopteridaceae consist of terrestrial, epipetric, or epiphytic perennials. The **stems** are rhizomatous, the rhizomes dictyostelic, creeping, ascending, erect, or scandent to climbing, shoot apices with non-clathrate scales. The **leaves** are usually monomorphic, rarely dimorphic, simple or pinnate to decompound; veins pinnate or forking, free to anastomosing, with or without included veinlets. **Sori** are usually round, indusiate or exindusiate, acrostichoid in some taxa. **Indusia**, when present, are round-reniform or peltate. **Sporangia** are mixed, sporangial stalks in 3 rows, short to long. **Spores** are reniform and monolete. Chromosome numbers: x = 41 (40).

Distribution of the Dryopteridaceae is pantropical to temperate. Economic importance includes numerous cultivated

Figure 4.40 Leptosporangiate Ferns–CYATHEALES. Cyatheaceae. **A,B.** *Cyathea medullaris*, black tree fern, one of the tallest in the world. **C,D.** *Cyathea cunninghamii*. **C.** Fern, from below; note trunk covered with remains of leaf bases. **D.** Pinna, showing dark sori. **E,F.** *Cyathea dealbata*, silver fern. **E.** Leaf. **F.** Close-up of pinna, abaxial surface, showing sori. **G–K.** *Cyathea australis*. **G.** Close-up of pinna, abaxial surface, showing sori. **H.** Cut petiole base, showing discontinuous pneumathodes. **I.** Sorus close-up, prior to sporangial dehiscence. **J.** Sorus after sporangial dehiscence. Note complex trichomes. **K.** Single leptosporangium after dehiscence.

Figure 4.41 Leptosporangiate Ferns–POLYPODIALES. Aspleniaceae. **A–G.** *Asplenium nidus*, bird's nest fern. **A.** Epiphytic plant on high branch of tree (Australia). **B.** Shoot apical region, the "bird's nest," **C,D.** Cross sections of petiole base. **C.** Midrib. **D.** Vascular transition. **E.** Close-up of abaxial blade surface, showing linear sori and indusia. **F,G.** *Asplenium bulbiferum*, mother fern. **F.** Plantlet ("bulbil"), a vegetative propagule growing from leaf. **G.** Leaf, abaxial surface, showing linear sori/indusia. **H,I.** *Asplenium oblongifolium*, a species with pinnate leaves. **J.** *Asplenium rhizophyllum*, showing plantlet formed by leaf tip after contacting a substrate.

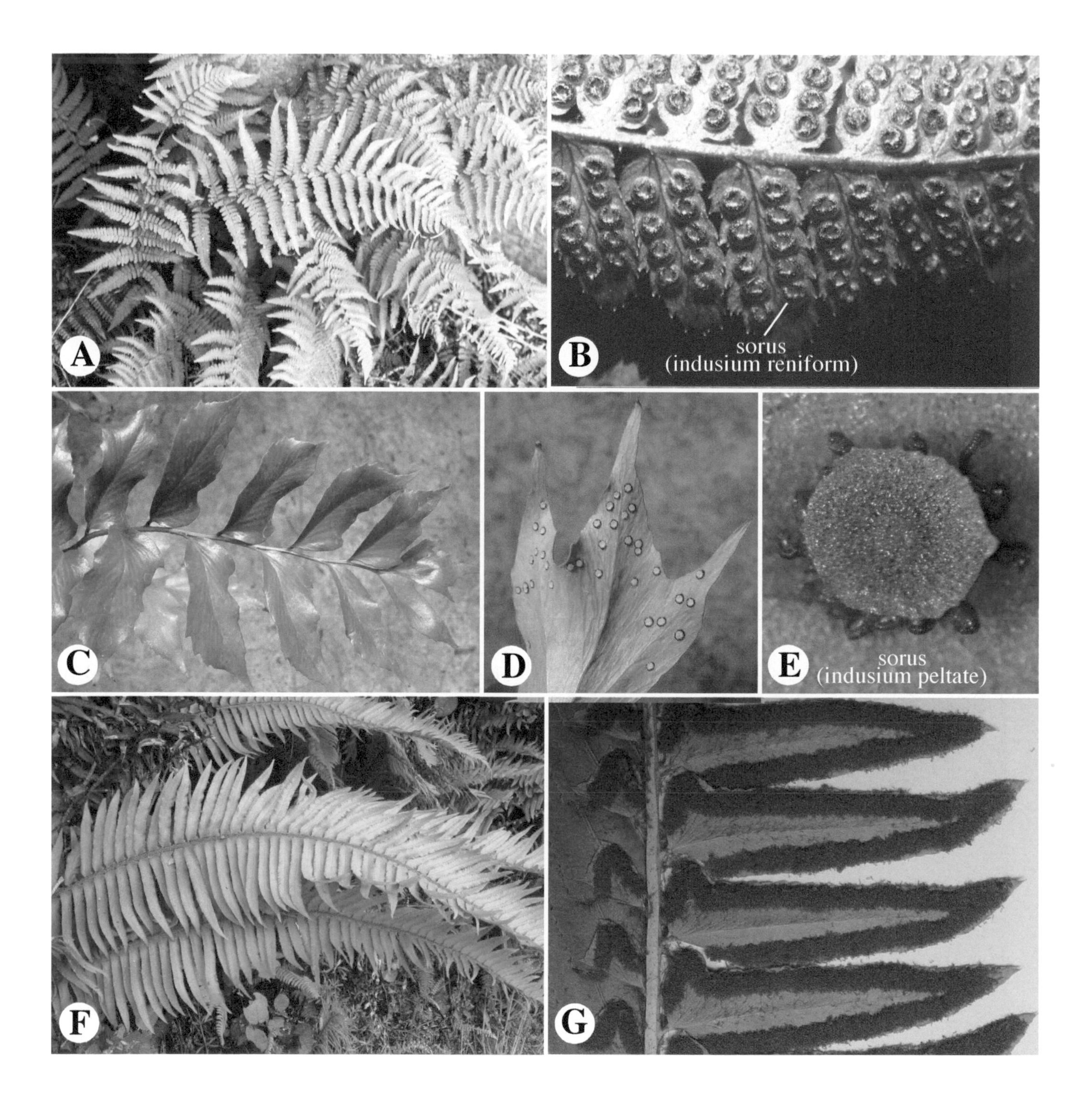

Figure 4.42 Leptosporangiate Ferns–POLYPODIALES. Dryopteridaceae. **A,B.** *Dryopteris arguta*, wood fern. **A.** Whole plant with bipinnatifid leaves arising from underground rhizomes. **B.** Close-up of pinnules, abaxial surface, showing sori with reniform indusia. **C,E.** *Cyrtomium falcatum*, holly fern. **C.** Pinnately compound leaf. **E.** Pinule (leaflet), showing abaxial, circular, indusiate sori. **E.** Sorus close-up, showing peltate indusium. **F–G.** *Polystichum imbricans*, sword fern. **F.** Pinnately compound leaves, from underground rhizomes. **G.** Close-up of pinnules (leaflets), abaxial surface, showing subrows of sporangia near the margin.

ornamentals, such as *Ctenitis pentangularis*, *Dryopteris* spp., *Rumohra adiantiformis* (leatherleaf fern), and *Polystichum* spp. See Kramer et al. (1990) for general information and Little and Barrington (2003) and Skog et al. (2004) for phylogenetic analyses of the family and complex.

The Dryopteridaceae are distinctive within the Polypodiales in being rhizomatous, creeping to climbing plants, the shoot apices with ***non-clathrate scales***, sori ***exindusiate*** (acrostichoid in some taxa) or ***indusiate*** with the indusia ***round-reniform or peltate***.

Polypodiaceae–Polygram/Polypody family (type *Polypodium*, from Greek *polys*, "many" + *pous*, "foot," in reference to knob-like petiole bases left after leaf abscission). ca. 65 genera/ca. 1,652 species (Figure 4.43).

The Polypodiaceae consist of epiphytic (usually), epipetric, or terrestrial perennials. The **stems** are rhizomatous, rhizomes dictyostelic, long to short-creeping, bearing scales. The **leaves** are simple (unlobed to pinnatifid) to 1-pinnate (rarely more), monomorphic or dimorphic; blades glabrous or with hairs or scales, in many taxa abscising near base, leaving short petiole bases (termed "phyllopodia"), veins often anastomosing or reticulate, sometimes with included veins, or free. **Sori** are abaxial (rarely marginal), round, oblong, or elliptic, rarely elongate or acrostichoid, the receptacle often with paraphyses, exindusiate (in some taxa covered by caducous scales when young). **Sporangia** are mixed, sporangial stalks in 1–3 rows, often long. **Spores** are hyaline to yellowish, reniform and monolete (non-grammitids) or greenish and globose-tetrahedral and trilete (most grammatids). Chromosome numbers: x = 35, 36, 37 (25, etc.).

Distribution of the Polypodiaceae is pantropical to temperate. Economic importance includes edible, medicinal, or flavoring plants (e.g., *Polypodium* spp.) and a number of ornamental cultivars, including species of *Aglaomorpha*, *Drynaria* and *Platycerium* (basket/staghorn ferns, both with dimorphic leaves, the basal leaves sterile, clasping, and humus-collecting), and various species of *Polypodium*, e.g., *P. aureum* (hare's foot fern), *P. glycyrrhiza* (licorice fern), and *P. vulgare*.

The Polypodiaceae, as treated by Smith et al. (2006), include of the so-called grammitid ferns, about 20 genera (incl. *Grammatis*) and 600 species of mostly small, tropical epiphytes with simple leaves. The grammatid ferns are often treated as the family Grammitidaceae but are nested within the Polypodiaceae. See Hennipman et al. (1990) for general information and Schneider et al. (2004b) for a recent phylogenetic analysis of the family.

The Polypodiaceae are distinctive within the Polypodiales in being ***exindusiate***, ***mostly epiphytic*** ferns; sori usually round, oblong, or elliptic, rarely elongate or acrostichoid.

Pteridaceae–Pteroid Fern family (type *Pteris*, from Greek *pteris*, "fern"). ca. 53 genera/ca. 1,211 species (Figure 4.44).

The Pteridaceae consist of terrestrial, epipetric, or epiphytic plants, rarely floating aquatics (*Ceratopteris* spp.). The **stems** are rhizomatous, the rhizomes creeping to erect, bearing scales or hairs. The **leaves** are simple, pinnate, pedate, or decompound, veins free or anastomosing. **Sori** are exindusiate, either marginal with a false indusium formed by a reflexed marginal flap or intramarginal in lines along veins, the receptacle generally not raised. **Sporangia** are mixed, sporangial stalks 1–3 cells thick, often long. **Spores** are globose or tetrahedral, trilete, and ornamented. Chromosome numbers: x = 29, 30.

Distribution of the Pteridaceae is subcosmopolitan, mostly in tropical and arid regions. Economic importance include many cultivated ornamentals, such as *Acrostichum, Adiantum* (maidenhair ferns), *Cheilanthes*, *Cryptogramma*, *Pellaea, Pentagramma, Platyzoma, Pteris*, and *Vittaria*. *Pteris vittata* has recently been used to remove arsenic from toxic landfills. The Pteridaceae consist of comprises five monophyletic groups: Cheilanthoideae (including *Cheilanthes, Dryopteris, Gaga*, *Pellaea*, and *Pentagramma*); Cryptogrammoideae (*Coniogramme, Cryptogramma*, and *Llavea*); Parkerioideae (*Acrostichum* and *Ceratopteris*); Pteridoideae (*Pteris, Taenitis*, and 11 other genera); and Vittarioideae (*Adiantum* and 11 other genera, incl. *Vittaria*). See Tryon et al. (1990) for general information and Zhang et al. (2005) for a phylogenetic study of the complex.

The Pteridaceae are distinctive within the Polypodiales in having ***exindusiate sori***, either ***marginal with false indusia***, or ***intramarginal in lines along veins***.

Figure 4.43 Leptosporangiate Fern–POLYPODIALES. Polypodiaceae. **A,B.** *Polypodium californicum*, California Polypody. **A.** Leaf, abaxial surface, showing circular, exindusiate sori. **B.** Close-up of sorus. **C,D.** *Polypodium scouleri*. **C.** Whole leaf. **D.** Close-up of leaf pinna, abaxial surface, showing exindusiate sori. **E–G.** *Platycerium grande*, staghorn fern. **E.** Epiphytic plant. **F.** Close-up of plant base showing dimorphic leaves, basal leaves clasping and turning brown at maturity and forming a shield that collects plant debris, apical leaves erect, photosynthetic. **G.** Close-up of erect leaf, abaxial surface, showing acrosticoid leptosporangia.

Figure 4.44 Leptosporangiate Ferns–POLYPODIALES. Pteridaceae. **A.** *Adiantum aleuticum*, showing pedate leaf. **B.** *Adiantum* sp., showing false indusium. **C.** *Adiantum jordanii*, pinnule, abaxial surface, showing false indusium. **D.** *Adiantum capillis-veneris*. Close-up of pinnule, abaxial surface, showing false indusia. **E,F.** *Pellaea rotundifolia*. **E.** Whole plant, with pinnately compound leaves. **F.** Close-up of pinnules, abaxial surface, showing intramarginal leptosporangia. **G–I.** *Pteris vittata*. **G.** Leaf, pinnately compound, abaxial view. **H.** Close-up of pinnule abaxial surface, showing intramarginal leptosporangia. **I.** Leptosporangia, close-up.

REVIEW QUESTIONS

VASCULAR PLANT APOMORPHIES

1. What is the formal, scientific name for the vascular plants?
2. Name the major apomorphies of the vascular plants.
3. What two features of the sporophyte are apomorphic for vascular plants, distinguishing them from liverworts, mosses, and hornworts?
4. What are two early evolving branching patterns in the vascular plants?
5. How was the evolution of lignin a major adaptive feature of the vascular plants?
6. What is the difference between a primary and secondary cell wall in terms of time of deposition and chemistry?
7. What is a pit? What is a primary pit field?
8. Is the secondary cell wall formed inside or outside the plasma membrane? Is it formed inside or outside the primary cell wall?
9. What are the general characteristics of sclerenchyma cells?
10. Name the two types of sclerenchyma and state how they differ.
11. What is the function and structure of tracheary elements?
12. What is xylem?
13. Name the two types of tracheary elements and cite how they differ structurally.
14. In what taxa are vessels found?
15. What is the function and structure of sieve elements?
16. What is phloem?
17. What is a sieve area and what compound is associated with them?
18. What is the difference, in morphology and taxonomic group where found, between a sieve cell and a sieve tube member?
19. What is the endodermis and Casparian strip, and what is the function of each?
20. What is the function of roots?
21. What is the name of the region of actively dividing cells in the root?
22. Name five diagnostic features of roots and their function, if known.
23. What are mycorrhizae and what is their function in vascular plants?

VASCULAR PLANT DIVERSITY

24. What are the characteristics of the rhyniophytes in terms of sporophyte morphology and stem anatomy?

LYCOPODIOPHYTA–LYCOPHYTES

25. Name and give the features of a (paraphyletic) fossil group that diverged along the immediate lineage to the lycophytes.
26. Name a fossil lycophyte that was a large tree in the Carboniferous and now makes up a large percentage of coal deposits.
27. What are the major apomorphies of the lycophytes?
28. What are the features of a lycophyll (microphyll)? An enation?
29. How are lycophylls thought to have evolved?
30. What is homospory? Name the order and family of lycophytes that have this condition.
31. What is a sporophyll? A strobilus?
32. Name three diagnostic features of the Lycopodiaceae.
33. What is a ligule?
34. Define: endospory, heterospory, megasporangium, megaspore, microsporangium, microspore.
35. Draw the life cycle of a heterosporous land plant, listing all structures, ploidy levels, and processes.
36. What order and two families of lycophytes have ligulate leaves and heterospory?
37. Describe the basic morphology of members of the Isoetaceae.
38. Name and define the two types of leaf morphology in *Selaginella* species.

EUPHYLLOPHYTA–EUPHYLLOPHYTES

39. Name the apomorphies of the euphyllophytes, and list the two major, vascular plant groups included.
40. How do euphylls differ from lycophylls?

41. What (paraphyletic) fossil group diverged along the immediate lineage to euphyllophytes?
42. Describe a widely accepted hypothesis regarding the evolution of the euphyll.
43. What is a shoot?
44. What is the name of the region of actively dividing cells in a shoot, and how does this differ among vascular plants?
45. Define node; internode.
46. What is the general morphology and function of leaves?
47. What is a vein?
48. What are the internal, chlorophyllous cells of a leaf called? Into which two layers may these cells be organized?
49. What is the definition of a bud and where are they typically located?
50. What is monopodial growth?

MONILOPHYA–MONILOPHYTES

51. Name the putative apomorphies of the monilophytes, and list the five major groups contained within it.
52. What is a siphonostele? Name the types of siphonosteles.

EQUISETOPSIDA–EQUISETOPHYTES

53. What is fossil member of the equisetophytes makes up a component of coal deposits?
54. Name the major apomorphies of the equisetophytes.
55. What is the only extant genus and family of this group?
56. What do equisetophytes have as a component of the cell wall?
57. What is the difference between a scouring rush and a horsetail? Into what two subgenera are these classified?
58. Describe the morphology of the strobilus (cone), sporangiophore, and sporangia of *Equisetum*.
59. What is unique about the spores of *Equisetum*? What is the function of this novelty?

PSILOLOPSIDA–PSILOTOPHYTES

60. What features about the roots and gametophytes are presumed apomorphies for the Psilopsida?
61. What is a eusporangium?
62. What is distinctive about the leaves of the Ophioglossales/Ophioglossaceae, the ophioglossoid ferns?
63. What is a synangium?
64. What is distinctive (and apomorphic) about the roots, leaves, and sporangia of the Psilotaceae, the whisk ferns?
65. What are the two genera of the Psilotaceae? What species is a commonly cultivated ornamental?

MARATTIOPSIDA–MARATTIOID FERNS

66. Name and describe the diagnostic features and a putative apomorphy of the marattioid ferns.
67. How do the gametophytes, leaf type, and leaf development of the marattioid ferns resemble the leptosporangiate ferns?
68. What type of sporangium is found in the marattioid ferns?

POLYPODIOPSIDA–LEPTOSPORANGIATE FERNS

69. Name three stem types/habits that occur in the leptosporangiate ferns.
70. What is circinate vernation? What terms are used for immature fern leaves that exhibit this?
71. Define frond, stipe, pinna, pinnule.
72. What aspects of venation and scale morphology are useful in leptosporangiate fern classification?
73. What is the major apomorphy of the Polypodiopsida? Describe its development and morphology.
74. Define annulus, sorus, indusium, false indusium, acrostichoid.
75. Name aspects of sorus morphology, indusium morphology, sporangium development, sporangium morphology, and spore type used in fern classification.
76. In a fern gametophyte, what is the name of the male gametangium? The female gametangium? What do they look like?
77. What is unique about the leaf morphology and sporangium annulus of the Osmundaceae?
78. Name the common name of the Hymenophyllaceae. What is unique about its indusium, receptacle, and leaf anatomy?

79. Describe the leaf morphology of the Gleicheniaceae.
80. What is unique about the leaf morphology, sorus, indusium, and sporangium annulus of the Lygodiaceae?
81. What is distinctive and apomorphic about the life cycle of the Salviniales? What is a sporocarp and what is its function?
82. Name the two families of the Salviniales and describe how they differ. Name two genera in each family.
83. How is *Azolla* of great economic importance?
84. What is distinctive about the sorus and indusium of the Aspleniaceae?
85. Name a diagnostic feature of the sorus and indusium of the Dryopteridaceae, Polypodiaceae, and Pteridaceae.

EXERCISES

1. Peruse the most recent literature on vascular plant phylogenetic relationships. Are there any differences from Figure 4.1?
2. Peruse botanical journals and find a systematic article on a nonseed vascular plant (e.g., a leptosporangiate fern or fern group). What is the objective of the article and what techniques were used to address it? What types of morphological characters are discussed by the author(s)?
3. Identify lycophytes, equisetophytes, psilotophytes, ophioglossoid ferns, or leptosporangiate ferns from live collections or specimens. What diagnostic features or apomorphies distinguish these groups?
4. Collect a leptosporangiate fern from one family and describe the following features: stem type, presence of scales and trichomes, leaf type/division, and sorus/indusium morphology. From hand-sections and wet mounts (see Chapter 10, Plant Anatomy Technique), observe under the microscope the cross-sectional stem anatomy, scale or trichome anatomy, leptosporangium morphology, and spore morphology.

REFERENCES FOR FURTHER STUDY

Banks, H. P., S. Leclercq, and F. M. Hueber. 1975. Anatomy and morphology of *Psilophyton dawsonii* sp. n. from the late Devonian of Quebec (Gaspé), and Ontario, Canada. Palaeontographica Americana 8: 77-126.

Bierhorst, David W. 1971. Morphology of Vascular Plants. Macmillan, New York.

Camus, J. M. 1990. Marattiaceae. In: K. U. Kramer and P. S. Green (eds.). The Families and Genera of Vascular Plants. I. Pteridophytes and Gymnosperms. Pp. 174-180. Springer-Verlag, Berlin.

Cantino, P. D., J. A. Doyle, S. W. Graham, W. S. Judd, R. G. Olmstead, D. E. Soltis, P. S. Soltis, and M. J. Donoghue. 2007. Towards a phylogenetic nomenclature of Tracheophyta. Taxon 56: 822-846.

Christenhusz, M. J. M. and M. W. Chase. 2018. PPG recognises too many fern genera. Taxon 67:481-487.

Cracraft, J., and M. J. Donoghue. 2004. Assembling the Tree of Life. Oxford University Press, New York.

Des Marais, D. L., A. R. Smith, D. M. Britton, and K. M. Pryer. 2003. Phylogenetic relationships and evolution of extant horsetails, *Equisetum*, based on chloroplast DNA sequence data (rbcL and trnL-F). International Journal of Plant Sciences 164: 737-751.

Ebihara, A., J.-Y. Dubuisson, K. Iwatsuki, S. Hennequin, and M. Ito. 2006. A taxonomic revision of Hymenophyllaceae. Blumea 51: 221-280.

Flora of North America Editorial Committee. 1993+. Pteridophytes and Gymnosperms. Volume 2, in Flora of North America North of Mexico. 7+ vols. New York and Oxford.

Foster, A. S., and E. M. Gifford. 1974. Comparative Morphology of Vascular Plants, 2nd edition. W. H. Freeman, San Francisco.

Gensel, P. G., and C. M. Berry. 2001. Early lycophyte evolution. American Fern Journal 91: 74–98.

Grewe, F., W. Guo, E. A. Gubbels, A. K. Hansen, and J. P. Mower. 2013. Complete plastid genomes from Ophioglossum californicum, Psilotum nudum, and Equisetum hyemale reveal an ancestral land plant genome structure and resolve the position of Equisetales among monilophytes. BMC Evolutionary Biology 13:8.

Gifford, E. M., and A. S. Foster. 1989. Morphology and Evolution of Vascular Plants, 3rd edition. W. H. Freeman and Co., New York.

Hauk, W. D., C. R. Parks, and M. W. Chase. 2003. Phylogenetic studies of Ophioglossaceae: evidence from rbcL and trnL-F plastid DNA sequences and morphology. Molecular Phylogenetics and Evolution 28: 131-151.

Hauke, R. L. 1990. Equisetaceae. In: K. U. Kramer and P. S. Green (eds.). The Families and Genera of Vascular Plants. I. Pteridophytes and Gymnosperms. Pp. 46-48. Springer-Verlag, Berlin.

Hennequin, S., A. Ebihara, M. Ito, K. Iwatsuki, and J.-Y. Dubuisson. 2003. Molecular systematics of the fern genus Hymenophyllum s.l. (Hymenophyllaceae) based on chloroplastic coding and noncoding regions. Molecular Phylogenetics and Evolution 27: 283–301.

Hennipman, E., P. Veldhoen, K. U. Kramer, and M. G. Price. 1990. Polypodiaceae. In: K. U. Kramer and P. S. Green (eds.). The Families and Genera of Vascular Plants. I. Pteridophytes and Gymnosperms. Pp. 203-230. Springer-Verlag, Berlin.

Hoot, S. B., W. C. Taylor, and N. S. Napier. 2006. Phylogeny and Biogeography of Isoëtes (Isoëtaceae) Based on Nuclear and Chloroplast DNA Sequence Data. Systematic Botany 31:449-460.

Hoshizaki, B. J., and R. C. Moran. 2001. Fern Grower's Manual. Timber Press, Portland, Oregon.

Iwatsuki, K. 1990. Hymenophyllaceae. In: K. U. Kramer and P. S. Green (eds.). The Families and Genera of Vascular Plants. I. Pteridophytes and Gymnosperms. Pp. 157-163. Springer Verlag, Berlin.

Jermy, A. C. 1990a. Isoetaceae. In: K. U. Kramer and P. S. Green (eds.). The Families and Genera of Vascular Plants. I. Pteridophytes and Gymnosperms. Pp. 26-31. Springer-Verlag, Berlin.

Jermy, A. C. 1990b. Selaginellaceae. In: K. U. Kramer and P. S. Green (eds.). The Families and Genera of Vascular Plants. I. Pteridophytes and Gymnosperms. Pp. 39-45. Springer-Verlag, Berlin.

Kenrick, P., and P. R. Crane. 1997. The Origin and Early Diversification of Land Plants: a Cladistic Study. Smithsonian Institution Press, Washington, DC.

Korall, P. and P. Kenrick. 2002. Phylogenetic relationships in Selaginellaceae based on rbcl sequences. American Journal of Botany 89: 506–517.

Korall, P. and P. Kenrick. 2004. The phylogenetic history of Selaginellaceae based on DNA sequences from the plastid and nucleus: extreme substitution rates and rate heterogeneity. Molecular Phylogenetics and Evolution 31: 852-864.

Korall, P., D. S. Conant, J. S. Metzgar, H. Schneider, and K. M. Pryer. 2007. A molecular phylogeny of scaly tree ferns (Cyatheaceae). American Journal of Botany 94: 873–886.

Kramer, K. U. 1990a. Cyatheaceae. In: K. U. Kramer and P. S. Green (eds.). The Families and Genera of Vascular Plants. I. Pteridophytes and Gymnosperms. Pp. 69-74. Springer-Verlag, Berlin.

Kramer, K. U. 1990b. Gleicheniacee. In: K. U. Kramer and P. S. Green (eds.). The Families and Genera of Vascular Plants. I. Pteridophytes and Gymnosperms. Pp. 145-152. Springer-Verlag, Berlin.

Kramer, K. U. 1990c. Marsileaceae. In: K. U. Kramer and P. S. Green (eds.). The Families and Genera of Vascular Plants. I. Pteridophytes and Gymnosperms. Pp. 180-183. Springer-Verlag, Berlin.

Kramer, K. U. 1990d. Notes on the Higher Classification of the Recent Ferns. In: K. U. Kramer and P. S. Green (eds.). The Families and Genera of Vascular Plants. I. Pteridophytes and Gymnosperms. Pp. 49-52. Springer-Verlag, Berlin.

Kramer, K. U. 1990e. Psilotaceae. In: K. U. Kramer and P. S. Green (eds.). The Families and Genera of Vascular Plants. I. Pteridophytes and Gymnosperms. Pp. 22-25. Springer-Verlag, Berlin.

Kramer, K. U. 1990f. Schizaeaceae. In: K. U. Kramer and P. S. Green (eds.). The Families and Genera of Vascular Plants. I. Pteridophytes and Gymnosperms. Pp. 258-263. Springer-Verlag, Berlin.

Kramer, K. U., and P. S. Green. 1990. The Families and Genera of Vascular Plants. I. Pteridophytes and Gymnosperms. Springer-Verlag, Berlin.

Kramer, K. U., and R. Viane. 1990a. Aspleniaceae. In: K. U. Kramer and P. S. Green (eds.). The Families and Genera of Vascular Plants. I. Pteridophytes and Gymnosperms. Pp. 52-57. Springer-Verlag, Berlin.

Kramer, K. U., and R. Viane. 1990b. Osmundaceae. In: K. U. Kramer and P. S. Green (eds.). The Families and Genera of Vascular Plants. I. Pteridophytes and Gymnosperms. Pp. 197-200. Springer-Verlag, Berlin.

Kramer, K. U., R. E. Holttum, R. C. Moran, and A. R. Smith. 1990. Dryopteridaceae. In: K. U. Kramer, and P. S. Green (eds.). The Families and Genera of Vascular Plants. I. Pteridophytes and Gymnosperms. Pp. 101-144. Springer-Verlag, Berlin.

Lellinger, D. B. 1985. A Field Manual of the Ferns and Fern-Allies of the United States and Canada. Smithsonian Institution Press, Washington, DC.

Lellinger, D. B. 2002. A Modern Multilingual Glossary for Taxonomic Pteridology. American Fern Society, Inc.

Little, D. P. and D. S. Barrington. 2003. Major evolutionary events in the origin and diversification of the fern genus *Polystichum* (Dryopteridaceae). American Journal of Botany 90: 508-514.

Mickel, John T. 1979. How to Know the Ferns and Fern Allies. Wm. C. Brown, Dubuque, IA.

Murdock, A. G. 2008a. Phylogeny of marattioid ferns (Marattiaceae): inferring a root in the absence of a closely related outgroup. American Journal of Botany 95: 626–641.

Murdock, A. G. 2008b. A taxonomic revision of the eusporangiate fern family Marattiaceae, with description of a new genus Ptisana. Taxon 57: 737-755.

Øllgaard, B. 1990. Lycopodiaceae. In: K. U. Kramer and P. S. Green (eds.). The Families and Genera of Vascular Plants. I. Pteridophytes and Gymnosperms. Pp. 31-39. Springer-Verlag, Berlin.

Perrie, L. R. and P. J. Brownsey. 2005. Insights into the biogeography and polyploid evolution of New Zealand *Asplenium* from chloroplast DNA sequence data. American Fern Journal 95: 1–21.

PPGI. 2016. A community-derived classification for extant lycophytes and ferns. Journal of Systematics and Evolution 54:563-603.

Pryer, K. M. 1999. Phylogeny of marsileaceous ferns and relationships of the fossil *Hydropteris pinnata* reconsidered. International Journal of Plant Sciences 160: 931–954.

Pryer, K. M., A. R. Smith, and J. E. Skog. 1995. Phylogenetic relationships of extant ferns based on evidence from morphology and *rbc*L sequences. American Fern Journal 85: 205–282

Pryer, K. M., H. Schneider, A. R. Smith, R. Cranfill, P. G. Wolf, J. S. Hunt, and S. D. Sipes. 2001a. Horsetails and ferns are a monophyletic group and the closest living relatives to seed plants. Nature 409: 618–622.

Pryer, K. M., A. R. Smith, J. S. Hunt, and J.-Y. Dubuisson. 2001b. *Rbc*L data reveal two monophyletic groups of filmy ferns (Filicopsida: Hymenophyllaceae). American Journal of Botany 88: 1118–1130.

Pryer, K. M., E. Schuettpelz, P. G. Wolf, H. Schneider, A. R. Smith, and R. Cranfill. 2004a. Phylogeny and evolution of ferns (monilophytes) with a focus on the early leptosporangiate divergences. American Journal of Botany 91: 1582–1598.

Pryer, K. M., H. Schneider, and S. Magallón. 2004b. The radiation of vascular plants. In J. Cracraft and M. J. Donoghue (eds.). Assembling the Tree of Life, pp. 138–153. Oxford University Press, London.

Qiu, Y.-L., L. Li, B. Wang, Z. Chen, V. Knoop, M. Groth-Malonek, O. Dombrovska, J. Lee, L. Kent, J. S. Rest, G. F. Estabrook, T. A. Hendry, D. W. Taylor, C. M. Testa, M. Ambros, B. Crandall-Stotler, R. J. Duff, M. Stech, W. Frey, D. Quandt, and C. C. Davis. 2006. The deepest divergences in land plants inferred from phylogenomic evidence. Proceedings of the National Academy of Sciences U.S.A. 103: 15511-15516.

Qiu, Y-L., L. Li, B. Wang, Z. Chen, O. Dombrovska, J. Lee, L. Kent, R. Li, R. W. Jobson, T. A. Hendry, D. W. Taylor, C. M. Testa, and M. Ambros. 2007. A nonflowering land plant phylogeny inferred from nucleotide sequences of seven chloroplast, mitochondrial, and nuclear genes. International Journal of Plant Sciences 168: 691-708.

Reid, J. D., G. M. Plunkett, and G. A. Peters. 2006. Phylogenetic relationships in the heterosporous fern genus *Azolla* (Azollaceae) based on DNA sequence data from three noncoding regions. International Journal of Plant Sciences 167: 529–538.

Rothfels, C. J., F.-W. Li, E. M. Sigel, L. Huiet, A. Larsson, D. O. Burge, M. Ruhsam, M. Deyholos, D. E. Soltis, C. N. Stewart, S. W. Shaw, L. Pokorny, T. Chen, C. dePamphilis, L. DeGironimo, L. Chen, X. Wei, X. Sun, P. Korall, D. W. Stevenson, S. W. Graham, G. K. S. Wong, and K. M. Pryer. 2015. The evolutionary history of ferns inferred from 25 low-copy nuclear genes. American Journal of Botany 102:1089-1107.

Rothwell, G. W., and K. C. Nixon. 2006. How Does the inclusion of fossil data change our conclusions about the phylogenetic history of Euphyllophytes?. International Journal of Plant Sciences 167: 737-749.

Ruhfel, B. R., M. A. Gitzendanner, P. S. Soltis, D. E. Soltis, and J. G. Burleigh. 2014. From algae to angiosperms - inferring the phylogeny of green plants (Viridiplantae) from 360 plastid genomes. BMC Evolutionary Biology 14:23.

Rydin, C. and N. Wikstrom. 2002. Phylogeny of Isoëtes (Lycopsida): resolving basal relationships using *rbc*L sequences. Taxon 51: 83–89.

Schneider, H. & Pryer, K. M. 2002. Structure and function of spores in the aquatic heterosporous fern family Marsileaceae. International Journal of Plant Sciences 163: 485–505.

Schneider, H., A. R. Smith, R. Cranfill, T. E. Hildebrand, C. H. Haufler, and T. A. Ranker. 2004b. Unraveling the phylogeny of polygrammoid ferns (Polypodiaceae and Grammitidaceae): exploring aspects of the diversification of epiphytic plants. Molecular Phylogeny and Evolution 31: 1041–1063.

Schneider, H., E. Schuettpelz, K. M. Pryer, R. Cranfill, S. Magallon, and R. Lupia. 2004. Ferns diversified in the shadow of angiosperms. Nature 428: 553–557.

Schneider, H., K. M. Pryer, R. Cranfill, A. R. Smith, and P. G. Wolf. 2002. Evolution of vascular plant body plans: a phylogenetic perspective. Systematics Association special volume 65: 330–364.

Schneider, H., S. J. Russell, C. J. Cox, F. Bakker, S. Henderson, M. Gibby, and J. C. Vogel. 2004a. Chloroplast phylogeny of asplenioid ferns based on *rbc*L and *trn*L-F spacer sequences (Polypodiidae, Aspleniaceae) and its implications for the biogeography. Systematic Botany 29: 260–274.

Schneller, J. J. 1990a. Azollaceae. In: K. U. Kramer and P. S. Green (eds.). The Families and Genera of Vascular Plants. I. Pteridophytes and Gymnosperms. Pp. 57-60. Springer-Verlag, Berlin.

Schneller, J. J. 1990b. Salviniaceae. In: K. U. Kramer and P. S. Green (eds.). The Families and Genera of Vascular Plants. I. Pteridophytes and Gymnosperms. Pp. 256-258. Springer-Verlag, Berlin.

Schuettpelz, E., P. Korall, and K. M. Pryer. 2006. Plastid *atp*A data provide improved support for deep relationships among ferns. Taxon 55: 897–906.

Schuettpelz, E., G. Rouhan, K. M. Pryer, C. J. Rothfels, J. Prado, M. A. Sundue, M. D. Windham, R. C. Moran, and A. R. Smith. 2018. Are there too many fern genera? Taxon 67:473-480.

Skog, J. E., E. Zimmer, and J. T. Mickel. 2002. Additional support for two subgenera of Anemia (Schizaeaceae) from data for the chloroplast intergenic spacer region trnL-F and morphology. American Fern Journal 92: 119–130.

Skog, J. E., J. T. Mickel, R. C. Moran, M. Volovsek, and E. A. Zimmer. 2004. Molecular studies of representative species in the fern genus Elaphoglossum (Dryopteridaceae) based on cpDNA sequences rbcL, trnL-F, and rps4-TRNS. International Journal of Plant Sciences 165: 1063-1075.

Smith, G. M. 1955. Cryptogamic Botany, 2nd edition. McGraw-Hill Book Company, Inc., New York.

Smith, A. R., K. M. Pryer, E. Schuettpelz, P. Korall, H. Schneider, and P. G. Wolf. 2006. A classification for extant ferns. Taxon 55: 705-731.

Smith, A. R., K. M. Pryer, E. Schuettpelz, P. Korall, H. Schneider, and P. G. Wolf. 2008. Fern classification. Pp. 417–467 in T. A. Ranker and C. H. Haufler (eds.), Biology and Evolution of Ferns and Lycophytes. Cambridge University Press, New York, USA.

Stewart, W. N., and G. W. Rothwell. 1993. Paleobotany and the Evolution of Plants, 2nd edition. Cambridge University Press, Cambridge, UK.

Tryon, R. M., and A. F. Tryon. 1982. Ferns and Allied Plants, with Special Reference to Tropical America. Springer-Verlag, New York.

Tryon, R. M., A. F. Tryon, and K. U. Kramer. 1990. Pteridaceae. In: K. U. Kramer and P. S. Green (eds.). The Families and Genera of Vascular Plants. I. Pteridophytes and Gymnosperms. Pp. 230-256. Springer-Verlag, Berlin.

Wagner, W. H., Jr. 1990. Ophioglossaceae. In: K. U. Kramer and P. S. Green (eds.). The Families and Genera of Vascular Plants. I. Pteridophytes and Gymnosperms. Pp. 193-197. Springer-Verlag, Berlin.

Weststrand, S. and P. Korall. 2016. Phylogeny of Selaginellaceae: There is value in morphology after all! American Journal of Botany 103:2136-2159.

Wikström, N., and P. Kenrick. 2000a. Phylogeny of epiphytic *Huperzia* (Lycopodiaceae): Paleotropical and neotropical clades corroborated by plastid rbcL sequences. Nordic Journal of Botany 20: 165–171.

Wikström, N., and P. Kenrick. 2000b. Relationships of *Lycopodium* and *Lycopodiella* based on combined plastid rbcL gene and trnL intron sequence data. Systematic Botany 25: 495–510.

Wikström, N. 2001. Diversification and relationships of extant homosporous lycopods. American Fern Journal 91: 150–165.

Wikström, N. and P. Kenrick. 2001. Evolution of Lycopodiaceae (Lycopsida): Estimating Divergence Times from rbcL Gene Sequences by Use of Nonparametric Rate Smoothing. Molecular Phylogenetics and Evolution 19: 177-186.

Wolf, P. G., K. M. Pryer, A. R. Smith, and M. Hasebe. 1998. Phylogenetic studies of extant pteridophytes. In: Molecular Systematics of Plants, 2nd ed. P. S. Soltis, D. E. Soltis, and J. J. Doyle (eds.). Chapman and Hall, New York.

Wolf, P. G., S. D. Sipes, M. R. White, M. L. Martines, K. M. Pryer, A. R. Smith, and K. Ueda. 1999. Phylogenetic relationships of the enigmatic fern families Hymenophyllopsidaceae and Lophosoriaceae: evidence from *rbc*L nucleotide sequences. Plant Systematics and Evolution 219: 263–270.

Yatabe, Y., H. Nishida, and N. Murakami. 1999. Phylogeny of Osmundaceae inferred from *rbc*L nucleotide sequences and comparison to the fossil evidence. Journal of Plant Research 112: 397–404.

Zhang, G., X. Zhang, and Z. Chen. 2005. Phylogeny of cryptogrammoid ferns and related taxa based on *rbc*L sequences. Nordic Journal of Botany 23: 485–493.

WEB SITE

American Fern Society web site <http://amerfernsoc.org> lists many resources, including publications, references, local, national, and international fern societies, education sites, commercial fern sites, and fern databases.

5

EVOLUTION AND DIVERSITY OF WOODY AND SEED PLANTS

LIGNOPHYTES—WOODY PLANTS

The lignophytes, or woody plants (also called Lignophyta), are a monophyletic lineage of euphyllous vascular plants that share the derived features of a **vascular cambium**, which gives rise to **wood**, and a **cork cambium,** which produces **cork** (Figures 5.1, 5.2) derived independently in these taxa, being unifacial. Growth of the vascular and cork cambia is called **secondary growth** because it initiates *after* the vertical extension of stems and roots due to cell expansion (primary growth). A vascular cambium is a sheath, or hollow cylinder, of cells that develops within the stems and roots as a continuous layer, between the xylem and phloem in extant, eustelic spermatophytes (see later discussion). The cells of the vascular cambium divide mostly *tangentially* (parallel to a tangential plane), resulting initially in two concentric layers of cells (Figure 5.3A). One of these layers remains as the vascular cambium and continues to divide indefinitely; the other layer eventually differentiates into either **secondary xylem = wood**, if produced to the inside of the cambium, or **secondary phloem**, if produced to the outside (Figure 5.3A,B). Because layers of cells are produced both to the inside and outside of a continuously generated cambium, this type of growth is termed **bifacial**. Generally, much more secondary xylem is produced than secondary phloem. [Note that a secondary cambium independently evolved in fossil lineages within the lycophytes (e.g., *Lepidodendron*) and equisetophytes (e.g., *Calamites*), but this cambium was **unifacial**, producing secondary xylem (wood) to the inside but no outer secondary phloem, likely limiting in terms of an adaptive feature.]

Secondary growth results in an increase of the width or girth of stems and roots (Figures 5.3B, 5.4). This occurs both by expansion of the new cells generated by the cambium and by accompanying *radial* divisions, increasing the number of cells within a given growth ring. Many woody plants have regular growth periods, e.g., forming annual rings of wood (Figure 5.4).

A cork cambium is similar to a vascular cambium, differing in differentiating near the periphery of the stem or root axis. The cork cambium and its derivatives constitute the **periderm** (referred to as the outer bark). The outermost layer of the periderm is **cork** (Figure 5.3B). Cork cells contain a

https://doi.org/10.1016/B978-0-12-812628-8.50005-5,

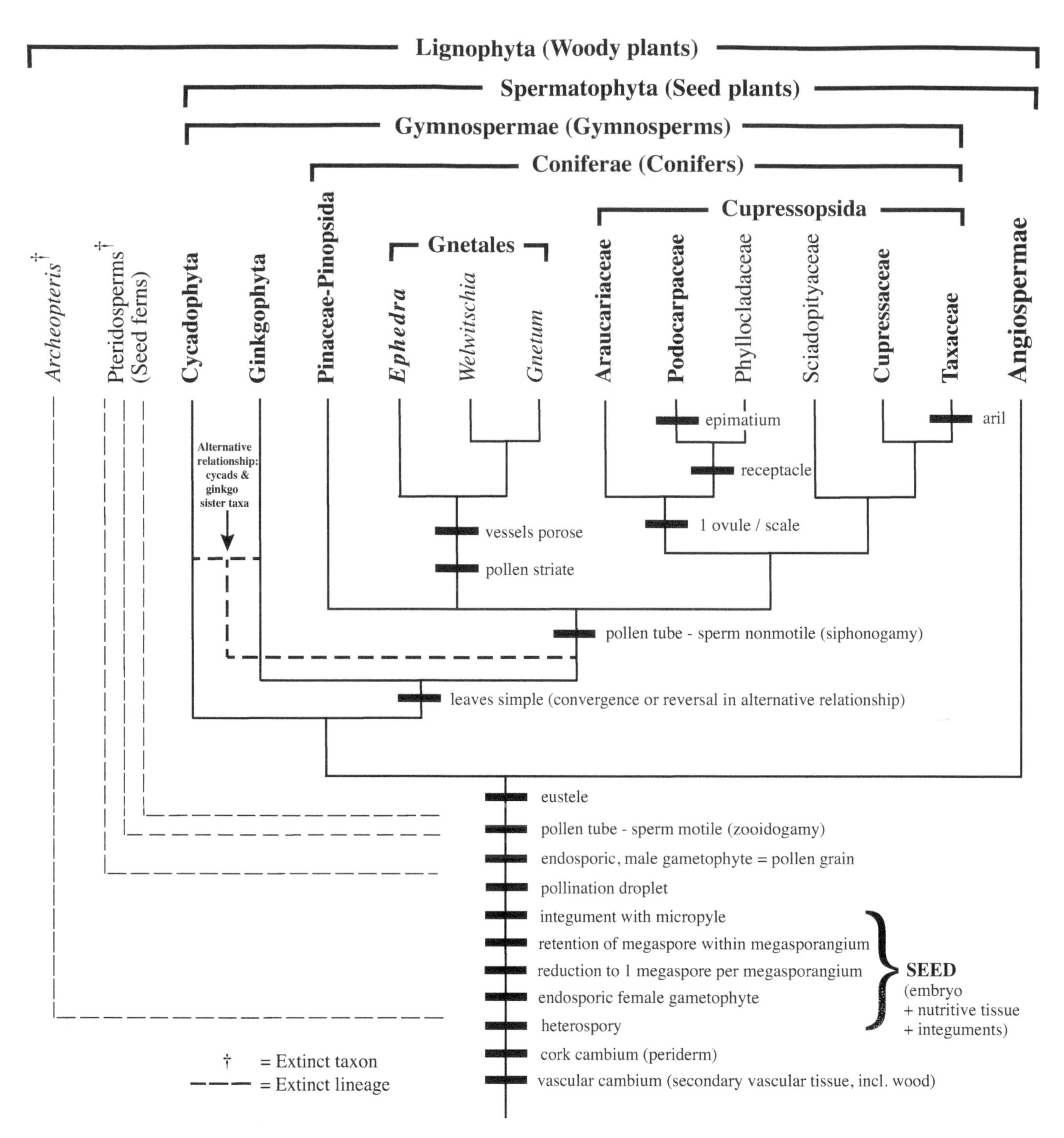

Figure 5.1 Cladogram of the woody and seed plants. Major apomorphies are indicated beside a thick hash mark. Families in bold are described in detail. Modified from Bowe et al. (2000), Chaw et al. (2000), Frohlich et al. (2000), and Samigullin et al. (1999).

waxy polymer called **suberin** (similar to cutin) that is quite resistant to water loss (see Chapter 10).

The vascular cambium and cork cambium were a major evolutionary novelty. Secondary xylem, or wood, functions in structural support, enabling the plant to grow tall and acquire massive systems of lateral branches. Thus, the vascular cambium was a precursor to the formation of intricately branched shrubs or trees with tall overstory canopies (e.g., Figure 5.2), a significant ecological adaptation. Cork produced by the cork cambium functions as a thick layer of cells that protects the delicate vascular cambium and secondary phloem from mechanical damage, predation, and desiccation.

Wood anatomy can be quite complex. The details of cellular structure are important characters used in the classifica-

Figure 5.2 Composite photograph of *Sequoiadendron giganteum*, a woody conifer that is the most massive, non-clonal organism on Earth, and among the tallest of trees.

tion and identification of woody plants. Wood anatomical features may also be used to study the past, a specialty known as **dendrochronology** (see Chapter 10).

Another feature of lignophytes is that they possess **monopodial growth**, in which a single main shoot develops branches from lateral (usually axillary) buds (see Chapters 4, 9). Although monopodial growth is presumed to have arisen prior to the monilophyte–lignophyte split, it enabled woody plants in particular the capability of forming extensive (sometimes massive) woody branching systems, permitting them to survive and reproduce more effectively.

SPERMATOPHYTES—SEED PLANTS

The Spermatophyta, commonly called spermatophytes or seed plants, are a monophyletic lineage within the lignophytes (Figure 5.1). The major evolutionary novelty that unites this group is the **seed**. A seed is defined as an **embryo**, which is an immature diploid sporophyte that develops from the zygote, surrounded by **nutritive tissue**, and enveloped by a **seed coat** (Figure 5.5). The embryo generally consists of an immature root called the **radicle**, a shoot apical meristem called the **epicotyl**, and one or more young seed leaves called **cotyledons**; the transition region between root and stem is called the **hypocotyl** (Figures 5.5, 5.10). An immature seed, prior to fertilization, is known as an **ovule**.

SEED EVOLUTION

The evolution of the seed involved several steps. The exact sequence of these is not certain, and two or more "steps" in seed evolution may have occurred concomitantly and be functionally correlated. The probable steps in seed evolution are as follows (Figure 5.6):

1. **Heterospory.** Heterospory is the formation of *two* types of haploid spores within *two* types of sporangia. Large, fewer numbered **megaspores** develop via meiosis from a **megasporocyte** in the **megasporangium**, this process known as **megasporogenesis**. Small, more numerous **microspores** are the products of meiosis of a **microsporocyte** in the **microsporangium**, the process termed **microsporogeneis** (Figures 5.6, 5.7). The ancestral condition, in which a single spore type forms, is called "homospory." Each megaspore develops into a **female gametophyte** that bears only archegonia; a microspore develops into a **male gametophyte**, bearing only antheridia. (Note that archegonia and antheridia were each lost in the angiosperms; see Chapter 6.) Although heterospory was prerequisite to seed evolution, there are fossil plants that were heterosporous but had not evolved seeds, among these being species of *Archaeopteris* (Figures 5.1, 5.13A; see later discussion). Note that heterospory has evolved *independently* in

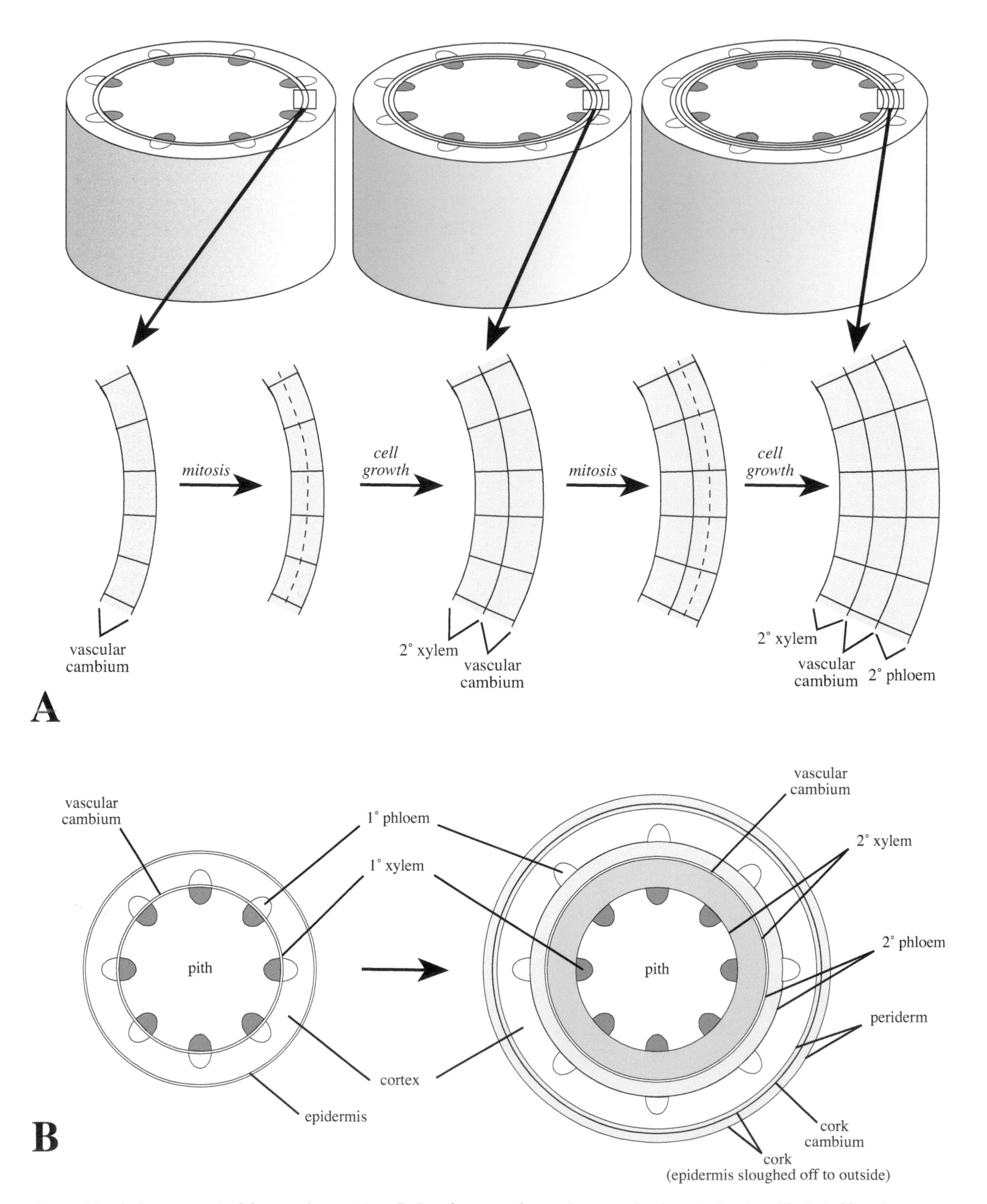

Figure 5.3 **A.** Development of the vascular cambium. **B.** Development of secondary vascular tissue in the stem, illustrated here for a eustelic stem.

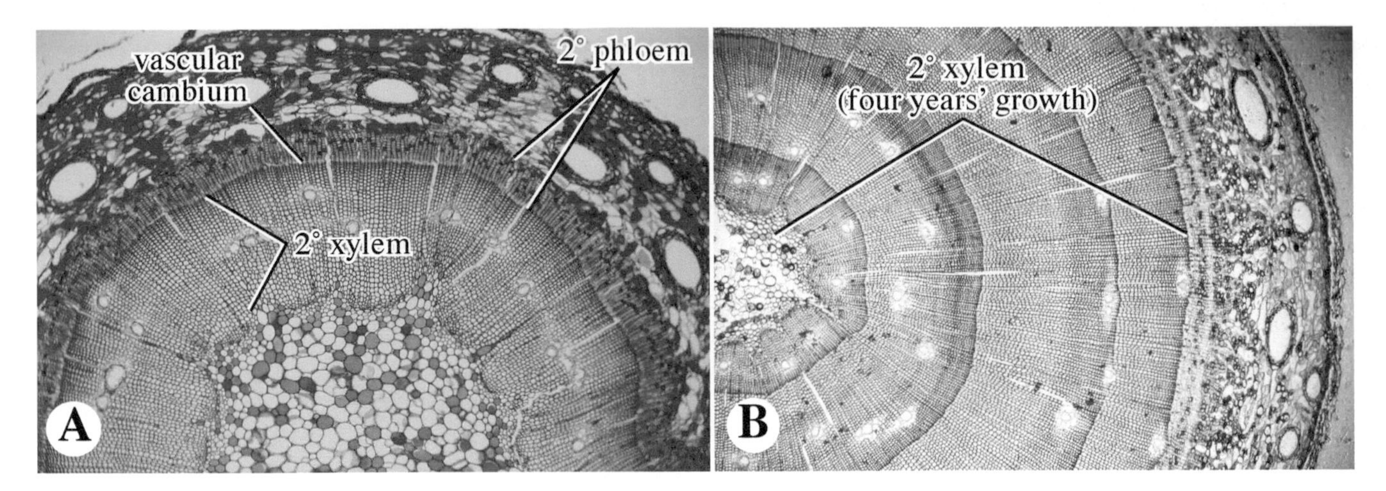

Figure 5.4 Woody stem cross section, *Pinus*. sp. **A.** One year's growth. **B.** Four years' growth.

other, nonseed plants, e.g., in the extant lycophytes *Selaginella* and *Isoetes* and in the water ferns (Chapter 4).

2. **Endospory.** Endospory is the complete development of, in this case, the female gametophyte *within the original spore wall* (Figure 5.6). The ancestral condition, in which the spore germinates and grows as an external gametophyte, is called **exospory**. The evolution of endosporic female gametophytes was correlated with that of endosporic male gametophytes (pollen grains); see later discussion.
3. **Reduction of megaspore number to one.** Reduction of megaspore number occurred in two ways. First, there evolved a reduction in the number of cells within the megasporangium that undergo meiosis (each termed a **megasporocyte** or **megaspore mother cell**) was reduced to one (Figure 5.6). After meiosis, the single diploid megasporocyte gives rise to four haploid megaspores. Second, of the four haploid megaspores produced by meiosis, *three consistently abort*, leaving only one functional megaspore. This single megaspore also undergoes a great increase in size, correlated with the increased availability of space and resources in the megasporangium.

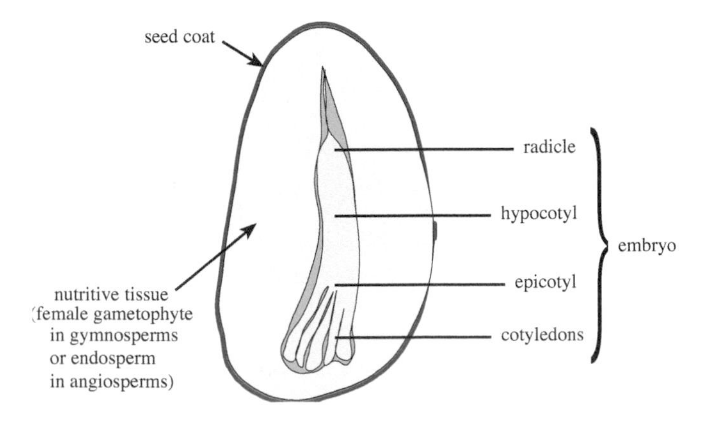

Figure 5.5 Morphology of a seed. *Pinus* sp. illustrated here.

4. **Retention of the megaspore.** Instead of the megaspore being released from the sporangium (the ancestral condition, as occurs in all homosporous nonseed plants), in seed plants it is retained within the megasporangium (Figure 5.6). This was accompanied by a reduction in thickness of the megaspore wall.
5. **Evolution of the integument & micropyle.** The final event in seed evolution was the envelopment of the megasporangium by a layer of tissue, called the **integument** (Figure 5.6). The integument grows from the base of the **megasporangium** (which is often called a **nucellus** when surrounded by an integument) and surrounds it, except at the distal end. Fossil evidence suggests that integuments likely evolved from separate lobes derived **telomes** (ancestral branches) that surrounded the megasporangium. These "preovules," i.e., ovules prior to the evolution of integuments, possessed a rim or ring of tissue at the apex of the megasporangium, the **lagenostome**, which functioned to funnel pollen grains to a pollination chamber. (See, e.g., Stewart and Rothwell 1993 for details.) The epitome of seed evolution occurred with the evolutionary "fusion" of the telomes to form the **integument**, a continuous sheath that completely surrounds the nucellus. The integument of all extant seed plants has a small pore at the distal end called the **micropyle**. The micropyle replaced the ancestral lagenostome as the site of entry of pollen grains (or in angiosperms, of pollen tubes). The micropyle also functions in the mechanics of pollination droplet formation and resorption (see below). Note that a single integument is the ancestral condition of spermatophytes; in angiosperms a second integument layer evolved later (Chapter 6).

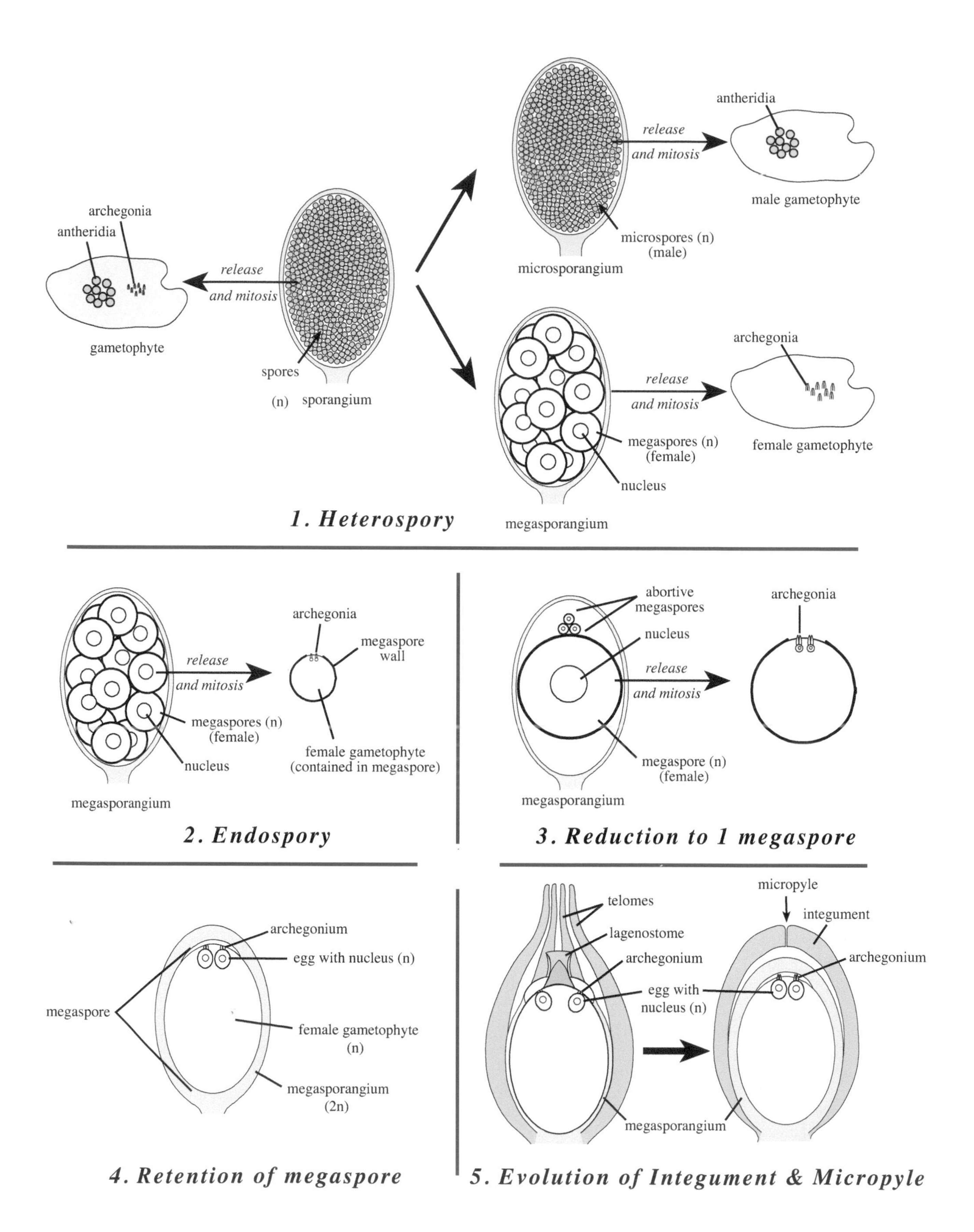

Figure 5.6 Ovule and seed evolution in the spermatophytes (hypothetical, for purpose of illustration).

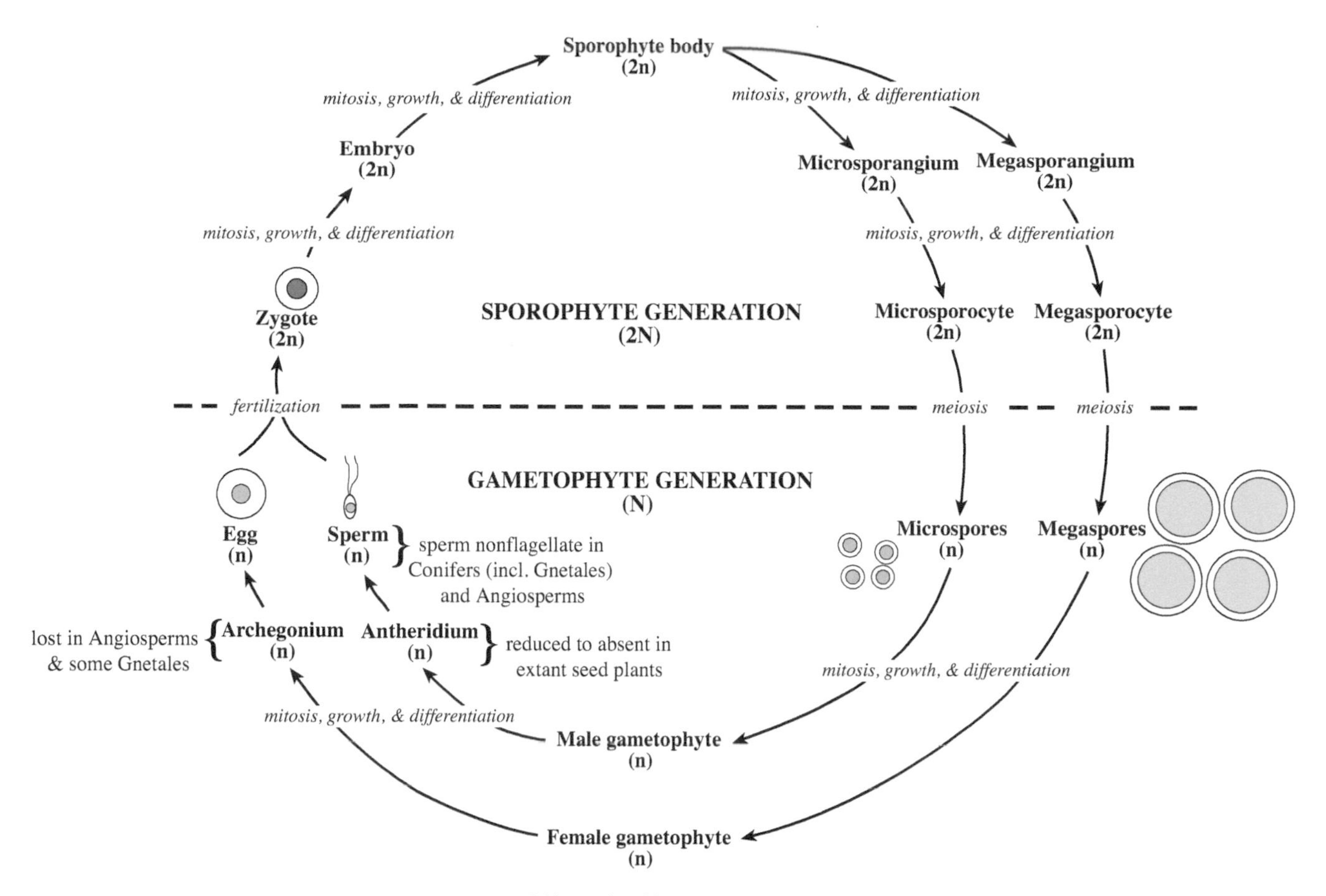

Figure 5.7 Life cycle of heterosporous seed plants.

POLLINATION DROPLET

One possible evolutionary novelty associated with seed evolution is the **pollination droplet**. This is a droplet of liquid that is secreted by the young ovule through the micropyle (Figures 5.10A, 5.17I). This droplet is mostly water plus some sugars or amino acids and is formed by the breakdown of cells at the distal end of the megasporangium (nucellus). The cavity formed by this breakdown of cells is called the **pollination chamber** (Figure 5.10A). The pollination droplet functions in transporting pollen grains through the micropyle. This occurs by resorption of the droplet, which "pulls" pollen grains that have contacted the droplet into the pollination chamber. It is unknown whether a pollination droplet was present in the earliest seed plants. However, the presence of a pollination droplet in many nonflowering seed plants suggests that its occurrence may be apomorphic for at least the extant seed plant lineages. Note that the ovules of angiosperms lack pollination droplets or pollination chambers, as flowering plants have evolved a different mechanism of pollen grain transfer (see Chapter 6).

POLLEN GRAINS

Concomitant with the evolution of the seed was the evolution of **pollen grains** (Figure 5.8). A pollen grain is, technically, an immature, endosporic male gametophyte. **Endospory** in pollen grain evolution was similar to the same process in seed evolution, involving the development of the male gametophyte within the original spore wall. Pollen grains of seed plants are extremely reduced male gametophytes, consisting of only a few cells. They are termed "immature" male gametophytes because, at the time of their release, they have not fully differentiated.

After being released from the microsporangium, pollen must be transported to the micropyle of the ovule (or, in angiosperms, to the stigmatic tissue of the carpel; see Chapter 6) in order to ultimately effect fertilization. Wind dispersal, in combination with an ovule pollination droplet (see later subsection), was probably the ancestral means of pollen transport. After being transported to the ovule (or stigmatic tissue), the male gametophyte completes development by undergoing additional mitotic divisions and differentiation. The male gametophyte grows an exosporic **pollen tube**, which functions as a haustorial organ, obtaining nutrition by absorp-

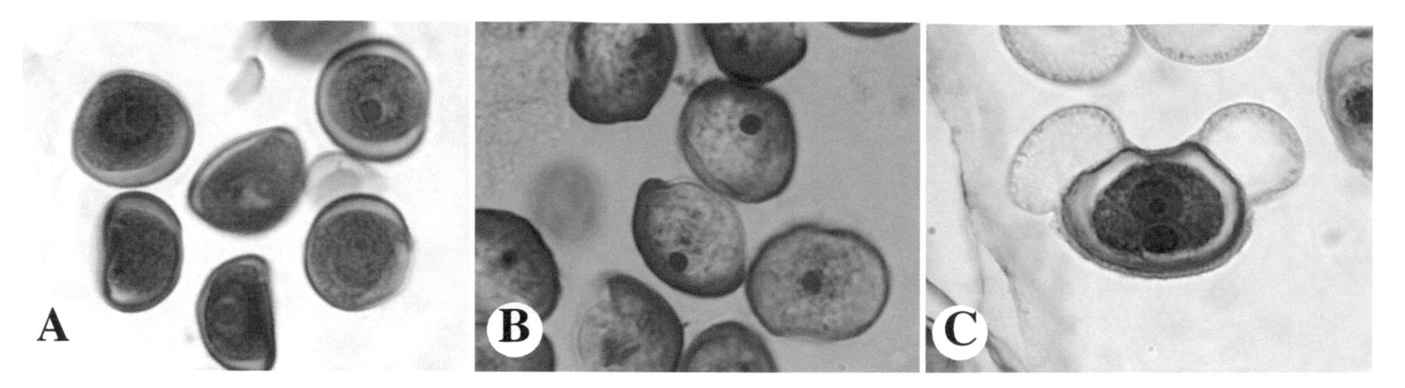

Figure 5.8 Pollen grains–immature male gametophytes of seed plants. **A.** *Zamia* sp., a cycad. **B.** *Ginkgo biloba*. **C.** *Pinus* sp., a conifer.

tion from the surrounding sporophytic tissue (Figure 5.9; see **Pollen Tube**).

POLLEN TUBE

The male gametophytes of all extant seed plants form a pollen tube (Figure 5.9) soon after the pollen grains make contact with the megasporangial (nucellar) tissue of the ovule. In extant seed plants the ancestral type of pollen type (found in cycads and ginkgophytes) was haustorial, in which the male gametophyte feeds (like a parasite) off the tissues of the nucellus. Motile sperm is delivered from this male gametophyte into a fertilization chamber, where the sperm swims to the archegonium containing the egg, a process known as **zooidogamy** (*zooin*, animal + *gamos*, marriage). In the conifers (including Gnetales), pollen tubes are also haustorial, but deliver non-motile sperm cells to the archegonium or egg, a process known as **siphonogamy** (*siphono*, tube + *gamos*, marriage). A type of siphonogamy evolved independently in the angiosperms. In angiosperms, however, the pollen tubes grow through stylar tissue prior to delivering the sperm to the egg of a female gametophyte (see Chapter 6).

OVULE AND SEED DEVELOPMENT

After pollination, the **megasporocyte** develops within the megasporangium of the ovule (Figures 5.10A, 5.11A). The megasporocyte is a single cell that undergoes meiosis, producing a tetrad of four haploid megaspores (**megasporogenesis**). In most extant seed plants the four megaspores are arranged in a straight line, or linearly (Figure 5.10A). The three megaspores that are distal (away from the ovule base) abort; only the proximal megaspore (near the ovule base) continues to develop. In the pollination chamber, the resorbed pollen grains (Figures 5.10A, 5.11A) develop into mature male gametophytes and form pollen tubes, which grow into the tissue of the megasporangium

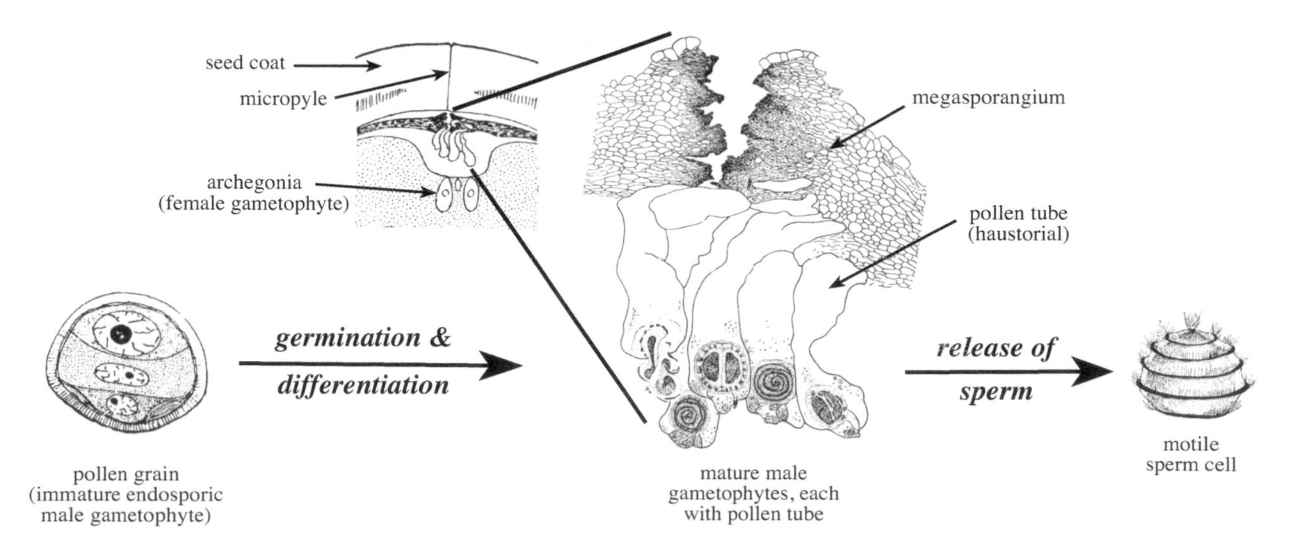

Figure 5.9 Male gametophyte morphology and development in the nonflowering spermatophytes; *Cycas* sp., illustrated. (Reproduced and modified from Swamy, B. G. L. 1948. American Journal of Botany 35: 77–88, by permission.)

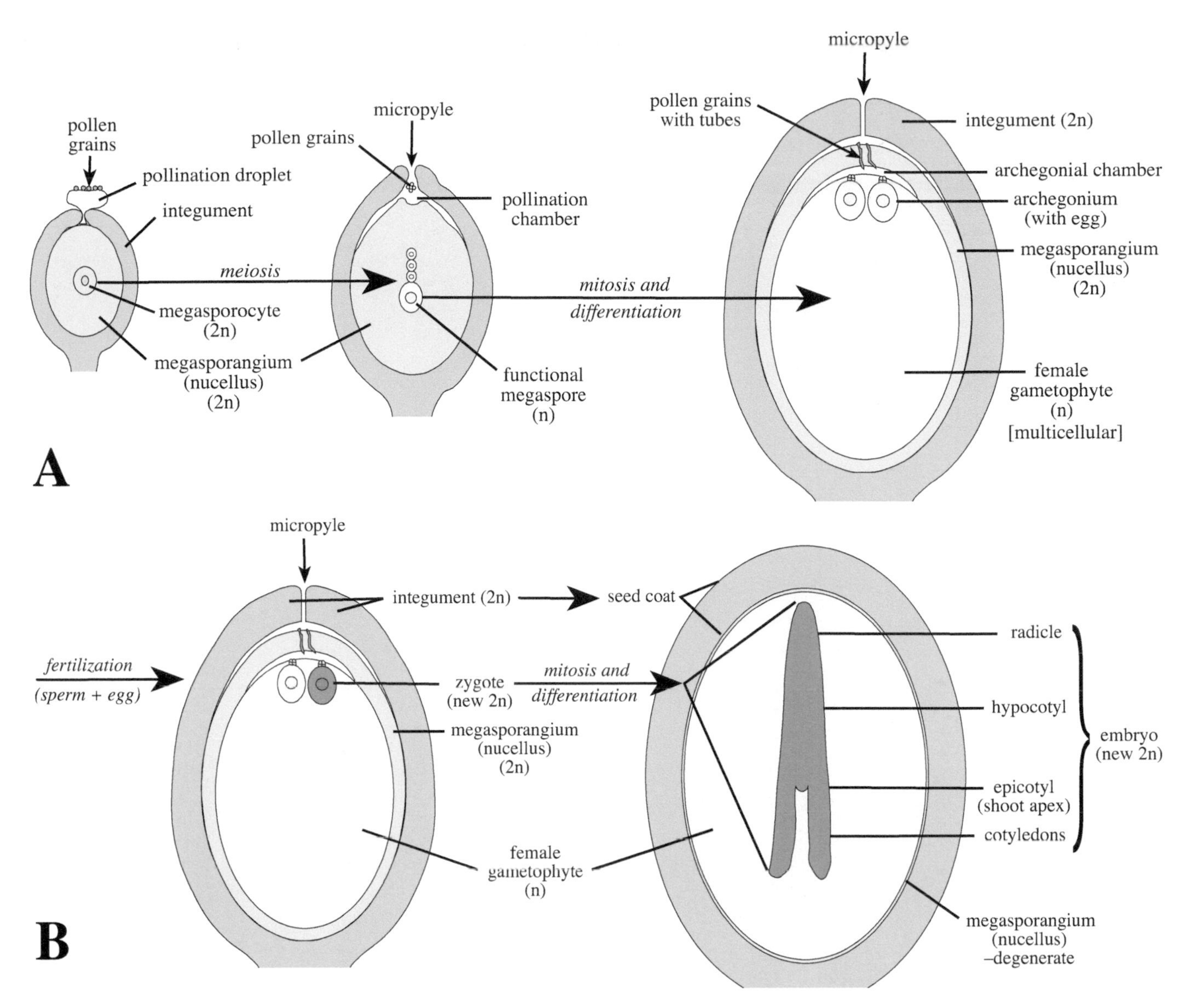

Figure 5.10 **A.** Ovule development in the in the nonflowering spermatophytes. **B.** Seed development.

(Figures 5.10A, 5.11B). In gymnosperms these male gametophytes may live in the megasporangial tissue for some time, generally several months to a year.

The functional megaspore greatly expands, accompanied by numerous mitotic divisions, to form the endosporic female gametophyte (Figures 5.10A, 5.11B,C), this process known as **megagametogenesis**. In the seeds of gymnosperms, archegonia differentiate at the apex of the female gametophyte (Figure 5.11C,D). As in the nonseed land plants, each archegonium has a large egg cell and a short line of neck cells (plus typically a ventral canal cell or nucleus). Eventually, the male gametophytes either release motile sperm cells (in cycads and *Ginkgo*) into a cavity between the megasporangium and female gametophyte (known as the **archegonial chamber**; Figure 5.10A), or the pollen tube of the male gametophyte delivers sperm cells directly into the archegonial neck (in conifers). (Note that the ovules of some Gnetales and all angiosperms lack archegonia.) The end result is that a sperm cell from the male gametophyte fertilizes the egg of the female gametophyte. A long period of time (perhaps a year or more) may ensue between **pollination**, which is delivery of the pollen grains to the ovule, and **fertilization**, actual union of sperm and egg. Note: This is not true for the flowering plants, in which fertilization generally occurs very soon after pollination (see Chapter 6).

The resulting diploid zygote, once formed, undergoes considerable mitotic divisions and differentiation, eventually maturing into the **embryo**, the immature sporophyte (Figures 5.10B, 5.11E). The tissue of the female gametophyte continues to surround the embryo (Figure 5.11E) and serves as

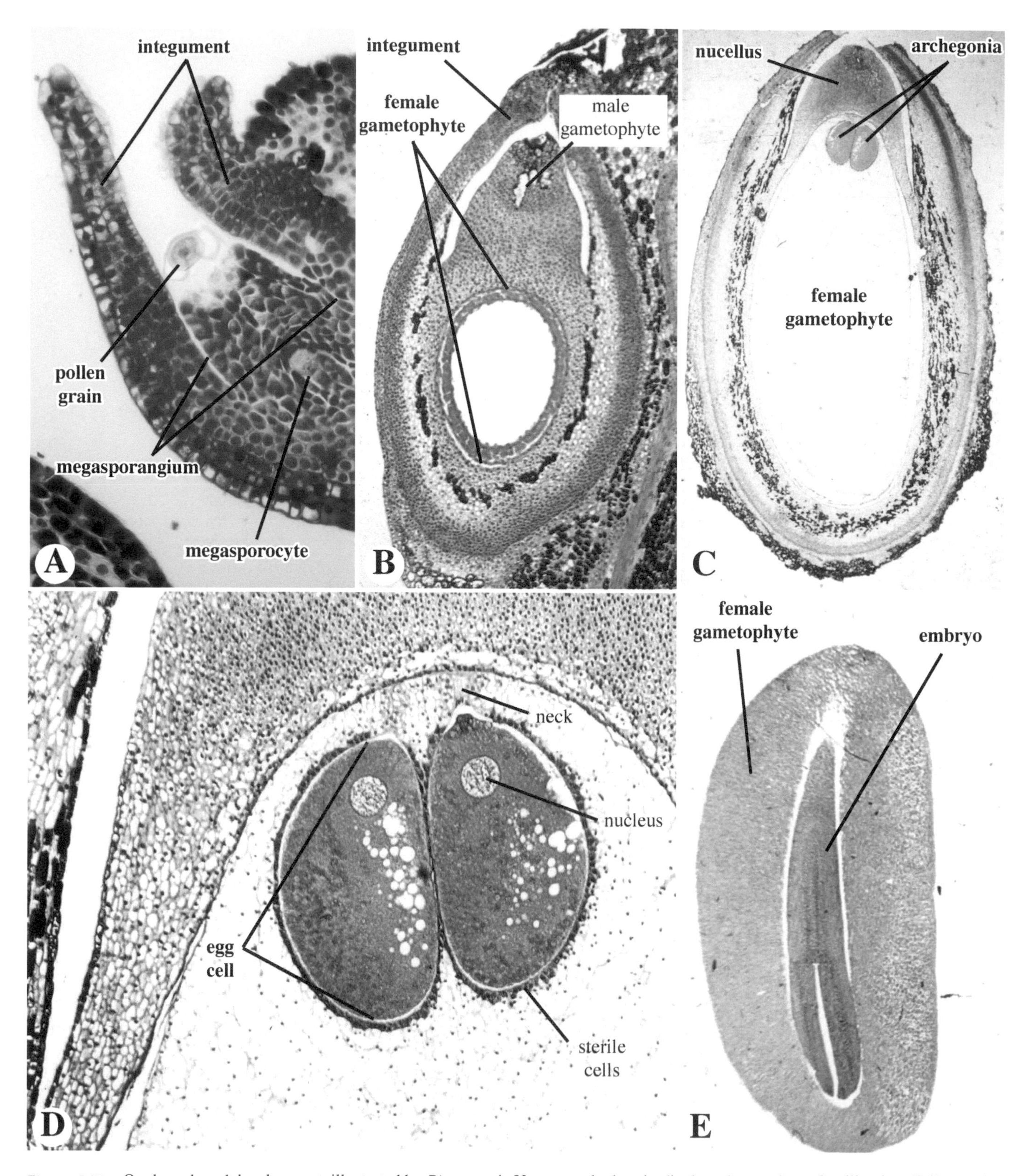

Figure 5.11 Ovule and seed development, illustrated by *Pinus* sp. **A.** Young ovule, longitudinal section, at time of pollination. Pollen grains are pulled into micropyle by resorption of pollination droplet. Meiosis of the megasporocyte has yet to occur. **B.** Postpollination, showing development of the female gametophyte and haustorial pollen tube growth of the male gametophytes within tissue of megasporangium (nucellus). **C.** Mature ovule, showing two functional archegonia within female gametophyte. **D.** Close-up of archegonia, each containing a large egg cell with a surrounding layer of sterile cells and apical neck. **E.** Seed longitudinal section, seed coat removed, showing embryo and surrounding nutritive layer of female gametophytic tissue.

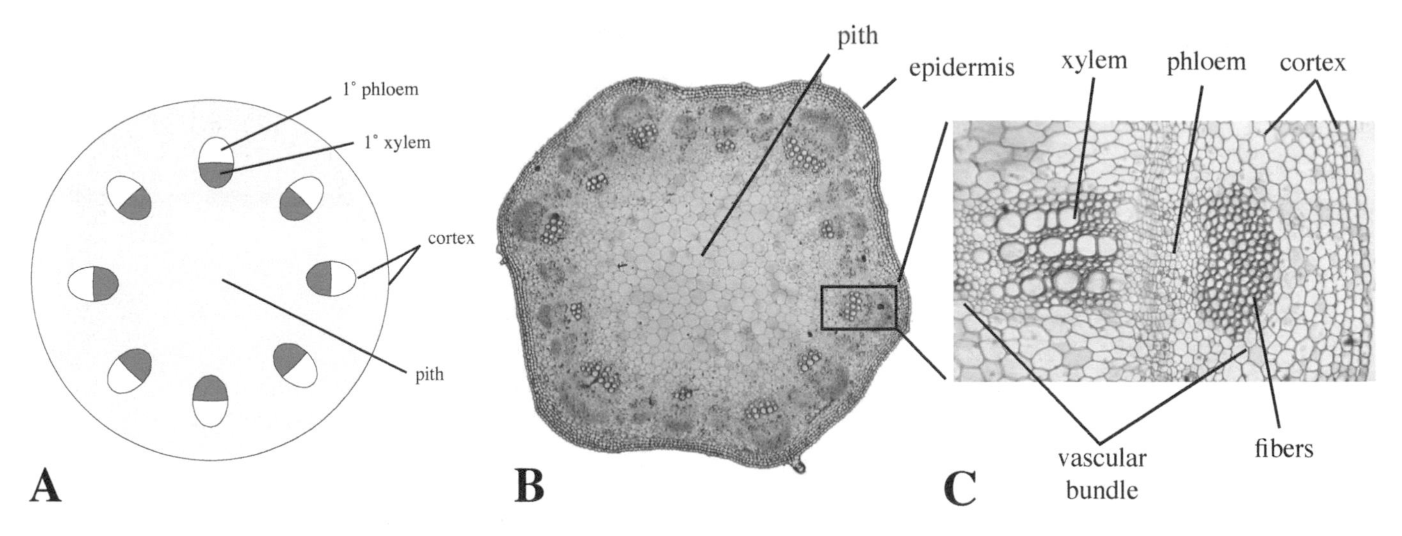

Figure 5.12 Eustele. **A.** Diagram of eustele. Note single ring of vascular bundles, with xylem inside, phloem outside. **B.** *Helianthus* stem cross section, an example of a eustele. **C.** Close-up of vascular bundle, showing xylem, phloem, and associated fibers.

nutritive tissue for the embryo upon seed germination (except in the flowering plants; see Chapter 6). The megasporangium (nucellus) eventually degenerates. The integument matures into a peripheral **seed coat**, which may differentiate into various hard and/or fleshy layers.

SEED ADAPTATIONS

The adaptive significance of the seed is unquestioned. First, seeds provide *protection*, mostly by means of the seed coat, from mechanical damage, desiccation, and often predation. Second, seeds function as the *dispersal unit* of sexual reproduction. In many plants the seed has become specially modified for dispersal. For example, a fleshy outer seed coat layer may function to aid in animal dispersal. In fact, in some plants the seeds are eaten by animals, the outer fleshy layer is digested, and the remainder of the seed (including the embryo protected by an inner, hard seed coat layer) passes harmlessly through the gut of the animal, ready to germinate with a built-in supply of fertilizer. In other plants, differentiation of the seed coat into one or more wings functions in seed dispersal by wind. Third, the seed coat may have *dormancy mechanisms* that ensure germination of the seed only under ideal conditions of temperature, sunlight, or moisture. Fourth, upon germination, the *nutritive tissue* surrounding the embryo provides *energy* for the young seedling, aiding in successful establishment.

Interestingly, in seed plants the female gametophyte (which develops within the megaspore) remains attached to and nutritionally dependent upon the sporophyte. This is exactly the reverse condition as is found in the liverworts, hornworts, and mosses (Chapter 3).

EUSTELE

In addition to the seed, an apomorphy for spermatophytes is the **eustele** (Figure 5.12). A eustele is a primary stem vasculature ("primary" meaning prior to any secondary growth) that consists of a single ring of discrete vascular bundles. Each vascular bundle contains an internal strand of xylem and an external strand of phloem that are radially oriented, i.e., positioned along a radius (Figure 5.12).

The protoxylem of the vascular bundles of a eustele is **endarch** in position, i.e., toward the center of the stem. This is distinct from the exarch protoxylem of the lycophytes and the mesarch protoxylem of most monilophytes (Chapter 4) and of some fossil relatives that diverged prior to the seed plants.

DIVERSITY OF WOODY AND SEED PLANTS

ARCHEOPTERIS

A well-known lignophyte that lacked seeds was the fossil plant ***Archeopteris*** (not to be confused with the very famous fossil, reptilian bird *Archeopteryx*). *Archeopteris* was a large tree, with wood like a conifer but leaves like a fern (Figure 5.13A,B). Sporangia, producing spores, were born on fertile branch systems. Some species of *Archeopteris* were heterosporous.

"PTERIDOSPERMS"—SEED FERNS

The "pteridosperms," or seed ferns, are almost certainly a nonnatural, paraphyletic group of fossil plants that had fern-like foliage, yet bore seeds. ***Medullosa*** is a well-known example of a seed fern (Figure 5.13C–E). As in many fossil

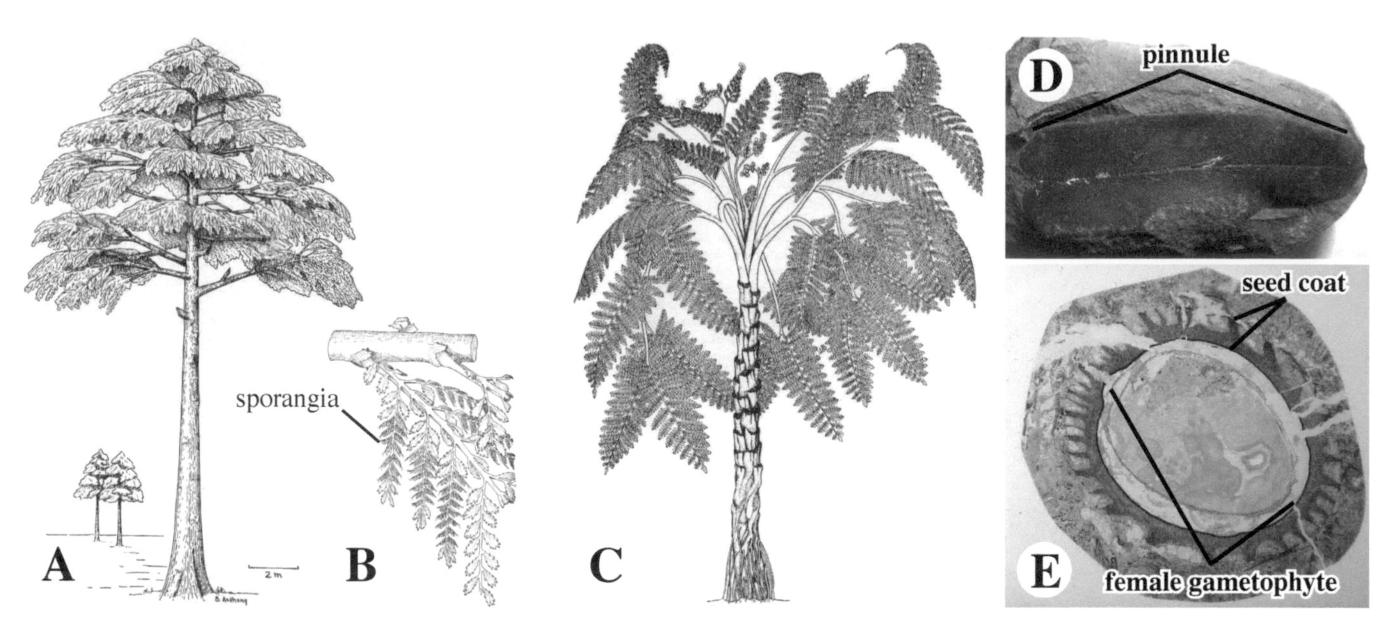

Figure 5.13 **A,B.** *Archeopteris*, an extinct lignophyte. [Reproduced from: Beck, C. B. 1962. American Journal of Botany 49: 373–382, by permission.] **A.** Reconstruction of plant. **B.** Branch system, showing leaves and sporangia. **C–E.** *Medullosa* (and form genera), an extinct "seed fern." **C.** Reconstruction of plant. [Reproduced from: Stewart, W. N., and T. Delevoryas. 1956. Botanical Review 22: 45–80, by permission.] **D.** Fossil leaf impression (*Neuropteris*). **E.** Seed longitudinal section (*Pachytesta*).

plants, different organs of *Medullosa* are placed in separate "form genera." For example, the fernlike leaves of *Medullosa* are in the form genera *Alethopteris* and *Neuropteris*. *Dolerotheca*, which had huge pollen grains, refers to the pollen-bearing organs of *Medullosa*, and seeds of *Medullosa* are placed in the form genus *Pachytesta*.

The relationships of various "pteridosperms" to extant seed plants are unclear. Some are basal to the extant seed plants (Figure 5.1; see Doyle 2006).

GYMNOSPERMAE—GYMNOSPERMS

Recent cladistic analyses using multiple gene sequences have provided strong evidence that the Spermatophyta (seed plants) are composed of two sister groups: Gymnospermae and Angiospermae (Figure 5.1), the common ancestor of which is estimated to have arisen some 310–350 million years ago. [Note that the extant gymnosperms are sometimes referred to as *Acrogymnospermae*.] The Gymnospermae, or gymnosperms (after *gymnos*, naked + *sperm*, seed), are called so because the ovules are not enclosed by surrounding carpel layer (thus, being "naked") at the time of pollination. (Note that the developing *seeds* often are enclosed, e.g., by megasporophylls or ovuliferous scales, after pollination.) Gymnosperms are essentially non-flowering seed plants. Hypotheses of relationships within the gymnosperms are contested, with some studies showing cycads (Cycadophyta) as sister to the remaining gymnosperms (e.g., Bowe et al. 2000), others with cycads and ginkgo (Ginkgophyta) forming a clade (e.g., Xi et al. 2013; Ruhfel et al. 2014), this clade sister to all other gymnosperms (Figure 5.1). In addition, most recent analyses place the Gnetales either as sister to the Pinopsida (the "gnepine" hypothesis; e.g., Mathews et al. 2010; Zhong et al. 2011); or to the Cupressopsida (the "gnecup" hypothesis; e.g., Ruhfel et al. 2014; see Figure 5.1).

Table 5.1 lists the classification system used here for the Gymnospermae. See Bierhorst (1971) and Foster and Gifford (1974) for general vascular plant morphology and Hill (2005) and Rydin et al. (2002) for studies of the gymnosperms.

CYCADOPHYTA—CYCADS

The Cycadophyta (also known as Cycadales), or cycads, are a relatively ancient group of plants that were once much more common than today and served as fodder for plant-eating nonavian dinosaurs. Extant cycads are now fairly restricted in distribution, consisting of approximately 210–250 species in 11 genera. Cycads are found in southeastern North America, Mexico, Central America, some Caribbean islands, South America, eastern and southeastern Asia, Australia, and parts of Africa. Many cycads throughout the world are of economic importance in being used as a source of food starch (sometimes termed "sago"), typically collected from the apex of the trunk just prior to a flush of leaves or reproductive structures. Some cycads, especially *Cycas revoluta*, the "sago palm," are planted horticulturally.

Cycads are an apparently monophyletic lineage consisting of plants with a mostly short, erect stem or trunk, rarely tall and palmlike (as in the strangely named *Microcycas*). The trunk bears spirally arranged, mostly pinnately compound

SPERMATOPHYTA
- **GYMNOSPERMAE (*ACROGYMNOSPERMAE*)**
  - **CYCADOPHYTA**
    - **Cycadaceae** (1/100–110)
    - **Zamiaceae** (10/220–230)
  - **GINKGOPHYTA**
    - **Ginkgoaceae** (1/1)
  - **CONIFERAE**
    - **PINOPSIDA**
      - **Pinaceae** (12/225)
    - **CUPRESSOPSIDA**
      - **Araucariaceae** (3/32)
      - **Cupressaceae** (32/130)
      - Phyllocladaceae (1/5)
      - **Podocarpaceae** (17/167)
      - Sciadopityaceae (1/1)
      - **Taxaceae** (incl. Cephalotaxaceae) (6/28)
    - **GNETALES**
      - **Ephedraceae** (1/40)
      - **Gnetaceae** (1/30)
      - **Welwitschiaceae** (1/1)
- **ANGIOSPERMAE** (See Chapters 6–8)

Table 5.1 Classification of the Spermatophyta, emphasis on the Gymnospermae. Some higher groups after Cantino et al. (2007). Families in bold are described in detail. Number of genera and species are indicated in parentheses, separated by slash mark.

leaves (Figures 5.14A, 5.15B-E). Only the genus *Bowenia* has bipinnately compound leaves (Figure 5.15A). The trunks of cycads do not usually exhibit lateral (axillary) branching; thus, the **loss of axillary branching** on the aerial trunk is diagnostic for the cycads (Figure 5.1). Interestingly, cycad pinnae (*Cycas*) or leaves (some Zamiaceae; e.g., *Bowenia*) exhibit **circinate vernation** (Figure 5.14B) as in ferns, perhaps a primitive retention that was lost in other seed plants. Reproductively, all cycad individuals are either male or female; this plant sex is termed **dioecious** (see Chapter 9).

All cycads have **pollen cones** or **strobili** (also called male cones/strobili). Recall that cones are determinate shoot systems, consisting of a single axis that bears **sporophylls**, modified leaves with attached sporangia. Pollen cones (Figures 5.14A,C; 5.16A,B,G) consist of an axis bearing **microsporophylls**, each of which bears numerous microsporangia (Figure 5.14D). These **microsporangia** produce great numbers of haploid microspores, each of which develops into a pollen grain, an immature, endosporic, male gametophyte (Figure 5.9). Interestingly, the pollen of all cycads (like the Ginkgophyta, to be discussed) release **motile sperm cells** (Figure 5.9) within the ovule of a seed cone, a vestige of an ancestrally aquatic condition.

Recent evidence (e.g., Rai et al., 2003) suggests that cycads are best grouped as two families: Cycadaceae and Zamiaceae, differing primarily in the absence of seed cones in the former. In the Cycadaceae seeds are produced on the margins of numerous **megasporophylls**, which are aggregated not in cones but at the trunk apex in dense masses (Figure 5.14E–G). In contrast, all members of the Zamiaceae have **seed [ovulate] cones** or **strobili** (also called female cones/strobili). Seed cones (Figures 5.15C–E, 5.16C,D,F–I) consist of an axis bearing **megasporophylls**, each bearing two seeds (Figure 5.16E,H,I). There is variation in the size and shape of the seed cones, megasporophylls, and seeds within groups.

See Johnson (1990c) for general information and Rai et al. (2003) for a phylogenetic analysis of the cycads.

Cycadaceae–Cycad family (type *Cycas*, from Greek *koikas* or *kykas*, for a kind of palm). 1 genus (*Cycas*, incl. *Epicycas*)/ 100–110 species (Figure 5.14).

The Cycadaceae consist of dioecious trees to perennial herbs. The **roots** are often vesicular-arbuscular mycorrhizal; some adventitious roots are "coralloid," being ageotropic (growing upward), branched and shaped like coral, and containing symbiotic, nitrogen-fixing cyanobacteria in the outer tissues. The **stem** is unbranched or dichotomously branched, either an aerial trunk, covered with persistent leaf bases, or subterranean from adventitious buds, the stem apex at ground level. The **leaves** are spiral, petiolate (petiole margins with prickles), pinnately compound, evergreen, and coriaceous, forming by means of circinate vernation, in which leaflets are coiled early in development; mature leaflets have a single midvein; non-photosynthetic, rigid cataphylls are typically produced in flushes alternately with photosynthetic leaves. The **pollen cones** are large, terminal from the trunk apex, with numerous microsporophylls, each abaxially bearing numerous, spherical microsporangia. The **seed-bearing** reproductive structures are not organized in determinate cones, consisting of numerous stalked, apically toothed to pinnately divided megasporophylls surrounding the trunk apex. The **seeds** are large, [1] 2–8 born marginally on each megasporophyll; the **embryo** has 2 cotyledons.

Members of the Cycadaceae are distributed in e. Africa, e. and s.e. Asia, and n. Australia. Economic importance includes cultivated ornamentals (esp. *Cycas revoluta*, sago-palm), food derived from the pith of the trunk (known as "sago," made into a flour, bread, that of some spp. toxic/carcinogenic), and edible seeds (after removal of toxins; e.g., *C. media*, of Australia, New Guinea). See Norstog and Nicholls (1997), Hill (1998 onwards), Johnson (1990b), and Jones (2002) for general information; Hill (2004) and Walters and Osborne (2004) for classification and nomenclature; and Hill et al. (2004) for a phylogenetic analysis.

The Cycadaceae are readily distinguished in consisting of ***dioecious*** trees or perennial herbs, having ***trunks or subter-***

Figure 5.14 CYCADOPHYTA. Cycadaceae (*Cycas*). **A–F.** *Cycas revoluta*, sago palm. **A.** Male individual with pollen cone. **B.** Leaves with circinate vernation. **C.** Close-up of pollen cone. **D.** Microsporophyll with sporangia. **E.** Female individual, showing aggregate of megasporophylls (cones lacking). **F.** Megasporophyll with marginal, immature seeds. **G.** *Cycas circinalis*, female, showing mature megasporophylls with seeds (cones lacking).

ranean stems, with large, ***coriaceous, evergreen, pinnate*** leaves (vernation involute ***circinate***), large, determinate ***pollen cones***, the ovulate reproductive structures ***not organized as cones***, consisting of numerous ***toothed to divided megasporophylls*** arising from apex of trunk, each bearing ***one or more marginal*** ovules.

Zamiaceae (incl. Boweniaceae, Stangeriaceae) family (type *Zamia*, from Latin *zamiae*, after *azaniae*, meaning pine cones). 10 genera/220–230 species (Figures 5.15, 5.16).

The Zamiaceae consist of dioecious trees to perennial herbs. The **roots** are often vesicular-arbuscular mycorrhizal; some adventitious roots are "coralloid," being ageotropic (growing upward), branched and shaped like coral, and containing sym-

Figure 5.15 CYCADOPHYTA. Zamiaceae. **A.** *Bowenia spectabilis* (Australia), showing a single, bipinnately compound leaf. **B.** *Macrozamia moorei* (Australia). **C.** *Encephalartos lebomboens* (Africa). **D.** *Lepidozamia peroffskyana* (eastern Australia). **E.** *Stangeria eriopus* (Africa).

biotic, nitrogen-fixing cyanobacteria in the outer tissues. The **stem** is unbranched or irregularly branched, either an aerial trunk, covered with persistent leaf bases, or subterranean, the stem apex at ground-level. The **leaves** are usu. pinnately compound (bipinnate in *Bowenia*), spiral, petiolate, stipulate or exstipulate, often large, forming by means of circinate vernation; leaflets usu. have several parallel veins (pinnately veined with dichotomous laterals in *Stangeria*); non-photosynthetic cataphylls are typically present. The **pollen cones** have numerous microsporophylls, each bearing numerous, spherical microsporangia abaxially. The **ovulate** reproductive structures are in determinate cones, which have numerous, spiral, usually peltate megasporophylls, each bearing 2 [3] inverted ovules on the adaxial margin. Plants are wind or insect (beetle or *Trigona* bee) pollinated; in some taxa, cones self-generate heat, effecting insect pollination by increasing odor or insect activity.

The genera of the Zamiaceae are: *Bowenia* (2 spp., n.e. Australia), *Ceratozamia* (26 spp., Mexico to Central America), *Chigua* (2 spp., Colombia), *Dioon* (14 spp., Mexico & Central America), *Encephalartos* (65 spp., tropical and s. Africa), *Lepidozamia* (2 spp., n.e. Australia), *Macrozamia* (41 spp., Australia), *Microcycas* (1 sp., *M. calocoma*, w. Cuba), *Stangeria* (1 sp., *S. eriopus*, S. Africa), and *Zamia* (68 spp., tropical/warm N. & S. Am.). See Norstog and Nicholls (1997), Hill (1998 onwards), and Jones (2002) for general information; Hill (2004) and Walters and Osborne (204) for classification and nomenclature; and Hill et al. (2004) for a phylogenetic analysis.

Members of the Zamiaceae are distributed in central to southern Africa, tropical N. & S. America, and Australia. Economic importance includes cultivated ornamentals (e.g., *Ceratozamia, Encephalartos,* and *Zamia* spp.), food (edible seeds and "sago" starch from stem pith). *Lepidozamia hopei*

Figure 5.16 CYCADOPHYTA. Zamiaceae. **A,B.** *Encephalartos altensteinii*, pollen cones. **C.** *Encephalartos arenarius*, female with cone. **D.** *Encephalartos ferox*, female, with bright red cone. **E.** *Encephalartos manikensis*, megasporophyll with two attached seeds. **F.** *Ceratozamia mexicana*, seed cone. **G.** *Zamia* sp., pollen and seed cones. **H,I.** *Ceratozamia* sp., seed cone and megasporophyll with ovules.

of n.e. Australia is the tallest cycad known (up to 20 m tall). See Johnson and Wilson 1990a,d,e for general information on the family.

The Zamiaceae are distinctive in being ***dioecious*** trees or perennial herbs, having ***trunks or subterranean stems***, the leaves ***pinnate*** [***rarely bipinnate***], with ***determinate pollen and seed cones***, the seed cones with usually ***peltate megasporophylls***, each bearing ***2 [3] adaxially marginal ovules/seeds***.

GINKGOPHYTA—GINKGOPHYTES

The Ginkgophyta, or ginkgophytes, have an extensive fossil record but contain only one extant species, *Ginkgo biloba*. This species is native only to certain remote regions of China but has now been planted worldwide as a popular street tree. *Ginkgo biloba*, unlike the cycads (and similar to conifers, discussed next), is a highly branched, woody tree. It can be recognized by the fact that it has short shoots in addition to long shoots, and by the distinctive obtriangular (fan-shaped), often two-lobed leaves with dichotomous venation (Figure 5.17A–C). *Ginkgo*, like the cycads, is **dioecious** and has ancestrally motile sperm.

Male *Ginkgo* trees bear reproductive structures that are called "cones" but that do not bear structures that resemble sporophylls. These pollen cones consist of a central axis with lateral stalks (Figure 5.17D–E), each of which bears two microsporangia (Figure 5.17F,G). The microsporangia dehisce longitudinally, releasing pollen grains. Female *Ginkgo* trees do not bear cones. The female reproductive structures each consist of an axis having two terminal ovules (Figure 5.17H,I).

Ginkgoaceae–Maidenhair Tree family (type *Ginkgo*, Japanese *gin*, silver, + *kyo*, apricot). 1 extant genus and species (*Ginkgo biloba*) (Figure 5.17).

The Ginkgoaceae consist of resinous, dioecious trees. The **roots** are mycorrhizal. The **stem** trunk is tall, with numerous lateral branches. The **leaves** are simple, a possible synapomorphy for the Ginkgophyta, spiral, petiolate, deciduous, and flabelliform, with open dichotomous venation; leaves are born on both long shoots and stout short shoots. The **male** reproductive shoots, which may be interpreted as modified cones, are catkin-like, consisting of an axis bearing numerous, spiral, stalk-like microsporophylls, each of which bears an apical pair of elongate, pendulous microsporangia. The **female** reproductive shoots consist of a stalk-like peduncle (arising from axils of the leaves of short shoots) that bear 2 [–4] erect ovules, each with a basal collar. The **seed** (usu. one per peduncle) is drupe-like, the integument with two layers: an inner, hard layer and an outer, fleshy layer, the latter becoming fetid at maturity (containing butyric acid and related compounds); the **embryo** has 2 cotyledons. Plants are wind pollinated.

Members of the family have a very limited, natural distribution in east China, where old, very large trees at temples are relictual. Now extinct members of the family had a worldwide distribution during the Mesozoic era. The family is economically important primarily as a cultivated ornamental, commonly used as a street tree because of its beautiful foliage (leaves turning yellow in the fall) and resistance to air pollution and pests. In addition, *Ginkgo* yields edible seeds (roasted female gametophytes, but toxic in large quantities), is a source of oil, and is used as an herbal remedy, including widespread claims that it aids in memory retention (but I forget why). See Page (1990f) for general information and Rydin et al. (2002) for information on relationships to other seed plants.

The Ginkgoaceae are distinctive in being ***dioecious trees*** with stout ***short shoots*** and simple, spiral, ***flabelliform (fan-shaped) leaves*** with ***open dichotomous venation***, the male reproductive structure ***catkin-like***, each consisting of an axis bearing microsporophylls with ***paired, pendant microsporangia***, the female consisting of an axis bearing gen. ***two, erect ovules each with a basal collar***, the seed with an outer fleshy and inner hard integument layer.

CONIFERAE—CONIFERS

The Coniferae, or conifers (also known as Coniferophyta or Pinophyta), are an ancient group of land plants that were once dominant in most plant communities worldwide. Today, they have largely been replaced by angiosperms, but still constitute the primary biomass of various "coniferous" forests.

Conifers comprise a monophyletic group of highly branched trees or shrubs with **simple leaves**, the latter a possible apomorphy shared with the Ginkgophytes. Leaves of conifers are often linear, acicular (needle-like), or subulate (awl-shaped; see Chapter 9), although they are sometimes broad and large. In some conifers the leaves are clustered into **short shoots**, in which adjacent internodes are very short in length. An extreme of this is the **fascicle**, e.g., in species of *Pinus*, the pines. A fascicle is a specialized short shoot consisting of stem tissue, one or more needle-shaped leaves, and persistent basal bud scales (Figure 5.18A,B; Chapter 9).

A second, apparent apomorphy of the conifers, including the Gnetales (discussed later), is the **loss of sperm cell motility** and **siphonogamy** (Figure 5.1). This distinguishes the conifers from the cycads and ginkgophytes, which have flagellated sperm cells. Conifers, like all extant seed plants, have pollen tubes, within which the male gametophytes develop. As in cycads and *Ginkgo*, these pollen tubes are haustorial, consuming the tissues of the nucellus (megasporangial tissue) for up to a year or so after pollination. One difference, however, (likely correlated with sperm nonmotility) is that the male gametophyte of conifers delivers the sperm cells more directly to the egg by the growth of the pollen tube into the archegonial neck, and release of non-motile sperm cells near the egg. This type of pollen tube and sperm transfer in conifers is known as **siphonogamy**, as opposed to zooidogamy. (Because there is more than one archegonium per seed, multiple fertilization events may occur, resulting in multiple young embryos, but usually only one survives in the mature seed.)

Reproductively, conifers produce pollen cones and seed cones, either on the same individual (**monoecy**) or, less commonly, on different individuals (**dioecy**). As with all vascular plants, cones consist of an axis that bears sporophylls. As in cycads, pollen cones (Figure 5.18C,D) consist on an axis with **microsporophylls** (Figure 5.18E,F). The microsporophylls bear **microsporangia**, which produce **pollen grains** (Figure 5.18E–G). The pollen grains of some (but not all) conifers are interesting in being bi-saccate, in which two bladder-like structures develop from the pollen grain wall (Figure 5.8C). These saccate structures, like air bladders,

Figure 5.17 GINKGOPHYTA. Ginkgoaceae. *Ginkgo biloba*. **A,B.** Vegetative growth. Note fan-shaped leaves, clustered into short shoots. **C.** Leaf close-up, showing dichotomous venation. **D.** Male tree bearing pollen cones. **E.** Pollen cone. **F,G.** Close-up of microsporangia, born in pairs on stalk arising from central axis of pollen cone. **H.** Female plant bearing stalk with pair of ovules. **I.** Close-up of ovule pair. Note pollination droplet from micropyle.

Figure 5.18 CONIFERAE. Pinaceae. **A–G.** *Pinus* spp. **A.** Shoot with young fascicles. **B.** Branch, showing scale leaves and fascicles. **C.** Apex of branch with fascicles and pollen cones. **D.** Pollen cones, close-up. **E.** Microsporophylls of pollen cones, each with two microsporangia. **F.** Male strobilus, longitudinal section, showing microsporangia and subtending microsporophylls. **G.** Close-up of microsporangium, full of mature pollen grains.

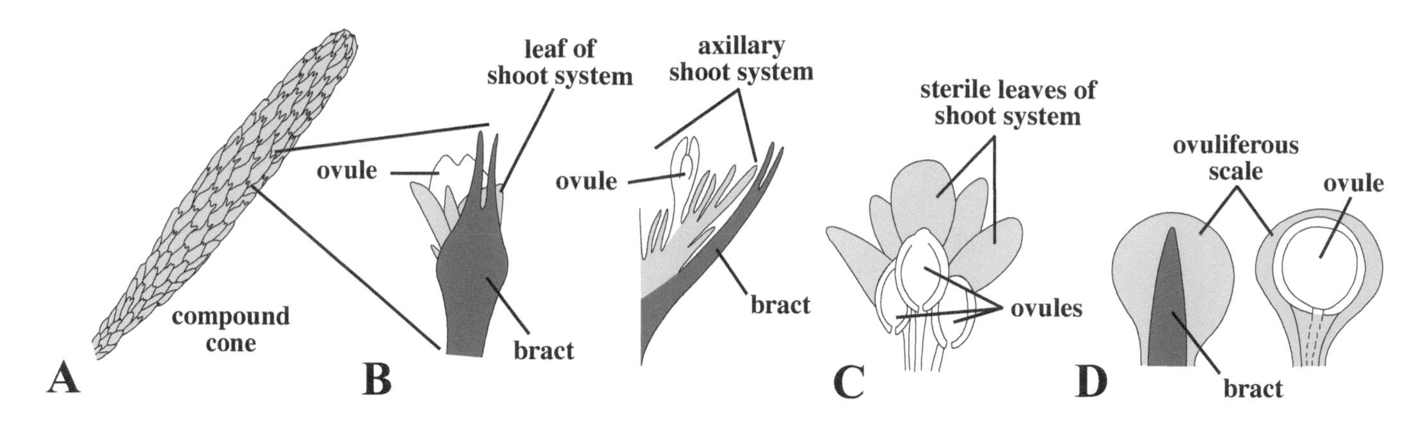

Figure 5.19 Evolution of the compound conifer cone. **A,B.** CORDAITALES, *Lebachia*. **A.** Compound cone, bearing numerous, spirally arranged bracts, each subtending a fertile short shoot. **B.** Close-up of bract and axillary fertile shoot system (left=abaxial view, right=side view), bearing spiral leaves and single ovule. **C,D.** FOSSIL CONIFERS. **C.** *Voltzia*, shoot system of compound cone, with 5 sterile leaves and 3 ovules. **D.** *Ulmannia*, shoot system of compound cone (left=abaxial view, right=adaxial view), showing outer bract and inner fused shoot system (ovuliferous scale) and ovule. (Redrawn from Florin, 1951.)

may function to transport the pollen more efficiently by wind. They may also function as flotation devices, to aid in the capture and transport of pollen grains by a pollination droplet formed in the nonflowering seed plants.

Seed cones of conifers are different from those of other seed plants in having what is known as a **compound** structure, consisting of an axis that bears modified leaves called **bracts**, each of which subtends a separate seed-bearing structure called an **ovuliferous scale** (Figures 5.19, 5.20). The ovuliferous scale is actually a *modified lateral shoot system*, as supported by the fossil record. Members of a fossil group known as the Cordaitales and relatives had seed cones that, like extant conifers, consisted of an axis bearing bracts. However, in the axils of the bracts was a shoot (branch) system, which consisted of an axis bearing sterile leaves and one to many ovules (each usually subtended by a "fertile" leaf). The Swedish botanist Rudolf Florin speculated that this lateral shoot system was evolutionarily modified into the single, dorsi-ventrally flattened ovuliferous scale of extant conifers. Transitions can be seen between fossil Cordaitales in which the axillary shoot system bore numerous, spiral (mostly sterile) leaves (Figure 5.19A,B), fossil conifers in which the leaves of the shoot system were flattened (Figure 5.19C), and other conifers in which the shoot system was flattened and fused into a single structure, the ovuliferous scale (Figure 5.19D). Corroborating evidence for the compound cone hypothesis is that veins of the ovuliferous scales of extant conifers are inverted (upside down), with respect to the bract, indicating that these veins may have been derived from ancestral leaves oriented 180° relative to the bract.

In the seed cones of many conifers, the ovuliferous scales are much larger than the small bracts (Figure 5.20D–F). In a few conifers, e.g., *Pseudotsuga* (Douglas-fir), the bracts are elongated and can be seen on the outside of the ovuliferous scales (Figure 5.20G). Each ovuliferous scale bears from one to many seeds on the upper (adaxial) surface (Figure 5.20H). Mature seeds are often winged (Figure 5.20H,I), an adaptation for seed dispersal by wind. In some conifers, e.g., Podocarpaceae and Taxaceae, the seed cones are greatly reduced such that bracts and ovuliferous scales are only present as developmental remnants or modified to form a fleshy structure, with the entire cone bearing only one ovule.

The conifers are divided here into three groups: Pinopsida (composed solely of the Pinaceae), Cupressopsida (composed of six families) and the Gnetales (composed of three families; see Figure 5.1). The Gnetales and Pinaceae are often sister taxa in molecular analyses, but to date the support for that relationship is not robust. See Page (1990d,e,g,m) for general information on conifers.

PINOPSIDA—PINOPHYTES

The Pinopsida, or pinophytes, consist of a single family.

Pinaceae–Pine family (type *Pinus*, Latin name for pine). 12 genera/225 species (Figures 5.18, 5.20, 5.21).

The Pinaceae consist of resinous, monoecious trees [rarely shrubs]. The **roots** are ectomycorrhizal. The **leaves** are simple, spiral, sessile or short-petiolate, usu. evergreen [deciduous in *Larix* and *Pseudolarix*], linear to long-acicular; photosynthetic leaves in specialized, short shoots in some taxa (*Cedrus, Larix*, and *Pseudolarix*; modified as indeterminate fascicles in *Pinus*), with non-photosynthetic, scale-like leaves sometimes borne on long shoots. The **pollen cones** are small, solitary or clustered, the microsporophylls spiral, each with two abaxial microsporangia. the **pollen** usu. 2-saccate. The **seed cones** are lateral or terminal, usu. woody, sometimes serotinous (not opening at maturity, seed release sometimes induced by fire), the ovuliferous scales spiral, each with usu. two, adaxial

Figure 5.20 CONIFERAE. Pinaceae. **A–F.** *Pinus* spp. **A.** Young seed cone, at time of pollination. **B.** Close-up, showing ovuliferous scales and bracts. Note pollen grains. **C.** One-year-old seed cone. **D.** *Pinus coulteri*, coulter pine, mature seed cone (most massive of any species). **E.** Female pine cones, right in section. **F.** Close-up of longitudinal section, showing bract and ovuliferous scale. **G.** *Pseudotsuga* sp. (Douglas-fir) seed cone. Note elongate bracts and wide ovuliferous scales. **H.** Immature ovuliferous scale, top view, showing two winged seeds. **I.** *Pinus*, mature winged seed.

Figure 5.21 CONIFERAE. Pinaceae. **A,B.** *Abies bracteata*, bristlecone fir. **C.** *Abies concolor*, white fir. **D.** *Abies fabri*, Faber's fir, seed cone. **E.** *Abies grandis*, grand fir, vegetative shoot. **F.** *Cedrus* sp., cedar, with short shoots and erect seed cones. **G.** *Picea orientalis*, oriental spruce. **H–J.** *P. sitchensis*, sitka spruce. **H.** Shoot with stiff, pungent leaves. **I.** Twig with persistent, knob-like, leaf bases (distinctive of spruces). **J.** Seed cone. **K–L.** *Pinus coulteri*, coulter pine, having most massive pine cones. **M.** *Pinus muricata*, with serotinous cones, opening only after fire. **N.** *Pinus lambertiana*, sugar pine, the tallest pine, with longest cones. **O.** *Pinus sabiniana*, gray pine. **P.** *Pseudotsuga menziesii*, douglas-fir, with bracts elongated from ovuliferous scales. **R,S.** *Tsuga mertensiana*, mountain hemlock, with pendant seed cones.

ovules, the subtending bracts free from the ovuliferous scale, bracts sometimes elongate (e.g., *Pseudotsuga*). The **seeds** are usually 2 per ovuliferous scale, inverted, usually winged, the **embryo** with multiple cotyledons; germination is epigeal [rarely hypogeal]. One copy of the inverted repeat of the chloroplast DNA is missing in Pinaceae.

The twelve genera of the Pinaceae are: *Abies* (fir, 46 sp., n. temperate, se. Asia, C. America), *Cathaya* [*Tsuga*] (1 sp., *C. argyrophylla*, China), *Cedrus* (cedar, 2–4 spp., n. Africa to Asia), *Hesperopeuce* [*Tsuga*] (1 sp., *H. mertensiana*, w. North America), *Keteleeria* (3 spp., s. China, Taiwan, s.e. Asia), *Larix* (larch, 10 spp., cool n. hemisphere), *Nothotsuga* [*Tsuga*] (1 sp., *N. longibracteata*, China), *Picea* (spruce, 34 spp., cool n. hemisphere), *Pinus* (pine, 110 spp., n. temperate to South America, Indonesia), *Pseudolarix* (golden-larch, 1 sp., *P. amabilis*, China), *Pseudotsuga* (douglas-fir, 4 spp., e. Asia and w. North America), and *Tsuga* (hemlock, 9 spp., temperate North America and e. Asia).

The Pinaceae is distributed in mostly temperate regions of the northern hemisphere (one *Pinus* spp. entering the s. hemisphere)–in most of North America, West Indies, northern Africa, and much of Eurasia. The family is of great economic importance, including very important lumber/timber trees (uses for electrical/telegraph/telephone poles, many used traditionally in wood ships) and wood pulp trees (used, e. g., in paper production), sources of turpentine, gums, resin (e.g., *Abies balsamea*, balsam fir), oils (used for scent and medicinally), food (seeds of *Pinus* spp., piñon/pignolias), and many other products (often used industrially), plus numerous cultivated ornamentals (including Christmas trees). Certain *Pinus* spp., originally introduced for timber or pulp, have become serious weeds in some areas. *Pinus longaeva*, the bristlecone pine, includes the oldest, single (non-clonal) organisms on earth, some over 5,000 years old. See Page (1990i) for general information, Gernandt et al. (2008) for a phylogenetic analysis, and Grenandt et al. (2005) for a study of the largest genus, *Pinus*.

The Pinaceae are distinctive in being trees [very rarely shrubs] with simple, ***linear to acicular***, ***spiral*** leaves, relatively small pollen cones, with ***two abaxial microsporangia*** per microsporophyll, and seed cones with ***woody, ovuliferous scales***, each usually bearing ***two adaxial, inverted ovules/seeds***, the seeds ***usually winged***, embryos with ***multiple cotyledons***.

CUPRESSOPSIDA—CUPRESSOPHYTES

The Cupressopsida, or cupressophytes (also termed the Cupressophyta) consist here of six families: **Araucariaceae**, **Cupressaceae**, **Phyllocladaceae** (Figure 5.24M–P; sometimes included in Podocarpaceae), **Podocarpaceae**, **Sciadopityaceae**, and **Taxaceae**, four of which are described in detail. The interrelationships of these families and selected apomorphies are seen in Figure 5.1.

Araucariaceae–Araucaria family (type *Araucaria*, after the *Araucani* Indians of Chile, where the type species, *Araucaria araucana*, occurs). 3 genera/32 species (Figure 5.22).

The Araucariaceae consist of monoecious or dioecious trees. The **roots** are endomycorrhizal. The **leaves** are evergreen, simple, spiral or opposite, and broad to acicular. The **pollen cones** are relatively large, with numerous microsporophylls, each with 5–20, pendant (inverted) microsporangia; **pollen** is not saccate. The **seed cones** are large, usually erect, globose to ovoid, falling and disintegrating when seeds mature; ovuliferous scales bear a single ovule, with the bract adnate to the scale. The **seeds** separate from the ovuliferous scale at maturity; the **embryo** has 2–4 cotyledons.

The Araucariaceae consist of the genera *Agathis* (ca. 13 spp., Australasia), *Araucaria* (18 spp., Australasia, S. America), and the monospecific *Wollemia* (*W. nobilis*, Australia).

The Araucariaceae are found mostly in the southern hemisphere, distributed in s. South America, Australasia, and s.e. Asia. Economic importance includes timber and canoe-building trees (e.g., *Agathis australis*, kauri, New Zealand), and several cultivated ornamentals (including *Araucaria araucana*, monkey-puzzle, *A. bidwillii*, bunya-bunya, and *A. heterophylla*, Norfolk Island-pine). Trees of *Araucaria columnaris*, endemic to New Caledonia but planted worldwide as a cultivar, lean to the south in the northern hemisphere and to the north in the southern hemisphere (Johns et al. 2017). Fossilized resin of *Agathis australis* is mined as a copal. *Wollemia nobilis* of New South Wales, Australia is famous as a rare "living fossil." See Page (1990a) for general information and Setoguchi et al. (1998) for a phylogenetic analysis of the family.

The Araucariaceae are distinctive in being ***dioecious or monoecious*** trees with ***broad to acicular*** leaves, the pollen cones large, with ***many (5–20) inverted microsporangia*** per microsporophyll, the seed cones ***large***, ***disintegrating when mature***, each ovuliferous scale bearing a ***single, median*** ovule/seed.

Cupressaceae [including Taxodiaceae]–Cypress family (type *Cupressus*, Latin name for *C. sempervirens*, Italian cypress). ca. 32 genera/130 species (Figure 5.23).

The Cupressaceae consist of resinous, monoecious or dioecious trees or shrubs. The **roots** are vesicular-arbuscular mycorrhizal and bear stout, above ground "knees" in *Taxodium*. The **stems** bear lateral branches that are strongly dorsi-ventrally flattened in some taxa (e.g., *Calocedrus, Libocedrus, Thuja*), less strongly so in others; deciduous leaf-bearing branchlets (resembling pinnate leaves) occur in some taxa (*Glyptostrobus, Metasequoia, Taxodium*). The **leaves** are simple, sessile, petiolate, or decurrent, usu. evergreen, sometimes dimorphic, spiral, opposite-decussate, or in whorls of 3–4, the shape linear,

Figure 5.22 CONIFERAE. Araucariaceae. **A–D.** *Agathis australis*, kauri. **A.** "Tane Mahuta," New Zealand, largest kauri in world (51.5 meters / 169 feet tall); note people below. **B.** Crown of kauri tree, with numerous epiphytes. **C.** Kauri leaves near tree base, lanceolate. **D.** Male and seed cones. (Image contributed by John Braggins.). **E–G.** *Araucaria bidwillii*, bunya-bunya. **E.** Vegetative shoot, with coriaceous, spinose leaves. **F.** Large, dried seed cone. **G.** Ovuliferous scale from seed cone, bearing single seed. **H.** *Araucaria cunninghamii*, vegetative shoot, with small, dense, subulate leaves. **I–L.** *Araucaria heterophylla*, Norfolk Island-pine, showing narrow, ultimate branches bearing linear leaves.

Figure 5.23 CONIFERAE. Cupressaceae. **A.** *Cupressus macrocarpa*, Monterey cypress. Inset: seed cones. **B,C.** *Cupressus sempervirens*, Italian cypress. **B.** Shoot with seed cones. **C.** Detached ovuliferous scale bearing numerous seeds. **D,E.** *Juniperus osterosperma*, Utah juniper. **D.** Vegetative shoot, showing decussate, scale-like leaves. **E.** Seed cones, with fleshy ovuliferous scales. **F.** *Taxodium distichum*, bald cypress, with deciduous branch systems and seed cones. **G,H.** *Metasequoia glyptostroboides*, dawn-redwood, showing seed cone and deciduous branch system with opposite, simple leaves. **I–K.** *Sequoia sempervirens*, redwood, with evergreen, flattened branch system bearing linear leaves and mature seed cone. **L–O.** *Sequoiadendron giganteum*, giant sequoia. **L.** Tree. **M.** Branches with subulate leaves. **N.** Mature seed cones. **O.** Ovuliferous scale, adaxial surface facing, bearing several winged seeds.

acicular, or deltoid-subulate (scale-like in mature plants, often completely covering younger shoots). The **pollen cones** are terminal, solitary [rarely in clusters], with 2–10 abaxial microsporangia per microsporophyll; the **pollen** is not saccate. The **seed cones** are usu. terminal, solitary [rarely clustered], woody, coriaceous, or fleshy (e.g., "berries" of *Juniperus* spp.); the ovuliferous scales are spiral, opposite, or in whorls of 3, each with usu. several (2–20) adaxial ovules, the bract adnate to the ovuliferous scale. The **seeds** are erect or inverted, winged or not, the **embryo** usu. with 2 [–15] cotyledons.

Notable among the 32 genera in the family are *Hesperocyparis* [*Cupressus*] (ca. 16 species, w. hemisphere), *Calocedrus* (incense-cedars, 3 spp., s.e. Asia and w. North America), *Chamaecyparis* (white cedars, 5 spp., e. Asia and North America), *Cupressus* (ca. 12 species, e. hemisphere), *Juniperus* (junipers, ca. 50 species, N. Hemisphere and tropical Africa mountains), *Metasequoia* (monospecific, *M. glyptostroboides*, dawn-redwood, China), *Sequoia* (monospecific, *S. sempervirens*, redwood, w. North America), *Sequoiadendron* (monospecific, *S. giganteum*, giant sequoia, California, USA), *Taxodium* (bald-cypress, 2 spp., e. North America, Mexico), and *Thuja* (arbor vitae/cedar, 5 spp., e. Asia and North America). *Sequoia sempervirens*, the redwood of California and Oregon (USA), includes the tallest living tree in the world (tallest cited as ca. 116 meters/380 feet). *Sequoiadendron giganteum*, the giant sequioia of California (USA), includes the largest, single (non-clonal) organism in the world (cited as 1,487 cubic meters/52,513 cubic feet).

Members of the Cupressaceae have a worldwide distrbution. Economic importance includes important timber trees (many taxa having rot- and termite-resistant wood), resin and flavoring plants, and numerous cultivated ornamentals, especially *Juniperus* (junipers) and *Cupressus* spp. (cypresses, e.g., *C. sempervirens*, Italian Cypress).

The Taxodiaceae (leaves mostly alternate) were formerly segregated from the Cupressaceae s.s. (leaves decussate or whorled). However, the Taxodiaceae, if treated separately, is paraphyletic, with the Cupressaceae s.s. nested within it; combining the two results in a larger, monophyletic Cupressaceae s.l. See Page (1990c,l) for general information, Farjon (2005) for a monographic treatment and Brunsfeld et al. (1994), Gadek et al. (2000), and Kusumi et al. (2000) for phylogenetic studies of the complex.

The Cupressaceae are distinguished in being monoecious or dioecious trees or shrubs with spiral, decussate, or whorled ***deltoid-subulate, linear, or acicular*** leaves (in flattened or deciduous branchlets in some taxa), the pollen cones usu. with ***multiple (2–10) microsporangia*** per microsporophyll, ovuliferous scales ***opposite or in whorls of 3***, ovules usu. ***several per scale***, embryos usu. with 2 cotyledons.

Podocarpaceae–Podocarp family (type *Podocarpus*, Greek *podos*, foot, + *karpos*, fruit, from the fleshy "receptacle" subtending the seed). 17 genera/ca. 167 species (Figure 5.24A–L).

The Podocarpaceae consist of resinous, usu. dioecious [rarely monoecious], trees [rarely shrubs]. The **leaves** are simple, spiral [rarely decussate or subopposite], linear, elliptic, or subulate to scale-like. The **pollen cones** are terminal or axillary, solitary or clustered, often "catkin-like" with an elongate axis bearing numerous microsporophylls, each with two microsporangia; **pollen** is usu. 2- [0,3-] saccate. The **seed cones** are terminal or axillary, cone-like or highly reduced, usu. fleshy [rarely dry], in some taxa subtended by a stalk ("peduncle") that may fuse with bracts, forming a fleshy "receptacle," ovuliferous scales 1–∞, each bearing a single ovule, in some members a single scale modified as a protective covering around the ovule, the "epimatium." The **seeds** are 1 [–∞], inverted or erect, often protruding, in some taxa with a fleshy, sometimes colorful epimatium (modified ovuliferous scale) and bract (termed the "carpidium"), and basal stalk-like region ("receptacle"), the fleshy tissue functioning in bird dispersal; **embryo** with 2 cotyledons.

Notable among the genera of the Podocarpaceae are *Dacrydium* (25 spp., s.e. Asia to New Zealand) and *Podocarpus* (94 spp., s. temperate, e. Asia, West Indies). The genus *Phyllocladus* (5 spp., s.e. Asia, Australasia), with interesting phyllodes/phylloclades (Figure 5.24M–P), is often included within the Podocarpaceae, but is here treated as the monogeneric Phyllocladaceae, the sister group to the Podocarpaceae.

Members of the Podocarpaceae are predominately distributed in the southern hemisphere (mainly Australasia to s.e. Asia, ranging to Japan), but are also found in Central and South America and tropical montane Africa. Economic importance includes several important timber or pulp trees (some traditionally used to make canoes), edible seed cones, and cultivated ornamentals (e.g., *Dacrydium, Podocarpus*). *Parasitaxus usta* of New Caledonia is the only known parasitic gymnosperm, being a root parasite on another family member (*Falcatifolium taxoides*). See Page (1990h,j) for general information and Kelch (1998) for a phylogenetic treatment of the family.

The Podocarpaceae are distinctive in being usu. ***dioecious*** trees, with linear, elliptic, or subulate to scale-like leaves, the seed cones with ***ovuliferous scales bearing 1 ovule***, entire cone often reduced to ***one seed***, sometimes borne on a ***fleshy receptacle***, the seed often enveloped by a ***fleshy epimatium and carpidium***, the embryo with two cotyledons.

Taxaceae [including Cephalotaxaceae]–Yew family (type *Taxus*, Latin for yew). 6 genera/28 species (Figure 5.25).

The Taxaceae consist of resinous or non-resinous, dioecious or monoecious, trees or shrubs. The **leaves** are simple, spiral

Figure 5.24 CONIFERAE. **A–L.** Podocarpaceae. **A–C.** *Dacrydium cupressinum*, rimu, with pendulous branches covered with simple leaves. **D,E.** *Dacrycarpus dacrydioides*, kahikatea. **D.** Pollen cones, with numerous, spiral microphylls. **E.** Seed cones, each bearing one purple seed atop fleshy, red receptacle. **F.** *Halocarpus bidwillii*. bog pine, showing seed, covered by black epimatium and carpidium, atop white receptacle. **G,H.** *Podocarpus gracilior.* **G.** Pollen cones. **H.** Seed cone with single seed enclosed by epimatium. **I.** *Podocarpus macrophyllus*, seed cones. **J–L.** *Podocarpus totara*. **J.** Trunk. **K.** Branch. **L.** Seed cones, red epimatium below seed. **M–P.** Phyllocladaceae, with flattened branch systems (phyllodes), resembling leaves. **M,N.** *Phyllocladus glaucus*. **O,P.** *Phyllocladus trichomanoides*. (Images D–F, L courtesy of John Braggins.)

Figure 5.25 CONIFERAE. Taxaceae, *Taxus* sp., yew. **A.** Branch with bearing pollen cones. **B.** Branch with seed cone of single seed surrounded by red, fleshy aril (open at apex). **C.** *Torreya californica*, California-nutmeg, seed surrounded by totally enclosing aril.

(often twisted to appear 2-ranked) to decussate, evergreen, linear to acicular, typically decurrent. The **pollen cones** are small, solitary or clustered, microsporophylls peltate, each bearing 2–16 microsporangia, the **pollen** not saccate. The **seed cones** are reduced to one ovule (in *Cephalotaxus*, seed cones with decussate bracts, each subtending 2 ovules). The **seed** is erect, 1 per cone, unwinged, partly or entirely enclosed by a fleshy, usu. brightly colored aril, the **embryo** with 2 cotyledons.

The six genera of the Taxaceae are: *Amentotaxus* (4 spp., China), *Austrotaxus* (1 sp., New Caledonia), *Cephalotaxus* (6 spp., Asia), *Pseudotaxus* (1 sp., China), *Taxus* (yews, 9 spp., n. temperate, s.e. Asia), and *Torreya* (7 spp., e. Asia, USA).

Members of the Taxaceae are distributed in North America, n. Africa, and Eurasia to s.e. Asia. Economic importance includes timber trees, some local medicinal plants, and cultivated ornamentals (e.g., *Taxus* spp., yew); taxol, derived from *Taxus brevifolia*, is used to treat ovarian cancer. *Cephalotaxus*, included in the Taxaceae here, is often treated as a separate, monogeneric family. See Page (1990b,k) for general information and Hao et al. (2008) for a phylogenetic analysis of the family.

The Taxaceae are distinctive in being ***evergreen*** trees or shrubs with linear to acicular, spiral to decussate leaves, the pollen cones with ***numerous microsporangiophores*** each bearing ***many (2–16) microsporangia***, the mature seed cones ***reduced to one seed***, the seed ***arillate***, cotyledons two.

GNETALES—GNETOPHYTES

The Gnetales or gnetophytes, also referred to as the Gnetopsida or Gnetophyta, are an interesting group containing three extant families: Ephedraceae (consisting solely of *Ephedra*, with ca. 40 species), Gnetaceae (consisting solely of *Gnetum* [including *Vinkiella*], with ca. 30 species), and the Welwitschiaceae (monospecific, consisting of *Welwitschia mirabilis*). The Gnetales has often been thought to be the sister group to the angiosperms, the two groups united by some obscure features, possibly including whorled, somewhat "perianth-like" microsporophylls in structures that may resemble flowers (see Chapter 6). However, as reviewed earlier, recent molecular studies have placed the Gnetales within the conifers, sister either to the Pinopsida or to the Cupressopsida (Figure 5.1).

The Gnetales are united by (among other things) the occurrence of (1) striate pollen (Figure 5.26A); and (2) vessels with porose (porelike) perforation plates (Figure 5.26B), as opposed to scalariform (barlike) perforation plates in basal angiosperms (see Chapter 6). The vessels of Gnetales were derived independently from those of angiosperms. The reproductive structures in various Gnetales show some parallels to the flowers of angiosperms.

Species of *Gnetum* of the Gnetaceae are tropical vines (rarely trees or shrubs) with opposite (decussate), simple leaves (Figure 5.27A), looking like an angiosperm but, of course, lacking true flowers. *Welwitschia mirabilis* of the Welwitschiaceae is a strange plant native to deserts of Namibia in southwestern Africa. An underground caudex bears only two leaves (Figure 5.27B), these becoming quite long and lacerated in old individuals. Male and seed cones are born on axes arising from the apex of the caudex (Figure 5.27C–G). *Ephedra* of the Ephedraceae is a rather common desert shrub (Figure 5.28) and can be recognized by the photosynthetic, striate stems and the very reduced scale-like leaves, only two or three per node. Pollen or seed cones may be found in the axils of the leaves (Figure 5.24; see family description). See Kubitzki (1990a,b,c,d) for information on the Gnetales.

Recently, the occurrence of a type of **double fertilization** was verified in species of the Gnetales. Double fertilization in *Ephedra* entails the fusion of each of two sperm cells from a

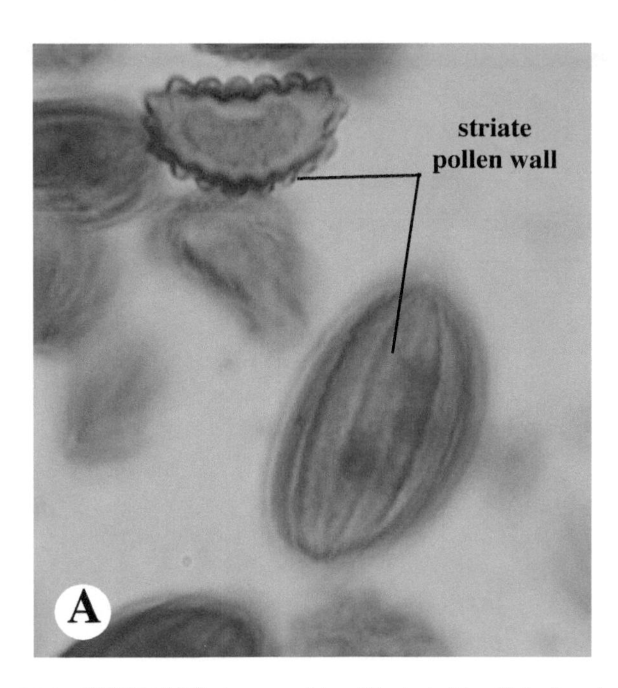

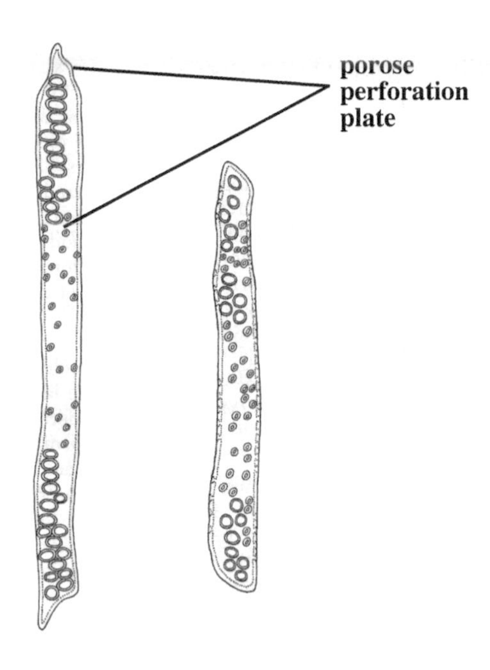

Figure 5.26 GNETALES. Apomorphies, illustrated by *Ephedra*. **A.** Striate pollen grains, face view below, cross section above. **B.** Vessels with porose perforation plates. (**B** reproduced from Evert, R. F. 2006. Esau's Plant Anatomy. 3rd ed. John Wiley and Sons, by permission.)

male gametophyte with nuclei in the archegonium of the female gametophyte. One sperm fuses with the egg nucleus and the other fuses with the ventral canal nucleus. In fact, the fusion product of sperm and ventral canal cell may even divide a few times mitotically, resembling angiospermous endosperm (Chapter 6), but this does not persist. Thus, double fertilization, which has long been viewed as a defining characteristic of the angiosperms alone, was recently interpreted as a possible apomorphy of the Gnetales and angiosperms together (formerly called the "Anthophytes"). This notion is rejected with the current acceptance of seed plant relationships as seen in Figure 5.1, in which the Gnetales are nested within the conifers. Thus, double fertilization in the Gnetales and angiosperms presumably evolved independently.

Ephedraceae– Mormon Tea family (type *Ephedra*, after Greek name for *Hippuris*, which resembles *Ephedra*). 1 genus/35–45 species (Figure 5.28).

The Ephedraceae consist of xeromorphic, usu. dioecious, shrubs, vines, or small trees. The **stems** underground are often rhizomatous; young aerial stems are narrow, striate, and photosynthetic. The **leaves** are reduced, being sessile, simple, opposite or in whorls of 3 [4], deltoid to subulate (usu. scale-like, becoming non-photosynthetic). The **pollen cones** are axillary on aerial shoots, each consisting of an axis bearing several pairs of decussate bracts (lowermost bracts usu. sterile); most upper bracts subtend a stalk-like microsporangiophore (also termed a microsporophyll) bearing 2–8 apical synangia. Each synangium contains 2 [4] poricidally dehiscent microsporangia; within the bract of the cone, the microsporangiophores are basally enclosed by two, connate bracteoles (sometimes termed a "perianth"). **Pollen** is striate, not saccate; the exine is shed after pollination (so male gametophytes are "naked"). The **seed cones** are axillary on aerial shoots, each consisting of an axis with 2–8 pairs of bracts (the lowermost bracts sterile, sometimes fleshy). The cones bear 1–3 ovules, one in the axil of each of the upper bracts. Each ovule is enveloped by two tissue layers: an outer layer (sometimes termed an "outer envelope"), usually interpreted as a pair of connate bracteoles, and an inner layer, the integument, which forms an apical "micropylar tube" that protrudes from the outer layer and receives the pollen. The mature **seeds** are gen. 1 or 2 per cone, either dry and winged or fleshy and colored; the **embryo** has 2 cotyledons. Plants are wind pollinated, although some are insect visited, obtaining a nectar-like secretion from the micropyle. Calcium oxalate crystals occur in intercellular spaces of the wood.

Members of the Ephedraceae are distributed in s.w. North America, w. South America, n. Africa, and Eurasia. Economic importance includes a traditional use as a tea (Mormon tea) in s.w. North America. The seeds of some species are edible. Some species are used medicinally for various ailments; "ma huang" (from *E. sinica*) has been used in China for many centuries. The alkaloid ephedrine has (among others) appetite supression, anti-asthma, and stimulant properties and has been used in weight loss products (but now largely banned because of harmful side effects). Some species are used as cultivated

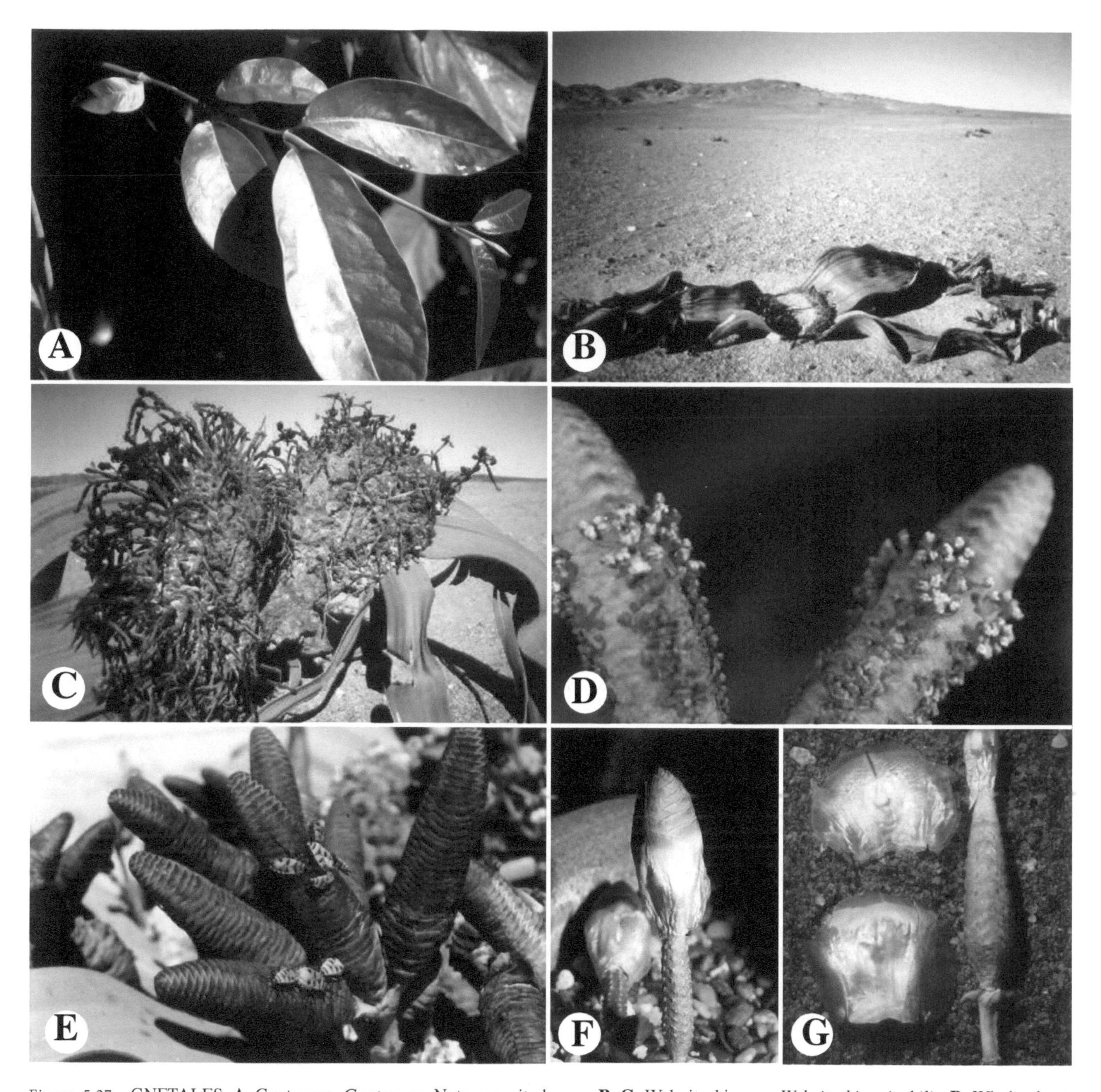

Figure 5.27 GNETALES. **A.** Gnetaceae. *Gnetum* sp. Note opposite leaves. **B–G.** Welwitschiaceae. *Welwitschia mirabilis*. **B.** Whole plant, showing two elongate opposite leaves. **C.** Close-up of central region of plant. **D.** Pollen cones. **E–G.** Seed cones. [**B–E.** contributed by Mark Olson]

ornamentals. See Kubitzki (1990a), Price (1996), and Rydin et al. (2004) for information on the morphology and phylogeny of the group.

The Ephedraceae are distinguished in being mostly ***dioecious shrubs, vines, or small trees*** with ***narrow, striate, photosynthetic*** aerial stems, the leaves ***scale-like, opposite or whorled***, the pollen cones with decussate bracts subtending **microsporangiophores**, each bearing apical ***synangia*** and subtended by an ***outer bract and two, inner connate bracteoles***, the seed cones bearing 1–3 ovules, each ovule subtended by a bract and enclosed by an ***outer layer ("envelope") of connate bracteoles*** and an ***inner integument***, the latter forming a protruding ***pollination tube***, the seeds winged or fleshy.

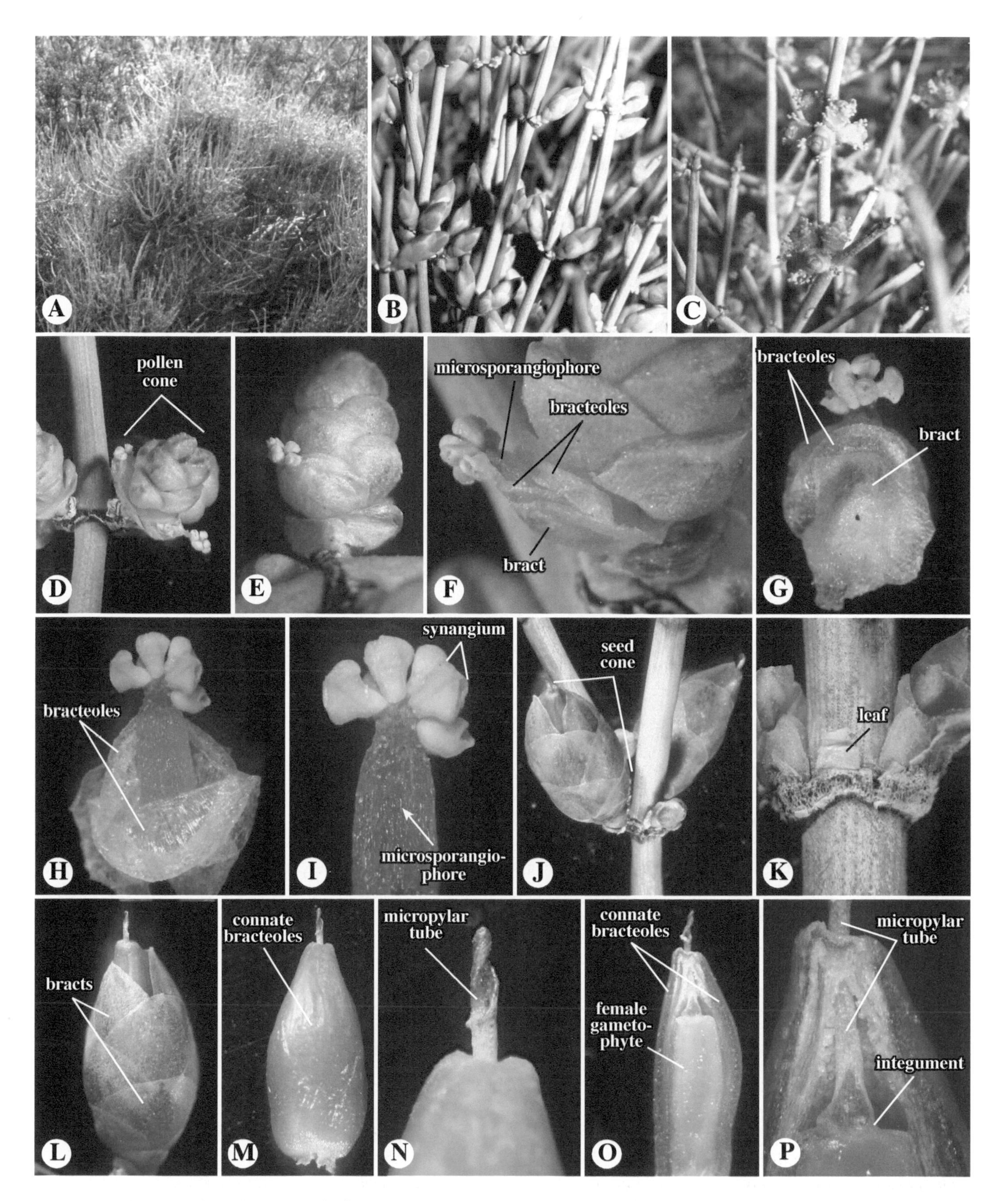

Figure 5.28 GNETALES. Ephedraceae. **A–C.** *Ephedra* sp. **A.** Whole plant. **B.** Female plant with cones. **C.** Male plant with cones. **D–P.** *Ephedra aspera*. **D–I.** Pollen cones, showing bracts and bracteoles subtending microsporangiophore, bearing synangia. **J–L.** Seed cones. Note bracts. **M–P.** Seed morphology. Note seed subtended by connate bracteoles and bearing extended micropylar tube.

REVIEW QUESTIONS

LIGNOPHYTES—WOODY PLANTS

1. What are the major evolutionary novelties for the lignophytes?
2. Describe the cell divisions of a vascular cambium during secondary growth.
3. What are the products of secondary growth of the vascular cambium? the cork cambium?

SPERMATOPHYTES—SEED PLANTS

4. Define seed and ovule.
5. Including heterospory, name and describe the steps that were involved in the evolution of the seed.
6. Based on fossil evidence, what was the precursor of the integument and micropyle?
7. Define and state the significance of the pollination droplet.
8. What is the definition of a pollen grain? From what does it develop?
9. What is a pollen tube and how does it function?
10. What is the difference between zooidogamy and siphonogamy, and for what groups were these evolutionary novelties?
11. Review the stages of ovule and seed development, and explain the lag period between pollination and fertilization.
12. Name four ways that seeds are adaptive.
13. Name and describe the stem stelar type that is an apomorphy for all extant seed plants.

SEED PLANT DIVERSITY

14. What were the basic features of *Archeopteris*?
15. What is a "pteridosperm" (seed fern)? Name a genus of the "seed ferns."
16. What is a gymnosperm, why are they called that, and what major groups are included within extant gymnosperms?
17. What group of seed plants is characterized by generally short trunks, pinnate (rarely bipinnate), coriaceous leaves with circinate vernation, dioecy, and motile sperm?
18. What is the definition of a cone (strobilus)? What are the parts of a seed cone? A pollen cone?
19. What are the diagnostic features of the Cycadaceae? Name an economically imporant member and its use.
20. How does the Zamiaceae differ with regard to reproductive morphology and genus diversity from the Cycadaceae?
21. What group/species is a dioecious tree with short shoots, obtriangular leaves with dichotomous venation?
22. What is the definition of a pine fascicle?
23. What is the morphology of some conifer pollen grains? What is the possible function of this morphology?
24. What is the name of the structure in a female pine cone that directly bears the ovules/seeds? From what was it evolutionarily derived? What subtends this structure?
25. What are the diagnostic features of the Pinaceae? Name several genera (scientific and common names) in the family.
26. How do the Araucariaceae and Cupressaceae vary with respect to leaf morphology, microsporangia number, and ovule number per scale? Name two important species of each family?
27. What is distinctive about the Podocarpaceae with respect to seed cone structure?
28. What layer covers the seeds of the Taxaceae and what is its function?
29. Name two apomorphies for the Gnetales.
30. Review the diagnostic features of the Ephedraceae. What is the family's common name? Economic importance?
31. Name the other two families and genera of the Gnetales. What do they look like and where do they occur?

EXERCISES

1. Peruse the most recent literature on phylogenetic relationships of the seed plants. Are there any differences relative to Figure 5.1?
2. Peruse botanical journals and find a systematic article on a member of the Cycadophyta, Ginkgophyta, or Coniferae (including Gnetales). What is the objective of the article and what techniques were used to address it?
3. Collect and identify several local conifers. What features are used to distinguish between families, genera, or species?

REFERENCES FOR FURTHER STUDY

Bierhorst, David W. 1971. Morphology of Vascular Plants. Macmillan, New York.

Bowe, L. Michelle, Gwénaële Coat, and Claude W. dePamphilis. 2000. Phylogeny of seed plants based on all three genomic compartments: Extant gymnosperms are monophyletic and Gnetales' closest relatives are conifers. Proceedings of the National Academy of Sciences of the United States of America 97: 4092–4097.

Brunsfeld, S. J., P. S. Soltis, D. E. Soltis, P. A. Gadek, C. J. Quinn, D. D. Strenge, and T. A. Ranker. 1994. Phylogenetic relationships among the genera of Taxodiaceae and Cupressaceae: Evidence from *rbc*L sequences. Systematic Botany 19: 253-262.

Cantino, P. D., J. A. Doyle, S. W. Graham, W. S. Judd, R. G. Olmstead, D. E. Soltis, P. S. Soltis, and M. J. Donoghue. 2007. Towards a phylogenetic nomenclature of Tracheophyta. Taxon 56: 822-846.

Chamberlain, C. J. 1935. Gymnosperms: Structure and Evolution. University of Chicago Press, Chicago.

Chaw, Shu-Miaw, Christopher L. Parkinson, Yuchang Cheng, Thomas M. Vincent, and Jeffrey D. Palmer. 2000. Seed plant phylogeny inferred from all three plant genomes: Monophyly of extant gymnosperms and origin of Gnetales from conifers. Proceedings of the National Academy of Sciences of the United States of America 97: 4086–4091.

Crane, Peter. 1985. Phylogenetic relationships in seed plants. Cladistics 1(4): 329–348.

Doyle, James A., and Michael J. Donoghue. 1986. Seed plant phylogeny and the origin of angiosperms: an experimental cladistic approach. The Botanical Review 52(4): 321–431.

Doyle, J. A. 2006. Seed ferns and the origin of angiosperms. Journal of the Torrey Botanical Society 133: 169-209.

Doyle, J. A. 2012. Molecular and fossil evidence on the origin of angiosperms. Annual Review of Earth and Planetary Science 40:301-326.

Farjon, A. 2005. A Monograph of Cupressaceae and Sciadopitys. Royal Botanic Gardens, Kew.

Florin, R. 1951. Evolution in Cordaites and Conifers. Acta Horti Bergiani 15:285-388.

Foster, A. S., and E. M. Gifford. 1974. Comparative Morphology of Vascular Plants, 2nd edition. W. H. Freeman, San Francisco.

Frohlich, Michael W., and David S. Parker. 2000. The mostly male theory of flower evolutionary origins: from genes to fossils. Syst. Bot. 25(2): 155-170.

Gadek, P. A., D. L. Alpers, M. M. Heslewood, and C. J. Quinn. 2000. Relationships within the Cupressaceae sensu lato: A combined morphological and molecular approach. American Journal of Botany 87: 1044-1057.

Gernandt, D. S., G. G. López, S. O. García, and A. Liston. 2005. Phylogeny and classification of *Pinus*. Taxon 54:29–42.

Gernandt, D. S., S. Magallón, G. G. López, O. Z. Flores, A. Willyard, and A. Liston. 2008. Use of simultaneous analyses to guide fossil-based calibrations of Pinaceae phylogeny. International Journal of Plant Sciences 169: 1986-1099.

Gifford, E. M., and A. S. Foster. 1989. Morphology and evolution of vascular plants, 3rd edition. W.H. Freeman and Co., New York.

Gugerli, F., C. Sperisen, U. Buchler, I. Brunner, S. Brodbeck, J. D. Palmer, and Y.-L. Qiu. 2001. The evolutionary split of Pinaceae from other conifers: evidence from an intron loss and a multigene phylogeny. Molecular Phylogenetics and Evolution 21: 167–175.

Hao, D. C., P. G. Xiao, B. L. Huang, G. B. Ge, and L. Yang. 2008. Interspecific relationships and origins of Taxaceae and Cephalotaxaceae revealed by partitioned Bayesian analyses of chloroplast and nuclear DNA sequences. Plant Systematics and Evolution 276: 89-104.

Hill, K. D., M. W. Chase, D. W. Stevenson, H. G. Hills, and B. Schutzman. 2003. The families and genera of cycads: a molecular phylogenetic analysis of Cycadophyta based on nuclear and plastid DNA sequences. International Journal of Plant Sciences 164: 933–948.

Hill, K. D., D. W. Stevenson, and R. Osborne. 2004. The world list of cycads. Botanical Review 70: 274-298.

Hill, K. D. 2005. Diversity and evolution of gymnosperms. In: Henry, R. J. (ed.), Plant Diversity and Evolution: Genotypic and Phenotypic Variation in Higher Plants. Pp. 25-44. CABI International, Wallingford.

Jiao, Y., N. J. Wickett, S. Ayyampalayam, A. S. Chanderbali, L. Landherr, P. E. Ralph, L. P. Tomsho, Y. Hu, H. Liang, P. S. Soltis, D. E. Soltis, S. W. Clifton, S. E. Schlarbaum, S. C. Schuster, H. Ma, J. Leebens-Mack, and C. W. dePamphilis. 2011. Ancestral polyploidy in seed plants and angiosperms. Nature 473:97.

Johns, J. W., J. M. Yost, D. Nicolle, B. Igic, and M. K. Ritter. 2017. Worldwide hemisphere-dependent lean in Cook pines. Ecology 98:2482-2484.

Johnson, L. A. S., and K. L. Wilson. 1990a. Boweniaceae. In: K. U. Kramer and P. S. Green (eds.). The Families and Genera of Vascular Plants. I. Pteridophytes and Gymnosperms. Pp. 369-370. Springer-Verlag, Berlin.

Johnson, L. A. S., and K. L. Wilson. 1990b. Cycadaceae. In: K. U. Kramer and P. S. Green (eds.). The Families and Genera of Vascular Plants. I. Pteridophytes and Gymnosperms. Pp. 370. Springer-Verlag, Berlin.

Johnson, L. A. S., and K. L. Wilson. 1990c. General Traits of the Cycadales. In: K. U. Kramer and P. S. Green (eds.). The Families and Genera of Vascular Plants. I. Pteridophytes and Gymnosperms. Pp. 363-368. Springer-Verlag, Berlin.

Johnson, L. A. S., and K. L. Wilson. 1990d. Stangeriaceae. In: K. U. Kramer and P. S. Green (eds.). The Families and Genera of Vascular Plants. I. Pteridophytes and Gymnosperms. Pp. 370-371. Springer-Verlag, Berlin.

Johnson, L. A. S., and K. L. Wilson. 1990e. Zamiaceae. In: K. U. Kramer and P. S. Green (eds.). The Families and Genera of Vascular Plants. I. Pteridophytes and Gymnosperms. Pp. 371-377. Springer-Verlag, Berlin.

Jones, David L. 1993. Cycads of the World: Ancient Plants in Today's Landscape. Smithsonian Institution Press, Washington, DC.

Kelch, D. G. 1998. Phylogeny of Podocarpaceae: Comparison of evidence from morphology and 18S rDNA. American Journal of Botany 85: 986-996.

Kubitzki, K. 1990a. Ephedraceae. In: K. U. Kramer and P. S. Green (eds.). The Families and Genera of Vascular Plants. I. Pteridophytes and Gymnosperms. Pp. 379-382. Springer-Verlag, Berlin.

Kubitzki, K. 1990b. General Traits of the Gnetales. In: K. U. Kramer and P. S. Green (eds.). The Families and Genera of Vascular Plants. I. Pteridophytes and Gymnosperms. Pp. 378-379. Springer-Verlag, Berlin.

Kubitzki, K. 1990c. Gnetaceae. In: K. U. Kramer and P. S. Green (eds.). The Families and Genera of Vascular Plants. I. Pteridophytes and Gymnosperms. Pp. 383-386. Springer-Verlag, Berlin.

Kubitzki, K. 1990d. Welwitschiaceae. In: K. U. Kramer and P. S. Green (eds.). The Families and Genera of Vascular Plants. I. Pteridophytes and Gymnosperms. Pp. 387-391. Springer-Verlag, Berlin.

Kusumi, J., Y. Tsumura, H. Yoshimaru, and H. Tachida. 2000. Phylogenetic relationships in Taxodiaceae and Cupressaceae sensu stricto based on *mat*K gene, *chl*L gene, *trn*L-*trn*F IGS region, and *trn*L intron sequences. American Journal of Botany 87:1480-1488.

Li, Z., A. E. Baniaga, E. B. Sessa, M. Scascitelli, S. W. Graham, L. H. Rieseberg, and M. S. Barker. 2015. Early genome duplications in conifers and other seed plants. Science Advances 1:e1501084.

Magallón, S., K. W. Hilu, and D. Quandt. 2013. Land plant evolutionary timeline: Gene effects are secondary to fossil constraints in relaxed clock estimation of age and substitution rates. American Journal of Botany 100:556-573.

Mathews, S. 2009. Phylogenetic relationships among seed plants: persistent questions and the limits of molecular data. American Journal of Botany 96: 228–236.

Mathews, S., D. Clements Mark, and A. Beilstein Mark. 2010. A duplicate gene rooting of seed plants and the phylogenetic position of flowering plants. Philosophical Transactions of the Royal Society B: Biological Sciences 365:383-395.

Nixon, Kevin C., William L. Crepet, Dennis Stevenson, and Else Marie Friis. 1994. A reevaluation of seed plant phylogeny. Annals of the Missouri Botanical Garden 81: 484–533.

Norstog, K. J. and T. J. Nicholls. 1997. The Biology of the Cycads. Cornell University Press, Ithaca.

Page, C. N. 1990a. Araucariaceae. In: K. U. Kramer and P. S. Green (eds.). The Families and Genera of Vascular Plants. I. Pteridophytes and Gymnosperms. Pp. 294-299. Springer-Verlag, Berlin.

Page, C. N. 1990b. Cephalotaxaceae. In: K. U. Kramer and P. S. Green (eds.). The Families and Genera of Vascular Plants. I. Pteridophytes and Gymnosperms. Pp. 299-302. Springer-Verlag, Berlin.

Page, C. N. 1990c. Cupressaceae. In: K. U. Kramer and P. S. Green (eds.). The Families and Genera of Vascular Plants. I. Pteridophytes and Gymnosperms. Pp. 302-316. Springer-Verlag, Berlin.

Page, C. N. 1990d. Economic Importance and Conifer Conservation. In: K. U. Kramer and P. S. Green (eds.). The Families and Genera of Vascular Plants. I. Pteridophytes and Gymnosperms. Pp. 293-294. Springer-Verlag, Berlin.

Page, C. N. 1990e. General Traits of Conifers. In: K. U. Kramer and P. S. Green (eds.). The Families and Genera of Vascular Plants. I. Pteridophytes and Gymnosperms. Pp. 290-292. Springer-Verlag, Berlin.

Page, C. N. 1990f. Ginkgoaceae. In: K. U. Kramer and P. S. Green (eds.). The Families and Genera of Vascular Plants. I. Pteridophytes and Gymnosperms. Pp. 284-289. Springer-Verlag, Berlin.

Page, C. N. 1990g. Interrelationships Between Families of Conifers. In: K. U. Kramer and P. S. Green (eds.). The Families and Genera of Vascular Plants. I. Pteridophytes and Gymnosperms. Pp. 292-293. Springer-Verlag, Berlin.

Page, C. N. 1990h. Phyllocladaceae. In: K. U. Kramer and P. S. Green (eds.). The Families and Genera of Vascular Plants. I. Pteridophytes and Gymnosperms. Pp. 317-319. Springer-Verlag, Berlin.

Page, C. N. 1990i. Pinaceae. In: K. U. Kramer and P. S. Green (eds.). The Families and Genera of Vascular Plants. I. Pteridophytes and Gymnosperms. Pp. 319-331. Springer-Verlag, Berlin.

Page, C. N. 1990j. Podocarpaceae. In: K. U. Kramer and P. S. Green (eds.). The Families and Genera of Vascular Plants. I. Pteridophytes and Gymnosperms. Pp. 332-346. Springer-Verlag, Berlin.

Page, C. N. 1990k. Taxaceae. In: K. U. Kramer and P. S. Green (eds.). The Families and Genera of Vascular Plants. I. Pteridophytes and Gymnosperms. Pp. 348-353. Springer-Verlag, Berlin.

Page, C. N. 1990l. Taxodiaceae. In: K. U. Kramer and P. S. Green (eds.). The Families and Genera of Vascular Plants. I. Pteridophytes and Gymnosperms. Pp. 353-361. Springer-Verlag, Berlin.

Page, C. N. 1990m. Taxonomic Concepts in Conifers and Ginkgoids. In: K. U. Kramer and P. S. Green (eds.). The Families and Genera of Vascular Plants. I. Pteridophytes and Gymnosperms. Pp. 282. Springer-Verlag, Berlin.

Penny, D., O. Deusch, R. A. Atherton, B. Zhong, P. J. Lockhart, S. V. Nikiforova, V. V. Goremykin, and P. J. Biggs. 2011. Systematic Error in Seed Plant Phylogenomics. Genome Biology and Evolution 3:1340-1348.

Price, R. A. 1996. Systematics of the Gnetales: A review of morphological and molecular evidence. International Journal of Plant Sciences 157 (6, suppl.): S40-S49.

Rai, H. S., H. E. O'Brien, P. A. Reeves, R. G. Olmstead, and S. W. Graham. 2003. Inference of higher-order relationships in the cycads from a large chloroplast data set. Molecular Phylogenetics and Evolution 29: 350–359.

Rothwell, Gar W., and Rudolph Serbet. 1994. Lignophyte phylogeny and the evolution of spermatophytes: a numerical cladistic analysis. Systematic Botany 19: 443–482.

Ruhfel, B. R., M. A. Gitzendanner, P. S. Soltis, D. E. Soltis, and J. G. Burleigh. 2014. From algae to angiosperms - inferring the phylogeny of green plants (Viridiplantae) from 360 plastid genomes. BMC Evolutionary Biology 14:23.

Rydin, C., M. Källersjö, and E. M. Friis. 2002. Seed plant relationships and the systematic position of Gnetales based on nuclear and chloroplast data: Conflicting data, rooting problems, and the monophyly of conifers. International Journal of Plant Sciences 163: 197-214.

Rydin, C., K. R. Pedersen, and E. M. Friis. 2004. On the evolutionary history of *Ephedra*: Cretaceous fossils and extant molecules. Proceedings of the National Academy of Sciences U.S.A. 101: 16571-16576.

Samigullin, Tagir Kh., William F. Martin, Aleksey V. Troitsky, Andrey S. Antonov. 1999. Molecular data from the chloroplast *rpo*C1 gene suggest a deep and distinct dichotomy of contemporary spermatophytes into two monophyla: gymnosperms (including Gnetales) and angiosperms. Journal of Molecular Evolution 49: 310–315.

Setoguchi, H., T. A. Osawa, J.-C. Pintaud, T. Jaffré, and J.-M. Veillon. 1998. Phylogenetic relationships within Araucariaceae based on rbcL gene sequences. American Journal of Botany 85: 1507-1516.

Stewart, W. N., and G. W. Rothwell. 1993. Paleobotany and the evolution of plants. Cambridge University Press, New York.

Walters, T. and R. Osborne (eds). 2004. Cycad Classification—Concepts and Recommendations. CAB International.

Xi, Z., J. S. Rest, and C. C. Davis. 2013. Phylogenomics and coalescent analyses resolve extant seed plant relationships. PLoS ONE 8:e80870.

WEB SITES

Earle, C. J. 2007 onwards. The Gymnosperm Database. http://www.conifers.org/cy/index.htm

Hill, K. D. 1998 onwards. The Cycad Pages. http://plantnet.rbgsyd.nsw.gov.au/PlantNet/cycad

6

EVOLUTION OF FLOWERING PLANTS

The flowering plants, or angiosperms (Greek *angio*, vessel + *sperm*, seed; i.e., seeds enclosed by a vessel), variously named Angiospermae, Magnoliophyta, or Anthophyta, are a monophyletic group currently thought to be the sister group to the gymnosperms (Chapter 5). Angiosperms are by far the most numerous, diverse, and "successful" extant plant group, containing well over 95% of all land plant species alive today. Flowering plants grow in virtually every habitable region and are dominant in some aquatic and most terrestrial ecosystems, the notable exception to the latter being coniferous forests. Angiosperms comprise the great bulk of our economically important plants, including our most valuable food crops (Chapter 1).

Several apomorphies distinguish the angiosperms from all other land plants, summarized as follows (Figure 6.1): (1) the flower, usually with an associated perianth; (2) stamens with two lateral thecae, each composed of two microsporangia; (3) a reduced, 3-nucleate male gametophyte; (4) carpels and fruit formation; (5) ovules with two integuments; (6) a reduced, 8-nucleate female gametophyte; (7) endosperm formation; and (8) sieve tube members. Some of these features, the product of a unique evolutionary event, have become further modified in particular lineages of angiosperms (see Chapters 7, 8).

Figure 6.1 shows a simplified cladogram of the major groups of angiosperms. The diversity and classification of these groups are discussed in Chapter 7 (Amborellales, Nymphaeales, Austrobaileyales, Chloranthales, Magnoliids, Monocots, and Ceratophyllales) and Chapter 8 (Eudicots). The following is a review of flowering plant apomorphies and general evolutionary history.

ANGIOSPERM APOMORPHIES

FLOWER

Perhaps the most obvious distinguishing feature of angiosperms is the **flower** (Figure 6.2; see Chapter 9 for detailed terminology of flower parts). A flower can be defined as a *modified*, *determinate shoot* system bearing one or more **stamens**, collectively called the **androecium**, and/or one or more **carpels** (making up one or more **pistils**), collectively called the **gynoecium** (see later discussion). Most angiosperm flowers are **bisexual** (**perfect**), containing both stamens and carpels, but some are **unisexual** (**imperfect**), having only stamens or carpels. In addition, most (but not all) flowers have a **perianth**, consisting of modified leaves at the base of the shoot system.

The perianth of a flower both protects the other floral parts during floral development and functions as an attractant for pollination (see later discussion and Chapter 13). Most flowers have a perianth of *two* discrete whorls or series of parts: an outer **calyx** and an inner **corolla** (Figure 6.3A). The calyx is generally green and photosynthetic, composed of leaf-like **sepals** or (if these are fused) of **calyx lobes**. The corolla is

PLANT SYSTEMATICS
https://doi.org/10.1016/B978-0-12-812628-8.50006-7,

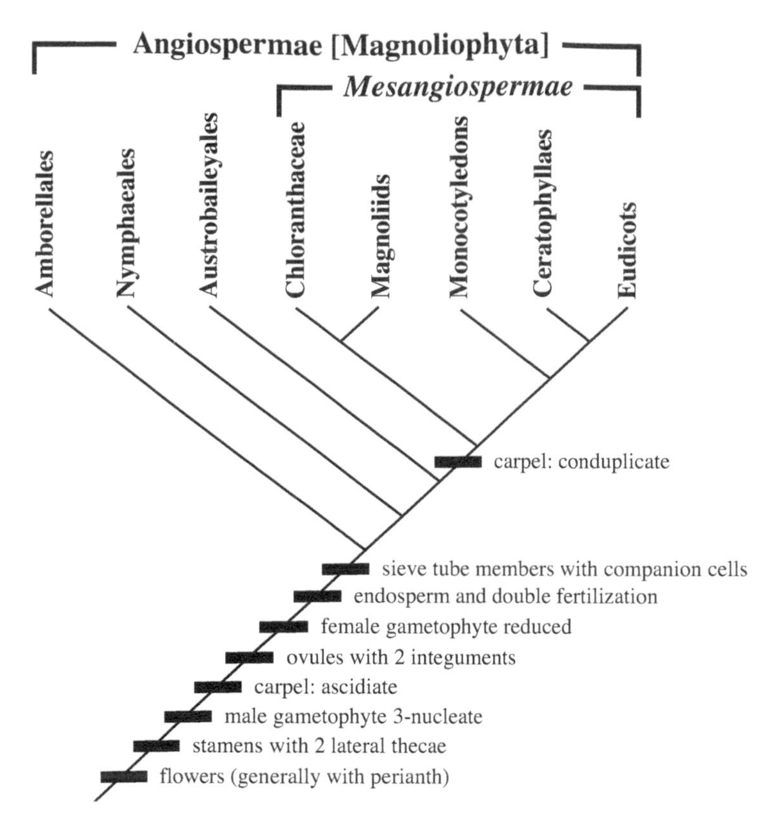

Figure 6.1 Cladogram of the angiosperms, showing apomorphies and major taxonomic groups, the latter largely after APG IV (2016). Carpel apomorphies after Endress and Doyle (2009) and included references.

typically colorful, showy, and odoriferous and is composed of individual **petals** or (if these are fused) of **corolla lobes**. However, in some flowering plants, there are two whorls of parts, but the outer and inner whorls of perianth parts are not otherwise differentiated, resembling one another in color and texture. The term **tepal** is often used for such similar perianth parts, and one may refer to **outer tepals** and **inner tepals** for the two whorls (Figure 6.3B). More rarely, the perianth may consist of a single whorl (this is usually called the calyx, by tradition) or of three or more discrete whorls (see Chapter 9). Finally, the perianth of some flowers consists of spirally arranged units that grade from sepal-like structures on the outside to petal-like structures on the inside, but with no clear point of differentiation between them; in this case, the units may be termed **tepals**, **perianth parts**, or **perianth segments** (Figure 6.3C).

The components of a flower develop in a manner very similar to leaves. In early floral development actively dividing regions of cells grow, forming bumplike mounds of tissue, the primordia. Typically, the primordia develop in whorls from outside to inside, in sequence as sepal (or outer tepal) primordia first, petal (or inner tepal) primordia second, stamen primordia third (often in two or more whorls), and carpel primordia last (Figure 6.4A–C). Each primordium typically becomes innervated by one or more vascular bundles (veins), primordia may also transform into a flattened, or "dorsiventral" (having a dorsal and ventral side) shape, resembling leaves. Fusion of floral parts may occur after they form, termed "postgenital fusion." Alternatively, floral parts may appear to be fused at maturity but may actually develop as a single structure. For example, the basal tube of a corolla in which the petals are fused (known as a "sympetalous" corolla; see Chapter 9) may form by vertical expansion of a ring of actively dividing tissue that develops beneath discrete primordia; only the upper corolla lobes may develop from discrete primordia. Overall, the resemblance of floral organs to leaves, in terms of initiating like leaf primordia of a vegetative shoot, being innervated by veins, and often having a dorsiventral shape, is why these organs — sepals, petals, stamens, and carpels — are thought to be "homologous" to leaves (Chapter 2).

Ongoing studies of the molecular basis of development in plants, originally in the species *Arabidopsis thaliana* and now in a range of others, have helped to elucidate the genetic basis of floral development and the nature of these presumed homologies. Research in this field is summarized in the "ABCE" model of floral development, in which gene products of the so-called A, B, and C classes combine to produce the four major floral organs: sepals, petals, stamens, and carpels (Figure 6.5; see Soltis et al. 2007 and Theissen and Melzer 2007 for summaries). In this model, sepals are expressed by A activity alone; petals by a combination of A

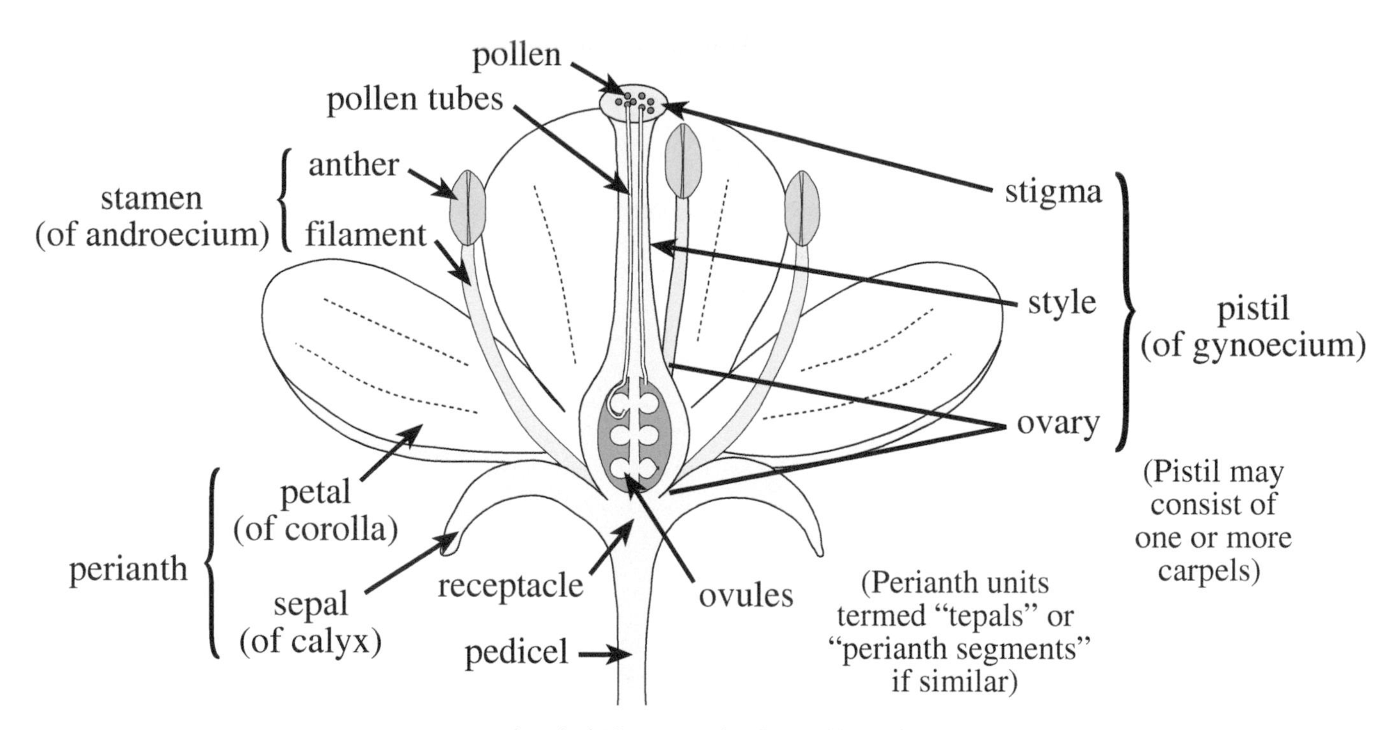

Figure 6.2 A typical (diagrammatic) flower, illustrating the parts.

and B activities, stamens by a combination of B and C activities, and carpels by C activity alone (Figure 6.5). In addition, genes of the so-called E class are needed in combination with those of the A, B, and C classes to effect proper floral organ identity (Figure 6.5). All of these floral organ identity genes work by producing "transcription factors" in the proper location of the flower, i.e., in the outermost, second, third, and innermost floral whorls. The transcription factors induce the expression of other genes that bring about the development of the four floral organs. Developmental studies like these, in a wide breath of angiosperms, will help to understand both the molecular basis of homology and the mechanisms of evolution that have given rise to the rich diversity of floral forms, including the origin of the flower.

The flower, with its typically showy and often scented perianth, evidently evolved in response to selective pressure for the transfer of pollen by animals. Animal pollination appears to be the primitive condition in the angiosperms, separating them from the predominantly wind-pollinated gymnosperms (Chapter 5). Numerous, intricate pollination mechanisms have evolved in various angiosperm lineages. These pollination mechanisms have largely driven the evolution of innumerable floral forms, accounting in large part for the distinctiveness of many angiosperm families (see Chapter

Figure 6.3 Various perianth types in flowers. **A.** Perianth of two whorls, dissimilar in appearance: a calyx of sepals and a corolla of petals (*Ruta*). **B.** Perianth of two whorls, similar in appearance: outer tepals and inner tepals (*Lilium*). **C.** Perianth undifferentiated, spiral (*Nymphaea*).

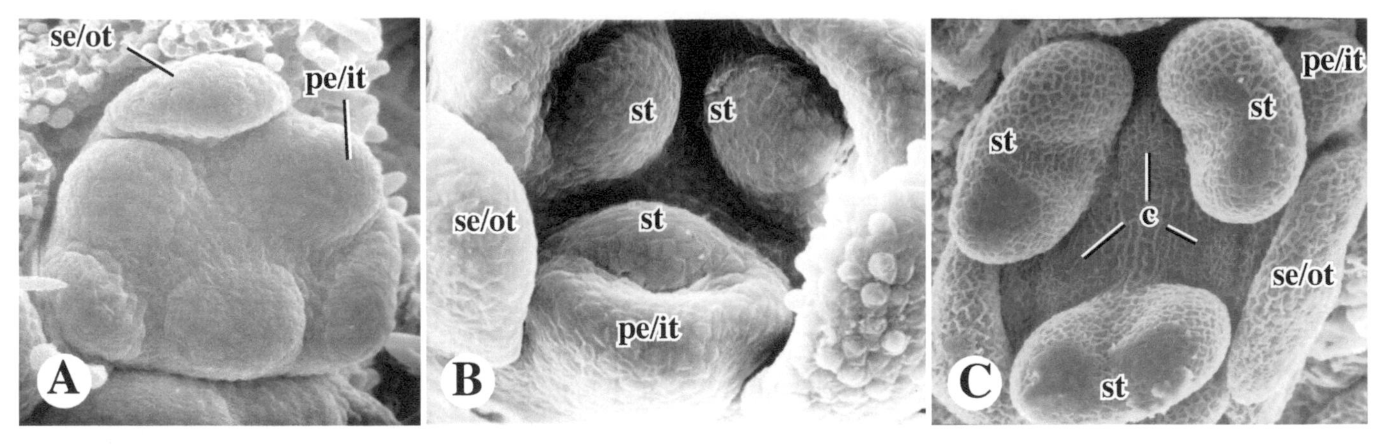

Figure 6.4 Flower development. **A.** Early development of sepal/outer tepal (se/ot) primordia and petal/inner tepal (pe/it) primordia. **B.** Later formation of stamen (st) primordia. **C.** More mature stamens and early initiation of carpel (c) primordia.

13 for floral "syndromes" related to pollination biology). Animal pollinators may include bees (Figure 6.6A), butterflies and moths (Figure 6.6B), flies (Figure 6.6C), bats (Figure 6.6D,E), and birds (Figure 6.6F). However, flowers of many groups are quite reduced in size or structural complexity, often lacking a perianth altogether; these may be water pollinated (Figure 6.6G) or wind pollinated (Figure 6.6H).

STAMENS

A distinctive apomorphy for the angiosperms is the **stamen**, the male reproductive organ of a flower. Stamens are interpreted as modified microsporophylls, modified leaves that bear microsporangia (see Chapter 5). Microsporangia produce microspores, which develop into pollen grains (Chapter 5; see later discussion). Some stamens have a laminar (leaf-like) structure, to which the anther is attached or embedded (Figure 6.7A). However, the stamens of most flowering plants have two parts: a stalk, known as a **filament**, and the pollen bearing part, known as the **anther** (Figure 6.7B). Some stamens lack a filament (or lamina), in which case the anther is **sessile**, directly attached to the rest of the flower.

The angiosperm anther is a type of synangium, a fusion product of sporangia. Anthers are unique in (ancestrally) containing two pairs of microsporangia arranged in a bilateral symmetry (i.e., having two mirror image halves). Each pair of microsporangia is typically located within a discrete half of the anther called a **theca** (plural, **thecae**; Figure 6.7C). Thus, such an anther consists of two thecae (termed **bithecal**), each theca having two microsporangia for a total of four (termed **tetrasporangiate**; Figure 6.7D). At maturity, the two microsporangia of a theca typically coalesce into a single, contiguous chamber, called the **anther locule**; each theca then opens to the outside by a specific dehiscence mechanism, releasing the pollen (Figure 6.7E). (Note that anthers of some angiosperms are secondarily reduced to a single theca, known as **monothecal** or **bisporangiate**, a distinctive systematic character; see Chapters 7–9.)

The adaptive value of the stamens of angiosperms over the microsporophylls of gymnosperms is likely connected with selective pressures for the flower itself. Stamens are generally smaller and lighter than gymnosperm microsporophylls, and stamens generally occur in bisexual flowers, rather than in more massive, unisexual cones. Modifications of the stamen have enabled the evolution of specialized pollination mechanisms, such as those involving stamens of the proper length or orientation to transfer pollen to a specific pollinator, flower heteromorphism (associated with stamens at different levels in the flower relative to differing style/stigma lengths), trigger devices, and very modified stamens such as pollinia (see Chapters 12 and 13 for more details).

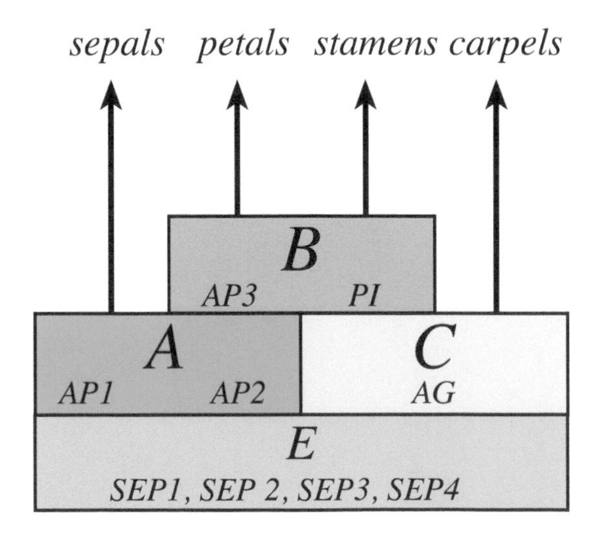

Figure 6.5 The "ABC" model of floral development. Within each gene class are specific genes (*AP1, AP2, AP3, AG, PI, SEP1, SEP2, SEP3, SEP4*), identified in mutant forms in *Arabidopsis thalliana*. (Diagram modifed from Jack, 2001.)

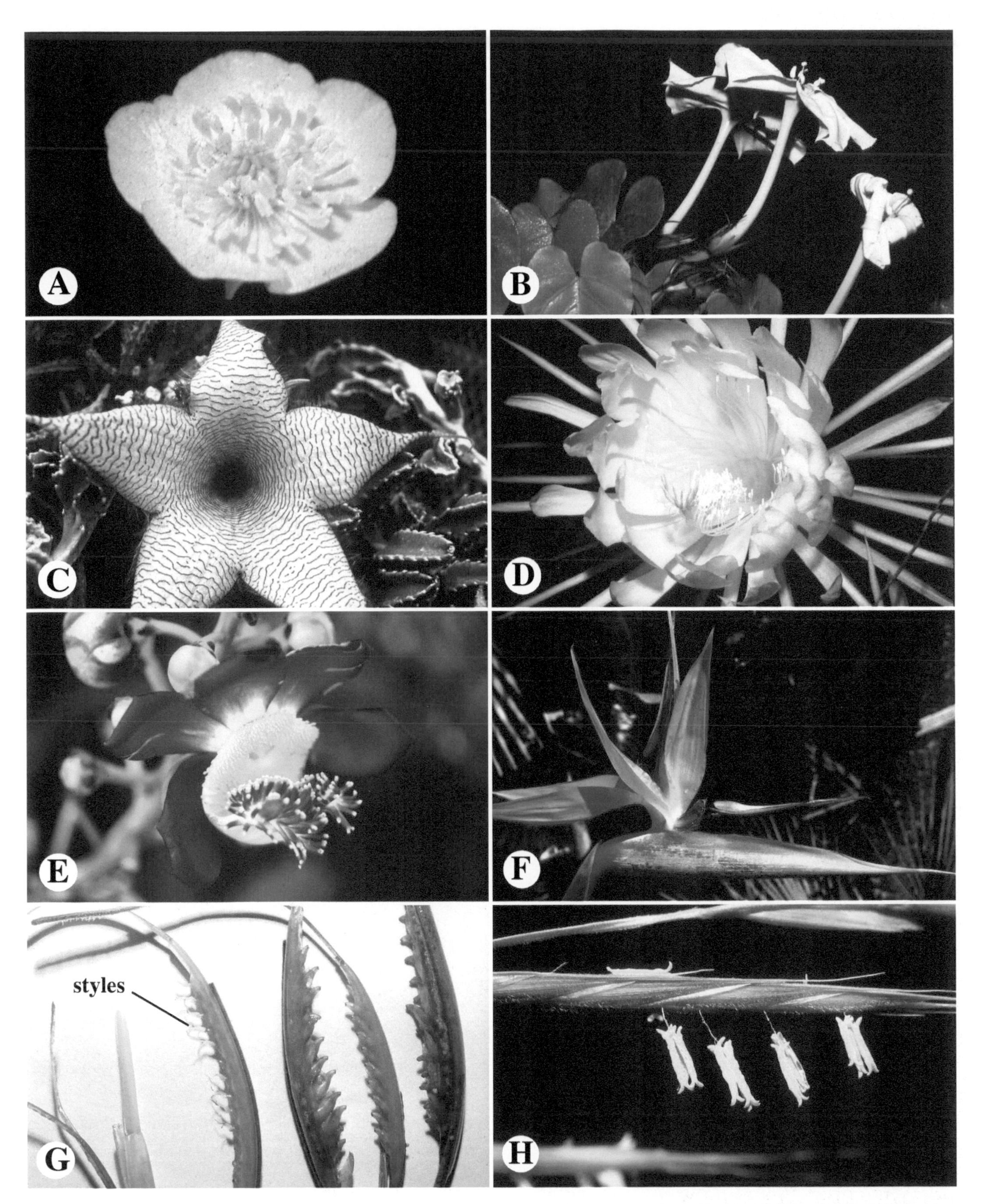

Figure 6.6 Flower modifications. **A.** *Ranunculus* sp., buttercup, insect-pollinated. **B.** *Calonyction* sp., moon flower, moth-pollinated. **C.** *Stapelia* sp., star flower, fly-pollinated. **D.** *Selenicereus*, night-blooming cereus, bat-pollinated. **E.** *Couroupida guianensis*, cannonball tree, bat-pollinated. **F.** *Strelitzia reginae*, bird of paradise, bird-pollinated. **G.** *Phyllospadix torreyi*, surf grass, water-pollinated. **H.** Grass, wind-pollinated.

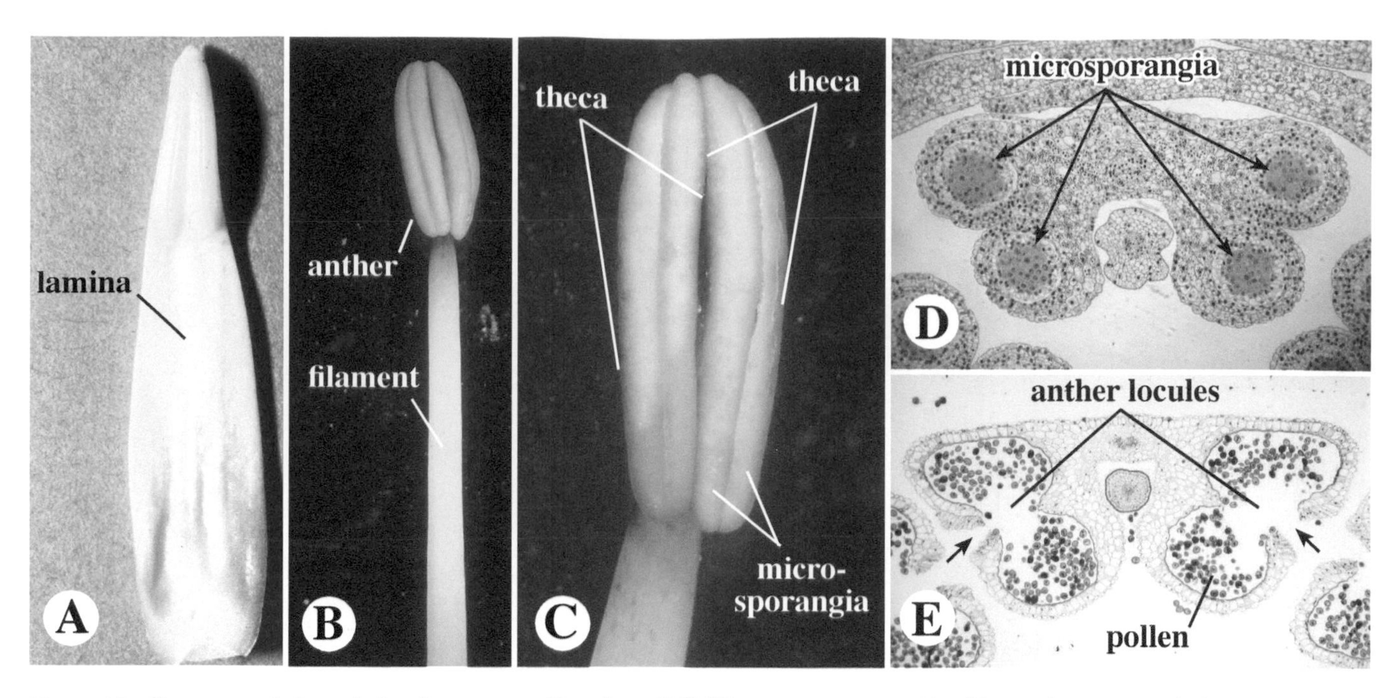

Figure 6.7 Stamen morphology. **A.** Laminar stamen, *Nymphaea*. **B,C.** Filamentous stamen, *Aloe*. Note anther composed of two thecae, each with two microsporangia. **D.** Young anther in cross section, showing four microsporangia. **E.** Cross section of older anther at time of dehiscence. Note that walls between adjacent microsporangia of each theca have broken down. Dehiscence line indicated by arrows.

REDUCED MALE GAMETOPHYTE

Another apomorphy for the angiosperms is a **reduced, three-celled male gametophyte** (Figure 6.8). No other plant group has a male gametophyte so reduced in cell number. Within the **microsporangium**, diploid microsporocytes undergo meiosis, each giving rise to four haploid **microspores**, this process known as **microsporogenesis** (Figure 6.8A). The process of **microgametogenesis** is initiated when a microspore nucleus undergoes a mitotic division forming two haploid cells within the microspore wall: a **tube cell** and a **generative cell** (Figure 6.8A,B). When this happens, the microspore is transformed into a **pollen grain**, defined as an immature, endosporic male gametophyte (see Chapter 5). The generative cell divides one time, producing **two sperm cells** (Figure 6.8A). Pollen grains of angiosperms are shed in either a two- or three-celled condition, depending on whether the generative cell division happens before or after the pollen grains are released. If pollen is released as two-celled, then the generative cell divides within the pollen tube as it travels down the style (Figure 6.8A). Whether pollen grains are 2- or 3-nucleate at release can be an important taxonomic character (Chapter 11).

The pollen grains of angiosperms, like those of gymnosperms, "germinate" during development, meaning that an elongate **pollen tube** grows out of the pollen grain wall, a condition known as **siphonogamy** (Figure 6.8A,C,D). In gymnosperms the pollen tube develops after the pollen grains enter the micropyle of the ovule and functions as a haustorial device (feeding from the tissues of the nucellus) for a long period of time (see Chapter 5). In contrast, the pollen tube of angiosperms forms immediately after transfer of pollen to the stigma. The angiosperm pollen tube elongates through (and feeds upon) the tissues of the stigma and style of the carpel and soon (generally within a day or so) reaches the ovule, where it penetrates the micropyle and transports the two sperm cells directly to the female gametophyte (see later discussion). The sperm cells of angiosperms lack flagella or cilia and are thus non-motile, a derived condition among the land plants. The loss of motility may be a function of the direct transport of the sperm cells to the micropyle of the ovule. The only other land plants with nonmotile sperm cells are the gymnospermous conifers (including the Gnetales), which lost sperm motility independently of flowering plants.

The adaptive significance of the reduced male gametophytes of angiosperms is probably correlated with the evolution of a reduced female gametophyte and relatively rapid seed development (discussed later). In gymnosperms, fertilization of sperm and egg occurs long after pollination, sometimes as long as a year or more; the male gametophytes must persist during this long period, feeding like a minature parasite off the tissues of the nucellus. In angiosperms, however, fertilization occurs very soon after pollination. Thus, angiospermous male gametophytes are "lean," apparently requiring a minimum number of cells and nuclei; they function to deliver sperm cells to the female gametophyte and effect fertilization very rapidly compared with gymnosperms.

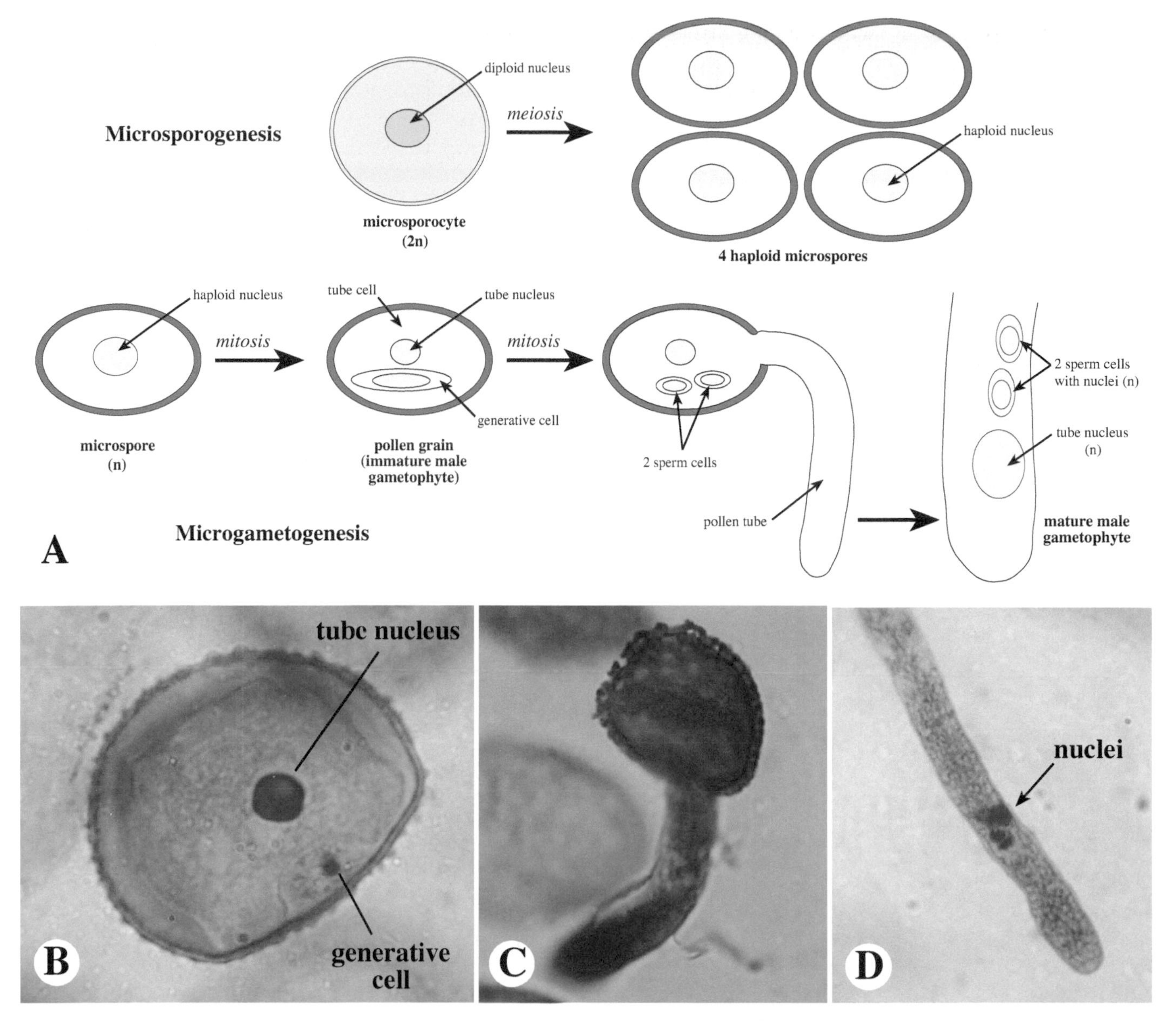

Figure 6.8 Angiosperm male gametophyte. **A.** Microsporogenesis, the development of haploid microspores via meiosis of a microsporocyte, and microgametogenesis, development of a reduced 3-nucleate male gametophyte (pollen grain) from a microspore. **B.** Mature binucleate pollen grain, with tube nucleus and generative cell. **C.** Germinating pollen grain, forming pollen tube. **D.** Tip of pollen tube, housing nuclear material (nuclei types unclear in this image).

CARPEL

A major apomorphy of angiosperms is the **carpel**. The traditional view of the carpel is that "angiospermy," the enclosure of ovules and seeds, arose via the evolution of a modified, conduplicate megasporophyll bearing two adaxial rows of ovules (Figure 6.9D). Recall that a "megasporophyll" is a modified leaf that bears megasporangia, which in the seed plants are components of the ovules and seeds (see Chapter 5); "conduplicate" (or plicate) means inwardly folded longitudinally and along the central margin (see Chapter 9). This megasporophyll is modified in that the margins — by virtue of the conduplicate folding — come together and fuse "postgenitally" (i.e., after being formed; Figures 6.9A–D, 6.10A), with certain parts differentiating into tissue for pollen reception and pollen tube growth, typically forming an apical stigma and style (Figure 6.9D). At maturity the carpel body completely encloses the ovules and seeds.

The sporophyll-like nature of the carpel is evident in in most flowering plants in that (1) it may develop like a leaf, having an initially flattened, dorsiventral shape, with an adaxial (toward the top center of the flower) and abaxial (away from the top center of the flower) surface; and (2) it has veins, typically one in the middle termed the **dorsal** (median) vein or bundle, corresponding to the midvein of a leaf, and two others near the two carpel margins termed the **ventral** (lateral or placental) veins/bundles (Figures 6.9D, 6.10A). Additional veins often occur between the dorsal and ventral bundles (e.g., Figure 6.10B), and veins will sometimes "fuse" together. The veins of a carpel are typically collateral (see Chapter 10), with xylem on the adaxial side and phloem on the abaxial side. The ventral veins become inverted in orientation after carpel formation, with the xylem and phloem disposed 180° from their original orientation, i.e., prior to conduplicate folding (Figure 6.9D).

However, the carpels of some angiosperm taxa show no evidence of a conduplicate, leaflike nature during develop-

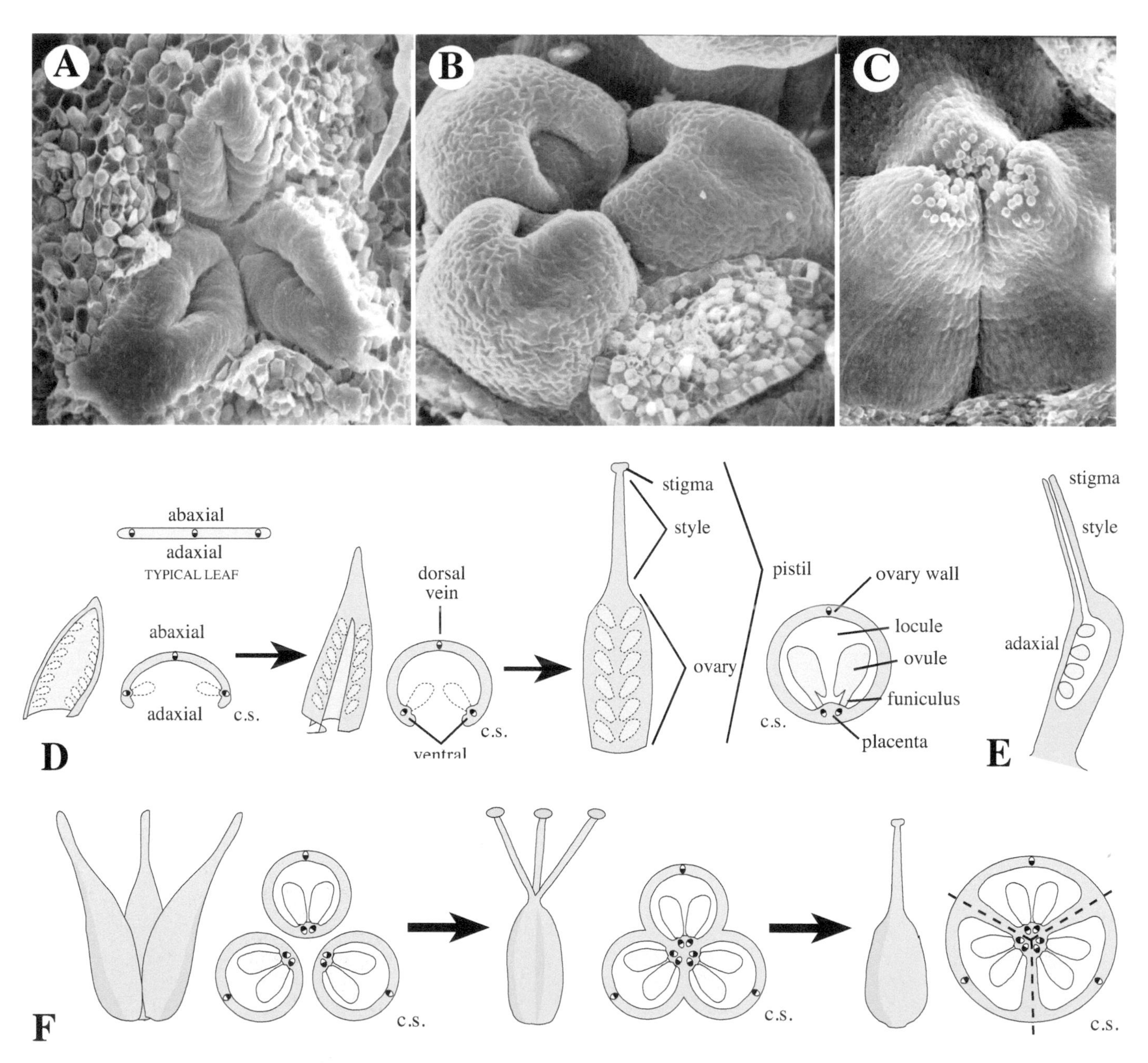

Figure 6.9 The carpel, an apomorphy of the angiosperms. **A–C.** Scanning electron micrographs of carpel development. **A.** Early formation of three carpels, showing conduplicate formation. **B.** Intermediate developmental stage. Note lateral contact of the three carpels. **C.** Mature stage, in which carpel margins have closed in and adjacent carpels have fused into a syncarpous gynoecium (compound pistil). **D.** Diagram of carpel development from early stages to mature ovary, adaxial side below. Note dorsal and ventral veins (black=xylem; white=phloem), the latter becoming inverted. **E.** Diagram (median longitudinal section) of pistil of *Austrobaileya* (Austrobaileyaceae), an ascidiate carpel type, the ancestral type. **F.** Diagram illustrating evolutionary sequence of carpel fusion (dashed lines=carpel boundaries).

ment, developing as a ring of tissue that grows upward, assuming a tube-like or urn-like form, this known as an **ascidiate** carpel (Figure 6.9E). Ascidiate carpels are likely the ancestral type among extant angiosperms (Figure 6.1; see Endress and Igersheim 2000); these were sealed (enclosing the ovules) not by post-genital fusion, but by secretion of a liquid, this secreted liquid serving as the contact zone for pollen germination (see below). The conduplicate type of carpel may constitute an apomorphy for the Mesangiospermae, although several reversals to an ascidiate carpel type have occurred in the flowering plants, with fusion of the carpel occurring in various groups by secretion and/or post-genital fusion (Endress and Doyle 2009).

A given flower can have one to many carpels. If two or more carpels are present, they may be separate from one another (distinct), termed **apocarpous**, or fused together (connate), termed **syncarpous**. Because of the frequent fusion of carpels, additional terms are useful in describing the female parts of a flower. The term **gynoecium** is the totality of female reproductive structures in a flower, regardless of their structure. Thus, a carpel may be alternatively defined as a unit of the gynoecium. The gynoecium is composed of one or more **pistils**. Each pistil consists of a basal **ovary**, an apical **style** (or styles), which may be absent, and one or more **stigmas**, the tissue receptive to pollen grains (Figure 6.9D). A pistil may be equivalent to one carpel (in which case, it may be termed a **simple pistil**) or composed of two or more, fused carpels (termed a **compound pistil**; Figures 6.9F, 6.10B). (The position of one or more ovules and the fusion of one or more carpels determine various placentation types; see Chapter 9 for complete terminology.)

The evolution of the carpel had considerable adaptive significance. First, because carpels are the receivers of pollen, they may function to selectively control fertilization. The transfer of pollen to the carpels is followed by germination of the pollen grain to form a pollen tube, which grows to the micropyle of the ovule, either through the tissue of the stigma and style, through a secretion liquid, or through a stylar canal. However, chemicals that are present in the stigma and style may inhibit either pollen germination or pollen tube growth; this is known as an **incompatibility reaction**, mediated by incompatibility genes (see Chapter 13). This type of chemical incompatibility often occurs between the pollen and stigmatic regions of *different* species. However, it may also occur between individuals of the *same* species, notably between individuals that are genetically similar and possess the same incompatibility alleles. Thus, incompatibility reactions may inhibit inbreeding, allowing for reproduction only between genetically dissimilar individuals of the species (i.e., promoting out-crossing; see Chapter 13 for more details). Thus, the carpel may ultimately provide some selective control as to which pollen grains contribute the sperm cells that fertilize the egg.

A second major adaptive function of the carpel pertains to fruit formation and seed dispersal. A **fruit** is the mature ovary or ovaries (made up of one or more carpels) plus any acces-

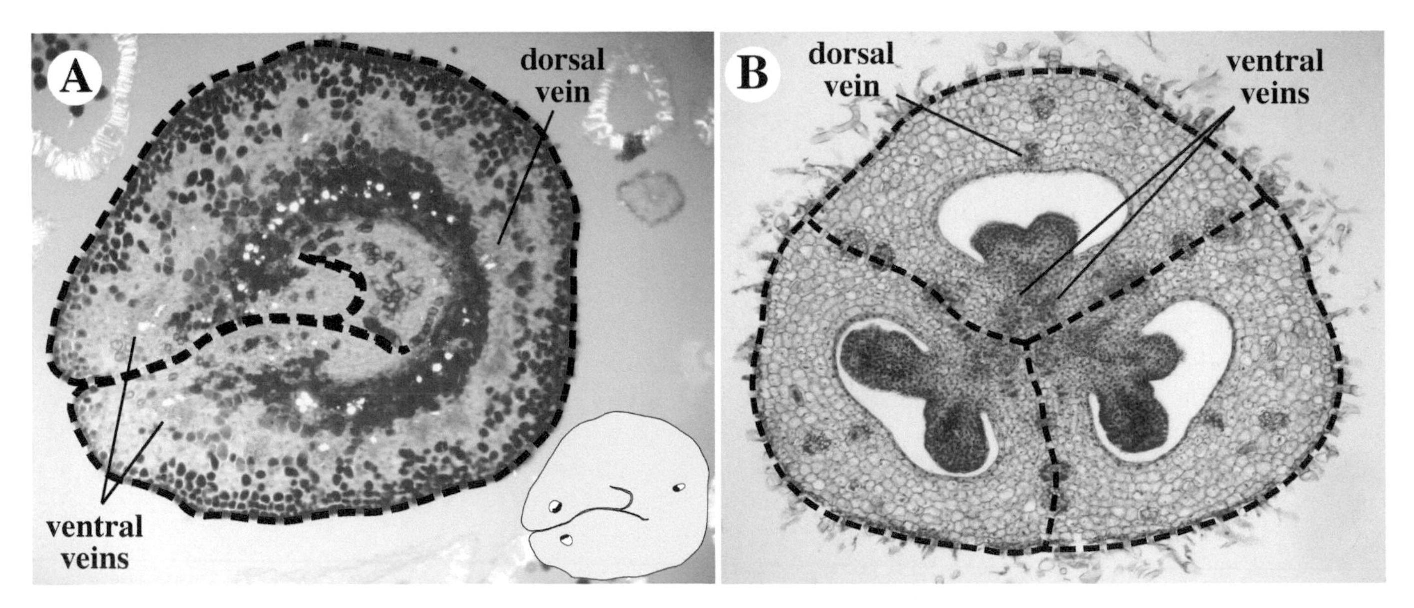

Figure 6.10 **A.** Ovary cross section of a taxon with a single carpel per flower (unicarpellate gynoecium). Note outline of carpel boundary (dashed line). Inset diagram: note orientation of xylem (black) and phloem (white) of veins. **B.** Ovary cross section of a taxon with a 3-carpellate, syncarpous pistil (carpels outlined by dashed lines), showing dorsal and ventral veins. (Note: vascular bundle outside dorsal vein supplies perianth and stamens, this ovary being inferior.)

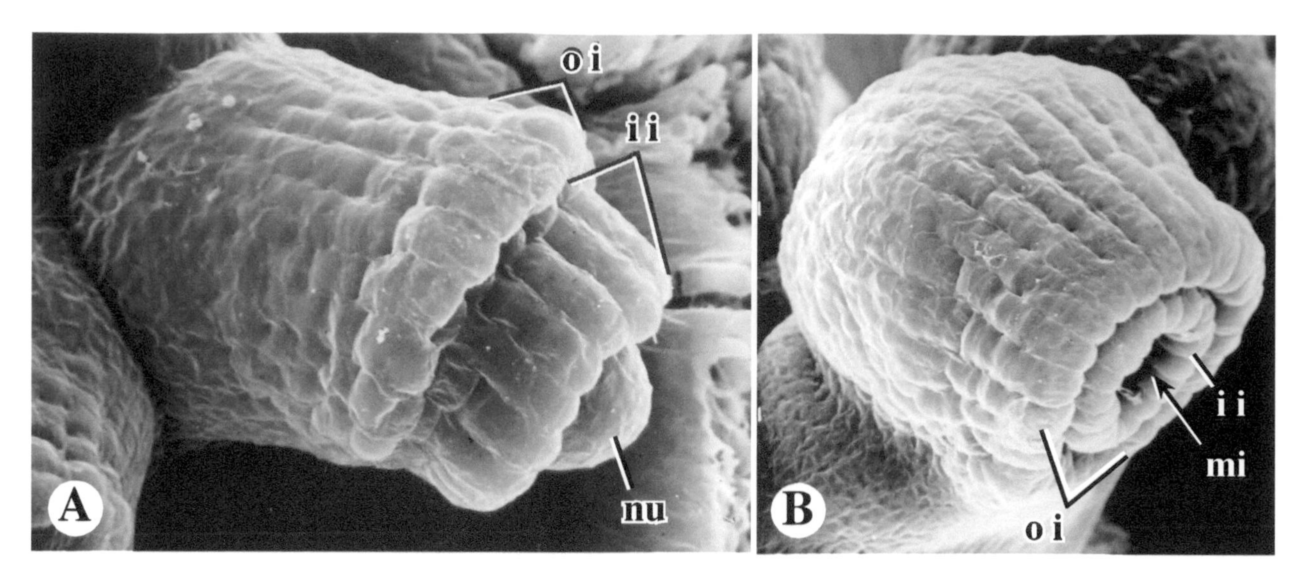

Figure 6.11 Bitegmic ovule, the ancestral condition of the angiosperms. **A.** Young ovule, showing intiation of inner integument (ii) and outer integument (oi), both growing around the nucellus (nu). **B.** Older ovule, in which inner and outer integuments have enveloped the nucellus, forming a micropyle (mi).

sory tissue that might be present (see Chapter 9). Fruits generally do not mature from ovaries if fertilization of the seed(s) does not occur. The mature ovary wall, termed the **pericarp**, may be highly modified. These modifications generally function in a tremendous variety of dispersal mechanisms (Chapter 9). In general, if the pericarp is fleshy, fruits are dispersed by animals. In these fleshy, animal-dispersed fruits, the seeds are transported either by passing through the gut of the animal unharmed (with only the pericarp being digested) or by being spilled during a sloppy eating session. Dry fruits may also be dispersed by animals, but typically via external barbs or prickles that catch on skin, fur, or feathers. Last, fruits may be dispersed by wind (aided by the development of wings or trichomes), water (via various flotation devices), or mechanically (by various explosive, hygroscopic, or catapulting methods).

TWO INTEGUMENTS

A unique apomorphy of angiosperms is the growth of **two integuments** during ovule development, the ovules known as **bitegmic** (Figure 6.11). All nonflowering seed plants have ovules with a single integument, termed **unitegmic**. The two integuments of angiosperms usually completely surround the nucellus, forming a small pore at the distal end; this opening, the **micropyle**, is the site of pollen tube entrance. Both of the integuments of angiosperm ovules contribute to the seed coat. The two integuments typically coalesce during seed coat development, but may form anatomically different layers.

The possible adaptive significance of two integuments, if any, is not clear, but may have enabled the evolution of specialized seed coat layers, although differential seed coat layers are found in several gymnosperm taxa as well. Interestingly, several angiosperm lineages have secondarily lost an integument, and are thus unitegmic. Notable unitegmic groups are many Poales of the Monocots (Chapter 7) and most of the Asterids of the Eudicots (Chapter 8).

REDUCED FEMALE GAMETOPHYTE

Several novelties of the angiosperms have to do with the evolution of a specialized type of ovule and seed. A major apomorphy of angiosperms is a **reduced female gametophyte**. As in other seed plants, a single megasporocyte within the megasporangium (nucellus) divides meiotically to form four haploid megaspores (Figure 6.12). The female gametophyte typically generates from only one of these megaspores (Figure 6.12), with a few exceptions in which others may contribute (see Chapter 11). In the great majority of angiosperms the megaspore divides in a sequence of three mitotic divisions, resulting in a total of eight haploid nuclei. Further differentiation usually results in an arrangement of these eight nuclei into seven cells, a pattern known as the *Polygonum* type (Figures 6.12, 6.13A; see Chapter 11). In the micropylar region three cells develop: an **egg** cell flanked by two **synergid cells**. Egg plus synergids is sometimes called the "egg apparatus." In the **chalazal** region, which is opposite to the micropyle, three **antipodal cells** form. The remaining volume of the female gametophyte is technically a single cell, called the **central cell**, which contains two **polar nuclei**. Archegonia do not form within the female gametophyte of angiosperms as they do in virtually all other seed plants. The female gametophyte in various angiospermous taxa may become further modified from the ancestral type described here by variations

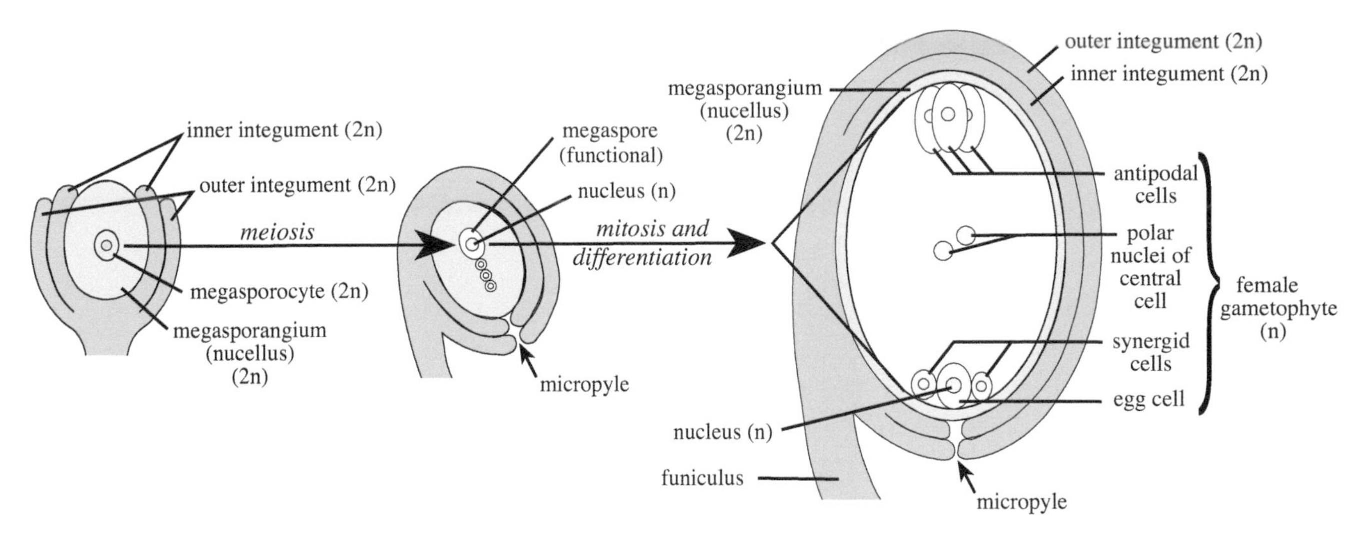

Figure 6.12 Angiosperm ovule development and morphology. Note meiosis of megasporocyte, producing four haploid megaspores, one of which undergoes mitotic divisions and differentiation, resulting in an 8-nucleate female gametophyte.

in cells divisions, nuclear fusions, and cell formations (see Chapter 11). (**Note:** The female gametophyte of angiosperms is often called an "embryo sac"; this terminology, although often used, is to be avoided, as it fails to denote the homology with the female gametophyte of other seed plants.)

A recent theory of female gametophyte evolution suggests that the ancestral condition of the angiosperms was not the common monosporic, 8-nucleate, 7-celled *Polygonum* type, but was instead a monosporic, 4-nucleate and celled condition found in virtually all Nymphaeales and Austrobaileyales, termed the *Nuphar/Schisandra* type (Figure 6.14A). This 4-nucleate condition, having one polar nucleus in a central cell and 3 cells (the egg apparatus) at the micropylar end could represent an ancestral *module*. This module would subsequently have been doubled (a third sequence of mitotic divisions) to yield the common *Polygonum* type (Figure 6.14A) or quadrupled to yield something like the 16-nucleate *Penaea* type (Figure 6.14A). In fact, the most basal angiosperm, *Amborella trichopoda* (Chapter 7), has a modified type of female gametophyte, being 9-nucleate and 8-celled via an extra mitotic division in the egg apparatus producing a third synergid cell; this type has been termed the *Amborella* type and it thought to have evolved independently of the common *Polygonum* type. A simplified cladogram of angiosperm relationships (Figure 6.14B) shows this scenario, with the 4-nucleate condition *Nuphar/Schisandra* type primitive

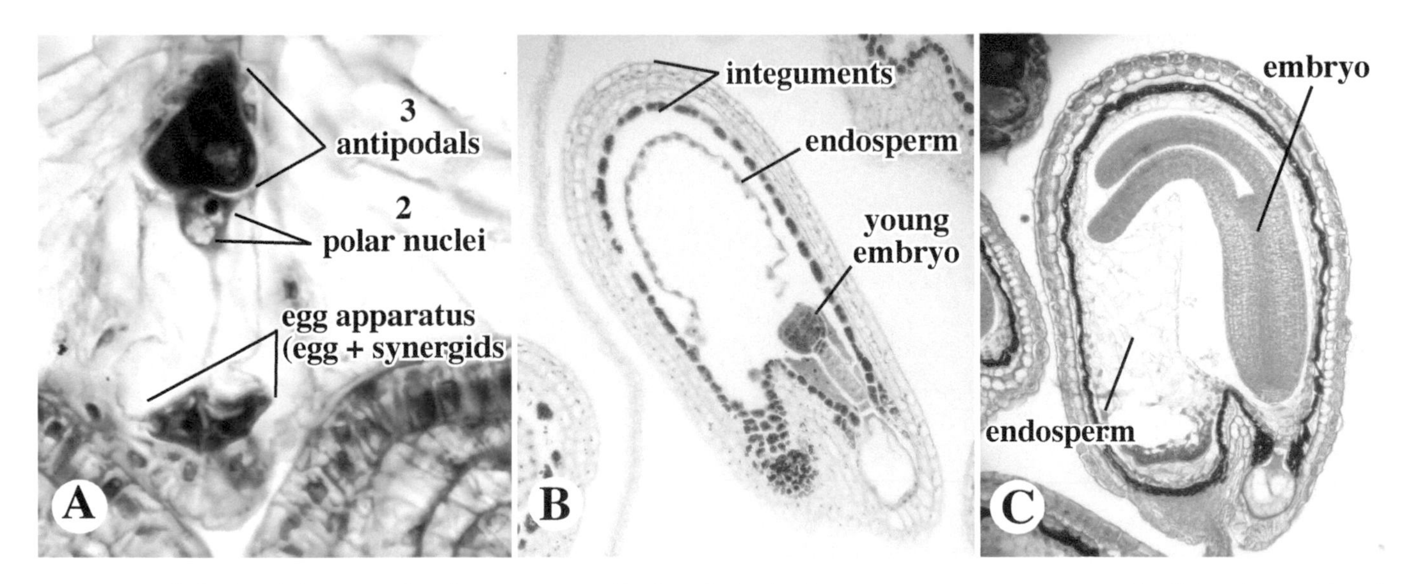

Figure 6.13 **A.** Reduced, 8-nucleate female gametophyte (*Lachnanthes*), showing egg apparatus (egg + synergid cells), polar nuclei, and antipodals. **B,C.** Endosperm formation (*Capsella*). **B.** Early stage. **C.** Later stage, forming seed.

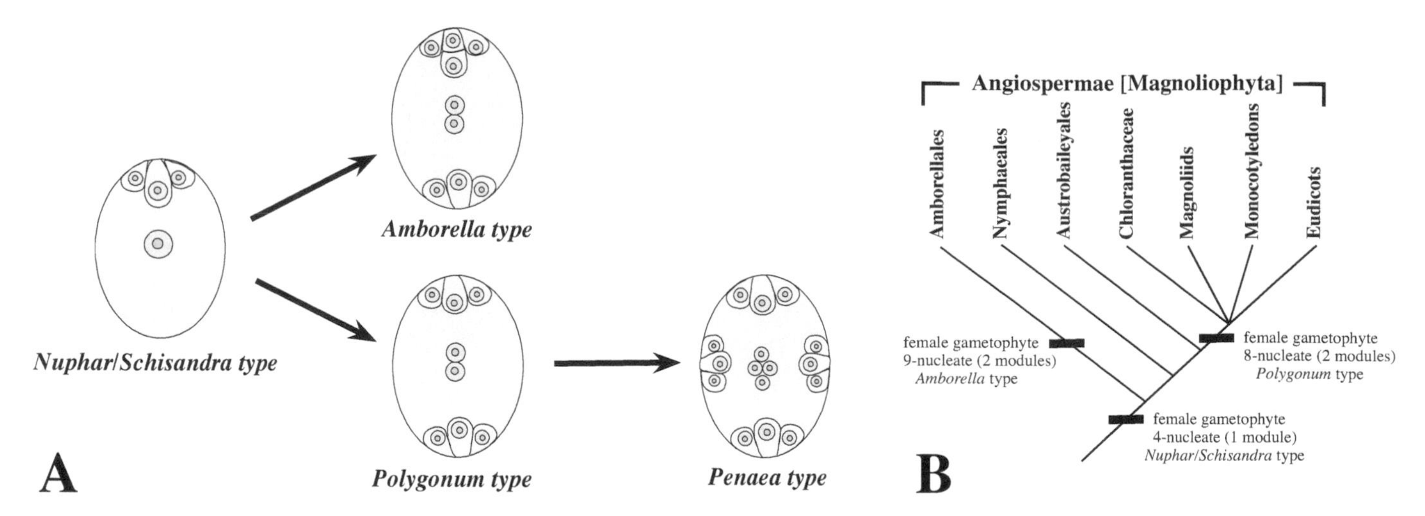

Figure 6.14 **A.** Modular hypothesis of female gametophyte evolution. The monosporic, 4-nucleate *Nuphar/Schisandra* type may represent the ancestral condition in the angiosperms, independently giving rise to the *Amborella* type and *Polygonum* type by duplication of the basic 4-nucleate module. **B.** Cladogram of angiosperms, showing evolutionary changes according to this modular hypothesis. Note that the *Polygonum* type is derived within the angiosperms. (After Friedman and Williams, 2004.)

and the 8-nucleate *Amborella* and *Polygonum* types derived. See Friedman and Williams (2004) and Friedman and Ryerson (2009) for more information on this idea.

The significance of a reduced female gametophyte in flowering plants is likely correlated with developmental timing. Fertilization in angiosperms occurs very shortly after pollination, unlike that of the gymnosperms, in which a long period of time may ensue between the two events. Thus, angiosperms have the capacity to more quickly generate seeds. This feature may be of tremendous adaptive value, enabling, for example, the evolution of rapidly spreading annual herbs.

ENDOSPERM FORMATION

Another major apomorphy of the angiosperms is the presence of **endosperm**. Endosperm is the product of **double fertilization**. When the pollen tube enters the micropyle of the ovule, it penetrates one of the synergid cells and releases the two sperm cells into the central cell of the female gametophyte

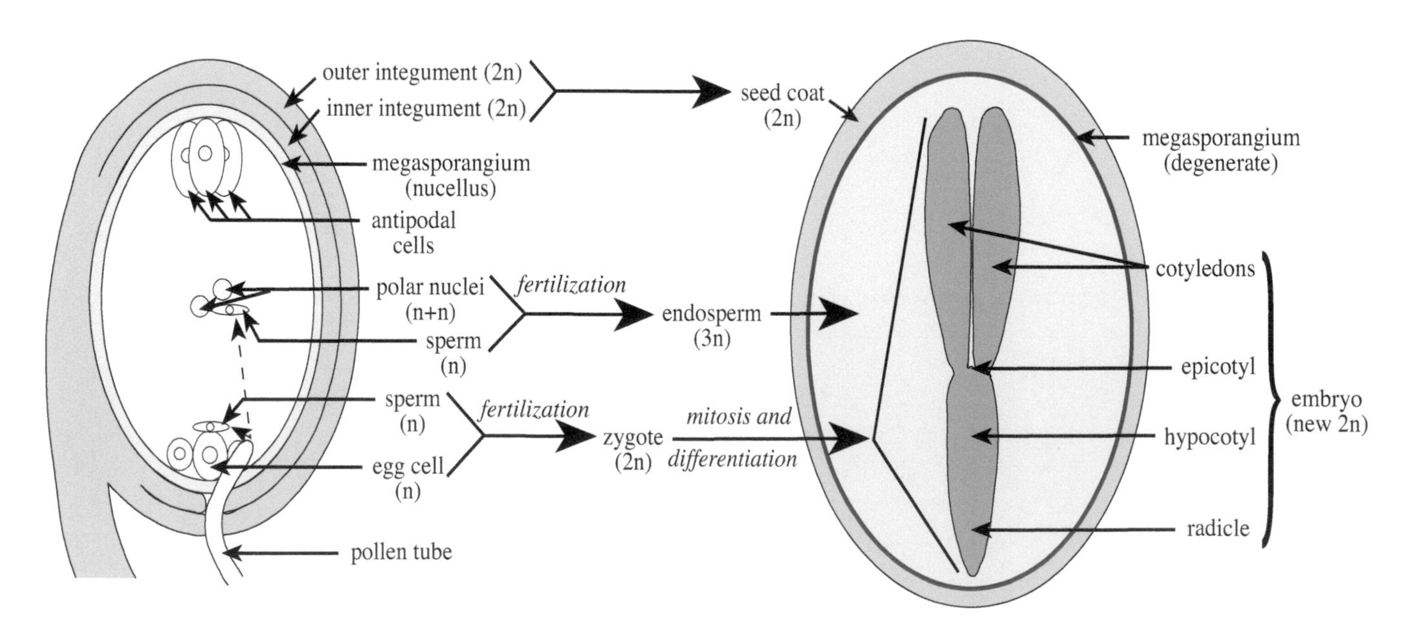

Figure 6.15 Angiosperm seed development and morphology. Note fertilization of egg, forming zygote and embryo, and fertilization of polar nuclei, forming triploid endosperm.

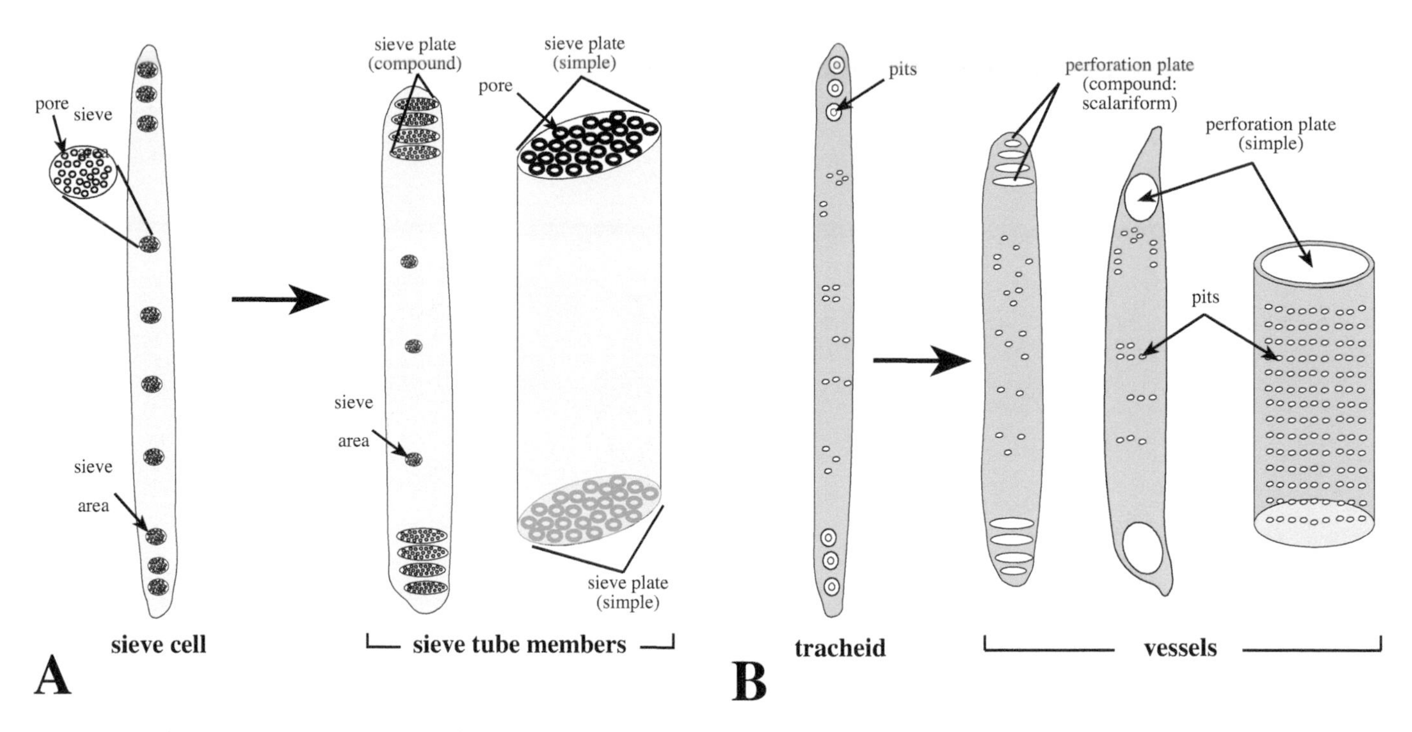

Figure 6.16 **A.** Evolutionary change from sieve cells (left) to sieve tube members, the latter an apomorphy of the angiosperms. **B.** Evolution of vessels in the angiosperms. Note transformation from imperforate tracheid to vessels with perforation plates. Trends within the angiosperms include change from elongate vessels with scalariform perforation plate to short vessels with simple perforation plates.

(Figure 6.15). One sperm cell migrates toward and fuses with the egg cell to produce a diploid **zygote**. As in other land plants, the zygote matures into an embryo, with structures similar to those in other seed plants (Figure 6.13). The other sperm cell fuses with the two polar nuclei to produce a triploid, or 3n, **endosperm cell**. This endosperm cell then repeatedly divides by mitosis, eventually forming the **endosperm**, a mass of tissue that generally envelopes the embryo of the seed (Figures 6.13B,C, 6.15). Endosperm replaces the female gametophyte as the primary nutritive tissue for the embryo in virtually all angiosperms, containing cells rich in carbohydrates, oil, or protein.

The adaptive significance of endosperm is, like that of the reduced female gametophyte, possibly correlated with developmental timing. The endospermous nutritive tissue of angiosperms does not begin to develop until *after* fertilization is achieved. This is in contrast with gymnospermous seed plants, in which considerable female gametophytic nutritive tissue is deposited after pollination, even if the ovules are never ultimately fertilized. Thus, a major selective pressure for the evolution of endosperm may have been conservation of resources, such that seed storage compounds are not formed unless fertilization is assured. An additional, functional feature of endosperm derives from the tissue being triploid. Having three sets of chromosomes (one from the male and two from the female) may enable the endosperm to develop more rapidly (correlated with rapid overall seed development) and may also provide greater potential for chemical variation in nutritive contents.

SIEVE TUBE MEMBERS

Angiosperms are unique (with minor exceptions) in having **sieve tube members** as the specialized sugar-conducting cells (Figure 6.16A). Sieve cells (and associated albuminous cells) are the primitive sugar-conducting cells and are found in all nonflowering vascular plants (see Chapter 4). Sieve tube members (and associated companion cells) were evolutionarily modified from sieve cells and are found only in flowering plants. Sieve tube members differ from the ancestral sieve cells in that the pores at the end walls are differentiated, being much larger than those on the side walls. These collections of differentiated pores at the end walls are called **sieve plates**. Sieve plates may be either compound (composed of two or more aggregations of pores) or simple (composed of one pore region). Parenchyma cells associated with sieve tube members are called **companion cells**. Companion cells function to load and unload sugars into the cavity of sieve tube members. Unlike the similar albuminous cells of gymnosperms, companion cells are derived from the same parent cell as the conductive sieve tube members.

The adaptive significance of sieve tube members over sieve cells is not clear, though they may provide more efficient sugar conduction.

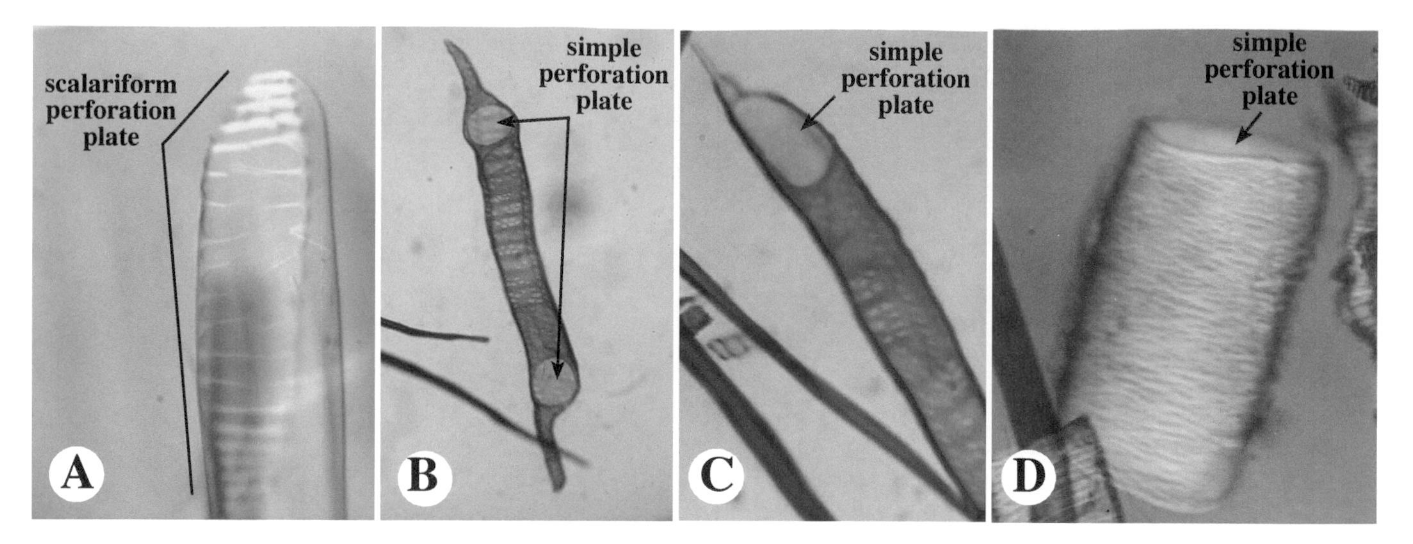

Figure 6.17 Variation in vessel anatomy in the angiosperms. **A.** *Liriodendron tulipifera*, with scalariform perforation plates. **B,C.** *Quercus* sp., elongate with simple perforation plates. **D.** *Cucurbita* sp., short and cylindrical, with simple perforation plates.

ANGIOSPERM SPECIALIZATIONS

Angiosperms are a tremendously diverse group of seed plants and have evolved a great number of novel structural features. Various lineages of angiosperms have acquired an amazing variety of specialized roots, stems, and leaf types not found in any other land plant taxa (see Chapters 7–9). And, as mentioned earlier, angiosperms have a number of specialized pollination systems and fruit/seed dispersal mechanisms, by-products of the evolution of flowers and fruits (see Chapter 13).

VESSELS

One angiosperm specialization concerns water and mineral conductive cells. The great majority of angiosperms have **vessels**, in which the two ends of the cells have openings, termed perforation plates (Figure 6.16B; see Chapters 4, 10). Vessels constituted a major evolutionary innovation within the angiosperms. Not all angiosperms have vessels, however, and some basal flowering plant groups (e.g., Amborellales, some Nymphaeales; see Figure 6.1, Chapter 7) are vessel-less, having only tracheids (which lack perforation plates). Thus, vessels apparently do not constitute an apomorphy for the flowering plants as a whole, and likely arose or were lost independently in more than one angiosperm lineage.

The tracheids of basal, vessel-less angiosperms characteristically have numerous transversely elongated pits (called "scalariform" pitting), especially at the tapering end walls where they join other tracheid cells. Tracheids with scalariform pitting may be the ancestral tracheary element for the angiosperms. In general, primitive vessels resemble tracheids in having scalariform perforation plates (Figure 6.17A) in which the openings consist of numerous, transversely oriented pits. Specializations of vessels (Figure 6.16B) include (1) modification of the perforation plate from scalariform to one with fewer, less transversely oriented openings, to a simple perforation plate (having a single opening; e.g., Figure 6.17B,C); (2) modification from tapering end walls to perpendicular ones; and (3) modification from long, narrow cells to short, wide cells (Figure 6.17D).

The adaptive advantage of vessels over tracheids is presumably more efficient solute conduction due to greater rate and lower flow resistance. However, under conditions of low water potential or freezing temperatures, wider vessels may increase the chance of cavitation (formation of a gas bubble in the water stream), which interrupts water flow and can result in plant death if not self-repaired. This may explain why tracheid-bearing conifers dominate over angiosperms in cold climatic regions.

ORIGIN OF ANGIOSPERMS

As is often stated, Charles Darwin described the relatively rapid diversification of the "higher plants" (presumed to mean angiosperms) as an "abominable mystery." The earliest definitive fossils of flowering plants are dispersed pollen grains from the earliest Cretaceous period, approximately 140 million years ago. The earliest definitive flowers occur slightly later in the fossil record, as early as 130 million years ago. These early flowering plant fossils can largely be assigned to recognizable, extant groups. Molecular phylogenetic studies estimate the age of origin of extant angiosperms from a comparable time of about 140 million years ago to considerably earlier (Magallón et al. 2015; see Coiro et al. 2019 and Li et al. 2019 for a discussion of the discrepancy between

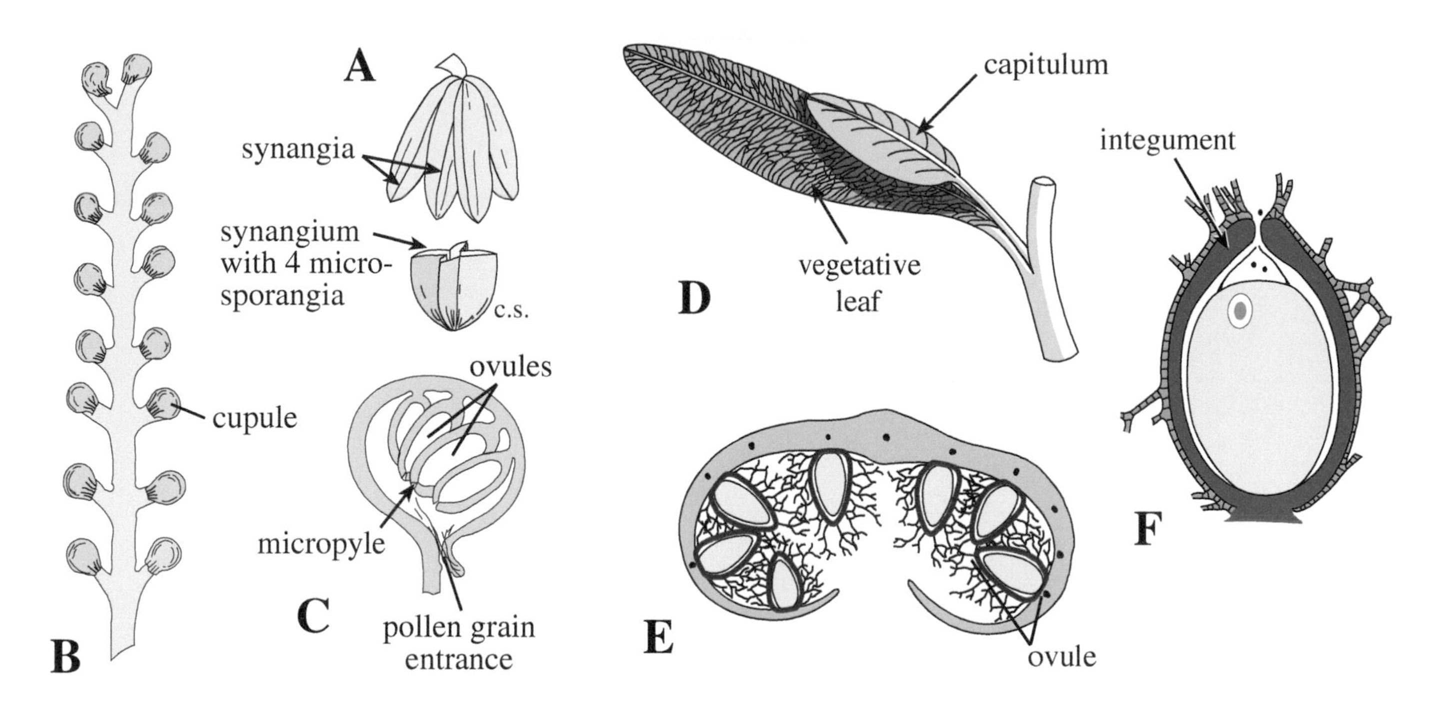

Figure 6.18 **A–C.** *Caytonia*, diagrams redrawn from Thomas (1925). **A.** Cluster of male reproductive units, each a radially symmetrical synangium of 4 microsporangia. (Cross section=c.s.) **B.** Reproductive axis, bearing two rows of cupules. **C.** Cupule, in sagittal section, showing four ovules and opening at base. **D–F.** *Glossopteris*, diagrams redrawn from Gould and Delevoryas (1977), by permission. **D.** Reproductive structure, showing vegetative leaf with reticulate venation and basally adnate reproductive structure, the capitulum. **E.** Cross section of capitulum, showing several ovules (each bearing branched trichomes) on lower surface. Note incurved margins. **F.** Ovule longitudinal section, showing single integument with micropyle.

fossil and molecular estimates of the origin of angiosperms). Once angiosperms arose, they radiated rapidly into several, distinct lineages and gradually replaced gymnosperms as the dominant plant life form on the earth.

However, the details of angiosperm evolution from a gymnosperm precursor are not clear. One problem is what to call an angiosperm. Many angiosperm features cited earlier, such as a reduced male gametophyte, reduced female gametophyte, and double fertilization with triploid endosperm, are microscopic and cytological and would be unlikely to be preserved in the fossil record. Cladistic analyses of extant angiosperms may help elucidate the features possessed by the common ancestor of the flowering plants. Given this, we might expect to find at least some of these features in the closest fossil relatives of the angiosperms. Based on recent cladistic studies, *Amborella trichopoda* of the Amborellales (Figure 6.1) is accepted as the best hypothesis for the most basal angiosperm lineage (see Chapter 7). *Amborella* lacks vessels and has unisexual flowers with a spiral perianth, laminar stamens, and separate carpels. However, other lineages of flowering plants vary in these features, making an assessment of the characteristics of the common ancestor of the angiosperms unclear.

Molecular phylogenetic studies, taking into account all angiosperm lineages, infer that the ancestral angiosperm flower was bisexual and radially symmetric, having more than two whorls of three, undifferentiated perianth parts, more than two whorls of three, unfused stamens, and more than five spirally arranged, unfused, ascidiate carpels closed by secretion, with other features uncertain (although differing in some results, see Endress and Doyle 2009 and Sauquet et al. 2017). This was followed in various angiosperms clades by successive reduction in the number of whorls of perianth parts and stamens, either by reduction or by merging of whorls. A reduced number of whorls made possible the evolution of fusion of parts and zygomorphy, features exhibited in some of the most speciose groups of angiosperms.

An ongoing hypothesis on the origin of angiosperms is that they were derived by modification of some member of the group known as "pteridosperms" (mentioned in Chapter 5), a paraphyletic assemblage of extinct plants that possessed seeds and had generally fernlike foliage. Some pteridosperms may represent possible angiosperm progenitors. (See Doyle 2006.) One fossil taxon that exemplifies a putative transition to angiosperms is *Caytonia* of the Caytoniales (Figure 6.18A–C). *Caytonia* possessed reproductive structures simi-

Figure 6.19 *Archaefructus*. **A.** Reconstruction of *Archaefructus sinensis*, showing reproductive axis bearing stamens proximally and carpels distally. (Contributed by K. Simons and David Dilcher (©).) **B.** Fossil impression of carpel units of *Archaefructus lianogensis*. (Contributed by David Dilcher (©) and Ge Sun.)

lar to those of the angiosperms. The male reproductive structures resemble anthers in consisting of a fusion product (synangium) of three or four microsporangia; however, these differ from angiosperm anthers in being radially (not bilaterally) symmetric (Figure 6.18A). The female reproductive structures of *Caytonia* consist of a spikelike arrangement of units that have been termed **cupules** (Figure 6.18B,C). Each cupule encloses a cluster of unitegmic ovules/seeds, with a small opening in the cupule near the proximal end (Figure 6.18C). The cupule has been hypothesized as being homologous with the angiosperm carpel. However, the cupule of *Caytonia* is different from what is presumed to be the ancestral carpel morphology, a conduplicate megasporophyll bearing ovules along two margins. In addition, (monosulcate) pollen grains have been discovered at the micropyle of *Caytonia* ovules, evidence that the pollen grains were transported directly to the ovules (perhaps by means of a pollination droplet, as occurs in extant gymnosperms), rather than to a stigmatic region where pollen tubes formed. Thus, the cupule apparently did not function as a carpel in terms of a site for pollen germination. Another interpretation of the cupule of *Caytonia* is that it is the homologue of the second integument apomorphic of all angiosperms, evolving by the reduction of the number of ovules within the cupule to one. In summary, the homology of the reproductive structure in *Caytonia* is difficult to decipher, and no other pteridosperm is clearly an angiosperm progenitor. However, some pteridosperms, like *Caytonia*, may still be more closely related to the angiosperms than to the extant gymnosperms (see Doyle 2006).

Another "pteridosperm" group that has been hypothesized as a close relative to the angiosperms are the Glossopteridales or glossopterids. Glossopterids were trees with simple leaves having a midrib giving rise to an extensive reticulate venation system (Figure 6.19D). Reproductive structures of some glossopterids consist of an appendage adnate to the leaf base. This appendage, termed a "capitulum," is often interpreted as a megaphyll partially enclosing ovules on the lower surface (Figure 16.18E). The ovules are orthotropous (Figure 16.18F) and unitegmic, with a single integument. Thus, the open, leaf-like capitulum may represent a partially closed megasporophyll, transitional to an angiospermous carpel in which the megasporophyll encloses ovules. The bitegmic ovule characteristic of flowering plants must necessarily have evolved independently in glossopterids. See Retallack and Dilcher (1981) and Doyle (2006) for more information on the origin of angiosperms from pteridosperm ancestors.

An example of a fossil that may help elucidate early angiosperm evolution is the genus *Archaefructus*, fairly recently collected from China, and evidently now dated to no earlier than 130 million years ago of the early Cretaceous.

Archaefructus (with two described species) was apparently an aquatic plant, having dissected leaves and elongate reproductive axes, each of the latter with paired stamens below and several-seeded carpels above (Figure 6.19). Although *Archaefructus* appears to have bona fide carpels, its relationship to extant angiosperms is debatable. By one hypothesis the reproductive axis is interpreted as an entire, perianth-less flower (with stamens below and carpels above), the axis perhaps homologous to an elongate receptacle reminiscent of some Magnoliaceae (see Chapter 7). By this interpretation, this reproductive structure might represent an ancestral flower (or flower precursor), and *Archaefructus* might be sister to the extant angiosperms. An alternative hypothesis views the reproductive axis of *Archaefructus* not as a single, achetypal flower, but as an inflorescence of individual, reduced male and female flowers, as seen in some aquatic angiosperms today. By this viewpoint, *Archaefructus* may just as likely represent an extinct off-shoot of an extant lineage within the angiosperms (such as the Nymphaeales). See Doyle (2008), Endress and Doyle (2009), and Sauquet et al. (2017) for analyses on the origin of the flower.

In summary, it seems that more fossils may need to be discovered and described (or reinvestigated with new techniques) before the abominable mystery can be satisfactorily solved. Cladistic analyses help, but there is always the problem of homology assessment with structures that are vastly different from contemporary forms. Despite the fact that the relationships among extant flowering plants are much better known with advanced molecular techniques (see Chapter 7), fossils will be key to understanding their origin. Paleobotanical work should be continuously emphasized as of the utmost importance in understanding plant relationships.

REVIEW QUESTIONS

ANGIOSPERM APOMORPHIES

1. What is another name for the flowering plants?
2. Name the apomorphies of the flowering plants.
3. What is the definition of a flower?
4. Name the major components of a typical flower.
5. Describe the morphology and adaptive significance of the perianth.
6. What is the "ABC" model of floral development, and what species served as the original exemplar for this?
7. What was a major selective pressure that resulted in the evolution of specialized types of flowers?
8. What is unique about the angiosperm stamen, and what are the types and parts of a stamen?
9. What is a theca and of what is it composed?
10. What about the male gametophyte of flowering plants is unique?
11. Describe the structure and function of a mature male gametophyte in the flowering plants.
12. What is the definition of a carpel?
13. What is the difference between carpel, pistil, and gynoecium?
14. Name and describe two major adaptive features of the carpel.
15. Contrast integument number in gymnosperms versus that in angiosperms.
16. Draw and label a mature female gametophyte in the flowering plants.
17. How many cells and nuclei are present in a typical, mature, female gametophyte of the flowering plants?
18. How might the reduced angiospermous female gametophyte be adaptive?
19. What is endosperm and what is its function?
20. What is the difference between a sieve cell and a sieve tube member? In what groups are each found?
21. What type of tracheary element do most angiosperms have, and what is its adaptive significance?

ANGIOSPERM ORIGINS

22. When are the earliest definitive angiosperm fossils found?
23. Describe the example of *Caytonia* and glossopterids as putative angiosperm progenitors, citing evidence for or against this idea.
24. Describe the reproductive structure of *Archaefructus* and indicate two competing hypotheses for its homology.

EXERCISES

1. Collect and observe a flowering plant. Looking at specific parts of the plant, go over in your mind the apomorphies (both macroscopic and microscopic) that have enabled the angiosperms to dominate the world's vegetation. Especially review all parts of a flower, citing the adaptive significance of each component.
2. Place various angiospermous pollen grains on a microscope slide, stain (e.g., with toluidine blue), and observe these reduced male gametophytes under a microscope. Look for the cells and nuclei inside. Are the pollen grains two-celled or three-celled at maturity?
3. Observe an angiosperm ovule in sagittal section under the microscope. Look for the two integuments and the (typically) eight nuclei and seven cells of the female gametophyte.
4. Contrast popcorn (an angiosperm) with pine nuts (a gymnosperm) in terms of the ploidy level and development of the nutritive tissue. Cite the selective advantage that flowering plant seeds might have in this regard.

REFERENCES FOR FURTHER STUDY

Andrews, H. N. 1961. Studies in Paleobotany. Wiley, New York.

APG III. 2009. An update of the Angiosperm Phylogeny Group classification for the orders and families of flowering plants: APG III. Botanical Journal of the Linnean Society 161: 105-121.

Coiro, M., J. A. Doyle, and J. Hilton. 2019. How deep is the conflict between molecular and fossil evidence on the age of angiosperms? New Phytologist: 1-17.

Crane, P. R., E. M. Friis, and K. Pedersen. 1995. The origin and early diversification of angiosperms. Nature 374: 27.

Crepet, W. L. 1998. The abominable mystery. Science 282: 1653–1654.

Cronquist, A. 1981. An integrated system of classification of flowering plants. Columbia University Press, New York.

Davies, T. J., T. G. Barraclough, M. W. Chase, P. S. Soltis, D. E. Soltis, and V. Savolainen. 2004. Darwin's abominable mystery: insights from a supertree of the angiosperms. Proceedings of the National Academy of Sciences of the United States of America 101: 1904–1909.

Doyle, J. A. 2006. Seed ferns and the origin of angiosperms. Journal of the Torrey Botanical Society 133: 169-209.

Doyle, J. A. 2008. Integrating molecular phylogenetic and paleobotanical evidence on origin of the flower. International Journal of Plant Sciences 169: 816-843.

Endress, P. K. and J. A. Doyle. 2009. Reconstructing the ancestral angiosperm flower and its initial specializations. American Journal of Botany 96: 22-66.

Eames, A. J. 1961. Morphology of the angiosperms. McGraw Hill, New York.

Friedman, W. E. and J. H. Williams. 2004. Developmental evolution of the sexual process in ancient flowering plant lineages. The Plant Cell 16, S119–S132, Supplement.

Friedman, W. E. and K. C. Ryerson. 2009. Reconstructing the ancestral female gametophyte of angiosperms: insights from Amborella and other ancient lineages of flowering plants. American Journal of Botany 96: 129–143.

Friis, E. M., J. A. Doyle, P. K. Endress, and Q. Leng. 2003. *Archaefructus*: Angiosperm precursor or specialized early angiosperm? Trends in Plant Science 8: 369–373.

Friis, E. M., K. R. Pedersen, and P. R. Crane. 2000. Reproductive structure and organization of basal angiosperms from the Early Cretaceous (Barremian or Aptian) of western Portugal. International Journal of Plant Sciences 161: S169–S182.

Gould, R. E. and T. Delevoryas. 1977. The biology of *Glossopteris*: evidence from petrified seed-bearing and pollen-bearing organs. Alcheringa 1: 387-399.

Li, H.-T., T.-S. Yi, L.-M. Gao, P.-F. Ma, T. Zhang, J.-B. Yang, M. A. Gitzendanner, P. W. Fritsch, J. Cai, Y. Luo, H. Wang, M. van der Bank, S.-D. Zhang, Q.-F. Wang, J. Wang, Z.-R. Zhang, C.-N. Fu, J. Yang, P. M. Hollingsworth, M. W. Chase, D. E. Soltis, P. S. Soltis, and D.-Z. Li. 2019. Origin of angiosperms and the puzzle of the Jurassic gap. Nature Plants 5:461-470.

Jack, T. 2001. Relearning our ABCs: new twists on an old model. Trends in Plant Science 6: 310–316.

Jenik, P. D., and V. F. Irish. 2000. Regulation of cell proliferation patterns by homeotic genes during *Arabidopsis* floral development. Development 127: 1267–1276.

Magallón, S., S. Gómez-Acevedo, L. L. Sánchez-Reyes, and T. Hernández-Hernández. 2015. A metacalibrated time-tree documents the early rise of flowering plant phylogenetic diversity. New Phytologist 207:437-453.

Retallack, G. and D. L. Dilcher. 1981. Arguments for a glossopterid ancestry of angiosperms. Paleobiology 7: 54-67.

Sauquet, H., M. v. Balthazar, S. Magallón, J. A. Doyle, P. K. Endress, E. J. Bailes, E. B. d. Morais, K. Bull-Hereñu, L. Carrive, M. Chartier, G. Chomicki, M. Coiro, R. Cornette, J. H. L. E. Ottra, C. Epicoco, C. S. P. Foster, F. Jabbour, A. Haevermans, T. Haevermans, R. Hernández, S. A. Little, S. Löfstrand, J. A. Luna, J. Massoni, S. Nadot, S. Pamperl, C. Prieu, E. Reyes, P. d. S. K. M. Schoonderwoerd, S. Sontag, A. Soulebeau, Y. Staedler, G. F. Tschan, A. W.-S. Leung, and J. Schönenberger. 2017. The ancestral flower of angiosperms and its early diversification. Nature Communications 8: 16047.

Soltis, D. E., P. S. Soltis, P. K. Endress, and M. W. Chase. 2005. Angiosperm phylogeny and evolution. Sinauer, Sunderland, MA.

Soltis, D. E., A. S. Chanderbali, S. Kim, M. Buzgo, and P. S. Soltis. 2007. The ABC model and its applicability to basal angiosperms. Annals of Botany 100:155-163.

Stebbins, G. L. 1974. Flowering Plants: Evolution above the Species Level. Belknap Press of Harvard University Press, Cambridge, MA.

Sun, G., D. L. Dilcher, S. Zheng, and Z. Zhou. 1998. In search of the first flower: a Jurassic angiosperm, *Archaefructus*, from Northeast China. Science 282: 1692–1695.

Sun, G., Q. Ji, D. L. Dilcher, S. Zheng, K. C. Nixon, and X. Wang. 2002. Archaefructaceae, a new basal angiosperm family. Science 296: 899–904.

Stewart, W. N., and G. W. Rothwell. 1993. Paleobotany and the evolution of plants. Cambridge University Press, New York.

Takhtajan, A. L. 1991. Evolutionary Trends in Flowering Plants. Columbia University Press, New York.

Theissen, G. and R. Melzer. 2007. Molecular mechanisms underlying origin and diversification of the angiosperm flower. Annals of Botany 100:603-619.

Thomas, H. H. 1925. The Caytoniales, a new group of angiospermous plants from the Jurassic rocks of Yorkshire. Philosophical Transactions of the Royal Society of London 213: 299–363.

Veit, B., R. J. Schmidt, S. Hake, and M. F. Yanofsky. 1993. Maize floral development: new genes and old mutants. The Plant Cell 5: 1205–1215.

Zanis, M. J., P. S. Soltis, Y. L. Qiu, E. Zimmer, and D. E. Soltis. 2003. Phylogenetic analyses and perianth evolution in basal angiosperms. Annals of the Missouri Botanical Garden 90: 129–150.

7

DIVERSITY AND CLASSIFICATION OF FLOWERING PLANTS:

AMBORELLALES, NYMPHAEALES, AUSTROBAILEYALES, MAGNOLIIDS, MONOCOTS, AND CERATOPHYLLALES

PLANT SYSTEMATICS
https://doi.org/10.1016/B978-0-12-812628-8.50007-9,

INTRODUCTION

The phylogenetic relationships within the angiosperms have been and continue to be a field of active research in plant systematics. Much progress has been made with the use of phylogenetic methodologies and algorithms and the elucidation of morphological, anatomical, embryological, palynological, karyological, chemical, and molecular evolution (see Chapters 9–14). The more recent use of entire plastome and cistron genomes and multiple nuclear genes has been particularly useful in assessing higher level angiosperm relationships. Whole genomes have been obtained for a number of plants, data that may be commonplace in the near future. But, research continues, especially with respect to certain recalcitrant groups. For a more precise understanding of relationships within a particular group, there is no substitute for consulting the most recent, primary scientific literature.

MAJOR ANGIOSPERM CLADES

Portrayal of the relationships of major angiosperm groups is modeled, with some deviations (see Tables 7.1–7.3; Cole 2015) after the system of the Angiosperm Phylogeny Group, 2016 (referred to as APG IV 2016), which supersedes APG (1998), APG II (2003), and APG III (2009). The APG IV system is based on published cladistic analyses primarily utilizing molecular data (see references cited within). The great majority of classifications of APG IV have been carried over from previous APG editions. In the APG systems, only those angiosperm families that are monophyletic are recognized. In many cases, angiosperm families have been redefined from their past, traditional circumscription, either being split into separate groups, e.g., traditional "Liliaceae" and "Scrophulariaceae," or united into one family, e.g., the Bombaceae, Malvaceae, Sterculiaceae, and Tiliaceae united into one family, Malvaceae, s.l. (see Chapter 8). The APG systems have classified one to several families into orders (these having the ending "-ales"; see Chapters 1, 16), where strong evidence suggests that the order is monophyletic. It must be understood, however, that the designated orders are not comparable evolutionary units and are not indicative of a hierarchical classification system (see Chapter 2). For example, a single "order" may be sister to a monophyletic group containing several orders. The orders can be viewed simply as convenient placeholders for one or more families that appear to comprise a monophyletic group with relatively high certainty. Some monophyletic groups containing several orders are given names, such as Mesangiosperms (*Mesangiospermae*), Magnoliids (Magnoliidae), Monocotyledoneae (monocots), Commelinids (Commelinidae), eu-dicots (*Eudicotyledoneae*), Superrosids (*Superrosidae*), Rosids (Rodidae), Superasterids (*Superasteridae*), Asterids (Asteridae) (Figure 7.1). Many of these have been formally named according to a phylogenetic system of nomenclature (see Cantino et al., 2007).

The precise interrelationships of the major groups of angiosperms show some differences among even recent analyses, but most relationships have generally converged. Figure 7.1 illustrates higher level phylogenetic relationships from Moore et al. (2011), Soltis et al. (2011), APG IV (2016), Sun et al. (2016), and Givnish et al. (2018). Note that some polytomies persist. The elucidation of lineages and clades arising from the common ancestor of the angiosperms and varioius groups within had yielded insight into their evolution, including timing of origin, character shifts, and habitat radiations.

As seen in Figure 7.1, the angiosperms can be broadly delimited into several groups: the Amborellales, Nymphaeales, Austrobaileyales, Chloranthales, Magnoliids (consisting of Laurales, Magnoliales, Canellales, and Piperales), monocotyledons or monocots (Monocotyledoneae), Ceratophyllales, and the eudicots. Of these major groups, the current chapter deals with all but the eudicots, which are covered in Chapter 8. Those angiosperm groups other than the eudicots are sometimes referred to as "basal" flowering plants because

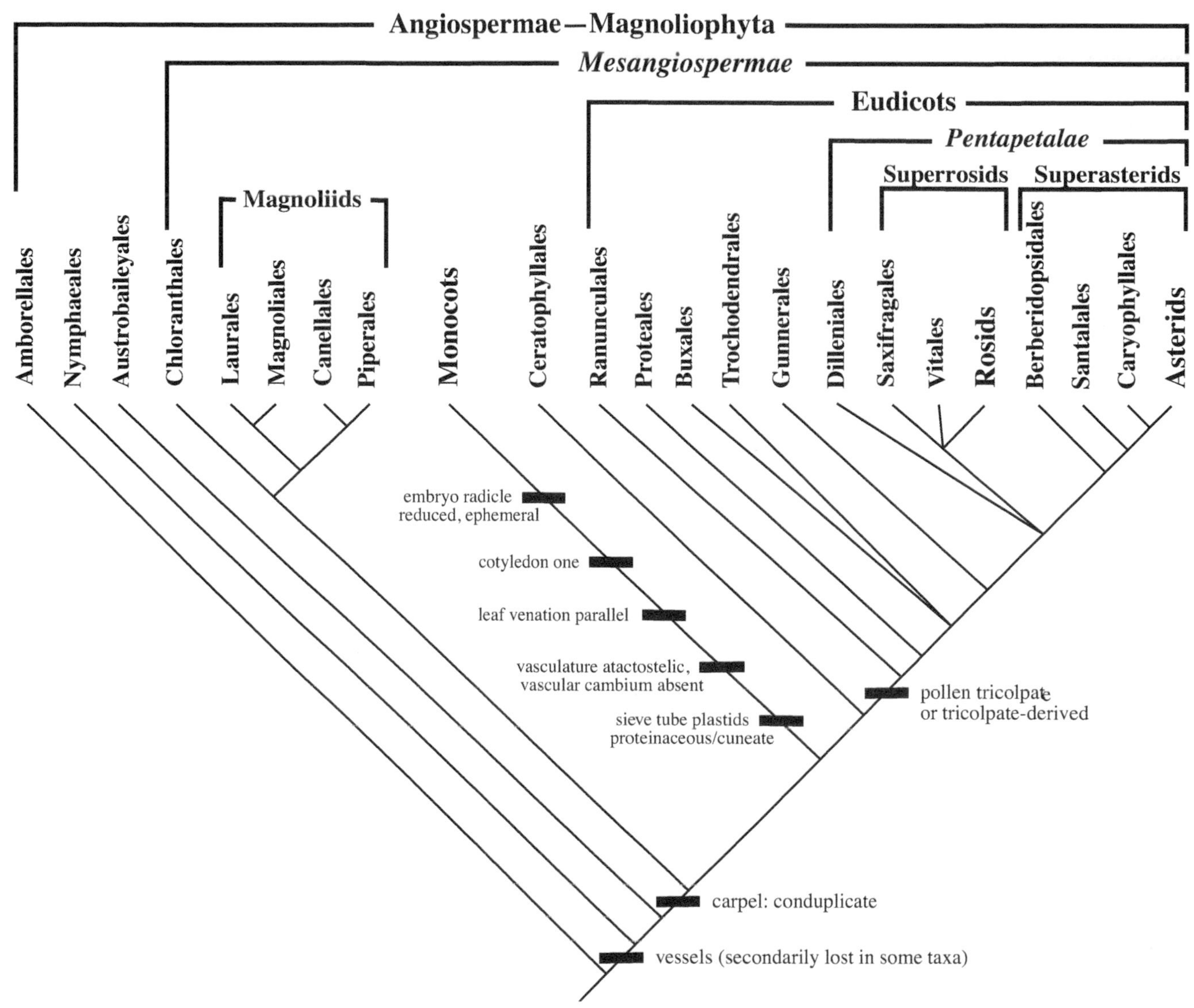

Figure 7.1 Phylogenetic relationships of major angiosperm clades, after Moore et al. (2011), Soltis et al. (2011), APG IV (2016), Sun et al. (2016), and Givnish et al. (2018), with selected apomorphies.

they include the first lineages that diverged from the common ancestor of the angiosperms. (Although difficult to do, the terms "basal" and even "early diverging" should be avoided.) The Amborellales, Nymphaeales, and Austrobaileyales are often referred at the "ANITA" grade (=those other than the *Mesangiospermae*), a paraphyletic assemblage. The families within the orders are listed in Table 7.1 (all except the monocots, including the Ceratophyllales), Tables 7.2 (non-commelinid monocots) and 7.3 (commelinid monocots); eudicot families are listed in Tables 8.1–8.3 of Chapter 8.

The great bulk of the angiosperms in terms of species diversity are contained within the monocots and eudicots. The monocotyledons are a large group, containing approximately 22% of all angiosperms (see later discussion). The eudicots comprise a very large group, including approximately 75% of all angiosperms, and will be treated separately in Chapter 8.

The traditionally defined group "Dicotyledonae," the dicotyledons or dicots, have been defined in the past by their possession of embryos with two cotyledons. It is now accepted that the possession of two cotyledons is an ancestral feature for the taxa of the flowering plants and not an apomorphy for any group within. Thus, "dicots" as traditionally delimited (all angiosperms other than monocots), are paraphyletic and must be abandoned as a formal taxonomic unit.

In the descriptions in this chapter and in Chapter 8, exemplars are used for each order or other major group. The choice of these exemplars is very limited in the context of the huge diversity of the angiosperms. These treatments are not designed as a substitute for the many fine references on flowering plant family characteristics (see the references at the end of this chapter), but are intended as an introduction to some of the common or important groups for the beginning student.

TABLE 7.1 Major groups of the angiosperms, listing the orders and their included families (after APG IV, 2016) for groups other than monocots (see Tables 7.2, 7.3) and eudicots (see Chapter 8). Families in **bold** are described in detail. An asterisk denotes a deviation from APG IV, with brackets indicating the more inclusive family recommended by APG IV.

ANGIOSPERMS		
AMBORELLALES	MAGNOLIIDS	MAGNOLIIDS (continued)
Amborellaceae	LAURALES	CANELLALES
NYMPHAEALES	Atherospermataceae	Canellaceae
Cabombaceae	Calycanthaceae	**Winteraceae**
Hydatellaceae	Gomortegaceae	PIPERALES
Nymphaeaceae	Hernandiaceae	**Aristolochiaceae** (incl. Lactoridac.)
AUSTROBAILEYALES	**Lauraceae**	Hydnoraceae* [Aristolochiaceae]
Austrobaileyaceae	Monimiaceae	**Piperaceae**
Illiciaceae* [Schisandraceae]	Siparunaceae	**Saururaceae**
Schisandraceae	MAGNOLIALES	**MONOCOTS** (see Table 7.2, p. 210; Table 7.3, p. 238)
Trimeniaceae	**Annonaceae**	CERATOPHYLLALES (p. 272)
CHLORANTHALES	Degeneriaceae	**Ceratophyllaceae**
Chloranthaceae	Eupomatiaceae	**EUDICOTS** (see Chapter 8)
	Himantandraceae	
	Magnoliaceae	
	Myristicaceae	

Taxa at the traditional rank of family are utilized as exemplar units; in a few cases subfamilies or tribes are described. Only major, general features of commonly encountered plant families are presented, with examples cited to show diagnostic features. More thorough descriptions and illustrations of angiosperm families may be obtained from references cited in the family descriptions and listed at the end of the chapter.

FAMILY DESCRIPTIONS

The family descriptions that follow use technical terms that are defined and illustrated in Chapter 9 and listed in the Glossary; some embryological or anatomical terms are defined in Chapters 10 and 11. The descriptions begin with a heading that lists the family name (scientific and common), the etymology of the type genus for the family if known (often from Greek or Latin), and the number of genera and species. The first paragraph is a description of plant characteristics of the family members, starting with plant habit and vegetative features, in the order of **root**, **stem**, and **leaf**. This is followed by reproductive features, in the order of **inflorescence**, **flower**, **perianth** (if undifferentiated) or **calyx** and **corolla** (if differentiated), **androecium**, **gynoecium**, **fruit**, and **seed**. Important anatomical or chemical characteristics are occasionally listed as well. The second paragraph lists infrafamilial classification (where pertinent), distribution, economically important members of the family, and citations of recent morphological, phylogenetic, or biogeographical studies. The third paragraph lists the diagnostic features of the family, i.e., how the family can be distinguished from other, related families. This is to aid the beginning student in recognizing the family at a glance; the most important diagnostic features are shown in ***boldface-italics***. Features thought to represent apomorphies for the family or groups within the family are cited as such. Finally, the family descriptions end with a floral formula.

The **floral formulas** are used to summarize the number and fusion of floral parts. In these formulas, **P** refers to perianth parts and is used where the perianth is uniseriate or undifferentiated into a typical outer calyx and inner corolla (e.g., being homochlamydeous, or having outer, calyx-like series and inner corolla-like series that intergrade). If the perianth is differentiated into a distinct calyx and corolla, **K** represents the number of sepals or calyx lobes and **C** the number of petals or corolla lobes. The androecium is denoted by **A** and represents the number of stamens; staminodes may also be tabulated, but are indicated as such in the formula. The gynoecium is denoted by **G**, showing the number of carpels in the gynoecium, followed by "superior" or "inferior" to denote ovary position. Connation, the fusion of similar parts, is illustrated with parentheses "( )" that enclose the number. Separate, discrete whorls of parts are separated by the "+" sign, delimiting the number of parts per whorl; the outermost whorl is indicated by the first number, the innermost whorl by the last number. Numbers that are enclosed by brackets "[ ]" represent a less common or rare condition. If there are more than about 10–12 parts, the "∞" sign is used for "numerous." (See Prenner et al. 2010 for more detail.)

The floral formulas used here summarize the variation that occurs *within the family as a whole*, not necessarily that for a single species. However, floral formulas certainly may also

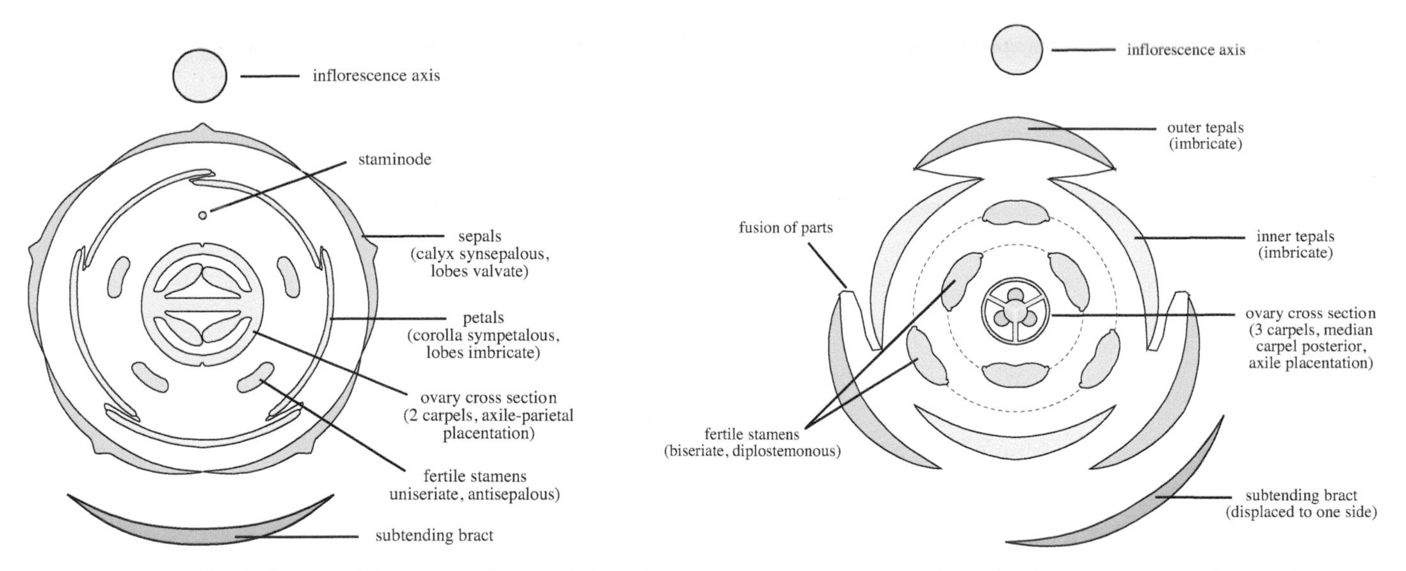

Figure 7.2 Floral diagrams, illustrating relative relationships among the components of the perianth, androecium, and gynoecium.

be used to summarize the floral characteristics of a single species. Some hypothetical examples of floral formulas are:

K (5) [(4)] **C** 5 [4] **A** 5+5 [4+4] **G** 5 [4], superior: represents a flower having a synsepalous calyx with five [rarely four] lobes, an apopetalous corolla of five [rarely four] petals, an androecium with ten distinct (not fused to one another) stamens in two whorls of five each [rarely eight stamens in two whorls of four], and an apocarpous gynoecium with five [rarely four], superior-ovaried carpels.

P (3+3) **A** 3+3 **G** (3), inferior: represents a flower with a homochlamydeous perianth (i.e., one not delimited into calyx and corolla) having connate, outer and inner whorls of three tepals each, six distinct stamens in two whorls of three each, and a syncarpous, inferior-ovaried gynoecium with three carpels.

Family descriptions are accompanied by figures of photographs and line drawings of exemplars. An effort is made to illustrate both diagnostic and apomorphic features. **Floral diagrams** are sometimes illustrated. These represent a diagrammatic cross-sectional view of a flower bud, showing the relative relationship of perianth, androecial, and gynoecial components (examples in Figure 7.2). Floral diagrams may show fusion of floral parts as well as things such as stamen position, placentation, and perianth, calyx, or corolla aestivation (see Chapter 9). They are very useful in visualizing floral structure, and, along with floral formulas, are a succinct summary of the characteristics of the group.

The following are detailed descriptions of selected families (shown in **bold** in Table 7.1) from these major groups. Those selected families were done so largely because live material is more likely to be available for classroom examination and dissection or because of their tremendous importance ecologically or with respect to biodiversity. An attempt was made to describe only information that can be generally seen by the student, unless the characters are of significant diagnostic significance. The source of data for family descriptions was largely taken from *Mabberley's Plant-Book* (Mabberley 2008), an excellent compendium of descriptions of vascular plant families and genera, which I highly recommend as a general reference. Other references used were Cronquist (1981) and Heywood et al. (2007). Very good recent family descriptions are found in the ongoing series *The Families and Genera of Flowering Plants*: Kubitzki et al. (1993, 1998a,b); Kubitzki and Bayer (2002); and Kubitzki (2004). Another good source is *Plants of the World: An Illustrated Encyclopedia of Vascular Plant Families* by Christenhusz et al. (2017). Many families have undergone major changes in circumscription in the APG system. Refer to the references cited earlier and at the end of the chapter for additional information and for descriptions of families not treated here. The *Angiosperm Phylogeny Website* (Stevens, 2001 onward) is an excellent, up-to-date resource for cladograms, classification, references, and apomorphies.

AMBORELLALES

This order comprises one family and one species (below). The Amborellaceae continues, in most molecular studies, to constitute the flowering plant group that is sister to all other angiosperms, although some studies suggest other possibilities (notably that Amborellaceae + Nymphaeaceae together are sister to the rest of the angiosperms; see Figure 7.1). See Mathews and Donoghue (1999, 2000), Qiu et al. (1999, 2000), Graham and Olmstead (2000a,b), Parkinson et al. (1999), Barkman et al. (2000), Zanis et al. (2002), Borsch et al. (2003), Kim et al. (2004), Moore et al. (2007), Moore et al. (2011), Soltis et al. (2011), and Sun et al. (2016) for studies on relationships of *Amborella* within the angiosperms. See Doyle and Endress (2000) and Zanis et al. (2003) for a discussion of character evolution in the basal angiosperms.

The absence of vessels in the order, which is rare in angiosperms, is likely an ancestral condition, and the absence of aromatic (ethereal) oil cells is significant in light of other groups that have them.

Amborellaceae—Amborella family. (type *Amborella*, L. for "around a little mouth," perhaps in reference to the flower). 1 genus and species (Figure 7.3).

The Amborellaceae comprises the single species *Amborella trichopoda*, a dioecious, tropical shrub. The **leaves** are alternate, spiral to distichous, undivided, exstipulate, evergreen, and simple. The **inflorescence** is an axillary cyme. The **flowers** are unisexual, actinomorphic, and hypogynous to perigynous. The **perianth** consists of 5–8, spiral, distinct to basally connate perianth parts (termed sepals by default). The **stamens** of male flowers are ∞, and somewhat laminar. **Anthers** are longitudinal in dehiscence. The **gynoecium** of female flowers is apocarpous, comprising 5–6 superior-ovaried pistils that are apically open. **Placentation** is marginal; the **ovule** is solitary in each pistil. The **fruit** is a drupecetum. Vessels and ethereal oil cells are lacking.

Amborella trichopoda, the single species of the Amborellaceae, is native only to New Caledonia. There are no economic uses, other than being a cultivar sought because of its distinctive, basal position in the angiosperms. See Thien et al. (2003) for a study of population structure and floral biology and Anger et al. (2017) for a study of plant sex and breeding systems of *Amborella*.

Figure 7.3 AMBORELLALES. Amborellaceae, *Amborella trichopoda*. **A.** Whole plant, in cultivation. **B.** Close-up of leaves. **C.** Male flowers, showing laminar stamens. **D.** Female flower close-up, showing spiral perianth and apocarpous gynoecium. (A and C, courtesy of Stephen McCabe; B, courtesy of the Arboretum at University of California-Santa Cruz; D, courtesy of Sandra K. Floyd.)

The Amborellaceae are distinctive in being ***vessel-less***, ***evergreen*** shrubs with ***unisexual*** flowers having an ***undifferentiated***, ***spiral*** perianth, ***numerous***, ***laminar*** stamens, and an ***apocarpous***, ***apically-open*** gynoecium, with ***1-ovuled*** carpels.

Male flowers: **P** 5–8 **A** ∞.
Female flowers: **P** 5–8 **G** 5–6, superior.

NYMPHAEALES

This order consists of three families. The Hydatellaceae were formerly placed in the monocotyledons (order Poales), but recent evidence places it basal in the Nymphaeales (see Saarela et al. 2007). The Nymphaeaceae and Cabombaceae are sometimes treated together (e.g., as subfamilies) in a broader Nymphaeaceae s.l. See Les et al. (1999), Borsch et al. (2008), Borsch and Soltis (2008), and Löhne et al. (2007, 2008) for studies on the phylogeny, classification, and biogeography of the Nymphaeales and included families.

Cabombaceae—Fanwort family (type *Cabomba*, Spanish for a South American aquatic plant). 2 genera (*Brasenia* and *Cabomba*)/6 species (Figure 7.4).

The Cabombaceae consist of aquatic herbs. The underground **stems** are rhizomatous, which give rise to elongate leafy shoots. The stem vasculature is atactostelic. The **leaves** are dimorphic, floating or submersed, exstipulate, spiral, opposite, or whorled, simple and undivided or highly divided into numerous segments. The **inflorescence** consists of a solitary, emergent flower. **Flowers** are bisexual, actinomorphic, and hypogynous. The **perianth** is dichlamydeous (differentiated into calyx and corolla), the parts whorled. The **calyx** consists of 3 [2 or 4] aposepalous sepals. The **corolla** consists of 3 [2 or 4] apopetalous petals. **Stamens** are 3 or 6 (in *Cabomba*) or 12–many (in *Brasenia*); the filaments are somewhat laminar. The **gynoecium** is apocarpous, with a superior ovary, and 2–18 [1] carpels; placentation is parietal; ovules are anatropous, bitegmic, and 2–3 [1] per carpel; styles are terminal or decurrent along the carpel. The **fruit** unit is a coriaceous follicle.

The Cabombaceae are distributed in tropical to temperate areas. *Cabomba* is found in the tropical Americas, whereas the monotypic *Brasenia* (*B. schreberi*) is distributed in tropical to temperate regions of the Americas, Africa, and Australasia. The Cabombaceae are sometimes treated as a subfamily (Cabomboideae) of the Nymphaeaceae, being different from the latter in having a trimerous [2 or 4] number of non-spirally arranged sepals and petals. See Williamson and Schneider (1993) for general information on the family.

The Cabombaceae are distinguished in being ***aquatic herbs*** with ***atactostelic*** stems (resembling those of monocots); dimorphic, floating or submersed, ***undivided or highly divided*** leaves; a perianth with ***3 [2,4] sepals and petals***, and an ***apocarpous*** gynoecium.

K 3 [2,4] **C** 3 [2,4] **A** 3, 6 or 12–∞ **G** 2–18 [1], superior.

Nymphaeaceae—Water-Lily family (type *Nymphaea*, meaning a water nymph). 5–6 genera/60–95 species (Figure 7.5).

The Nymphaeaceae consist of aquatic, annual, or perennial herbs, with a milky latex often present. The underground **stems** are rhizomatous or tuberous. The stem vasculature is a eustele, that may be modified as concentric vascular bundles. The **leaves** are simple, often peltate, stipulate or exstipulate, floating, spiral, usually orbicular in shape. The **inflorescence** consists of a solitary, floating or emergent flower. **Flowers** are bisexual, actinomorphic, and hypogynous or epigynous, with long peduncles arising from the underground stem. The **perianth** is usually differentiated into calyx and corolla, the parts spirally arranged. The **calyx** consists of 4–6 [up to 14],

Figure 7.4 NYMPHAEALES. Cabombaceae. *Brasenia schreberi*. **A.** Floating leaves and emergent flowers. **B,C.** Flower (protogynous) close-up. **B.** Pistils mature. **C.** Stamens mature. All photos courtesy of Jeffrey M. Osborn and Mackenzie L. Taylor.

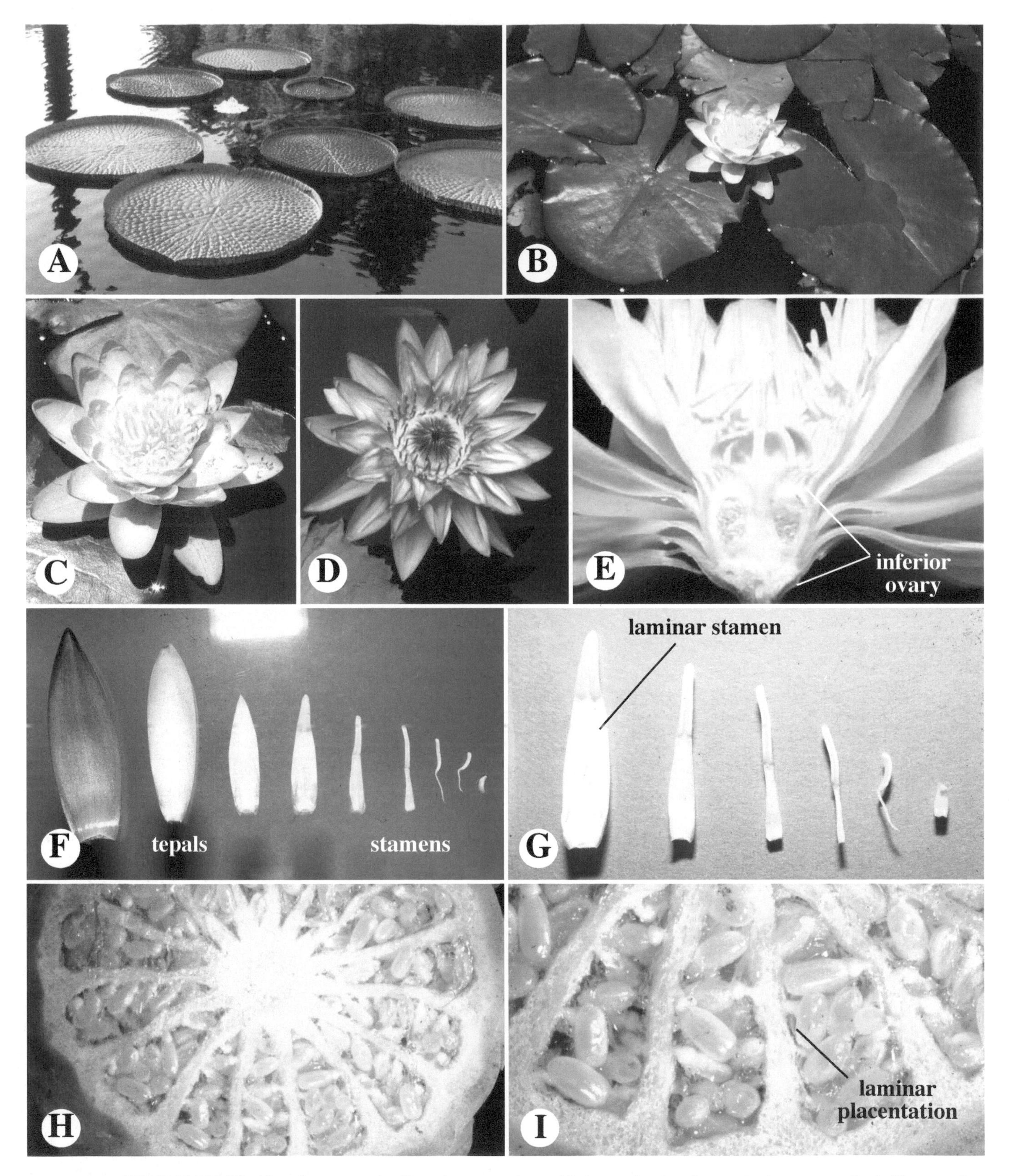

Figure 7.5 NYMPHAEALES. Nymphaeaceae. **A.** *Victoria amazonica*, with large, floating leaves having upturned, rimlike margins. **B–I.** *Nymphaea* spp. **B.** Whole plant, showing floating leaves and solitary flower. **C,D.** Close-up of flower. Note numerous, spiral perianth parts and stamens. **E.** Flower in longitudinal section, showing perianth series, inferior ovary, and numerous stamens. **F.** Removed floral parts (outer to inner = left to right), showing gradation from sepal-like structures (left) to petal like structures (second and third from left) to stamens (right). **G.** Close-up of stamens, showing gradation from outer, laminar stamens (left) to filamentous stamens (middle) to sub-sessile stamens (right). **H.** Ovary cross section. **I.** Close up of ovary cross section, showing laminar placentation, i.e., attachment of ovules to inner surface of septae.

aposepalous sepals. The **corolla** consists of 8–many [0], apopetalous petals, the inner of which grade into laminar stamens. **Stamens** are numerous, spiral, apostemonous; the filaments are laminar to the outside, grading morphologically into petals, to terete toward the flower center; anthers are longitudinal in dehiscence, dithecal, with thecae and connective often extending beyond the anther. The **gynoecium** is syncarpous, with a superior or inferior ovary, and 3–many carpels; placentation is laminar or parietal; ovules are anatropous, bitegmic, and numerous per carpel. The **fruit** is a berry.

The Nymphaeaceae has in the past included the subfamilies Cabomboideae and Nelumboideae, but these are treated here as separate families: Cabombaceae and Nelumbonaceae (the latter distantly grouped within the eudicots; see Chapter 8). Members of the Nymphaeaceae are distributed worldwide. Economic uses include species with edible rhizomes and seeds; many species are used as ornamental cultivars, especially *Nuphar* (cow-lily), *Nymphaea* (water-lily), and *Victoria* (giant water-lily), the last having huge, peltate, floating leaves with upturned, ridged margins. See Schneider and Williamson (1993) for general information and Borsch and Soltis (2008) and Löhne et al. (2007, 2008) for phylogenetic studies.

The Nymphaeaceae are distinguished from related families in being ***aquatic*** herbs with ***floating*** leaves and ***solitary***, ***floating to emergent*** flowers with mostly ***spiral*** floral parts and petals ***grading into usually laminar*** stamens.

K 4–6 [–14] **C** 8-∞ [0] **A** ∞ **G** (3–∞), superior or inferior.

AUSTROBAILEYALES

The Austrobaileyales consist of 3 or 4 families (Table 7.1). Only the Illiciaceae (treated as a component of the Schisandraceae in APG IV, 2016) are described here.

Illiciaceae [Schisandraceae]—Star-Anise family (type *Illicium*, L. for alluring). 1 genus/42 species (Figure 7.6).

The Illiciaceae consist of trees and shrubs with aromatic (ethereal) oil cells. The **leaves** are simple, spiral (often appearing whorled), pellucid-punctate, exstipulate, evergreen, and glabrous. The **inflorescence** is an axillary or supra-axillary, solitary flower or group of 2 or 3 flowers. The **flowers** are small, bisexual, actinomorphic, and hypogynous. The **perianth** consists of numerous (7–33), distinct tepals, typically spirally arranged, the outer sepal-like parts grading into inner petallike parts, which grade into central anther-like parts. The **stamens** are few to numerous (4–ca. 50), in one or more spiral series, and apostemonous; filaments are short and thick. **Anthers** are longitudinal in dehiscence, with an extended connective. The **gynoecium** is apocarpous, with numerous (5–21), superior, unilocular carpels in a single whorl. The **style** is open. **Placentation** is ventrally sub-basal; **ovules** are anatropous, 1 per carpel. The **fruit** is an aggregate of follicles (follicetum). The **seeds** are endospermous, the endosperm oil-rich. Flowers are beetle-pollinated.

The Illiciaceae are distributed in southeastern Asia and southeastern U.S. and Caribbean. Economic importance includes *Illicium anisatum*, Japanese anise, used to kill fish and used medicinally and in religious rites, and *Illicium verum*, star anise, used as a spice, e.g., in liqueurs (Figure 7.7). See Hao et al. (2000) and Liu et al (2006) for phylogenetic studies of the Illiciaceae and related family Schisandraceae.

The Illiciaceae are distinctive in being ***evergreen*** trees or shrubs having ***aromatic oil cells***, with glabrous, spiral, ***pellucid-punctate***, ***exstipulate*** leaves, the flowers with ***numerous***, ***spiral*** tepals (outer sepal-like, inner petal-like), few-numerous stamens, and few-numerous, ***one-seeded***, ***apocarpous*** pistils ***in a single whorl***, the fruit a ***follicetum***.

P ∞ [7–33] **A** ∞ [4–50] **G** ∞ [5–21], superior.

Figure 7.6 AUSTROBAILEYALES. Illiciaceae. **A.** *Illicium floridanum* flower, face view. (Photo courtesy of Jack Scheper, Floridata.com LC) **B–D.** *Illicium parviflorum*. **B.** Branch, showing simple, glabrous, evergreen leaves and one- to few-flowered inflorescences. **C.** Flower, face view, showing grading perianth. **D.** Flower, longitudinal section, showing encircling stamens and apocarpous gynoecium.

Figure 7.7 AUSTROBAILEYALES. Illiciaceae. *Illicium verum*. Fruits of star anise.

MAGNOLIIDS

The Magnoliids, formally Magnoliidae, recognized by APG IV (2016), contain the four orders Canellales, Laurales, Magnoliales, and Piperales. See Qiu et al. (2005) and Soltis et al. (2007, 2011), Massoni et al. (2014, 2015a,b), and Sun et al. (2016) for recent phylogenetic studies of the complex.

LAURALES

The Laurales, sensu APG IV (2016), contain seven families (Table 7.1), only one of which, the Lauraceae, are described here. Notable among the other families are the **Atherospermataceae** (6–7 genera, 16–25 spp., Australasia/Chile; Figure 7.8A), with opposite, simple, serrate leaves, inner staminodes, and plumose unit fruits; **Calycanthaceae** (3–5 genera, 10–11 spp., n.e. Australia, China, N. America), with opposite, simple, entire leaves and numerous, undifferentiated perianth parts and numerous stamens and carpels within a hollow receptacle (Figure 7.8B,C); and **Monimiaceae** (22–24 genera, 200 spp., mostly tropical S. Hemisphere), with opposite, simple, glandular leaves and small, usu. unisexual flowers. See Endress and Igersheim (1997), Renner (1999), and Renner and Chanderbali (2000) for further information on the order and Soltis et al. (2007, 2011) and Massoni et al. (2014, 2015a,b) for phylogenetic analyses.

Lauraceae—Laurel family (type *Laurus*, L. for laurel or bay). ca. 50 genera/ca. 2,500+ species (Figure 7.9).

The Lauraceae consist of mostly trees or shrubs (except *Cassytha*, a parasitic vine) with aromatic oil glands. The **leaves** are evergreen, simple, exstipulate, spiral, rarely whorled or opposite, undivided or lobed, pinnate-netted, usually punctate. The **inflorescence** is an axillary cyme or raceme, rarely a solitary flower. **Flowers** are small, bisexual or unisexual, actinomorphic, perigynous or epiperigynous, the subtending receptacle often enlarging in fruit. The **perianth** is 1–3-whorled, usu. 3+3 [6, 2+2, or 3+3+3], apotepalous, hypanthium present. **Stamens** are 3–12 or more, with staminodes often present as an inner whorl; filaments often have a pair of basal, nectar-bearing appendages; anthers are valvular, with 2–4 [1] valves per anther opening from the base, introrse or extrorse in dehiscence, dithecal [monothecal], tetrasporangiate [bi- or monosporangiate]. The **gynoecium** consists of a single superior [inferior in *Hypodaphnis*] ovary, unicarpellous, [up to 3], with 1 locule, 1 terminal style, and 1–3 stigmas; placentation is apical; ovules are anatropous, bitegmic, 1 per carpel. The **fruit** is a berry, drupe, or is dry and indehiscent, often with an enlarged receptacle and accrescent calyx; seeds are exalbuminous.

The Lauraceae are distributed in tropical to warm temperate regions, esp. S.E. Asia and tropical America. Economic importance includes several timber trees, spice, and other

Figure 7.8 LAURALES. **A.** Atherospermataceae, *Doryphora sassafras*. Note valvular anthers. **B,C.** Calycanthaceae, *Calycanthus occidentalis*, spicebush. Note numerous, undifferentiated perianth parts and numerous stamens and carpels.

Figure 7.9 LAURALES. Lauraceae. **A.** *Sassafras albidum*, sassafras. **B–D.** *Laurus nobilis*, laurel. **B.** Branch, in flower. **C.** Whole flower, showing tepals and multiple whorls of stamens. **D.** Anther, showing valvular dehiscence from base of anther, one valve per theca. **E–I.** *Persea americana*, avocado. **E.** Shoot and inflorescence. **F.** Flower. **G.** Anther, showing valvular dehiscence from base of anther, two valves per theca. **H.** Pistil, ovary in longitudinal section, showing single, apical ovule. **I.** Mature fruit, a drupe. **J–M.** *Cinnamomum camphora*, camphor tree. **J.** Shoot, with inflorescence. **K.** Flower. **L.** Drupes, with enlarged receptacle at base. **M.** Drupe longitudinal section, showing hard endocarp.

flavoring plants (including the bark of *Cinnamomum cassia*, cassia, and *C. zeylanicum*, cinnamon; oils derived from *C. camphora*, camphor; and the leaves of *Laurus nobilis*, laurel or bay), and food plants, especially avocado, *Persea americana*. See Rohwer (1993) for general information and Rohwer and Rudolf (2005) for a phylogenetic study.

The Lauraceae are distinguished in being ***perennial trees or shrubs*** [rarely vines] with ***aromatic oil glands***, ***evergreen*** leaves, an ***undifferentiated*** perianth, ***valvular*** anther dehiscence, and a ***single***, ***superior*** ovary having one ovule per carpel with ***apical*** placentation, seeds lacking endosperm.
P 3+3 [6, 2+2, or 3+3+3] **A** 3–12+ **G** 1 [-(3)], superior, rarely inferior, hypanthium present.

MAGNOLIALES

The Magnoliales, sensu APG IV (2016), contain six families (Table 7.1), two of which are described here. Notable among the others are the **Myristicaceae**, containing *Myristica fragrans*, from which are derived nutmeg and mace (from the seeds and aril, respectively). See Sauquet et al. (2003) and general references on angiosperm phylogeny (e.g., Soltis et al. 2007, 2011), Massoni et al. (2014).

Annonaceae—Custard-Apple family (type *Annona*, from a Haitian name). ca. 106 genera/2,400 species (Figure 7.10).

The Annonaceae consist of trees, shrubs, or woody vines (lianas). The **leaves** are usually distichous, simple, and exstipulate. The **inflorescence** is a solitary flower or cyme. The **flowers** are bisexual [unisexual] and hypogynous. The **perianth** is triseriate, usu. 3+3+3, hypanthium absent. The **stamens** are numerous, usually spiral, apostemonous, rarely basally connate. **Anthers** are longitudinally dehiscent. The **pollen** is released as monads, tetrads, or polyads. The **gynoecium** consists of numerous carpels with superior ovaries, either apocarpous with usually spiral carpels, or rarely syncarpous with whorled carpels. **Placentation** is variable; **ovules** are anatropous or campylotropous, bitegmic or rarely tritegmic, 1-numerous per carpel. The **fruit** is an aggregate of berries or dry and indehiscent units, or a syncarp in which the unit berries fuse to a fleshy receptacular axis. The **seeds** are endospermous, the endosperm ruminate (having an uneven, coarsely wrinkled texture), oily, sometimes starchy. Resin canals and a septate pith are usually present.

The Annonaceae have a mainly tropical distribution. Economic importance includes *Annona* spp. (e.g., *Annona cherimola*, cherimoya/custard-apple) grown for their edible fruits, species used for scent (e.g., *Cananga odorata* (ylang-ylang) used in perfumes) or timber, and some cultivated ornamentals, e.g., *Polyalthia*. See Doyle and Le Thomas (1994, 1996) for analyses emphasizing pollen evolution and Pirie et al. (2006, 2007) and Guo et al. (2017) for phylogenetic studies.

The Annonaceae are distinctive in being trees, shrubs, or woody vines with ***simple***, usually ***distichous*** leaves, a ***trimerous*** perianth, ***numerous***, ***usually spiral*** stamens and pistils (apocarpous or syncarpous), and seeds with ***ruminate endosperm***.
P 3+3+3 **A** ∞ **G** ∞, superior.

Magnoliaceae—Magnolia family (type *Magnolia*, after Pierre *Magnol*, 1638–1715). 7 genera/200 species (Figure 7.11, 7.12).

The Magnoliaceae consist of species of trees or shrubs. The **leaves** are simple, spiral, pinnate-netted, and stipulate, with caducous stipules enclosing the buds, leaving encircling stipule scars. The **inflorescence** is a terminal solitary flower. **Flowers** are large, bisexual (rarely unisexual), actinomorphic, hypogynous; the receptacle grows into an elongate axis (called a torus or androgynophore), which bears the androecium and gynoecium. The **perianth** is multiwhorled or spiral, apotepalous. **Stamens** are numerous, spiral, apostemonous; filaments are thickened to laminar; anthers are longitudinal in dehiscence (variable in direction), tetrasporangiate, dithecal, the paired sporangia sometimes appearing embedded, with a connective often extending beyond thecae. The **gynoecium** is apocarpous, with [2-] numerous, superior, spirally arranged ovaries/carpels, each unilocular, with one terminal style, and one stigma; placentation is marginal; ovules are anatropous and bitegmic, 2–numerous per carpel. The **fruit** is an aggregate of follicles, samaras, or berry-like units; seeds are endospermous, rich in oils and protein with a sarcotesta (fleshy seed coat resembling an aril) often present.

The Magnoliaceae are distributed in tropical to warm temperate regions, especially in the northern hemisphere. Economic importance includes ornamental cultivars and some important timber trees, e.g., *Liriodendron*, *Magnolia*, and *Michelia*. See Kim et al. (2001) for a detailed treatment and Sauquet et al. (2003), Azuma et al. (2001), and Nie et al. (2008) for phylogenetic studies.

The Magnoliaceae are distinguished in being trees and shrubs with simple ***stipulate*** leaves (twigs with ***encircling stipule scars***), ***solitary*** flowers, a usually ***undifferentiated petaloid*** perianth with ***numerous*** tepals, and ***numerous***, ***spiral*** stamens and an ***apocarpous gynoecium*** of ***numerous***, ***spiral*** pistils born on ***elongate receptacular axis (torus or androgynophore)***, the fruit an ***aggregate*** of follicles, berries, or samaras, seeds usually with a ***sarcotesta***.
P ∞ **A** ∞ **G** ∞ [2–∞], superior.

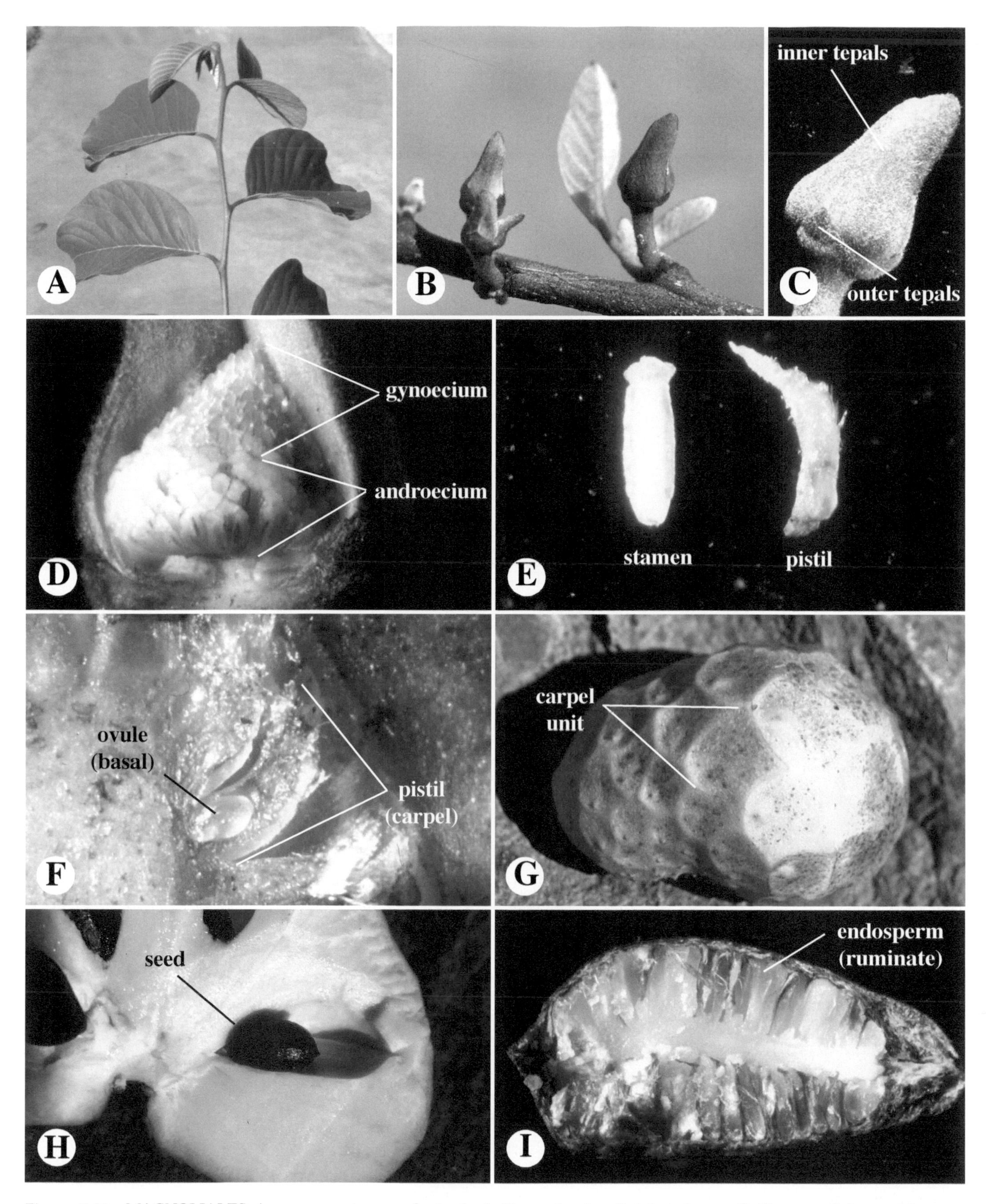

Figure 7.10 MAGNOLIALES. Annonaceae. *Annona cherimola*. **A.** Shoot, showing distichous leaves. **B.** Close-up of shoot, with young leaves and flowers. **C.** Flower close-up, showing undifferentiated perianth. **D.** Flower, perianth removed. Note basal androecium of numerous stamens and apical, apocarpous gynoecium of numerous pistils. **E.** Close-up of removed stamen and pistil. **F.** Pistil, longitudinal section, showing single ovule with basal placentation. **G.** Fruit, a syncarp of laterally fused carpel units. **H.** Fruit in section, showing dark seeds and surrounding fleshy tissue. **I.** Seed in longitudinal section, showing characteristic ruminate endosperm.

Figure 7.11 MAGNOLIALES. Magnoliaceae. *Liriodendron tulipifera*, tulip tree. **A.** Shoot, showing simple leaves and solitary flower. **B.** Twig, with encircling stipule scars. **C.** Flower, with numerous stamens and elongate receptacle. **D.** Flower, side view, showing 3 whorls of perianth parts (3 per whorl). **E.** Fruit, a samaracetum (aggregate fruit of samaras); note receptacle axis. **F.** Unit samaras, each 1-seeded.

CANELLALES

The Canellales, sensu APG IV (2016), contains two families (Table 7.1), one of which is described here. The other family, **Canellaceae**, contains plants of the eastern African and eastern American/Caribbean tropics, some used as flavorings, medicinals, and ornamentals, including *Canella winterana*, wild-cinnamon. See Massoni et al. (2014) for phylogenetic information on the order.

Winteraceae—Wintera family (type *Wintera* [=*Drimys*], after John Winter, 1540–1596, who traveled with Sir Frances Drake). 5 genera/105 species (Figure 7.13).

The Winteraceae consist of vessel-less, evergreen, hermaphroditic or dioecious, aromatic shrubs and trees, insect or wind-pollinated. The **leaves** are simple, spiral, exstipulate, entire, often abaxially glaucous. The **inflorescence** is of solitary flowers or cymose clusters. The **flowers** are bisexual [unisexual], actinomorphic or zygomorphic, hypogynous. The **calyx** consists of 2–4[6] valvate sepals, either aposepalous or synsepalous and basally connate or completely connate and calyptrate. The **corolla** consists of (2–)5–∞, petals, in 2 or more whorls. The **stamens** are 3-∞, spiral, apostemonous, with often laminar filaments. The **pollen** is usually released in tetrads. The **gynoecium** is apocarpous [unicarpellous] of 1–∞ carpels, the carpels not completely closed in some taxa, the ovary superior and unilocular. The **style** is terminal or positioned along the margin of the conduplicate carpels walls. **Placentation** is marginal or laminar; **ovules** are anatropous, bitegmic, ∞ per carpel. The **fruit** is an aggregate of berries or follicles, rarely a capsule or syncarp

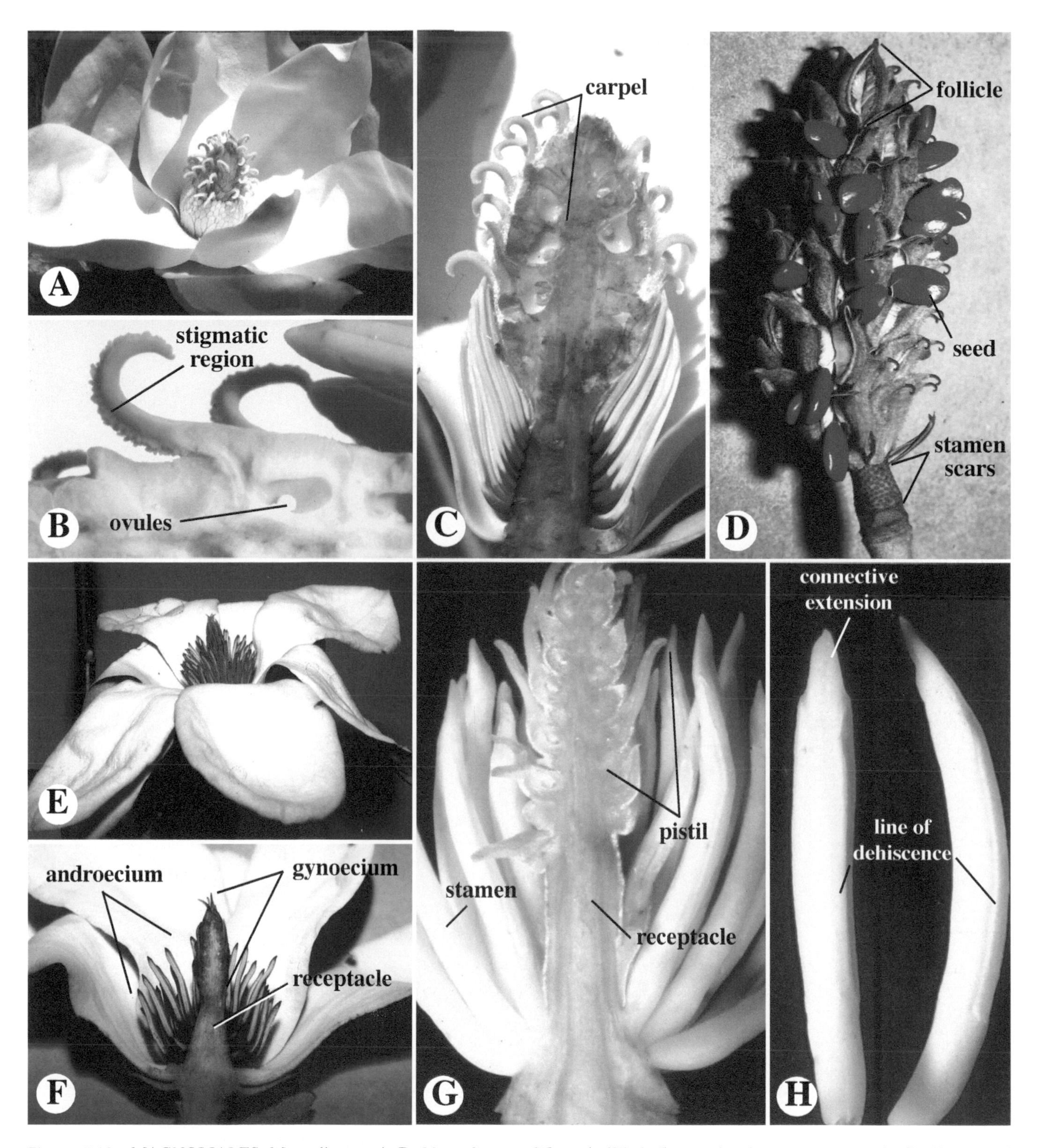

Figure 7.12 MAGNOLIALES. Magnoliaceae. **A–D.** *Magnolia grandiflora*. **A.** Whole flower, showing numerous tepals. **B.** Close-up of pistil. Note marginal placentation. **C.** Flower l.s., showing pistils. **D.** Fruit, an aggregate of follicles. Note seeds, having fleshy (red) sarcotesta. **E,F.** *Magnolia stellata*. **E.** Whole flower. **F.** Flower l.s. Note elongate, central receptacle (torus, androgynophore). **G,H.** *Michelia doltsopa*. **G.** Flower l.s., close-up, showing androecium (below) and receptacle bearing pistils. **H.** Stamens, adaxial (left) and side (right) views. Note lack of differentiation between filament and anther.

Figure 7.13 CANNELALES. Winteraceae. *Drimys winteri*. **A.** Plant, a tree. **B.** Inflorescence, a cymose cluster. **C.** Flower, showing multiseriate corolla. **D.** Flower close-up, showing (slightly) laminar stamens and apocarpous gynoecium. **E.** Flower back side, showing two sepals, a common condition. **F.** Leaves, abaxially glaucous, common in the family.

(by post-genital fusion of carpels). The **seeds** are oily endospermous.

The Winteraceae are often classified into two subfamilies: Taktajanioideae (*Taktajania*, Madagascar) and Winteroideae (*Drimys*, *Pseudowintera*, *Tasmannia*, and *Zygogynum* [*Bubbia*, *Exospermum*, *Belliolum*]). Members of the family are distributed in Australia, New Guinea, Madagascar, montane South America, and the Pacific southwest. Economic importance includes some cultivated ornamentals and medicinal plants, e.g., *Drimys*. See Suh et al. (1993), Endress et al. (2000), Karol et al. (2000), Doust (2003), Doust and Drinnan (2004), and Marquínez et al. (2009) for general studies of the family. See Marquínez et al. (2009) and Thomas et al. (2014) for phylogenetic and phylogeographical analyses of the family.

The Winteraceae are distinctive in being ***vessel-less***, ***evergreen***, ***aromatic*** shrubs or trees with exstipulate, entire, simple, spiral, often abaxially glaucous leaves, the flowers bisexual [unisexual] with valvate or basally connate to calyptrate sepals, ***often laminar stamens***, often ***tetradhedral pollen***, an ***apocarpus [unicarpellous] gynoecium***, the fruit an ***aggregate*** of berries or follicles [capsule or syncarp].

K 2–4[6] or (2–4[6]) C [2–]5-∞ A 3-∞ G 1–∞, superior.

PIPERALES

The Piperales, sensu APG IV (2016), contain 3–4 families (Table 7.1), three of which are described here. Notable in the order is the achlorophyllous, parasitic **Hydnoraceae**. See Nickrent et al. (2002) and Wanke (2007a) for a phylogenetic analysis of the order.

Aristolochiaceae—Birthwort family (type *Aristolochia*, Gr. *aristos*, best + *lochia*, childbirth, from resemblance of a species of the genus to the correct fetal position). 5–8 genera/465–480 species (Figure 7.14).

The Aristolochiaceae consist of hermaphroditic [gynomonoecious in *Lactoris*] shrubs, vines, or rhizomatous herbs, usually climbing. The **leaves** are simple, petiolate, spiral or distichous, and usually exstipulate (stipules present, large in *Lactoris*). The **inflorescence** consists of a solitary flower or of terminal or lateral racemes or cymes. **Flowers** are bisexual, actinomorphic or zygomorphic (in *Aristolochia*), generally epigynous. The **perianth** consists of a three-lobed, synsepalous, petaloid calyx. The **corolla** is absent, of 3 petals (*Saruma*), or reduced to 3 minute petal-like structures (in *Asarum*). **Stamens** are 6–ca. 40 (staminodes sometimes present), free or fused with the style forming a gynostemium (also called a column or gynostegium); filaments, when present, are short and thick; anthers are longitudinal and extrorse [introrse] in dehiscence, dithecal. The **gynoecium** is syncarpous, with an inferior or half-inferior [superior in *Lactoris*] ovary, with 3–6 carpels, 3–6 locules, one style, and 3–6 stigmas; placentation is axile [parietal in *Lactoris*]; ovules are usually anatropous, bitegmic, many per carpel. The **fruit** is usually a capsule, less commonly a schizocarp of follicles or indehiscent; seeds are oily to starchy endospermous.

Members of the family have distributions in tropical and warm temperate regions, esp. in the Americas. Economic importance includes cultivated ornamentals, e.g., *Aristolochia* (Dutchman's-pipe, pelican flower, birthwort) and *Asarum* (wild ginger), with some species used medicinally (*Aristolochia*, *Thottea*), some to cure snakebites. The monospecific Lactoridacae (*Lactoris fernandeziana*) of the Juan Fernandez Islands, Chile, is sometimes included within the Aristolochiaceae, as are the two genera of the parasitic group, Hydnoraceae. See Kelly and Gonzalez (2003), Neinhuis et al. (2005), Ohi-Toma et al. (2006), Wanke et al. (2006a), and Naumann et al. (2013) for phylogenetic studies of the family.

The Aristolochiaceae are distinguished in being usually ***climbing*** plants, having an ***enlarged***, ***petaloid*** calyx, an ***absent to reduced*** corolla, often adnate stamens (forming a ***gynostemium***), and an ***inferior to superior***, ***3–6-carpeled and loculed*** ovary.

K (3) **C** 0 [3] **A** 6–∞, usu. adnate to style **G** (3–6), inferior, half-inferior, or superior.

Piperaceae—Pepper family (type *Piper*, from an Indian name for pepper). 5 genera/ca. 3600 species (Figure 7.15).

The Piperaceae consist of herbs, shrubs, vines, or trees. The **leaves** are spiral, simple, stipulate (the stipules adnate to the petiole) or exstipulate. The **inflorescence** is a spike or spadix. The **flowers** are very small, bisexual or unisexual, actinomorphic, bracteate, with bracts peltate, and hypogynous. The **perianth** is absent. The **stamens** are 3+3 [1–10]. **Anthers** are longitudinally dehiscent, dithecal (sometimes appearing monothecal by fusion of thecae). The **gynoecium** consists of a single pistil with a superior ovary, having 1 or 3–4 carpels, and one locule. The **style** is absent or solitary; **stigma(s)** are 1 or 3–4, being brushlike and lateral in *Peperomia*. **Placentation** is basal; **ovules** are orthotropous, bitegmic or (in *Peperomia*) unitegmic, one per ovary. The **fruit** is a 1-seeded berry or drupe. The **seeds** have a starchy perisperm (the endosperm scanty). Plants have spherical, aromatic (ethereal) oil cells in the parenchyma and an atactostele-like vasculature (but with an outer cambium).

Members of the family have distributions in tropical regions. Economic importance includes *Piper nigrum*, the source of black and white pepper; other species are used for flavoring, medicinal plants, euphoric plants (e.g., *Piper methysticum*,

Figure 7.14 PIPERALES. Aristolochiaceae. **A.** *Asarum canadense*, wild ginger. **A.** Whole plant, showing flower with trimerous calyx. **B,C.** *Hexastylis minor*. **B.** Whole plant. **C.** Close-up of flower, showing trimerous, petaloid calyx. **D,E.** *Aristolochia elegans*. **D.** Flower bud, just prior to opening. **E.** Mature flower, face view. **F,G.** *Aristolochia macrophylla*. **F.** Vine, with large, cordate leaves and flower (circled). **G.** Flower longitudinal section. Note synsepalous perianth and inferior ovary. **H–M.** *Aristolochia trilobata*. **H.** Flower base longitudinal section, showing inferior ovary and gynostemium (column or androgynophore). **I.** Gynostemium close-up. Note anther thecae and stigmatic surface. **J.** Gynostemium, upper view, showing six lobes, corresponding to carpels. **K.** Gynostemium cross section. **L.** Close-up of K, showing thecae of anthers. **M.** Ovary cross section. Note six carpels and axile placentation.

Figure 7.15 PIPERALES. Piperaceae. **A.** *Piper nigrum*, pepper. Vegetative morphology. **B.** *Peperomia argyreia*, watermelon peperomia. Spadix inflorescence. **C.** *Peperomia* sp. Close-up of inflorescence, showing numerous small, bracteate flowers. Note absence of perianth. **D–G.** *Macropiper excelsum*. **D.** Whole plant with spadix. **E.** Immature male flowers. **F.** Mature male flowers, anthers dehiscing. **G.** Inflorescence cross section, showing thick, fleshy axis.

kava), and cultivated ornamentals, e.g., *Peperomia* spp. See Wanke et al. (2006b, 2007a,b), Jaramillo et al. (2008), and Massoni et al. (2014, 2015a,b) for phylogenetic studies.

The Piperaceae are distinctive in having an ***atactostelic*** stem, a ***spike or spadix*** with ***numerous***, ***very small***, ***unisexual or bisexual*** flowers ***lacking*** a perianth, the ovary solitary, 1-ovulate, the fruit a ***1-seeded berry or drupe***.
P 0 **A** 3+3 [1–10] **G** 1 or (3,4), superior.

Saururaceae—Lizard's-Tail family (type *Saururus*, Gr. *saur*, lizard + *our*, tail, after the tail-shaped inflorescence of *Saururus cernuus*). 4 genera/6 species (Figure 7.16).

The Saururaceae consist of perennial herbs. The **leaves** are spiral, simple, and stipulate, the stipules adnate to the petiole. The **inflorescence** is a bracteate spike or raceme, with involucrate bracts enlarged and petal like in some taxa. The **flowers** are bisexual, hypogynous. The **perianth** is absent. The **stamens** are 3, 3+3, or 4+4, apostemonous, adnate to base of the gynoecium in some taxa. **Anthers** are longitudinal in dehiscence. The **gynoecium** is syncarpous or apically apocarpous, with a superior ovary, 3–5 carpels, and one locule. The **styles** are 3–5. **Placentation** is parietal (to marginal in *Saururus*); **ovules** are orthotropous to hemitropous, bitegmic, 1–10 per ovary. The **fruit** is an apically dehiscent capsule. The **seeds** are perispermous. Stems have 1 or 2 vascular bundle rings.

Members of the family have distributions in eastern Asia and N. America. Economic importance includes some cultivated ornamentals. See Meng et al. (2003) and Neinhuis et al. (2005) for a recent phylogenetic studies of the family.

The Saururaceae are distinctive in being ***perennial herbs*** with a ***bracteate spike or raceme*** and with flowers ***lacking*** a perianth, the ovary solitary, many-ovulate, the fruit a ***capsule***.
P 0 **A** 3, 3+3, or 4+4 **G** (3–5), superior.

MONOCOTYLEDONS—MONOCOTS

The monocotyledons, or monocots (also known as the Monocotyledoneae or Liliidae), have long been recognized as a major and distinct group, comprising roughly 56,000 species, 22% of all angiosperms. All recent studies, including several molecular ones, agree with the notion that monocots are monophyletic (Figure 7.1). Monocots include the well-known aroids, arrowleaves, lilies, gingers, orchids, irises, palms, and grasses. Grasses are perhaps the most economically important of all plants, as they include grain crops such as rice, wheat, corn, barley, and rye.

Traditionally, monocots have been defined in part by the occurrence of floral parts in multiples of three. However, this feature is now thought to represent an ancestral condition, one present or common in several non-monocot lineages of flowering plants such as the Laurales, Magnoliales, and Piperales.

The phylogenetic relationships of the major groups of monocots, as summarized from recent studies, are seen in Figure 7.17. The monophyly of monocots is corroborated by several major morphological, anatomical, and ultrastructural apomorphies. These apomorphies will be discussed first, followed by a treatment of the major groups and exemplar families.

MONOCOT APOMORPHIES

First, all monocots have **sieve tube plastids** with **cuneate** (wedge-shaped) **proteinaceous inclusions** (Figure 7.18A) of the "P2 type" (also found in *Asarum* of the Aristolochiaceae; see Behnke 2000). This sieve tube plastid type, which can only be resolved with transmission electron microscopy, is found in all investigated monocotyledons, with some variation in form (Behnke 2000). Thus, it is likely that the cuneate, proteinaceous plastid type constitutes an apomorphy for the monocots (Figures 7.1, 7.17). The adaptive significance of this plastid type in monocots (if any) is unknown.

Second, all monocots have an atactostele stem vasculature, an apparent apomorphy for the group. An **atactostele** (Figure 7.18B) consists of numerous discrete vascular bundles that, in cross section, consist of two or more rings or (more commonly) appear to be rather randomly organized (but which actually have a high complexity of organization). In addition, no monocot has a true vascular cambium that produces true wood (Chapter 5); this feature is likely correlated with the evolution of the atactostele. Thus, for example, tall palm trees have no wood, relying on the deposition and expansion of cells during primary growth for support. Some monocots (e.g., members of the Agavaceae and Asphodelaceae) do have secondary growth by means of so-called "anomalous" cambia, but these do not develop as a single continuous cylinder that deposit rings of secondary tissue, as in plants that produce true wood. A few eudicots (e.g., some Nelumbonaceae) have evolved an atactostele, but this was most likely a secondary innovation. Atactosteles may have evolved in response to selective pressure for adaptation to an aquatic habitat, but this is not clear.

Third, *most* monocots have **parallel leaf venation** (Figure 7.18C–E), another apomorphy for the group. In leaves with parallel venation, the veins are either strictly parallel (as in most grasses), curved and approximately parallel,

Figure 7.16 PIPERALES. Saururaceae. **A–D.** *Anemopsis californica*, yerba santa. **A.** Whole plant, showing basal leaves and scapose spikes. **B.** Inflorescence, close-up, with showy bracts. **C.** Close-up of inflorescence. Note tightly clustered flowers (one flower circled). **D.** Inflorescence in longitudinal section, showing partially embedded gynoecium with parietal placentation. **E,F.** *Saururus cernuus*, lizard's tail. **E.** Whole plant, showing cauline leaves and elongate raceme. **F.** Close-up of inflorescence. **G,H.** *Houttuynia cordata*.

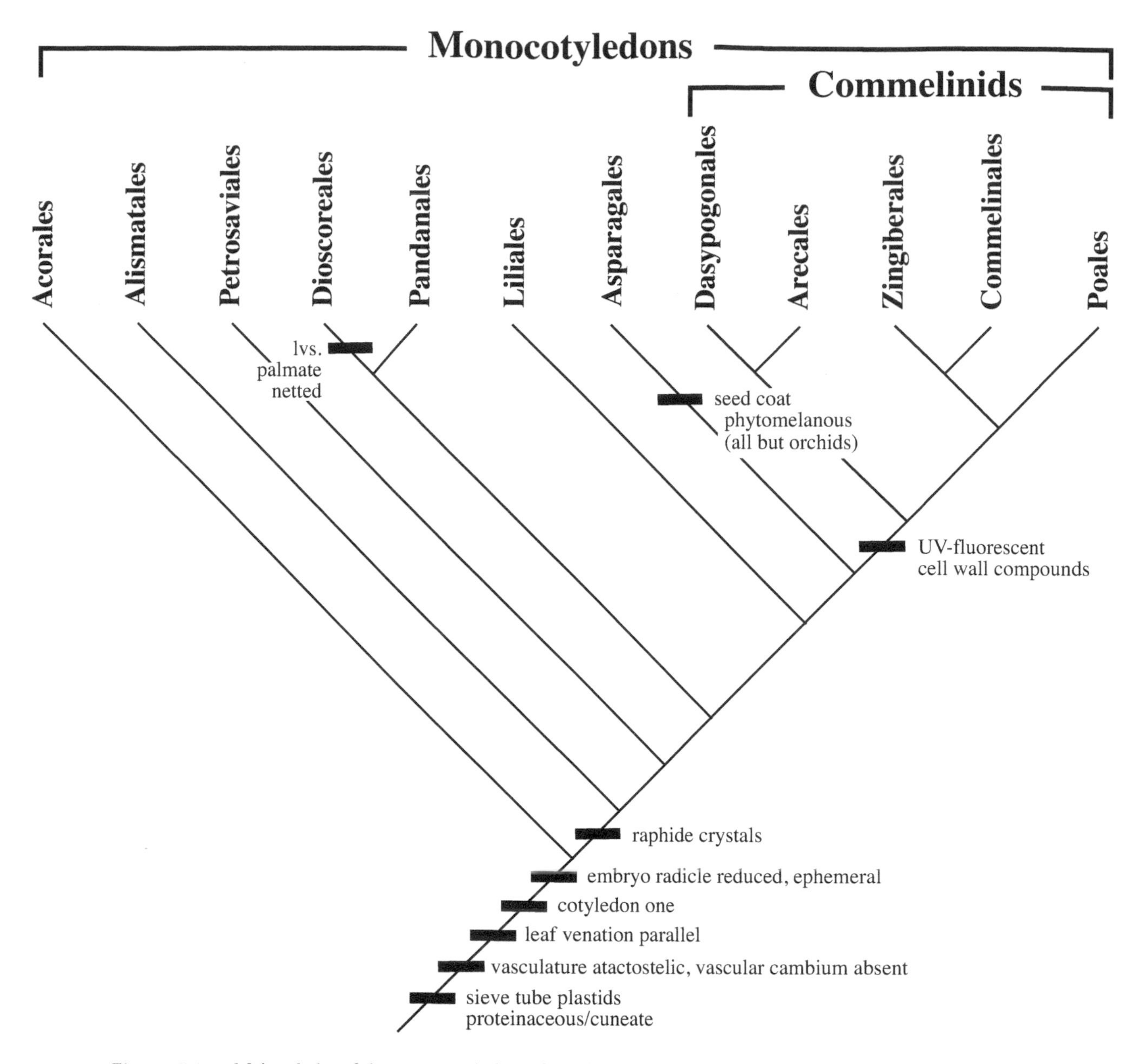

Figure 7.17 Major clades of the monocotyledon, after Givnish et al. (2018), with selected apomorphies shown.

or penni-parallel (= pinnate-parallel). A penni-parallel leaf has a central midrib with secondary veins that are essentially parallel to one another (Figure 7.18E). In all types of parallel venation, the ultimate veinlets connecting the major parallel veins are transverse and do not form a netlike **reticulate venation** (see Chapter 9) as found in almost all nonmonocotyledonous flowering plants. Parallel leaf venation is not a characteristic of all monocots. Numerous monocot taxa, for example some Araceae, the Dioscoreaceae (yam family), Smilacaceae (green briar family), and many others, have a reticulate leaf venation similar to that found in nonmonocots. However, the evidence supports the notion that a reticulate venation evolved in these monocot taxa secondarily, after the common evolution of parallel veins.

Fourth, all monocots have a **single cotyledon** (Figure 7.19), the feature responsible for the name *monocot*. A single cotyledon appears to be a valid apomorphy for all monocots. Its adaptive significance, if any, is unknown. Some of the angiosperm lineages closely related to monocots may have a reduced second cotyledon, but these are almost certainly not homologous.

And, fifth, monocots have a **reduced, ephemeral radicle** (Figure 7.19), the root of the embryo. Subsequent roots of monocot taxa are adventitious (see Chapter 9).

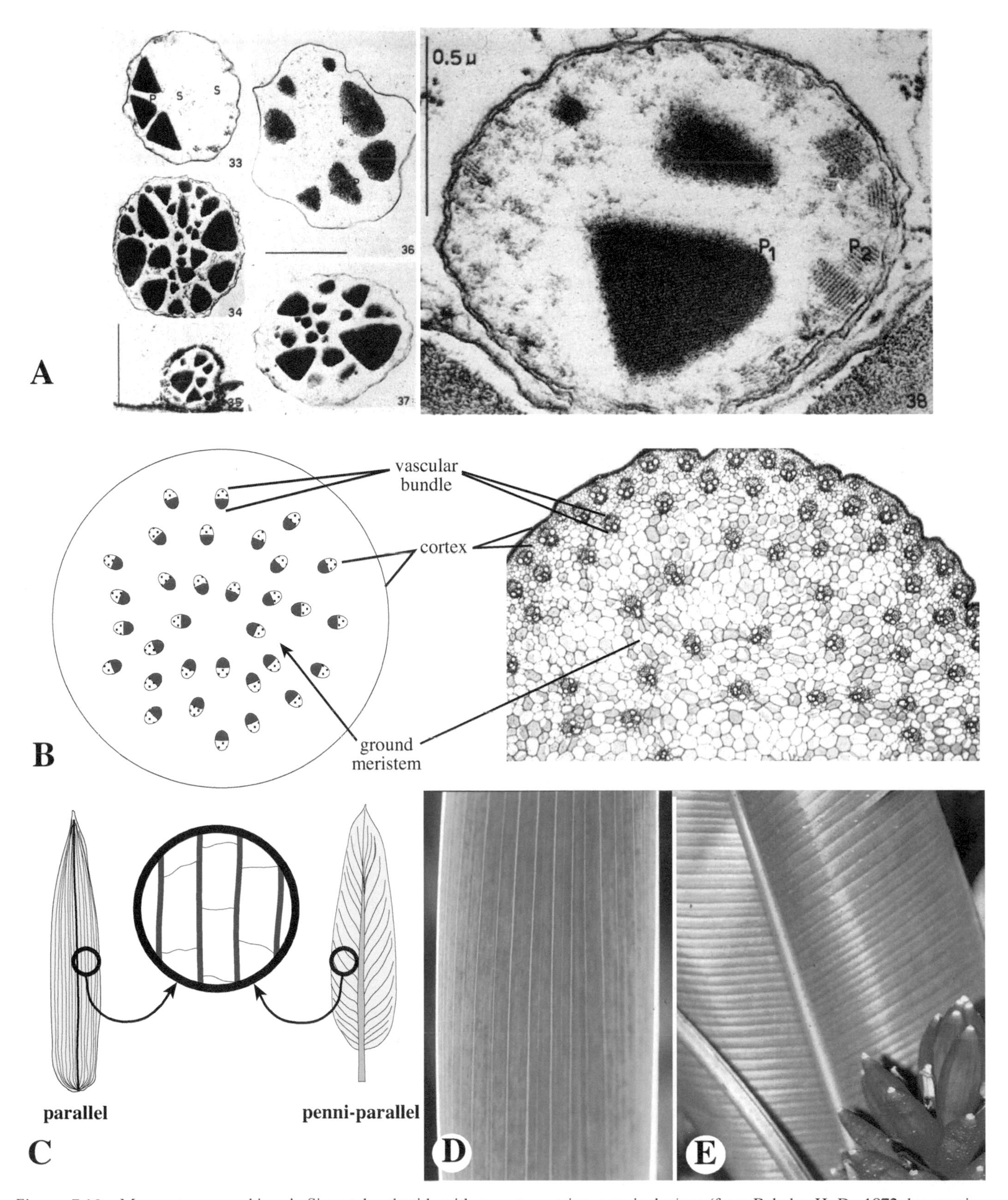

Figure 7.18 Monocot apomorphies. **A.** Sieve tube plastids with cuneate proteinaceous inclusions (from Behnke, H.-D., 1972, by permission). **B.** The atactostele. Note numerous vascular bundles; at left: xylem = dark; phloem = stippled. **C–E.** Parallel venation. **C.** Parallel venation (left) and penni-parallel venation (right), a modification of parallel venation. **D.** *Elymus condensatus* (Poaceae), an example of parallel venation. **E.** *Musa coccinea* (Musaceae), exemplifying penni-parallel venation.

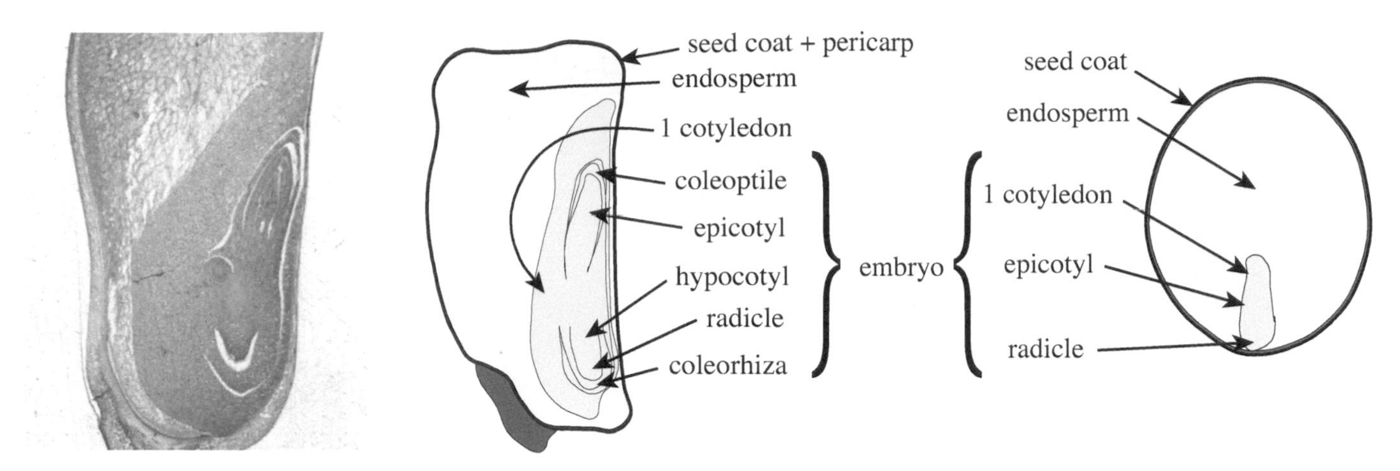

Figure 7.19 A single cotyledon, an apomorphy of the monocotyledons. Left, *Zea mays* (Poaceae). Right, *Xiphidium caeruleum* (Haemodoraceae).

CLASSIFICATION OF THE MONOCOTYLEDONS

The orders of monocots and their included families are listed in Tables 7.2 and (for the Commelinid monocots) 7.3. The Acorales, which consists of the single family Acoraceae and the single genus *Acorus*, continues to be supported as sister to all remaining monocots. The Alismatales, including the family Araceae, is sister to all other monocots except the Acorales. Subsequent, sequential lineages include Petrosaviales, a Dioscoreales-Pandanales clade, Liliales, Asparagales, and the Commelinids. Commelinid monocots form a well-resolved clade that consists of the Dasypogonaceae, the Arecales (the sole member being the Arecaceae, or Palmae, the palms), the Commelinales, the Zingiberales (ginger group), and the Poales (grasses and their

TABLE 7.2 Orders and included families of Monocotyledons (excluding Commelinids, see Table 7.3), based on APG IV, 2016, modified in part by Givnish et al., 2018. Families in **bold** are described in detail. An asterisk denotes a deviation from APG IV, with brackets indicating the more inclusive family recommended by APG IV; s.l. = sensu lato, in the broad sense. Subfamilies of synonymous names after Stevens (2001, onwards).

MONOCOTYLEDONS		
ACORALES	PANDANALES	ASPARAGALES (continued)
Acoraceae	Cyclanthaceae	Asparagaceae s.s.*
ALISMATALES	**Pandanaceae**	**Asphodelaceae s.s.***
Alismataceae	Stemonaceae	Asteliaceae
Aponogetonaceae	Triuridaceae	Blandfordiaceae
Araceae	Velloziaceae	Boryaceae
Butomaceae	LILIALES	Doryanthaceae
Cymodoceaceae	Alstroemeriaceae	Hemerocallidaceae*
Hydrocharitaceae	Campynemataceae	[Hemerocallidoideae, Asphodelaceae s.l.]
Juncaginaceae	Colchicaceae	Hyacinthaceae* [Scilloideae, Asparagaceae s.l.]
Maundiaceae	Corsiaceae	Hypoxidaceae
Posidoniaceae	**Liliaceae**	**Iridaceae**
Potamogetonaceae	Luzuriagaceae*	Ixioliriaceae
Ruppiaceae	Melanthiaceae	Lanariaceae
Scheuchzeriaceae	Petermanniaceae	Laxmanniaceae*
Tofieldiaceae	Philesiaceae	[Lomandroideae, Asparagaceae s.l.]
Zosteraceae	Ripogonaceae	**Orchidaceae**
PETROSAVIALES	Smilacaceae	Ruscaceae* [Nolinoideae, Asparagaceae s.l.]
Petrosaviaceae	ASPARAGALES	Tecophilaeaceae
DIOSCOREALES	Agapanthaceae* [Agapanthoideae, Amaryllidaceae s.l.]	**Themidaceae***
Burmanniaceae	**Agavaceae*** [Agavoideae, Asparagaceae s.l.]	[Brodiaeoideae, Asparagaceae s.l.]
Dioscoreaceae	**Alliaceae*** [Alloideae, Amaryllidaceae s.l.]	Xanthorrhoeaceae*
Nartheciaceae	**Amaryllidaceae s.s.***	[Xanthorrhoeoideae, Asphodelaceae s.l.]
Taccaceae*	Aphyllanthaceae*	Xeronemataceae
Thismiaceae*	[Aphyllanthoideae, Asparagaceae s.l.]	**COMMELINIDS** (see Table 7.3, p. 2380)

close relatives). See Chase et al. (2000a, 2006), Stevenson et al. (2000), Hertweck et al. (2015), and Givnish et al. (2018) for phylogenetic analyses of the monocots. See Rudall et al. (1995), Wilson and Morrison (2000), and Columbus et al. (2006) for collections of papers from monocot symposia.

ACORALES

The Acorales contain only one family, one genus, with 2–4 species. In molecular analyses, it arises as sister to the remaining monocots. See general angiosperm phylogeny references, Duvall et al. (1993), Chase et al. (2000a), Chen et al. (2002), Moore et al. (2011), Soltis et al. (2011), Sun et al. (2016), and Givnish et al. (2018).

Acoraceae—Sweet Flag family (type *Acorus*, meaning "without pupil," originally in reference to a species of *Iris* used to treat cataracts). 1 genus/2–4 species (Figure 7.20).

The Acoraceae consist of perennial herbs found in marshy habitats. The **stems** are rhizomatous. The **leaves** are ensiform, unifacial, distichous, sheathing, simple, undivided, exstipulate, and parallel veined, with intravaginal (axillary) squamules present. The **inflorescence** is a terminal spadix borne on a leaf-like peduncle and subtended by a long, linear spathe. The **flowers** are bisexual, actinomorphic, ebracteate, sessile, and hypogynous. The **perianth** is biseriate, of 3+3 distinct tepals. The **stamens** are biseriate, 3+3, apostemonous, with flattened filaments. **Anthers** are longitudinal and introrse in dehiscence. The **gynoecium** is syncarpous, with a superior ovary, 2–3 carpels, 2–3 locules, and a minute stigma. **Placentation** is apical-axile; **ovules** are ∞ per carpel, pendent. The **fruit** is a 1–5 [–9] seeded berry, with a persistent perianth. The **seeds** are perispermous and endospermous. Aromatic ethereal oil cells are present. Raphide crystals are absent.

The Acoraceae are similar to the family Araceae (discussed later) in having a spadix and spathe, but is clearly separated from that family (within which it used to be placed) based on morphology and analyses of DNA sequence data. The Acoraceae differs from the Araceae in having ensiform, unifacial leaves, perispermous/endospermous seeds, and aromatic (ethereal) oil cells, and in lacking raphide crystals. Members of the Acoraceae are distributed in the Old World and North America. Economic importance includes *Acorus calamus* used medicinally (e.g., as "calamus oil"), in religious rituals, as an insecticide, and as a perfume and flavoring plant (e.g., in liqueurs). See Grayum (1990), Bogner and Mayo (1998), Govaerts and Frodin (2002), and Bogner (2011) for more information on the family.

The Acoraceae are distinctive in being ***marsh*** plants with a ***spadix*** and ***spathe*** (resembling Araceae) but having ***distichous***, ***ensiform***, ***unifacial*** leaves, ***perispermous and endospermous*** seeds, and ***ethereal oil cells***, and in ***lacking raphide crystals***.

P 3+3 **A** 3+3 **G** (2–3) superior.

Figure 7.20 ACORALES. Acoraceae. *Acorus calamus*. **A.** Whole plant. **B.** Close-up of spadix inflorescence.

ALISMATALES

The Alismatales, sensu APG IV (2016), contain 14 families, only two of which are described here. The family Araceae appears to be sister to the remainder of those in the order (Soltis et al. 2011, Givnish et al. 2018). Notable among the families of the order that are not described here (see Figure 7.21) are the terrestrial **Tofieldiaceae** and a number of aquatic groups, including the **Aponogetonaceae** (e.g., *Aponogeton distachyon*), **Cymodoceaceae** (several marine sea-grasses), **Hydrocharitaceae** (Figure 7.21A,B, including marine sea-grasses such as *Halophila* and *Thallasia*, fresh water aquatics such as the aquarium plants *Elodea* and *Vallisneria,* and problematic weedy species of *Elodea*, *Hydrilla*, and *Lagarosiphon*), **Juncaginaceae** (Figure 7.21C,D), **Posidoniaceae** (*Posidonia* spp., marine sea-grasses), **Potamogetonaceae** (freshwater aquatics, Figure 7.21E,F), **Ruppiaceae** (*Ruppia* spp., fresh to brackish water plants), and **Zosteraceae** (including deep, marine sea-grass species such as *Phyllospadix*, Figure 7.21G–J). See Les and Haynes (1995), Les et al. (1997), and Ross et al. (2016) for more information on the order.

Trichomes located in the axils of sheathing leaves, known as **intravaginal squamules** (see Chapter 9), are common in many Alismatales (also found in the Acorales). The evolution

Figure 7.21 ALISMATALES exemplars. **A,B.** *Najas guadalupensis*, Hydrocharitaceae. **C.** *Triglochin scilloides*, Juncaginaceae. **D.** *Triglochin concinnus*, Juncaginaceae. **E,F.** *Potamogeton foliosus*, Potamogetonaceae. **G–J.** *Phyllospadix torreyi*, Zosteraceae.

of raphide crystals (see Chapter 10) may constitute an apomorphy for the Alismatales plus all other monocots except the Acorales (Figure 7.17). However, if so, they have been secondarily lost in a number of monocot lineages, including many Poales, Zingiberales, and most of the Alismatales themselves (except for the Araceae).

Araceae — Arum family (type *Arum*, from a name used by Theophrastus). 123 genera/ca. 4,430 species (Figures 7.22, 7.23).

The Araceae consist of terrestrial or aquatic shrubs, vines, or herbs (the vegetative body reduced and globose to thalloid in the Lemnoideae). The **roots** are often mycorrhizal, without root hairs. The **stems** are rhizomatous, cormose, tuberous, or reduced. The **leaves** are simple, bifacial, spiral, or distichous, sometimes highly divided or fenestrate (often exhibiting heteroblasty), with parallel, penni-parallel, or netted venation. The **inflorescence** is a terminal, many-flowered spadix (with a sterile apical portion in some), usually subtended by a prominent, often colored spathe, or reduced to 1–4 flowers in a small pouch in the Lemnoideae. **Flowers** are small, bisexual or unisexual (female flowers often proximal, and the male distal on a spadix), actinomorphic, sessile, ebracteate, hypogynous, sometimes foul-smelling. The **perianth** is biseriate and 2+2 or 3+3 [4+4] or absent, apotepalous or basally syntepalous, a hypanthium absent. **Stamens** are 4,6, or 8 [1–12], distinct or connate, antitepalous in bisexual flowers; anthers are poricidal, longitudinal, or transverse in dehiscence. The **gynoecium** is syncarpous, with a superior ovary, 3 [1–ca. 50] carpels, usu. as many locules as carpels, style and stigma one and short or absent; placentation is variable; ovules are usu. anatropous and bitegmic, 1–∞ per carpel. The **fruit** is typically a multiple of berries, less often dry, e.g., of utricles. **Seeds** are oily (sometimes also starchy) endospermous (rarely endosperm absent) with a sometimes fleshy seed coat. Some have cyanogenic compounds or alkaloids. Raphides are present and laticifers are common.

The Araceae are divided into several subfamilies; the traditional Lemnaceae (small, thalloid to globose aquatics with very reduced flowers; Figure 7.23E–G) are now known to be nested within the Araceae as subfamily Lemnoideae. Members of the family have distributions in tropical and subtropical regions. Economic importance includes many taxa that are important food sources (from rootstocks, leaves, seeds, or fruits) in the tropics, e.g., *Alocasia*, *Amorphophallus*, *Colocasia esculenta* (taro), *Monstera*, *Xanthosoma sagittifolium*; indigenous medicinal, fiber (from roots), or arrow-poison plants; and numerous cultivated ornamentals, such as *Aglaonema*, *Anthurium*, *Caladium* (elephant's ear), *Dieffenbachia* (dumb cane), *Epipremnum*, *Monstera*, *Philodendron*, *Spathiphyllum*, *Syngonium*, and *Zantedeschia* (calla-lily). *Amorphophallus titanum* (Figure 7.23C) is a novelty in having a massive, rapidly developing inflorescence. *Wolffia* spp. (Figure 7.23F,G) are unique in having the smallest flowers of any angiosperm. See Grayum (1990), French et al. (1995), Mayo et al. (1998, 2014), Rothwell et al. (2004), and Cabrera et al. (2008) for more information and phylogenetic studies.

The Araceae are distinguished from related families in having ***bifacial*** leaves with ***parallel or netted*** venation, usually a ***spadix*** of numerous, small flowers with a subtending ***spathe***, ***endospermous*** seeds, and ***raphide crystals***.

P 2+2,3+3,(2+2),(3+3) or 0 [4+4,(4+4)] **A** 4,6,8 or (4,6,8) [1–12] **G** (3) [1–(∞)] superior.

Alismataceae—Water-Plantain family (type *Alisma*, after a name used by Dioscorides for a plantain-leaved aquatic plant). 15 genera/ca. 88+ species (Figure 7.24).

The Alismataceae consist of perennial [or annual], monoecious, dioecious, or polygamous, floating to emergent, aerenchymatous, aquatic, or marsh herbs. The **stem** is a corm or rhizome, the latter sometimes bearing tubers. The **leaves** are basal, simple, petiolate [rarely sessile], sheathing, spiral, and often dimorphic (the juvenile linear, adult leaves linear to ovate to triangular sagittate or hastate), parallel, or reticulate in venation. The **inflorescence** is a scapose raceme or panicle [sometimes umbel-like] with flowers or flower axes whorled or flowers solitary, spathe absent. **Flowers** are bisexual or unisexual, actinomorphic, subsessile to pedicellate, bracteate, hypogynous; the receptacle is flat or expanded and convex. The **perianth** is biseriate and dichlamydeous, trimerous, hypanthium absent. The **calyx** consists of 3, aposepalous sepals. The **corolla** consists of 3, apopetalous, caducous petals. **Stamens** are 6, 9, or ∞ [3], whorled, distinct and free or connate in bundles, uniseriate or biseriate (often in pairs); anthers are longitudinal, and extrorse or latrorse in dehiscence. The **gynoecium** is apocarpous, with a superior ovary, 3–∞ carpels, and 1 terminal style and stigma; placentation is basal [rarely marginal]; ovules are anatropous or campylotropous, bitegmic, 1 [∞] per carpel. The **fruit** is an aggregate of achenes or basally dehiscing follicles. **Seeds** are exalbuminous.

The Alismataceae have a worldwide distribution, esp. in N. temperate regions. Economic importance includes taxa used as food by indigenous people, others used as aquatic, cultivated ornamentals. See Haynes et al. (1998), Lehtonen (2009), and Chen et al. (2012) for more information on and phylogenetic studies of the family.

The Alismataceae are distinguished from related families in consisting of ***aquatic or marsh herbs***, with solitary or

Figure 7.22 ALISMATALES. Araceae. **A–D.** *Xanthosoma sagittifolium*. **A.** Whole plant, with large, sagittate leaves. **B.** Inflorescence, a spadix and surrounding spathe. **C.** Close-up of distal male flowers. **D.** Close-up of proximal female flowers. **E.** *Anthurium* sp., multiple fruit of berries. **F.** *Gymnostachys anceps*, inflorescences. **G.** *Aglaonema modestum*, inflorescence. **H.** *Arisaema triphyllum* (jack-in-the-pulpit), inflorescence and leaf. **I.** *Symplocarpus foetidus* (skunk weed), inflorescence. **J.** *Monstera deliciosa*, flowers (bisexual), showing outer face of hexagonal pistil and peripheral stamens. **K–Q.** *Zantedeschia aethiopica* (calla lily). **K.** Sagittate leaves. **L,M.** Inflorescence. **N.** Female flowers, face view. **O.** Female flowers, pistil longitudinal section, showing basal placentation. **P.** Ovary cross section, showing three carpels and locules. **Q.** Anther, with poricidal dehiscence.

Figure 7.23 ALISMATALES. Araceae diversity. **A.** *Pothos* sp., a vine. **B.** *Philodendron selloum*, a rhizomatous shrub. **C.** *Amorphophallus titanum*, having among the largest inflorescences of any flowering plant. **D.** *Pistia stratiotes,* water-lettuce, a floating aquatic. **E.** *Lemna* sp., duckweed, a floating aquatic. **F.** *Wolffia* sp., water-meal, a floating aquatic, having the smallest flower of any flowering plant. **G.** *Wolffia borealis* in flower, with no perianth, one anther, and one pistil (with circular stigma) inside dorsal pouch. Note tip of sewing needle for size. (C, courtesy of Constance Gramlich; G, courtesy of Wayne Armstrong.)

often ***whorled*** flowers or flower axes, and ***dichlamydeous*** flowers with an ***apocarpous*** gynoecium having ***basal placentation***, the fruit an ***aggregate of achenes or follicles***.
K 3 **C** 3 **A** 6,9–∞ [3] **G** 3–∞ superior.

PETROSAVIALES

The Petrosaviales, sensu APG IV (2016), consists of a single family, **Petrosaviaceae**, containing 4 species in 2 genera (*Japonolirion* and *Petrosavia*), native to eastern Asia. Recent phylogenetic analyses have placed this group as sister to the monocots other than Acorales and Alismatales (Figure 7.17).

DIOSCOREALES

This order contains 3–5 families, with APG IV (2016): recognizing only three, but others (e.g., Stevens, 2001 onwards; Givnish et al., 2018) recognizing five (Table 7.2). Only the Dioscoreaceae are described here. Notable among the others are the **Taccaceae** (Figure 7.25A–C), a pantropical group of one genus (*Tacca*) and ca. 12 species, and the **Thismiaceae** (ca. 5 genera/85 species), which are achlorophyllous and mycoheterotrophic. See Caddick et al. (2002a,b) and Givnish et al. (2018) for phylogenetic studies of the group.

Dioscoreaceae—Yam family (type *Dioscorea*, after Dioscorides, Greek herbalist and physician of 1st century A.D.). 4 genera/ca. 870 species (Figure 7.25D–I).

Figure 7.24 ALISMATALES. Alismataceae. **A–C.** *Sagittaria montevidensis*, arrowhead. **A.** Emergent, aquatic plant, with sagittate leaves. **B.** Inflorescence. **C.** Male flower close-up, showing dichlamydeous perianth. **D–F.** *Sagittaria* spp. **D.** Female flower longitudinal section, showing expanded receptacle and numerous pistils. **E.** Close-up of pistils, each with a single ovule having basal placentation. **F.** Leaf and inflorescence, the latter a raceme of whorled flowers. **G–I.** *Echinodorus berteroi*, burhead. **G.** Leaf. **H.** Flower (bisexual) close-up. **I.** Maturing fruit, an achenecetum, with persistent sepals.

Figure 7.25 DIOSCORALES. **A–C.** Taccaceae. *Tacca chantrieri*. **A.** Whole plant; note net-veined leaves. **B.** Flower close-up. **C.** Fruits. **D–G.** Dioscoreaceae, yam family. *Dioscorea* spp. **D.** Shoot, leaves palmate-netted; inset showing close-up of reticulate venation. Leaf surface close-up, showing reticulate venation. **E.** Basal tubers. **F–H.** *Dioscorea saxatilis*. **F.** Male flowers. **G.** Female flower; note inferior ovary. **H.** Pendant inflorescence of immature, winged fruits. **I.** Fruit close-up.

The Dioscoreaceae consist of dioecious or hermaphroditic, perennial herbs. The **stems** are rhizomatous or tuberous, often with climbing aerial stems, secondary growth present in some taxa. The **leaves** are spiral, opposite, or whorled, petiolate (typically with a pulvinus at proximal and distal ends), simple to palmate, undivided to palmately lobed, stipulate or not, with parallel or often net (reticulate) venation, the primary veins arising from the leaf base. The **inflorescence** is an axillary panicle, raceme, umbel, or spike of monochasial units (reduced to single flowers), with prominent involucral bracts in *Tacca*. The **flowers** are bisexual or unisexual, actinomorphic, pedicellate, bracteate or not, and epigynous. The **perianth** is biseriate, homochlamydeous, 3+3, a hypanthium absent or present. The **stamens** are 3+3 or 3+0, whorled, diplostemonous or antisepalous, distinct or monadelphous, free or epitepalous. **Anthers** are longitudinal and introrse or extrorse in dehiscence, tetrasporangiate, dithecal. The **gynoecium** is syncarpous, with an inferior ovary, 3 carpels, and 3 locules. The **style(s)** are 3 or 1 and terminal; **stigmas** are 3. **Placentation** is axile or parietal; **ovules** are 1–2 [∞] per carpel. The **fruit** is a capsule or berry, often winged, 1–3 locular at maturity. **Seeds** are exalbuminous.

Members of the Dioscoreaceae have a mostly pantropical distribution. The family as most recently circumscribed contains 4 genera: *Dioscorea*, *Stenomeris*, *Tacca* (previously classified in Taccaceae), and *Trichopus* (sometimes classified in Trichopodaceae). Several segregate genera have been merged into *Dioscorea* (Caddick et al. 2002). Economic importance includes various species of *Dioscorea*, the true yam, which are very important food sources in many tropical regions and which are also a source of steroidal saponins, used pharmaceutically in semisynthetic corticosteroid and sex hormones (especially birth control products) and used indigenously as a poison or soap. See Huber (1998a,b), Caddick et al. (2002), and Viruel et al. (2018) for studies.

The Dioscoreaceae are distinctive in being perennial, ***hermaphroditic or dioecious***, ***rhizomatous or tuberous*** herbs with ***simple to palmate leaves*** having ***net venation*** and ***epigynous***, trimerous flowers.
P 3+3 **A** 3+3 or 3+0 **G** (3), inferior, hypanthium absent or present.

PANDANALES

This order contains five families in APG IV (2016), only one of which is described here. Notable among the other four is the **Cyclanthaceae**, containing *Carludovica palmata*, source of fiber, e.g., for Panama hats. See Caddick et al. (2002a), Rudall and Bateman (2006), and Givnish et al. (2018) for more information on the order.

Pandanaceae—Screw-Pine family (type *Pandanus*, after a Malayan name for screw-pines). 5 genera/ca. 900 species (Figure 7.26).

The Pandanaceae consist of perennial, dioecious, woody trees, shrubs, or vines. The adventitious **roots** are often branched, prop roots. The **stems** are sympodially branched, with prominent, encircling leaf scars. The **leaves** are acrocaulis, 3- or 4-ranked, appearing spiral because of twisting of the stem, sheathing, simple, undivided, linear to ensiform, parallel veined, the margin and adaxial midrib typically with prickles. The **inflorescence** is a terminal, rarely axillary, panicle, spike, or raceme or a pseudo-umbel of spikes or spadices subtended by spathes. The **flowers** are minute, usually unisexual, often with pistillodes or staminodes present, pedicellate, bracteate, hypogynous. The **perianth** is absent or an obscure 3–4-lobed, cuplike structure. The **stamens** are ∞; filaments are fleshy. The **gynoecium** is syncarpous, with a superior ovary and 1–∞ carpels and locules. **Ovules** are anatropous, bitegmic, 1–∞. The **fruit** is a berry or drupe, forming multiple fruits in some taxa.

Members of the Pandanaceae are distributed from western Africa east to the Pacific islands. Economic importance includes use as ornamentals in some taxa and uses by indigenous people for thatch (for roofing), weaving, fiber, food (fruits and stems), spices, and perfumes. See Cox et al. (1995) Stone et al. (1998), and Buerki et al. (2017) for more information on the family.

The Pandanaceae are distinctive in being mostly ***dioecious***, ***sympodially branched***, woody plants with ***prop roots***, ***3- or 4-ranked***, simple, ***acrocaulis***, linear to ensiform leaves (appearing spiral), and small, usually ***unisexual*** flowers of variable morphology, the fruit a ***berry or drupe***, multiple in some.
P (3–4) or 0 **A** ∞ (male) **G** 1(–∞) (female), superior.

LILIALES

The Liliales is a fairly large group of monocotyledons that include 10–11 families (Table 7.2; Figure 7.27). As with the Asparagales, family delimitations of the Liliales have undergone a number of changes in recent years. Only the Liliaceae are described here. Notable among the other families are the **Alstroemeriaceae** (Figure 7.27A–D), *Alstroemeria* being a commonly cultivated ornamental, having interesting resupinate leaves; **Colchicaceae** (Figure 7.27E), containing *Colchicum autumnale*, autumn-crocus, source of colchicine used medicinally (e.g., formerly to treat gout) and in plant breeding (inducing chromosome doubling); **Melanthiaceae** (Figure 7.27G–I); **Philesiaceae** (Figure 7.27F); and **Smilacaceae** (Figure 7.27J), including *Smilax*, the green-briers, species of which are of economic importance as the source of sarsaparilla. See Rudall et al. (2000), Fay et al. (2006), Kim et al. (2013), and Givnish et al. (2016a) for phylogenetic analyses of the order.

Liliaceae [including Calochortaceae]—Lily family (type *Lilium*, fr. Greek *lirion* for lily). ca. 15 genera/ca. 600-700 species (Figure 7.28).

The Liliaceae consist of perennial herbs. The **roots** are typically contractile. The **stems** are usually bulbous, rhizomatous in some. The **leaves** are basal or cauline, spiral or (in *Lilium* and *Fritillaria* spp.) whorled, usually sheathing,

Figure 7.26 PANDANALES. Pandanceae, Screw-pine family. *Pandanus* sp. **A–C.** Whole plant, showing acrocaulis, narrow leaves. **D.** Fruit, a multiple fruit of drupes. **E.** Base of stem with prop roots. **F.** Male inflorescence. **G.** Male flowers, close-up.

Figure 7.27 LILIALES, exemplars. **A–D.** Alstroemeriaceae. **A,B.** *Alstroemeria* sp. **C,D.** *Bomarea* sp. **E.** Colchicaceae, *Burchardia umbellata*. **F.** Philesiaceae, *Geitonoplesium* sp. **G–I.** Melanthiaceae. **G.** *Trillium grandiflorum*. **H.** *Trillium erectum*. **I.** *Toxicoscordion fremontii*. **J.** Smilacaceae. *Smilax glyciphylla*.

Figure 7.28 LILIALES. Liliaceae, Lily family. **A.** *Lilium pardalinum* subsp. *shastense*, with pendant flower. **B.** *Lilium* sp., with erect flower. Note nectary at base of tepal. **C.** *Tulipa* sp., tulip. **D,E.** *Erythronium americanum*, trout lily. **D.** Flower close-up, some tepals removed. **E.** Ovary cross section, showing three carpels and locules. **F,G.** *Medeola virginiana*, Indian cucumber-root. **F.** Plant, with whorled leaves. **G.** Flower close-up. **H.** *Scoliopus bigelovii*, with fungus gnat visitors, photo by Benjamin Lowe. **I.** *Fritillaria biflora*, chocolate lily. **J.** *Calochortus splendens*. **K.** *Calochortus venustus*. **L,M.** *Calochortus weedii*. **L.** Whole flower. **M.** Basal perigonal gland (at inner tepal).

rarely petiolate, simple, and parallel veined [rarely net-veined]. The **inflorescence** is a terminal raceme, of a solitary flower, or rarely an umbel. The **flowers** are bisexual, actinomorphic or zygomorphic, pedicellate, bracteate or not, hypogynous. The **perianth** is biseriate and 3+3, homochlamydeous or dichlamydeous, apotepalous, perianth parts sometimes spotted or striate. The **stamens** are 3+3, whorled, diplostemonous, distinct, and free. **Anthers** are peltately attached to the filament or pseudo-basifixed (the filament tip surrounded by but not adnate to connective tissue), and longitudinally dehiscent. The **gynoecium** is syncarpous, with a superior ovary, 3 carpels, and 3 locules. The **style** is solitary; **stigmas** are 3, trilobed or with 3 crests. **Placentation** is axile. "Perigonal" nectaries are present, at the tepal bases. The female gametophyte is of the monosporic, *Polygonum* type or the tetrasporic, *Fritillaria* type. The **fruit** is a loculicidal, septicidal, or irregularly dehiscent capsule or a berry. The **seeds** are flat and discoid or ellipsoid, the endosperm with aleurone and fatty oils, but no starch. Raphide crystals and chelidonic acid are lacking. Allyl sulfide compounds are absent.

The Liliaceae in the past has been treated as a large assemblage (Liliaceae sensu lato), which has more recently been broken up into numerous segregate families. Members of the family grow in mostly steppes and mountain meadows of the northern hemisphere, with the center of diversity in S.W. Asia to China. Economic importance includes several taxa of value as ornamental cultivars, including lilies, *Lilium*, and tulips, *Tulipa*. See Tamura (1998a,b), Hayashi and Kawano (2000), Hayashi and Kawano (2000), Patterson and Givnish (2002), and Rønsted et al. (2005).

The Liliaceae are characterized in being ***perennial***, ***usually bulbous herbs***, ***lacking an onion-like odor***, with basal or cauline leaves, the inflorescence a ***raceme***, ***umbel, or of solitary flowers*** with a **superior** ovary.

P 3+3 **A** 3 **G** (3), superior.

ASPARAGALES

The Asparagales include 14 families, sensu of APG IV (2016), although 24 families are recognized here. This order encompasses a large and diverse number of taxa (Table 7.2). Based on recent phylogenetic studies, it is likely that an apomorphy previously thought to unite the Asparagales, the presence of seeds having a seed coat containing a black substance called **phytomelan** (Figure 7.29), may actually be apomorphic for the all except the Orchidaceae, which is sister to all other members of the order (Figure 7.30). The phytomelaniferous seeds of the Asparagales were lost in some lineages, particularly those that have evolved fleshy fruits.

Figure 7.29 ASPARAGALES. Seeds of *Agapanthus* (left) and *Yucca* (right), both with black, phytomelan-encrusted seed coat, a general apomorphy for the order.

The phylogenetic relationships of families in the Asparagales are seen in Figure 7.30. Apomorphies for the order may include simultaneous microsporogenesis (see Chapter 11) and an inferior ovary; if so, several reversals in these features occurred in various lineages (Figure 7.30). Family delimitations of the Asparagales have undergone a number of changes in recent years. The treatment listed here recognizes several families that are united in APG IV (2016), notably Alliaceae and Agapanthaceae [=Amaryllidaceae s.l. in APG IV], Agavaceae, Aphyllanthaceae, Hyacinthaceae, Laxmanniaceae, Ruscaceae, and Themidaceae [=Asparagaceae s.l. in APG IV], and Hemerocallidaceae and Xanthorrhoeaceae [=Asphodelaceae s.l. in APG IV]. See Table 7.2, Figure 7.30.

Seven families of the Asparagales are described here. Notable among the others are the **Agapanthaceae**, with *Agapanthus* spp. being common cultivars (Figure 7.31A ,B); **Asparagaceae**, including the vegetable, *Asparagus officinalis*, and several ornamental species, such as *A. setaceus*, "asparagus fern"; **Blandfordiaceae** (Figure 7.31C); **Doryanthaceae** (Figure 7.31J,K); **Hemerocallidaceae** (Figure 7.31E–G), including *Hemerocallis fulva*, day-lily; **Hyacinthaceae**, including several ornamental cultivars; **Hypoxidaceae** (Figure 7.31H); **Laxmanniaceae** (Figure 7.31D,I); and **Xanthorrhoeaceae**, the "grass trees" Figure 7.31L,M). See Fay et al. (2000), Rudall (2003), Chase et al. (2006), Graham et al. (2006), and Pires et al. (2006) for recent phylogenetic and morphological studies of the Asparagales.

Agavaceae—Agave family (type *Agave*, meaning "admirable, noble"). [Asparagaceae s.l.] Ca. 23 genera/640 species (Figures 7.32, 7.33).

The Agavaceae consist of perennial subshrubs, shrubs, trees, or possibly herbs. The **stems** are a acaulescent caudex, rhizome, bulb, or are arborescent, sympodial in taxa with branched

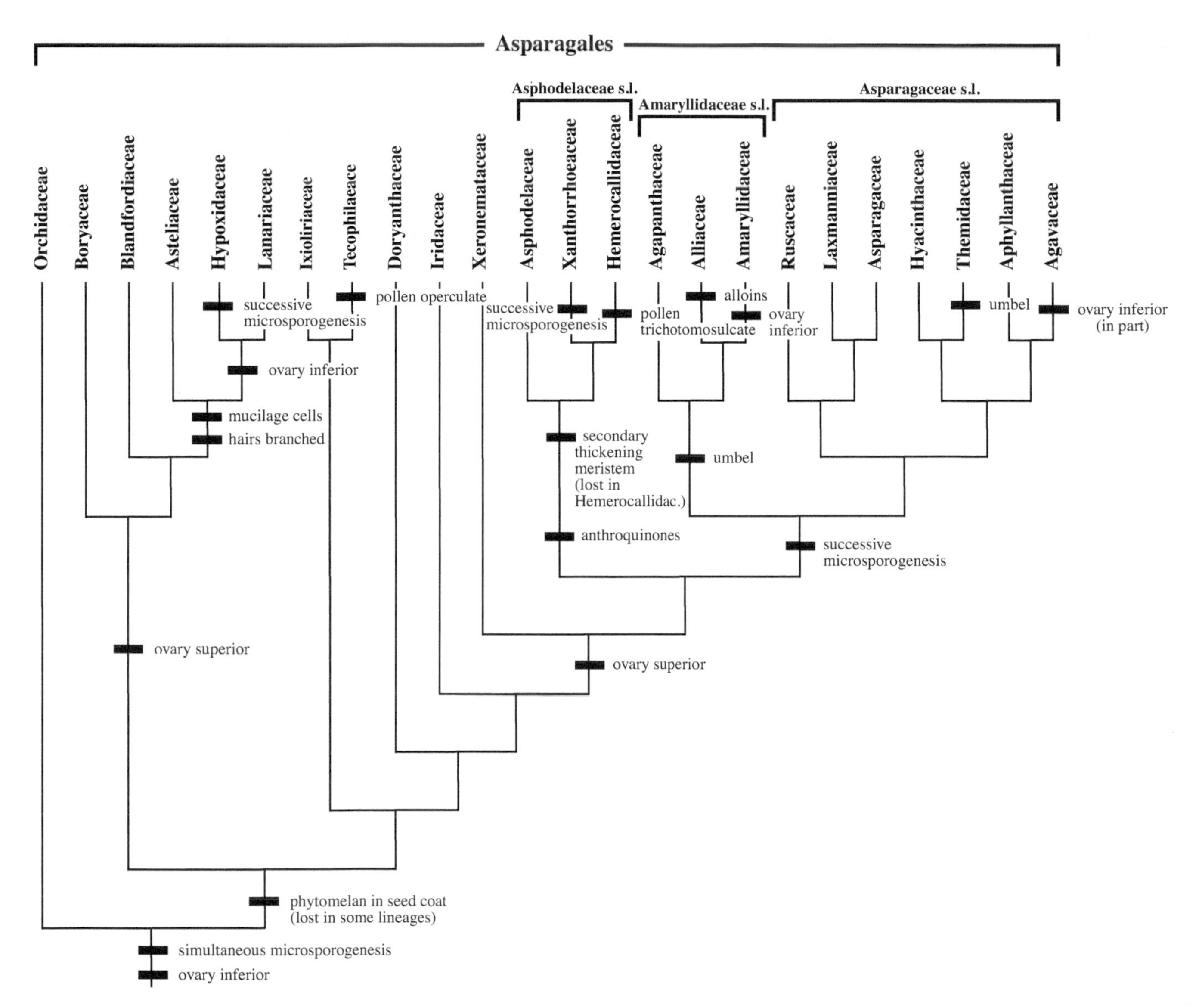

Figure 7.30 ASPARAGALES cladogram with selected apomorphies, collated from Chase et al. (2006), Graham et al. (2006), Pires et al. (2006), and Givnish et al. (2018). Note alternative classification of APG IV (2016), as Amaryllidaceae s.l., Asparagaceae s.l., and Asphodelaceae s.l. The Aphyllanthaceae (with asterisk) varies significantly in its placement in different analyses. An inferior ovary is assumed to be apomorphic for the order, with several reversals, but other optimizations are possible. The group including Amaryllidaceae s.l. and Asparagaceae s.l. have been called "higher" Asparagoids.

stems, some species with anomalous secondary growth. The **leaves** are parallel veined, often large, xeromorphic, fibrous or rarely succulent, basal and rosulate or acrocaulis, spiral, simple, undivided, the apex or margin sometimes toothed or spined. The **inflorescence** is a panicle, raceme, or spike in some producing vegetative plantlets. The **flowers** are bisexual, actinomorphic or zygomorphic, bracteate, hypogynous or epigynous. The **perianth** is biseriate, homochlamydeous of 3+3 tepals, apotepalous or syntepalous, a hypanthium present in some. The **stamens** are 6, distinct, the filaments long and thin to short and thick. **Anthers** are dorsifixed, versatile, longitudinal and introrse in dehiscence, tetrasporangiate, dithecal.The **gynoecium** is syncarpous, with a superior or inferior ovary and 3 carpels and locules. The **style** is solitary; **stigmas** are solitary or 3-lobed. **Placentation** is axile; **ovules** are anatropous, bitegmic, ∞ and in 2 rows per carpel. Septal **nectaries** are present. The **fruit** is a loculicidal or septicidal capsule or indehiscent (dry or fleshy). The **seeds** are black, phytomelanous, and flattened. Flowers are pollinated by bats, bees, hummingbirds, or moths; *Tegiticula* moths have a symbiotic relationship with *Yucca* species, the female moths transferring pollen and ovipositing the ovaries (the

Figure 7.31 ASPARAGALES exemplars. **A,B.** *Agapanthus orientalis*, Agapanthaceae. **C.** *Blandfordia nobilis*, Blandfordiaceae. **D.** *Thysanotus* sp., Laxmanniaceae. **E–G.** Hemerocallidaceae. **E.** *Dianella laevis*. **F.** *Hemerocallis fulva*, day-lily. **G.** *Johnsonia* sp. **H.** *Hypoxis* sp., Hypoxidaceae. **I.** *Cordyline* sp., Laxmanniaceae. **J,K.** *Doryanthes excelsa*, Doryanthaceae. **L,M.** *Xanthorrhoea* spp., Xanthorrhoeaceae.

Figure 7.32 ASPARAGALES. Agavaceae. *Agave deserti*. **A.** Whole plant, with basal, fibrous leaves and terminal panicle. **B.** Fruit, a loculicidal capsule; note phytomelan-encrusted seeds. **C.** Flower close-up, showing homochlamydeous perianth and inferior ovary.

developing larvae feeding on some of the seeds). The chromosomes are dimorphic in size, 5 long and 25 short.

Members of the Agavaceae occur in xeric to mesic habitats, with many found in dry areas, and often have CAM photosynthesis. The family is distributed in the New World, ranging from the central U.S. to Panama, Caribbean islands, and northern South America. Economic importance includes use by indigenous cultures as a source of fiber, food, beverages, soap, and medicinals. The leaves of *Agave sisalana* are the source of sisal fiber and *A. fourcroydes* of henequen. The fermented and distilled young flowering shoots of *Agave tequilana* are the primary source of tequila.

Pires et al. (2004) and Bogler et al. (2006) suggest expanding the Agavaceae to include at least four other genera, *Camassia, Chlorogalum, Hesperocallis,* and *Hosta,* with additional genera likely to be added. Many of these are herbaceous, and all seem to have dimorphic chromosomes as occur in traditional family members. See also Bogler and Simpson (1995, 1996), Verhoek (1998), and Archibald et al. (2015). APG IV (2016) places Agavaceae within an expanded Asparagaceae.

The Agavaceae are distinctive in being ***perennial*** subshrubs to branched trees with spiral, ***xeromorphic***, ***generally fibrous*** leaves, trimerous ***hypogynous to epigynous*** flowers, and possibly apomorphic ***dimorphic chromosomes*** (base number with 5 long and 25 short chromosomes).

P 3+3 **A** 6 **G** (3), superior or inferior, hypanthium in some.

Alliaceae—Onion family (type *Allium*, Latin name for garlic). [Amaryllidaceae s.l.] ca. 13 genera/800 species (Figure 7.34).

The Alliaceae consist of biennial or perennial herbs, usually with a distinctive onion-like (alliaceous) odor. The **stems** are acaulescent and usually a bulb, rarely a short rhizome or corm, typically enveloped by membranous scale leaves or leaf bases. The **leaves** are simple, basal, spiral, closed sheathing, acicular, linear, or lanceolate [rarely ovate], parallel veined. The **inflorescence** is a terminal, scapose umbel (derived from condensed, monochasial cymes, sometimes

Figure 7.33 ASPARAGALES. Agavaceae. **A.** *Yucca schidigera*, showing arborescent habit with acrocaulis leaves. **B.** *Hesperoyucca whipplei*. Note trimerous, homochlamydeous flowers with superior ovaries. **C.** *Yucca* sp. anther. **D.** *Yucca* sp. ovary. **E,F.** *Yucca brevifolia*, Joshua tree. **E.** Whole plant, arborescent with acrocaulis leaves and terminal panicles. **F.** Close-up of leaves and panicle. **G–J.** Basal members of expanded Agavaceae s.l. **G.** *Camassia scilloides*. **H.** *Hooveria parviflora*. **I,J.** *Hesperocallis undulata*, desert-lily, a bulbous, perennial herb.

Figure 7.34 ASPARAGALES. Alliaceae. **A.** *Allium praecox*, showing basal leaves, scape, and simple umbel. **B.** *Allium peninsulare*, showing close-up of umbel with subtending spathe-like bract. **C.** *Allium praecox*, flower close-up, showing biseriate, homochlamydeous perianth, six stamens, and superior ovary. **D.** *Allium cepa*, onion, bulb longitudinal section. **E,F.** *Tulbaghia violacea*.

termed a "pseudo-umbel"), rarely a spike or of solitary flowers, with membranous and spathelike bracts. The **flowers** are bisexual, actinomorphic, pedicellate (pedicels sometimes apically articulate), membranous-bracteate, and hypogynous. The **perianth** is biseriate, homochlamydeous, campanulate to tubular, hypanthium absent, with 3 outer and 3 inner, distinct to connate tepals, a corona sometimes present. The **stamens** are 3+3 [rarely 3 or 2 with staminodes], whorled, diplostemonous, biseriate, unfused or epitepalous; the filaments are generally flat. **Anthers** are versatile, longitudinal and introrse in dehiscence. The **gynoecium** is syncarpous, with a superior [rarely half-inferior] ovary, 3 carpels, and 3 locules. The **style** is solitary, terminal or gynobasic; the **stigma** is solitary, trilobed to capitate, dry to wet. **Placentation** is axile; **ovules** are campylotropous to anatropous, 2–∞ per carpel. Septal **nectaries** are present. The **fruit** is a loculicidal capsule. The **seeds** are black, phytomelanous, ovoid, ellipsoid or subglobose, endospermous, the endosperm rich in oils and aleurone. Family members contain alliin, which is enzymatically converted by wounding to allyl sulfide compounds, the latter imparting the distinctive onion-like odor and taste.

The Alliaceae have a mostly worldwide distribution, mainly northern hemisphere, S. American, and S. African. Economic importance includes important food and flavoring plants, including onion (*Allium cepa*), garlic (*A. sativum*), leek (*A. ampeloprasum*), chive (*A. schoenoprasum*), and other *Allium* species. Garlic also has documented medicinal properties. Several taxa are used as ornamental cultivars, e.g., *Ipheion, Leucocoryne,* and *Tulbaghia* spp. See Fay and Chase (1996) for information on phylogeny, Rahn (1998a) for a family treatment, and Nguyen et al. (2008) for a phylogenetic study of *Allium*. APG IV (2016) places Alliaceae within an expanded (and conserved) Amaryllidaceae.

The Alliaceae are distinctive in being generally ***bulbous herbs***, with ***basal***, usually ***narrow*** leaves, an ***umbellate*** inflorescence, and a usually ***superior ovary***.

P 3+3 **A** 3+3 [3,2] **G** (3), superior [rarely half-inferior].

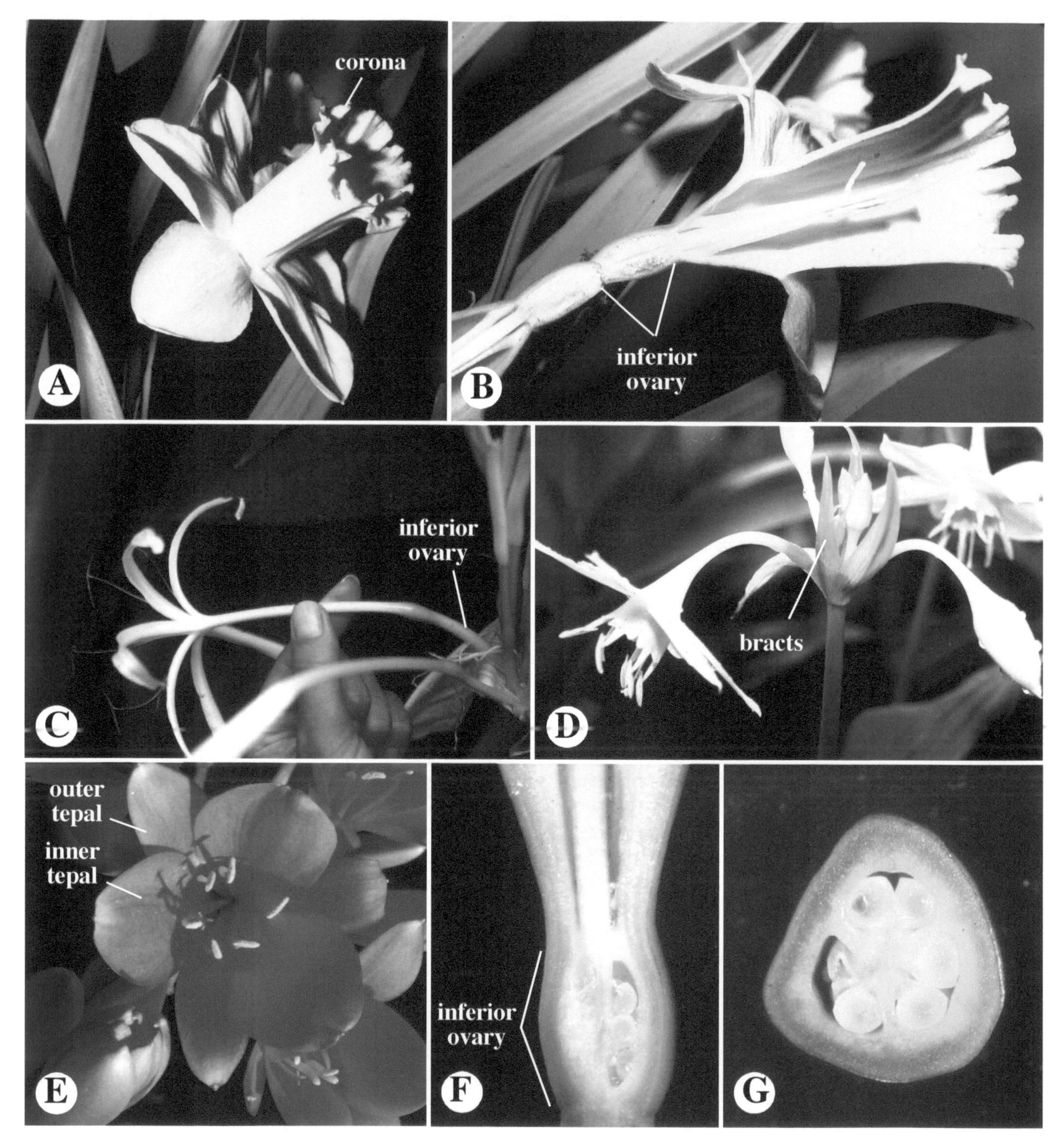

Figure 7.35 ASPARAGALES. Amaryllidaceae. **A,B.** *Narcissus pseudonarcissus*. **A.** Flower, showing elongate, tubular corona. **B.** Flower, longitudinal section. Note inferior ovary. **C.** *Crinum* sp., showing inferior ovary. **D.** *Eucharis grandiflora*. Note spathaceous bracts subtending flowers. **E–G.** *Clivia miniata*. **E.** Flower, face view. **F.** Ovary longitudinal section. **G.** Ovary cross section. Note three carpels and locules with axile placentation.

Amaryllidaceae (s.s.) —Amaryllis family (type *Amaryllis*, Latin name after a beautiful shepherdess). ca. 60 genera/ca. 800 species (Figure 7.35).

The Amaryllidaceae consist of terrestrial, rarely aquatic or epiphytic, perennial herbs. The **stems** are bulbs, covered by membranous leaf bases, the "tunica." The **leaves** are simple, undivided, spiral or distichous, sheathing or not, sessile or petiolate, and parallel veined. The **inflorescence** is a terminal, scapose umbel (derived from condensed, monochasial cymes, sometimes termed a "pseudo-umbel"), rarely of solitary flowers, with bracts present, enclosing the flower buds. The **flowers** are bisexual, actinomorphic or zygomorphic, pedicellate or sessile, bracteate, epigynous to epiperigynous. The **perianth** is biseriate, homochlamydeous, trimerous, apotepalous or syntepalous, and forming a short to long hypanthial tube, sometimes with a perianth corona (e.g., *Narcissus*). The **stamens** are generally biseriate, 3+3 [3–18], distinct or connate, forming a staminal corona in some (e.g., *Hymenocallis*). **Anthers** are usually dorsifixed, longitudinal [rarely poricidal], and introrse in dehiscence. The **gynoecium** is syncarpous, with an inferior ovary, 3 carpels, and 3 [1] locules. **Placentation** is axile or basal; ovules are anatropous, bitegmic, unitegmic, or ategmic. The **fruit** is a loculicidal capsule or rarely a berry. The **seeds** are phytomelaniferous.

The Amaryllidaceae have a worldwide distribution, being especially concentrated in South America and South Africa. Economic importance is primarily as innumerable cultivated ornamentals, such as *Amaryllis* (belladonna-lily), *Crinum*, *Galanthus* (snowdrop), *Hippeastrum* (amaryllis), *Leucojum* (snowflake), *Lycoris* (spider-lily), and *Narcissus* (daffodil); several taxa are used by indigenous peoples for medicinal, flavoring, psychotropic, or other purposes. APG IV (2016) includes Alliaceae and Agapanthaceae in an expanded Amaryllidaceae s.l. See Meerow and Snijman (1998), Meerow et al. (1999, 2000, 2006), Meerow and Snijman (2006), and Rønsted et al. (2012) for phylogenetic studies.

The Amaryllidaceae are distinctive in being perennial, ***bulbous herbs*** with an ***umbellate*** inflorescence and an ***inferior ovary***.

P 3+3 or (3+3) **A** 3+3 or (3+3) [3–18] **G** (3), inferior, hypanthium present.

Asphodelaceae (s.s.) —Asphodel or Aloe family. (type *Asphodelus*, after an ancient Greek name.) ca. 19 genera/ca. 790-940 species (Figure 7.36).

The Asphodelaceae consist of herbs to [rarely] pachycaulous trees. **Roots** are often succulent, with a velamen in some taxa. The **stems** exhibit anomalous secondary growth in some taxa, as in *Aloe*. The **leaves** are usually succulent, simple, spiral to distichous, undivided, parallel-veined, and dorsiventral to terete, the margins entire to toothed or spinose. The **inflorescence** is a raceme or panicle. The **flowers** are bisexual, actinomorphic or zygomorphic, pedicellate, bracteate or not, hypogynous. The **perianth** is biseriate, homochlamydeous, 3+3, apotepalous or syntepalous. The **stamens** are 3+3, distinct. **Anthers** are dorsifixed to basifixed, longitudinal and introrse in dehiscence. The **gynoecium** is syncarpous, with a superior ovary, 3 carpels, and 3 locules. **Placentation** is axile; **ovules** are 2–∞ per carpel. Septal nectaries are present. The **fruit** is a loculicidal capsule or (rarely) berry. The **seeds** have an aril present.

Members of the Asphodelaceae grow in temperate and subtropical Africa, particularly southern Africa. Economic importance includes *Aloe* spp. (esp. *A. vera* and *A. ferox*, from which aloin is derived), which have important uses medicinally (e.g., as laxatives and treatment of burns) as well as in skin, hair, and health products; many family members are important as cultivated ornamentals, e.g., *Aloe*, *Asphodelus*, *Gasteria*, *Haworthia*, *Kniphofia*. See Smith and v. Wyk (1998) for a general treatment and Chase et al. (2000b) and Devey et al. (2006) for phylogenetic analyses. APG IV (2016) includes Xanthorrhoeaceae and Hemerocallidaceae in an expanded (and conserved) Asphodelaceae s.l.

The Asphodelaceae are distinguished from related taxa in being ***herbs*** or ***pachycaulous trees*** with leaves usually ***succulent***, flowers trimerous with a superior ovary, and the seeds ***arillate***.

P 3+3 or (3+3) **A** 3+3 **G** (3), superior.

Iridaceae—Iris family (type *Iris*, the mythical goddess of the rainbow). ca. 66 genera/ca. 2,120 species (Figure 7.37).

The Iridaceae consist of perennial [rarely annual] herbs or shrubs with anomalous secondary growth, achlorophyllous and saprophytic in *Geosiris*. The **stems** are rhizomatous, cormose, bulbous, or a woody caudex. The **leaves** are unifacial (with leaf plane parallel to stem) or terete, simple, narrow and generally ensiform, sheathing, often equitant, distichous, and parallel-veined [scalelike and achlorophyllous in *Geosiris*]. The **inflorescence** is a terminal spike, solitary flower, or a spike or panicle of clusters of 1–many monochasial cymes (often rhipidia), typically subtended by two spathelike bracts; inflorescence subterranean in *Geosiris*. **Flowers** are bisexual, actinomorphic or zygomorphic, pedicellate or sessile, bracteate, epigynous or rarely hypogynous (*Isophysis*). The **perianth** is biseriate, homochlamydeous, 3+3, apotepalous or syntepalous (forming a prominent tube in Ixioideae), a hypanthium present or absent. **Stamens** are 3, opposite the outer tepals, distinct or monadelphous; anthers are longitudinally extrorse or poricidal in dehiscence. The **gynoecium** is syncarpous, with an inferior (superior in *Isophysis* only) ovary, 3 carpels and locules, style(s) terminal, petaloid in many Iridoideae; placentation is axile (rarely parietal); ovules are anatropous,

Figure 7.36 ASPARAGALES. Asphodelaceae. **A.** *Aloe marlothii*, rosette of succulent, spinose leaves. **B.** *Aloe* sp., showing zygomorphic flowers with tubular perianth. **C.** *Kniphofia* sp., red-hot poker. **D.** *Aloe* sp., cut succulent leaf. **E,F.** *Gasteria trigona*. **E.** Flowers. **F.** Ovary cross section. **G.** *Asphodelus fistulosus*, flower. **H.** *Bulbine* sp., showing actinomorphic flower with apotepalous perianth and pilose stamen filaments. **I.** *Haworthia limifolia*, basal rosette of leaves. **J–L.** *Haworthia cooperi*. **J.** Inflorescence, a raceme. **K.** Flower longitudinal section, showing syntepalous perianth. **L.** Close-up of ovary longitudinal section and epitepalous stamens.

Figure 7.37 ASPARAGALES. Iridaceae. **A.** *Iris* sp., showing unifacial leaves that are equitant and distichous. **B.** *Dietes* sp., showing the three outer tepals, three inner tepals, and petaloid styles (corresponding in position to the three carpels). **C.** *Iris* sp., with petaloid style pulled back to show stamen opposite outer tepal. **D.** *Crocus* sp. **E.** *Chasmanthe aethiopica*, an example of a zygomorphic member of the family. **F.** *Iris* sp., inferior ovary cross section, showing axile placentation. **G,H.** *Sisyrinchium bellum*. **G.** Whole flower. **H.** Close-up of flower, showing central connate stamens. **I.** Close-up of connate anthers. **J.** *Melasphaerula ramosa*. **K.** *Pillansia templemannii*. **L.** *Moraea fugax*. **M.** *Tritoniopsis*.

bitegmic, 1–∞ per carpel. The **fruit** is a loculicidal capsule; seeds are endospermous with a dry or fleshy seed coat.

The Iridaceae has recently been classified as seven subfamilies: Isophysidoideae, consisting of only one species, *Isophysis tasmanica*, the only superior-ovaried family member; Patersonioideae; Geosiridoideae, consisting solely of *Geosiris aphylla* of Madagascar, an achlorophyllous saprophyte; Aristeoideae; Nivenioideae; Crocoideae; and Iridoideae. (See Goldblatt et al., 2008.) Members of the family have a worldwide distribution, being especially diverse in southern Africa. Economic importance includes extensive use as ornamental cultivars, e.g., as cut flowers, especially species of *Iris*, *Gladiolus*, *Freesia*, and *Crocus*; the styles and stigmas of *Crocus sativus* are the source of the spice saffron; corms of some species are eaten by indigenous people. See Goldblatt et al. (1998, 2001, 2008), Reeves et al. (2001), and Goldblatt and Manning (2008) for morphological and phylogenetic studies of the Iridaceae.

The Iridaceae are distinguished from related families in being usually ***perennial herbs*** with generally ***ensiform, unifacial*** leaves, a bracteate ***spike or panicle*** of ***solitary flowers*** or ***monochasial cyme (rhipidia) clusters***, and flowers with ***three stamens opposite outer tepals***.

P 3+3 or (3+3) **A** 3 or (3) **G** (3), inferior (superior in *Isophysis*).

Orchidaceae—Orchid family (type *Orchis*, meaning testicle, from the shape of the root tubers). ca. 800 genera/ca. 28,000 species (Figures 7.38–7.40).

The Orchidaceae consist of terrestrial or epiphytic, perennial [rarely annual] herbs [rarely vines]. The **roots** are often tuberous (in terrestrial species) or aerial (in epiphytic species), typically with a multilayered velamen. The **stems** are rhizomatous or cormose in terrestrial species, the epiphytic species often with pseudobulbs. The **leaves** are spiral, distichous, or whorled, usually sheathing, simple, and parallel veined. The **inflorescence** is a raceme, panicle, spike, or a solitary flower. The **flowers** are bisexual, rarely unisexual, zygomorphic, usually resupinate, resulting in a 180° shift of floral parts (Figure 7.42C), epigynous. The **perianth** is biseriate, homochlamydeous (although outer and inner whorls are often differentiated), 3+3, apotepalous or basally syntepalous, extremely variable in shape and color, sometimes spurred or with enlarged saclike tepal. The inner median, anterior tepal (when resupinate; actually posterior early in development) is termed the "labellum," which is typically enlarged, sculptured, or colorful and often functions as a landing platform for pollinators. The **stamen** in most species is solitary, derived from the median stamen of the ancestral outer whorl, often with two vestigial staminodes derived from the lateral stamens of an ancestral inner whorl; in Apostasioideae or Cypripedioideae, there are two or three fertile stamens, when two, derived from the two lateral stamens of the ancestral inner whorl, when three, derived from these plus the median stamen of the outer whorl; the androecium is fused with the style and stigma to form the **gynostemium** (also called the **column** or **gynostegium**). **Anthers** are longitudinally or modified in dehiscence, bisporangiate, dithecal; in all but the Apostasioideae and most Cypripedioideae, the pollen is agglutinated into 1–12 (typically 2 or 4) discrete masses, each termed a "pollinium" (derived from individual anther microsporangia or from fusion products or subdivisions of the microsporangia); the pollinia plus a sticky stalk (derived from either the anther or stigma) are together termed a "pollinarium," the unit of transport during pollination, the anther connective often modified into an "operculum" (anther cap) that covers the anther(s) prior to pollination. The **pollen** consists of tetrad units in most family members, but may be massulae or monads in various groups (see Chapter 12, Palynology). The **gynoecium** is syncarpous, with an inferior ovary, 3 carpels, and 1–3 locules. The style is solitary and terminal and is the major component of the gynostemium; a single, enlarged lobe, termed the "rostellum" and interpreted as part of the stigma(s), is positioned above the stigmatic region; the rostellum typically is adherent to the pollinarium stalk, the tip of which derives a sticky substance from the surface of the rostellum (this sticky region termed the "viscidium"). **Placentation** is parietal or axile; **ovules** are anatropous, usually bitegmic, very many per carpel (sometimes on the order of a million). Nectaries are typically present, variable in position and type. The **fruit** is a loculicidal capsule or rarely a berry. The **seeds** are often membranous-winged, possibly functioning in wind dispersal, and exalbuminous, the endosperm abortive early in development. **Pollination** is effected by various insects (often one species having a specific association with one orchid species), birds, bats, or frogs. The transfer of pollen grains together within the pollinia is an apparent adaptation for ensuring fertilization of many of the tremendous number of ovules. Some species have remarkable adaptations for pollination. Among the more remarkable are several species with visual and chemical mimicry, fooling a male insect into perceiving the flower as a potential mate. The bucket orchid, *Coryanthes*, has an pouch-like labellum that fills with a fluid secreted from the gynostemium; a bee, falling into this fluid, travels through a tunnel, forcing deposition of the pollinarium on its body.

The Orchidaceae are recently classified into five subfamilies: Apostasioideae (2–3 stamens, axile placentation, lacking pollinia), Vanilloideae (1 stamen, parietal placentation), Cypripedioideae (2 stamens, parietal placentation, lacking pollinia), and the Orchidoideae and Epidendroideae (1 stamen, parietal placentation, pollinia), the latter designated by Cameron et al. (2004, 2006) as a paraphyletic "Lower Epidendroid" and a monophyletic Higher Epidendroids

Figure 7.38 ASPARAGALES. Orchidaceae. **A.** *Oncidium lanceanum*. **B.** *Vanilla planifolia*, vanilla. **C.** *Zygopetalum* sp. **D,E.** *Stanhopea tigrina*, with pendant flowers. **F.** *Caladenia fuscata*. **G.** *Calopogon* sp., a nonresupinate species.

(Figure 7.39A,B). The single stamen of the Vanilloideae were hypothesized by Cameron et al. (2006) to have evolved independently of that in the Orchidoideae-Epidendroideae (Figure 7.39A,B). Members of the family are distributed worldwide. Economic importance is largely as cultivated ornamentals, including some quite monetarily valuable in the horticultural trade. The fermented capsules of *Vanilla planifolia* (Figure 7.40B) are the source of vanilla food flavoring. *Angraecum sesquipedale* Thouars (Madagascar) is known for its long spur (up to 45 cm long); this orchid is pollinated by a moth with a proboscis of that spur length, a fact that Charles Darwin predicted prior to the discovery of the moth (and recent observation that it is indeed the pollinator). See Cameron et al. (1999), Cameron and Chase (2000), Cameron (2004, 2006), Górniak et al. (2010), and Givnish et al. (2015, 2016b) for phylogenetic and biogeographic studies of the orchids.

The Orchidaceae are distinctive in consisting of ***mycorrhizal***, mostly perennial, ***terrestrial or epiphytic*** herbs having trimerous, often ***resupinate*** flowers with a showy ***labellum***, the ***androecium and gynoecium adnate*** (termed a ***column***,

Figure 7.39 ASPARAGALES. Orchidaceae (following page). **A.** *Cattleya* sp., showing basic structure of a resupinate orchid flower. Note enlarged and colorful inner, median tepal, the "labellum." **B–F.** *Cymbidium* sp. **B.** Whole flower, illustrating prominent gynostemium. **C.** Flower longitudinal section. **D.** Close-up of gynostemium apex. Note operculum covering anther. **E.** Gynostemium apex with operculum removed. **F.** Close-up of pollinarium with two pollinia. **G.** *Ludisia* sp., cross section of inferior ovary, showing parietal placentation. **H.** *Thelymitra antennifera*, an orchid mimicking an insect. **I,J.** *Epidendrum* sp. **I.** Close-up of gynostemium, which is adnate to the labellum. **J.** Flower longitudinal section. **K,L.** *Orchis spectabilis*. **K.** Whole flower. **L.** Close-up view of gynostemium. **M,N.** *Encyclia cochleata*. **M.** Pseudobulb, found in many epiphytic orchids. **N.** Flower, showing rare nonresupinate orientation. **O.** *Dendrobium* sp. **P.** *Cypripedium* sp., lady's slippers. Note enlarged, swollen labellum. **Q.** *Paphiopedilum* sp.

(*See Figure in next page*)

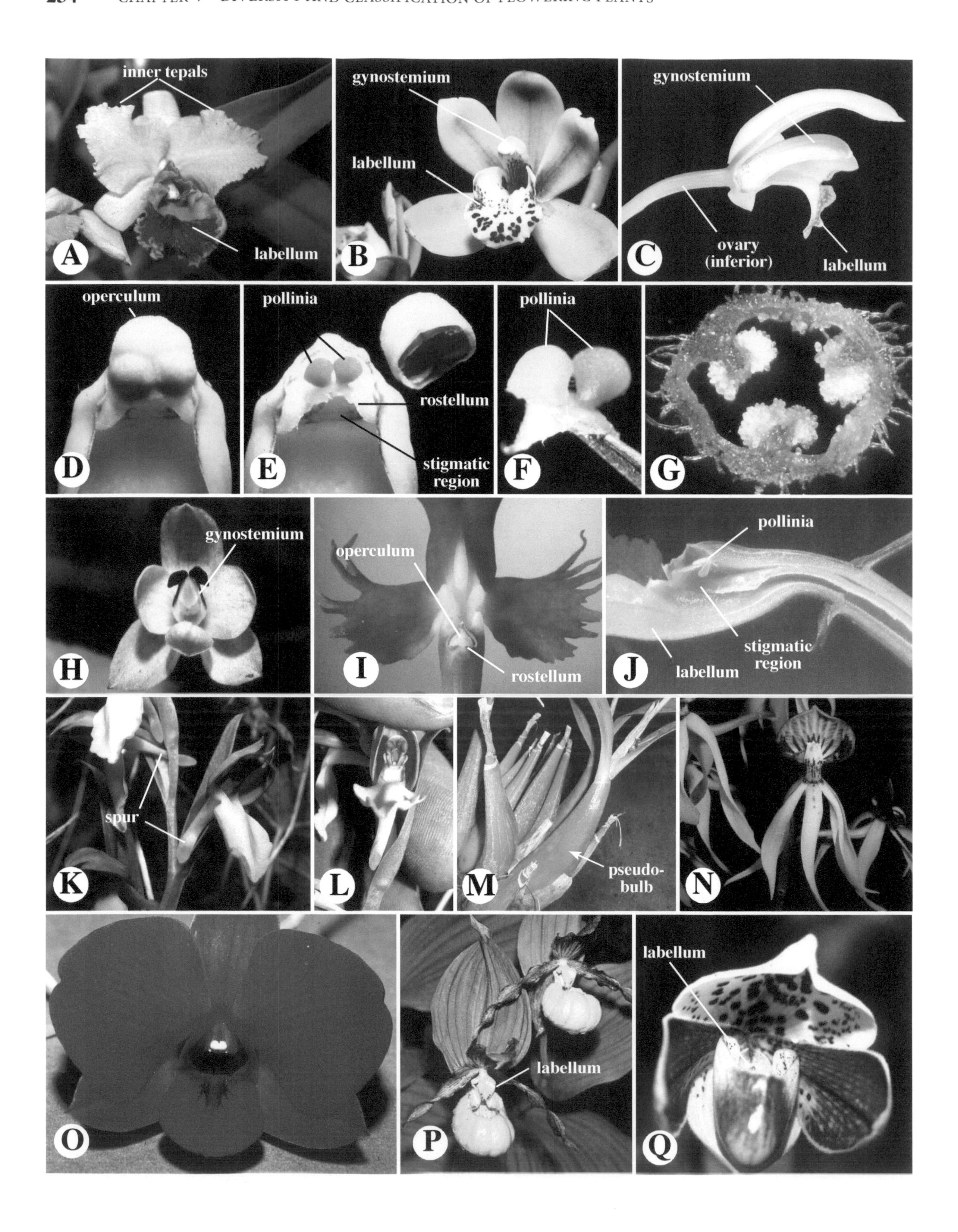
inner tepals
labellum
A
gynostemium
labellum
B
gynostemium
ovary
(inferior)
labellum
C
operculum
D
pollinia
rostellum
stigmatic
region
E
pollinia
F
G
gynostemium
H
operculum
rostellum
I
pollinia
stigmatic
region
labellum
J
spur
K
L
pseudo-
bulb
M
N
O
labellum
P
labellum
Q

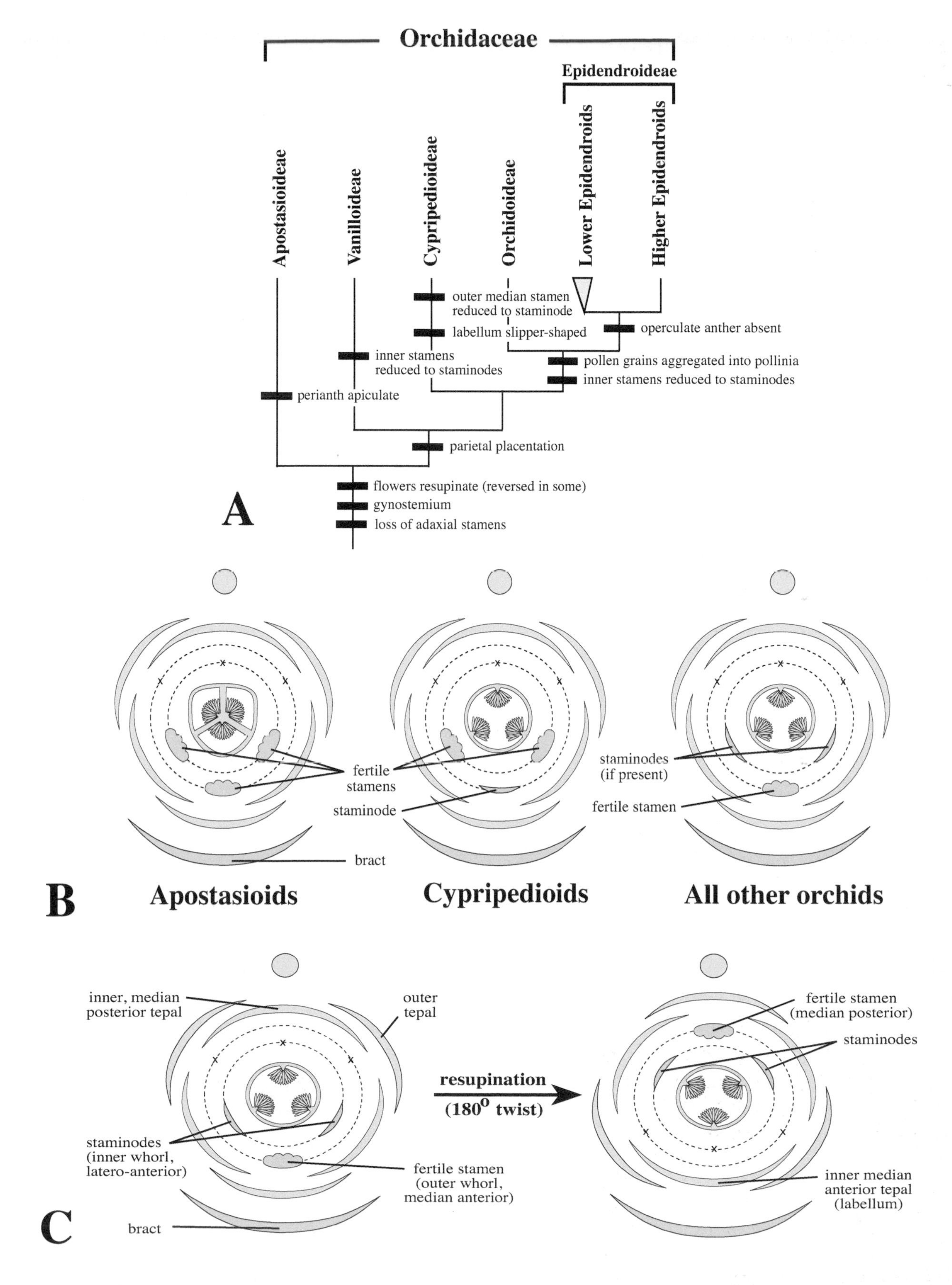

Figure 7.40 ASPARAGALES. Orchidaceae. **A.** Cladogram of major orchid groups, after Cameron et al. (2006), with putative, selected apomorphies. **B.** Floral diagrams of Apostasioids, Cypripedioids, and all other orchids (lower), after Dahlgren et al. (1985). **C.** Floral diagram before (left) and after (right) resupination.

gynostegium, or ***gynostemium***), the pollen grains often fused into 1–several masses (***pollinia***), bearing a sticky-tipped stalk, pollinia and stalk termed a ***pollinarium***, which is the unit of pollen dispersal during pollination.

P (3+3) **A** 1-3, when 1 a pollinarium **G** (3), inferior, with gynostemium.

Themidaceae—The Brodiaea family (type *Themis* [=*Triteleia*]) [Asparagaceae s.l.]. ca. 12 genera/ca. 62 species (Figure 7.41).

The Themidaceae consist of perennial herbs. The **stems** are corms, typically with a membranous to fibrous covering from previous leaf bases, termed a "tunica." **Leaves** are simple, closed-sheathing, flat, terete, or fistulose, acicular, linear, or lanceolate in outline. The **inflorescence** consists of a terminal scapose umbel. **Flowers** are bisexual, actinomorphic, and hypogynous. The **perianth** is biseriate and homochlamydeous, tepals 3+3, connate below or distinct. **Stamens** are 6 (3+3) or 3 (3 outer staminodes + 3 fertile, or 3 fertile in the position of the inner whorl), whorled, diplostemonous or antipetalous, usually distinct. The **gynoecium** is syncarpous; the ovary is superior, with 3 carpels, 3 locules, and 1 terminal style. **Placentation** is axile with 2–many ovules per carpel. The **fruit** is a loculicidal capsule. **Seeds** are ovoid, ellipsoid, or subglobose, endospermous, rich in oils and aleurone. An onionlike (alliaceous) odor is absent.

Members of the Themidaceae are distributed in North America from S.W. Canada to Central America. There are no economic uses other than a few being used in cultivation. See Fay and Chase (1996) regarding the "resurrection" of the Themidaceae, Rahn (1998b) for detailed information on the family, and Pires et al. (2001) and Pires and Sytsma (2002) for phylogenetic analyses. APG IV (2016) places Themidaceae within an expanded Asparagaceae.

The Themidaceae are distinctive in being perennial, ***cormose herbs***, ***lacking an onionlike odor***, and having an ***umbellate*** inflorescence.

P 3+3 **A** 3+3, 3+3 staminodes, or 0+3 **G** (3), superior.

Figure 7.41 ASPARAGALES. Themidaceae. **A.** *Dichelostemma capitatum*, showing basal leaves, scape, and simple umbel. **B.** *Bloomeria crocea*, showing close-up of umbel with scape. **C.** *Brodiaea elegans*, flower close-up, showing biseriate, homochlamydeous perianth, three central, fertile stamens, and three staminodes. **D.** *Dichelostemma capitatum*, corm in longitudinal section. **E,F.** *Brodiaea orcuttii*. **E.** Flower close-up in longitudinal section, showing three fertile stamens and superior ovary. **F.** Ovary cross section, showing axile placentation.

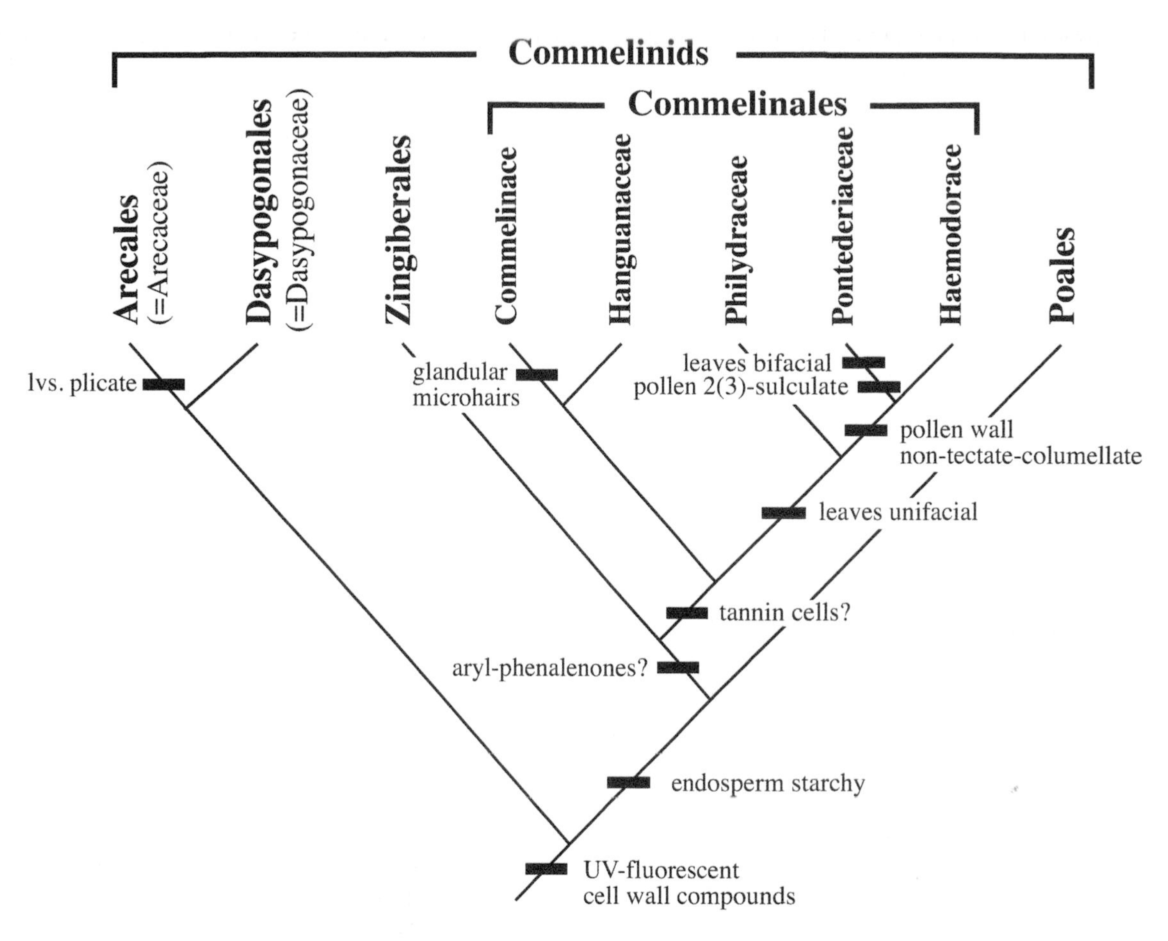

Figure 7.42 Major clades of the commelinid monocotyledons, modified from APG IV (2016) and Givnish et al. (2018), with selected apomorphies shown.

COMMELINIDS

The Commelinids, or Commelinidae, are a monophyletic assemblage of monocots, as evidenced by morphological and molecular data (Figure 7.42). The commelinids are characterized by an apparent chemical apomorphy, the presence of a class of organic acids (including coumaric, diferulic, and ferulic acid) that impregnate the cell walls. These acids can be identified microscopically in being UV-fluorescent (Figure 7.43). The orders and families of the Commelinids (after APG IV, 2016) are listed in Table 7.3).

The Commelinids include a number of economically important plants, including the palms (Arecaceae), gingers and bananas (Zingiberales), and grasses (Poaceae). The grass family in particular is perhaps the most important family of plants, as grasses include the grain crops. As can be seen from Figure 7.42, the monofamilial Dasypogonales and Arecacales form a clade sister to the rest of the commelinid monocots. See Graham et al. (2006), Chase et al. (2006), and Givnish et al. (2018) for recent analyses.

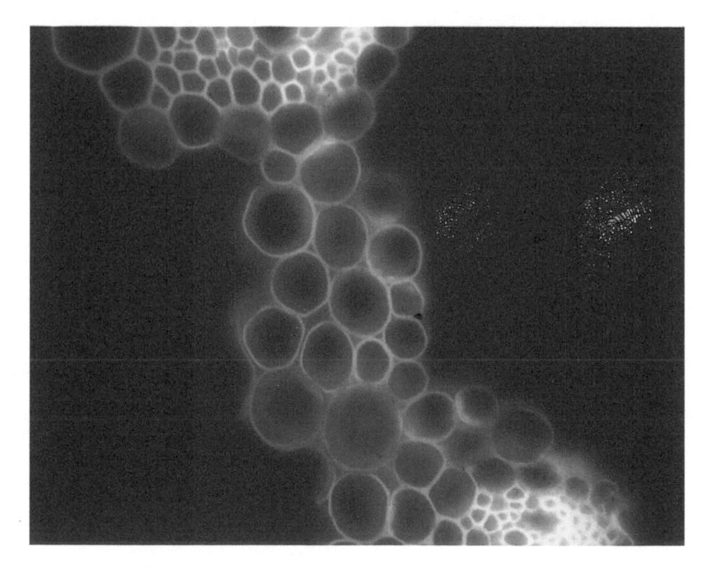

Figure 7.43 Leaf cross section of *Lachnanthes caroliniana* (Haemodoraceae), showing the UV fluorescence of nonlignified cell walls (center). This fluorescence is indicative of the presence of certain organic acids, apomorphic for the Commelinid monocots.

TABLE 7.3 Orders and included families of the Commelinid Monocotyledons, after APG IV (2016). Families in **bold** are described in detail. An asterisk denotes a deviation from APG IV, with brackets indicating the more inclusive family recommended by APG IV.

COMMELINIDS			
DASYPOGONALES*	ZINGIBERALES	POALES	POALES (continued)
Dasypogonaceae	**Cannaceae**	Anarthriaceae* [Restionaceae]	Mayacaceae
ARECALES	Costaceae	**Bromeliaceae**	**Poaceae (Gramineae)**
Arecaceae (Palmae)	Heliconiaceae	Centrolepidaceae* [Restionaceae]	Rapateaceae
COMMELINALES	Lowiaceae	**Cyperaceae**	**Restionaceae**
Commelinaceae	Marantaceae	Ecdeiocoleaceae	**Sparganiaceae*** [Typhaceae]
Haemodoraceae	**Musaceae**	**Eriocaulaceae**	Thurniaceae
Hanguanaceae	**Strelitziaceae**	Flagellariaceae	(including Prioniaceae)
Philydraceae	**Zingiberaceae**	Joinvilleaceae	**Typhaceae***
Pontederiaceae		**Juncaceae**	**Xyridaceae**

Figure 7.44 COMMELINIDS. Dasypogonales, Dasypogonaceae. *Dasypogon bromeliaefolius*, native to southwestern Australia.

DASYPOGONALES

This order contains the single family, Dasypogonaceae. Its position has varied in past analyses, but was firmly placed sister to the Arecales by Givnish et al. (2018). It is treated as part of the Arecales by APG IV (2016). The Dasypogonaceae consist of four genera—*Baxteria, Calectasia, Dasypogon,* and *Kingia*—native to southern and southwestern Australia (Figure 7.44).

ARECALES

This order contains the single family Arecaceae. See Dransfield and Uhl (1998), Asmussen et al. (2000, 2006), Hahn (2002), and Lewis and Doyle (2001) for information and phylogenetic analyses of the palms.

Arecaceae (Palmae)—Palm family (type *Areca*, after Portuguese name for the betel palm). ca. 190 genera/ca. 2,600 species (Figures 7.45, 7.46).

The Arecaceae consist of perennial trees, large rhizomatous herbs, or lianas. Plant sex is variable, and secondary growth is absent. The **roots** are mycorrhizal, lacking root hairs. The **stem** is usually arborescent, consisting of a single, unbranched trunk [dichotomously branched in *Hyphaene*], or a cespitose cluster of erect stems, or a stout, dichotomously branched rhizome (*Nypa*), or an elongate liana with long internodes (rattan palms). The **leaves** are typically quite large, generally terminal (acrocaulis), spiral [rarely distichous or tristichous], with a sheathing base and an elongate, stout petiole (sometimes referred to as "pseudopetiole") between the sheath apex and blade. In arborescent taxa the sheathing bases of adjacent leaves may overlap one another, forming a distinctive "crownshaft" at the trunk apex. Leaves are simple, pinnate, bipinnate, costapalmate, or palmate; if simple, the leaves are often pinnately or palmately divided, sometimes bifid, with leaflet spines present in some taxa. Leaves are typically ligulate (with an appendage, the ligule, at the inner junction of blade and petiole); in taxa with palmate leaves, another distinctive process, called the hastula, may be present at the junction of the petiole and blade. The leaf blade is characteristically plicate (pleated), with the leaflets or blade divisions in cross section either induplicate (V-shaped, with the point of the fold below, or abaxial) or reduplicate (Λ-shaped, with the point of the fold above, or adaxial). Venation is pinnate- or palmate-parallel. The **inflorescence** is typically an axillary, bracteate panicle or spike of solitary flowers or of cyme units, the inflorescence arising either below (infrafoliar) or among (interfoliar) or above (suprafoliar) the leaves of the crownshaft. The peduncle is subtended by an often large prophyll and 1–∞ spathes. The **flowers** are unisexual or bisexual, actinomorphic, sessile, and hypogynous. The **perianth** is usually biseriate and homochlamydeous, 3+3 [0, 2+2, or ∞], apotepalous. The **stamens** are 3+3 [3 or ∞], distinct or connate, epitepalous in

Figure 7.45 ARECALES. Arecaceae. **A.** *Archontophoenix cunninghamiana*, king palm, showing single, unbranched trunk with acrocaulis "crown" of pinnately compound leaves and lateral inflorescences below crownshaft (infrafoliar). **B.** *Phoenix dactylifera*, date palm, with several inflorescences arising within crownshaft (interfoliar). **C.** *Syagrus romanzoffiana*, queen palm, with pinnate leaves. **D.** *Washingtonia robusta*, with palmately divided leaves. **E.** *Licuala peltata*, with palmately lobed leaves. **F.** *Livistona drudei* leaf close-up, showing ligula at junction of petiole and blade. **G.** *Jubaea chilensis* leaf close-up, showing plicate posture of pinnate leaves. **H.** Reduplicate (*Syagrus romanzoffiana*) and induplicate (*Phoenix dactylifera*) leaf posture. Adaxial side of leaflet blade is at top.

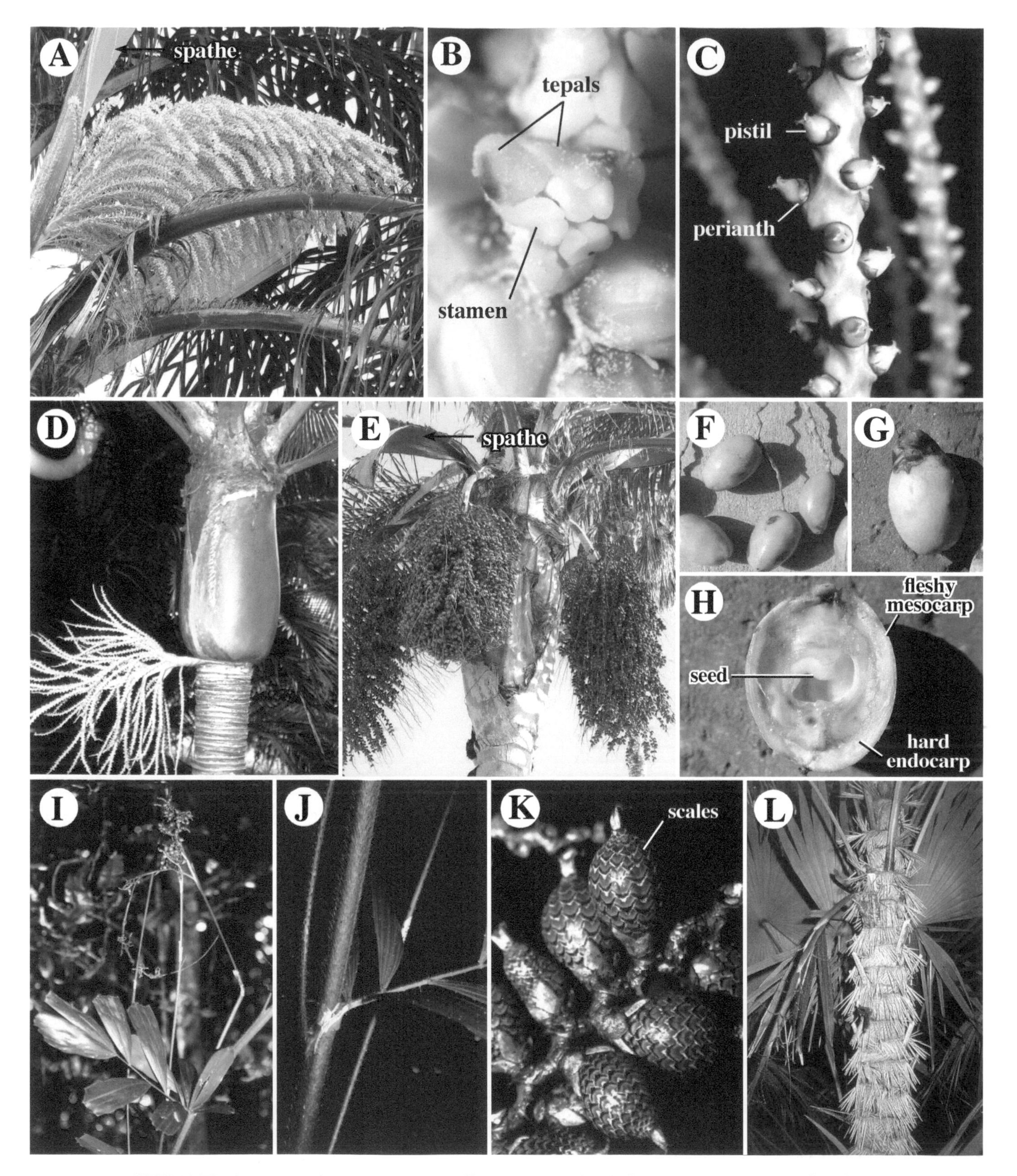

Figure 7.46 ARECALES. Arecaceae. **A.** *Syagrus romanzoffiana*, queen palm, inflorescence, with spathe. **B.** *Chamaerops humilis*, Mediterranean palm. Close-up of flowers, showing trimerous perianth and androecium. **C,D.** *Rhopalostylis sapida*. **C.** Close-up of female flowers, with reduced, scalelike perianth and superior, 3-carpeled ovary. **D.** Close-up of sheathing leaf bases forming crownshaft; note lateral, infrapetiolar inflorescence. **E.** *Syagrus romanzoffiana*, infructescences, with spathes. **F.** *Phoenix dactylifera*, date palm, drupes. **G,H.** *Syagrus romanzoffiana*, drupes. **G.** Whole fruit. **H.** Drupe longitudinal section, showing pericarp layers (hard endocarp and fleshy mescocarp). **I–K.** *Calamus* sp. (rattan palm). **I.** Whole plant, showing pinnate leaf. **J.** Leaf base close-up; note sheath and long internodes. **K.** Fruit (drupe) close-up, showing retrorse scales typical of the rattan palms. **L.** *Zombia antillarum*, a palm with numerous spines.

some spp., staminodes present in some spp. **Anthers** are longitudinal, rarely poricidal, in dehiscence. The **gynoecium** is syncarpous or apocarpous, with a superior ovary, usually 3 [1, 2, 4–∞] carpels, and 3 or 1 [∞] locules. The styles, if present, are distinct or connate; stigmas are sessile or at tip of styles. **Placentation** is variable; ovules are variable in type, bitegmic, and 1 per locule. Septal nectaries are present in some taxa. The **fruit** is fleshy or fibrous, usually a drupe [rarely dehiscent or a pyrene], some with outer scales (Calamoideae), hairs, prickles, or other processes. **Seeds** are usually 1 [–10] per fruit and have an oil or hemicellulose-rich, sometimes ruminate, endosperm; starch is absent.

The Arecaceae have distributions mostly in warm, tropical regions and are often ecologically important where they occur. The family has of late been classified into five subfamilies: Arecoideae, Calamoideae, Ceroxyloideae, Coryphoideae, and Nypoideae (Asmussen et al. 2006). The lianous Calamoideae (*Calamus* and relatives) is sister to the rest of the family; the rhizomatous Nypoideae (*Nypa fruticans* only) are sister to the remainder; and the Coryphoideae is sister to the Ceroxyloideae and Arecoideae. The plicate leaf is a probable apomorphy for the family as a whole, as is the drupaceous fruit. (The family Cyclanthaceae and a few other scattered monocot taxa also have plicate leaves, but these are thought to have evolved independently.) The palms are of great economic importance, including uses as fruits (e.g., *Cocos nucifera*, coconut palm, *Phoenix dactylifera*, date palm), furniture/canes (rattan palms), fibers (e.g., "coir" from the mesocarp of *Cocos nucifera*), oils (e.g., *Elaeis oleifera*, oil palm), starch (e.g., *Metroxylon* spp., sago palms), waxes (e.g., *Copernicia cerifera*, wax palm), and many species used indigenously as timber or in building construction; fruits of *Areca catechu*, betel palm, are chewed in India (with *Piper betle* leaves and lime) as a stimulant. See Barrett et al. (2015) and Faurby et al. (2016) for additional phylogenetic studies.

The Arecaceae are distinctive in having a rhizomatous, lianous, or usually ***arborescent*** stem, with large, sheathing, ***plicate*** leaves, a ***fleshy, usually drupaceous*** fruit, and seeds lacking starch. The plicate leaf posture and drupaceous fruit are likely apomorphies for the family.

P 3+3 [0,2+2,∞] **A** 3+3 or (3+3) [3,∞; 0 in female fls.] **G** 3 or (3) [1,2,4–∞; 0 in male fls.], superior.

COMMELINALES, ZINGIBERALES, AND POALES

The taxa of the Commelinid monocots other than the Dasypogonales and Arecales are classified into the three orders Zingiberales, Commelinales, and Poales. All of these have seeds that contain endosperm rich in starch, an apparent apomorphy for the three orders (Figure 7.42). In contrast, the palms have seeds rich in oils and hemicellulose and lacking in starch.

The Commelinales and Zingiberales are sister taxa according to recent phylogenetic analyses (see APG IV, 2016; Givnish et al., 2018). A possible apomorphy uniting them is the presence of arylphenalenone chemical compounds (Figure 7.42), which are common in the Haemodoraceae and have been discovered also in some Pontederiaceae and Zingiberales.

COMMELINALES

The Commelinales, sensu APG IV (2016) consist of five families (Table 7.3), three of which are described here. Family interrelationships (after Givnish et al., 2018) and putative apomorphies are portrayed in Figure 7.42. The family Hanguanaceae has now been placed here. The Commelinales is not well defined morphologically, although floral tannin cells may constitute an apomorphy (Figure 7.42). The Haemodoraceae and Philydraceae have unifacial leaves, a likely apomorphy, but this would necessitate the reversal to bifacial leaves in the aquatic Pontederiaceae (Figure 7.42). The Haemodoraceae and Pontederiaceae may be united by the apomorphy of non-tectate-columellate pollen wall structure (Figure 7.42). See Givnish et al. (1999, 2018), Davis et al. (2004), and Saarela et al. (2008) for phylogenetic analyses of the order.

Commelinaceae—Spiderwort family (type *Commelina*, presumed to be named after the two Dutch botanists, Jan Commelin (1629–1692) and his nephew Caspar Commelin (1667–1731). ca. 40 genera/ca. 700 species (Figure 7.47).

The Commelinaceae consist of mostly perennial herbs. The **stems** typically have swollen nodes. The **leaves** are spiral, sheathing (sheath closed), simple, undivided, with each half of the blade rolled adaxially toward the midrib early in development. The **inflorescence** is a cyme, rarely a raceme or of solitary flowers, the flowers often piercing the subtending bract. The **flowers** are usually bisexual, actinomorphic or zygomorphic, and hypogynous. The **perianth** is biseriate, usually dichlamydeous. The **calyx** consists of 3 distinct or basally fused sepals or lobes. The **corolla** contains 3, equal or unequal (anterior petal smaller), distinct or basally connate [sometimes clawed] petals or lobes, which are characteristically ephemeral. The **stamens** are usually 3+3, sometimes with 3 fertile and 3 staminodes [rarely of 1 fertile stamen], apostemonous, the filaments often with pilose trichomes, fertile stamens sometimes dimorphic. **Anthers** are basifixed, versatile, longitudinally dehiscent [rarely pori-

Figure 7.47 COMMELINALES. Commelinaceae. **A–C.** *Commelina* sp. **A.** Inflorescence apex, showing closed sheath of leaves and spathaceous inflorescence bracts. **B.** Flower, face view. Note staminodes and fertile stamen dimorphism. **C.** Tricellular microhair, a putative apomorphy of the Commelinaceae. **D.** *Dichorisandra reginae*. **E.** *Cyanotis somaliensis*. **F–H.** *Tradescantia*. **F.** *T. hirsuta*. **G.** *T. virginiana*. whole flower. **H.** *Tradescantia* sp. Close-up of androecium, showing pilose filaments.

cidal apically and basally], with the connective often extended; prominent, antherodes (sterile anthers) present on staminodes. The **gynoecium** is syncarpous, with a superior ovary, 3 carpels (the median carpel anterior), locules 3 or 1 at the apex only or 1–2 (the other locule(s) undeveloped or absent). **Placentation** is axile; **ovules** are orthotropous to anatropous, bitegmic, 1–∞ in number. The **fruit** is a loculicidal capsule, rarely an indehiscent capsule or berry. The **seeds** are rarely winged or arillate, having a starchy endosperm. Plant surfaces typically bear 3-celled, glandular "microhairs," a putative apomorphy for at least the great bulk of the family (Figure 7.49C), and tissues often have raphide-containing mucilage cells.

Members of the Commelinaceae have distributions in most tropical to subtemperate regions worldwide. Economic importance includes ornamental cultivars, such as *Rhoeo*, *Tradescantia*, and *Zebrina*, and some local medicinal and edible species. See Faden (1998), Evans et al. (2003), and Pellegrini (2017a) for treatments and phylogenetic analyses.

The Commelinaceae are distinctive in being mostly perennial ***herbs*** with ***closed sheathed*** leaves and a trimerous, hypogynous flower with an ***ephemeral*** corolla, staminodia in some, most species with characteristic ***3-celled glandular microhairs***, the latter a probable apomorphy for the family (Figure 7.44).

K 3 or (3) **C** 3 or (3) **A** 3 or 3 + 3 staminodes or 1 **G** (3), superior

Haemodoraceae—Bloodwort family (type *Haemodorum*, from Greek *haimo*, blood, in reference to red pigmentation in roots and rootstocks of that and other family members). 13 (–16) genera/ca. 115 species (Figures 7.48, 7.49).

The Haemodoraceae consist of perennial herbs. The **stems** are rhizomatous, stoloniferous or cormose. The **leaves** are simple, unifacial, mostly basal, distichous, sheathing and often equitant, undivided, narrow, flat or terete, and parallel veined. The **inflorescence** is a terminal thyrse or corymb of single or 2–3 branched monochasial cyme units, a simple raceme, or rarely reduced to a single flower. The **flowers** are bisexual, actinomorphic or zygomorphic, pedicellate, bracteate, hypogynous, epigynous, or epiperigynous, glabrous to densely tomentose, trichomes tapering, pilate-glandular, or branched, often brightly colored; the receptacle is extended proximally in some taxa (e.g., *Wachendorfia*). The **perianth** is apotepalous or syntepalous, biseriate, imbricate, and homochlamydeous with 3 outer and 3 inner tepals (median outer tepal posterior in zygomorphic flowers) or syntepalous and uniseriate with 6 valvate tepal lobes; tepals are red, red-orange, yellow-orange, yellow, to green, white, or black in color, hypanthium present or absent. The **stamens** are 6, 3, or 1, whorled, diplostemonous or antitepalous in taxa with 6 stamens, or antipetalous in taxa with 3 or 1 stamens, unfused or epitepalous, with staminodes present in some taxa. **Anthers** are basifixed, longitudinal and introrse in dehiscence, tetrasporangiate, dithecal, with thecae and connective having appendages in *Tribonanthes*. The **pollen** is monosulcate or 2–many porate and binucleate at release. The **gynoecium** is syncarpous, with an inferior or superior ovary, 3 carpels (the median carpel posterior), and 3 (rarely 1 at ovary apex) locules. The **style** and **stigma** are solitary, the latter often 3-lobed. **Placentation** is axile; **ovules** are anatropous, bitegmic, 1, 2, 5–7, or ∞ per carpel. Septal nectaries occur in most taxa. The **fruit** is a capsule or rarely a schizocarp. The **seeds** are globose, ellipsoid and ridged, or flattened and marginally winged, with starchy endosperm. Distinctive arylphenalenone chemicals are found in all investigated family members, comprising a reddish pigmentation in the roots and rhizomes of some taxa (hence the name "Bloodwort").

The Haemodoraceae contain two monophyletic groups: Haemodoroideae [Haemodoreae], with unbranched, pilate or tapering trichomes, 3 (1) stamens, and monosulcate pollen, and Conostylidoideae [Conostylideae], with branched to dendritic trichomes, 6 stamens, and porate pollen. Members of the family grow in seasonally wet habitats with distributions in S.W. and E. Australia, New Guinea, S. South Africa, N. South America, Central America and S. Mexico, Cuba, or E. to S.E. North America. Economic importance includes ornamental cultivars, especially *Anigozanthos* spp. (kangaroo paws), and historical uses by native people for food and as euphorics. See Simpson (1990, 1998) and Hopper et al. (1999, 2009) for recent treatments of the family.

The Haemodoraceae are distinctive in being perennial herbs with ***arylphenalenone*** compounds (imparting a reddish coloration to stems and roots in almost all Haemodoroideae), ***unifacial*** leaves, and variable flowers.

P 3+3 or (3+3) or (6) **A** 1,3,6 **G** (3), inferior or superior, hypanthium present or absent.

Pontederiaceae—Pickerel-Weed family (type *Pontederia*, after Buillo Pontedera, former Professor of Botany at Padua, 1688–1757). 2 genera/ca. 33 species (Figures 7.50, 7.51).

The Pontederiaceae consist of perennial or rarely annual, emergent or free-floating, aquatic herbs. The **stems** are rhizomatous or stoloniferous. The **leaves** are bifacial, ligulate, mostly basal, distichous or spiral, basally sheathing and petiolate (the petiole swollen in *Pontederia crassipes* [*Eichhornia crassipes*]), simple, undivided, narrow to broad, flat, and parallel curved-convergent veined (filiform in some *Heteranthera* s.l.). The **inflorescence** is a terminal or axillary raceme, spike, thryse, or of solitary flowers, with spathe-like bract present. The **flowers** are bisexual, zygomorphic or actinomorphic, hypogynous, glabrous or with scattered pilate-glandular trichomes on the outer perianth, filaments, or style. The

Figure 7.48 COMMELINALES. Haemodoraceae. **A.** *Haemodorum laxum*, showing cormlike rootstock with red pigmentation. **B.** *Haemodorum spicatum*, opened flower. **C.** *Lachnanthes caroliniana*. **D.** *Dilatris viscosa*, showing three, dimorphic stamens. **E.** *Dilatris corymbosa*, flower longitudinal section. **F.** *Wachendorfia thyrsiflora*, showing zygomorphic, enantiostylous flower. **G.** *Xiphidium caeruleum*, showing unifacial, ensiform leaves. **H.** *Xiphidium caeruleum*, actinomorphic flower with three stamens and superior ovary. **I.** *Tribonanthes uniflora*, with inferior ovary. **J,K.** *Conostylis juncea*. **J.** Whole plant. **K.** Flower close-up, showing six stamens. **L.** *Phlebocarya ciliata*, with actinomorphic flowers. **M,N.** *Anigozanthos manglesii*. **M.** Inflorescence, the unit a monochasium. **N.** Zygomorphic flower with six stamens and inferior ovary.

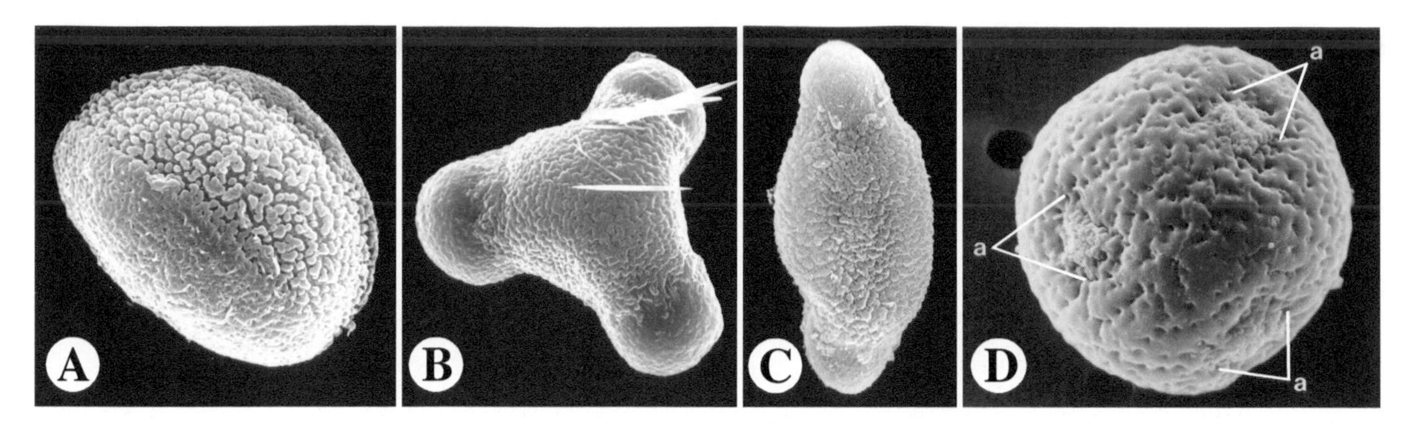

Figure 7.49 COMMELINALES. Haemodoraceae. Pollen diversity. **A.** Ellipsoid and monosulcate, *Schiekia orinocensis*. **B.** Triangular and triporate, *Conostylis aurea*. **C.** Fusiform and diporate, *Macropidia fuliginosa*. **D.** Spherical and oligoporate, *Tribonanthes australis*. Apertures = "a."

perianth is biseriate and homochlamydeous with 3 [4] outer and 3 [4] inner imbricate tepals, with median inner tepal posterior in zygomorphic flowers; tepals basally connate, and blue, lilac, white, or yellow, with nectar guide on median tepal in zygomorphic flowers, hypanthium present. The **stamens** are six (3+3) or three (+ staminodes) or 1 (+ 2 staminodes), whorled, of different lengths in some taxa (often associated with trystyly), diplostemonous (in taxa with 6 stamens) or antipetalous (in taxa with 3 or 1 stamens), epitepalous, filaments with appendages in some taxa. **Anthers** are basifixed, introrse, longitudinal or (in *Pontederia* subg. *Monochoria*) poricidal in dehiscence, tetrasporangiate and dithecal. The **pollen** is di- (tri-) sulculate and trinucleate at release. The **gynoecium** is syncarpous, with a superior ovary, 3 carpels (2 carpels reduced and abortive in *Pontederia*), and 1 to 3 locules. The **styles** are heteromorphic or enantiostylous in some taxa; the **stigma** is solitary, often 3-lobed. **Placentation** is apical, axile, parietal, or axile below and parietal above; **ovules** are anatropous, bitegmic, 1–∞ per carpel. Septal nectaries are present or absent. The **fruit** is a loculicidal capsule or nut/utricle (e.g., *Pontederia*). The **seeds** are longitudinally ribbed, with a starchy endosperm.

Members of the Pontederiaceae have distributions in tropical to north temperate regions in Africa, Asia, and esp. the Americas. Economic importance includes species that are serious weeds (especially *Pontederia crassipes* [*Eichhornia crassipes*], water hyacinth, which clogs waterways), some species with edible parts, and cultivated ornamentals (e.g., *Heteranthera* and *Pontederia*). See Barrett and Graham (1997), Cook (1998), Graham et al. (1998), and Pellegrini (2017b) and Pellegrini et al. (2018) for treatments and phylogenetic analyses. The last two studies proposed merging formerly seven genera into two, *Heteranthera* and *Pontederia*.

The Pontederiaceae are distinctive in being ***emergent to free-floating aquatic herbs*** with simple, sheathing, ***bifacial*** leaves, actinomorphic or zygomorphic flowers, and ***di-(tri-)sulculate*** pollen. The bifacial leaves and sulculate pollen are probable apomorphies for the family (Figure 7.42).
P (3+3) or (4+4) **A** 3+3 or 3+staminodes or 1+2 staminodes **G** (3), superior, hypanthium present.

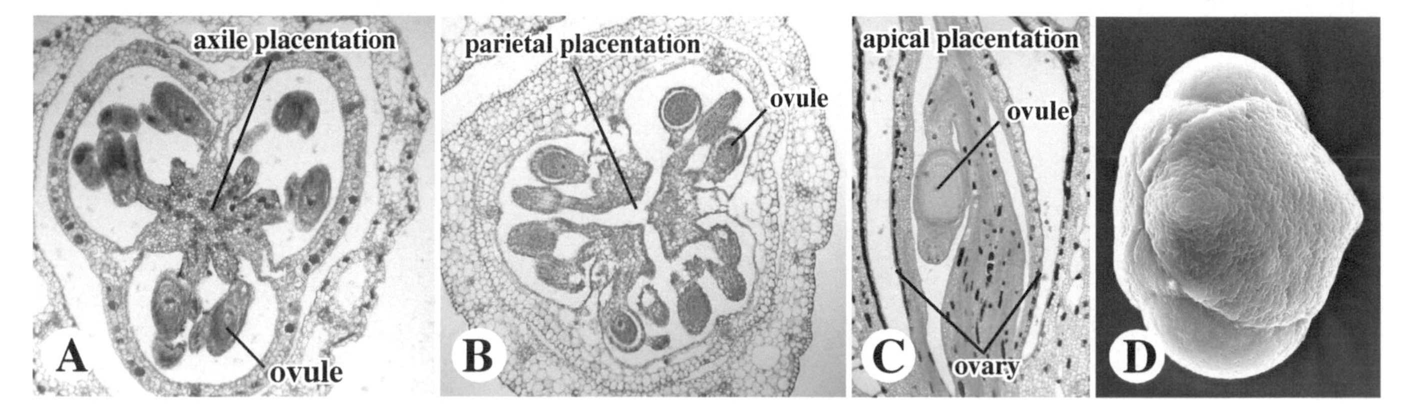

Figure 7.50 COMMELINALES. Pontederiaceae. **A.** *Pontederia diversifolia* [*Eichornia diversifolia*], ovary cross section, showing axile placentation. **B.** *Heteranthera reniformis*, ovary cross section, with parietal placentation. **C.** *Pontederia cordata*, ovary longitudinal section, showing apical placentation. **D.** Pollen of *Pontederia cordata*, showing disulculate apertures.

Figure 7.51 COMMELINALES. Pontederiaceae. **A–F.** *Pontederia crassipes* [*Eichhornia crassipes*]. **A.** Habit, showing masses of floating clonal plants clogging lake. **B.** Plant with inflorescence. **C.** Close-up of leaves with swollen petioles. **D.** Inflorescence. **E.** Close-up of zygomorphic flower, showing median posterior inner tepal with nectar guide. Note dimorphic stamens: three long and three short. **F.** Flower close-up, showing three long stamens and style (above). **G–I.** *Pontederia cordata*. **G.** Whole plant, in flower. **H.** Inflorescence. **I.** Flower close-up, showing zygomorphic perianth with nectar guides. **J,K.** *Heteranthera reniformis*. **J.** Whole plant, with inflorescence. **K.** Close-up of tubular, actinomorphic flowers.

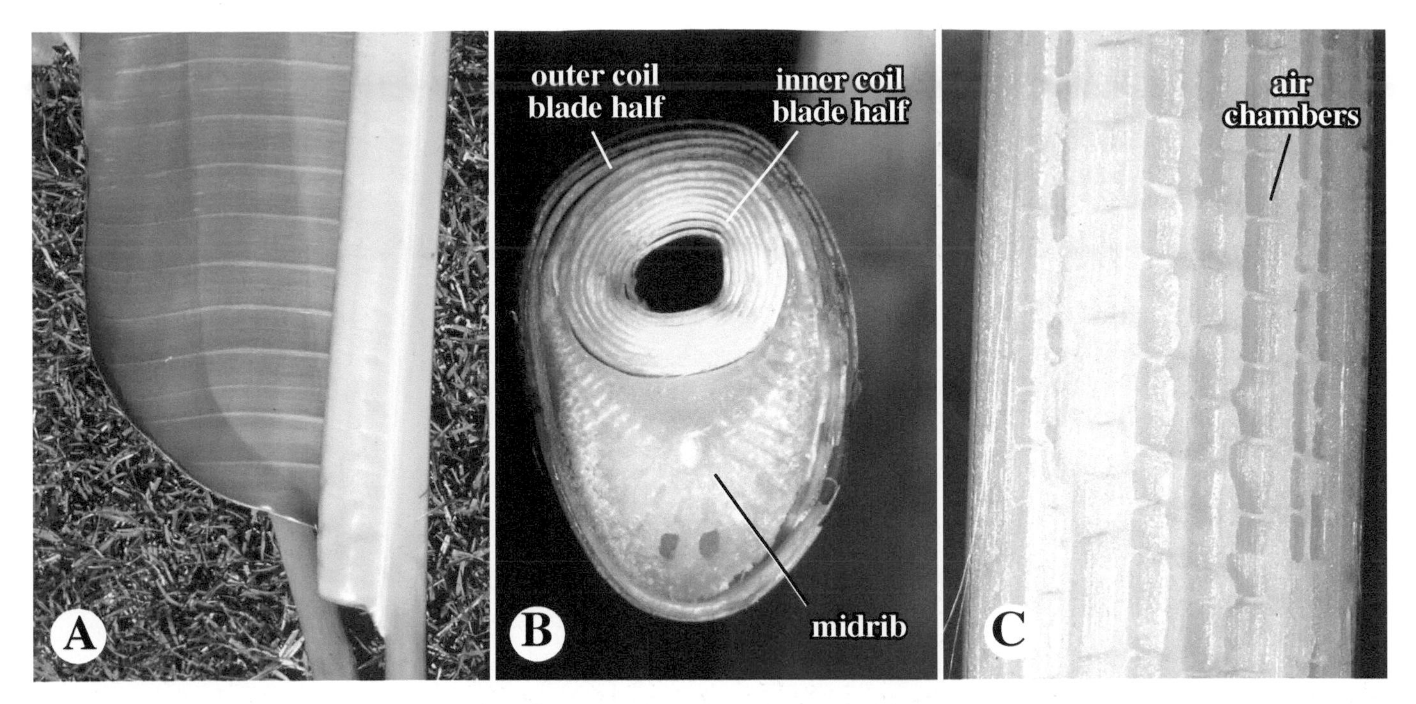

Figure 7.52 Apomorphies of the Zingiberales. **A–B.** Supervolute ptyxis. **A.** *Strelitzia reginae* (bird-of-paradise) young leaf, showing right half still partially coiled. **B.** *Musa acuminata* (banana), immature leaf cross section, showing rolled left and right leaf blade halves. **C.** Diaphragmed air chambers, seen here in a leaf petiole longitudinal section of *Musa acuminata* (banana).

ZINGIBERALES

The Zingiberales, commonly called the gingers and bananas, are a well-defined, monophyletic group of eight families (Table 7.3). Several apomorphies unite the ginger group. One obvious apomorphy is the occurrence of leaves with **penni-parallel venation** (Figures 7.52A, 7.55A). In addition, virtually all members of the Zingiberales have a **ptyxis** (the posture of immature leaves or leaf parts; see Chapter 9) that is **supervolute**, in which the two opposing (left and right) halves of the blade are rolled along a longitudinal axis, one half being rolled completely within the other (Figure 7.52A,B). Leaves and stems of all members of the order have **diaphragmed air chambers** (Figure 7.52C) and possess **silica cells** (although the latter is not apomorphic for this order alone). Lastly, all Zingiberales have an **inferior ovary** (Figures 7.58D; see also Figure 7.53A).

The four terminal families of the Zingiberales make up a monophyletic assemblage, commonly known as the "ginger group," their phylogenetic interrelationships well accepted (Figure 7.53A–C). Relationships of the other four families, referred to as the "bananas," have varied in different analyses. The phylogenies in Figures 7.53A/7.54 (after Kress et al., 2001) and 7.53B (after Sass et al., 2016) portray different paraphyletic banana groups, but that of Figure 7.53C (after Givnish et al., 2018) shows a monophyletic banana group. See these studies and Johansen (2005) for phylogenetic analyses of the order.

Four of the eight families of the Zingiberales are described here. Among those not treated are the **Heliconiaceae** (Figure 7.55A–C), **Marantaceae** (Figure 7.55D–F), and **Costaceae** (Figure 7.55G–I). Descriptions of families below are in approximate phylogenetic order.

Musaceae—Banana family (type *Musa*, after Antonia Musa, physician to Emperor Augustus 63–14 BC). 2–3 genera (*Ensete*, *Musa*, *Musella* [*Ensete*])/ca. 40 species (Figure 7.56).

The Musaceae consist of monoecious, perennial herbs. The **stems** are subterranean, sympodial, rhizomatous to cormose, and hapaxanthic. The **leaves** are large, basal, spiral, sheathing (with the long, sheathing leaf bases overlapping, forming a pseudo-stem), a petiole (sometimes termed a "pseudo-petiole") present in *Musa*, lacking in *Ensete*, simple (often tearing in several places perpendicular to the midrib), and penni-parallel-veined. The **inflorescence** (which arises from the apical meristem of the corm and grows inside the rolled leaf sheaths) is a terminal thyrse, equivalent to a raceme of spirally arranged, fasciculate, monochasial cymes (commonly called "banana hands"), bracteate, the bracts large, coriaceous, each enclosing a fasciculate unit cyme, female cymes proximal, male cymes distal. **Flowers** are ebracteate, unisexual, zygomorphic, epigynous. The **perianth** is biseriate and homochlamydeous, 3+3, syntepalous (the inner, adaxial tepal usually distinct). **Stamens** are apostemonous, 5 or 6, the missing stamen or staminode opposite the inner, median,

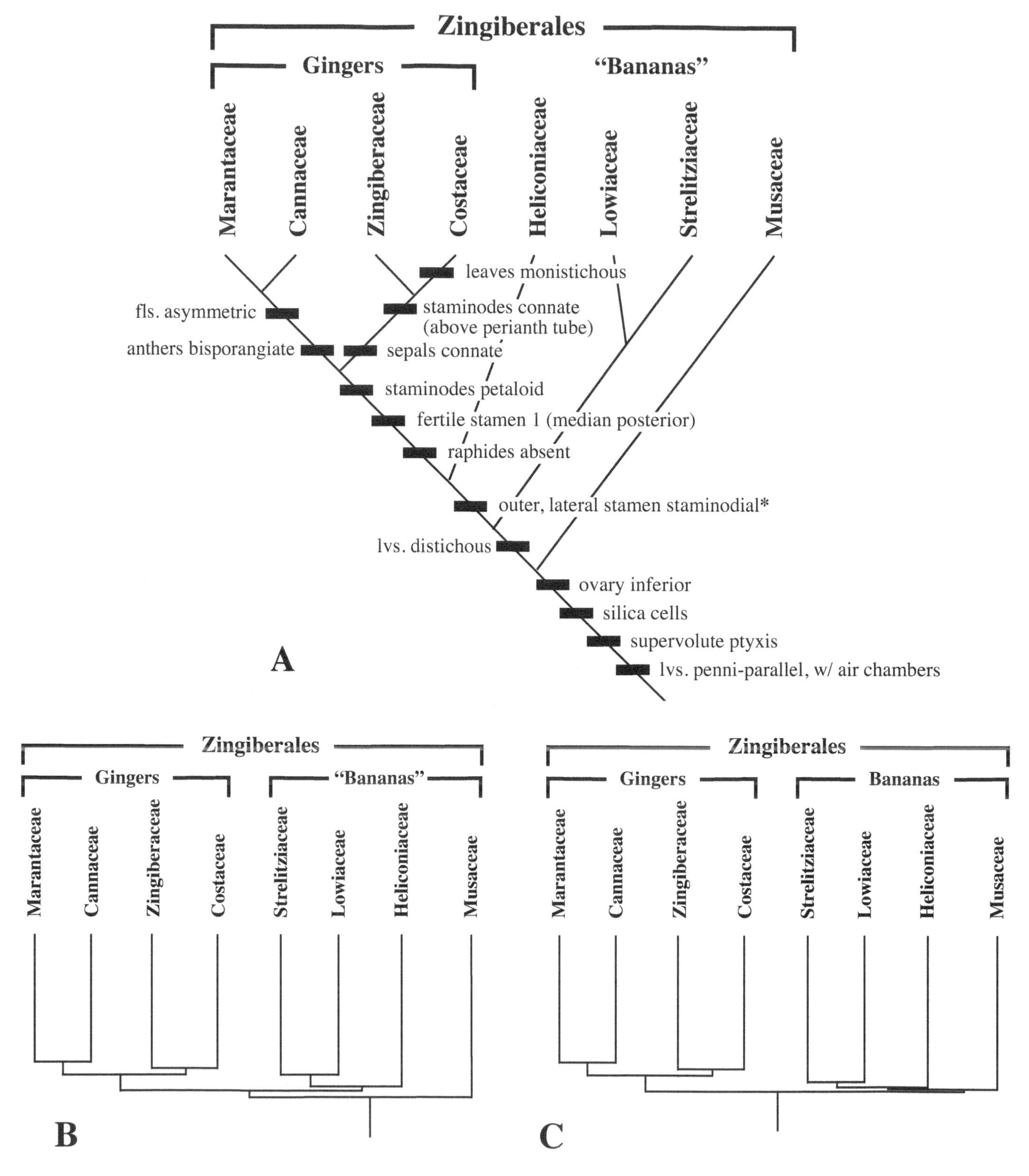

Figure 7.53 **A.** Cladogram of the Zingiberales (after Kress et al. 2001), with selected apomorphies, that at "*" after Kirchoff (2003). Note the monophyletic ginger group but banana group forming a grade, with Musaceae sister to all others of the order. **B.** Phylogram, redrawn after Sass et al. (2016), showing alternative relationship, in which Musaceae remains sister to the rest of the order, but three families of the "banana" group form a clade, the latter sister to the ginger clade. **C.** Phylogram, redrawn after Givnish et al. (2018), showing yet another relationship, in which the banana group is a clade. In "**B**" and "**C**," however, note short branches of some banana group lineages.

Figure 7.54 Cladogram of the Zingiberales, after Kress et al. (2001). Note relationships in the banana group different from that of Sass et al. (2016) or Givnish et al. (2018). (See Figure 7.53.) Artwork by Ida Lopez, by permission of W. J. Kress.

adaxial tepal; anthers are longitudinal in dehiscence, dithecal. The **gynoecium** is syncarpous, with an inferior ovary, 3 carpels (the median carpel anterior), and 3 locules; the styles are terminal; placentation is axile; ovules are anatropous, bitegmic, ∞ per carpel. Septal nectaries are present and occur above the locules. The **fruit** is a berry; seeds are endospermous, with rudimentary arils. Pollinated by bats or birds in the wild.

Members of the family have Old World distributions in tropical Africa and southeast Asia to northern Australia. Economic importance includes use of fruits of *Musa* spp. as a food source (esp. triploid forms of *Musa acuminata* and the triploid hybrid *Musa ×paradisiaca*); *Musa textilis* (Manila-hemp, abacá) and *Musa basjoo* are used as a fiber source for twine, textiles, and building materials. See Andersson (1998a) and Liu et al. (2010) for a recent treatments of the Musaceae.

The Musaceae are distinguished from related families of the Zingiberales in having a ***spiral*** leaf arrangement and ***monoecious*** plant sex.

P (3+3) **A** 5–6 **G** (3), inferior.

Strelitziaceae—Bird-of-paradise family (type *Strelitzia*, after Charlotte of Mecklenburg-Strelitz, wife of King George III). 3 genera (*Phenakospermum, Ravenala, Strelitzia*)/7 species (Figures 7.57A, 7.58).

The Strelitziaceae consist of perennial herbs or trees. The underground **stems** are rhizomatous (dichotomously branching in at least some), the aerial stems decumbent and herbaceous or arborescent and woody-textured. The **leaves** are distichous, sheathing, petiolate, simple, and penni-parallel-veined (veins marginally fused). The **inflorescence** is a terminal or axillary

Figure 7.55 ZINGIBERALES, exemplars of three other families, not described. **A–C.** Heliconiaceae, *Heliconia* spp., all with showy bracts. **A,B.** Taxa with erect inflorescences. **C.** Taxa with pendant inflorescences. **D–F.** Marantaceae. **D,E.** *Calathea louisae*. **F.** *Thalia geniculata*. **G–I.** Costaceae, *Costus* spp. **G.** Leaves spiral, monistichous (borne along one row). **H.** Inflorescence. **I.** Flowers.

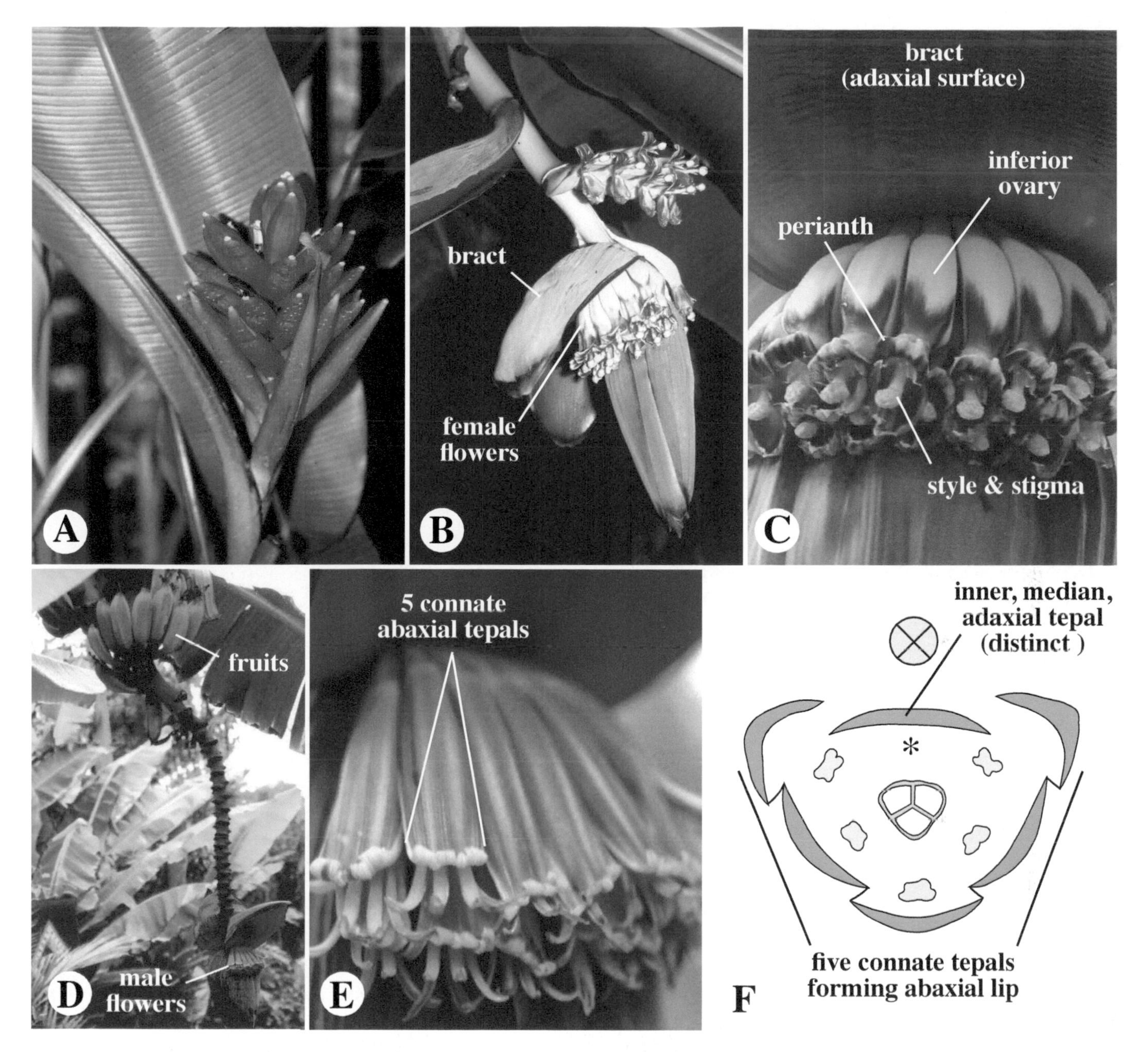

Figure 7.56 ZINGIBERALES. Musaceae. **A.** *Musa coccinea*, showing leaf with penni-parallel venation and terminal inflorescence (of bright red flowers and bracts). **B–F.** *Musa acuminata*, cultivated banana. **B.** Young inflorescence, with proximal cyme unit ("banana hand") of female flowers, subtended by large bract. **C.** Cyme of female flowers. **D.** Inflorescence, which grew through pseudostem, having proximal female flowers (in fruit) and distal male flowers (below). **E.** Close-up of male flowers. **F.** Floral diagram (combining male and female flowers); * = missing stamen.

thryse of 1–many monochasial cymes, each cyme subtended by a large, spathaceous bract. **Flowers** are bisexual, zygomorphic, bracteate, epigynous. The **perianth** is biseriate and homochlamydeous, 3+3, syntepalous, the median inner tepal smaller than the lateral, sometimes connivent inner tepals. **Stamens** are 5 or 6; anthers are basifixed, longitudinal in dehiscence, and bithecal. The **gynoecium** is syncarpous, with an inferior ovary, 3 carpels (the median carpel anterior), and 3 locules; the style is terminal and filiform; placentation is axile; ovules are anatropous, bitegmic, and ∞ per carpel. Septal nectaries are present. The **fruit** is a loculicidal capsule; seeds are arillate, with a starch-rich endosperm and starch-less perisperm. Pollinated by insects or birds.

Members of the family have distributions in tropical South America, Southern Africa, and Madagascar. Economic importance includes some species used as ornamental culti-

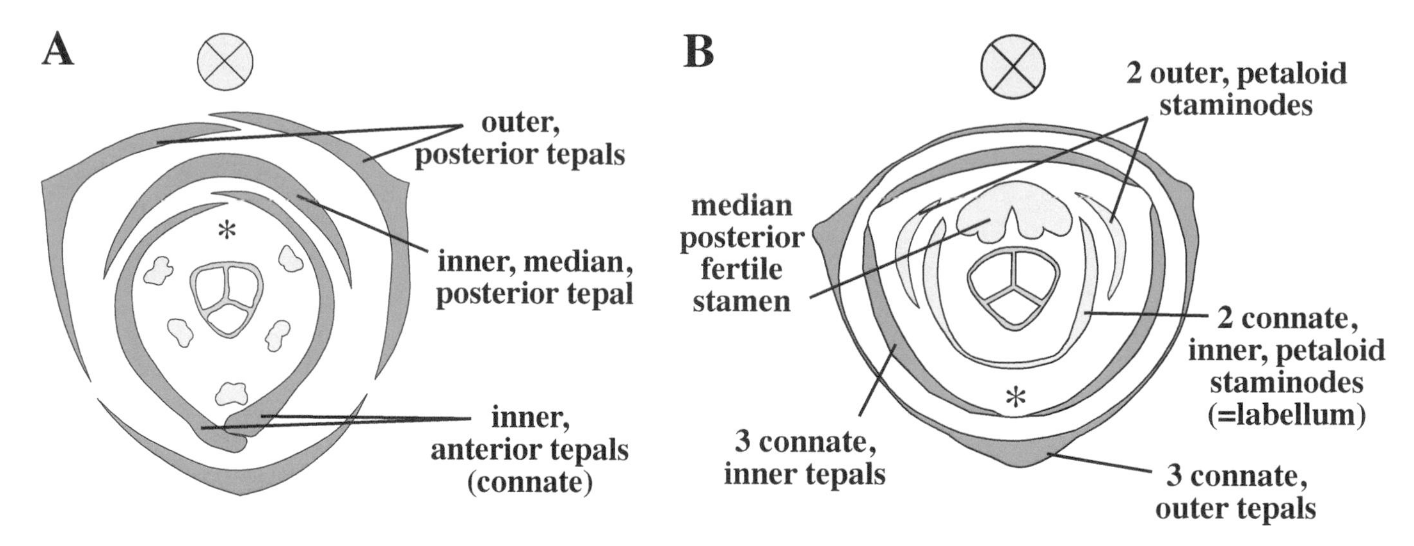

Figure 7.57 Floral diagrams. **A.** Strelitziaceae. **B.** Zingiberaceae. * = missing stamen.

vars, e.g., *Strelitzia reginae* (bird-of-paradise) and *S. nicolai* (tree bird-of-paradise). See Andersson (1998b) and Cron et al. (2012) for a treatments and analyses of the Strelitziaceae.

The Strelitziaceae are distinguished from related families of the Zingiberales in having ***rhizomatous and decumbent*** or ***erect, arborescent*** stems with ***distichous*** leaves and flowers having ***5–6 stamens***.

P (3+3) **A** 5 or 6 **G** (3), inferior.

Zingiberaceae—Ginger family (type *Zingiber*, from a pre-Greek name, possibly from India). ca. 56 genera/ca. 1,100-1,600 species (Figures 7.57B, 7.59).

The Zingiberaceae consist of perennial herbs. The **stems** are rhizomatous and sympodial. The **leaves** are distichous, simple, sheathing (sheaths forming a pseudo-stem in some), petiolate, usually ligulate, penni-parallel-veined, a pulvinus present in *Zingiber*. The **inflorescence** is a bracteate spike, raceme, thyrse, or of solitary flowers. **Flowers** are bisexual, zygomorphic, bracteate, and epigynous. The **perianth** is biseriate and homochlamydeous, 3+3, syntepalous, each whorl 3-lobed. **Stamens** are 1 fertile (median posterior in position); the anther is longitudinal or poricidal in dehiscence, dithecal. Staminodes are 4, petaloid, the two in the inner whorl connate, forming an anterior labellum, the two in the outer whorl distinct above the floral tube or fused to labellum (the third member of the outer whorl absent). The **gynoecium** is syncarpous, with an inferior ovary, 3 carpels (the median carpel anterior), and 1 or 3 locules; the style is terminal and positioned in the furrow of the filament and between the anther thecae; placentation is axile or parietal; ovules are anatropous, bitegmic, and ∞ per carpel. Septal nectaries are absent and replaced by two epigynous nectaries. The **fruit** is a dry or fleshy loculicidal or indehiscent capsule; seeds are arillate, with a starch-rich endosperm and perisperm. Plants are insect-pollinated.

The Zingiberaceae are a large family, classified (sensu Kress 2002) into four subfamilies: Siphonochiloideae; Tamijioideae; Alpinioideae (with tribes Riedelieae and Alpinieae), distinctive in having lateral staminodes small or absent, never petaloid; and Zingiberoideae (with tribes Zingibereae and Globbeae), distinctive in having the distichous leaf plane parallel to the rhizome. Members of the family have distributions in the tropics of south and southeastern Asia, especially Indo-malaysia. Economic importance includes the source of important spice plants, e.g., *Curcuma* spp., including *C. domestica* (turmeric), *Elettaria cardamomum* (cardamom), and *Zingiber* spp., including *Z. officinale* (ginger); some species are grown as cultivated ornamentals, e.g., *Alpinia* and *Hedychium*. See Larsen et al. (1998), Kress et al. (2002), and Kress and Specht (2006) for recent studies of the Zingiberaceae.

The Zingiberaceae are distinguished from related families of the Zingiberales in having ***distichous, usually ligulate*** leaves with a ***single, dithecal*** stamen and a ***petaloid labellum derived from two staminodes***.

P (3+3) **A** 1 fertile + 2 + (2) petaloid staminodes **G** (3), inferior.

Cannaceae—Canna-Lily family (type *Canna*, after Greek name for reed). 1 genus (*Canna*)/ca. 10–25 species (Figure 7.60).

The Cannaceae consist of perennial herbs. The **stems** are rhizomatous and sympodial. The **leaves** are distichous [to spiral], sheathing, petiolate, simple, and penni-parallel-

Figure 7.58 ZINGIBERALES. Strelitziaceae. **A–E.** *Strelitzia reginae*, bird of paradise. **A.** Whole plant, showing (basal) leaves (arising from rhizome) and lateral, erect inflorescence. **B.** Close-up of inflorescence. Note large, subtending spathe and two visible flowers. **C.** Inflorescence cross section, showing spathe and monochasium of flowers (ovaries seen in cross section), each flower subtended by a bract. **D.** Flower close-up, showing inner and outer tepals and inferior ovary. **E.** Inner anterior tepals pulled back, exposing the enclosed five stamens and central style/stigma. **F–H.** *Strelitzia nicolai*, giant bird of paradise. **F.** Whole plant. Note distichous, sheathing, cauline leaves. **G.** Inflorescence. **H.** Loculicidal capsular fruits. Note black seeds covered at base with (orange) arils.

Figure 7.59 ZINGIBERALES. Zingiberaceae. **A–C.** *Alpinia* sp. (shell ginger). **A.** Whole plant, erect aerial stem with inflorescence. **B.** Flower close-up, showing tepals and petaloid staminodes, forming an anterior labellum. **C.** Flower close-up. Note single anther of stamen posterior to and partially enclosing style. **D–F.** *Hedychium* sp. (ginger-lily). **D.** Whole plant. Note distichous leaves. **E.** Inflorescence, showing single stamen of flower. **F.** Flower close-up (removed), showing inferior ovary, outer and inner tepals, and showy, petaloid staminodes.

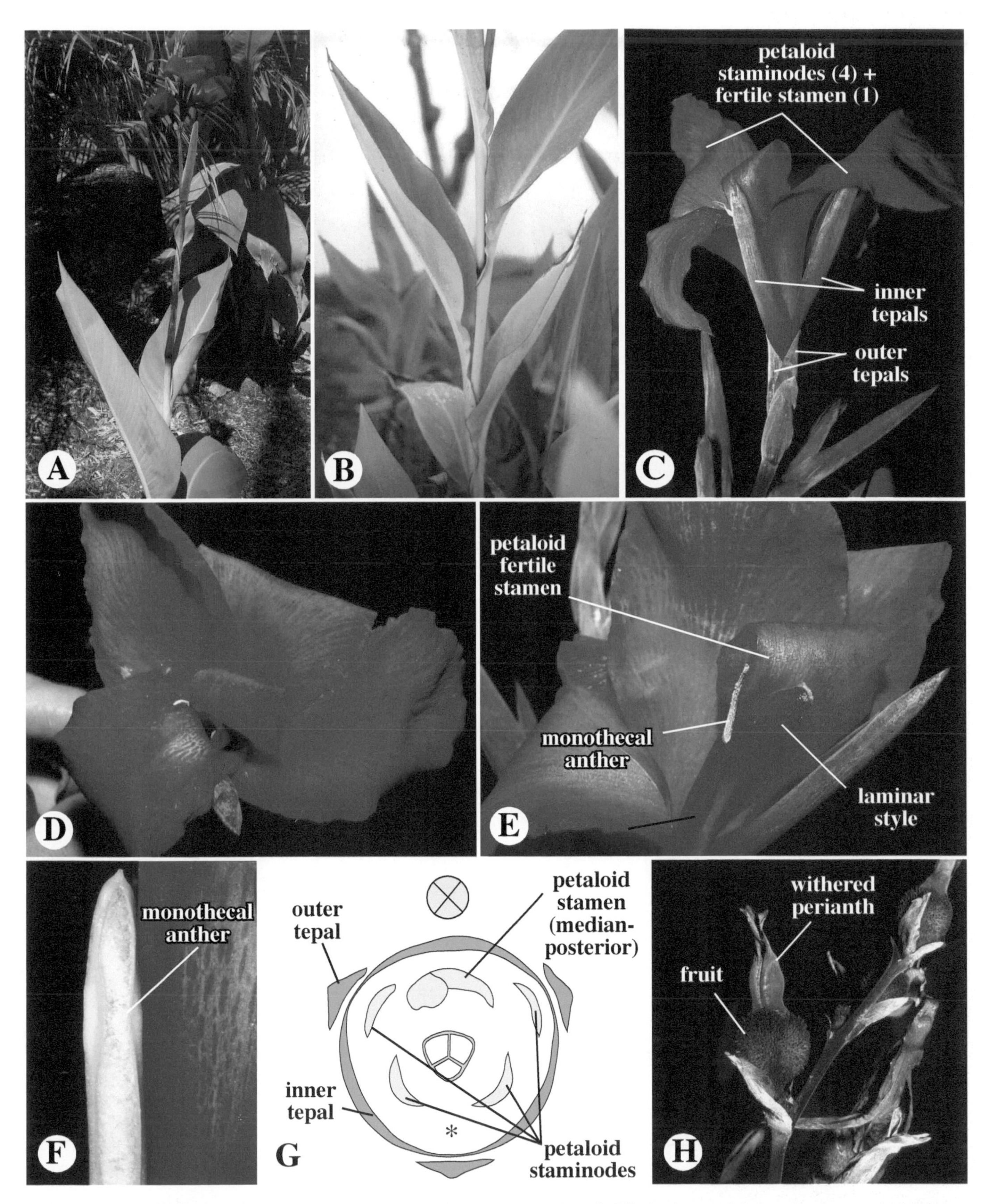

Figure 7.60 ZINGIBERALES. Cannaceae. **A–F.** *Canna ×generalis* (canna-lily). **A.** Whole plant, aerial shoot bearing terminal inflorescence. **B.** Close-up of shoot, showing sheathing leaves. **C.** Flower. Note reduced outer tepals, narrow, showy inner tepals, and large, showy, petaloid staminodes and stamen. **D.** Top view of flower, showing petaloid staminodes. **E.** Flower close-up, showing petaloid fertile stamen with laterally adnate, monothecal anther. Note laminar style. **F.** Monothecal anther, close-up. **G.** Floral diagram. * = missing stamen. **H.** Close-up of fruit, a capsule.

veined. The **inflorescence** is a bracteate thyrse consisting of a spike or raceme of 2-flowered cymes (or reduced to a raceme). **Flowers** are bisexual, asymmetric, and epigynous. The **perianth** is biseriate and homochlamydeous, 3+3, and apotepalous. **Stamens** are 1 fertile, (median posterior in position), the fertile stamen petaloid. Staminodes are 1–4[5], large, petaloid, resembling the fertile stamen; the anther is laterally subapically positioned on the petaloid stamen, longitudinal in dehiscence, bisporangiate, and monothecal. The **gynoecium** is syncarpous, with an inferior ovary, 3 carpels (the median carpel anterior), and 3 locules; the style is terminal and laminar; placentation is axile; ovules are anatropous, bitegmic, ∞ per carpel. The **fruit** is a usually a capsule; seeds are exarillate, with a starch-rich endosperm and perisperm.

Members of the Cannaceae have distributions in the warm American tropics. Economic importance includes ornamental cultivars of *Canna* spp. (canna lily) and a source of starch (from rhizome of *Canna edulis*). See Kubitzki (1998c), Maas-van de Kamer and Maas (2008), and Prince (2010) for family treatments and phylogenetic studies.

The Cannaceae are distinguished from related families of the Zingiberales in having usually ***distichous*** leaves and flowers with ***one petaloid, monothecal*** stamen associated with ***1–4[5] petaloid staminodes***.

P 3+3 **A** 1, petaloid & monothecal + 1–4 petaloid staminodes **G** (3), inferior.

POALES

The Poales is a large group of 16–17 families, nine of which are described here. One hypothesis of relationships of the families is seen in Figure 7.61. The order includes several basal groups with showy, insect-pollinated flowers. Many members of the Poales have small, reduced, typically wind pollinated flowers. See Bremer (2002) and Michelangeli et al. (2003) and general studies of the angiosperms (e.g., Soltis et al. 2007, 2011) and monocots (e.g., Chase et al. 2000a, 2006; Davis et al. 2004; Graham et al. 2006; McKain et al. 2016; and Givnish et al. 2010, 2018) for studies on phylogenetic relationships in the order.

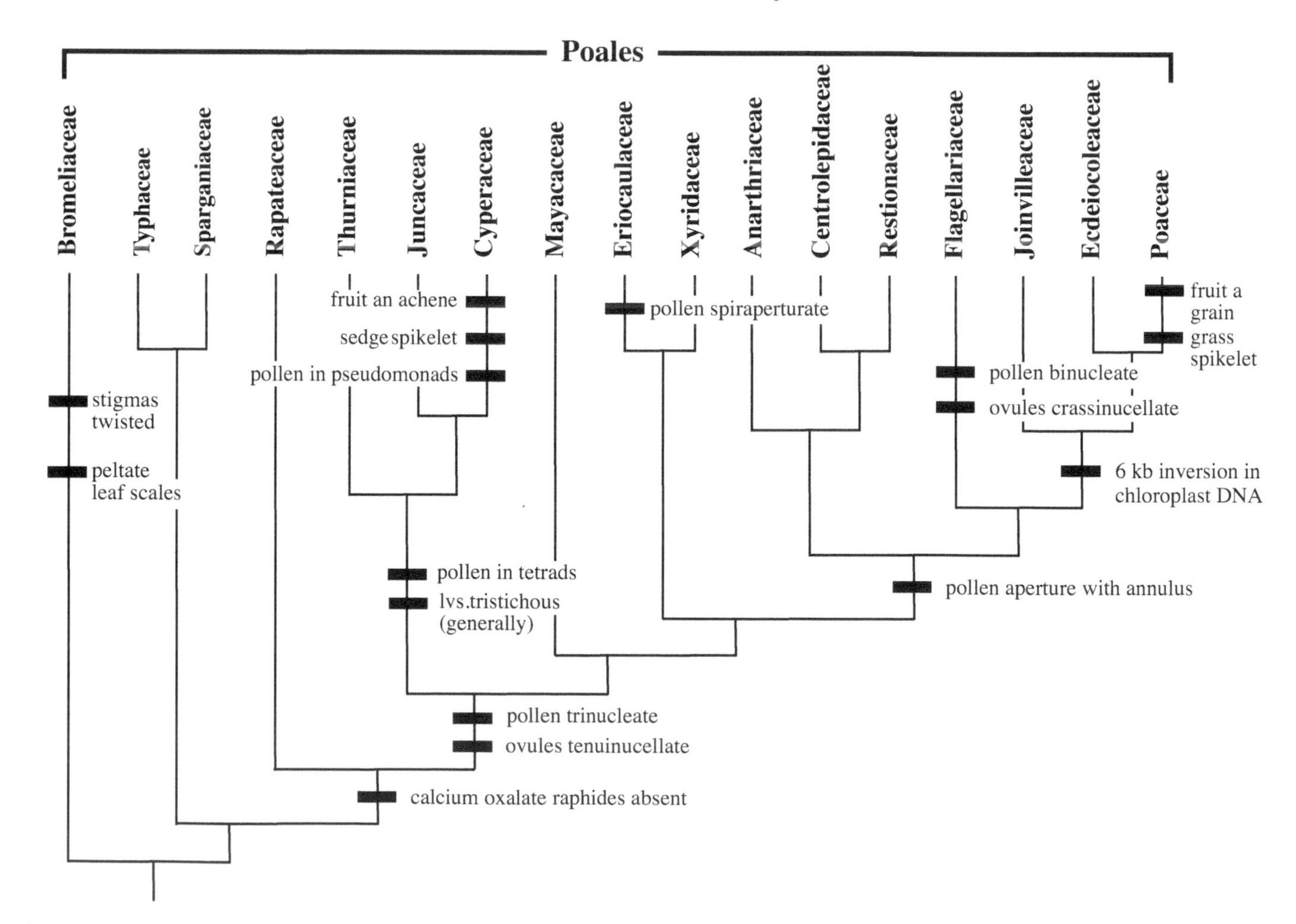

Figure 7.61 Cladogram of the families of the Poales, after Givnish et al. (2018), with selected apomorphies shown.

Bromeliaceae—Bromeliad family (type *Bromelia*, after Swedish medical doctor and botanist O. Bromell, 1639–1705). ca. 69 genera/ca. 3,400 species (Figure 7.62).

The Bromeliaceae consist of terrestrial or epiphytic, perennial herbs to rosette trees. **Roots** are absorbing or function as holdfasts or rarely absent. The **stem** is a caudex or rarely arborescent, often sympodially branched. The **leaves** are spiral, simple, often adaxially concave, sheathing, in some (the "tank" bromeliads) tightly overlapping and channeling rain water and runoff to storage in a central cavity, the margins entire or serrate-spinose, the surface usually bearing (at least when young) absorptive, usually peltate, scale-like trichomes, functioning in water and mineral uptake. The **inflorescence** is a terminal, bracteate, spike, raceme, or head, the bracts often brightly colored. The **flowers** are bisexual [rarely unisexual], actinomorphic or slightly zygomorphic, bracteate, hypogynous or epigynous. The **perianth** is biseriate and dichlamydeous to homochlamydeous, 3+3, the perianth parts distinct to basally connate, the petals/inner tepals often with one or more basal scale- or flap-like appendages. The **stamens** are diplostemonous, 3+3, distinct or connate, often epipetalous. **Anthers** are longitudinally and introrsely dehiscent. The **gynoecium** is syncarpous, with a superior or inferior ovary, 3 carpels, and 3 locules. The **style** is solitary, with 3 typically twisted **stigmas**. **Placentation** is axile; **ovules** are mostly anatropous, bitegmic, few–∞ per carpel. Septal nectaries are present. The **fruit** is a septicidal capsule or berry, rarely a sorosis (*Ananas*). The **seeds** are winged, plumose, or glabrous. Pollinated by birds, insects, bats, wind [rarely], or flowers cleistogamous.

The Bromeliaceae were traditionally classified into three subfamilies: Pitcairnioideae, with superior (to half-inferior) ovaries, forming capsules with winged seeds; Tillandsioideae, with superior ovaries forming capsules with plumose seeds; and Bromelioideae, with inferior ovaries forming berries with unappendaged seeds. Givnish et al. (2008) proposed a classification into eight monophyletic subfamilies: Brocchinioideae, Lindmanioideae, Hechtioideae, Tillandsioideae, Navioideae, Pitcairnioideae, Puyoideae, and Bromelioideae. Their Tillandsioideae and Bromelioideae correspond to the traditional subfamilies; the other six were formerly classified together (as a now paraphyletic Pitcairnioideae, s.l.). Members of the Bromeliaceae are distributed almost entirely in the American tropics. Economic importance includes uses as fruit plants (e.g., *Ananas comosus*, pineapple), fiber plants, and cultivated ornamentals. See Smith and Till (1998) for a general description and Givnish et al. (2007) and Escobedo-Sarti et al. (2013) for phylogenetic analyses of the family.

The Bromeliaceae are distinctive in being perennial ***terrestrial or epiphytic*** herbs or shrubs with ***absorptive, peltate trichomes***, often ***colorful*** bracts, and trimerous flowers, the petals/outer tepals often with ***basal scales or appendages***, stigmas typically ***twisted***.
P 3+3 or (3)+(3) **A** 3+3 **G** (3), superior or inferior.

Cyperaceae—Sedge family (type *Cyperus*, Greek for several species of that genus). ca. 98 genera/ca. 5,700 species (Figures 7.63, 7.64).

The Cyperaceae consist of perennial or annual herbs, rarely shrubs or lianas. The **stems** of perennials are rhizomes, stolons, bulbs, or caudices bearing aerial culms that are often tufted (cespitose), usually 3-sided, with a solid pith. The **leaves** are bifacial, spiral, and usually tristichous [rarely distichous], sheathing (sheath usually closed), simple, undivided, narrow, flat, and parallel veined, a ligule present or absent; lower leaves (or in some taxa all leaves) reduced to sheaths. The **inflorescence** consists of one or more bisexual or unisexual "sedge" spikelets (either solitary or in various types of secondary inflorescences), each spikelet consisting of a central axis (the rachilla), bearing spiral or distichous bracts (also called scales or glumes), each (except sometimes the lower) subtending a single flower. The **flowers** are small, unisexual or bisexual, actinomorphic, hypogynous. The **perianth** is absent or 6-merous [1–∞], of reduced, distinct bristles or scalelike tepals. **Stamens** are 3 [1–6+], anthers introrse and longitudinal in dehiscence, filaments elongating during anthesis. The **pollen** is released as "pseudomonads," in which 3 of the 4 nuclei of the microspore tetrad degenerate after microsporogenesis. The **gynoecium** is syncarpous, with a superior ovary, 2 or 3 [rarely 4] carpels, and 1 locule; the gynoecium of *Carex* and relatives is surrounded by an inflated bract, known as the perigynium, at the apex of which the style protrudes. The **styles** are usually 2 or 3. **Placentation** is basal; **ovules** are anatropous, bitegmic, 1 per ovary. **Nectaries** are absent. The **fruit** is a lenticular (2-sided) or trigonous (3-sided), 1-seeded achene (also called a nutlet), rarely a drupe or berry. Plants are generally wind pollinated.

Members of the Cyperaceae have a worldwide distribution, especially in temperate regions. The (paraphyletic) genus *Carex* is especially diverse with ca. 2000 species, important in a number of ecosystems. Economic importance is limited, with some species used as mats, thatch, weaving material, or writing material (*Cyperus papyrus*, papyrus, the culm pith of which was historically used to make paperlike scrolls), a few used as ornamental cultivars (e.g., *Cyperus involucratus*, umbrella plant), and some species, such as the nutsedges, being noxious weeds. See Goetghebeur (1998), Simpson et al. (2003, 2007), Muasya et al. (2009), Naczi (2009), Escudero and Hipp (2013), and Hinchcliff and Roalson (2013) for recent descriptions and phylogenetic analyses.

Figure 7.62 POALES. Bromeliaceae. **A–D.** Puyoideae. **A.** *Puya venusta*, whole plants. **B.** *Puya alpestris*, flower. **C.** *Puya* sp., showing superior ovary. **D.** Pitcairnioideae. *Dyckia dawsonii*, flower. **E–H.** Tillandsioideae. **E.** *Tillandsia fasciculata*. **F–H.** *Tillandsia usneoides*. **F.** Whole plant. **G.** Close-up of stem, covered with absorptive, scale-like trichomes. **H.** Close-up of absorptive, peltate trichome, an apomorphy for the family. **I–P.** Bromelioideae. **I,J.** *Neoregelia* sp. **I.** Whole plant, a "tank" bromeliad. **J.** Inflorescence emerging from apex. **K–N.** *Aechmea* sp. **K.** Inflorescence. **L.** Style/stigma close-up. **M.** Flower, longitudinal section, showing inferior ovary. **N.** Ovary cross section. Note septal nectaries. **O.** *Billbergia* sp., flower longitudinal section. Note inferior ovary and scales at base of perianth. **P.** *Ananas comosus*, pineapple. Classification of subfamilies after Givnish et al. (2007).

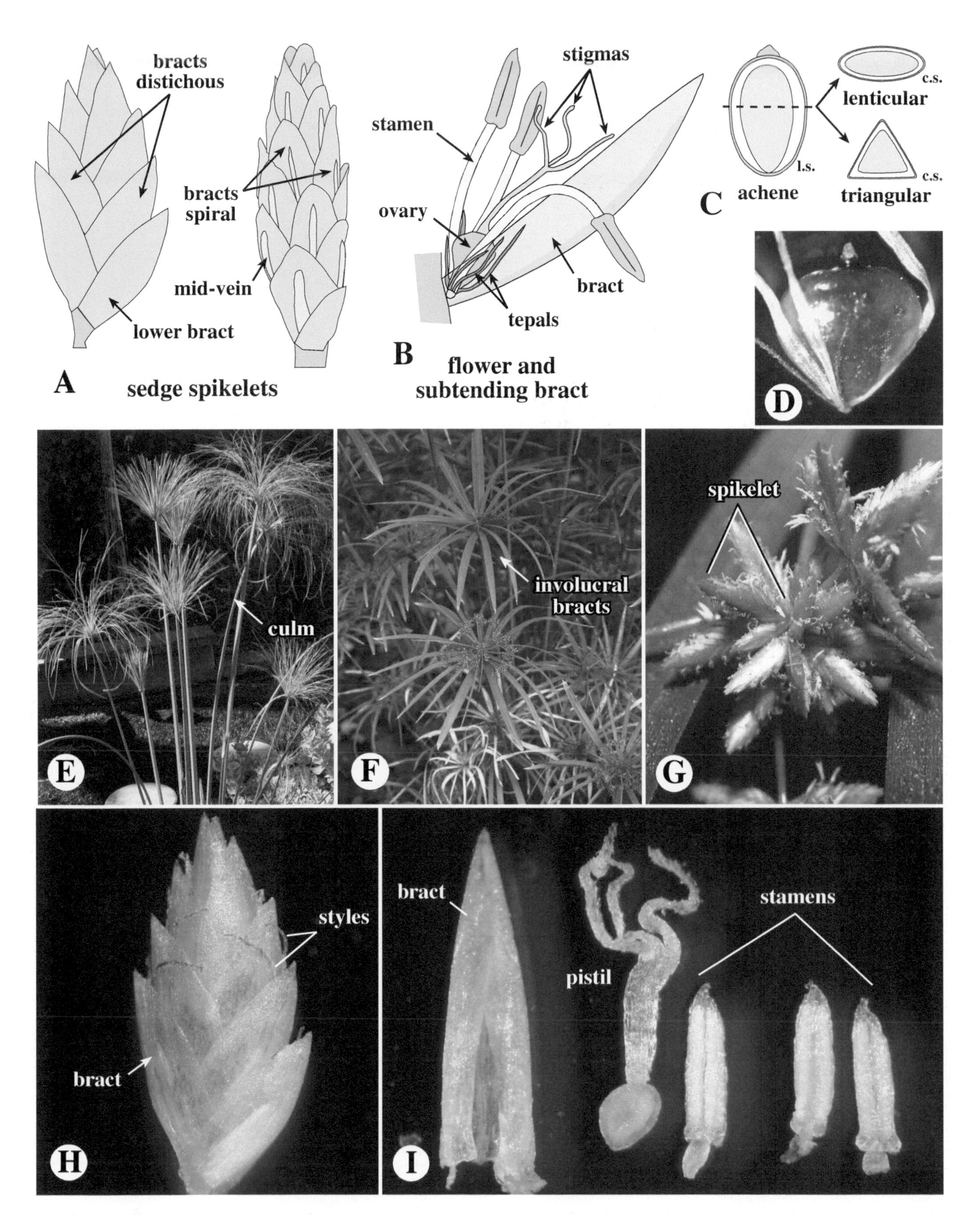

Figure 7.63 POALES. Cyperaceae. **A.** Diagram of sedge spikelets with distichous (left) and spiral (right) bracts. **B.** Diagram of sedge flower. **C.** Diagram of achene, illustrating two shape types; l.s. = longitudinal section; c.s. = cross section. **D.** *Schoenoplectus* sp., mature achene. **E.** *Cyperus papyrus*, papyrus. **F–I.** *Cyperus involucratus*. **F.** Whole plant showing prominent inflorescence bracts. **G.** Close-up of inflorescence, a glomerule of spikelets. **H.** Spikelet, close-up. **I.** Subtending bract (left) and dissected flower components (right).

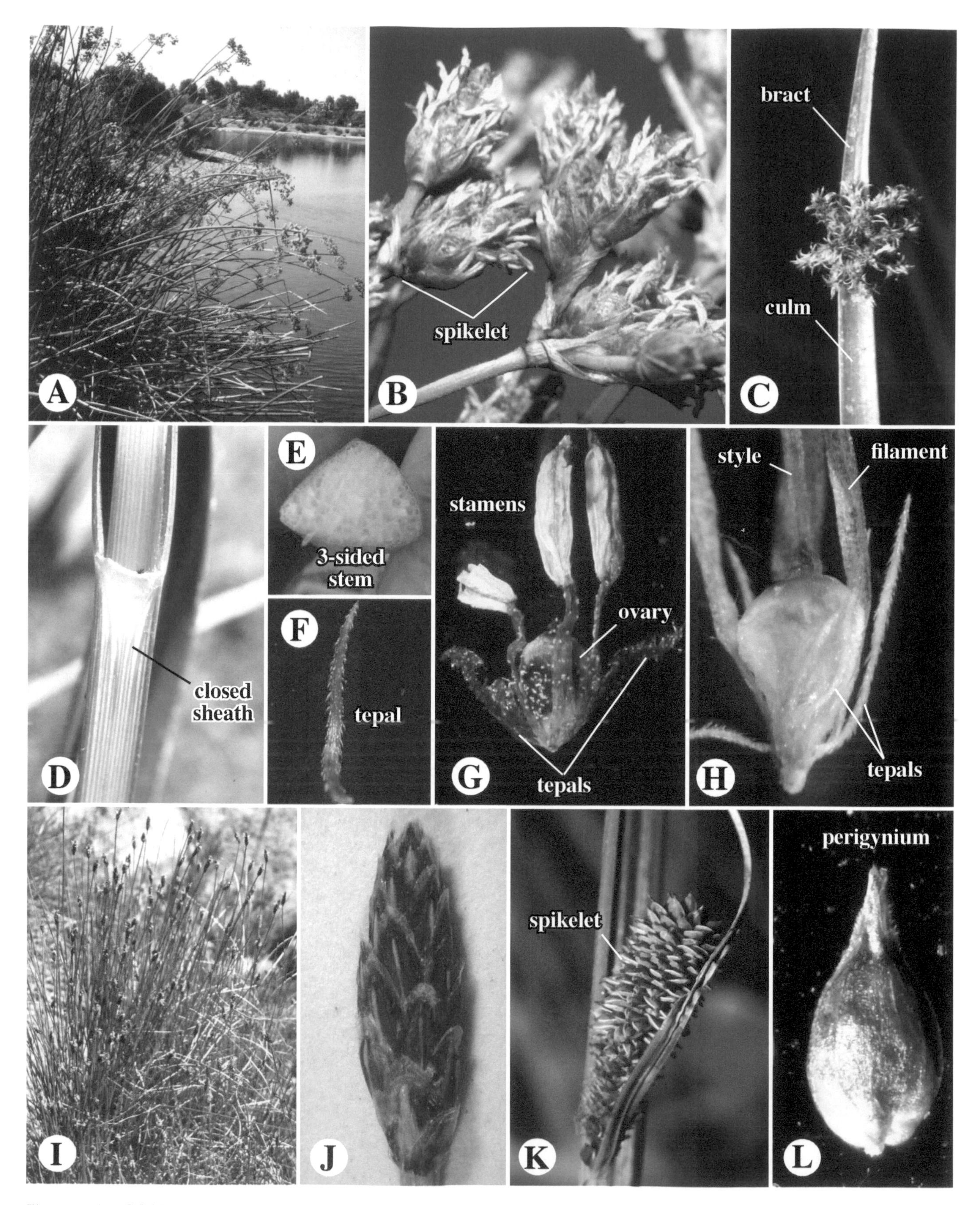

Figure 7.64 POALES. Cyperaceae. **A,B.** *Schoenoplectus californicus*. **A.** Plants growing along pond edge. **B.** Close-up of inflorescence, a dense panicle of spikelets. **C.** *Schoenoplectus americanus*, inflorescence arising from culm. Note stemlike bract at apex. **D,E.** *Bolboschoenus maritimus*. **D.** Closed leaf sheath. **E.** Stem cross section, showing three-sided shape. **F–H.** *Schoenoplectus* spp. **F.** Tepal, detached. **G.** Flower. **H.** Base of flower, showing tepals. **I,J.** *Eleocharis montevidensis*. **I.** Whole plant; note cespitose culms. **J.** Spikelet, close-up. **K.** *Carex barbarae*, spike of female spikelets. **L.** *Carex praegracilis*, perigynium, covering pistil (inside).

The Cyperaceae are distinctive in being herbs with usually ***3-sided, solid-pithed*** stems, ***closed-sheathed, often tristichous*** leaves, the inflorescence a "***sedge spikelet,***" consisting of a central axis bearing many sessile, ***distichous or spiral bracts***, each subtending a single, reduced unisexual or bisexual flower, with perianth ***absent or reduced to bristles or scales***, usually 3 stamens, and a 2–3-carpellate ovary, the fruit a 2- or 3-sided ***achene***.

P 6 or 0 [1–∞] **A** 3 [1–6+] **G** (2–3)[(4)], superior.

Eriocaulaceae—Pipewort family (type *Eriocaulon*, Greek for woolly stem). 7 genera/ca. 1,200 species (Figure 7.65).

The Eriocaulaceae consist of monoecious [rarely dioecious], perennial or annual herbs. The **stems** are rhizomatous, basal shoots often tufted (cespitose). The **leaves** are basal, often rosulate, spiral [rarely distichous], basally sheathing, simple, usually narrow (flat or terete, canaliculate in some), and parallel veined. The **inflorescence** is a scapose head with subtending imbricate bracts (phyllaries), the compound receptacle often with trichomes or chaffy bracts, in monoecious species the male and female flowers mixed or females marginal. The **flowers** are small, whitish, unisexual, actinomorphic or zygomorphic, sessile or short-pedicellate, subtended by chaffy bracts or bractless. The **perianth** is biseriate, dichlamydeous, transparent, whitish, or variously colored. The **calyx** is distinct or basally fused into a tube with 2 or 3 parts. The **corolla** is also distinct or basally fused into a tube and with 2 or 3 parts [corolla rarely absent]. The **stamens** are 2 or 4 in dimerous flowers, 3, 6, or 1 in trimerous flowers, antipetalous (when 2 or 3), epipetalous or arising from a stalk-like "anthophore" (or "androphore"), apically bearing petals and stamens. **Anthers** are longitudinal and introrse in dehiscence, bi- or tetrasporangiate. The **pollen** is spheroidal, usually spiraperturate-spinulose. The **gynoecium** is syncarpous, with a superior ovary, 2 or 3 carpels, and 2 or 3 locules. The **style** is solitary, sometimes style-like appendages also present; **stigmas** are 2–3, dry. **Placentation** is apical, ventral-pendulous; **ovules** are orthotropous, bitegmic, 1 per carpel. **Nectaries** are absent except for glands at tepal tips in some taxa. The **fruit** is a loculicidal capsule. The **seeds** are ellipsoidal, endospermous, starchy. Flowers are wind or insect pollinated.

The Eriocaulaceae consist of the principal genera *Eriocaulon*, *Leiothrix*, *Paepalanthus*, and *Syngonanthus*. Members of the family grow in wet areas with distributions in tropical to subtropical warm regions, especially the Americas, a few northern temperate. Economic importance includes inflorescences of *Syngonanthus* used in the floral trade as "everlastings." See Stützel (1998) for a detailed family description and Gomes de Andrade et al. (2010) and Giulietti et al. (2012) for phylogenetic analyses.

The Eriocaulaceae are distinctive in being perennial or annual herbs with ***basal, often rosulate*** leaves and a ***scapose head*** of very small, ***unisexual*** usually ***white*** flowers.

K 2–3 or (2–3) **C** 2–3 or (2–3) **A** 2, 3, 2+2 or 3+3 **G** (2–3), superior.

Juncaceae—Rush family (type *Juncus*, from the Latin for binder, in reference to use in weaving and basketry). 7 genera/ca. 430 species. (Figure 7.66)

The Juncaceae consist of perennial, rarely annual, herbs. The **stems** of perennials are usually rhizomatous. The **leaves** are simple, parallel veined, undivided, bifacial or unifacial, mostly basal, spiral, usually tristichous [rarely distichous], sheathing, usually with auricles and ligulate, flat or terete. The **inflorescence** is of solitary flowers or compound of 1–many cymes, glomerules, or heads. The **flowers** are bisexual, rarely unisexual, actinomorphic, bracteate, hypogynous. The **perianth** is usually scarious, biseriate, homochlamydeous, rarely uniseriate, 3+3 [2+2 or 3], apotepalous, with hypanthium absent. The **outer** and **inner tepals** are distinct, each whorl of 3 [2] parts. The **stamens** are 3+3 [3+0 or 2+2], whorled, diplostemonous when biseriate, unfused. **Anthers** are basifixed, longitudinally dehiscent. The **pollen** is released as tetrads. The **gynoecium** is syncarpous, with a superior ovary, 3 carpels, and 3 or 1 locules. The **style** is usually 3-branched, **stigmas** sometimes twisted. **Placentation** is axile, basal, or parietal; **ovules** are anatropous, bitegmic, 1–∞ per carpel. The **fruit** is a loculicidal capsule, rarely indehiscent. **Seeds** are starchy endospermous. Flowers are wind or insect pollinated.

The Juncaceae, Cyperaceae, and another family (Thurniaceae, including Prioniaceae) probably share two major apomorphies: tristichous leaves and pollen in tetrads (Figure 7.63). Members of the Juncaceae have a worldwide distribution, generally in temperate and cool regions. Economic importance is limited, some used as ornamental cultivars, *Juncus* spp. used indigenously to make matting, bowls, or other products; cushion-forming *Distichia* is used as fuel in Peru. See Balslev (1998), Záveská Drábková et al. (2003), Roalson (2005), and Záveská Drábková (2010) for recent treatments and phylogenetic analyses of the Juncaceae.

The Juncaceae are distinctive in being usually ***perennial herbs*** with ***spiral, sheathing***, bifacial or unifacial leaves, trimerous, actinomorphic flowers with a typically ***scarious*** perianth and a ***loculicidal capsule***.

P 3+3 [2+2 or 3] **A** 3+3 [3+0 or 2+2] **G** (3), superior.

Figure 7.65 POALES. Eriocaulaceae. *Eriocaulon* sp. **A.** Whole plant. **B.** Inflorescence, a scapose head. **C.** Head, face view, showing dark anthers of male flowers. **D.** Male flower. **E.** Female flower. **F.** Ovary of female flower, tepals removed. **G.** Spiraperturate pollen. Abbreviations: c.l. = corolla lobe; k.l. = calyx lobe.

Poaceae (Gramineae)—Grass family (type *Poa*, Greek name for a grass). 707 genera/ca. 11,300 species (Figures 7.67–7.70).

The Poaceae consist of perennial or annual, hermaphroditic, monoecious, or dioecious herbs or (in the bamboos) trees. The **roots** are adventitious, often endomycorrhizal. The underground **stems** of perennials are rhizomes or stolons, the erect stems (termed "culms") are hollow (solid at the nodes), often cespitose, woody-textured in some (e.g., bamboos). The **leaves** are simple, basal or cauline, distichous, rarely spiral, with a usually open, basal sheath; the leaf blade is bifacial, parallel-veined, often auriculate at base, and typically ligulate, with a ligule at junction of sheath and blade (resembling a sheathlike structure or tuft of trichomes); in the bamboos the first leaves are scalelike and sheathing, followed by branches that bear photosynthetic leaves; in the bamboos and other taxa, a stalklike "pseudo-petiole" is present between the sheath and blade. The **inflorescence** consists of terminal or axillary spikelets (more properly termed "grass spikelets"), these aggregated in secondary inflorescences of spikes, racemes, panicles, or glomerules; the spikelets are sessile or stalked (the spikelet stalk termed a "pedicel"), and are whorled, opposite, or distichous (on 1 or 2 sides) on the inflorescence axes; the grass spikelet itself consists of an axis (termed the "rachilla") bearing distichous parts: two basal bracts (termed "glumes," the lower one called the "first glume," the upper the "second glume," sometimes modified or absent) and one or more "florets"; each floret consists of a minute lateral axis with two additional bracts (termed the "lemma" and "palea") and a flower; the lemma is the lower and larger bract, typically with an odd number of veins (nerves); the palea is the upper, smaller bract, which has 2 veins and is partially enveloped or enclosed by the lemma. A bristlelike **awn** may be present at the apex of glumes or lemmas. The **flowers** are bisexual or unisexual, sessile, and hypogynous. The **perianth** is absent or modified into 2 or 3 **lodicules** (located on the

Figure 7.66 POALES. Juncaceae. **A.** *Juncus bufonius*, whole plant. **B.** *Juncus dubius*, flower close-up. Note six, imbricate, scarious tepals. **C,D.** *Juncus phaeocephalus*. **C.** Close-up of unifacial leaves. **D.** Flowers, showing exserted styles. **E–F.** *Juncus acutus*. **E.** Whole plant. **F.** Close-up of infructescence of capsules (left) and inflorescence (right).

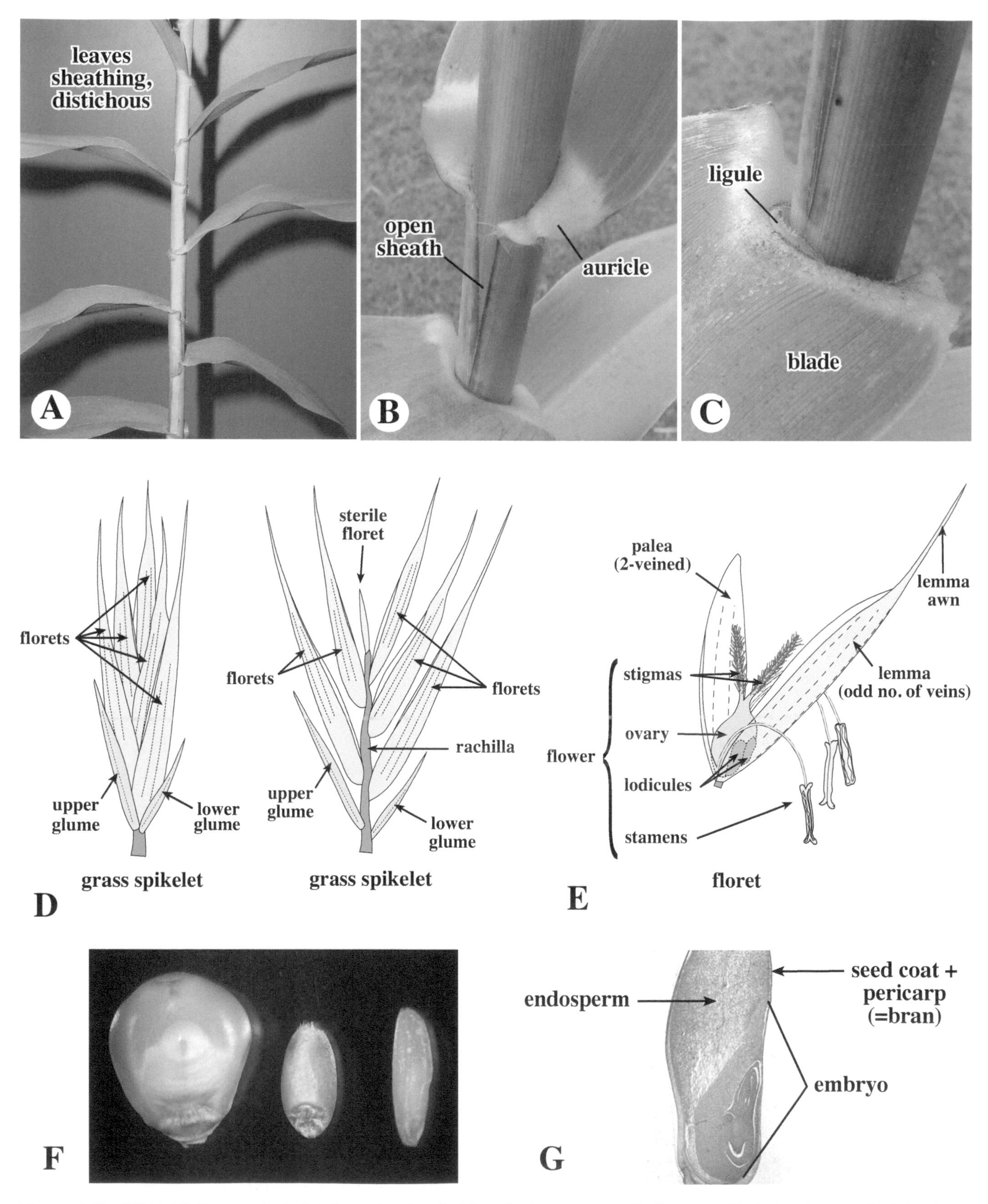

Figure 7.67 POALES. Poaceae. **A.** Aerial shoot, showing distichous leaf arrangement. **B.** Close-up of leaf, showing open leaf sheath and basal auricles of blade. **C.** Ligule, at adaxial junction of sheath and blade. **D.** Diagram of grass spikelets, consisting of an axis (rachilla) bearing two, basal glumes (one or both absent or modified in some taxa) plus 1–∞ florets. **E.** Floret, consisting of a short, lateral branch bearing two bracts, lemma and palea, plus a flower; a given floret may be sterile or unisexual in some taxa. **F.** Grains of (left to right) *Zea mays* (corn), *Triticum aestivum* (bread wheat), and *Oryza sativa* (rice). **G.** Grain of *Zea mays* in longitudinal section, showing embryo, endosperm, and fusion product of seed coat and pericarp.

Figure 7.68 POALES. Poaceae. Bamboos, with stout, woody, aerial stems that bear large, nonphotosynthetic, scale leaves and upper, lateral branches with photosynthetic leaves. **A,B.** *Dendrocalamus giganteus*. **C.** Bamboo showing lateral branches with photosynthetic leaves.

lower side, toward the lemma), which upon swelling function to open the floret by separating the lemma from palea. The **stamens** are 2 or 3. **Anthers** are basifixed-versatile, usually sagittate at the base, generally pendulous on elongate filaments, dithecal, and longitudinal in dehiscence. The **pollen** is monoporate. The **gynoecium** is syncarpous, with a superior ovary, 2–3 carpels, and 1 locule. The **stigmas** are 2 or 3, usually plumose. **Placentation** is basal; **ovules** are orthotropous to anatropous, usually bitegmic, 1 per ovary. **Nectaries** are absent. The **fruit** is a caryopsis (grain). The **seeds** are endospermous. Plants are wind pollinated.

The Poaceae are worldwide in distribution. The grasses are perhaps the most economically important group of plants, containing the agricultural grains (vital food and alchoholic beverage sources), including barley (*Hordeum*), corn (*Zea*), oats (*Avena*), rice (*Oryza*), rye (*Secale*), wheat (*Triticum*), and others, as well as important forage and grazing plants. Members of the family are also important components of many ecosystems, such as grasslands and savannahs.

Recent molecular studies of the Poaceae provide the basis for its classification into twelve subfamilies. See Grass Phylogeny Working Group (2001), Duvall et al. (2007, 2010), Hodkinson et al. (2007), Simon (2007), Grass Phylogeny Working Group II (2012), Kellogg (2015) for information on character evolution, phylogeny, and classification.

The Poaceae are distinctive in being herbs (trees in the bamboos) with ***hollow-pithed stems*** and ***open- (rarely closed-) sheathed, distichous leaves*** with a ***ligule*** at inner junction with blade; the inflorescence is a ***grass spikelet***, typically with 2 basal bracts (***glumes***) on a central axis and 1–∞ ***florets***, each consisting of a short lateral axis with 2 bracts (a lower, odd-veined ***lemma*** and an upper, 2-veined ***palea***) and a flower, the flower with perianth reduced to usually 2–3 ***lodicules***, usually 2–3 pendulous stamens, and a single 2–3-carpellate, 1-ovuled ovary with 2–3 ***plumose stigmas***, the fruit a ***caryopsis*** (***grain***).

P 2–3 [-6+] lodicules **A** 2–3 [1] **G** (2–3), superior.

Restionaceae—Restio family (type *Restio*, from the Latin for rope or cord, in reference to the cordlike stems). 64 genera/ca. 550 species (Figure 7.71).

The Restionaceae consist of dioecious [rarely monoecious or hermaphroditic], evergreen, perennial herbs. The underground **stems** are rhizomatous or stoloniferous, the erect culms photosynthetic, hollow or solid. The **leaves** are simple, unifacial, spiral, with a usually open sheath, usually eligulate, often reduced to sheaths in mature plants, sometimes caducous. The **inflorescence** is a solitary flower or an aggregate of spikelets, in variously branched groups, each group sometimes subtended by bracts (spathes). The spikelets consist of an axis bearing 1–∞ flowers, each flower subtended by 1 [2] bract, lowermost bracts often sterile; male and female spikelets may be similar or dimorphic. The **flowers** are small, unisexual, actinomorphic, hypogynous. The **perianth** is biseriate, homochlamydeous, 3+3 [0–2+0–2], apotepalous. The **tepals** are membranous to indurate. The **stamens** are 3 [1–4],

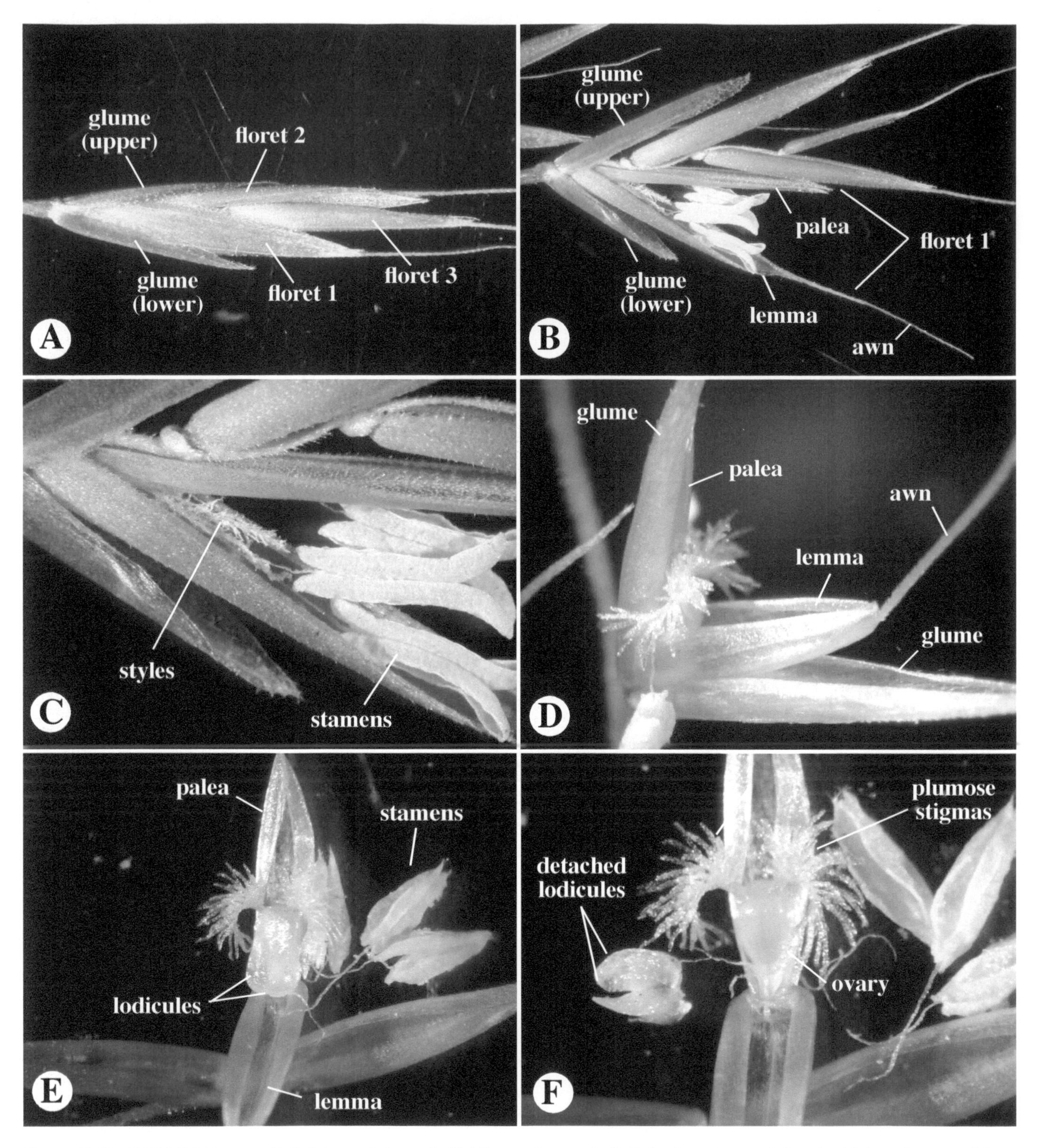

Figure 7.69 POALES. Poaceae. Spikelet morphology. **A–C.** *Elymus glaucus*. **A.** Spikelet, immature and closed, with two glumes and three florets. **B.** Mature spikelet, showing palea and awned lemma of open floret. **C.** Close-up of flower, showing three stamens and styles of ovary. **D–F.** *Stipa miliacea*. **D.** Open, mature spikelet, showing two glumes and palea and lemma of single floret. **E.** Floret dissected open, showing three stamens and two lodicules on lemma side of ovary. **F.** Lodicules removed; note ovary with two plumose styles.

Figure 7.70 POALES. Poaceae. Spikelet diversity. **A.** *Cynodon dactylon*, bermuda grass, close-up of spikelets in two rows, each bearing exserted, pendulous anthers and red, fimbriate styles. **B.** *Stipa pulchra*, needle grass, having one floret per spikelet. **C.** *Sorghum bicolor*, in which two, reduced, male spikelets are grouped with a single, bisexual spikelet. **D.** *Phalaris minor,* spikelet with prominent glumes and one floret. **E.** *Avena barbata*, pendulous spikelet. **F.** *Distichlis spicata*, inflorescence of female plant, a condensed panicle of spikelets. **G.** *Brachypodium distachyon*, spikelet with numerous, awned florets. **H.** *Festuca perennis*, inflorescence a spike of distichously arranged spikelets. **I.** *Oryza sativa*, rice.

Figure 7.71 POALES. Restionaceae. **A.** *Chondropetalum mucronatum*, whole plant. **B.** *Elegia* sp., close-up of male and female inflorescences. **C,D.** *Empodisma minus*. **C.** Whole plants. **D.** Close-up of male and female inflorescences. **E.** *Lyginia barbata*, female inflorescence, fruit at left. **F,G.** *Leptocarpus aristatus*. **F.** Inflorescence. **G.** Female flowers, whole flower (left) and dissected gynoecium and tepals (right).

antipetalous, distinct [rarely connate]. **Anthers** are usually unilocular, bisporangiate, and monothecal, longitudinally and usually introrse in dehiscence. The **pollen** is monoulcerate at release. The **gynoecium** is syncarpous or unicarpelous, with a superior ovary, 3 [1 or 2] carpels, and 3 [1 or 2] locules. The **style(s)** are 1–3. **Placentation** is apical-axile; **ovules** are orthotropous, bitegmic, solitary. **Nectaries** are absent. The **fruit** is an achene, nut, or capsule. The **seeds** are endospermous and sometimes have an elaiosome, functioning in ant dispersal. Flowers are wind pollinated.

Members of the Restionaceae are a major component of fynbos or heath vegetation with distributions in the southern hemisphere, especially South Africa and Australia. Economic importance includes local use as thatching and brooms. See Linder et al. (1998, 2000), Hardy et al. (2008), Briggs and Linder (2009), and Briggs et al. (2000, 2010, 2014) for phylogenetic analyses and treatments of the family.

The Restionaceae are distinctive in being ***perennial, rhizomatous***, mostly ***dioecious herbs*** with photosynthetic erect stems, ***leaves reduced to sheaths***, an inflorescence of solitary flowers or variously branched ***spikelets***, and ***small, unisexual, wind-pollinated*** flowers with usually ***monothecal, bisporangiate*** anthers.

P 3+3 [0–2+0–2] **A** 3 [1–4] **G** (3) [1–(2)], superior.

Sparganiaceae—Bur-Reed family [Typhaceae, sensu APG IV, 2016] (type *Sparganium*, from the Greek for a band used to wrap or bind, after the long, narrow leaves). 1 genus (*Sparganium*)/14 species (Figure 7.72A-F).

The Sparganiaceae consist of emergent, aquatic, monoecious, perennial herbs. The **stems** are rhizomatous. The **leaves** are bifacial, distichous, sheathing, simple, undivided, flat, elongate and narrow, and parallel veined. The **inflorescence** is compound, of globose, bracteate, unisexual heads, male heads above, female below. The **flowers** are small, unisexual, actinomorphic, sessile, the female flowers hypogynous. The **perianth** is apparently bracteate in female flowers, the scale-like tepals 1–6 in males, 3–4 [2–5] in females. The **stamens** are 1–8, antitepalous, distinct or basally connate. The **gynoecium** is unicarpellous or syncarpous, with a superior ovary, 1 [2–3] carpel(s), and 1 [2–3] locule(s). **Placentation** is apical; **ovules** are anatropous, bitegmic, 1 per carpel. **Nectaries** are absent. The **fruit** is dry and drupelike with a persistent perianth and style. The **seeds** are endospermous. Flowers are wind pollinated.

Members of the Sparganiaceae have a worldwide distribution. Taxa are of no significant economic importance. The family is often united in the Typhaceae; see Kubitzki (1998d).

The Sparganiaceae are distinctive in being ***perennial, rhizomatous, monoecious, emergent aquatics*** with ***distichous, bifacial*** leaves and ***unisexual, globose heads*** (male heads above, female below) of numerous, minute, wind-pollinated flowers having a ***scalelike*** perianth, and a ***drupelike*** fruit with ***persistent style***.

Male flowers: **P** 1–6 **A** 1–8 or (1–8).
Female flowers: **P** 3–4 [2–5] **G** 1 [(2–3)], superior.

Typhaceae—Cattail family (Gr. for various plants). 1 genus (*Typha*)/8–13 species (Figure 7.72G-I).

The Typhaceae consist of emergent, aquatic, monoecious, perennial herbs. The **stems** are rhizomatous. The **leaves** are bifacial, mostly basal, distichous, sheathing, simple, undivided, flat, elongate and narrow, and parallel veined, with spongy parenchyma. The **inflorescence** is a terminal, cylindrical spike of very dense flowers, male above and female below. The **flowers** are very small, unisexual, actinomorphic, female flowers hypogynous. The **perianth** consists of 0–3 [–8] bristle-like tepals in male flowers, ∞ bristle or scale-like tepals (in 1–4 whorls) in female flowers. The **stamens** are 3 [1–8], apostemonous. **Anthers** are basifixed, with connective broad, extended beyond thecae. The **pollen** is released as tetrads or monads. The **gynoecium** is unicarpellous, with a superior ovary. The **style** is accrescent. **Placentation** is apical; the **ovule** is solitary, anatropous, bitegmic. **Nectaries** are absent. The **fruit** is a dehiscent, achenelike fruit, with an accrescent gynophore (stipe) and style and persistent perianth parts, aiding in wind dispersal. The **seeds** are starchy endospermous. Flowers are wind pollinated.

Members of the Typhaceae grow as emergents in ponds, ditches, and marshes with worldwide distributions. Economic importance includes local uses as food (pollen or starchy rhizome), matting (leaves), paper, or as ornamental cultivars. See Kubitzki (1998d) for a description of the Typhaceae and Kim and Choi (2011) and Zhou et al. (2018) for phylogenies.

The Typhaceae are distinctive in being ***perennial, rhizomatous, monoecious, emergent aquatics*** with ***distichous, bifacial*** leaves, a ***spike*** of numerous, minute, ***wind-pollinated*** flowers (male above and female below) having a ***bristlelike or scalelike*** perianth, and an ***achenelike, dehiscent*** fruit with an ***accrescent stipe and style***.

Male flowers: **P** 0–3 [–8] **A** 3 [1–8].
Female flowers: **P** ∞ **G** 1, superior.

Xyridaceae—Yellow-eyed-grass family (type *Xyris*, after Greek name for plant with razorlike leaves). 5 genera/ca. 250–400 species (Figure 7.73).

The Xyridaceae consist of perennial or annual herbs. The **stem** of perennials is a caudex, less commonly a rhizome or corm. The **leaves** are bifacial or unifacial-ensiform, usually basal and rosulate, alternate distichous or spiral, sheathing with sheaths often persistent, simple, ligulate in some, narrow, flat or terete, and parallel veined. The **inflorescence**

Figure 7.72 POALES. **A–F.** Sparganiaceae. *Sparganium* sp. **A.** Whole plant. **B.** Inflorescence, a panicle of globose heads. **C.** Shoot cross section, showing distichous leaf arrangement. **D.** Inflorescence, female heads below, male above. **E.** Female heads. **F.** Male heads. **G–I.** Typhaceae. **G,H.** *Typha latifolia*. **G.** Whole plants, emergent, rhizomatous herbs. **H.** Inflorescence. **I.** *Typha domingensis*, close-up of inflorescence, female flowers below, male above.

Figure 7.73 POALES. Xyridaceae. *Xyris* sp. **A.** Whole plant. **B.** Inflorescence, a scapose, bracteate spike. **C.** Flower, face view; note three stamens and three staminodes. **D.** Fruit, with persistent sepals and bract. **E,F.** Pollen grain in midsagittal (**E**) and surface (**F**) views, the latter showing sulcate aperture type. **G.** Orthotropous ovules.

is a terminal, scapose, usually solitary spike or head with bracts subtending single or (in *Achlyphila*) 2 or 3 flowers. The **flowers** are bisexual, hypogynous, actinomorphic or slightly zygomorphic, sessile or pedicellate, subtended by imbricate, indurate bracts. The **perianth** is biseriate, dichlamydeous. The **calyx** is aposepalous with 3 sepals, the anterior one reduced to absent. The **corolla** is ephemeral, of three, distinct or basally connate petals/corolla lobes, usually yellow, rarely white, blue, or magenta. The **stamens** are 3 [6], whorled, biseriate or uniseriate, or with 3 staminodes and 3 fertile stamens. **Anthers** are longitudinal in dehiscence. The **pollen** is sulcate or inaperturate. The **gynoecium** is syncarpous, with a superior ovary, 3 carpels and 1 or 3 locules. The **style** is solitary; **stigmas** are 3 [1]. **Placentation** is axile, basal, free-central, or parietal; **ovules** are orthotropous or anatropous, bitegmic, few–∞ per carpel. The **fruit** is a loculicidal or irregularly dehiscent capsule, sometimes enclosed by persistent sepals and bracts. The **seeds** are small, endospermous (starchy and proteinaceous, sometimes oily).

Three genera of the family, *Abolboda*, *Aratitiyopea*, and *Orectanthe* (sometimes classified in the family Abolbodaceae) have spiral, bifacial leaves, spinose pollen, highly connate petals, and asymmetric, appendaged styles. The other two genera, *Achlyphila* and *Xyris*, have distichous, unifacial leaves, nonspinose pollen, slightly connate or distinct petals, and symmetric, unappendaged styles. Members of the Xyridaceae grow in wet areas, such as marshy savannas, and have a worldwide distribution in tropical and warm and some temperate regions; three genera are restricted to northern South America. Economic importance includes *Xyris* spp. used occasionally ornamentally and medicinally. See Kral (1998) for a detailed family treatment.

The Xyridaceae are distinctive in being perennial or annual herbs with a terminal, ***scapose bracteate head or dense spike***, bracts subtending showy flowers with ***ephemeral, usually yellow petals***.

K 3 **C** 3 or (3) **A** 3 or 3+3 or 3+3 staminodes **G** (3), superior.

Figure 7.74 CERATOPHYLLALES. Ceratophyllaceae. *Ceratophyllum demersum*. **A.** Whole plant. **B.** Close-up, showing whorled, dichotomously forked leaves.

CERATOPHYLLALES

This order, containing one family and genus (APG IV 2016; Table 7.1), has been placed in different positions in various phylogenetic analyses. Recent studies (Moore et al. 2011; Soltis et al. 2011; Sun et al. 2016), including the synthesis of APG IV (2016), have agreed on its position as sister to the eudicots (Figure 7.1; see Chapter 8).

Ceratophyllaceae—Hornwort family (type *Ceratophyllum*, Gr. *cerato*, horn + *phyllum*, leaf, from the forked leaves resembling horns). 1 genus/2–30 species (depending on treatment) (Figure 7.74).

The Ceratophyllaceae consist of monoecious, floating or submerged, aquatic, perennial herbs with rootlike anchoring branches. The **leaves** are exstipulate, whorled, 3–10 per node, 1–4 dichotomously divided, and marginally serrulate. The **inflorescence** consists of solitary and axillary flowers, male and female usually on alternate nodes. **Flowers** are unisexual. The **perianth** is uniseriate and consists of 8–12, basally fused, linear tepals. **Stamens** are generally numerous (5–27), spirally arranged on a flat receptacle; filaments are not clearly distinct from anthers, the thecae and connective apically two-pointed. The **gynoecium** is unicarpellous, with a superior ovary, 1 carpel, and 1 locule; placentation is marginal with a solitary anatropous or orthotropous, unitegmic ovule. The **fruit** is an achene, with a persistent, spiny style; seeds are exalbuminous.

Members of the family are worldwide in distribution. Economically, *Ceratophyllum demersum* is used as an aquarium plant and as a protective cover in fisheries. See Les (1993) for more information on the family.

The Ceratophyllaceae are distinguished from related families in being ***monoecious***, ***aquatic herbs*** with ***whorled***, ***dichotomously branched***, ***serrulate*** leaves, and ***solitary***, ***unisexual*** flowers.

P (8–12) **A** 5–27 **G** 1, superior.

REVIEW QUESTIONS

GENERAL

1. What is the Angiosperm Phylogeny Group system of classification and what higher taxonomic rank does it utilize?
2. What are the major groups of non-eudicot angiosperms?
3. Why have the traditional "dicots" been abandoned as a taxonomic group?
4. What is a floral formula? What are the symbols used in floral formulas?
5. What is a floral diagram and what does it represent?

NONMONOCOT GROUPS

6. Name the family and species of what is thought to represent the most basal lineage of angiosperms.
7. Name the diagnostic characteristics of the Amborellaceae. Do these necessarily represent ancestral angiosperm features?
8. How does the Nymphaeales compare with the Amborellaceae in: plant habit, flower sex, perianth arrangement, stamen number and type, gynoecial fusion type, and ovary position?
9. What anatomical feature is characteristic of the family Illiciaceae?
10. How is the Illiciaceae different from and similar to the Amborellaceae and Nymphaeales?
11. What distinctive anther dehiscence occurs in the Lauraceae?
12. Name two economically important members of the Lauraceae.
13. Name at least two families of the Magnoliales.
14. For the Annonaceae, what is distinctive about the leaf arrangement and endosperm structure?
15. What is distinctive about the receptacle and gynoecial fusion of the Magnoliaceae?
16. What is the fruit type of the Magnoliaceae?
17. What is distinctive about the anatomy, stamen structure, pollen morphology, and gynoecial fusion of the Winteraceae?
18. Name at least three families of the Piperales.
19. What is the etymology of "*Aristolochia*"? To what does it refer?
20. What are the diagnostic features of the Aristolochiaceae?
21. How does the Piperaceae differ from the Aristolochiaceae?
22. What is an economically important member of the Piperaceae?
23. How does the Saururaceae differ from the Piperaceae?

MONOCOTS: BASAL LINEAGES

24. Name and describe the major apomorphies of the monocots.
25. Name the order, family, and genus of the most basal lineage of monocots.
26. How does *Acorus* differ from the Araceae in: leaf structure; seed nutritive tissue; crystal type?
27. What is the leaf venation of members of the Araceae?
28. What is the inflorescence type of the Araceae?
29. Name an economically important member of the Araceae.
30. What are two putative apomorphies of the Asparagales? Of the Asparagales minus the Orchidaceae?
31. What is a cytological apomorphy of the Agavaceae?
32. What is the ovary position of the Agavaceae?
33. What is a chemical apomorphy of the Alliaceae?
34. Name two economically important members of the Alliaceae.
35. How are members of the Asphodelaceae distinguished? What is their distribution?
36. Name and define the leaf structure of the Iridaceae.
37. What is the range of inflorescence morphology of the Iridaceae?
38. What is the floral formula of the Iridaceae?
39. How many species occur in the orchid family?
40. For the Orchidaceae, name the ovary position, placentation, and name for specialized androecium.
41. What is a gynostemium? What are other names for this structure?
42. What orchid is used as a food flavoring and what part of the plant is utilized?

MONOCOTS: COMMELINIDS

43. Name and describe the major chemical apomorphy of the Commelinid monocots.
44. What family (order) is now considered to be sister to the palm family?
45. Name an apomorphy of the Arecaceae, the palms.
46. What are the two acceptable scientific names of the palm family?
47. For the Arecaceae, what is the: flower sex, ovary position, fruit type?
48. What is the seed nutrition of the Commelinid monocots, minus the Arecales and Dasypogonaceae?
49. Name three apomorphies of the Zingiberales.

50. What is the leaf arrangement and plant sex of the the Musaceae?
51. What is the scientific name of banana?
52. What leaf arrangement apomorphy unites all of the Zingiberales, minus the Musaceae?
53. Name two apomorphies that unite the clade Cannaceae + Costaceae + Marantaceae + Zingiberaceae.
54. What, in reality, are the showy petaloid structures of this group?
55. Name two economically important members of the Zingiberaceae.
56. What is the specialized anther type of the Cannaceae and Marantaceae?
57. Name three families of the Commelinales.
58. Name one or more apomorphies for these families.
59. Name two apomorphies of the Bromeliaceae.
60. What is the ovary position of the Bromeliaceae?
61. What special ecological adaptations do various bromeliads have?
62. Name an economically important bromeliad.
63. What pollen and leaf apomorphies may unite the Cyperaceae, Thurniaceae (including Prionaceae) and Juncaceae?
64. How do the Cyperaceae and Juncaceae families differ in inflorescence, perianth, and fruit type?
65. How do the Eriocaulaceae and Xyridaceae differ in flower sex? pollen aperture type?
66. What are the two scientific names for the grass family?
67. How does the leaf arrangement and attachment of grasses differ from that of sedges?
68. What is the structural difference between a "sedge spikelet" and a "grass spikelet"?
69. What is a: grass spikelet? Floret? Glume? Lemma? Palea? Lodicule?
70. What is the fruit type of the grasses?
71. Name the generic and common names of five economically important grain crops.
72. Name the characteristics and distribution of the Restionaceae.
73. What is the flower sex of the Sparganiaceae and Typhaceae?
74. How do these two families differ?
75. For the Ceratophyllaceae name the plant habitat, plant habit, leaf arrangement and morphology, and economic importance.

EXERCISES

1. Select a family of angiosperms treated in this chapter and learn everything you can about it. Perform a literature search (e.g., family name + "systematics") on journal articles published in the last 5 years. Consult family descriptions, recent data on phylogenetic relationships, and information on intrafamilial groupings.
2. From this same family, collect living material of an exemplar. Describe this species in detail, using the character list of Appendix 1 as a guide (see Chapter 9). Illustrate the vegetative and reproductive parts (see Appendix 2).
3. Assimilate all of your information in a written report and computerized slide presentation to present to an audience.
4. For the Asparagales, Commelinids, Zingiberales, or Poales, study competing phylogenetic analyses and compare them with respect to methods and results. What morphological apomorphies occur in lineages of the group?

REFERENCES FOR FURTHER STUDY

GENERAL REFERENCES ON ANGIOSPERM RELATIONSHIPS AND EVOLUTION

Christenhusz, M. J. M., M. F. Fay, and M. W. Chase. 2017. Plants of the World: An Illustrated Encyclopedia of Vascular Plant Families. Kew Publishing, Royal Botanic Gardens, Kew, and The University of Chicago Press.

Heywood, V. H., R. K. Brummitt, A. Culham, and O. Seberg. 2007. Flowering Plants of the World. Firefly Books, London.

Mabberley, D. J. 2008. Mabberley's Plant-Book: A Portable Dictionary of the Higher Plants, Their Classification and Uses. 3rd edition. Cambridge University Press, Cambridge..

Soltis, D., P. Soltis, P. Endress, M. W. Chase, S. Manchester, W. Judd, L. Majure, and E. Mavrodiev. 2018. Phylogeny and Evolution of the Angiosperms: Revised and Updated Edition. The University of Chicago Press, Chicago, London.

GENERAL REFERENCES ON ANGIOSPERM PHYLOGENY AND ON SPECIFIC GROUPS

Andersson, L. 1998a. Musaceae. In: Kubitzki, K. (ed.), The Families and Genera of Vascular Plants. IV. Flowering Plants. Monocotyledons. Alismatanae and Commelinanae (except Gramineae). Pp. 296–300. Springer, Berlin.

Andersson, L. 1998b. Strelitziaceae. In: Kubitzki, K. (ed.), The Families and Genera of Vascular Plants. IV. Flowering Plants. Monocotyledons. Alismatanae and Commelinanae (except Gramineae). Pp. 451–454. Springer, Berlin.

Anger, N., B. Fogliani, C. P. Scutt, and G. Gâteblé. 2017. Dioecy in Amborella trichopoda: Evidence for genetically based sex determination and its consequences for inferences of the breeding system in early angiosperms. Annals of Botany 119:591-597.

Angiosperm Phylogeny Group (APG). 1998. An ordinal classification for the families of flowering plants. Annals of the Missouri Botanical Garden 85: 531–553.

APG II. 2003. An update of the Angiosperm Phylogeny Group classification for the orders and families of flowering plants: APG II. Botanical Journal of the Linnean Society 141: 399–436.

APG III. 2009. An update of the Angiosperm Phylogeny Group classification for the orders and families of flowering plants: APG III. Botanical Journal of the Linnean Society 161: 105-121.

APG IV. 2016. An update of the Angiosperm Phylogeny Group classification for the orders and families of flowering plants: APG IV. Botanical Journal of the Linnean Society 181: 1-20.

Arber, A. 1925. Monocotyledons; A Morphological Study. The University Press, Cambridge, England.

Archibald, J. K., S. R. Kephart, K. E. Theiss, A. L. Petrosky, and T. M. Culley. 2015. Multilocus phylogenetic inference in subfamily Chlorogaloideae and related genera of Agavaceae - informing questions in taxonomy at multiple ranks. Molecular Phylogenetics and Evolution 84:266-283.

Asmussen, C. B., Baker, W. J., and Dransfield, J. 2000. Phylogeny of the palm family (Arecaceae) based on rps16 intron and trnL-trnF plastid DNA sequences. In: Wilson, K. L., and D. A. Morrison (eds.), Monocots: Systematics and Evolution. Pp. 525–535. CSIRO, Collingwood.

Asmussen, C. B., J. Dransfield, V. Deickmann, A. S. Barfod, J.-C. Pintaud, and W. J. Baker. 2006. A new subfamily classification of the palm family (Arecaceae): Evidence from plastid DNA phylogeny. Botanical Journal of the Linnean Society 151: 15-38.

Azuma, H., J. G. Garciá-Franco, V. Rico-Gray, and L. B. Thien. 2001. Molecular phylogeny of the Magnoliaceae: The biogeography of tropical and temperate disjunctions. American Journal of Botany 88: 2275-2285.

Balslev, H. 1998. Juncaceae. In: Kubitzki, K. (ed.), The Families and Genera of Vascular Plants. IV. Flowering Plants. Monocotyledons. Alismatanae and Commelinanae (except Gramineae). Pp. 252–259. Springer, Berlin.

Barkman, T. J., G. Chenery, J. R. McNeal, J. Lyons-Weiler, W. J. Ellisens, G. Moore, A. D. Wolfe, and C. W. dePamphilis. 2000. Independent and combined analyses of sequences from all three genomic compartments converge on the root of flowering plant phylogeny. Proceedings of the National Academy of Sciences of the United States of America 97: 13166–13171.

Barrett, C. F., W. J. Baker, J. R. Comer, J. G. Conran, S. C. Lahmeyer, J. H. Leebens-Mack, J. Li, G. S. Lim, D. R. Mayfield-Jones, L. Perez, J. Medina, J. C. Pires, C. Santos, D. W. Stevenson, W. B. Zomlefer, and J. I. Davis. 2015. Plastid genomes reveal support for deep phylogenetic relationships and extensive rate variation among palms and other commelinid monocots. New Phytologist 209:855-870.

Barrett, S. C. H., and S. W. Graham. 1997. Adaptive radiation in the aquatic plant family Pontederiaceae: Insights from phylogenetic analysis. In: Givnish, T. J., and K. J. Sytsma (eds.), Molecular Evolution and Adaptive Radiation. Pp. 225–258. Cambridge University Press, Cambridge, UK.

Behnke, H.-D. 1972. Sieve-tube plastids in relation to Angiosperm systematics—an attempt towards a classification by ultrastructural analysis. Botanical Review 38: 155–197.

Behnke, H.-D. 2000. Forms and sizes of sieve-element plastids and evolution of the monocotyledons. In: Wilson, K. L., and D. A. Morrison (eds.), Monocots: Systematics and Evolution. Pp. 163–188. CSIRO, Collingwood.

Bell, C. D., D. E. Soltis, and P. S. Soltis. 2010. The age and diversification of the angiosperms re-revisited. American Journal of Botany 97:1296–1303.

Bogler, D. J., and B. B. Simpson. 1995. A Chloroplast DNA Study of the Agavaceae. Systematic Botany 20: 191.

Bogler, D. J., and B. B. Simpson. 1996. Phylogeny of Agavaceae based on ITS rDNA sequence variation. American Journal of Botany 83: 1225.

Bogler, D. J., C. Pires, and J. Francisco-Ortega. 2006. Phylogeny of Agavaceae based on ndhF, rbcL, and ITS sequences: implications of molecular data for classification. In: Columbus, J. T., E. A. Friar, J. M. Porter, L. M. Prince, and M. G. Simpson (eds.), Monocots: Comparative Biology and Evolution. Pp. 313-328. Rancho Santa Ana Botanic Garden, Claremont, California. [Aliso 22: 313-328.]

Bogner, J. 2011. Acoraceae. Pp. in H. P. Noteboom (ed.), Flora malesiana. Ser. 1, vol. 20. Nationaal Herbarium Nederland, Leiden.

Bogner, J., and S. J. Mayo. 1998. Acoraceae. In: Kubitzki, K., H. Huber, P. J. Rudall, P. S. Stevens, and T. Stutzel (eds.). The Families and Genera of Vascular Plants IV: Flowering Plants Monocotyledons Alismatanae and Commelinanae (except Gramineae). Springer-Verlag, New York.

Borsch, T., and P. S. Soltis. 2008. Nymphaeales - the first globally diverse clade? Taxon 57: 1051.

Borsch, T., C. Löhne, and J. H. Wiersema. 2008. Phylogeny and evolutionary patterns in Nymphaeales: Integrating genes, genomes and morphology. Taxon 57: 1052-1081.

Borsch, T., K. W. Hilu, D. Quandt, V. Wilde, C. Neinhuis, and W. Barthlott. 2003. Noncoding plastid trnT-trnF sequences reveal a well resolved phylogeny of basal angiosperms. Journal of Evolutionary Biology 16: 558–576.

Bremer, K. 2002. Gondwanan evolution of the grass alliance of families (Poales). Evolution 56: 1374–1387.

Briggs, B. G. and H. P. Linder. 2009. A new subfamilial and tribal classification of Restionaceae (Poales). Telopea 12:333-345.

Briggs, B. G., A. D. Marchant, and A. J. Perkins. 2010. Phylogeny and features in Restionaceae, Centrolepidaceae and Anarthriaceae (restiid clade of Poales). Pp. 357-388 in O. Seberg, Petersen, G., Barfod, A. S., & Davis, J. I. (ed.), Diversity, Phylogeny, and Evolution in the Monocotyledons. Åarhus University Press, Århus.

Briggs, B. G., A. D. Marchant, and A. J. Perkins. 2014. Phylogeny of the restiid clade (Poales) and implications for the classification of Anarthriaceae, Centrolepidaceae and Australian Restionaceae. Taxon 63:24-46.

Briggs, B. G., A. D. Marchant, and A. J. Perkins. 2014. Phylogeny of the restiid clade (Poales) and implications for the classification of Anarthriaceae, Centrolepidaceae and Australian Restionaceae. Taxon 63:24-46.

Briggs, B. G., A. D. Marchant, S. Gilmore, and C. L. Porter. 2000. A molecular phylogeny of Restionaceae and allies. Pp. 661-671 in K. L. Wilson, & Morrison, D. A. (ed.), Monocots: Systematics and Evolution. CSIRO, Collingwood.

Buerki, S., T. Gallaher, T. Booth, G. Brewer, F. Forest, J. T. Pereira, and M. W. Callmander. 2017. Biogeography and evolution of the screw-pine genus Benstonea Callm. & Buerki (Pandanaceae). Candollea 71:217-229.

Cabrera, L. I., G. A. Salazar, M. W. Chase, S. J. Mayo, J. Bogner, and P. Dávila. 2008. Phylogenetic relationships of aroids and duckweeds (Araceae) inferred from coding and noncoding plastid DNA. American Journal of Botany 95:1153-1165.

Cabrera, L. I., G. A. Salazar, M. W. Chase, S. J. Mayo, J. Bogner, and P. Dávila. 2008. Phylogenetic relationships of aroids and duckweeds (Araceae) inferred from coding and noncoding plastid DNA. American Journal of Botany 95:1153-1165.

Caddick, L. R., P. J. Rudall, P. Wilkin, T. A. J. Hedderson, and M. W. Chase. 2002a. Phylogenetics of Dioscoreales based on combined analyses of morphological and molecular data. Botanical Journal of the Linnean Society 138: 123–144.

Caddick, L. R., P. Wilkin, P. J. Rudall, T. A. J. Hedderson, and M. W. Chase. 2002b. Yams reclassified: A recircumscription of Dioscoreaceae and Dioscoreales. Taxon 51: 103–114.

Cameron, K. M. 2004. Utility of plastid psaB gene sequences for investigating intrafamilial relationships within Orchidaceae. Molecular Phylogenetics and Evolution 31: 1157–1180.

Cameron, K. M. 2006. A comparison and combination of plastid atpB and rbcL gene sequences for inferring phylogenetic relationships within Orchidaceae. In: Columbus, J. T., E. A. Friar, J. M. Porter, L. M. Prince, and M. G. Simpson (eds). Pp. 447-464. Monocots: Comparative Biology and Evolution. Excluding Poales. Rancho Santa Ana Botanical Garden, Claremont, Ca. [Aliso 22: 447-464.]

Cameron, K. M., and M. W. Chase. 2000. Nuclear 18S rDNA sequences of Orchidaceae confirm the subfamilial status and circumscription of Vanilloideae. In: Wilson, K. L., and D. A. Morrison (eds.), Monocots: Systematics and Evolution. Pp. 457–464. CSIRO, Collingwood.

Cameron, K. M., M. W. Chase, W. M. Whitten, P. J. Kores, D. C. Jarrell, V. A. Albert, T. Yukawa, H. G. Hills, and D. H. Goldman. 1999. A phylogenetic analysis of the Orchidaceae: evidence from rbcL nucleotide. American Journal of Botany 86: 208–224.

Cantino, P. D., J. A. Doyle, S. W. Graham, W. S. Judd, R. G. Olmstead, D. E. Soltis, P. S. Soltis, and M. J. Donoghue. 2007. Towards a phylogenetic nomenclature of Tracheophyta. Taxon 56: 822-846.

Chase, M. W. 2004. Monocot relationships: an overview. American Journal of Botany 91:1645-1655.

Chase, M. W., A. Y. De Bruijn, A. V. Cox, G. Reeves, P. J. Rudall, M. A. T. Johnson, and L. E. Eguiarte. 2000b. Phylogenetics of Asphodelaceae (Asparagales): an analysis of plastid rbcL and trnL-F DNA sequences. Annals of Botany 86: 935–952.

Chase, M. W., D. E. Soltis, P. S. Soltis, P. J. Rudall, M. F. Fay, W. J. Hahn, S. Sullivan, J. Joseph, M. Molvray, P. J. Kores, T. J. Givnish, K. J. Sytsma, and J. C. Pires. 2000a. Higher-level systematics of the monocotyledons: an assessment of current knowledge and a new classification. In: Wilson, K. L., and D. A. Morrison (eds.), Monocots: Systematics and Evolution. CSIRO, Collingwood.

Chase, M. W., D. E. Soltis, R. G. Olmstead, D. Morgan, D. H. Les, B. D. Mishler, M. R. Duvall, R. A. Price, H. G. Hills, Y.-L. Qiu, K. A. Kron, J. H. Rettig, E. Conti, J. D. Palmer, J. R. Manhart, K. J. Sytsma, H. J. Michaels, W. J. Kress, K. G. Karol, W. D. Clark, M. Hedrén, B. S. Gaut, R. K. Jansen, K.-J. Kim, C. F. Wimpee, J. F. Smith, G. R. Furnier, S. H. Strauss, Q.-Y. Xiang, G. M. Plunkett, P. S. Soltis, S. M. Swensen, S. E. Williams, P. A. Gadek, C. J. Quinn, L. E. Eguiarte, E. Golenberg, G. H. Learn Jr., S. W. Graham, S. C. H. Barrett, S. Dayanandan, and V. A. Albert. 1993. Phylogenetics of seed plants: an analysis of nucleotide sequences from the plastid gene rbcL. Annals of the Missouri Botanical Garden 80: 528–580.

Chase, M. W., M. F. Fay, and V. Savolainen. 2000. Plant Systematics: A half-century of progress (1950–2000) and future challenges—Higher-level classification in the angiosperms: New insights from the perspective of DNA sequence data. Taxon 49: 685–704.

Chase, M. W., M. F. Fay, D. S. Devey, O. Maurin, N. Rønsted, T. J. Davies, Y. Pillon, G. Petersen, O. Seberg, M. N. Tamura, C. B. Asmussen, K. Hilu, T. Borsch, J. I. Davis, D. W. Stevenson, J. C. Pires, T. J. Givnish, K. J. Sytsma, M. A. Mcpherson, S. W. Graham, and H. S. Rai. 2006. Multigene analyses of monocot relationships: a summary. In: J. T. Columbus, E. A. Friar, J. M. Porter, L. M. Prince, and M. G. Simpson (eds.). Monocots: comparative biology and evolution, 2 vols. Aliso 22: 63-75. Rancho Santa Ana Botanic Garden, Claremont, California, USA.

Chen, L.-Y., J.-M. Chen, R. W. Gituru, T. D. Temam, and Q.-F. Wang. 2012. Generic phylogeny and historical biogeography of Alismataceae, inferred from multiple DNA sequences. Molecular Phylogenetics and Evolution 63:407-416.

Chen, Y.-Y., D.-Z. Li, and H. Wang. 2002. Infrageneric phylogeny and systematic position of the Acoraceae inferred from ITS, 18S and rbcL sequences. Acta Botanica Yunnanica 24: 699–706.

Cole, T. C. H. 2015. Angiosperm Phylogeny Group (APG) in jeopardy – Where have the flowers gone? PeerJ PrePrints 3:e1238v1

Columbus, J. T., E. A. Friar, J. M. Porter, L. M. Prince, and M. G. Simpson (eds.). 2006. Monocots: Comparative Biology and Evolution. 2 vols. Rancho Santa Ana Botanic Garden, Claremont, CA.

Cook, C. D. K. 1998. Pontederiaceae. In: Kubitzki, K. (ed.), The Families and Genera of Vascular Plants. IV. Flowering Plants. Monocotyledons. Alismatanae and Commelinanae (except Gramineae). Pp. 395–403. Springer, Berlin.

Cox, P. A., K.-L. Huynh, and B. C. Stone. 1995. Evolution and systematics of Pandanaceae. In: Rudall, P. J., P. J. Cribb, D. F. Cutler, and C. J. Humphries (eds.), Monocotyledons: Systematics and Evolution. Pp. 663–684. Royal Botanic Gardens, Kew.

Cron, G. V., C. Pirone, M. Bartlett, W. J. Kress, and C. Specht. 2012. Phylogenetic relationships and evolution in the Strelitziaceae (Zingiberales). Systematic Botany 37:606-619.

Cronquist, A. 1981. An Integrated System of Classification of Flowering Plants. Columbia University Press, New York.

Dahlgren, R., H. T. Clifford, and P. F. Yeo. 1985. The Families of the Monocotyledons: Structure, Evolution, and Taxonomy. Springer-Verlag, New York.

Davis, J. I., D. W. Stevenson, G. Petersen, O. Seberg, L. M. Campbell, J. V. Freudenstein, D. H. Goldman, C. R. Hardy, F. A. Michelangeli, and M. P. Simmons. 2004. A phylogeny of the monocots, as inferred from rbcL and atpA sequence variation, and a comparison of methods for calculating jackknife and bootstrap values. Systematic Botany 29: 467–510.

Devey, D. S., I. Leitch, P. J. Rudall, J. C. Pires, Y. Pillon, and M. W. Chase. 2006. Systematics of Xanthorrhoeaceae sensu lato, with an emphasis on Bulbine. In: Columbus, J. T., E. A. Friar, J. M. Porter, L. M. Prince, and M. G. Simpson (eds). Monocots: Comparative Biology and Evolution. Excluding Poales. Pp. 345-351. Rancho Santa Ana Botanical Garden, Claremont, Ca. [Aliso 22: 345-351.]

Doyle, J. A., and A. Le Thomas. 1994. Cladistic analysis and pollen evolution in Annonaceae. Acta Botanica Gallica 141: 149–170.

Doyle, J. A., and A. Le Thomas. 1996. Phylogenetic analysis and character evolution in Annonaceae. Bulletin of the Museum of Natural History. Paris, 4e sér. [sect. B, Adansonia] 18: 279–334.

Doyle, J. A., and P. K. Endress. 2000. Morphological phylogenetic analysis of basal angiosperms: Comparison and combination with molecular data. International Journal of Plant Sciences 161: S121–S153.

Dransfield, J. and N. W. Uhl. 1998. Palmae. In: Kubitzki, K. (ed.), The Families and Genera of Vascular Plants. IV. Flowering Plants. Monocotyledons. Alismatanae and Commelinanae (except Gramineae). Pp. 306–388. Springer, Berlin.

Drew, B. T., B. R. Ruhfel, S. A. Smith, M. J. Moore, B. G. Briggs, M. A. Gitzendanner, P. A. Soltis, and D. E. Soltis. 2014. Another look at the root of angiosperms reveals a familiar tale. Systematic Biology 63:368-382.

Duvall, M. R., C. H. Leseberg, C. P. Grennan, and L. M. Morris. 2010. Molecular evolution and phylogenetics of complete chloroplast genomes in Poaceae. Pp. 437-450 in O. Seberg, Petersen, G., Barfod, A. S., & Davis, J. I. (ed.), Diversity, Phylogeny, and Evolution in the Monocotyledons. Aarhus, Aarhus University Press.

Duvall, M. R., J. I. Davis, L. G. Clark, J. D. Noll, D. H. Goldman, and G. Sánchez-Ken. 2007. Phylogeny of the grasses (Poaceae) revisited. In: Columbus, J. T., E. A. Friar, J. M. Porter, L. M. Prince, and M. G. Simpson. (eds). Monocots: Comparative Biology and Evolution. Poales. Pp. 237-247. Rancho Santa Ana Botanical Garden, Claremont, Ca. [Aliso 23: 237-247.]

Duvall, M. R., M. T. Clegg, M. W. Chase, W. D. Clark, W. J. Kress, H. G. Hills, L. E. Eguiarte, J. F. Smith, B. S. Gaut, E. A. Zimmer, and G. H. Learn Jr. 1993. Phylogenetic hypotheses for the monocotyledons constructed from rbcL sequence data. Annals of the Missouri Botanical Garden 80: 607–619.

Eiserhardt, W. L., A. Antonelli, D. J. Bennett, L. R. Botigué, J. G. Burleigh, S. Dodsworth, B. J. Enquist, F. Forest, J. T. Kim, A. M. Kozlov, I. J. Leitch, B. S. Maitner, S. Mirarab, W. H. Piel, O. A. Pérez-Escobar, L. Pokorny, C. Rahbek, B. Sandel, S. A. Smith, A. Stamatakis, R. A. Vos, T. Warnow, and W. J. Baker1. 2018. A roadmap for global synthesis of the plant tree of life. American Journal of Botany 105:614-622.

Endress, P. K., and A. Igersheim. 1997. Gynoecium diversity and systematics of the Laurales. Botanical Journal of the Linnean Society 125: 93–168.

Escobedo-Sarti, J., I. Ramírez, C. Leopardi, G. Carnevali, S. Magallón, R. Duno, and D. Mondragón. 2013. A phylogeny of Bromeliaceae (Poales, Monocotyledoneae) derived from an evaluation of nine supertree methods. Journal of Systematics and Evolution 51:743–757.

Escudero, M. and A. L. Hipp. 2013. Escudero, M., & Hipp, A. L. 2013. Shifts in diversification rates and clade ages explain species richness in higher-level sedge taxa (Cyperaceae). American J. Bot. 100: 2403-2411. American Journal of Botany 100:2403-2411.

Evans, T. M., K. J. Sytsma, R. B. Faden, and T. J. Givnish. 2003. Phylogenetic relationships in the Commelinaceae: II. A cladistic analysis of rbcL sequences and morphology. Systematic Botany 28: 270–292.

Faden, R. B. 1998. Commelinaceae. In: Kubitzki, K. (ed.), The Families and Genera of Vascular Plants. IV. Flowering Plants. Monocotyledons. Alismatanae and Commelinanae (except Gramineae). Pp. 109–127. Springer, Berlin.

Faurby, S., W. L. Eiserhardt, W. J. Baker, and J.-C. Svenning. 2016. An all-evidence species-level supertree for the palms (Arecaceae). Molecular Phylogenetics and Evolution 100:57-69.

Fay, M. F., and M. W. Chase. 1996. Resurrection of Themidaceae for the Brodiaea alliance, and recircumscription of Alliaceae, Amaryllidaceae and Agapanthoideae. Taxon 45: 441–451.

Fay, M. F., M. W. Chase, N. Rønsted, D. S. Devey, Y. Pillon, J. C. Pires, G. Petersen, O. Seberg, and J. I. Davis. 2006. Phylogenetics of Liliales: Summarized evidence from combined analyses of five plastid and one mitochondrial loci. In: Columbus, J. T., E. A. Friar, J. M. Porter, L. M. Prince, and M. G. Simpson (eds). Monocots: Comparative Biology and Evolution. Excluding Poales. Pp. 559-565. Rancho Santa Ana Botanical Garden, Claremont, Ca. [Aliso 22: 559-565.]

Fay, M. F., P. J. Rudall, S. Sullivan, K. L. Stobart, A. Y. de Bruijn, G. Reeves, F. Qamaruz-Zaman, W.-P. Hong, J. Joseph, W. J. Hahn, J. G. Conran, and M. W. Chase. 2000. Phylogenetic studies of Asparagales based on four plastid DNA regions. In: Wilson, K. L., and D. A. Morrison (eds.), Monocots: Systematics and Evolution. Pp. 360–371. CSIRO, Collingwood.

French, J. C., M. Chung, and Y. Hur. 1995. Chloroplast DNA phylogeny of Ariflorae. In: Rudall, P. J., P. J. Cribb, D. F. Cutler, and C. J. Humphries (eds.), Monocotyledons: Systematics and Evolution. Pp. 255–275. Royal Botanic Gardens, Kew.

Giulietti, A. M., M. J. G. Andrade, V. L. Scatena, M. Trovó, A. L. Coan, P. T. Sano, F. A. R. Santos, R. L. B. Borges, and C. van den Berg. 2012. Molecular phylogeny, morphology and their implications for the taxonomy of Eriocaulaceae. Rodriguésia 63:1-19.

Givnish, T. J., A. Zuluaga, D. Spalink, M. S. Gomez, V. K. Y. Lam, J. M. Saarela, C. Sass, W. J. D. Iles, D. J. L. d. Sousa, J. Leebens-Mack, J. C. Pires, W. B. Zomlefer, M. A. Gandolfo, J. I. Davis, D. W. Stevenson, C. dePamphilis, C. D. Specht, S. W. Graham, C. F. Barrett, and C. Ané. 2018. Monocot plastid phylogenomics, timeline, net rates of species diversification, the power of multi-gene analyses, and a functional model for the origin of monocots. American Journal of Botany 105:1-23.
Givnish, T. J., A. Zuluaga, I. Marques, V. K. Y. Lam, M. S. Gomez, W. J. D. Iles, M. Ames, D. Spalink, J. R. Moeller, B. G. Briggs, S. P. Lyon, D. W. Stevenson, W. Zomlefer, and S. W. Graham. 2016a. Phylogenomics and historical biogeography of the monocot order Liliales: Out of Australia and through Antarctica. Cladistics 32:581-605.
Givnish, T. J., D. Spalink, M. Ames, S. P. Lyon, S. J. Hunter, A. Zuluaga, A. Doucette, G. C. Caro, J. McDaniel, M. A. Clements, M. T. K. Arroyo, L. Endara, R. Kriebel, N. H. Williams, and K. M. Cameron. 2016b. Orchid historical biogeography, diversification, Antarctica and the paradox of orchid dispersal. Journal of Biogeography 43:1905-1916.
Givnish, T. J., D. Spalink, M. Ames, S. P. Lyon, S. J. Hunter, A. Zuluaga, W. J. D. Iles, M. A. Clements, M. T. K. Arroya, J. Leebens-Mack, L. Endara, R. Kriebel, K. M. Neubig, W. M. Whitten, N. H. Williams, and K. M. Cameron. 2015. Orchid phylogenomics and multiple drivers of their extraordinary diversification. Proceedings of the Royal Society of London, Series B: Biological Sciences 282:20151553.
Givnish, T. J., J. C. Pires, S. W. Graham, M. A. McPherson, L. M. Prince, and T. B. Patterson. 2007. Phylogeny, biogeography, and ecological evolution in Bromeliaceae: Insights from ndhF sequences. In: Columbus, J. T., E. A. Friar, J. M. Porter, L. M. Prince, and M. G. Simpson (eds). Monocots: Comparative Biology and Evolution. Poales. Pp. 3-26. Rancho Santa Ana Botanical Garden, Claremont, Ca. [Aliso 23: 3-26.]
Givnish, T. J., M. S. Ames, J. R. McNeal, M. R. McKain, P. R. Steele, C. W. dePamphilis, S. W. Graham, J. C. Pires, D. W. Stevenson, W. B. Zomlefer, B. G. Briggs, M. R. Duvall, M. J. Moore, J. M. Heaney, D. E. Soltis, P. S. Soltis, K. Thiele, and J. H. Leebens-Mack. 2010. Assembling the tree of the monocotyledons: Plastome sequence phylogeny and evolution of Poales. Annals of the Missouri Botanical Garden 97:584-616.
Givnish, T. J., T. M. Evans, J. C. Pires, and K. J. Sytsma. 1999. Polyphyly and convergent morphological evolution in Commelinales and Commelinidae: Evidence from rbcL sequence data. Molecular Phylogenetics and Evolution 12: 360–385.
Goetghebeur, P. 1998. Cyperaceae. In: Kubitzki, K. (ed.), The Families and Genera of Vascular Plants. IV. Flowering Plants. Monocotyledons. Alismatanae and Commelinanae (except Gramineae). Pp. 141–189. Springer, Berlin.
Goldblatt, P., A. Rodriguez, M. P. Powell, T. J. Davies, J. C. Manning, M. van der Bank, and V. Savolainen. 2008. Iridaceae 'out of Australia'? Phylogeny, biogeography and divergence times based on plastid DNA sequences. Systematic Botany 33: 495-508.
Goldblatt, P., and J. Manning. 2008. The Iris Family. Natural history and Classification. Timber Press, Portland, Ore.
Goldblatt, P., J. C. Manning, and P. Rudall. 1998. Iridaceae. In: Kubitzki, K. (ed.), The Families and Genera of Vascular Plants. III. Flowering Plants. Monocotyledons. Lilianae (except Orchidaceae). Pp. 295–333. Springer, Berlin.
Gomes de Andrade, M. J., A. M. Giuletti, A. Rapini, L. P. de Queiroz, A. Concição de Souza, P. R. Macahado de Almeida, and C. van den Berg. 2010. A comprehenesive phylogenetic analysis of Eriocaulaceae: Evidence from nuclear (ITS) and plastid psbA-trnH and trnL-F) DNA sequences. Taxon 59:379-388.
Górniak, M., O. Paun, and M. W. Chase. 2010. Phylogenetic relationships with Orchidaceae based on a low-copy nuclear-coding gene, Xdh: Congruence with organellar and nuclear ribosomal DNA results. Molecular Phylogenetics and Evolution 56:784-795.
Govaerts, R. and D. G. Frodin. 2002. World Checklist and Bibliography of Araceae (and Acoraceae). Royal Botanic Gardens, Kew.
Graham, S. W. and R. G. Olmstead. 2000a. Evolutionary significance of an unusual chloroplast DNA inversion found in two basal angiosperm lineages. Current Genetics 37: 183–188.
Graham, S. W., and R. G. Olmstead. 2000b. Utility of 17 chloroplast genes for inferring the phylogeny of the basal angiosperms. American Journal of Botany 87: 1712–1730.
Graham, S. W., J. M. Zgurski, M. A. Mcpherson, D. M. Cherniawsky, J. M. Saarela, E. F. C. Horne, S. Y. Smith, W. A. Wong, H. E. O'Brien, V. L. Biron, J. C. Pires, R. G. Olmstead, M. W. Chase, and H. S. Rai. 2006. Robust inference of monocot deep phylogeny using an expanded multigene plastid data set. In: J. T. Columbus, E. A. Friar, J. M. Porter, L. M. Prince, and M. G. Simpson (eds.). Monocots: comparative biology and evolution, 2 vols. Aliso 22: 3-21. Rancho Santa Ana Botanic Garden, Claremont, California, USA.
Graham, S. W., J. R. Kohn, B. R. Morton, J. Eckenwalder, and S. C. H. Barrett. 1998. Phylogenetic congruence and discordance among one morphological and three molecular data sets from Pontederiaceae. Systematic Biology 47: 545-567.
Grass Phylogeny Working Group. 2001. Phylogeny and subfamilial classification of the grasses (Poaceae). Annals of the Missouri Botanical Garden 88: 373–457.
GrassPhylogenyWorkingGroupII. 2012. New grass phylogeny resolves deep evolutionary relationships and discovers C4 origins. New Phytologist 193:304-312.
Grayum, M. H. 1990. Evolution and phylogeny of the Araceae. Annals of the Missouri Botanical Garden 77: 628–697.
Guo, X., C. C. Tang, D. C. Thomas, T. L. P. Couvreur, and R. M. K. Saunders. 2017. A megaphylogeny of the Annonaceae: Taxonomic placement of five enigmatic genera and support for a new tribe, Phoenicantheae. Scientific Reports 7:7323.
Hahn, W. J. 2002. A molecular phylogenetic study of the Palmae (Arecaceae) based on atpB, rbcL and 18s nrDNA sequences. Systematic Biology 51: 92–112.
Hao, G., R. M. K. Saunders, and M.-L. Chye. 2000. A phylogenetic analysis of the Illiciaceae based on sequences on internal transcribed spacers (ITS) of nuclear ribosomal DNA. Plant Systematics and Evolution 223: 81-90.
Hardy, C. R., P. Moline, and H. P. Linder. 2008. A phylogeny for the African Restionaceae and new perspectives on morphology's role in generating complete species phylogenies for large clades. International Journal of Plant Sciences 169: 377-390.

Hayashi, K. and S. Kawano. 2000. Molecular systematics of Lilium and allied genera (Liliaceae): Phylogenetic relationships among Lilium and related genera based on the rbcL and matK gene sequence data. Plant Species Biology 15:73-93.
Hayashi, K., and Kawano, S. 2000. Molecular systematics of Lilium and allied genera (Liliaceae): Phylogenetic relationships among Lilium and related genera based on the rbcL and matK gene sequence data. Plant Species Biology 15: 73–93.
Haynes, R. R., D. H. Les, and L. B. Holm-Nielsen. 1998. Alismataceae. In: Kubitzki, K. (ed.), The Families and Genera of Vascular Plants. IV. Flowering Plants. Monocotyledons. Alismatanae and Commelinanae (except Gramineae). Pp. 11–18. Springer, Berlin.
Hertweck, K. L., M. S. Kinney, S. A. Stuart, O. Maurin, S. Mathews, M. W. Chase, M. A. Gandolfo, and J. C. Pires. 2015. Phylogenetics, divergence times and diversification from three genomic partitions in monocots. Botanical Journal of the Linnean Society 178:375-393.
Heywood, V. H., R. K. Brummitt, A. Culham, and O. Seberg. 2007. Flowering Plants of the World. Firefly Books, London.
Hilu, K. W., T. Borsch, K. Muller, D. E. Soltis, P. S. Soltis, V. Savolainen, M. W. Chase, M. P. Powell, L. A. Alice, R. Evans, H. Sauquet, C. Neinhuis, T. A. B. Slotta, J. G. Rohwer, C. S. Campbell, and L. W. Chatrou. 2003. Angiosperm phylogeny based on matK sequence information. American Journal of Botany 90: 1758–1766.
Hinchcliff, C. E. and E. H. Roalson. 2013. Using supermatrices for phylogenetic enquiry: An example using the sedges. Systematic Biology 62:205-219.
Hodkinson, T. R., N. Salamin, M. W. Chase, Y. Bouchenak-Khelladi, S. A. Renvoize, and V. Savolainen. 2007. Large trees, supertreess, and diversification of the grass family. In: Columbus, J. T., E. A. Friar, J. M. Porter, L. M. Prince, and M. G. Simpson. (eds). Monocots: Comparative Biology and Evolution. Poales. Pp. 248-258. Rancho Santa Ana Botanical Garden, Claremont, Ca. [Aliso 23: 248-258.]
Hopper, S. D., M. G. Fay, M. Rossetto, and M. W. Chase. 1999. A molecular phylogenetic analysis of the bloodroot and kangaroo paw family, Haemodoraceae: taxonomic, biogeographic and conservation implications. Botanical Journal of the Linnean Society 131: 285–299.
Hopper, S. D., R. J. Smith, M. F. Fay, J. C. Manning, and M. F. Chase. 2009. Molecular phylogenetics of Haemodoraceae in the greater cape and southwest Australian floristic regions. Molecular Phylogenetics and Evolution 51: 19-30.
Huber, H. 1998a. Dioscoreaceae. Pp. 216–235. In: The Families and Genera of Flowering Plants. IV. Flowering Plants, Monocotyledons: Lilianae (except Orchidaceae) (Kubitski, K. (ed.), Springer-Verlag, Berlin, Heidelberg, New York.
Huber, H. 1998b. Trichopodaceae. Pp. 441–444. In: The Families and Genera of Flowering Plants. IV. Flowering Plants, Monocotyledons: Lilianae (except Orchidaceae) (Kubitski, K. (ed.), Springer-Verlag, Berlin, Heidelberg, New York.
Hutchinson, J. 1973. The Families of Flowering Plants; Arranged According to a New System Based on their Probable Phylogeny, 3rd edition. Clarendon Press, Oxford.
Jansen, R. K., Z. Cai, L. A. Raubeson, H. Daniell, C. W. dePamphilis, J. Leebens-Mack, K. F. Müller, M. Guisinger-Bellian, R. C. Haberle, A. K. Hansen, T. W. Chumley, S.-B. Lee, R. Peery, J. R. McNeal, J. V. Kuehl, and J. L. Boore. 2007. Analysis of 81 genes from 64 plastid genomes resolves relationships in angiosperms and identifies genome-scale evolutionary patterns. Proceedings of the National Academy of Sciences U.S.A. 104: 19369-19374.
Jaramillo, M. A., R. Callejas, C. Davidson, J. F. Smith, A. C. Stevens, and E. J. Tepe. 2008. A phylogeny of the tropical genus Piper using ITS and the chloroplast intron psbJ–petA. Systematic Botany 33: 647-660.
Johansen, L. B. 2005. Phylogeny of Orchidantha (Lowiaceae) and the Zingiberales based on six DNA regions. Systematic Botany 30: 106-117.
Kellogg, E. A. 2015. Poaceae. Pp. 1-416 in K. Kubitzki (ed.), The Families and Genera of Vascular Plants. XIII. Springer, Berlin.
Kelly, L. M., and F. Gonzalez. 2003. Phylogenetic relationships in Aristolochiaceae. Systematic Botany 28: 236–249.
Kim, C. and H.-K. Choi. 2011. Molecular systematics and character evolution of Typha (Typhaceae) inferred from nuclear and plastid DNA sequence data. Taxon 60:1417-1428.
Kim, J. S., J.-K. Hong, M. W. Chase, M. F. Fay, and J.-H. Kim. 2013. Familial relationships of the monocot order Liliales based on a molecular phylogenetic analysis using four plastid loci: matK, rbcL, atpB and atpF-H. Botanical Journal of the Linnean Society 172:5-21.
Kim, S., C.-W. Park, Y.-D. Kim, and Y. Suh. 2001. Phylogenetic relationships in family Magnoliaceae inferred from ndhf sequences. American Journal of Botany 88: 717–728.
Kim, S., D. E. Soltis, P. S. Soltis, M. J. Zanis, and Y. Suh. 2004. Phylogenetic relationships among early-diverging eudicots based on four genes: Were the eudicots ancestrally woody? Molecular Phylogenetics and Evolution 31: 16–30.
Kirchoff, B. K. 2003. Shape matters: Hofmeister's rule, primordium shape, and flower orientation. International Journal of Plant Sciences 164: 505–517.
Kral, R. 1998. Xyridaceae. In: Kubitzki, K. (ed.), The Families and Genera of Vascular Plants. IV. Flowering Plants. Monocotyledons. Alismatanae and Commelinanae (except Gramineae). Pp. 461–469. Springer, Berlin.
Kress, W. J. and C. D. Specht. 2006. The evolutionary and biogeographic origin and diversification on the tropical monocot order Zingiberales. Pp. in J. T. Columbus, Friar, E. A., Porter, J. M., Prince, L. M., & Simpson, M. G. (ed.), Monocots: Comparative Biology and Evolution. Excluding Poales. Claremont, California, Rancho Santa Ana Botanical Garden [Aliso 22: 621-632.].
Kress, W. J., L. M. Prince, and K. J. Williams. 2002. The phylogeny and a new classification of gingers: evidence from molecular data. American Journal of Botany 89: 1682–1696.
Kress, W. J., L. M. Prince, W. J. Hahn, and E. A. Zimmer. 2001. Unraveling the evolutionary radiation of the families of the Zingiberales using morphological and molecular evidence. Systematic Biology 50: 926–944.
Kubitzki, K. 1998c. Cannaceae. In: Kubitzki, K. (ed.), The Families and Genera of Vascular Plants. IV. Flowering Plants. Monocotyledons. Alismatanae and Commelinanae (except Gramineae). Pp. 103–105. Springer, Berlin.

Kubitzki, K. 1998d. Typhaceae. In: Kubitzki, K. (ed.), The Families and Genera of Vascular Plants. IV. Flowering Plants. Monocotyledons. Alismatanae and Commelinanae (except Gramineae). Pp. 457–460. Springer, Berlin.

Kubitzki, K., H. Huber, P. J. Rudall, P. S. Stevens, and T. Stutzel (eds.). 1998a. The Families and Genera of Vascular Plants III: Flowering Plants Monocotyledons Lilianae (except Orchidaceae). Springer-Verlag, Berlin.

Kubitzki, K., H. Huber, P. J. Rudall, P. S. Stevens, and T. Stutzel (eds.). 1998b. The Families and Genera of Vascular Plants IV: Flowering Plants Monocotyledons Alismatanae and Commelinanae (except Gramineae). Springer-Verlag, Berlin.

Kubitzki, K., J. G. Rohwer, and V. Bittrich (eds). 1993. The Families and Genera of Vascular Plants. II. Flowering Plants: Dicotyledons, Magnoliid, Hamamelid and Caryophyllid Families. Springer, Berlin.

Larsen, K., J. M. Lock, H. Maas, and P. J. M. Maas. 1998. Zingiberaceae. In: Kubitzki, K. (ed.), The Families and Genera of Vascular Plants. IV. Flowering Plants. Monocotyledons. Alismatanae and Commelinanae (except Gramineae). Pp. 474–495. Springer, Berlin.

Lehtonen, S. 2009. Systematics of the Alismataceae - a morphological evaluation. Aquatic Botany 91:279-290.

Les, D. H. 1993. Ceratophyllaceae. In: Kubitzki, K., J. G. Rohwer, and V. Bittrich (eds.), The Families and Genera of Vascular Plants. II. Flowering Plants: Dicotyledons, Magnoliid, Hamamelid and Caryophyllid Families. Pp. 246–249. Springer, Berlin.

Les, D. H., and Haynes, R. R. 1995. Systematics of subclass Alismatidae: a synthesis of approaches. In: Rudall, P. J., P. J. Cribb, D. F. Cutler, and C. J. Humphries (eds.), Monocotyledons: Systematics and Evolution, Vol. 2. Pp. 353–377. Royal Botanic Gardens, Kew.

Les, D. H., E. L. Schneider, D. J. Padgett, P. S. Soltis, D. E. Soltis, and M. Zanis. 1999. Phylogeny, classification and floral evolution of water lilies (Nymphaeaceae; Nymphaeales): a synthesis of non-molecular, rbcL, matK, and 18s rDNA data. Systematic Botany 24: 28–46.

Les, D. H., M. A. Cleland, and M. Waycott. 1997. Phylogenetic studies in Alismatidae, II: evolution of marine angiosperms (seagrasses) and hydrophily. Systematic Botany 22: 443–463.

Lewis, C. E., and J. J. Doyle. 2001. Phylogenetic utility of the nuclear gene malate synthase in the palm family (Arecaceae). Molecular Phylogenetics and Evolution 19: 409–420.

Linder, H. P., B. G. Briggs, and L. A. S. Johnson. 1998. Restionaceae. In: Kubitzki, K. (ed.), The Families and Genera of Vascular Plants. IV. Flowering Plants. Monocotyledons. Alismatanae and Commelinanae (except Gramineae). Pp. 425–444. Springer, Berlin.

Linder, H. P., B. G. Briggs, and L. A. S. Johnson. 2000. Restionaceae: A morphological phylogeny. In: Wilson, K. L., and D. A. Morrison (eds.), Monocots: Systematics and Evolution. Pp. 653–660. CSIRO, Collingwood.

Liu, A.-Z., W. J. Kress, and D.-Z. Li. 2010. Phylogenetic analyses of the banana family (Musaceae) based on nuclear ribosomal (ITS) and chloroplast (trnL-F) evidence. Taxon 59:20-28.

Liu, Z., G. Hao, Y.-B. Luo, L. B. Thien, S. W. Rosso, A.-M. Lu, and Z.-D. Chen. 2006. Phylogeny and androecial evolution in Schisandraceae, inferred from sequences of nuclear ribosomal DNA ITS and chloroplast DNA trnL-F regions. International Journal of Plant Sciences 167: 539-550.

Löhne, C., M.-J. Yoo, T. Borsch, J. H. Wiersema, V. Wilde, C. D. Bell, W. Barthlott, D. E. Soltis, and P. S. Soltis. 2008. Biogeography of Nymphaeales: Extant patterns and historical events. Taxon 57: 1123-1146.

Löhne, C., T. Borsch, and J. H. Wiersema, J. H. 2007. Phylogenetic analysis of Nymphaeales using fast-evolving and noncoding chloroplast markers. Botanical Journal of the Linnean Society 154: 141-163.

Maas-van de Kamer, H., and P. J. M. Maas. 2008. The Cannaceae of the world. Blumea 53: 247-318.

Mabberley, D. J. 2008. Mabberley's Plant-Book: A Portable Dictionary of the Higher Plants, Their Classification and Uses. 3rd edition. Cambridge University Press, Cambridge..

Marquínez, X., L. G. Lohmann, M. L. F. Salatino, A. Salatino, and F. González. 2009. Generic relationships and dating of lineages in Winteraceae based on nuclear (ITS) and plastid (rpS16 and psbA-trnH) sequence data. Molecular Phylogenetics and Evolution 53:435-449.

Massoni, J., F. Forest, and H. Sauquet. 2014. Increased sampling of both genes and taxa improves resolution of phylogenetic relationships within Magnoliidae, a large and early-diverging clade of angiosperms. Molecular Phylogenetics and Evolution 70:84-93.

Massoni, J., J. Doyle, and H. Sauquet. 2015b. Fossil calibration of Magnoliidae, an ancient lineage of angiosperms. Palaeontologia Electronica 18.1.2FC.

Massoni, J., T. L. P. Couvreur, and H. Sauquet. 2015a. Five major shifts in diversification through the long evolutionary history of Magnoliidae (angiosperms). BMC Evolutionary Biology 15:49.

Mathews, S., and M. J. Donoghue. 1999. The root of angiosperm phylogeny inferred from duplicate phytochrome genes. Science 286: 947–950.

Mathews, S., and M. J. Donoghue. 2000. Basal angiosperm phylogeny inferred from duplicate phytochromes A and C. International Journal of Plant Sciences 161: S41–S55.

Mayo, S. J., J. Bogner, and N. Cusimano. 2013. Recent progress in the phylogenetics and classification of Araceae. Pp. 208-242 in P. Wilkin, & Mayo, S. J. (ed.), Early Events in Monocot Evolution. Cambridge University Press, Cambridge.

Mayo, S. J., J. Bogner, and P. C. Boyce. 1998. Araceae. In: Kubitzki, K. (ed.), The Families and Genera of Vascular Plants. III. Flowering Plants. Monocotyledons. Lilianae (except Orchidaceae). Pp. 26–73. Springer, Berlin.

McKain, M. R., H. Tang, J. R. McNeal, S. Ayyampalayam, J. I. Davis, C. W. DePamphilis, T. J. Givnish, J. C. Pires, D. W. Stevenson, and J. Leebens-Mack. 2016. A phylogenomic assessment of ancient polyploidy and genome evolution across the Poales. Genome Biology and Evolution 8:1150-1164.

Meerow, A. W., and D. A. Snijman. 1998. Amaryllidaceae. In: Kubitzki, K. (ed.), The Families and Genera of Vascular Plants. III. Flowering Plants. Monocotyledons. Lilianae (except Orchidaceae). Pp. 83–110. Springer, Berlin.

Meerow, A. W., and D. A. Snijman. 2006. The never-ending story: Multigene approaches to the phylogeny of Amaryllidaceae. In: Columbus, J. T., E. A. Friar, J. M. Porter, L. M. Prince, and M. G. Simpson (eds). Monocots: Comparative Biology and Evolution. Excluding Poales. Pp. 355-366. Rancho Santa Ana Botanical Garden, Claremont, Ca. [Aliso 22: 355-366.]

Meerow, A. W., M. F. Fay, C. L. Guy, Q.-B. Li, F. Q. Zaman, and M. W. Chase. 1999. Systematics of Amaryllidaceae based on cladistic analysis of plastid rbcL and trnL-F sequence data. American Journal of Botany 86: 1325–1345.

Meerow, A. W., M. F. Fay, M. W. Chase, C. L. Guy, Q.-B. Li, D. A. Snijman, and S. L. Yang. 2000. Phylogeny of Amaryllidaceae: Molecules and morphology. In: Wilson, K. L., and D. A. Morrison (eds.), Monocots: Systematics and Evolution. Pp. 372–386. CSIRO, Collingwood.

Meng, S.-W., A. W. Douglas, D.-Z. Li, Z.-D. Chen, H.-X. Liang, and J.-B. Yang. 2003. Phylogeny of Saururaceae based on morphology and five regions from three plant genomes. Annals of the Missouri Botanical Garden 90: 592–602.

Merckx, V. S. F. T., S. Huysmans, and E. F. Smets. 2010. Cretaceous origins of mycoheterotrophic lineages in Dioscoreales. Pp. 39–53 in O. Seberg, G. Petersen, A. S. Barfod, and J. I. Davis (eds.), Diversity, phylogeny and evolution in the monocotyledons. Aarhus University Press, Aarhus, Denmark.

Michelangeli, F. A., J. I. Davis, and D. W. Stevenson. 2003. Phylogenetic relationships among Poaceae and related families as inferred from morphology, inversions in the plastid genome, and sequence data from the mitochondrial and plastid genomes. American Journal of Botany 90: 93–106.

Moore, M. J., C. D. Bell, P. S. Soltis, and D. E. Soltis. 2007. Using plastid genome-scale data to resolve enigmatic relationships among basal angiosperms. Proceedings of the National Academy of Sciences U.S.A. 104: 19363-19368.

Moore, M. J., N. Hassan, M. A. Gitzendanner, R. A. Bruenn, M. Croley, A. Vandeventer, J. W. Horn, A. Dhingra, S. F. Brockington, M. Latvis, J. Ramdial, R. Alexandre, A. Piedrahita, Z. Xi, C. C. Davis, P. S. Soltis, and D. E. Soltis. 2011. Phylogenetic analysis of the plastid inverted repeat for 244 species: Insights into deeper-level angiosperm relationships from a long, slowly evolving sequence region. International Journal of Plant Sciences 172:541-558.

Muasya, A. M., D. A. Simpson, G. A. Verboom, P. Goetghebeur, R. F. C. Naczi, M. W. Chase, and E. Smets. 2009. Phylogeny of Cyperaceae based on DNA sequence data: Current progress and future prospects. Botanical Review 75: 2-21.

Naczi, R. F. C. 2009. Insights on using morphologic data for phylogenetic analysis in sedges (Cyperaceae). Botanical Review 75: 67-95.

Nandi, O. I., M. W. Chase, and P. K. Endress. 1998. A combined cladistic analysis of angiosperms using rbcL and non-molecular data sets. Annals of the Missouri Botanical Garden 85: 137–212.

Naumann, J., K. Salomo, J. P. Der, E. K. Wafula, J. F. Bolin, E. Maass, L. Frenzke, M.-S. Samain, C. Neinhuis, C. W. dePamphilis, and S. Wanke. 2013. Single-copy nuclear genes place haustorial Hydnoraceae within Piperales and reveal a Cretaceous origin of multiple parasitic Angiosperm lineages. PLoS ONE 8:e79204.

Neinhuis, C., S. Wanke, K. W. Hilu, K. Müller, and T. Borsch. 2005. Phylogeny of Aristolochiaceae based on parsimony, likelihood, and Bayesian analyses of trnL-trnF sequences. Plant Systematics and Evolution 250: 7-26.

Nguyen, N. H., H. E. Driscoll, and C. D. Specht. 2008. A molecular phylogeny of the wild onions (Allium; Alliaceae) with a focus on the western North American center of diversity. Molecular Phylogenetics and Evolution 47: 1157-1172.

Nickrent, D. L., A. Blarer, Y.-L. Qiu, D. E. Soltis, P. S. Soltis, and M. Zanis. 2002. Molecular data place Hydnoraceae with Aristolochiaceae. American Journal of Botany 89: 1809–1817.

Nie, Z.-L., J. Wen, H. Azuma, Y.-L. Qiu, H. Sun, Y. Meng, W.-B. Sun, and E. A. Zimmer. 2008. Phylogeny and biogeographic complexity of Magnoliaceae in the Northern Hemisphere inferred from three nuclear data sets. Molecular Phylogenetics and Evolution 48: 1027-1040.

Ohi-Toma, T., T. Sugawara, H. Murata, S. Wanke, C. Neinhuis, and J. Murata. 2006. Molecular phylogeny of Aristolochia sensu lato (Aristolochiaceae) based on sequences of rbcL, matK, and phyA genes, with special reference to differentiation in chromosome numbers. Systematic Botany 31: 481-492.

Parkinson, C. L., K. L. Adams, and J. D. Palmer. 1999. Multigene analyses identify the three earliest lineages of extant flowering plants. Current Biology 9: 1485–1488.

Patterson, T. B. and T. J. Givnish. 2002. Phylogeny, concerted convergence, and phylogenetic niche conservatism in the core Liliales: Insights from rbcL and ndhF sequence data. Evolution 56: 233–252.

Pellegrini, M. O. O. 2017a. Morphological phylogeny of Tradescantia L. (Commelinaceae) sheds light on a new infrageneric classification for the genus and novelties on the systematics of subtribe Tradescantiinae. PhytoKeys 89:11-72.

Pellegrini, M. O. O. 2017b. Two new synonyms for Heteranthera (Pontederiaceae, Commelinales). Nordic Journal of Botany 35:124-128.

Pellegrini, M. O. O., C. N. Horn, and R. F. Almeida. 2018. Total evidence phylogeny of Pontederiaceae (Commelinales) sheds light on the necessity of its recircumscription and synopsis of Pontederia L. PhytoKeys 108:25-83.

Pires, J. C., and K. J. Sytsma. 2002. A phylogenetic evaluation of a biosystematic framework: Brodiaea and related petaloid monocots (Themidaceae). American Journal of Botany 89: 1342–1359.

Pires, J. C., I. J. Maureira, J. P. Rebman, G. A. Salazar, L. I. Cabrera, M. F. Fay, and M. W. Chase. 2004. Molecular data confirm the phylogenetic placement of the enigmatic Hesperocallis (Hesperocallidaceae) with Agave. Madroño 51:307-311.

Pires, J. C., I. J. Maureira, T. J. Givnish, K. J. Sytsma, O. Seberg, G. Petersen, J. I. Davis, D. W. Stevenson, P. J. Rudall, M. F. Fay, and M. W. Chase. 2006. Phylogeny, genome size, and chromosome evolution of Asparagales. In: Columbus, J. T., E. A. Friar, J. M. Porter, L. M. Prince, and M. G. Simpson (eds). Monocots: Comparative Biology and Evolution. Excluding Poales. Pp. 287-304. Rancho Santa Ana Botanical Garden, Claremont, Ca. [Aliso 22: 287-304.]

Pires, J. C., M. F. Fay, W. S. Davis, L. Hufford, J. Rova, M. W. Chase, and K. J. Sytsma. 2001. Molecular and phylogenetic analyses of Themidaceae (Asparagales). Kew Bulletin 56: 601-626.

Pirie, M. D., L. W. Chatrou, J. B. Mols, R. H. J. Erkens, and J. Oosterhof. 2006. 'Andean centred' genera in the short branch clade of Annonaceae: testing biogeographic hypotheses using phylogeny reconstruction and molecular dating. Journal of Biogeography 33: 31-46.

Pirie, M. D., M. P. B. Vargas, M. Botermans, F. T. Bakker, and L. W. Chatrou. 2007. Ancient paralogy in the cpDNA trnL-F region in Annonaceae: Implications for plant molecular systematics. American Journal of Botany 94: 1003-1016.

Prenner, G., R. M. Bateman, and P. J. Rudall. 2010. Floral formulae updated for routine inclusion in formal taxonomic descriptions. Taxon 59:241–250.

Prince, L. M. 2010. Phylogenetic relationships and species delimitation in Canna (Cannaceae). Pp. 307-331 in O. Seberg, Petersen, G., Barfod, A. S., & Davis, J. I. (ed.), Diversity, Phylogeny, and Evolution in the Monocotyledons. Aarhus University Press, Århus.

Qiu, Y.-L., J. Lee, F. Bernasconi-Quadroni, D. E. Soltis, P. S. Soltis, M. Zanis, E. A. Zimmer, Z. Chen, V. Savolainen, and M. W. Chase. 1999. The earliest angiosperms: evidence from mitochondrial, plastid and nuclear genomes. Nature 402: 404–407.

Qiu, Y.-L., J. Lee, F. Bernasconi-Quadroni, D. E. Soltis, P. S. Soltis, M. Zanis, E. A. Zimmer, Z. Chen, V. Savolainen, and M. W. Chase. 2000. Phylogeny of basal angiosperms: analyses of five genes from three genomes. International Journal of Plant Sciences 161: S3–S27.

Qiu, Y.-L., O. Dombrovska, J. Lee, L. Li, B. A. Whitlock, F. Bernasconi-Quadroni, J. S. Rest, C. C. Davis, T. Borsch, K. W. Hilu, S. S. Renner, D. E. Soltis, P. S. Soltis, M. J. Zanis, J. J. Cannone, R. R. Gutell, M. Powell, V. Savolainen, L. W. Chatrou, and M. W. Chase, M. W. 2005. Phylogenetic analysis of basal angiosperms based on nine plastid mitochondrial and nuclear genes. International Journal of Plant Sciences 166: 815-842.

Rahn, K. 1998a. Alliaceae. In: Kubitzki, K. (ed.), The Families and Genera of Vascular Plants. III. Flowering Plants. Monocotyledons. Lilianae (except Orchidaceae). Pp. 70–78. Springer, Berlin.

Rahn, K. 1998b. Themidaceae. In: Kubitzki, K. (ed.), The Families and Genera of Vascular Plants. III. Flowering Plants. Monocotyledons. Lilianae (except Orchidaceae). Pp. 436–440. Springer, Berlin.

Reeves, G., M. W. Chase, P. Goldblatt, P. Rudall, M. F. Fay, A. V. Cox, B. Lejeune, and T. Souza-Chies. 2001. Molecular systematics of Iridaceae: evidence from four plastid DNA regions. American Journal of Botany 88: 2074–2087.

Renner, S. S. 1999. Circumscription and phylogeny of the Laurales: Evidence from molecular and morphological data. American Journal of Botany 86: 1301–1315.

Renner, S. S., and A. S. Chanderbali. 2000. What is the relationship among Hernandiaceae, Lauraceae, and Monimiaceae, and why is this question so difficult to answer? International Journal of Plant Sciences 161: S109–S119.

Roalson, E. H. 2005. Phylogenetic relationships in the Juncaceae inferred from nuclear ribosomal DNA internal transcribed spacer sequence data. International Journal of Plant Sciences 166: 397-413.

Rohwer, J. G. 1993. Lauraceae. In: Kubitzki, K., J. G. Rohwer, and V. Bittrich (eds.), The Families and Genera of Vascular Plants. II. Flowering Plants: Dicotyledons, Magnoliid, Hamamelid and Caryophyllid Families. Pp. 366–390. Springer, Berlin.

Rohwer, J. G., and B. Rudolph. 2005. Jumping genera: The phylogenetic positions of Cassytha, Hypodaphnis, and Neocinnamomum (Lauraceae) based on different analyses of trnK intron sequences. Annals of the Missouri Botanical Garden 92: 153-178.

Ronse Decraene, L.-P., P. S. Soltis, and D. E. Soltis. 2003. Evolution of floral structures in basal angiosperms. International Journal of Plant Sciences 164: S329–S363.

Rønsted, N., M. R. E. Symonds, T. Birkholm, S. Brøgger Christensen, A. W. Meerow, M. Molander, P. Mølgaard, G. Petersen, N. Rasmussen, J. van Staden, G. I. Stafford, and A. K. Jäger. 2012. Can phylogeny predict chemical diversity and potential medicinal activity of plants? A case study of Amaryllidaceae. BMC Evolutionary Biology 12:182.

Rønsted, N., S. Law, H. Thornton, M. F. Fay, and M. W. Chase. 2005. Molecular phylogenetic evidence for the monophyly of Fritillaria and Lilium (Liliaceae; Liliales) and the infrageneric classification of Fritillaria. Molecular Phylogenetics and Evolution 35:509-527.

Ross, T. G., C. F. Barrett, M. S. Gomez, V. K. Y. Lam, C. L. Henriquez, D. H. Les, J. I. Davis, A. Cuenca, G. Petersen, O. Seberg, M. Thadeo, T. J. Givnish, J. Conran, D. W. Stevenson, and S. W. Graham. 2016. Plastid phylogenomics and molecular evolution of Alismatales. Cladistics 32:160-178.

Rothwell, G. W., M. R. Van Atta, H. E. J. Ballard, and R. A. Stockey. 2004. Molecular phylogenetic relationships among Lemnaceae and Araceae using the chloroplast trnL-trnF intergenic spacer. Molecular Phylogenetics and Evolution 30:378-385.

Rudall, P. 2003. Unique floral structures and iterative evolutionary themes in Asparagales: insights from a morphological cladistic analysis. Botanical Review 68: 488–509.

Rudall, P. J. and R. M. Bateman. 2006. Morphological phylogenetic analysis of Pandanales: Testing contrasting hypotheses of floral evolution. Systematic Botany 31:223-238.

Rudall, P., K. L. Stobart, W.-P. Hong, J. G. Conran, C. A. Furness, G. C. Kite, and M. W. Chase. 2000. Consider the lilies: Systematics of Liliales. In: Wilson, K. L., and D. A. Morrison (eds.), Monocots: Systematics and Evolution. Pp. 347–359. CSIRO, Collingwood.

Rudall, P., P. J. Cribb, D. F. Cutler, and C. Humphries (eds). 1995. Monocotyledons: Systematics and Evolution. Royal Botanic Gardens Kew [London].

Saarela, J. M., H. S. Rai, J. A. Doyle, P. K. Endress, S. Mathews, A. D. Marchant, B. Briggs, and S. W. Graham. 2007. Hydatellaceae identified as a new branch near the base of the angiosperm phylogenetic tree. Nature 446: 312-315.

Saarela, J. M., P. J. Prentis, H. S. Rai, and S. W. Graham. 2008. Phylogenetic relationships in the monocot order Commelinales, with a focus on Philydraceae. Botany 86:719-731.

Saarela, J. M., P. J. Prentis, H. S. Rai, and S. W. Graham. 2008. Phylogenetic relatiionships in the monocot order Commelinales, with a focus on Philydraceae. Botany 86: 719-731. doi:10.1139/B08-063

Sass, C., W. J. D. Iles, C. F. Barrett, S. Y. Smith, and C. D. Specht. 2016. Revisiting the Zingiberales: Using multiplexed exon capture to resolve ancient and recent phylogenetic splits in a charismatic plant lineage. PeerJ 4:31584.

Sauquet, H., J. A. Doyle, T. Scharaschkin, T. Borsch, K. W. Hilu, L. Chatrou, and A. Le Thomas. 2003. Phylogenetic analysis of Magnoliaceae and Myristicaceae based on multiple data sets: implications for character evolution. Botanical Journal of the Linnean Society 142: 125–186.

Sauquet, H., M. v. Balthazar, S. Magallón, J. A. Doyle, P. K. Endress, E. J. Bailes, E. B. d. Morais, K. Bull-Hereñu, L. Carrive, M. Chartier, G. Chomicki, M. Coiro, R. Cornette, J. H. L. E. Ottra, C. Epicoco, C. S. P. Foster, F. Jabbour, A. Haevermans, T. Haevermans, R. Hernández, S. A. Little, S. Löfstrand, J. A. Luna, J. Massoni, S. Nadot, S. Pamperl, C. Prieu, E. Reyes, P. d. S. K. M. Schoonderwoerd, S. Sontag, A. Soulebeau, Y. Staedler, G. F. Tschan, A. W.-S. Leung, and J. Schönenberger. 2017. The ancestral flower of angiosperms and its early diversification. Nature Communications 8:16047.

Savolainen, V., M. W. Chase, S. B. Hoot, C. M. Morton, D. E. Soltis, C. Bayer, M. F. Fay, A. Y. de Bruijn, S. Sulllivan, and Y.-L. Qiu. 2000. Phylogenetics of flowering plants based on combined analysis of plastid atpB and rbcL sequences. Systematic Biology 49: 306–362.

Schneider, E. L. and P. S. Williamson. 1993. Nymphaeaceae. In: Kubitzki, K., J. G. Rohwer, and V. Bittrich (eds.), The Families and Genera of Vascular Plants. II. Flowering Plants: Dicotyledons, Magnoliid, Hamamelid and Caryophyllid Families. Pp. 486–493. Springer, Berlin.

Simon, B. K. 2007. Grass phylogeny and classification: Conflict of morphology and molecules. In: Columbus, J. T., E. A. Friar, J. M. Porter, L. M. Prince, and M. G. Simpson. (eds). Monocots: Comparative Biology and Evolution. Poales. Pp. 259-266. Rancho Santa Ana Botanical Garden, Claremont, Ca. [Aliso 23: 259-266.]

Simpson, D. A., A. M. Muasya, M. V. Alves, J. J. Bruhl, S. Dhooge, M. W. Chase, C. A. Furness, K. Ghamkhar, P. Goetghebeur, T. R. Hodkinson, A. D. Marchant, A. A. Reznicek, R. Nieuwborg, E. H. Roalson, E. Smets, J. R. Starr, W. W. Thomas, K. L. Wilson, and X. Zhang. 2007. Phylogeny of Cyperaceae based on DNA sequence data - a new rbcL analysis. In: Columbus, J. T., E. A. Friar, J. M. Porter, L. M. Prince, and M. G. Simpson. (eds). Monocots: Comparative Biology and Evolution. Poales. Pp. 72-83. Rancho Santa Ana Botanical Garden, Claremont, Ca. [Aliso 23: 72-83.]

Simpson, D. A., C. A. Furness, T. R. Hodkinson, A. M. Muasya, and M. W. Chase. 2003. Phylogenetic relationships in Cyperaceae subfamily Mapanioideae inferred from pollen and plastid DNA sequence data. American Journal of Botany 90: 1071–1086.

Simpson, M. G. 1990. Phylogeny and classification of the Haemodoraceae. Annals of the Missouri Botanical Garden 77: 722–784.

Simpson, M. G. 1998. Haemodoraceae. In: Kubitzki, K. (ed.), The Families and Genera of Vascular Plants. IV. Flowering Plants. Monocotyledons. Alismatanae and Commelinanae (except Gramineae). Pp. 212–222. Springer, Berlin.

Smith, G. F., and B.-E. v. Wyk. 1998. Asphodelaceae. In: Kubitzki, K. (ed.), The Families and Genera of Vascular Plants. III. Flowering Plants. Monocotyledons. Lilianae (except Orchidaceae). Pp. 130–140 Springer, Berlin.

Smith, L. B., and W. Till. 1998. Bromeliaceae. In: Kubitzki, K. (ed.), The Families and Genera of Vascular Plants. IV. Flowering Plants. Monocotyledons. Alismatanae and Commelinanae (except Gramineae). Pp. 74–99. Springer, Berlin.

Soltis, D. E., M. A. Gitzendanner, and P. S. Soltisy. 2007. A 567-taxon data set for angiosperms: the challenges posed by Bayesian analyses of large data sets. International Journal of Plant Sciences 168:137–157.

Soltis, D. E., P. S. Soltis, D. L. Nickrent, L. A. Johnson, W. J. Hahn, S. B. Hoot, J. A. Sweere, R. K. Kuzoff, K. A. Kron, M. W. Chase, S. M. Swensen, E. A. Zimmer, S.-M. Chaw, L. J. Gillespie, W. J. Kress, and K. J. Sytsma. 1997. Angiosperm phylogeny inferred from 18S ribosomal DNA sequences. Annals of the Missouri Botanical Garden 84: 1–49.

Soltis, D. E., P. S. Soltis, M. W. Chase, M. E. Mort, D. C. Albach, M. Zanis, V. Savolainen. W. H. Hahn, S. B. Hoot, M. F. Fay, M. Axtell, S. M. Swensen, L. M. Prince, W. J. Kress, K. C. Nixon, J. S. Farris. 2000. Angiosperm phylogeny inferred from 18S rDNA, rbcL, and atpB sequences. Botanical Journal of the Linnean Society 133(4): 381–461.

Soltis, D. E., P. S. Soltis, P. K. Endress, and M. W. Chase. 2005. Angiosperm phylogeny and evolution. Sinauer, Sunderland, MA.

Soltis, D. E., S. A. Smith, N. Cellinese, K. J. Wurdack, D. C. Tank, S. F. Brockington, N. F. Refulio-Rodriguez, J. B. Walker, M. J. Moore, B. S. Carlsward, C. D. Bell, M. Latvis, S. Crawley, C. Black, D. Diouf, Z. Xi, C. A. Rushworth, M. A. Gitzendanner, K. J. Sytsma, Y.-L. Qiu, K. W. Hilu, C. C. Davis, M. J. Sanderson, R. S. Beaman, R. G. Olmstead, W. S. Judd, M. J. Donoghue, and P. S. Soltis. 2011. Angiosperm phylogeny: 17 genes, 640 taxa. American Journal of Botany 98:704-730.

Stevenson, D. W., J. I. Davis, J. V. Freudenstein, C. R. Hardy, M. P. Simmons, and C. D. Specht. 2000. A phylogenetic analysis of the monocotyledons based on morphological and molecular character sets, with comments on the placement of Acorus and Hydatellaceae. In: Wilson, K. L., and D. A. Morrison (eds.), Monocots: Systematics and Evolution, CSIRO, Collingwood.

Stone, B. C., K.-L. Huynh, and H.-H. Poppendieck. 1998. Pandanaceae. In: Kubitzki, K. (ed.), The Families and Genera of Vascular Plants. III. Flowering Plants. Monocotyledons. Lilianae (except Orchidaceae). Pp. 397–403. Springer, Berlin.

Stützel, T. 1998. Eriocaulaceae. In: Kubitzki, K. (ed.), The Families and Genera of Vascular Plants. IV. Flowering Plants. Monocotyledons. Alismatanae and Commelinanae (except Gramineae). Pp. 197–207. Springer, Berlin.

Sun, Y., M. Moore, S. Zhang, P. Soltis, D. Soltis, T. Zhao, A. Meng, L. Xiaodong, J. Li, and H. Wang. 2016. Phylogenomic and structural analyses of 18 complete plastomes across nearly all families of early-diverging eudicots, including an angiosperm-wide analysis of IR gene content evolution. Molecular Phylogenetics and Evolution 96:93-101.

Takhtajan, A. L. 1991. Evolutionary Trends in Flowering Plants. Columbia University Press, New York.

Takhtajan, A. L. 1997. Diversity and Classification of Flowering Plants. Columbia University Press, New York.

Tamura, M. N. 1998a. Calochortaceae. In: Kubitzki, K. (ed.), The Families and Genera of Vascular Plants. III. Flowering Plants. Monocotyledons. Lilianae (except Orchidaceae). Pp. 164–172. Springer, Berlin.

Tamura, M. N. 1998b. Liliaceae. In: Kubitzki, K. (ed.), The Families and Genera of Vascular Plants. III. Flowering Plants. Monocotyledons. Lilianae (except Orchidaceae). Pp. 343–353. Springer, Berlin.

Thien, L. B., T. L. Sage, T. Jaffre, P. Bernhardt, V. Pontieri, P. H. Weston, D. Malloch, H. Azuma, S. W. Graham, M. A. McPherson, H. S. Rai, R. F. Sage, and J.-L. Dupre. 2003. The population structure and floral biology of Amborella trichopoda (Amiborellaceae). Annals of the Missouri Botanical Garden 90: 466–490.

Thomas, N., J. J. Bruhl, A. Ford, and P. H. Weston. 2014. Molecular dating of Winteraceae reveals a complex biogeographical history involving both ancient Gondwanan vicariance and long-distance dispersal. Journal of Biogeography 41:894-904.

Thorne, Robert F. 2000. The classification and geography of the flowering plants: Dicotyledons of the class Angiospermae: (Subclasses Magnoliidae, Ranunculidae, Caryophyllidae, Dilleniidae, Rosidae, Asteridae, and Lamiidae). Botanical Review 66(4): 441–647.

Verhoek, S. 1998. Agavaceae. In: Kubitzki, K. (ed.), The Families and Genera of Vascular Plants. III. Flowering Plants. Monocotyledons. Lilianae (except Orchidaceae). Pp. 60–70. Springer, Berlin.

Viruel, J., F. Forest, O. Paun, M. W. Chase, D. Devey, R. S. Couo, J. G. Segarra-Moragues, P. Catalán, and P. Wilkin. 2018. A nuclear Xdh phylogenetic analysis of yams (Dioscorea, Dioscoreaceae) congruent with plastid trees reveals a new Neotropical lineage. Botanical Journal of the Linnean Society 187:232-246.

Wanke, S., F. González, and C. Neinhuis. 2006a. Systematics of pipevines: Combining morphological and fast-evolving molecular characters to investigate the relationships within subfamily Aristolochioideae (Aristolochiaceae). International Journal of Plant Sciences 167: 1215-1227.

Wanke, S., L. Vanderschaeve, G. Mathieu, C. Neinhuis , P. Goetghebeur and M. S. Samain. 2007b. From Forgotten Taxon to a Missing Link? The Position of the Genus Verhuellia (Piperaceae) Revealed by Molecules. Annals of Botany 99: 1231–1238.

Wanke, S., M. A. Jaramillo, T. Borsch, M.-S. Samain, D. Quandt, and C. Neinhuis. 2007a. Evolution of Piperales - matK gene and trnK intron sequence data reveal lineage specific resolution contrast. Molecular Phylogenetics and Evolution 42: 477-497.

Wanke, S., M.-S. Samain, L. Vanderschaeve, G. Mathieu, P. Goetghebeur, and C. Neinhuis. 2006b. Phylogeny of the genus Peperomia (Piperaceae) inferred from the trnK/matK region (cpDNA). Plant Biology 8: 93-102.

Williamson, P. S. and E. L. Schneider. 1993. Cabombaceae. In: Kubitzki, K., J. G. Rohwer, and V. Bittrich (eds.), The Families and Genera of Vascular Plants. II. Flowering Plants: Dicotyledons, Magnoliid, Hamamelid and Caryophyllid Families. Pp. 157–160. Springer, Berlin.

Wilson, K. L., and D. A. Morrison (eds). 2000. Monocots: Systematics and Evolution. CSIRO, Collingwood, VIC, Australia.

Zanis, M. J., D. E. Soltis, P. S. Soltis, S. Mathews, and M. J. Donoghue. 2002. The root of the angiosperms revisited. Proceedings of the National Academy of Sciences of the United States of America 99: 6848–6853.

Zanis, M. J., P. S. Soltis, Y. L. Qiu, E. Zimmer, and D. E. Soltis. 2003. Phylogenetic analyses and perianth evolution in basal angiosperms. Annals of the Missouri Botanical Garden 90: 129–150.

Záveská Drábková, L. 2010. Phylogenetic relationships within Juncaceae: Evidence from five regions of plastid, mitochondrial and nuclear ribosomal DNA, with notes on morphology. Pp. 389-416 in O. Seberg, G. Petersen, A. S. Barfod, and J. I. Davis (eds.), Diversity, Phylogeny, and Evolution in the Monocotyledons. Aarhus University Press, Aarhus, Denmark.

Záveská Drábková, L., J. Kirschner, O. Seberg, G. Petersen, and C. Vlcek. 2003. Phylogeny of the Juncaceae based on rbcL sequences, with special emphasis on Luzula DC. and Juncus L. Plant systematics and evolution = Entwicklungsgeschichte und Systematik der Pflanzen 240: 133–148.

Zhou, B., T. Tu, F. Kong, J. Wen, and X. Xu. 2018. Revised phylogeny and historical biogeography of the cosmopolitan aquatic plant genus Typha (Typhaceae). Science Reports 8:6613.

Zomlefer, W. B. 1994. Guide to Flowering Plant Families. University of North Carolina Press, Chapel Hill.

WEB SITES

POWO. 2019. Plants of the World Online. Facilitated by the Royal Botanic Gardens, Kew. Published on the Internet. *http://plantsoftheworldonline.org*.

Reveal, J. L. (2001 onwards). Indices Nominum Supragenericorum Plantarum Vascularium. Alphabetical Listing by Family of Validly Published Suprageneric Names. *http://www.plantsystematics.org/reveal/PBIO/fam/allspgnames.html*
An index of accepted names above the rank of family.

Royal Botanic Gardens, Kew, Vascular Plant Families and Genera Database *http://data.kew.org/vpfg1992/vascplnt.html*
A searchable database taken from Vascular Plant Families and Genera, compiled by R. K. Brummitt and published by the Royal Botanic Gardens, Kew, in 1992.

Stevens, P. F. (2001 onwards). Angiosperm Phylogeny Website. *http://www.mobot.org/MOBOT/research/APweb*
An excellent graphical representation and update of the Angiosperm Phylogeny Group classification, cladograms, family characteristics, references, and apomorphies.

World Checklist of Monocotyledeons. (2009 onwards). The Board of Trustees of the Royal Botanic Gardens, Kew. Published on the Internet. *http://www.kew.org/monocotChecklist*
A Checklist of accepted scientific names and synonyms of all monocot families, including search functions for all scientific names and distribution of monocot taxa.

8

DIVERSITY AND CLASSIFICATION OF FLOWERING PLANTS: EUDICOTS

(Continued)

https://doi.org/10.1016/B978-0-12-812628-8.50008-0,

EUDICOTS

The **eudicots** (*Eudicotyledoneae*, sensu Cantino et al., 2007) are a large, monophyletic assemblage of angiosperms, comprising roughly 190,000 described species, or 75% of all angiosperms. The monophyly of eudicots is well supported from molecular data and delimited by at least one palynological apomorphy: a **tricolpate or tricolpate-derived** pollen grain (Figure 8.1). A tricolpate pollen grain is one that has three apertures, equally spaced and approximately parallel to the polar axis of the grain (Figure 8.2; see Chapter 12). Apertures are differentiated regions of the pollen grain wall that may function as the site of pollen tube exitus as well as to allow for expansion and contraction of the pollen grain with changes in humidity (Chapter 12). Tricolpate pollen grains evolved from a **monosulcate** type (having a single distal aperture; Figure 8.2), which is considered to be ancestral in the angiosperms, as well as for many seed plant clades. Many eudicots have pollen grains with more than three apertures, of a great variety of numbers, shapes, and position (constituting important taxonomic characters; see Chapter 12). These are all thought to have been derived from a tricolpate type.

The orders of the eudicots and their included families (mostly after APG IV 2016, with some deviations; see Cole 2015) are listed in Tables 8.1–8.3. Table 8.1 lists the non-superrosid and non-superaserid groups, including the families of the Ranunculales, Proteales, Buxales, Trochodendrales, Gunnerales, and Dilleniales. Table 8.2 lists orders and families within the Superrosids inclusive of Rosids, and Table 8.3 lists those of the Superasterids inclusive of Asterids. For detailed information on eudicot relationships, see a recent analysis of general angiosperm relationships (e.g., Moore et al. 2011; Soltis et al. 2011; Sun et al. 2016a) and earlier studies (see Literature). Note that several Phylocode names for clades within the Eudicots were introduced by Cantino et al. (2007). These include, from groups used here, the *Pentapetalae* (Dilleniales + Superrosids + Superasterids; Figure 8.1) and *Gunneridae* (Gunnerales + *Pentapetalae*).

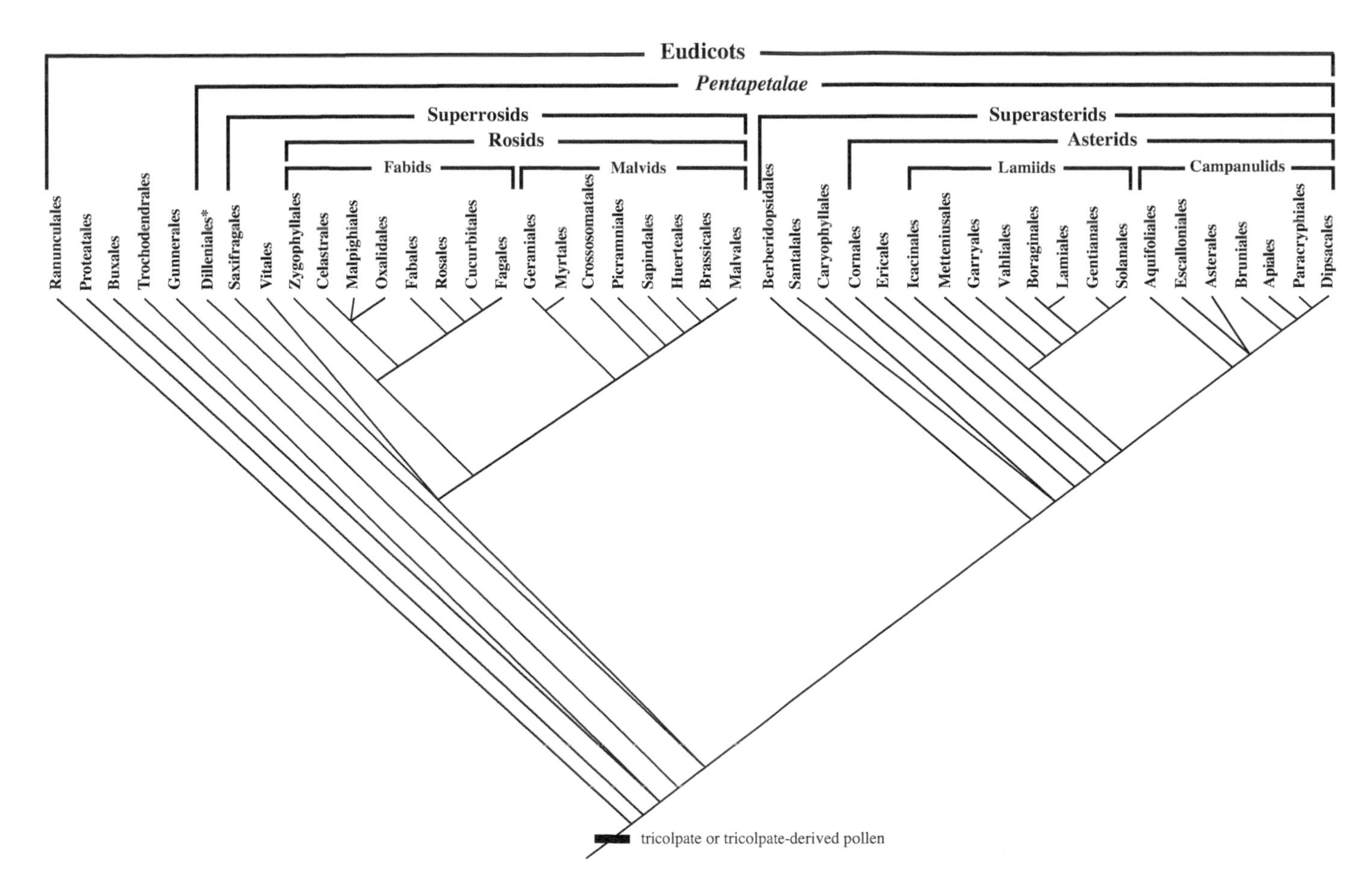

FIGURE 8.1 Cladogram of the orders of the Eudicots, after Moore et al. (2011), Soltis et al. (2011), APGIV (2016), and Sun et al. (2016). *=Position of Dilleniales uncertain.

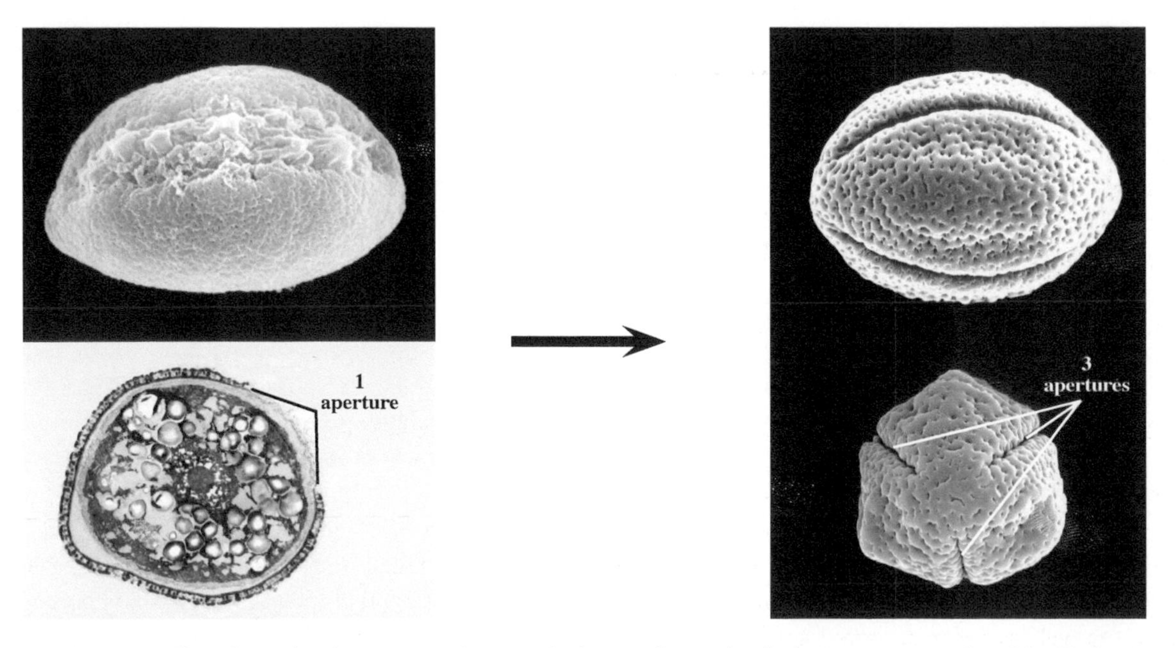

FIGURE 8.2 Transformation from monosulcate to tricolpate pollen grain, the latter an apomorphy of the Eudicots.

TABLE 8.1 Orders and families of the Eudicots, largely after APG IV (2016). Families in bold are described in detail. * = Position uncertain (Dilleniales). See Tables 8.2 and 8.3 for listings of the Superrosids/Rosids and Superasterids/Asterids.

EUDICOTS		
RANUNCULALES	PROTEALES	GUNNERALES
Berberidaceae	**Nelumbonaceae**	**Gunneraceae**
Circaeasteraceae	**Platanaceae**	Myrothamnaceae
Eupteleaceae	**Proteaceae**	DILLENIALES*
Lardizabalaceae	Sabiaceae	Dilleniaceae
Menispermaceae	BUXALES	**SUPERROSIDS & ROSIDS**
Papaveraceae	Buxaceae	(see Table 8.2, p. 298)
Ranunculaceae	TROCHODENDRALES	**SUPERASTERIDS & ASTERIDS**
	Trochodendraceae	(see Table 8.3, p. 365)

RANUNCULALES

The Ranunculales contain seven families, three of which are described here. Among those not described are the **Menispermaceae**, of which *Chondrodendron tomentosum*, curare, is used as an arrow poison by native Amazonians, and from which is derived tubocurarine used medicinally as a muscle relaxant in surgery and to treat diseases. See Wang, et al. (2009) and Lane et al. (2018) for phylogenetic analyses.

Berberidaceae — Barberry family (type *Berberis*, after Barbary, an ancient Arabic name for N. Africa). 14 genera/ca. 700 species (Figure 8.3).

The Berberidaceae consist of perennial trees, shrubs, or herbs. The **leaves** are spiral [rarely opposite], petiolate with the petiole often flared basally, and either pinnate, ternate, simple, or unifoliolate. The **inflorescence** is a raceme, spike, panicle, cyme, or a solitary, axillary flower. The **flowers** are bisexual, actinomorphic, and hypogynous. The **perianth** is 6–7-seriate with 3 [2,4] parts per whorl, the outer 2 whorls sepaloid, the inner 4–5 whorls petaloid, with the innermost 2–3 of these nectariferous (sometimes interpreted as staminodes). The **stamens** are 6 [4–18], mostly in two whorls, opposite the inner most whorls of petals. **Anthers** are valvular (opening from the base) or longitudinal in dehiscence, and are tetrasporangiate and dithecal. The **gynoecium** is unicarpellous (sometimes interpreted as derived from 2–3 carpels), with a superior ovary, and 1 [2] locules. **Placentation** is marginal; **ovules** are ∞ [1,2] per ovary. The **fruit** is a berry [rarely dry].

Members of the family have a worldwide distribution, especially in north-temperate regions. Economic importance includes cultivated ornamentals, such as *Berberis* and *Mahonia*; *Podophyllum* is reported to be used against testicular cancer; *Berberis vulgaris*, the common barberry, is the alternate host of stem rust of wheat. See Wang, W. et al. (2007, 2009) and Sun et al. (2018) for phylogenetic analyses.

The Berberidaceae are distinctive in having flowers with a ***multiseriate*** perianth (possibly apomorphic for the family) differentiated into ***outer sepaloid*** and ***inner petaloid*** parts (the innermost nectariferous), a ***biseriate androecium***, and a **single**, apparently ***unicarpellate*** pistil.

P ∞, 6–7-seriate in whorls of 3 [2,4] each **A** 6 [4–18] **G** 1, superior.

Papaveraceae (including Fumariaceae & Pteridophyllaceae) — Poppy family (type *Papaver*, Latin for poppy). Ca. 44 genera/ca. 825 species (Figure 8.4).

The Papaveraceae consist of annual or perennial herbs, shrubs, or small (sometimes pachycaulous) trees, with milky latex from articulated laticifers in some taxa. The **leaves** are spiral to subopposite, usually lobed to divided or dissected, exstipulate. The **inflorescence** is a solitary flower or cyme. The **flowers** are bisexual, actinomorphic, zygomorphic, or biradial, hypogynous [rarely perigynous]. The **perianth** is dichlamydeous, in 3 [2,4] series, a hypanthium absent [rarely present]. The **calyx** is uniseriate and aposepalous to basally synsepalous, with 2 [3], usually caducous sepals. The **corolla** is biseriate, apopetalous, of 2+2 or 3+3 [–16] petals, sometimes imbricate and crumpled in bud, the outer petals with a spur or sac in some taxa. The **stamens** are usually numerous [4–6], centripetal, sometimes in two or three bundles. **Anthers** are tetrasporangiate or bisporangiate, dithecal or monothecal (sometimes in the same flower). The **gynoecium** is syncarpous, with a superior ovary, 2 [to several] carpels, and 1 [2–several] locule(s). The **style** and **stigma** are usually solitary, the latter sometimes connate to form a disklike structure. **Placentation** is parietal [rarely axile]; **ovules** are anatropous to campylotropous, bitegmic, 1–∞ per carpel. **Nectaries** are sometimes present at base of stamens. The **fruit** is a longitudinally dehiscent or poricidal capsule, sometimes a schizocarp or nut. The **seeds** are oily endospermous, arillate in some taxa.

FIGURE 8.3 RANUNCULALES. Berberidaceae. **A–E.** *Berberis* sp., barberry. **A.** Whole plant, with pinnately compound leaves. **B.** Flower bud, showing numerous tepals. **C.** Open flower. **D.** Tepals with opposing stamens. **E.** Pistil, longitudinal section, showing marginal placentation. **F.** *Diphylleia cymosa*, umbrella leaf; whole plant with peltate, palmately cleft leaves. **G,H.** *Podophyllum peltatum*, mayflower. **G.** Plant, showing peltate leaves. **H.** Solitary flower. **I.** *Caulophyllum thalictroides*, blue cohosh; cymose inflorescence. **J–M.** *Nandina domestica*. **J.** Flower bud, showing multiseriate perianth. **K.** Open flower. **L.** Flower close-up, showing pistil and stamens. **M.** Ovary longitudinal section, showing marginal placentation.

FIGURE 8.4 RANUNCULALES. Papaveraceae. **A.** *Ehrendorferia chrysantha*, with biradial, saccate flowers. **B.** *Dendromecon rigida*, tree poppy, showing biseriate (2+2) corolla (sepals caducous). **C.** *Eschscholzia californica*, California poppy, ovary cross section showing parietal placentation. **D.** *Papaver californicum*, fire poppy. **E.** *Papaver somniferum*, mature poricidal capsule. **F.** *Platystemon californicus*, cream cups. **G.** *Romneya trichocalyx*, Matilija poppy. **H,I.** *Sanguinaria canadensis*, bloodroot, having several petals per flower.

Members of the Papaveraceae are distributed in mostly north temperate regions. Economic importance includes many cultivated ornamentals and taxa used as oil seeds. *Papaver somniferum*, opium poppy, is an addictive narcotic plant, the source of heroin (which has shaped human history) and very important medicinally, e.g., as the source of the analgesic morphine and other alkaloids. The Papaveraceae are now treated as 3 subfamilies: Pteridophylloideae, Fumarioideae, and Papaveroideae. Members of the Fumarioideae (formerly Fumariaceae) differ in having biradial or bilateral flowers, the outer whorl of petals usually with a spur or sac. See Hoot et al. (1997), Kadereit et al. (1994, 1995), and Wang, W. et al. (2009) for detailed phylogenetic studies of the family.

The Papaveraceae are distinctive in being herbs, shrubs, or small trees (some with milky sap), with a ***dichlamydeous, triseriate*** perianth (the corolla ***biseriate***), usually ***numerous*** stamens, and a superior, compound ovary having ***parietal*** placentation, the fruit usually a ***loculicidal or poricidal*** capsule.

K 2 [3] or (2 [3]) **C** 2+2 or 3+3 [–16] **A** ∞ [4–6] **G** (2) [–several], superior.

Ranunculaceae — Buttercup family (type *Ranunculus*, meaning "little frog," after the amphibious habit of many species). ca. 62 genera/ca. 2,500 species (Figure 8.5).

The Ranunculaceae consist of terrestrial or aquatic, perennial or annual shrubs, herbs, or lianas. The **leaves** are spiral, simple to compound, stipulate or exstipulate. The **inflorescence** is a cyme or a solitary flower. The **flowers** are bisexual, rarely unisexual, actinomorphic or zygomorphic, hypogynous; the receptacle is somewhat elongate. The **perianth** is dichlamydeous, a hypanthium absent. The **calyx** is aposepalous with 5–8 [3], often petaloid sepals, sometimes spurred or cucullate. The **corolla** is apopetalous with few–∞ [rarely 0] petals, sometimes spurred. The **stamens** are apostemonous, usually ∞, spiral and usually centripetal, sometimes multiwhorled. **Anthers** are longitudinal in dehiscence, tetrasporangiate, and dithecal. The **gynoecium** is apocarpous [rarely syncarpous], usually of ∞ [1–few] pistils/carpels, each unilocular, ovaries superior. **Placentation** is marginal, apical, or basal, axile in the syncarpous taxa; **ovules** are anatropous or hemitropous, bitegmic or unitegmic, several–∞ per carpel. **Nectaries** are often perigonal (at the base of staminode-like petals). The **fruit** is an aggregate of follicles, achenes, or berries. Flowers are insect- or wind-pollinated.

Members of the Ranunculaceae have distributions mainly in temperate and boreal regions. Economic importance includes cultivated ornamentals, medicinal plants (such as *Hydrastis canadensis*, goldenseal), poisonous plants, and weeds. See Hoot (1991, 1995) and Wang, W. et al. (2009) for phylogenetic analyses of the family.

The Ranunculaceae are distinctive in being herbs, shrubs, or lianas having flowers with ***spirally arranged*** perianth parts, usually in two series, ***numerous*** stamens, and an ***apocarpous*** gynoecium.

K 5–8 [3] **C** few–∞ [0] **A** ∞ **G** ∞ [1–few], superior.

PROTEALES

The Proteales contain four very different families, classified together only rather recently. All but the Sabiaceae are similar in having mostly 1 ovule per ovary. The **Nelumbonaceae** and **Platanaceae** are similar in having apical placentation. The **Platanaceae** and **Proteaceae** have similarities in wood anatomy. See Sun et al. (2016) for a phylogenetic analysis including relationships in the order.

Nelumbonaceae — Water Lotus family (type *Nelumbo*, a Sinhalese name). 1 genus (*Nelumbo*)/1–2 species (Figure 8.6).

The Nelumbonaceae consist of aquatic, perennial herbs, with milky latex present from articulated laticifers. The **stems** are rhizomatous. The **leaves** are spiral, peltate, petiolate (petiole emergent), simple, undivided, orbicular, concave above, and net-veined. The **inflorescence** is an axillary (from a scale leaf of the rhizome), emergent, solitary flower. The **flowers** are large, bisexual, actinomorphic, long-pedunculate, ebracteate, hypogynous; the receptacle is enlarged and spongy, with numerous sunken cavities containing individual pistils. The **perianth** is approximately 3–seriate, with an outermost whorl of 2 green, sepaloid tepals, and two inner whorls of numerous, yellow or red, petaloid tepals, all tepals distinct. The **stamens** are numerous, spiral, apostemonous; filaments are narrow. **Anthers** are latrorse to introrse and longitudinal in dehiscence, tetrasporangiate, with a laminar connective. The **gynoecium** is apocarpous, with 12–40 one-loculed superior ovaries. The **style** and **stigma** are solitary and terminal. **Placentation** is apical; **ovules** are anatropous and bitegmic, 1 per ovary. The **fruit** is an aggregate of nuts, each sunken in an accrescent receptacle. The **seeds** are exalbuminous. Flowers are beetle-pollinated. The stems have a eustele or atactostele-like vasculature.

The Nelumbonaceae were formerly treated as a subfamily (Nelumboideae) of the Nymphaeaceae. The two species that make up the Nelumbonaceae are distributed from eastern North America to northern South America (*N. lutea*) and Asia to northern Australia (*N. nucifera*). Economic importance includes use as cultivated ornamentals, edible rhizomes and seeds, medicine, and in religious rites (the sacred lotus); *N. nucifera* is famous for having long-lived seeds, some discovered 3000 years old. See Williamson and Schneider (1993) for more detailed information.

The Nelumbonaceae are distinctive in being ***aquatic herbs*** with often atactostelic stems, ***emergent concave-peltate*** leaves, and ***emergent, solitary*** flowers with numerous tepals, numerous stamens, and an ***apocarpous*** gynoecium having pistils ***partially embedded within an expanded receptacle***; the fruit is an ***aggregate of nuts*** within an ***accrescent receptacle***.

P 2+∞+∞ **A** ∞ **G** ∞, superior.

Platanaceae — Plane Tree or Sycamore family (type *Platanus*, Greek for broad, referring to the leaf). 1 genus (*Platanus*)/ca. 10 species (Figure 8.7).

The Platanaceae consist of monoecious trees with exfoliating bark. The **leaves** are alternate (the petiole enclosing an infrapetiolar bud), simple, palmately lobed, stipulate (stipules usually encircling twig), deciduous, with stellate trichomes, and usually palmately netted venation. The **inflorescence** is a terminal, pendant spike of unisexual heads. The **flowers** are small, unisexual. The **perianth** is biseriate, hypanthium absent. The **calyx** is aposepalous to basally synsepalous, of 3–4 [7] sepals. The **corolla** is apopetalous, of 3–4 [7]

FIGURE 8.5 RANUNCULALES. Ranunculacaeae. **A,B.** *Caltha leptosepala*, white marsh-marigold, showing numerous stamens and pistils. **C.** *Aconitum columbianum*, monkshood, with cucullate (hooded) calyx. **D.** *Aquilegia canadensis*, columbine, corolla with prominent spurs. **E,F.** *Clematis pauciflora*, virgin's bower. **G,H.** *Hepatica americana*, liverleaf. **I,J.** *Delphinium* sp., larkspur. **I.** Flower, showing calyx spur. **J.** Perianth opened to show numerous stamens and pistils. **K.** *Ranunculus californica*, buttercup. **L,M.** *Thalictrum fendleri*, meadow-rue. **L.** Male flowers, pendant. **M.** Female flowers, erect, each with several pistils. **N.** *Myosurus minimus*, mouse tail, with elongate receptacle. **O.** *Xanthorhiza simplicissima*, yellow root.

FIGURE 8.6 PROTEALES. Nelumbonaceae. *Nelumbo nucifera*, Indian/Chinese lotus. **A.** Whole plants, showing emergent, peltate/concave leaves and flowers. **B.** Whole flower, showing numerous tepals. **C.** Flower close-up; note numerous stamens and enlarged receptacle. **D.** Receptacle longitudinal section, with sunken pistils at apex. **E.** Close-up of pistils in longitudinal section, showing apical placentation. **F.** Fruit, an aggregate of nuts, sunken in accrescent receptacle. **G.** Close-up of fruit, longitudinal section, showing nuts.

(usually 0 in females) petals. The **stamens** are 3–4 [7], antisepalous, apostemonous, staminodes present in some female flowers. **Anthers** are sessile, longitudinal in dehiscence, the connective with an apical, peltate appendage. The **gynoecium** is apocarpous, with a superior ovary and 5–8 [3–9] carpels and locules. **Placentation** is apical; **ovules** are orthotropous, bitegmic, 1 [2] per ovary. **Nectaries** are absent. The **fruit** is a multiple of achenes, with an accrescent, bristly perianth (functioning in wind dispersal) and persistent style. The **seeds** are endospermous. Flowers are wind-pollinated.

Members of the Platanaceae are distributed in the northern hemisphere. Economic importance includes timber and use as cultivated ornamentals. See Kubitzki et al. (1993b) for a treatment of the family and Grimm and Denk (2008) for a phylogenetic analysis.

The Platanaceae are distinctive in being ***monoecious trees*** with ***encircling stipules, infrapetiolar buds,*** usually ***palmately lobed and veined leaves***, and a ***pendant spike of heads*** bearing ***unisexual*** flowers, with a ***multiple fruit of bristly achenes.***

Male: K 3–4 [7] or (3–4 [7]) **C** 3–4 [7] **A** 3–4 [7].
Female: K 3–4 [7] or (3–4 [7]) **C** 0 **G** 5–8 [3–9], superior.

Proteaceae — Protea family (type *Protea*, after Proteus, the sea god, for his versatility in changing form). Ca. 80 genera/ca. 1800 species (Figure 8.8).

The Proteaceae consist of shrubs and trees. The **roots** are without mycorrhizae, often with short, lateral "proteoid" roots. The **leaves** are usually spiral and simple, pinnate or bipinnate, evergreen, and coriaceous. The **inflorescence** is a bracteate raceme, umbel, involucrate head, or of solitary or paired flowers. The **flowers** are bisexual or unisexual, actinomorphic or zygomorphic. The **perianth** is uniseriate. The **calyx** is valvate, consisting of 4 distinct or connate sepals. The **corolla** is absent (or interpreted as modified into 4-lobed nectariferous disk or minute scales). The **stamens** are 4, antisepalous, usually episepalous. **Anthers** are longitudinal in dehiscence, tetrasporangiate or bisporangiate, dithecal or monothecal, with thecae and connective often extended as an appendage. The **gynoecium** is unicarpellous, with a superior ovary and 1 locule. **Placentation** is marginal or appearing basal; **ovules** are variable in type, bitegmic, 1–2 [–∞] per ovary. The **fruit** is a follicle, nut, achene, or drupe. **Seeds** are exalbuminous.

Members of the Proteaceae occur in rain forest to xeric habitats with distributions in tropical and subtropical regions,

FIGURE 8.7 PROTEALES. Platanaceae, *Platanus racemosa*. **A.** Tree, early in spring. **B.** Inflorescence (maturing to fruit), a spike of heads. **C.** Close-up of inflorescence. **D.** Close-up of head. **E.** Male head, longitudinal section. **F.** Close-up of female head, showing pistils and perianth of ill-defined flowers. **G.** Stamen of male flower. Note apical extension of connective. **H.** Female head, longitudinal section. Note distinct pistils (apocarpous gynoecium) and bristlelike perianth. **I.** Ovary, longitudinal section, showing single, apical ovule. **J.** Fruit, a multiple of achenes. Note persistent styles. **K.** Individual achenes, with accrescent, bristly perianth.

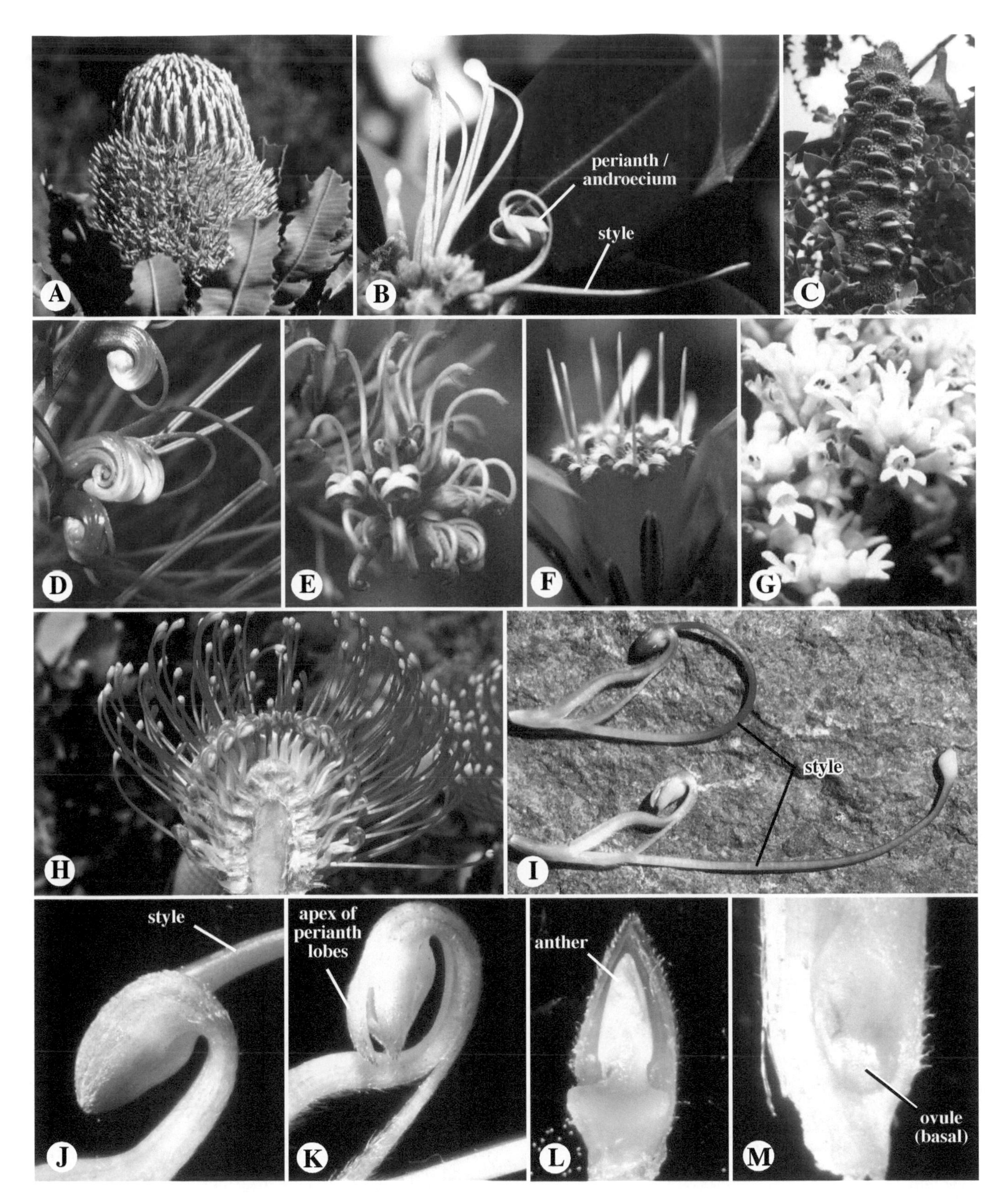

FIGURE 8.8 PROTEALES. Proteaceae. **A–C.** *Banksia* spp. **A.** *Banksia menziesii*, headlike inflorescence. **B.** *Banksia* sp., showing partially opened (above) and open (below) flowers. **C.** *Banksia grandis*, multiple fruit of follicles. **D.** *Grevillea longistyla*, flowers. **E.** *Grevillea sericea*, inflorescence. **F.** *Lambertia formosa*, inflorescence. **G.** *Conospermum taxifolium*, flowers. **H–M.** *Leucospermum* sp., pincushions. **H.** Headlike inflorescence in longitudinal section. **I.** Flowers removed, before (above) and after (below) anthesis. **J.** Close-up of flower tip before anthesis, showing four coherent perianth lobe apices enclosing style tip. **K.** Flower tip, after release of style and stigma. **L.** Close-up of perianth lobe apex, adaxial surface, showing sessile anther. **M.** Ovary longitudinal section, showing single, basal ovule.

FIGURE 8.9 GUNNERALES. Gunneraceae, *Gunnera manicata*. **A.** Whole plant, in cultivation, showing large size (one of the largest herbs). **B.** Petiolate, peltate leaf, seen from underneath, showing palmate venation. **C.** Plant base. **D.** Inflorescence, panicle-like. **E.** Inflorescence apex, showing floral bracts. **F.** Bisexual or male flowers; note stamens with short filaments. **G.** Maturing female flowers, showing small, vestigial sepals and inferior ovaries. **H.** Infructescence of drupes.

especially Australia and South Africa. Economic importance includes cultivated ornamentals (e.g., *Banksia*, *Grevillea*, *Protea*), important timber trees, and species with edible seeds (e.g., *Macadamia*). See Hoot and Douglas (1998) and Weston and Barker (2006) for phylogenetic studies.

The Proteaceae are distinctive in having flowers with a ***uniseriate perianth*** of ***4 sepals***, ***4 antisepalous stamens***, and a ***unicarpellous***, superior ovary.

K 4 or (4) **C** 0 **A** 4 **G** 1, superior.

GUNNERALES

The Gunnerales (sensu Stevens 2001 onwards) contain two families, **Gunneraceae** and **Myrothamnaceae**; only the former is described here.

Gunneraceae — Gunnera family (type *Gunnera*, after Johan Ernst Gunnerus, 1718–1773, Norwegian botanist). 1 genus (*Gunnera*)/ca. 60 species (Figure 8.9).

The Gunneraceae consist of terrestrial, perennial [annual in *G. herteri*], polygamous or dioecious herbs, most quite large, with channels in stems and adventitious roots housing symbiotic cyanobacteria *Nostoc* colonies. The **stems** are usually rhizomatous, pachycaulous in some, anatomy polystelic. The **leaves** are small to extremely large (up to 3 m in diameter), simple, palmately veined, spiral, petiolate, often peltate, orbicular to ovate, with colleters and scales in or near leaf axiles. The **inflorescence** is an axillary or pseudo-terminal raceme, compound raceme, or spike. The **flowers** are small, unisexual or bisexual, bracteate or not, epigynous. The **perianth** is biseriate. The **calyx** has usu. 2 valvate sepals, these sometimes vestigial. The **corolla** has 2 [0] petals, often cauducous. The **stamens** are 2 [1], with short filaments. **Anthers** are small, dithecal, tetrasporangiate, longitudinally dehiscent. Pistillodes may be present in male flowers. The **gynoecium** is syncarpous, with an inferior ovary, 2 carpels, and 1 locule. The **styles** are 2, terminal. **Placentation** is apical; **ovules** are anatropous, bitegmic, pendulous, 1 [2] per ovary. The **fruit** is a drupe. The

FIGURE 8.10 DILLENIALES. Dilleniaceae. **A.** *Hibbertia scandens*. **B.** *Hibbertia* sp. Both showing actinomorphic flower with numerous, centrifugal stamens and numerous pistils.

seed is solitary, oily endospermous.

The Gunneraceae are distributed mostly in tropical regions of the southern hemisphere. Economic importance includes some cultivated ornamentals (e.g., *Gunnera manicata*), soil -improving crop pants (from the nitrogen-fixing, symbiotic cyanobacteria), and local medicinal, tanning, or edible plants. See Wanntorp and Wanntorp (2003) and Wanntorp (2006) for reviews of the morphology, phylogeny, and biogeography of the family. Interestingly, the small, annual *Gunnera herteri* is sister to all other species, which are large-leaved perennials.

The Gunneraceae are distinctive in being ***terrestrial herbs***, the stems and adventitious roots containing the symbiotic ***cyanobacterium*** **Nostoc**, leaves ***often very large***, spiral, orbicular to ovate***, with palmate venation***, the inflorescence of ***unisexual or bisexual*** flowers with usu. ***2 sepals, 2 [0], often caducous petals, 2 [1] anthers***, and an ***inferior, 2-carpellate*** ovary with a gen. ***single, apical ovule***, the fruit a 1-seeded ***drupe***.
K 2 **C 2 [0] A** 2 [1] **G** 2, inferior.

DILLENIALES

The Dilleniales contain the single pantropical/Australasian family **Dilleniaceae**, with simple, spiral, conduplicate leaves (secondary veins typically terminating at teeth) and actinomorphic flowers with ***numerous centrifugal stamens***, often in ***bundles***, the gynoecium ***apocarpous*** [rarely syncarpous] (Figure 8.10). The phylogenetic relationship of the Dilleniales varies in different Eudicot analyses, some placing it with the Superrosids, others with the Superasterids; see Figure 8.1.

SUPERROSIDS

The so-called Superrosids or *Superrosidae* (after Soltis et al., 2011) is a clade that includes the orders Saxifragales, Vitales (this sometimes placed in the Rosids), plus a large group, the Rosids (see Figure 8.1, Table 8.2).

SAXIFRAGALES

The Saxifragales include 15 families (Table 8.2), of which three are described here. Notable among the families not treated are the **Altingiaceae** (including *Liquidambar*, sweetgum; Figure 8.11A–C), **Cercidiphyllaceae** (only 2 species, used as timber trees, much more widespread in the past), **Grossulariaceae** (including *Ribes*, the currants and gooseberries, with edible fruits; Figure 8.11F–J), **Haloragaceae** (including aquarium aquatics such as *Myriophyllum*), and **Paeoniaceae** (peonies; Figure 8.11D,E). See Soltis and Soltis (1997), Qiu et al. (1998), Fishbein et al. (2001), Jian et al. (2008), and Soltis et al. (2016) for information on the order and relatives.

Crassulaceae — Stonecrop family (type *Crassula*, meaning “thick or succulent little plant”). Ca. 34 genera/ca. 1,400 species (Figure 8.12).

The Crassulaceae consist of herbs, shrubs, or rarely trees. The **leaves** are spiral, opposite, or whorled, simple, exstipulate, and characteristically succulent. The **inflorescence** is a branched cyme or of solitary flowers. The **flowers** are usually bisexual, actinomorphic, pedicellate or sessile, hypogynous or slightly perigynous. The **perianth** is biseriate and

TABLE 8.2 Orders and families of the Superrosids and (nested within that group) the Rosids, after APG IV (2016). Families in **bold** are described in detail. An asterisk indicates a classification deviating from APGIV (2016), following Zhand and Simmons (2006) for Celastrales and Jeiter et al. (2016) for Geraniales. See Table 8.3 for listing of orders and families in the Superasterids and Asterids.

SUPERROSIDS & ROSIDS

- SAXIFRAGALES
  - Altingiaceae
  - Aphanopetalaceae
  - Cercidiphyllaceae
  - **Crassulaceae**
  - Cynomoriaceae
  - Daphniphyllaceae
  - Grossulariaceae
  - Haloragaceae
  - **Hamamelidaceae**
  - Iteaceae
  - Paeoniaceae
  - Penthoraceae
  - Peridiscaceae
  - **Saxifragaceae**
  - Tetracarpaeaceae
- VITALES (within Rosids; APG IV)
  - **Vitaceae**

ROSIDS

- FABIDS
  - ZYGOPHYLLALES
    - **Krameriaceae**
    - **Zygophyllaceae**
  - CELASTRALES
    - Celastraceae
    - Lepidobotryaceae
    - Parnassiaceae*
  - MALPIGHIALES
    - Achariaceae
    - Balanopaceae
    - Bonnetiaceae
    - Callophyllaceae
    - Caryocaraceae
    - Centroplacaceae
    - Chrysobalanaceae
    - Clusiaceae (Guttiferae)
    - Ctenolophonaceae
    - Dichapetalaceae
    - Elatinaceae
    - Erythroxylaceae
    - **Euphorbiaceae**
    - Euphroniaceae
    - Goupiaceae
    - Humiriaceae
    - **Hypericaceae**
    - Irvingiaceae
    - Ixonanthaceae
    - Lacistemataceae
    - Linaceae
    - Lophopyxidaceae
    - **Malpighiaceae**
    - Ochnaceae
    - Pandaceae
    - **Passifloraceae**
    - Peraceae
    - Phyllanthaceae
    - Picrodendraceae
    - Podostemaceae
    - Putranjivaceae
    - Rafflesiaceae
    - Rhizophoraceae
    - **Salicaceae**
    - Trigoniaceae
    - **Violaceae**
  - OXALIDALES
    - Brunelliaceae
    - Cephalotaceae
    - Connaraceae
    - Cunoniaceae
    - Elaeocarpaceae
    - Huaceae
    - **Oxalidaceae**
  - FABALES
    - **Fabaceae (Leguminosae)**
    - **Polygalaceae**
    - Quillajaceae
    - Surianaceae
  - ROSALES
    - Barbeyaceae
    - Cannabaceae
    - Dirachmaceae
    - Elaeagnaceae
    - **Moraceae**
    - **Rhamnaceae**
    - **Rosaceae**
    - **Ulmaceae**
    - **Urticaceae**
  - CUCURBITALES
    - Apodanthaceae
    - Anisophylleaceae
    - **Begoniaceae**
    - Coriariaceae
    - Corynocarpaceae
    - **Cucurbitaceae**
    - Datiscaceae
    - Tetramelaceae
  - FAGALES
    - **Betulaceae**
    - Casuarinaceae
    - **Fagaceae**
    - **Juglandaceae**
    - Myricaceae
    - Nothofagaceae
    - Ticodendraceae
- MALVIDS
  - CROSSOSOMATALES
    - Aphloiaceae
    - Crossosomataceae
    - Geissolomataceae
    - Guamatelaceae
    - Stachyuraceae
    - Staphyleaceae
    - Strasburgeriaceae
  - GERANIALES
    - Francoaceae
    - **Geraniaceae**
    - Hypseocharitaceae*
    - Melianthaceae*
    - Vivianiaceae*
  - MYRTALES
    - Alzateaceae
    - Combretaceae
    - Cryteroniaceae
    - **Lythraceae**
    - **Melastomataceae**
    - **Myrtaceae**
    - **Onagraceae**
    - Penaeaceae
    - Vochysiaceae
  - PICRAMNIALES
    - Picramniaceae
  - HUERTEALES
    - Dipentodontaceae
    - Gerrardinaceae
    - Tapisciaceae
  - BRASSICALES
    - Akaniaceae
    - Bataceae
    - **Brassicaceae (Cruciferae)**
    - Capparaceae
    - Caricaceae
    - Cleomaceae
    - Emblingiaceae
    - Gyrostemonaceae
    - Koeberliniaceae
    - Limnanthaceae
    - Moringaceae
    - Pentadiplandraceae
    - Resedaceae
    - Salvadoraceae
    - Setchellanthaceae
    - Tovariaceae
    - **Tropaeolaceae**
  - MALVALES
    - Bixaceae
    - Cistaceae
    - Cytinaceae
    - Dipterocarpaceae
    - **Malvaceae**
    - Mutingiaceae
    - Neuradaceae
    - Sarcolaenaceae
    - Sphaerosepalaceae
    - Thymelaeaceae
  - SAPINDALES
    - **Anacardiaceae**
    - Biebersteiniaceae
    - Burseraceae
    - Kirkiaceae
    - Meliaceae
    - Nitrariaceae
    - **Rutaceae**
    - **Sapindaceae**
    - Simaroubaceae

FIGURE 8.11 SAXIFRAGALES. **A–C.** Altingiaceae, *Liquidambar styraciflua*, sweetgum. **A.** Shoot with leaves, inflorescences, and last year's fruit (multiple fruit of capsules). **B.** Pendant female inflorescence. **C.** Erect male inflorescence. **D–E.** Paeoniaceae, *Paeonia californica*, peony. **F–J.** Grossulariaceae. **F.** *Ribes indecorum*, white-flowered currant; inflorescence. **G.** *Ribes montigenum*, mountain gooseberry. **H.** *Ribes nevadense*, mountain pink currant. **I–J.** *Ribes speciosum*, fuchsia-flower gooseberry. Note inferior ovary and hypanthium.

dichlamydeous. The **calyx** is usually aposepalous, with 5 [rarely 3–6+] sepals. The **corolla** is apopetalous to basally sympetalous, with 5 [rarely 3–6+] petals. The **stamens** are 1–2× the sepal or petal number, biseriate and obdiplostemonous or uniseriate and antisepalous, free or epipetalous. **Anthers** are longitudinal in dehiscence, tetrasporangiate, dithecal. The **gynoecium** is apocarpous, with a superior ovary and 5 [3–6 or more] carpels. **Placentation** is marginal; **ovules** are anatropous, bitegmic, and numerous [rarely few or 1]. **Nectaries** are present, consisting of scale-like structures at the base of and opposite the carpels. The **fruit** is a follicetum, rarely a capsule. The **seeds** are endospermous (oily and proteinaceous). The stem xylem is usually in a continuous cylinder; leaves often have Kranz anatomy, with Crassulacean Acid Metabolism (CAM) photosynthesis (see Chapter 10).

The Crassulaceae is traditionally treated in six subfamilies, but recent studies discount the monophyly of most of these. Members of the family often grow in arid environments and also occur in mesic or moist habitats with distributions worldwide, except Australia and Pacific islands; species are most diverse in S. Africa and mtns. of Mexico and Asia. Economic importance includes cultivated ornamentals, especially *Aeonium*, *Crassula*, *Echeveria*, *Kalanchöe*, and *Sedum*. See Thiede and Eggli (2006) for general information and Ham and 't Hart (1998) and Mort et al. (2001, 2010) for recent phylogenetic analyses and summaries.

The Crassulaceae are distinctive in being herbs, shrubs, or rarely trees, with simple, ***succulent leaves*** having ***CAM photosynthesis***, a cymose inflorescence with bisexual, actinomorphic, dichlamydeous flowers, obdiplostemonous or uniseriate stamens, and an ***apocarpous gynoecium*** with opposed, ***scale-like nectaries***, the fruit a ***follicetum***.

K 5 [3–6+] **C** 5 [3–6+] **A** 5+5 or 5 [3–6+] **G** 5 [3–6+], superior.

Hamamelidaceae — Witch-Hazel family (type *Hamamelis*, Greek for a plant with pyriform fruits). Ca. 27 genera/ca. 82 species (Figure 8.13).

The Hamamelidaceae consist of trees or shrubs, often with stellate trichomes. The **leaves** are simple, unlobed to palmately lobed, spiral [rarely distichous], usually stipulate. The **inflorescence** is a usually a spike or head. The **flowers** are unisexual or bisexual. The **perianth** is dichlamydeous . The **calyx** is aposepalous or synsepalous with 4–5 [0–7] sepals/lobes. The

FIGURE 8.12 SAXIFRAGALES. Crassulaceae. **A–D.** Vegetative morphology. **A.** *Kalanchöe beharensis*, large shrub with thick, tomentose leaves. **B.** *Graptopetalum paraguayense*, shrub with succulent, glaucous leaves. **C.** *Crassula argentea*, shrub with succulent, glabrous leaves. **D.** *Sedum smallii*, small annual, with minute, succulent leaves. **E–J.** Reproductive morphology. **E.** *Dudleya pulverulenta*, showing branched, scorpioid cyme; note bracts displaced ca. 90° from pedicel. **F.** *Bryophyllum pinnatum*, with pendulous, tubular flowers on large, paniculate inflorescence. **G.** *Crassula argentea*. Note flower with apocarpous gynoecium and scalelike nectaries at base of and opposite carpels. **H,I.** *Graptopetalum paraguayense*. **H.** View of biseriate, obdiplostemonous androecium. **I.** Carpel (ovary) cross section, showing marginal placentation. **J.** *Pachyphytum hookeri*, mature follicetum fruit.

corolla is apopetalous with 4–5 [0] usu. narrow petals. The **stamens** are 4–5 [1–24], alternipetalous. **Anthers** are valvular or longitudinal in dehiscence, with connective often extended beyond thecae. The **gynoecium** is syncarpous (at least basally), with a superior to inferior ovary, 2–3 carpels, and 2–3 locules. The **styles** are 2–3. **Placentation** is axile; **ovules** are anatropous or orthotropous, bitegmic, 1–∞, apical. The **fruit** is a woody capsule. The **seeds** are endospermous, often with a hard, shiny seed coat. Plants are wind or insect pollinated.

Members of the Hamamelidaceae are distributed from eastern North America to northern South America, southeast Africa, and eastern Asia. Economic importance includes extraction of essential oils used for scent and medicinally, timber trees, and ornamental cultivars. *Hamamelis virginiana*, witch-hazel, is the source of the similarly named medicinal astringent. The Altinginaceae (e.g., *Liquidambar*) is often treated as part of the family. See Li and Bogle (2001) and Li (2008) for phylogenetic analyses of the family.

The Hamamelidaceae are distinctive in being ***trees or shrubs***, with ***stellate trichomes*** and simple, ***stipulate***, usu. spiral leaves, the inflorescence usu. a ***spike or head***, flowers ***unisexual or bisexual***, dichlamydeous with usu. 4–5 sepals and petals, the ovary superior to inferior, 2–3-carpellate, the fruit a ***woody capsule***, seeds with a hard seed coat.
K 4–5 [0–7] **C** 4–5 [0] **A** 4–5 [1–24] **G** (2–3), superior to inferior.

Saxifragaceae — Saxifrage family (type *Saxifraga*, Latin for "rock breaking"). Ca. 33 genera/600 species (Figure 8.14).

The Saxifragaceae consist of perennial herbs or subshrubs. The **leaves** are usually spiral, often rosulate, simple, pinnate, or palmate, usually exstipulate, sometimes succulent. The **inflorescence** is a cyme, raceme, or of solitary flowers. The **flowers** are bisexual, actinomorphic, perigynous to epigynous. The **perianth** is biseriate, dichlamydeous, with hypanthium. The **calyx** is aposepalous with 5 [3–10] sepals; the **corolla** is apopetalous with 5 [0, 3–10] corolla lobes, these sometimes lobed. The **stamens** are variable in number, uniseriate or biseriate, with staminodes present in some. The **gynoecium** is syncarpous and often lobed, rarely apocarpous, with a superior to inferior ovary, and 2–4 [7] carpels. **Placentation** is marginal, axile, or parietal; **ovules** are anatropous, uni- or bitegmic, usually numerous. A **nectariferous** ovary disk is often present. The **fruit** is a (usually) septicidal capsule. **Seeds** are (oily) endospermous.

The Saxifragaceae are widely distributed, especially in northern temperate and cold regions. Economic importance is primarily as cultivated ornamentals. See Soltis and Soltis (1997), Soltis et al. (2001), Xiang et al. (2012), and Deng et al. (2015) for phylogenetic analyses of the family.

The Saxifragaceae are distinctive in being ***perennial herbs, rarely subshrubs***, with spiral, sometimes succulent leaves (often in rosettes), flowers usually with 5 [3–10] distinct sepals and petals [0], and 1–2 whorls of stamens, the ***gynoecium usually syncarpous and lobed***, the ***ovary superior to inferior***, with numerous ovules, fruit a ***septicidal capsule***.
K 5 [3–10] **C** 5 [0, 3–10] **A** 5 or 5+5 [variable] **G** (2–4) [(–7)], superior to inferior, hypanthium present.

VITALES

The Vitales contain only the family **Vitaceae**, described here. The APG IV (2016) system tentatively places the Vitales/Vitaceae within the Rosids.

Vitaceae [incl. Leeaceae] — Grape family (type *Vitis*, Latin name for cultivated grape). Ca. 17 genera/ca. 955 species (Figure 8.15).

The Vitaceae consist of lianas, herbs, or trees, some succulent or pachycaulous. The **stems** often have tendrils opposite the leaves. The **leaves** are simple (these often palmately lobed and veined), palmate, or pinnate, spiral, opposite, or distichous, stipulate (stipules caducous). The **inflorescence** is a cyme, corymb, or panicle, terminal from shoots opposite leaves [rarely axillary]. The **flowers** are small, bisexual or unisexual, actinomorphic. The **perianth** is biseriate. The **calyx** is aposepalous, reduced to a collar in some, with 4–5 [3–7] sepals. The **corolla** is usu. apopetalous, forming a dehiscent calyptra in *Vitis*, with 4–5 [3–7], valvate petals. The **stamens** are 4–5 [3–7], antipetalous. The **gynoecium** is syncarpous, with a superior [half-inferior] ovary, 2 [3–4] carpels, and gen. 2 locules (some taxa with false partitions). The **style** is solitary. **Placentation** is +/– axile; **ovules** are anatropous, bitegmic, 2 per carpel. **Nectaries** are 5-lobed and alternate with the stamens or discoid around ovary base. The **fruit** is a berry. The **seeds** are endospermous.

The Vitaceae are distributed from eastern North America to South America, from central to southern Africa, and in southeast Asia/Australasia. Economic importance is predominantly grapes (the cultivated *Vitis vinifera* and other native species) used as food and to make wine, plus some cultivated ornamentals, e.g., *Cissus*, *Parthenocissus*, and *Rhoicissus*. The Vitaceae can be classified into two subfamilies: Leeoideae [Leeaceae] and Vitoideae. See Wen (2006) and Wen et al. (2018) for information on phylogeny and classification of the Vitaceae.

The Vitaceae are distinctive in being ***lianas*** (rarely herbs or pachycauls) with ***tendrils opposite leaves***, leaves simple, palmate, or pinnate, flowers small, with a ***valvate apopetalous or calyptrate*** corolla, ***antipetalous*** stamens, and ***lobed or discoid nectary*** at ovary base, the fruit a berry, 1–2 seeds per locule.
K 4–5 [3–7] **C** 4–5 [3–7] **A** 4–5 [3–7] **G** 2 [3–4], superior [rarely half-inferior].

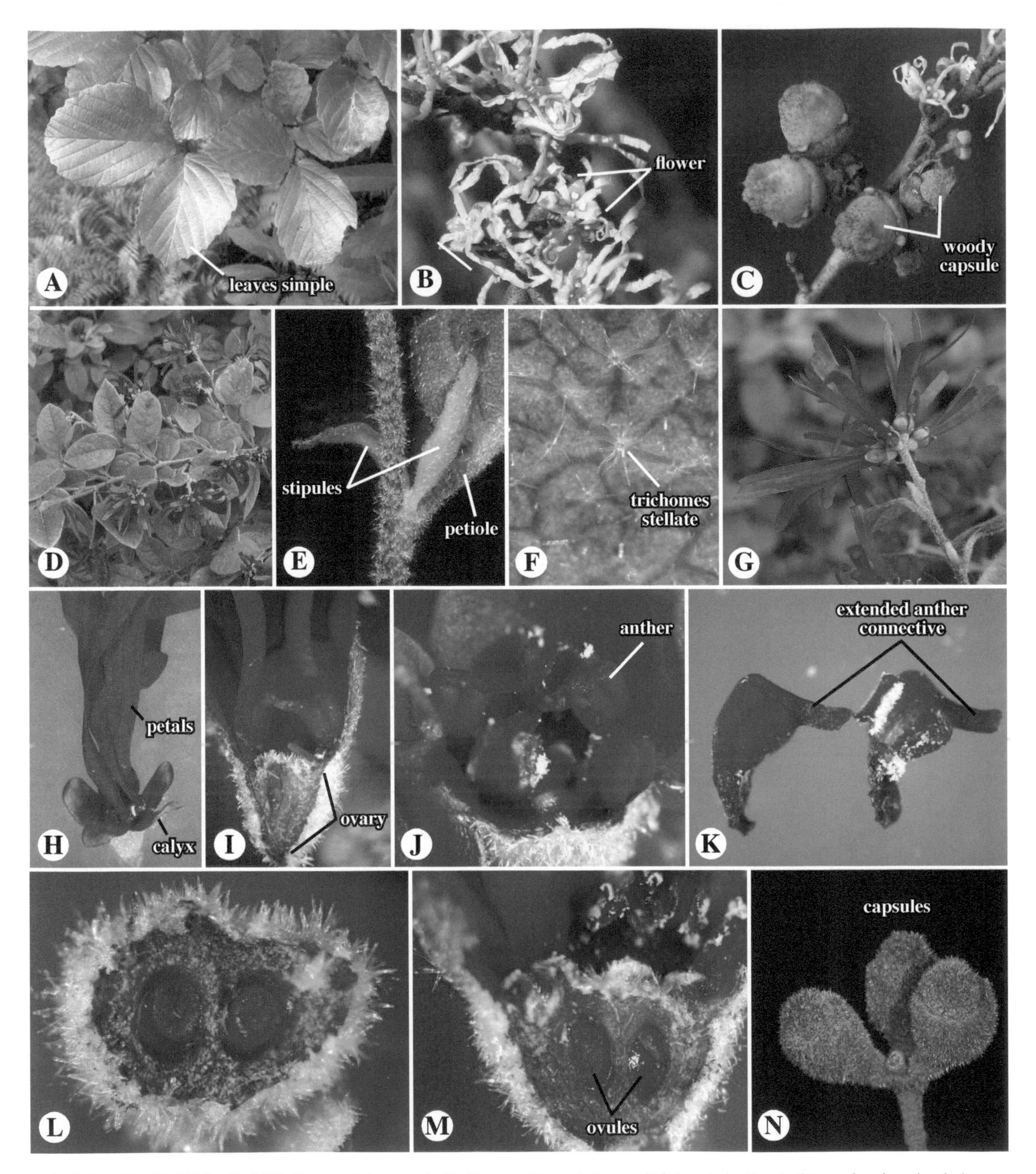

FIGURE 8.13 SAXIFRAGALES. Hamamelidaceae. **A–C.** *Hamamelis virginiana*, witch-hazel. **A.** Shoot of tree, showing simple leaves. **B.** Inflorescence. **C.** Fruits, woody, dehiscent capsules. **D.** *Loropetalum chinense*. **D.** Shoot of shrub. **E.** Close-up of leaf node, showing stipules. **F.** Leaf abaxial surface, showing stellate trichomes. **G.** Inflorescence, a head-like cluster of flowers. **H.** Flower, showing calyx and elongate petals of corolla. **I.** Flower longitudinal section, showing mostly inferior ovary. **J.** Close-up of flower center, showing five, alternipetalous stamens. **K.** Stamens, with short filaments and connectives extending from anther apex. **L.** Ovary cross section, showing two carpels and locules. **M.** Ovary longitudinal section; placentation axile. **N.** Fruits, capsules, prior to dehiscence. [Images at **A–C** courtesy of Will Cook.]

FIGURE 8.14 SAXIFRAGALES. Saxifragaceae. **A,B.** *Heuchera hirsutissima*, a perennial herb, with a basal rosette of leaves. Inset (above): flower longitudinal section, showing half-inferior ovary. **C.** *Heuchera sanguinea*, showing inferior ovary and hypanthium. **D.** *Jepsonia aphylla*, inflorescence. **E.** *Lithophragma affine*, woodland star, showing dichlamydeous perianth. **F.** *Mitella diphylla*, with dissected petals. **G.** *Saxifraga bronchialis*, flower close-up. **H.** *Micranthes californica*, flower close-up, with half-inferior ovary. **I.** *Tellima grandiflora*, flower close-up, showing lobed petals. **J,K.** *Tiarella cordifolia*, foamflower.

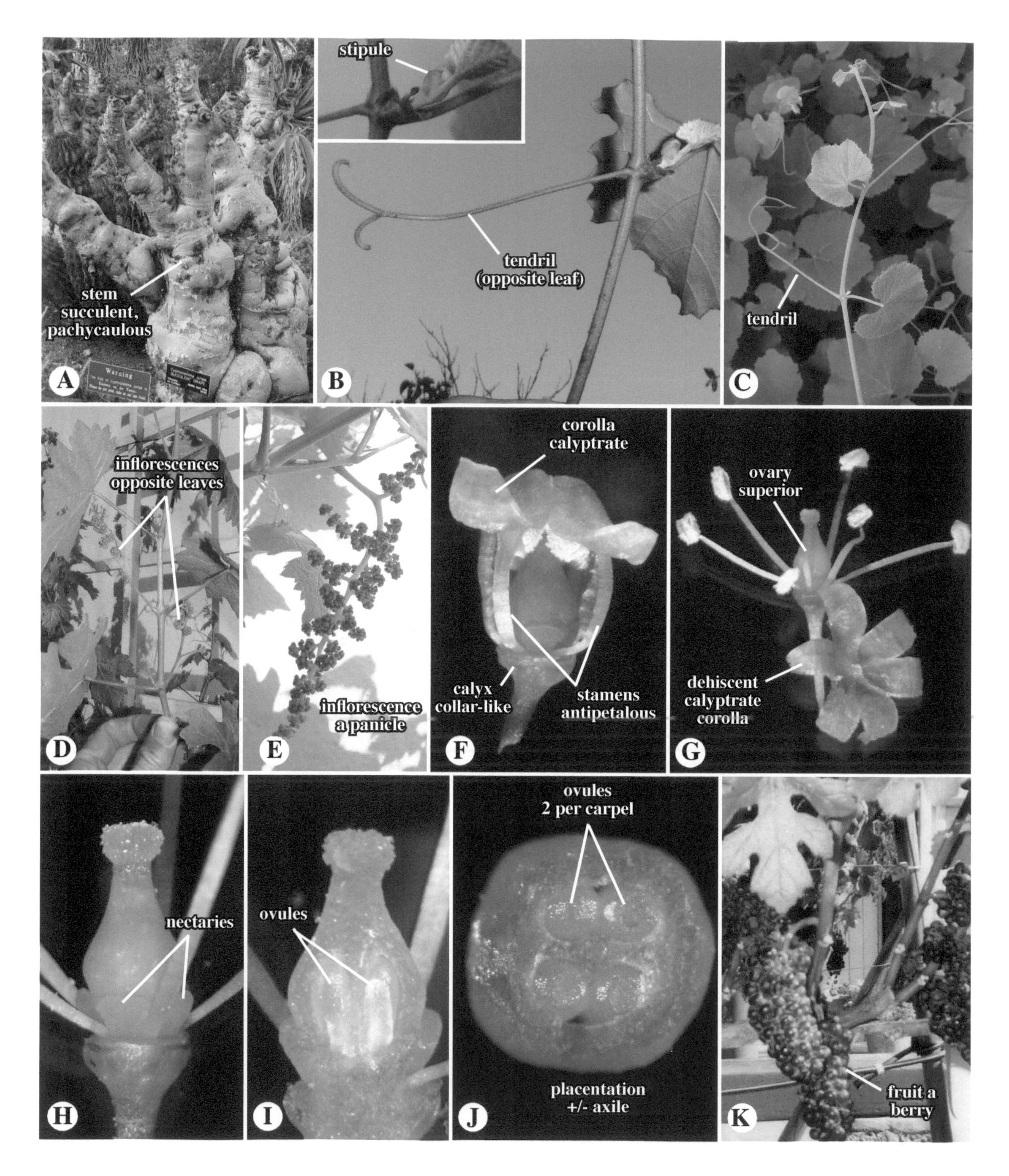

FIGURE 8.15 VITALES. Vitaceae. **A.** *Cyphostemma juttae*, a small tree with succulent, pachycaulous stems. **B,C.** *Rhoicissus capensis*, cape grape, showing stipules and tendril opposite leaves, typical of lianaceous taxa. **D-J.** *Vitis girdiana*, a wild grape. **D.** Shoot, with opposite tendril. **E.** Inflorescence, a panicle. **F.** Flower at anthesis, showing collar-like calyx a–d calyptra like corolla, the latter dehiscing from base. **G.** Flower, later stage, with corolla almost detached. **H.** Flower, showing nectaries alternating with stamens. **I.** Ovary longitudinal section, showing ovules. **J.** Ovary cross section, showing two locules, two ovules per carpel. **K.** *Vitis vinifera*, cultivated grape, with ripe berries.

ROSIDS

The Rosids (Rosidae) comprise a very large group of eudicots. Recent molecular studies verify the monophyly of this group. Rosids are largely equivalent to the subclass Rosidae of Cronquist (1981), but contain several taxa that various authors placed in other groups (particularly in the subclass Dilleniidae of Cronquist).

No clear nonmolecular apomorphies unite the Rosids. Members tend to have perianths with unfused parts and a stamen merosity greater than that of the calyx or corolla, but there are many exceptions. Generally, Rosids have bitegmic, crassinucellate ovules, distinguishing them from the Asterids, which largely have unitegmic, tenuinucellate ovules.

The Rosids are here delimited into 16 orders, which are split between the Fabids, known previously as Eurosids I, and the Malvids, known previously as Eurosids II (Figures 8.1, 8.16). A listing of families placed within these orders is seen in Table 8.2 (after APG IV 2016). See Soltis et al. (2007), Zhu et al. (2007),Wang, H. et al. (2009), Sun et al. (2016b), and Folk et al. (2018) for analyses of the Rosids.

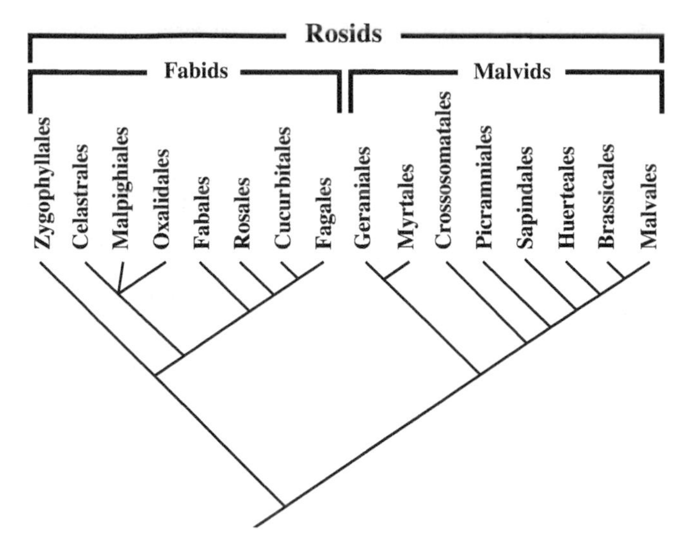

FIGURE 8.16 Cladogram of the orders of the Rosids, after Moore et al. (2011), Soltis et al. (2011), APGIV (2016), and Sun et al. (2016). Note Vitales are included in Rosids in some studies.

FABIDS

The Fabids (*Fabidae*, sensu Cantino et al. 2007, also known as Eurosids I) is a subgroup of Rosids containing 8 orders (Table 8.2, Figures 8.1, 8.16). Some of the orders are quite large, in terms of number of families, genera, and species. Among Fabids, are taxa of great agricultural importance, such as members of the **Cucurbitaceae** (squash family), **Fabaceae** (bean/pea family), **Rosaceae** (rose family), and **Euphorbiaceae** (spurge family). Others are of great ecological or industrial significance, such as the oaks (**Fagaceae**).

ZYGOPHYLLALES

This order, sensu APG IV (2016), contains two families (Table 8.2), both described here.

Krameriaceae — Rhatany family (type *Krameria*, after Johann G. H. Kramer, 1684–1744, Austrian Army physician and botanist). 1 genus (*Krameria*)/ca. 18 species (Figure 8.17).

The Krameriaceae consist of hemiparasitic shrubs, trees, or herbs. The **roots** are haustorial. The **stems** are thorny in some taxa. The **leaves** are simple or ternate, spiral, exstipulate. The **inflorescence** is of solitary, axillary flowers or a terminal raceme. The **flowers** are bisexual, zygomorphic, hypogynous, apparently resupinate. The **perianth** is biseriate. The **calyx** is aposepalous with 5 [4], imbricate, sepals, the median sepal anterior, adaxial surface of sepals petaloid. The **corolla** is pentamerous, the 3 posterior petals elongate, clawed, often sympetalous, the 2 anterior petals modified into lipid-secreting glands. The **stamens** are 4 [5], uniseriate, alternipetalous, apostemonous, the median anterior stamen missing or staminodial, filaments sometimes basally connate and adnate to petals. **Anthers** are longitudinal or poricidal in dehiscence. The **gynoecium** is syncarpous, with a superior ovary and 2 carpels, one of these sterile. **Placentation** is apical-pendulous, with two, anatropous, bitegmic **ovules**. **Nectaries** are absent. The **fruit** is a 1seeded, dry, prickly nut, prickles often with retrorse barbs. Plants are visited by female *Centris* bees, which collect oils from the lipid-secreting glands for their developing larvae.

Members of the Krameriaceae are distributed in warm, usually dry regions of the Americas. Economic importance includes local uses as medicines or dyes. See Simpson et al. (2004) for phylogenetic relationships in the family and Simpson (2006) for general information.

The Krameriaceae are distinctive in being ***hemiparasitic*** shrubs, trees, or herbs with biseriate flowers, the corolla with the ***3 posterior petals elongate and clawed***, the ***2 anterior petals modified as lipid-secreting glands***, the gynoecium 2-carpellate with ***1 carpel abortive***, the fruit a 1-seeded, ***prickly nut or capsule,*** the ***prickles often with retrorse barbs***.

K 5 [4] **C** (3)+2 **A** 4 or (4) [5] **G** (2), one carpel abortive, superior.

Zygophyllaceae - Caltrop family (type *Zygophyllum*, Greek for "yoke" "leaf," in reference to paired leaflets in the type genus). Ca. 22 genera/ca. 285–325 species (Figure 8.18).

The Zygophyllaceae consist of trees, shrubs, or herbs. The

FIGURE 8.17 ZYGOPHYLLALES. Krameriaceae. **A–G.** *Krameria bicolor*. **A.** Shrub. **B.** Flower, face view. **C–D.** Flower, side view. Note glandular, lipid-secreting anterior petals. **E.** Flower longitudinal section. Note apical-pendulous placentation. **F.** Ovary longitudinal section, close-up. **G.** Fruit, showing prickles with apically retrorse (glochidiate) barbs. **H–I.** *Krameria erecta*. **H.** Flowers. **I.** Flower and fruit.

stems often have swollen nodes. The **leaves** are paripinnate, geminate, trifoliolate, or sometimes simple, mostly opposite (less often spiral), often resinous, usually stipulate, the stipules often persistent, spiny in some taxa. The **inflorescence** is a cyme, raceme, or of solitary flowers. The **flowers** are usually bisexual and actinomorphic. The **perianth** is usually biseriate, dichlamydeous. The **calyx** is aposepalous or basally synsepalous with 5 [4,6], imbricate or valvate, sepals/lobes. The **corolla** is apopetalous with 5 [4,6,0], imbricate, convolute, or valvate petals. The **stamens** are usually 10 [4–18], usually biseriate [1-,3-seriate], the filaments often with basal glands or scale-like appendages. **Anthers** are longitudinal in dehiscence. The **gynoecium** is syncarpous, with a superior ovary (gynophore/stipe present in some taxa) and 5 [2,4,6] carpels and locules. The **style** is solitary. **Placentation** is axile or apical-axile; **ovules** are bitegmic, of variable type, 1-many per carpel. **Nectaries** are usually present and intrastaminal and/or extrastaminal. The **fruit** is an often lobed capsule or schizocarp (each carpels split into two mericarps in some taxa), rarely a drupe or berry. The **seeds** have an oily or absent endosperm.

FIGURE 8.19 CELASTRALES. **A–C.** Celastraceae. **A.** *Catha edulis*, khat. **B,C.** *Euonymus occidentalis*, burning bush. **D.** Parnassiaceae. *Parnassia californica*, grass of Parnassus.

The Zygophyllaceae are distributed in Old and New World tropical and arid regions. Economic importance includes local uses as timber (esp. *Guiacum* spp., lignum-vitae, having a very hard, heavy wood, used in machinery), medicinal resins, and waxes (e.g., *Bulnesia* spp.), edible fruits (e.g., *Balanites* spp.), dyes and other medicines (e.g., *Peganum harmala*, harmal, source of the dye "turkey red"). *Tribulus terrestris*, caltrop, can be a noxious weed, with dangerous spiny fruits that can cause sickness in livestock when ingested. *Larrea* spp., creosote bush, is an ecologically important shrub, dominant in many desert ecosystems of North and South America. The Zygophyllaceae have recently been classified into five subfamilies. See Sheahan (2006) for general information and Sheahan and Chase (1996, 2000) for phylogenetic analyses of the family.

The Zygophyllaceae are distinctive in being trees, shrubs, or trees, with leaves usually opposite and ***paripinnate, geminate, or trifoliolate*** [simple], the flowers ***usually dichlamydeous***, nectariferous, with a 5 [4–6]-merous perianth, stamens ***usually 2 [1,3]-seriate***, the ***filaments basally glandular or with scale-like appendages***, gynoecium syncarpous with 5 [1,4,6] carpels, the fruit ***usually a lobed capsule or schizocarp*** (rarely a drupe or berry).

K 5 [4,6] or (5) [(4,6)] **C** 5 [4,6,0] **A** 5+5 [4–18] **G** (5) [(2,4,6)], superior.

CELASTRALES

This order, here treated, contains 3 families (Table 8.2). Notable among members of the **Celastraceae**, *Catha edulis*, khat (Figure 8.19A), the leaves of which are chewed as a stimulant, and *Euonymus*, burning bush (Figure 8.19B,C); and **Parnassiaceae** (included within the Celastraceae in APG IV), containing *Parnassia*, grass of Parnassus (Figure 8.19D). See Zhang and Simmons (2006) and Simmons et al. (2012) for a treatment of the Celastrales.

MALPIGHIALES

The Malpighiales are a very large order of 36 families (Table 8.2), six of which are described here. Notable among the other families are the **Achariaceae**, containing *Hydnocarpus* spp. and relatives, which produce cyclopentenoid fatty acids—e.g., hydnocarpic and chaulmugric acid—used in treating leprosy; **Chrysobalanaceae**, including *Chrysobalanus icaco*, cocoplum (Figure 8.20A); **Clusiaceae** (Figure 8.20B), sometimes inclusive of Hypericaceae; **Elatinaceae**, the aquatic waterworts (Figure 8.20C); **Erythroxylaceae**, containing *Erythoxylum* spp., the source of the alkaloid cocaine; **Linaceae** (Figure 8.20D), including *Linum usitatissimum*, flax, the source of linen cloth; **Rafflesiaceae**, renowned for containing the largest flower in the world, *Rafflesia arnoldii*; **Phyllanthaceae** (Figure 8.20E); and **Rhizophoraceae** (Figure 8.20F), the ecologically important mangrove family. See Chase et al. (2002), Wurdack and Davis (2009), Korotkova et al. (2009), and Xi et al. (2012) for phylogenetic analyses.

Euphorbiaceae — Spurge family (type *Euphorbia*, after Euphorbus, physician to the king of Mauritania, 1st century). Ca. 218 genera/ca. 6,300–6,700 species (Figure 8.21).

The Euphorbiaceae consist of monoecious or dioecious, herbs, shrubs, vines, or trees, latex present in some major groups. The **stems** are succulent and cactuslike in some (e.g., some Euphorbias). The **leaves** are simple, rarely trifoliolate or palmate, spiral, opposite, or whorled, stipules generally present, these sometimes modified as glands or spines (e.g., many succulent Euphorbias). The **inflorescence** is generally a cyme, modified as a cyathium in some Euphorbioideae. The **flowers** are unisexual, actinomorphic, rarely zygomorphic, bracteate in some, hypogynous. The **perianth** is biseriate, uniseriate, or absent, generally 5-merous. The **calyx** is usu-

FIGURE 8.18 ZYGOPHYLLALES. Zygophyllaceae. **A–G.** *Larrea tridentata*, creosote bush. **A.** Whole plant, a desert shrub. **B.** Geminate leaves. **C.** Flower, face view. **D.** Flower close-up. **E.** Stamens, removed, showing scale-like appendages. **F.** Ovary cross section. **G.** Fruit, a hirsute schizocarp. **H.** *Fagonia laevis*, a small shrub with trifoliolate leaves and spinose stipules; inset showing 5-lobed, capsular fruit. **I–O.** *Tribulus terrestris*, caltrop. **I.** Flowering shoot. **J.** Fruiting shoot, showing branching pattern. **K.** Node close-up, showing stipules of adjacent leaves. **L.** Flower close-up; note 10 stamens and superior ovary. **M.** Flower, with faacing petal removed, showing antisepalous glands, 5 intrastaminal and 5 extrastaminal. **N.** Fruit, a schizocarp with stout prickles. **O.** *Viscainoa geniculata*, flowers and fruits.

FIGURE 8.20 MALPIGHIALES. **A.** Chrysobalanaceae, *Chrysobalanus icaco*, shoot, in fruit; inset, flower close-up. **B.** Clusiaceae, *Clusia* sp., flowering shoot; inset, flower close-up. **C.** Elatainaceae, *Elatine brachysperma*, a small herb, showing flower (left) and fruit (right). **D.** Linaceae, *Linum lewisii*, flower. **E.** Phyllanthaceae, *Phyllanthus* sp. **F.** Rhizophoraceae, *Ceriops* sp., immature fruits.

ally aposepalous with 5 [rarely 0] sepals. The **corolla** is usually apopetalous with 5 [rarely 0] valvate or imbricate petals. The **stamens** are 1–∞, distinct or connate. **Anthers** are longitudinal, poricidal, or transverse in dehiscence. The **gynoecium** is syncarpous, with a superior ovary, 3 [2–∞] carpels and locules. The **styles** are as many as carpels, each style sometimes 2-branched. **Placentation** is apical-axile and pendulous, with an obturator (a protuberance from the funiculus or placenta at the base of the ovule); **ovules** are anatropous or hemitropous, bitegmic, 1 per carpel. **Nectaries** are often present. The **fruit** is a schizocarp, drupe, berry, or samara.

The Euphorbiaceae traditionally were classified into five subfamilies: Phyllanthoideae, Oldfieldioideae, Acalyphoideae, Crotonoideae, and Euphorbioideae. The former two subfamilies have two ovules per carpel (biovulate) and have been elevated to family rank, Phyllanthaceae and Picrodendraceae, respectively (in the Malpighiales). Within the Euphorbiaceae as delimited here, Wurdack et al. (2005) recognized nine monophyletic infrafamilial groups, including the newly recognized subfamilies Peroideae and Cheilosoideae and a monophyletic Euphorbioideae; the traditional Acalyphoideae and Crotonoideae are paraphyletic. Members of the Euphorbioideae have latex and reduced staminal flowers, culminating in the highly specialized and characteristic cyathium in some members of the subfamily. See Wurdack and Chase (2002) and Wurdack et al. (2005) for phylogenetic analyses.

Members of the Euphorbiaceae have worldwide distributions. Economic importance includes *Aleurites fordii*, tung oil and *A. moluccana*, candlenut oil; *Hevea brasiliensis*, the major source of natural rubber; *Manihot esculentus*, cassava/manioc, a very important food crop and the source of tapioca; *Ricinus communis*, the source of castor bean oil and the deadly poison ricin; and various oil, timber, medicinal, dye, and ornamental plants. Succulent *Euphorbia* species are major components of plant communities in southern Africa, as well as important ornamental cultivars.

The Euphorbiaceae are distinctive in having ***unisexual*** flowers with a superior, ***usually 3-carpellate*** ovary with ***1 ovule per carpel***, ***apical-axile*** in placentation, many taxa with ***red, yellow, or usually white (milky) latex,*** the Euphorbioideae alone with reduced staminal flowers, some with a characteristic ***cyathium*** inflorescence.

K 5 [0] **C** 5 [0] **A** 1–∞ **G** (3) [(2–∞)], superior.

Hypericaceae [Guttiferae] — St. John's Wort family (type *Hypericum*, after the Greek *hyper*, above, + *eikon*, picture, perhaps in reference to flowers of this group being placed above pictures at Walpurgisnacht (later the feast of St. John, 24 June) to ward off evil spirits.). 9 genera/ca. 480-590 species (Figure 8.22).

FIGURE 8.21 MALPIGHIALES. Euphorbiaceae. **A.** *Euphorbia maculata*, close-up of leaves and cyathia. **B.** *Euphorbia ingens*, a stem-succulent plant. **C.** *Euphorbia baioensis*, showing paired stipular spines. **D.** *Euphorbia obesa*. **E.** *Euphorbia milii*, crown-of-thorns, aerial shoot of plant. **F.** Close-up of cyathia, having red bracts. **G,H.** *Euphorbia grandicornis*. **G.** Whole plant, showing enlarged stipular spines. **H.** Cyathia. **I–K.** *Euphorbia* sp. **I.** Cyathium, showing bracts and male and female flowers. **J.** Female and male flower, removed. **K.** Close-up of male flower, showing junction between filament and pedicel. **L.** *Euphorbia ingens*, simple dichasium of cyathia, the central one in fruit. **M.** *Manihot esculenta*, cassava. **N.** *Aleurites moluccana*, candle-nut tree.

FIGURE 8.22 MALPIGHIALES. Hypericaceae, *Hypericum canariense* illustrated. **A.** Whole plant, showing terminal, corymbose cymes, from primary and lateral shoots. **B.** Vegetative shoot, showing simple, entire, opposite leaves. **C.** Leaf close-up, showing characteristic dotted glands. **D.** Flower, with pentamerous perianth, numerous stamens, and 3-carpellate pistil. **E.** Flower close-up, showing basally synsepalous calyx. (Petals apopetalous.) **F.** Close-up of stamen fascicle, characteristic of the family. **G.** Stamen fascicle removed from flower. **H.** Close-up of two anthers, showing longitudinal dehiscence. **I.** Flower, dissected, showing pistil with superior ovary and three styles. **J.** Ovary cross section, showing 3 carpels with parietal placentation, the placentae appressed in the center.

The Hypericaceae consist of trees, shrubs, or herbs. The **leaves** are simple, cauline, opposite or whorled, exstipulate, entire, and (often pellucid) gland dotted or lined. The **inflorescence** is a terminal or axillary cyme, corymb, or of solitary flowers. The **flowers** are bisexual, actinomorphic, hypogynous. The **calyx** is usually basally synsepalous with 4–5, sepals/lobes. The **corolla** is apopetalous with 4–5, often contorted, petals. The **stamens** are [5–] ∞, centrifugal, often fasciculate in 3–5 bundles. **Anthers** are dithecal, longitudinally dehiscent, often with glands. The **gynoecium** is syncarpous, with a superior ovary, 3–5 carpels, and 1–3 locules. The **style(s)** are 1 or more in number. **Placentation** is axile or parietal. Extrafloral **nectaries** are found in some taxa. The **fruit** is (usu.) a capsule, berry, or drupe. The **seeds** are small, often numerous.

The Hypericaceae have a cosmopolitan distribution. Economic importance includes cultivated ornamentals, medicinal plants, and wood plants; some *Hypericum* spp. are toxic to livestock when ingested, the poisons photo-activated. *Hypericum perforatum* (klamath weed, St. John's wort) is popular as an herbal remedy to help counter depression. See Stevens 2006 for a general treatment of the family and Nürk et al. (2015) and Ruhfel et al. (2016) for phylogenetic analyses.

The Hypericaceae are distinctive in having ***opposite, simple, entire*** leaves with ***glandular dots or lines***, flowers with a 4–5-merous perianth, usu. ***numerous, centrifugal*** stamens ***often fasciculate***, and a superior, syncarpous, ***3–5 carpellate*** ovary, the fruit a ***capsule (usu.), berry, or drupe***.

K 4–5 or (4–5) **C** 4–5 **A** 5–∞ **G** (3–5), superior.

Malpighiaceae — Barbados-Cherry family (type *Malpighia*, after Marcello Malpighi (1628–1694), Italian physician and anatomist). Ca. 68 genera/ca. 1,250 species (Figure 8.23).

The Malpighiaceae consist of vines, shrubs, or trees, often with anomalous secondary growth. The surfaces of various organs have characteristic "malpighian" trichomes, which are unicellular with two opposite, tapering arms perpendicular to the short attachment point. The **leaves** are simple, mostly undivided, pinnately veined, usually opposite [rarely subopposite or ternate], petiolate, usually stipulate (stipules large or connate in some taxa), with two fleshy glands near the junction of petiole and blade. The **inflorescence** is a terminal or axillary raceme, panicle, or cyme. The **flowers** are perfect [rarely unisexual], actinomorphic to zygomorphic, pedicellate with jointed pedicels, bracteate and 2-bracteolate, and hypogynous. The **perianth** is dichlamydeous and pentamerous. The **calyx** is aposepalous or basally synsepalous with 5 sepals/calyx lobes, each with 2 fleshy, basally abaxial glands, sometimes reduced to 1 gland or absent on one sepal, rarely absent entirely. The **corolla** is apopetalous, often with crumpled aestivation, consisting of 5 clawed (unguiculate) petals, these typically marginally ciliate, toothed, or fringed, the adaxial-lateral petal often different from the others. The **stamens** are 10 [2–15], biseriate, rarely 1- or 3-seriate, apostemonous or basally fused into a tube, staminodes or stamens with partially sterile anthers often present, sometimes within an entire whorl. **Anthers** are longitudinal [rarely poricidal] in dehiscence, dithecal or rarely monothecal, with thecae and connective enlarged and glandular in some taxa. The **gynoecium** is usually syncarpous, but essentially apocarpous in some taxa, with a superior ovary, and 3 [2–5] carpels and locules. The **styles** are 3, sometimes fused; **stigmas** are 3. **Placentation** is apical-axile; **ovules** are anatropous or hemitropous, bitegmic, 1 per carpel, epitropous-ventral. In New World taxa, the sepal glands (termed "elaiophores") secrete an oil, which functions as the attractant/reward for anthrophorid bees; in some Old World taxa, these sepal glands secrete sugar-rich nectar. The **fruit** is a drupe, nut, or samara (wing morphology variable), often schizocarpic, the calyx accrescent in some taxa. **Seeds** are exalbuminous.

The Malpighiaceae are distributed in Old and New World tropics and subtropics, mostly in South America. Economic importance includes uses as fruits, e.g., *Malpighia glabra* (Barbados-cherry, rich in vitamin C), and *Bunchosia* spp. (marmelo). The liana *Banisteriopsis caapi* (Ayahuasca) yields (from bark) hallucinogenic alkaloids used by some South American natives in religious/spiritual rites. Intrafamilial classification of the Malpighiaceae (which traditionally has been based mainly on fruit characteristics) is still not firmly established, although many lineages have been identified from recent molecular studies. See Cameron et al. (2001), Davis et al. (2001), and Davis and Anderson (2010).

The Malpighiaceae are distinctive in being vines, shrubs, or trees with simple, usually opposite and stipulate leaves, a ***pair of glands present near junction of petiole with blade***, the flowers usually bisexual, perianth pentamerous, the ***sepals usually with 2 [1] fleshy, abaxial glands*** (secreting oil or nectar), the ***petals clawed and marginally ciliate, toothed, or fringed***, the androecium 1–3-seriate, ***staminodes often present***, the ovary superior and ***3 [2–5] carpellate and loculate***, with ***1, apical-axile ovule per carpel***, the fruit a drupe, nut, or samara, ***often schizocarpic***.

K 5 or (5) **C** 5 **A** 10 [2–15] **G** (3) [(2–5)], superior.

Passifloraceae (including Malesherbiaceae) — Passion Flower family (type *Passiflora*, Latin for "passion flower," after events of the Christian Passion, signified in floral parts). Ca. 27 genera/ca. 1,035 species (Figure 8.24).

The Passifloraceae consist of hermaphroditic to dioecious lianas, shrubs, or trees. The **stems** have axillary tendrils in lianous species. The **leaves** are simple, rarely palmately compound, often palmately lobed, spirally arranged, stipulate or exstipulate, the petioles often with extrafloral nectaries. The

FIGURE 8.23 MALPIGHIALES. Malpighiaceae. **A–N.** *Stigmaphyllon ciliatum*. **A.** Vine, with axillary inflorescences. **B.** Leaf margin. **C.** Junction of petiole with blade, showing paired leaf glands. **D.** Inflorescence base, showing jointed pedicels and bract glands. **E.** Malpighian trichome. **F.** Flower, face view, showing characteristic petals. **G.** Close-up of flower center. Note incurved sepals. **H.** Flower center, side view. **I.** Flower base from below, showing paired sepal glands. **J.** Dissected style/stigma, fertile stamen, and staminodes. **K.** Fertile stamen, with swollen connective. **L.** Ovary cross section, with one ovule per carpel (appearing as two by curvature of anatropous ovule). **M.** Ovary longitudinal section, showing apical-axile placentation. **N.** Ovule clearing (apex of flower above), showing epitropous ovule position. **O–Q.** *Malpighia coccigera*. **O.** Flower, face view. **P.** Flower, side view, note paired sepal glands. **Q.** Fruit, a drupe, with persistent sepal glands.

FIGURE 8.24 MALPIGHIALES. Passifloraceae. **A–C.** *Passiflora* sp. **A.** Flower. **B.** Close-up, showing androgynophore. Note five stamens and three-branched style. **C.** Perianth, back view, showing three bracts, five sepals, and five petals. **D.** *Passiflora* sp., flower. **E.** *Passiflora* sp., ovary cross section, showing parietal placentation with three carpels. **F.** *Passiflora foetida*, fruit. **G,H.** *Passiflora edulis*, fruit, a berry.

inflorescence is a cyme or a solitary flower. The **flowers** are bisexual or unisexual, bracteate, actinomorphic, usually perigynous, rarely hypogynous. The **calyx** is aposepalous or basally synsepalous, with 5 [3–8] sepals or calyx lobes. The **corolla** is apopetalous or basally sympetalous, with 5 [3–8] petals or corolla lobes, and a corona of 1 or more whorls of filamentous or scalelike structures between the perianth and androecium. The **stamens** are 5 [4–∞], alternipetalous, whorled, uniseriate, distinct or connate, free or adnate to an androgynophore. **Anthers** are longitudinal in dehiscence.

The **gynoecium** is syncarpous, with a superior ovary, 3 [2–5] carpels, and 1 locule, usually arising from a prominent androgynophore, rarely from a gynophore or sessile. The **styles** and **stigma** are as many as carpels and usually basally connate. **Placentation** is parietal; **ovules** are anatropous, bitegmic, and numerous per carpel. The **fruit** is a berry or capsule. The **seeds** are endospermous.

The Passifloraceae have a worldwide distribution in tropical and subtropical regions. Economic importance includes use as cultivated ornamentals, e.g. *Adenia* and *Passiflora* spp., and as edible fruits, esp. *Passiflora edulis*, passion fruit. See Feuillet and Macdougal (2006) for general information. The families Malesherbiaceae and Turneraceae are now merged with the Passifloraceae (APG IV, 2016). See Tokuoka (2012) for a recent phylogenetic analysis of the family.

The Passifloraceae are distinctive in being lianas, shrubs, or trees with actinomorphic, usually bisexual flowers, having ***one or more whorls of coronal appendages*** between perianth and androecium, an ***androgynophore*** usually present, and typically ***5 stamens and 3 carpels***, with ***parietal placentation***, the fruit a ***berry or capsule***.

K 5 [3–8] or (5) [(3–8)] **C** 5 [3–8] or (5) [(3–8)] **A** 5 [4–∞] or (5) [(4–∞)] **G** (3) [(2–5)], superior, hypanthium present in most species.

Salicaceae [including Flacourtiaceae] — Willow family (type *Salix*, Latin for willow). 54 genera/1200 species (Figure 8.25).

The Salicaceae consist of hermaphroditic to dioecious, trees or shrubs. The roots of *Salix* and *Populus* often are ectomycorrhizal. The **leaves** are simple, spiral [rarely opposite or subopposite], usually stipulate (stipules often caducous), often with "salicoid teeth," in which a vein traverses into the tooth apex, expanding into a usually pigmented, rounded gland or a stout trichome. The **inflorescence** is a terminal or axillary raceme, spike, catkin (*Salix, Populus*), corymb, cyme, glomerule, or of solitary flowers. The **flowers** are often small, bisexual or unisexual, bracteate in some, hypogynous. The **perianth** is lacking, uniseriate, or biseriate. The **calyx** is aposepalous or synsepalous with 0–8 [–15], sepals/lobes, adnate to ovary base in some taxa. The **corolla** consists of 0–8 [–15] petals, absent or equal in number to calyx. The **stamens** are 2–∞ [1], when ∞ centrifugal in development, with antipetalous clusters in some taxa, filaments rarely connate. **Anthers** are usually small, longitudinally dehiscent. The **gynoecium** is syncarpous, with a superior [rarely semi-inferior] ovary, 2–∞ carpels, and 1 locule. The **styles** are 1–8, **stigmas** 2–∞ in number. **Placentation** is parietal to basal [rarely axile]; **ovules** are anatropous to orthotropous, unitegmic or bitegmic, 2–∞. **Nectaries** are typically present, as an intrastaminal, extrastaminal disk, or extragynoecial disk, or of discrete, sometimes protruding, glands. The **fruit** is a berry, drupe, 3 [2–6]-valved capsule [samara]. The **seeds** are arillate or comose in some taxa. Plants are wind or insect pollinated.

The Salicaceae as recognized currently are much expanded from the traditional family delimitation and encompasses the former Flacourtiaceae. Chase et al. (2002) studied phylogenetic relationships within the complex and proposed a tentative classification in which the family is split into nine tribes, with the traditional Salicaceae s.s. being equivalent to tribe Saliceae.

Members of the Salicaceae are distributed mostly worldwide, from tropical to cold-temperate climates. Economic importance includes timber trees, fruit trees, and ornamental cultivars, including *Azara*, *Idesia*, *Olmediella* (Guatemalan-holly), *Populus* (cottonwoods or poplars), and *Salix* (willows); the bark of willows (*Salix*) was the original source of salicin, the chemical modification of which is aspirin (acetyl-salicylic acid). The Samydaceae and Scyphostegiaceae have been merged into the Salicaceae in APG IV (2016), treated as subfamilies along with Salicoideae in Sun et al. (2016b).

The Salicaceae, s.l. are ***dioecious to hermaphroditic trees or shrubs*** with ***simple, usually spiral, stipulate*** leaves, often with ***salicoid teeth***, flowers are ***often small***, perianth variable, sometimes one or both whorls absent, stamens 2–∞, ovaries with ***parietal to basal*** placentation, the fruit usually a ***capsule, berry, or drupe***.

K 0–8 [–15] **C** 0–8 [–15] **A** 2–∞ **G** 2–∞, superior.

Violaceae — Violet family (type *Viola*, Latin for various fragrant plants). Ca. 34 genera/ca. 985 species (Figure 8.26).

The Violaceae consist of herbs, shrubs, trees, or lianas. The **leaves** are simple, undivided to divided, usually spiral, and stipulate. The **inflorescence** is of solitary, axillary flowers or in heads, panicles, or racemes. The **flowers** are usually bisexual (cleistogamous in some, e.g., *Viola* spp.), actinomorphic or zygomorphic. The **calyx** is aposepalous with 5 imbricate, often persistent sepals. The **corolla** is apopetalous with 5 imbricate or convolute petals, the anterior petal in zygomorphic flowers often with a nectar-storing spur. The **stamens** are 5 [3], apostemonous or monadelphous, usually connivent around the ovary. The **gynoecium** is syncarpous, with a superior ovary, 3 [2–5] carpels, and 1 locule. The **style** is solitary. **Placentation** is parietal; **ovules** are anatropous, bitegmic, 1–∞ per carpel. **Nectaries** are present on the dorsal side of stamens. The **fruit** is a berry or loculicidal capsule, rarely a nut. The **seeds** are endospermous and often with an aril or caruncle.

The Violaceae have a mostly worldwide distribution. Economic importance includes some plants of medicinal value, cultivated ornamentals, e.g., *Viola* ×*wittrockiana* (pansies), and use for extraction of volatile oils. See Ballard et al. (2013) for general information and Tokuoka (2008) and Wahlert et al. (2014) for a phylogenetic studies of the family.

FIGURE 8.25 MALPIGHIALES. Salicaceae, s.l. **A.** *Populus fremontii*, female, shoot with catkins. **B.** *Populus deltoides*, male with pendant catkins. **C.** *Salix laevigata*, male flowering shoot. **D–M.** *Salix lasiolepis*. **D.** Tree. **E.** Male catkin. **F.** Male flower, showing bract and basal, adaxial gland. **G.** Female catkin. **H.** Female flower, with bract and gland. **I.** Ovary longitudinal section, showing basal to parietal placentation. **J.** Ovary cross section, showing two placental regions. **K.** Fruiting catkin. **L.** Valvular capsule after dehiscence. **M.** Comose seed. **N.** *Dovyalis caffra*, with showy, bisexual flower having tricarpellate gynoecium. **O–P.** *Xylosma congestum*, with numerous centrifugal stamens and extrastaminal glands.

FIGURE 8.26 MALPIGHIALES. Violaceae. **A.** *Hybanthus calycinus*, flower. Note large anterior petal. **B.** *Hybanthus concolor*, with reduced flowers. **C.** *Viola pedata*. **D.** *Viola purpurea*, a perennial herb. **E,F.** *Viola papilionacea*. **E.** Flower, face view. **F.** Flower longitudinal section. Note connivent stamens and nectary inside corolla spur. **G–K.** *Viola* ×*wittrockiana*, pansy. **G.** Flower, face view. **H–J.** Flower, longitudinal sections. **H.** Corolla spur with recessed nectaries. **I.** Connivent stamens appressed to pistil. **J.** Ovary, style, and stigma. **K.** Ovary cross section, showing three carpels with parietal placentation, placental regions at arrows.

FIGURE 8.27 OXALIDALES. **A.** Cephalotaceae, *Cephalotus follicularis*, the Western Australian Pitcher Plant. **B–D.** Cunoniaceae. **B.** *Bauera rubioides*. **C.** *Ceratopetalum gummiferum*. **D.** *Cunonia capensis*. **E.** Elaeocarpaceae, *Tetratheca* sp.

The Violaceae are distinctive in being ***herbs, shrubs, trees, or lianas*** with simple, undivided or divided leaves, actinomorphic or zygomorphic flowers with a 5-merous perianth, usually ***5, connivent stamens***, and a superior, usually ***3-carpellate ovary*** with ***parietal placentation***, the seeds often with a ***caruncle or aril***.

K 5 **C** 5 **A** 5 [3] or (5) [(3)] **G** (3) [(2–5)], superior.

OXALIDALES

The Oxalidales, sensu APG IV (2016), contain seven families (Table 8.2), only one of which is described here. Notable among the others are the **Cephlotaceae**, composed solely of *Cephalotus follicularis*, the insectivorous Western Australia pitcher plant (Figure 8.27A); the **Cunoniaceae**, mostly Southern hemisphere trees and shrubs with compound leaves (Figure 8.27B–D); and the **Elaeocarpaceae**, simple-leaved tropical trees or shrubs with poricidal anthers (Figure 8.27E). See Matthews and Endress (2002) and Kubitzki (2004) for recent information about this order.

Oxalidaceae — Oxalis family (type *Oxalis*, Greek for sour, from accumulation of oxalic acid in the tissues). Ca. 6 genera/ ca. 570-770 species (Figure 8.28).

The Oxalidaceae consist of herbs, shrubs, or small trees. The **stems** are bulbs or tubers in some herbaceous taxa. The **leaves** are pinnate, palmate, often trifoliolate, rarely unifoliolate or of phyllodes, spirally arranged, usually exstipulate, the leaflets often folded at night. The **inflorescence** is a cyme. The **flowers** are bisexual, actinomorphic, and hypogynous. The **calyx** is aposepalous with 5 imbricate sepals. The **corolla** is apopetalous or basally sympetalous with 5 [0], convolute to imbricate petals. The **stamens** are biseriate, 5+5, whorled, the outer fertile stamens shorter, filaments basally connate, with outer whorl of staminodes sometimes present. **Anthers** are longitudinally dehiscent. The **gynoecium** is syncarpous, with a superior ovary, 5 [3] carpels, and 5 [3] locules. The **styles** are 5 [3,1]. **Placentation** is axile; **ovules** are anatropous or hemitropous, bitegmic, 2–∞ [1] per carpel. **Nectaries** are often present at base of outer stamens. The **fruit** is a loculicidal capsule or berry. The **seeds** are endospermous, often with basal aril.

The Oxalidaceae are mostly worldwide in distribution. Economic importance includes fruit trees, e.g., *Averrhoa carambola* (star fruit), tuber plants, e.g., *Oxalis tuberosa*, and ornamental cultivars, e.g., *Oxalis* spp. See Oberlander et al. (2004) and Heibl and Renner (2012) phylogenetic studies.

The Oxalidaceae are distinctive in being herbs, shrubs, or small trees, usually with ***pinnate or palmate*** (***often trifoliolate***) leaves (***leaflets often folding*** at night), flowers bisexual, actinomorphic, pentamerous, the ***stamens usually biseriate***, with ***outer, basal nectaries***, the fruit a loculicidal capsule or berry.

K 5 **C** 5 or (5) [0] **A** (5+5) [(+5 staminodes)] **G** (5) [(3)], superior.

FABALES

The Fabales contain four families (Table 8.2), two of which are described in detail here. The **Quillajaceae** of this order, formerly classified in the Rosaceae, has flowers with diplostemonous stamens, clawed petals, and a strongly lobed ovary (Figure 8.29), the fruit bearing winged seeds. See Bello et al. (2009, 2012), and Sun et al. (2016) for studies of the order.

FIGURE 8.28 OXALIDALES. Oxalidaceae. **A,B.** *Averrhoa carambola*. **A.** Plant in flower. **B.** Winged fruit. **C,D.** *Oxalis gigantea*, a shrub. **E.** *Oxalis oregana*, redwood sorrel, a perennial herb. **F.** *Oxalis rosea*, flower close-up. **G–I.** *Oxalis pes-caprae*. **G.** Flower, dissected, showing biseriate androecium, plus an outer whorl of staminodes. **H.** Central ovary, with apical, divergent styles and stigmas. **I.** Ovary cross-section, showing five carpels with axile placentation.

FIGURE 8.29 FABALES. Quillajaceae. *Quillaja saponaria*. **A.** Whole flower, showing calyx, corolla, biseriate stamens, and pistil. **B.** Close-up of 5-lobed pistil and inner stamens.

Fabaceae (Leguminosae) — Bean/Pea family (type *Faba* [=*Vicia*], after the Latin name for broad bean). Ca. 770 genera/19,500 species (Figures 8.30–8.32).

The Fabaceae consist of herbs, shrubs, trees, or vines, with spines sometimes present. The **roots** of many members have a symbiotic association with nitrogen-fixing bacteria (*Rhizobium* spp.), which induce formation of root nodules (this especially common in the Faboideae). The **leaves** are usually compound (pinnate, bipinnate, trifoliolate, rarely palmate), sometimes simple or unifoliolate, usually spiral, basal pulvini often present, sometimes functioning in tactile ("thigmonastic"), leaflet folding responses (e.g., *Mimosa* spp.), generally stipulate, sometimes stipellate, stipules spinose in some. The **inflorescence** is variable, typically bracteate. The **flowers** are usually bisexual, sometimes unisexual, actinomorphic or zygomorphic, pedicellate or sessile, hypogynous or perigynous. The **perianth** is biseriate, dichlamydeous, with a hypanthium sometimes present. The **calyx** is aposepalous or synsepalous with 5 [3–6] sepals. The **corolla** is apopetalous or sympetalous, with 5 [rarely 0] valvate or imbricate petals. The **stamens** are 5 or 10–∞, distinct or connate. **Anthers** are longitudinal, rarely poricidal in dehiscence. The **gynoecium** is unicarpellous, with a superior ovary, 1 [rarely 2 or more in some Mimosoids] carpel, and 1 locule. The **style** and **stigma** are solitary. **Placentation** is marginal; **ovules** are anatropous or campylotropous, bitegmic, 2–∞ [1] per carpel. **Nectaries** are often present as a ring at the base of the ovary. The **fruit** is generally a legume, sometimes indehiscent (e.g., *Arachis*, peanut), winged (a samara), drupe-like, or divided into transverse partitions (a loment).

The Fabaceae are now classified into six, monophyletic subfamilies, some of which have been recircumscribed (The Legume Phylogeny Working Group, 2017): **Caesalpinioideae** (148 genera, pantropical to temperate, including *Ceratonia*, *Senna*; inclusive of the "Mimosoid clade," this including *Acacia*, *Mimosa*), **Cercidoideae** (12 genera, including *Bauhinia*, *Cercis*), **Dialioideae** (17 genera, pan-tropical trees and shrubs), **Detarioideae** (84 genera, mostly tropical trees and shrubs, including *Tamarindus*), **Duparquetioideae** (monospecific, containing only the African *Duparquetia orchidacea*), and **Faboideae=Papilionoideae** (503 genera, cosmopolitan).

The **Cercidioideae** (Figure 8.30) are distinctive in having unifoliolate (blade unlobed or, e.g., in *Bauhinia*, bilobed with central apical process) or bifoliolate (geminate) leaves, flowers with imbricate petals, the posterior inner to the laterals, 10 or fewer distinct stamens, and seeds with an apical crescent -shaped hilum.

The **Caesalpinioideae** (Figure 8.31) are distinctive in having extrafloral nectaries often present on the leaf petiole, rachis, or rachillae; flowers bilateral or radial, corolla aestivation imbricate (then posterior petal inner to lateral) or valvate, stamens [3] 5–∞, distinct (staminodes present in some); anthers often with an apical gland, and pollen commonly in tetrads or polyads. Nested within the Caesalpinioideae, the **Mimosoid Clade** (formerly subfamily Mimosoideae) is distinctive in having radial flowers with a valvate corolla of distinct or basally fused petals, hypanthium sometimes present, and usually numerous, distinct or basally fused stamens.

The **Faboideae** (**Papilionoideae**; Figure 8.32) are distinctive in having ***"papilionaceous" flowers***, which are ***zygomorphic,*** with ***connate stamens*** (either ***10 monadelphous or 9+1 diadelphous***) and a corolla (imbricate in bud) with five petals consisting of a ***large, median, usually posterior petal*** (the "***banner***" or "***standard***"), which is ***outer*** to (overlapping) the adjacent petals, ***two lateral "wing" petals*** (overlapped by the banner), and ***two anterior, distally fused "keel" petals***. The flowers are resupinate in some species, e.g., *Clitoria*, in which the banner is anterior in position.

The Fabaceae are a very large group with a worldwide distribution. Members of the family are dominant species in some ecosystems (e.g., *Acacia* spp. in parts of Africa and Australia) and ecologically important for containing nitrogen-fixing rhizobial nodules in many species. Economically, legumes are one of the important plant groups, being the source of numerous pulses (such as *Arachis hypogaea*, peanut; *Glycine max*, soybeans; *Lens culinaris*, lentil; *Phaseolus* spp., beans; *Pisum sativum*, peas); flavoring plants (such as *Ceratonia siliqua*, carob), fodder and soil rotation plants (such as *Medicago sativa*, alfalfa, or *Trifolium* spp., clovers), oils, timber trees, gums, dyes, and insecticides. See The Legume Phylogeny Working Group (2017).

The Fabaceae are distinctive in being trees, shrubs, vines, or herbs, with ***stipulate***, often ***compound*** leaves and typically pentamerous flowers, usually with a ***single, unicarpellous***

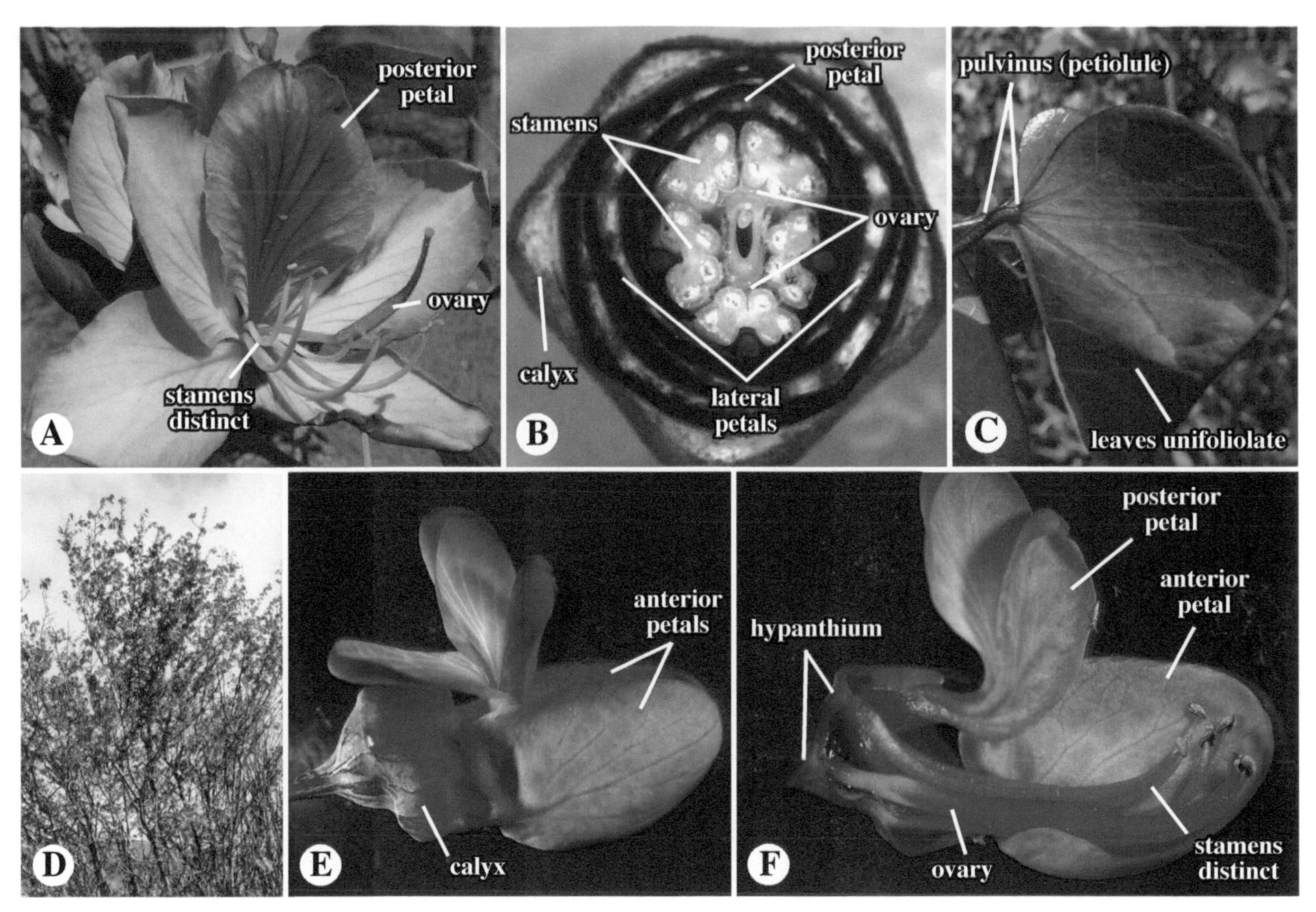

FIGURE 8.30 FABALES. Fabaceae. Cercidioideae. **A–B.** *Bauhinia variegata,* orchid tree. **A.** Flower. **B.** Flower bud cross section. Note medial posterior petal inner to laterals and distinct stamens. **C–F.** *Cercis occidentalis,* red bud. C. Unifoliolate leaf, characteristic of subfamily. Note pulvinus, thickening at distal end of petiole, interpreted as a vestigial petiolule. **D.** Tree. **E.** Whole flower. **F.** Flower longitudinal section, showing hypanthium. Note medial posterior petal inner to laterals and distinct stamens.

pistil with ***marginal placentation***, the fruit a ***legume*** (or modified legume).

K 5 or (5) [(3–6)] **C** 5 or (5) [0,1–6, or (1–6)] **A** 5, 10, or (∞) [1–∞] **G** 1 [2–16], superior, hypanthium sometimes present.

Polygalaceae—Milkwort family (type *Polygala*, Greek for "much milk," in the belief that some species consumed by cows increase milk flow). Ca. 21 genera/ca. 965 species (Figure 8.33).

The Polygalaceae consist of trees, shrubs, lianas, or herbs. The **leaves** are simple, spiral, usually exstipulate (modified as a pair of glands or spines in some). The **inflorescence** is a spike, raceme, or panicle. The **flowers** are bisexual, zygomorphic [rarely almost actinomorphic], hypogynous to perigynous, and subtended by a pair of bracteoles. The **perianth** is biseriate, a hypanthium present in some. The **calyx** is usually aposepalous, sepals 5, the two inner, latero-posterior sepals often petaloid (resembling wing petals), rarely all or the two anterior sepals basally connate. The **corolla** is often adnate to the androecium forming a tube, petals 5 or 3 (the latter by suppression or loss of two lateral petals), when 3, the median-anterior (lower) petal apically fringed and boat-shaped. The **stamens** are 4+4, 10, or 3–7, usually basally connate forming a staminal tube. **Anthers** are longitudinal or apically poricidal in dehiscence. The **gynoecium** is syncarpous with a superior ovary, 2–5 [–8] carpels, and locules [locule rarely 1]. The **style** is often curved, often 2-lobed, one lobe stigmatic, the other sterile and comose. **Placentation** is usually apical-axile; **ovules** are pendulous, epitropous, anatropous to hemitropous, bitegmic, 1 per carpel [rarely 1 per ovary]. **Nectaries** consist of a nectariferous disk surrounding the base of the ovary; extrafloral nectaries are present in many species. The **fruit** is a loculicidal capsule, nut, samara, or drupe. The **seeds** are arillate (with caruncle) and endospermous (proteinaceous).

Members of the Polygalaceae have a mostly worldwide distribution. Economic importance includes some ornamental cultivars and plants of local medicinal value. See Eriksen

FIGURE 8.31 FABALES. Fabaceae. **A–H.** Caesalpinioideae. **A.** *Balsamocarpon brevifolium* flower, showing medial posterior petal inner to lateral petals. **B.** *Senna cumingii* flower, with medial posterior petal inner to lateral petals. Note distinct, heteromorphic (trimorphic) stamens. **C–I.** Mimosoid clade. **C–E.** *Acacia longifolia*. **C.** Leaves (phyllodes) and spicate inflorescences. **D.** Close-up of flowers of spike. **E.** Individual flower. Note inconspicuous perianth, numerous stamens, and ovary (removed from another flower). **F.** *Acacia urophylla*, with globose heads of flowers. **G.** *Pithecellobium unguis-cati*, cat claw, with heads of flowers. **H–I.** *Calliandra haematocephala*. **H.** Inflorescence heads, the flowers with very long, showy stamens. **I.** Shoot with fruits (legumes) and immature heads.

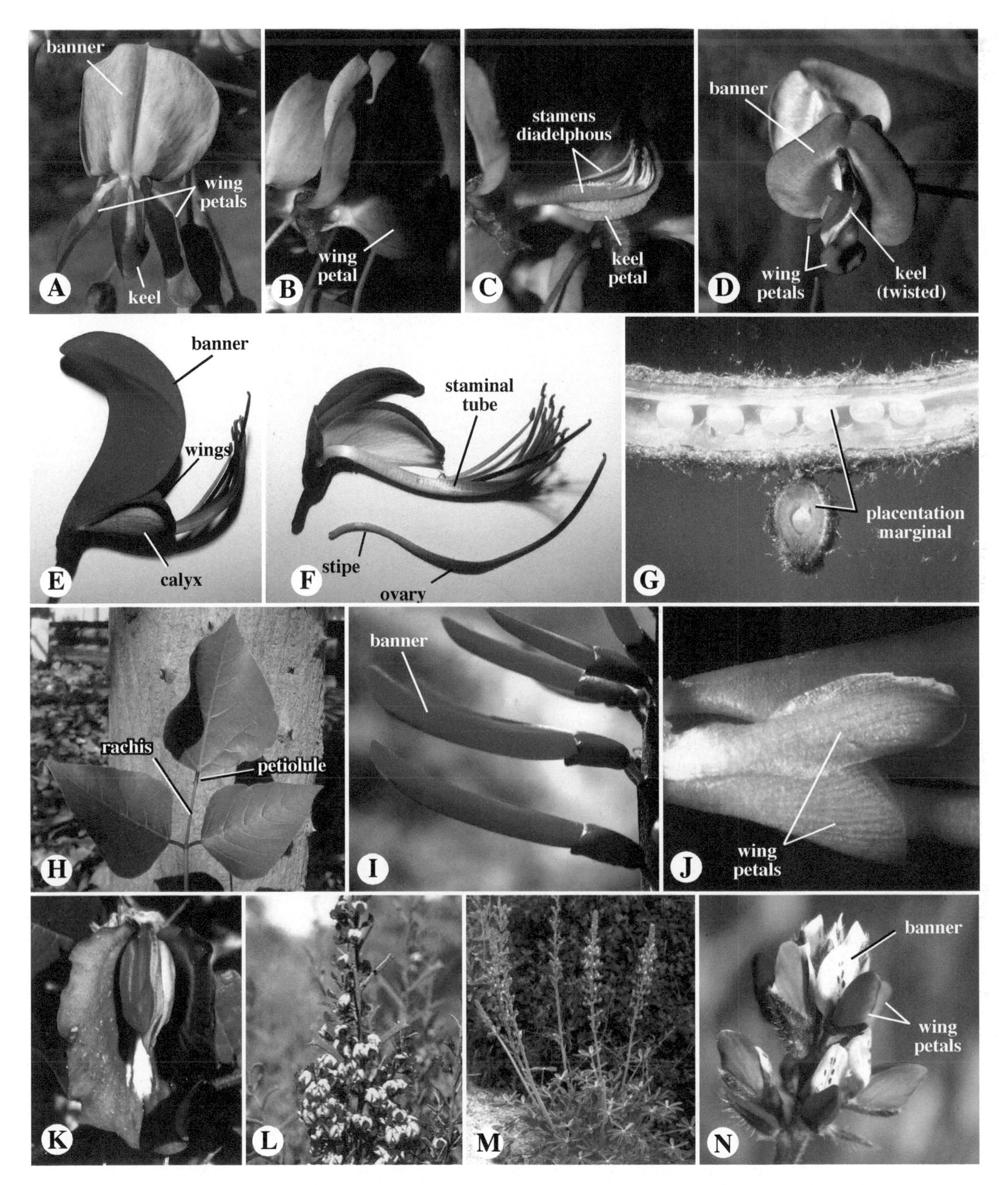

FIGURE 8.32 FABALES. Fabaceae, Faboideae (Papilionoideae). **A–C.** *Wisteria sinensis*. **A.** Papilionaceous lower, face view. Note large banner, outer to wing petals. **B.** Side view, showing banner and one of two wing petals. **C.** Side view, one wing and keel petal removed, showing diadelphous stamen fusion. **D.** *Strophostyles umbellata*, with asymmetric flowers. **E–G.** *Erythrina caffra*. **E.** Flower, side view, showing banner and reduced wing and keel petals. **F.** Dissected flower, showing staminal tube (monadelphous stamen fusion) and removed stipitate ovary. **G.** Immature legume in longitudinal (above) and cross- (below) section, showing marginal placentation. **H–J.** *Erythrina coralloides*, coral bean tree. **H.** Pinnate-ternate leaf. **I.** Flowers, side view, with elongate, semitubular banner petals. **J.** Close-up of reduced wing petals. **K.** *Clitoria mariana*, a resupinate papilionoid, with banner below, keel above. **L.** *Daviesia* sp., a papilionoid. **M.** *Lupinus excubitus*, with palmate leaves. **N.** *Lupinus bicolor*, flower close-ups, with contrasting, spotted banner.

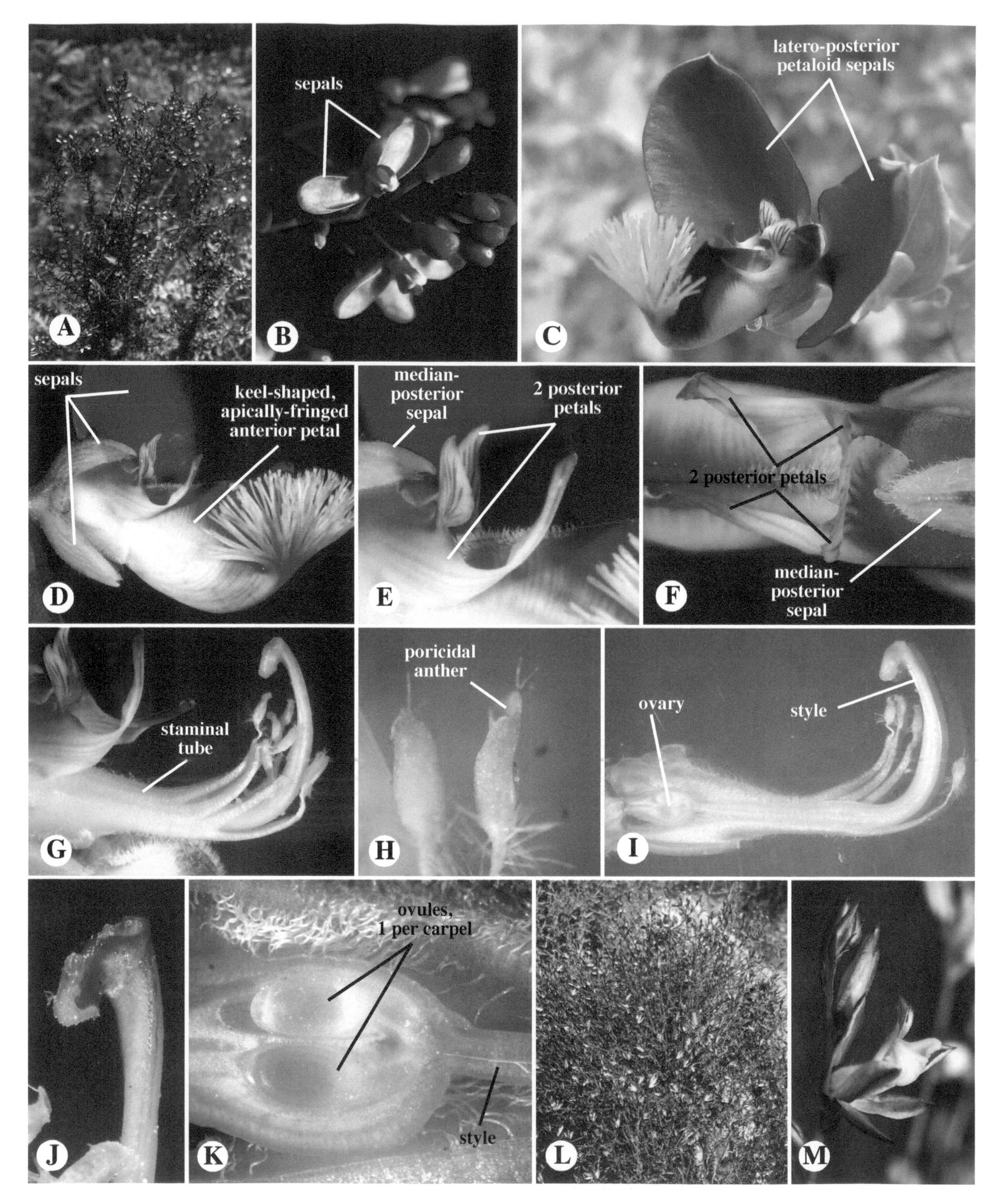

FIGURE 8.33 FABALES. Polygalaceae. **A,B.** *Comesperma ericinum*. **A.** Shrub. **B.** Flowers, showing winglike, petaloid sepals. **C–J.** *Polygala* ×*dalmaisiana*. **C.** Flower, showing two latero-posterior, petaloid sepals. **D.** Flower close-up, showing three of five sepals (one petaloid sepal removed) and petals of perianth, the anterior petal keel-shaped and apically fringed. **E.** Side view of two posterior petals, each two-lobed. **F.** Top view of two posterior petals. **G.** Androecium, the filaments basally fused into a staminal tube. **H.** Poricidal anthers. **I.** View of pistil, with style extending through middle of staminal tube. **J.** Style tip, showing two-lobed stigma. **K.** Ovary longitudinal section, showing ovules with apical-axile placentation. **L,M.** *Rhinotropis desertorum*, shrub and flower close-up.

and Persson (2006) for general information and Persson (2001) and Abbott (2011) for phylogenetic studies.

The Polygalaceae are distinctive in being trees, shrubs, lianas, or herbs, with ***simple***, ***spiral***, usually exstipulate leaves, the flowers bisexual, the perianth biseriate, with the ***2 inner (of 5) sepals often petaloid*** (resembling wing petals), petals 3–5, when 3, the anterior petal often ***apically fringed*** and ***boat-shaped***, the anthers ***poricidal or longitudinally dehiscent***, the style ***often 2-lobed with one lobe stigmatic, the other sterile***, ***ovule 1 per carpel***, the seeds ***arillate*** (with caruncle).
K 5 or (5) or (2)+3 **C** 3 or 5 **A** 4+4, 10, 3-7 **G** (2–5) [(–8)], superior.

ROSALES

The Rosales, sensu APG IV (2016), contain nine families (Table 8.2), five of which are described here. Of these, the large family Rosaceae is of particular economic importance. Four families of the Rosales — Cannabaceae, Moraceae, Ulmaceae, and Urticaceae — comprise a monophyletic group, sometimes referred to as the Urticalean Rosids (descriptions following Rosaceae treatment). See Kubitzki (2004) and Sun et al. (2016) for studies of the order.

Rhamnaceae — Buckthorn family (type *Rhamnus*, Greek name for buckthorn or other thorny shrubs). 52–57 genera/ca. 950 species (Figure 8.34).

The Rhamnaceae consist of trees, shrubs, lianas, or rarely herbs. The **roots** of some taxa are associated with nitrogen -fixing Actinomycetes bacteria. The **stems** are sometimes modified as thorns, tendrils, or "hooks." The **leaves** are simple, sometimes rudimentary, pinnately or palmately veined, spiral or opposite, stipulate or exstipulate, with stipular spines present in some taxa. The **inflorescence** is a cyme, thyrse, fascicle, or rarely a solitary flower. The **flowers** are unisexual or bisexual, actinomorphic, perigynous to epiperigynous. The **perianth** typically has a hypanthium, sometimes adnate to staminal disk. The **calyx** is aposepalous with 4–5 sepals. The **corolla** is apopetalous with 4–5 petals [absent in some taxa], these often clawed, concave, and cucullate (hooded). The **stamens** are 4–5, whorled, alternisepalous, and apostemonous. **Anthers** are longitudinal in dehiscence. The **gynoecium** is syncarpous, with a superior to inferior ovary and 2–3 [–5] carpels and locules. **Placentation** is apical-axile; **ovules** are anatropous, bitegmic, 1 [2] per carpel. **Nectaries** consist of a staminal disk, often fused to the hypanthium and/or (inferior) ovary. The **fruit** is a drupe with 1–many endocarps, a circumscissile capsule, or a schizocarp of mericarps. The **seeds** are exalbuminous.

The Rhamnaceae have a mostly worldwide distribution, especially in tropics. Economic importance includes edible fruits (e.g., *Ziziphus jujuba*, jujube, or *Z. lotus*, lotus fruit), ornamental cultivars, and dye, medicinal, soap, timber and varnish plants; *Ziziphus spina-christi* is purported to be the true Christ's crown of thorns. See Medan and Schirarend (2004) for general information and Richardson et al. (2000) and Hauenschild et al. (2016) for phylogenetic studies.

The Rhamnaceae are distinctive in being trees, shrubs, lianas, or rarely herbs with simple, spiral or opposite leaves, unisexual or bisexual, ***perigynous to epiperigynous flowers***, the perianth/androecium ***4–5-merous***, petals sometimes absent, ***stamens alternisepalous***, a ***nectariferous disk usually adnate to hypanthium***, the fruit a ***drupe, circumscissile capsule, or schizocarp***.
K 4–5 **C** 4–5 [0] **A** 4–5 **G** (2–3) [(–5)], superior to inferior, hypanthium usually present.

Rosaceae — Rose family (type *Rosa*, Latin for various roses). 85–90 genera/2,500–3,000 species (Figures 8.35–8.36).

The Rosaceae consist of trees, shrubs, or herbs. The **leaves** are spiral (rarely opposite), simple or compound, undivided to divided, usually stipulate (lost in some taxa), the stipules often adnate to the petiole base. The **inflorescence** is variable. The **flowers** are bisexual (usually), actinomorphic, perigynous or epiperigynous; the receptacle is sometimes expanded or sunken. The **perianth** is biseriate and dichlamydeous, usually pentamerous, imbricate, a hypanthium present. The **calyx** is aposepalous with 5 [3–10] sepals. The **corolla** is apopetalous with 5 [0, 3–10] petals. The **stamens** are 20–∞ [1,5], whorled, arising centripetally, usu. apostemonous. **Anthers** are longitudinal or rarely poricidal in dehiscence and dithecal. The **gynoecium** is syncarpous or apocarpous, with a superior or inferior ovary, 1–∞ carpels, and 1–∞ locules. The **style(s)** are terminal or lateral. **Placentation** is axile, basal, or marginal; **ovules** are 1–∞. **Nectaries** are often present on the hypanthium. The **fruit** is a drupe, pome, hip, follicetum, achenecetum, or capsule. The **seeds** are usually without endosperm.

The Rosaceae have been classified into three subfamilies (Potter et al. 2007; see also Zhang et al. 2017): Rosoideae, having an apocarpous or unicarpellous gynoecium forming an achene, achenecetum, or drupecetum, including taxa in which the receptacle is expanded and fleshy (e.g., *Fragaria*) or sunken (e.g., the hips of *Rosa*); Dryadoideae, having a drupecetum or achene, but all having a symbiotic relationship with the nitrogen fixing actinomycete *Frankia*; and Spiraeoideae. The Spiraedoideae show complicated variation in ovary/fruit morphology, encompassing taxa with an apocarpous gynoecium forming a follicetum (*Spiraea* and

FIGURE 8.34 ROSALES. Rhamnaceae. **A,B.** *Ceanothus crassifolius*. **C.** *Ceanothus tomentosus*, inflorescence. **D.** *Ceanothus verrucosus*, flower close-up. **E.** *Ceanothus griseus*, flower close-up. Note antipetalous stamens. **F,G.** *Ceanothus tomentosus*. Flowers, showing nectariferous disk, ovary, and hypanthium. **H.** *Frangula californica*. **I,J.** *Rhamnus crocea*. **I.** Flower. Note absence of petals and alternisepalous stamens. **J.** Shoot bearing fruits. **K.** *Colletia cruciata*, with thick, spinose leaves. **L.** *Colletia paradoxa*. **M.** *Ziziphus* sp.

The Rose Family

The rose is a rose,
And was always a rose.
But the theory now goes
That the apple's a rose,
And the pear is, and so's
The plum, I suppose.
The dear only knows
What will next prove a rose.
You, of course, are a rose—
But were always a rose.

Robert Frost (1874–1963)

FIGURE 8.35 ROSALES. Rosaceae. **A.** *Rosa* sp., one of many forms of cultivated roses. **B.** *The Rose Family*, by Robert Frost. (Frost, R., and J. J. Lankes. 1928. West-Running Brook. H. Holt, New York.)

relatives, classified as Spiraeeae), taxa with a single, superior-ovaried pistil bearing one ovule, the fruit a drupe (*Prunus* and relatives, classified as the tribe Amygdaleae, formerly Prunoideae), and taxa with an inferior ovary, forming a pome (*Malus* and relatives, classified as the Pyrinae, formerly Maloideae).

Members of the family have mostly worldwide distributions, but are more concentrated in north temperate regions. The family is very economically important as the source of many cultivated fruits, including *Fragaria* (strawberry), *Malus* (apples), *Prunus* (almond, apricot, cherry, peach, plum), *Pyrus* (pear), and *Rubus* (blackberry, raspberry), as well as essential oils (e.g., *Rosa*), and numerous ornamental cultivars, such as *Cotoneaster*, *Photinia*, *Prunus* (cherries), *Pyracantha*, *Rosa* (roses), and *Spiraea*. See Potter et al. (2002, 2007) and Zhang et al. (2017) for phylogenetic studies of the family.

The Rosaceae are distinctive in having usually ***stipulate*** leaves (often adnate to petiole) and an actinomorphic, generally ***pentamerous*** flower with ***hypanthium present***, variable in gynoecial fusion, ovary position, and fruit type.

K 5[3–10] **C** 5[0,3–10] **A** 20–∞[1,5] **G** 1(–∞), superior or inferior, hypanthium present.

URTICALEAN ROSIDS

Nested within the Rosales are four families, which in the past have been variously delimited in an order "Urticales." This group, informally termed Urticalean Rosids, include taxa with small, unisexual, wind-pollinated flowers, an apparent apomorphy for the complex (Figure 8.37). Additional apomorphies for lineages in the complex are seen in Figure 8.50. Three of the Urticalean Rosid families — **Moraceae**, **Ulmaceae**, and **Urticaceae** — are described in detail below. The fourth family, **Cannabaceae**, includes the economically and culturally important *Cannabis sativa* (used as a bast fiber plant, hemp, and the euphoric and medicinal plant, marijuana; see Chapter 1) and *Humulus lupulus* (hops, the flavoring plant of beer production), as well as taxa previously classified in the Celtidaceae, such as *Celtis*, hackberry (Figure 8.37). See Sytsma et al. (2002) for a phylogenetic and character analysis of the complex and Sun et al. (2016b) and Zhang et al. (2017) for phylogenetic analyses.

Moraceae — Mulberry family (type *Morus*, Latin name for mulberry). Ca. 39 genera/ca. 1,125 species (Figure 8.38).

The Moraceae consist of monoecious or dioecious trees, shrubs, lianas, and herbs, often with laticifers bearing a milky latex. The **roots** are prop or buttress in some taxa. The **leaves** are simple [rarely compound], spiral or opposite, stipulate. The **inflorescence** is axillary and variable in morphology, consisting of a spike (catkinlike in *Morus*), raceme, head (in some taxa with flowers borne upon the surface of an

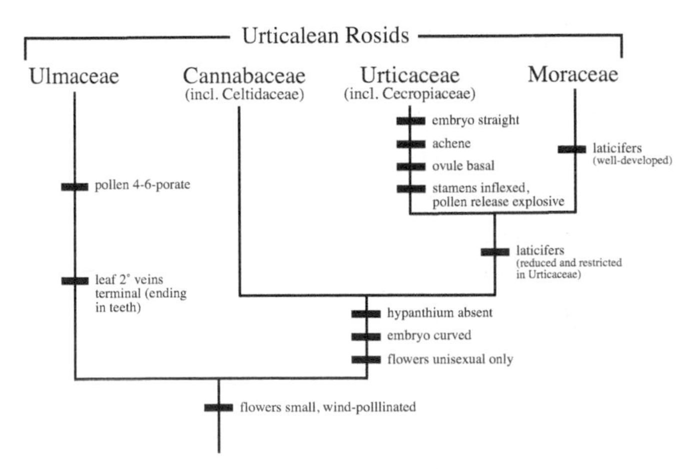

FIGURE 8.37 Cladogram of "Urticalean Rosids, after Sytsma (2002).

FIGURE 8.36 ROSALES. Rosaceae. **A–E.** Rosoideae. **A–C.** *Rosa californica*. **A.** Flower. **B.** Sunken receptacle of flower. **C.** Fruit, a hip. **D,E.** *Drymocallis glandulosa*, with numerous pistils of an apocarpous gynoecium. **F–M.** Spiraeoideae. **F–H.** Pyrinae. **F.** *Pyrus kawakamii*, flower, face view. **G,H.** *Malus domestica*, apple. **G.** Flower longitudinal section, showing inferior ovary, characteristic of the subfamily. **H.** Apple fruit, a pome, comprised mostly of hypanthial tissue. **I,J.** Amygdaleae, *Prunus* sp., cherry, with a superior ovary. **K–M.** Spiraeeae, *Spiraea vanhoutii*. **K.** Flower close-up, showing 5 pistils inside hypanthium. **L.** Flower longitudinal section. **M.** Ovary longitudinal-section, with marginal placentation.

invaginated compound receptacle), or in *Ficus* an enclosed hypanthodium. The **flowers** are unisexual, small, actinomorphic, hypogynous or epigynous. The **perianth** is uniseriate [rarely biseriate], 0–10, the perianth parts (often termed a calyx) connate, at least basally. The **stamens** are 1–6, opposite and usually as many as the perianth parts; anthers are dithecal or (in *Ficus* spp.) monothecal. The **gynoecium** is syncarpous, with a superior or inferior ovary, 2 [3] carpels, and 1 [2–3] locules. The **styles** are typically 2. **Placentation** is apical (to subapical); **ovules** are solitary, anatropous to campylotropous, and bitegmic. The **fruit** is a multiple of achenes, each unit achene often surrounded by the accrescent, fleshy perianth (thus resembling a drupe) or borne on a fleshy compound receptacle, in *Ficus* borne on the inner surface of an enclosed syconium. The **seeds** are 1 per unit fruit, albuminous or exalbuminous. Plants are wind or insect pollinated, in *Ficus* spp. having an intricate pollination mechanism with wasps.

The Moraceae have a worldwide distribution, from tropical to temperate regions. Economic importance includes fruit trees, such as *Artocarpus altilis* (breadfruit), *Ficus carica* (edible fig), and *Morus* spp. (mulberry); paper, rubber, and timber trees; and some cultivated ornamentals, especially *Ficus* spp., figs; the leaves of *Morus alba* are the food source of silkworm moth larvae. See Datwyler and Weiblen (2004) and Zerega et al. (2005, 2010), and Clement and Weiblen (2009) and Gardner et al. (2017) for studies of the Moraceae.

The Moraceae are distinctive in being ***monoecious or dioecious*** trees, shrubs, lianas, or herbs with a ***milky latex***, ***stipulate***, simple leaves, and ***unisexual flowers***, the female with a usually ***2-carpellate (2 styled) pistil*** and a ***single, apical to subapical ovule***, the fruit a ***multiple of achenes***, in some taxa with an enlarged compound receptacle or syconium. **P** (0–10) **A** 1–6 **G** (2) [(3)], superior or inferior.

Ulmaceae — Elm family (type *Ulmus*, Latin for elm). 6 genera/35 species (Figure 8.39).

The Ulmaceae consist of monoecious or hermaphroditic trees, often with mucilage canals. The **leaves** are simple, with secondary veins terminating in teeth (except *Ampelocera*), usu. spiral but usually becoming distichous, stipulate, serrate, often basally oblique. The **inflorescence** is axillary, of solitary flowers or a cyme or panicle. The **flowers** are small, unisexual or bisexual, actinomorphic. The **perianth** is of 5 [2–9], spiral tepals, apotepalous or basally connate. The **stamens** are typically numerous. **Anthers** are dorsifixed-versatile. The **pollen** is 4–7-porate. The **gynoecium** is syncarpous, with a superior ovary, 2–3 carpels and locules. The **styles** are 2–3. **Placentation** is apical-pendulous; **ovules** are anatropous to amphitropous, bitegmic, 1 per locule. The **fruit** is a dry, often a flattened samara.

The Ulmaceae are distributed in temperate N. America and Eurasia, tropical America, tropical Africa, and s.e. Asia. Economic importance includes timber trees (esp. *Ulmus*, elm, plus *Holoptelea, Phyllostylon, Zelkova*), fiber, fodder, and medicinal plants, and cultivated ornamentals. Species of *Ulmus* (e.g., *U. americana*) are susceptible to Dutch Elm disease (*Ophiostoma* spp.), an ascomycete fungus, decimating entire forests. See Sytsma et al. (2002) for a phylogenetic study.

The Ulmaceae are distinctive in being ***trees*** with simple, often ***distichous, serrate, oblique*** leaves, the secondary veins ***terminating at teeth***, the flowers small, ***unisexual or bisexual***, wind-pollinated, the ovary superior, 2–3-loculed with 1 pendulous ovule per locule, the fruit ***dry,*** often a ***flattened samara***. **P** 5 [2–9] **A** ∞ **G** (2–3), superior, unisexual or bisexual.

Urticaceae [including Cecropiaceae] — Nettle family (type *Urtica*, Latin for "to burn," after the stinging trichomes of some family members). Ca. 54-55 genera/1,600-2,600 species (Figure 8.40).

The Urticaceae consist of dioecious or monoecious [rarely hermaphroditic] herbs (annual or perennial), shrubs, trees, or lianas, some epiphytic with aerial or prop roots. Siliceous stinging (urent) trichomes are found in some taxa; the trichome tip breaks off upon contact, injecting toxins (reported to be 5-hydroxytryptamine, serotonin, and acetylcholine, or an unidentified neurotoxin, apparently *not* formic acid) from a basal reservoir of the trichome. Crystalline, calcium carbonate cystoliths (within "lithocyst" cells) are often present and of taxonomic importance. Latex channels are sometimes present, but do not produce a milky sap. The **stems** of branches are specialized in *Cecropia* in being hollow between nodes, often functioning as a home for guard *Azteca* ants. The **leaves** are simple, unlobed to deeply palmately divided (e.g., *Cecropia*), spiral or opposite, usually stipulate (stipules connate in some taxa), with pinnate, palmate, or unequally ternate venation; pad-like food (Müllerian) bodies at petiole bases of *Cecropia* function as a reward for guard ants. The **inflorescence** is an axillary cyme, globose head, spike, or of solitary flowers, inflorescence bracts connate or forming an involucre in some taxa. The **flowers** are small, usually unisexual, mostly actinomorphic, hypogynous. The **perianth** is a uniseriate calyx, apo- or synsepalous, mostly 4–5-merous [1–6, 0 in female flowers of some taxa]. The **stamens** are 4–5 [1–6], uniseriate, generally equal in number to sepals, antisepalous, apostemonous; filaments are straight or inflexed, the latter under tensile pressure and elastically reflexing at anthesis, functioning to catapult the pollen (at the highest speed every measured for a plant). **Anthers** are dithecal and longitudinally dehiscent. A **pistillode** is present in male flowers of some taxa. The **gynoecium** of female flowers consists of a single pistil with a superior ovary with locule and carpel 1 (by reduction, often termed "pseu-

FIGURE 8.38 ROSALES. Moraceae. **A.** *Artocarpus altilis*, breadfruit, a tree with pinnately cleft leaves and heads of unisexual flowers. **B–D.** *Broussonetia papyrifera*, paper mulberry. **B.** Inflorescence, a head. **C.** Flower, showing fleshy perianth surrounding pistil. **D.** Achene longitudinal section, showing single seed. **E,F.** *Dorstenia* sp., with headlike, compound receptacle bearing minute flowers. **G,H.** *Ficus benghalensis*. **G.** Prop roots. **H.** Habit, a large tree, here covering more than an acre of ground. **I–K.** *Ficus carica*, edible fig. **I.** Shoot, with fruits (syconia). **K.** Mature syconium, longitudinal section. **K.** Syconium apex, showing scales surrounding pore and wasp. **L.** *Maclura pomifera*, osage-orange, showing globose, multiple fruit of drupelike achenes. **M–P.** *Morus* sp., mulberry. **M.** Female inflorescence, a head of flowers. **N.** Flower close-up, showing outer perianth, which forms a fleshy layer in fruit. **O.** Ovary longitudinal section, showing subapical placentation. **P.** Fruit, a multiple of achenes, each surrounded by a fleshy perianth.

FIGURE 8.39 ROSALES. Ulmaceae, **A–F.** *Ulmus pumila*. **A.** Shoot, showing distichous leaf arrangement. **B.** Node close-up, showing stipule scar and oblique leaf base. **C.** Shoot with fruit. **D–E.** Leaf venation, showing veinlet terminating at tooth. **F.** Fruits, flattened samaras. **G.** *Ulmus alata*, pistillate flowers, maturing to fruits. (Image courtesy of W. Cook.) **H.** *Ulmus thomasii*, vegetative shoot. **I.** *Ulmus americana*, infructescence of flattened samaras.

FIGURE 8.40 ROSALES. Urticaceae. **A–C.** *Cecropia* sp. **A–B.** Tree, with deeply pamlately divided leaves. **C.** Stem close-up, showing guard ants. **D.** *Boehmeria cylindrica*. **E,F.** *Parietaria hespera*. **E.** Close-up of cyme; achene at inset. **F.** Leaf, showing spherical cystoliths. **G.** *Soleirolia soleirolii*, leaf face view, showing cystolith crystal of lithocyst cell. **H–K.** *Hesperocnide tenella*. **H.** Whole plant, with opposite leaves and urent trichomes. **I.** Inflorescence of axillary cymes. **J.** Cyme close-up. **K.** Stinging (urent) trichome. **L–P.** *Urtica urens*, a stinging nettle. **L.** Male flower, showing 4 sepals and inflexed stamens. **M.** Male flower after filaments have reflexed. **N.** Mature female flower, with achene. **O.** Achene, longitudinal section, showing straight embryo. **P.** Embryo, face view, dissected from seed.

FIGURE 8.41 CUCURBITALES. **A–C.** Apodanthaceae, a family of endo-parasites. *Pilostyles thurberi*. **A.** Flowers of parasite erupting from bark of host (*Psorothamnus thurberi*, Fabaceae). **B,C.** Close-up of flowers. **D–F.** Datiscaceae, *Datisca glomerata*. **D.** Plant, a perennial herb. Inset: leaf, pinnately divided. **E.** Bisexual flower. **F.** Fruit, a capsule.

domonomerous"; rarely 2). The **style** is solitary and terminal, **stigma** 1 [rarely 2], "penicillate" (paintbrush-shaped) in some taxa. **Staminodes** sometimes present. **Placentation** is basal, with a solitary, orthotropous [hemitropous], bitegmic **ovule**. **Nectaries** are absent. The **fruit** is a achene, nut, or drupe, often with accrescent perianth; in some taxa the perianth is fleshy and envelopes the achene, forming a "pseduodrupe"; the achenes of some taxa are ejected by elastically reflexing staminodes of the female flowers. The **seeds** have an oily or starchy endosperm with a straight embryo. Flowers are wind pollinated.

The Urticaceae are distributed worldwide in temperate and tropical regions. Economic importance includes several taxa used as fiber plants (e.g., *Boehmeria nivea*, ramie), cultivated ornamentals (e.g., *Pilea* spp., *Elatostema* [incl. *Pellionia*] spp., and *Soleirolia soleirolii*), and some leaf vegetable crops. *Cecropia* spp. are ecologically important pioneer trees in the neotropics. Trichomes of the "stinging nettles" (tribe Urticeae) cause contact dermatitis; some of these (e.g., *Dendrocnide moroides*, the "Gympie Gympie" of Australia) have a very painful and long-lasting sting, apparently caused by an unidentified neurotoxin, that is quite dangerous or even fatal to domesticated animals and humans. The former Cecropiaceae, consisting of woody, tropical plants, are non-monophyletic and nested within the Urticaceae s.l. See Friis (1993) and Kubitzki (1993a) for general information about the family and Systma et al. (2002), Hadiah et al. (2008), and Zerega et al. (2005) for recent phylogenetic analyses of the family and close relatives.

The Urticaceae are distinctive in being usually ***monoecious or dioecious*** herbs, shrubs, trees, or lianas, often with calcium carbonate ***cystoliths***, some taxa having ***stinging trichomes***, the flowers ***small, wind-pollinated***, with a ***uniseriate perianth*** [rarely absent], the male flowers with straight or, in many taxa, ***inflexed, pollen-catapulting filaments***, female flowers with a ***unilocular,*** *usu.* ***unicarpellate (pseudomonomerous) ovary*** having a ***single, basal, orthotropous ovule***, the fruit an ***achene, nut, or drupe***, often attached to an ***accrescent perianth***.

Male: P 4–5 or (4–5) [1–6] **A** 4–5 [1–6].
Female: P 4–5 or (4–5) [0] **G** 1 [(2)], superior.

CUCURBITALES

The Cucurbitales contain eight families (Table 8.2), two of which are described here in detail. Notable of the families not described are the **Apodanthaceae**, endo-parasitic plants, bursting from the bark of their host when flowering (Figure 8.41A–C), and the **Datiscaceae** (Figure 8.41D–F), the two species found in w. North America and Asia, respectively. See Zhang and Renner (2003) and Zhang et al. (2006) for studies of the order.

Begoniaceae — Begonia family (type *Begonia*, after Michel Bégon (1638–1710), governor of French Canada). 2 genera (*Begonia* and *Hillebrandia*)/ca.1,870 species (Figure 8.42).

The Begoniaceae consist of usually monoecious herbs, shrubs, or vines. The **roots** are fibrous or tuberous. The **stems** are generally succulent, with rhizomes or pachycauls in some taxa. The **leaves** are alternate, spiral, or distichous, often oblique-asymmetrical, usually simple [palmately lobed or compound in some], usually palmately veined, stipulate, the stipules often large. The **inflorescence** is an axillary cyme. The **flowers** are unisexual, actinomorphic, biradial, or bilateral, epigynous. The **perianth** is homochlamydeous, tepals petaloid, uniseriate or biseriate, distinct or basally connate, male flowers usually 2+2 [5+5] and valvate, female flowers usually 5

FIGURE 8.42 CUCURBITALES. Begoniaceae. **A–B.** *Hillebrandia sandwicensis*. [Images courtesy of Gerald Carr.] **A.** Whole plant. **B.** Flower close-ups. Note half-inferior ovary and 5 carpels/styles of female flowers. **C–M.** *Begonia* sp. **C.** Whole plant, showing alternate leaves with characteristic oblique base. **D.** Stem cross section, showing succulent cortex. **E.** Close-up of stipules. **F.** Cymose inflorescence. **G.** Male flower, with biseriate perianth. **H.** Androecium. **I.** Anther, dithecal, longitudinally dehiscent. **J–M.** Female flowers. **J.** Face view. Note 5, imbricate perianth parts. **K.** Three styles, each bifurcate and coiled. **L.** Side view, showing inferior, winged ovary. **M.** Ovary cross section, showing axile placentation, placentae bifurcate. **N–S.** *Begonia* cultivars, illustrating beautiful variation in leaves and flowers.

[5+5], and imbricate. The **stamens** are 4–∞, whorled or secund, distinct. **Anthers** are longitudinal or poricidal in dehiscence, dithecal, with connective often elongate. The **gynoecium** is syncarpous, with an often winged, inferior [half-inferior in *Hillebrandia*] ovary having 2–3 [–6] carpels and locules. The **styles** are as many as carpels, **stigmas** bifid and twisted in some taxa. **Placentation** is axile or parietal and almost touching, placentae unbranched or bifid; **ovules** are anatropous, bitegmic, numerous. The **fruit** is usually a loculicidally or septicidally dehiscent capsule, rarely a berry. The **seeds** are numerous. Plants often have CAM photosynthesis.

The Begoniaceae are distributed worldwide in tropical or warm regions. Economic importance is primarily as cultivated ornamentals, the genus *Begonia* having thousands of hybrids and cultivars. The monotypic genus *Hillebrandia* (*H. sandwicensis*, of Hawaii) is sister to *Begonia* (Clement et al. 2004). See Clement et al. (2004) for a phylogenetic analysis of the family and close relatives and Forrest and Hollingsworth (2003), Forrest et al. (2005), Plana (2003), and Moolight et al. (2018) for phylogenetic analyses of *Begonia* and relatives.

The Begoniaceae are distinctive in being mostly ***monoecious*** herbs or shrubs with tuberous to fibrous roots, often succulent stems, and often ***oblique-asymmetrical, simple or palmately lobed to compound*** leaves, the inflorescence axillary and cymose, the males flowers typically with ***2+2, valvate tepals***, female flowers typically with ***5 [5+5], imbricate tepals***, the ***ovary inferior [half-inferior], often winged***, with ***2–3 [–6] carpels and locules*** and numerous ovules/seeds, the fruit a ***capsule or berry***.

Male: **P** 2+2 [5+5] **A** 4–∞.
Female: **P** 5 [5+5] **G** (2–3) [(–6)], inferior.

Cucurbitaceae — Cucumber/Gourd family (type *Cucurbita*, Latin for gourd). Ca. 98 genera/1,000 species (Figure 8.43).

The Cucurbitaceae consist of monoecious or dioecious [rarely hermaphroditic] vines [rarely tree-like], usually with one tendril per node. The **leaves** are simple, palmately veined and often palmately lobed, spiral, and exstipulate. The **inflorescence** is axillary, variable in type or with flowers solitary. The **flowers** are usually unisexual, actinomorphic, the female flowers epiperigynous. The **perianth** is biseriate and dichlamydeous, with hypanthium present. The **calyx** is aposepalous with 5 [3–6] imbricate sepals. The **corolla** is apopetalous or sympetalous with 5 [3–6] valvate petals. The **stamens** are 3–5, alternipetalous, distinct or connate. **Anthers** are longitudinal in dehiscence, dithecal or monothecal. The **gynoecium** is syncarpous, with an inferior ovary, 3 [1–5] carpels, and 1 locule [locules rarely as many as carpels]. The **styles** are 1–3; **stigmas** are 1–2. **Placentation** is parietal, rarely axile; **ovules** are anatropous, bitegmic, generally ∞ [rarely 1–few]. Extrafloral **nectaries** are often present. The **fruit** is a berry, pepo, capsule, or samara. The **seeds** are exalbuminous. Stem anatomy is typically bicyclic, with bicollateral vascular bundles.

The Cucurbitaceae have largely worldwide distributions, but occur mostly in tropical regions. Economic importance includes important food crops such as *Citrullus lanatus* (watermelon), *Cucumis melo* (melons), *Cucumis sativa* (cucumber), *Cucurbita pepo* and other spp. (squashes, pumpkins) and a number of other taxa; the dried fruits of several species are used as gourds, those of *Luffa* (luffa) are used as a sponge; some taxa have medicinal or horticultural uses. See Renner et al. (2002), Schaefer et al. (2008), and Schaefer and Renner (2011a,b) for recent studies of the family.

The Cucurbitaceae are distinctive in being mostly ***monoecious or dioecious vines*** with ***simple, palmately veined and/or lobed leaves***, usually with ***tendrils***, the ***female flowers epiperigynous***, with usually ***parietal placentation and three carpels***, the fruit a berry, pepo, capsule, or samara.

K 5 [3–6] **C** 5 [3–6] or (5) [(3–6)] **A** 3–5 or (3–5) **G** (3) [(2–5)], inferior, hypanthium present.

FAGALES

The Fagales contain eight families (Table 8.2), members of which are largely monoecious and wind pollinated. Three families are described here. Notable among the others are the **Casuarinaceae**, including *Casuarina*, the Australian-pine (Figure 8.44A,B); **Myricaceae**, the wax-myrtle family (Figure 8.44C); **Nothofagaceae**, the southern-beech family, which include important timber trees (Figure 8.44D); and **Ticodendraceae**, composed of one species native to central America (Figure 8.44E). See Li et al. (2002, 2004, 2016) and Manos and Steele (1997) for detailed treatments of the order.

Betulaceae—Birch family (type *Betula*, Latin name for birch or for pitch, derived from bark). 6 genera/ca. 145 species (Figure 8.45).

The Betulaceae consist of monoecious trees or shrubs. The **leaves** are simple, deciduous, usually spiral, caducous -stipulate, with the margin usually toothed. The **inflorescences** are unisexual, the male inflorescence is a pendulous catkin, the female inflorescence a short, pendulous or erect catkin, both bearing numerous, 1–3-flowered, bracteate, simple dichasia. The **flowers** are unisexual, hypogynous or epigynous. The **perianth** is uniseriate (by default termed a **calyx**), of 1–6 [0], scale-like sepals/lobes. A **corolla** is absent. The **stamens** are 1–∞, generally the same number as the perianth parts, antisepalous. **Anthers** have thecae either divided along the connective or connate. The **gynoecium** is syncarpous, with an inferior or superior ovary (the latter sometimes termed "nude" because of lacking a perianth and therefore a

FIGURE 8.43 CUCURBITALES. Cucurbitaceae. **A.** *Cucumis melo*, cantaloupe. **B.** *Cucurbita foetidissima*, calabazilla. **C.** *Cucurbita pepo*, a pumpkin. **D–F.** *Cucurbita pepo*, a squash. **D,E.** Female flower. **F.** Male flower. **G–L.** *Marah macrocarpus*, manroot. **G.** Whole plant. Note vine habit, simple, palmately lobed leaves, and tendrils. **H.** Male flower. **I.** Close-up of connate anthers of male flower. **J.** Female flower, showing inferior ovary. **K.** Ovary cross section, showing parietal placentation (only two carpels). **L.** Immature, prickly fruit.

FIGURE 8.44 FAGALES. **A,B.** Casuarinaceae, *Casuarina equisetifolia*. **A.** Flowering branch. Inset: inflorescence (above), infructescence (below). **B.** Close-up of lateral, photosynthetic branches, with striate stems and whorled, scale leaves. **C.** Myricaceae, *Myrica californica*. **D.** Nothofagaceae, *Nothofagus fuscus*, in fruit. **E.** Ticodendraceae, *Ticodendron incognitum*.

point of reference for ovary position), and 2 [3] carpels; locules are 2 [3] below, 1 above. The **styles** are 2–3, distinct. **Placentation** is apical-axile, the ovules pendulous from the apex of the septa; **ovules** are anatropous, unitegmic or bitegmic, 1–2 per locule. The **fruit** is a nut or 2-winged samara, subtended by woody bracts of a cone-like infructescence or partially enclosed by leafy bracts. The **seeds** are with or without endosperm. Plants are wind pollinated.

The Betulaceae are distributed in northern temperate and mountainous, tropical regions. The family is usually divided into two subfamilies: Betuloideae (including *Alnus*, alders, and *Betula*, birches) with male flowers in groups of 3 (a full, simple dichasium), and the Coryloideae (including *Corylus*, hazels/filberts, *Carpinus*, ironwood, and *Ostrya*, hornbeam) with male flower units reduced to 1. Economic importance of the family includes lumber trees (some woods very dense), chemical derivatives, nuts (e.g., *Corylus*, hazels/filberts), cultivated ornamentals, and numerous uses by aboriginal people. See Forest et al. (2005), Yoo and Wen (2008), and Grimm and Renner (2013) for phylogenetic studies of the family.

The Betulaceae are distinctive in being ***monoecious trees or shrubs*** with simple, toothed leaves, and bearing ***pendulous, elongate male catkins*** and ***pendulous to erect female catkins***, each with numerous, ***bracteate dichasia***, the fruit a ***nut or 2-winged samara***.

P 1–6 [0] **A** 1–∞ **G** (2) [(3)], superior (nude) or inferior.

Fagaceae — Oak family (type *Fagus*, Latin for the beech tree). 7 genera/ca. 730 species (Figure 8.46).

The Fagaceae consist of monoecious, rarely dioecious, trees or shrubs. The **leaves** are simple, undivided to divided, usually spiral, rarely opposite or whorled, stipulate, the stipules deciduous. The **inflorescence** is usually unisexual, the male inflorescence a catkin, spike, or head of reduced dichasia, the female flowers located at the base of male inflorescences or solitary. The **flowers** are small, unisexual, actinomorphic, the female flowers epigynous and involucrate, the involucral bracts often fused forming a cupule (e.g., acorn cup). The **perianth** is composed of 6 [4–9] tepals. The **stamens** are 6–12 [4–90], distinct. **Anthers** are longitudinal in dehiscence. The **gynoecium** is syncarpous, with an inferior ovary, and usually 3 or 6 [2,7–12] carpels, the locules as many as carpels basally, opening to one locule apically. The **styles** are as many in number as carpels. **Placentation** is basally axile; **ovules** are anatropous, bitegmic, 2 per carpel. The **fruit** is a nut (sometimes termed a "glans") with a usually hard pericarp, subtended by a 2 or more valved cupule (e.g., the acorn "cup") of sometimes spiny appendages. The **seeds** are exalbuminous. Plants are usually wind pollinated, although they are insect pollinated in *Castanea*.

The Fagaceae have a mostly worldwide distribution in nontropical regions. Economic importance includes important lumber trees, such as *Quercus* (oak), *Fagus* (beech), and *Castanea* (chestnut); the outer bark of *Quercus suber* is the source of commercial cork; the seeds of various species have been a traditionally important source of food for humans and other animals. See Manos et al. (2001) and Oh and Manos (2008) for phylogenetic studies of the Fagaceae.

The Fagaceae are distinctive in being ***monoecious*** (rarely dioecious) ***trees or shrubs*** with ***simple leaves*** (sometimes divided), the ***flowers unisexual and small***, the male flowers in ***catkins or heads of reduced dichasia***, the female at base of male inflorescences or solitary, with an inferior, multicarpellate ovary, the fruit a ***nut with subtending 2-many valved cupule bearing appendages***.

Male: **P** 6 [4–9] **A** 6–12 [4–90]

Female: **P** 6 [4–9] **G** (3,6) [(2,7–12)], inferior.

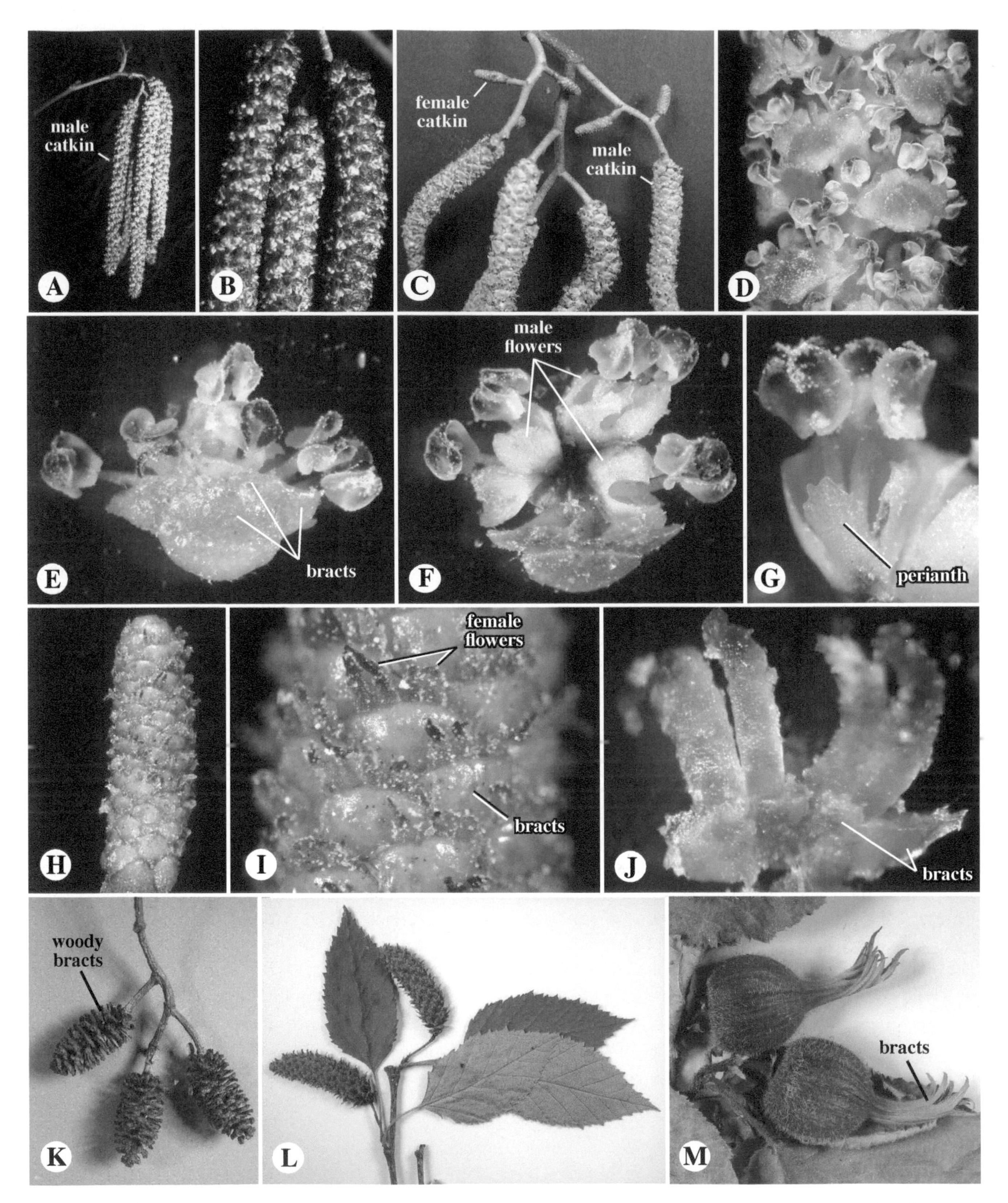

Figure 8.45 FAGALES. Betulaceae. **A,B.** *Alnus serrulata*, showing male catkins. **C–K.** *Alnus rhombifolia*, white alder. **C.** Male and female catkins. **D.** Close-up of male catkin. **E,F.** Male simple dichasium unit, three flowers with subtending bracts. **G.** Male flower, with perianth. **H.** Female catkin. **I.** Catkin, close-up, showing styles exserted from bracts. **J.** Female flowers with subtending bracts. **K.** Infructescence with woody textured bracts. **L.** *Betula papyrifera*, paper birch, with infructescence. **M.** Fruits of *Corylus cornuta*, beaked hazelnut, with subtending bracts.

Figure 8.46 FAGALES. Fagaceae. **A,B.** *Castanea dentata*, chestnut. **A.** Male spikes. **B.** Fruits, showing prickly involucral bracts. **C–E.** *Quercus suber*, cork oak. **C.** Outer bark, from which commercial cork is derived. **D.** Male flowers of catkin. **E.** Female flower, showing styles and young involucral bracts. **F.** *Quercus* sp., showing catkins and female flowers. **G.** *Quercus chrysolepis*, with mature acorns in which involucral bracts are fused into acorn cup. **H.** *Quercus acutissima*, with prominent, relatively distinct involucral bracts.

Juglandaceae — Walnut family (type *Juglans*, after Latin *Jovis*, Jupiter + *glans*, acorn). Ca. 9 genera/ca. 51 species (Figure 8.47).

The Juglandaceae consist of monoecious or dioecious trees or shrubs. The **twigs** of some taxa have superposed buds and a diaphragmed pith. The **leaves** are pinnate or trifoliolate, often with aromatic glands, usu. spiral [opposite], exstipulate, the leaflet margin often serrate. The male **inflorescence** is a catkin or panicle, the female inflorescence is a solitary flower or small groups of flowers at shoot tips. The **flowers** are small, unisexual, bracteate. The **perianth** is uniseriate [absent]. The **calyx** consists of 4 [0–5] sepals, which are adnate to bracts and bracteoles of an involucre, the fusion product in female flowers often developing into a fruit husk adnate to the ovary. The **corolla** is absent. The **stamens** of male flowers are 2–∞, uniseriate or biseriate, distinct. **Anthers** are longitudinally dehiscent. The **gynoecium** of female flowers is syncarpous, with an inferior ovary, 2 [3] carpels, 2 [3] locules below, 1 locule

Figure 8.47 FAGALES. Juglandaceae. **A.** *Carya* sp., hickory/pecan. Note pinnate leaves and pendant, male catkins. **B,C.** *Juglans regia*, English walnut, showing characteristic superposed buds and diaphragmed pith. **D–H.** *Julgans hindsii*. **D.** Tree. **E.** Shoot, showing pinnate leaves and cluster of male catkins. **F.** Close-up of male flowers of catkins. Note subtending calyx, adnate to bracts. **G.** Female flower, with inferior ovary surrounded by adnate involucral bracts. Note epigynous calyx and two styles/stigmas. **H.** Immature fruit. **I,J.** *Juglans regia*. **I.** Fruit (at left), a pseudodrupe, with outer fleshy layer derived from bracts adnate to ovary. Pericarp (center and right), containing a single seed. **J.** Pseudodrupe with half of fleshy involucral layer removed. [Images at D–H courtesy of Gerald Carr; that at J courtesy of The Rampant Gardener.]

Figure 8.48 CROSSOSOMATLES. Crossosomataceae. **A.** *Crossosoma bigelovii*, flower, showing pentamerous perianth, numerous stamens, and apocarpous gynoecium. **B,C.** *Crossosoma californicum*. **B.** Flower, showing numerous stamens and apocarpous gynoecium. **C.** Immature fruits. **D.** Mature, dehiscent fruits, each a follicle. [Image at A, courtesy of Jerry Green; those at B–D, of Reid Moran.]

above (extra locules sometimes forming by false partitions). The **styles** are 2–3. **Placentation** is apical; the solitary **ovule** is orthotropous and unitegmic. The **fruit** is a nut, samara, tryma (in which the enveloping involucre dehisces at maturity), or pseudodrupe (in which the involucre is fleshy and indehiscent). The **seeds** are exalbuminous, cotyledons 4-lobed. Plants are wind-pollinated.

The Juglandaceae is distributed in the North America, South America, Europe, and Asia (esp. e. and s.e. Asia). Economic importance includes uses for timber and furniture (especially *Carya,* hickory, *Engelhardtia*, and *Juglans,* walnut), nut trees (e.g., *Carya illinoinensis*, pecan, and various *Juglans* spp., walnuts), and cultivated ornamentals (e.g., *Carya, Juglans, Pterocarya*, and *Platycarya*). See Manos and Stone (2001) and Manos et al. (2007) for phylogenetic studies of the family.

The Juglandaceae are distinctive in being ***monoecious or dioecious*** trees or shrubs with ***pinnate or trifoliolate*** leaves, flowers ***small, corolla absent***, male flowers gen. in ***catkins***, female flowers at tips of shoots, the ovary ***inferior*** with ***2–3 carpels and locules*** (***1 loculate above***), the subtending ***bracts fusing and forming an outer husk*** in many taxa, the fruit a 1-seeded ***nut, samara, tryma, or pseudodrupe***.

Male flowers: **K** 4 [1–5] **C** 0 **A** 2–∞.
Female flowers: **K** 4 or 0 **C** 0 **G** (2) [(3)], inferior.

MALVIDS

Malvids (*Malvidae*, sensu Cantion et al. 2007; formerly called Eurosids II), the second major subgroup of the Rosids, include 8 orders (Table 8.2, Figure 8.16). Some of the orders of the this complex are quite large, in terms of both number of families and species. Among the Malvids are taxa of great agricultural importance, such as members of the **Anacardiaceae** (cashew family), **Brassicaceae** (mustard family), **Malvaceae** (mallow family), **Myrtaceae** (myrtle family), and **Rutaceae** (citrus family).

CROSSOSOMATLES

The Crossosomatales are composed of 7 families (APG IV 2016; Table 8.2). Only the **Crossosomataceae** are described. See Oh (2010) for a phylogenetic study of the order.

Crossosomataceae — Crossosoma family (type the Greek *krossoi*, "fringe," and *soma*, "body," referring to the fimbriate seed aril). 4 genera/12 species (Figure 8.48).

The Crossosomataceae consist of shrubs or small trees. The **leaves** are simple, spiral, rarely opposite, minutely stipulate or exstipulate, deciduous or marcesent. The **inflorescence** is of solitary flowers. The **flowers** are bisexual or unisexual, actinomorphic, perigynous. The **perianth** is biseriate, dichlamydeous, hypanthium short. The **calyx** is aposepalous with 4–5 [3,6] sepals. The **corolla** is apopetalous with 4–5 [3,6], imbricate petals. **Stamens** are ca. 50, in 1–4 whorls, basally connate in ca. 10 bundles. **Anthers** are longitudinal in dehiscence. The **gynoecium** is apocarpous, the ovaries superior, of 1–5 [–9] unilocular carpels. **Placentation** is marginal; ovules are amphitropous or campylotropous, bitegmic, 2–∞ [1] per carpel. **Nectaries** are present, as a disk, either adnate to or inner to stamens. The **fruit** is a follicetum. The **seeds** are fimbriate-arillate, with an (oily) endosperm.

The Crossosomataceae are found in southwestern North America. See Sosa and Chase(2003) for a phylogenetic study.

The Crossosomataceae are distinctive in being ***shrubs or small trees*** with simple, spiral leaves, the flowers with ***numerous stamens in ca. 10 trunk bundles***, the gynoecium ***apocarpous***, forming a ***follicetum***.

K 4–5 [3–6] **C** 4–5 [3–6] **A** ∞, in ca. 10 trunk bundles **G** 1–5 [–9], superior, hypanthium present, short.

Figure 8.49 GERANIALES. Geraniaceae. **A,B.** *Erodium botrys*, storksbill. **A.** Whole plant in flower. **B.** Base of mature schizocarpic fruit. **C,D.** *Erodium moschatum*. **C.** Flowers. **D.** Young fruits, showing accrescent styles, forming beak. **E.** *Geranium carolinianum*, in flower and immature fruit. **F.** *Geranium* sp., base of flower with sepals and petals removed, showing alternipetalous glands. **G–L.** *Pelargonium* spp. **G.** Flower, face view, slightly zygomorphic. **H.** Inflorescence. **I.** Androecium and gynoecium, showing style branches. **J.** Cross section of ovary, strongly 5-lobed. **K.** Ovary lobe longitudinal section, showing apical-axile placentation. **L.** Fruit, a beaked schizocarp of mericarps.

GERANIALES

The Geraniales contain 5 families (Table 8.2), one of which, the **Geraniaceae**, is described here. See Palazzasi et al. (2012) for a detailed analysis of the order.

Geraniaceae — Geranium family (type *Geranium*, Greek for crane, from accrescent styles resembling a long bird's beak). 7 genera/ca. 866 species (Figure 8.49).

The Geraniaceae consist of herbs or shrubs. The **stems** are a pachycaul in some taxa. The **leaves** are simple or compound, if simple, usually pinnately or palmately lobed to divided, spiral, rarely opposite, usually stipulate, leaves modified as spines, with axillary fascicles, in *Monsonia*. The **inflorescence** is a cyme or a solitary, axillary flower. The **flowers** are bisexual, actinomorphic (zygomorphic in *Pelargonium*), hypogynous, often bracteate, an epicalyx present in some. The **perianth** is biseriate and dichlamydeous. The **calyx** is aposepalous or synsepalous with 5 [4], imbricate or valvate

Figure 8.50 MYRTALES. Combretaceae. **A.** *Bucida buceras*. **B.** *Conocarpus erectus*. **C.** *Quisqualis* sp. **D–F.** *Terminalis catappa*.

sepals, the adaxial sepal a nectariferous spur in *Pelargonium*. The **corolla** is apopetalous with 5 [0,4,8] imbricate, rarely convolute petals. The **stamens** are usually 10 [rarely 8 or 15], in two whorls, basally connate, with staminodes present in the outer whorl of some. **Anthers** are longitudinal in dehiscence. The **gynoecium** is syncarpous, with a superior ovary, 5 [rarely 2, 3, or 8] carpels, and as many locules as carpels. The **style** is usually solitary; **stigmas** are as many as carpels. **Placentation** is apical-axile; **ovules** are anatropous to campylotropous, bitegmic, usually two per carpel. **Nectaries** are present, between petals and stamens (except in *Pelargonium*). The **fruit** is a loculicidal capsule or a schizocarp of mericarps or follicles, usually separating from a persistent beak arising from an accrescent style. Multicellular, capitate, glandular trichomes are often present, usually with aromatic oils in trichome glands.

The genera *Geranium*, *Pelargonium*, and *Erodium* (stork's bill, crane's bill, or filaree) are distinctive in having a schizocarpic fruit with an elongate, persistent beak, possibly apomorphic for the family. Members of the family are distributed in mostly temperate, some tropical, regions. Economic importance includes the use of taxa as cultivated ornamentals (such as *Geranium* and *Pelargonium*), forage plants (such as *Alfilaria* in w. U.S.), and in essential oil extraction (e.g., *Pelargonium*). See Lis-Balchin (2002) for general information and Fiz et al. (2008), Palazzasi et al. (2012), and Zhang et al. (2015) for a phylogenetic analyses.

The Geraniaceae are distinctive in being herbs or shrubs with generally pentamerous, dichlamydeous flowers usually having ***nectariferous glands alternating with the petals*** and generally ***two or more whorls of stamens***, ***staminodes*** often present; the tribe Geranieae is distinctive in having ***beaked, schizocarpic fruits***.

K 5 [4] **C** 5 [0,4,8] **A** 5+5 [8,15] **G** (5) [(2,3,8)], superior.

MYRTALES

The Myrtales contain ca. 12 families (Table 8.2), of which four are described here. Notable among the other families are the **Combretaceae**, including savannah and mangrove species, timber and dye plants, and ornamental cultivars (Figure 8.50). A possible apomorphy for many families in the order may be an inferior ovary with a hypanthium (epiperigynous perianth/androecial position). See Conti et al. (1996, 1998), Wilson et al. (2004), and Sytsma et al. (2005).

Lythraceae — Loosestrife family (type *Lythrum*, after Greek *lythron*, "blood," either in reference to flower color of family members or styptic [bleeding inhibiting] qualities of some family members). 31 genera/ca. 605 species. (Figures 8.51, 8.52).

The Lythraceae s.l. consist of terrestrial, mangrove (*Sonneratia*), or aquatic (*Trapa*), annual or perennial herbs, shrubs, or trees. The **roots** of *Trapa* are adventitious at stem nodes; in the mangrove *Sonneratia* roots give rise to pneumatophores. The **stems** are often 4-sided, often with internal phloem; thorns are present in *Punica*; in the aquatic *Trapa* under-water stems bear opposite, elongate, filiform-dissected, photosynthetic organs of uncertain origin. The **leaves** are simple, not glandular-punctate, usually entire, opposite [rarely whorled or spiral], exstipulate or with vestigial or caducous stipules, the aerial, rosulate leaves of *Trapa* with aerenchymatous, floating petioles. The **inflorescence** is a raceme, panicle, fascicle, dichasium, or of solitary flowers. **Flowers** are bisexual, actinomorphic or zygomorphic, perigynous to epiperigynous, an epicalyx sometimes present. The **perianth** is mostly 4–8-merous [rarely 3–16-merous], a prominent hypanthium present, usually ribbed, often colored, with a basal spur in some taxa, in *Trapa* with a basal, lobed, cup-like structure. The **calyx** is synsepalous with valvate lobes extending from the apex of the hypanthium, the lobes often alternating with outer hypanthial appendages at calyx sinuses (calyx lobes in *Trapa*

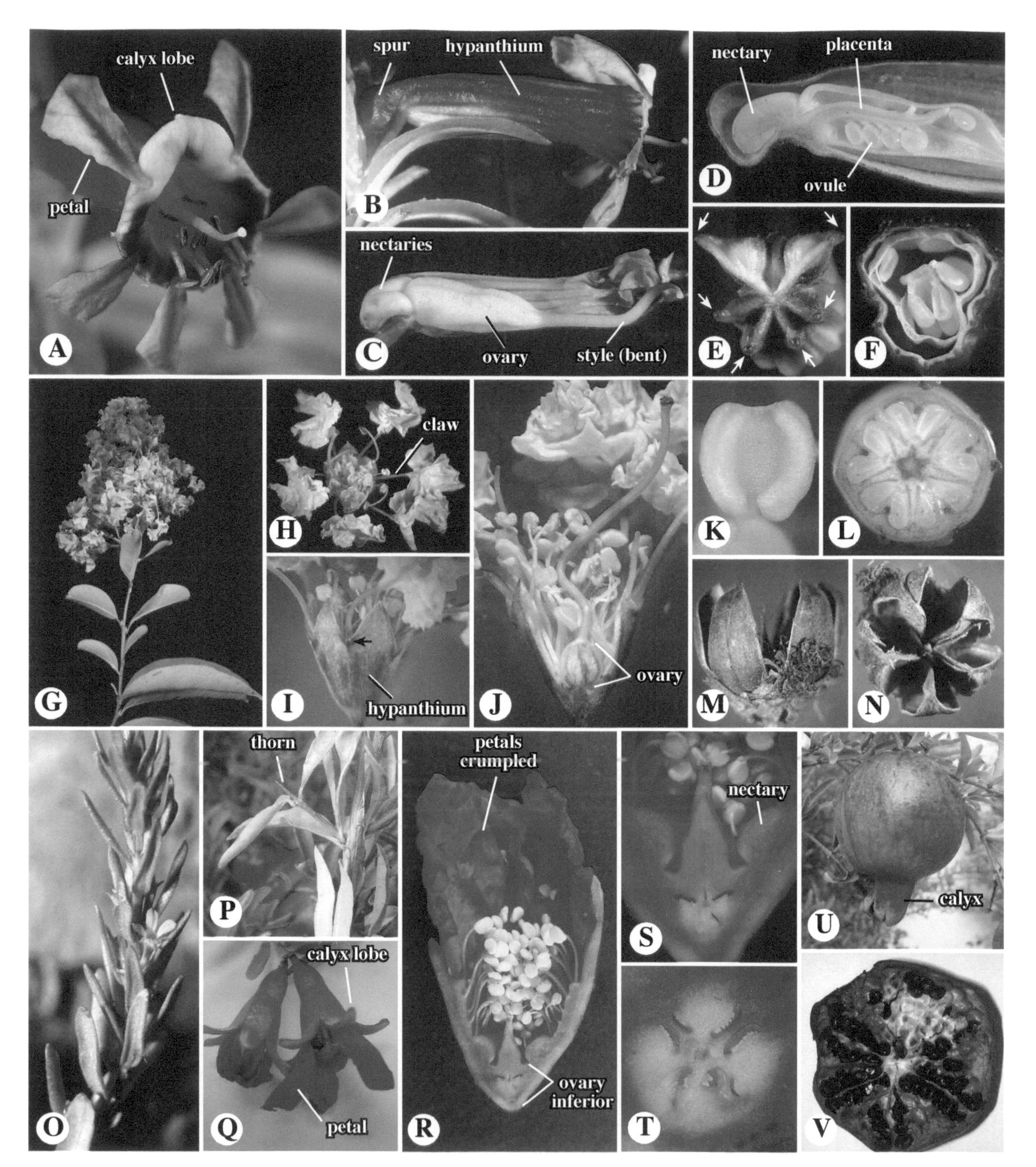

Figure 8.51 MYRTALES. Lythraceae. **A–F.** *Cuphea* sp. **A–B.** Flower, face and side views. **C.** Flower longitudinal-section. **D.** Base of flower in longitudinal section. **E.** Flower bud, face view, showing valvate calyx. Note hypanthial appendages (arrows) between calyx lobes. **F.** Ovary cross section, showing 2 carpels/locules. Note hypanthial tube outside ovary wall. **G–N.** *Lagerstroemia indica*, crape-myrtle. **G.** Inflorescence, a panicle. **H.** Flower close-up, showing clawed petals. **I.** Flower side view. **J.** Flower longitudinal section, showing superior ovary and numerous stamens inserted on inner hypanthial wall. Note 5, imbricate perianth parts. **K.** Anther, dithecal. **L.** Ovary cross section, showing axile placentation of 5 carpels. **M–N.** Fruit, a loculicidal capsule. **O.** *Lythrum hyssopifolium*, flowering shoot. **P–V.** *Punica granatum*, pomegranate. **P.** Thorn close-up; note opposite leaves. **Q.** Flower fascicle, showing biseriate perianth. **R,S.** Flower longitudinal section, showing inferior ovary and numerous stamens. **T.** Ovary cross section, showing multi-leveled placentae. **U.** Fruit, with persistent calyx. **V.** Fruit in section, showing seeds with edible sarcotestas.

Figure 8.52 MYRTALES. Lythraceae. **A.** *Cuphea ignea*, cigar plant, a cultivar. **B.** *Heimia salicifolia*, siniciche, a hallucinogen used by Mexican shamans. **C.** *Sonneratia* sp., a mangrove species. **D.** *Trapa bispinosa*, water-chestnut, with unusual horned fruits.

accrescent, becoming spine- or horn-like in fruit). The **corolla** is imbricate-crumpled in bud, apopetalous, with usually clawed petals arising between calyx lobes near the apex and from the inner surface of the hypanthium. The **stamens** are 8–∞ [rarely solitary], apostemonous, often twice the number of calyx or corolla parts, if numerous then centrifugal in development, usually biseriate, inserted from inner surface of hypanthium below apex often at different levels, filaments are elongate and often of different lengths. **Anthers** are longitudinal in dehiscence, tetrasporangiate, dithecal. The **gynoecium** is syncarpous, with a superior, half-inferior (in *Trapa*, becoming inferior in fruit), or inferior (*Punica*) ovary, 2–4 [–20] carpels, and 2–4 [–20], locules, apically unilocular in some taxa, rarely unilocular throughout. The **style** is solitary and terminal, usually filiform (subulate in *Trapa*), sometimes bent, often heterostylous; the **stigma** is often capitate. **Placentation** is axile, rarely basal; placentae occur in 2–3 layers in *Punica granatum* with upper layers parietal; in *Trapa* each of two locules has one apical-axile ovule; **ovules** are anatropous, bitegmic, usually ∞, rarely 1 or 2+. **Nectaries** consist of an annular disk surrounding ovary or nectaries present in hypanthial spur; superior part of ovary of *Trapa* surrounded by lobed, cup-like structure. The **fruit** is either a capsule of variable dehiscence, an indehiscent capsule with leathery exocarp (*Punica*), a drupe-like, unilocular, 1-seeded fruit with 2–4 accrescent, horn-like calyx lobes (*Trapa*), or a berry (*Sonneratia*). The **seeds** are exalbuminous, sometimes winged, in many taxa possessing distinctive mucilaginous seed coat trichomes; in *Punica* the seed coat is fleshy (a "sarcotesta"). The large flowers of *Duabanga* and *Sonneratia* are bat-pollinated.

The Lythraceae have a worldwide distribution, mostly in tropical regions. Economic importance includes timber plants, dye plants (especially *Lawsonia inermis*, henna), weeds (e.g., *Trapa*), and numerous ornamental cultivars such as *Cuphea* spp., *Lagerstroemia indica* (crape-myrtle), and *Lythrum* spp. (loosestrife); *Punica granatum* (pomegranate) is used medicinally and is an important fruit tree, the seeds having a fleshy, edible seed coat, yielding the fermentation product grenadine, and the persistent, crown-like calyx thought to be original model for the crowns of royalty; the fruits of *Trapa* (water chestnut) are an important food plant locally, especially in oriental cooking. See Graham et al. (2005, 2011) and Narzary et al. (2016) for phylogenetic analyses.

The Lythraceae s.l. are distinctive in being herbs (aquatic in *Trapa*), shrubs, or trees, usually opposite leaves ***lacking punctate glands***, the flowers with a ***prominent, usually ribbed and often colored hypanthium***, calyx lobes ***valvate*** from hypanthial apex, petals ***imbricate-crumpled in bud***, stamens ***often numerous*** and of different lengths, the ovary superior to inferior, the fruit a capsule, a berry, or drupe-like.

K 4–8 [3–16] **C** 4–8 [3–16] **A** 8–∞ [1] **G** 2–4 [–20], superior to inferior, hypanthium present.

Melastomataceae — Melastome family (type *Melastoma*, Greek for "black mouth," from fruits that stain). Ca. 188 genera/ca. 4,960 species (Figure 8.53).

The Melastomataceae consist of shrubs, herbs, rarely trees or lianas. The **stems** are often 4-sided. The **leaves** are simple, opposite, rarely whorled, usually exstipulate, often with 3–9 subparallel, major veins. The **inflorescence** is a cyme. The **flowers** are bisexual, epiperigynous or perigynous, with perianth mostly actinomorphic, the androecium zygomorphic. The **perianth** has a hypanthium usually present. The **calyx** consists of 4–5 [3–10], valvate or calyptrate sepals. The **corolla** is apopetalous, with 4–5 [3–10] convolute petals. The **stamens** are usually 8 or 10, biseriate, often dimorphic; filaments are often twisted during anthesis, positioning the anthers to one side of flower. **Anthers** are poricidal with 1–2 pores per anther, rarely longitudinal in dehiscence, the anther connective often appendaged. The **gynoecium** is syncarpous, with an inferior or superior ovary, 3–5 [2–15] carpels, and locules, or unilocular by formation of incom-

Figure 8.53 MYRTALES. Melastomataceae. **A,B.** *Medinilla myriantha*. **A.** Flowering shoot, showing subparallel major veins. **B.** Flower close-up. **C,D.** *Rhexia mariana*. **C.** Flower. **D.** Urn-shaped fruit; note inferior ovary. **E.** *Tetrazygia bicolor*, inflorescence. **F–H.** *Tibouchina urvilleana*. **F.** Flowering shoot. **G.** Flower, face view. **H.** Flower, longitudinal section, showing inferior ovary and stamen connective appendages. **I.** *Tibouchina multijuga*, bud cross section, showing ovary with axile placentation and recessed anthers of stamens.

plete septa. **Placentation** is usually axile; **ovules** are anatropous to campylotropous, bitegmic, with "zig-zag" micropyle, ∞ [1] per carpel. The **fruit** is a loculicidal or septifragal capsule or a berry. The **seeds** are exalbuminous.

The Melastomataceae have distributions in tropical regions, especially South America. Economic importance includes timber trees, edible fruit plants, dye plants, and several ornamental cultivars, such as *Tibouchina*. See Clausing and Renner (2001) and Morley and Dick (2003).

The Melastomataceae are distinctive in being shrubs, herbs, rarely trees or lianas, the ***stems often 4-sided***, with ***simple, opposite*** (rarely whorled) ***leaves***, usually with ***3–9 subparallel major veins***, the inflorescence a cyme, ***flowers epiperigynous or perigynous***, the perianth usually 4–5-merous, the ***stamens biseriate and dimorphic***, anthers oriented to one side of flower by filament twisting, ***connective often appendaged and dehiscence usually poricidal***, the inferior ovary with usually axile placentation, and fruit a capsule or berry.
K 4–5 [3–10] **C** 4–5 [3–10] **A** 4+4 or 5+5 **G** (3–5) [(2–15)], inferior, hypanthium usually present.

Myrtaceae — Myrtle family (type *Myrtus*, Greek name for myrtle). 131 genera/ca. 5900 species (Figure 8.54).

The Myrtaceae consist of hermaphroditic [dioecious in *Psiloxylon*] trees and shrubs. The **roots** possess ectotrophic mycorrhizae. The **stems** have secretory cavities and internal phloem in the pith. **Leaves** are opposite (usually) or spiral, rarely whorled, simple, glandular-punctate or pellucid, and often coriaceous, with stipules present and small or absent. The **inflorescence** is variable. The **flowers** are bisexual, actinomorphic, bracteate, epiperigynous, rarely perigynous. The **perianth** is biseriate, perianth segments distinct or connate, fused into a lidlike calyptra (operculum) in some (e.g., *Eucalyptus*). The **calyx** consists of 4–5 [3,6], imbricate sepals. The **corolla** consists of 4–5 [3,6] petals. The **stamens** are ∞, centripetal, distinct or connate into 4 or 5 groups. **Anthers** are loculicidal or poricidal in dehiscence. The **gynoecium** is syncarpous, with an inferior [rarely half-inferior or superior] ovary, 2–5 [–16] carpels, and 2–5 [–16, 1] locules. The **style** is terminal; the **stigma** is capitate or lobed. **Placentation** is axile (being basal-axile to apical-axile) [parietal if unilocular]; **ovules** are anatropous or campylotropous, bitegmic or unitegmic, 2–∞ per locule. **Nectaries** are present, as a disk atop the ovary or on the inner hypanthium. The **fruit** is a berry or loculicidal capsule [rarely a drupe or nut].

The Myrtaceae are currently classified into two subfamilies (with several tribes): Myrtoideae, with 17 tribes, and Psiloxyloideae. Members of the family have distributions in warm tropics and temperate Australia. Economic importance includes important timber trees, especially *Eucalyptus* spp., edible fruits (e.g., *Psidium guajava*, guava), spices (e.g., *Syzygium aromaticum*, cloves, *Pimenta dioica*, allspice), oils (e.g., *Eucalyptus* spp.), and cultivated ornamentals such as *Callistemon* (bottlebrush), *Chamelaucium* (wax-flower), *Eucalyptus* spp., *Leptospermum* (tea tree), and *Myrtus* (myrtle). See Wilson et al. (2001, 2005) and Biffin et al. (2007) for phylogenetic studies of the family.

The Myrtaceae are distinctive in being trees and shrubs with ***glandular-punctate or pellucid leaves*** and usually ***epiperigynous flowers*** with ***numerous stamens***.
K 4–5 [3,6] **C** 4–5 [3,6] **A** ∞ **G** (2–5) [(–16)], inferior [rarely half-inferior or superior], with hypanthium.

Onagraceae — Evening-Primrose family (after *Onagra* [=*Oenothera*], Greek for oleander, an unrelated plant). Ca. 22 genera/ca. 660 species (Figure 8.55).

The Onagraceae consist of terrestrial or aquatic herbs and shrubs, rarely trees. The **stems** have internal phloem, often with epidermal oil cells. The **leaves** are spiral, opposite, or whorled, simple, undivided to pinnatifid, and exstipulate or stipulate. The **inflorescence** is a spike, panicle, or of solitary flowers. The **flowers** are usually bisexual, actinomorphic or zygomorphic, usually epiperigynous, with an elongate hypanthium in some taxa. The **perianth** is biseriate and dichlamydeous. The **calyx** is valvate, aposepalous, consisting of usually 4 [rarely 2–6] sepals. The **corolla** is valvate or imbricate, apopetalous, consisting of usually 4 [rarely 2–6 or 0] petals. The **stamens** are 4+4 [rarely 2–6]. **Anthers** are longitudinal in dehiscence, tetrasporangiate and dithecal, with cross-partitions in some species. The **pollen** is shed in monads or tetrads, often with viscin threads, which function to adhere grains together. The **gynoecium** is syncarpous, with an inferior ovary, and usually 4 [rarely 2–6] carpels and locules. **Placentation** is axile or parietal; **ovules** are anatropous, bitegmic, usually ∞ per locule, with a monosporic, 4-nucleate (*Oenothera* type) female gametophyte. The **fruit** is a capsule, berry, or nut. **Seeds** are (oily) endospermous.

The Onagraceae have a worldwide distribution. Economic importance includes several cultivated ornamentals, such as species of *Clarkia*, *Fuchsia*, and *Oenothera*. See Levin et al. (2003), Ford and Gottlieb (2007), and Wagner et al. (2007) for phylogenetic studies and infrafamilial classifications.

The Onagraceae are distinctive in being herbs and shrubs (rarely trees) with usually ***4-merous [2–6-merous], epiperigynous flowers*** with usually ***4+4 stamens*** and a ***monosporic, 4-nucleate female gametophyte***, the latter a possible apomorphy for the family.
K 4 [2–6] **C** 4 [2–6,0] **A** 4+4 [2–6] **G** (4) [(2–6)], inferior, with hypanthium.

Figure 8.54 MYRTALES. Myrtaceae. **A.** *Actinodium cunninghamii*, having a capitulum resembling that of the Asteraceae. **B,C.** *Callistemon citrinus*. **B.** Note capsules and flower spike, which continues apical vegetative growth. **C.** Close-up of flower, showing reduced perianth and numerous, showy stamen filaments. **D.** *Calothamnus sanguineus*, with connate, zygomorphic stamens (in bundles). **E.** *Darwinia fascicularis*, having heads and flowers with elongate styles. **F.** *Eucalyptus obtusiflora*, showing capsules and flowers. **G.** *Eucalyptus torquata*. Note calyptra and incurved stamens (prior to unfurling). **H–J.** *Eucalyptus sideroxylon*. **H.** Flower longitudinal section, showing inferior ovary and hypanthium. **I.** Style and stigma close-up. **J.** Ovary cross section, showing axile placentation. **K.** *Eucalpytus diversicolor*, karri, a tall tree species. **L.** *Leptospermum laevigatum*. **M.** *Myrtus communis*, leaf close-up showing internal, pellucid glands. **N.** *Lophostemon confertus*, Brisbane box, flower with fascicles of stamens. **O.** *Verticordia grandiflora*, with fringed petals.

Figure 8.55 MYRTALES. Onagraceae. **A,B.** *Chylismia claviformis*, with capitate stigma. **C.** *Clarkia delicata*. **D.** *Clarkia purpurea*. **E,F.** *Clarkia unguiculata*. **E.** Flower. **F.** Ovary cross section, with axile placentation. **G,H.** *Fuchsia* ×*hybrida*. **G.** Pendant flower. **H.** Flower in longitudinal section, showing epiperigynous structure. **I.** *Epilobium canum*, with zygomorphic flowers. **J.** *Ludwigia hexapetala*, with pentamerous flowers. **K.** *Ludwigia sedoides*, a floating aquatic. **L.** *Oenothera deltoides*, with four linear stigmas. **M.** *Oenothera speciosa*. **N.** *Oenothera elata*, showing four-valved capsule.

BRASSICALES

The Brassicales are composed of ca. 17 families (Table 8.2; Figure 8.56A), two of which are described here. The order is generally united in having indolic **glucosinolate** secondary compounds (Figure 8.56A,B). These function to deter herbivory and parasitism and also serve as flavoring agents in the commercially important members of the Brassicaceae, such as broccoli, cauliflower, and mustard. Some well-known families of the order not described here include the **Bataceae** (saltworts; Figure 8.57A,B), **Capparaceae** (including capers, *Capparis spinosa*), **Caricaceae** (including the fruit tree *Carica papaya*, papaya; Figure 8.57E–H), **Cleomaceae** (Figure 8.58), **Koeberliniaceae** (crown of thorns; Figure 8.57I–L), **Limnanthaceae** (meadowfoam; Figure 8.57C–D), and **Moringaceae** (Figure 8.57M–Q). See Rodman et al. (1997, 1998), Kubitzki and Bayer (2002), and Edger et al. (2018) for more information on the order.

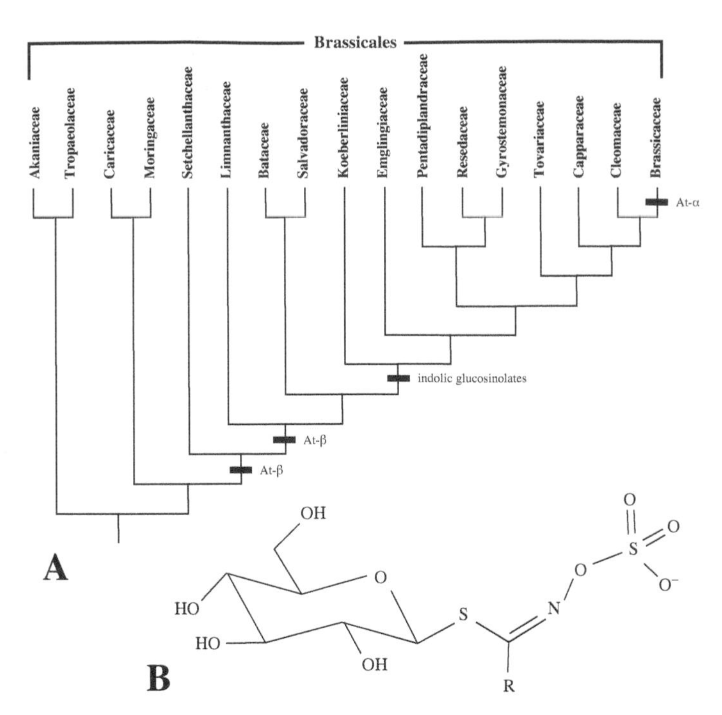

Figure 8.56 A. Cladogram of 17 families of Brassicales, after Edger et al. (2018). At-α and At-β are whole genome duplications involved in the glucosinolate pathway. B. General structure of indole glucosinolates, an apomorphy of most Brassicales.

Brassicaceae (Cruciferae) — Mustard family (type *Brassica*, after name used by Pliny for cabbagelike plants). 321–338 genera/3400–3700 species (Figures 8.59).

The Brassicaceae consist of usually hermaphroditic herbs, rarely shrubs (pachycaulous in some). The **leaves** are simple [rarely compound], often lobed to divided, spiral [rarely opposite], exstipulate. The **inflorescence** is usually a raceme, rarely of solitary, axillary flowers. The **flowers** are bisexual, rarely unisexual, usually actinomorphic, pedicellate, ebracteate, hypogynous; the receptacle is rarely elongate into a gynophore. The **perianth** is dichlamydeous, cruciate. The **calyx** is aposepalous [rarely synsepalous] with 2+2, decussate outer sepals, often basally gibbous. The **corolla** is apopetalous, rarely basally connate, with 4 [rarely absent] petals, which are often clawed. The **stamens** are apostemonous, biseriate, 2+4 tetradynamous [rarely 2 or 4 or up to 16], the outer 2 shorter, antisepalous, the inner 4 longer, of two pairs, each pair (from a single primordium) flanking adjacent petals. **Anthers** are longitudinal in dehiscence. The **gynoecium** is syncarpous, with a superior ovary, 2 carpels, and 2 locules. The **style** is 1 or absent. **Placentation** is axile -parietal, each carpel with two rows of ovules, the placentae at junction of septum and ovary wall; **ovules** are anatropous or campylotropous, 1–∞ per carpel. **Nectaries** are discrete or ringlike around stamens or pistil. The **fruit** is a specialized capsule, called a **silique** (>3× longer than broad) or **silicle** (<3× longer than broad), that usually dehisces by the two valves falling entire (rarely transversely dehiscent or indeshiscent) and leaving a persistent cross-wall consisting of a peripheral rim, termed the **replum**, and membranous intervening tissue spanning the replum, termed a **false septum**. The **seeds** are usually exalbuminous.

The Brassicaceae, sensu APG IV (2016) are now separated from the traditional families Capparaceae and the Cleomaceae (the last two in the past treated as subfamilies of the Capparaceae). Each of these three families appears to be monophyletic (Table 8.2). The Capparaceae differ from the Brassicaceae largely in having a woody habit, an elongate gynophore or androgynophore, a generally greater number of stamens, a unilocular ovary with parietal placentation, and an indehiscent fruit type lacking a replum. The Cleomaceae (Figure 8.58) resemble the Capparaceae but are largely herbaceous and have a dehiscent fruit with a replum (but lacking a complete partition, thus the ovary unilocular). See Al-Shehbaz et al. (2006), Bailey et al. (2006) and Beilstein et al. (2008) for phylogenetic analyses of the group.

The Brassicaceae have a worldwide distribution. Economic importance encompasses numerous vegetable and flavoring plants (notably the crucifers or mustard plants), including horse-radish (*Armoracia rusticana*), arugula (*Eruca vesicaria*), broccoli, brussels sprouts, cauliflower, cabbage, collards, kale (all cultivars of *Brassica oleracea*), rutabaga and canola oil (*B. napus*), mustard (*B. nigra*), turnip (*B. rapa*), wasabi (*Eutrema japonicum*); radish (*Raphanus sativus*), and many more; plus numerous cultivated ornamentals, dye plants (*Isatis tinctoria*, woad), and some noxious weeds; *Arabidopsis thalliana* is noted as a model for detailed molecular studies.

Figure 8.57 BRASSICALES. **A,B.** Bataceae, *Batis maritima*, saltwort. **C,D.** Limnanthaceae, *Limnanthes gracilis*, meadowfoam. **E–H.** Caricaceae, *Carica papaya*, papaya. **I–L.** Koeberliniaceae, *Koeberlinia spinosa*, crown of thorns. **M–Q.** Moringaceae. **M.** *Moringa drouhardii*. **N.** *Moringa borziana*. **O.** *Moringa oleifera*. **P,Q.** *Moringa longituba*. [Images at J–L courtesy of Reid Moran, those at M–Q courtesy of Mark Olson.]

Figure 8.58 BRASSICALES. **A,B.** Cleomaceae, *Peritoma arborea*. **A.** Flowers. **B.** Fruit. **C–E.** *Cleome bassleriana*. **C.** Inflorescence. **D.** Palmate leaves. **E.** Flower, showing elongate gynophores.

The Brassicaceae as treated here are distinctive in being herbs, rarely shrubs, with ***glucosinolates*** (mustard oil glucosides), the ***perianth cruciate*** (petals usually clawed), the androecium with usually ***2+4, tetradynamous stamens***, the gynoecium with a superior, ***2-carpellate/loculate ovary***, with ***axile-parietal placentation*** and a usually ***2-valved, dehiscent fruit with a replum (silique or silicle).***
K 2+2 **C** 4 **A** 2+4 [2,4-16] **G** (2), superior.

Tropaeolaceae — Nasturtium family (type *Tropaeolum*, after Greek *tropaion*, monument of war victory, for resemblance of leaves to shields used to construct monument). 1 genus (*Tropaeolum*)/90-95 species (Figure 8.60).

The Tropaeloaceae consist of herbaceous vines, often climbing by means of twining leaf petioles. The **roots** are tuberous in some taxa. The **leaves** are simple, unlobed or palmately lobed to divided, spiral, palmately-veined, peltate in some, stipulate or exstipulate at maturity. The **inflorescence** is of solitary, axillary flowers. The **flowers** are bisexual, zygomorphic, pedicellate, perigynous. The **perianth** is dichlamydeous, with hypanthium present. The **calyx** is generally aposepalous, of 5, imbricate sepals, with the 3 adaxial sepals forming a spur. The **corolla** is apopetalous, of 5, clawed petals. The **stamens** are 8 (4+4), apostemonous. **Anthers** are longitudinal in dehiscence. The **gynoecium** is syncarpous, with a superior ovary, 3 carpels (median carpel adaxial), and 3 locules. The **style** is 3-branched. **Placentation** is apical-axile; **ovules** are bitegmic, 2 per carpel (1 ovule aborting). **Nectaries** are present in the calyx spur. The **fruit** is a schizocarp of 3 [1] fleshy, dry, or winged, 1-seeded mericarps. The **seeds** are exalbuminous.

The Tropaeloaceae is distributed in Central and South America. Economic importance includes cultivated ornamentals (especially *Tropaeolum majus*, nasturtium) and local medicinal, and food plants. See Andersson and Andersson (2000) for a molecular phylogeny of the family and Bayer and Appel (2002) for a general family treatment.

The Tropaeloaceae are distinctive in being ***climbing herbs*** with ***peltate or palmately lobed-divided*** leaves, solitary, axillary flowers with a ***spurred calyx*** and ***clawed petals***, ***8 stamens***, and ***3 carpels***, the fruit a ***schizocarp of 1-seeded mericarps***.
K 5 **C** 5 **A** 8 **G** (3), superior.

Figure 8.59 BRASSICALES. Brassicaceae. **A–E.** *Brassica rapa*. **A–C.** Flower. **A.** Side view. **B.** Top view. **C.** Perianth removed, showing tetradynamous stamens. **D.** Silique. **E.** Silque after dehiscence. **F,G.** *Lepidium nitidum*, pepper-grass. **F.** Silicle prior to dehiscence. **G.** Silicles after dehiscence. **H,I.** *Cardamine californica*, milkmaids. **J–K.** *Rorippa nasturtium-aquaticum*, watercress, flowers and fruits. **L.** *Raphanus sativus*, radish, flowers. **M.** Silique of *Brassica nigra*, mustard. **N.** Silicle of *Thysanocarpus laciniatus*. **O.** Transversely dehiscent capsule of *Cakile maritima*.

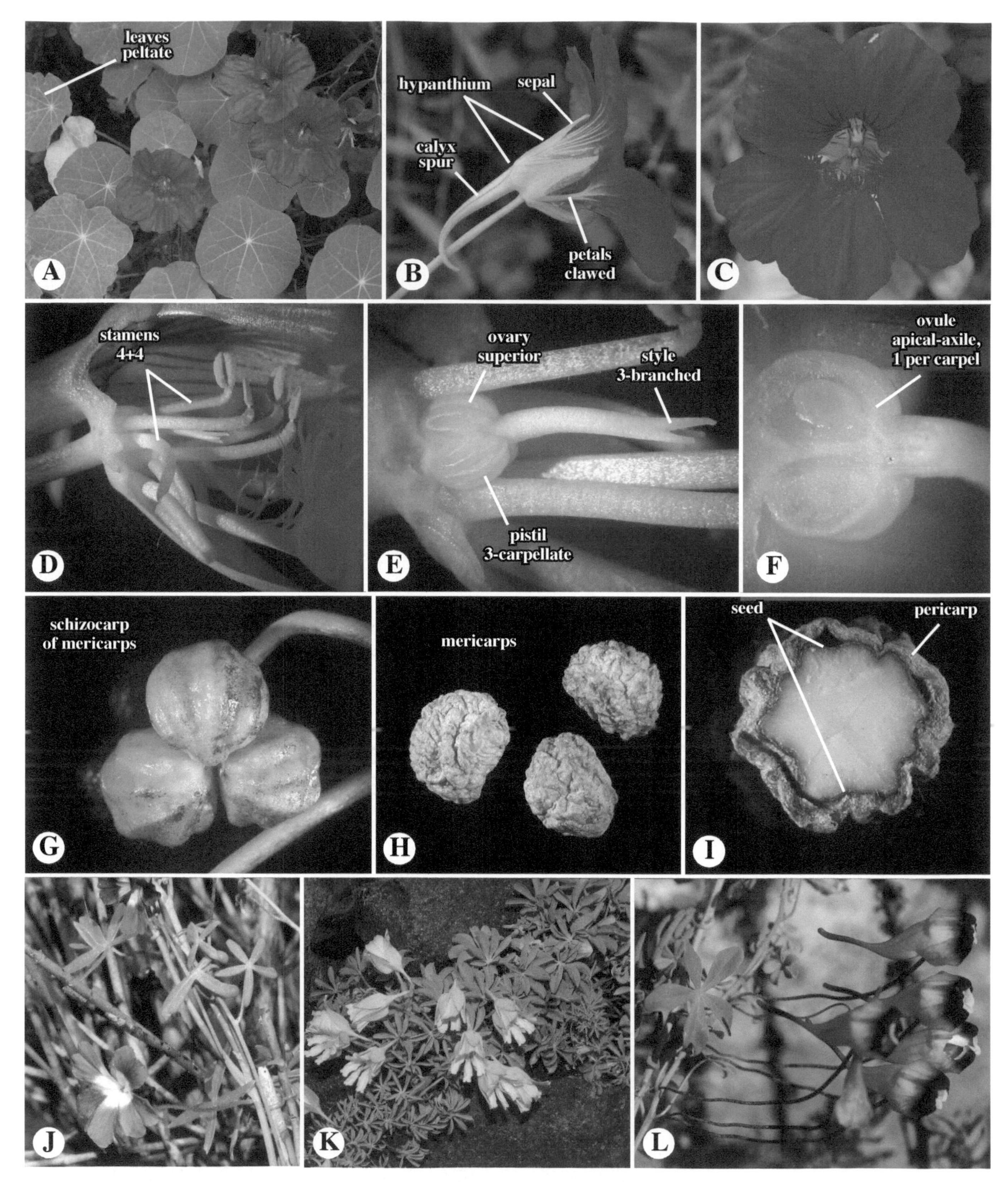

Figure 8.60 BRASSICALES. Tropaeolaceae. *Tropaeolum majus*, nasturtium. **A.** Whole plant, with peltate leaves and solitary, axillary flowers. **B,C.** Flower (zygomorphic), showing calyx spur and 5 clawed petals. **D,E.** Flower longtidtual section, showing 4+4 stamens and 3-branched style. **F.** Ovary longitudinal section, showing apical-axile placentation. **G.** Fruit, a schizocarp of mericarps, prior to splitting. **H.** Mericarps, after dehiscence. **I.** Mericarp cross section, showing single seed. **J.** *Tropaeolum azureum*. **K.** *T. polycephalus*. **L.** *T. tricolor*.

Figure 8.61 MALVALES. **A,B.** Bixaceae, *Bixa orellana*, annato. Note reddish seeds, source of plant coloring. **C–E.** Cistaceae. **C.** *Cistus* sp. **D,E.** *Crocanthemum scoparium*. Note 3 larger and 2 smaller sepals at "E." **F–H.** Thymeleaceae. **F.** *Pimelia ferruginea*. **G.** *Pimelia* sp. **H.** *Pimelia physodes*.

MALVALES

The Malvales include 10 families (Table 8.2). Of these, only the **Malvaceae** (s.l.) are covered here. More well known among the others are the **Bixaceae,** containing *Bixa orellana*, anatto, commonly used as a natural food coloring (Figure 8.61A,B); **Cistaceae**, the rock-rose family (Figure 8.61C–E); **Dipterocarpaceae**, the dipterocarps of s.e. Asia, source of important hardwood timber trees and gum/resin plants; and **Thymelaeaceae** (Figure 8.61F–H). The order as a whole may have chemical and anatomical apomorphies, including the presence of lysigenous mucilage canals in most members. See Kubitzki and Chase (2002) and Le Péchon and Gigord (2014) for more information.

Malvaceae [including Bombaceae, Sterculiaceae, and Tiliaceae] — Mallow family (type *Malva*, after name used by Pliny, for "soft"). Ca. 250 genera/ca. 4200 species (Figures 8.62–8.65).

The Malvaceae, sensu APG IV (2016), consist of usually hermaphroditic, rarely monoecious or polygamous trees, shrubs, or herbs, often with either stellate trichomes or peltate scales. The **leaves** are simple or palmately compound, sometimes lobed to divided, palmately or pinnately veined, usually spiral and stipulate, the stipules often caducous. The **inflorescence** is of solitary or paired flowers or cymelike, sometimes complex. The **flowers** are bisexual [rarely unisexual], mostly actinomorphic, an epicalyx typically present, hypogynous, rarely perigynous. The **perianth** is biseriate or uniseriate. The **calyx** is aposepalous or basally synsepalous with 5 [less often 3–4], valvate sepals. The **corolla** is apopetalous [sometimes adnate to the base of an androecium tube; absent in some], when present of 5 [3–4], sometimes clawed, convolute, valvate, or imbricate petals. The **stamens** are 5–∞, the filaments usually connate, either as a tube surrounding the ovary, or as 5–15 bundles of stamens or a tube bearing bundles. **Anthers** are longitudinal or poricidal in dehiscence. The **pollen** is spinulose or smooth. The **gynoecium** is syncarpous, rarely apocarpous or with carpels fused only apically, with a superior [rarely inferior] ovary, 2–∞ carpels, and 2–∞ [1] locules. The **style** is unlobed, lobed, or branched at the apex. **Placentation** is usually axile, rarely marginal; **ovules** are 2–∞ [1] per carpel. **Nectaries** consist of glandular trichomes typically present at the adaxial base of the calyx. The **fruit** is a loculicidal, septicidal, or indehiscent capsule, a schizocarp of mericarps, rarely a berry or samara. **Seeds** are exalbuminous or endospermous (oily or starchy).

The Malvaceae s.l. as treated here were formerly (and still in some references) divided into four families: Malvaceae s.s., Bombacaceae, the Bombax family, Sterculiaceae, the chocolate family, and Tiliaceae, the Linden family. Recent morphological and molecular analyses indicate that these groups are largely nonmonophyletic and best classified together. Bayer et al. (1999) tentatively recognized nine subfamilies, some putative apomorphies shown in Figure 8.64 (but see also Nyffeler et al., 2005 and Richardson et al., 2015). The Malvaceae s.l. as a whole may be united by an inflorescence apomorphy, the occurrence of a "bicolor unit" (Bayer, 1999, the term derived from *Theobroma bicolor*, where it was first observed), consisting of a modified, 3-bracted cyme, the trimerous epicalyx of family members possibly derived from these 3 bracts. Other possible apomorphies of the family are a valvate calyx, stellate or lepidote trichomes, and dilated secondary tissue rays (Figures 8.64, 8.65).

Figure 8.62 MALVALES. Malvaceae. **A,B.** *Theobroma cacao*, cacao (Byttnerioideae). **A.** Flower, showing cucullate (hooded) petals. **B.** Fruit, a berry, source of chocolate. **C.** *Guichenotia ledifolia* (Byttnerioideae). **D.** *Grewia occidentalis* (Grewioideae). **E.** *Tilia americana* (Tilioideae), showing prominent inflorescence bract. **F.** *Durio zibethinus*, durian (Helicteroideae), fruit in longitudinal section. **G,H.** *Dombeya* spp. (Dombeyoideae). **I.** *Brachychiton discolor* (Sterculioideae), with uniseriate perianth (corolla absent). **J.** *Heritiera littoralis* (Sterculioideae), showing androgynophore. **K,L.** *Pseudobombax ellipticum* (Bombacoideae). **K.** Flower with numerous stamens. **L.** Close-up of monothecal anthers. Subfamilies after Bayer et al. (1999).

Figure 8.63 MALVALES. Malavaceae (continued). **A–C.** *Chorisia speciosa* (Bombacoideae). **A.** Whole flower, showing staminal tube. **B.** Close-up of staminal tube bearing monothecal anthers. **C.** Bud cross section, showing convolute corolla. **D,E.** *Chiranthodendron pentadactylon* (Bombacoideae). **D.** Flower, abaxial view, showing five connate stamens. **E.** Fruit, a capsule. **F.** *Fremontodendron mexicanum* (Bombacoideae). **G–K.** *Hibiscus* sp. (Malvoideae). **G.** Whole flower, showing convolute corolla and staminal tube. **H.** Nectariferous adaxial surface of calyx. **I.** Close-up of style branches and anthers of surrounding staminal tube. **J.** Close-up of monothecal anther. **K.** Ovary cross-section, showing axile placentation. **L.** *Alyogyne huegelii* (Malvoideae), flower longitudinal section, showing corolla adnate to androecium, staminal tube, and superior ovary. Subfamilies after Bayer et al. (1999).

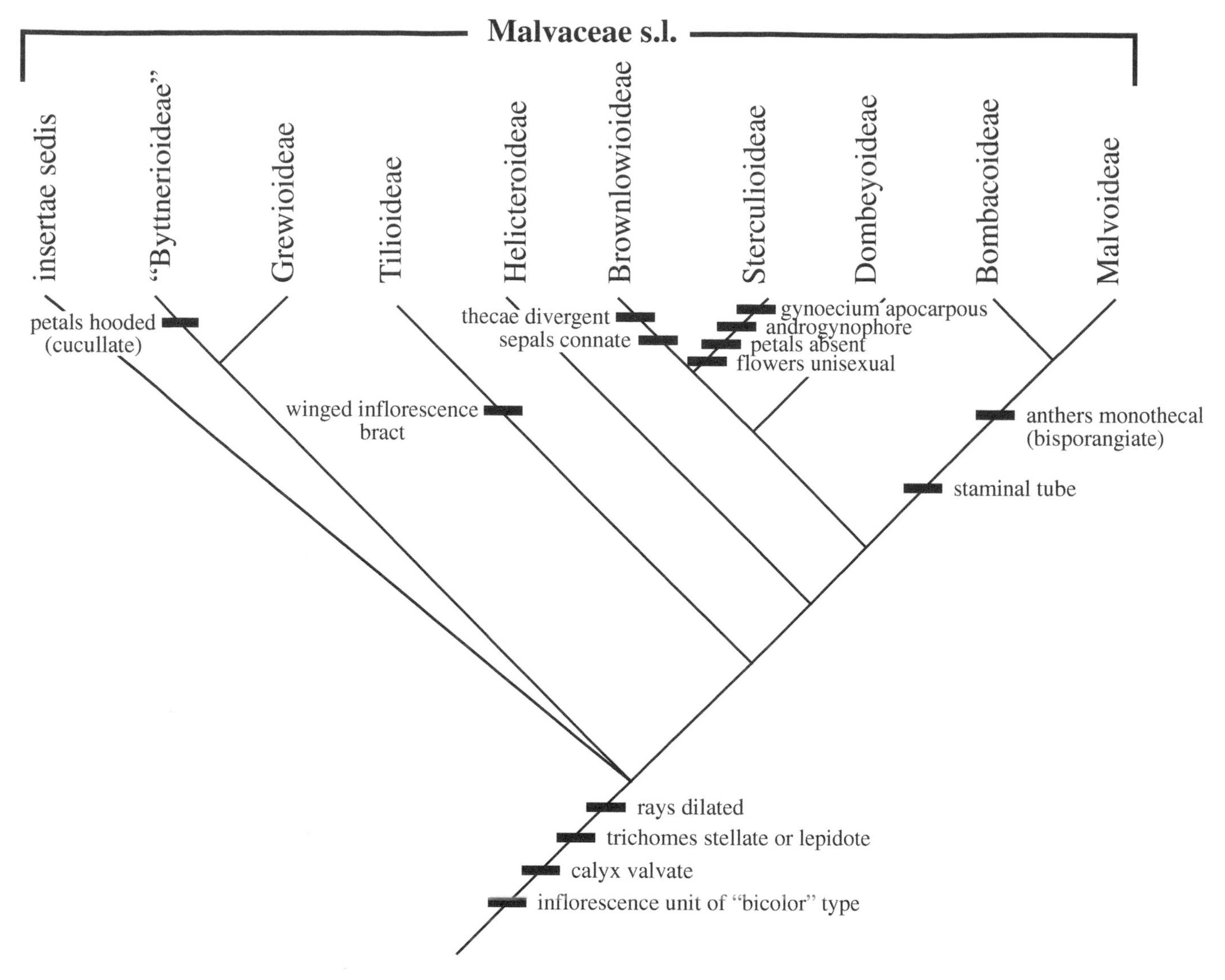

Figure 8.64 MALVALES. Cladogram of the Malavaceae s.l., after Bayer et al. (1999). See also Nyffeler et al. (2005) and Richardson et al. (2015). Apomorphies listed are suggestive and may need further study for verification.

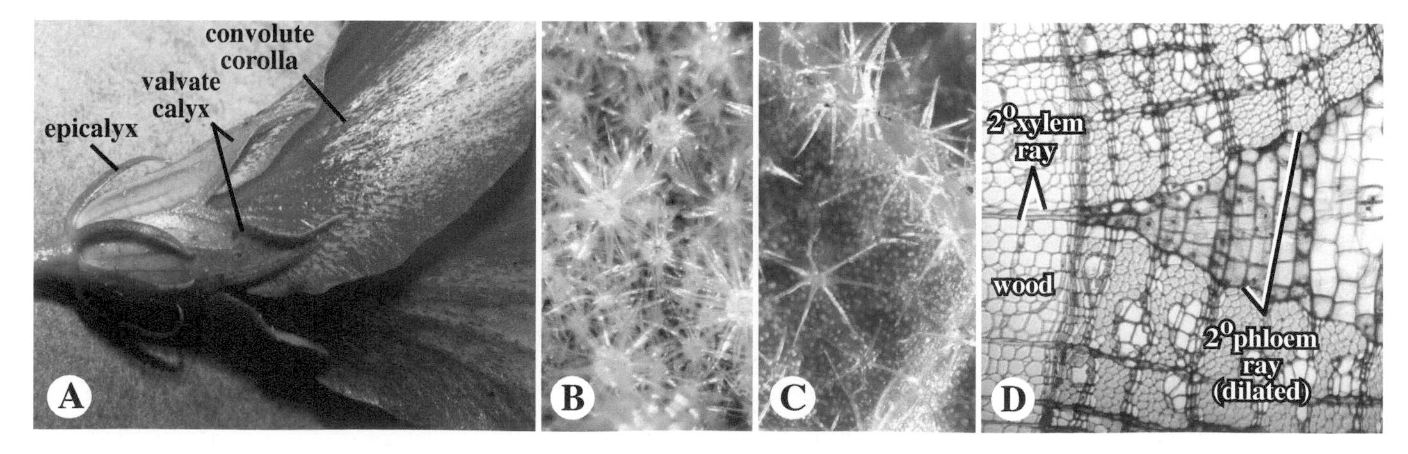

Figure 8.65 MALVALES. Possible apomorphies of the Malavaceae s.l. **A.** Valvate calyx (*Hibiscus*); also note epicalyx and convolute corolla. **B,C.** Stellate trichomes (*Alyogyne* and *Fremontodendron*, respectively). **D.** Dilated wood rays (*Tilia*).

Members of the Malvaceae are distributed worldwide, especially in tropical regions. Economic importance includes medicinal plants; several fiber plants, especially *Gossypium* spp. (cotton, the world's most important fiber plant) and *Ceiba pentandra* (kapok), in both of which the seed trichomes are utilized, and *Corchorus* spp. (jute), a bast fiber plant and source of burlap; food and flavoring plants, such as *Theobroma cacao* (cacao, the source of chocolate), *Cola nitida* (cola), *Abelmoschus* (okra), and *Durio zibethinus* (durian); wood, such as *Ochroma pyramidale* (balsa) and *Pachira aquatica*; and numerous ornamental cultivars, such as *Brachychiton*, *Chorisia* (floss-silk tree), *Dombeya*, *Fremontodendron*, *Hibiscus* (mallows), and *Tilia* (linden tree). Many others, such as *Adansonia digitata* (baobab, tropical Africa) are of great local economic or ecological importance. See Bayer et al. (1999), Nyffeler et al. (2005), and Richardson et al. (2015) for phylogenetic analyses.

The Malvaceae are distinctive in being herbs, shrubs, or trees, often with ***stellate trichomes***, typically with an ***epicalyx***, the ***calyx valvate***, the ***corolla often convolute*** [sometimes valvate or imbricate] the ***stamens connate*** as a tube or 5–∞ bundles, with ***monothecal or dithecal anthers***, gynoecium syncarpous [rarely apocarpous], ovary superior [rarely inferior], ovules axile or marginal, the fruit a capsule, schizocarp of mericarps, berry, or samara.
K 3–5 or (3–5) **C** 3–5 [0] **A** 5–∞ **G** 2–∞ [1], superior [rarely inferior].

SAPINDALES

This order contains nine families, three of which are described here. Among the others, the **Burseraceae** is notable as the source of frankincense (*Boswellia* spp., esp. *B. sacra*) and myrrh (*Commiphora* spp., esp. *C. myrrha*). See Gadek et al. (1996) and Muellner-Riehl et al. (2016) for a phylogenetic analyses of the order.

Anacardiaceae—Cashew family (type *Anacardium*, Greek for heart-shaped, after swollen, red pedicel in cashew fruit). Ca. 80 genera/ca. 870 species (Figure 8.66).

The Anacardiaceae consist of trees, shrubs, lianas, or rarely perennial herbs, tissues of plant organs with resin ducts or laticifers, the resin allergenic in some taxa. The **leaves** are pinnate, trifoliolate, or simple, spiral, rarely opposite or whorled, exstipulate or stipules vestigial. The **inflorescence** is a terminal or axillary thyrse. The **flowers** are bisexual or unisexual, actinomorphic, usually hypogynous; the receptacle is swollen and fleshy in some taxa (e.g., *Anacardium*). The **perianth** is biseriate and dichlamydeous, parts valvate or imbricate. The **calyx** is usually basally synsepalous with usually 5 sepals or lobes. The **corolla** is apopetalous with usually 5 [0] petals. The **stamens** are 5–10 [1, ∞], apostemonous or rarely basally connate. The **gynoecium** is syncarpous [rarely apocarpous], with a superior [rarely inferior] ovary, 1–3, or 5 [rarely 12] carpels, and usually 1 [sometimes as many as carpels] locule. **Placentation** is apical/pendulous or basal; **ovules** are anatropous, bitegmic or unitegmic, 1 per carpel. **Nectaries** are present as a staminal, intrastaminal, or extrastaminal nectariferous disk. The **fruit** is a drupe, with the mesocarp usually resinous. The **seeds** have endosperm absent or scanty.

The Anacardiaceae have a broad distribution in tropical to temperate regions. Economic importance includes ornamental cultivars (e.g., *Cotinus*, *Schinus* spp.), fruit, seed, and spice trees, such as *Pistacia vera* (pistachio), *Rhus coriaria* (Sicilian sumac), *Anacardium occidentale* (cashew), and *Mangifera indica* (mango), plus dye, tannin, timber, and lacquer trees. *Toxicodendron* spp. (poison oak, poison ivy) and related taxa cause contact dermatitis, and fruits/seeds can be allergenic in sensitive individuals. See Wannan (2006) and Aguilar-Ortigoza and Sosa (2004) phylogenetic studies of the family and Pell et al. (2011) for general information.

The Anacardiaceae are distinctive in being trees, shrubs, lianas, or perennial herbs with ***resin ducts or laticifers*** (some species causing allergenic responses), flowers generally 5-merous, with a ***nectariferous disk*** and ***single ovule per carpel***, the fruit a ***drupe*** with a resinous mesocarp.
K usu. 5 or (5) **C** usu. 5 [0] **A** 5–10 [1, ∞] **G** (1–3,5) [(12)], superior, rarely inferior.

Rutaceae [incl. Cneoraceae] — Rue/Citrus family (type *Ruta*, Latin for rue). Ca. 161 genera/ca. 2100 species (Figure 8.67).

The Rutaceae consist of trees, shrubs, lianas, or rarely herbs. The **stems** of some taxa have thorns. The **leaves** are simple, trifoliolate, or pinnate, sometimes pinnatifid, exstipulate, usually with pellucid or punctate glands. The **inflorescence** is a cyme or raceme, rarely of solitary flowers. The **flowers** are usually bisexual and actinomorphic, hypogynous, rarely epigynous. The **calyx** is aposepalous or synsepalous with 4–5 [2–3] sepals or lobes. The **corolla** is apopetalous or sympetalous with 4–5 [0, 2–3], imbricate or valvate petals or lobes. The **stamens** are 8–10–∞, usually diplostemonous, in 2 [1–4] whorls, with staminodes present in some taxa; filaments are often basally connate. **Anthers** are longitudinal in dehiscence. The **gynoecium** is syncarpous, rarely apocarpous, with a superior ovary, 4–5 [1–∞] carpels, and 4–5 [1–∞] locules. **Placentation** is axile; **ovules** are anatropous or hemitropous, bitegmic, 2 [1–∞] per carpel. **Nectaries** are usually present as an annular disk at the base of the ovary. The **fruit** is a schizocarp, berry, drupe, or hesperidium (the last with internal, swollen trichomes termed juice

Figure 8.66 SAPINDALES. Anacardiaceae. **A–C.** *Mangifera indica*, mango. **A.** Inflorescence and leaves. **B.** Flower close-up. **C.** Fruit (drupe) longitudinal section, showing single, apical seed and endocarp. **D.** *Rhus integrifolia*, shoot with drupes. **E.** *Schinus terebinthifolius*, Brazilian pepper, shoot with drupes. **F.** *Toxicodendron diversilobum*, poison oak. **G–K.** *Malosma laurina*, laurel sumac. **G.** Vegetative shoot. **H.** Inflorescence. **I,J.** Flower close-ups. **K.** Flower longitudinal section, showing nectaries and ovary.

Figure 8.67 SAPINDALES. Rutaceae. **A–C.** *Calodendrum capense*. **A.** Flower. **B.** Fruit. **C.** Leaf close-up, showing internal, pellucid glands. **D.** *Cneoridium dumosum*. Note glands on fruits. **E,F.** *Eriostemon myoporoides*. **G.** *Eriostemon spicatum*. **H.** *Eriostemon buxifolius*. **I–K.** *Ruta graveolens*, rue. **I.** Flower, face view. **J.** Ovary close-up, showing glandular surface. **K.** Flower longitudinal section. **L.** *Philotheca salsolifolia*. **M.** *Citrus aurantiacus*, navel orange, fruit (seedless cultivar), a hesperidium.

sacs). Secretory cavities containing ethereal oils are present in many tissues, including the leaves and pericarp.

Members of the Rutaceae have a worldwide distribution, especially in tropical regions. Economic importance includes many important fruits, among them *Citrus* spp. (oranges, grapefruits, lemons, limes, etc.), herbs such as *Ruta graveolens* (rue), timber trees, medicinal plants, and a number of ornamental cultivars. See Groppo et al. (2008, 2012), Bayer et al. (2009), and Morton and Telmer (2014) for phylogenetic analyses, the latter proposing four subfamilies.

The Rutaceae are distinctive in being trees, shrubs, lianas, or herbs, with simple to compound leaves and usually bisexual, actinomorphic, hypogynous, 4–5-merous flowers, typically with an ***annular, nectariferous disk***, the fruit a schizocarp, berry, hesperidium, or drupe; ***secretory glands containing ethereal oils*** occur in many tissues, appearing as ***pellucid -punctate glands*** in the leaves and pericarp.

K 4–5 [2–3] **C** 4–5 [0, 2–3] **A** 8–10–∞ **G** (4–5) [(1–∞)], superior.

Sapindaceae [including Aceraceae & Hippocastanaceae]—Soapberry/Maple/Buckeye family (type *Sapindus*, name meaning "Indian soap," from the use of soapberry). Ca. 144 genera/ca. 1900 species (Figures 8.68, 8.69).

The Sapindaceae consist of trees, shrubs, lianas, or herbaceous vines, tendrils present in viney species. The **leaves** are simple, palmate, trifoliolate, pinnate, or bipinnate, usually spiral, opposite in Hippocastanoideae (incl. *Acer* & *Aesculus*), usually exstipulate. The **inflorescence** is a cyme, thyrse, or raceme, rarely of solitary, axillary flowers, sometimes umbel-like or corymb-like. The **flowers** are unisexual or bisexual, actinomorphic or zygomorphic, hypogynous, rarely perigynous, hypanthium sometimes present. The **calyx** is aposepalous or basally synsepalous with 4–5 sepals/lobes. The **corolla** is apopetalous with 4–5 [3,6+,0] petals, often with a basal, scale-like appendage. The **stamens** are 4–∞, filaments often with trichomes. **Anthers** are longitudinal in dehiscence. The **gynoecium** is syncarpous, with a superior ovary, 2–3 [–6] carpels and locules, all but one aborting in some taxa. **Placentation** is variable; **ovules** are anatropous, orthotropous, to campylotropous, bitegmic, 1–∞ per carpel. **Nectaries** are a (usually) extrastaminal or intrastaminal annular disk or pad. The **fruit** is a variable, fleshy or dry at maturity, e.g., a schizocarp of samaras in *Acer* and *Dipteronia*, a loculicidal capsule in *Aesculus*. The **seeds** are often arillilate or with fleshy integuments.

Members of the Sapindaceae s.l. are distributed in tropical and temperate regions worldwide. Economic importance includes include edible fruits/seeds, such as *Blighia* (akee), *Dimocarpus* (longan), and *Litchi* (litchi nut); timber trees; oil seeds; medicinal plants; stimulating (caffeine-containing) beverages, especially *Paullinia cupana* (guarana, prepared as a soft drink, especially popular in parts of South America); ornamental cultivars, such as *Koelreuteria* (golden-rain tree), *Acer* (maple), and *Aesculus* (buckeye, horse-chestnut) spp.; arrow or fish poisons, e.g., *Jagera* and *Paullinia* spp.; and various species used locally as a soap. Harrington et al. (2005), in a phylogenetic analysis of the family, proposed classification into 4 subfamilies: Dodonaeoideae, Hippocastanoideae (incl. *Acer* & *Aesculus*), Sapindoideae, and the monogeneric Xanthoceroideae. See also Buerki et al. (2009, 2010) for a recent analyses and classification.

The Sapindaceae, s.l. are distinctive in being ***trees, shrubs, lianas, or herbaceous vines*** with ***simple, palmate, trifoliolate, or pinnate*** leaves, the flowers typically with a ***4–5-merous*** perianth, ***extrastaminal or intrastaminal nectariferous disk***, and superior, usually ***2–3-carpellate*** ovary, the seeds often with an aril or fleshy seed coat (except in *Acer, Aesculus*, and relatives), many taxa having soap-like saponins in tissues.

K 4–5 or (5) **C** 4–5 [3, 6+,0] **A** 5–8 [4–10+] **G** (2–3) [(–6)], superior, hypanthium present in some.

SUPERASTERIDS

This group, the *Superasteridae* sensu Soltis et al. (2011), has been identified from several recent molecular phylogenetic analyses (see Figure 1; Table 8.3). The Superasterids usually consist of the Santalales Caryophyllales, Berberidopsidales, plus the Asterids (Figure 1; Table 8.3). As noted earlier, the Dilleniales are sometimes placed within the Superasterids.

SANTALALES

The largely parasitic Santalales are classified into seven families in APG IV (2016) and twenty families by Nickrent et al. (2010) and Huei-Jiun et al. (2015), the latter accepted here; see Table 8.3. Only the Viscaceae are described here. Among the others, the **Balanophoraceae** (Figure 8.70A–C) consist of achlorophyllous root parasites, erupting form the soil at flowering; **Comandraceae**, including *Comandra umbellata*, the bastard toad-flax (Figure 8.70D–E), widespread in North America; **Santalaceae**, including *Santalum*, sandlewood, used for timber, oil, and incense; and **Thesiaceae** (Figure 8.70F), including the speciose genus *Thesium* of the Old World and South America. See Nickrent and Malécot (2001), Der and Nickrent (2008), Nickrent et al. (2010), and Huei-Jiun et al. (2015) for phylogenetic analyses and classification of the order. Santalaceae in the APG IV (2016) treatment encompass the Santalaceae, s.s. (sandalwoods) plus the families Amphorogynaceae, Cervantesiaceae, Comandraceae, Nanodeaceae, Thesiaceae, and Viscaceae (mistletoes) of Nickrent et al. (2010, 2019) and Huei-Jiun et al. (2015).

Figure 8.68 SAPINDALES. Sapindaceae. **A.** Dodonaeoideae. *Dodonaea triquetra*, with winged fruits. **B–N.** Sapindoideae. **B.** *Alectryon subcinereus*. **C.** *Cupaniopsis anacardioides*, capsule with arillate seeds. **D.** *Cardiospermum corindum*, with bladdery capsule. **E–K.** *Koelreuteria spp*. **E,F.** Flower close-ups. **G.** Tree. **H.** Ovary cross section, showing axile placentation and three carpels. **I.** Androecium. **J,K.** Immature fruit. **L,M.** *Litchi chinensis*, litchi, the fruit a nut with a single, arillate seed (the aril edible). **N.** *Sapindus saponaria*, soap tree.

Figure 8.69 SAPINDALES. Sapindaceae, Hippocastenoideae. **A–D.** *Acer macrophyllulm*, big-leaf maple. **A.** Shoot, with opposite leaves. **D.** Inflorescence, a raceme. **D.** Flower, with actinomorphic perianth. **D.** Fruit, a schizocarp of samaras. **E.** *Acer rubrum*, red maple, showing inflorescence (right) and young fruits (samaras, at left). **F–J.** *Aesculus californica*, buckeye. **F.** Flowering shoot, showing palmate leaves and paniculate inflorescence. **G.** Flower, side view, showing bilaeral symmetry. **H.** Flower, longitudinal section. **I.** Ovary cross section, 3-carpellate. **J–K.** *Aesculus* sp. **J.** Immature fruit. **K.** Mature fruit, a 3-valved capsule, and seeds.

TABLE 8.3 Orders and families of the Superasterids and Asterids, after APG IV (2016). Families in bold are described in detail. An asterisk denotes a deviation from APG IV, following Martins et al. (2003) for Primulaceae s.l., Nickrent et al. (2010) and Huei-Jiun (2015) for Santalales, Hernández-Ledesma et al. (2015) and Walker et al. (2018) for Caryophyllales, Leubert et al. (2016) for Boraginales, and Stevens (2001 onwards) for Peltantheraceae.

SUPERASTERIDS

- BERBERIDOPSIDALES
  - Aextoxicaceae
  - Berberidopsidaceae
- SANTALALES
  - Amphorogynaceae*
  - Aptandraceae*
  - Balanophoraceae
  - Cervantesiaceae*
  - Comandraceae*
  - Coulaceae*
  - Erythropalaceae*
  - Loranthaceae
  - Misodendraceae
  - Mystropetalaceae*
  - Nanodeaceae*
  - Olacaceae
  - Oktoknemaceae*
  - Opiliaceae
  - Santalaceae
  - Schoepfiaceae
  - Strombosiaceae*
  - Thesiaceae*
  - **Viscaceae***
  - Ximeniaceae*
- CARYOPHYLLALES
  - Achatocarpaceae
  - Agdestidaceae*
  - **Aizoaceae**
  - **Amaranthaceae**
  - Anacampserotaceae
  - Ancistrocladaceae
  - Asteropeiaceae
  - Barbeuiaceae
  - Basellaceae
  - **Cactaceae**
  - **Caryophyllaceae**
  - **Chenopodiaceae***
  - Corbichoniaceae*
  - Didiereaceae
  - Dioncophyllaceae
  - **Droseraceae**
  - Drosophyllaceae
  - Frankeniaceae
  - Gisekiaceae
  - Halophytaceae
  - Kewaceae
  - Limeaceae
  - Lophiocarpaceae
  - Macarthuriaceae
  - Microteaceae
  - Molluginaceae
  - Montiaceae
  - Nepenthaceae
  - **Nyctaginaceae**
  - Petiveriaceae
  - Physenaceae
  - Phytolaccaceae
  - **Plumbaginaceae**
  - **Polygonaceae**
  - Portulacaceae
  - Rhabdodendraceae
  - Sarcobataceae
  - Simmondsiaceae
  - Stegnospermataceae
  - Talinaceae
  - Tamaricaceae

ASTERIDS

- CORNALES
  - **Cornaceae**
  - Curtisiaceae
  - Grubbiaceae
  - Hydrangeaceae
  - Hydrostachyaceae
  - **Loasaceae**
  - Nyssaceae
- ERICALES
  - Actinidiaceae
  - **Balsaminaceae**
  - Clethraceae
  - Cyrillaceae
  - Diapensiaceae
  - Ebenaceae
  - **Ericaceae**
  - **Fouquieriaceae**
  - Lecythidaceae
  - Maesaceae*
  - Marcgraviaceae
  - Mitrastemonaceae
  - **Myrsinaceae***
  - Pentaphylacaceae
  - **Polemoniaceae**
  - **Primulaceae***
  - Roridulaceae
  - Sapotaceae
  - **Sarraceniaceae**
  - Sladeniaceae
  - Styracaceae
  - Symplocaceae
  - Tetrameristaceae
  - **Theaceae**
  - Theophrastaceae*
- LAMIIDS
  - GARRYALES
    - Eucommiaceae
    - **Garryaceae**
  - BORAGINALES
    - **Boraginaceae**
    - Codonaceae*
    - Coldeniaceae*
    - **Cordiaceae***
    - Ehretiaceae*
    - **Heliotropiaceae***
    - Hoplestigmataceae*
    - **Hydrophyllaceae***
    - Lennoaceae*
    - Namaceae*
    - Wellstediaceae*
  - ICACINALES
    - Icacinaceae
    - Oncothecaceae
  - METTENIUSALES
    - Metteniusaceae
  - VAHLIALES
    - Vahliaceae
  - GENTIANALES
    - **Apocynaceae**
    - Gelsemiaceae
    - **Gentianaceae**
    - Loganiaceae
    - **Rubiaceae**
  - LAMIALES
    - **Acanthaceae**
    - **Bignoniaceae**
    - Byblidaceae
    - Calceolariaceae
    - Carlemanniaceae
    - Gesneriaceae
    - **Lamiaceae (Labiatae)**
    - Lentibulariaceae
    - Linderniaceae
    - Martyniaceae
    - Mazaceae
    - **Oleaceae**
    - **Orobanchaceae**
    - Paulowniaceae
    - Pedaliaceae
    - Peltantheraceae*
    - **Phyrmaceae**
    - **Plantaginaceae**
    - Plocospermataceeae
    - Schlegeliaceae
    - **Scrophulariaceae**
    - Stilbaceae
    - Tetrachondraceae
    - Thomandersiaceae
    - **Verbenaceae**
  - SOLANALES
    - **Convolvulaceae**
    - Hydroleaceae
    - Montiniaceae
    - **Solanaceae**
    - Sphenocleaceae
- CAMPANULIDS
  - AQUIFOLIALES
    - **Aquifoliaceae**
    - Cardiopteridaceae
    - Helwingiaceae
    - Phyllonmaceae
    - Stemonuraceae
  - ESCALLONIALES
    - Escalloniaceae
  - BRUNIALES
    - Bruniaceae
    - Columelliaceae
  - APIALES
    - **Apiaceae (Umbelliferae)**
    - **Araliaceae**
    - Griseliniaceae
    - Myodocarpaceae
    - Pennantiaceae
    - **Pittosporaceae**
    - Torricelliaceae
  - PARACRYPHIALES
    - Paracryphiaceae
  - DIPSACALES
    - Adoxaceae
    - **Caprifoliaceae***
    - Diervillaceae*
    - **Dipsacaceae***
    - Linnaeaceae*
    - Morinaceae*
    - **Valerianaceae***
  - ASTERALES
    - Alseuosmiaceae
    - Argophyllaceae
    - **Asteraceae (Compositae)**
    - Calyceraceae
    - **Campanulaceae**
    - **Goodeniaceae**
    - Menyanthaceae
    - Pentaphragmataceae
    - Phellinaceae
    - Rousseaceae
    - Stylidiaceae

Viscaceae — Mistletoe family (type *Viscum*, Latin for mistletoe). 7 genera/ca. 520 species (Figure 8.70D–I).

The Viscaceae consist of dioecious, monoecious, or andromonoecious herbs or shrubs. The **roots** are haustorial. The **leaves** are simple, opposite-decussate, exstipulate, scale-like (squamate) in some. The **inflorescence** is an axillary or terminal usually compound of simple dichasia. The **flowers** are small, unisexual, actinomorphic, sessile. The **perianth** is uniseriate, 2–4-merous, often reduced, valvate, apotepalous or basally syntepalous, often yellow or green in color. The **stamens** are 2–4, antitepalous, free or epitepalous, filaments short or absent. **Anthers** are basifixed, porcidial or transverse in dehiscence, tetra-, bi-, or unisporangiate-columellate. The **gynoecium** is syncarpous, with an inferior ovary, 3–4 carpels, 1 locule. The style is solitary, terminal. **Placentation** is basal, ovules often absent with female gametophytes arising within a "placental nucellar complex" (termed *mamelon*). **Nectaries** are absent or reduced (female flowers) or well-developed glandular disks (male flowers). The **fruit** is a berry or (in Arceuthobium) an explosive capsule. **Seeds** have viscid from mesocarp tissue, functioning in bird dispersal, seed coat absent, endospermous, the endosperm chlorophyllous, starchy. Plants are wind- or insect-pollinated.

Members of the Viscaceae have a worldwide distribution. Economic importance includes mistletoes, esp. *Phoradendron* spp., used in decorative displays, and pests on crop trees (e.g., *Arceuthobium*). See Nickrent et al. (2010) for family diagnosis.

Figure 8.70 SANTALALES. **A–C.** Balanophoraceae, *Balanophora fungosa*, an achorophyllous, root parasitic plant. **D.** Comandraceae. *Comandra umbellata*, close-up of flowers; note uniseriate, valvate perianth. **E,F.** Thesiaceae. **H.** *Pyrularia pubera*, buffalo nut. **I.** *Thesium* sp. flower close-ups. **G–I.** Viscaceae. **G.** *Arceuthobium campylopodum*, parasitic on pine. **H,I.** *Phoradendron californicum*, desert mistletoe.

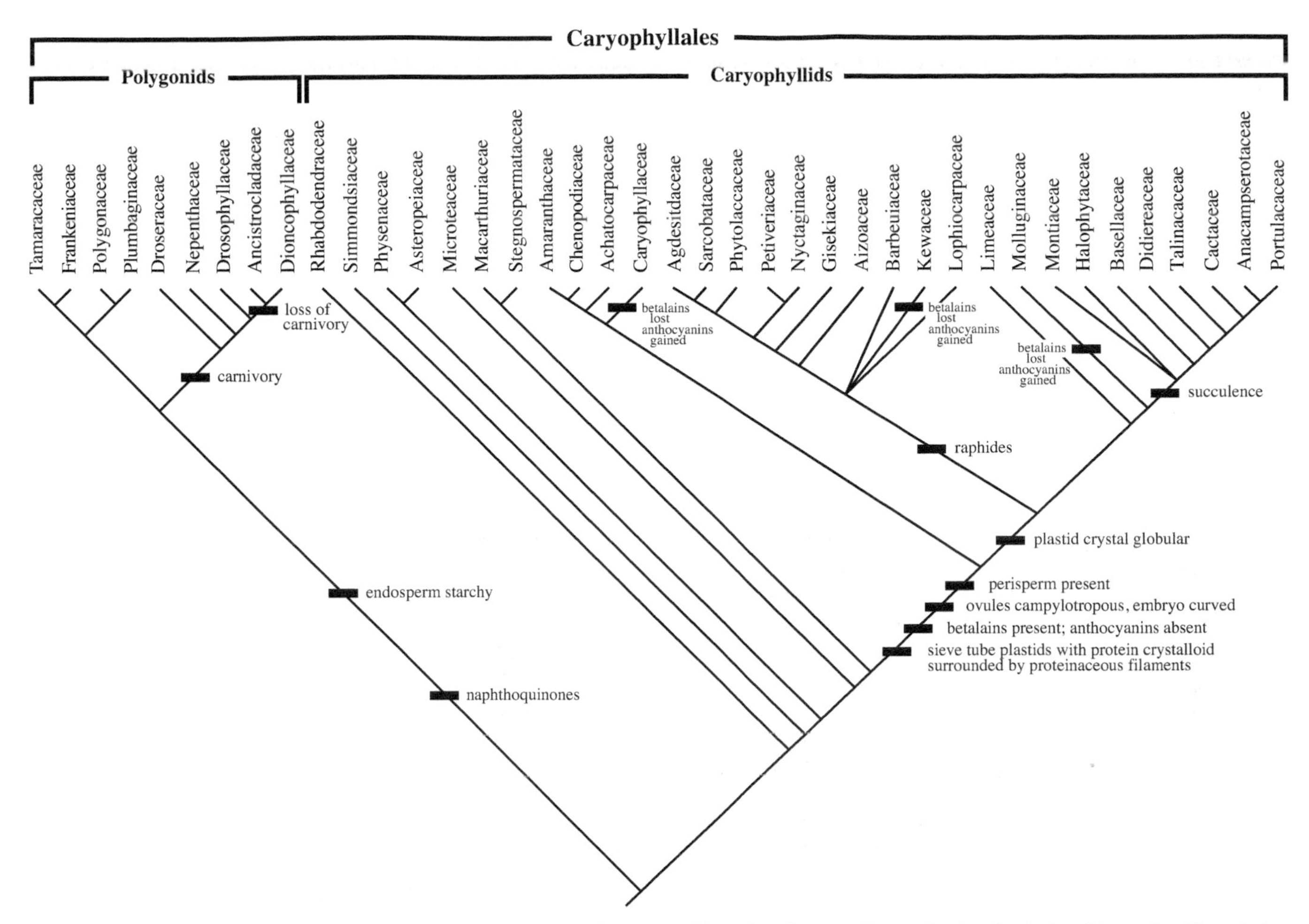

Figure 8.71 Cladogram of the Caryophyllales, with selected apomorphies, showing one hypoothesis of relationships, after Hernández-Ledesma et al. (2015), Yang et al. (2017), and Walker et al. (2018). Within one clade of the Polygonids note evolution and loss of carnivory. Within the Caryophyllids note the clade of ca. 24 families characterized by four ultrastructural, embryological, and chemical apomorphies, including the gain of betalain pigments, and the subsequent loss of these pigments in at least three lineages.

The Viscaceae are distinctive in being ***photosynthetic hemiparasites from stems*** of host, ***decussate leaves***, ***small, unisexual flowers*** with a ***uniseriate perianth***, ovules often absent with female gametophytes from a ***placental nucellar complex***, fruit a ***berry*** or ***explosive capsule***, and seeds with ***seed coat absent***, ***chlorophyllous endosperm***, and ***viscin***.
P 2–4 **A** 2–4 **G** (3), inferior.

CARYOPHYLLALES

The Caryophyllales contain 38–41 families (depending on the treatment; see Table 8.3) and about 12,500 species. The relationship of the Caryophyllales within the Eudicots has varied significantly in different analyses, but it appears to be more closely related to the Asterids than to the Rosids (see Moore et al. 2011; Soltis et al. 2011; Sun et al. 2016; Figure 8.1), and is included here within the Superasterids.

The Caryophyllales as currently defined include a number of families previously unsuspected of close relationships prior to molecular studies. The relationships of families portrayed in Figure 8.71 shows a consensus reached by numerous experts in the group (after Hernández-Ledesma et al. 2015, Yang et al. 2017, and Walker et al. 2018), with an informal classification (after Schäferhoff et al. 2009) into two sister clades: Polygonids (also called "non-core" Caryophyllales) and Caryophyllids (also called "core" or "expanded core" Caryophyllales). Several new families have been proposed of late. See Brockington et al. (2009), Schäferhoff et al. (2009), Hernández-Ledesma et al. (2015), Yang et al. (2017), and Walker et al. (2018) for recent analyses of the group, with Brockington et al. (2011, 2015) discussing the evolution of pigments, Devi et al. (2016) naphthoquinone chemistry, Wang et al. (2018) succulence, Yang et al. (2017) polyploidy, and Renner and Specht (2011, 2013) and Walker et al. (2017) carnivory in the complex.

The Caryophyllales encompass a traditional group formerly known as the Centrospermae (or Caryophyllidae, after Cronquist 1981). The traditional Centrospermae

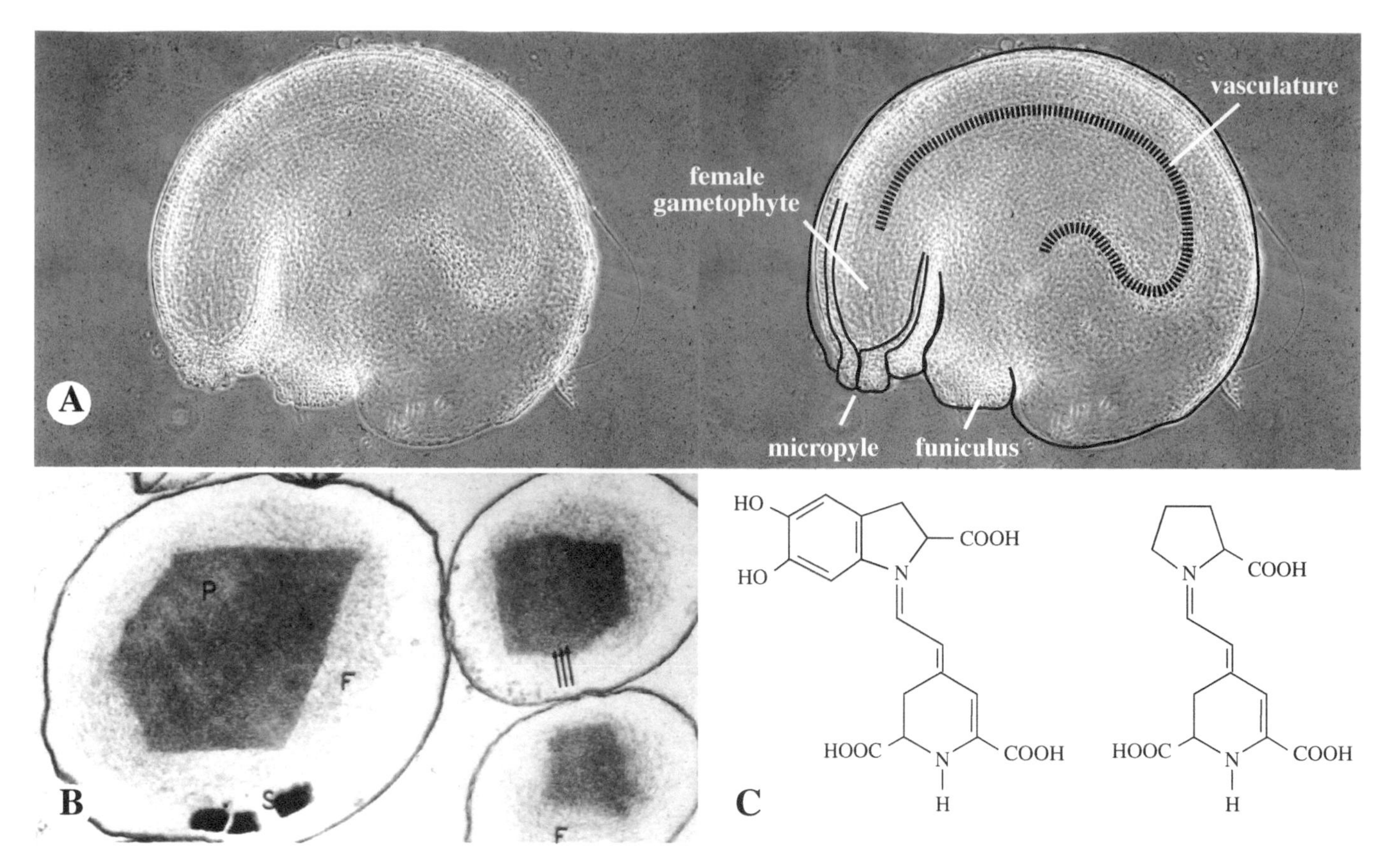

Figure 8.72 Caryophyllales apomorphies. **A.** Campylotropous ovule. **B.** Sieve tube plastid with proteinaceous filaments (F) encircling central crystalloid protein (P); from Benhke (1972), by permission. **C.** Betanidin, a betacyanin (left) and indicoxanthin, a betaxanthin, both examples of betalain pigments.

or Caryophyllidae are largely equivalent to the Caryophyllids, a complex of approximately 31 families (Figure 8.71), although the Polygonaceae, Plumbaginaceae, Frankeniaceae, and Tamaricaceae were often included in the past.

Many, though not all, members of the Caryophyllids possess pollen that is trinucleate upon being released from the anther, a relatively rare feature in angiosperms (most being binucleate at release). Also, many members of the Caryophyllids have either free-central (hence the name "Centrospermae; see below," from the seeds arising from a central column) or basal placentation (see Chapter 9). More clear-cut apomorphies for approximately 24 families of Caryophyllids are an ultrastructural feature—**sieve tube plastids with protein crystalloid inclusions surrounded by proteinaceous filaments** (Figure 8.72B)—and ovule/seed apomorphies — **ovules campylotropous** (Figure 8.72A) and **perispermous seeds** (see Chapter 11). In addition, these families are thought to share the evolution of **betalains** (Figure 8.72C), which are reddish, purplish, or yellowish pigmented compounds that functionally replace the anthocyanins found in other angiosperms. Note that the occurrence of betalains apparently reversed to anthocyanins in some family clades (Figure 8.71; see Brockington et al., 2011, 2015). Within the Caryophyllids is a sub-clade, composed of families that have a proteinaceous sieve tube plastid that is "globular" in shape (Figure 8.71). This "Globular Inclusion Clade" is composed of two sister groups, one with raphides as an apomorphy and the other composed mostly of succulents (Figure 8.71).

Nine of the 38–41 families of the order are described in detail here (Table 8.1). Notable among the families not described are the **Didieriaceae** (Figure 8.73A,B), including the pachycaulous xerophytes *Alluaudia* and *Portulacaria*; **Frankeniaceae** (Figure 8.73C–E), herbaceous or shrubby halophytes; **Molluginaceae** (Figure 8.73F), mostly herbs, many of which are weeds; **Montiaceae** [formerly in Portulacaceae] (Figure 8.73G,H), leaf-succulent herbs and shrubs; **Nepenthaceae** (Figure 8.73I–L), carnivorous plants with pitchers at the ends of photosynthetic leaves; **Phytolaccaceae** (Figure 8.74A–D), the pokeberry family; **Portulacaceae** (Figure 8.74E–H), the purslane family, mostly succulents with free-central placentation, circumscissile capsules, and seeds bearing strophioles; **Simmondsiaceae** (Figure 8.74I,J), consisting solely of *Simmondsia chinensis*, jojoba, the seed a source of oil-like wax with many uses, including cosmetics and skin/hair products; and **Tamaricaceae** (Figure 8.74K,L), including the cultivar and serious weed *Tamarix*.

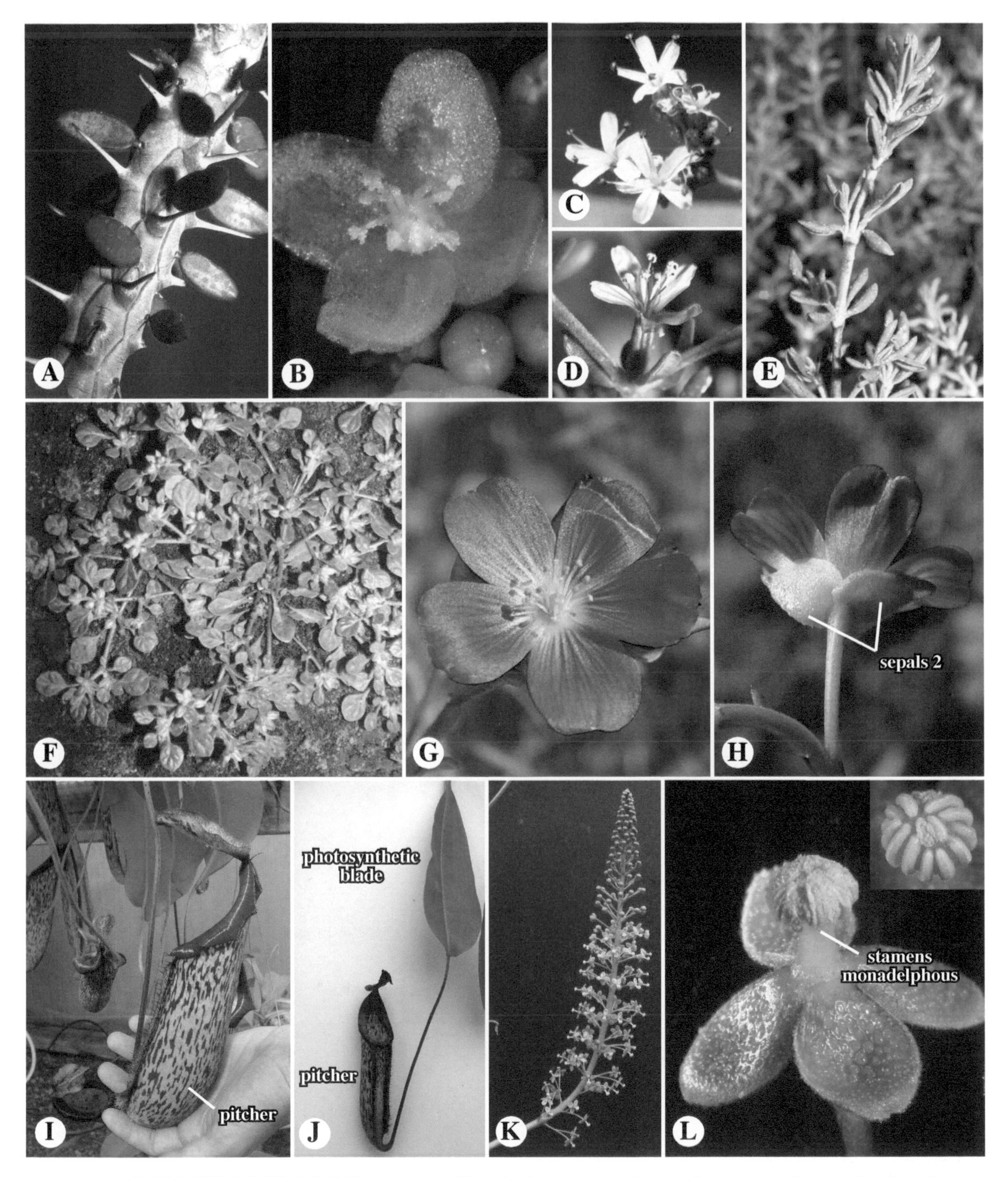

Figure 8.73 CARYOPHYLLALES. **A,B.** Didiereaceae. **A.** *Alluaudia dumosa*, stem with succulent leaves and thorns. **B.** *Alluaudia procera*, flower. **C–E.** Frankeniaceae, halophytes. **C.** *Frankenia palmeri*. **D,E.** *Frankenia salina*. **F.** Molluginaceae. *Glinus lotoides*, a weedy, annual herb. **G,H.** Montiaceae, *Calandrinia menziesii*. Note calyx with two sepals. **I–L.** Nepenthaceae, a carnivorous group. **I.** *Nepenthes rafflesiana*, showing pitcher from end of leaves. **J–L.** *Nepenthes* sp., showing pitcher leaf, inflorescence, and male flower (with connate, monadelphous stamens).

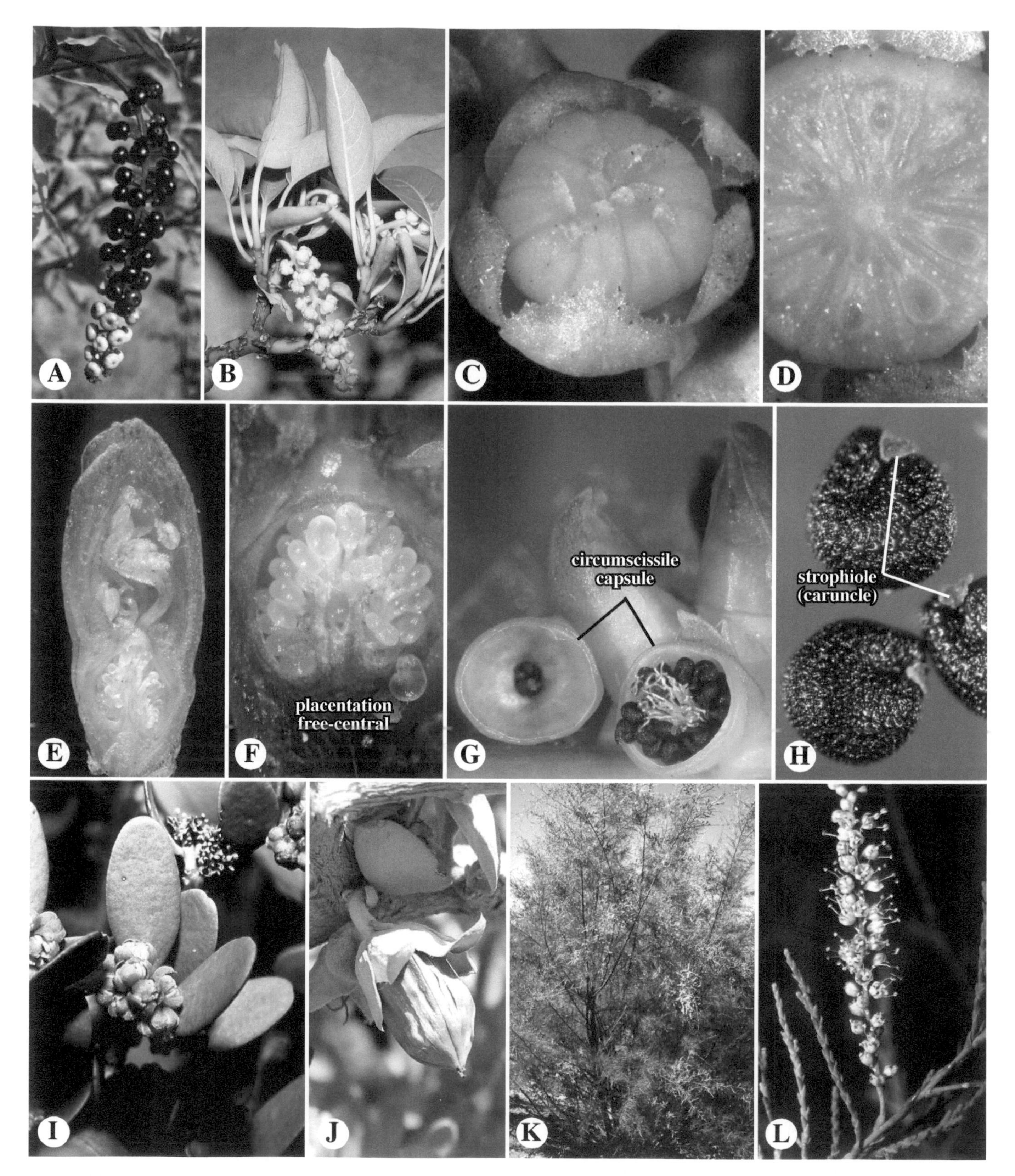

Figure 8.74 CARYOPHYLLALES. **A–D.** Phytolaccaceae. **A.** *Phytolacca americana*, pokeberry. **B–D.** *Phytolacca dioica*, ombu tree. **B.** Shoot, with inflorescence. **C.** Female flower. **D.** Ovary cross section, showing multiple carpels. **E–H.** Portulacaceae, *Portulaca oleracea*. **E.** Flower longitudinal section. **F.** Ovary longitudinal section, showing free-central placentation, typical of Caryophyllales. **G.** Fruit, a circumscissile capsule. **H.** Seeds, with stropiole (caruncle). **I,J.** Simmondsiaceae, *Simmondsia chinensis*. jojoba. **I.** Flowers of male individual. **J.** Fruit of female individual.**K–L.** Tamaricaceae, *Tamarix ramosissima*, tamarisk.

Aizoaceae — Mesembryanthemum or Vygie family (type *Aizoon*, meaning "always alive"). Ca. 124 genera/ca. 1,900 species (Figure 8.75).

The Aizoaceae consist of annual or perennial herbs, rarely shrubs or trees, rarely spiny. The **leaves** are often "centric," without a bifacial structure (some with apical "lens" through which light enters), opposite or whorled (rarely alternate), simple, undivided, usually exstipulate, succulent, often terete or angled. The **inflorescence** is terminal or axillary, of solitary flowers or cymes. The **flowers** are bisexual (rarely unisexual), actinomorphic, epiperigynous or perigynous. The **perianth** is uniseriate, a hypanthium present. The **calyx** is distinct with 5 [3–8] sepals. The **corolla** is absent, the petaloid structures of some taxa interpreted as petaloid staminodes. The **stamens** are [4] 5–∞, apostemonous or basally connate into bundles or monadelphous, with an outer whorl(s) of petaloid staminodes in some taxa. The **gynoecium** is syncarpous, with a superior or inferior ovary, 2–∞ carpels, and 2–∞ (rarely 1) locules. The **styles** are generally as many as carpels. **Placentation** is axile, parietal with septa, or basal; **ovules** are campylotropous to anatropous, bitegmic, [1–] ∞. **Nectaries** are present, inner to the insertion of the androecium. The **fruit** is a loculicidal capsule or berry. Betalain pigments are present, anthocyanins absent. Photosynthesis is often C4 or CAM (see Chapter 10).

Members of the Aizoaceae grow in tropical and subtropical regions, primarily in South Africa, less so in Australia. Economic importance includes mostly numerous ornamental cultivars, some (e.g., *Sceletium*) with medicinal properties, *Tetragonia* (New Zealand spinach) used as table greens. See Klak et al. (2003) and Klak (2010) for recent treatments of the Aizoaceae.

The Aizoaceae are distinctive in being herbs, rarely shrubs or trees, with generally ***opposite, succulent leaves*** (often with C4 or CAM photosynthesis) and solitary or cymose flowers with a ***uniseriate perianth*** (***outer petaloid staminodes*** present in many), usually ***numerous stamens***, and usually ***numerous ovules***, betalain pigments only present.

P 5 [3–8] **A** [4] 5–∞ **G** (2–∞), superior or inferior, hypanthium present.

Amaranthaceae — Amaranth family (type *Amaranthus*, meaning "unfading"). 71 genera/750 species. Inclusive of the Chenopodiaceae, sensu APG IV 2016 (Figure 8.76A–D).

The Amaranthaceae consist of annual or perennial, hermaphroditic, dioecious, monoecious, or polygamous, herbs, vines, shrubs, or rarely trees. The **leaves** are simple, spiral or opposite, usually entire. The **inflorescence** is of solitary flowers, cymes, or thyrses, with bracts and bracteoles often bristle-like and pigmented. The **flowers** are bisexual or unisexual, hypogynous. The **perianth** is uniseriate, 3–5 [0–2], apotepalous, rarely basally syntepalous. The **stamens** are 3–5 [0–2], generally the same number as perianth parts, antitepalous, basally connate, forming a tube. Anthers are dithecal or monothecal. The **gynoecium** is unicarpellous, with a superior ovary. The **style** is solitary. **Placentation** is basal; **ovules** are campylotropous, bitegmic, 1 [∞] per ovary. **Nectaries** are present, as an intrastaminal disk. The **fruit** is a irregularly dehiscing capsule or berry. The **seeds** are mostly starchy -perispermous. Betalain pigments are present, anthocyanins absent. Stems may form concentric rings of vascular bundles.

Members of the family are distributed worldwide, in tropical and some temperate regions. Economic importance includes weeds and pseudograin crops (*Amaranthus* spp.) and some cultivated ornamentals, such as *Celosia argentea*, cockscomb. See Kadereit et al. (2003) and Müller and Borsch (2005) for phylogenetic studies of the complex. Subfamily Polycnemoideae has shifted between the Amaranthaceae and Chenopodiaceae in different analyses.

The Amaranthaceae are distinctive in being herbs to trees with ***bristle-like bracts***, a ***uniseriate perianth*** of 3–5 [0–2] parts, basally connate stamens of same number and ***opposite perianth parts***, and a ***1-loculed, mostly 1-ovuled ovary with basal placentation***.

P 3–5 [0–2] **A** 3–5 [0–2] **G** 1, superior

Chenopodiaceae — Goosefoot family (type *Chenopodium*, after the Greek chen, "goose" plus podion, "little foot," in reference to the leaf shape of some species). Ca. 10 genera/ca. 500 species (Figure 8.76E–T).

The Chenopodiaceae consist of annual or perennial herbs or shrubs. The **stems** are sometimes jointed or succulent. The **leaves** are simple, exstipulate, succulent or reduced in some. The **inflorescence** is of solitary flowers or in spikes, panicles, or cymes. The **flowers** are small, bisexual or unisexual, usually actinomorphic, hypogynous. The **perianth** is uniseriate [1] 5 [6–8], apotepalous, rarely basally syntepalous. The **stamens** are [0–2] 3–5 [6–8], antitepalous, distinct or basally connate forming a tube. Anthers are longitudinal in dehiscence. The **gynoecium** is syncarpous, with a superior, rarely half-inferior ovary, 2–3 [5] carpels, and 1 locule. The styles are 1-several. **Placentation** is basal; **ovules** are campylotropous or amphitropous, bitegmic, 1 per ovary. **Nectaries** are present in some, as an annular disk. The **fruit** is a nutlet, rarely a circumscissile capsule. The **seeds** are starchy-perispermous. Betalain pigments are present, anthocyanins absent. Stems may form concentric rings of vascular bundles or alternating concentric rings of xylem and phloem. Plants often have C4 or CAM photosynthesis.

The Chenopodiaceae have a worldwide distribution. Economic importance includes the vegetable crops beet (*Beta vulgaris*) and spinach (*Spinacia oleracea*), the pseudograin

Figure 8.75 CARYOPHYLLALES. Aizoaceae. **A,B.** *Carpobrotus edulis*. **A.** Flower, top view, showing numerous, petaloid staminodes and numerous fertile stamens. **B.** Flower, longitudinal section. Note inferior ovary and hypanthium. **C–E.** *Aptenia cordifolia*. **C.** Flowering shoot. Note decussate leaf arrangement. **D.** Flower, longitudinal section, showing inferior ovary and hypanthium. **E.** Flower, cross section, showing axile placentation (four carpels and locules in this species). **F–H.** *Fenestera aurantiaca*. **F.** Leaves, with apical "lens." **G.** Leaf longitudinal section. **H.** Flower, longitudinal section, showing numerous petals and stamens and inferior ovary. **I.** *Lithops* sp., one of the "stone" plants, the leaves camouflaged as pebbles. **J.** *Faucaria tigrina*, tiger's jaw, flower and shoots.

Figure 8.76 CARYOPHYLLALES. **A–D.** Amaranthaceae. **A.** *Amaranthus caudatus*, tassel flower. **B–D.** *Amaranthus* sp. **B.** Plant with inflorescence. **C.** Female flower. **D.** Male flower. **E–T.** Chenopodiaceae. **E,F.** *Suaeda esteroa*. **E.** Whole plant, a succulent halophyte. **F.** Solitary flowers in axils of leaves; note antitepalous stamens. **G–J.** *Chenopodium* spp. **G.** Open bisexual flower, with five sepals, five antitepalous stamens, and single pistil. **H.** Fruit, surrounded by persistent calyx. **I.** Calyx removed to show fruit. **J.** Fruit sectioned, showing single seed with curved embryo. **K,L.** *Atriplex prostrata*. **K.** Whole plant. **L.** Inflorescence, a compound spike of cymes. **M.** *Atriplex canescens*, saltbush/shadscale, with accrescent bracts surrounding fruits. **N,O.** *Salsola tragus*. **N.** Whole plant, a tumbleweed. **O.** Flower close-up, showing spine-tipped bracts and prominent sepals. **P,Q.** *Allenrolfea occidentalis*, iodine bush. **P.** Estuarine community dominated by plant. **Q.** Stem close-up, showing jointed, succulent stems with reduced leaves. **R,S.** *Salicornia pacifica*, pickleweed. **R.** Shoot, with succulent, jointed stems and reduced leaves. **S.** Inflorescence, flowers inside bracts, styles exserted. **T.** *Arthroceras subterminale*, inflorescence with stamens exserted from bracts.

quinoa (*Chenopodium quinoa*), several detrimental weeds, and some local medicinal plants. Subfamily Polycnemoideae has shifted between the Amaranthaceae and Chenopodiaceae in different analyses. See Kadereit et al. (2003, 2010) for a recent phylogenetic analyses of the complex.

The Chenopodiaceae are distinctive in being herbs or shrubs with a ***uniseriate perianth*** of [1] 5 [6–8] parts, a ***superior, unilocular ovary*** with a ***single seed,*** and a ***nutlet or circumscissile capsule*** as a fruit.

P [0–2]3–5[6–8] **A** [0–2]3–5[6–8] **G** superior to half-inferior

Cactaceae — Cactus family (type *Cactus* [genus no longer recognized], Greek for a spiny plant). 111–118 genera/1,200–1,500 species (Figures 8.77, 8.78).

The Cactaceae consist of perennial shrubs or trees, with leaf spines arising from specialized axillary meristems termed **areoles**, which in some taxa may also bear small, trichome-like leaves with retrorse barbs called **glochidia**. The **stems** are typically succulent, and may be cylindrical and either short (sometimes shoots bunched) or tall-columnar or cladodes (e.g., prickly-pears), often with ribs or tubercles, radially plicate in some (e.g., barrel cacti), sometimes bearing dense masses of trichomes. The **leaves** are simple, often succulent, spiral, usually caducous (persistent and well-developed in Pereskioideae and Leuenbergerioideae and a few others). The **inflorescence** is of axillary, solitary flowers or rarely terminal cymes. The **flowers** are often large, bisexual (rarely unisexual), actinomorphic, epiperigynous. The **perianth** consists of numerous, distinct, spirally arranged tepals, grading from outer bractlike to inner petal-like structures, a hypanthium present. The **stamens** are numerous, spiral or in whorled clusters, and apostemonous. **Anthers** are longitudinal in dehiscence, tetrasporangiate, and dithecal. The **gynoecium** is syncarpous, with an inferior or half-inferior ovary, 3–many carpels, and 1 locule. The **style(s)** are single at the base, branched above. **Placentation** is parietal [basal]; **ovules** are campylotropous [anatropous], bitegmic, numerous per carpel. **Nectaries** are present within the hypanthium. The **fruit** is a berry, rarely dry and indehiscent. Bracts, which may sprout axillary flower or branch buds, commonly occur on the outer surface of the ovary and fruit. The **seeds** are arillate in some, exalbuminous, perispermous in some, embryos straight to curved. Flowers are pollinated by bees, moths, hummingbirds, or bats. Betalain pigments are present, anthocyanins absent. Photosynthesis is CAM (Crassulacean Acid Metabolism), in which stomata are opened at night (when carbon dioxide is fixed and stored), closed during the day to conserve water (see Chapter 10).

The Cactaceae are recently classified into up to five subfamilies: Pereskioideae, with persistent, broad vegetative leaves, glochidia absent, and seeds exarillate; Leuenbergerioideae, resembling Pereskioideae but forming a distinct clade; Maihuenioideae, cushion- or mat-forming shrubs with persistent leaves, of the southern Andes and Patagonia; Opuntioideae, with specialized glochidia and arillate seeds, many with terete, caducous leaves; and Cactoideae, with leaves and glochidia absent and seeds exarillate. Some workers propose that the monotypic South American *Blossfeldia* (*B. liliputana*, claimed as the smallest cactus in the world) should be placed in its own subfamily, Blossfeldioideae. Members of the Cactaceae grow mostly in desert regions, but many in temperate regions, with distributions in the New World (except for *Rhipsalis* in Africa). Economic importance includes many local, indigenous uses for medicine, food, and dyes (e.g., carmine from the cochineal insect), and innumerable cultivated ornamentals; *Opuntia* spp. are eaten for their fruits (prickly-pears) and stems (nopales), but can be serious, introduced weeds; *Lophophora williamsii* (peyote) is used as a hallucinogen and in religious ceremonies (e.g., Religion of the Native American Church). See Nyffeler (2002), Wallace and Gibson (2002), Butterworth (2006), Hernández-Hernández et al. (2011), and Guerrero et al. (2018) for phylogenetic studies of Cactaceae.

The Cactaceae are distinctive in being typically ***stem-succulent***, CAM shrubs or trees, with ***leaves usually reduced or absent***, the axillary meristems modified into specialized ***areoles*** bearing ***leaf spines***, the flowers ***epiperigynous*** with ***spiral perianth parts*** intergrading from outer bractlike to inner petal-like parts, having ***numerous stamens*** and an ***inferior [half-inferior] ovary*** with numerous ovules and ***parietal*** placentation, betalain pigments only present.

P ∞ **A** ∞ **G** (3–∞), inferior [half-inferior], with hypanthium.

Caryophyllaceae — Carnation family (type *Caryophyllus* [genus no longer recognized], meaning "clove-leaved"). Ca. 101 genera/ca. 2200 species (Figure 8.79).

The Caryophyllaceae consist of annual or perennial herbs, rarely shrubs, lianas, or trees. The **stems** often have swollen nodes. The **leaves** are opposite (rarely spiral), simple, usually exstipulate. The **inflorescence** is of dichasial cymes or solitary flowers. The **flowers** are bisexual or unisexual, actinomorphic, hypogynous, rarely perigynous. The **perianth** is biseriate, dichlamydeous, hypanthium absent [rarely present]. The **calyx** is synsepalous or aposepalous with 5 [4] sepals. The **corolla** is apopetalous and often unguiculate (clawed), with 5 [0,4] petals. The **stamens** are 5–10 [1–4], uniseriate or biseriate, apostemonous, epipetalous, or episepalous, basally epipetalous and forming a tube in some species. **Anthers** are longitudinal in dehiscence. The **gynoecium** is syncarpous, with a superior ovary (often with a stipe/gynophore), 2–5+ carpels, and 1 locule, often with basal septa. The **style(s)** are terminal, single below, often branched above.

Figure 8.77 CARYOPHYLLALES. Cactaceae. **A–D.** Pereskioideae. *Pereskia grandiflora*. **A.** Vegetative shoot, showing persistent, photosynthetic leaves and axillary areoles. **B.** Flower longitudinal section, showing inferior ovary and hypanthium. **C.** Ovary cross section, showing parietal placentation. **D.** Seed, longitudinal section, with curved embryos. **E–N.** Opuntioideae. **E.** *Cylindropuntia ganderi*, a cholla with cylindrical stems. **F.** *Cylindropuntia prolifera*, flower. Note tubercles bearing areoles along outter surface of inferior ovary. **G–I.** *Opuntia littoralis*. **G.** Vegetative shoots, with cladodes, flattened photosynthetic stems. **H.** Mass of glochidia from areole. **I.** Single glochidium with retrorse barbs (high magnification, scanning electron microscopy). **J.** *Opuntia basilaris* open flower, showing numerous tepals and stamens. **K–M.** *Cumulopuntia ignescens*. **K.** Habit, a cushion plant of the high-elevation altiplano of Chile. **L.** Flower, nested within mass of cushion. **M.** Stem, showing tubercles with terminal areoles bearing glochidia. **N.** *Quiabentia zehntneri*, an Opuntioid with persistent, photosynthetic leaves.

Figure 8.78 CARYOPHYLLALES. Cactaceae, Cactoideae. **A.** *Carnegiea gigantea*, saguaro cactus, a columnar cactus with a ridged, pleated stem. **B–C.** *Echinocereus engelmannii*, hedgehog cactus. **B.** Habit, with umerous, erect, clumped stems. **C.** Flower close-up. **D.** *Echinopsis chiloensis*. **E–F.** *Ferocactus cylindraceus*, a barrel cactus with pleated stems. **G.** *Ferocactus* sp., close-up of areole with radiating leaf spines. **H–I.** *Oreocereus leucotrichus,* a columnar cactus with dense stem trichomes. **J–K.** *Mammillaria dioica*. **J.** Flower. **K.** Fruits, fleshy red berries. **L.** *Mammillaria longimama*, having elongate stem tubercles. **M.** *Pachycereus pringlei*, cardon, fruits dry. **N,O.** *Selenicereus* sp., night-blooming cereus. **N.** Open, zygomorphic flower, with numerous tepals and stamens. **O.** Side view, with transition between bracts and inner petaloid tepals. **P.** *Zygocactus* sp., a cactus with cladodes bearing marginal flowers.

Figure 8.79 CARYOPHYLLALES. Caryophyllaceae. **A,B.** *Arenaria caroliniana*. **A.** Whole plants, an annual herb. **B.** Flower close-up. **C.** *Cerastium glomeratum*, capsules, with apical lobes ("teeth"). **D–H.** *Lychnis coronaria*. **D.** Flower, face view. **E.** Calyx, side view, showing synsepalous fusion. **F.** Single petal, showing basal claw. **G.** Ovary, with five styles. **H.** Ovary cross section, showing free-central placentation (lacking septa). **I.** *Silene gallica*, side view of flower and immature fruits. **J.** *Silene laciniata*, flower with strongly lobed petals. **K.** *Spergularia villosa*, flower close-up. **L,M.** *Spergula arvensis*. **L.** Swollen node of stem, bearing lobed leaves. **M.** Fruit, a five-valved capsule. **N,O.** *Stellaria media*. **N.** Shoot, showing swollen node and opposite leaves. **O.** Flower, with five bifid petals.

Placentation is free-central at least above, often axile below, or basal; **ovules** are campylotropous to hemitropous, bitegmic, 1-many per ovary. **Nectaries** occur as a nectariferous disk in some. The **fruit** is an achene or capsule, with valves or teeth 1–2× the carpel number. The **seeds** are perispermous, often with sculptured seed coat. Anthocyanin pigments are present, betalains absent. Anomalous secondary growth with concentric rings of vascular tissue found in some taxa.

The Caryophyllaceae have a worldwide distribution, especially in the northern hemisphere. Economic importance includes several ornamental cultivars, such as *Dianthus*, carnation. See Smissen et al. (2002), Fior et al. (2006), Harbaugh et al. (2010), and Greenberg and Donoghue (2011) for phylogenetic analyses of the Caryophyllaceae.

The Caryophyllaceae are distinctive in having ***nodes often swollen***, with ***simple, opposite leaves***, an inflorescence of ***solitary flowers or dichasial cymes***, and biseriate, actinomorphic, usually pentamerous flowers with ***distinct, clawed petals***, a superior ovary with distally ***free-central or basal placentation***, and a ***capsular fruit***, anthocyanin pigments only present.

K 5[4] or (5[4]) **C** 5[0,4] **A** 5 or 5+5 [1–4] **G** (2–5+), superior, hypanthium usually absent.

Droseraceae — Sundew family (type *Drosera*, after Greek *droseros*, dewy, in reference to glandular hairs). 3 genera (*Aldrovanda, Dionaea, Drosera*)/ca. 205 species (Figure 8.80).

The Droseraceae consist of carnivorous (usu. insectivorous), terrestrial or (in *Aldrovandra*) aquatic herbs. The **leaves** are simple (divided in some *Drosera* spp.), often in basal rosettes, spiral or whorled, usually stipulate, adaxially circinate or folded in development, of two structural types: "snap-trap," with two halves of blade closing rapidly — triggered by hairs — to entrap prey (*Aldrovanda* and *Dionaea*), or tentacular, with gland-tipped trichomes snaring prey (*Drosera*), all with sessile digestive glands. The **inflorescence** is terminal, composed of circinate monochasia or a solitary flower (*Aldrovandra*). The **flowers** are bisexual, actinomorphic, hypogynous. The **perianth** is dichlamydeous. The **calyx** is basally synsepalous with 4–8, imbricate, sepals/lobes. The **corolla** is apopetalous, of 4–8 convolute petals. The **stamens** are 5 [4–20], antisepalous, apostemonous. The **gynoecium** is syncarpous, with a superior ovary, 3 [5] carpels, and a unilocular locule. The **styles** are 3 [5], often bifid. **Placentation** is parietal or basal; **ovules** are anatropous, bitegmic, ∞ [3+]. The **fruit** is a loculicidal [rarely indehiscent] capsule. The **seeds** are endospermous.

The Droseraceae have a cosmopolitan distribution, often growing in wet areas. Economic importance includes ornamental cultivars, esp. *Drosera* spp., the sundews; the monotypic *Aldrovandra vesiculosa* and *Dionaea muscipula* (Venus fly-trap) are cultivated as curiosities. The Droseraceae are sister to a clade of four other families in most recent analyses (Figure 8.71). See Cameron et al. (2002) and Rivadavia et al. (2003) for phylogenetic analyses. See also Ellison and Adamec (2018).

The Droseraceae are distinctive in being ***carnivorous herbs*** with ***trap or tentacular leaves*** often in a basal rosette, the inflorescence of ***circinate monochasia*** or a solitary flower, flowers bisexual with a superior ovary having ***3 [5] carpels*** with ***parietal placentation***, the fruit a loculicidal capsule.

K (5–8) [(4)] **C** (5–8) [(4)] **A** 4–20 **G** (3) [(5)], superior.

Nyctaginaceae — Four-o'clock family (type *Nyctaginia*, after Greek *nyct*, night, in reference to some night-flowering species). Ca. 31 genera/ca. 405 species (Figure 8.81).

The Nyctaginaceae consist of hermaphroditic or dioecious trees, shrubs, lianas, or herbs. The **stems** are often with anomalous secondary growth, with concentric rings of vascular bundles or alternating xylem and phloem layers. The **leaves** are simple, opposite [rarely alternate], exstipulate. The **inflorescence** is usually a cyme, sometimes umbellate or of solitary flowers, an involucre present in some, with involucral bracts petaloid or connate and resembling a calyx. The **flowers** are bisexual [sometimes unisexual], bracteate in some, hypogynous. The **perianth** is uniseriate (termed calyx by default). The **calyx** is synsepalous and petaloid, usually salverform or infundibular, constricted above the ovary, with 5 [3–8] induplicate-valvate or contorted lobes. The **corolla** is absent. The **stamens** are 1-∞ and are apostemonous or basally connate, of the same or differing lengths. **Anthers** are longitudinally dehiscent. The **gynoecium** is unicarpellous, with a superior ovary and 1 locule. The **style** is terminal, mostly filiform. **Placentation** is basal; the **ovule** is solitary, usually campylotropous. **Nectaries** consist of an annular disk surrounding the ovary. The **fruit** is an achene or nut, often surrounded by the persistent and accrescent base of a coriaceous or fleshy calyx (often glandular, ribbed, winged, and/or grooved), forming an accessory fruit (anthocarp or diclesium), which functions in animal dispersal. **Seeds** are perispermous, embryos straight or curved. Betalain pigments are present.

Members of the Nyctaginaceae are distributed in tropical or warm temperate regions. Economic importance includes local edible and medicinal plants and cultivated ornamentals, such as *Bougainvillea* spp. and *Mirabilis japonica*, the four-o'clock (named because of flowers opening late in the afternoon). See Douglas and Manos (2007) for a phylogenetic analysis, Bittrich and Kühn (1993) for a detailed description, and Douglas and Spellenberg (2010) for a classification of the family.

The Nyctaginaceae are distinctive in being trees, shrubs, or herbs with ***opposite leaves***, the flower(s) subtended by a ***calyx-like involucre*** in some, having a ***uniseriate perianth*** (***calyx, often petaloid***), an annular, nectariferous disk, and a

Figure 8.80 CARYOPHYLLALES. Droseraceae. **A,B.** *Dionaea muscipula*, venus flytrap. Note conduplicate trap leaves, which rapidly close after being triggered. **C.** *Drosera spathulata*, a sundew with basal rosette of tentacular leaves. **D–E.** *Drosera capensis*. **D.** Elongate, tentacular leaves, showing glandular trichomes. **E.** Flower. **F–J.** *Drosera adelae*. **F.** Bud, with imbricate sepals. **G.** Flower, dichlamydeous, actinomorphic. **H.** Flower close-up, showing five, T-shaped stamens and pistil with superior ovary and bifid styles. **I.** Stamen; note poricidal dehiscence. **J.** Ovary cross section, showing parietal placentation with three carpels (arrows at placentae).

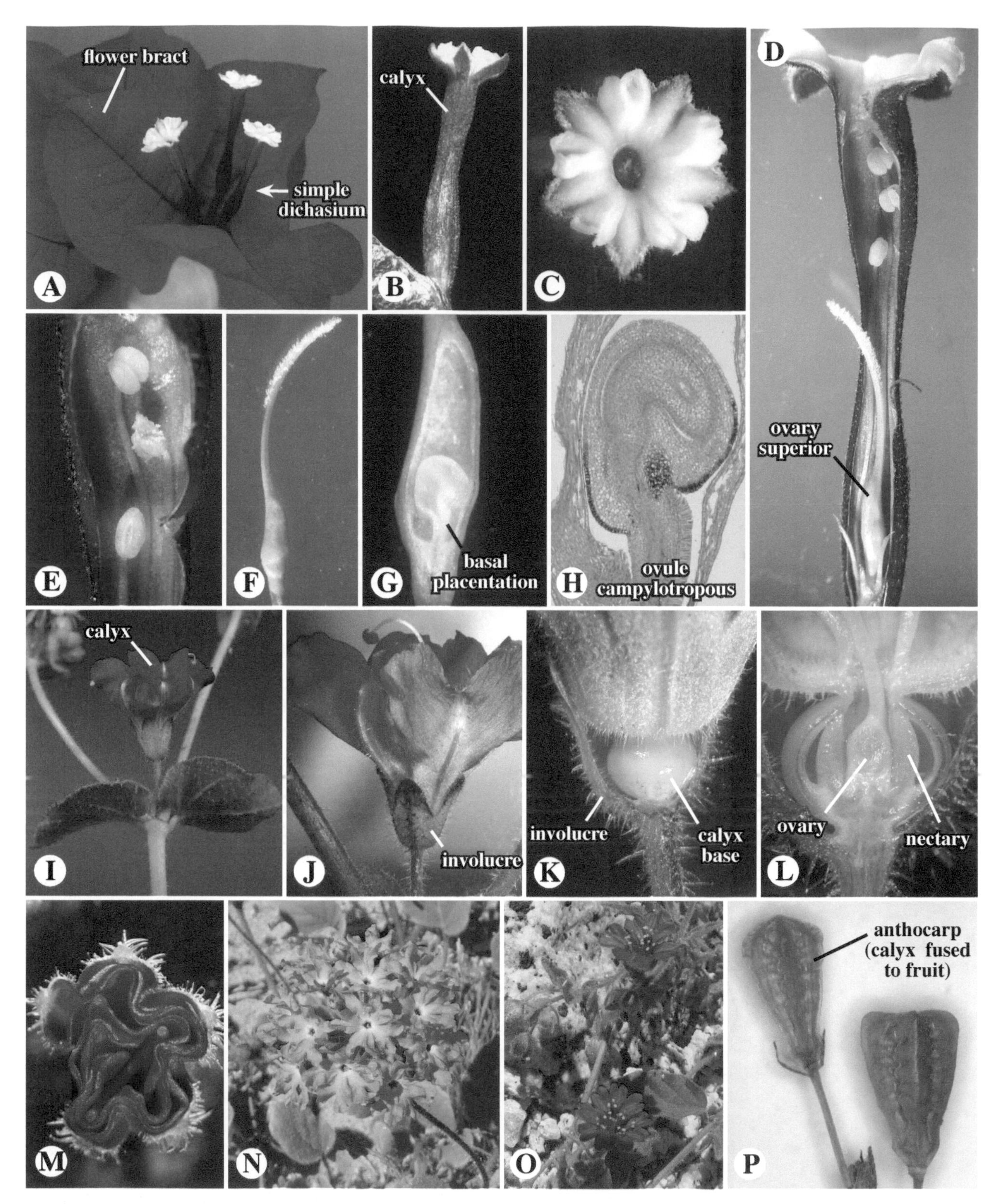

FIGURE 8.81 CARYOPHYLLALES. Nyctaginaceae. **A–H.** *Bougainvillea spectabilis*. **A.** Dichasium, with showy floral bracts. **B.** Flower, perianth uniseriate. **C.** Perianth (calyx), face view. **D.** Flower, longitudinal section; note superior ovary. **E.** Stamens; note differing lengths. **F.** Pistil. **G.** Ovary longitudinal section, showing basal placentation. **H.** Campylotropous ovule, sagittal-section. **I–M.** *Mirabilis laevis*, wishbone plant, showing terminal flower of cymose inflorescence. **J.** Flower, with subtending involucre (resembling calyx). **K.** Flower base, involucre partially removed, showing calyx base. **L.** Longitudinal section of flower base; note superior ovary and surrounding nectary. **M.** Perianth, cross section, with contorted aestivation. **N.** *Abronia villosa*, sand-verbena, inflorescence an umbel. **O.** *Allionia incarnata*, trailing windmills. **P.** *Boerhavia triquetra*, ringstem, with anthocarp fruits

unicarpellous ovary with a ***single, basal, usually campylotropous*** ovule, the fruit an achene or nut often surrounded by ***persistent, accrescent calyx***, forming an ***anthocarp***.
K (5) [(3–8)] **C** 0 **A** 1–∞ **G** 1, superior.

Plumbaginaceae — Leadwort family (type *Plumbago*, after Latin *plumbum*, lead, + *ago*, resemblance). Ca. 29 genera/ca. 730-840 species (Figure 8.82).

The Plumbaginaceae consist of perennial shrubs, lianas, or herbs. The **leaves** are simple, spiral, usu. exstipulate, often with calcium salt secreting glands. The **inflorescence** is a spike, raceme, head, or thyrse with circinate units. The **flowers** are bisexual, actinomorphic, hypogynous. The **perianth** is dichlamydeous. The **calyx** is synsepalous with 5 lobes and 5 or 10 ribs, often petaloid, glandular or scarious in some taxa. The **corolla** is sympetalous with 5, convolute lobes. The **stamens** are 5, antipetalous, epipetalous in some. **Anthers** are longitudinally dehiscent. The **gynoecium** is syncarpous, with a superior ovary, 5 carpels, and 1 locule. The **style(s)** are terminal, either 5 or 1 and 5-lobed; flowers are often heterostylous. **Placentation** is basal; **ovules** are anatropous and bitegmic, the funiculus often slender. The **fruit** is a achene or circumscissile capsule (with apical valves in some), the scarious or glandular calyx aiding in dispersal in some taxa. The **seeds** are endospermous or exalbuminous. Anthocyanin pigments are present, betalains absent; glycine betaines, functioning in salt secretion, are present in some taxa.

The Plumbaginaceae have a cosmopolitan distribution, but often grow in salt marshes or coastal habitats. Economic importance includes some local medicinal uses, but mostly ornamental cultivars, such as *Armeria* (thrift, sea pink), *Ceratostigma*, *Limonium* (statice, sea lavender), and *Plumbago* (esp. *P. auriculata*, cape plumbago). The family is usually divided into two subfamilies: Plumbaginoideae, with a spike, raceme, or head inflorescence, and Staticoideae, with a thyrse of circinate units. See Lledó et al. (1998, 2001, 2005) and Moharrek et al. (2014) for phylogenetic studies of the family.

The Plumbaginaceae are distinctive in being ***perennial shrubs, lianas, or herbs*** with simple, spiral leaves and ***pentamerous flowers*** with an actinomorphic calyx and corolla, the ovary ***unilocular*** with a ***single, basal, anatropous ovule***, the fruit an ***achene or circumscissile capsule***.
K (5) **C** (5) **A** 5 **G** (5), superior.

Polygonaceae — Buckwheat family (type *Polygonum*, meaning "many knees," from swollen nodes found in some species). Ca. 55 genera/ca. 1,110 species (Figure 8.83).

The Polygonaceae consist of annual or perennial herbs, shrubs, lianas, vines, or trees. The **stems** often have swollen nodes. The **leaves** are usually spiral, simple, stipulate or exstipulate, when stipulate, stipules are typically connate into a scarious, appressed sheath extending above the node, termed an "ocrea." The **inflorescence** consists of involucrate fasciculate units, the fascicles arranged in various branched or unbranched secondary inflorescences, cymose in some. The **flowers** are hypogynous, small, bisexual or unisexual, actinomorphic, pedicellate, the pedicels often articulated (jointed) above, an "ocreola" often subtending individual flowers. The **perianth** is uniseriate or appearing biseriate (actually spiral), homochlamydeous, usually 3+3, or 5 and quincuncial (rarely 2+2), the tepals (perianth parts) basally connate, hypanthium absent or present. The **stamens** are 3+3 or 8 [2,9+], often of two lengths, generally antitepalous, apostemonous to basally connate. **Anthers** are versatile, longitudinal and introrse in dehiscence. The **gynoecium** is syncarpous, with a superior ovary, 3 [2,4] carpels, and 1 locule. The **styles** are distinct or basally connate. **Placentation** is basal; **ovules** are orthotropous, bi- or unitegmic, solitary. **Nectaries** are often present, consisting of a nectariferous disk or nectary pads at base of stamens. The **fruit** is usually a 3-sided achene or nutlet, sometimes with an accrescent perianth or hypanthium. The **seeds** are endospermous, oily and starchy. Anthocyanin pigments are present, betalains absent. The vasculature is often anomalous.

The Polygonaceae are typically classified into two subfamilies: Polygonoideae, with ocrea present, and Eriogonoideae, with ocrea absent, but current classification supports the notion that the loss of an ocrea is not an apomorphy of the Eriogonoideae but arose within (for tribe Eriogoneae). Members of the family have a worldwide distribution, especially in the northern temperate hemisphere. Economic importance includes edible plants, such as *Fagopyrum esculentum*, buckwheat, and *Rheum rhabarbarum*, rhubarb; medicinal plants; timber, charcoal, and tanning plants; and a number of cultivated ornamentals, such as *Antigonon leptopus*, coral vine, *Muehlenbeckia*, and *Polygonum*. See Sanchez and Kron (2008), Burke et al. (2010), Sanchez et al. (2011), and Schuster et al. (2015) for phylogenetic analyses of the family. The ocrea, a 5-merous, quincuncial perianth, orthotropic ovules, and achenes are likely apomorphies for the family (see Sanchez and Kron 2008).

The Polygonaceae are distinctive in having simple, spiral leaves, with or without a stipular ***ocrea***, an inflorescence of ***fasciculate units***, small actinomorphic flowers usually with ***3+3 or 5 (quincuncial) connate tepals***, a 3 [2,4] carpellate ovary with a ***single, basal, mostly orthotropous ovule***, and a usually ***3-sided achene or nutlet***, anthocyanin pigments only present.
P (3+3) or (5) [(2+2)] **A** 3+3, 8 [2,9+] **G** (3) [(2,4)], superior, hypanthium absent or present.

Figure 8.82 CARYOPHYLLALES. Plumbaginaceae. **A,B.** *Armeria martitima*, thrift, sea-pink. Note basal leaves and head inflorescence. **C–E.** *Plumbago auriculata*, Cape plumbago. **C.** Inflorescence, a spike. **D.** Corolla, face-view. **E.** Flower, side view. Note glandular calyx. **F.** *Limonium californicum*, a salt marsh plant. **G.** *Limonium sinuatuam*. **H–M.** *Limonium perezii*. **H–I.** Inflorescence, showing circinate units. **J.** Flower, side view, showing dichlamydeous perianth. **K.** Flower, longitudinal section. Note superior ovary. **L.** Ovary, removed, showing 5 styles. **M.** Ovary, unilocular, with single, anatropous ovule pulled out. Note basal placentation and slender funiculus.

Figure 8.83 CARYOPHYLLALES. Polygonaceae. **A–C.** *Polygonum amphibium*. **A.** Shoot with spikelike inflorescence. **B.** Close-up of node, showing ocrea. **C.** Close-up of flowers. **D.** *Polygonum* sp., flower close-up, showing quincuncial aestivation (outlined). **E.** *Eriogonum fasciculatum*, ovary longitudinal section, showing single basal, orthotropous ovule. **F.** *Dedekera eurekensis*, ovary with three styles. **G.** *Rumex crispus*, fruits, surrounded by accrescent calyx. **H.** *Fagopyrum esculentum*, buckwheat, trigonous achenes. **I.** *Cocoloba diversifolia*, sea-grape. **J.** *Chorizanthe fimbriata*, spine-flower, showing 3+3 tepals. **K.** *Antigonon leptopus*, coral vine. **L,M.** *Eriogonum fasciculatum*, California buckwheat. **L.** Whole plant. **M.** Fascicle with involucre. **N.** Flower close-up, showing 3+3 tepals, androecium, and superior ovary.

ASTERIDS

The Asterids (Asteridae) are a major group of eudicots, comprising a large percentage of angiosperms in total. The Asterids are divided into 17 orders here (APG IV 2016; Figure 8.84, Table 8.3). The Asterids include well-known and economically important plants, such as dogwoods, hydrangeas, blueberries, phlox, tea, borage, gentians, mints, snapdragons, tomatoes/potatoes, carrots, scheffleras, hollies, bluebells, daisies, and a host of others.

Asterids appear to have three major apomorphies (Stull et al. 2015): (1) the presence of iridoid chemical compounds (Figure 8.85A); (2) unitegmic, tenuinucellate ovules (Figure 8.85B); and (3) cellular endosperm (see Chapter 11 for an explanation of the latter two). In addition, most Asterids have a sympetalous corolla (Figure 8.85C), but this likely has undergone considerable homoplasy.

The following are representative family descriptions of Asterids from most of the orders listed in Table 8.3. See Albach et al. (2001), Bremer et al. (2001, 2002), Soltis et al. (2007), Moore et al. (2011), Soltis et al. (2011), Oxelman et al. (2004), Stull et al. (2015), and Sun et al. (2016a) for information about relationships and evolution within the Asterids.

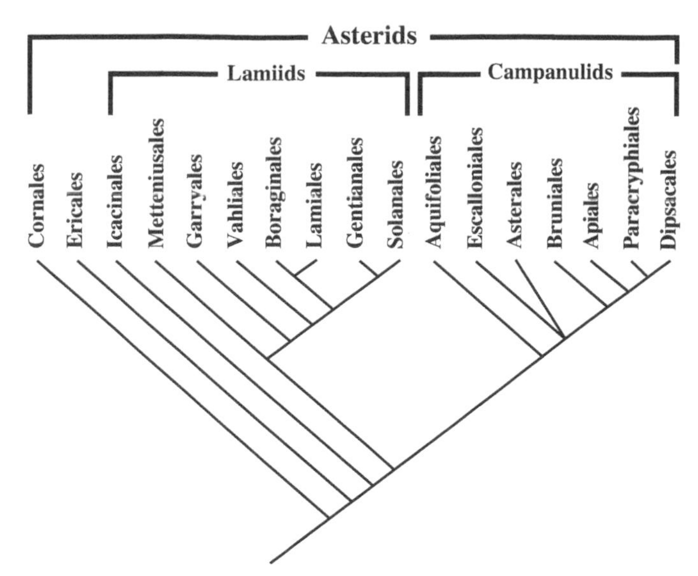

Figure 8.84 Cladogram of the major Asterid orders, after APG IV (2016).

CORNALES

The Cornales consists of seven families (Table 8.3), of which two are described in detail here. Notable among the other families is the **Hydrangeaceae** (Figure 8.86), including important cultivated ornamentals, such as *Hydrangea* and *Philadelphus*. See Xiang et al. (2002, 2011), Kubitzki (2004), and references therein for information about relationships and families within the order.

Cornaceae — Dogwood family (type *Cornus*, Latin for horn, after the hard wood). Ca. 2 genera (*Alangium, Cornus*)/ca. 85 species (Figure 8.87).

The Cornaceae consist of trees, shrubs, or rarely perennial, rhizomatous herbs. The **leaves** are simple, usually undivided [rarely pinnatifid], usually opposite [rarely spiral], and usually exstipulate. The **inflorescence** is a cyme or head of cymes, rarely a raceme, with showy, petaloid inflorescence bracts in some taxa. The **flowers** are bisexual [rarely unisexual], actinomorphic, and epigynous. The **perianth** is biseriate. The **calyx** is aposepalous, synsepalous and tubular in some unisexual male flowers, with 4 [5–7,0] sepals or calyx lobes. The **corolla** is apopetalous with 4–5 [–10, 0 in uni-

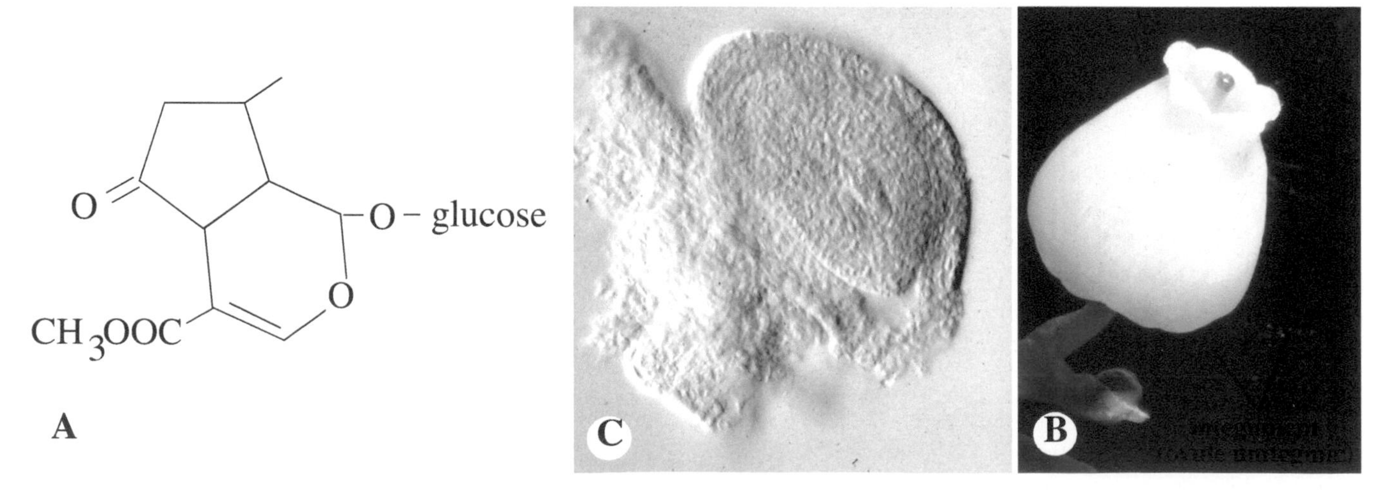

Figure 8.85 Putative apomorphies of the Asterids. **A.** Cornin, an iridoid compound. **B.** Unitegmic, tenuinucellate ovule. **C.** Sympetalous corolla.

Figure 8.86 CORNALES. Hydrangeaceae. **A.** *Carpenteria californica*. **B–C.** *Hydrangea* sp. **B.** Inflorescence, with peripheral sterile flowers bearing large calyces. **C.** Fertile flowers. **D.** *Whipplea modesta*.

sexual female flowers] valvate or imbricate petals. The **stamens** are 4–5 [10], usually alternipetalous, uniseriate, rarely biseriate, and apostemonous. **Anthers** are longitudinal in dehiscence. The **gynoecium** is syncarpous, with an inferior ovary, 2–4 [–9] carpels, and as many locules as carpels [rarely 1]. The **style** is solitary or as many as there are carpels. **Placentation** is apical and pendulous; **ovules** are anatropous, bitegmic [rarely unitegmic], 1 per locule. **Nectaries** usually consist of an infrastaminal annular disk. The **fruit** is a usually a drupe, the endocarp grooved, 1–5-locular. The **seeds** are endospermous.

The Cornaceae are generally distributed in northern temperate regions. Economic importance includes cultivated ornamentals, such as *Cornus* (e.g., *C. florida*, flowering dogwood) and some timber and edible fruit trees. The Nyssaceae are included in this family in APG IV (2016). See Xiang et al. (1993, 1997, 2006), Xiang and Thomas (2008), and Feng et al. (2009) for more information.

The Cornaceae are distinctive in being trees, shrubs, or rhizomatous herbs, with ***simple, usually opposite and undivided leaves***, a usually ***cymose inflorescence***, sometimes in ***heads***, with ***showy, petaloid bracts*** in some taxa, the flowers usually biseriate and bisexual, ***epigynous***, generally 4–5-merous, with an infrastaminal annual disk and ***inferior ovary***, the fruit usually a multi-locular ***drupe***.

K 4 [5–7, 0, connate in male flowers] **C** 4–5 [10, 0 in female flowers] **A** 4–5 [10] **G** (2–4) [(–9)], inferior.

Loasaceae — Loasa/Chili Nettle family (type *Loasa*, etymology unknown). Ca. 21 genera/350 species (Figure 8.88).

The Loasaceae consist of herbs, shrubs, or small trees, with silicified, often glochidiate, urent (stinging), or apically glandular trichomes. The **leaves** are simple, often lobed, spiral or opposite, exstipulate. The **inflorescence** is a thyrse or of solitary flowers. The **flowers** are bisexual, actinomorphic, epigynous or hypogynous. The **perianth** is dichlamydeous. The **calyx** is aposepalous with 5 [4–8] convolute or imbricate, persistent (often accrescent) sepals. The **corolla** is apopetalous or basally sympetalous, with 5 [4–8] petals, cybiform (boat-shaped) in some taxa. The **stamens** are 5–∞, antipetalous, distinct or basally connate, sometimes in fascicles, with petaloid or nectariferous staminodes present in some. **Anthers** are longitudinally dehiscent, sometimes subsessile from corolla tube. The **gynoecium** is syncarpous, with a superior or inferior ovary, 3–5 [7] carpels, and usually one locule. **Placentation** is usually parietal, rarely axile or apical; **ovules** are hemitropous, unitegmic. The **fruit** is a dehiscent or rarely indehiscent capsule. The **seeds** are with or without endosperm.

The Loasaceae are distributed in the Americas, southwestern Africa, and Arabia. Economic importance includes some cultivated ornamentals. See Moody et al. (2001), Hufford et al. (2003), Weigend (2004), and Acuña et al. (2017) for phylogenetic analyses and general information.

The Loasaceae are distinctive in being herbs, shrubs, or small trees with ***silicified, glochidiate, urent, or apically glandular*** trichomes, the flowers dichlamydeous with a ***persistent or accrescent*** calyx, ovary superior or inferior, ***usu. unilocular with parietal placentation***, the fruit a ***capsule***.

K 5 [4–8] **C** 5 or (5) [4–8] **A** 5, 10–∞ **G** (3–7), superior or inferior.

ERICALES

The Ericales is a large group of 22–25 families (Table 8.3), eight of which are treated here. Notable among the families not described are the **Actinidiaceae** (containing *Actinidia chinensis*, kiwi fruit), **Ebenaceae** (including *Diospyros* spp., persimmon/ebony, and *Euclea pseudebenus*, black ebony), and **Sapotaceae** (source of important fruit and timber trees, as well as of *Palaquium gutta*, gutta-percha, and *Manilkara*

Figure 8.87 CORNALES. Cornaceae. **A,B.** *Cornus canadensis*, a subshrub. **A.** Inflorescence with showy bracts subtending flowers. **B.** Infructescence, of berries. **C.** *Cornus florida*, a tree with large showy inflorescence bracts. **D–I.** *Cornus sericea*, a small tree. **D.** Inflorescence, a corymb lacking showy bracts. **E.** Leaves, opposite in arrangement. **F.** Flower, face view, showing petals, alternipetalous stamens, nectariferous disk, and style/stigma. **G.** Flower, longitudinal section, showing inferior ovary and nectary. **H.** Ovary, longitudinal section, showing apical-axile placentation. **I.** Ovary, cross section, 2-carpellate and 2-loculate.

zapota, chicle, the original chewing gum). See Geuten et al. (2004), Kubitzki (2004), and Rose et al. (2018) for information about relationships within the order.

Balsaminaceae — Jewel-Weed / Touch-Me-Not family (type *Balsamina* [=*Impatiens*], from Greek *balsamos*, balsam). 2 genera (*Hydrocera*, monotypic, & *Impatiens*)/ca. 1,000 species (Figure 8.89).

The Balsaminaceae consist of herbs, rarely subshrubs or very small trees, with raphide crystals. The **stems** underground are tubers or rhizomes, aerial stems are distinctively fleshy and translucent, rarely forming a storage pachycaul. The **leaves** are simple, usually spiral [opposite, or whorled], exstipulate or glandular-stipulate. The **inflorescence** is a cyme or solitary flower. The **flowers** are bisexual, zygomorphic, usually resupinate, hypogynous. The **calyx** is aposepalous

Figure 8.88 CORNALES. Loasaceae. **A.** *Eucnide urens*, rock nettle, flower. **B.** *Loasa tricolor*, ortiga brava, flower. **C–F.** *Mentzelia albicaulis*. **C.** Flower, side view, showing hispid trichomes and inferior ovary. **D.** Perianth, face view. Note 5 sepals and petals, numerous stamens. **E.** Flower longitudinal section, showing inferior ovary, fascicled stamens. **F.** Ovary cross section. Note parietal placentation and single locule. **G.** *Mentzelia involucrata*, flower side view. **H.** *Mentzelia laevicaulis*, blazing star, flower face view, showing outer, petaloid staminodes and numerous fascicled stamens. **I.** *Scyphanthus elegans*, open flower, face view; inset, side view. [Image at A contributed by Reid Moran.]

Figure 8.89 ERICALES. Balsaminaceae. **A–C.** *Impatiens niamniamensis*. Note simple, toothed leaves, and zygomorphic flower with spurred sepal and projecting androecium (covering gynoecium). **D–P.** *Impatiens wallerana*. **D.** Plant, in flower. **E.** Leaves, simple, toothed, usually spiral. **F.** Translucent stem. **G.** Bud, showing lower sepal spur. **H.** Flower, face view, showing 5 petals. **I,J.** Flower, back view, showing 3 sepals, 1 spurred. **K.** Close-up of staminal tube, forming calyptra around pistil. **L.** Longitudinal section of androecium and gynoecium, showing pistil nested within staminal tube. **M.** Pistil, after staminal tube fallen off. **N.** Ovary cross section, showing axile placentation; septae thin-walled (at arrows). **O.** Fruit, an explosively dehiscent capsule. **P.** Remains of fruit valves after dehiscence.

with 3 [5] often petaloid sepals, lowermost forming slender, necariferous spur. The **corolla** is sympetalous (*Impatiens*) or apopetalous (*Hydrocera*), with 5 petals. The **stamens** are 5, whorled, monadelphous (upper filaments and anthers connate), the fused anthers forming a calyptra covering stigma(s). **Anthers** are 3–4-sporangiate, the sporangia with thread-like "trabeculae" that partition sporogenous tissue and hold pollen together at dehiscence. The **gynoecium** is syncarpous, with a superior ovary, 5 [4] carpels and locules. The **stigma(s)** are 1 or 5. **Placentation** is axile; **ovules** are anatropous, bitegmic or unitegmic, 1 (*Hydrocera*) or ∞ (*Impatiens*) per carpel. Extrafloral **nectaries** are present in some. The **fruit** is a drupe with 5, schizocarpic pyrenes (*Hydrocera*) or a touch-sensitive, explosively dehiscent capsule with fleshy, twisted, enrolling valves, which separate rapidly from base and scatter seeds (*Impatiens*). Plants are insect or bird pollinated.

The Balsaminaceae are distributed in eastern and northern North America, central/southern Africa, and Europe to southeastern Asia. Economic importance includes uses as cosmetic dyes, local medicines, edible plants, and popular ornamental cultivars; some *Impatiens* spp. are naturalized weeds. The monotypic *Hydrocera* (*H. triflora*) and the large genus *Impatiens* are sister taxa. See Yuan et al. (2004) and Janssens et al. (2006) for recent phylogenetic studies of the family.

The Balsaminaceae are distinctive in being mostly herbs with ***translucent*** aerial stems and simple leaves, the flowers ***zygomorphic***, usu. ***resupinate***, with a ***lower, slender sepal spur***, the 5 stamens ***monadelphous***, with connate anthers forming ***calyptra over pistil***, anther sporogenous tissue with ***trabeculae***, the fruit a ***drupe with pyrenes*** or a touch-sensitive, ***explosively dehiscent capsule***.

K 3 [5] **C** 5 or (5) **A** (5) **G** (5), superior.

Ericaceae [including Empetraceae, Epacridaceae, Monotropaceae, Pyrolaceae] — Heath family (type *Erica*, a name used by Pliny, adapted from Theophrastus). Ca. 126 genera/ca. 4,000-4,400 species (Figures 8.90, 8.91).

The Ericaceae consist of perennial, hermaphroditic or dioecious, shrubs and small trees, rarely lianas. Some taxa are achlorophyllous and mycotrophic ("fungus feeding," i.e., obtaining nutrition from mycorrhizal fungi in the soil, which in turn obtain nutrition from roots of vascular plants). The **roots** have endotrophic mycorrhizae. The **leaves** are simple, exstipulate, spiral, opposite, or whorled, sometimes with a basal pulvinus, often evergreen, generally coriaceous, variable in shape, ranging from linear and strongly abaxially concave-revolute ("ericoid") to broad and relatively flat, with pinnate-netted venation. The **inflorescence** is a raceme, fascicle, headlike cluster, or of solitary flowers. The **flowers** are bisexual [rarely unisexual], actinomorphic, pedicellate, bracteate (with two bracteoles), hypogynous or epigynous. The **perianth** is biseriate and dichlamydeous or uniseriate, urceolate, tubular, or campanulate. The **calyx** is aposepalous with 5 [2–7], valvate or imbricate sepals. The **corolla** is sympetalous, rarely apopetalous, with 5 [2–7 or 0 in some], convolute or imbricate lobes or petals. The **stamens** are 5+5 [rarely >10, 2–4], whorled, usually obdiplostemonous and biseriate [rarely uniseriate], usually distinct and epipetalous. **Anthers** of all but basal members are developmentally inverted such that the anther base assumes an apical position, dehiscence poricidal or longitudinal and introrse at maturity by inversion, connective appendages present in some taxa. The **pollen** is typically shed in tetrahedral tetrads [monads in basal members]. The **gynoecium** is syncarpous, with a superior or less often inferior (e.g., *Vaccinium*) ovary, 5 [2–10] carpels, and 5 [2–10] locules. The **style** is solitary, terminal, hollow, fluted. **Placentation** is axile, apical-axile, basal-axile, or parietal; **ovules** are anatropous to campylotropous, unitegmic, 1–∞ per carpel. **Nectaries** are present, as an intrastaminal disk. The **fruit** is a capsule, berry, or drupe. The **seeds** are endospermous (oily and proteinaceous).

The Ericaceae are now circumscribed to include the former families Empetraceae, Epacridaceae, Monotropaceae, Pyrolaceae, and others. The Ericaceae have been classified into 8 subfamlilies: Enkianthoideae, Monotropoideae (including some achlorophyllous and parastitic members), Arbutoideae (including *Arbutus* and *Arctostaphylos*, the manzanitas), Cassiopoideae, Ericoideae (including *Erica*, *Rhododendron*, and the former Empetraceae), Harrimanelloideae, Styphelioideae (including the former Epacridaceae), and Vaccinioideae (including *Vaccinium*, blueberries, and relatives; see Kron et al. 2002). Members of the family grow in acid soils typically; various species (e.g., *Erica*) are dominants in bog, moorland, or heathland communities with worldwide distributions in temperate and tropical (mostly montane) regions. Economic importance includes cultivated ornamentals, especially *Rhododendron* [*Azalea*] and *Erica*; *Vaccinium* species (including blueberry and cranberry) are important fruit plants. See Kron et al. (2002) and and Schwery et al. (2014) for information on phylogenetic relationships and evolution within the Ericaceae.

The Ericaceae are distinctive in being usually ***evergreen shrubs and trees*** (some achlorophyllous and mycotrophic), with ***coriaceous, linear-revolute to broad-flat*** leaves, a sympetalous corolla, stamens usually developing by ***anther inversion***, dehiscence often ***poricidal*** and an intrastaminal disk.

K 5 [2–7] **C** (5) [(2–7), 0] **A** 5+5 [2–4, ∞] **G** (5) [(2–10)], superior (usually) or inferior.

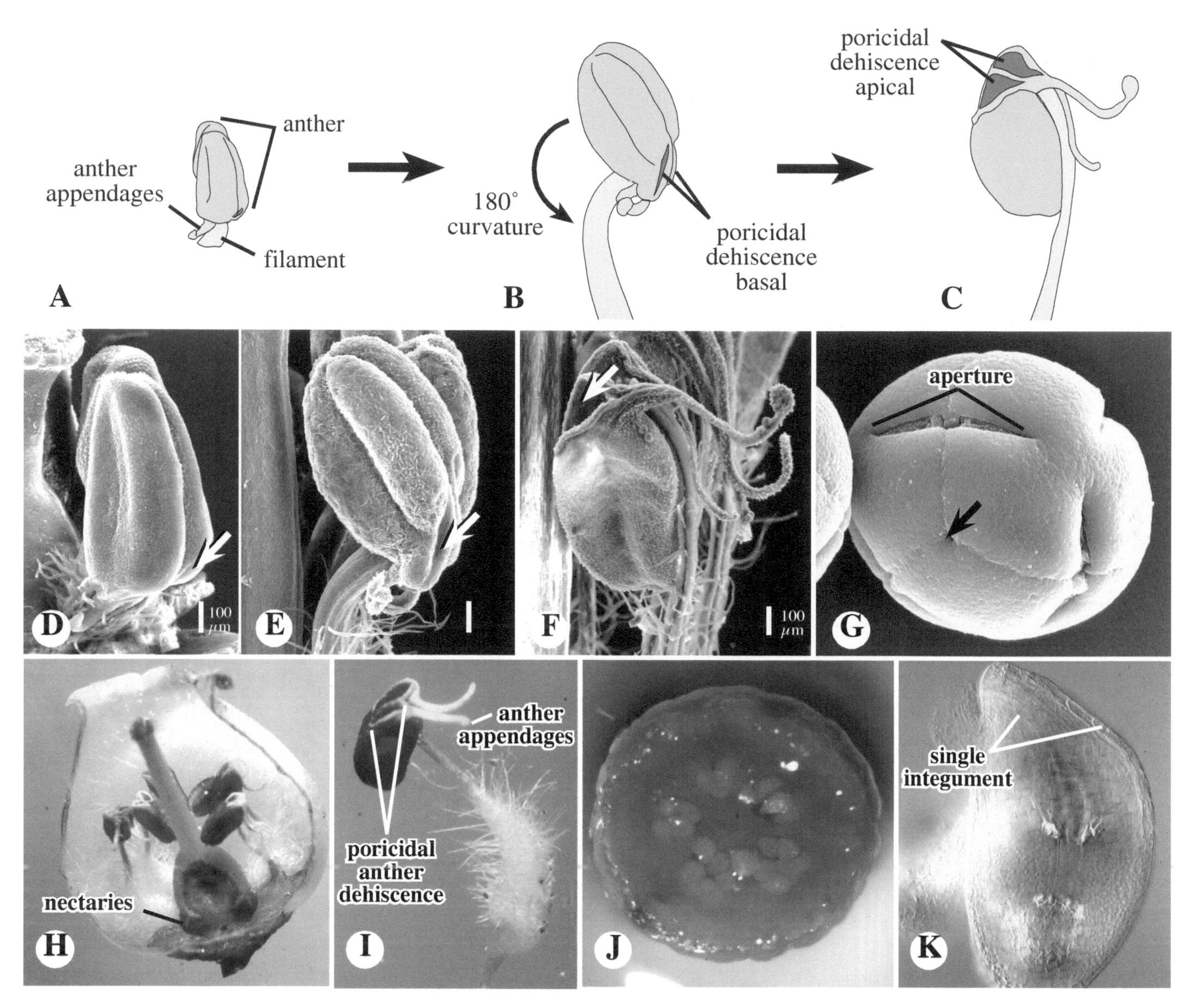

Figure 8.90 ERICALES. Ericaceae. **A–C.** Diagrammatic development of inverted anthers, characteristic of the family; note 180° inversion of anther base, in this case associated with anther appendages. **D–F.** *Xylococcus bicolor*, development of inverted anthers, showing inversion of anther base (arrow). **G–K.** *Arbutus unedo*. **G.** Pollen grain tetrad, showing colporate apertures; note junction of three (of four) grains (arrow). **H.** Flower in longitudinal section; note urceolate corolla, superior ovary, and basal nectaries. **I.** Stamen, showing poricidal dehiscence and anther appendages. **J.** Ovary cross section, showing five carpels and locules and axile placentation. **K.** Unitegmic ovule, typical of the Ericaceae and other Asterids.

Fouquieriaceae — Ocotillo family (type *Fouquieria*, after the Frenchman P. E. Fouquier). 1 genus/11 species (Figure 8.92).

The Fouquieriaceae consist of xeromorphic shrubs or trees. The **stems** are woody to succulent, ridged. The **leaves** are simple, undivided, spiral, those of long shoots forming petiolar spines, these with axillary fascicles of drought-deciduous nonspiny leaves. The **inflorescence** is a terminal spike, raceme, or panicle. The **flowers** are bisexual, actinomorphic, and hypogynous. The **calyx** is aposepalous with 5 imbricate sepals. The **corolla** is sympetalous and with 5, tubular to salverform, imbricate lobes. The **stamens** are 10–18 [23], uniseriate, apostemonous. **Anthers** are longitudinal in dehiscence. The **gynoecium** is syncarpous, with a superior ovary, 3 carpels, with 3 locules at base, 1 at apex. The **style** is solitary and 3-branched. **Placentation** is axile below, parietal above; **ovules** are ∞, anatropous, bitegmic. The **fruit** is a loculicidal capsule. The **seeds** are endospermous.

The Fouquieriaceae are distributed in southwestern North America. Economic importance is limited; *Fouquieria*

Figure 8.91 ERICALES. Ericaceae. **A.** *Vaccinium erythrocarpum*. **B.** *Xylococcus bicolor*, with urceolate corollas, typical of many Ericaceae. **C.** *Epacris longiflora*, with long tubular, reddish flowers and short, flat, coriaceous leaves. **D.** *Leucopogon ericoides*. **E.** *Pyrola rotundifolia*, an achlorophyllous, mycotrophic species. **F.** *Monotropa uniflora*, another achlorophyllous, mycotrophic species. **G.** *Rhododendron maximum*, with zygomorphic flowers and large, evergreen, coriaceous leaves. **H,I.** *Rhododendron occidentale*. **H.** Zygomorphic flowers with long, exserted stamens and style. **I.** Close-up of style and stamens, the latter with poricidal anthers. **J.** *Corema conradii*, close-up of low, decumbent shrub; note reduced, wind-pollinated flowers and "ericoid" leaves with abaxial cleft (arrow). **K.** *Erica centranthoides*, with red, tubular corollas and small, linear, "ericoid" leaves. **L.** *Kalmia latifolia*, in which stamens are bent (arrow) into pockets that catapult pollen with release of anthers.

Figure 8.92 ERICALES. Fouquieriaceae. **A–C.** *Fouquieria columnaris*, boojum or cirio, a pachycaulous plant native to deserts of Baja California, Mexico. **A,B.** Plant habit. **C.** Close-up of leaf of long shoot, the blade senescing, leaving a petiolar spine. **D.** *Fouquieria fasciculata*, a caudiciform species. **E–K.** *Fouquieria splendens*, ocotillo. **E.** Whole plant, a tall shrub ca. 4 m tall. **F.** Inflorescence. **G.** Close-up of shoot, showing petiolar spines (derived from original leaves of long shoot) plus fascicles (short shoots), having drought deciduous leaves. **H.** Flowers, with exserted stamens. **I.** Flower base, showing calyx. **J.** Ovary longitudinal section. **K.** Ovary cross section; note three carpels with axile placentation.

splendens (ocotillo) is planted locally as a fence or hedge; *Fouquieria columnaris* (boojum, cirio) is a spectacular pachycaul of Mexican deserts. See Schultheis and Baldwin (1999) and De-Nova et al. (2018) for more information about the phylogeny and evolution of the family.

The Fouquieriaceae are distinctive in being ***xeromorphic, sometimes succulent*** shrubs or trees, bearing ***long shoot leaves with petiolar spines***, in axils of which develop ***fascicles of drought-deciduous, nonspiny leaves***, the flowers mostly pentamerous, with a sympetalous corolla and superior, ***tricarpellate ovary*** having axile-parietal placentation.
K 5 **C** (5) **A** 10-18 [23] **G** (3), superior.

Myrsinaceae — Myrsine family [=Primulaceae, sensu APG IV, 2016] (type *Myrsine*, Greek for myrtle). Ca. 41 genera/ ca. 1,440 species (Figures 8.93, 8.94).

The Myrsinaceae consist of trees, shrubs, lianas, or herbs, often with resin ducts or cavities on leaves, flowers, or fruits appearing as yellowish to black, dot- or dash-shaped glands, some taxa with glandular trichomes. The **leaves** are simple, spiral or opposite, exstipulate. The **inflorescence** is an ebracteate fascicle or corymb. The **flowers** are small, bisexual, actinomorphic, hypogynous. The **perianth** is biseriate and dichlamydeous. The **calyx** is usually basally connate with 4–5 [3–7] lobes. The **corolla** is usually sympetalous with 4–5 [3–7] imbricate, convolute, or valvate lobes. The **stamens** are 4–5 [3–7], antipetalous, epipetalous on the corolla tube, monadelphous in some taxa. **Anthers** are longitudinal or poricidal in dehiscence. The **gynoecium** is syncarpous with a superior ovary, 3–5 [6] carpels, and 1 locule. The **style** is solitary and terminal. **Placentation** is generally free-central; **ovules** are anatropous to campylotropous, bitegmic (rarely unitegmic), few to ∞. The **fruit** is a 1–∞-seeded berry, drupe, or capsule. The **seeds** are (oily) endospermous, rarely exalbuminous.

The Myrsinaceae are distributed worldwide, but concentrated in Old World tropics. Economic importance includes cultivated ornamentals [e.g., *Ardisia crenata* (Christmas berry), *Cyclamen*, *Lysimachia* (loosestrife), and *Myrsine africana* (African-boxwood)] and locally used medicinal plants; the viviparous *Aegiceras* is an important component of mangrove forests. This family is part of a monophyletic complex of four—Maesaceae, Theophrastaceae, Primulaceae, and Myrsinaceae—that are similar in having antipetalous stamens and generally free-central placentation. For phylogenetic analyses of the complex, see Källersjö et al. (2000), Martins et al. (2003), Hao et al. (2004), Anderberg et al. (2007), and Yesson et al. (2009). Ståhl and Anderberg (2004) provide a general treatment of the Myrsinaceae, s.s.

The Myrsinaceae are distinctive in being trees, shrubs, lianas, or herbs, tissues often with ***pigmented dot- or dash-shaped glands*** or with ***glandular trichomes***, leaves simple and spiral or opposite, the flowers small, actinomorphic, dichlamydeous, 4–5 [3–6]-merous with ***antipetalous stamens***, the ovary superior, ***free-central***, the fruit a ***berry, drupe, or capsule***.
K 4–5 [3–7] **C** 4–5 [3–7] **A** 4–5 [3–7] **G** 3–5 [–6], superior.

Polemoniaceae — Phlox family (type *Polemonium*, associated with the Greek herbalist Polemon). Ca. 18 genera/385 species (Figure 8.95).

The Polemoniaceae consist of annual or perennial herbs, shrubs, lianas, or small trees. The **leaves** are simple or pinnate, divided in some, spiral, opposite, or whorled, exstipulate. The **inflorescence** is a head, cyme, or of solitary flowers. The **flowers** are bisexual, actinomorphic or zygomorphic, hypogynous. The **perianth** is biseriate and dichlamydeous. The **calyx** is usually synsepalous with five [4–6] lobes. The **corolla** is sympetalous and rotate, salverform, or bilabiate, with five [4–6] lobes, convolute in bud. The **stamens** are five [4–6], whorled, alternipetalous, uniseriate, epipetalous (inserted at different levels on corolla tube in some taxa). **Anthers** are longitudinal in dehiscence. The **gynoecium** is syncarpous, with a superior ovary, 3 [2,4] carpels, and 3 [2,4] locules. The **style** branches and **stigmas** are 3 [2,4]. **Placentation** is axile; **ovules** are anatropous to hemitropous, unitegmic, 1–∞. **Nectaries** consist of an annular nectariferous disk around ovary base. The **fruit** is a capsule, usually longitudinally dehiscent. The **seeds** are endospermous (oily). The stem xylem typically occurs in a continuous ring.

Members of the Polemoniaceae have distributions in the Americas (especially western North America) and Eurasia. Economic importance includes numerous cultivated ornamentals, such as *Cobaea*, *Gilia*, *Ipomopsis*, *Phlox*, and *Polemonium*. See Wilken (2004) for a general description and Porter (1997), Porter and Johnson (1998), Prather et al. (2000), and Johnson et al. (2008) phylogenetic relationships.

The Polemoniaceae are distinctive in being herbs, shrubs, or small trees with simple (divided in some) or pinnate leaves, a typically ***5–[4,6] merous, synsepalous, sympetalous perianth***, ***5 [4,6] epipetalous stamens***, and a typically ***3–[2,4] carpellate***, superior ovary, the fruit a ***loculicidal capsule***.
K (5) [4,6] **C** (5) [4,6] **A** 5 [4,6] **G** (3) [(2),(4)], superior.

Primulaceae (s.s.) — Primrose family (type *Primula*, Latin diminutive of "first," for early flowering). Ca. 9 genera/900 species (Figure 8.96).

The Primulaceae consist of annual or perennial herbs, rarely subshrubs. The **leaves** are usually simple, often basal and rosulate, exstipulate, spiral, opposite, or whorled. The **inflorescence** is a scapose head, panicle, umbel, or of solitary flowers. The **flowers** are bracteate, bisexual, actinomorphic

Figure 8.93 ERICALES. Myrsinaceae. **A–B.** *Ardisia crenata*. **A.** Flowering shoot, showing resionous glands. **B.** Flower, showing glands and antipetalous stamens. **C.** *Ardisia escallonioides*, with fleshy fruits (drupes). [Images at A,B contributed by Gerald Carr.]

[rarely zygomorphic]. The **perianth** is biseriate and dichlamydeous. The **calyx** is synsepalous with five [3–9] lobes. The **corolla** is sympetalous with five [3–9, 0] lobes, these convolute or imbricate. The **stamens** are five [3–9] (as many as petals or sepals), whorled, uniseriate (inserted at two levels in different individuals of heterostylous taxa), antipetalous, epipetalous from corolla tube, with antisepalous staminodes present in some taxa. **Anthers** are dithecal, introrse and longitudinal or poricidal in dehiscence. The **gynoecium** is syncarpous, with a superior [half-inferior in *Samolus*] ovary, with 1 locule and usually 5 carpels, often with partial septa at the ovary base, heterostylous in some taxa. The **style** is solitary and terminal. The **stigma** is typically capitate. **Placentation** is free-central; **ovules** are anatropous or campylotropous, bitegmic, ∞ [5+]. The **fruit** is a valvular, circumscissile, or indehiscent capsule. **Seeds** are [1–] usually numerous, with a non-starchy endosperm.

The Primulaceae are primarily distributed in the northern hemisphere. Economic importance includes numerous cultivated ornamentals, such as *Androsace* (rock-jasmine) and *Primula* (primrose). *Primula* spp. and other family members are heterostylous, with "pin" (styles long, anthers low in corolla tube) and "thrum" (styles short, anthers high in corolla tube) forms that promote cross-pollination. This family is part of a monophyletic complex of four—Maesaceae, Theophrastaceae, Primulaceae, and Myrsinaceae—that are similar in having antipetalous stamens and generally free-central placentation (the four combined as Primulaceae s.l. in APG IV, 2016 and treated as subfamilies in, e.g., Stevens, 2001 onwards). For a recent phylogenetic analysis of the complex, see Källersjö et al. (2000) and Martins et al. (2003), who propose transferring some members of a paraphyletic Primulaceae s.l. to the Myrsinaceae. See Anderberg (2004) for a general treatment of the Primulaceae s.l.

The Primulaceae are distinctive in being herbs or subshrubs with dichlamydeous, ***usually pentamerous*** [3–9] flowers, sepals and petals connate, the ***stamens antipetalous***, the ovary superior [rarely half-inferior] with ***free-central placentation***, the fruit a capsule.

K (5) [(3–9)] **C** (5) [(3–9), 0] **A** 5 [3–9] **G** (5), superior, rarely half-inferior.

Sarraceniaceae — Pitcher Plant family (type *Sarracenia*, after Michel *Sarrazin*, French-Canadian physician and naturalist, 1659–1734). 3 genera/ca. 32 species (Figure 8.97).

The Sarraceniaceae consist of perennial herbs [rarely subshrubs]. The **stems** are typically rhizomatous. The **leaves** are simple, exstipulate, usually in a basal rosette, with a "pitcher" or "pitfall trap" structural type, being tubular (by conduplication and fusion of margins during development), with an apical hood ("nectar spoon" in *Heliamphora*; "operculum" in others), the tube becoming fluid-filled, often with digestive glands on inner surface; insects and other small animals are attracted to pitcher mouth visually, by scent, or by nectary-like glands (located on nectar spoon in *Heliamphora*); prey animals are directed downward by introrse trichomes at pitcher's upper end and slide downward on slightly lower, slick surface into fluid, where they drown and are digested by means of secreted digestive enzymes or symbiotic bacteria. The **inflorescence** is a solitary, scapose flower or a short raceme. The **flowers** are large, bisexual, actinomorphic, nodding, hypogynous. The **perianth** is dichlamydeous. The **calyx** is aposepalous with 5 [3–6], imbricate, petaloid, persistent sepals. The **corolla** is apopetalous with 5 [0], imbricate, deciduous petals. The **stamens** are 10–∞. **Anthers** are basifixed. The **gynoecium** is syncarpous, with a superior ovary and 3–5 carpels, unilocular at ovary apex, 3–5-locular below. The **style** is 5-branched with terminal stigmas (*Darlingtonia*, *Heliamphora*) or apically peltate with stigmatic regions underneath (*Sarracenia*). **Placentation** is axile below, parietal from protruding placentae above; **ovules** are anatropous, unitegmic or bitegmic, numerous per carpel. The **fruit** is a loculicidal capsule.

Figure 8.94 ERICALES. Myrsinaceae. **A–G.** *Lysimachia arvensis*, scarlet pimpernel. **A.** Flower, face view. **B.** Flower, bottom view. **C.** Flower center, close-up; note antipetalous stamens. **D.** Ovary cross section. **E.** Flower longitudinal section, showing superior ovary. **F.** *Lysimachia quadrifolia*, whorled loostrife. **G.** *Lysimachia terrestris*, swamp loostrife. **H–N.** *Cyclamen persicum*. **H.** Whole plant, a perennial herb, with solitary flowers. **I.** Recurved flower bud, with convolute corolla lobes. **J.** Flower, oblique view. **K.** Close-up of flower throat, showing 5 connivent, antipetalous stamens. **L.** Flower longitudinal section. **M.** Stamens, abaxial (left) and adaxial (right) views. **N.** Ovary cross section, showing free-central placentation.

Figure 8.95 ERICALES. Polemoniaceae. **A.** *Gilia latifolia*, whole plant. **B.** *Gilia* sp., flower close-up, showing actinomorphic, sympetalous corolla. **C.** *Acanthogilia gloriosa*, a spiny shrubby species. **D.** *Langloisia setosissima* subsp. *punctata*, corolla with prominent nectar guides. **E.** *Langloisia setosissima* subsp. *s.* **F.** *Leptosiphon aureus*. **G.** *Linanthus dianthiflorus*, with fringed corolla lobes. **H.** *Leptosiphon floribundus*, showing cyme inflorescence. **I–K.** *Leptosiphon parviflorus*. **I.** Ovary cross section, showing three carpels/locules and axile placentation. **J.** Whole plant, showing long corolla tube. **K.** Style, apically three-branched. **L.** *Loeseliastrum schottii*, with zygomorphic flowers.

Figure 8.96 ERICALES. Primulaceae. **A–G.** *Primula* [*Dodecatheon*] *clevelandii*, shooting stars. **A.** Inflorescence, an umbel. **B.** Flower, petals reflexed. **C.** Flower, face view, showing antipetalous stamens. **D.** Androecium, side view, anthers introrse. **E.** Flower longitudinal section; note superior ovary and reflexed perianth. **F.** Anther, adaxial side facing. **G.** Ovary cross section.

The Sarraceniaceae are distributed in eastern and western North America and the Guayana Highland of South America, typically growing in acidic boggy habitats low in nitrogen availability. Economic importance is as cultivated ornamentals. See Bayer et al. (1996), Neyland and Merchant (2006), and Ellison et al. (2012) for phylogenetic and biogeographic studies. See also Ellison and Adamec (2018).

The Sarraceniaceae are distinctive in being ***carnivorous***, perennial herbs with ***pitcher, apically hooded, fluid-filled*** leaves that drown and digest insects and other small animals, flowers relatively ***large, nodding***, with ***persistent sepals*** and ***deciduous petals***, the ovary ***axile below, parietal above***, the fruit a ***loculicidal capsule***.

K 5 [3–6] **C** 5 **A** 10–∞ **G** (3–5), superior.

Theaceae — Tea family (after *Thea* [=*Camellia*], a Titaness in Greek mythology). Ca. 9 genera/ca. 190-460 species (Figure 8.98).

The Theaceae consist of trees and shrubs, rarely lianas, usually with unicellular trichomes. The **leaves** are simple, entire or toothed, usually spiral, exstipulate, evergreen. The **inflorescence** is of solitary, axillary flowers or a raceme or panicle. The **flowers** are usually large and bisexual, actinomorphic, hypogynous or rarely epigynous, with 2–∞ bracteoles. The **perianth** is biseriate and usually dichlamydeous (calyx and corolla intergrading in some taxa). The **calyx** is usually basally synsepalous with 5 [4–7], imbricate lobes, sometimes persistent and/or accrescent. The **corolla** is apopetalous or basally sympetalous with 5 [4–∞], imbricate or convolute petals or lobes. The **stamens** are usually numerous, developing centrifugally (from inside out), apostemonous or basally connate, in some taxa connate in 5 bundles opposite petals. **Anthers** are longitudinal, rarely poricidal in dehiscence. The **gynoecium** is syncarpous, with a superior or rarely inferior ovary, 3–5 [2–10] carpels, and as many locules as carpels. The **styles** are as many as carpels, sometimes basally connate. **Placentation** is axile; **ovules** are anatropous or campylotropous, bitegmic, 2 [1–∞] per carpel. **Nectaries** are often present at stamens

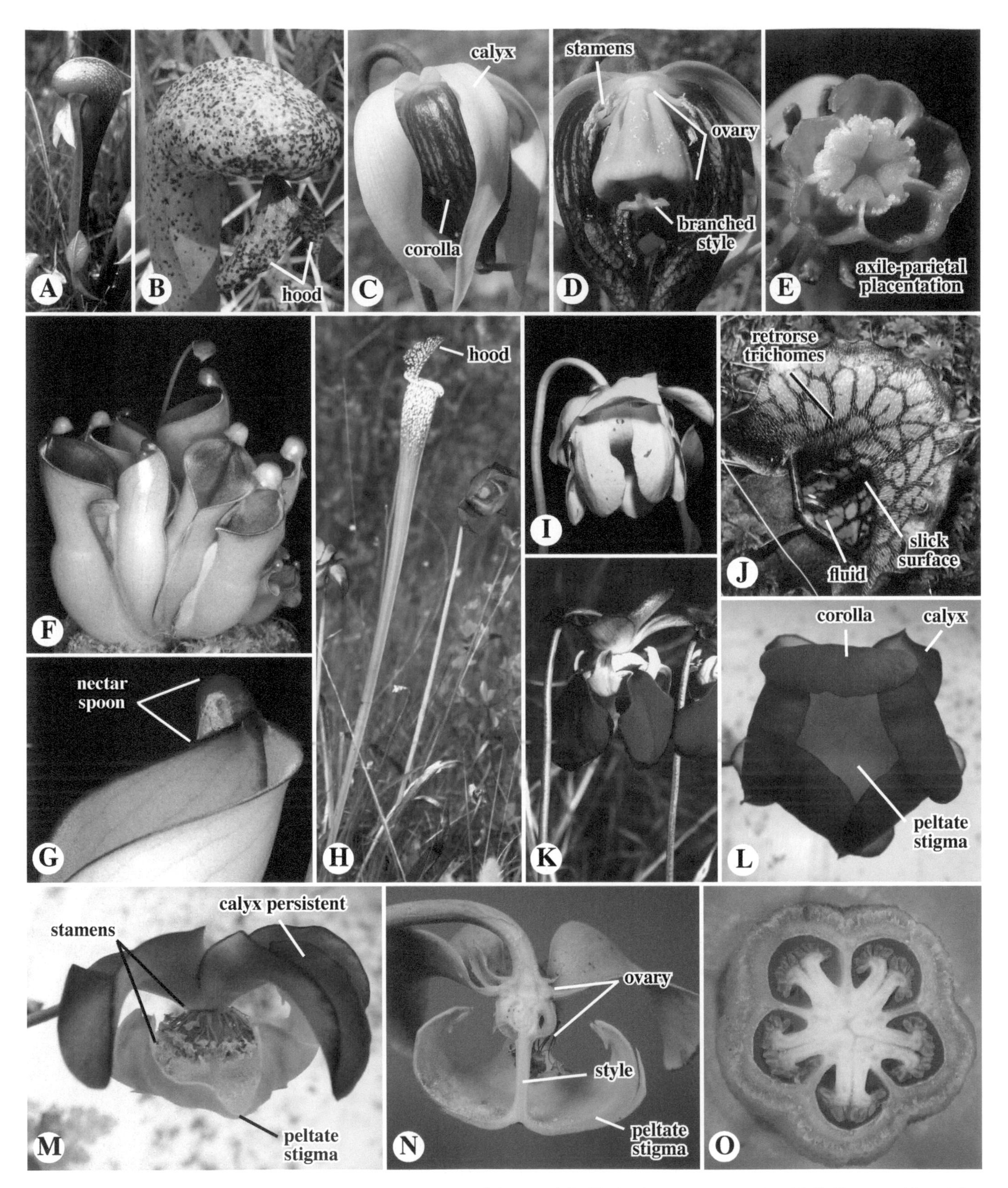

Figure 8.97 ERICALES. Sarraceniaceae. **A–E.** *Darlingtonia californica.* **A,B.** Pitcher leaf. Note bifid "hood." **C.** Flower, with nodding orientation. **D.** Flower, perianth partially removed, showing superior ovary with 5-branched style. **E.** Ovary cross section, showing apically parietal placentation (axile below). **F,G.** *Heliamphora* sp. Note slit along upper part of pitcher leaf and apical "nectar spoon." **H.** *Sarracenia leucophylla*, with tall pitcher leaves and solitary flowers. **I.** *S. flava* flower. Note pendant petals. **J.** *S. purpurea* leaf, showing retrorse trichomes above slick surface and fluid-filled pitcher. **K–M.** *S. rubra.* **K.** Flower, side view. **L.** Flower, face view, with 5 petals and peltate stigma. **M.** Flower side view, petals off, showing persistent calyx and peltate stigma. **N,O.** *S. leucophylla.* **N.** Flower longitudinal section, showing style and peltate stigma. **O.** Ovary base cross section, showing axile placentation. [Images at F,G courtesy of Steven Swartz.]

Figure 8.98 THEALES. Theaceae. **A–H.** *Camellia* sp. **A.** Flower, face view. **B.** Flower center, showing androecium, numerous, centrifugal stamens, and central pistil. **C.** Stamen, with dithecal, longitudinally dehiscent anthers. **D.** Flower longitudinal section, showing superior ovary. **E.** Ovary longitudinal section. **F.** Ovary cross section, showing axile placentation. **G.** Leaf margin. **H.** Node of stem, leave exstipulate. **I.** *Gordonia* sp., flower, with numerous petals and stamens.

bases. The **fruit** is a loculicidal [rarely septicidal] capsule, rarely indehiscent or fleshy. The **seeds** are mostly exalbuminous, the middle seed coat layer (mesotesta) typically lignified.

Members of the Theaceae are distributed worldwide in tropical and warm temperate regions. Economic importance includes *Camellia sinensis*, leaves of which are used to make common tea (green=unfermented, black=fermented), plus several cultivated ornamentals, including *Camellia* spp. and cultivars, *Franklinia*, *Gordonia*, and *Stewartia*. See Prince and Parks (2001) and Yang et al. (2004, 2006) for phylogenetic analyses and Stevens et al. (2004) for a general treatment.

The Theaceae are distinctive in being ***trees or shrubs*** (rarely lianas) with ***simple, spiral, evergreen*** leaves, flowers ***usually solitary***, biseriate and dichlamydeous, usually with ***numerous, centrifugal stamens***, the ovary superior (rarely inferior) with axile placentation, the fruit a capsule, indehiscent, or fleshy, seed ***mesotesta lignified***.

K (5) [(4–7)] **C** 5 [4–∞] or (5) [(4–∞)] **A** ∞ or (∞) **G** (3–5) [(2–10)], superior or rarely inferior.

LAMIIDS

The Lamiids (Bremmer et al. 2002; *Lamiidae*, sensu Cantino et al. 2007; formerly called Euasterids I) are a monophyletic group of eight orders (APG IV 2016; Table 8.3; Figure 8.84), five of which are treated here. See Refulio-Rodriguez and Olmstead (2014) and Stull et al. (2015) for phylogenetic analyses of the Lamiids.

GARRYALES

The Garryales, sensu APG IV (2016) contain two families (Table 8.3), Eucommiaceae and Garryaceae, the latter described here.

Garryaceae — Silk-Tassel family (type *Garrya*, after N. Garry, 1782–1856). 2 genera (*Aucuba*, *Garrya*)/17–19 species (Figure 8.99).

The Garryaceae consist of dioecious trees and shrubs. The **leaves** are simple, opposite decussate, and exstipulate. The **inflorescence** is an axillary, pendant, thyrsoid catkin, with decussate, often connate bracts, each bract subtending 1–3 flowers (if 3 in a simple dichasium). The **flowers** are small, unisexual, actinomorphic, pedicellate to sessile. The **perianth** of male flowers is uniseriate, 4-merous, valvate, the perianth parts (sepals) apically connate; that of female flowers absent or reduced to 2 appendages near styles. The **stamens** are 4, alternisepalous. **Anthers** are longitudinal in dehiscence, dithecal. The **gynoecium** has an inferior ovary (by interpretation), 2(–3) carpels, and 1 locule. The **styles/stigmas** are 2–3. **Placentation** is apical; **ovules** are anatropous, unitegmic, 2(–3) per pistil. The **fruit** is a berry, dry at maturity. The **seeds** are 1–2 per fruit, oily-endospermous. Plants contain toxic alkaloids.

The Garryaceae are distributed in western North America and Central America. Economic importance includes local medicinal uses and occasional cultivation as an ornamental, such as *Aucuba japonica*. See Liston (2003) for more information.

The Garryaceae are distinctive in being ***dioecious shrubs or trees*** with ***pendant catkins***, the flowers of males ***4-merous*** with the ***perianth uniseriate***, female flowers with perianth absent or reduced and with a ***2–3-carpellate, unilocular*** ovary, the fruit a 1–2-seeded, dry berry.

Male: **P** 4 **A** 4

Female: **P** 0–2 **G** 2(3), inferior.

BORAGINALES

This order, sensu APG IV (2016), contains only the family Boraginaceae. A recent study accepted here (Luebert et al., 2016) treats the order as containing 11 families (see Table 8.3), although some of these might warrant merging (see Stevens, 2001 onwards). Notable among the families not treated in detail are: **Ehretiaceae** (Figure 8.100A–C), mostly trees, ovary ***4-lobed*** with a ***terminal, 2-branched style***, the fruit a ***drupe with 2 (2-seeded) or 4 (1-seeded) pyrenes***; **Lennoaceae** (Figure 8.100D–G), ***achlorophyllous, root-parasitic*** herbs with a fleshy, ***circumscissile capsule***, and **Namaceae** (Figure 8.100H–K; a family of now three genera now segregated from the Hydrophyllaceae), ***shrubs, small trees, or herbs*** (*Nama*), ovary ***unlobed*** giving rise to ***two terminal styles*** (stylodia) or a style ***united 3/4 of its length***.

Boraginaceae — Borage family (type *Borago*, possibly meaning "shaggy coat," in reference to the leaves). 90 genera/1,600-1,700 species (Figure 8.101).

The Boraginaceae consist of annual or perennial herbs, rarely shrubs or trees, often with hirsute or hispid vestiture. The **roots** are taprooted. The **stems** are sometimes rhizomatous. The **leaves** are simple, entire, spiral, sessile or petiolate, exstipulate. The **inflorescence** is a terminal or axillary, monochasial (usually scorpioid) or dichasial cyme, rarely of solitary, axillary flowers. The **flowers** are bisexual, actinomorphic, hypogynous. The **perianth** is biseriate and dichlamydeous. The **calyx** is apo- or synsepalous 5-merous, valvate, often covered with trichomes. The **corolla** is sympetalous, 5-merous, rotate, salverform, infundibular, or campanulate, white, yellow, pink, or blue, with a convolute or imbricate aestivation. The **stamens** are 5 [4–6], whorled, alternipetalous, inserted or exserted, uniseriate, epipetalous. **Anthers** are longitudinal in dehiscence. The **gynoecium** is syncarpous, with a superior ovary, 2 carpels, and 4 locules by formation of false septa dividing each carpel, the mature ovary deeply 4-lobed. The

Figure 8.99 GARRYALES. Garryaceae. **A–H.** *Garrya veatchii*. **A–E.** Male plant. **A.** Catkins of male flowers. **B.** Paired, opposite simple dichasia, apical view (axis removed). **C.** Male flower, sepals apically connate. **D.** Male flower, opened, showing 4 sepals (perianth uniseriate) and 4 alternisepalous stamens. **E.** Male catkin, distal end above. **F–H.** Female plant. **F.** Female catkins. **G.** Female flower, adaxial view, with bract behind. **H.** Ovary longitudinal section, showing 1 of 2 ovules, with apical placentation. **I.** *Garrya elliptica*, in fruit. **J.** *Garrya veatchii*, female catkin, distal end above.

Figure 8.100 Boraginales **A–C.** Ehretiaceae. **A,B.** *Tiquilia plicata*. **A.** Plant. **B.** Flower dissected, showing 2-branched style, diagnostic of family along with fruit characters. **C.** *Tiquilia palmeri*, close-up of flower. **D–G.** Lennoaceae. **D,E.** *Pholisma arenarium*, sand plant. **D.** Inflorescence (morel-shaped), arising from ground. **E.** Close-up of flowers. **F,G.** *Pholisma sonorae*, sand food. **F.** Inflorescence (mushroom-shaped). **G.** Inflorescence longitudinal section. Both species desert root-parasites with actinomorphic corollas. **H–K.** Namaceae. **H–I.** *Eriodictyon*, spp., shrubs. **H.** *Eriodictyon sessilifolium*, Baja California yerba santa. **I.** *Eriodictyon californicum*, California yerba santa. **J.** *Nama demissa*, purple mat, a prostrate to decumbent annual. **K.** *Wigandia urens*, stinging wigandia, a tree. Flower longitudinal section, showing superior, unlobed ovary and two terminal styles (stylodia) and stigmas. [Images at F,G contributed by Matt Guilliams.]

Figure 8.99 Boraginales, Boraginaceae. **A.** *Amsinckia intermedia*, showing salverform corolla and circinate, scorpioid inflorescence unit. **B–H.** *Borago officinalis*. **B.** Inflorescence, a non-circinate, scorpioid cyme. **C.** Flower close-up. **D.** Close-up of corona and staminal appendages. **E,F.** Gynoecium, showing four-lobed ovary and gynobasic style. **G.** Fruit close-up. **H.** Single nutlet, with caruncle. **I.** *Cryptantha intermedia*, inflorescence unit, a monochasial cyme. **J.** *Plagiobothrys acanthocarpus*, having uncinate nutlets. **K,L.** *Echium fastuosum*, with monochasial (helicoid) cyme unit. **M.** *Mertensia virginica*. **N–R.** Nutlet variation. Scale bar = 1 mm. **N.** *Amsinckia tessellata*. **O.** *Cryptantha pterocarya* var. *p*. **P.** *Eremocarya micrantha* var. *m*. **Q.** *Oreocarya bakeri*. **R.** *Plagiobothrys acanthocarpa*.

style is solitary, gynobasic, often with a pyramidal to subulate gynobase; stigmas are 1–2, minute, capitate, or bilobed. **Placentation** is basal; **ovules** are 2 per carpel. **Nectaries** are present in some taxa as a ring around the ovary base. The **fruit** is a schizocarp of 4 [1–3] nutlets, these smooth or with various surface protuberances or bristles (glochidiate or uncinate in some taxa), sometimes winged.

Members of the family are distributed worldwide. Economic importance is limited, some used as herbs (e.g., *Borago officinalis*), dyes, or cultivated ornamentals (e.g., e.g., *Echium*, pride of Madeira, or *Myosotis*, forget-me not). The Boraginaceae is currently classified into three subfamilies and 11 tribes (Chacon et al., 2016). See Weigend et al. (2016) for a thorough summary of the family.

The Boraginaceae are distinctive in being ***herbs*** (rarely shrubs or trees) with simple, spiral leaves, the inflorescence usually monochasial of ***scorpioid cymes***, with actinomorphic, sympetalous flowers having a 4-ovuled, ***deeply 4-lobed ovary*** (by development of "false septa") the style ***gynobasic***, the fruit a ***schizocarp of 4 (1–3) nutlets***.
K 5 or (5) **C** (5) **A** 5 **G** (2), superior, hypanthium absent.

Cordiaceae — Cordia family (type *Cordia*, after the Germans Euricius Cordus and his son Valerius Cordus). 2 genera (*Cordia* & *Varronia*)/ca. 400 species (Figure 8.102A-E).

The Cordiaceae consist of perennial, hermaphroditic or dioecious, trees, shrubs, or rarely lianas, often pubescent or hispid. The **leaves** are simple, entire, spiral, rarely sub-opposite, petiolate, exstipulate. The **inflorescence** is terminal, of monochasial or dichasial units, sometimes congested. The **flowers** are bisexual or unisexual, gen. actinomorphic, hypogynous. The **perianth** is biseriate and dichlamydeous, hypanthium absent. The **calyx** is synsepalous with 5 [4], valvate lobes. The **corolla** is sympetalous with 5 [4] lobes, tubular, rotate, or campanulate, white (rarely yellow). The **stamens** are 5 [4–15], usually exserted, epipetalous (sometimes basally). The **gynoecium** is syncarpous, with a superior ovary, 2 carpels, and 4 locules by formation of false septa dividing each carpel, the ovary not 4-lobed at maturity. The **style** is terminal; **stigmas** 4-branched, clavate to capitate. **Ovules** are 2 per carpel. **Nectaries** are usually present as a ring around the ovary base. The **fruit** is a drupe of 4 seeds, sometimes 1 seeded by abortion.

The Cordiaceae are distributed worldwide, especially in tropics and subtropics. Economic importance includes some cultivated ornamentals. See Miller and Gottschling (2007) and Gottschling et al. (2016, with Cordiaceae included within Ehretiaceae) for treatments.

The Cordiaceae are distinctive in being ***shrubs or trees***, rarely lianas, with simple, spiral leaves, an inflorescence with ***dichasial or monochasial*** (sometimes congested) units, the perianth 5(4)-merous, calyx and corolla of fused parts, the ovary superior, 2-carpeled, 4-loculed, the ***stigmas 4-branched***, and the fruit a ***drupe of 4 [1] seeds***.
K (5) [(4)] **C** (5) [(4)] **A** (5) [(4)] **G** (2), superior, hypanthium absent.

Heliotropiaceae — Heliotrope family (type *Heliotropium*, after Greek *helios*, "sun," and *trope*, "turning," perhaps from a mistaken belief that the flowers are indeed heliotropic). 4 genera (*Heliotropium* [incl. *Tournefortia*], *Euploca*, *Ixorhea*, and *Myriopus*/ca. 450 species (Figure 8.102I–L) .

The Heliotropiaceae consist of annual or perennial herbs, subshrubs, shrubs, lianas, or small trees, often strigose, sericeous, or glandular. The **leaves** are simple, entire [very rarely crenate or dentate], often revolute, spiral, petiolate or sessile, exstipulate. The **inflorescence** is terminal or axillary, of scorpioid cyme units, often thyrsoid. The **flowers** are bisexual, rarely unisexual, gen. actinomorphic, hypogynous. The **perianth** is biseriate and dichlamydeous. The **calyx** is usually basally synsepalous with 5, valvate, often pubescent lobes. The **corolla** is sympetalous, 5-lobed, valvate or imbricate, white, yellow, blue, pink, or orange. The **stamens** are 5, whorled, alternipetalous, inserted, uniseriate, epipetalous. The **gynoecium** is syncarpous, the ovary superior, with 2 carpels, 4 locules by formation of false septa dividing each carpel, the mature ovary not 4-lobed. The **style** is solitary, terminal; the stigmatic region is conical of a distinctive "stigmatic head," consisting of a basal ring-shaped stigma and a sterile, sometimes 2-lobed apex; ovules are 2 per carpel. **Nectaries** are present as disc around ovary base. The **fruit** is a dry or fleshy, 4 [1–2] seeded, units separating into 1–4 nutlets with 1–2 seeds each.

The Heliotropiaceae are distributed worldwide, especially in the tropics and subtropics. Economic importance includes some horticultural members, esp. *Heliotropium arborescens*, plus some weeds and toxic members. See Craven (2005), Luebert et al. (2011), and Luebert and Diane et al. (2016) for treatments of the Heliotropiaceae.

The Heliotropiaceae are distinctive in being annual or perennial ***herbs, subshrubs, shrubs, lianas or, small trees*** with petamerous flowers, the calyx and corolla of fused parts, the ovary superior, 2-carpeled and 4-loculed, the ***style terminal*** with a ***conical stigmatic head***, the latter a likely apomorphy for the family.
K 5 **C** 5 **A** 5 **G** (2), superior, hypanthium absent.

Hydrophyllaceae (not including Namaceae) — Waterleaf family (type *Hydrophyllum*, after *hydro*, water, + *phyllum*, leaf). 12 genera/240–280 species (Figure 8.103).

The Hydrophyllaceae consist of annual, biennial, or perennial herbs, often hirsute/hispid, and/or glandular. The **leaves** are simple and undivided to divided, or pinnate to bipinnate, rarely palmate. The **inflorescence** is terminal or axillary, of

Figure 8.102 Boraginales. **A–E.** Cordiaceae. *Cordia* sp. **A,B.** Plant, with actinomorphic, sympetalous flowers. **C.** Close-up of 4-lobed style, diagnostic of subfamily. **D,E.** Fruit, a drupe with 4-loculed endocarp, also diagnostic of the subfamily. **F–I.** Heliotropiaceae. **F.** *Heliotropium arborescens*, a shrub. **G–I.** *H. curassavicum*, a perennial herb. **G.** Circinate, scorpioid cyme unit. **H.** Flower, showing actinomorphic, pentamerous rotate-salverform corolla. **I.** Ovary, with terminal, unbranched style and conic stigma.

scorpiod cyme units, sometimes thyrsoid or congested. The **flowers** are bisexual, actinomorphic to slightly zygomorphic, hypogynous. The **perianth** is biserate and dichlamydeous, hypanthium absent. The **calyx** is usually synsepalous (only basally connate in some taxa) with 5 [4] valvate lobes. The **corolla** is sympetalous, convolute, with 5 [4] lobes, often blue to purple, also white, pink, or yellow. The **stamens** are 5 [4], whorled, alternipetalous, uniseriate, epipetalous, inserted at the same or different levels, often with flanking scale-like structures at junction with corolla tube. The **gynoecium** is syncarpous, with a superior ovary, 2 carpels, and 1 or 2 locules, sometimes appearing more by intrusion of placentae. The **style** is one, terminal; **stigmas** are 2-branched, capitate. **Placentation** is parietal or axile, the placentae sometimes enlarged; ovules are 2–∞ per carpel. **Nectaries** are present as a disk or glands around the ovary base. The **fruit** is a 2-valved capsule, sometimes irregularly or not dehiscent.

The Hydrophyllaceae are distributed in North and Central America and western South America. Economic importance includes some cultivated ornamentals. The family has recently been separated from *Eriodictyon*, *Nama*, and *Wigandia* of the Namaceae (Luebert et al., 2016). See Ferguson (1999) and Walden and Patterson (2012) for phylogenetic analyses within the family and Hofmann et al. (2016) for a thorough summary.

The Hydrophyllaceae are distinctive in being ***herbs*** with ***simple (undivided or divided), or pinnate to bipinnate***, spiral leaves, the inflorescence of scorpioid cymes, with actinomorphic, sympetalous flowers, often having a ***pair of scales*** at junction of stamen filament with the corolla tube, the ovary superior, 2-carpeled, 2–∞ ovuled, unlobed, the fruit a ***2-valved capsule***.

K 5 [4] **C** 5 [4] **A** 5 [4] **G** (2), superior, hypanthium absent.

Figure 8.103 Boraginales, Hydrophyllaceae. **A.** *Emmenanthe penduliflora*, showing monochasial cyme. **B.** *Eucrypta chrysanthemifolia*, inflorescence unit. **C.** *Nemophila menziesii*, baby blue eyes. **D.** *Phacelia cicutaria*, two valves of capsule. **E.** *Phacelia minor*, flowers. **F,G.** *Phacelia parryi*. **F.** Flower. **G.** Corolla, basal adaxial side, showing scales at base of stamen filaments. **H.** *Phacelia pedicellata*, with helicoid cyme units. **I.** *Pholistoma auritum*, flower, side view, showing sympetalous, rotate corolla.

GENTIANALES

The Gentianales, sensu APG IV (2016), contain five families (Table 8.3), three of which are treated here. See Struwe et al. (1995) and Refulio-Rodriguez and Olmstead (2014) for phylogenetic analyses of the order.

Apocynaceae [including Asclepiadaceae] — Dogbane/Milkweed family (type *Apocynum*, Greek for "away from dog," in reference to some taxa used as dog poison). ca. 400 genera/4,555-5,100 species (Figures 8.104, 8.105).

The Apocynaceae consist of lianas, trees, shrubs, or herbs, with latex present in tissues. The **stems** are succulent in some taxa, e.g., the stapelioids. The **leaves** are simple, undivided, sometimes reduced, opposite, whorled, or rarely spiral, usually exstipulate. The **inflorescence** is a cyme (often umbelliform in Asclepiadoids), raceme, or of solitary flowers. The **flowers** are usually bisexual, actinomorphic, and hypogynous. The **calyx** is usually synsepalous (at least basally) with 5 imbricate or valvate lobes. The **corolla** is sympetalous with 5 convolute (rarely valvate or imbricate) lobes. The **stamens** are 5, alternipetalous, often epipetalous, apostemonous to

Figure 8.104 GENTIANALES. Apocynaceae. **A.** *Carissa grandiflora*, Natal plum; flower, face view. **B.** *Catharanthus roseus*, Madagascar periwinkle. **C–F.** *Nerium oleander*, oleander. **C.** Flower, face view. **D.** Flower, side view. **E,F.** Flower longitudinal section, showing style, stigma (E), and ovary (F). **G.** *Plumeria* sp., flower in face view. **H,I.** *Pachypodium lamerei*. **H.** Upper shoot. **I.** Spines of trunk. **J.** *Pachypodium namaquanum*. **K.** *Pachypodium bispinosum*. **L.** *Strophanthus speciosus*.

Figure 8.105 GENTIANALES. Apocynaceae, Asclepiadoids. **A–G.** *Asclepias tuberosa*. **A.** Inflorescence, an umbelliform cyme. **B.** Flower, side view. **C.** Gynostemium close-up, showing hood and horn appendages and one (of five) translator apparatus. **D.** Translator apparatus, inside sheath of tissue, with glandular corpusculum protruding from slit of sheath. **E.** Translator apparatus removed, showing corpusculum, translator arms (retinacula) and pollinia. **F.** Flower longitudinal section, showing two distinct ovaries and styles and a single stigma. **G.** Close-up of ovaries in longitudinal section. **H.** *Sarcostemma cynanchoides*, flowers. **I,J.** *Gomphocarpus physocarpus*. **I.** Umbelliform cyme. **J.** Fruits, bladderlike schizocarpic follicles. **K–M.** Succulent taxa. **K.** *Stapelia gigantea*, producing a foul odor and fly-pollinated. **L.** *Stapelia variegata*. **M.** *Huernia* sp.

monadelphous. In Asclepiadoids the stamens are connate to the stigma to form a **gynostegium**, often elaborated with appendages: "hoods" and "horns," functioning to contain nectar. In these taxa, **pollen** grains of each theca of an anther are fused into a waxy mass called a **pollinium**. The right pollinium of each anther is attached to the left pollinium of the adjacent anther by a **translator**, consisting of two hair-like **translator arms** (also termed **retinacula**) attached together to a two-parted, gland-like **corpusculum**. Thus, the unit of pollen dispersal in Asclepiadoids is the **translator apparatus**, made up of the yoke-shaped translator and the two, adnate pollinia (half-anthers) of adjacent stamens, and located within "pollination slits" on the sides of the gynostegium. The **gynoecium** is syncarpous, often only apically (with ovaries distinct), with a superior, rarely half-inferior ovary, 2 [–8] carpels, and 1–2 locules. The **stigma** is capitate or lobed, in Asclepiadoids consisting of a single, broad fusion product with the receptive regions within the lateral pollination slits. **Placentation** is apical and pendulous or marginal; **ovules** are anatropous, unitegmic, ∞ [1–] per carpel. **Nectaries** are sometimes with 5 nectar glands or a disk at ovary base. The **fruit** is a variable and can be a berry, drupe, or follicle; in Asclepiadoids the fruit is a schizocarp of two follicles (one often not developing). The **seeds** are endospermous. Plants typically contain various glycosides and alkaloids. In Asclepiadoids, the gland-like corpusculum of the translator apparatus becomes attached to the leg of an insect, which pulls the apparatus from the pollination slit within a sheath-like membrane on the side of the gynostegium; pollination is effected when the translator apparatus is re-inserted by the insect into an empty pollination slit of another flower.

The Apocynaceae were treated as two families, Apocynaceae (Dogbane family) and Asclepiadaceae (Milkweed family), the latter now classified as subfamily Asclepiadoideae (the Asclepiadoids). Members of the Apocynaceae produce a number of secondary chemicals, functioning in herbivory deterrence. The family has a largely worldwide distribution, mostly in tropical regions. Economic importance includes uses as cultivated ornamentals, such as *Nerium* (oleander), *Plumeria*, *Stapelia*, and *Vinca* (periwinkle); medicinal uses, such as *Catharanthus roseus* (Madagascar periwinkle), from which vincritine/vinblastine used to treat childhood leukemia, and *Rauvolfia serpentina*, from which the drug reserpine is derived; and uses as timber, fiber, rubber, dye, and poison plants. See Endress and Bruyns (2000), Sennblad and Bremer (1996, 2002), Fishbein et al. (2018), and Livshultz et al. (2018) for phylogenetic analyses, classification, and chemical evolution of the family.

The Apocynaceae are distinctive in being lianas, trees, shrubs, or herbs with a 5-merous perianth/androecium, the gynoecium usually with 2 carpels, the ***ovaries distinct in some taxa with stigmas connate*** (in Asclepiadoids ***androecium adnate to single stigma forming a gynostegium*** and pollen fused to form ***pollinia***, each half derived from an adjacent anther), the fruits variable, but a ***schizocarp of follicles*** in the Asclepiadoids.
K (5) **C** (5) **A** 5 or (5) **G** (2) [(–8)], superior, rarely half-inferior.

Gentianaceae — Gentian family (type *Gentiana*, after Gentius, king of Illyria). Ca. 91 genera/1,700 species (Figure 8.106).

The Gentianaceae consist of trees, shrubs, or usually herbs. The **stems** of tree species are sometimes pachycaulous, and some taxa are achlorophyllous and mycotrophic (obtaining nutrition from mycorrhizal fungi in the soil). The **leaves** are simple (leaves scalelike in mycotrophic species), opposite (rarely whorled or spiral), and exstipulate. The **inflorescence** is a cyme, raceme, or of solitary flowers. The **flowers** are usually bisexual and actinomorphic, and hypogynous. The **calyx** is usually synsepalous with 4–5 [–12], imbricate or valvate lobes. The **corolla** is sympetalous, rotate to salverform, with 4–5 [–12], usually convolute lobes. The **stamens** are 4–5 [–12], alternipetalous, epipetalous, with staminodes present in some. **Anthers** are usually longitudinal (rarely poricidal) in dehiscence. The **gynoecium** is syncarpous, with a superior ovary, 2 carpels, and 1, rarely 2 locules. The **style** is solitary and terminal; the **stigma** is solitary, 2-lobed, or decurrent. **Placentation** is parietal with placentae sometimes protruding and branched, rarely axile or free-central; **ovules** are anatropous, unitegmic, numerous. **Nectaries** are often present as pits on corolloa lobes, with a nectariferous disk or glands usually at the ovary base. The **fruit** is a septicidal capsule, rarely a berry. The **seeds** are oily endospermous.

The Gentianaceae have a mostly worldwide distribution. Economic importance consists primarily of ornamental cultivars such as *Eustoma*, *Exacum*, *Gentiana* (gentian), and *Sabatia*; some species are used medicinally or as timber. See Struwe et al. (2002), Merckx et al. (2013), and Struwe (2014) for information on and phylogenetic studies of the family.

The Gentianaceae are distinctive in being trees, shrubs, or ***usually herbs*** (sometimes achlorophyllous and mycotrophic), with a usually ***4–5 [–12] -merous perianth/androecium***, often with ***nectary pits on corolla lobes*** and a ***disk or glands at ovary base***, the ovary superior, 2-carpellate, with ***usually parietal*** (sometimes axile or free-central) placentation, the fruit a ***septicidal capsule or berry***.
K (4–5) [(–12)] **C** (4–5) [(–12)] **A** 4–5 [–12] **G** (2), superior.

Rubiaceae — Coffee family (type *Rubia*, name used by Pliny for madder). ca. 614 genera/13,240 species (Figure 8.107).

The Rubiaceae consist of terrestrial (rarely epiphytic or aquatic) trees, shrubs, lianas, or herbs. The **leaves** are simple, undivided and entire, opposite or decussate, rarely whorled

Figure 8.106 GENTIANALES. Gentianaceae. **A–F.** *Zeltnera venusta*. **A,B.** Whole plants. **C.** Flower close-up. **D.** Flower in longitudinal section. **E.** Close-up of coiled stamens and style. **F.** Ovary cross section, showing parietal placentation. **G,H.** *Exacum affine*. **I.** *Obolaria virginica*, an herb. **J,K.** *Swertia parryi*. **J.** Whole plant. **K.** Flower close-up, showing corolla nectaries. **L.** *Sabatia dodecandra*, a species with multiple perianth and androecial parts.

Figure 8.107 GENTIANALES. Rubiaceae. **A–C.** *Coffea arabica*, coffee. **A.** Flowers. **B.** Leaves; note opposite arrangement. **C.** Fruit, a drupe. **D.** *Pentas* sp. **E.** *Galium aparine*, bedstraw. **F,G.** *Psychotria kirkii*. **H–M.** *Psychotria viridiflora*, flower morphology. **H.** Simple dichasium. **I.** Flower longitudinal section, showing inferior ovary. **J.** Ovary, showing two ovules, with axile placentation. **K.** Apex of corolla tube, showing style. **L.** Stigmas. **M.** Ovary cross section; note two carpels and locules.

Figure 8.108 LAMIALES. **A.** Calceolariaceae, *Calceolaria* sp., pocketbook plant. **B.** Gesneriaceae, *Syringa leucotricha*. **C.** *Pinguicula* sp., a carnivorous plant with "flytrap" basal leaves.

(sometimes interpreted as opposite leaves with expanded stipules), or spiral by suppression, stipulate, stipules of opposite leaves connate, often bearing structures termed colleters, which produce mucilaginous compounds protecting the young shoot. The **inflorescence** is a cyme, rarely of solitary flowers. The **flowers** are usually bisexual and usually epigynous. The **perianth** is usually biseriate, although the calyx is lost in some taxa. The **calyx** is synsepalous with 4–5 or 0 lobes. The **corolla** is sympetalous with 4–5 [rarely 3 or 8–10], actinomorphic or bilabiate lobes. The **stamens** are 4–5 [rarely 3 or 8–10], alternipetalous and epipetalous. **Anthers** are longitudinal in dehiscence. The **gynoecium** is syncarpous, with a usually inferior [rarely superior] ovary, 2 [3–5+] carpels, and 1–2 [3–5+] locules. **Placentation** is axile, rarely parietal; **ovules** are anatropous to hemitropous, unitegmic, with a funicular obturator, 1–∞ per carpel. **Nectaries** are often present as a nectariferous disk atop ovary. The **fruit** is a berry, capsule, drupe, or schizocarp. The **seeds** are usually endospermous.

The Rubiaceae have a mostly worldwide distribution, more concentrated in tropical regions. Economic importance includes *Cinchona*, the source of quinine used to treat malaria, *Coffea arabica* and other species, the source of coffee, *Pausinystalia johimbe*, the source of the sexual stimulant yohimbine, some timber trees, fruiting plants, dye plants (such as *Rubia*, madder), and ornamental cultivars (e.g., *Pentas*, among others). See Robbrecht and Manen (2006), Bremer (2009), Bremer and Eriksson (2009), and Rydin et al. (2017) for phylogenetic studies of the Rubiaceae.

The Rubiaceae are distinctive in being trees, shrubs, lianas, or herbs with ***simple, entire, usually opposite or decussate leaves*** and ***connate stipules***, the stipules often with mucilage -secreting ***colleters***, the ***inflorescence usually a cyme***, flowers usually bisexual, the perianth dichlamydeous, perianth and androecium often 4–5-merous (calyx absent in some), the ***ovary usually inferior*** (rarely superior), often with an apical ***nectariferous disk***, ovules with a ***funicular obturator***, the fruit a berry, capsule, drupe, or schizocarp.

K (4–5) [0] **C** (4–5) [(3,8–10)] **A** 4–5 [3,8–10] **G** (2) [(3–5+)], usually inferior, rarely superior.

LAMIALES

The Lamiales, sensu APG IV (2016), contain 24–25 families (Table 8.3), many of which have undergone considerable changes in classification (e.g., see Scrophulariaceae, discussed later). Nine families are described here. Among those not described are the **Calceolariaceae** (Figure 8.108A) including *Calceolaria* spp., the pocketbook plant; **Gesneriaceae** (Figure 8.108B), a large family including many cultivated ornamentals, such as *Saintpaulia*, African violets, and *Sinningia*, gloxinia; **Lentibulariaceae**, including the interesting carnivorous plants *Genlisea*, with subterranean "eel-trap" leaves, *Pinguicula* (butterwort), with sticky "fly-paper," insect-trapping leaves (Figure 8.108C), and *Utricularia* (bladderwort), an aquatic with leaf bladders that "suction-trap" small aquatic animals; and **Pedaliaceae**, containing *Sesamum indicum*, sesame. See Kadereit (2004) information and Schäferhoff et al. (2010) for phylogenetic analyses of the order.

Acanthaceae — Acanthus family (type *Acanthus*, prickly-one). Ca. 220 genera/4,000 species (Figure 8.109).

The Acanthaceae consist of terrestrial or aquatic herbs, shrubs, or rarely trees. The **leaves** are opposite (usually) and simple. The **inflorescence** is a cyme, raceme, or of solitary flowers. The **flowers** are bisexual, zygomorphic, bracteate and bracteolate (the bracts often colored), and hypogynous. The **perianth** is biseriate and dichlamydeous, with hypanthium absent. The **calyx** is synsepalous with 5 [4,6], imbricate or valvate lobes. The **corolla** is sympetalous and usually

Figure 8.109 LAMIALES. Acanthaceae. **A–D.** *Acanthus mollis*. **A.** Inflorescence. **B.** Flower close-up, showing reduced corolla. **C.** Anther close-up. **D.** Ovule and jaculator. **E.** *Hypoestes aristata*, inflorescences. **F,G.** *Justicia brandegeana*, shrimp plant. **F.** Inflorescence, with showy bracts. **G.** Flower close-up, showing bilabiate corolla. **H,I.** *Justicia californica*. **H.** Inflorescence. **I.** Open fruit, showing peglike jaculators. **J.** *Ruellia graecizans*. **K–M.** *Thunbergia gregorii*. **K.** Flower, face view. **L.** Flower, side view, showing basal, spathelike bracts. **M.** Open flower. Note calyx, stamens, and style/stigma.

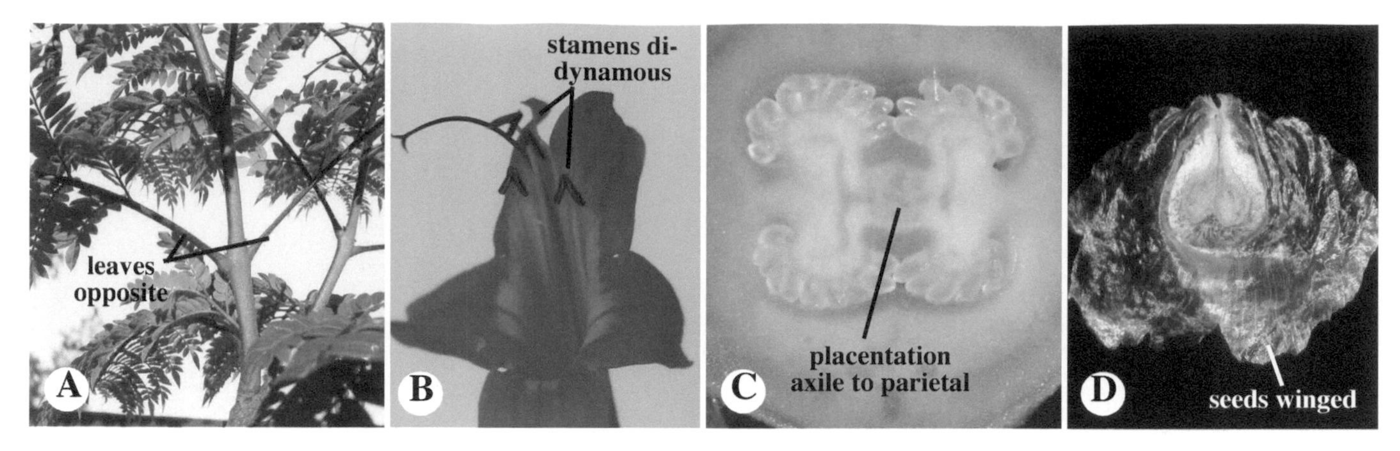

Figure 8.110 LAMIALES. Bignoniaceae. Diagnostic characteristics. **A.** Opposite leaf arrangement (*Jacaranda mimosifolia*). **B.** Didynamous stamen arrangement (*Tecoma capensis*). **C.** Axile to parietal placentation (*Spathodea campanulata*). **D.** Winged seeds (*S. campanulata*).

bilabiate (the upper lip suppressed in some species) with 4–5 imbricate or convolute lobes. The **stamens** are 2, 4, or rarely 5, with staminodes present in some. **Anthers** are tetrasporangiate or bisporangiate, dithecal or monothecal, with parallel or divergent thecae. The **pollen** is tricolpate, triporate, diporate, pantoporate, or inaperturate. The **gynoecium** is syncarpous, with a superior ovary, 2 carpels, and 2 locules. The **style** is solitary and terminal. **Placentation** is axile; **ovules** are variable in type, 2–∞ per carpel. Nectaries are usually present as a disk at the ovary base. The **fruit** is often an explosively dehiscent, loculicidal capsule. The seeds have funiculi that are modified into rigid, often hook-shaped structures that function to catapult the seeds; these are termed funicular retinacula or jaculators. Cystoliths are characteristic of some taxa, appearing as streaks in the leaves.

Members of the Acanthaceae are distributed from the tropics to temperate regions. Economic importance includes several cultivated ornamentals, such as *Acanthus mollis*, *Aphelandra,* and *Justicia* [*Beloperone*]. *Avicennia* spp., the black mangroves, are now classified as a subfamily of the Acanthaceae (Schwarzbach and McDade, 2002). See McDade et al. (2000, 2008) for studies of phylogeny and classification of the family.

The Acanthaceae are distinctive in having simple, **opposite leaves** with **zygomorphic**, **bracteate**, usually **bilabiate flowers**, the fruit of many members an explosively dehiscent, loculicidal capsule with distinctive **funicular retinacula** (jaculators) that function in seed dispersal.

K (5) [(4,6)] **C** (4–5) **A** 2,4,or 5 **G** (2), superior.

Bignoniaceae — Bignonia family (type *Bignonia*, after Abbé Jean-Paul Bignon, 1662–1743, court librarian at Paris, friend of Tournefort). Ca. 110 genera/800 species (Figures 8.110, 8.111).

The Bignoniaceae consist of trees, shrubs, and lianas, rarely herbs. The **leaves** are usually pinnate or ternate, less often simple or palmate, usually opposite, sometimes whorled, rarely simple, exstipulate, the terminal leaflets modified as tendrils in some taxa. The **inflorescence** is a cyme, raceme, or of solitary flowers. The **flowers** are bisexual, zygomorphic [corolla rarely actinomorphic], hypogynous. The **calyx** is synsepalous with 5 zygomorphic, often bilabiate lobes [sometimes unlobed or spathaceous, rarely calyptrate]. The **corolla** is sympetalous with 5, usually bilabiate lobes [rarely actinomorphic]. The **stamens** are alternipetalous, whorled, usually didynamous, 2+2 or 2+2 + 1 staminode [rarely 2 + 3 staminodes]. The **gynoecium** is syncarpous, with a superior ovary, 2 carpels, and 1, 2, or 4 locules. The **style** is solitary and terminal, with two **stigmas**. **Placentation** is axile or parietal with intruding septae [false septa dividing each carpel into two locules in some]; **ovules** are anatropous or hemitropous, unitegmic, numerous. **Nectaries** are usually present as a ring or cup-shaped structure around ovary base. The **fruit** is a two-valved capsule, rarely fleshy or fibrous and indehiscent. The **seeds** are usually flat and winged in taxa with capsules, exalbuminous. The stem anatomy of members of the family with lianas is unique.

Members of the Bignoniaceae are distributed primarily in tropical [some temperate] regions. Economic importance includes important timber trees and many ornamental cultivars (e.g., *Jacaranda*, *Spathodea*). See Spangler and Olmstead (1999) and Olmstead et al. (2009) for recent phylogenetic studies of the family.

The Bignoniaceae are distinctive in being trees, shrubs, or vines with ***opposite leaves*** and usually ***zygomorphic***, often ***bilabiate,*** flowers with ***didynamous*** stamens, a superior, 2-carpellate ovary having ***axile or parietal placentation*** with numerous ovules, the fruit a ***capsule*** [rarely indehiscent] with usually ***flat, winged, exalbuminous*** seeds.

K (5) **C** (5) **A** 2+2 [+1 staminode in some; rarely 2 fertile +3 staminodes] **G** (2), superior.

Figure 8.111 LAMIALES. Bignoniaceae. **A.** *Campsis radicans*, a vine. **B.** *Chilopsis linearis*, desert-willow, flowers. **C,D.** *Distictis buccinatoria*, a vine. **C.** Flowers. **D.** Corolla, opened to reveal didynamous stamens and style/stigmas. **E–G.** *Kigelia africana*, African sausage tree. **E.** Flower. **F.** Flower with corolla removed and calyx dissected, showing large nectaries. **G.** Fruit, a rare indehiscent capsule. **H,I.** *Jacaranda mimosifolia*, a tree. **H.** Opened fruit, showing winged seeds. **I.** Flower, close-up, with bilabiate corolla. **J–L.** *Markhamia* sp., a vine. **J.** Flower. **K.** Dehiscent fruit, with persistent septum (replum). **L.** Winged seeds. **M,N.** *Spathodea campanulata*, a tree. **M.** Inflorescence. **N.** Flower close-up, showing spathaceous calyx and large corolla. **O.** *Pyrostegia venusta*, a vine.

Lamiaceae (Labiatae) — Mint family (type *Lamium*, possibly from gullet, after the shape of the corolla tube). Ca. 236 genera/7,280 species (Figures 8.112, 8.113).

The Lamiaceae consist of hermaphroditic, sometimes gynodioecious, herbs, shrubs, or rarely trees, often with short-stalked glandular trichomes producing aromatic ethereal oils. The **stems** are usually 4-sided (square in cross section), at least when young. The **leaves** are simple [rarely pinnate], opposite, sometimes whorled [rarely spiral], exstipulate. The **inflorescence** consists of lateral cyme units in a verticillaster or thyrse, or of solitary, axillary flowers. The **flowers** are bisexual [rarely unisexual], mostly zygomorphic, bracteate and bracteolate, hypogynous. The **perianth** is biseriate and dichlamydeous, the corolla usually bilabiate, sometimes actinomorphic, hypanthium absent. The **calyx** is synsepalous of 5 zygomorphic, sometimes bilabiate, lobes. The **corolla** is sympetalous, of 4 or 5 corolla lobes (if 4, by fusion of two lobes). The **stamens** are 2, 4, or 2 fertile + 2 staminodes, whorled, epipetalous (adnate to corolla tube). **Anthers** are longitudinal in dehiscence, with connective split in *Salvia* and relatives, separating the thecae of anthers (one theca lost in some taxa). The **gynoecium** is syncarpous, with a superior ovary, 2 carpels, and 4 locules. The **style** is solitary, often apically 2-branched, terminal or gynobasic with the ovary deeply 4-lobed by formation of false septa dividing each carpel; **stigmas** are usually 2. **Placentation** is basal; **ovules** are anatropous to hemitropous, unitegmic, 2 per carpel, 1 per locule. **Nectaries** are usually present as a disk or pad of tissue at base of ovary. The **fruit** is a schizocarp of usually four [1–3] nutlets, a drupe, or a berry. Plants often have ethereal oils and the carbohydrate stachyose (a tetrasaccharide).

The Lamiaceae have a mostly worldwide distribution. Economic importance includes medicinal plants, culinary herbs (e.g., *Mentha*, mint; *Ocimum*, basil; *Salvia rosmarinus*, rosemary; *Salvia* spp., sage; *Thymus*, thyme), fragrance plants (e.g., *Lavandula*, lavender; *Pogostemon*, patchouli), food (e.g., *Stachys affinis*, Chinese artichoke), and a plethora of cultivated ornamentals. See Wagstaff et al. (1998) and Harley et al. (2004) for information on phylogeny and classification.

The Lamiaceae are distinctive in being herbs or shrubs, ***often aromatic with ethereal oils***, with usually ***4-sided stems***, ***opposite*** [or whorled] leaves, a ***verticillaster or thyrse*** inflorescence [flowers solitary and axillary in some], and zygomorphic [rarely actinomorphic], ***usually bilabiate*** flowers having a superior ovary, often ***deeply 4-lobed*** (by formation of "false septa") with a ***gynobasic style***, the fruit a ***schizocarp of usually 4 nutlets*** or a berry or drupe.

K (5) **C** (5) [(4)] **A** 4 or 2 [+2 staminodes] **G** (2), superior.

Oleaceae — Olive family (type *Olea*, L. for olive, oil). 24 genera/615-800 species (Figure 8.114).

The Oleaceae consist of trees, shrubs, or rarely lianas, characteristically with peltate, secretory trichomes. The **leaves** are opposite [rarely spiral], exstipulate, and either simple, pinnate, unifoliolate, or ternate. The **inflorescence** is a cyme or of solitary flowers. The **flowers** are bisexual [rarely unisexual], actinomorphic, hypogynous. The **perianth** is usually dichlamydeous. The **calyx** is synsepalous with 4 [–15] valvate lobes [rarely absent]. The **corolla** is sympetalous [rarely apopetalous or absent], with 4 [–12], lobes that may be convolute, imbricate, or valvate. The **stamens** are 2 [4], epipetalous. **Anthers** are dithecal, longitudinal in dehiscence. The **gynoecium** is syncarpous, with a superior ovary, 2 carpels, and 2 locules. The **style** is terminal. **Placentation** is apical-axile; **ovules** are anatropous or amphitropous, unitegmic, usually 2 [1–4, ∞] per carpel. **Nectaries** consist of an annular disk around the ovary in some taxa. The **fruit** is a berry, capsule, drupe (e.g., *Olea*, olive), or samara (e.g., *Fraxinus*, ash). The **seeds** have an endosperm that is oily or absent.

Members of the Oleaceae are mostly worldwide in distribution. Economic importance includes timber trees (e.g., *Fraxinus*, ash, the source of wooden baseball bats), food and oil plants (*Olea europaea,* olive), scent/perfume plants (e.g., *Jasminum*, jasmine, and *Syringa*, lilac), and numerous ornamental cultivars, such as *Chionanthus*, *Forsythia*, *Fraxinus*, *Ligustrum* (privet), and *Osmanthus*. See Wallander and Albert (2000) and Kim and Kim (2011) for a phylogenetic studies of the family and Green (2004) for a detailed description.

The Oleaceae are distinctive in being ***trees or shrubs*** (rarely lianous), usually with ***peltate secretory trichomes*** and ***opposite leaves***, the inflorescence a ***cyme or solitary-flowered***, the flowers usually bisexual, actinomorphic, the perianth mostly dichlamydeous and ***tetramerous***, with ***2*** [4] ***stamens***, a superior ovary with 2 carpels and locules, and apical-axile placentation, the fruit a berry, capsule, drupe, or samara.

K (4) [(–15), 0] **C** (4) [(–12), 0] **A** 2 [4] **G** (2), superior.

Orobanchaceae — Broom-Rape family (type *Orobanche*, from Greek *orobos,* a legume, + *anche*, strangle, after the parasitic habit). Ca. 99 genera/2,060-2,100 species (Figure 8.115).

The Orobanchaceae consist of achlorophyllous or chlorophyllous, mostly parasitic to hemiparasitic herbs [rarely shrubs or lianas], plants often turn black after drying (on herbarium sheets). The **roots** are usually haustorial, parasitizing roots of a host plant. The **stems** of achlorophyllous parasites are often fleshy. The **leaves** are simple, spiral or opposite, exstipulate, scale-like in achlorophyllous taxa. The **inflorescence** is a raceme, spike, or of solitary flowers. The **flowers** are bisexual,

Figure 8.112 LAMIALES. Lamiaceae, diagnostic features. **A.** *Salvia leucantha*, showing opposite leaf arrangement. **B.** *Marrubium vulgare*, unit inflorescence of verticillaster. **C.** *Orthosiphon*, showing thryrse. **D.** *Salvia rosmarinus*, flower with zygomorphic, bilabiate corolla. **E.** *Hemiandra pungens*, showing four-lobed ovary and gynobasic style. **F,G.** *Salvia rosmarinus*. **F.** Style, 2-branched at apex. **G.** Fruit, showing two of four schizocarpic nutlets. **H.** *Acanthomintha ilicifolia*, an endangered species, with solitary, axillary flowers. **I.** *Hemiandra pungens*, corolla. **J.** *Hyptis emoryi*, inflorescence. **K,L.** *Monarda fistulosa*, congested headlike verticillaster, with showy bracts. **M.** *Monarda didyma*, bee-balm.

Figure 8.113 LAMIALES. Lamiaceae, diversity. **A.** *Monardella macrantha*. **B.** *Pogogyne abramsii*, with thyrse reduced to solitary, axillary flowers. **C.** *Prunella vulgaris*. **D.** *Salvia apiana*, having enantiostyly. **E.** *Salvia clevelandii*. **F.** *Salvia leucophylla*. **G–I.** *Salvia mellifera*. **G.** Flower, side view. **H.** Flower, longitudinal section, showing gynoecium and epipetalous stamens, with unusual, extended connective and monothecal anthers. **I.** Ovary, four-lobed with gynobasic style. **J,K.** *Scutellaria tuberosa*. **L.** *Stachys ajugoides*, with strongly bilabiate corolla.

Figure 8.114 LAMIALES. Oleaceae. **A–G.** *Olea europaea*, olive. **A.** Inflorescences, each an axillary thyrse. Note opposite leaves. **B–C.** Leaf close-up, abaxial (B) and adaxial (C) surface; note scale-like, peltate trichomes. **D.** Dichasial inflorescence unit . **E.** Flower, face view, showing 4 petals and 2 stamens. **F.** Flower longitudinal section, showing superior ovary. **G.** Ovary longitudinal section; note apical-axile placentation. **H.** *Chionanthus* sp., fringe tree. Note 4 petals. **I–M.** *Fraxinus* sp., ash, female plant. **I.** Dichasial inflorescence unit. **J.** Female flower. **K.** Ovary longitudinal section, showing apical-axile placentation. **L.** Immature fruit. **M.** Mature fruits, samaras. **N–R.** *Ligustrum japonicum*, privet. **N.** Terminal thyrse. Note opposite leaves. **O.** Young corolla, with valvate aestivation. **P.** Mature flower, with 4 calyx/corolla lobes, 2 stamens. **Q.** Ovary cross section, with 4 ovules. **R.** Ovary longitudinal section, with apical-axile placentation.

Figure 8.115 LAMIALES. Orobanchaceae. **A.** *Epifagus virginiana*, beech-drops. **B,C.** *Kopsiopsis strobilacea*. **B.** Whole plant (dug up). **C.** Flower close-ups. **D.** *Aphyllon fasciculatum*. **E,F.** *Pedicularis canadensis*. **E.** Whole plant. **F.** Flower close-up, showing bilabiate corolla. **G.** *Pedicularis densiflora*, flowers. **H.** *Pedicularis groenlandica*, elephant heads. **I–L.** *Castilleja* spp., Indian paintbrush, hemiparasites with showy bracts. **I.** *C. densiflora*. **J.** *C. exserta*. **K,L.** *C. foliolosus*.

bracteate, hypogynous. The **perianth** is dichlamydeous. The **calyx** is synsepalous with 4–5 [0,1–3], usu. valvate lobes. The **corolla** is sympetalous with 5, usu. bilabiate, imbricate lobes, the abaxial lobe outer. The **stamens** are 4 [2], alternipetalous, epipetalous, with an adaxial staminode present in some. **Anthers** are longitudinally dehiscent. The **gynoecium** is syncarpous, with a superior ovary, 2 [3] carpels and 1–2 locules, each carpel typically with 2 placentae. The **style** is soliatry, terminal; **stigmas** are 2–4. **Placentation** is axile or parietal; **ovules** are anatropous, unitegmic, 1–∞ per carpel. The **fruit** is a loculicidal capsule [rarely a drupe], each valve usu. with 2 placentae. The **seeds** are oily endospermous or exalbuminous.

The Orobanchaceae are distributed mostly in the northern hemisphere. Economic importance includes some cultivated ornamentals or weeds and locally used medicinal or food plants. Many taxa, especially the photosynthetic members of the family, were formerly (and are still often) classified in the Scrophulariaceae s.l. See Olmstead et al. (2001) and Tank et al. (2006) for general placement of the Orobanchaceae and Tank et al. (2009) and McNeal et al. (2013) for relationships within the family.

The Orobanchaceae are distinctive in being ***achlorophyllous or chlorophyllous root parasites***, with spiral or opposite leaves (stems ***fleshy*** and leaves ***scale-like*** in achlorophyllous species), the flowers with a ***sympetalous, bilabiate*** corolla, typically with 4 [2] stamens, a posterior staminode present in some, the ovary superior, 2[3]-carpellate, with parietal or axile placentation, each carpel often with ***two placentae***, the fruit a loculicidal capsule with numerous, minute seeds.

K (4–5) [0, (1–3)] **C** (5) **A** 4 [2] **G** (2) [(3)], superior.

Phrymaceae — Lopseed family (type *Phryma*, after an early genus name). Ca. 13 genera/ca. 200 species (Figure 8.115).

The Phrymaceae consist of annual or perennial herbs or shrubs. The **stems** are 4-angled when young. The **leaves** are simple, opposite, exstipulate, toothed. The **inflorescence** is a spike, raceme, or of solitary, axillary flowers. The **flowers** are bisexual, zygomorphic or actinomorphic, ebracteate, hypogynous. The **perianth** is dichlamydeous. The **calyx** is synsepalous with 5 lobes. The **corolla** is sympetalous, with 5 lobes, usually bilabiate, the upper lip 2-lobed. The **stamens** are 2–4, alternipetalous, the anterior pair longer than the posterior pair, epipetalous, staminodes sometimes present. **Anthers** are longitudinally dehiscent. The **gynoecium** is syncarpous, with a superior ovary, 2 carpels (appearing as 1 in *Phryma*, but stigma 2-lobed), and 1 locule. The **style** is solitary and terminal; **stigma(s)** are 2, bilamellate and thigmonastic in some taxa, associated with pollination mechanism. **Placentation** is axile, parietal, or basal; **ovules** are anatropous, unitegmic, 1–∞. The **fruit** is a a dehiscent capsule (opening late in some taxa), berry, or reflexed achene with persistent calyx (*Phryma*), the latter accounting for the common name "lopseed."

The Phrymaceae have recently been expanded from one (*Phryma leptostachya*) to approximately 200 species, including the large genus *Mimulus*. Members of the family have a mostly worldwide distribution, especially western North America. Economic importance includes some ornamental cultivars, e.g., *Mazus* and *Mimulus* spp. See Beardsley and Olmstead (2002), Cantino (2004), Beardsley et al. (2004), Tank et al. (2006), and Barker et al. (2012) for phylogenetic analyses of the complex and for information on interfamilial relationships.

The Phrymaceae are distinctive in being herbs or shrubs with ***simple, opposite*** leaves, and ***zygomorphic synsepalous and sympetalous*** flowers with a ***bilabiate*** corolla and 2–4 stamens, the superior ovary 2-carpellate, the fruit a ***capsule, achene, or berry***.

K (5) **C** (5) **A** 2–4 **G** (2), superior.

Plantaginaceae (including Callitrichaceae, Globulariaceae, Plantaginaceae s.s., and many members of the traditional Scrophulariaceae s.l.) — Plantain/Speedwell family (type *Plantago*, Latin for "sole of the foot" or "footprint," after resemblance of the leaves of some taxa lying flat on the ground). Ca. 92 genera/2,000 species (Figure 8.116).

The Plantaginaceae consist of terrestrial or aquatic, hermaphroditic, monoecious, or gynomonoecious herbs or shrubs. The **leaves** are usually simple, spiral, opposite, or whorled, exstipulate, the major veins sometimes parallel. The **inflorescence** is a head, raceme, thyrse, or of solitary flowers. The **flowers** are usually bisexual, zygomorphic or actinomorphic, bracteate, hypogynous or epigynous. The **perianth** is membranous to scarious. The **calyx** is synsepalous with 3–5, imbricate lobes. The **corolla** is sympetalous, bilabiate in some, with 3–5 imbricate lobes. The **stamens** are 1-8, alternipetalous and epipetalous. **Anthers** are versatile, longitudinal or transverse in dehiscence. The **gynoecium** is syncarpous, with a superior or inferior ovary and 1–2 carpels and locules. The **style** is solitary; **stigma(s)** are 2-lobed or capitate. **Placentation** is axile or basal; **ovules** are anatropous to hemitropous, unitegmic, 1–40 per carpel. The **fruit** is a capsule (circumscissile or septicidal), achene, berry, or schizocarp of nutlets. The **seeds** are endospermous. Plants are insect- or wind-pollinated.

The Plantaginaceae have a worldwide distribution. As delimited here (see Olmstead et al. 2001) the Plantaginaceae contains many taxa that have traditionally been placed in the Scrophulariaceae s.l., including *Antirrhinum* (snapdragons), *Chelone* (turtleheads), *Collinsia* (Chinese houses), and *Digitalis* (foxglove). Many members of the Plantaginaceae are important in the horticultural trade, e.g., *Antirrhinum*. *Digitalis* is both horticulturally and medicinally important, being the source of the "cardiac" glycoside digitoxin and others, used to treat heart ailments. Other family members are important

Figure 8.116 LAMIALES. Phrymaceae. **A.** *Erythranthe guttata*, seep monkey flower. **B–H.** *Diplacus puniceus*, coast monkey flower. **B.** Axillary, opposite flowers, face view. **C.** Winged calyx, side view. **D.** Androecium (didynamous) and style/stigma. **E.** Cross section at base of flower, showing perianth and ovary. **F.** Ovary cross section, with parietal placentation (septum incomplete). **G.** Bilamellate stigmas, open. **H.** Stigmas closed, 5 seconds after physical contact. **I,J.** *Phryma leptostachya*, lopseed; images courtesy of David G. Smith. **I.** Flower close-up, showing bilabiate corolla; note depressed orientation after flowering (below). **J.** Inflorescence, a spike.

weeds or food plants. See Olmstead and Reeves (1995), Beardsley and Olmstead (2002), Schwarzbach (2004), Oxelman et al. (2005), and Albach et al. (2005) for general information and studies of phylogenetic relationships in the family. The latter study tentatively grouped family members into 12 tribes.

The Plantaginaceae as a whole cannot be readily delimited from other Scrophulariaceae s.l. based on morphological characteristics. Within the Plantaginaceae are clades that have previously been treated as the separate families Callitrichaceae, Globulariaceae, Hippuridaceae, and Plantaginaceae s.s. Two of these former families, the Callitrichaceae and Plantaginaceae s.s., are briefly described here, listing them as tribes, sensu Albach et al. (2005).

The Callitricheae (sensu Albach et al. 2005), commonly known as Starworts, contain 1 genus, *Callitriche*, and approximately 17 species. The Callitricheae are distinctive in being

Figure 8.117 LAMIALES. Plantaginaceae, showing tribal classification of Albach et al. (2005). **A–C.** Antirrhineae. **A.** *Antirrhinum majus*, snapdragons. **B.** *Sairocarpus nuttallianus*. **C.** *Nuttallanthus texanus* [*Linaria canadensis*]. **D.** Gratioleae. **C.** *Gratiola amphiantha* [*Amphianthus pusillus*], an aquatic species. **E,F.** Digitalideae. *Digitalis purpurea*, foxglove, source of digitoxin. **G.** Cheloneae. *Collinsia heterophylla*, Chinese houses. **H,I.** Veroniceae. *Veronica anagallis-aquatica*. **J–M.** Callitricheae. **J,K.** *Callitriche heterophylla*. **J.** Note dimorphic leaves, floating and submerged. **K.** Fruit close-up. **L,M.** *Callitriche marginata*. **L.** Floating shoots. **M.** Flower close-up. **N–Q.** Plantagineae. **N,O.** *Plantago erecta*. **N.** Whole plant. **O.** Inflorescence. **P.** *Plantago lanceolata*, inflorescence, with exeserted stamens. **Q.** *Plantago major*, whole plant.

aquatic, submerged to floating herbs with decussate leaves and unisexual flowers lacking a perianth, the stamens usually solitary, the gynoecium of 2 carpels, each with 2 locules, and the fruit a schizocarp of 4 nutlets. The floral formula is: **P** 0 **A** 1 [2,3] **G** (2), superior. Economic importance includes some aquarium plants.

The Plantagineae (sensu Albach et al., 2005) contain 3 genera, *Bougueria, Littorella*, and *Plantago*, and approximately 275 species, almost all of these in the genus *Plantago*. The Plantagineae are distinctive in being herbs, rarely shrubs, the leaves spiral with parallel major veins, the flowers bisexual or unisexual, with 4 [3] connate, membranous to scarious sepals and petals, usually 4 [1–3] stamens, and a superior, 2-carpellate ovary, the fruit a circumscissile capsule or achene. The floral formula is **K** (4) [(3)] **C** (4) [(3)] **A** 4 [1–3] **G** (2), superior.

The Plantaginaceae are distinctive in being herbs, rarely shrubs, the leaves spiral to whorled, the flowers ***bisexual or unisexual***, with ***3–5 connate sepals and petals***, 1–8 stamens, and a superior to inferior, 1-2 carpellate ovary, the fruit a ***capsule, achene, berry, or shizocarp of nutlets***.
K (3–5) **C** (3–5) **A** 1-8 **G** (1–2), superior or inferior.

Scrophulariaceae (including Buddlejaceae, Myoporaceae) — Figwort family (type *Scrophularia*, from Latin *scrofule*, scrophula, tubercular lymph nodes, alluding to resemblance to rhizome thickenings or to curing properties). Ca. 60 genera /ca. 1,900 species (Figure 8.117).

The Scrophulariaceae consist of terrestrial or aquatic, trees, shrubs, or herbs. The **leaves** are simple, spiral or opposite, exstipulate. The **inflorescence** is a spike, raceme, or thyrse. The **flowers** are bisexual, zygomorphic [rarely actinomorphic], hypogynous. The **perianth** is dichlamydeous. The **calyx** is synsepalous with 4–5 [2], valvate or imbricate, lobes. The **corolla** is sympetalous, usually bilabiate, sometimes spurred, with 5 [0, 4–8] valvate or imbricate lobes. The **stamens** are 2–5, staminoides sometimes present. **Anthers** are longitudinally dehiscent, typically with confluent locules. The **gynoecium** is syncarpous, with a superior ovary and 2 [3] carpels and locules. The **style** is solitary, terminal. **Placentation** is axile; **ovules** are anatropous to campylotropous, unitegmic, 1–∞ per carpel. **Nectaries** are often present as a disk around the ovary base. The **fruit** is a capsule (septicidal, loculicidal, or poricidal), drupe, or berry. The **seeds** are endospermous.

The Scrophulariaceae have a worldwide distribution. Economic importance includes local medicinal plants, timber plants, and numerous cultivated ornamentals (e.g., *Buddleja, Diascia, Myoporum, Verbascum*). The Scrophulariaceae as treated here are a small subset of the family as formerly circumscribed. Molecular studies have shown that the traditionally defined Scrophulariaceae are not monophyletic. Several major clades that include at least some taxa that were formerly placed in the Scrophulariaceae s.l., are now recognized at the family rank (Olmstead et al. 2001; Oxelman et al. 2005; see Stevens 2001 onwards): Calceolariaceae, Gratiolaceae, Linderniaceae, Orobanchaceae, Pawloniaceae, Phrymaceae, Plantaginaceae, Scrophulariaceae s.s., and Stilbaceae (see Table 8.3). These families are difficult to diagnose and differentiate from one another morphologically. See Fischer (2004) for a general treatment and Kornhall et al. (2001), Oxelman et al. (2005), Tank et al. (2006) for phylogenetic analyses of the complex.

The Scrophulariaceae are characterized as trees, shrubs or herbs with opposite or spiral leaves and usually ***zygomorphic*** flowers with a ***superior, 2 [3]-carpellate*** ovary having ***axile*** placentation with ***usually numerous*** ovules, the fruit a capsule, berry, or drupe.
K (2–5) C (5) [0,(4–8)] A 2–5 G (2) [(3)], superior.

Verbenaceae — Verbena family (type *Verbena*, Latin name for plants used medicinaly and in religious ceremonies). Ca. 31 genera/ca. 920 species (Figure 8.118).

The Verbenaceae consist of hermaphroditic [rarely dioecious], aromatic trees, shrubs, lianas, or herbs. The **stems** are usually 4-sided, at least when young, the nodes often ridged, thorns present in some taxa. The **leaves** are usu. simple, serrate, opposite [rarely whorled], and exstipulate. The **inflorescence** is a cyme, panicle, raceme, or head, in some taxa with an involucre of colored bracts. The **flowers** are hypogynous, usu. bisexual. The **perianth** is dichlamydeous. The **calyx** is synsepalous, actinomorphic or zygomorphic, with 5 [4] imbricate lobes. The **corolla** is sympetalous, zygomorphic (often weakly so), often salverform with a narrow tube, sometimes bilabiate, with 5 [4], imbricate lobes. The **stamens** are 4 [5], alternipetalous, inserted, epipetalous, staminodes sometimes present. **Anthers** are longitudinally dehiscent. The **gynoecium** is syncarpous, with a superior ovary, usu. 2 carpels, and 2 or 4 locules, the latter by development of false septae. The **style** is terminal; **stigmatic regions** are swollen, glandular. **Ovules** are 4 per pistil (2 per carpel), anatropous, unitegmic. The **fruit** is a schizocarp of 1-seeded mericarps or a drupe of 1–4 pyrenes. The **seeds** are exalbuminous.

Members of the Verbenaceae are distributed in mostly tropical regions (esp. South America), some taxa temperate. Economic importance includes uses as timber (e.g., *Citharexylum* spp., fiddlewood), flavoring, medicinal, and tea plants, numerous cultivated ornamentals (e.g., *Aloysia, Duranta, Lantana, Verbena*), and weeds. See Wagstaff and Olmstead (1997), Oxelman et al. (2005), Marx et al. (2010), and Frost et al. (2017) for phylogenetic studies.

The Verbenaceae are distinctive in being trees, shrubs, lianas, or herbs with ***4-sided stems***, leaves ***opposite, simple, exstipulate,*** *usu. serrate*, flowers 4–5-merous, mostly zygo-

Figure 8.118 LAMIALES. Scrophulariaceae s.s. **A–D.** *Scrophularia californica*. **A.** Whole plant. **B.** Flower close-up, showing bilabiate corolla. **C.** Flower longitudinal section; note superior ovary. **D.** Ovary cross section, with axile placentation. **E–G.** Myoporoids. *Myoporum laetum*. **E.** Branch of tree, in fruit (a berry). **F.** Leaf close-up showing glands. **G.** Flower close-up. **H–J.** *Verbascum thapsus*. **H.** Inflorescence, a spike. **I.** Flower close-up. **J.** Whole plant.

morphic, with gynoecium 2-carpellate with a ***terminal style***, the fruit a ***schizocarp of 1-seeded mericarps*** or a ***drupe of pyrenes***. K (5) [(4)] C (5) [(4)] A 4 [5] G (2), superior.

SOLANALES

The Solanales, sensu APG IV (2016), contain 5 families (Table 8.3). Of these, two families are described here. See Soltis et al. (2011), and Refulio-Rodriguez and Olmstead (2014) for phylogenetic analyses.

Convolvulaceae — Morning Glory / Bindweed family (type *Convolvulus*, Latin for "interwoven"). Ca. 59 genera/ca, 1,900 species (Figure 8.119).

The Convolvulaceae consist of herbaceous to woody vines, less commonly herbs, shrubs, or rarely trees. Some family members are achlorophyllous and parasitic (e.g., *Cuscuta*). The **roots** are haustorial in parasitic taxa. The **stems** of viney members are dextrorse (twining clockwise when moving away, like the grooves of a typical "right-handed" screw). The **leaves** are simple, undivided to divided, spiral, exstipu-

Figure 8.119 LAMIALES. Verbenaceae. **A–B.** *Aloysia triphylla*, lemon-verbena. **C.** *Duranta repens*. Note orange drupes. **D.** *Holmsiolodia sanguinea*, with showy calyces. **E–I.** *Lantana camara*. **E.** Shoot, showing exstipulate, opposite leaves. **F.** Node, showing ridge between petioles. **G.** Head inflorescences. Note salverform corollas of flowers. **H.** Fruits, drupes. **I.** Drupe cross section, showing hard endocarp. **J–M.** *Verbena rigida*. **J.** Plant, showing opposite leaves and corymb of heads. **K.** Head, close-up. Note slightly zygomorphic corollas. **L.** Flower longitudinal section, with salverform corolla, epipetalous stamens, and superior ovary. **M.** Pistil, with asymmetric stigmatic region.

Figure 8.120 SOLANALES. Convolvulaceae. **A–D.** *Convolvulus* [*Calystegia*] sp. **A.** Whole vine, in flower. **B.** Young flower, showing involute aestivation. **C.** Flower, side view. **D.** Bracts (epicalyx) subtending flower. **E.** *Ipomoea alba*, moon flower, with long, narrow corolla tube (moth pollinated). **F–J.** *Ipomoea* sp., morning glory. **F–G.** Open flowers, with infundibular corolla. **H.** Flower base longitudinal section, showing basal placentation. **I.** Ovary cross section, showing four ovules. **J.** Style, with two style branches, and adjacent anthers. **K.** *Dichondra occidentalis*. **L–N.** *Cressa truxillensis*. **O,P.** *Cuscuta gronovii*, an achlorophyllous, parasitic vine with haustorial roots.

late, reduced and scalelike in *Cuscuta*. The **inflorescence** is a head, dichasium, or of solitary flowers, bracteate, of usually two, often accrescent bracts. The **flowers** are bisexual, actinomorphic, and hypogynous. The **perianth** is dichlamydeous. The **calyx** has 5 [3,4], sepals or lobes. The **corolla** is sympetalous, often infundibular, with 5 [3,4] lobes, with usually involute (plicate) aestivation (imbricate in *Cuscuta*). The **stamens** are 5 [3,4], filaments often unequal in length, the stamens epipetalous. **Anthers** are longitudinal in dehiscence. The **gynoecium** is syncarpous with a superior ovary, 2 [3–5] carpels, and 1–several locules (as many as carpels). The **style(s)** are solitary to as many as carpels. **Placentation** is basal; **ovules** are anatropous, unitegmic, 2 per carpel [rarely ∞]. **Nectaries** are present, consisting of an annular disk around base of ovary. The **fruit** is a capsule (loculicidal, circumscissile, or irregularly dehiscing), berry, drupe, or nut. The **seeds** are endospermous. Internal phloem (inner to the xylem) is present in many family members.

The Convolvulaceae have a mostly worldwide distribution. Economic importance includes cultivated ornamentals such as *Convolvulus*, *Ipomoea*, and *Jacquemontia*; *Ipomoea batatas* is the sweet potato, source of the starchy storage root. See Staples and Brummit (2007) for general information and Stefanovic et al. (2002, 2003), Stefanovic and Olmstead (2004), and Garcia et al. (2014) for phylogenetic studies.

The Convolvulaceae are distinctive in being often ***dextrorse -twining vines***, less commonly shrubs or trees, usually with ***internal phloem***, with simple, spiral leaves, and ***actinomorphic, sympetalous*** flowers, corollas typically with ***involute aestivation*** and often ***infundibular***.

K (5) [(3,4)] **C** (5) [(3,4)] **A** (5) [3,4] **G** (2) [(3–5)], superior.

Solanaceae — Nightshade family (type *Solanum*, Latin for sleeping or comforter, after narcotic properties of some taxa). Ca. 102 genera/ca. 2,460 species (Figure 8.121).

The Solanaceae consist of herbs, shrubs, trees, or lianas, with prickles present in some taxa, many with stellate trichomes. The **leaves** are simple, pinnate, or ternate, usually spiral and exstipulate. The **inflorescence** is of solitary flowers or cyme units. The **flowers** are bisexual, actinomorphic, rarely zygomorphic. The **perianth** is biseriate, dichlamydeous, usually tubular, rotate, or salverform, hypanthium absent. The **calyx** is synsepalous, persistent, sometimes accrescent, with 5 calyx lobes. The **corolla** is sympetalous and with 5 [4,6] convolute, imbricate, or valvate lobes, with usually involute (plicate) aestivation. The **stamens** are 5 [rarely 4 or 2 + 2 staminodes], antisepalous and epipetalous, the anthers often connivent, with staminodes rarely present. **Anthers** are longitudinal or poricidal in dehiscence. The **gynoecium** is syncarpous, with a superior ovary, 2 [rarely 3–5] carpels, and 2 [rarely 1 or 4–5] locules. **Placentation** is axile, rarely basal; **ovules** are variable in type, unitegmic, ∞ [rarely 1–few] per carpel. The **fruit** is a berry, drupe, or capsule (often septicidal). The **seeds** are endospermous. Alkaloids and internal phloem (inner to the xylem, surrounding pith) are present in many family members.

Members of the Solanaceae have mostly worldwide distributions, concentrated in South America. Economic importance includes many edible plants, such as *Capsicum* (peppers), *Physalis philadelphica* (tomatillo), *Solanum* [*Lycopersicon*] *esculentum* (tomato), and *Solanum tuberosum* (potato), and the infamous fumatory *Nicotiana tabacum* (tobacco). Alkaloids from various taxa have medicinal properties (e.g., atropine from *Atropa belladonna*), hallucinogenic properties (e.g., *Datura* spp., angel trumpet), or are deadly poisons (e.g., *Datura* spp., e.g., Jimson weed, *Solanum* spp., nightshades) or known carcinogens (*Nicotiana tabacum*); some are used as ornamental cultivars, others are noxious weeds. See Olmstead et al. (2008) and Särkinen et al. (2013) for phylogenetic studies and Barboza et al. (2016) for a general treatment.

The Solanaceae are distinctive in being herbs, shrubs, trees, or lianas with ***internal phloem***, spiral leaves, a usually ***actinomorphic [zygomorphic], 5-merous*** perianth and androecium (***corolla involute in aestivation***), a usually ***bicarpellate, syncarpous*** gynoecium, and usually ***numerous ovules*** per carpel, the fruit a berry, drupe, or capsule.

K (5) **C** (5) [(4),(6)] **A** 5 [4 or 2+2 staminodes] **G** (2) [(3–5)], superior.

CAMPANULIDS

The Campanulids (Bremmer et al. 2002; *Campanulidae*, sensu Cantino et al., 2007; formerly called Euasterids II) are a monophyletic group of seven orders within the Asterids of APG IV, 2016 (Table 8.3). See Tank and Donoghue (2010) for a recent study of the group and Winkworth et al. (2008) and Beaulieu and O'Meara (2018) for a study on the efficacy of phylogenetic studies of the group.

AQUIFOLIALES

The Aquifoliales contain 5 families (APG IV, 2016; Table 8.3), of which only the Aquifoliaceae are described here. See Chandler and Plunkett (2004) for detailed information on the order and Kårehed (2002) for phylogenetic studies.

Aquifoliaceae — Holly family (type *Aquifolium* [=*Ilex*], after the classical name for holly). 1 genus (*Ilex*)/400-500 species (Figure 8.122).

The Aquifoliaceae consist of usually dioecious trees or shrubs. The **leaves** are usually spiral, rarely opposite, stipulate or exstipulate, usually evergreen, spine-margined in some taxa

Figure 8.121 SOLANALES. Solanaceae. **A,B.** *Brugmansia* sp. Note actinomorphic, plicate corolla. **C.** *Datura wrightii*, jimson weed, a poisonous/hallucinogenic plant. **D–H.** *Nicotiana glauca*, tree tobacco. **D.** Corolla. **E.** Style and stigma. **F.** Ovary in longitudinal section. **G.** Ovary cross section, showing bicarpellate axile placentation with large placentae. **H.** Fruit, a capsule. **I.** *Iochroma cyaneum*, with tubular corollas. **J.** *Physalis ixocarpa*, tomatillo, having an acrescent calyx. **K.** *Solandra maxima*. **O.** *Schizanthus littoralis*, having zygomorphic flowers. **M,N.** *Solanum douglasii*, one of the nightshades. **M.** Flower, with connivent stamens. **N.** Stamen, showing poricidal dehiscence. **O.** *Solanum parishii*, with prominent nectar guides.

Figure 8.122 AQUIFOLIALES. Aquifoliaceae. **A–J.** *Ilex opaca*, American holly. **A.** Inflorescence. **B.** Flower, face view. **C.** Flower, back (abaxial) view. **D.** Flower, side view, showing superior ovary and nectar. **E.** Ovary longitudinal section, showing apical-axile placentation. **F.** Ovary cross section, showing endocarp of pyrenes surrounding each ovule. **G.** Shoot in fruit. **H.** Berry. **I.** Berry opened, revealing pyrene units. **J.** Pyrene cross section. **K–L.** *Ilex paraguarensis*, yerba mate, used to make a stimulating tea.

(many *Ilex* spp.). The **inflorescence** is an axillary or supra-axillary cyme, fascicle, or raceme. The **flowers** are small, usually unisexual, actinomorphic, hypogynous. The **perianth** is dichlamydeous. The **calyx** is synsepalous with 4 [–8,0], imbricate lobes. The **corolla** is basally sympetalous [rarely apopetalous] with 4 [–8,0] imbricate petals/lobes. The **stamens** are 4 [–8], alternipetalous, often epipetalous. The **gynoecium** is syncarpous, with a superior ovary and 4-6 [2–24] carpels and locules. **Placentation** is apical-axile; **ovules** are anatropous to campylotropous, unitegmic, 1 [rarely 2] per carpel. A **nectariferous** disk is absent. The embryo is small. The **fruit** is a drupe with pyrenes. Plants are often with resins or latex in leaf mesophyll cells.

The Aquifoliaceae have a mostly worldwide distribution. Economic importance includes uses for wood, stimulating teas (esp. *Ilex paraguariensis*, yerba mate), and numerous cultivars, especially those of *Ilex*, holly. See Cuénoud et al. 2000, Powell et al. (2000), Manen et al. (2002, 2010), and Selbach-Schnadelbach et al. (2009) for phylogenetic studies.

The Aquifoliaceae are distinctive in being usually ***evergreen, dioecious trees or shrubs***, the flowers ***actinomorphic***, dichlamydeous, hypogynous, with usually ***1 ovule per carpel***, the fruit a ***drupe of pyrenes***.
K (4) [(–8),0] **C** (4) [(–8),0] **A** 4 [–8] **G** (4–6) [(2–24)], superior.

APIALES

The Apiales contain seven families (APG IV, 2016; Table 8.3), of which three are described here. See Chandler and Plunkett (2004), Tank and Donoghue (2010), and Nicolas and Plunkett (2014) for more information on the order.

Apiaceae (Umbelliferae) — Carrot family (type *Apium*, a name used by Pliny for a celerylike plant). Ca. 434 genera/ca. 3,780 species (Figures 8.123, 8.124).

The Apiaceae consist of herbs, less often shrubs or trees. The **leaves** are usually pinnate, ternate, or decompound [rarely simple, palmate, or phyllodinous], spiral, with a broad sheathing base, stipular flanges sometimes present. The **inflorescence** is usually a compound umbel often with subtending involucral bracts, sometimes a head or simple umbel or reduced to a single flower or dichasium. The **flowers** are small, bisexual [marginal flowers sometimes sterile], actinomorphic, epigynous. The **perianth** is biseriate and dichlamydeous or uniseriate by loss of the calyx. The **calyx** is aposepalous with 5 lobes, which may be reduced or absent. The **corolla** is apopetalous and with 5 [rarely 0], valvate petals. The **stamens** are 5, whorled, alternipetalous, and apostemonous. The **gynoecium** is syncarpous, with an inferior ovary, 2 carpels, and 2 [rarely 1] locules, often with a stylopodium at apex of ovary. **Placentation** is apical-axile; **ovules** are anatropous, pendulous, unitegmic, 1 per carpel. The **fruit** is a schizocarp of mericarps, supported by carpophores upon splitting. The **seeds** are endospermous, endosperm oily. Some taxa have anomalous secondary thickening.

The Apiaceae have a worldwide distribution. Economically important members include a number of food, herb, and spice plants, such as *Anethum*, dill; *Apium*, celery; *Carum*, caraway; *Coriandrum*, coriander; *Cuminum*, cumin; *Daucus*, carrot; *Foeniculum*, fennel; and *Petroselinum*, parsley; some species are poisonous, such as *Conium maculatum*, poison-hemlock (an extract of which Socrates drank in execution); others are used as ornamental cultivars. See Plunkett et al. (1997, 2004a) and Nicolas and Plunkett (2009) for phylogeny and classification of the family.

The Apiaceae are distinctive in being ***herbs***, with ***sheathing leaves*** (compound or simple, often decompound), the inflorescence usually an ***involucrate compound umbel*** [rarely a head, simple umbel, or reduced] with actinomorphic flowers having a 2-carpellate and 2-loculate, ***inferior ovary***, each carpel with ***one, axile-apical, pendulous ovule***, the fruit a ***schizocarp of mericarps***.
K 5 or 0 **C** 5 [0] **A** 5 **G** (2), inferior.

Araliaceae — Ginseng family (type *Aralia*, possibly from French Canadian Aralie). Ca. 43 genera/ca. 1,450 species (Figure 8.125).

The Araliaceae consist of trees, shrubs, lianas, or herbs. The **stems** of trees are often pachycaulous. The **leaves** are palmate, pinnate, or simple (these often divided), usually spiral, rarely opposite or whorled, usually stipulate. The **inflorescence** is a usually terminal umbel, head, or secondary inflorescence (e.g., panicle) of umbels, rarely of solitary flowers. The **flowers** are usually bisexual, actinomorphic, epigynous (rarely hypogynous). The **calyx** is aposepalous with 5 [3–∞] sepals, often reduced or absent. The **corolla** is apopetalous, rarely basally sympetalous or calyptrate, with 5 [rarely 3–12], valvate or imbricate petals or lobes. The **stamens** are 5–10 [rarely 3–∞]. **Anthers** are longitudinal in dehiscence. The **gynoecium** is syncarpous, with an inferior (rarely superior and secondarily derived) ovary, 2–5 [–∞] carpels, and 1–∞ locules. The **styles** are 1–∞. **Placentation** is apical-axile; **ovules** are anatropous, unitegmic, 1 [2] per carpel. **Nectaries** are sometimes present as an epigynous disk. The **fruit** is a drupe with multiple endocarps, berry, or schizocarp with carpophore. The **seeds** are oily endospermous. Plant tissues usually have secretory canals; leaf nodes are multilacunar.

The Araliaceae have a mostly worldwide distribution, mainly in tropical regions. Economic importance includes medicinally important plants, especially *Panax* (ginseng);

Figure 8.123 APIALES. Apiaceae. **A–E.** *Apium graveolens*, wild celery. **A.** Flower, immature, with incurved petals. **B.** Flower at anthesis, showing stylopodium and alternipetalous stamens. **C.** Flower in side view; note inferior ovary. **D.** Leaf. **E.** Inflorescence, a compound umbel. **F.** *Conium maculatum*, poison-hemlock, of Socrates fame. **G.** *Daucus carota*, wild carrot or Queen Anne's lace, inflorescence. **H,I.** *Daucus pusillus*. **H.** Inflorescence, showing subtending involucre. **I.** Infructescence, with bristly fruits. **J–L.** *Foeniculum vulgare*, wild fennel. **J.** Compound umbel, showing rays. **K.** Close-up of simple umbel unit. **L.** Ovary longitudinal section, showing two carpels and apical-axile placentation.

Figure 8.124 APIALES. Apiaceae. **A–F.** *Foeniculum vulgare*. **A.** Umbel unit of infructescence. **B.** Fruit, a schizocarp of mericarps, showing supporting carpophores. **C.** Fruit in cross section, showing two carpels, oil canals, and seeds. **D.** Leaf, with prominent, adnate stipules. **E,F.** Florence fennel (var. *dulce*), bred for succulent stems and petioles, used as a vegetable. **G.** *Heracleum lanatum*, showing flowers of umbel unit. **H.** *Lomatium dasycarpum*, with immature fruits. **I–J.** *Lomatium lucidum*, having winged fruits. **K.** *Sanicula bipinnatifidum*, with an umbel of headlike units. **L.** *Sanicula tuberosa*, showing short, bisexual flowers and long-pedicellate male flowers. **M.** *Xanthosia* sp., southern cross, with unusual bracts and winged rays.

Figure 8.125 APIALES. Araliaceae. **A.** *Fatsia japonica*, the inflorescence a panicle of umbels. **B,C.** *Hedera canariensis*, Algerian ivy. **B.** Reproductive shoot, with a raceme of umbels. **C.** Close-up of unit umbels. **D.** *Hydrocotyle* sp., pennywort. **E–J.** *Schefflera actinophylla*, octopus tree. **E.** Whole plant, showing palmate leaves and large, terminal panicle of umbels. **F.** Close up of inflorescence axis. **G.** Unit umbel. Note open flower, petals caducous. **H.** Mature bud in longitudinal section. **I.** Carpel close-up, showing apical-axile placentation. **J.** Ovary cross section, 10-carpellate. **K–N.** *Schefflera* sp. **K.** Whole plant. **L.** Close-up of stem nodes, showing broad point of attachment. **M.** Inflorescence umbel unit. **N.** Fruits, drupes.

Tetrapanax papyrifer, used as Chinese rice paper; some timber plants; and several cultivated ornamentals such as *Fatsia*, *Hedera* (ivy), and *Schefflera*. See Wen et al. (2000), Plunkett et al. (2004b), and Nicolas and Plunkett (2009) for studies of the phylogeny and classification of the Araliaceae.

The Araliaceae are distinctive in being ***mostly tropical*** trees, shrubs, lianas, or herbs with ***palmate or pinnate*** (rarely simple, then usually divided) leaves, an ***inflorescence of heads, umbels, or with umbel units***, the flowers with often reduced calyx, apopetalous to sympetalous corolla, and a ***1–∞-carpellate inferior ovary*** with usually ***apical-axile placentation***, the fruit a berry, drupe, or schizocarp.

K 5 [0, 3–∞] **C** 5 [3–12] **A** 5–10 [3–∞] **G** (2–5) [(–∞)], inferior, rarely superior.

Pittosporaceae — Parchment Bark/Australian-Laurel family (type *Pittosporum*, after Greek *pitta*, pitch, + *sporum*, seed, from resinous covering of seeds). Ca. 8 genera/ca. 200 species (Figure 8.126).

The Pittosporaceae consist of trees, shrub, or lianas, some members with thorns (e.g., *Bursaria*). The **leaves** are evergreen, simple, spiral (sometimes whorled at shoot apices), exstipulate, and coriaceous. The **inflorescence** is an axillary or terminal cyme, corymb, thyrse, or of solitary flowers. The **flowers** are mostly bisexual (some functionally unisexual), actinomorphic (rarely slightly zygomorphic), and bracteate, with two bracteoles. The **perianth** is dichlamydeous, often showy and fragrant. The **calyx** is aposepalous or basally synsepalous with 5 deciduous sepals or calyx lobes. The **corolla** is usu. sympetalous with a basal tube and 5, imbricate lobes. The **stamens** are 5, apostemonous or basally connate, whorled, alternipetalous. **Anthers** are dithecal, longitudinal or poricidal in dehiscence. The **gynoecium** is syncarpous, with a superior ovary, 2 [3–5] carpels, and usu. 1 [2–5] locule. **Placentation** is usu. parietal; **ovules** are anatropous to campylotropous, unitegmic, usu. ∞ per carpel. The **fruit** is a loculicidal capsule or berry with a viscid, resinous pulp. **Seeds** are arillate (winged in *Hymenosporum*), with an oily, proteinaceous endosperm.

Members of the Pittosporaceae grow in the tropical and subtropical Old World, especially in Australasia. Economic importance includes local uses as medicine, fuel oil, lumber, and food, plus several ornamental cultivars, including *Billardiera, Bursaria, Hymenosporum flavum*, and numerous *Pittosporum* species. See Cayzer et al. (2004) and Chandler et al. 2007 for phylogenetic studies of the Pittosporaceae.

The Pittosporaceae are distinctive in being trees, shrubs, or lianas with a ***pentamerous perianth and androecium***, ***superior ovary*** with ***2 [3–5] carpels***, usu. ***parietal*** placentation, the fruit a ***capsule or berry with a resinous pulp*** and ***arillate [winged] seeds***.

K 5 or (5) **C** (5) **A** 5 **G** (2) [(3–5)], superior.

DIPSACALES

The Dipsacales is treated here as containing seven families (congruent, e.g., with Xiang et al. 2019), although APG IV (2016) treats the Caprifoliaceae as inclusive of all but the Adoxaceae (Table 8.3). Three families are described here in detail; these families are largely equivalent to subfamilies of the Caprifoliaceae s.l. in some references (e.g., Stevens 2001, onwards). Among groups not described, the Adoxaceae includes *Viburnum*, with several species used as ornamentals, and *Sambucus*, elderberry, used as fruit and wine plants (Figure 8.127A,B). The Linnaeaceae [Caprifoliaceae, Linnaeoideae], the twinberry family, (Figure 8.127C,D) is renowned for being named after Carolus Linnaeus, the "father of taxonomy" (see Chapter 16). See Donoghue et al. (2001), Zhang et al. (2003), Moore and Donoghue (2007), and Xiang et al. (2019) for recent phylogenetic analyses of the Dipsacales.

Caprifoliaceae s.s. [Caprifoliaceae s.l., Caprifolioideae] — Honeysuckle family (type *Caprifolium* [=*Lonicera*], Latin for "goat leaf"). 5 genera/220 species (Figure 8.127E–I).

The Caprifoliaceae consist of shrubs, trees, lianas, or herbs. The **leaves** are simple [rarely pinnate], opposite, exstipulate or reduced-stipulate. The **inflorescence** is usually a cyme. The **flowers** are bisexual, epigynous, rarely epihypogynous. The **calyx** is synsepalous with 5 [4] imbricate or open lobes, the calyx often constricted below lobes. The **corolla** is sympetalous with 5 [4], imbricate or valvate lobes. The **stamens** are 5 [2,4], alternipetalous, and epipetalous. **Anthers** are longitudinal in dehiscence. The **gynoecium** is syncarpous, with an inferior ovary, 2–5 [–8] carpels (not all carpels fertile in some taxa), and 1–5 [–8] locules. The **style** is solitary or absent. **Placentation** is axile or parietal; **ovules** are anatropous, unitegmic, 1–∞ per carpel. **Nectaries** are often present on inner corolla tube, with extrafloral nectaries present in some taxa. The **fruit** is a capsule or berry. The **seeds** are oily endospermous.

The Caprifoliaceae have a mostly worldwide distribution. Economic importance includes several ornamental cultivars, such as *Lonicera* (honeysuckles). See Pyck et al. (1999) for a classification system and Theis et al. (2008) for a phylogenetic analysis of the group.

The Caprifoliaceae are distinctive in being trees, shrubs, herbs, or lianas with ***opposite, usually simple*** (rarely pinnate) ***leaves***, a ***cymose inflorescence***, the flowers usually ***epigynous***, with a ***4–5-merous perianth***, 5 [2,4] stamens, and 2–5 [–8] carpels, ***not all fertile in some taxa***, the fruit a berry, capsule, or drupe.

K (5) [(4)] **C** (5) [(4)] **A** 5 [2,4] **G** (2–5) [(–8)], inferior.

Figure 8.126 APIALES. Pittosporaceae. **A.** *Bursaria spinosa*. Note thorns. **B–G.** *Hymenosporum flavum*. **B.** Inflorescence at terminus of shoot. **C.** Inflorescence unit, a cyme. Note deciduous sepals, characteristic of family. **D.** Flower, actinomorphic, with aposepalous calyx and sympetalous corolla. **E.** Flower longitudinal section, showing superior ovary. **F.** Two-valved, loculidal capsule. **G.** Winged seeds, found only in *Hymenosporum flavum*. **H.** *Pittosporum kirkii*. **I–L.** *Pittosporum tobira*. **I.** Inflorescence, a corymb. **J.** Flower, with five, alternipetalous stamens. **K.** Flower, two petals removed, showing superior ovary. **L.** Ovary cross section. Note parietal placentation. **M.** *Pittosporum turneri*. **N–P.** *Pittosporum undulatum*. **N.** Inflorescence. **O.** Two-valved, loculicidal capsule. **P.** Seeds, arillate and viscid-resinous.

Figure 8.127 DIPSACALES. **A,B.** Adoxaceae, *Sambucus mexicana*, one of the elderberries. **C–D.** Linnaeaceae, *Linnaea borealis*, twinberry, named after Carolus Linnaeus (images courtesy of Michael Silveira). **D.** *Abelia ×grandiflora*. **E–I.** Caprifoliaceae s.s. **E–G.** *Symphoricarpos albus*, snow berry. **E.** Flowering shoot; note opposite leaves. **F.** Flower longitudinal-section. **G.** Ovary longitudinal section. **H,I.** *Lonicera japonica*, honeysuckle.

Dipsacaceae [Caprifoliaceae, Dispsacoideae] — Teasel family (type *Dipsacus*, from *dipsa*, thirst, relative to "water-collecting," in reference to the water collecting leaf bases). 11 genera/290 species (Figure 8.128).

The Dipsacaceae consist of herbs or shrubs. The **leaves** are simple, opposite or whorled, and exstipulate. The **inflorescence** is a head or raceme of cyme units, subtended by an involucre. The **flowers** are bisexual, usually bracteate, forming an epicalyx in some. The **calyx** is synsepalous or bristle-like with 4–5 [0,10] lobes or elements. The **corolla** is sympetalous with 4–5 imbricate lobes. The **stamens** are 4, alternipetalous, and epipetalous. **Anthers** are longitudinally dehiscent. The **gynoecium** is syncarpous, with an inferior ovary, 2 carpels, and 1 functional locule. The **style** is solitary. **Placentation** is apical; the solitary **ovule** is anatropous and unitegmic. A **nectary** is present, consisting of an annular disk at base of style. The **fruit** is an achene, usually enclosed by the epicalyx and with a terminal, persistent calyx. The **seeds** have an oily endosperm.

Figure 8.128 DIPSACALES. Dipsacaceae. **A–D.** *Dipsacus sativus*, teasel. **A.** Plant habit, an herb. **B.** Opposite, connate-perfoliate, water-collecting leaves, functioning to stop crawling insects. **C.** Inflorescence, an involucrate head. **D.** Infructescence, with persistent, woody flower bracts. **E–L.** *Scabiosa* sp. **E.** Inflorescence, a head. **F.** Bottom of inflorescence, showing involucral bracts. **G.** Inflorescence longitudinal section, showing compound receptacle. **H.** Flower, face view. **I.** Flower, showing epicalyx, calyx, and corolla. **J.** Style and stigmatic region. **K.** Ovary longitudinal section, showing inferior position and apical ovule. **L.** Fruit, with persistent epicalyx and filiform calyx.

The Dipsacaceae are distributed in Eurasia and Africa. Economic importance includes some cultivated ornamentals. See Pyck and Smets (2004) for a phylogenetic analysis.

The Dipsacaceae are distinctive in being herbs or shrubs with ***opposite or whorled*** leaves, an ***involucrate head of cyme units***, the flowers usually ***bracteate*** with an ***epicalyx***, the ovary ***inferior with 2 carpels*** and ***1 locule*** (1 locule abortive) and a ***single, apical ovule***, the fruit an ***achene*** usually with a ***persistent epicalyx and calyx***.

K (4–5) [0, 10] **C** (4–5) **A** 4 **G** (2), inferior.

Valerianaceae [Caprifoliaceae, Valerianoideae] — Valerian family (type *Valeriana*, after Latin *valere*, to be strong, or Valeria, a Roman province where plant found, or perhaps after Valerianus, Roman emperor of 3rd century A.D.). 17 genera/315 species (Figure 8.129).

The Valerianaceae consist of herbs, rarely shrubs. The **leaves** are simple or pinnate, opposite, and exstipulate. The **inflorescence** is composed of cyme units. The **flowers** are usually bisexual, bracteate. The **perianth** is biseriate or uniseriate. The **calyx** is absent, composed of teeth, or of

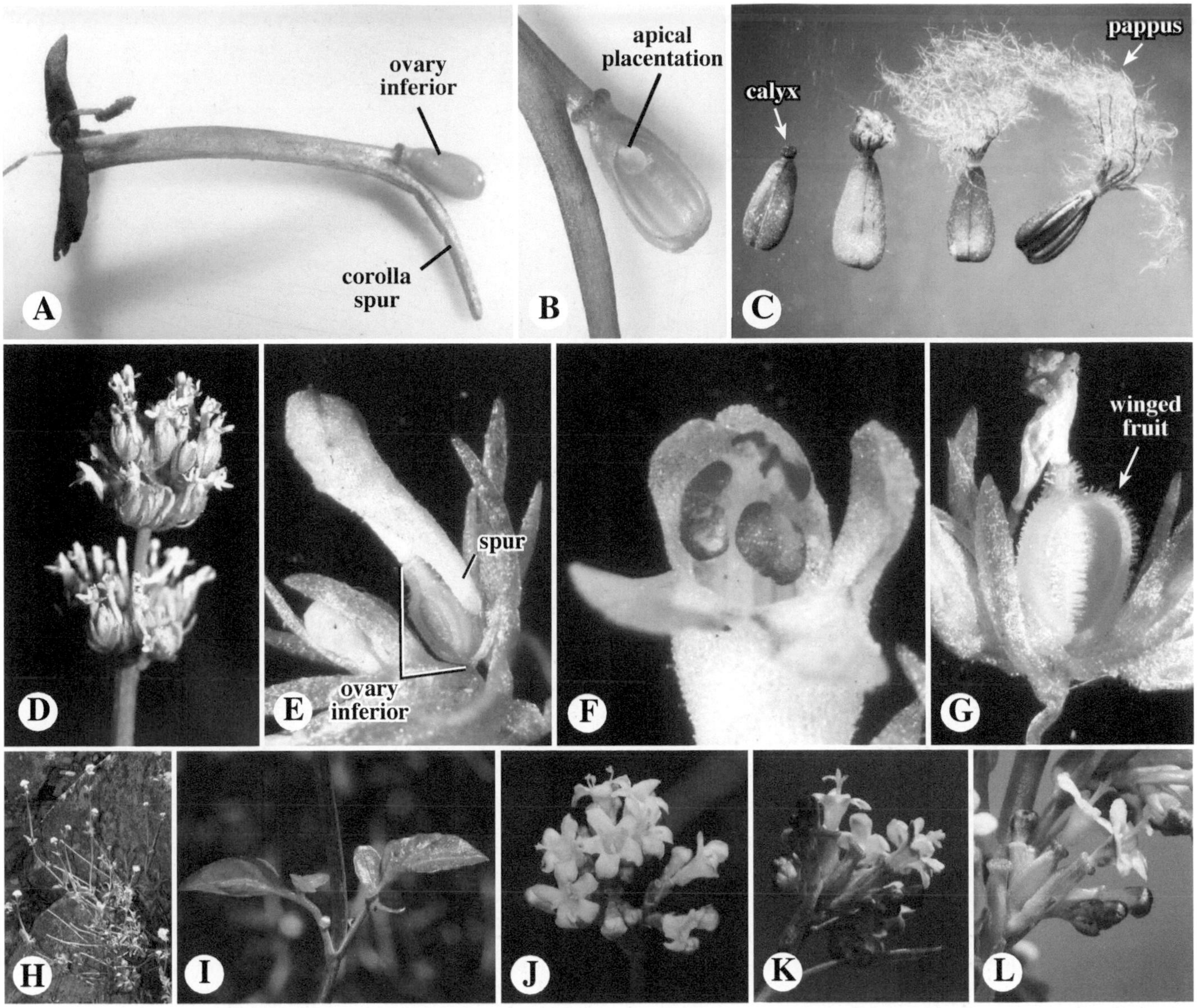

Figure 8.129 DIPSACALES. Valerianaceae. **A–C.** *Centranthus ruber*, red valerian. **A.** Whole flower, with sympetalous corolla and spur, solitary stamen, and inferior ovary. **B.** Ovary longitudinal section, showing apical placentation. **C.** Series of fruit development (left to right), showing maturation of pappus from calyx. **D–G.** *Plectritis ciliosa*. **D.** Inflorescence. **E.** Flower with inferior ovary and corolla spur. **F.** Apex of five-lobed corolla, showing three epipetalous stamens. **G.** Winged fruit, beneath withering perianth/androecium. **H–L.** *Valeriana* sp., vegetative and floral morphology.

5 sepals, forming an accrescent pappus in some taxa. The **corolla** is sympetalous with 5 [3,4] imbricate lobes and a basal, nectariferous spur, actinomorphic or zygomorphic, bilabiate in some. The **stamens** are 1–4 (less than the number of corolla lobes), whorled, epipetalous. **Anthers** are versatile and longitudinally dehiscent. The **gynoecium** is syncarpous, with an inferior ovary, 3 carpels, and 1 functional locule (2 locules abortive). The **style** is solitary. **Placentation** is apical; the **ovule** is solitary, anatropous, and unitegmic. The **fruit** is an achene, sometimes winged, with a plumose, pappuslike calyx in some. The **seeds** have an oily endosperm.

The Valerianaceae are mostly worldwide in distribution. Economic importance includes cultivated ornamentals (e.g., *Centranthus, Valeriana*) and edible, medicinal (e.g., *Valeriana officinalis*, garden valerian), or essential oil plants.

The Valerianaceae are distinctive in being herbs, rarely shrubs, with ***opposite*** leaves, a ***sympetalous, spurred*** corolla, 1–4 stamens, and a ***tricarpellate, inferior*** ovary with ***1 functional locule*** (2 locules abortive) and a ***single, apical ovule***, the fruit an ***achene***, with a ***pappuslike calyx*** in some members.

K 0–5 **C** (5) [(3,4)] **A** 1–4 **G** (3), inferior.

ASTERALES

The Asterales, sensu APG IV (2016), contain 11 families (Table 8.3), three of which are described here. Notable among the other families are the **Stylidiaceae** (Figure 8.130), the trigger plants, having an interesting "trigger" pollination mechanism (see Chapter 13). See Lundberg and Bremer (2003) and Soltis et al. (2007, 2011) for phylogenetic studies of the Asterales.

Asteraceae (Compositae)—Sunflower family (type *Aster*, meaning star). Ca. 1,900 genera/ca. 33,000 species (Figures 8.131–8.135).

The Asteraceae consist of herbs, shrubs, trees, or vines, with laticifers or resin ducts present in some taxa. The **leaves** are simple or compound, spiral or opposite [rarely whorled], exstipulate. The **inflorescence** consists of one or more heads (capitula) arranged in various secondary inflorescences, each head consisting of a flat to conical compound receptacle that bears one to many flowers (developing centripetally) and is subtended by one or more series of bracts, the **phyllaries** (collectively termed the **involucre**), any outer often reduced bracts known as the **calyculus**; heads of six general types: (1) **discoid**, with only disk flowers, all bisexual; (2) **disciform**, with only disk flowers, a mixture of pistillate and sterile with bisexual and staminate, in the same or different heads; (3) **radiate**, with central (bisexual or male) disk flowers and peripheral (female or sterile) ray flowers; (4) **radiant**, with only disk flowers, but those on the periphery expanded and often somewhat zygomorphic; (5) **ligulate**, with all ray flowers, these typically with 5-toothed corolla apices; and (6) **bilabiate**, with all bilabiate flowers. The **flowers** are epigynous, bisexual or unisexual, subtended in some taxa by bracts, known as **chaff** or **paleae**; if present head termed "paleate," if absent "epaleate." The **perianth** is biseriate or uniseriate with hypanthium absent. The **calyx**, known as the **pappus**, is modified as 2–∞ (sometimes connate) awns, scales, or capillary bristles (typically barbed or plumose), pappus absent in some taxa. The **corolla** is sympetalous with 5 [4] lobes (reduced to 3 marginal teeth in some), of three structural types (also called "flower types"): (1) **bilabiate**, corolla zygomorphic with a short tube having upper and lower lips; (2) **disk**, corolla actinomorphic with short to elongate tube bearing 5 [4] teethlike or elongate lobes; or (3) **ray** or **ligulate**, corolla zygomorphic with generally short tube having elongate, flat, extension bearing 3–5 apical teeth. The **stamens** are 5 [4], whorled, alternipetalous, usually syngenesious, the anthers fused into a tube through which the style grows. **Anthers** are basifixed, with apical extensions and sometimes basal lobes, longitudinal and introrse in dehiscence. The **gynoecium** is syncarpous, with an inferior ovary, 2 carpels, and 1 locule. The **style** is solitary and apically two-branched; **stigmas** are two, occurring as stigmatic lines on the adaxial surface of style branches. **Placentation** is basal; **ovules** are anatropous, unitegmic, 1 per ovary. **Nectaries** are usually present ovary apex. The **fruit** is an achene (or "cypsela," an achene derived from an inferior ovary), typically in a multiple fruit, an elongate beak forming between fruit and pappus in some. The **seeds** are exalbuminous.

The Asteraceae are now considered to be the largest of the angiosperms in terms of species diversity. It has recently been classified into at least twelve subfamilies (Panero and Funk, 2008; Panero et al., 2014). Members of the family have a worldwide distribution. Economic importance includes food plants (e.g., *Cynara scolymus*, artichoke, *Helianthus annuus*, sunflower, and *Lactuca sativa*, lettuce), a number of

Figure 8.130 ASTERALES. Stylidiaceae. *Stylidium* sp., triggerplant. **A.** Plant. **B.** Inflorescence. **C.** Demonstration of "trigger" mechanism, in which an insect visitation (simulated by pencil here), causes stamen to catapult forward, dusting insect with pollen.

Figure 8.131 ASTERALES. Asteraceae. **A–C.** *Gazania* sp., head morphology. **A.** Radiate head, face view. **B.** Head, side view, showing involucre. **C.** Head, longitudinal section, showing compound receptacle, central disk flowers, and peripheral ray flowers. **D–G.** *Encelia californica*. **D.** Disk flower, showing inferior ovary and corolla. **E.** Flower apex. Note two-branched style and subtending anthers. **F,G.** Dissected flower, showing syngenesious androecium, with central style. **H–J.** Disk flower morphology. **H.** *Glebionis coronaria*. **I.** *Bahiopsis laciniata*. **J.** *Carduus pycnocephalus*. **K.** Bilabiate flower morphology, *Trixis californica*. **L–N.** Ray flower morphology, *Sonchus oleraceus*.

Figure 8.132 ASTERALES. Asteraceae. **A–D.** Involucre morphology. **A.** Imbricate, multiseriate, isomorphic phyllaries in *Encelia californica*. **B.** Imbricate, multiseriate, dimorphic phyllaries in *Ursinia* sp. **C.** Mostly uniseriate phyllaries (but with tiny outer bracts below) in *Senecio vulgaris*. **D.** Decussate phyllaries in *Jaumea carnosa*. **E–G.** Secondary inflorescence types. **E.** Glomerule of heads in *Carduus pycnocephalus* (also having spine-tipped phyllaries). **F.** Corymb of heads in *Achillea* sp. **G.** Secund, narrow panicle of heads in *Solidago pinetorum*. **H–J.** Radiate heads, having inner disk flowers and peripheral ray flowers. **H.** *Aster* sp. **I.** *Layia platyglossa*. **J.** *Tithonia rotundifolia*. **K–M.** Ligulate heads, with all ray flowers. **K.** *Malacothrix californica*. **L.** *Cichorium intybus*, chicory. **M.** *Rafinesquia neomexicana*.

Figure 8.133 ASTERALES. Asteraceae. **A,B.** Bilabiate heads, i.e., having bilabiate flowers. **A.** *Polyachyrus* sp. **B.** *Acourtia microcephala*. **C–E.** Discoid heads, with all disk flowers. **C.** *Psathyrotes ramosissima*. **D.** *Cirsium vulgare*, a thistle. **E.** *Carduus pycnocephalus*. **F,G.** Disciform heads, with inner staminate or bisexual and outer pistillate flowers, *Cotula coronopifolia*, brass buttons. Note reduced, 4-merous corollas and androecium. **H.** Discoid head, *Palafoxia arida*. **I.** Radiant head, with enlarged, somewhat zygomorphic, peripheral disk flowers. **J–L.** Disciform heads, having unisexual flowers. **J.** *Ambrosia chamissonis*, male (staminate) heads above, female (pistillate) below. **K,L.** *Ambrosia salsola*. **K.** Male heads. **L.** Female heads.

Figure 8.134 ASTERALES. Asteraceae. **A–H.** Pappus morphology. **A.** Pappus absent, *Glebionis coronaria*. **B.** Pappus of capillary bristles, *Carduus pycnocephalus*. **C.** Capillary bristles antrorsely barbellate, *Isocoma menziesii*. **D.** Fruit, with beak and distal pappus of capillary bristles, *Lactuca serriola*. **E.** Pappus of plumose capillary bristles, *Cirsium mohavense*. **F.** Pappus of awns, *Palafoxia arida*. **G.** Pappus of two awns and several scales, *Bahiopsis laciniata*. **H.** Pappus of flat awns, *Uropappus lindleyi*. **I.** Fruits (achenes), *Sonchus oleraceus*. **J.** Bur fruits, achenes with a spiny involucre as accessory tissue, *Xanthium strumarium*. **K.** Compound receptacle bearing numerous bristles, *Carduus pycnocephalus*. **L,M.** Heads with chaff subtending flowers. **L.** *Bahiopsis laciniata*. **M.** *Encelia californica*.

Figure 8.135 ASTERALES. Asteraceae. *Argyroxiphium sandwicense*, silversword. **A.** Whole plant at time of flowering. **B.** Close-up of basal leaves, showing silver, UV-reflectant, sericeous trichome layer. **C.** Close-up of head (capitulum), characteristic of the Asteraceae.

ornamental cultivars, and various species used locally or industrially; the prickly fruits of *Arctium lappa* (burdock) are purported to have been the model for invention of velcro. See Bremer (1994, 1996), Panero and Funk (2002, 2008), and Panero et al. (2014) for more detailed information on relationships in the Asteraceae.

The Asteraceae are distinctive in being herbs, shrubs, vines, or trees, the inflorescence a ***head*** (***capitulum***) subtended by an ***involucre*** of ***phyllaries***, flowers either ***bilabiate***, ***disk***, or ***ray/ligulate*** (heads of many taxa a mixture of central disk flowers and peripheral ray flowers), with the calyx, termed a ***pappus***, modified as scales, awns, or capillary bristles (or absent), the androecium ***syngenesious***, and with an ***inferior ovary*** with a ***single, basal ovule***, the fruit a multiple of ***achenes***.

K 0-∞ (pappus) **C** (5) [(4)] or (3) in some ray flowers **A** (5) [(4)] **G** (2), inferior.

Campanulaceae — Bluebell family (type *Campanula*, after Latin *campana*, bell, after the corolla shape). Ca. 84 genera /ca. 2,380 species (Figure 8.136).

The Campanulaceae consist of hermaphroditic [dioecious], herbs [shrubs, trees]. The **stems** are rarely tuberous, trees often pachycaulous. The **leaves** are simple, spiral [opposite, whorled], exstipulate. The **inflorescence** unit is a raceme or cyme [rarely epiphyllous]. The **flowers** are bisexual [unisexual], actinomorphic or zygomorphic, resupinate or not, epigynous [perigynous]. The **perianth** is dichlamydeous. The **calyx** is synsepalous with 5 [3–10], imbricate or valvate, persistent lobes. The **corolla** is sympetalous [apopetalous], with 5 [3–10] petals/lobes, bilabiate in Lobelioideae. The **stamens** are 5, whorled, alternipetalous, connivent or connate with a staminal tube, epipetalous or not. **Anthers** are longitudinally dehiscent. The **gynoecium** is syncarpous, with an inferior [superior] ovary, 2–5 carpels, and 1–10 locules. **Placentation**

Figure 8.136 ASTERALES. Campanulaeae. **A.** *Adenophora* sp., ladybells. **B,C.** *Brighamia insignis*, olulu. **D**. *Campanula* sp., bluebells. **E–G.** *Campanula muralis*. **E–F.** Flower with corolla partially removed, showing inferior ovary and basally fused fused and dilated filaments. **G.** Ovary cross section, **H.** *Canarina canariensis*, canary bellflower. **I.** *Downingia concolor*, flower in face view, showing bilabiate corolla. **J–M.** *Lobelia eriantha*. **J.** Flower, oblique view. Note bilabiate corolla. **K.** Flower longitudinal section, showing inferior ovary and staminal tube terminating in connate anthers. **L.** Close-up of apex of staminal tube. Note protruding stigma. **M.** Longitudinal section of staminal tube, showing connate anthers, style, and terminal stigma. **N–O.** *Lobelia cardinalis*, cardinal flower. **P–Q.** *Nemacladus* spp., threadplants.

is axile [parietal]; **ovules** are anatropous, unitegmic, numerous per carpel. **Nectaries** are present in some taxa as a nectariferous disk at ovary apex. The **fruit** is a berry or capsule. The **seeds** are oily endospermous.

The Campanulaceae have a worldwide distribution. Economic importance includes local medicinal uses and cultivated ornamentals (e.g., *Adenophora, Campanula, Lobelia*). The remarkable giant lobelias of montane Africa are purported to have evolved from herbaceous ancestors. See Eddie et al. (2003), Roquet et al. (2008), and Crowl et al. (2014, 2016) for phylogenetic and biogeographic studies of the family.

The Campanulaceae are distinctive in being herbs, less often shrubs or trees, the flowers ***actinomorphic or zygomorphic***, perianth/androecium 5 [3–10], the stamens ***connivent or connate***, the ovary ***inferior*** [rarely superior] with ***2–5 carpels***, the fruit a ***berry or capsule***.

K (5) [(3–10)] **C** (5) [(3–10)] **A** (5) [(3–10)] **G** (2–5), inferior [superior].

Goodeniaceae — Goodenia family. (type *Goodenia*, after Samuel *Gooden*ough, English bishop and botanical writer) Ca. 12 genera/ca. 430 species (Figure 8.137).

The Goodeniaceae consist of shrubs, herbs, or rarely trees. The **leaves** are simple, alternate, spiral [rarely opposite or whorled], exstipulate. The **inflorescence** is of solitary flowers or a head, raceme, or cyme. **Flowers** are bisexual, zygomorphic, epigynous [to hypogynous]. The **perianth** is dichlamydeous. The **calyx** is synsepalous with 5 [3] lobes, reduced in some taxa. The **corolla** is valvate, sympetalous with 5 lobes, bilabiate or unilabiate with the 5 lobes anterior. The **stamens** are 5, alternipetalous, epipetalous or free. **Anthers** are connivent or connate, forming a tube, longitudinal introrse in dehiscence. The **gynoecium** is syncarpous, with an inferior [rarely half-inferior or superior] ovary, 2 carpels, and 2 [rarely 1 or 4] locules. The **style** is solitary, growing through the tube formed by the anthers, the style having, below the stigma, a cupular indusium (having marginal hairs) that collects pollen and presents to a visting insect. **Placentation** is axile; **ovules** are anatropous, unitegmic, 1–∞ per carpel. Intrastaminal **nectaries** are present in some taxa. The **fruit** is a capsule, rarely a nut or drupe. The **seeds** are flat, winged in some taxa, and oily endospermous.

The Goodeniaceae are mostly Australian in distribution. Economic importance includes some taxa used as ornamental cultivars. See Carolin et al. (2006) for general information on the family and Jabaily et al. (2012, 2014) and Gardner et al. (2016) for phylogenetic and biogeographic analyses.

The Goodeniaceae are distinctive in being herbs, shrubs, rarely trees, the flowers ***5-merous***, with sympetalous, ***bilabiate or often unilabiate (with the 5 lobes anterior)*** corollas, ***stamens forming a tube***, the style ***growing through the connivent or connate anthers*** with a ***cupular indusium*** that collects pollen, and a ***usually inferior ovary***, the fruit a capsule, rarely a drupe or nut.

K (5) [(3)] **C** (5) **A** (5) or 5 **G** (2), inferior [rarely half-inferior or superior].

Figure 8.137 ASTERALES. Goodeniaceae. **A–G.** *Scaevola* spp. **A.** Flower, face view, showing unilabiate corolla and style along cleft of corolla tube. **B.** Close-up of style, with cupular indusium. Note lower stamens. **C.** Bud stage, showing style during elongation between stamens, the cupular indusium collecting pollen grains. **D.** Flower base in longitudinal section, showing inferior ovary. **E.** Ovary cross section, showing two carpels and locules. **F.** Fruits (berries). **G.** Flower, face view. **H.** *Dampiera* sp.

REVIEW QUESTIONS

GENERAL FEATURES AND EUDICOT ORDERS RANUNCULALES, PROTEALES, GUNNERALES, AND DILLENIALES

1. Name and describe the major apomorphy of the eudicots.
2. Name three families in the order Ranunculales.
3. What is distinctive about the perianth of the Berberidaceae?
4. What is the corolla cycly and placentation of the Papaveraceae?
5. What economically important member of the Papaveraceae has shaped human history?
6. What is the etymology of the root name for *Ranunculus*, the type genus of Ranunculaceae?
7. What is the gynoecial fusion of the Ranunculaceae?
8. Name three families of the Proteales.
9. For the Nelumbonaceae what is the family common name, plant habitat, leaf base/shape, floral formula, and placentation?
10. What is the fruit type of the Nelumbonaceae and what accessory tissue is part of this fruit?
11. What is the single genus and common name of the Platanaceae?
12. What is the stipule type, bud type, inflorescence type, flower sex, and fruit type of the Platanaceae?
13. How are the Nelumbonaceae and Platanaceae similar with regard to placentation?
14. For the Proteaceae, name the perianth cycly, perianth merosity, stamen number, ovary position, and placentation.
15. In what two regions of the world are most Proteaceae found?
16. What is distinctive about the Gunneraceae in terms of habit, root symbiosis, leaf morphology, and fruit type?
17. How are the Dilleniales/Dilleniaceae distinctive with regard to stamen development and gynoecial fusion?

SUPERROSIDS

18. What is the leaf arrangement and typical leaf texture of the Crassulaceae?
19. For the Crassulaceae, give the photosynthetic mechanism, inflorescence type, perianth cycly, gynoecium fusion, and fruit type.
20. Name a few common cultivars of the Crassulaceae.
21. What is the plant habit, trichome type, flower sex, and fruit type of the Hamamelidaceae?
22. For the Saxifragaceae, what is the leaf arrangement, and what is the variation of gynoecial fusion and ovary position?
23. Name the diagnostic features of the Vitaceae with respect to plant habit, stem morphology, perianth morphology, and fruit type.
24. What is the most economically important member of the Vitaceae (common and scientific name)?

ROSIDS: FABIDS

25. What are some common features of members of the Rosids?
26. Name 4 families of great economic or ecological importance in the Fabids.
27. What family are hemiparasites and have flowers with lipid-secreting anterior petals (functioning in pollination) and prickly fruits?
28. For the Zygophyllaceae name two diagnostic features and three economically or ecologically important members.
29. Name at least five families of the large order Malpighiales.
30. What is the plant sex of members of the Euphorbiaceae?
31. What two groups (typically treated as subfamilies) of the Euphorbiaceae yield a latex?
32. What is the typical carpel and locule number and inflorescence type in the Euphorbiaceae?
33. What is the highly specialized inflorescence type found in many Euphorbioideae?
34. Name three (scientific and common name) economically important members of the Euphorbiaceae.
35. Which Euphorbiaceae are important members of plant communities in southern Africa?
36. Characterize the Hypericaceae with respect to leaf structure and stamen number, development, and fusion.
37. What is distinctive about the Malpighiaceae with respect to leaf morphology, calyx structure, petal structure, and stamen structure.
38. Name two economically important members of the Malpighiaceae.
39. What is the common name of the Passifloraceae?

40. What two floral features are distinctive for the Passifloraceae?
41. What is the common name and medicinal importance of the Salicaceae?
42. How are the Salicaceae characteristic with respect to plant sex, leaf tooth morphology, and placentation?
43. What is the typical floral formula of the Violaceae?
44. Members of the Violaceae with zygomorphic flowers may have the anterior petal modified as what structure and function?
45. What is the anther orientation and placentation of the Violaceae?
46. What are the diagnostic features of the Oxalidaceae?
47. What is the alternate, traditional/classical name for the Fabaceae family?
48. What is distinctive about the typical gynoecial fusion, carpel number, placentation, and fruit type of the Fabaceae?
49. How many subfamilies of the Fabaceae are accepted by the Legume Phylogeny Working Group?
50. Within what subfamily are the "Mimosoids" nested? How do they generally differ from other members of that subfamily?
51. Describe in detail a papilionaceous flower and indicate the subfamily (giving both acceptable names) having this type of flower.
52. Name several economically important members of the Fabaceae.
53. What symbiotic bacterium is found in root nodules of many members of the Fabaceae?
54. What is distinctive about the calyx of the Polygalaceae, and what other family/subfamily do the flowers superficially resemble?
55. What is unusual about the roots in some members of the family Rhamnaceae?
56. What is the significance of *Ziziphus spina-christi*?
57. What is the characteristic perianth/androecial position and stamen position of the Rhamnaceae?
58. What are the three, currently recognized subfamilies of the Rosaceae and how do they differ?
59. Name several economically important members (scientific and common names) of the Rosaceae.
60. What four families are "Urticalean Rosids"?
61. Name two economically important members of the Cannabaceae (scientific and common names).
62. Name three diagnostic features of the Moraceae.
63. What are three economically important members of the Moraceae?
64. For the Ulmaceae name the family common name, plant habit, leaf arrangement, venation structure, and fruit type.
65. Name the following for the Urticaceae: common name, plant sex, pollination mechanism, perianth cycly, stamen posture and pollen transfer mechanism, ovary structure, and ovule number, position, and type.
66. What is unusual about the trichome structure and internal anatomy of some members of the Urticaceae?
67. What is the plant sex of the Begoniaceae? How do male and female flowers differ in perianth structure?
68. What is the leaf base shape, ovary position, and ovary/fruit shape of the Begoniaceae?
69. What is the typical plant habit, plant sex, and leaf morphology of the Cucurbitaceae.
70. Name the typical ovary position, perianth/androecial position, carpel number, and placentation of the Cucurbitaceae.
71. Name three economically important members of the Cucurbitaceae.
72. What is the common name, plant sex, inflorescence type, and fruit type of the Betulaceae?
73. How does the Fagaceae differ from the Betulaceae?
74. For the Fagaceae what is distinctive about the (a) male inflorescence? (b) fruit accessory part?
75. What are the plant sex, male inflorescence type, ovary position, and fruit structure of the Juglandaceae.
76. Name two economically important members of the Juglandaceae.

ROSIDS: MALVIDS

77. Name a few features of the Crossosomataceae.
78. In the Geraniaceae, what is distinctive about the nectary position?
79. What is distinctive about the fruit type of the tribe Geranieae of the Geraniaceae?
80. For the Lythraceae, what is distinctive about the leaf structure, calyx aestivation, and corolla aestivation?
81. What is distinctive about the leaf arrangement and venation in the Melastomataceae?
82. What is the ovary position, anther dehiscence, and anther connective form of the Melastomataceae?
83. How can the Myrtaceae often be recognized with respect to leaf structure?
84. What is the perianth/androecial position and stamen number of the Myrtaceae?
85. Name two spices, one fruit tree, and an important timber/pulp genus of the Myrtaceae.
86. What is the common name of the Onagraceae family?

87. Name the typical floral formula (including ovary position) of the Onagraceae.
88. What is an apomorphy for most members of the Brassicales?
89. What is the alternate, traditional/classical name for the Brassicaceae family? The common name?
90. What is the typical corolla type for the Brassicaceae? Stamen arrangement?
91. Give the typical floral formula for the Brassicaceae.
92. How do the Brassicaceae differ from the Capparaceae and Cleomaceae with in placentation and fruit morphology?
93. Name several economically important members of the Brassicaceae (scientific and common names).
94. How are the Tropaeolaceae distinctive with respect to plant habit and perianth structure?
95. What is the floral formula of the Tropaeolaceae?
96. Name four putative apomorphies for the Malvaceae, as accepted today.
97. What former three families are now included as part of the family Malvaceae?
98. For the Malvoideae and Bombacoideae together, name the (a) stamen fusion; (b) anther type.
99. What is the name of the specialized bracts that subtend the calyx in many Malvaceae?
100. What are the common names of *Gossypium* spp., *Theobroma cacao*, and *Cola nitida*?
101. Name the common name of and several economically important members of the family Anacardiaceae.
102. How is the Anacardiaceae distinctive with regard to (a) nectaries; (b) anatomy?
103. Name several economically important members of the family Rutaceae.
104. How is the Rutaceae distinctive with regard to (a) nectaries; (b) glandular secretions?
105. How are many Sapindaceae distinctive with regard to (a) leaf morphology; (b) nectaries; (c) chemistry?
106. What former family is now included within the Sapindaceae?

SUPERASTERIDS

107. What is the general plant nutritional syndrome found in the order Santalales.
108. What is the term for the root type of the Viscaceae?
109. Name an economically important member of the Viscaceae.
110. What carnivorous plant families are found in the Caryophyllales?
111. What are the two major (sister) groups of the Caryophyllales?
112. Name the apomorphies for the terminal members (including ca. 24 families) of the Core Caryophyllales.
113. What are betalains and what is their function?
114. For the Aizoaceae, what is the leaf texture and typical leaf arrangement?
115. What is the perianth cycly of the Aizoaceae, and what structures appear like petals in many members?
116. What types of photosynthesis are found in many Aizoaceae?
117. What is distinctive about the perianth cycly, locule number, and placentation of the Amaranthaceae?
118. How do the Amaranthaceae and Chenopodiaceae differ?
119. For the Cactaceae, what is the plant habit, stem texture, and geographic distribution?
120. What are the specialized axillary meristems of cacti termed, and what do these produce?
121. What is unusual about the perianth of the Cactaceae?
122. What type of photosynthesis is found in cacti and what is its physiological significance?
123. Give the floral formula, perianth/androecial position, and placentation for the Cactaceae.
124. What is the common name of the Caryophyllaceae, and what genus denotes this common name?
125. For the Caryophyllaceae, what is distinctive about stem nodes, leaf arrangement, corolla (petal) type, and placentation?
126. What is unusual about the plant nutrition of the Droseraceae?
127. Name and describe the two leaf structural types of the Droseraceae.
128. What is the common name and "claim to fame" of *Dionaea muscipula*?
129. What is distinctive about the perianth cycly, carpel number, ovule position and type, and fruit type of the Nyctaginaceae?
130. For the Plumbaginaceae what is the carpel number, locule number, and ovule position and type?
131. What is the common name of the Polygonaceae, and what species denotes this common name?
132. What two perianth morphologies occur in the Polygonaceae? Which of these is now considered to be ancestral?
133. What is the name of the distinctive stipular structures found in some Polygonaceae?
134. What is the ovule position and type of the Polygonaceae?

ASTERIDS

135. Name two apomorphies of the Asterids and a third common feature.
136. What is the common name of the Cornaceae, and how is the family distinctive with regard to (a) leaf arrangement; (b) inflorescence types; (c) ovary position; and (d) fruit type?
137. How are the Loasaceae distinctive with respect to trichome anatomy, calyx duration, and placentation?
138. Name three families of the Ericales of economic importance.
139. How is the Balsaminaceae distinctive with respect to calyx morphology, stamen fusion/structure, anther anatomy, and fruit type?
140. What is the common name of the Ericaceae, and how are many family members distinctive with regard to (a) associated soil chemistry; (b) leaf morphology; and (c) leaf duration?
141. What is the anther dehiscence of many Ericaceae and how do these anthers develop?
142. Name some economically important members of the Ericaceae.
143. What is the common name of the Fouquieriaceae, and what is the (a) perianth cycly; (b) corolla fusion; (c) carpel number?
144. Describe the shoot and leaf morphology of the Fouquieriaceae. How are these adaptive?
145. How is the Myrsinaceae distinctive with respect to leaf structure, stamen arrangement, and placentation?
146. What is the common name and floral formula of the Polemoniaceae?
147. What is the common name of the Primulaceae, and how are they similar to the Myrsinaceae?
148. What is the specialized leaf of the Sarraceniaceae termed? Describe its structure and function.
149. Name the three genera of the Sarraceniaceae, their distribution, and how they differ in leaf and stigma morphology.
150. How are the Sarraceniaceae characterized with respect to flower orientation, calyx/corolla duration, and placentation?
151. What is the common name of the Theaceae and the scientific name of its most economically important member?
152. Characterize the Theaceae with respect to plant habit, leaf type/arrangement/duration, and stamen number/development.

ASTERIDS: LAMIIDS

153. What is distinctive about the Garryaceae with regard to plant sex, inflorescence type/orientation, and perianth cycly?
154. How do recent systems of classification of the family composition of the Boraginales differ from that of APG IV (2016)?
155. Describe the gynoecial morphology (including carpel, locule, and ovule number), style position, and fruit type for the Boraginaceae.
156. How does the Hydrophyllaceae resemble the Boraginaceae in corolla symmetry, stamen merosity, ovary position, and carpel number?
157. How is the Hydrophyllaceae different from the Boraginaceae in ovary shape, style position, and fruit type?
158. Name one or more diagnostic features of the Cordiaceae and Heliotropiaceae.
159. Which group (family) of the Boraginales is achlorophyllous/parasitic?
160. What class of chemical compounds are found in the tissues of the Apocynaceae?
161. What is unusual about the gynoecial fusion in many members of the Apocynaceae?
162. Describe the distinctive androecium and pollen fusion type found in the Asclepiadoids (milkweeds).
163. Name two medicinally important members of the Apocynaceae, including the compounds used and diseases these are used to treat.
164. Review the diagnostic features of the gentian family (Gentianceae), noting the floral nectaries and glands.
165. Describe the leaf arrangement, stipular morphology, and ovary position of the Rubiaceae.
166. Name economically important members of the Rubiaceae with respect to uses as a medicine, beverage, and sexual stimulant.
167. For the Acanthaceae, describe the leaf arrangement, flower symmetry, and modified funiculus.
168. For the Bignoniaceae, describe the plant habit, leaf arrangement, flower symmetry, and seed morphology and nutrition.
169. How can the Lamiaceae be recognized with respect to (a) plant chemistry; (b) stem shape; (c) leaf type and arrangement; (d) corolla type; (e) style position (in most); and (f) fruit type (in most)?
170. What is the common name of the Oleaceae and how are they distinctive in trichome structure, leaf arrangement, and stamen number?
171. What is distinctive about the Orobanchaceae with respect to plant nutrition and physiology?
172. What is the common name of the Phyrmaceae and why is it called that?

173. What relatively large, mostly western North American genus is now classified in the Phyrmaceae?
174. What traditionally defined families, or portions of these families, are now included within the (expanded) Plantaginaceae?
175. What are the general floral characteristics of the Scrophulariaceae?
176. For the Verbenaceae, what is the stem shape, leaf type/arrangement, corolla symmetry, style position, and fruit type?
177. How are the Convolvulaceae and Solanaceae similar and how different?
178. Name an important agricultural species in the Convolvulaceae.
179. For the Solanaceae, what is the (a) flower symmetry; (b) corolla aestivation (in bud); (c) stamen number; (d) ovary position; (e) carpel number; and (f) ovule number (per carpel)?
180. Name three members of the Solanaceae of great economic importance.

ASTERIDS: CAMPANULIDS

181. Give the common name and list two economically important members of the Aquifoliaceae.
182. What is the plant sex, leaf duration, and fruit type of the Aquifoliaceae?
183. Give the common name of the Apiaceae and list three economically important members of the family.
184. For the Apiaceae, what is the (a) leaf base; (b) inflorescence type; (c) ovary position; (d) fruit type?
185. How does the Araliaceae resemble the Apiaceae? How does it differ?
186. How are the fruits and seeds of the Pittosporaceae distinctive?
187. Name two alternative ways to classify the Dipsacales.
188. What is the common name of the Caprifoliaceae?
189. For the Caprifoliaceae, name the leaf arrangement, inflorescence type, and ovary position.
190. What is distinctive about the carpel number, locule number, and fruit type in the Dipsacaceae?
191. What is distinctive about the carpel number, locule number, and corolla type in the Valerianaceae?
192. What is the alternate traditional/classical name for the Asteraceae?
193. Name and describe the three corolla types of the Asteraceae.
194. Name and define the inflorescence unit of the Asteraceae. What are forms of this inflorescence and how do they differ?
195. Define (a) involucre; (b) phyllary; (c) chaff; (d) pappus.
196. For the Asteraceae, what is the (a) stamen fusion; (b) ovary position; (c) fruit type?
197. Name two economically important members of the Asteraceae used for food.
198. How are the Campanulaceae distinctive with regard to flower symmetry, stamen fusion, ovary position, and fruit type?
199. Name the diagnostic features of the Goodeniacae.

EXERCISES

1. Select a family of eudicots and learn everything you can about it. Perform a literature search (e.g., family name + "systematics" or "phylogeny") on journal articles published in the last five years. Consult family descriptions, recent data on phylogenetic relationships, and information on intrafamilial groupings.
2. From this same family, collect living material of an exemplar. Describe this species in detail, using the character list of Appendix 1 as a guide (see Chapter 9). Illustrate the vegetative and reproductive parts (see Appendix 2).
3. Assimilate all of your information in a written report and computerized slide show and present it.

REFERENCES FOR FURTHER STUDY

GENERAL ANGIOSPERM FAMILY CLASSIFICATION, DESCRIPTIONS, AND IMAGES

Christenhusz, M. J. M., M. F. Fay, and M. W. Chase. 2017. Plants of the World: An Illustrated Encyclopedia of Vascular Plant Families. Kew Publishing, Royal Botanic Gardens, Kew, and The University of Chicago Press.
Heywood, V. H., R. K. Brummitt, A. Culham, and O. Seberg. 2007. Flowering Plants of the World. Firefly Books, London.
Mabberley, D. J. 2008. Mabberley's Plant-Book: A Portable Dictionary of the Higher Plants, Their Classification and Uses. 3rd edition. Cambridge University Press, Cambridge..
Soltis, D., P. Soltis, P. Endress, M. W. Chase, S. Manchester, W. Judd, L. Majure, and E. Mavrodiev. 2018. Phylogeny and Evolution of the Angiosperms: Revised and Updated Edition. The University of Chicago Press, Chicago, London.

GENERAL REFERENCES ON ANGIOSPERM PHYLOGENY AND ON SPECIFIC EUDICOT GROUPS

Abbott, J. R. 2011. Notes on the disintegration of *Polygala* (Polygalaceae), with four new genera for the Flora of North America. Journal of the Botanical Research Institute of Texas 5: 125-137.

Acuña, R., S. Fliesswasser, M. Ackermann, T. Henning, F. Luebert, and M. Weigend. 2017. Phylogenetic relationships and generic re-arrangements in "South Andean loasas" (Loasaceae). Taxon 66: 385-378.

Aguilar-Ortigoza, C. J. and V. Sosa. 2004. The evolution of toxic phenolic compounds in a group of Anacardiaceae genera. Taxon 53: 357-364.

Al-Shehbaz, I. A., M. A. Beilstein, M. A., and E. A. Kellogg. 2006. Systematics and phylogeny of the Brassicaceae (Cruciferae): An overview. Plant Systematics and Evolution 259: 89-120

Albach, D. C., P. S. Soltis, D. E. Soltis, and R. G. Olmstead. 2001. Phylogenetic analysis of Asterids based on sequences of four genes. Annals of the Missouri Botanical Garden 88: 163–212.

Albach, D. C., H. M. Meudt, and B. Oxelman. 2005. Piecing together the "new" Plantaginaceae. American Journal of Botany 92: 297-315.

Anderberg, A. A. 2004. Primulaceae. In: Kubitzki, K. (ed.), The Families and Genera of Vascular Plants. VI. Flowering Plants. Dicotyledons. Celastrales, Oxalidales, Rosales, Cornales, Ericales. Pp. 313-319. Springer, Berlin.

Anderberg, A. A., U. Manns, and M. Källerjö. 2007. Phylogeny and floral evolution of the Lysimachieae (Ericales, Myrsinaceae): Evidence from ndhF sequence data. Willdenowia 37: 407-421.

Andersson, L., and S. Andersson. 2000. A molecular phylogeny of Tropaeolaceae and its systematic implications. Taxon 49: 721-736.

Angiosperm Phylogeny Group (APG). 1998. An ordinal classification for the families of flowering plants. Annals of the Missouri Botanical Garden 85: 531–553.

APG II. 2003. An update of the Angiosperm Phylogeny Group classification for the orders and families of flowering plants: APG II. Botanical Journal of the Linnean Society 141: 399–436.

APG III. 2009. An update of the Angiosperm Phylogeny Group classification for the orders and families of flowering plants: APG III. Botanical Journal of the Linnean Society 161: 105-121.

APG IV. 2016. An update of the Angiosperm Phylogeny Group classification for the orders and families of flowering plants: APG IV. Botanical Journal of the Linnean Society 181: 1-20.

Bailey, C. D., M. A. Koch, M. Mayer, K. Mummenhorf, S. L. Okane, Jr., S. I. Warwick, M. D. Windham, and I. A. Al-Shehbaz. 2006. Toward a global phylogeny of the Brassicaceae. Molecular Biology and Evolution 23: 2142-2160.

Ballard, H. E., J. de Paula-Souza, and G. A. Wahlert. 2013. Violaceae. Pp. 303-322 in K. Kubitzki (ed.), The Families and Genera of Flowering Plants. XI. Flowering Plants: Eudicots. Malpighiales. Springer, Berlin.

Barboza, G. E., A. T. Hunziker, G. Bernardello, A. A. Cocucci, C. Carrizo Garcia, V. Fuentes, M. O. Dillon, V. Bittrich, M. T. Cosa, R. Subils, A. Romanutti, S. Arroyo, and A. Anton. 2016. Solanaceae. Pp. in J. W. Kadereit and V. Bittrich (eds.), The Families and Genera of Vascular Plants, Volume 14: Flowering Plants: Eudicots - Aquifoliales, Boraginales, Bruniales, Dipsacales, Escalloniales, Garryales, Paracryphiales, Solanales (except Convolvulaceae), Icacinaceae, Metteniusaceae, Vahliaceae. Springer, Berlin.

Barker, W. R., G. l. Nesom, P. M. Beardsley, and N. S. Fraga. 2012. A taxonomic conspectus of Phrymaceae: A narrowed circumscription for Mimulus, new and resurrected genera, and new names and combinations. Phytoneuron 2012-39: 1–60.

Barkman, T. J., G. Chenery, J. R. McNeal, J. Lyons-Weiler, W. J. Ellisens, G. Moore, A. D. Wolfe, and C. W. dePamphilis. 2000. Independent and combined analyses of sequences from all three genomic compartments converge on the root of flowering plant phylogeny. Proceedings of the National Academy of Sciences of the United States of America 97: 13166–13171.

Bayer, C. 1999. The bicolor unit-homology and transformation of an inflorescence structure unique to core Malvales. Plant Systematics and Evolution 214: 187–198.

Bayer, C., M. F. Fay, A. Y. D. Bruijn, V. Savolainen, C. M. Morton, K. Kubitzki, W. S. Alverson, and M. W. Chase. 1999. Support for an expanded family concept of Malvaceae within a recircumscribed order Malvales: a combined analysis of plastid atpB and rbcL DNA sequences. Botanical Journal of the Linnaean Society 129:267–303.

Bayer, C., and O. Appel. 2002. Tropaeolaceae. In: Kubitzki, K. (ed.), The Families and Genera of Vascular Plants. V. Flowering Plants. Dicotyledons. Malvales, Capparales and Non-betalain Caryophyllales. Pp. 400-404. Springer, Berlin.

Bayer, R. J., L. Hufford, and D. E. Soltis. 1996. Phylogenetic relationships in Sarraceniaceae based on rbcL and ITS sequences. Systematic Botany 21: 121-134.

Bayer, R. J., D. J. Mabberley, C. Morton, C. H. Miller, I. K. Sharma, B. E. Pfeil, S. Rich, R. Hitchcock, and S. Syskes. 2009. A molecular phylogeny of the orange subfamily (Rutaceae: Aurantioideae) using nine cpDNA sequences. American Journal of Botany 96: 668-685.

Beardsley, P. M., and R. G. Olmstead. 2002. Redefining Phrymaceae: the placement of *Mimulus*, tribe Mimuleae, and Phryma. American Journal of Botany 89: 1093–1102.

Beardsley, P. M., S. Schoenig, J. B. Whittall, and R. G. Olmstead. 2004. Patterns of evolution in western North American Mimulus (Phrymaceae). American Journal of Botany 91: 474-489.

Beaulieu, J. M. and B. C. O'Meara. 2018. Can we build it? Yes we can, but should we use it? Assessing the quality and value of a very large phylogeny of campanulid angiosperms. American Journal of Botany 105: 417–432.

Behnke, H. D. 1972. Sieve-tube plastids in relation to Angiosperm systematics—an attempt towards a classification by ultrastructural analysis. Botanical Review 38: 155–197.

Beilstein, M. A., I. A. Al-Shehbaz, S. Mathews, and E. A. Kellogg. 2008. Brassicaceae phylogeny inferred from phytochrome A and ndhF sequence data: Tribes and trichomes revisited. American Journal of Botany 95: 1307-1327.

Bello, M. A., A. Bruneau, F. Forest, and J. A. Hawkins. 2009. Elusive relationships within order Fabales: phylogenetic analyses using matK and rbcL sequence data. Systematic Botany 34: 102-114.

Bello, M. A., P. J. Rudall, and J. A. Hawkins. 2012. Combined phylogenetic analyses reveal interfamilial relationships and patterns of floral evolution in the eudicot order Fabales. Cladistics 28: 393-421.

Biffin, E., M. G. Harrington, M. D. Crisp, L. A. Craven, and P. A. Gadek. 2007. Structural partitioning, paired-sites models and evolution of the ITS transcript in Syzygium and Myrtaceae. Molecular Phylogenetics and Evolution 43: 124-139.

Bittrich, V. and Kühn, U. 1993. Nyctaginaceae. In: Kubitzki, K., J. G. Rohwer, and V. Bittrich (eds.), The Families and Genera of Vascular Plants. II. Flowering Plants: Dicotyledons, Magnoliid, Hamamelid and Caryophyllid Families. Pp. 473-485. Springer, Berlin.
Borsch, T., K. W. Hilu, D. Quandt, V. Wilde, C. Neinhuis, and W. Barthlott. 2003. Noncoding plastid trnT-trnF sequences reveal a well resolved phylogeny of basal angiosperms. Journal of Evolutionary Biology 16: 558–576.
Bremer, B., K. Bremer, N. Heidari, P. Erixon, R. G. Olmstead, A. A. Anderberg, M. Källersjö, and E. Barkhordarian. 2002. Phylogenetics of asterids based on 3 coding and 3 non-coding chloroplast DNA markers and the utility of non-coding DNA at higher taxonomic levels. Molecular Phylogenetics and Evolution 24: 274–301.
Bremer, B. 2009. A review of molecular phylogenetic studies of Rubiaceae. Annals of the Missouri Botanical Garden 96: 4-26.
Bremer, B. and O. Eriksson. 2009. Time tree of Rubiaceae: Phylogeny and dating the family, subfamilies and tribes. International Journal of Plant Sciences 170: 766-793.
Bremer, K. 1994. Asteraceae: Cladistics and Classification. Timber Press, Portland, OR.
Bremer, K. 1996. Major clades and grades of the Asteraceae. In: Hind, D. J. N., and H. J. Beentje (eds.), Compositae: Systematics. Pp. 1–7. Proceedings of the International Compositae Conference, Kew. Royal Botanic Gadens, Kew.
Bremer, K., A. Backlund, B. Sennblad, U. Swenson, K. Andreasen, M. Hjertson, J. Lundberg, M. Backlund, and B. Bremer. 2001. A phylogenetic analysis of 100+ genera and 50+ families of euasterids based on morphological and molecular data with notes on possible higher level morphological synapomorphies. Plant Systematics and Evolution 229: 137–169.
Brockington, S. F., R. Alexandre, J. Ramdial, M. J. Moore, S. Crawley, A. Dhingra, K. Hilu, D. E. Soltis, and P. S. Soltis. 2009. Phylogeny of the Caryophyllales Sensu Lato: Revisiting Hypotheses on Pollination Biology and Perianth Differentiation in the Core Caryophyllales. International Journal of Plant Sciences 170:627–643.
Brockington, S. F., R. H. Walker, B. J. Glover, P. S. Soltis, and D. E. Soltis. 2011. Complex pigment evolution in the Caryophyllales. New Phytologist 190:854–864.
Brockington, S. F., Y. Yang, F. Gandia-Herrero, S. Covshoff, J. M. Hibberd, R. F. Sage, G. K. S. Wong, M. J. Moore, and S. A. Smith. 2015. Lineage-specific gene radiations underlie the evolution of novel betalain pigmentation in Caryophyllales. New Phytologist 207:1170-1180.
Buerki, S., F. Forest, P. Acevedo-Rodríguez, M. W. Callmander, J. A. A. Nylander, M. Harrington, I. Sanmartín, F. Küpfer, and N. Alvarez. 2009. Plastid and nuclear DNA markers reveal intricate relationships at subfamilial and tribal levels in the soapberry family (Sapindaceae). Molecular Phylogenetics and Evolution 51: 238-258.
Buerki, S., P. P. I. Lowry, N. Alvarez, S. G. Razafimandimbison, P. Küpfer, and M. W. Callmander. 2010. Phylogeny and circumscription of Sapindaceae revisited: Molecular sequence data, morphology, and biogeography support recognition of a new family, Xanthoceraceae. Plant Ecology and Evolution 143: 148-159.
Burke, J. M., A. Sanchez, K. A. Kron, and M. Luckow. 2010. Placing the woody tropical genera of Polygonaceae: A hypothesis of character evolution and phylogeny. American J. Bot. 97: 1377-1390. American Journal of Botany 97: 1377-1390.
Butterworth, C. A. 2006. Molecular phylogenetics of Cactaceae Jussieu - a review. In: Sharma A. K., & Sharma, A. (eds), Plant Genome Biodiversity and Evolution. Volume 1, Part C. Phanerogams (Angiosperm-Dicotyledons). Pp. 489-524. Science Publishers, Enfield.
Cameron, K. M., M. W. Chase, W. R. Anderson, and H. G. Hills. 2001. Molecular systematics of Malpighiaceae: Evidence from plastid rbcL and matK sequences. American Journal of Botany 88: 1847-1862.
Cameron, K. M., K. J. Wurdack, and R. W. Jobson. 2002. Molecular evidence for the common origin of snap-traps among carnivorous plants. American Journal of Botany 89: 1503-1509.
Cantino, P. D. 1992. Evidence for a polyphyletic origin of the Labiatae. Annals of the Missouri Botanical Garden 79: 361–379.
Cantino, P. D. 2004. Phrymaceae. In: Kadereit, J. (ed.), The Families and Genera of Vascular Plants. VII. Flowering Plants: Dicotyledons: Lamiales (except Acanthaceae including Avicenniaceae). Pp. 323–326. Springer, Berlin.
Cantino, P. D., J. A. Doyle, S. W. Graham, W. S. Judd, R. G. Olmstead, D. E. Soltis, P. S. Soltis, and M. J. Donoghue. 2007. Towards a phylogenetic nomenclature of Tracheophyta. Taxon 56: 822-846.
Carolin, R. C. 2006. Goodeniaceae. In: Kadereit, J. W. and C. Jeffrey (eds), The Families and Genera of Vascular Plants. Volume VIII. Flowering Plants. Eudicots. Asterales. Pp. 589-598. Springer, Berlin.
Cayzer, L. W., M. D. Crisp, and I. R. H. Telford. 2004. Cladistic analysis and revision of *Billardiera* (Pittosporaceae). Australian Systematic Botany 17: 83-125.
Chacón, J., F. Luebert, H. H. Hilger, S. Ovcinnikova, F. Selvi, L. Cecchi, C. M. Guilliams, K. Hasenstab-Lehman, K. Sutorý, M. G. Simpson, and M. Weigend. 2016. The borage family (Boraginaceae s.str.): A revised infrafamilial classification based on new phylogenetic evidence, with emphasis on the placement of some enigmatic genera. Taxon 65:523–546.
Chandler, G. T., and G. M. Plunkett. 2004. Evolution in Apiales: nuclear and chloroplast markers together in (almost) perfect harmony. Botanical Journal of the Linnean Society 144: 123–147.
Chandler, G. T., G. M. Plunkett, S. M. Pinney, L. W. Cayzer, and C. E. C. Gemmill. 2007. Molecular and morphological agreement in Pittosporaceae: Phylogenetic analysis with nuclear ITS and plastid trnL-trnF sequence data. Australian Systematic Botany 20: 390-401.
Chase, M. W., et al. 1993. Phylogenetics of seed plants: an analysis of nucleotide sequences from the plastid gene rbcL. Annals of the Missouri Botanical Garden 80: 528–580.
Chase, M. W., C. M. Morton, and J. A. Kallunki. 1999. Phylogenetic relationships of Rutaceae: a cladistic analysis of the subfamilies using evidence from rbcL and atpB sequence variation. American Journal of Botany 86: 1191–1199.
Chase, M. W., M. F. Fay, and V. Savolainen. 2000. Plant systematics: a half-century of progress (1950–2000) and future challenges—higher -level classification in the angiosperms: new insights from the perspective of DNA sequence data. Taxon 49: 685–704.
Chase, M. W., S. Zmarzty, M. D. Lledó, K. J. Wurdack, S. M. Swensen, and M. F. Fay. 2002. When in doubt, put it in Flacourtiaceae: A molecular phylogenetic analysis based on plastid L DNA sequences. Kew Bulletin 57: 141–181.
Clausing, G., and S. S. Renner. 2001. Molecular phylogenetics of Melastomataceae and Memecylaceae: implications for character evolution. American Journal of Botany 88: 486–498.
Clement, W. L., M. C. Tebbitt, L. L. Forrest, J. E. Blair, L. Brouillet, T. Eriksson, and S. M. Swensen. 2004. Phylogenetic position and biogeography of *Hillebrandia sandwicensis* (Begoniaceae): a rare Hawaiian relict. American Journal of Botany 91:905-917.

Clement, W. L. and G. D. Weiblen. 2009. Morphological evolution in the mulberry family (Moraceae). Systematic Botany 34: 530-552.
Cole, T. C. H. 2015. Angiosperm Phylogeny Group (APG) in jeopardy – Where have the flowers gone? PeerJ PrePrints 3:e1238v1
Conti, E., A. Litt, and K. J. Sytsma. 1996. Circumscription of Myrtales and their relationships to other Rosids: evidence from rbcL sequence data. American Journal of Botany 83: 221–233.
Conti, E., A. Litt, P. G. Wilson, S. A. Graham, B. G. Briggs, L. A. S. Johnson, and K. J. Sytsma. 1998. Interfamilial relationships in Myrtales: molecular phylogeny and patterns of morphological evolution. Systematic Botany 22: 629–647.
Craven, L. A. 2005. Malesian and Australian *Tournefortia* transferred to *Heliotropium* and notes on delimitation of Boraginaceae. Blumea 50: 375-381.
Crowl, A. A., E. Mavrodiev, G. Mansion, R. Haberle, A. Pistarino, G. Kamari, D. Phitos, T. Borsch, and N. Cellinese. 2014. Phylogeny of Campanuloideae (Campanulaceae) with emphasis on the utility of nuclear pentatricopeptide repeat (PPR) genes. PLoS ONE 9: e94199.
Crowl, A. A., N. W. Miles, C. J. Visger, K. Hansen, T. Ayers, R. Haberle, and N. Cellinese. 2016. A global perspective on Campanulaceae: Biogeographic, genomic and floral evolution. American Journal of Botany 103: 233-245.
Cuénoud, P., M. A. Del Pero Martinez, P.-A. Loizeau, R. Spichiger, S. Andrews, and J.-F. Manen. 2000. Molecular phylogeny and biogeography of the genus *Ilex* L. (Aquifoliaceae). Annals of Botany 85: 111-122.
Cuénoud, P., V. Savolainen, L. W. Chatrou, M. Powell, R. J. Grayer, and M. W. Chase. 2002. Molecular phylogenetics of Caryophyllales based on nuclear 18S rDNA and plastid rbcL, atpB, and matK DNA sequences. American Journal of Botany 89:132-144.
Datwyler, S. L. and G. Weiblen. 2004. On the origin of the fig: Phylogenetic relationships of Moraceae from ndhF sequences. American Journal of Botany 91: 767-777.
Davis, C. C., W. R. Anderson, and M. J. Donoghue. 2001. Phylogeny of Malpighiaceae: Evidence from chloroplast ndhF and trnL-F nucleotide sequences. American Journal of Botany 88: 1830-1846.
Davis, C. C. and W. R. Anderson. 2010. A complete generic phylogeny of Malpighiaceae inferred from nucleotide sequence data and morphology. American Journal of Botany 97: 2031-2046.
De-Nova, J. A., L. L. Sánchez-Reyes, L. E. Eguiarte, and S. Magallón. 2018. Recent radiation and dispersal of an ancient lineage: The case of Fouquieria (Fouquieriaceae, Ericales) in North American deserts. Molecular Phylogenetics and Evolution 126: 92-104.
Deng, J.-b., B. T. Drew, E. V. Mavrodiev, M. A. Gitzendanner, P. S. Soltis, and D. E. Soltis. 2015. Phylogeny, divergence times, and historical biogeography of the angiosperm family Saxifragaceae. Molecular Phylogenetics and Evolution 83: 86-98.
Der, J., and D. Nickrent. 2008. A molecular phylogeny of Santalaceae (Santalales). Systematic Botany 33: 107-116.
Devi, S. P., S. Kumaria, S. R. Rao, and P. Tandon. 2016. Carnivorous plants as a source of potent bioactive compound: Naphthoquinones. Tropical Plant Biology 9: 267-279.
Diane, N., H. H. Hilger, H. Förther, M. Weigend, and F. Luebert. 2016. Heliotropiaceae. Pp. 203-211 in J. W. Kadereit and V. Bittrich (eds.), The Families and Genera of Vascular Plants. Flowering Plants. Eudicots. 14. Springer International Publishing, Switzerland.
Donoghue, M. J., T. Eriksson, P. A. Reeves, and R. G. Olmstead. 2001. Phylogeny and phylogenetic taxonomy of Dipsacales, with special reference to Sinadoxa and Tetradoxa (Adoxaceae). Harvard Papers in Botany 6:459-479.
Douglas, N. A., and P. S. Manos. 2007. Molecular phylogeny of Nyctaginaceae: taxonomy, biogeography, and characters associated with a radiation of xerophytic genera in North America. American Journal of Botany 94: 856-872.
Douglas, N. and R. Spellenberg. 2010. A new tribal classification of Nyctaginaceae. Taxon 59: 905-910.
Eddie, W. M., T. Shulkina, J. Gaskin, R. C. Haberle, and R. K. Jansen. 2003. Phylogeny of Campanulaceae s. str. inferred from ITS sequences of nuclear ribosomal DNA. Annals of the Missouri Botanical Garden 90: 554-576.
Edger, P. P., J. C. Hall, A. Harkess, M. Tang, J. Coombs, S. Mohammadin, M. E. Schranz, Z. Xiong, J. Leebens-Mack, B. C. Meyers, K. J. Sytsma, M. A. Koch, I. A. Al-Shehbaz, and J. C. Pires. 2018. Brassicales phylogeny inferred from 72 plastid genes: A reanalysis of the phylogenetic localization of two paleopolyploid events and origin of novel chemical defenses. American Journal of Botany 105: 463–469.
Eiserhardt, W. L., A. Antonelli, D. J. Bennett, L. R. Botigué, J. G. Burleigh, S. Dodsworth, B. J. Enquist, F. Forest, J. T. Kim, A. M. Kozlov, I. J. Leitch, B. S. Maitner, S. Mirarab, W. H. Piel, O. A. Pérez-Escobar, L. Pokorny, C. Rahbek, B. Sandel, S. A. Smith, A. Stamatakis, R. A. Vos, T. Warnow, and W. J. Baker1. 2018. A roadmap for global synthesis of the plant tree of life. American Journal of Botany 105: 614-622.
Ellison, A. M., E. D. Butler, E. J. Hicks, R. F. C. Naczi, P. J. Calie, C. D. Bell, and C. C. Davis. 2012. Phylogeny and biogeography of the carnivorous plant family Sarraceniaceae. PLoS ONE 7: e39291.
Ellison, A. M. and L. Adamec. 2018. Carnivorous Plants: Physiology, Ecology, and Evolution. Oxford University Press, Oxford.
Endress, M. E., and P. V. Bruyns. 2000. A revised classification of the Apocynaceae s. l. Botanical Review 66: 1–56.
Eriksen, B., and C. Persson. 2006. Polygalaceae. In: Kubitzki, K. (ed.), The Families and Genera of Vascular Plants. IX. Flowering Plants. Eudicots. Berberidopsidales, Buxales, Crossosomatales, Fabales p.p., Geraniales, Gunnerales, Myrtales p.p., Proteales, Saxifragales, Vitales, Zygophyllales, Clusiaceae Alliance, Passifloraceae Alliance, Dilleniaceae, Huaceae, Picramniaceae, Sabiaceae Series. Pp. 345-363. Springer, Berlin.
Feng, C.-M., S. R. Manchester, and Q.-Y. J. Xiang. 2009. Phylogeny and biogeography of Alangiaceae (Cornales) inferred from DNA sequences, morphology, and fossils. Molecular Phylogenetics and Evolution 51: 201-214.
Feuillet, C., and J. M. Macdougal. 2006. Passifloraceae. In: Kubitzki, K. (ed.), The Families and Genera of Vascular Plants. IX. Flowering Plants. Eudicots. Berberidopsidales, Buxales, Crossosomatales, Fabales p.p., Geraniales, Gunnerales, Myrtales p.p., Proteales, Saxifragales, Vitales, Zygophyllales, Clusiaceae Alliance, Passifloraceae Alliance, Dilleniaceae, Huaceae, Picramniaceae, Sabiaceae Series. Pp. 270-281. Springer, Berlin.
Fior, S., P. O. Karis, G. Casazza, L. Minuto, and F. Sala. 2006. Molecular phylogeny of the Caryophyllaceae (Caryophyllales) inferred from chloroplast matK and nuclear rDNA ITS sequences. American Journal of Botany 93: 399-411.
Fischer, E. 2004. Scrophulariaceae. In: Kadereit, J. (ed.), The Families and Genera of Vascular Plants. VII. Flowering Plants. Dicotylendons. Lamiales (except Acanthaceae including Avicenniaceae). Pp. 333–432. Springer, Berlin.
Fishbein, M., C. Hibsch-Jetter, D. E. Soltis, and L. Hufford. 2001. Phylogeny of Saxifragales (Angiosperms, Eudicots): analysis of a rapid, ancient radiation. Systematic Biology 50: 817–847.

Fishbein, M., T. Livshultz, S. C. K. Straub, A. O. Simões, J. Boutte, A. McDonnell, and A. Foote. 2018. Evolution on the backbone: Apocynaceae phylogenomics and new perspectives on growth forms, flowers, and fruits. American Journal of Botany 105: 495-513.

Fiz, O., P. Vargas, M. L. Alarcón, M. Aedo, J. L. Garcia, and J. J. Aldasoro. 2008. Phylogeny and historical biogeography of Geraniaceae in relation to climate changes and pollination ecology. Systematic Botany 33: 326-342.

Folk, R. A., M. Sun, P. S. Soltis, S. A. Smith, D. E. Soltis, and R. P. Guralnick. 2018. Challenges of comprehensive taxon sampling in comparative biology: Wrestling with rosids. American Journal of Botany 105: 433–445.

Ford, V. S, and L. D. Gottlieb. 2007. Tribal relationships within Onagraceae inferred from PgiC sequences. Systematic Botany 32: 348-356.

Forest, F., V. Savolainen, M. W. Chase, R. Lupia, A. Bruneau, and P. R. Crane. 2005. Teasing apart molecular- versus fossil-based error estimates when dating phylogenetic trees: A case study in the birch family (Betulaceae). Systematic Botany 30: 118-133.

Forrest, L. L. and P. M. Hollingsworth. 2003. A recircumscription of *Begonia* based on nuclear ribosomal sequences. Plant Systematics and Evolution 241: 193-211.

Forrest, L. L., M. Hughes, and P. M. Hollingsworth. 2005. A phylogeny of *Begonia* using nuclear ribosomal sequence data and morphological characters. Systematic Botany 30: 671-682.

Friis, I. 1993. Urticaceae. In: Kubitzki, K., J. G. Rohwer, and V. Bittrich. (eds), The Families and Genera of Vascular Plants. II. Flowering Plants: Dicotyledons, Magnoliid, Hamamelid and Caryophyllid Families. Pp. 612-629. Springer, Berlin.

Frost, L. A., S. M. Tyson, P. Lu-Irving, N. O'Leary, and R. G. Olmstead. 2017. Origins of North American arid-land Verbenaceae: More than one way to skin a cat. American Journal of Botany 104: 1708-1716.

Gadek, P. A., E. S. Fernando, C. J. Quinn, S. B. Hoot, T. Terrazas, M. C. Sheahan, and M. W. Chase. 1996. Sapindales: molecular delimitation and infraordinal groups. American Journal of Botany 83: 802–811.

García, M. A., M. Costea, M. Kuzmina, and S. Stefanovic. 2014. Phylogeny, character evolution, and biogeography of *Cuscuta* (dodders; Convolvulaceae) inferred from coding plastid and nuclear sequences. American Journal of Botany 101: 670-690.

Gardner, A. G., E. B. Sessa, P. Michener, E. Johnson, K. A. Shepherd, D. G. Howarth, and R. S. Jabaily. 2016. Utilizing next-generation sequencing to resolve the backbone of the Core Goodeniaceae and inform future taxonomic and floral form studies. Molecular Phylogenetics and Evolution 94: 605-617.

Geuten, K., E. Smets, P. Schols, Y.-M. Yuan, S. Janssens, P. Küpfer, and N. Pyck. 2004. Conflicting phylogenies of balsamoid families and the polytomy in Ericales: Combining data in a Bayesian framework. Molecular Phylogenetics and Evolution 31: 711–729.

Gottschling, M., H. H. Hilger, M. Wolf, and N. Diane. 2001. Secondary structure of the ITS1 transcript and its application in a reconstruction of the phylogeny of Boraginales. Plant Biology 3: 629–636.

Gottschling, M., M. Weigend, and H. H. Hilger. 2016. Ehretiaceae. Pp. 165-178 in J. W. Kadereit and V. Bittrich (eds.), The Families and Genera of Vascular Plants. Flowering Plants. Eudicots. 14. Springer International Publishing, Switzerland.

Graham, S. A., J. Hall, K. Sytsma, and S.-H. Shi. 2005. Phylogenetic analysis of the Lythraceae based on four gene regions and morphology. International Journal of Plant Sciences 166: 995-1017.

Graham, S. A., M. Diazgranados, and J. C. Barber. 2011. Relationships among the confounding genera *Ammannia, Hionanthera, Nesaea*, and *Rotala* (Lythraceae). Botanical Journal of the Linnean Society 166: 1-19.

Graham, S. W., and R. G. Olmstead. 2000. Utility of 17 chloroplast genes for inferring the phylogeny of the basal angiosperms. American Journal of Botany 87: 1712–1730.

Green, P. S. 2004. Oleaceae. Pp. 296-306, in Kadereit, J. (ed)., The Families and Genera of Vascular Plants. VII. Flowering Plants. Dicotyledons. Lamiales (except Acanthaceae including Avicenniaceae). Springer, Berlin.

Greenberg, A. K. and M. J. Donoghue. 2011. Molecular systematics and character evolution in Caryophyllaceae. Taxon 60: 1637-1652.

Grimm, G. and S. S. Renner. 2013. Harvesting Betulaceae sequences from GenBank to generate a new chronogram for the family. Botanical Journal of the Linnean Society 172: 465-477.

Grimm, G. W., and T. Denk. 2008. ITS evolution in *Platanus* (Platanaceae): Homoeologues, pseudogenes and ancient hybridization. Annals of Botany 101: 403-419.

Groppo, M., J. R. Pirani, M. L. F. Salatino, S. R. Blanco, and J. A. Kallunki. 2008. Phylogeny of Rutaceae based on two non-coding regions from cpDNA. American Journal of Botany 95: 985-1005.

Groppo, M., J. A. Kallunki, J. R. Pirani, and A. Antonelli. 2012. Chilean *Pitavia* more closely related to Oceania and Old World Rutaceae than to Neotropical groups: Evidence from two cpDNA non-coding regions, with a new subfamilial classification of the family. PhytoKeys 19: 9-29.

Guerrero, P. C., L. C. Majure, A. Cornejo-Romero, and T. Hernández-Hernández. 2018. Phylogenetic relationships and evolutionary trends in the cactus family. Journal of Heredity 110: 4-21.

Hadiah, J. T., B. J. Conn, and C. J. Quinn. 2008. Infra-familial phylogeny of Urticaceae, using chloroplast sequence data. Australian Systematic Botany 21: 375-385.

Hall, J. C., K. J. Sytsma, and H. H. Iltis. 2002. Phylogeny of Capparaceae and Brassicaceae based on chloroplast sequence data. American Journal of Botany 89: 1826–1842.

Ham, R. C., H. J. van, and H. 't Hart. 1998. Phylogenetic relationships in the Crassulaceae inferred from chloroplast DNA restriction-site variation. American Journal of Botany 85: 123–134.

Hao, G., Y.-M. Yuan, C.-M. Hu, X.-J. Ge, and N.-X. Zhao. 2004. Molecular phylogeny of *Lysimachia* (Myrsinaceae) based on chloroplast trnL-F and nucclear ribosomal ITS sequences. Molecular Phylogenetics and Evolution 31: 323-339.

Harbaugh, D. T., M. Nepokroeff, R. K. Rabeler, J. Mc Neill, E. A. Zimmer, and W. L. Wagner. 2010. A new lineage-based tribal classification of the family Caryophyllaceae. Internat. J. Plant Sci. 171: 185-198. International Journal of Plant Sciences 171: 185-198.

Harley, R., S. Atkins, A. L. Budantsev, P. D. Cantino, B. J. Conn, M. Grayer, M. M. Harley, R. De Kok, T. Krestovskaya, R. Moralaes, A. J. Paton, O. Ryding, and T. Upson. 2004. Labiatae. In: Kadereit, J. (ed.), The Families and Genera of Vascular Plants. VII. Flowering Plants. Dicotyledons. Lamiales (except Acanthaceae including Avicenniaceae). Pp. 167–275. Springer, Berlin.

Harrington, M. G., K. J. Edwards, S. A. Johnson, M. W. Chase, and P. A. Gadek. 2005. Phylogenetic inference in Sapindaceae sensu lato using plastid matK and rbcL DNA sequences. Systematic Botany 30: 366-382.

Hauenschild, F., S. Matuszak, A. N. Muellner-Riehl, and A. Favre. 2016. Phylogenetic relationships within the cosmopolitan buckthorn family (Rhamnaceae) support the resurrection of Sarcomphalus and the description of Pseudoziziphus gen. nov. Taxon 65: 47-64.

Heibl, C. and S. S. Renner. 2012. Distribution models and a dated phylogeny for Chilean *Oxalis* species reveal occupation of new habitats by different lineages, not rapid adaptive radiation. Systematic Biology 61: 823-834.

Hernández-Hernández, T., H. M. Hernández, J. A. De-Nova, R. Puente, L. E. Eguiarte, and S. Magallón. 2011. Phylogenetic relationships and evolution of growth form in Cactaceae (Caryophyllales, Eudicotyledoneae). American Journal of Botany 98:44-61.

Hernández-Ledesma, P., W. G. Berendsohn, T. Borsch, S. V. Mering, H. Akhani, S. Arias, I. Castañeda-Noa, U. Eggli, R. Eriksson, H. Flores-Olvera, S. Fuentes-Bazán, G. Kadereit, C. Klak, N. Korotkova, R. Nyffeler, G. Ocampo, H. Ochoterena, B. Oxelman, R. K. Rabeler, A. Sanchez, B. O. Schlumpberger, and P. Uotila. 2015. A taxonomic backbone for the global synthesis of species diversity in the angiosperm order Caryophyllales. Willdenowia 45:281-383.

Hilu, K. W., Borsch, T., Muller, K., Soltis, D. E., Soltis, P. S., Savolainen, V., Chase, M. W., Powell, M. P., Alice, L. A., Evans, R., Sauquet, H., Neinhuis, C., Slotta, T. A. B., Rohwer, J. G., Campbell, C. S., and Chatrou, L. W. 2003. Angiosperm phylogeny based on matK sequence information. American Journal of Botany 90: 1758–1766.

Hofmann, M., G. K. Walden, H. H. Hilger, and M. Weigend. 2016. Hydrophyllaceae. Pp. 221-238 in J. W. Kadereit and V. Bittrich (eds.), The Families and Genera of Vascular Plants. Flowering Plants. Eudicots. 14. Springer International Publishing, Switzerland.

Hoot, S. B. 1991. Phylogeny of the Ranunculaceae based on epidermal characters and micromorphology. Systematic Botany 16: 741–755.

Hoot, S. B., and P. R., Crane. 1995. Inter-familial relationships in the Ranunculiidae based on molecular systematics. In: Jensen, U., and J. W. Kadereit (eds.), Systematics and Evolution of the Ranunculiflorae. Springer, Vienna. Pp. 119–131 [Plant Syst. Evol. Suppl. 9.]

Hoot, S. B. 1995. Phylogeny of the Ranunculaceae based on preliminary atpB, rbcL and 18S ribosomal DNA sequence data. In: Jensen, U., and J. W. Kadereit (eds.), Systematics and Evolution of the Ranunculiflorae. Pp. 241–251. Springer, Vienna. [Plant Syst. Evol. Suppl. 9.]

Hoot, S. B., J. W. Kadereit, F. R. Blattner, K. B. Jork, A. E. Schwarzbach, and P. R. Crane. 1997. Data congruence and phylogeny of the Papaveraceae s.l. based on four data sets: atpB and rbcL sequences, trnK restriction sites, and morphological characters. Systematic Botany 22:575–590.

Hoot, S. B., and A. W. Douglas. 1998. Phylogeny of the Proteaceae based on atpB and atpB-rbcL intergenic spacer regions. Australian Systematic Botany 11: 301–320.

Hoot, S. B., S. Magallon-Puebla, and P. R. Crane. 1999. Phylogeny of basal eudicots based on three molecular data sets: atpB, rbcL and 18S nuclear ribosomal DNA sequences. Annals of the Missouri Botanical Garden 86: 119–131.

Hufford, L., M. M. McMahon, A. M. Sherwood, G. Reeves, and M. W. Chase. 2003. The major clades of Loasaceae: Phylogenetic analysis using the plastid matK and trnL-trnF regions. American Journal of Botany 90: 1215-1228.

Jabaily, R. S., K. A. Shepherd, M. H. G. Gustafsson, L. W. Sage, S. L. Krauss, D. G. Howarth, and T. J. Motley. 2012. Systematics of the Austral-Pacific family Goodeniaceae: Establishing a taxonomic and evolutionary framework. Taxon 61: 419-436.

Jabaily, R. S., K. A. Shepherd, A. G. Gardner, M. H. G. Gustafsson, D. G. Howarth, and T. J. Motley. 2014. Historical biogeography of the predominantly Australian plant family Goodeniaceae. Journal of Biogeography 41: 2057-2067.

Jansen, R. K., Z. Cai, L. A. Raubeson, H. Daniell, C. W. dePamphilis, J. Leebens-Mack, K. F. Müller, M. Guisinger-Bellian, R. C. Haberle, A. K. Hansen, T. W. Chumley, S.-B. Lee, R. Peery, J. R. McNeal, J. V. Kuehl, and J. L. Boore. 2007. Analysis of 81 genes from 64 plastid genomes resolves relationships in angiosperms and identifies genome-scale evolutionary patterns. Proceedings of the National Academy of Sciences U.S.A. 104: 19369-19374.

Janssens, S. B., K. Geuten, Y. M. Yuan, Y. Song, P. Küpfer, and E. F. Smets. 2006. Phylogenetics of *Impatiens* and *Hydrocera* (Balsaminaceae) using chloroplast atpB-rbcL spacer sequences. Systematic Botany 31: 171-180.

Jeiter, J., M. Weigend, and H. H. Hilger. 2016. Geraniales flowers revisited: Evolutionary trends in floral nectaries. Annals of Botany 119:395–408.

Jian, S., P. S. Soltis, M. A. Gitzendanner, M. J. Moore, R. Li, T. A. Hendry, Y.-L.Qiu, A. Dhingra, C. Bell, and D. E. Soltis. 2008. Resolving an ancient, rapid radiation in Saxifragales. Systematic Biology 57: 38-57.

Johnson, L. A., L. M. Chan, T. L. Weese, L. D. Busby, and S. McMurry. 2008. Nuclear and cpDNA sequences combined provide strong inference of higher phylogenetic relationships in the phlox family (Polemoniaceae). Molecular Phylogenetics and Evolution 48: 997-1012.

Kadereit, G., T. Borsch, K. Weising, and H. Freitag. 2003. Phylogeny of Amaranthaceae and Chenopodiaceae and the evolution of C4 photosynthesis. International Journal of Plant Sciences 164:959-986.

Kadereit, G., E. V. Mavrodiev, E. H. Zacharias, and A. P. Sukhorukov. 2010. Molecular phylogeny of Atripliceae (Chenopodioideae, Chenopodiaceae): Implications for systematics, biogeography, flower and fruit evolution, and the origin of C4 photosynthesis. American Journal of Botany 97:1664–1687.

Kadereit, J. (ed.). 2004. The Families and Genera of Vascular Plants. VII. Flowering Plants. Dicotyledons. Lamiales (except Acanthaceae including Avicenniaceae). Springer, Berlin.

Kadereit, J. W., F. R. Blattner, K. B. Jork, and A. Schwarzbach. 1994. Phylogenetic analysis of the Papaveraceae s.l. (inc. Fumariaceae, Hypecoaceae, and Pteridophyllum) based on morphological characters. Botanische Jahrbücher für Systematik 116: 361–390.

Kadereit, J. W., F. R. Blattner, K. B. Jork, and A. Schwarzbach. 1995. The phylogeny of the Papaveraceae sensu lato: morphological, geographical, and ecological implications. In: Jensen, U., and J. W. Kadereit (eds.), Systematics and Evolution of the Ranunculiflorae. Pp. 133-145. Springer, Vienna. [Plant Syst. Evol. Suppl. 9.]

Källersjö, M., G. Bergqvist, and A. A. Anderberg. 2000. Generic realignment in primuloid families of the Ericales s.l.: A phylogenetic analysis based on DNA sequences from three chloroplast genes and morphology. American Journal of Botany 87: 1325-1341.

Kårehed, J. 2002. Not just hollies - the expansion of Aquifoliales. Pp. 1-14 Evolutionary Studies in Asterids Emphasising Euasterids II. Acta Universitatis Upsaliensis, Uppsala.

Kim, D.-K. and J.-H. Kim. 2011. Molecular phylogeny of the tribe Forsythieae (Oleaceae) based on nuclear ribosomal DNA internal transcribed spacers and plastid DNA trnL-F and matK gene sequences. J. Plant Res. 124: 339-347. Journal of Plant Research 124: 339-347.

Klak, C., A. Khunou, G. Reeves, and T. Hedderson. 2003. A phylogenetic hypothesis for the Aizoaceae (Caryophyllales) based on four plastid DNA regions. American Journal of Botany 90: 1433–1445.

Klak, C. 2010. Phylogeny and diversification of Aizoaceae: Progress and prospects. Schumannia 6, Biodiversity & Ecology 3: 87-107.

Kornhall, P., N. Heidari, and B. Bremer. 2001. Selagineae and Manuleeae, two tribes or one? Phylogenetic studies in the Scrophulariaceae. Plant Systematics and Evolution 288: 199-218.

Korotkova, N., J. V. Schneider, D. Quandt, A. Worberg, G. Zizka, and T. Borsch. 2009. Phylogeny of the eudicot order Malpighiales - analysis of a recalcitrant clade with sequences of the petD group II intron. Plant Systematics and Evolution 282: 201-228.

Kron, K. A., W. S. Judd, P. F. Stevens, D. M. Crayn, A. A. Anderberg, P. A. Gadek, C. J. Quinn, and J. L. Luteyn. 2002. A phylogenetic classification of Ericaceae: molecular and morphological evidence. Botanical Review 68: 335–423.

Kubitzki, K., Rohwer, J. G., and Bittrich, V. 1993. The Families and Genera of Vascular Plants. II. Flowering Plants: Dicotyledons, Magnoliid, Hamamelid and Caryophyllid Families. Springer-Verlag, Berlin.

Kubitzki, K. 1993a. Cecropiaceae. In: Kubitzki, K., J. G. Rohwer, and V. Bittrich (eds.), The Families and Genera of Vascular Plants. II. Flowering Plants: Dicotyledons, Magnoliid, Hamamelid and Caryophyllid Families. Pp. 243-245. Springer, Berlin.

Kubitzki, K. 1993b. Platanaceae. In: Kubitzki, K., J. G. Rohwer, and V. Bittrich (eds.). The Families and Genera of Vascular Plants. II. Flowering plants:dicotylendons, magnoliid, hamamelid, and caryophyllid families. Pp. 521–522. Springer-Verlag, Berlin, New York.

Kubitzki, K., and C. Bayer. 2002. The Families and Genera of Vascular Plants. V. Flowering Plants. Dicotyledons: Malvales, Capparales, and non-betalain Caryophyllales. Springer, Berlin.

Kubitzki, K., and M. W. Chase. 2002. Introduction to Malvales. In: K. Kubitzki and C. Bayer (eds), The Families and Genera of Vascular Plants. V. Flowering Plants. Dicotyledons. Malvales, Capparales and Non-betalain Caryophallales. Pp. 12–16. Springer, Berlin.

Kubitzki, K. 2004. The Families and Genera of Vascular Plants. VI. Flowering Plants. Dicotyledons. Ceslastrales, Oxalidales, Rosales, Cornales, Ericales. Springer, Berlin.

Lamb Frye, A. S., and K. A. Kron. 2003. RbcL phylogeny and character evolution in Polygonaceae. Systematic Botany 28:326–332.

Lane, A. K., M. M. Augustin, S. Ayyampalayam, A. Plant, S. Gleissberg, V. S. di Stilio, C. W. Depamphilis, G. K.-S. Wong, T. M. Kutchan, and J. H. Leebens-Mack. 2018. Phylogenomic analysis of Ranunculales resolves branching events across the order. Botanical Journal of the Linnean Society 187: 157-166.

Langstrom, E. and M. W. Chase. 2002. Tribes of Boraginoideae (Boraginaceae) and placement of *Antiphytum, Echiochilon, Ogastemma*, and *Sericostoma*: A phylogenetic analysis based on atpB plastid DNA sequence data. Plant Systematics and Evolution 234: 137-153.

Le Péchon, T. and L. D. B. Gigord. 2014. On the relevance of molecular tools for taxonomic revision in Malvales, Malvaceae s.l., and Dombeyoideae. Pp. 337-363 in P. Besse (ed.), Molecular Plant Taxonomy Methods and Protocols. Humana Press, Springer, New York.

Levin, R. A., W. L. Wagner, P. C. Hoch, M. Nepokroeff, J. C. Pires, E. A. Zimmer, and K. J. Sytsma. 2003. Family-level relationships of Onagraceae based on chloroplast rbcL and ndhF data. American Journal of Botany 90: 107–115.

Li, H.-L., W. Wang, R.-Q. Li, J.-B. Zhang, M. Sun, R. Naeem, J.-X. Su, X.-G. Xiang, P. E. Mortimer, D.-Z. Li, K. D. Hyde, J.-C. Xu, D. E. Soltis, P. S. Soltis, J. Li, S.-Z. Zhang, H. Wu, Z.-D. Chen, and A.-M. Lu. 2016. Global versus Chinese perspectives on the phylogeny of the N-fixing clade. Journal of Systematics and Evolution 54: 392–399.

Li, J. and A. L. Bogle. 2001. A new suprageneric classification system of Hamamelidoideae based on morphology and sequences of nuclear and chloroplast DNA. Harvard Papers in Botany 5: 499-515.

Li, J. 2008. Molecular phylogenetics of Hamamelidaceae: Evidence from DNA sequences of nuclear and chloroplast genomes. Pp. 227-250 in A. K. Sharma and A. Sharma (eds.), Plant Genome Biodiversity and Evolution. Volume 1. Part E. Phanerogams - Angiosperm. Science Publishers, Delhi.

Li, R.-Q., Z.-D. Chen, Y.-P. Hong, and A.-M. Lu. 2002. Phylogenetic relationships of the "higher" hamamelids based on chloroplast trnL-F sequences. Acta Botanica. Sinica 44: 1462–1468.

Li, R.-Q., Z.-D. Chen, A.-M. Lu, D. E. Soltis, P. S. Soltis, and P. S. Manos. 2004. Phylogenetic relationships in Fagales based on DNA sequences from three genomes. International Journal of Plant Sciences 165: 311-324.

Lis-Balchin, M. (ed.). 2002. Geranium and Pelargonium: The Genera Geranium and Pelargonium. Taylor and Francis, London.

Liston, A. 2003. A new interpretation of floral morphology in *Garrya* (Garryaceae). Taxon 52: 271-276.

Livshultz, T., E. Kaltenegger, S. C. K. Straub, K. Wietemier, E. Hirsch, K. Koval, L. Mema, and A. Liston. 2018. Evolution of pyrrolizine alkaloid biosynthesis in Apocynaceae: Revisiting the defence de-escalation hypothesis. New Phytologist 218: 762-773.

Lledó, M. D., M. B. Crespo, K. M. Cameron, M. F. Fay, and M. W. Chase. 1998. Systematics of Plumbaginaceae based on cladistic analysis of rbcL sequence data. Systematic Botany 23: 21-29.

Lledó, M. D., P. O. Karis, M. B. Crespo, M. F. Fay, and M. W. Chase. 2001. Phylogenetic position and taxonomic status of the genus Aegialitis and subfamilies Staticoideae and Plumbaginoideae (Plumbaginaceae): Evidence from plastid DNA sequences and morphology. Plant Systematics and Evolution 229: 107-124.

Lledó, M. D., M. B. Crespo, M. F. Fay, and M. W. Chase. 2005. Molecular phylogenetics of *Limonium* and related genera (Plumbaginaceae): Biogeographical and systematic implications. American Journal of Botany 92: 1189-1198.

Luebert, F., G. Brokamp, J. Wen, M. Weigend, and H. H. Hilger. 2011. Phylogenetic relationships and morphological diversity in neotropical *Heliotropium* (Heliotropiaceae). Taxon 60: 663–680.

Luebert, F., L. Cecchi, M. W. Frohlich, M. Gottschling, C. M. Guilliams, K. E. Hasenstab-Lehman, H. H. Hilger, J. S. Miller, M. Mittelbach, M. Nazaire, M. Nepi, D. Nocentini, D. Ober, R. G. Olmstead, F. Selvi, M. G. Simpson, K. Sutorý, B. Valdés, G. K. Walden, and M. Weigend. 2016. Familial classification of the Boraginales. Taxon 65: 502–522.

Lundberg, J. and K. Bremer. 2003. A phylogenetic study of the order Asterales using one morphological and three molecular data sets. International Journal of Plant Sciences 164: 553–578.

Manen, J.-F., M. C. Boulter, and Y. Naciri-Graven. 2002. The complex history of the genus *Ilex* L. (Aquifoliaceae): evidence from the comparison of plastid and nuclear DNA sequences and from fossil data. Plant Sytematics and Evolution 235: 79-98.

Manen, J.-F., G. Barriera, P.-A. Loizeau, and Y. Naciri. 2010. The history of extant *Ilex* species (Aquifoliaceae): Evidence of hybridization within a Mioece radiation. Molecular Phylogenetics and Evolution 57: 961-977.

Manos, P. S., and K. P. Steele. 1997. Phylogenetic analysis of "higher" Hamamelididae based on plastid sequence data. American Journal of Botany 84: 1407–1419.

Manos, P. S., and D. E. Stone. 2001. Evolution, phylogeny and systematics of the Juglandaceae. Annals of the Missouri Botanical Garden 88: 231-269.
Manos, P. S., Z.-K. Zhou, and C. H. Cannon. 2001. Systematics of Fagaceae: Phylogenetic tests of reproductive trait evolution. International Journal of Plant Sciences 162: 1361–1379.
Manos, P. S., P. S. Soltis, D. E. Soltis, S. R. Manchester, S.-H. Oh, C. D. Bell, D. L. Dilcher, and D. E. Stone. 2007. Phylogeny of extant and fossil Juglandaceae inferred from the integration of molecular and morphological data sets. Systematic Biology 56: 412-430.
Martins, L., C. Oberprieler, and F. H. Hellwig. 2003. A phylogenetic analysis of Primulaceae s.l. based on internal transcribed spacer (ITS) DNA sequence data. Plant Systematics and Evolution 237:75-85.
Marx, H., N. O'Leary, Y.-W. Yuan, P. Lu-Irving, D. C. Tank, M. E. Múlgura, and R. Olmstead. 2010. A molecular phylogeny and classification of Verbenaceae. American Journal of Botany 97: 1647-1663.
Matthews, M. L., and P. K. Endress. 2002. Comparative floral morphology and systematics in Oxalidales (Oxalidaceae, Connaraceae, Brunelliaceae, Cephalotaceae, Cunoniaceae, Elaeocarpaceae, Tremandraceae). Botanical Journal of the Linnean Society 140: 321–381.
Matthews, M. L., and P. K. Endress. 2005. Comparative floral structure and systematics in Crossosomataceae, Stachyuraceae, Staphyleaceae, Aphloiaceae, Geissolomataceae, Ixerbaceae, Strasburgeriaceae). Botanical Journal of the Linnean Society 147: 1–46.
Mayta, L. and E. A. Molinari-Novoa. 2015. L'intégration du genre Leuenbergeria Lodé dans sa propre sous-famille, Leuenbergerioideae Mayta & Mol. Nov., subfam. nov. Succulentopi 12:6-7.
McDade, L. A., S. E. Masta, M. L. Moody, and E. Waters. 2000. Phylogenetic relationships among Acanthaceae: Evidence from two genomes. Systematic Botany 25: 106–121.
McDade, L. A., T. F. Daniel, and C. A. Kiel. 2008. Toward a comprehensive understanding of phylogenetic relationships among lineages of Acanthaceae s.l. (Lamiales). American Journal of Botany 95: 1136-1152.
McNeal, J. R., J. R. Bennett, A. D. Wolfe, and S. Mathews. 2013. Phylogeny and origins of holoparasitism in Orobanchaceae. American Journal of Botany 100: 971-983.
Medan, D., and C. Schirarend. 2004. Rhamnaceae. In: Kubitzki, K. (ed.), The Families and Genera of Vascular Plants. VI. Flowering Plants. Dicotyledons. Celastrales, Oxalidales, Rosales, Cornales, Ericales. Pp. 320-338. Springer, Berlin.
Meimberg, H., P. Dittrich, G. Bringmann, J. Schlauer, and G. Heubl. 2000. Molecular phylogeny of Caryophyllidae s.l. based on matK sequences with special emphasis on carnivorous taxa. Plant Biology 2: 218–228.
Merckx, V. S. F. T., J. Kissling, H. Hentrich, S. B. Janssens, C. B. Mennes, C. D. Specht, and E. F. Smets. 2013. Phylogenetic relationships of the mycoheterotrophic genus *Voyria* and the implications for the biogeographic history of Gentianaceae. American Journal of Botany 100: 712-721.
Miller, J. S. 2003. Classification of Boraginaceae subfam. Ehretioideae: Resurrection of the genus Hilsenbergia Tausch ex Meisn. Adansonia Series 3, 25: 151-189.
Miller, J. S. and M. Gottschling. 2007. Generic classification in the Cordiaceae (Boraginales): Resurrection of the genus *Varronia* P.Br. Taxon 56:163-169.
Moharrek, F., S. Kasempour Osaloo, and M. Assadi. 2014. Molecular phylogeny of Plumbaginaceae with emphasis on *Acantholimon* Boiss. based on nuclear and plastid DNA sequences in Iran. Biochemical Systematics and Ecology 57: 117-127.
Monro, A. K. 2006. The revision of species-rich genera: A phylogenetic framework for the strategic revision of Pilea (Urticaceae) based on cpDNA, nrDNA, and morphology. American Journal of Botany 93: 426-441.
Moonlight, P. W., W. H. Ardi, L. A. Padilla, K.-F. Chung, D. Fuller, D. Girmansyah, R. Hollands, A. Jara-Muñoz, R. Kiew, W.-C. Leong, Y. Liu, A. Mahardika, L. D. K. Marasinghe, M. O'Connor, C.-I. Peng, Á. J. Pérez, T. Phutthai, M. Pullan, S. Rajbhandary, C. Reynel, R. R. Rubite, J. Sang, D. Scherberich, Y.-M. Shui, M. C. Tebbitt, D. C. Thomas, H. P. Wilson, N. H. Zaini, and M. Hughes. 2018. Dividing and conquering the fastest-growing genus: Towards a natural sectional classification of the mega-diverse genus *Begonia* (Begoniaceae). Taxon 27: 276-323.
Moore, B. R. and M. J. Donoghue. 2007. Correlates of diversification in the plant clade Dipsacales: Geographic movement and evolutionary innovations. The American Naturalist 170:S28–S55.
Moore, M. J., N. Hassan, M. A. Gitzendanner, R. A. Bruenn, M. Croley, A. Vandeventer, J. W. Horn, A. Dhingra, S. F. Brockington, M. Latvis, J. Ramdial, R. Alexandre, A. Piedrahita, Z. Xi, C. C. Davis, P. S. Soltis, and D. E. Soltis. 2011. Phylogenetic analysis of the plastid inverted repeat for 244 species: Insights into deeper-level angiosperm relationships from a long, slowly evolving sequence region. International Journal of Plant Sciences 172:541-558.
Morley, R. J. and C. W. Dick. 2003. Missing fossils, molecular clocks, and the origin of the Melastomataceae. American Journal of Botany 90: 1638-1645.
Mort, M. E., D. E. Soltis, P. S. Soltis, J. Francisco-Ortega, and A. Santos-Guerra. 2001. Phylogenetic relationships and evolution of Crassulaceae inferred from matK sequence data. American Journal of Botany 88: 76–91.
Mort, M. E., T. R. O'Leary, P. Carillo-Reyes, T. Nowell, J. K. Archibald, and C. P. Randle. 2010. Phylogeny and evolution of Crassulaceae: Past, present and future. Schumannia 6: 69-86.
Morton, C. M. and C. Telmer. 2014. New subfamily classification for Rutaceae. Annals of the Missouri Botanical Garden 99: 620-641.
Muellner-Riehl, A. N., A. Weeks, J. W. Clayton, S. Buerki, L. Nauheimer, Y.-C. Chiang, S. Cody, and S. K. Pell. 2016. Molecular phylogenetics and molecular clock dating of Sapindales based on plastid rbcL, atpB and trnL-trnF DNA sequences. Taxon 65: 1019-1036.
Müller, K., and T. Borsch. 2005. Phylogenetics of Amaranthaceae based on matK/trnK sequence data - evidence from parsimony, likelihood, and Bayesian analysis. Annals of the Missouri Botanical Garden 92: 66-102.
Nandi, O. I., M. W. Chase, and P. K. Endress. 1998. A combined cladistic analysis of angiosperms using rbcL and non-molecular data sets. Annals of the Missouri Botanical Garden 85: 137–212.
Narzary, D., S. A. Ranade, P. K. Divakar, and T. S. Rana. 2016. Molecular differentiation and phylogenetic relationship of the genus Punica (Punicaceae) with other taxa of the order Myrtales. Rheedea 26: 37-51.
Neyland, R. and M. Merchant. 2006. Systematic relationships of Sarraceniaceae inferred from nuclear ribosomal DNA sequences. Madroño 53: 223-232.

Nickrent, D. L., and V. Malécot. 2001. A molecular phylogeny of Santalales. In: Fer, A., P. Thalouarn, D. M. Joel, L. J. Musselman, C. Parker, and J. A. C. Verkleij (eds.), Proceedings of the 7th International Parasitic Weed Symposium. Faculté des Sciences, Université de Nantes, Nantes, France. See http://www.science.siu.edu/parasitic-plants/Santalales.IPWC/Sants.IPWC.html for an on-line version of the paper.

Nickrent, D. L., V. Malécot, R. Vidal-Russell, and J. P. Der. 2010. A revised classification of Santalales. Taxon 59:538-558.

Nickrent, D. L., F. Anderson, and J. Kuijt. 2019. Inflorescence evolution in Santalales: integrating morphological characters and molecular phylogenetics. American Journal of Botany 3: 1-13.

Nicolas, A. N. and G. M. Plunkett. 2009. The demise of subfamily Hydrocotyloideae (Apiaceae) and the re-alignment of its genera across the whole order Apiales. Molecular Phylogenetics and Evolution 53: 134-151.

Nicolas, A. N. and G. M. Plunkett. 2014. Diversification times and biogeographic patterns in Apiales. Botanical Review 80: 30-58.

Nürk, N. M., S. Uribe-Convers, B. Gehrke, D. C. Tank, and F. R. Blattner. 2015. Oligocene niche shift, Miocene diversification - cold tolerance and accelerated speciation rates in the St John's worts (Hypericum, Hypericaceae). BMC Evolutionary Biology 15: 80.

Nyffeler, R. 2002. Phylogenetic relationships in the cactus family (Cactaceae) based on evidence from trnK/matK and trnL-trnF sequences. American Journal of Botany 89: 312–326.

Nyffeler, R., C. Bayer, W. S. Alverson, A. Yen, B. A. Whitlock, M. W. Chase, and D. A. Baum. 2005. Phylogenetic analysis of the Malvadendrina clade (Malvaceae s.l.) based on plastid DNA sequences. Organisms Diversity & Evolution 5: 109-123.

Oh, S.-H., and P. S. Manos. 2008. Molecular phylogenetics and cupule evolution in Fagaceae as inferred from nuclear CRABS CLAW sequences. Taxon 57: 434-451.

Oh, S.-H. 2010. Phylogeny and systematics of Crossosomatales as inferred from chloroplast atpB, matK, and rbcL sequences. Korean Journal of Plant Taxonomy 40: 208-217.

Olmstead, R. G., and P. A. Reeves. 1995. Evidence for the polyphyly of the Scrophulariaceae based on chloroplast rbcL and ndhF sequences. Annals of the Missouri Botanical Garden 82: 176–193.

Olmstead, R. G., J. A. Sweere, R. E. Spangler, L. Bohs, and J. D. Palmer. 1999. Phylogeny and provisional clasification of the Solanaceae based on chloroplast DNA. In: Nee, M., D. Symon, R. N. Lester, and J. P. Jessop (eds.), Solanaceae IV: Advances in Biology and Utilization. Pp. 111–137. Royal Botanic Gardens, Kew.

Olmstead, R. G., C. W. DePamphilis, Andrea D. Wolfe, Nelson D. Young, Wayne J. Elisons, and Patrick A. Reeves. 2001. Disintegration of the Scrophulariaceae. American Journal of Botany 88: 348.

Olmstead, R. G., L. Bohs, H. A. Migid, E. Santiago-Valentin, V. F. Garcia, and S. M. Collier. 2008. A molecular phylogeny of the Solanaceae. Taxon 57: 1159-1181.

Olmstead, R. G., M. L. Zjhra, L. G. Lohmann, S. O. Grose, and A. J. Eckert. 2009. A molecular phylogeny and classification of Bignoniaceae. American Journal of Botany 96: 1731-1743.

Oxelman, B., N. Yoshikawa, B. L. McConaughy, J. Luo, A. L. Denton, and B. D. Hall. 2004. RPB2 gene phylogeny in flowering plants, with particular emphasis on asterids. Molecular Phylogenetics and Evolution 32: 462-479.

Oxelman, B., P. Kornhall, R. G. Olmstead, and B. Bremer. 2005. Further disintegration of Scrophulariaceae. Taxon 54: 411-425.

Palazzesi, L., M. Gottschling, V. Barreda, and M. Weigend. 2012. First Miocene fossils of Vivianiaceae shed new light on phylogeny, divergence times, and historical biogeography of Geraniales. Biol. Biological Journal of the Linnean Society 107: 67-85.

Panero, J. L. and V. A. Funk. 2002. Toward a phylogenetic classification for the Compositae (Asteraceae). Proceedings of the Biological Society of Washington 115: 909–922.

Panero, J. L. and V. A. Funk. 2008. The value of sampling anomalous taxa in phylogenetic studies: Major clades of the Asteraceae revisited. Molecular Phylogenetics and Evolution 47: 757-782.

Panero, J. L., S. E. Freire, L. A. Espinar, B. S. Crozier, G. E. Barboza, and J. J. Cantero. 2014. Resolution of deep nodes yields an improved backbone phylogeny and a new basal lineage to study early evolution in Asteraceae. Molecular Phylogenetics and Evolution 80:43-53.

Pell, S. K., J. D. Mitchell, A. J. Miller, and T. A. Lobova. 2011. Anacardiaceae. Pp. 7-50 in K. Kubitzki (ed.), The Families and Genera of Flowering Plants. X. Flowering Plants: Eudicots. Sapindales, Cucurbitales, Myrtaceae. Springer, Berlin.

Persson, C. 2001. Phylogenetic relationships in Polygalaceae based on plastid DNA sequences from the trnL-F region. Taxon 50: 763–779.

Plana, V. 2003. Phylogenetic relationships of the Afro-Malagasy members of the large genus Begonia inferred from trnL intron sequences. Systematic Botany 28: 693-704.

Plunkett, G. M., D. E. Soltis, and P. S. Soltis. 1997. Evolutionary patterns in Apiaceae: Inferences based on matK sequence data. Systematic Botany 21: 477–495.

Plunkett, G. M., G. T. Chandler, P. P. Lowry II, S. M. Pinney, and T. S. Sprenkle. 2004a. Recent advances in understanding Apiales and a revised classification. South African Journal of Botany 70: 371-381.

Plunkett, G. M., J. Wen, and P. P. Lowry II. 2004b. Infrafamilial classifications and characters in Araliaceae: Insights from the phylogenetic analysis of nuclear (ITS) and plastid (trnL-trnF) sequence data. Plant Systematics and Evolution 245: 1-39.

Porter, J. M. 1997. Phylogeny of Polemoniaceae based on nuclear ribosomal internal transcribed DNA sequences. Aliso 15: 57–77.

Porter, J. M., and L. A. Johnson. 1998. Phylogenetic relationships of Polemoniaceae: inferences from mitochondrial nad1b intron sequences. Aliso 17: 157–188.

Potter, D., F. Gao, P. E. Bortiri, S.-H. Oh, and S. Baggett. 2002. Phylogenetic relationships in Rosaceae inferred from chloroplast matK and trnL-trnF nucleotide sequence data. Plant Systematics and Evolution 231: 77–89.

Potter, D., T. Eriksson, R. C. Evans, S. Oh, J. E. E. Smedmark, D. R. Morgan, M. Kerr, K. R. Robertson, M. Arsenault, T. A. Dickinson, and C. S. Campbell. 2007. Phylogeny and classification of Rosaceae. Plant Systematics and Evolution 266: 5-43.

Powell, M., V. Savolainen, P. Cuénoud, J.-F. Manen, and S. Andrews. 2000. The mountain holly (*Nemopanthus mucronatus*: Aquifoliaceae) revisited with molecular data. Kew Bulletin 55: 341-347.

Prather, C. A., C. J. Ferguson, and R. K. Jansen. 2000. Polemoniaceae phylogeny and classification: Implications of sequence data from the chloroplast gene ndhF. American Journal of Botany 87: 1300–1308.

Prince, L. M. and C. R. Parks. 2001. Phylogenetic relationships of Theaceae inferred from chloroplast DNA sequence data. American Journal of Botany 88: 2309-2320.

Pyck, N., P. Roels, and E. Smets. 1999. Tribal relationships of Caprifoliaceae: Evidence from a cladistic analysis using ndhF sequences. Systematics and Geography of Plants 69: 145-159.

Pyck, N., and E. Smets. 2004. On the systematic position of Triplostegia (Dipsacales): A combined molecular and morphological approach. Belgian Journal of Botany 137: 125-139.

Qui, Y.-L., M. W. Chase, S. B. Hoot, E. Conti, P. R. Crane, K. J. Sytsma, and C. R. Parks. 1998. Phylogenetics of Hamamelidae and their allies: parsimony analyses of nucleotide sequences of the plastid gene rbcL. International Journal of Plant Sciences 159: 891–905.

Reeves, P. A., and R. G. Olmstead. 1998. Evolution of novel morphological, ecological, and reproductive traits in a clade containing Antirrhinum. American Journal of Botany 85: 1047–1056.

Refulio-Rodriguez, N. F. and R. G. Olmstead. 2014. Phylogeny of Lamiidae. American Journal of Botany 101: 287-299.

Renner, S. S., A. Weerasooriya, and M. E. Olson. 2002. Phylogeny of Cucurbitaceae inferred from multiple chloroplast loci. In: Botany 2002: Botany in the Curriculum, Abstracts. Madison, WI. (http://www.botany2002.org)

Renner, T. and C. D. Specht. 2011. A sticky situation: assessing adaptations for plant carnivory in the Caryophyllales by means of stochastic character mapping. International Journal of Plant Sciences 172:889-901.

Renner, T. and C. D. Specht. 2013. Inside the trap: gland morphologies, digestive enzymes, and the evolution of plant carnivory in the Caryophyllales. Current Opinion in Plant Biology 16:436-442.

Richardson, J. E., M. E. Fay, Q. C. B. Cronk, and M. W. Chase. 2000. A revision of the tribal classification of Rhamnaceae. Kew Bulletin 55: 311–340.

Richardson, J. E., B. A. Whitlock, A. W. Meerow, and S. Madriñán. 2015. The age of chocolate: A diversification history of *Theobroma* and Malvaceae. Frontiers in Ecology and Evolution 3: 120.

Rivadavia, F., K. Kondo, M. Kato, and M. Hasebe. 2003. Phylogeny of the sundews, Drosera (Droseraceae), based on chloroplast rbcL and nuclear 18S ribosomal DNA sequences. American Journal of Botany 90: 123-130.

Robbrecht, E. and J. F. Manen. 2006. The major evolutionary lineages of the coffee family (Rubiaceae, angiosperms). Combined analysis (nDNA and cpDNA) to infer the position of Coptosapelta and Luculia, and supertree construction based on rbcL, rps16, trnL-trnF and atpB-rbcL data. A new classification in two subfamilies, Cinchonioideae and Rubioideae. Systematics and Geography of Plants 76: 85-146.

Rodman, J. E., K. G. Karol, R. A. Price, and K. J. Sytsma. 1997. Molecules, morphology, and Dahlgren's expanded order Capparales. Systematic Biology 21: 289–307.

Rodman, J. E., P. S. Soltis, D. E. Soltis, K. J. Sytsma, and K. G. Karol. 1998. Parallel evolution of glucosinolate biosynthesis inferred from congruent nuclear and plastid gene phylogenies. American Journal of Botany 85: 997–1007.

Roquet, C., L. Sáez, J. J. Aldasoro, A. Susanna, M. L. Alarcón, and N. Garcia-Jacas. 2008. Natural delineation, molecular phylogeny and floral evolution in Campanula. Systematic Botany 33: 203-217.

Rose, J. P., T. J. Kleist, S. D. Löfstrand, B. T. Drew, J. Schönenberger, and K. J. Sytsma. 2018. Phylogeny, historical biogeography, and diversification of angiosperm order Ericales suggest ancient Neotropical and East Asian connections. Molecular Phylogenetics and Evolution 122: 59-79.

Ruhfel, B. R., C. P. Bove, C. T. Philbrick, and C. C. Davis. 2016. Dispersal largely explains the Gondwanan distribution of the ancient tropical clusioid plant clade. American Journal of Botany 103: 1117-1128.

Rydin, C., N. Wikström, and B. Bremer. 2017. Conflicting results from mitochondrial genomic data challenge current views of Rubiaceae phylogeny. American Journal of Botany 104: 1522-1532.

Salomo, K., J. F. Smith, T. S. Feild, M.-S. Samain, L. Bond, C. Davidson, J. Zimmers, C. Neinhuis, and S. Wanke. 2017. The emergence of earliest angiosperms may be earlier than fossil evidence indicates. Systematic Botany 42:607-619.

Sanchez, A., and K. A. Kron. 2008. Phylogenetics of Polygonaceae with an emphasis on the evolution of Eriogonoideae. Systematic Botany 33: 87-96.

Sanchez, A., T. M. Schuster, J. M. Burke, and K. A. Kron. 2011. Taxonomy of Polygonoideae (Polygonaceae): A new tribal classification. Taxon 60: 151-160.

Särkinen, T., L. Bohs, R. G. Olmstead, and S. Knapp. 2013. A phylogenetic framework for evolutionary study of the nightshades (Solanaceae): A dated 1000-tip tree. BMC Evolutionary Biology 13: 214.

Savolainen, V., Chase, M. W., Hoot, S. B., Morton, C. M., Soltis, D. E., Bayer, C., Fay, M. F., de Bruijn, A. Y., Sulllivan, S., and Qiu, Y.-L. 2000a. Phylogenetics of flowering plants based on combined analysis of plastid atpB and rbcL sequences. Systematic Biology 49: 306–362.

Savolainen, V., M. F. Fay, D. C. Albach, A. Backlund, M. v. d. Bank, K. M. Cameron, S. A. Johnson, M. D. Lledo, J. C. Pintaud, M. Powell, M. C. Sheahan, D. E. Soltis, P. S. Soltis, P. Weston, W. M. Whitten, K. J. Wurdack, and M. W. Chase. 2000b. Phylogeny of the eudicots: a nearly complete familial analysis based on rbcL gene sequences. Kew Bulletin 55: 257–310.

Scatigna, A. V., P. W. Fritsch, V. C. Souza, and A. O. Simões. 2018. Phylogenetic relationships and morphological evolution in the carnivorous genus Philcoxia (Plantaginaceae, Gratioleae). Systematic Botany 43: 910–919.

Schaefer, H., C. Heibl, and S. S. Renner. 2008. Gourds afloat: A dated phylogeny reveals an Asian origin of the gourd family (Cucurbitaceae) and numerous oversea dispersal events. Proceedings of the Royal Society B, 276: 843-851.

Schaefer, H. and S. S. Renner. 2011a. Cucurbitaceae. Pp. 112-174 in K. Kubitzki (ed.), The Families and Genera of Flowering Plants. X. Flowering Plants: Eudicots. Sapindales, Cucurbitales, Myrtaceae. Springer, Berlin.

Schaefer, H. and S. S. Renner. 2011b. Phylogenetic relationships in the order Cucurbitales and a new classification of the gourd family (Cucurbitaceae). Taxon 60: 122-138.

Schäferhoff, B., K. F. Müller, and T. Borsch. 2009. Caryophyllales phylogenetics: Disentangling Phytolaccaceae and Molluginaceae and description of Microteaceae as a new isolated family. Willdenowia 39:209-228.

Schäferhoff, B., A. Fleischmann, E. Fischer, D. C. Albach, T. Borsch, G. Heubl, and K. F. Müller. 2010. Towards resolving Lamiales relationships: insights from rapidly evolving chloroplast sequences. BMC Evolutionary Biology 10: 352.

Schönenberger, J. 2009. Comparative floral structure and systematics of Fouquieriaceae and Polemoniaceae (Ericales). International Journal of Plant Sciences 170:1132-1167.

Schultheis, L., and B. G. Baldwin. 1999. Molecular phylogenetics of Fouquieriaceae: evidence from nuclear rDNA ITS studies. American Journal of Botany 86: 578–589.
Schuster, T. M., J. L. Reveal, M. J. Bayly, and K. A. Kron. 2015. An updated molecular phylogeny of Polygonoideae (Polygonaceae): Relationships of *Oxygonum, Pteroxygonum*, and *Rumex*, and a new circumscription of *Koenigia*. Taxon 64: 1188-1208.
Schwarzbach, A. E. and L. A. McDade. 2002. Phylogenetic relationships of the mangrove family Avicenniaceae based on chloroplast and nuclear ribosomal DNA sequences. Systematic Botany 27: 84-98.
Schwarzbach, A. E. 2004. Plantaginaceae. In: Kadereit, J. (ed.), The Families and Genera of Vascular Plants. VII. Flowering Plants. Dicotyledons. Lamiales (except Acanthaceae including Avicenniaceae). Pp. 327–329. Springer, Berlin.
Schwery, O., R. E. Onstein, Y. Bouchenak-Khelladi, Y. Xing, R. J. Carter, and H. P. Linder. 2014. As old as the mountains: The radiations of the Ericaceae. New Phytologist 207: 355-367.
Selbach-Schnadelbach, A., S. Cavalli, J.-F. Mannen, G. C. Coelho, and T. Teixeira de Souza-Chies. 2009. New information for Ilex phylogenetics based on the plastid psbA-trnH intergenic spacer (Aquifoliaceae). Botanical Journal of the Linnean Society 159: 182-193.
Sennblad, B., and B. Bremer. 1996. The familial and subfamilial relationships of Apocynaceae and Asclepiadaceae evaluated with rbcl data. Plant Systematics and Evolution 202: 155-175.
Sennblad, B., and B. Bremer. 2002. Classification of Apocynaceae s.l. according to a new approach combining Linnaean and phylogenetic taxonomy. Systematic Biology 51: 389-409.
Sheahan, M. C., and Chase, M. W. 1996. A phylogenetic analysis of Zygophyllaceae R. Br. based on morphological, anatomical and rbcL sequence data. Botanical Journal of the Linnean Society 122: 279-300.
Sheahan, M. C., and Chase, M. W. 2000. Phylogenetic relationships within Zygophyllaceae based on DNA sequences of three plastid regions, with special emphasis on Zygophylloideae. Systematic Botany 25: 371-384.
Sheahan, M. C. 2006. Zygophyllaceae. In: Kubitzki, K. (ed.), The Families and Genera of Vascular Plants. IX. Flowering Plants. Eudicots. Berberidopsidales, Buxales, Crossosomatales, Fabales p.p., Geraniales, Gunnerales, Myrtales p.p., Proteales, Saxifragales, Vitales, Zygophyllales, Clusiaceae Alliance, Passifloraceae Alliance, Dilleniaceae, Huaceae, Picramniaceae, Sabiaceae Series. Pp. 488-500. Springer, Berlin.
Simmons, M., M. McKenna, C. Bacon, K. Yakobson, J. Cappa, R. Archer, and A. Ford. 2012. Phylogeny of Celastraceae tribe Euonymeae inferred from morphological characters and nuclear and plastid genes. Molecular phylogenetics and evolution 62:9-20.
Simpson, B. B., A. Weeks, D. M. Helfgott, and L. L. Larkin. 2004. Species relationships in Krameria (Krameriaceae) based on ITS sequences and morphology: implications for character utility and biogeography. Systematic Botany 29:97-108.
Simpson, B. B. 2006. Krameriaceae. In: Kubitzki, K. (ed.), The Families and Genera of Vascular Plants. IX. Flowering Plants. Eudicots. Berberidopsidales, Buxales, Crossosomatales, Fabales p.p., Geraniales, Gunnerales, Myrtales p.p., Proteales, Saxifragales, Vitales, Zygophyllales, Clusiaceae Alliance, Passifloraceae Alliance, Dilleniaceae, Huaceae, Picramniaceae, Sabiaceae Series. Pp. 208-212. Springer, Berlin.
Smissen, R. D., J. C. Clement, P. J. Garnock-Jones, and G. K. Chambers. 2002. Subfamilial relationships in Caryophyllaceae as inferred from 5' ndhF sequences. American Journal of Botany 89: 1336–1341.
Soejima, A., and J. Wen. 2006. Phylogenetic analysis of the grape family (Vitaceae) based on three chloroplast markers. American Journal of Botany 93: 278-287.
Soltis, D. E., and P. S. Soltis. 1997. Phylogenetic relationships in Saxifragaceae sensu lato: a comparison of topologies based on 18S rDNA and rbcL sequences. American Journal of Botany 84: 504–522.
Soltis, D. E., P. S. Soltis, M. W. Chase, M. E. Mort, D. C. Albach, M. Zanis, V. Savolainen. W. H. Hahn, S. B. Hoot, M. F. Fay, M. Axtell, S. M. Swensen, L. M. Prince, W. J. Kress, K. C. Nixon, and J. S. Farris. 2000. Angiosperm phylogeny inferred from 18S rDNA, rbcL, and atpB sequences. Botanical Journal of the Linnean Society 133(4): 381–461.
Soltis, D. E., R. K. Kuzoff, M. E. Mort, M. Zanis, M. Fishbein, L. Hufford, J. Koontz, and M. K. Arroyo. 2001. Elucidating deep-level phylogenetic relationships in Saxifragaceae using sequences for six chloroplastic and nuclear DNA regions. Annals of the Missouri Botanical Garden 88: 669–693.
Soltis, D. E., P. S. Soltis, P. K. Endress, and M. W. Chase. 2005. Angiosperm phylogeny and evolution. Sinauer, Sunderland, MA.
Soltis, D. E., M. A. Gitzendanner, and P. S. Soltisy. 2007. A 567-taxon data set for angiosperms: the challenges posed by Bayesian analyses of large data sets. International Journal of Plant Sciences 168:137–157.
Soltis, D. E., S. A. Smith, N. Cellinese, K. J. Wurdack, D. C. Tank, S. F. Brockington, N. F. Refulio-Rodriguez, J. B. Walker, M. J. Moore, B. S. Carlsward, C. D. Bell, M. Latvis, S. Crawley, C. Black, D. Diouf, Z. Xi, C. A. Rushworth, M. A. Gitzendanner, K. J. Sytsma, Y.-L. Qiu, K. W. Hilu, C. C. Davis, M. J. Sanderson, R. S. Beaman, R. G. Olmstead, W. S. Judd, M. J. Donoghue, and P. S. Soltis. 2011. Angiosperm phylogeny: 17 genes, 640 taxa. American Journal of Botany 98:704-730.
Soltis, D. E., M. E. Mort, M. Latvis, E. V. Mavrodiev, B. C. O'meara, P. S. Soltis, J. G. Burleigh, and R. R. D. Casas. 2013. Phylogenetic relationships and character evolution analysis of Saxifragales using a supermatrix approach. American Journal of Botany 100.
Sosa, V. and M. W. Chase. 2003. Phylogenetics of Crossosomataceae based on rbcL sequence data. Systematic Botany 28:96-105.
Sosa, V. 2006. Crossosomataceae. Pp. 119-122 in K. Kubitzki (ed.), The Families and Genera of Vascular Plants. Volume 9: Flowering Plants: Eudicots: Berberidopsidales, Buxales, Crossosomatales. Springer, Berlin.
Spangler, R. E., and R. G. Olmstead. 1999. Phylogenetic analysis of the Bignoniaceae based on the cpDNA gene sequence rbcL and ndhF. Annals of the Missouri Botanical Garden 86: 33–46.
Ståhl, B. and A. A. Anderberg. 2004. Myrsinaceae. In: Kubitzki, K. (ed.), The Families and Genera of Vascular Plants. VI. Flowering Plants. Dicotyledons. Celastrales, Oxalidales, Rosales, Cornales, Ericales. Pp. 266-281. Springer, Berlin.
Staples, G. W., and R. K. Brummitt. 2007. Convolvulaceae. In: Heywood, V. H., R. K. Brummitt, A. Culham, and O. Seberg. (eds). Flowering Plant Families of the World. Pp. 108-110. Royal Botanic Gardens, Kew.
Stefanovic, S., L. Krueger, and R. G. Olmstead. 2002. Monophyly of the Convolvulaceae and circumscription of their major lineages based on DNA sequences of multiple chloroplast loci. American Journal of Botany 89: 1510-1522.

Stefanovic, S., D. F. Austin, D. F., and R. G. Olmstead. 2003. Classification of Convolvulaceae: A phylogenetic approach. Systematic Botany 28: 791-806.

Stefanovic, S. and R. G. Olmstead. 2004. Testing the phylogenetic position of a parasitic plant (Cuscuta, Convolvulaceae, Asteridae): Bayesian inference and the parametric bootstrap on data drawn from three genomes. Systematic Biology 53: 384-399.

Stevens, P. F., S. Dressler, and A. L. Weitzman. 2004. Theaceae. In: Kubitzki, K. (ed.), The Families and Genera of Vascular Plants. VI. Flowering Plants. Dicotyledons. Celastrales, Oxalidales, Rosales, Cornales, Ericales. Pp. 463-471. Springer, Berlin.

Stevens, P. F. 2006. Hypericaceae. In: Kubitzki, K. (ed.), The Families and Genera of Vascular Plants. Volume IX. Flowering plants, Eudicots: Berberidopsidales, Buxales, Crossosomatales, Fabales p.p., Geraniales, Gunnerales, Myrtales pp., Proteales, Saxifragales, Vitales, Zygophyllales, Clusiaceae Alliance, Passifloraceae Alliance, Dilleniaceae, Huaceae, Picramniaceae, Sabiaceae. Pp. 194-201. Springer, Berlin.

Struwe, L., V. A. Albert, and B. Bremer. 1995. Cladistics and family-level classification of the Gentianales. Cladistics 10: 175-206.

Struwe, L., J. W. Kadereit, J. Klackenberg, S. Nilsson, M. Thiv, K. B. von Hagen, and V. A. Albert. 2002. Systematics, character evolution, and biogeography of Gentianaceae, including a new tribal and subtribal classification. In: Struwe, L., & Albert, V. A. (eds), Gentianaceae: Systematics and Natural History. Pp. 21-209. Cambridge University Press, Cambridge.

Struwe, L. 2014. Classification and evolution of the family Gentianaceae. Pp. 13-35 in J. J. Rybczynski, M. R. Davey, and A. Mikula (eds.), The Gentianaceae = Volume 1: Characterization and Ecology. Springer, Heidelberg.

Stull, G. W., R. D. d. Stefano, D. E. Soltis, and P. S. Soltis. 2015. Resolving basal lamiid phylogeny and the circumscription of Icacinaceae with a plastome-scale data set. American Journal of Botany 102:1-20.

Stull, G. W., M. Schori, D. E. Soltis, and P. S. Soltis. 2018. Character evolution and missing (morphological) data across Asteridae. American Journal of Botany 105: 470–479.

Su, H.-J., J.-M. Hu, F. E. Anderson, J. P. Der, and D. L. Nickrent. 2015. Phylogenetic relationships of Santalales with insights into the origins of holoparasitic Balanophoraceae. Taxon 64:491-506.

Sun, Y., M. J. Moore, S. Zhang, P. S. Soltis, D. E. Soltis, T. Zhao, A. Meng, X. Li, J. Li, and H. Wang. 2015. Phylogenomic and structural analyses of 18 complete plastomes across all families of early-diverging eudicots, including an angiosperm-wide analysis of IR gene content evolution. Molecular Phylogenetics and Evolution 96:93-101.

Sun, Y., M. Moore, S. Zhang, P. Soltis, D. Soltis, T. Zhao, A. Meng, L. Xiaodong, J. Li, and H. Wang. 2016a. Phylogenomic and structural analyses of 18 complete plastomes across nearly all families of early-diverging eudicots, including an angiosperm-wide analysis of IR gene content evolution. Molecular Phylogenetics and Evolution 96:93-101.

Sun, M., R. Naaem, J.-X. Su, Z.-Y. Cao, J. G. Burleigh, P. S. Soltis, D. E. Soltis, and Z.-D. Chen. 2016b. Phylogeny of the Rosidae: A dense taxon sampling analysis. Journal of Systematics and Evolution 54: 363-391.

Sun, Y., M. J. Moore, J. B. Landis, N. Lin, L. Chen, T. Deng, J. Zhang, A. Meng, S. Zhang, K. S. Tojibaev, H. Sun, and H. Wang. 2018. Plastome phylogenomics of the early-diverging eudicot family Berberidaceae. Molecular Phylogenetics and Evolution 128: 203-211.

Sytsma, K. J., J. Morawetz, J. C. Pires, M. Nepokroeff, E. Conti, M. Zjhra, J. C. Hall, and M. W. Chase. 2002. Urticalean rosids: Circumscription, rosid ancestry, and phylogenetics based on rbcL, trnL-F, and ndhF sequences. American Journal of Botany 89: 1531-1546.

Sytsma, K. J., A. Litt, M. L. Zjhra, C. Pires, M. Nepokroeff, E. Conti, J. Walker, and P. G. Wilson. 2004. Clades, clocks, and continents: Historical and biogeographical analysis of Myrtaceae, Vochysiaceae, and relatives in the southern hemisphere. International Journal of Plant Sciences 165: S85-S105.

Takhtajan, A. L. 1991. Evolutionary Trends in Flowering Plants. Columbia University Press, New York.

Tank, D. C., P. M. Beardsley, S. A. Kelchner, and R. G. Olmstead. 2006. Review of the systematics of Scrophulariaceae s.l. and their current disposition. Australian Systematic Botany 19: 289-307.

Tank, D. C., J. M. Egger, and R. G. Olmstead. 2009. Phylogenetic classification of subtribe Castillejinae (Orobanchaceae). Systematic Botany 34: 182-197.

Tank, D. C. and M. J. Donoghue. 2010. Phylogeny and phylogenetic nomenclature of the Campanulidae based on an expanded sample of genes and taxa. Systematic Botany 35:435-441.

The Legume Phylogeny Working Group (LPWG). 2017. A new subfamily classification of the Leguminosae based on a taxonomically comprehensive phylogeny. Taxon 66: 44–77.

Theis, N., M. J. Donoghue, and J. Li. 2008. Phylogenetics of the Caprifolieae and Lonicera (Dipsacales) based on nuclear and chloroplast DNA sequences. Systematic Botany 33: 776-783.

Thiede, J., and U. Eggli. 2006. Crassulaceae. In: Kubitzki, K. (ed.), The Families and Genera of Vascular Plants. IX. Flowering Plants. Eudicots. Berberidopsidales, Buxales, Crossosomatales, Fabales p.p., Geraniales, Gunnerales, Myrtales p.p., Proteales, Saxifragales, Vitales, Zygophyllales, Clusiaceae Alliance, Passifloraceae Alliance, Dilleniaceae, Huaceae, Picramniaceae, Sabiaceae Series. Pp. 83-118. Springer, Berlin.

Thorne, Robert F. 2000. The classification and geography of the flowering plants: Dicotyledons of the class Angiospermae: (Subclasses Magnoliidae, Ranunculidae, Caryophyllidae, Dilleniidae, Rosidae, Asteridae, and Lamiidae). Botanical Review 66(4): 441–647.

Thulin, M., A. J. Moore, H. El-Seedi, A. Larsson, P. A. Christin, and E. J. Edwards. 2016. Phylogeny and generic delimitation in Molluginaceae, new pigment data in Caryophyllales, and the new family Corbichoniaceae. Taxon 65:775-793.

Tokuoka, T. 2008. Molecular phylogenetic analysis of Violaceae (Malpighiales) based on plastid and nuclear DNA sequences. Journal of Plant Research 121: 253-260.

Tokuoka, T. 2012. Molecular phylogenetic analysis of Passifloraceae sensu lato (Malpighiales) based on plastid and nuclear DNA sequences. Journal of Plant Research 125: 489-497.

Wagner, W. L., P. C. Hoch, and P. H. Raven. 2007. Revised classification of the Onagraceae. Systematic Botany Monographs 83: 1-240.

Wagstaff, S. J., and R. G. Olmstead. 1997. Phylogeny of Labiatae and Verbenaceae inferred from rbcL sequences. Systematic Botany 22: 165-179.

Wagstaff, S. J., L. Hickerson, R. Spangler, P. A. Reeves, R. G. Olmstead. 1998. Phylogeny of Labiatae s.l., inferred from cpDNA sequences. Plant Systematics and Evolution 209: 265-274.

Wahlert, G. A., T. Marcussen, J. de Paula-Souza, M. Feng, and H. E. J. Ballard. 2014. A phylogeny of the Violaceae (Malpighiales) inferred from plastid DNA sequences: Implications for generic diversity and intrafamilial classification. Systematic Botany 39: 239-252.

Walden, G. K. and R. Patterson. 2012. Nomenclature of subdivisions within *Phacelia* (Boraginaceae: Hydrophylloideae). Madroño 59:211-222.

Walker, J. F., Y. Yang, M. J. Moore, J. Mikenas, A. Timoneda, S. F. Brockington, and S. A. Smith. 2017. Widespread paleopolyploidy, gene tree conflict, and recalcitrant relationships among the carnivorous Caryophyllales. American Journal of Botany 104:858 – 867.

Walker, J. F., Y. Yang, T. Feng, A. Timoneda, J. Mikenas, V. Hutchison, C. Edwards, N. Wang, S. Ahluwalia, J. Olivieri, N. Walker-Hale, L. C. Majure, R. Puente, G. Kadereit, M. Lauterbach, U. Eggli, H. Flores-Olvera, H. Ochoterena, S. F. Brockington, M. J. Moore, and S. A. Smith. 2018. From cacti to carnivores: Improved phylotranscriptomic sampling and hierarchical homology inference provide further insight into the evolution of Caryophyllales American Journal of Botany 105:446–462.

Wallace, R. S., and A. C. Gibson. 2002. Evolution and systematics. In: Nobel, P. S. (ed.), Cacti: Biology and Uses. Pp. 1–21. University of California Press, Berkeley.

Wallander, E., and Albert, V. A. 2000. Phylogeny and classification of Oleaceae based on rps16 and trnL-F sequence data. American Journal of Botany 87: 1827-1841.

Wang, H., M. J. Moore, P. S. Soltis, C. D. Bell, S. F. Brockington, R. Alexandre, C. C. Davis, M. Latvis, S. R. Manchester, and D. E. Soltis. 2009. Rosid radiation and the rapid rise of angiosperm-dominated forests. Proceedings of the National Academy of Sciences 106: 3853-3858.

Wang, N., Y. Yang, M. J. Moore, S. F. Brockington, J. F. Walker, J. W. Brown, and e. a. B. Liang. 2018. Evolution of Portulacineae marked by gene tree conflict and gene family expansion associated with adaptation to harsh environments. BioArxiv Issue 294549.

Wang, W., A.-M. Lu, Y. Ren, M. E. Endress, and Z.-D. Chen. 2009. Phylogeny and classification of Ranunculales: Evidence from four molecular loci and morphological data. Perspectives in Plant Ecology, Evolution and Systematics 11: 81-110.

Wang, W., D. L. Dilcher, G. Sun, H.-S. Wang, and Z.-D. Chen. 2016. Accelerated evolution in the early angiosperms: Evidence from ranunculalean phylogeny by integrating living and fossil data. J. Syst. Evol. 54:336-341.

Wang, W., Z.-D. Chen, R.-Q. Li., and J.-H. Li. 2007. Phylogenetic and biogeographic diversification of Berberidaceae in the northern hemisphere. Systematic Botany 32: 731-742.

Wang, X. Z., D. Ren, and Y. F. Wang. 2000. First discovery of angiospermous pollen from Yixian Formation in Western Liaoning. Acta Geologica Sinica 74.

Wannan, B. S. 2006. Analysis of generic relationships in Anacardiaceae. Blumea 51: 165-195.

Wanntorp, L. 2006. Molecular systematics and evolution of the genus Gunnera. Pp. 419-435 in A. K. Sharma and A. Sharma (eds.), Plant Genome Biodiversity and Evolution. Volume 1, Part C Phanerogams (Angiosperm-Dicotyledons). Science Publishers, Enfield, NH.

Wanntorp, L. and H.-E. Wanntorp. 2003. The biogeography of *Gunnera* L.: vicariance and dispersal. Journal of Biogeography 30: 979-987.

Weigend, M., F. Selvi, D. C. Thomas, and H. H. Hilger. 2016. Boraginaceae. Pp. 41-102 in K. Kubitzki, J. W. Kadereit, and V. Bittrich (eds.), The Families and Genera of Vascular Plants. Flowering Plants, Eudicots. 14. Springer International Publishing, Switzerland.

Wen, J. 2006. Leeaceae and Vitaceae. In: Kubitzki, K. (ed.), The Families and Genera of Vascular Plants. IX. Flowering Plants. Eudicots. Berberidopsidales, Buxales, Crossosomatales, Fabales p.p., Geraniales, Gunnerales, Myrtales p.p., Proteales, Saxifragales, Vitales, Zygophyllales, Clusiaceae Alliance, Passifloraceae Alliance, Dilleniaceae, Huaceae, Picramniaceae, Sabiaceae Series. Pp. 221-225 and 467-479. Springer, Berlin.

Wen, J., G. M. Plunkett, A. D. Mitchell, and S. J. Wagstaff. 2000. The evolution of Araliaceae: a phylogenetic analysis based on ITS sequences of nuclear ribosomal DNA. Systematic Botany 26: 144–167.

Wen, J., Z.-L. Nie, X.-Q. Liu, N. Zhang, S. Ickert-Bond, J. Gerrath, S. R. Manchester, J. Boggan, and Z.-D. Chen. 2018. A new phylogenetic classification of the grape family (Vitaceae). Journal of Systematics and Evolution 56: 262-272.

Weston, P. H., and N. P. Barker. 2006. A new generic classification of the Proteaceae with an annotated checklist of genera. Telopea 11: 314-344.

Wilken, D. H. 2004. Polemoniaceae. In: Kubitzki, K. (ed.), The Families and Genera of Vascular Plants. VI. Flowering Plants. Dicotyledons. Celastrales, Oxalidales, Rosales, Cornales, Ericales. Pp. 300–312. Springer, Berlin.

Williamson, P. S., and E. L. Schneider. 1993. Nelumbonaceae. In: Kubitzki, K., Rohwer, J. G., and Bittrich, V. (eds.), The Families and Genera of Vascular Plants. II. Flowering Plants: Dicotyledons, Magnoliid, Hamamelid and Caryophyllid Families. Pp. 470–472. Springer, Berlin.

Wilson, P. G., M. M. O'Brien, P. A. Gadek, and C. J. Quinn. 2001. Myrtaceae revisited: A reassessment of infrafamilial groups. American Journal of Botany 88: 2013–2025.

Wilson, P. G., M. M. O'Brien, M. M. Heslewood, and C. J. Quinn. 2005. Relationships within Myrtaceae sensu lato based on a matK phylogeny. Plant Systematics and Evolution 251: 3-19.

Winkworth, R. C., J. Lundberg, and M. J. Donoghue. 2008. Towards a resolution of campanulid phylogeny, with special reference to the placement of Dipsacales. Taxon 57: 53-65.

Wurdack, K. J., and M. W. Chase. 2002. Phylogenetics of Euphorbiaceae s.s. using plastid (rbcL and trnL-F) DNA sequences. In: Botany 2002 Botany in the Curriculum: Integrating Research and Teaching. Abstract. Madison, WI. (http://www.botany2002.org)

Wurdack, K. J., P. Hoffmann, and M. W. Chase. 2005. Molecular phylogenetic analysis of uniovulate Euphorbiaceae (Euphorbiaceae sensu stricto) using plastid rbcL and trnL-F DNA sequences. American Journal of Botany 92: 1397-1420.

Wurdack, K. J., and C. C. Davis. 2009. Malpighiales phylogenetics: Gaining ground on one of the most recalcitrant clades in the angiosperm tree of life. American Journal of Botany 96:1551-1570.

Xi, Z., B. R. Ruhfel, H. Schaefer, A. M. Amorim, M. Sugumaran, K. J. Wurdack, P. K. Endress, M. Matthews, P. F. Stevens, S. Mathews, and C. C. Davis, III. 2012. Phylogenomics and a posteriori data partitioning resolve the Cretaceous angiosperm radiation in Malpighiales. Proceedings of the National Academy of Sciences 109: 17519-17524.

Xiang, C. L., M. A. Gitzendanner, D. E. Soltis, H. Peng, and L. G. Lei. 2012. Phylogenetic placement of the enigmatic and critically endangered genus Saniculiphyllum (Saxifragaceae) inferred from combined analysis of plastid and nuclear DNA sequences. Molecular Phylogenetics and Evolution 64: 357-367.

Xiang, C.-L., H.-J. Dong, S. Landrein, F. Zhao, W.-B. Yu, D. E. Soltis, P. S. Soltis, A. Backlund, H.-F. Wang, D.-Z. Li, and H. Peng. 2019. Revisiting the phylogeny of Dipsacales: new insights from phylogenomic analyses of complete plastomic sequences. Journal of Systematics and Evolution doi: 10.1111/jse.12526.

Xiang, Q.-Y., D. E. Soltis, D. R. Morgan, and P. S. Soltis. 1993. Phylogenetic relationships of Cornus L. sensu lato and putative relatives inferred from rbcL sequence data. Annals of the Missouri Botanical Garden 80: 723-734.

Xiang, Q.-Y., S. J. Brunsfeld, D. E. Soltis, and P. S. Soltis. 1997. Phylogenetic relationships in Cornus based on chloroplast DNA restriction sites: implications for biogeography and character evolution. Systematic Botany 21: 515–534.

Xiang, Q.-Y., S. J. Brunsfeld, and P. S. Soltis. 1998. Phylogenetic relationships of Cornaceae and close relatives inferred from matK and rbcL sequences. American Journal of Botany 85: 285–297.

Xiang, Q.-Y., M. Moody, D. E. Soltis, C.-Z. Fan, and P. S. Soltis. 2002. Relationships within Cornales and circumscription of Cornaceae—matK and rbcL sequence data and effects of outgroups and long branches. Molecular Phylogenetics and Evolution 24: 35–57.

Xiang, Q.-Y., D. T. Thomas, W. H. Zhang, S. R. Manchester, and Z. Murrell. 2006. Species level phylogeny of the genus Cornus (Cornaceae) based on molecular and morphological evidence - implications for taxonomy and Tertiary intercontinental migration. Taxon 55: 9-30.

Xiang, Q.-Y. and D. T. Thomas. 2008. Tracking character evolution and biogeographic history through time in Cornaceae - does choice of methods matter? Journal of Systematics and Evolution 46: 349-374.

Xiang, Q.-Y. J., D. T. Thomas, and Q. P. Xiang. 2011. Resolving and dating the phylogeny of Cornales - effects of taxon sampling, data partitions and fossil calibrations. Molecular Phylogenetics and Evolution 50: 123-138.

Yang, J.-B., S.-X. Yang, D.-Z. Li, L.-G. Lei, T. Ikeda, and H. Yoshino. 2006. Phylogenetic relationships of Theaceae inferred from mitochondrial matR gene sequence data. Acta Botanica Yunnanica 29: 29-36.

Yang, S.-X., J.-B. Yang, L.-G. Lei, D.-Z. Li, H. Yoshino, and T. Ikeda, T. 2004. Reassessing the relationships between Gordonia and Polyspora (Theaceae) based on the combined analysis of molecular data from the nuclear, plastid and mitochondrial genomes. Plant Systematics and Evolution 248: 45-55.

Yang, Y., M. J. Moore, S. F. Brockington, D. E. Soltis, G. K.-S. Wong, E. J. Carpenter, Y. Zhang, L. Chen, Z. Yan, Y. Xie, R. F. Sage, S. Covshoff, J. M. Hibberd, M. N. Nelson, and S. A. Smith. 2015. Dissecting Molecular Evolution in the Highly Diverse Plant Clade Caryophyllales Using Transcriptome Sequencing. Molecular Biology and Evolution 32:2001-2014.

Yang, Y., M. J. Moore, S. F. Brockington, J. Mikenas, J. Olivieri, J. F. Walker, and S. A. Smith. 2017. Improved transcriptome sampling pinpoints 26 ancient and more recent polyploidy events in Caryophyllales, including two allopolyploidy events. New Phytologist 217: 855-870.

Yesson, C., N. H. Toomey, and A. Culham. 2009. Cyclamen: Time, sea and speciation biogeography using a temporally calibrated phylogeny. Journal of Biogeography 36: 1234-1252.

Yoo, K.-O., and J. Wen. 2008. Phylogenetic relationships of Coryloideae based on waxy and atpB-rbcL sequences. Korean Journal of Plant Taxonomy 38: 371-388.

Yuan, Y.-M., Y. Song, K. Geuten, E. Rahelivololona, S. Wohlhauser, E. Fischer, E. Smets, and P. Küpfer. 2004. Phylogeny and biogeography of Balsaminaceae inferred from ITS sequences. Taxon 53: 391-403.

Zerega, N. J. C., M. N. Nur Supardi, and T. J. Motley. 2010. Phylogeny and recircumscription of Artocarpeae (Moraceae) with a focus on Artocarpus. Systematic Botany 35: 766-782.

Zerega, N. J. C., W. L. Clement, S. L. Datwyler, and G. D. Weiblen. 2005. Biogeography and divergence times in the mulberry family (Moraceae). Molecular Phylogenetics and Evolution 37: 402-416.

Zhang, J., T. A. Ruhlmann, J. Sabir, J. C. Blazier, and R. K. Jansen. 2015. Coordinated rates of evolution between interacting plastid and nuclear genes in Geraniaceae. Plant Cell 27: 563-573.

Zhang, L.-B., and S. S. Renner. 2003. Phylogeny of Cucurbitales inferred from seven chloroplast and mitochondrial loci. In: Botany 2003: Aquatic and Wetland Plants: Wet and Wild. Mobile, Alabama. (http://www.botany2003.org)

Zhang, L.-B. and M. Simmons. 2006. Phylogeny and delimitation of the Celastrales inferred from nuclear and plastid genes. Systematic Botany 31:122-137.

Zhang, L.-B., M. P. Simmons, A. Kocyan, and S. S. Renner. 2006. Phylogeny of the Cucurbitales based on DNA sequences of nine loci from three genomes: Implications for morphological and sexual system evolution. Molecular Phylogenetics and Evolution 39: 305-322.

Zhang, S.-D., J.-J. Jin, S.-Y. Chen, M. W. Chase, D. E. Soltis, H.-T. Li, J.-B. Yang, D.-Z. Li, and T.-S. Yi. 2017. Diversification of Rosaceae since the Late Cretaceous based on plastid phylogenomics. New Phytologist 214: 1355-1367.

Zhang, W.-H., H. Li, H.-B. Chen, and Y.-C. Tang. 2003. Phylogeny of the Dipsacales s.l. based on chloroplast trnL-F and ndhF sequences. Molecular Phylogenetics and Evolution 26:176–189.

WEB SITES

POWO. 2019. Plants of the World Online. Facilitated by the Royal Botanic Gardens, Kew. Published on the Internet. *http://plantsoftheworldonline.org*.

Reveal, J. L. (2001 onwards). Indices Nominum Supragenericorum Plantarum Vascularium. Alphabetical Listing by Family of Validly Published Suprageneric Names. *http://www.plantsystematics.org/reveal/PBIO/fam/allspgnames.html*
An index of accepted names above the rank of family.

Royal Botanic Gardens, Kew, Vascular Plant Families and Genera Database *http://data.kew.org/vpfg1992/vascplnt.html*
A searchable database taken from Vascular Plant Families and Genera, compiled by R. K. Brummitt and published by the Royal Botanic Gardens, Kew, in 1992.

Stevens, P. F. (2001 onwards). Angiosperm Phylogeny Website. *http://www.mobot.org/MOBOT/research/APweb*
An excellent graphical representation and update of the Angiosperm Phylogeny Group classification, cladograms, family characteristics, references, and apomorphies.

III

SYSTEMATIC EVIDENCE AND DESCRIPTIVE TERMINOLOGY

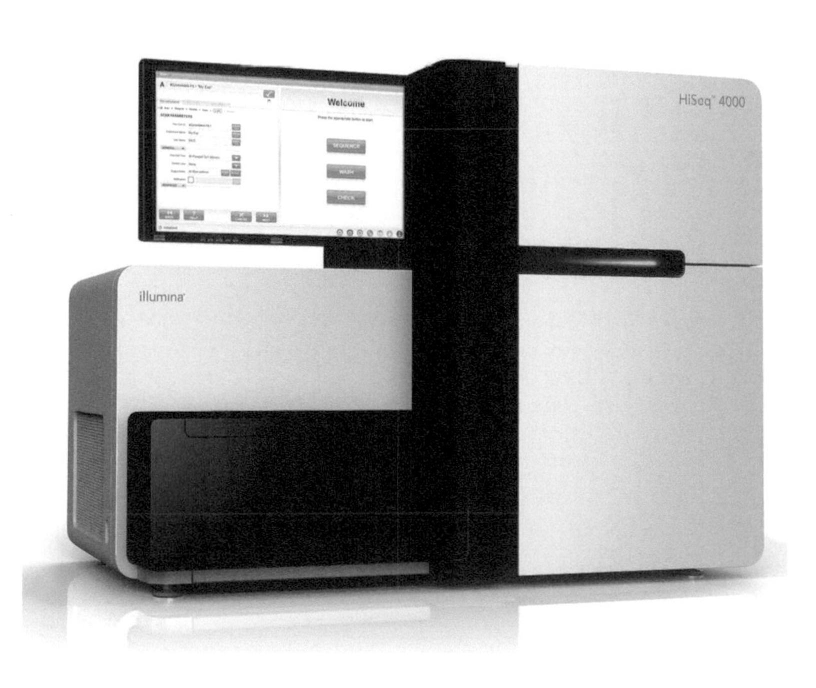

9

PLANT MORPHOLOGY

https://doi.org/10.1016/B978-0-12-812628-8.50009-2,

Plant morphology is a field of study dealing with the external and gross internal structure of plant organs. Morphology intergrades somewhat with plant anatomy, which is the study of tissue and cell structure of plant organs (see Chapter 10). Morphology forms the basis of taxonomic descriptions and generally constitutes the most important data in delimiting and circumscribing taxa.

The terms cited here are largely descended from those used by herbalists and botanists of the past, beginning perhaps with Theophrastus (370–ca. 285 BC), one of the first to write detailed plant descriptions using technical terminology (as in *Historia Plantarum*). The terms have evolved tremendously over the years, especially since the invention of microscopes, having become more detailed and specific. Many of these terms were borrowed from classical Latin (or Greek converted to Latin) and used in new meanings; some were modified from preexisting terms, many invented, and some discarded along the way.

As with all fields of evidence, the terms used in plant morphological descriptions may vary from one source to the next. In using a particular flora, for example, its glossary (if present) should be checked to verify usage of terms. The terms cited below are from a number of sources but are largely derived and classified, with some exceptions, from Radford et al. (1974), which is a precise and logical system of organizing morphological terms (see also Bell 1991). By this classification scheme, the section on **PLANT STRUCTURE** includes characters and character states for specific plant organs and parts. This is followed by a section on **GENERAL TERMINOLOGY**, which lists characters and states that can be used to describe a variety of plant organs.

Note that some terms may need to be explained using other terms (especially those from the general terminology section). Please refer to the Glossary if an unknown term is encountered.

PLANT STRUCTURE

PLANT ORGANS

The basic structural components, or organs, of plants are delimited by and strongly correlated with their specific functions. Among the liverworts, hornworts, and mosses (see Chapter 3), these organs are components of the haploid gametophyte. The gametophyte of these taxa contains **rhizoids**, which are uniseriate, filamentous chains of cells functioning in anchorage and water/mineral absorption. The basic body of the gametophyte can either be a flat mass of cells, termed a **thallus** (found in some liverworts and all hornworts) or a **shoot**, consisting of a generally cylindrical stem bearing leaves (found in some liverworts and all mosses; see Chapter 3). It should be noted that the shoot systems of liverworts and mosses are gametophytic tissue.

The major organs of vascular plants are sporophytic roots and shoots. **Roots** are present in almost all vascular plants and typically function in anchorage and absorption of water and minerals. Roots consist of an apical meristem that gives rise to a protective root cap, a central endodermis-bounded vascular cylinder, absorptive epidermal root hairs, and endogenously developed lateral roots (Figure 9.1).

The sporophytic **shoots** of vascular plants consist of stem plus leaves (Figure 9.1). Shoots contain an apical meristem of actively dividing cells that, through continued differentiation, result in the elongation of the stem and formation of leaves and buds (see later discussion). The **stem** is a generally cylindrical organ that bears the photosynthetic leaves. Stems typically function in conduction of water and minerals from the roots and in support and elevation of both leaves and reproductive structures, although some stems are highly modified for other functions (see later discussion). The **leaf** is that organ of the shoot that is generally dorsiventrally flattened and that usually functions in photosynthesis and transpiration. Leaves are derived from **leaf primordia** within the shoot apex and are often variously modified. In vascular plants, leaves contain one to many vascular bundles, the **veins**; in some mosses, the gametophytic leaves may contain a veinlike **costa**, consisting of specialized (although not truly vascular) conductive tissue. **Buds** are immature shoot systems, typically located in the axils of leaves. Buds may grow to form lateral vegetative branches or reproductive structures (see later discussion).

Among reproductive plant organs, the **sporangium** is the basic spore-producing part of all land plants. The sporangium of liverworts, hornworts, and mosses is known as a **capsule** and typically makes up most of the sporophyte (see Chapter 3). In heterosporous plants, including all of the seed plants, sporangia are of two types: male or microsporangium and female or megasporangium (see Chapter 4). A **cone**, also called a **strobilus**, is a modified, determinate, reproductive shoot system of many nonflowering vascular plants, consisting of a stem axis bearing either sporophylls, in "simple" cones, or modified shoot systems, in "compound" cones (see Chapter 5). An **ovule** is a megasporangium enveloped by one or more protective integuments. A **seed** is the mature ovule of the seed plants, consisting of an internal embryo surrounded by nutritive tissue (comprising female gametophyte or endosperm) and enveloped by a protective seed coat (Chapter 5). The reproductive organ of angiosperms is the **flower**, a modified, determinate shoot bearing sporophylls called stamens and carpels, with or without outer modified leaves, the perianth (see Chapter 6). An **inflorescence** is an aggregate of one or more flowers, the boundaries of which generally occur

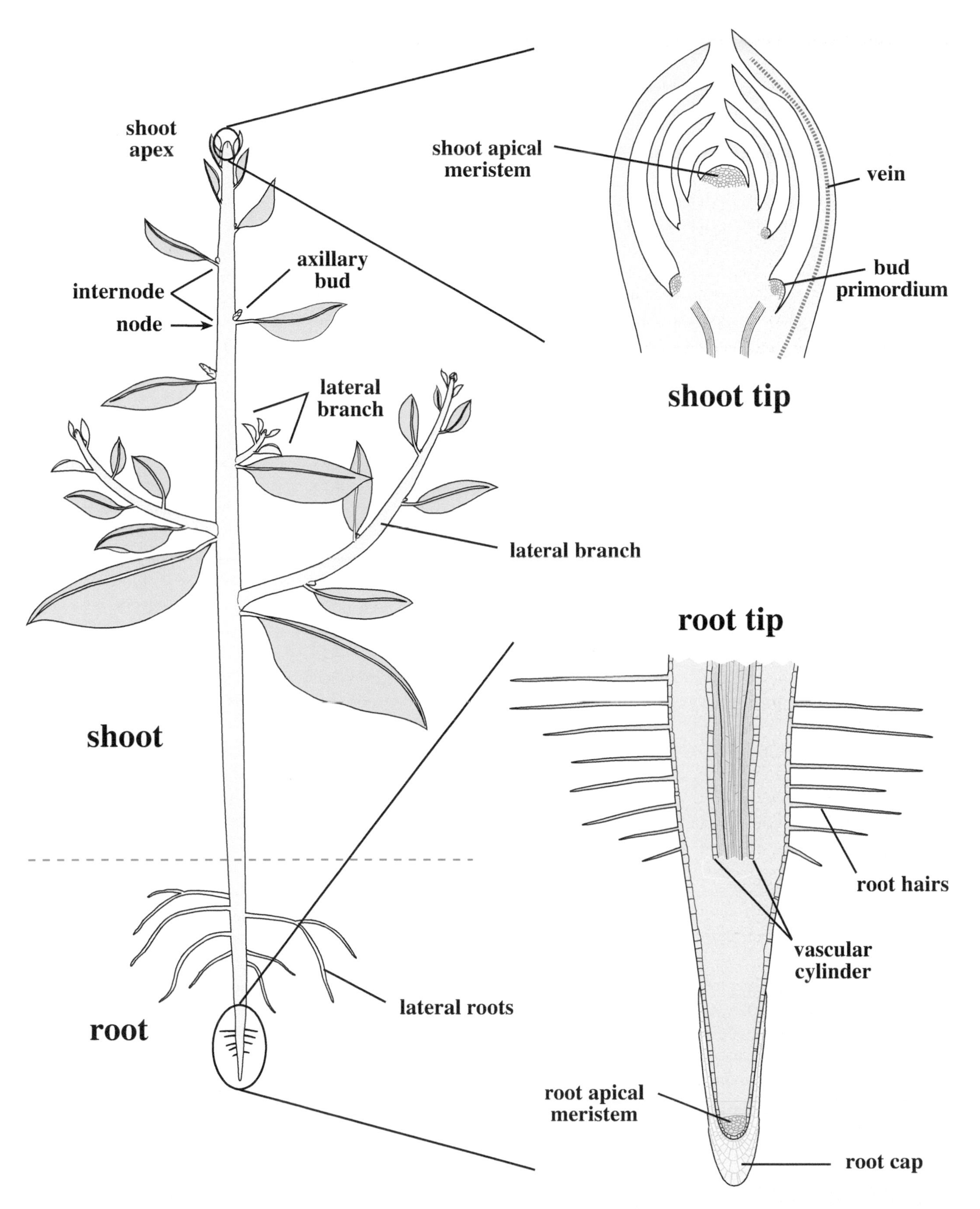

Figure 9.1 General plant structure, showing primary root and primary shoot. Note that all parts of both root and shoot are derived from cell divisions of the root or shoot apical meristem.

with the presence of vegetative leaves. A **fruit** is the mature ovary of flowering plants, consisting of the pericarp (mature ovary wall), seeds, and (if present) accessory parts.

Plant Habit

Plant habit refers to the general form of a plant, encompassing a variety of components such as stem duration and branching pattern, development, or texture. Most plants can be clearly designated as an herb, vine, liana, shrub, or tree (with some subcategories; see later discussion); however, some species are difficult to accommodate into these categories. An **herb** is a plant in which any aboveground shoots, whether vegetative or reproductive, die back at the end of an annual growth season. Although the aboveground shoots are annual, the herb itself may be annual, biennial, or perennial, the last by means of long-lived underground rootstocks. Such perennial herbs, having a bulb, corm, rhizome, or tuber as the underground stem, are termed **geophytes**. A **vine** is a plant with elongate, weak stems, that are generally supported by means of scrambling, twining, tendrils, or roots; vines may be annual or perennial, herbaceous or woody. A **liana** (also spelled *liane)* is a vine that is perennial and woody; lianas are major components in the tree canopy layer of some tropical forests. A **shrub** is a perennial, woody plant with several main stems arising at ground level. A **subshrub** is a short shrub that is woody only at the base and that seasonally bears new, nonwoody, annual shoots above. Finally, a **tree** is defined as a generally tall, perennial, woody plant having one main stem (the trunk) arising at ground level. (Some plant ecologists will sometimes distinguish between shrubs and trees based primarily on an arbitrary height.)

Plant Habitat

Plant habitat refers to the general environment where the plant is growing. General habitat terms include whether the plant is **terrestrial**, growing on land; **aquatic**, growing in water; or **epiphytic**, growing on another plant. If aquatic, a plant can be **submersed**, occurring under water; **floating**, occurring at the water surface; or **emergent**, having roots or stems anchored to the substrate under water and aerial shoots growing above water. A **rheophyte** is a plant found along (often swiftly flowing) streams and river banks. Other aspects of the habitat include the type of substrate that the plant is growing in (e.g., whether on sandy, loam, clay, gravelly, or rocky soil). **Saxicolous [epipetric]** refers to a plant growing on rocks or boulders, either in or on the surface (**lithophyte**) or in the cracks (**chasmophyte**). Other habitat features include slope, aspect, elevation, moisture regime, and surrounding vegetation, community, or ecosystem. (See Chapter 17, "Plant Collecting and Documentation.")

Plant Life Form

Plant life form denotes aspects of their structure, life cycle, and physiology. (See Raunkiaer 1934.) Life form types include: **chamaephyte**, an overwintering perennial with buds at or just below ground level; **epiphyte**, a plant growing on another plant, e.g., *Tillandsia* (Bromeliaceae); **geophyte**, a perennial herb with underground perennating rootstocks such as bulbs, corms, rhizomes; **halophyte**, a salt-adapted plant, e.g., *Salicornia* (Chenopodiaceae); **mycotroph** [myco-heterotroph], a usually achlorophyllous plant having an intimate contact with a fungus, from which it receives most of its nutrition; **phreatophyte**, a plant with a long taproot, in contact with ground water (e.g., mesquite, *Prosopis*); **succulent**, a plant with fleshy stems (**stem succulents**, e.g., cacti) or leaves (**leaf succulents**, e.g., members of Aizoaceae or Crassulaceae); **therophyte**, an annual plant; and **xerophyte**, a plant adapted to live in a dry, generally hot environment;

Plant life form types denoting nutritional physiology include: **saprophyte**, a heterotrophic plant living off dead organic matter; **mycotroph [myco-heterotroph]**, a usually achlorophyllous plant having an intimate contact with a fungus, from which it receives most of its nutrition; and **parasite**, a plant feeding on another plant, dependent on it for all or part of its nutrition. A parasite may be further classified into a **holoparasite**, one that lacks photosynthesis/chloroplasts and must attach to a host to survive and reproduce (e.g., *Cuscuta*, *Orobanche*, *Balanophora*), or a **hemiparasite**, a parasitic plant that is photosynthetic during at least part of its life cycle. A hemiparasite may be either an **obligate hemiparasite**, requiring a host to survive and reproduce (e.g., *Phoradendron*, mistletoe, Viscaceae) or a **facultative hemi-parasite**, not requiring a host connection to survive and reproduce (e.g., *Krameria*, Krameriaceae, or *Pedicularis*, lousewort, a photosynthetic member of the Orobanchaceae).

ROOTS

Roots are plant organs that function in anchorage and in absorption of water and minerals. Roots are found in all of the vascular land plants except for the psilotophytes—*Psilotum* and relatives. (As discussed earlier, nonvascular land plants have **rhizoids** that assume a similar function.)

Roots, like shoots, develop by the formation of new cells within the actively growing **apical meristem** of the root tip. The apical meristem is covered on the outside by a **rootcap**, functioning both to protect the root apical meristem and to provide lubrication as the root grows into the soil. The epidermal cells away from the root tip develop hairlike extensions called **root hairs**; these function in greatly increasing the surface area available for water and mineral absorption. Roots of many (if not most) species of plants have an interest-

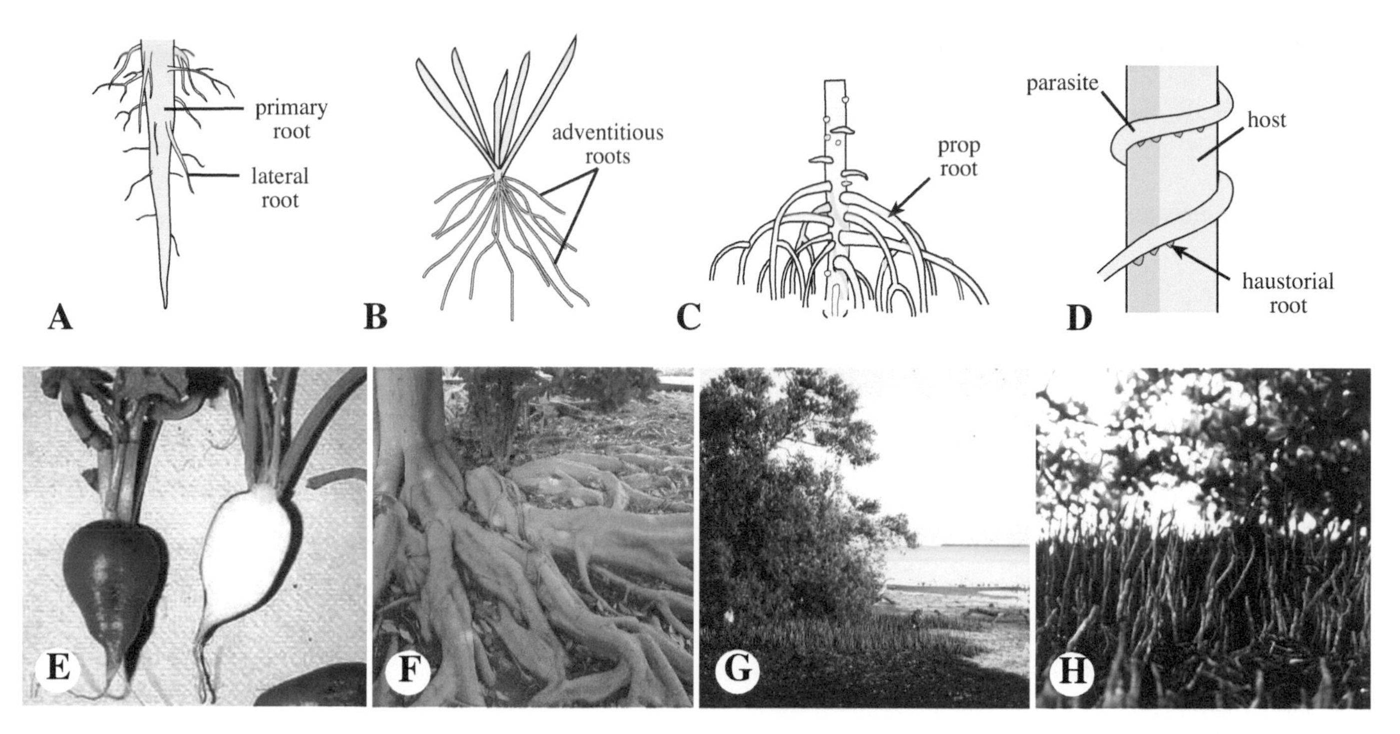

Figure 9.2 Root types. **A.** Tap root. **B.** Fibrous root system. **C.** Prop roots. **D.** Haustorial roots. **E.** Storage roots; *Raphanus sativus*, radish. **F.** Buttress roots; *Ficus rubiginosa*, fig. **G,H.** Pneumatophores; *Avicennia germinans*, black mangrove.

ing symbiotic interaction with a species of fungus, known as **mycorrhizae**. Although the exact function of mycorrhizae is often unclear, in some species at least the fungus host aids the plant both in increasing overall surface area for absorption and in increasing the efficiency of mineral uptake, particularly phosphorus. Roots have a central **vascular cylinder** of conductive cells, xylem and phloem. This vascular cylinder is surrounded by a special cylinder of cells known as the **endodermis**. Lateral roots develop by cell divisions within the **pericycle**, a cylindrical layer of parenchyma cells located just inside the endodermis itself. (See Chapter 10 for more details of root anatomy.)

The first root to develop in a vascular plant is the **radicle** of the embryo. If the radicle continues to develop after embryo growth, it is known as the **primary root**. Additional roots may arise from internal tissue of either another root, the stem/shoot (often near buds), or (rarely) a leaf. Roots that arise from other roots are called **lateral roots**. Roots that arise from a nonroot organ (stem or leaf) are **adventitious roots**.

Root Types (Figure 9.2)

Various modifications of roots have evolved. If the primary root becomes dominant, it is called a **taproot**, and the plant is described as having a **taproot system**. If the primary root soon withers and subsequent roots are adventitious, the plant has a **fibrous root system**. Several plant species, particularly those that are biennials, have **storage roots** in which the taproot has become greatly thickened, accumulating reservoirs of high-energy storage compounds (usually starch). Many plants that are epiphytic (grow on another plant), particularly tropical members of the monocot families Araceae and Orchidaceae, have **aerial roots**. These are adventitious roots that generally do not enter the soil and may absorb water and minerals from the air or from runoff from plants. Many plant species with bulbs or corms have **contractile roots**, roots that actually contract vertically, functioning to pull the rootstock further into the soil. Parasitic plants have specialized roots called **haustoria** that penetrate the tissues of a host plant. Some adventitious roots called **prop roots** grow from the base of the stem and function to further support the plant. Some plant species that grow in swamps or marshes have **pneumatophores**, roots that grow upwardly from soil to air that function to obtain additional oxygen. **Buttress roots** are enlarged, horizontally spreading and often vertically thickened roots at the base of trees that aid in mechanical support; they are found in certain tropical or marsh/swamp tree species.

STEMS AND SHOOTS

Stems function both as supportive organs (supporting and usually elevating leaves and reproductive organs) and as conductive organs (conducting both water/minerals and sugars through the vascular tissue between leaves, roots, and reproductive organs). Structurally, stems can be distinguished from roots based on several anatomical features (see Chapter 10). As mentioned earlier, a **shoot** is a stem plus its associated leaves. Sporophytic shoots that are branched and bear leaves are an apomorphy for all extant vascular plants; the leafy shootlike structures of mosses and some liverworts are gametophytic and not directly homologous with shoots of vascular plants.

The first shoot of a seed plant develops from the **epicotyl** of the embryo (see Seeds). The epicotyl elongates after embryo growth into an axis (the stem) that bears leaves from its tip, which contains the actively dividing cells of the shoot **apical meristem**. Further cell divisions and growth results in the formation of a mass of tissue that develops into the immature leaf, called a **leaf primordium** (Figure 9.1). The point of attachment of a leaf to a stem is called the **node**. The region between two adjacent nodes is the **internode** (Figure 9.1). A bit later in development, the tissue at the upper (adaxial) junction of leaf and stem (called the **axil**) begins to divide and differentiate into a **bud primordium**. As the shoot matures, the leaves fully differentiate into an amazing variety of forms. The bud primordium matures into a **bud**, defined as an immature shoot system, often surrounded by protective scale leaves (see Buds). Buds have an architecture identical to the original shoot. They may develop into a lateral branch or may terminate by developing into a flower or inflorescence. **Vascular strands** run between stem and leaf, providing a vascular connection, composed of xylem and phloem, for water, mineral, and sugar transport. The vascular strands of leaves are termed **veins**.

The mostly parenchymatous tissue external to the vascular (conductive) tissue of a stem is termed the **cortex**. The **pith** is the central, mostly parenchymatous tissue, internal to the stem vasculature (e.g., in siphonosteles and eusteles). In monocots, in which there are numerous, "scattered" vascular bundles (an atactostele), the intervening parenchymatous tissue is termed **ground meristem** (see Chapter 10).

The stems of some vascular plants, notably the conifers and nonmonocot flowering plants, contain **wood**, which technically is secondary xylem tissue, derived from a vascular cambium (see Chapter 10). In these woody plants **bark** refers to all the tissues external to the vascular cambium, consisting of secondary phloem (inner bark), leftover cortex, and derivatives of the cork cambium (the last comprising the outer bark, or periderm; see Chapter 10).

Stem/Shoot Types (Figure 9.3)

Various modifications of stems and shoots have evolved, many representing specific adaptations. For example, perennial and some biennial herbs have underground stems, which are generally known as **rootstocks**. Rootstocks function as storage and protective organs, remaining alive underground during harsh conditions of cold or drought. When environmental conditions improve, rootstocks serve as the site of new shoot growth, sending out new adventitious roots and new aerial shoots from the apical meristem or from previously dormant buds. Different types of rootstocks have evolved in various taxonomic groups. These include the following:

1. **Bulb**, in which the shoot consists of a small amount of vertical stem tissue (bearing roots below) and a massive quantity of thick, fleshy storage leaves (e.g., *Allium* spp., onions)
2. **Corm**, in which the shoot consists mostly of generally globose stem tissue surrounded by scanty, scale-like leaves (e.g., some *Iris* spp., irises)
3. **Caudex**, in which the rootstock consists of a relatively undifferentiated but vertically oriented stem
4. **Rhizome**, in which the stem is horizontal and underground, typically with short internodes (compare stolon, below) and bearing scale-like leaves (e.g., *Zingiber officinale*, ginger)
5. **Tuber**, which consists of a thick, underground storage stem, usually not upright, typically bearing outer buds and lacking surrounding storage leaves or protective scales (e.g., *Solanum tuberosum*, potato).

Rootstocks may function as reproductive structures in vegetative (clonal) propagation, either by splitting apart into separate plants or by forming proliferative structures that subsequently separate (and may even be dispersed by animals). For example, buds in the axils of the leaves of bulbs can develop into proliferative **bulbels** (e.g., garlic); some taxa (e.g., certain onions) can even form tiny, propagative bulbs within the aerial shoots or inflorescence of the plant, these termed **bulbils**. Cormose plants can, similarly from axillary buds, form proliferative corms, termed **cormels**. Tuberous plants typically form numerous tubers at the tips of elongate stems; these tubers can become easily separated, growing into an individual plant. Tubers can even form on aerial shoots (e.g., *Dioscorea*, true yams), ultimately falling off and growing into a new individual. Rhizomes frequently become highly branched; when older parts die or become broken, the separated rhizomes function as separate individuals.

A **stolon** or **runner** is a stem with long internodes that runs on or just below the surface of the ground, typically terminating in a new plantlet, as in *Fragaria* (strawberry). Because stolons can be underground, they are sometimes termed rootstocks and resemble narrow, elongate rhizomes. Stolons func-

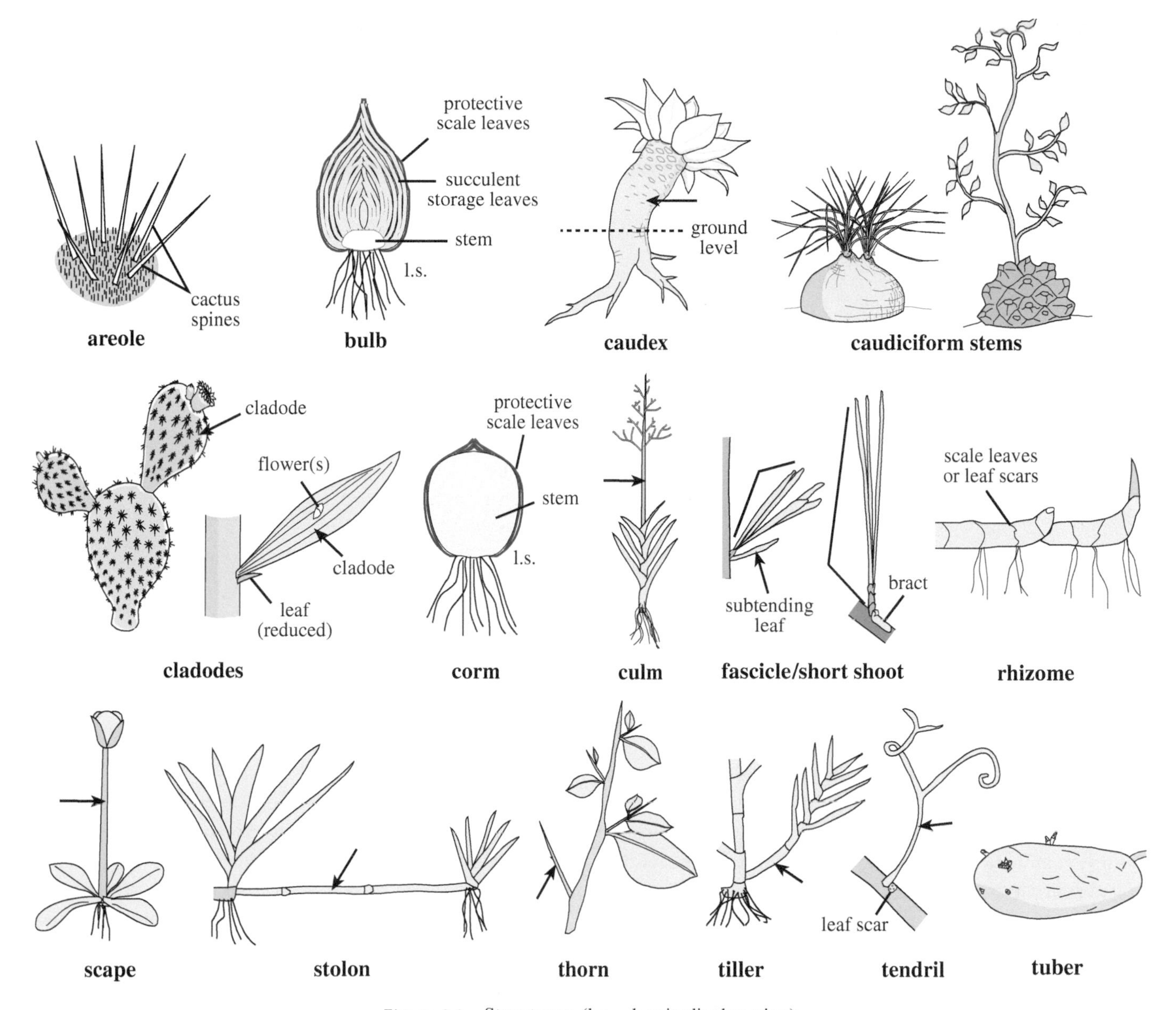

Figure 9.3 Stem types. (l.s. = longitudinal section)

tion specifically as vegetative propagative structures, however, as the terminal plantlet often becomes separated from the parent plant.

Many modified types of stems that are aerial (above-ground) also have specific functions. For example, a **cladode** (also termed a **cladophyll** or **phylloclade**) is a flattened, photosynthetic stem that may resemble and function as a leaf, found, e.g., in prickly-pear cacti, *Asparagus*, and *Ruscus*. Cladodes take over the primary photosynthetic function of leaves and may function to reduce water loss.

Some aerial stems may function for storage of food reserves or water. So-called succulent stems (the plants often referred to as "stem succulents") contain a high percentage of parenchyma tissue that may store great quantities of water, allowing the plant to survive subsequent drought periods. The cacti of the New World and the stem succulent euphorbs of South Africa are classic examples of plants with succulent stems. Some of these, most notably the barrel cacti and the large columnar cacti such as saguaro's or cardon's, have fluted trunks that can expand rapidly following a rain, enabling the plant to store more water. Other aerial, storage stems include:

1. A **caudiciform stem**, which is a low, swollen, perennial storage stem (at or above-ground level), from which arise annual or nonpersistent photosynthetic shoots (e.g., *Calibanus*, some *Dioscorea* spp.)

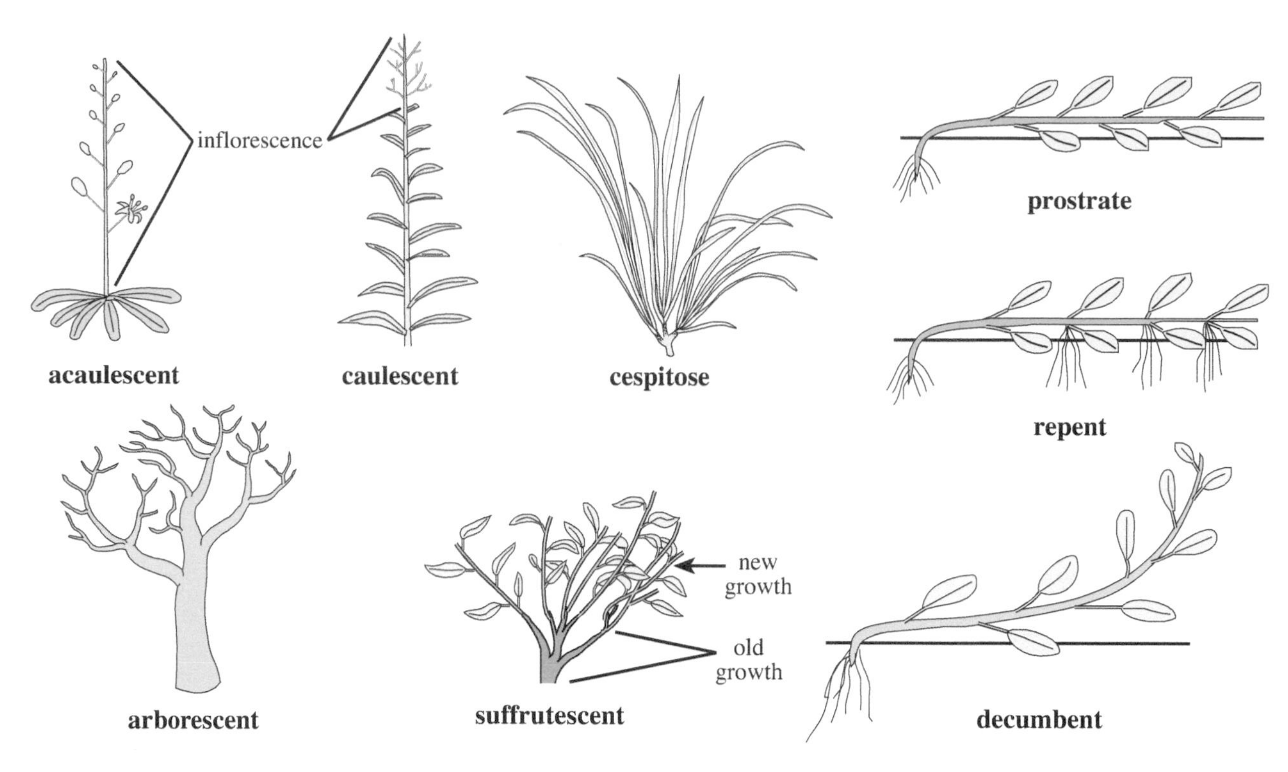

Figure 9.4 Stem habit.

2. A **pachycaul**, which is a woody, trunklike stem that is swollen basally, the swollen region functioning in storage (e.g., bottle trees, *Brachychiton* spp., and the boojum tree, *Fouquieria columnaris*).

Some stems or shoot types function as protective devices by deterring an herbivore from taking a bite of the plant. A **thorn** is a sharp-pointed stem or shoot. (A thorn is not to be confused with a spine, which is a sharp-pointed leaf or leaf part, or a prickle, which is a sharp-pointed epidermal structure found anywhere on the plant; see later discussion.) A very specialized type of shoot is the **areole**, a modified, reduced, nonelongating shoot apical meristem bearing leaf spines. Areoles are characteristic of the cactus family, Cactaceae.

Some stems are specialized for reproduction. For example, a **scape** is a "naked" (lacking vegetative leaves) peduncle (inflorescence axis), generally arising from a basal rosette of vegetative leaves and functioning to elevate flowers well above the ground. A **culm** refers to the flowering and fruiting stem(s) of grasses and sedges. A **tiller** is the general term for a proliferative grass shoot, typically growing in masses from axillary buds at the base of the stem.

Stems may have multiple or varied functions. A **lignotuber** or **burl** is largely a protective and regenerative stem following fires. Lignotubers or burls are typically swollen, woody stems, at or slightly below ground level, from which arise persistent, woody, aerial branches (e.g., some *Manzanita* spp.). A **pseudobulb** is a short, erect, aerial storage or propagative stem of certain epiphytic orchids. A **short shoot** or **fascicle** (also called a spur shoot or dwarf shoot) is a modified shoot with very short internodes from which flowers or leaves are borne. Short shoots enable the production of leaves or reproductive organs relatively quickly, with minimal stem (branch) tissue being formed. Short shoots may be found on so-called drought deciduous plants, which produce a quick flush of leaves from short shoots following a rain. Short shoots arise from the buds of more typical shoots (branches) with longer internodes, the latter termed, in contrast, *long shoots*. Finally, a **tendril** is a long, slender, coiling branch, adapted for climbing. Tendrils are typically found on weak -stemmed vines and function in support. (Note that most tendrils are leaves or leaf parts; see Leaf Structural Type.)

Stem Habit (Figure 9.4)

Stem habit is a character describing the relative position of the stem or shoot, but may also be based on stem structure, growth, and orientation. Stem habit features, like stem types, represent adaptations that enhance survival and reproduction. For example, a plant with an above-ground stem is **caulescent**;

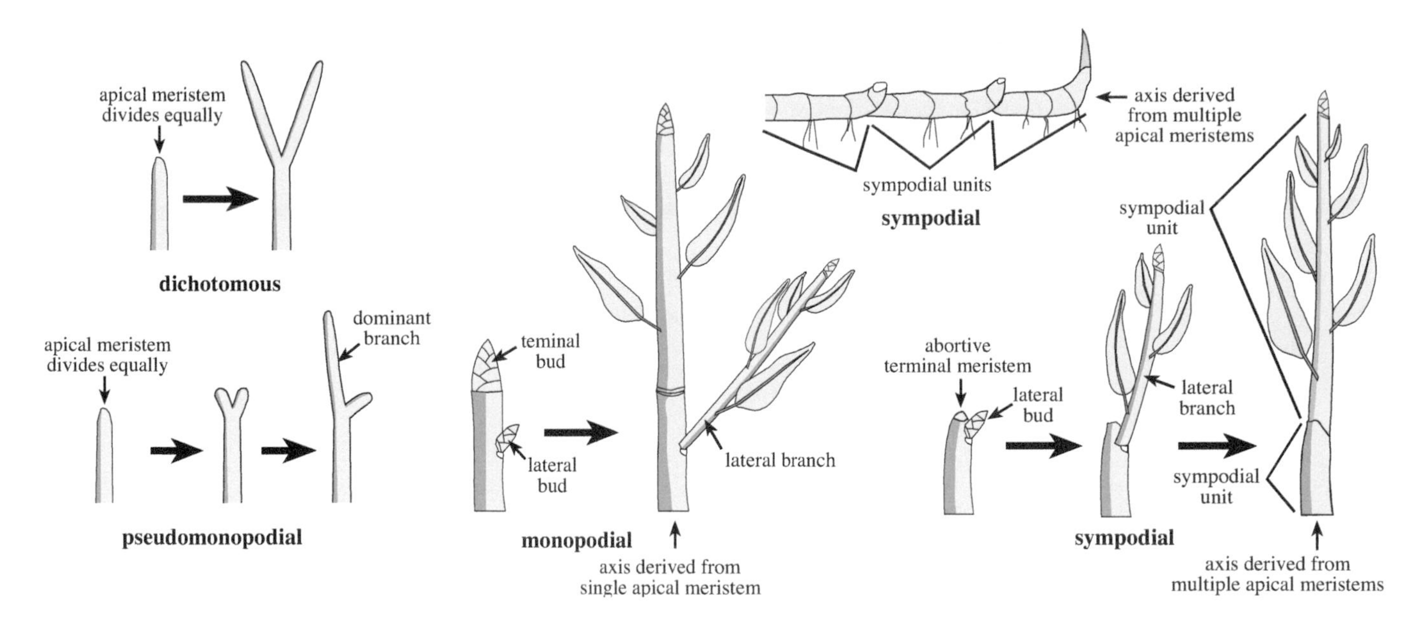

Figure 9.5 Stem branching patterns.

one that lacks an above-ground stem, other than the inflorescence axis, is termed **acaulescent**. Acaulescent plants bear major photosynthetic leaves only at ground level, often in a basal rosette, with the only shoot becoming aerial, being an inflorescence that eventually dies off. Acaulescent plants are often biennial herbs, in which a storage root develops in the first year and flowering (bolting) occurs in the second, or perennial herbs, in which the persistent stem remains underground and protected during extreme environmental conditions. Plants with caulescent stem habits include shrubs, trees, and herbs with aerial vegetative shoots and leaves. Some corresponding stem habit terms are **arborescent**, treelike in appearance and size; **frutescent**, having the habit of a shrub, with numerous, woody, aerial trunks; and **suffrutescent**, being basally woody and herbaceous apically, the habit of a subshrub. Vines are also types of caulescent plants. The stem habit of vines can be either **clambering** (also called **scandent**), sprawling across objects without specialized climbing structures, or **climbing**, growing upward by means of tendrils, petioles, or adventitious roots. Some plants are adapted to lying on the ground, at least in part. These include those that are **prostrate** (also called **procumbent**), trailing or lying flat, not rooting at the nodes; **repent**, creeping or lying flat but rooting at the nodes; or **decumbent**, being basally prostrate but apically ascending. Finally, some plants have a **cespitose** stem habit, in which multiple aerial but short-stemmed shoots arise from the base, forming a much-branched cushion. Many grasses are cespitose, these being the so-called "bunch" grasses.

Stem Branching Pattern (Figure 9.5)
The below- or above-ground stems or shoots of a plant often exhibit characteristic branching patterns. Branching pattern is determined by the relative activity of apical meristems, both the "original" shoot apical meristem derived from the seedling epicotyl and apical meristems subsequently derived from lateral buds. One major feature of branching pattern has to do with the duration of apical meristematic growth of a shoot. If a given shoot has the potential for unlimited growth, such that the apical meristem is continuously active, the growth is termed **indeterminate**. If instead a shoot terminates growth after a period of time, with either the abortion of the apical meristem or its conversion into a flower, inflorescence or specialized structure (such as a thorn or tendril), the growth is termed **determinate**. (Note that these same terms are used for inflorescence development; see later discussion.)

Other terms for branching pattern center on the developmental origin of a given branch or axis. A relatively rare type of branching is **dichotomous,** in which a single apical meristerm divides equally into branches, e.g., *Psilotum*. A variant of dichotomous is branching that is **pseudomonopodial**, in which one branch of the initial dichotomy overtops and becomes dominant over the other. Dichotomous and pseudomonopodial branching are found primarily in lycophytes. In contrast, if a given stem axis is derived from growth of a *single* apical meristem, the pattern is termed **monopodial**. The monopodial axis may grow indefinitely and thus be indeterminate. In contrast, if a given axis (which may appear to be a single, continuous structure) is made up of numerous units that are derived from *separate* apical meristems, the branching pattern is **sympodial**. These sympodial units arise from lateral buds that are proximal to the apical meristem of the original shoot. Many rhizomes have sympodial growth.

Stem Branching Models (Figure 9.6)
Aside from the general stem branching patterns, more specific models of tree branching pattern have been described (Hallé et al. 1978). These are used almost exclusively for trees, but may be used with herbs as well. The models are based first on whether the tree is **monoaxial**, unbranched with a single (vegetative) apical meristem, or **polyaxial**, branched with more than one vegetative apical meristem. Additional considerations are whether the shoots are **orthotropic**, erect and essentially radially symmetric, the branching three-dimensional, or **plagiotropic**, more or less horizontal with dorsiventral symmetry, the branching two -dimensional and leaves generally in one plane (either distichous or secund). Plagiotropy may occur in two general ways. **Plagiotropy by apposition** is that in which extension growth of the branch is taken over by an axillary meristem, but with the original branch terminal meristem continuing growth, usually as a short shoot. **Plagiotropy by substitution** is that in which the original branch terminal meristem aborts or converts into a terminal inflorescence or flower, extension growth of the branch being taken over by an axillary meristem. In addition, the timing of development of a shoot can be important in plant growth. **Syllepsis** (or sylleptic growth) is growth of an axillary bud into a shoot without a period of rest. **Prolepsis** (or proleptic growth) is growth of an axillary bud into a shoot only after a period of rest.

Other, related terms have to do with flowering. An indeterminate shoot that bears lateral flowers but that continues vegetative growth is termed **pleonanthic**. A plant with a determinate shoot that completely transforms into a flower or inflorescence is called **hapaxanthic**. If the entire plant flowers and fruits only once, and then dies, it is termed **monocarpic**; the plant itself can be an annual or perennial, but the term is usually used only for perennials, given that all annuals are monocarpic.

The following are tree growth models (after Hallé et al., 1978, illustrated in Figure 9.6), each of which is named after a botanist who contributed to our knowledge of that model. Only a very few examples of the models are listed, ones that might be familiar to the reader; see Hallé et al. (1978) for elaboration of the models and considerably more examples. An important concept, however, is that of **reiteration**, the growth of shoots not conforming to the parameters of the model, e.g., due to environmental stress, such as mechanical or animal damage, obscuring its normal expression. Although these models are rather specialized, they are useful in the study of tree architecture (and a challenging and intriguing exercise for the student to decipher). A given model may represent the end product of evolutionary adaptations to a given environment or life strategy and their elucidation may have taxonomic, ecological, or biomechanical significance.

Attim's Model. Polyaxial; with a monopodial trunk with continuous growth, bearing equivalent branches, flowers always lateral. E.g., *Avicennia germinans*, black mangrove (Acanthaceae), *Alnus incana* (Betulaceae), *Casuarina equisetifolia* (Casuarinaceae), *Euphorbia* spp. (Euphorbiaceae), *Eucalyptus* spp. (Myrtaceae), *Rhizophora mangle*, red mangrove (Rhizophoraceae).

Aubréville's Model. Polyaxial; rhythmically growing and branching, with rhythmic growth, having a monopodial trunk bearing modular whorls or pseudo-whorls of branch tiers, plagiotropic by apposition, all with a similar phyllotaxis, the inflorescences lateral. E.g., *Terminalia* spp. (Combretaceae), *Manilkara zapota* (Sapotaceae).

Chamberlain's Model. Polyaxial; having regular, sympodial branching, the modules usually orthotropic, each hypaxanthic by producing a terminal flower or inflorescence, linear growth continued by distal, lateral meristems. E.g., many cycads (such as male *Cycas* spp.), *Jatropha* spp. (Euphorbiaceae), *Dieffenbachia* spp., *Philodendron selloum* (Araceae).

Champagnat's Model. Polyaxial; with successive modular orthotropic axes with spiral leaf arrangement, each curving and becoming pendulous by its own weight, a new modular unit arising from the upper part of the curved axis. E.g., *Sambucus* spp. (Adoxaceae), *Crescentia cujete*, calabash tree (Bignoniaceae), *Caesalpinia pulcherrima* (Fabaceae), *Lagerstroemia indica*, crepe myrtle (Lythraceae), *Bougainvillea spectabilis* (Nyctaginaceae).

Corner's Model. Monoaxial, in which the inflorescences or sporophylls are lateral, the single stem capable of growth after flowering, not monocarpic. E.g., many tree ferns, cycads (such as female *Cycas* spp.), many palms (Arecaceae).

Fagerlind's Model. Polyaxial; with a monopodial, orthotropic trunk producing tiers of modular branches, each branch sympodial and plagiotropic by apposition, with spiral or decussate leaves. E.g., *Hymenosporum flavum* (Pittosporaceae); *Magnolia grandiflora*, flowering magnolia (Magnoliaceae).

Holttum's Model. Monoaxial, with the terminal meristem developing entirely into an inflorescence, the tree dying after fruit maturation, therefore monocarpic. E.g., many *Agave* spp. (Agavaceae), many palms (Arecaceae).

Koriba's Model. Polyaxial; having orthotropic modules, each of which is sympodial and aborts or produces a terminal inflorescence, the modules initially equivalent, but later one becoming dominant and erect as a trunk, the others developing into branches. E.g., *Ochrosia* spp. (Apocynaceae), *Catalpa* spp. (Bignoniaceae), *Phytolacca dioica* (Phytolaccaceae).

Leeuwenberg's Model. Polyaxial; having equivalent, orthotropic modules, each of which is sympodial and produces a terminal inflorescence, with two or more new modules arising below it. E.g., tree *Aloë* spp. (Asphodelaceae), *Dracaena draco* (Ruscaceae), *Nerium oleander*, *Pachypodium* spp. (Apocynaceae), *Pandanus* spp. (Pandanaceae).

Mangenot's Model. Polyaxial; with axes composed of modular units from a single apical meristem composed of an orthotropic proximal part (leaves often spiral), abruptly recurving into a plagiotropic distal part (the leaves often distichous), a new module orthotropically arising from the bend of the recurved section. E.g., *Vaccinium corymbosum*, blueberry (Ericaceae), *Strychnos* sp., strychnine (Loganiaceae), *Eugenia* sp. (Myrtaceae).

Massart's Model. Polyaxial; rhythmically growing and branching, with an orthotropic, monopodial trunk having rhythmic growth, producing regular tiers of lateral branches that are plagiotropic by leaf arrangement or symmetry, never by apposition, reproductive structure position variable. E.g., *Agathis* spp., *Araucaria* spp. (Araucariaceae), *Sequoia sempervirens*. redwood (Cupressaceae), *Diospyros* spp. (Ebenaceae), *Myristica fragrans* (Myristicaceae), *Abies* spp. (Pinaceae).

Nozeran's Model. Polyaxial; rhythmically growing and branching, with an orthotropic, sympodial trunk, each sympodial unit bearing a distal tier of monopodial or sympodial plagiotropic branches, the leaf arrangement of trunk and branches different. E.g., *Theobroma cacao*, chocolate (Malvaceae).

Petit's Model. Polyaxial; continuously growing and branching, with a monopodial, orthotropic trunk producing tiers of modular branches, each branch sympodial and plagiotropic by substitution (thus branch modules hapaxanthic), with spiral or decussate leaves. E.g., *Gossypium* spp., cotton (Malvaceae); *Morinda citrifolia* (Rubiaceae).

Prévost's Model. Polyaxial; rhythmically growing and branching, having two types of orthotropic modules forming trunk and branches from inception, the branch modules plagiotropic by apposition, arising sylleptically from subapical region of the trunk module, successive trunk modules proleptic and arising well below the tier of branch modules. E.g., *Cordia* spp. (Cordiaceae), *Euphorbia pulcherrima*, crucifixion thorn (Euphorbiaceae).

Rauh's Model. Polyaxial; with a monopodial trunk, rhythmically producing tiers of branches, each branch identical to trunk; flowers always lateral. E.g., *Ilex* spp. (Aquifoliaceae), *Araucaria* spp. (Araucariaceae); *Kalanchoe beharensis* (Crassulaceae), *Euphorbia* spp. (Euphorbiaceae), *Quercus* spp. (Fagaceae), *Cecropia* spp./*Ficus* spp. (Moraceae), *Fraxinus* spp. (Oleaceae), most *Pinus* spp. (Pinaceae); *Acer* spp. (Sapindaceae).

Roux's Model. Polyaxial; continuously growing and branching, with a monopodial, orthotropic trunk having spiral leaf arrangement, bearing lateral branches that are plagiotropic, but never by apposition, usually with distichous leaf arrangement, the branches inserted continuously on the trunk, reproductive structure position variable, but usually lateral on branches. E.g., *Polyalthia* spp. (Annonaceae), *Dipterocarpus* spp. (Dipterocarpaceae), *Bertholletia excelsa*, Brazil nut (Lecythidaceae), *Durio zibethinus*, durio (Malvaceae), *Coffea arabica* (Rubiaceae).

Scarrone's Model. Polyaxial; rhythmically growing and branching, with a monopodial trunk bearing tiers of branches, each branch sympodial by terminal flowering, but becoming orthotropic. E.g.,Anacardium occidentale, cashew, *Mangifera indica*, mango (Anacardiaceae), *Jacaranda mimosifolia* (Bignoniaceae), *Echium* spp. (Boraginaceae), *Aeonium* spp. (Crassulaceae), *Arbutus unedo* (Ericaceae), *Aesculus* spp., horse-chestnut (Sapindaceae).

Schoute's Model. Polyaxial; growth from meristems that produce orthotropic or plagiotropic trunks forking at regular, distinct intervals by equal dichotomy, otherwise with no lateral branches; inflorescences lateral. E.g., *Hyphaene thebaica*, *Nypa fruticans* (Arecaceae); *Flagellaria indica* (Flagellariaceae).

Stone's Model. Polyaxial; continuously growing and branching, with a orthotropic trunk that may flower terminally, bearing orthotropic branches, with additional branching occurring sympodially below terminal inflorescences. E.g., *Mikania cordata* (Asteraceae), *Pandanus* spp. (Pandanceae).

Tomlinson's Model. Polyaxial; vegetative axes all equivalent, orthotropic, with equivalent orthotopic modules developing from basal nodes in subsequent axes. E.g., many monocots (Bromeliaceae, Cyperaceae, Poaceae, Zingiberales), *Euphorbia* spp. (Euphorbiaceae), *Kalanchoë* spp. (Crassulaceae).

Troll's Model. Polyaxial; with all axes plagiotropic, the modules superposed upon one another, the proximal part of each module becoming erect in development, the distal part a branch. E.g., *Psidium guineense* (Myrtaceae), *Erythroxylum coca* (Erythroxylaceae), *Albizzia julibrissin*, *Bauhinia* spp. (Fabaceae), *Fagus grandifolia*, beech (Fagaceae), *Averrhoa carambola*, star-apple (Oxalidaceae).

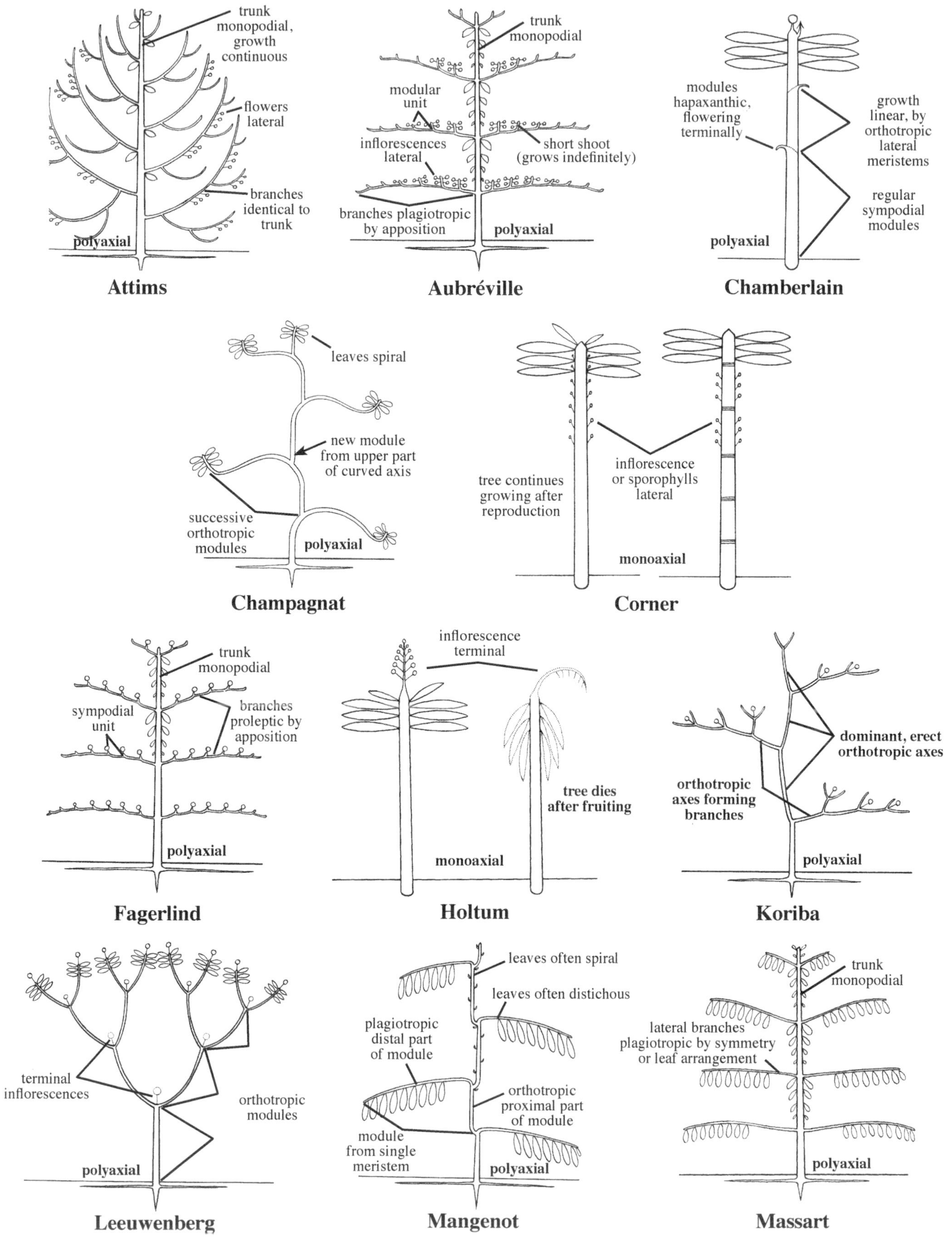

Figure 9.6 Stem branching pattern models, after Hallé et al. (1978). See text for explanation.

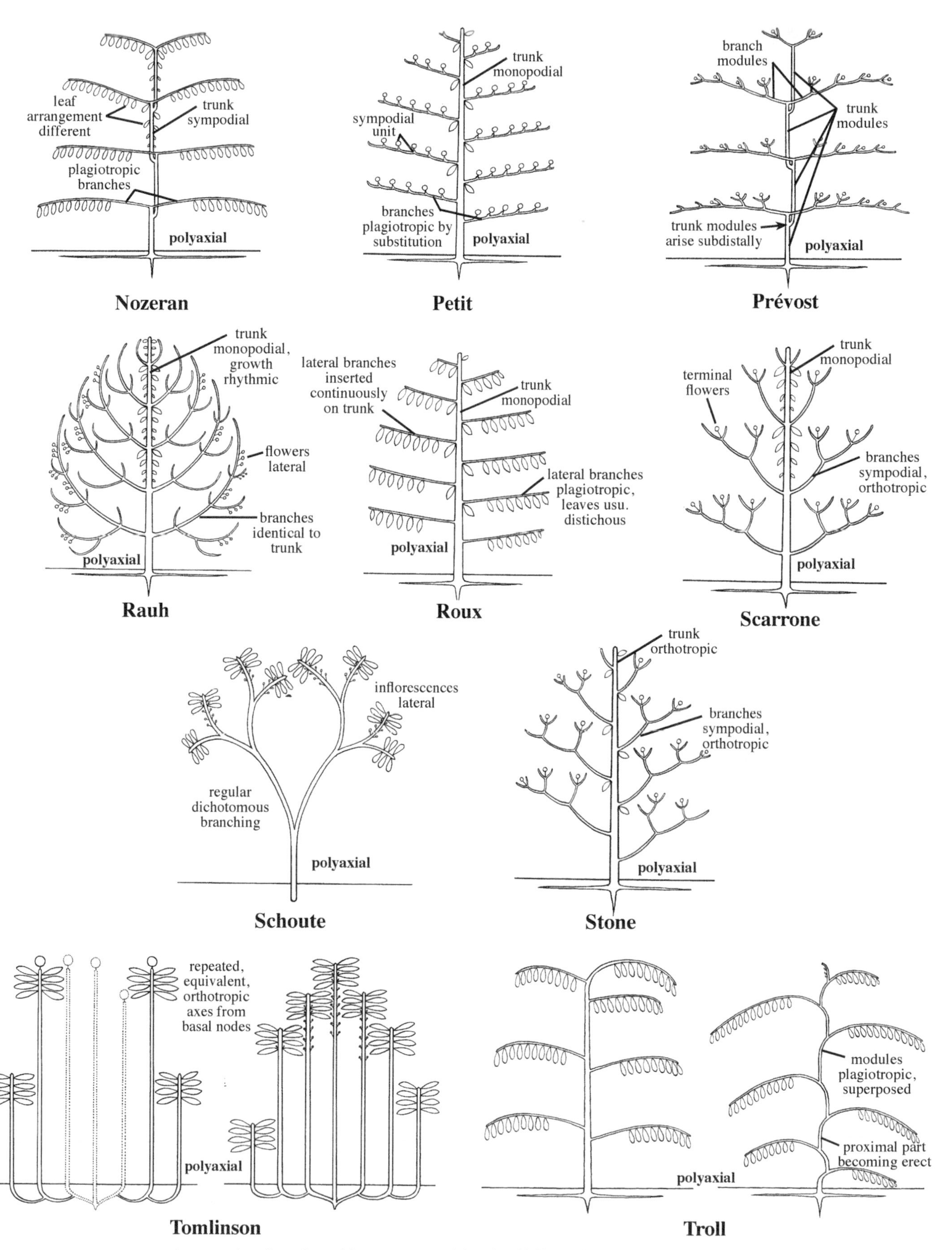

Figure 9.6 (continued) Stem branching pattern models, after Hallé et al. (1978). See text for explanation.

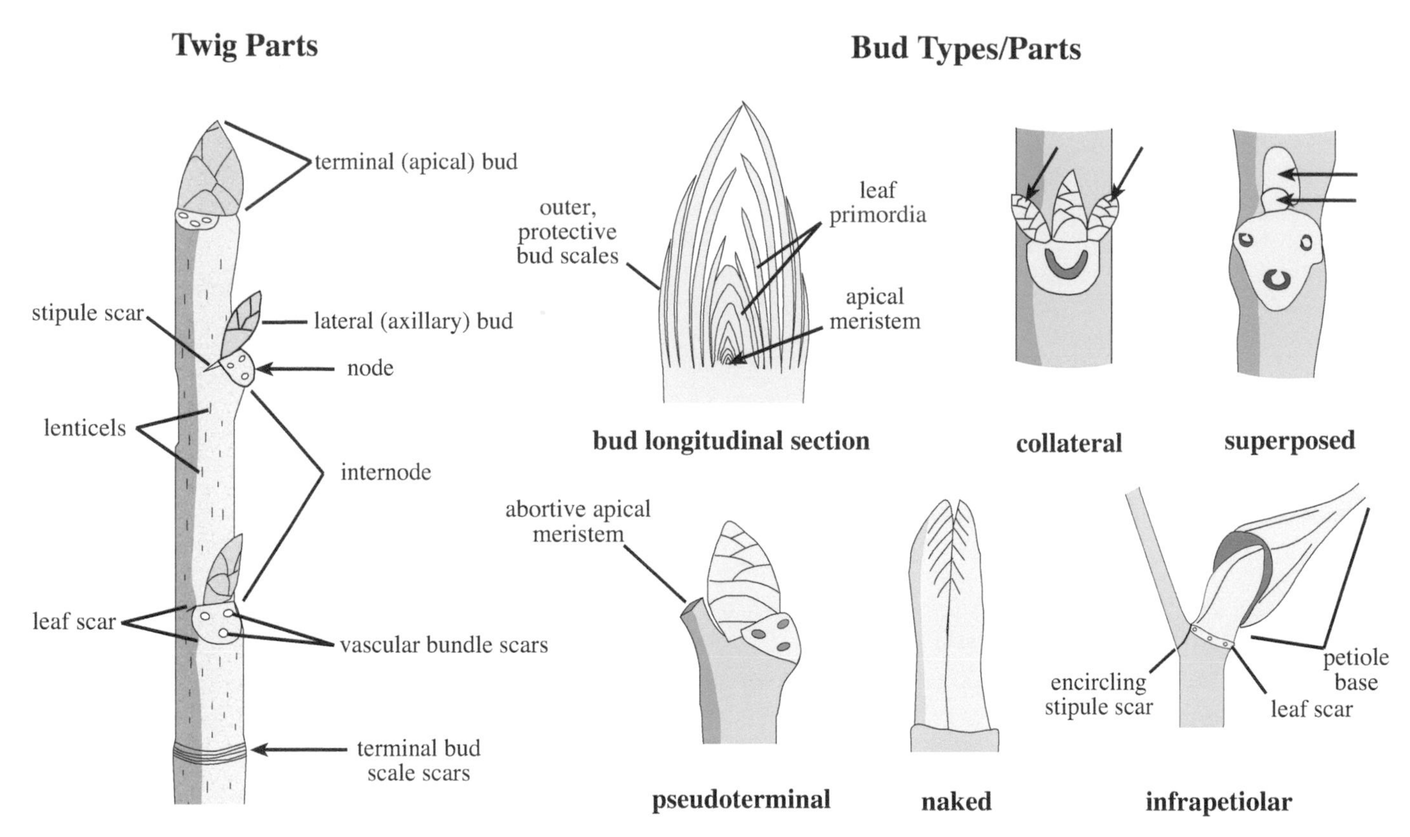

Figure 9.7 Twigs parts and bud types (l.s. = longitudinal section).

Twigs, Trunks, and Buds (Figure 9.7)

Twigs are the woody, recent-growth branches of trees or shrubs. **Buds** are immature shoot systems that develop from meristematic regions. In deciduous woody plants the leaves fall off at the end of the growing season and the outermost leaves of the buds may develop into protective **bracts** (modified leaves) known as **bud scales**. The bud of a twig that contains the original apical meristem of the shoot (which by later growth may result in further extension of the shoot) is called the **terminal** or **apical** bud. Buds formed in the axils of leaves are called **axillary** [**axial**] or **lateral** buds.

A given bud may be **vegetative**, if it develops into a vegetative shoot bearing leaves; **floral** or **inflorescence**, if it develops into a flower or inflorescence; or **mixed**, if it develops into both flower(s) and leaves. In some species more than one axillary bud forms per node. Two or more axillary buds that are oriented sideways are called **collateral buds**; two or more axillary buds oriented vertically are called **superposed buds**. If the original terminal apical meristem of a shoot aborts (e.g., by ceasing growth or maturing into a flower), then an axillary bud near the shoot apex may continue extension growth; because this axillary bud assumes the function of a terminal bud, it is called a **pseudoterminal bud**.

Several scars may be identified on a woody, deciduous twig. These include the **leaf scar**, **leaf vascular bundle scars**, **stipule scars** (if present), and **bud scale scars.** Bud scale scars represent the point of attachment of the bud scales of the original terminal bud after resumption of growth during the new season. Thus, bud scale scars represent the point where the branch ceased elongation the previous growing season; the region between adjacent bud scale scars represents a single year's growth in temperate climates, but could be shorter or longer in tropical climates.

Bark technically comprises all the tissue outside the vascular cambium of a plant with true wood (see Chapter 10). The outer bark, or periderm, are the tissues derived from the cork cambium itself. Morphologically, bark may refer to the outermost protective tissues of the stems or roots of a plant with some sort of secondary growth, whether derived from a true cork cambium or not. Bark types are often good identifying characteristics of plant taxa, particularly of deciduous trees during the time that the leaves have fallen. Various bark types include:

1. **Exfoliating**, a bark that cracks or splits into large sheets
2. **Fissured**, a bark split or cracked into vertical or horizontal grooves

3. **Plated**, a bark split or cracked, with flat plates between the fissures
4. **Shreddy**, bark coarsely fibrous
5. **Smooth**, a non-fibrous bark without fissures, fibers, plates, or exfoliating sheets.

LEAVES

Leaves are the primary photosynthetic organs of plants, functioning also as the main site of transpiration. Leaves are derived from leaf primordia of the shoot apex and are, at least early in development, generally "dorsiventrally" flattened (i.e., with "dorsal" and "ventral" sides; see Position). A leaf can be gametophytic, in the leafy liverworts and mosses, or sporophytic, in the vascular plants. As mentioned earlier, sporophytic leaves characteristically are associated with **buds**, immature shoot systems, typically located in the axils of leaves. Buds may grow to form lateral vegetative branches or reproductive structures (see later discussion).

Leaf Parts (Figures 9.8, 9.9, 9.10)

The expanded, flat portion of the leaf, which contains the bulk of the chloroplasts, is termed the **blade** or **lamina**. Many leaves also have a proximal stalk, the **petiole** or (e.g., in ferns) the **stipe**. A leaf or leaf part (typically at the base) that partially or fully clasps the stem above the node is a leaf **sheath**, such as in the Poaceae (grasses) and many Apiaceae. A **pseudopetiole** is a petiole-like structure arising between a leaf sheath and blade, found in several monocots, such as bananas and bamboos. As mentioned earlier, leaves contain one to many vascular bundles, the **veins** (also sometimes called **nerves**); similar specialized (although not truly vascular) conductive tissue is present in mosses.

Many leaves have **stipules**, a pair of leaflike appendages, which may be modified as spines or glands, at either side of the base of a leaf. If stipules are present, the leaves are **stipulate**; if absent, they are **exstipulate**. A specialized, scarious, sheathlike structure arising above the node in some members of the family Polygonaceae, interpreted as modified stipules, is termed an **ocrea** (see Polygonaceae treatment in Chapter 8). **Stipels** are paired leaflike structures, which may also be modified as spines or glands, at either side of the base of the leaflet of a compound leaf, as in some Fabaceae. If stipels are present, the leaves are **stipellate**; if absent, they are **exstipellate**. Stipules and stipels may, in some cases, function to protect the young, developing leaf primordia. They often are small and fall off (are "caducous") soon after leaf maturation. In some taxa, stipules or stipels may be highly modified into spines or glands. Extreme examples are some African acacias, in which the swollen stipular spines function as a home for protective populations of ants. In the Rubiaceae the inner surface of the connate stipules (from opposite leaves) bear **colleters**, structures that secrete mucilage (aiding to protect young, developing shoots).

Some leaves are compound (as discussed later), i.e., divided into discrete components called **leaflets**. The stalk of a leaflet is termed the **petiolule**. Some other specialized leaf parts, restricted to certain taxa, are:
1. **Hastula**, an appendage or projection at the junction of petiole and blade, as in some palms
2. **Ligule**, an outgrowth or projection from the inner, top of the sheath, at its junction with the blade, as in the Poaceae
3. **Pulvinus**, the swollen base of a petiole or petiolule, as in some Fabaceae.

Leaf Behavior

The pulvinus may, in some taxa, e.g., some Fabaceae (legumes), function in **thigmonasty** [**seismonasty**] which is movement (closing) of the leaflets of a compound leaf as a response to touch, vibration, or heat (e.g., as in *Mimosa pudica*, a sensitive plant); a similar physiological response due to darkness (in photoperiodism) is termed **nyctinasty**. These physiological responses may protect the leaf from mechanical damage or help to inhibit water loss.

Leaf Structural Type (Figures 9.8, 9.9)

Leaf structural type (in contrast to "leaf type," discussed later) deals with specialized modifications of leaves. One basic leaf structural type in vascular plants is whether the leaves are lycophyllous or euphyllous. **Lycophylls** are small, simple leaves with intercalary growth and a single, central vein that joins to the stem without a leaf gap (below). Lycophylls are found only in lycophytes and are similar to the type of leaf found in the earliest ancestors of vascular plants. **Euphylls** are larger, simple or compound leaves with marginal or apical growth, a leaf gap (region of parenchymatous tissue above the junction of the leaf and stem vasculature), and generally multiple veins. Euphylls are found in ferns (in the broad sense), gymnosperms, and angiosperms (see Chapter 4).

A leaf that is modified in shape and usually smaller than the major photosynthetic leaves is called a **bract**. In angiosperms bracts are typically associated with flowers (flower bracts) or the axes of inflorescences (inflorescence bracts). A **bractlet** or **bracteole** (also called a **prophyll** or **prophyllum**) is a smaller or secondary bract often borne on the side of a pedicel in flowering plants. The term *bract* is also used for the largely nonphotosynthetic leaves that subtend the ovuliferous scales in conifer cones or that subtend the fascicles or short shoots of members of the pine family (Pinaceae). The term **scale** is used for a small, non-green leaf, either of a bud (**bud scales**), functioning to protect the delicate apical meristem and leaf primordia, or of an underground rootstock,

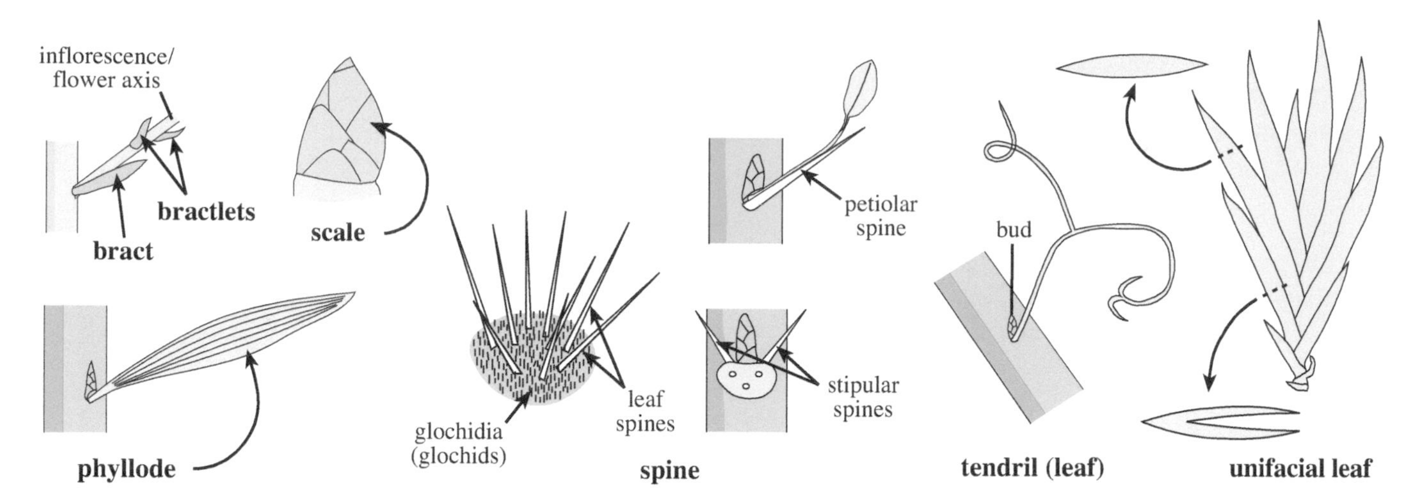

Figure 9.8 Leaf structural types.

e.g., along the internodes of a rhizome. **Squamate** means having or producing scales. Scales can also refer to the reduced bracts of sedge spikelets (Cyperaceae). The term **cataphyll** can be used to denote a scale-like, often non-green, protective leaf (e.g., in cycads or palms) or can refer to a rudimentary scale leaf found in usually hypogeous (cryptocotylar) seedlings.

Some bractlike leaves are found in specific taxonomic groups and are given specialized names. A group of bracts resembling sepals immediately below the true calyx is termed an **epicalyx** (**calyculus**) found, e.g., in many members of the Malvaceae. Bracts subtending individual flowers of composites (Asteraceae) are collectively termed **chaff** or **paleae** (singular, **palea**), e.g., as found in the tribe Heliantheae of that family. The specialized bracts of the grass (Poaceae) spikelet are given different terms: **glumes**, the two bracts occurring at the base of a grass spikelet; **lemma**, the outer and lower bract at the base of the grass floret; and **palea**, the inner and upper bract at the base of the grass floret (See **Inflorescence Type**, later, and treatment of Poaceae in Chapter 7.)

A **phyllary** is one of the involucral bracts subtending a head (see later discussion), as in the Asteraceae. A **spathe** is an enlarged, sometimes colored bract subtending and usually enclosing an inflorescence, e.g., that subtending the spadix of the Araceae.

Phyllodes are leaves that consist of a flattened, bladelike petiole. Phyllodes are found in a group of mostly Australian *Acacia* species (the phyllodinous Acacias) and are derived from ancestrally compound leaves by loss of the rachis and leaflets. A **tendril** is a coiled and twining leaf or leaf part, usually a modified rachis or leaflet. (*Tendril* can also refer to a modified, coiling stem; see Stem Type).

A **spine** is a sharp-pointed leaf or leaf part. The typical spines of cacti (Cactaceae) are **leaf spines**, as they develop from the entire leaf primordia. A very small, deciduous leaf spine with numerous, retrorse barbs along its length is a **glochidium** (plural, **glochidia** or **glochids**), as found in the areoles of opuntioid cacti. Some taxa have spines that develop from a petiole, midrib, or secondary vein of a leaf, e.g., the **petiolar spines** of *Fouquieria* spp. In some palms, e.g., *Phoenix*, the leaflets may be modified into sharp-pointed **leaflet spines**. Many plants, such as the stem-succulent *Euphorbias*, have **stipular spines**; these are typically paired, at the base of a leaf.

A **unifacial leaf** is *isobilateral*, i.e., flattened side-to-side and having a left and right side, except at the base, where they are often sheathing. Some monocots belonging to several different families have unifacial leaves, notably members of the Iridaceae, the Iris family. A **centric leaf** is one that is cylindrical in shape, e.g., *Fenestraria* of the Aizoaceae. Centric leaves are sometimes a subcategory of unifacial leaves.

Some leaves are very specialized adaptations of carnivorous plants. **Pitcher** (or "pitfall") leaves are those that are shaped like a container, which bears an internal fluid and functions in the capture and digestion of small animals. Several taxa have pitcher leaves, including among others *Nepenthes* (Nepenthaceae) and *Sarracenia* (Sarraceniaceae), the pitcher plants. **Tentacular** leaves (an "adhesive" or flypaper" type) are those bearing numerous, sticky, glandular hairs or bristles that function in capturing and digesting small animals; these are characteristics of *Drosera* spp. (Droseraceae), the sundews. **Snap-trap** leaves are those that mechanically move after being triggered, in the process capturing and digesting small animals, found in *Dionaea muscipula*, Venus fly trap, and *Aldrovandra vesiculosa* (both Droseraceae). Other specialized leaves include the "suction trap" leaves of *Utricularia* and the "eel trap" leaves of *Genlisia* (both

Figure 9.9 Leaf structural types. **A,B.** Phyllode, *Acacia longifolia*. **A.** Mature. **B.** Young, with vestigial, caducous rachillae, representative of ancestral condition. **C.** Tendril, *Lathyrus vestitus*. **D–F.** Spines. **D.** Stipular spines, *Euphorbia* sp. **E.** Petiolar spines, *Fouquieria splendens*. Note mature leaf (above), dehiscence of blade and upper tissue of petiole, leaving petiolar spine (below). **F.** Leaf spines, cactus areole. **G–K.** Leaf modifications of carnivorous plants. **G,H.** Pitcher (pitfall) leaves. **G.** *Nepenthes* sp. **H.** *Sarracenia purpurea*. **I.** Tentacular leaves of *Drosera capensis*. Note glandular trichomes (arrow). **J.** Snap-trap leaf, *Dionaea muscipula*. **K.** Showy flower bracts, *Bougainvillea*. **L.** Epicalyx, *Lavatera bicolor*. **M.** Bud scale, *Liquidambar styraciflua*. **N.** Unifacial leaf, *Juncus phaeocephalus*.

Lentibulariaceae); see Ellison and Adamec (2018) for more details about carnivorous plant adaptations.

Leaf Type (Figure 9.10)
The pattern of division of a leaf into discrete components or segments is termed **leaf type**. A **simple** leaf is one bearing a single, continuous blade. A **compound** leaf is one divided into two or more, discrete **leaflets**. Leaf type should not be confused with leaf division; a simple leaf may be highly divided, but as long as the divisions are not discrete leaflets, it is still technically a simple leaf; see General Terminology. For either compound or divided leaves of ferns, the first (largest) division of a leaf is termed a **pinna**; the ultimate divisions are termed **pinnules**. If the leaves are compound or divided into more than two orders, the terms "primary pinna," "secondary pinna," etc. can be used, with the ultimate divisions or leaflets always being pinnules.

Simple leaves were the ancestral condition in the vascular plants, as in the lycophylls of the lycopods. Simple leaves are also the norm among the psilotophytes, equisetophytes, *Ginkgo*, and conifers (including the Gnetales). Compound leaves are characteristic of many "ferns," and all of the cycads. Angiosperms have the greatest diversity of leaves, ranging from simple to highly compound.

Various types of compound leaves have evolved, perhaps as a means of increasing total blade area without sacrificing structural integrity. For example, the blade tissue of a compound leaf generally may have better structural support (e.g., under windy conditions) than that of a comparably sized simple leaf. Compound leaves tend to be more common in mesic to wet environments and simple leaves in dry environments, but there are many exceptions to this and no clear trends.

Compound leaves are defined based on the number and arrangement of leaflets. A **pinnately compound** or **pinnate** leaf is one with leaflets arranged (either oppositely or alternately) along a central axis, the **rachis**. If a pinnate leaf has a terminal leaflet (and typically an odd number of leaflets), it is **imparipinnate** or **odd-pinnate**; if it lacks a terminal leaflet (and has an even number of leaflets), it is **paripinnate** or **even-pinnate**. A **bipinnately compound** or **bipinnate** leaf is with two orders of axes, each of which is pinnate (equivalent to a compound leaf of compound leaves). The central axis of a bipinnate leaf is still termed the **rachis**; the lateral axes that bear leaflets are termed **rachillae** (singular **rachilla**). Similarly, a compound leaf with three orders of axes, each pinnate, is termed **tripinnately compound** or **tripinnate**, etc.

A compound leaf in which four or more leaflets arise from a common point, typically at the end of the petiole, is termed **palmately compound** or **palmate**. A **costapalmate** leaf type is one that is essentially palmately compound to divided, but has an elongate, rachislike extension of the petiole (termed the **costa**), as occurs in some palms.

A compound leaf with only three leaflets is termed **trifoliolate** or **ternately compound**. (A leaf with two orders of axes, each ternately compound, is termed **biternately compound**. Further orders, e.g., **triternately compound**, can also occur.) Most ternately compound leaves are **palmate-ternate**, in which the three leaflets join at a common point (whether petiolulate or sessile). Rarely, ternately compound leaves can be **pinnate-ternate**, in which the terminal leaflet arises from the tip of a rachis. Pinnate-ternate leaves are actually derived (by reduction) from an ancestral pinnately compound leaf; they are found, e.g., in some members of the Fabaceae.

Decompound is a general term for a leaf that is more than once compound, i.e., with two or more orders, being bi-, tri-, etc. pinnately, palmately, or ternately, compound. However, *decompound* is also used for a highly divided leaf; see Division).

A compound leaf consisting of only two leaflets is termed **geminate** (after Gemini, the twins, in Greek mythology). A compound leaf with two rachillae, each bearing two leaflets, is termed **bigeminate**. A compound leaf with two rachillae, each of these bearing a pinnate arrangement of leaflets, is termed **geminate-pinnate**. Finally, a very specialized type of leaf is one that appears superficially to be simple, but actually consists of a single leaflet attached to the apex of a petiole, the junction between them clearly defined. This leaf type, known as **unifoliolate**, is interpreted as being derived by reduction of an ancestrally compound leaf.

In some taxa, e.g., many Araceae, the leaves exhibit **heteroblasty** (adjective, **heteroblastic**), in which the juvenile leaves are distinctly different in size or shape from the adult leaves (making species identification difficult).

Leaf Attachment (Figure 9.11)
The nature of the joining of the leaf to the stem is termed leaf attachment (sometimes treated under "Base"; see **General Morphology**). In general, leaves may be **petiolate**, with a petiole, or **sessile**, without a petiole. Leaflets of a compound leaf are, correspondingly, either **petiolulate** or **sessile**. (The term **subsessile** is sometimes used for a leaf/leaflet with a small, rudimentary petiole/petiolule.) Sessile or petiolate leaves can also have a **sheathing** leaf attachment, in which a flattened leaf base (the *sheath*) partially or wholly clasps the stem, typical of the Poaceae (grasses) and many Apiaceae. If a leaf appears to extend down the stem from the point of attachment, as if fused to the stem, the leaf attachment is **decurrent** (e.g., as in many Cupressaceae). A decurrent leaf

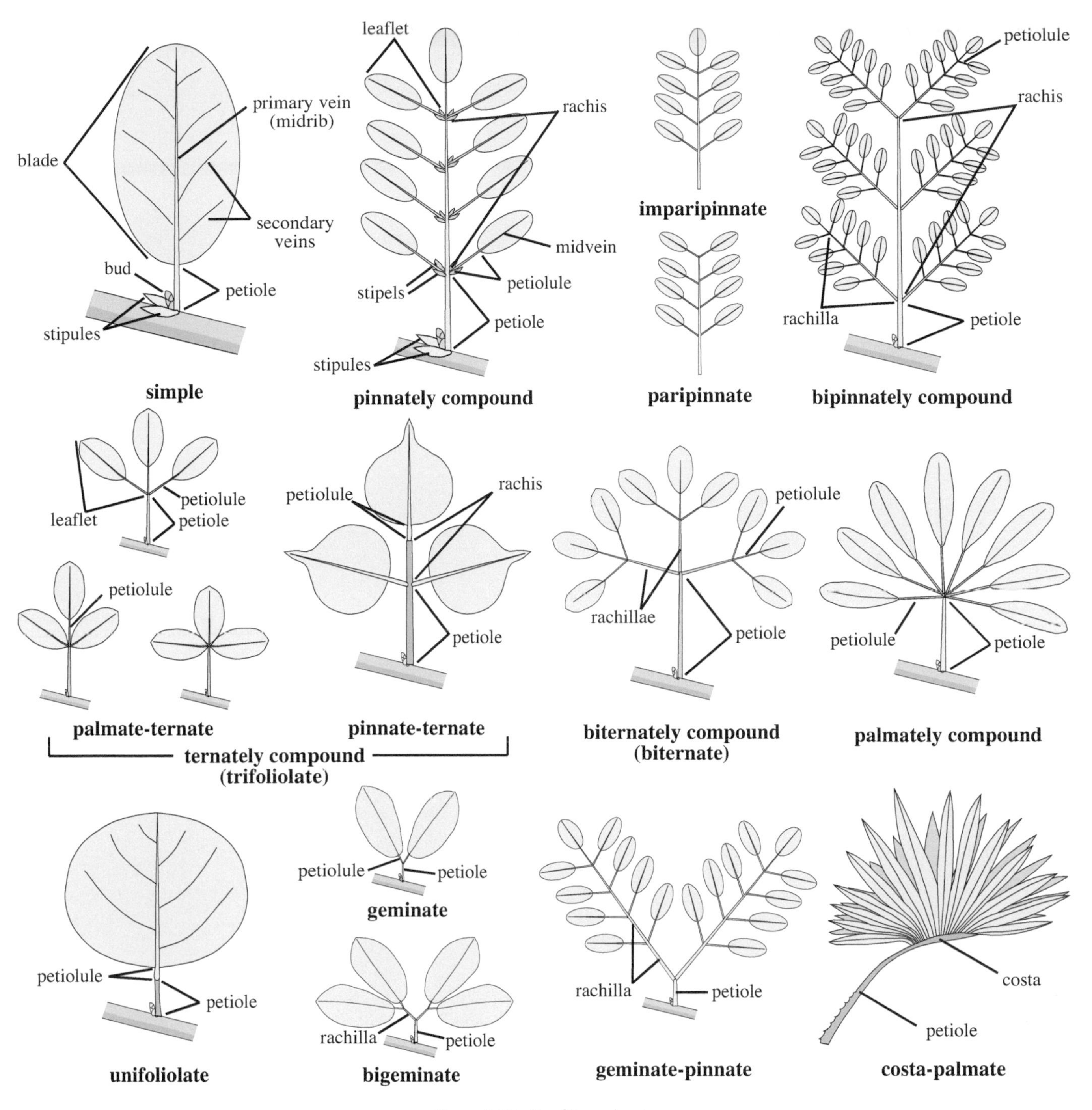

Figure 9.10 Leaf types/parts.

base is not actually caused by later fusion of the leaf to the stem, but by extension growth of actively dividing cells of the leaf primordium at the leaf–stem junction. Last, specializations of sessile leaves may occur. If a leaf is sessile and clasps the stem most, but not all, of its circumference, the attachment is termed **amplexicaul**. If the leaf is sessile with the base of the blade completely surrounding the stem, it is termed **perfoliolate**. A special case of the latter (involving fusion of leaves) is **connate-perfoliate**, whereby typically two opposite leaves fuse basally such that the blade bases of the fusion product completely surrounds the stem.

Leaf Venation (Figures 9.12, 9.13)

The sporophytic leaves of vascular plants contain vascular bundles, known as **veins**, which conduct water, minerals, and sugars between the leaf and the stem. The leaves of some

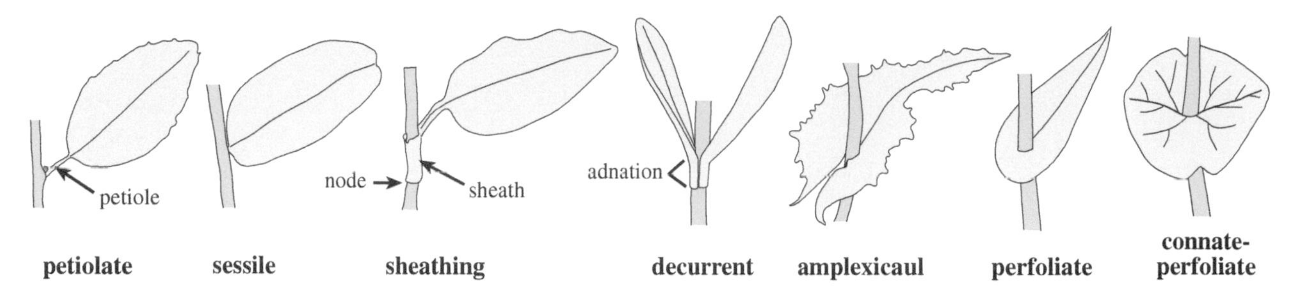

Figure 9.11 Leaf attachment.

vascular plants have only a single vein, but in most the veins are branched (termed "ramified" or "anastomosing"), sometimes in a very intricate pattern. **Venation** refers to this pattern of veins and vein branching. Although venation is usually described for vegetative leaves, it can also be assessed in other leaf homologues, such as bracts, sepals, petals, stamens, or carpels.

The major vein (or veins) of a leaf, with respect to size, is termed the **primary vein**. From the primary vein(s), smaller, lateral veins may "branch off," these known as **secondary veins**; from secondary veins, even smaller **tertiary veins** may arise, and so forth. [The distinctions between these vein classes can be difficult to determine in some taxa.] If a simple leaf has a single, primary vein, that vein is termed the **midrib** or **costa** (although *costa* may also be used for the nonvascularized conductive tissue found in the gametophytic leaves of mosses). The central, primary vein of the leaflet of a compound leaf is termed the **midvein**.

Venation patterns can be quite complex, and the terminology formidable (see later discussion). Four, very general venation classes are as follows (Figure 9.11):

1. **Uninervous**, in which there is a central midrib with no lateral veins, e.g., as in the lycophytes, psilotophytes, and equisetophytes, as well as many conifers
2. **Dichotomous**, in which veins successively branch distally into a pair of veins of equal size and orientation, e.g., in *Ginkgo biloba*, in which there is no actual midrib
3. **Parallel**, in which the primary and secondary veins are essentially parallel to one another, the ultimate veinlets being transverse (at right angles), e.g., in most monocots
4. **Netted** or **reticulate**, in which the ultimate veinlets form an interconnecting netlike pattern, e.g., most nonmonocot flowering plants.

Reticulate leaves can be **pinnately veined** (**pinnate-netted**), with secondary veins arising along length of a single primary vein (the midrib or, in a compound leaf, midvein); **palmately veined** (**palmate-netted**), with four or more primary veins arising from a common basal point; or **ternately veined** (**ternate-netted**), with three primary veins arising from a common basal point.

Similar to parallel venation in having transverse ultimate veinlets are **penni-parallel** (also called **pinnate-parallel**), with secondary veins arising from a single primary vein region, the former essentially parallel to one another (e.g., the Zingiberales); and **palmate-parallel**, with several primary veins (of leaflets or leaf lobes) arising from one point, the adjacent secondary veins parallel to these (e.g., "fan" palms).

A more detailed classification system of venation (and many other leaf features) is that of Hickey (1973) and Hickey and Wolf (1975). This system is based on the pattern of primary, secondary, and tertiary venation. The following is a summary of the terms used in this system, illustrated in Figure 9.12.

Three general venation categories are used for a basically pinnate venation: **craspedodromous**, in which secondary veins terminate at the leaf margin; **camptodromous**, in which secondary veins do not terminate at the margin; and **hyphodromous**, with only the primary midrib vein present or evident and secondary veins either absent, very reduced, or hidden within the leaf mesophyll.

Subcategories of craspedodromous venation include **simple craspedodromous**, in which all secondary veins terminate at the margin; **semicraspedodromous**, in which the secondary veins branch near the margin, one terminating at the margin, the other looping upward to join the next secondary vein; and **mixed craspedodromous** (not illustrated), with some secondary veins terminating at the margin, but with many terminating away from the margin.

Subtypes of camptodromous venation include **brochidodromous**, in which secondary veins form prominent upward loops near the margin, joining other, more distal, secondary veins; **eucamptodromous**, in which secondary veins curve upward near the margin but do not directly join adjacent secondaries; **cladodromous**, in which secondary veins branch

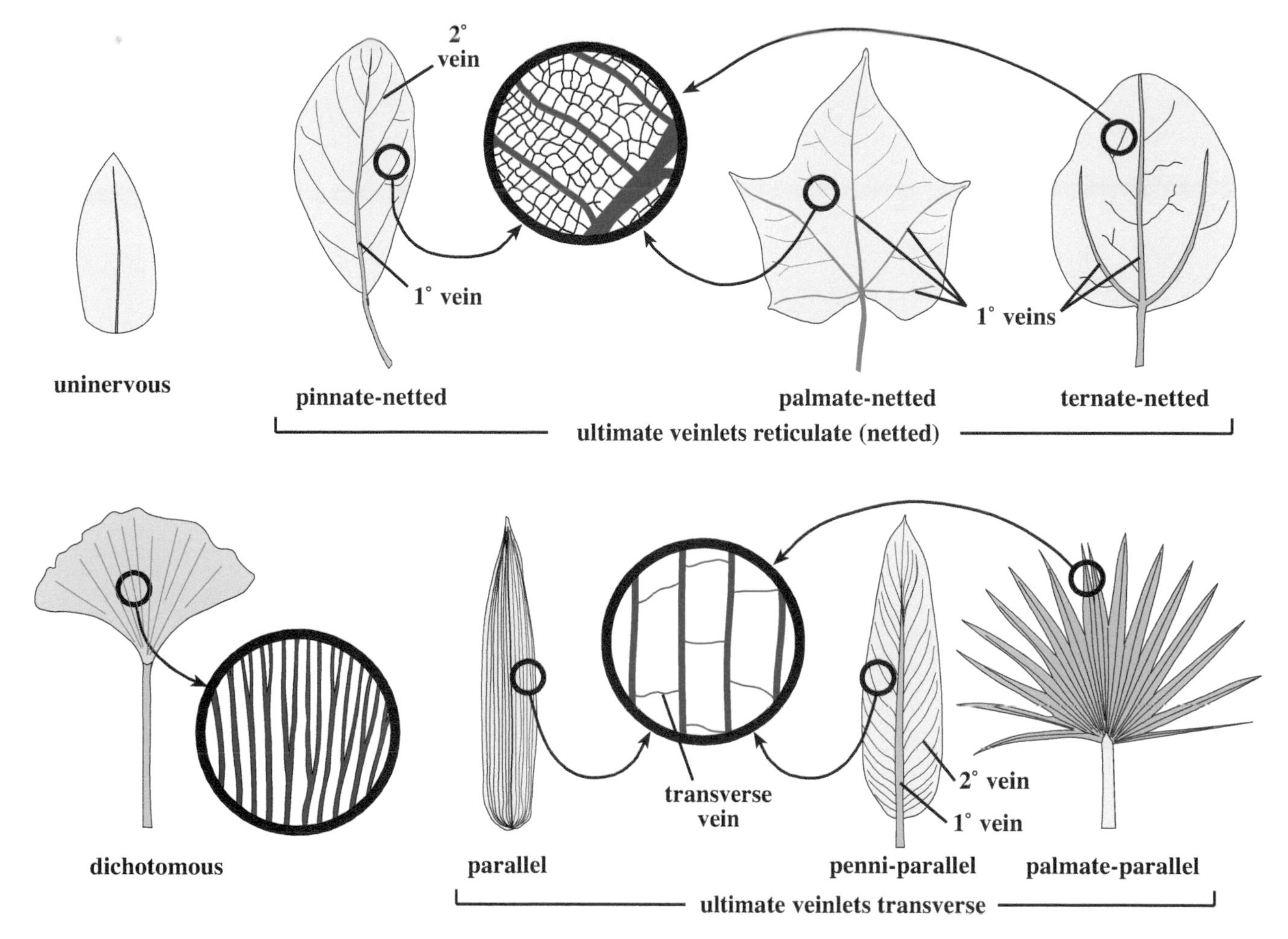

Figure 9.12 Leaf venation, generalized terminology.

toward the margin; and **reticulodromous**, in which secondary veins branch repeatedly, forming a very dense, netlike structure.

Parallelodromous venation is equivalent to parallel (defined earlier), in which two or more primary or secondary veins run parallel to one another, converging at the apex.

Venation is **actinodromous** if three or more primary veins diverge from one point (equivalent to ternate or palmate venation). **Palinactinodromous** is similar, but the primary veins have additional branching above the main point of divergence of the primaries.

For actinodromous and palinactinodromous types, the venation is **marginal** if the main, primary veins reach the blade margin, and **reticulate** (not to be confused with "reticulate" in the more general venation terminology) if they do not. **Flabellate** venation is that in which several equal, fine veins branch toward the apex of the leaf.

Campylodromous venation is that in which several primary veins run in prominent, recurved arches at the base, curving upward to converge at the leaf apex.

Finally, venation is **acrodromous**, if two or more primary veins (or strongly developed secondary veins) run in convergent arches toward the leaf apex (but are not recurved at the base, as in campylodromous).

For actinodromous, palinactinodromous, and acrodromous types, the venation is **basal** if the primaries are joined at the blade base, and **suprabasal** if the primaries diverge above the blade base. The venation is **perfect** if branching of the lateral primary veins and their branches cover at least two thirds of the leaf blade area (or reach at least two thirds of the distance toward the leaf apex), and **imperfect** if these veins cover less than two thirds of the leaf blade area (or reach less than two thirds of the way toward the leaf apex).

These complex venation types, along with many other details of the leaf, can be specific to certain taxonomic groups

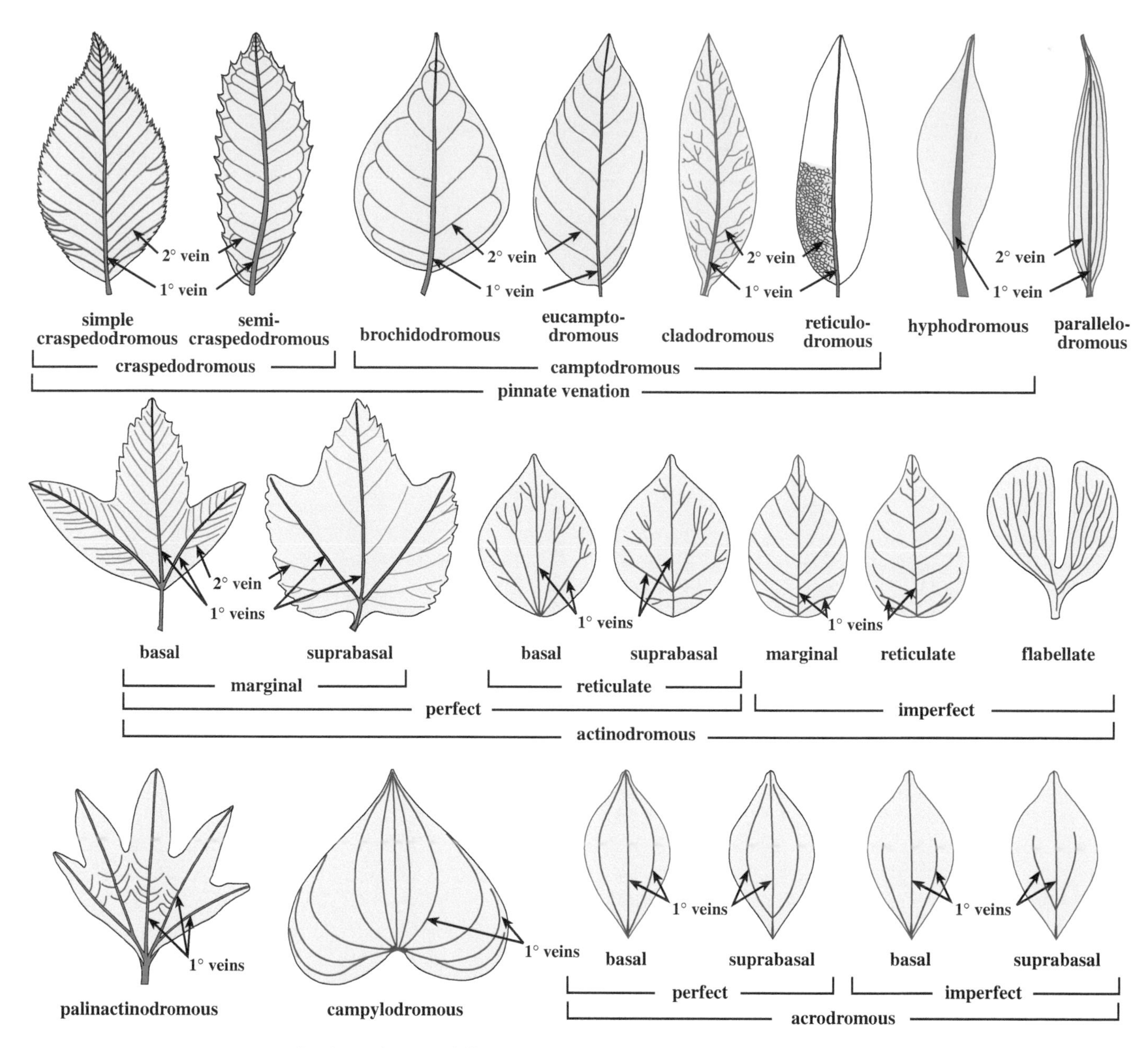

Figure 9.13 Leaf venation, specialized terminology, redrawn from Hickey (1973), by permission.

of plants. Although they are not widely used in standard morphological descriptions, their recognition can be important in identification (e.g., of many tropical and fossil plants) and classification (see Hickey and Wolf 1975).

FLOWERS

A major diagnostic feature of angiosperms is the flower. As discussed in Chapter 6, a **flower** is a modified reproductive shoot, basically a stem with an apical meristem that gives rise to leaf primordia. Unlike a typical vegetative shoot, however, the flower shoot is determinate, such that the apical meristem stops growing after the floral parts have formed. At least some of the leaf primordia of a flower are modified as reproductive sporophylls (leaves bearing sporangia). Flowers are unique, differing, e.g., from the cones of gymnosperms, in that the sporophylls develop either as stamens or carpels (see Chapter 6, and later discussion).

Flower Parts (Figure 9.14)

The basic parts of a flower, from the base to the apex, are as follows. The **pedicel** is the flower stalk. (If a pedicel is absent, the flower attachment is *sessile*.) Flowers may be subtended by a **bract**, a modified, generally reduced leaf; a smaller or secondary bract, often borne on the side of a pedicel, is

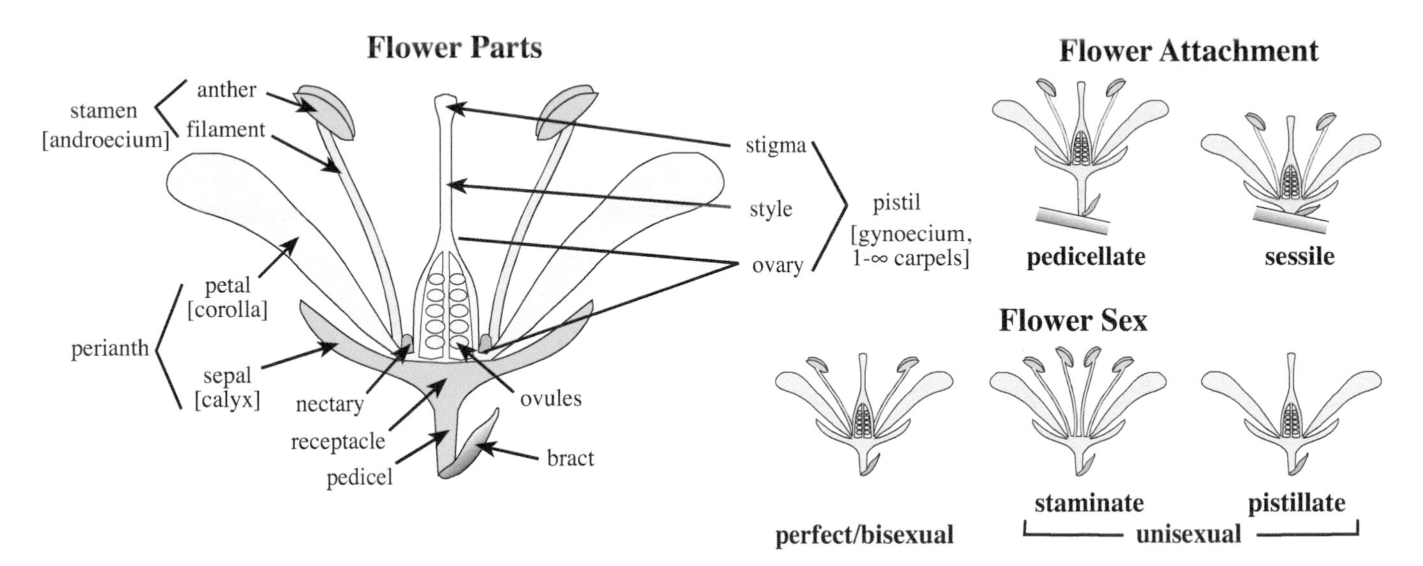

Figure 9.14 Flower parts, sex, and attachment.

termed a **bracteole** or **bractlet** (also called a **prophyll** or **prophyllum**). Bracteoles, where present, are typically paired. [In some taxa, a series of bracts, known as the **epicalyx**, immediately subtends the calyx (see later discussion), as in *Hibiscus* and other members of the Malvaceae.] The **receptacle** or **floral receptacle** (also termed a **torus**, although "torus" can also be used for a compound receptacle; see Inflorescence Parts) is the tissue or region of a flower to which the other floral parts are attached. The receptacle is typically at the very tip of the floral axis (derived from the original apical meristem). In some taxa the receptacle can grow significantly and assume an additional function. From the receptacle arises the basic floral parts. The **perianth** (also termed the **perigonium**) is the outermost, nonreproductive group of modified leaves of a flower. If the perianth is relatively undifferentiated, or if its components intergrade in form, the individual leaflike parts are termed **tepals**. In most flowers the perianth is differentiated into two groups. The **calyx** is the outermost series or whorl of modified leaves. Individual units of the calyx are **sepals**, which are typically green, leaflike, and function to protect the young flower. The **corolla** is the innermost series or whorl of modified leaves in the perianth. Individual units of the corolla are **petals**, which are typically colored (nongreen) and function as an attractant for pollination. Some flowers have a **hypanthium** (floral tube), a cup-like or tubular structure, around or atop the ovary, bearing along its margin the sepals, petals, and stamens.

Many flowers have a **nectary**, a specialized structure that secretes nectar. Nectaries may develop on the perianth parts, within the receptacle, on or within the androecium or gynoecium (below), or as a separate structure altogether. Some flowers have a **disk**, a discoid or doughnut-shaped structure arising from the receptacle. Disks can form at the outside and surrounding the stamens (termed an *extrastaminal disk*), at the base of the stamens (*staminal disk*), or at the inside of the stamens and/or base of the ovary (*intrastaminal disk*). Disks may be nectar-bearing, called a *nectariferous disk*.

The **androecium** refers to all of the male organs of a flower, collectively all the stamens. A **stamen** is a microsporophyll, which characteristically bears two thecae (each theca comprising a pair of microsporangia; see Chapter 6). Stamens can be leaflike (laminar), but typically develop as a stalklike **filament**, bearing the pollen-bearing **anther**, the latter generally equivalent to two fused thecae.

The **gynoecium** refers to all of the female organs of a flower, collectively all the carpels. A **carpel** is the unit of the gynoecium, at maturity enclosing one or more ovules. Carpels may form as a ring of tissue (ascidiate development) or as a modified, conduplicate female megasporophyll of a flower, (plicate or conduplicate development; see Chapter 6). A **pistil** is that part of the gynoecium composed of an **ovary**, one or more **styles** (which may be absent), and one or more **stigmas** (see later discussion).

In some taxa, e.g., Aristolochiaceae and Orchidaceae, the androecium and gynoecium are fused into a common structure, known variously as a **column**, **gynandrium**, **gynostegium**, or **gynostemium**. A stalk that bears the androecium and gynoecium is an **androgynophore**, e.g., Passifloraceae. A stalk-like structure that bears stamens alone is termed an **androphore** (e.g., some Eriocaulaceae); one that bearing one or more pistils is a **gynophore** or **stipe** (see Gynoecium, Carpel, and Pistil).

Flower Sex and Plant Sex (Figure 9.14)

Flower sex refers to the presence or absence of male and female parts within a flower. Most flowers are **perfect** or **bisexual [monoclinous]**, having both stamens and carpels. Bisexual flower sex is likely the ancestral condition in angiosperms.

Many angiosperm taxa, however, have **imperfect** or **unisexual [diclinous]** flower sex. In this case, flowers are either **pistillate/female**, in which only carpels develop, or **staminate/male**, in which only stamens develop.

Plant sex refers to the presence and distribution of perfect or imperfect flowers on individuals of a species. A **hermaphroditic** plant is one with only bisexual flowers. A **monoecious** (*mono*, one + *oikos*, house) plant is one with only unisexual flowers, both staminate and pistillate on the same individual plant; e.g., *Quercus* spp., oaks. A **dioecious** (*di*, two + *oikos*, house) plant is one with unisexual flowers, but with staminate and pistillate on separate individual plants (i.e., having separate male and female individuals; e.g., *Salix* spp., willows).

Plant sex can vary within individuals of a species, and there may also be a combination of perfect and imperfect flowers in different individuals. (These terms are confusing, but occasionally seen in the literature.) **Polygamous** is a general term for a plant with both bisexual and unisexual flowers. **Andromonoecious** refers to a plant with both staminate and perfect flowers on the same individual, and **gynomonoecious** is a plant with both pistillate and perfect flowers on the same individual. **Polygamomonoecious [Trimonoecious]** refers to a plant with pistillate, staminate, and perfect flowers on the same individual. **Androdioecious** refers to a plant with staminate flowers on some individuals and perfect flowers on other individuals. **Gynodioecious** refers to a plant with pistillate flowers on some individuals and perfect flowers on other individuals. **Polygamodioecious** is a plant with staminate and perfect flowers on some individuals, pistillate and perfect flowers on other individuals. **Trioecious** refers to a plant with pistillate, staminate, and perfect flowers on different individuals. All of these types of nonhermaphroditic plant sex may function as a mechanism of promoting increased outcrossing between individuals of a species. (However, many hermaphroditic plants can outcross by other means; see Chapter 13.)

Flower Attachment (Figure 9.14)

Flower attachment is **pedicellate**, having a pedicel; **sessile**, lacking a pedicel; or **subsessile**, having a short, rudimentary pedicel. The terms **bracteate**, with bracts, and **ebracteate**, lacking bracts, may also be used with respect to flower

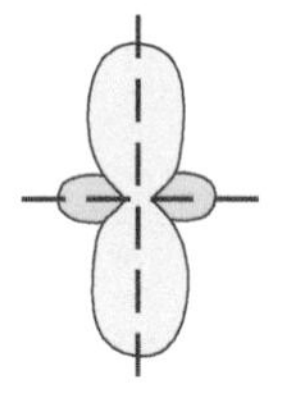
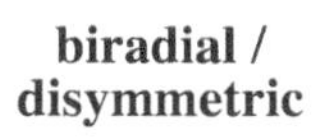

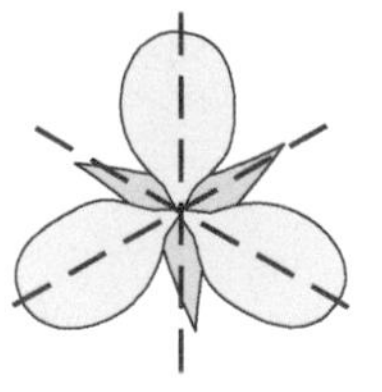

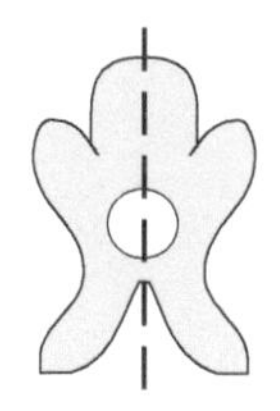

Figure 9.15 Flower symmetry types.

attachment. The adaptive significance of pedicels is likely correlated with the spatial positioning of flowers relative to pollination or eventual fruit or seed dispersal.

Flower Cycly

Flower cycly refers to the number of cycles (series or whorls) or floral parts. The two basic terms used are **complete**, for a flower having all four major series of parts (sepals, petals, stamens, and carpels) and **incomplete**, for a flower lacking one or more of the four major whorls of parts (e.g., any unisexual flower, or a bisexual flower lacking a corolla).

Flower Symmetry (Figures 9.15, 9.16)

Flower symmetry is an assessment of the presence and number of mirror-image planes of symmetry. **Actinomorphic** or **radial** symmetry [also called **polysymmetric** or **regular**] is that in which there are three or more planes of symmetry, such that there is a repeating structural morphology when rotated less than 360° about an axis. (A variant on actinomorphic symmetry is **haplomorphic**, appearing radially symmetric but not having strict mirror image halves because the parts are numerous and/or spirally inserted.) **Biradial symmetry** [also called **disymmetric**] means having two (and only two) planes of symmetry. (The difference between biradial and radial symmetry is sometimes not recognized, both being termed *radial symmetry* or *actinomorphy*; however, the distinction can be useful and is recognized here.) **Zygomorphic** or **bilateral** symmetry [also called **monosymmetric** or **irregular**] is that in which there is only one plane of symmetry. An **asymmetric** flower lacks any plane of symmetry, usually the result of twisting of parts. Flower symmetry can sometimes be subtle and can even vary within a flower; if so, it should be separately described for calyx, corolla, androecium, and gynoecium to avoid confusion.

Flower symmetry can be an important adaptation relative to pollination systems. Actinomorphic flower symmetry is likely the ancestral condition in angiosperms and is

Figure 9.16 Flower symmetry examples. **A,B.** Actinomorphic/radial symmetry. **A.** Five planes of symmetry. **B.** Six planes of symmetry. **C.** Slight zygomorphy (bilateral symmetry), vertical plane of symmetry. **D,E.** Strong zygomorphy of all floral parts. **F.** Asymmetry, caused by twisting of floral parts.

found in a large number of groups. Zygomorphy has evolved repeatedly in many groups, typically as a means of more efficiently transferring pollen to an animal (usually insect) pollinator. Zygomorphy is typically correlated with a more horizontal floral orientation, and there are many different ways that zygomorphy can come about developmentally and morphologically.

Flower Maturation

Flower maturation refers to the time of development of flowers or flower parts (see also General Terminology). **Anthesis** is the general time of flowering, the opening of flowers with parts available for pollination. The relative timing of development of male versus female flowers or floral parts can be an important feature in reproductive biology. **Protandrous** refers to stamens developing, or pollen release occurring, prior to the maturation of carpels or stigmas being receptive. **Protogynous** is the reverse, with carpels or stigmas developing before stamens mature or pollen is released. Both protandry and protogyny may function to promote outcrossing (and thus inhibit selfing) within individuals of a species.

Two flower maturation terms dealing with the relative direction of development of parts can be important in describing taxonomic groups. **Centrifugal** refers to developing from the center toward the outside or periphery, whereas **centripetal** is development from the outside or periphery toward the center region. Both *centrifugal* and *centripetal* can be applied to parts of the perianth, calyx, corolla, androecium, or gynoecium; the terms are often used to describe the direction of development of stamens in a multiwhorled androecium.

Finally, the term **cleistogamy** (adj. **cleistogamous**) refers to a flower in which the perianth remains closed, such that pollen produced from within the flower pollinates only the stigma(s) of that flower. **Chasmogamy** (adj. **chasmogamous**) is the normal situation, in which the perianth opens and pollen may be dispersed.

PERIANTH

The **perianth** (or *perigonium)* is the outermost, nonreproductive group of modified leaves of a flower. (The term *perianth* has also been used for components of the reproductive structures of various Gnetales, but these are not homologous.) A perianth is absent in some flowering plants, typically those taxa that have very small, reduced flowers. The perianth, where present, functions both to protect the young flowering parts and to aid in pollination.

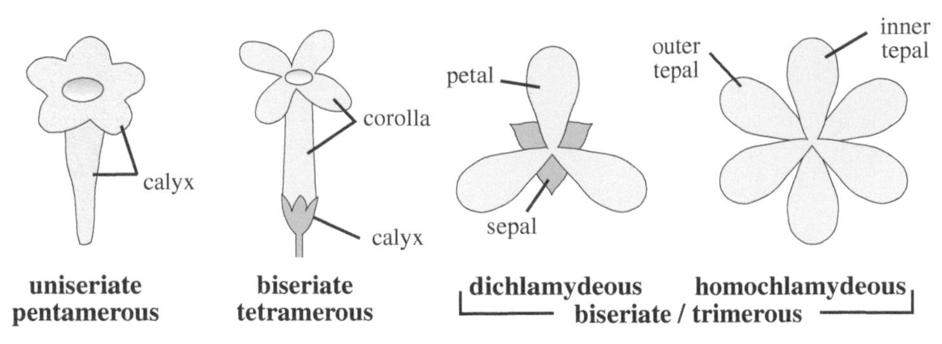

Figure 9.17 Perianth cycly and merosity.

The units of the perianth arise like leaves as primordia from the apical meristem of the flower. Typically, they may retain leaflike characters. Sepals, in fact, are usually green with stomata and veins; even petals will have veins and may have vestigial stomata. However, the perianth can undergo significant developmental changes and be highly modified (and unleaflike) at maturity.

Perianth Arrangement/Cycly/Merosity (Figures 9.17, 9.18)
A fundamental aspect of perianth structure is **perianth arrangement**, the position of perianth parts relative to one another. In some taxa, such as some magnolias and water lilies, the perianth parts have a **spiral** arrangement, i.e., spirally arranged with only one perianth part per node, not in distinct whorls. Typically, flowers with a spiral perianth arrangement have parts that are either undifferentiated (similar to one another) or that grade from an outer, sepal-like form to an inner petal-like form. In either case, the term **tepal** is used to describe undifferentiated or intergrading perianth parts. In most flowering plants the perianth parts have a **whorled** arrangement, in which the parts appear to arise from the same nodal region. (Note that, developmentally, the perianth parts may actually initiate as primordia at slightly different times and positions; however, at maturity, this is usually undetectable.)

Cycly refers to the number of whorls (cycles, series) of parts. (See General Terminology.) Thus, **perianth cycly** is the number of whorls of perianth parts. The most common type of perianth cycly by far is **biseriate** (also called *dicyclic*), in which there are two discrete whorls, an outer (= lower) and inner (= upper). A less common condition in flowering plants is a **uniseriate** perianth cycly, with perianth parts in a single whorl. Uniseriate perianths may arise by loss or reduction of one of the whorls of an ancestrally biseriate perianth. If it is known that the calyx was evolutionarily lost, what remains should be called a *corolla*; if the corolla was lost, what remains should be termed a *calyx*. If this directionality is not known, a uniseriate perianth is usually termed a calyx by tradition (although it may simply be called a *perianth*). Perianths may also rarely be **triseriate** (or "tricyclic") = three-whorled, **tetraseriate** (or "tetracyclic") = four-whorled, etc. The term **multiseriate** may be used to mean "composed

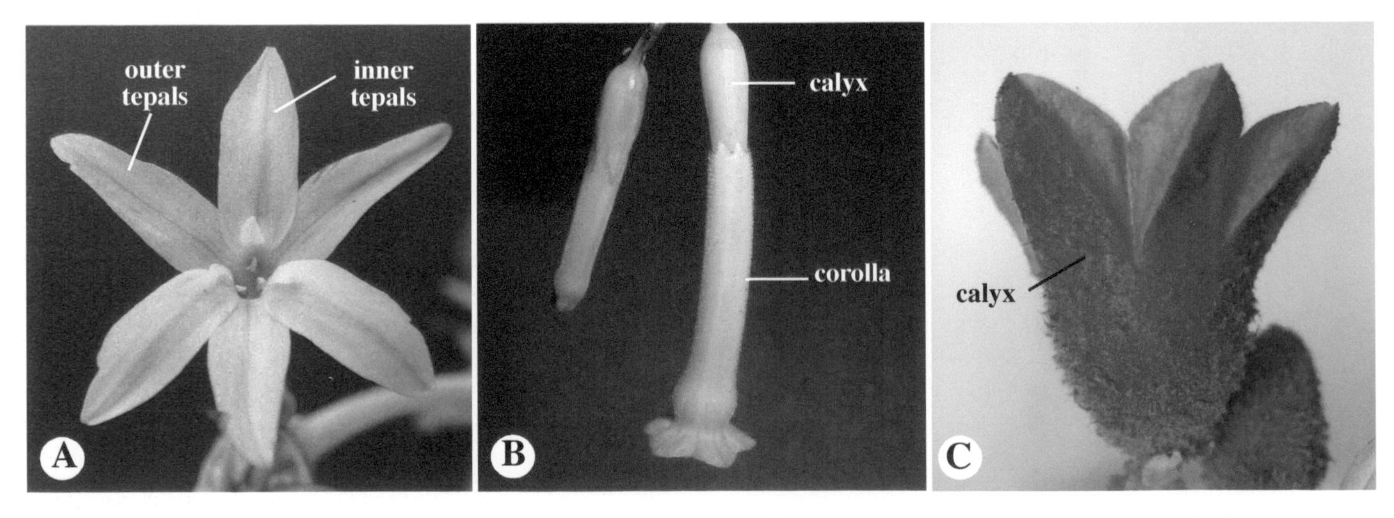

Figure 9.18 Perianth cycly. **A.** Biseriate, homochlamydeous, with outer and inner perianth parts (tepals) similar. **B.** Biseriate, dichlamydeous, with distinct calyx and corolla. **C.** Uniseriate, with a single whorl of perianth parts, by default termed a calyx.

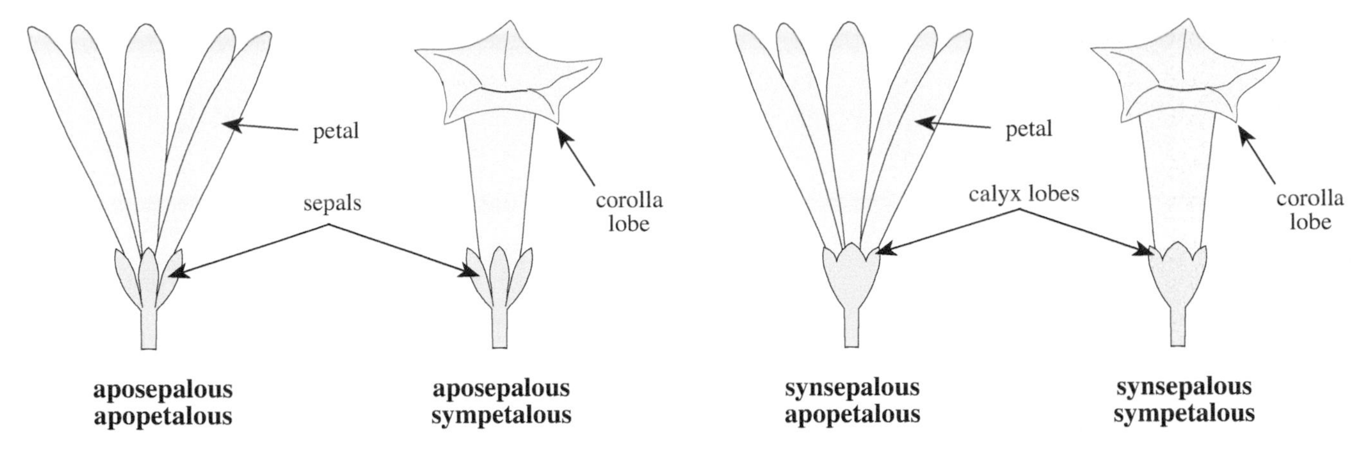

Figure 9.19 Perianth fusion.

of three or more whorls." Other cycly terms evaluate the similarity of the whorls of parts to one another. **Dichlamydeous** describes a perianth composed of a distinct outer calyx and inner corolla; in most cases, a dichlamydeous perianth is also biseriate, but it may be multiseriate (i.e., the calyx or corolla containing more than one whorl). **Homochlamydeous** refers to a perianth composed of similar parts, each part a tepal. Most monocots have a homochlamydeous perianth, whereas most eudicots have a dichlamydeous one. In some cases, the distinction between *dichlamydeous* and *homochlamydeous* can be difficult, as it may be difficult to assess whether outer and inner series are similar or different.

Merosity refers to the number of parts per whorl or cycle. (See General Terminology.) Thus, **perianth merosity** is the number of parts per whorl of the perianth. General terms for perianth merosity are **isomerous**, having the same number of members in different whorls (e.g., five sepals and five petals) and **anisomerous**, having a different number of members in different whorls (e.g., two sepals and five petals). Merosity may be described separately for each whorl of the perianth, e.g., **calyx merosity** and **corolla merosity**. It is assessed for numbers of discrete petals, sepals, and tepals, or, if perianth fusion occurs, for numbers of calyx, corolla, or perianth lobes (see later discussion). Perianth, calyx, or corolla merosity is usually designated as a simple number, although terms such as **bimerous** (a whorl with two members), **trimerous** (a whorl with three members), **tetramerous** (a whorl with four members), and **pentamerous** (a whorl with five members), etc., can be used. Terms for absence of parts include **achlamydeous**, lacking a perianth altogether, **apetalous**, having no petals or corolla, and **asepalous**, having no sepals or calyx.

Perianth Fusion (Figure 9.19)

The term **perianth fusion** deals with the apparent fusion of perianth parts to one another. (This character may be treated separately as *calyx* or *corolla fusion*.) If sepals, petals, or tepals are discrete and unfused, the respective terms **aposepalous [chorisepalous]**, **apopetalous [choripetalous]**, and **apotepalous [choritepalous]** may be used. If sepals, petals, or tepals appear to be fused (even slightly at the base), the respective terms **synsepalous [gamosepalous]**, **sympetalous [gamopetalous]**, and **syntepalous [gamotepalous]** are used. The "fusion" of perianth parts does not usually occur as a separate event, e.g., petals fusing together after they are individually formed. The "fusion" is apparent, and typically results by the growth of a common floral primordium at the base of the calyx, corolla, or peranth.

Perianth fusion results in the development of a tubelike or cuplike structure (the region of "fusion") in the calyx, corolla, or perianth. If little fusion occurs, the tubelike region occurs only at the base and gives rise to calyx, corolla, or perianth lobes.

Perianth Parts (Figure 9.20)

Various specialized terms are used for parts of the perianth. These include the following: **anterior** or **ventral**, referring to the lower, abaxial lobe(s) or side, toward a subtending bract; **beard**, a tuft, line, or zone of trichomes on a perianth or perianth part (see Vestiture); **claw**, an abruptly narrowed base of a sepal or petal; **corona**, a crownlike outgrowth between stamens and corolla, which may be petaline or staminal in origin; **faucal**, referring to the throat of a corolla; **hypanthium** or **floral cup**, a generally tubular or cup-shaped structure at the top rim of which are attached the calyx, corolla, and androecium; **labellum**, a modified, typically expanded, median petal, tepal, or perianth lobe, such as in the Orchidaceae; **limb**, the expanded portion of usually sympetalous corolla above the tube and throat; **lip**, either of two variously shaped parts into which a calyx or corolla is divided, usually into upper (posterior) and/or lower (anterior) lips, such as most Lamiaceae, Orchidaceae (Note: each lip may be composed of one or

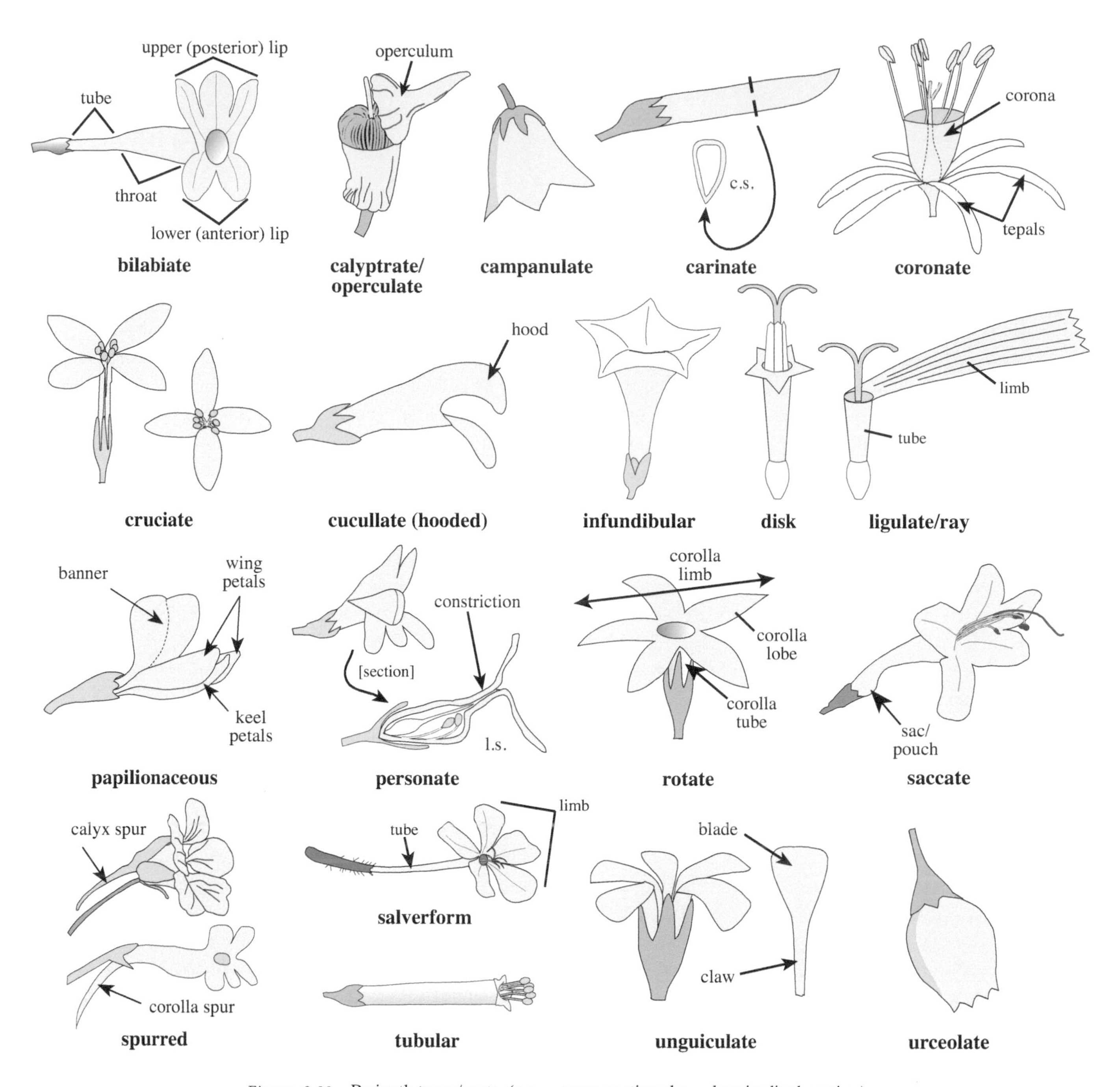

Figure 9.20 Perianth types/parts. (c.s. = cross section; l.s. = longitudinal section)

more lobes); **lobe**, a segment of a synsepalous calyx or sympetalous corolla; **petal**, a corolla member or segment; a unit of the corolla; **posterior** or **dorsal**, referring to the upper, adaxial lobe(s) or side, nearest to the axis, away from the subtending bract; **sepal**, a calyx member or segment, a unit of the calyx; **spur**, a tubular, rounded or pointed projection from the calyx or corolla, functioning to contain nectar; **tepal**, a perianth member or segment not differentiated into distinct sepals or petals; **throat**, an open, expanded region of a perianth, usually of a sympetalous corolla; **tube**, a cylindrically shaped perianth or region of the perianth, usually of a sympetalous corolla.

Perianth Type (Figure 9.20)

Perianth type can include aspects of the entire perianth; however it could include aspects of only the calyx, corolla, or hypanthium (if present). Generally, perianth type is based on the structure of the corolla alone, in which case it could logically be termed *corolla type*. The terminology for perianth type takes into account various aspects of shape, fusion,

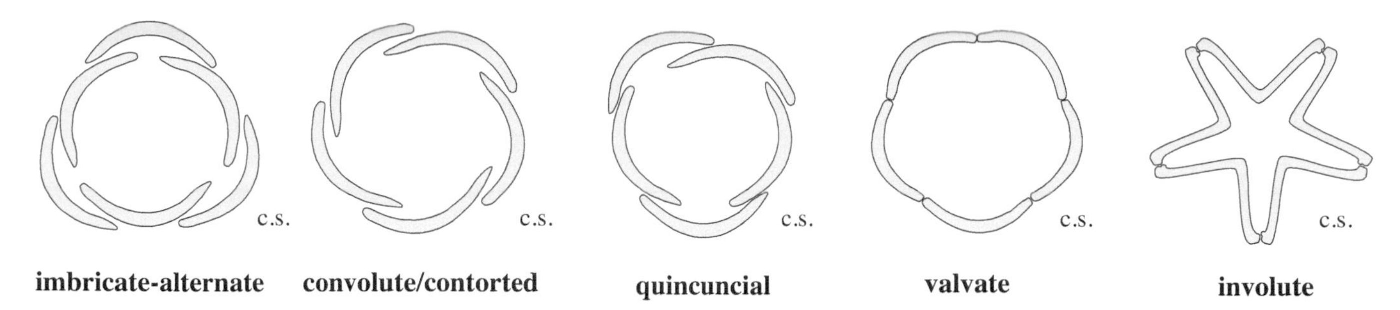

Figure 9.21 Perianth aestivation.

orientation, and merosity. Perianth type is often of systematic value and may be diagnostic for certain clades of angiosperms. The perianth type typically reflects adaptive features related to pollination biology, such as attracting a pollinator or better effecting the transfer of pollen. Some perianths are highly modified for other functions, such as the **lodicules** of grasses, which are reduced perianth parts that, upon swelling, open up the grass floret (see Inflorescence Type, later, and Poaceae of Chapter 7).

Specific perianth types include the following: **bilabiate**, two-lipped, with two, generally upper and lower segments, as in many Lamiaceae; **calyptrate/operculate**, having calyx and corolla fused into a cap that falls off as a unit, as in *Eucalyptus*; **campanulate**, bell-shaped, with a basally rounded flaring tube about as broad as long and flaring lobes, as in *Campanula* (may also be used for bell-shaped apopetalous corolla or apotepalous perianth); **carinate**, keeled, with a sharp median fold, usually on the abaxial side; **coronate**, with a tubular or flaring perianth or staminal outgrowth, as in *Narcissus*, *Asclepias* spp.; **cruciate**, with four distinct petals in cross form, as in many Brassicaceae; **cucullate/galeate**, hooded, with an abaxially concave posterior lip; **disk**, having an actinomorphic, tubular corolla with flaring lobes, as in some Asteraceae; **hypocrateriform**, a corolla with a tube having abruptly spreading lobes (encompasses both rotate and salverform, below); **infundibular**, funnel-shaped, with a tubular base and continuously expanded apex, as in *Ipomoea*, morning glory; **ligulate** or **ray**, having a short, tubular corolla with a single, elongate, strap-like apical extension, as in some Asteraceae; **papilionaceous**, with one large posterior petal (banner or standard), two inner, lateral petals (wings), and two usually apically connate lower petals (keel), the floral structure of the Faboideae (Fabaceae); **personate**, two -lipped, with the upper arched and the lower protruding into the corolla throat, as in *Antirrhinum*, snapdragon; **rotate**, with a short tube, typically no longer than the calyx, and wide limbs oriented at right angles to the tube, as in *Phlox*; **saccate**, having a pouchlike evagination; **salverform**, trumpet-shaped; with a long tube extending beyond the calyx and flaring limbs at right angles to tube; **tubular**, mostly cylindrical; **unguiculate**, clawed, as in many Brassicaceae, Caryophyllaceae; **unilabiate**, one-lipped, as in many Goodeniaceae; and **urceolate**, urn-shaped, expanded at base and constricted at apex, as in many Ericaceae.

Perianth Aestivation (Figure 9.21)

Perianth aestivation is defined by the position, arrangement, and overlapping of floral perianth parts. Aestivation can be an important systematic character for delimiting or diagnosing some flowering plant taxa. In practice, aestivation is best observed by making hand sections of mature flower buds, because after anthesis, the perianth aestivation may be obscured. For very small flowers, histological sectioning may be needed to clearly see the aestivation type.

Some standard perianth aestivation terms are as follows: **imbricate**, general term for overlapping perianth parts; **convolute** or **contorted**, imbricate with perianth parts of a single whorl overlapping at one margin, being overlapped at the other, as in the corolla of many Malvaceae; **crumpled**, having a wrinkled or crinkled appearance, particularly in bud; **imbricate-alternate**, imbricate with the outer whorl of perianth parts (sepals or outer tepals) alternating with (along different radii) the inner whorl of perianth parts (petals or inner tepals); **quincuncial**, imbricate with perianth parts of a single pentamerous whorl having two members overlapping at both margins, two being overlapped at both margins, and one overlapping only at one margin; **valvate**, with a whorl of perianth parts meeting at the margins, not overlapping; and **involute**, valvate with each perianth part induplicate (folded longitudinally inward along central axis).

ANDROECIUM

The androecium consists of all the floral male (pollen -producing) reproductive organs, the units of which are stamens. Stamens are interpreted as being modified, sporangia -bearing leaves or microsporophylls. Stamens initiate as primordia from the flower apical meristem, but at maturity are attached to the receptacle, corolla (having an epipetalous

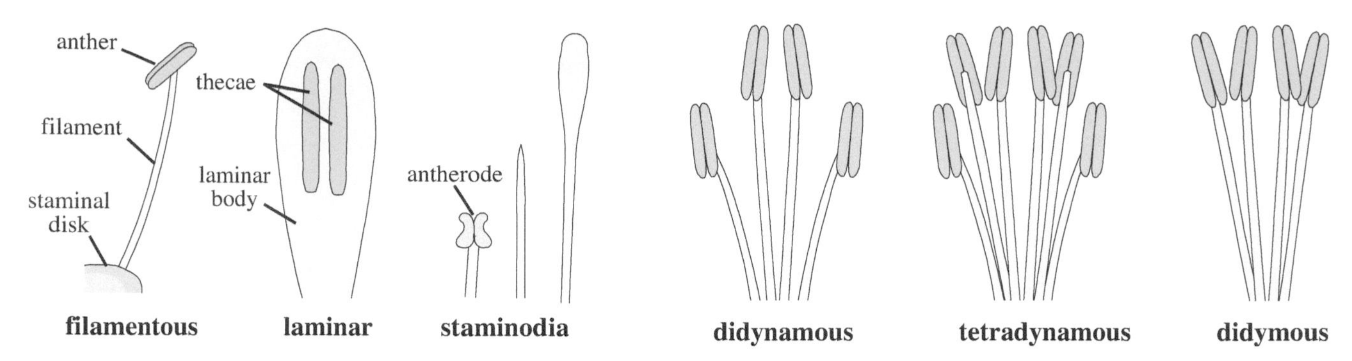

Figure 9.22 Stamen types and parts.

Figure 9.23 Stamen arrangement.

stamen fusion; see below), hypanthium rim, or **staminal disk**, a fleshy, elevated, often nectariferous cushion of tissue.

Stamen Type (Figure 9.22)

There are two basic **stamen types**: laminar and filamentous (although intermediates can occur). **Laminar** stamens possess a leaflike, dorsiventrally flattened structure bearing two **thecae** (pairs of microsporangia), these typically on the adaxial surface. Laminar stamens may represent the ancestral type in flowering plants, although they have evolved secondarily in some groups. **Filamentous** stamens are far more common, having a stalklike, generally terete **filament** with a discrete pollen-bearing part, the **anther**.

In some taxa one or more stamens will initially form but will be nonfertile. Such a sterile stamen is termed a **staminode** or **staminodium**. Staminodes may resemble the fertile stamens and can only be identified by determining if viable pollen is released. Other staminodes may be highly modified in structure, being petaloid, clavate (clublike), nectariferous, or very reduced and vestigial. Staminodes may or may not possess an **antherode**, a sterile antherlike structure.

Stamen Arrangement, Cycly, Position, and Number

Stamen arrangement (Figure 9.23) is the placement of stamens relative to one another (see General Terminology). Two basic stamen arrangements are **spiral**, with stamens arranged in a spiral, and **whorled**, with stamens in one or more discrete whorls or series. Additional stamen arrangement types consider the relative lengths of stamens to one another: **didymous**, with stamens in two equal pairs; **didynamous**, with stamens in two unequal pairs (as in many Bignoniaceae, Lamiaceae, Scophulariaceae, etc.); and **tetradynamous**, with stamens in two groups of four long and two short (typical of the Brassicaceae).

Stamen cycly (Figure 9.24) refers to the number of whorls or series of stamens present (applying only if the stamens are whorled to begin with). The two major types of stamen cycly are **uniseriate**, having a single whorl of stamens, and **biseriate**, with two whorls of stamens. If additional whorls are present, the terms *triseriate*, *tetraseriate*, etc., can be used.

Stamen position (Figure 9.23) is the placement of stamens relative to other, unlike floral parts, in particular to the sepals and petals. An **antisepalous** (also called *antesepalous*) stamen position is one in which the point of stamen attach-

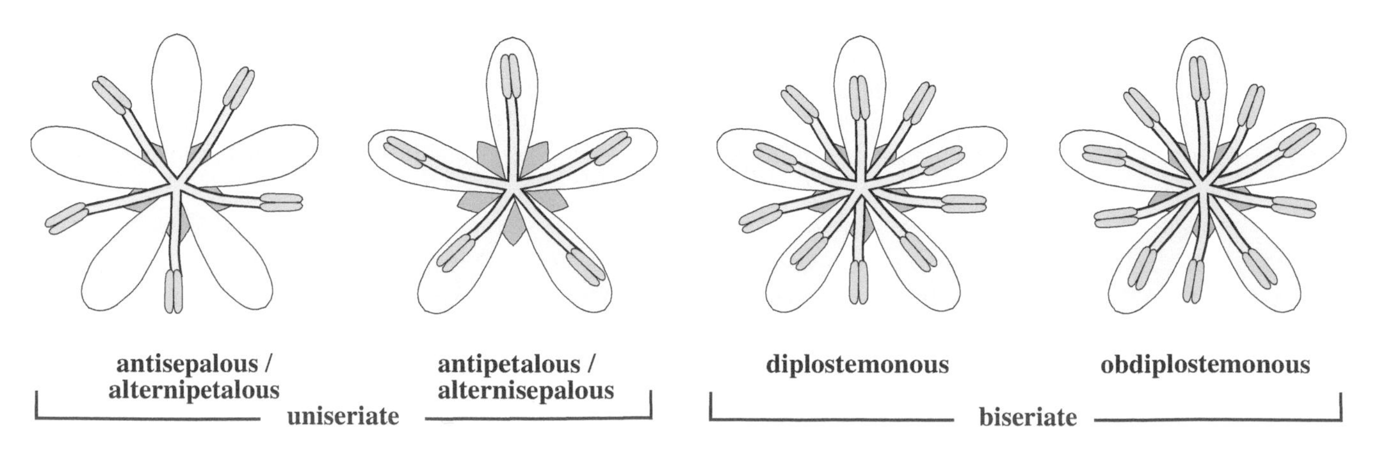

Figure 9.24 Stamen cycly (uniseriate or biseriate) and position.

ment is in line with (opposite) the sepals, calyx lobes, or outer whorl of tepals; similarly, **alternipetalous** means having the stamens positioned between the petals or corolla lobes. Antisepalous and alternipetalous are usually synonymous because (in a biseriate perianth) petals/corolla lobes are almost always inserted between sepals/calyx lobes; however, one should describe only what is evident, such that either or both terms may be used. Antisepalous or alternipetalous stamens are very common in taxa with uniseriate stamens.

An **antipetalous** (also called *antepetalous*) stamen position is one in which the point of attachment is in line with (opposite) the petals, corolla lobes, or inner whorl of tepals; **alternisepalous** means that the stamens are positioned between the sepals or calyx lobes. Antipetalous and alternisepalous are usually synonymous (for the same reason cited earlier). An antipetalous/alternisepalous stamen position is relatively rare and may be diagnostic for specific groups, such as the Primulaceae and Rhamnaceae.

Other stamen position terms, that also take into account stamen cycly and number are: **haplostemonous**, stamens uniseriate, equal in number to the petals, and opposite the sepals (antisepalous); **obhaplostemonous**, stamens uniseriate, equal in number to the petals, and opposite the petals (antipetalous); **diplostemonous**, stamens biseriate, the outer whorl opposite the sepals and the inner whorl opposite petals; and **obdiplostemonous**, stamens biseriate, the outer whorl opposite the petals, the inner opposite sepals. Among taxa with a uniseriate stamen cycly, a haplostemonous position is much more common. Among those with a biseriate stamen cycly, a diplostemonous position is much more common; obdiplostemonous stamens are relatively rare, being diagnostic, e.g., for some Crassulaceae.

Stamen number is typically simply expressed as just that, a number. The term **polystemonous** may be used for an androecium with numerous stamens, usually many more than the number of petals.

Stamen Attachment and Insertion (Figure 9.25)

Stamen attachment refers to the presence or absence of a stalk, being either **filamentous**, with a filament present, **sessile**, with filament absent, or **subsessile**, with filament very short and rudimentary. Laminar stamens are, by default, sessile.

Stamen insertion (Figure 9.25) can refer to either of two things. First, it can indicate whether stamens extend past the perianth or not, the two terms being **exserted** (also termed *phanerantherous*), with stamens protruding beyond the perianth, and **inserted** (also termed *cryptantherous*), with stamens included within the perianth. Insertion may also correspond to the *point of insertion*, which is the point of adnation of an epipetalous stamen to the corolla (see later discussion). Examples of

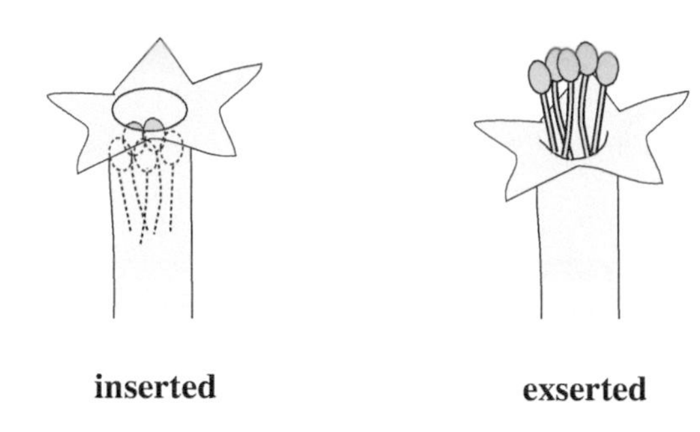

Figure 9.25 Stamen insertion.

the latter usage are "the stamens are inserted halfway up the corolla tube" or "stamens are inserted unequally" (meaning they are inserted at different levels along the length of, say, a corolla tube). Stamen insertion, by either usage, is generally indicative of an adaptation for some particular pollination mechanism, functioning to present the anthers effectively to an animal pollinator.

Stamen Fusion (Figure 9.26)

Stamen fusion refers to whether and how stamens are fused. The general terms *distinct* (unfused to one another), *connate* (fused to one another), *free* (unfused to a different structure), and *adnate* (fused to a different structure) may be used (see General Terminology). Common specialized terms are **apostemonous**, with stamens unfused (both distinct and free); **diadelphous**, with two groups of stamens, each connate by filaments only, as in many Faboideae (Fabaceae), which typically have nine stamens fused most of their length and one fused only at the base or not at all; **epipetalous** (also called *petalostemonous*), with stamens adnate to (inserted on) petals or the corolla (the terms *epitepalous* and *episepalous* can be used for adnation of stamens to tepals or sepals, respectively); **monadelphous**, with one group of stamens connate by their filaments, as in Malvaceae; and **syngenesious**, with anthers connate but filaments distinct, diagnostic of the Asteraceae. Stamen fusion, like stamen insertion, typically functions as a "presentation" mechanism for animal pollination.

Anther Parts, Type, and Attachment (Figure 9.27)

Anthers are discrete pollen containing units, found in the stamens of the great majority of angiosperms. Anthers typically consist of two compartments called **thecae** (singular *theca*), with each theca containing two microsporangia (the fusion product of which is a *locule*). (Thus, anthers are typically tetrasporangiate.) The tissue between and interconnecting the two thecae is termed the **connective**, to which the filament (if present) is attached. **Microsporangia** are the sites of production of pollen grains, the immature male gametophytes of seed plants.

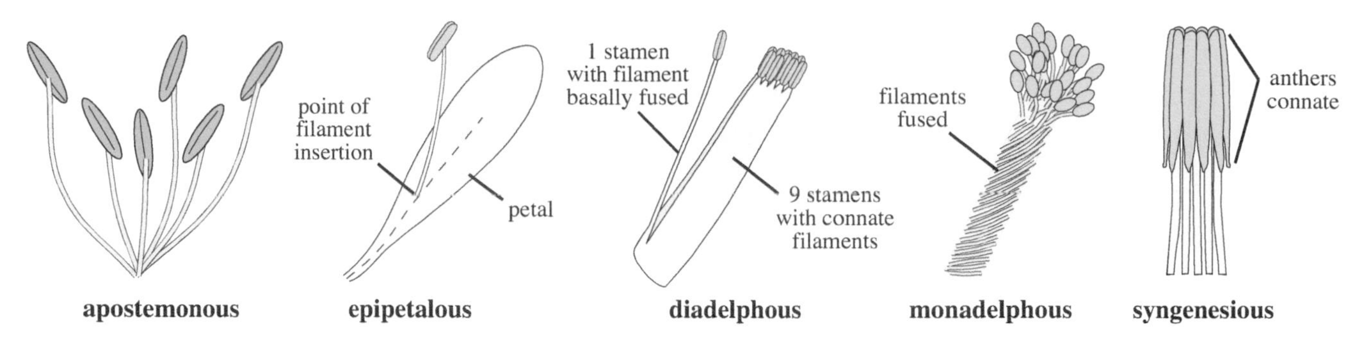

Figure 9.26 Stamen fusion.

Various anther types occur, as determined by their internal structure. The typical anther is **dithecal**, having two thecae with typically four microsporangia. In a very few taxa, such as the Cannaceae and Malvaceae, anthers are **monothecal**, having one theca with typically two microsporangia. Finally, an extreme type of anther is the **pollinium**, a typically dithecal anther in which all the pollen grains of both thecae (Orchidaceae) or of adjacent thecae (*Asclepias*) are fused together as a single mass. The pollinia of the Orchidaceae and *Asclepias* have different developmental origins and structures.

Anther attachment refers to the position or morphology of attachment of the filament to the anther. Standard anther attachment types are **basifixed**, anther attached at its base to apex of the filament; **dorsifixed**, anther attached dorsally and medially to the apex of the filament; and **subbasifixed**, anther attached near its base to the apex of the filament. A **versatile** anther attachment is one in which the anther freely pivots ("teeter -totters") at the point of attachment with the filament; versatile anthers may be dorsifixed, basifixed, or subbasifixed.

Anther Dehiscence (Figures 9.28, 9.29)
Anther dehiscence refers to the opening of the anther in releasing pollen grains. **Anther dehiscence type** (Figure 9.27) is the physical mechanism of anther dehiscence. The most common, and ancestral, anther dehiscence type is **longitudinal**, dehiscing along a suture parallel to the long axis of the thecae. Other types are rare and specific to given groups, including **poricidal**, dehiscing by a pore at one end of the thecae, such as the Ericaceae; **transverse**, dehiscing at right angles to the long axis of the theca; and **valvular**, dehiscing through a pore covered by a flap of tissue, as in the Lauraceae.

Anther dehiscence direction (Figure 9.28) indicates the position of the anther opening relative to the center of the flower or to the ground. Anther dehiscence direction is best detected when the anthers are immature (e.g., in bud) or just beginning to open. After dehiscence, the anthers usually shrivel and twist, obscuring the original direction in which they opened. Common types of dehiscence direction are: **extrorse**, dehiscing outward, away from the flower center; **introrse**, dehiscing inward, toward the flower center; and **latrorse**, dehiscing laterally, to the sides. In horizontally oriented flowers, anthers may face **upward** or **downward**, relative to the ground.

One fine point of anther dehiscence direction concerns some flowers, in which at least some of the stamens have one direction early in development but become reoriented to

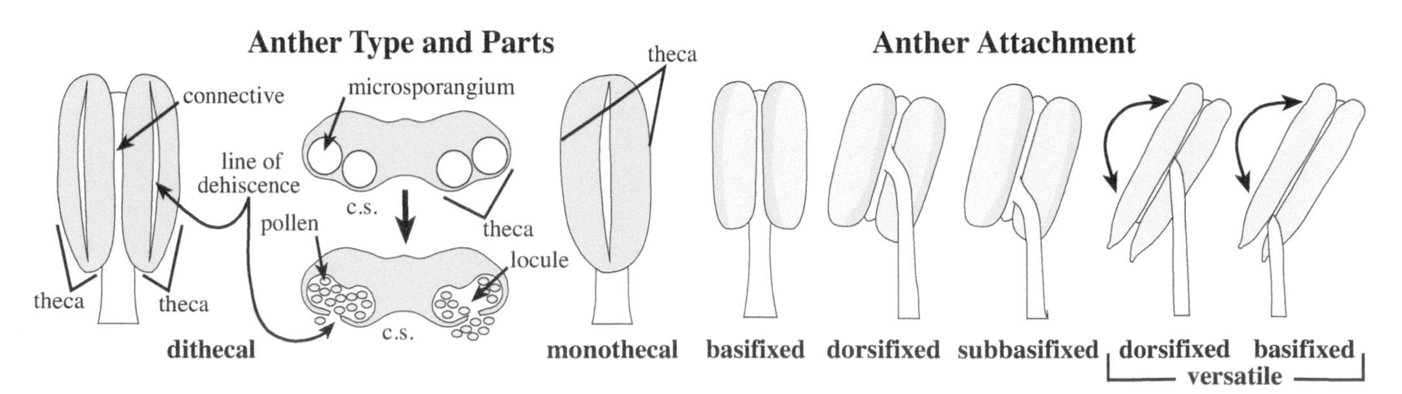

Figure 9.27 Anther types, parts, and attachment (c.s. = cross section)

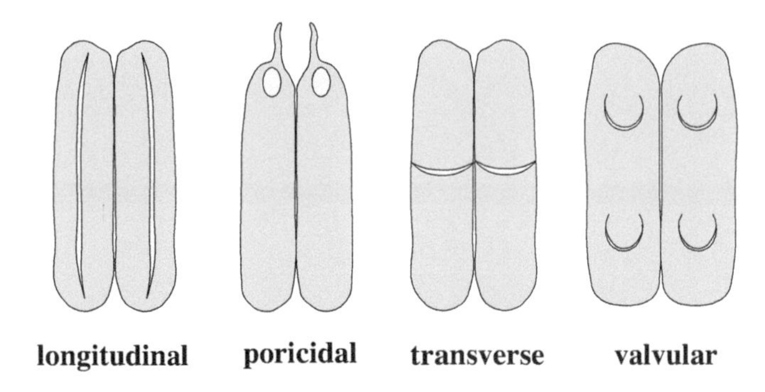

Figure 9.28 Anther dehiscence types.

another direction at maturity. In such a case, the dehiscence direction can be described both in the early developmental stage and in the mature stage. For example, a common condition is one in which the anthers are introrse early in development, but reorient to the top of the flower, with all the anthers facing downward. Such a dehiscence direction can be described as **introrse** early in development (based on observation of buds), and **downward** at maturity (see Figure 9.28). (In another example, the poricidal anthers of members of the Ericaceae are extrorse early in development, but introrse at maturity by inversion; see Chapter 8.)

NECTARIES

Nectaries are specialized nectar-producing structures of the flower (Figure 9.14). Nectar is a solution of one or more sugars and various other compounds and functions as an attractant (a "reward") to promote animal pollination. Nectaries may be **padlike**, developing as a discrete pad of tissue extending only part-way around the base of the flower. Commonly, a floral **disk**, consisting of a disk-like or dough-nut-shaped mass of tissue surrounding the ovary base or top, functions as a nectary. These **nectariferous disks** may be inner to (**intrastaminal**), beneath (**staminal**), or outer to (**extrastaminal**) the androecium. A **perigonal nectary** is one on the perianth, usually at the base of sepals, petals, or tepals. **Septal nectaries** are specialized tissues embedded *within* the septae of an ovary, secreting nectar via a pore at the ovary base or apex.

Note that other specialized glands may secrete non-sugar compounds that function as a pollination reward, such as waxes by members of the Krameriaceae. These are not termed nectaries, but are simply called glands, e.g., wax glands.

GYNOECIUM, CARPEL, AND PISTIL (Figure 9.30)

The **gynoecium** refers to all female organs of a flower (Figure 9.30). The unit of the gynoecium is the **carpel**,

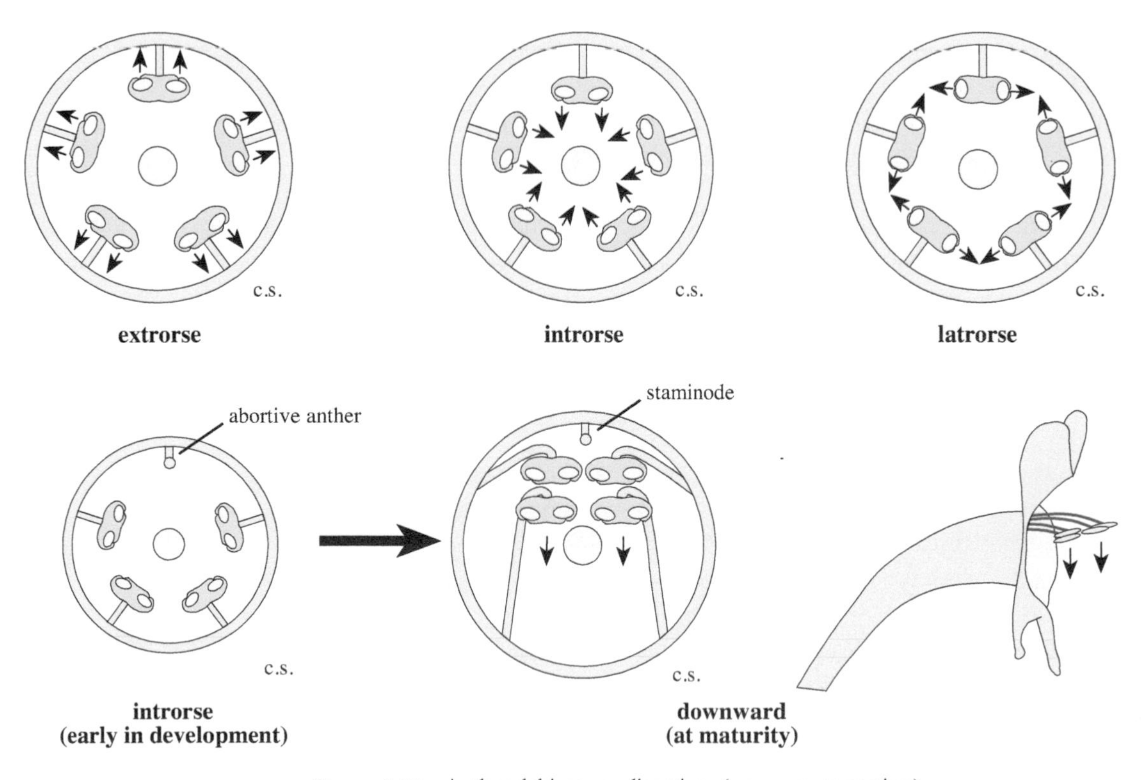

Figure 9.29 Anther dehiscence direction. (c.s. = cross section)

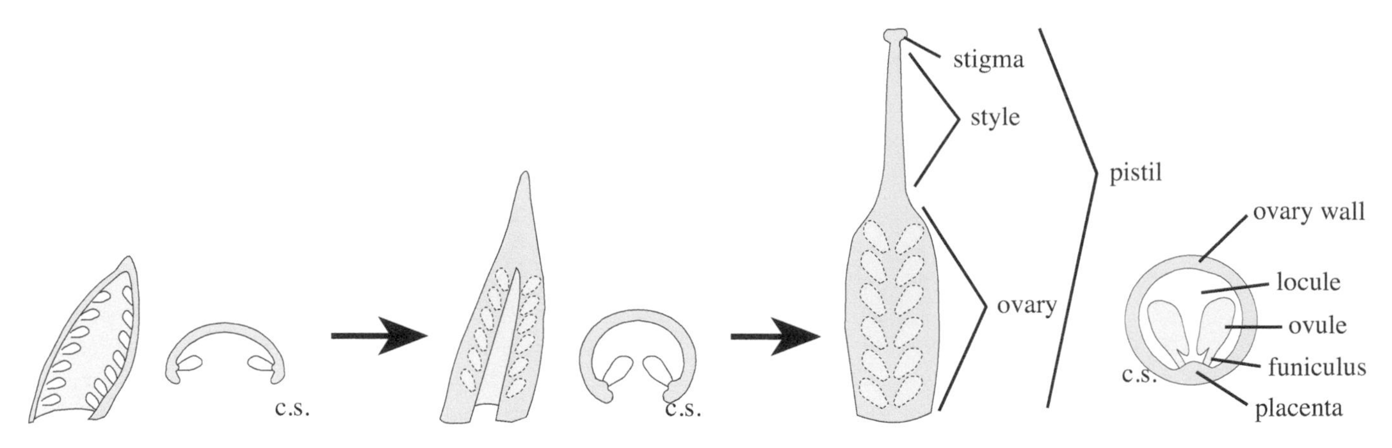

Figure 9.30 Gynoecium: carpel development. (c.s. = cross section)

defined as a modified, typically conduplicate megasporophyll that encloses one or more ovules (see Chapter 6). The carpel is one of the major features (apomorphies) that make angiosperms unique within the seed plants. Like all flower parts, a carpel is interpreted as a modified leaf, in this case a megasporophyll, defined as a reproductive leaf bearing megasporangia (which in seed plants are components of the ovules). Carpels, in fact, may develop as dorsiventrally flattened leaves that fold conduplicately, ultimately enclosing the ovules.

A **pistil** is that part of the gynoecium composed of an **ovary**, one or more **styles**, and/or one or more **stigmas** (see later discussion). The **ovary** is the part of the pistil containing the ovules. A **style** is a generally stalklike, non-ovule-bearing portion of the pistil between the stigma and ovary. Styles may be absent in some pistils. A **stigma** is the pollen-receptive portion of the pistil. Stigmas may be discrete structures or they may be a region (the *stigmatic region*) of a style or style branch, e.g., the stigmatic "lines" on the styles of Asteraceae pistils. Finally, the term **stipe** or **gynophore** is used for a basal stalk of the pistil; stipes are usually absent. [Note that *stipe* is also used as a synonym for a leaf petiole, especially that of ferns.]

Pistils or ovaries may be **simple**, composed of one carpel, or **compound**, composed of two or more carpels (see Carpel Number). By convention, if there is more than one ovary, style, or stigma, but if any of these appear fused in any way (e.g., three apparent ovaries fused at the base), they are all part of the same pistil. (One unique case are the Asclepiadoids, in which the gynoecium consists of two carpels made up of two distinct ovaries and styles but a single stigma joining the styles; because the stigmas of the two carpels are connate, the whole structure is termed a single pistil.)

Within the ovary, a **septum** (plural *septa*) is a partition or cross-wall. A **locule** is an ovary cavity, enclosed by the ovary walls and septa. **Placentae** (singular, **placenta**) are the tissues of the ovary that bear the ovules, the immature seeds. A **funiculus** is a stalk that may lead from the placenta to the ovule. A **column** is the central axis to which septae and/or placentae are attached in axile or free-central placentation (see later discussion).

Gynoecial Fusion (Figures 9.31, 9.32)

Fusion of carpels is a very important systematic character, the features of which are characteristic of major taxonomic groups. An **apocarpous** gynoecial fusion is one in which the carpels are distinct. An apocarpous gynoecium is generally thought to be the ancestral condition in the angiosperms. In contrast, a **syncarpous** gynoecial fusion is one in which carpels are connate (the pistil *compound*) and is the most common type in flowering plants. In a syncarpous gynoecium, the degree of carpel fusion can vary considerably; from connation only at the extreme base (having a strongly lobed ovary), to fusion into one, unlobed ovary but distinct styles and/or stigmas, to complete fusion with one ovary, style, and stigma. Fusion of carpels can determine the placentation type (Figure 9.32; see later discussion). Last, if the gynoecium is composed of a single carpel (in which fusion is really inapplicable), the term **unicarpellous** is used.

Carpel/Locule Number (Figure 9.31)

Carpel and locule number are important characters in angiosperm systematics. **Locule number** is generally easy to determine from ovary cross- and/or longitudinal sections, being equivalent to the number of wall-enclosed chambers within the ovary. In a general sense, ovaries may be **unilocular**, with a single locule, or **plurilocular**, having two or more locules. In some angiosperms, septa may divide the ovary into chambers in one region, such as the ovary base, but not in another region, such as the ovary apex; in such a case, the chambers

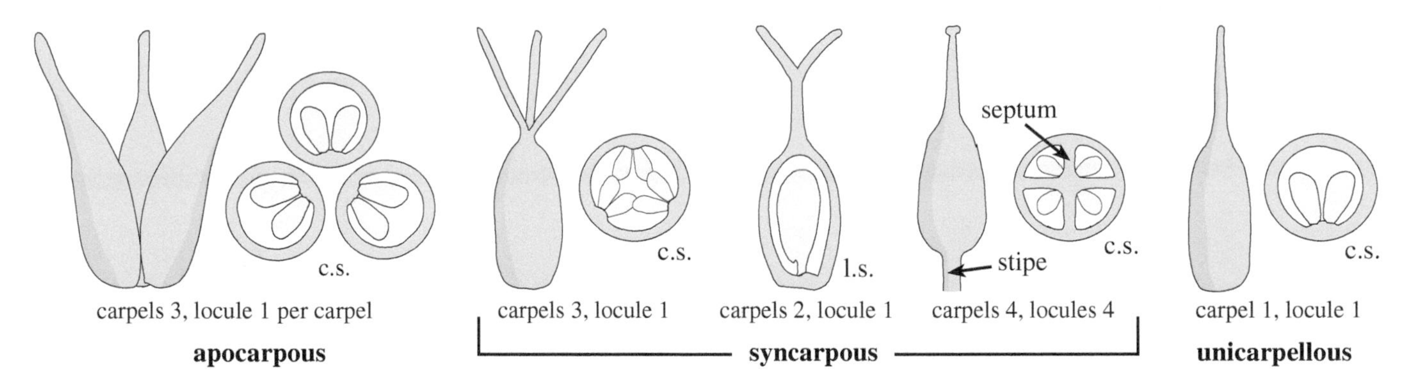

Figure 9.31 Gynoecial fusion, carpel number, and locule number. (c.s. = cross section, l.s. = longitudinal section)

below are continuous with one chamber above, and the locule number is technically 1, or unilocular.

Carpel number is often critical in classification and identification of flowering plants. It is determined as follows: If the gynoecium is apocarpous, the number of carpels is equal to the number of pistils; this is because each pistil is equivalent to a single carpel in any apocarpous gynoecium. If there is a single pistil, that pistil can be equivalent to one carpel (i.e., unicarpellous) or be composed of any number of fused carpels. For one pistil the carpel number is determined (in sequence) as follows: First, carpel number is equal to the number of styles or stigmas, if either of these is greater than 1. This is true regardless of the structure of the ovary because each of the styles or stigmas is a part of a carpel or is interpreted as a vestige of an ancestral carpel. (For example, pistils of all members of the Asteraceae have two styles and stigmas, and thus carpel number is interpreted as 2. This is true even though there is but one locule, ovule, and placenta; the two styles are interpreted as ancestral vestiges of a two -carpellate pistil, which became evolutionarily reduced to a single ovuled and loculed structure.) Second, if a single pistil has only one style and stigma, the ovary must be dissected to reveal the carpel number. If the ovary is plurilocular, then locule number is generally equal to the number of carpels. Each locule, in such a case, represents the chamber of the original ancestral or developmental carpel (except in some gynobasic taxa; see later discussion). Finally, if the ovary is unilocular, the number of carpels is equal to the number of placentae. For example, a violet, with one pistil, one style/ stigma, and one locule, has three carpels because of the three placentae (having parietal placentation). (Exceptions to the last two rules are the gynobasic taxa of the Lamiaceae and Boraginaceae, s.s. In both of these groups, each of the two carpels is bisected early in development by a so-called false

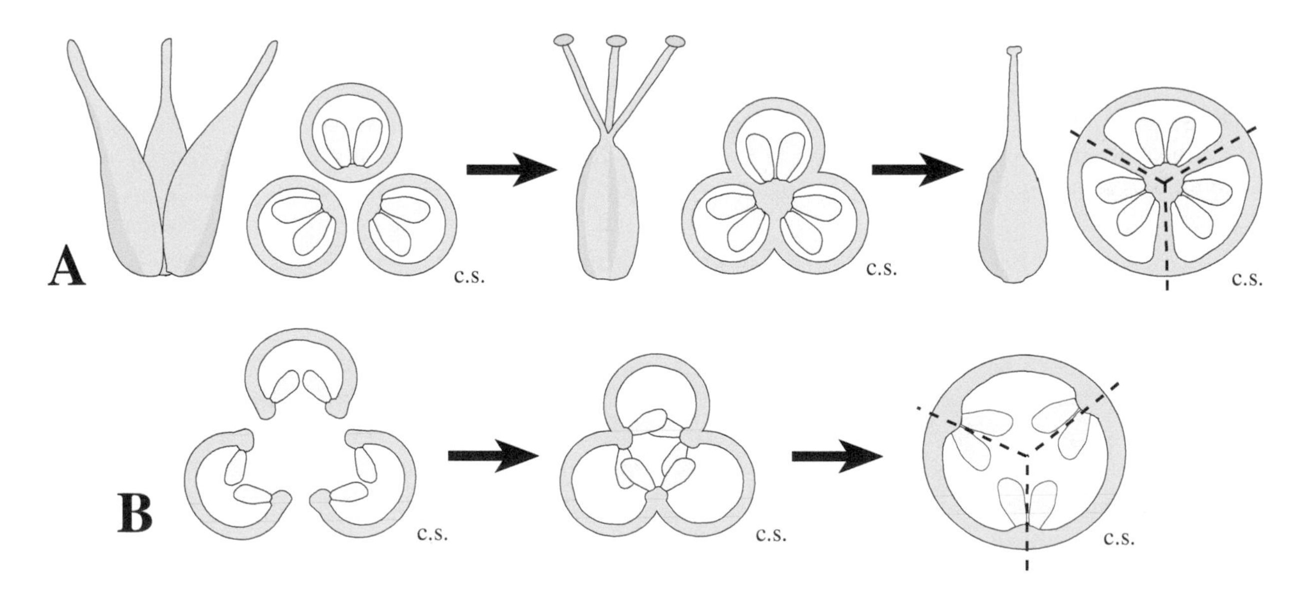

Figure 9.32 Evolutionary sequences of carpel fusion. **A.** Sequence leading to axile placentation. **B.** Sequence leading to parietal placentation. Carpel boundaries shown with dashed lines (c.s. = cross section).

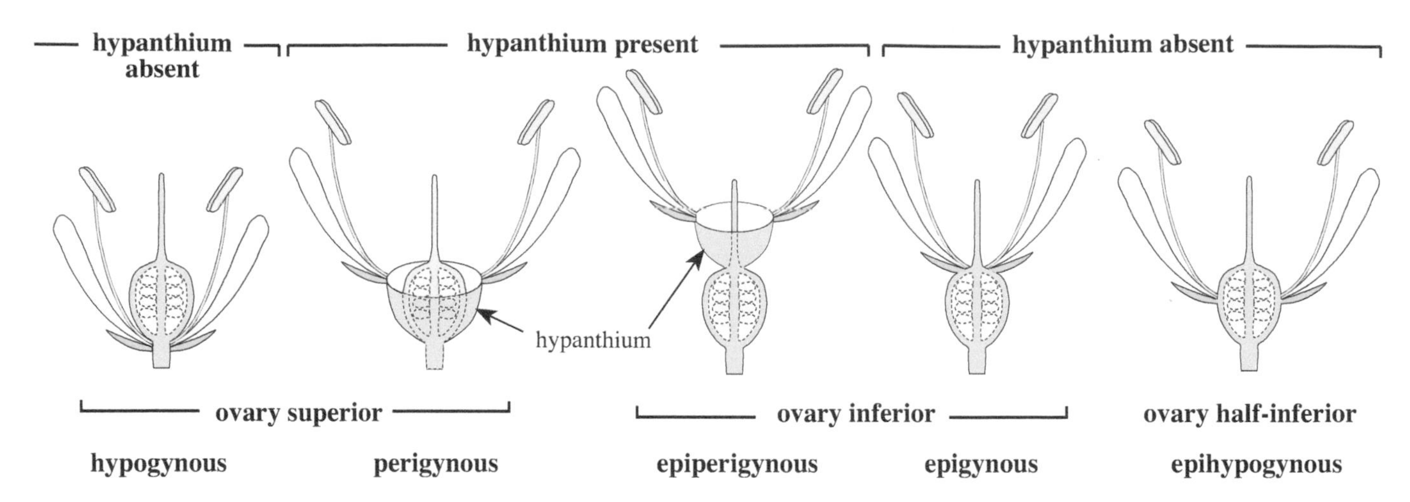

Figure 9.33 Ovary position and perianth/androecial position.

septum, such that the mature ovary typically has four locules, each with a single placenta and ovule. Thus, in this case, the number of locules and placentae, which is four, is twice that of the number of carpels.) The term **pseudomonomerous** is sometimes used for a gynoecium composed of more than one carpel but that appears to be unicarpellous.

Ovary Attachment and Position (Figures 9.31, 9.33)

Ovary attachment deals with the presence or absence of a basal stalk or **stipe**. A **sessile** ovary is one lacking a stipe and is by far the most common situation. A **stipitate** ovary is one having a **stipe** and is relatively rare (Figure 9.31).

Ovary position (Figure 9.33) assesses the position or placement of the ovary relative to the other floral parts: hypanthium, calyx, corolla, and androecium. A **superior** ovary is one with sepals, petals, and stamens, and/or hypanthium attached at the base of the ovary. An **inferior** ovary position has sepals, petals, stamens, and/or hypanthium attached at the ovary apex. A range of intermediates between superior and inferior ovaries can occur; the term **half-inferior** is used for sepals, petals, stamens, and/or hypanthium attached at the middle of the ovary.

Perianth/Androecial Position (Figure 9.33)

Perianth/androecial position describes placement of the perianth and androecium relative both to the ovary and to a hypanthium, if present. Although used widely, perianth/androecial position may be simply substituted with a description of both ovary position and hypanthium presence/absence.

Three perianth/androecial position terms describe a flower without a hypanthium (and are rather repetitious with ovary position). The term **hypogynous** is used for sepals, petals, and stamens attached at base of a superior ovary. **Epigynous** refers to the sepals, petals, and stamens attached at apex of an inferior ovary. **Epihypogynous** is used for sepals, petals, and stamens attached at middle of the ovary, the ovary being half-inferior.

Other perianth/androecial position terms denote the presence of a hypanthium, with the sepals, petals, and stamens attached to the hypanthium rim. **Perigynous** denotes a hypanthium attached at the base of a superior ovary. **Epiperigynous** denotes a hypanthium attached at the apex of an inferior ovary. (The awkward term **epihypoperigynous** may be used to describe a hypanthium attached at the middle of a half-inferior ovary.)

Placentation (Figure 9.34A)

Placentation refers to the positioning of the ovules and takes into account the number and position of placentae, septa, and locules. Determining placentation requires probing or making a cross and/or longitudinal section of the ovary.

Standard placentation types are **axile**, with the placentae arising from the column in a compound ovary with septa, common in many flowering plants such as the Liliaceae; **apical** or **pendulous**, with a placenta at the top of the ovary; **apical-axile**, with two or more placentae at the top of a septate ovary, as occurs in the Apiaceae; **basal**, with a placenta at the base of the ovary, as occurs in the Asteraceae and Poaceae; **free-central**, with the placentae along the column in a compound ovary without septa, such as in the Caryophyllaceae; **laminar**, with ovules arising from the surface of the septae; **marginal**, with the placentae along the margin of a unicarpellate (simple) ovary, as in the Fabaceae; **parietal**, with the placentae on the ovary walls or upon intruding partitions of a unilocular, compound ovary, such as in the Violaceae; **parietal-axile**, with the placentae at the junction of the septum and ovary wall of a two or more loculate ovary, such as in the Brassicaceae; and **parietal-septate**,

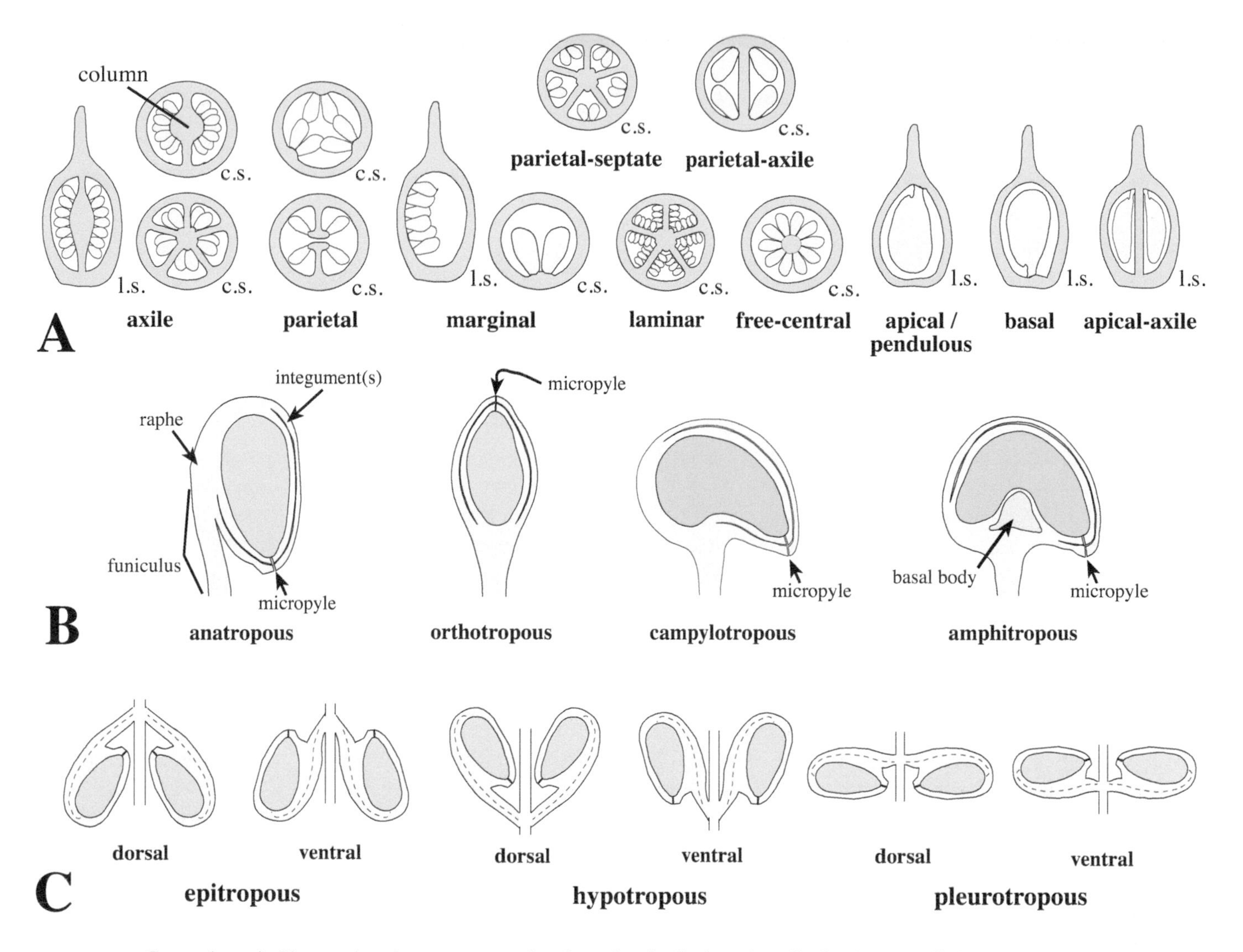

Figure 9.34 Gynoecium. **A.** Placentation. (c.s. = cross section; l.s. = longitudinal section). **B.** Ovule types. **C.** Ovule position, illustrated with anatrompous ovules.

with placentae on the inner ovary walls but within septate locules, as in some Aizoaceae.

Ovule Parts, Type, and Position (Figure 9.34B,C)
Ovules are immature seeds, technically consisting of a megasproangium enveloped by one or more integuments (Chapter 5). The basic parts of an ovule (see also **Seeds**) are the **nucellus** or **megasporangium**, within which the female gametophyte develops; **integument(s)**; **funiculus**, the stalk of the ovule; **micropyle**, the opening in the integument through which pollen or a pollen tube enters; and the **raphe**, a ridge on the seed coat often present, formed from an adnate funiculus.

Ovule type is based primarily on the curvature of the funiculus and nucellus/female gametophyte. An **anatropous** ovule is one in which curvature during development results in displacement of the micropyle to a position adjacent to the funiculus base; this is the most common ovule type of the angiosperms and is presumed to be ancestral. An **orthotropous** [**atropous**] ovule is one in which no curvature takes place during development; the micropyle is positioned opposite the funiculus base. (An ovule somewhat intermediate in curvature between anatropous and orthotropous is sometimes termed **hemitropous** or **hemianatropous**.) A **campylotropous** ovule type is one in which the nucellus is bent only along the lower side. An **amphitropous** ovule is one in which the nucellus is bent strongly along both upper and lower sides, with a lower "basal body." (See Chapter 11, for a more precise terminology of ovule types.)

Ovule position refers to the direction that an ovule faces relative to the floral axis, with the micropyle and raphe as regions of orientation. An **epitropous** ovule is one in which the micropyle points distally (toward the flower apex). This type can be further divided into **epitropous-dorsal**, in which the raphe is dorsal (abaxial, pointing away from the central floral or ovary axis) or **epitropous-ventral**, in which the raphe is ventral (adaxial, pointing toward the central floral or ovary axis). A **hypotropous** ovule is one in which the micropyle points proximally. This type can be further divided into **hypotropous-dorsal**, in which the raphe is dorsal (abaxial, pointing away from the central floral or ovary axis) or **hypotropous-ventral**, in which the raphe is ventral (adaxial, pointing toward the central floral or ovary axis). A final position type is a **pleurotropous** ovule, one in which the micropyle points to the side. This type can be further divided into **pleurotropous-dorsal**, in which the raphe is above or **pleurotropous-ventral**, in which the raphe is below. (A **heterotropous** ovule is one that varies in orientation.)

Style Position/Structural Type (Figure 9.35)

Style position is the placement of the style relative to the body of the ovary. A **terminal** or **apical** style position is one arising at the ovary apex; this is by far the most common type. A **subapical** style arises to one side, near and slightly below the ovary apex. A **lateral** style position is one arising at the side of an ovary, as in members of the Rosaceae, such as *Fragaria*. Finally, a **gynobasic** style arises from the base of the ovary. Gynobasic styles are characteristic of the Boraginaceae, s.s. and of most Lamiaceae, in which the style arises from the base and center of a strongly lobed ovary.

Styles may be structurally specialized in some taxa. One specialized **style structural type** is a stylar **beak**, a persistent, extended style or basal (to subbasal) stylar region. A beak is typically accrescent and elongates during fruit formation. Beaks function in fruit dispersal, as in members of the Asteraceae (e.g., *Taraxacum*, dandelion) or Geraniaceae (e.g., *Geranium*).

Stigma/Stigmatic Region Types (Figure 9.34)

The term **stigma** is used for a discrete structure that is receptive to pollen on the entire surface, whereas *stigmatic region* may be used for that portion of a larger structure (generally a style or style branch) that is receptive to pollen. General shape terms may be used to describe **stigma** or **stigmatic region types**. A few common stigma or stigmatic region types are **discoid**, with stigma(s) disk-shaped; **globose**, with stigma(s) spherical in shape; **linear**, with stigmas or stigmatic tissue long and narrow in shape; and **plumose**, stigmas with feathery, trichome-like extensions, often found in wind-pollinated taxa (e.g., in Cyperaceae, Poaceae).

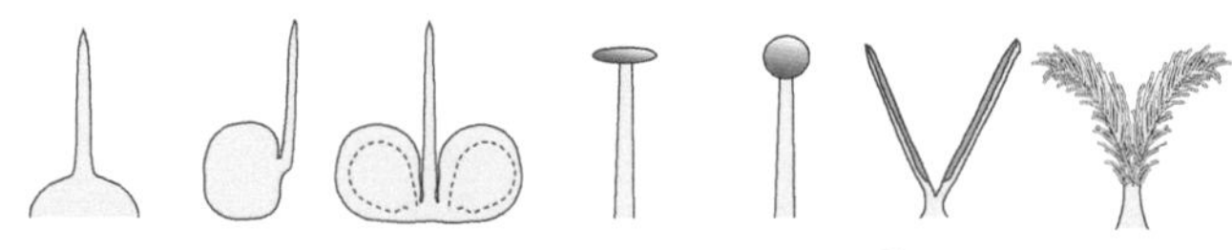

Figure 9.35 Gynoecium: style position (left) and stigma/stigmatic region type (right).

INFLORESCENCES

An inflorescence is a collection or aggregation of flowers on an individual plant. Inflorescences often function to enchance reproduction. For example, the aggregation of flowers in one location will make them visually more attractive to potential pollinators. Other inflorescences are related to very specialized reproductive mechanisms, examples being the spadices and associated spathes of some Araceae or the syconia of figs (see later discussion). The structure of an inflorescence can be complicated, requiring detailed developmental study.

Inflorescence Parts

Several terms deal with leaflike structures found in the inflorescence. An **inflorescence bract** is one that subtends not an individual flower but an inflorescence axis or a group of flowers. (Bracts that subtend an individual flower should be termed *floral bracts*; however, some sources do not make the distinction or will use *inflorescence bract* to refer to either.) A group or cluster of bracts subtending an entire inflorescence is termed an **involucre** (adjective *involucrate*); a similar group of bracts subtending a unit of the inflorescence is an **involucel**. A **spathe** (adjective *spathaceous*) is an enlarged, sometimes colored bract subtending and usually enclosing an inflorescence; many Araceae are good examples of spathes, which subtend the spadix inflorescence (see later discussion). An **awn** is a bristlelike, apical appendage on the glumes or lemmas of grass (Poaceae) spikelets.

Other terms deal with various (stem) axes in an inflorescence. A **peduncle** (adjective *pedunculate*) is the stalk of an entire inflorescence. A **compound receptacle** (also called a **torus**, although the latter term is also used for the floral receptacle; see Flower Parts) is a mass of tissue at the apex of a peduncle that bears more than one flower. A peduncle that lacks well-developed leaves, arising from a basal rosette of vegetative leaves is termed a **scape** (adjective *scapose*), the plant habit in such a case being acaulescent. A **rachis** is a major, central axis within an inflorescence. However, the central axis of a grass or sedge spikelet is a **rachilla**. Finally, a **ray** is a secondary axis of a compound umbel (see later discussion).

Inflorescence Position

There are three major **inflorescence positions**, defined based on where the inflorescence develops: (1) **axillary**, in which the entire inflorescence is positioned in the axil of the nearest vegetative leaf; (2) **terminal**, in which the inflorescence develops as part of a terminal shoot that gave rise to the nearest vegetative leaves; and (3) **cauliflorous**, in which the inflorescence grows directly from a woody trunk. Three specialized inflorescence position terms for palms are **infrafoliar**, in which the inflorescence arises below the crownshaft, **interfoliar**, in which it arises within the crownshaft, and **suprafoliar**, in which it arises above the leaves of the crownshaft.

Inflorescence Development

Inflorescence development is a major aspect of defining inflorescence type. The two major inflorescence developmental types are determinate and indeterminate. A **determinate** inflorescence is one in which the apical meristem of the primary inflorescence axis terminates in a flower; typically, the terminal flower matures first, with subsequent maturation occurring from apex to base. Determinate inflorescences are characteristic of cymes. An **indeterminate** inflorescence is one in which the apical meristem of the primary inflorescence axis does not develop into a flower; typically, the basal flower matures first, with maturation occurring from base to apex. Indeterminate inflorescences include a number of types, such as spikes, racemes, and panicles (see later discussion).

Inflorescence Type (Figures 9.36–9.38)

Inflorescences that have a common development and structure with respect to presence, number, arrangement, or orientation of bracts, axes, and certain specialized structures, define an **inflorescence type**. One difficulty with determining inflorescence type is simply delimiting its boundaries. Generally, an inflorescence is bounded by the lowest vegetative leaf. However, there may be a gradation between lower or basal vegetative leaves and small floral bracts, such that the delimitation of the inflorescence is somewhat arbitrary. (Note that if an inflorescence consists of a single flower, it is termed **solitary**; a **scapose** inflorescence is one with one or more flowers on an essentially leafless peduncle or **scape**, usually arising from a basal rosette.)

Inflorescence types are valuable characters in systematics and are often characteristic of specific groups, such as the compound umbels of the Apiaceae, heads of the Asteraceae, and helicoid or scorpioid cymes of the Boraginaceae. Some inflorescence types are quite specialized adaptations for reproduction, such as the cyathia of Euphorbioids.

The term **cyme** (Figure 9.36) can be used as a general term to denote a determinate inflorescence. One type of cyme is the **dichasium**, one that develops along two axes, forming one or more pairs of opposite, lateral axes. A **simple dichasium** is a three-flowered cyme, having a single terminal flower and two, opposite lateral flowers, the pedicels of all of equal length; bracts typically subtend the two lateral flowers, although the bracts may be absent. (The term **cymule** may be used for a small, simple dichasium.) A **compound dichasium** is a many-flowered cyme of repeatedly branching simple dichasia units. In a compound dichasium, the branches are typically decussately arranged and are thus in multiple planes. Finally, a **compound cyme** is a branched cyme, similar to a compound dichasium but lacking a consistent dichasial branching pattern. Some compound cymes actually have the same branching pattern as a compound dichasium but with certain internodal axes being reduced or missing, yielding a more congested appearance.

A **monochasium** (Figure 9.36) is a cyme that develops along one axis only. (The terminology for monochasial cymes can vary from author to author, the following being just one.) A **helicoid cyme** or **bostryx** is a monochasium in which the axes develop on only one side of each sequential axis, appearing coiled at least early in development. A **scorpioid cyme** or **cincinnus** is a monochasium in which the branches develop on alternating sides of each sequential axis, typically resulting in a geniculate (zig-zag) appearance. Both helicoid cymes and scorpioid cymes have branches or axes that are in more than one plane and can be viewed as being derived by reduction from the decussate branches of a compound dichasium. Two other monochasial cymes have, by definition, axes that are in one plane. A **drepanium** is a monochasium in which the axes develop on only one side of each sequential axis; like a helicoid cyme, drepania typically appear coiled at least early in development. (Drepania are treated as helicoid cymes in some terminology.) A **rhipidium** is a monochasium in which the branches develop on alternating sides of each sequential axis; like scorpioid cymes, rhipidia typically have a geniculate (zig-zag) appearance. (Rhipidia are treated as scorpioid cymes in some terminology.) In reality, these four monochasial structures may intergrade with one another. For example, a monochasium intermediate between a helicoid cyme and a drepanium may occur. Thus, simply using the term *monochasial cyme* may be best in lieu of more detailed observations and descriptions.

Several indeterminate inflorescence types are recognized (Figure 9.37). All of these generally lack a flower at the top of the main axis and develop from base to apex. A **spike** is an indeterminate inflorescence, consisting of a single axis bearing sessile flowers. Similarly, a **raceme** is an indeterminate

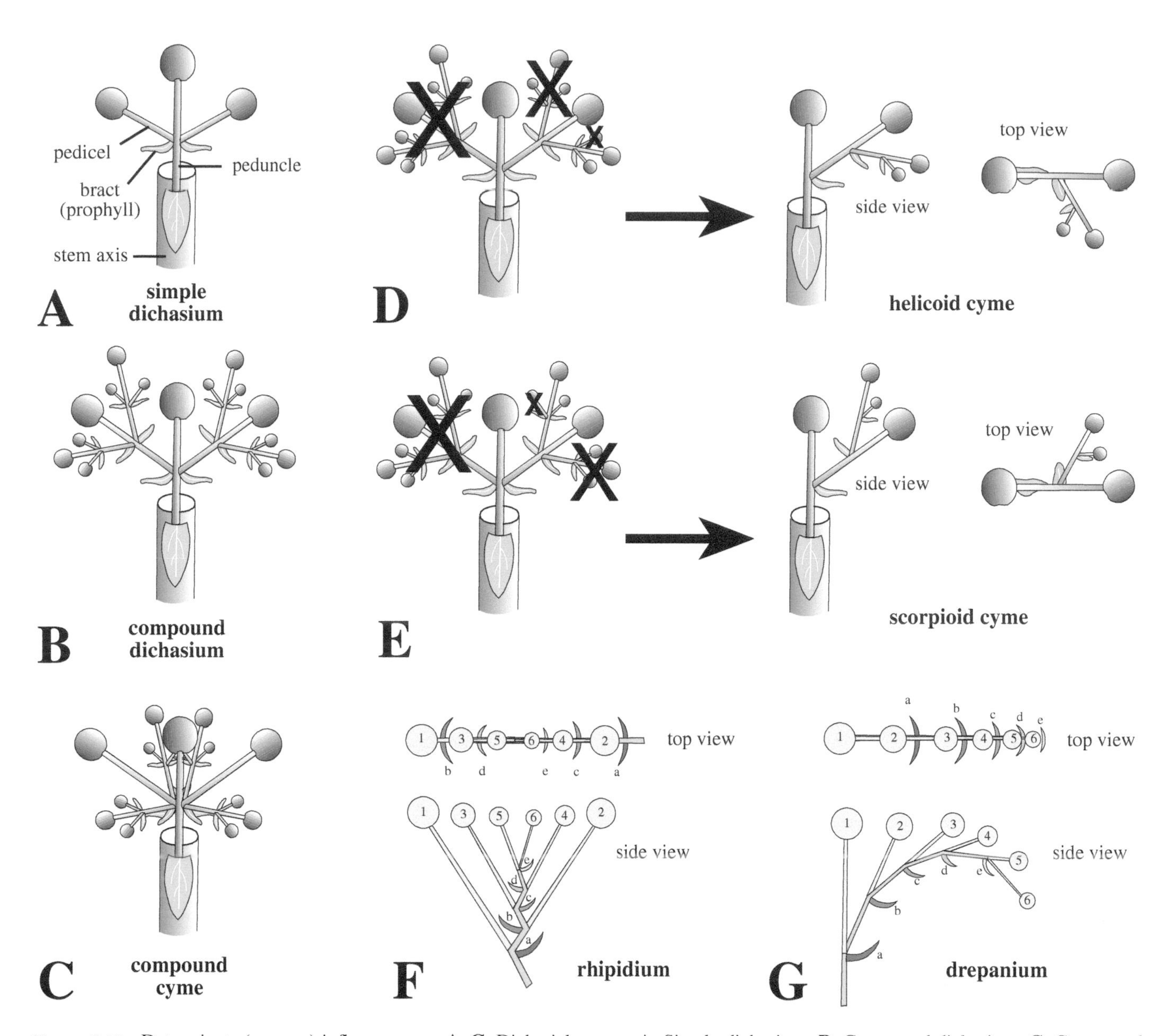

Figure 9.36 Determinate (cymose) inflorescences. **A–C.** Dichasial cymes. **A.** Simple dichasium. **B.** Compound dichasium. **C.** Compound cyme. **D–G.** Monochasial cymes. **D.** Helicoid cyme, showing derivation from compound dichasium by development of one axis on the side of the primary axis. **E.** Scorpioid cyme, showing derivation from compound dichasium by development of one axis on alternating sides of the primary axis. **F.** Rhipidium. **G.** Drepanium. (Terminology after and redrawn from Weberling 1989, by permission.)

inflorescence in which the single axis bears pedicellate flowers. A **panicle** is like a branched raceme, defined as an indeterminate inflorescence having several branched axes bearing pedicellate flowers. Finally, a **corymb** is an indeterminate inflorescence consisting of a single axis with lateral axes and/or pedicels bearing flat-topped or convex flowers. Corymbs can be either simple or compound. A **simple corymb** is unbranched, consisting of a central axis bearing pedicellate flowers, the collection of flowers being flat-topped or convex; simple corymbs are like racemes in which the lower pedicels are much more elongate than the upper. A **compound corymb** is branched, consisting of two or more orders of inflorescence axes bearing flat-topped or convex, pedicellate flowers; compound corymbs are like panicles in which the lower axes and pedicles are much more elongate than the upper.

Some inflorescences may be either determinate or indeterminate (Figure 9.38). A **simple umbel** is a determinate or indeterminate, flat-topped or convex inflorescence with

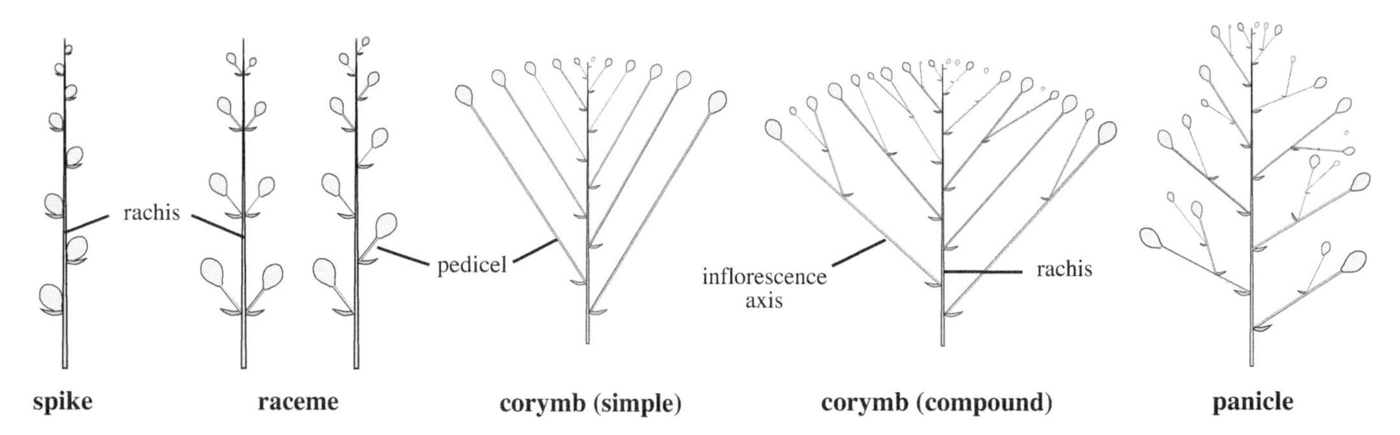

Figure 9.37 Indeterminate inflorescence types.

pedicels attached at one point to a peduncle. Two inflorescences in which the flowers at the point of attachment appear congested are the fascicle and glomerule. A **fascicle** is a racemelike or paniclelike inflorescence with pedicellate flowers in which internodes between flowers are very short. A **glomerule** is an inflorescence of sessile or subsessile flowers in which the internodes between flowers are very short.

In some taxa an inflorescence will appear to be one type, but (upon detailed examination) is actually a modification of another type. For example, the term **pseudoumbel** is used for an inflorescence appearing like a simple umbel, but actually composed of condensed, monochasial cymes, as in the Alliaceae and Amaryllidaceae.

Secondary Inflorescences (Figure 9.39)

Secondary inflorescences are defined as aggregates of unit inflorescences (also called "primary" or "partial" inflorescences); each **unit inflorescence** is a subunit of the secondary inflorescence that resembles an inflorescence type, per se. Examples of secondary inflorescences are "a panicle of spikelets," "a corymb of heads," or "a raceme of spikes." A **paracladium** is a unit inflorescence resembles the secondary inflorescence, e.g., an umbel of umbels (compound umbel).

Two specific types of secondary inflorescences are the thryse and verticillaster. A **thyrse** is essentially a raceme of cymes, in which the main axis is indeterminate but the opposite, lateral, unit inflorescences are pedicellate cymes, typically either simple dichasia, compound dichasia, or compound cymes, occasionally monochasial cymes. A **verticillaster** is essentially a "spike of opposite cymes," similar to a thyrse in having an indeterminate main axis but differing in that the lateral cymes have very reduced to absent internodal axes and pedicels, giving a congested appearance. Verticillasters are found in several members of the Lamiaceae, the mint family. A **compound umbel** is another secondary inflorescence in which the peduncle bears secondary axes called **rays** that are attached at one point and unit, simple umbels attached at the tip of the rays, as in many Apiaceae.

Specialized Inflorescences (Figure 9.40)

Some inflorescences are quite specialized and often restricted to certain taxonomic groups. A **catkin** (also called an *ament*) is a unisexual, typically male spike or elongate axis that falls as a unit after flowering or fruiting, as in *Quercus*, *Salix*. A **cyathium** is an inflorescence bearing small, unisexual flowers and subtended by an involucre (frequently with petaloid glands), the entire inflorescence resembling a single

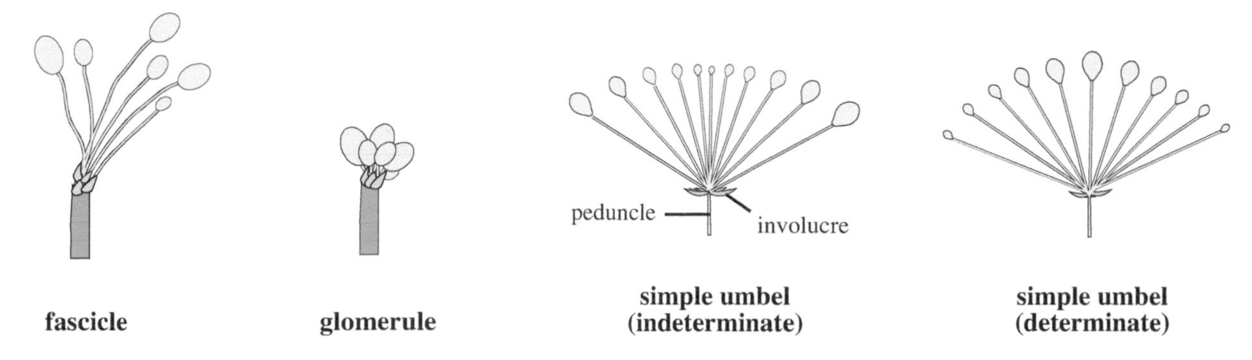

Figure 9.38 Indeterminate or determinate inflorescence types.

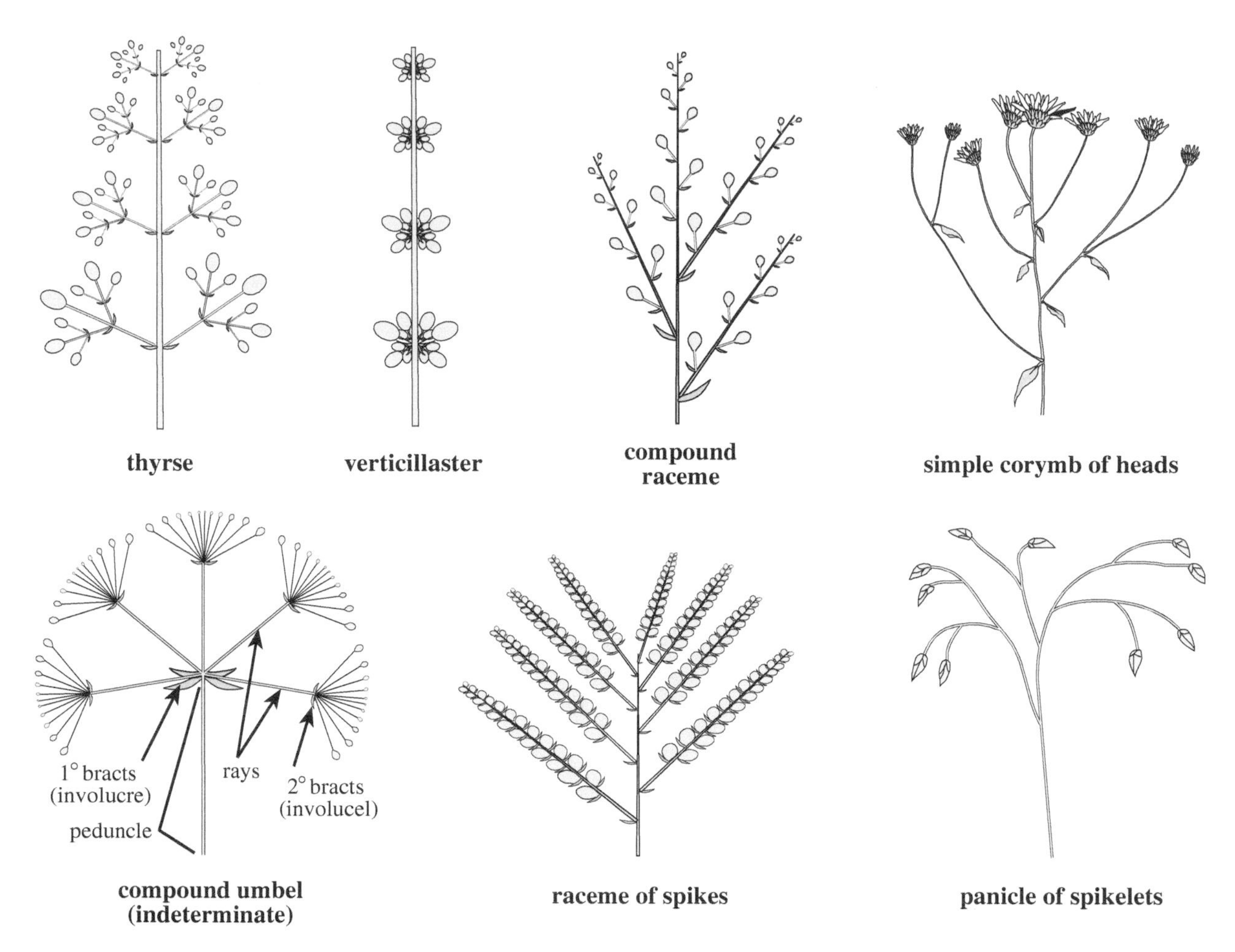

Figure 9.39 Secondary inflorescences.

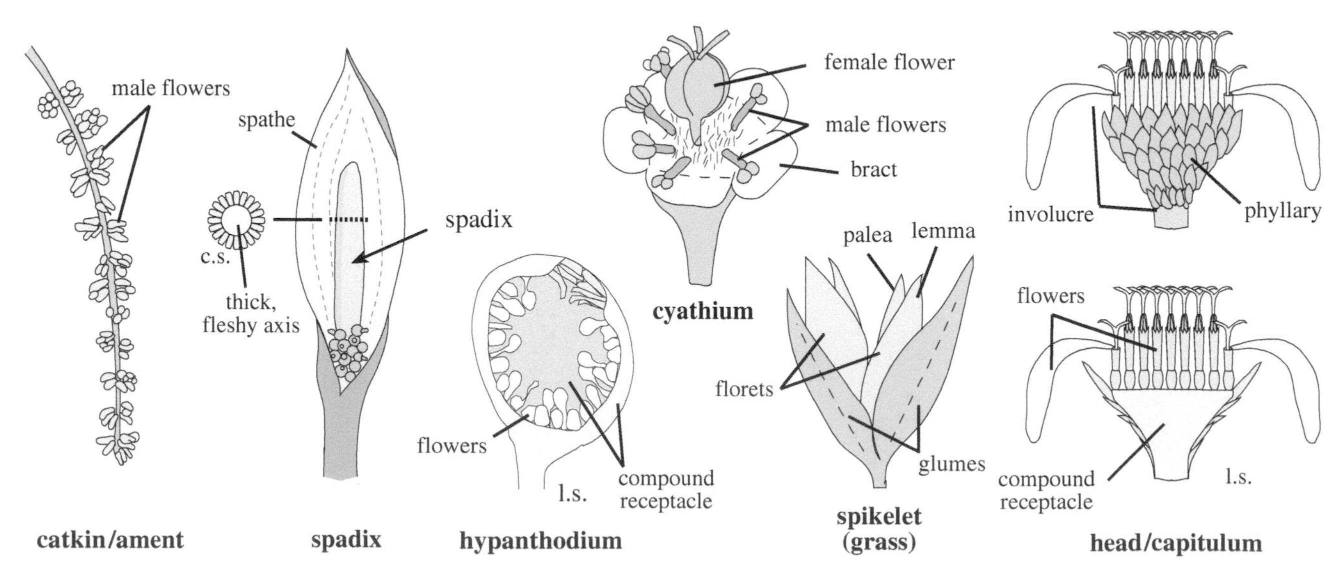

Figure 9.40 Specialized inflorescence types.

flower, as in *Euphorbia* and relatives. An inflorescence unit, such as a cyathium, that appears as and may function like a single flower is termed a **pseudanthium**, typically consisting of two or more flowers fused or tightly grouped together. A **head** or **capitulum** is a determinate or indeterminate, crowded group of sessile or subsessile flowers on a compound receptacle, often subtended by an **involucre**, composed of involucral bracts, or **phyllaries**; a **calyculus** refers to often reduced outer bracts in the heads of the Asteraceae. Heads are typical of the Asteraceae and some other groups. (Note that some inflorescences resemble a head but lack a compound receptacle; these can be termed *head-like*.) A **hypanthodium** is an inflorescence bearing numerous flowers on the inside of a convex or involuted compound receptacle, as in *Ficus*. A **spadix** is a spike with a thickened or fleshy central axis, typically with congested flowers and usually subtended by a spathe, as in the Araceae. A **spikelet** literally means a "small spike" and refers to the basic inflorescence unit in the Cyperaceae, the sedges, and Poaceae, the grasses. Sedge spikelets are like a small spike, with sessile (reduced) flowers on an axis (**rachilla**), each flower subtended by a bract (also called a *scale*). A grass spikelet consists of an axis (**rachilla**), typically bearing two basal bracts (*glumes*) and one or more short lateral branch units called **florets**, each of which bears two bracts (*lemma* and *palea*) that subtend a terminal, reduced flower. (See family treatments of Cyperaceae and Poaceae in Chapter 7.)

FRUITS

Fruits are the mature ovaries or pistils of flowering plants plus any associated accessory parts. **Accessory parts** are organs attached to a fruit but not derived directly from the ovary or ovaries, including the bracts, axes, receptacle, compound receptacle (in multiple fruits), hypanthium, or perianth. The term **pericarp** (rind, in the vernacular) is used for the fruit wall, derived from the mature ovary wall. The pericarp is sometimes divisible into layers: endocarp, mesocarp, and exocarp (see fleshy fruit types, discussed later).

Fruit types are based first on fruit development. The three major fruit developments are **simple** (derived from a single pistil of one flower), **aggregate** (derived from multiple pistils of a single flower, thus having an apocarpous gynoecium), or **multiple** (derived from many coalescent flowers; see later discussion). In aggregate or multiple fruits, the component derived from an individual pistil is called a **unit fruit**. The term **infructescence** denotes a mature inflorescence in fruit.

As mentioned in Chapter 6, the evolution of fruits was correlated with the evolution of carpels and is a significant adaptation for seed dispersal in the angiosperms.

Simple Fruit Types (Figures 9.41–9.43)

The simple fruit type, as well as unit fruit types of aggregate and multiple fruits, are classified based on a number of criteria, including (1) whether fleshy (succulent) or dry *at maturity*; (2) whether **indehiscent** (not splitting open at maturity) or **dehiscent** (splitting open along definite pores, slits, or sutures); (3) if dehiscent, the type (e.g., location, shape, and direction) of dehiscence; (4) carpel and locule number, including presence of septa; (5) seed/ovule number; (6) placentation; (7) structure of the pericarp wall; and (8) ovary position.

One class of simple fruits are those that are dry and indehiscent at maturity (Figure 9.41). An **achene** is a one-seeded, dry, indehiscent fruit with seed attached to the pericarp at one point only, such as the unit fruits of sunflowers. An **anthocarp** or **diclesium** is an achene or nut, surrounded by the persistent and accrescent perianth, as in *Pontederia* or the Nyctaginaceae. A **grain** or **caryopsis** is a one-seeded, dry, indehiscent fruit with the seed coat adnate to pericarp wall; grains are the fruit type of all Poaceae (grasses). (The embryo of grain crops is known as *germ*, as in "wheat germ"; the pericarp and seed coat together are the *bran*.) A **nut** is a one -seeded, dry indehiscent fruit with a hard pericarp, usually derived from a one-loculed ovary. (Nuts and achenes may intergrade; the terms are sometimes used interchangeably.) A **nutlet** is a small nutlike fruit; for example, the mericarps (see

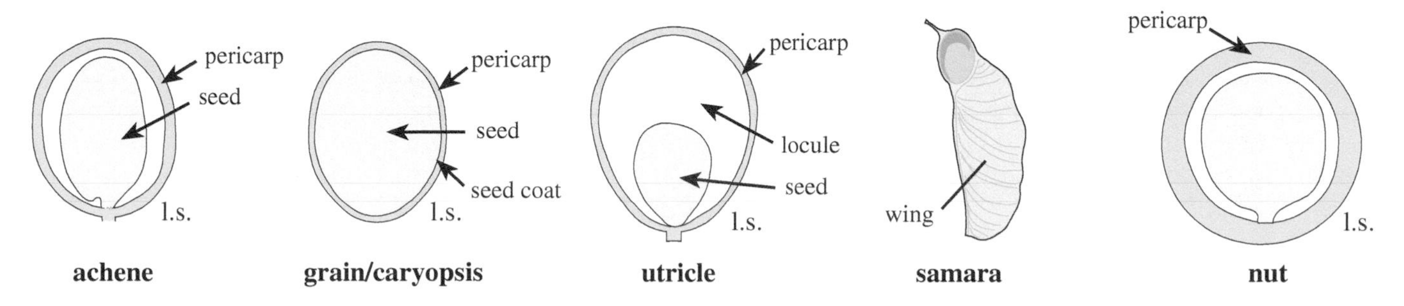

Figure 9.41 Fruits: simple, dry, and indehiscent fruit types (l.s. = longitudinal section).

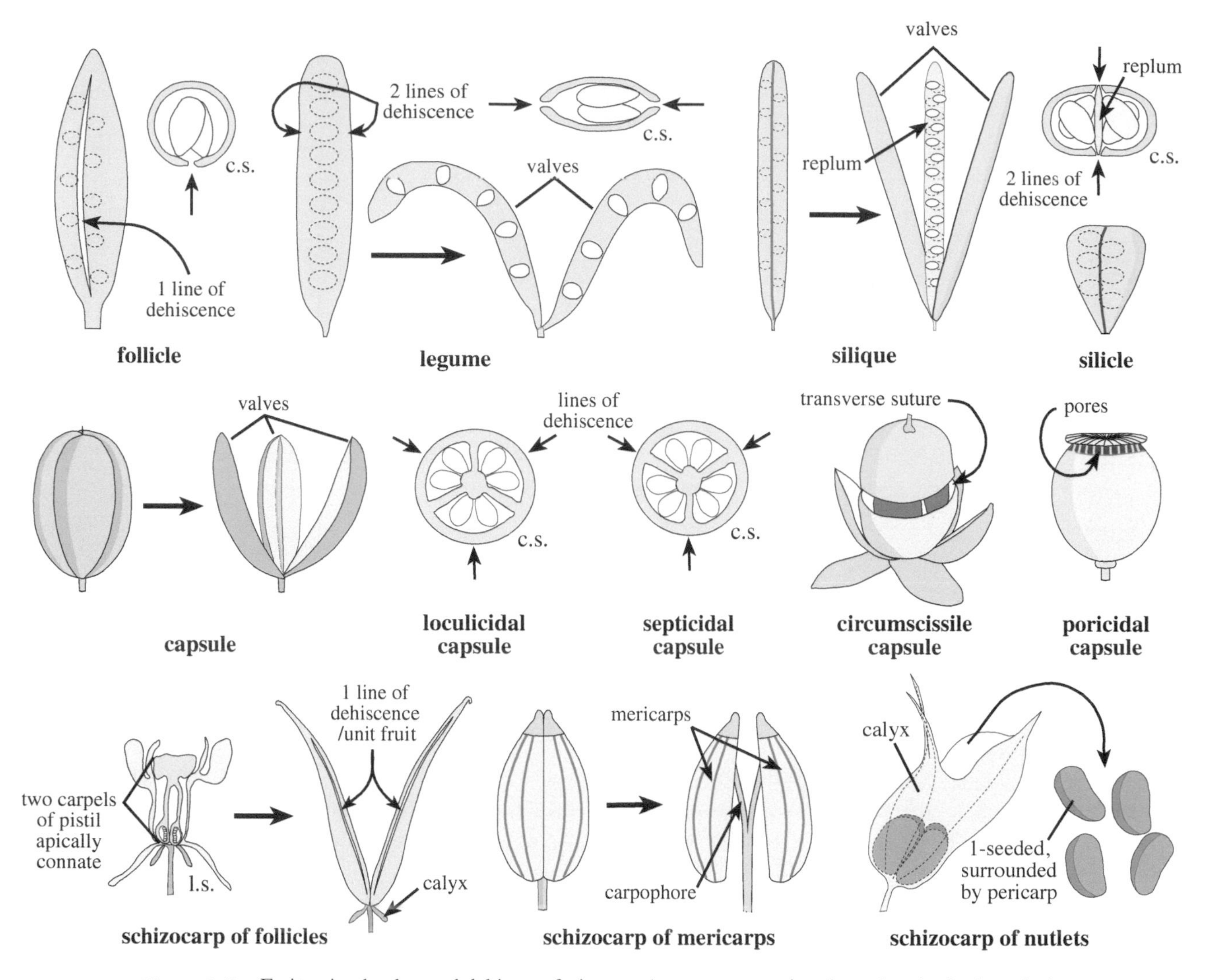

Figure 9.42 Fruits: simple, dry, and dehiscent fruit types (c.s. = cross section; l.s. = longitudinal section).

schizocarp) of the Boraginaceae and Lamiaceae are termed nutlets. A **samara** is a winged, dry, usually indehiscent fruit, as in *Acer* (maple) and *Ulmus* (elm). A **tryma** is a nut surrounded by an involucre that dehisces at maturity, such as in *Carya* (pecan). Finally, a **utricle** is a small, bladdery or inflated, one-seeded, dry fruit; utricles are essentially achenes in which the pericarp is significantly larger than the mature seed, as in *Atriplex* (salt bush).

Other simple fruits are dry and dehiscent at maturity (Figure 9.42). Most dry, dehiscent fruits open by means of a valve, pore, or mericarp (see later discussion). However, some, of various fruit types, are **explosively dehiscent**, i.e., will open with force (by various mechanisms), functioning to eject the seeds, e.g., *Ecballium elaterium*, the squirting cucumber, or *Impatiens*.

A general type of dry, dehiscent fruit is the capsule. **Capsules** are generally dry (rarely fleshy), dehiscent fruits derived from compound (multicarpeled) ovaries. Several types of capsules can be recognized based on the type or location of dehiscence. **Loculicidal capsules** have longitudinal lines of dehiscence radially aligned with the locules (or between the placentae, if septa are absent). **Septicidal capsules** have longitudinal lines of dehiscence radially aligned with the ovary septa (or with the placentae, if septa are absent). Both loculicidal and septicidal capsules split into **valves**, a portion of the pericarp wall that splits off, but does not enclose the seed(s); valves may remain attached to the fruit or may fall off, depending on the taxon. A **circumscissile capsule** (also called a **pyxis** or **pyxide**) has a transverse (as opposed to longitudinal) line of dehiscence, typically forming a terminal lid or operculum, as in *Plantago*. A **septifragal** or **valvular capsule** is one in which the valves break off from the septa, as in *Ipomoea*, morning glory. **Poricidal capsules** have dehiscence occurring by means of

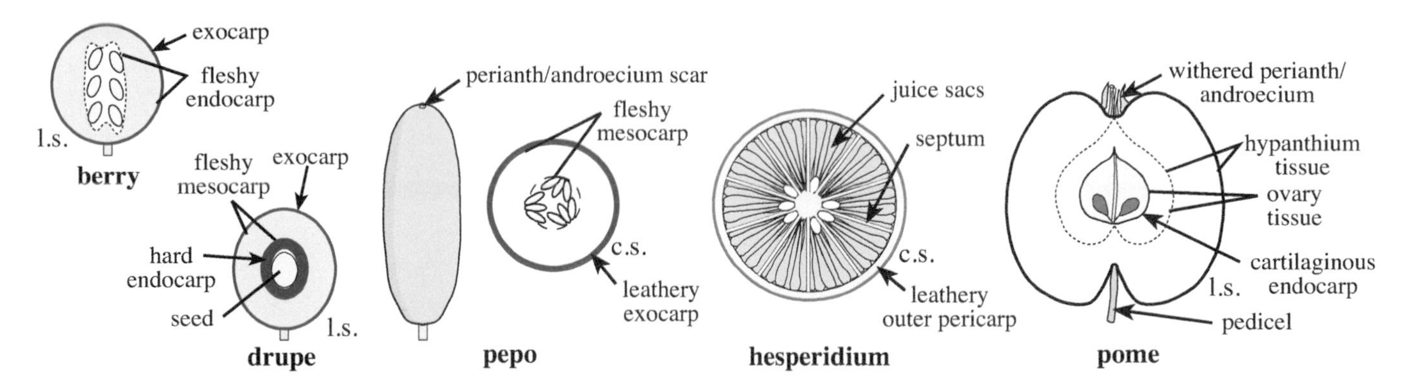

Figure 9.43 Fruits: simple, fleshy fruit types (c.s. = cross section; l.s. = longitudinal section).

pores, as in *Papaver*, poppy. Other capsules can be defined by the location of dehiscence, such as **acrocidal capsules**, dehiscing by means of apical slits, or **basicidal capsules**, dehiscing by means of basal slits, as in *Aristolochia* spp.

Some other dry, dehiscent fruit types are really just specialized capsulelike structures. A **follicle** is a dry, dehiscent fruit derived from one carpel that splits along one suture, such as in the unit fruits of *Magnolia*. A **legume** is a dry, dehiscent fruit derived from one carpel that splits along two longitudinal sutures; legumes are the diagnostic fruit type of the Fabaceae, the legume family. Some legumes retain the vestige of the two, longitudinal sutures, but have become secondarily modified, such as **loments**, which split transversely into one-seeded segments, and **indehiscent legumes**, which do not split open at all (e.g., peanut). **Silicles** and **siliques** are dry, dehiscent fruits derived from a two-carpeled ovary that dehisces along two sutures but that has a persistent partition, the **replum** (the mature septum, generally with attached seeds). The two fruit types differ is that a silicle is about as broad or broader than long, a silique is longer than broad; both are characteristic fruit types of the Brassicaceae, the mustard family.

Finally, a **schizocarp** is a dry, dehiscent fruit type derived from a two or more loculed compound ovary in which the locules separate at maturity. The individual unit fruits containing each locule can be defined based on other simple fruit types. For example, a **schizocarp of follicles** is a fruit in which the (generally two) carpels of a pistil split at maturity, each carpel developing into a unit follicle, as in *Asclepias*, milkweed. A **schizocarp of mericarps** is one in which the carpels of a single ovary split during fruit maturation, each carpel developing into a unit mericarp, as in the Apiaceae. **Mericarps** are portions of the fruit that separate from the ovary as a distinct unit completely enclosing the seed(s); in the Apiaceae the two mericarps are typically attached to one another via a stalklike structure called the **carpophore**. Lastly, a **schizocarp of nutlets** is distinct in that a single ovary becomes lobed during development, the lobes developing at maturity into nutlets, which split off. **Nutlets** here may be viewed as specialized types of mericarps. Schizocarpic nutlets are typical of the Boraginaceae and most Lamiaceae, which have gynobasic styles attached between adjacent ovary lobes. (Note that the term **eremocarp** may be used for one of the fruit units developing from ovary lobes that are separate from one another at their inception, as in the Boraginaceae and Lamiaceae. By this terminology, the term schizocarp is reserved for a fruit unit that is part of a larger, single structure at inception, splitting off only at fruit maturity, as with the schizocarp of mericarps in the Apiaceae; see Hilger 2014.)

Another class of simple fruits includes those that, at maturity, are fleshy or succulent (also termed baccate or carnose; see Texture) (Figure 9.43). Fleshy fruits are general adaptations for seed dispersal by animals, the succulent pericarp being the reward (with at least some seeds either falling out or passing through the animal's gut unharmed). Fleshy fruits are generally indehiscent, but may rarely be dehiscent, as in some *Yucca* spp. The pericarp of some fleshy fruits may be divided into layers. These pericarp wall layers, if present, are termed the **endocarp** (the innermost wall layer), **mesocarp** (the middle wall layer), and **exocarp** (the outermost wall layer); if only two layers are evident, the terms *endocarp* and *exocarp* alone are used. A **berry** is the general, unspecialized term for a fruit with a succulent pericarp, as in *Vitis*, grape. A **drupe** is a fruit with a hard, stony endocarp and a fleshy mesocarp, as in *Prunus* (peach, plum, cherry, etc.). The term **pyrene** can be used either for a fleshy fruit in which each of two or more seeds is enclosed by a usually bony-textured endocarp, or **pyrene** can refer to the seed covered by a hard endocarp unit itself, regardless of the number. A **hesperidium** is a septate fleshy fruit with a thick-skinned, leathery outer pericarp wall and fleshy modified trichomes (juice sacs) arising from the inner walls, as in *Citrus* (orange, lemon, grapefruit, etc.). A **pepo** is a

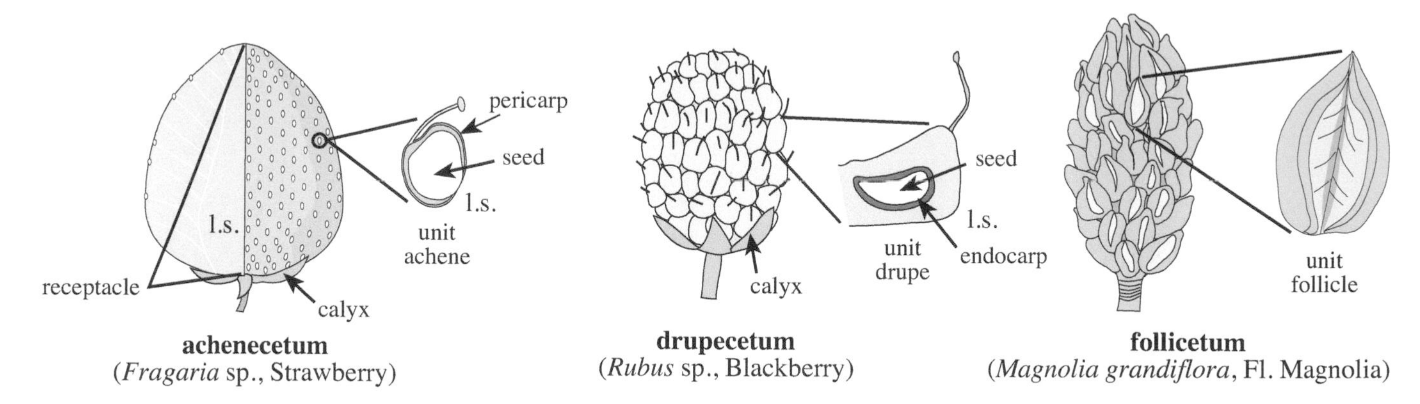

Figure 9.44 Fruits: aggregate fruit types (l.s. = longitudinal section).

nonseptate fleshy fruit with parietal placentation and a leathery exocarp derived from an inferior ovary, the fruit type of the Cucurbitaceae. A **pome** is a fleshy fruit with a cartilaginous endocarp derived from an inferior ovary, with the bulk of the fleshy tissue derived from the outer, adnate hypanthial tissue, as in *Malus* (apple) and *Pyrus* (pear). Finally, a **pseudodrupe** is a nut surrounded by a fleshy, indehiscent involucre, as in *Juglans* (walnut); thus, pseudodrupes have accessory tissue serving as the fleshy component.

Aggregate Fruit Types (Figure 9.44)

An **aggregate fruit** is one derived from two or more pistils (ovaries) of one flower. In determining the aggregate fruit type, one first identifies the **unit fruit** that corresponds to a single pistil. The aggregate fruit type is then indicated either as "aggregate fruit of" the particular unit fruits or by adding the suffix "-acetum" to the unit fruit term.

An **achenecetum** is an aggregate fruit of achenes. A common example is *Fragaria*, strawberry, in which the achenes are on the surface of accessory tissue, an enlarged, fleshy receptacle. A **drupecetum** is an aggregate fruit of drupes, as in *Rubus*, raspberry or blackberry. A **follicetum** is an aggregate fruit of follicles, as occurs in *Magnolia*. A **syncarp** is an aggregrate fruit, typically of berries, in which the unit fruits fuse together, as in *Annona*. (Note that syncarps may form at the floral stage or later during fruit development; if the latter, the fruit is sometimes called a **pseudosyncarp**.)

Multiple Fruit Types (Figure 9.45)

A **multiple fruit** is one derived from two or more flowers that coalesce. In determining the multiple fruit type, one may also identify the unit fruit corresponding to a single pistil of a single flower; the fruit type may be indicated as a "multiple fruit of" the particular unit fruit present.

Some specialized multiple fruit types are as follows: A **bur** is a multiple fruit of achenes or grains surrounded by a prickly involucre, such as in *Cenchrus*, sandbur (Poaceae), or *Xanthium*, cocklebur (Asteraceae). A **sorosis** is a multiple fruit in which the unit fruits are fleshy berries and are laterally fused along a central axis, as in *Ananas*, pineapple. A **syconium** is a multiple fruit in which the unit fruits are small achenes covering the surface of a fleshy, inverted compound receptacle (derived from a hypanthodium), as in *Ficus*, fig.

Seeds (Figure 9.46)

Aspects of seed morphology can be important systematic characters used in plant classification and identification. Some valuable aspects of seed morphology are size and shape, as well as the color and surface features of the **seed coat,** the outer protective covering of seed derived from the integument(s). The seed coat of angiosperms consists of two, postgenitally fused layers, an outer **testa** derived from the outer integument (itself sometimes divided into layers, an inner **endotesta**, middle **mesotesta**, and outer **exotesta**) and an inner **tegmen** derived from the inner integument (which can be divided into similar layers, the **endotegmen**, **mesotegmen**, and **exotegmen**). A seed coat that is fleshy at maturity may be termed a **sarcotesta** (although this may be confused with an *aril*, which is separate from the integuments; see later discussion). Also important in seed morphology are the shape, size, and color of the **hilum**, the scar of attachment of the funiculus on the seed coat, and of the **raphe**, a ridge on the seed coat formed from an adnate funiculus. Some seeds have an **aril** (adj. *arillate*), a fleshy outgrowth of the funiculus, raphe, or integuments (but separate from the integuments) that generally functions in animal seed dispersal. Arils may be characteristic of certain groups, such as the Sapindaceae. Similar to the aril is a **caruncle** (also called an **elaiosome** or **strophiole**), a fleshy outgrowth at the base of the seed; caruncles also function in animal seed dispersal, such as the caruncFulate seeds of *Viola*, violets, with regard to seed dispersal by ants.

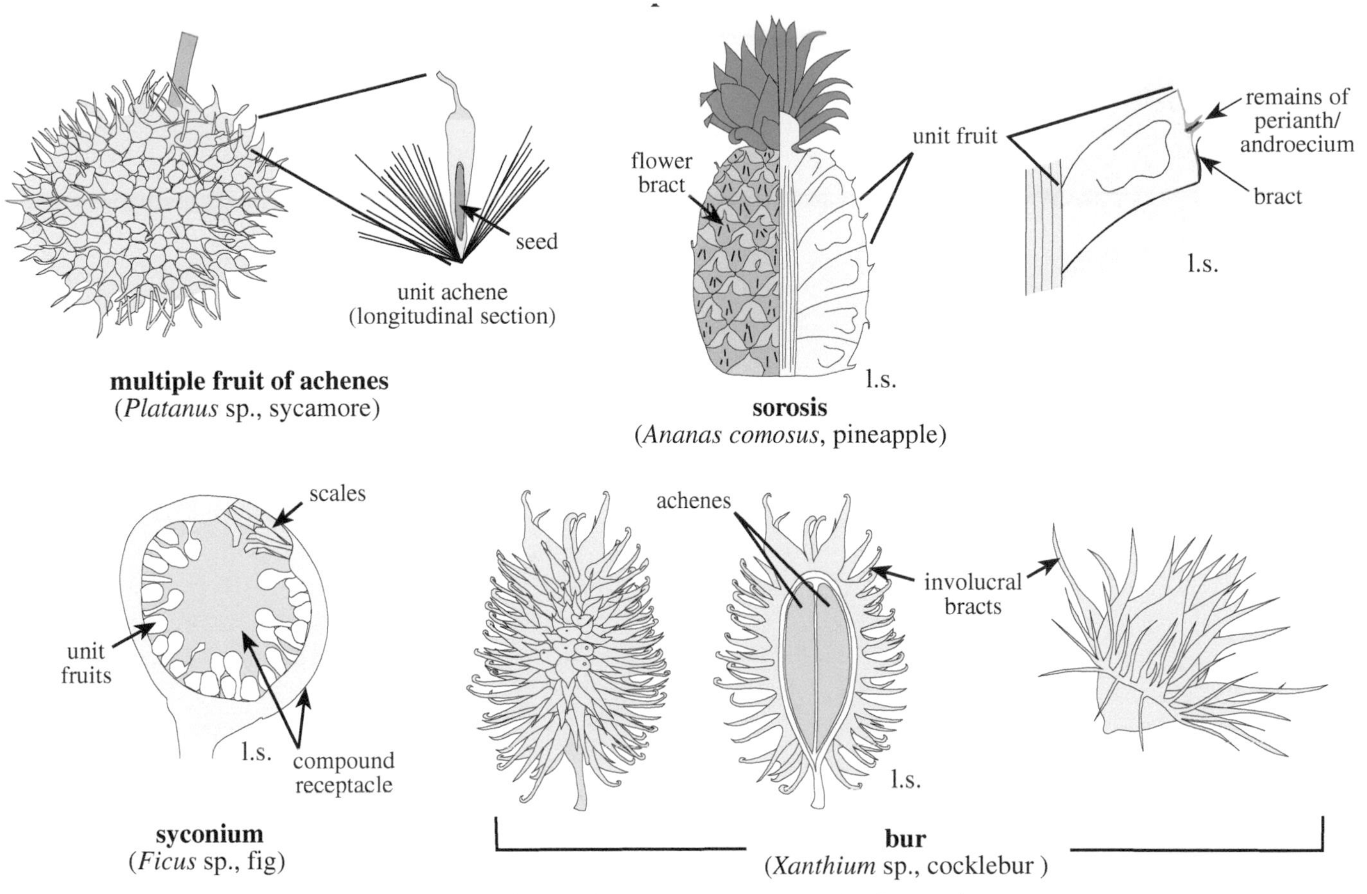

Figure 9.45 Fruits: multiple fruit types (l.s. = longitudinal section).

Specific details of the **embryo**, the immature sporophyte, can be studied. These include aspects of the **epicotyl** (the immature shoot), **radicle** (the immature root; not to be confused with a "radical" position; see later discussion), **hypocotyl** (the transition region between the root and epicotyl), and **cotyledon(s)** (the first leaf/leaves of the embryo, often functioning in storage of food reserves). Some members of the Poaceae, the grass family, have the epicotyl surrounded by a protective sheath known as the **coleoptile**, and the radicle surrounded by a protective sheath known as the **coleorhiza**. Cotyledon aestivation (or *ptyxis*) can be a valuable systematic feature.

Seed Endosperm Type

All angiosperms form endosperm, the food reserve tissue derived from fusion of sperm with the polar nuclei of the female gametophyte. The typical angiosperm seed is **albuminous** or **endospermous**, having endosperm as the food reserve in mature seeds. In some angiosperms endosperm develops, but very little to none is deposited in mature seeds, a feature termed **exalbuminous** or **nonendospermous**, as in orchid seeds. Finally, some flowering plants are **cotylespermous**, in which the main food reserve is stored in the cotyledons. Cotylespermous seeds are typical of beans and peas.

Seed Germination Type

Seed germination type requires observation of young seedlings during germination and describes positioning of the cotyledons. **Hypogeous [cryptocotylar]** refers to a type in which the cotyledon(s) remain in the ground during germination. **Epigeous [phanerocotylar]** has cotyledon(s) elevated above the ground during germination. **Vivipary** refers to a seed that germinates into a seedling before being shed from the parent plant, e.g., *Rhizophora*, red mangrove.

FRUIT AND SEED DISPERSAL

The dispersal unit, or **diaspore**, of a plant (seeds and/or fruits, including accessory parts) often exhibits specific adaptations for dispersal from the parent plant, giving it a selective advantage. These include: **anemochory**, dispersal by wind (e.g., dandelion fruits, with a wind-blown pappus); **autochory**, self-dispersal (including **ballochory**, dispersal by explosive dehiscence, **barochory**, dispersal by gravity, or **geocarpy**, dispersal in which the plant pushes fruits into the ground, e.g., *Arachis hypogaea*, peanut); **hydrochory**, dispersal by water (e.g., coconuts); **myrmecochory**, dispersal by ants (e.g., violet seeds); and **zoochory**, general dispersal by (larger) animals. Two important descriptors of zoochory

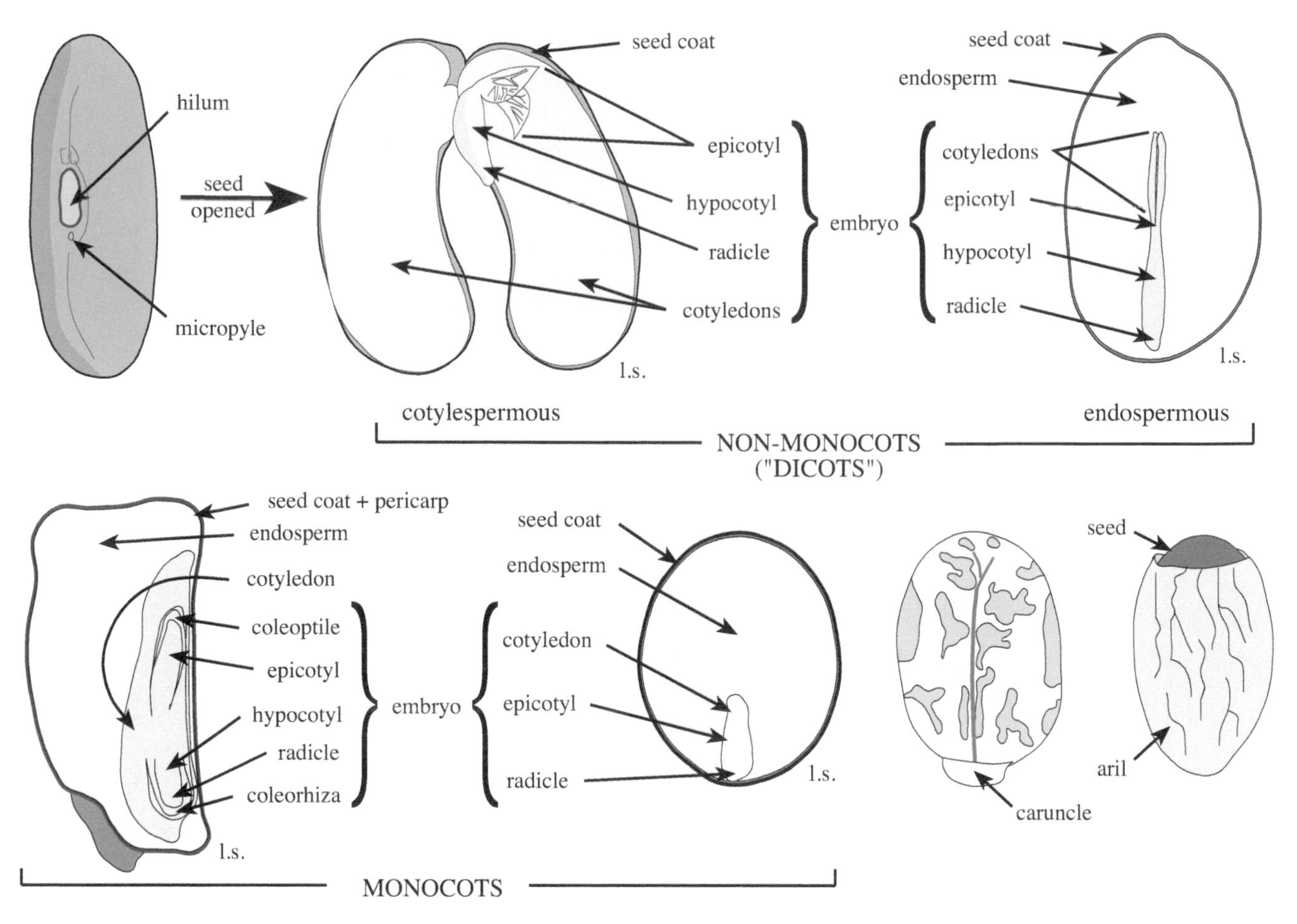

Figure 9.46 Seed parts, endosperm types (l.s. = longitudinal section).

are: **epizoochory** (**exozoochory**), in which a fruit or seed becomes attached to and is carried away by an animal (e.g., a burr becoming attached to an animal's fur or a human's socks) and **endozoochory**, in which a fruit or seed is eaten and passes out via the animal's feces unharmed. The absence of a specialized diaspore dispersal is termed **atelochory**.

GENERAL TERMINOLOGY

Many plant morphological terms can apply to a number of different plant organs (or even to features of other types of organisms). These general terms are defined below.

COLOR

Color is a measure of the wavelengths of light reflected from or transmitted through an object. When describing color, that of each component organ or part should be precisely designated. For example, instead of just stating "flowers yellow," describe as "corolla and filaments yellow, anthers maroon, pollen white, ovary green." Color itself may be defined in a very precise way, utilizing components of hue, value, and chroma. For precise designation of color, a color chart is invaluable (see Tucker et al. 1991).

Color pattern is a measure of the distribution of colors on an object. Common color pattern terms are **maculate**, spotted, with small spots on a more or less uniform background; **pellucid**, having translucent spots or patches; and **variegated**, with two or more colors occurring in various irregular patterns, generally used for leaves.

SIZE

Of course, measuring the size of plant organs and parts is important in description and identification. Generally, size of parts refers to linear measurements, as in "leaf length" or "corolla width." Metric units should be used throughout.

NUMBER

Number refers to a simple count of parts. Of course, number of parts can be very valuable information in systematic studies. With whorled structures, a distinction is made between cycly and merosity.

Cycly is the number of cycles or whorls of parts. It may simply be designated as a number, or terms may be used such as **monocyclic** or **uniseriate**, with a single whorl of parts; **dicyclic** or **biseriate**, with two whorls of parts; **tricyclic** or **triseriate**, with three whorls of parts; etc. Cycly is most commonly used for parts of the perianth or androecium (see earlier discussion).

Merosity is the number of parts per whorl or cycle. Merosity may also be designated as a simple number, or the terms **bimerous**, a whorl with two members, **trimerous**, a whorl with three members, **tetramerous**, a whorl with four members, **pentamerous**, a whorl with five members, etc., may be used. Two general merosity terms are **isomerous**, having the same number of members in different whorls, and **anisomerous**, having a different number of members in different whorls. In addition, the terms **polymerous**, having a larger than typical number of parts, and **oligomerous**, having a fewer than typical number of parts, are sometimes used. Merosity is most commonly designated for floral parts: the calyx, corolla, androecium, and gynoecium (equivalent to carpel number in a syncarpous gynoecium).

TEXTURE

Texture is the internal structural consistency of an object; some texture terms also take color into account. Texture is often described for leaves but can be used for any plant part, such as bracts or flower parts. Texture may be correlated with plant habitat and can be representative of the amount of water storage tissue (as in leaf or stem succulent plants), fibers, vascular bundles, lignin, suberin, or other internal anatomical features of a plant organ. Common texture terms include **cartilaginous**, with the texture of cartilage; hard and tough but flexible, usually whitish; **chartaceous**, opaque and of the texture of writing paper; **coriaceous**, thick and leathery, but somewhat flexible; **herbaceous**, having a soft or slightly succulent texture; **indurate**, hardened and inflexible; **membranous**, thin and somewhat translucent, membranelike; **mesophytic**, having an intermediate texture, between coriaceous and membranous (typical of many, common leaves); **ruminate,** unevenly textured, coarsely wrinkled, looking as if chewed (e.g., the endosperm of the Annonaceae); **scarious**, thin and appearing dry, usually whitish or brownish; **succulent** [*baccate* or *carnose*], fleshy and juicy; and **woody**, having a hard, woodlike texture.

FUSION

Fusion refers to the apparent joining (or lack of joining) of two or more discrete plant organs or parts. Entities that are "fused" may have developed separately and then come into contact and joined later. This process, known as *postnatal* or *postgenital fusion*, may happen, e.g., when organs (e.g., anthers) fuse after being separately formed. However, organs or plant parts that appear fused often actually develop from a common meristematic tissue early in development, a process known as *congenital fusion*. A typical example of congenital fusion is a sympetalous corolla, in which a common, ring-like meristem develops into the corolla tube. (See also Perianth Fusion, Stamen Fusion, and Gynoecial Fusion.)

Fusion terms are distinguished as to whether fusion is between like or unlike parts. **Connate** is integral fusion of *like* parts, such that the parts are not easily separable. **Adnate** is a similar integral fusion of *unlike* parts. Thus, saying "stamens are connate" means that they are fused to one another (e.g., monadelphous, diadelphous, syngenesious, etc.), whereas "stamens adnate" means they are fused to something else (e.g., to the corolla). Two similar terms to represent partial or incomplete fusion are **coherent**, with *like* parts joined but only superficially and easily separable; and **adherent**, with *unlike* parts joined, but likewise only superficially and easily separable.

Some terms designate lack of fusion. **Distinct** means with *like* parts unfused and separate. **Free** is with *unlike* parts unfused and separate. Lastly, **contiguous** means with parts touching but not connate, adnate, coherent, or adherent. Contiguous plant parts may appear fused, but are only in close contact.

SHAPE

Shape terms may be used for stems, leaves, leaflets or other leaf parts, bracts, sepals, petals, stamens, pistils, trichomes, or other plant parts. Shape is an important feature in plant description and identification. Shape may be classified as solid (three-dimensional) versus plane (two-dimensional). The latter, plane shape, may be divided into overall plane shape, base, margin, apex shape, apical process, and division.

Shape: Solid (Three-Dimensional) (Figure 9.47)

Several specific three-dimensional shapes are widely used. **Capitate** is head-shaped, spherical with a short basal stalk. The term for spherical is **globose**; that for half-sphere-shaped is **hemispheric**. An ellipsoid shape with the long axis parallel to the point of attachment is termed **prolate**; one extended perpendicular to the point of attachment is **oblate**. **Clavate** means club-shaped, cylindrical with a gradually tapering, thickened and rounded end. **Discoid** is disk-shaped, and **fusiform** is spindle-shaped, narrowly ellipsoid with two attenuate ends. **Cymbiform** means shaped like a boat, as the glumes of many grasses. **Filiform** means threadlike or filamentous, being long, thin, and typically flexuous. **Fistulose** or **fistular** means cylindrical and hollow within. **Lenticular** means lens-shaped, disk-shaped with two convex sides. **Ligulate** is tongue-shaped; flattened and somewhat oblong in shape, as

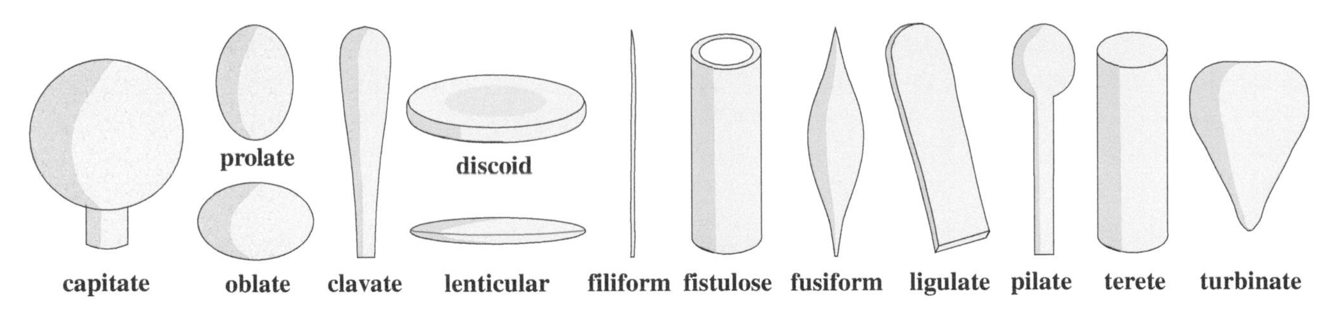

Figure 9.47 Solid (three-dimensional) shapes.

in the ligulate (ray) corollas of some Asteraceae. **Pilate** means with a long cylindrical stalk terminating in a globose or ellipsoid apical thickening, as in pilate-glandular trichomes (see later discussion). **Terete** is the general plant term meaning cylindrical. **Turbinate** means turban or top-shaped, as in turbinate heads or compound receptacles of the Asteraceae.

In addition to these specific terms, other three-dimensional shape terms can be derived from those for two-dimensional shapes (next) by adding the suffix "-oid;" as in "ellipsoid," "oblanceoloid," "ovoid."

Shape: Plane (Two-Dimensional) (Figure 9.48)

Overall plane or two-dimensional shape has been standardized (see Systematics Association Committee for Descriptive Terminology, 1962). These shape terms are based, in part, on the ratio of the length to the width of the shape outline, the common length width ratios being >12:1, 12:1–6:1, 6:1–3:1, 2:1–3:2, approximately 6:5, and approximately 1:1. (Note that the bases, apices, and details of the margin can vary in these general planar shape terms; see later discussion.)

Shapes in which the margins (sides) of the object are straight and approximately parallel are **acicular**, needle-like with length:width ratio greater than 12:1; **ensiform**, sword-shaped, with length:width ratio greater than 12:1, e.g., leaves of *Iris* spp.; **strap-shaped**, flat, not needle-like but with length:width ratio greater than 12:1; **linear**, length:width ratio between 12:1 and 6:1; **narrowly oblong**, length:width ratio between 6:1 and 3:1; and **oblong**, length:width ratio between 2:1 and 3:2.

Shapes in which the margins are symmetrically curved, with the widest point near the midpoint of the object, are **narrowly elliptic**, length:width ratio between 6:1 and 3:1; **elliptic**, length:width ratio between 2:1 and 3:2; **widely elliptic**, length:width ratio approximately 6:5; and **orbicular** (*circular*), length:width ratio approximately 1:1.

Shapes in which the margins are curved, with the widest point near the base, are **lanceolate**, length:width ratio between 6:1 and 3:1; **lance-ovate**, length:width ratio between 3:1 and 2:1; **ovate**, length:width ratio between 2:1 and 3:2; **widely ovate**, length:width ratio approximately 6:5; and **very widely ovate**, length:width ratio close to 1.

Shapes in which the margins are curved, with the widest point near the apex, are **oblanceolate**, length:width ratio between 6:1 and 3:1; **oblance-ovate**, length:width ratio between 3:1 and 2:1; **obovate**, length:width ratio between 2:1 and 3:2; **widely obovate**, length:width ratio approximately 6:5; and **very widely obovate**, length:width ratio close to 1.

Three-sided shapes, in which the sides are approximately straight, are **narrowly triangular**, length:width ratio between 6:1 and 3:1; **triangular**, length:width ratio between 2:1 and 3:2; **widely triangular**, length:width ratio approximately 6:5; and **deltate**, length:width ratio approximately 1.

Four-sided, parallelogram-like shapes are **rhombic**, widest near middle, length:width ratio between 2:1 and 3:2; and **trullate**, widest near base; length:width ratio between 2:1 and 3:2.

Finally, some specialized shapes are **alate**, winged; **cordate** (**cordiform**), shaped like an inverted Valentine heart, approximately ovate with a cordate base (see Base); **falcate** (**falciform**), scimitar-shaped, lanceolate to linear and curved to one side; **lyrate**, pinnatifid, but with a large terminal lobe and smaller basal and lateral lobes; **pandurate**, violin-shaped, obovate with the side margins concave; **reniform**, kidney-shaped, wider than long with a rounded apex and reniform base (see Base); **spatulate**, oblong, obovate, or oblanceolate with a long attenuate base; and **subulate**, awl-shaped, approximately narrowly oblong to narrowly triangular.

Base (Figure 9.49)

Base shapes in which the sides are incurved or are approximately straight are **attenuate**, basal margins abruptly incurved (concave), intersection angle less than 45°; **narrowly cuneate**, basal margins approximately straight, intersection angle less than 45°; **cuneate**, basal margins approximately straight, intersection angle 45°–90°; **obtuse**, basal margins approximately straight, intersection angle greater than 90°; and **truncate**, basal margin cut straight across, intersection angle approximately 180°.

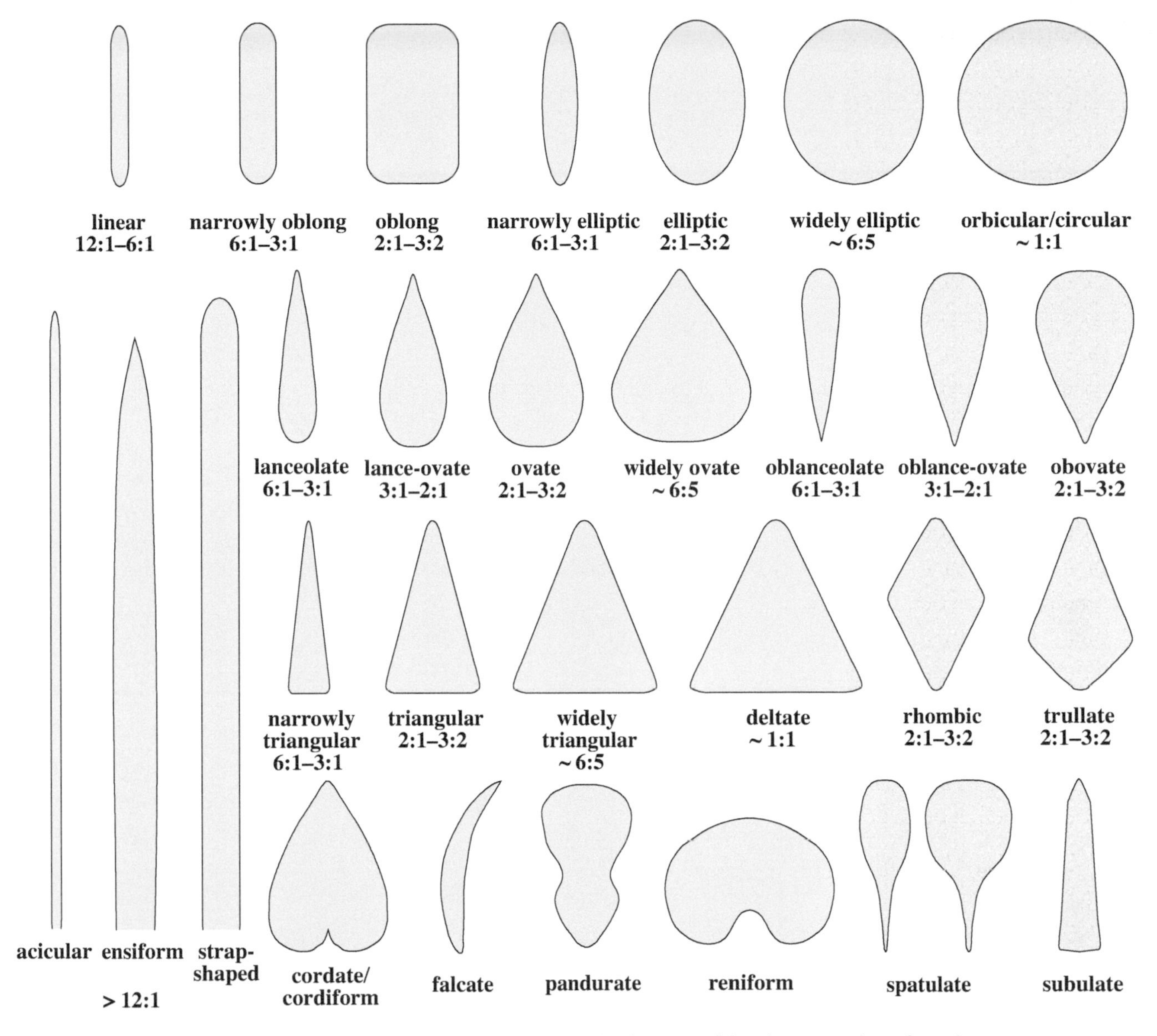

Figure 9.48 Planar (two-dimensional) shapes. Note that bottoms of drawings are points of attachment.

Base shapes in which the sides are curved are **rounded**, basal margins convex, forming a single, smooth arc; **cordate**, with two rounded, basal lobes intersecting at sharp angle, the margins above lobes smoothly rounded; and **reniform**, with two rounded, basal lobes, smoothly concave at intersection of lobes.

Bases in which there are two protruding lobes are **auriculate**, with two rounded, basal lobes, the margins above lobes concave; **hastate**, with two basal lobes, more or less pointed and oriented outwardly, approximately 90° relative to central axis; and **sagittate**, with two basal lobes, more or less pointed and oriented downward, away from the apex.

Finally, some other, specialized base shapes are **oblique**, having an asymmetrical base; **peltate**, with the petiole attached away from the margin, on the underside of the blade, as in *Tropaeolum*; and **sheathing**, having a basal, clasping leaf sheath. (Note: see also Leaf Attachment.)

APEX (Figures 9.50, 9.51)

Apex shape (Figure 9.49) refers to the shape of the apical region below the apical process or vein extension, if present (see later discussion). For a leaf or bract, this refers to the shape of the blade tissue at the apex.

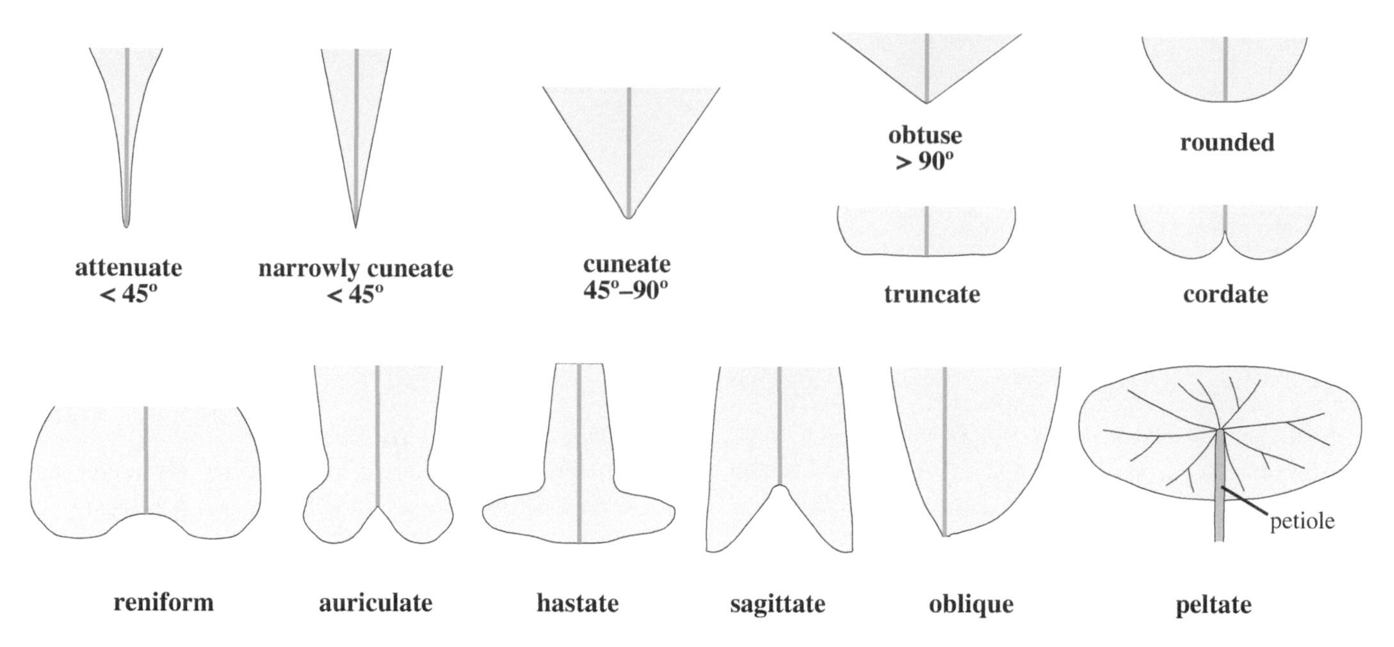

Figure 9.49 Bases. Note that bottoms of drawings are points of attachment.

An **acuminate** apex is one with the apical margins abruptly incurved (concave), the apical intersection angle <45°. Two other apex shapes are specialized variants of acuminate: **caudate**, abruptly acuminate into a long, narrowly triangular (tail-like) apical region; and **cuspidate**, abruptly acuminate into a triangular, stiff or sharp apex.

Four apex shapes have straight, not curved, sides. A **narrowly acute** apex is one with the margins approximately straight, the intersection angle less than 45°. (Thus, narrowly acute differs from acuminate, caudate, and cuspidate in part by having straight margins.) An **acute** apex also has more or less straight margins, with the intersection angle between 45° and 90°. An **obtuse** apex shape has apical margins approximately straight, the intersection angle greater than 90°. A **truncate** apex has the apical margin cut straight across, the angle approximately 180°.

A **rounded** apex has convex apical margins, forming a single, smooth arc. An **oblique** apex has an asymmetrical shape (see Base). Finally, two terms that describe an apical cleft (differing only in the depth of that cleft) are **emarginate**, having an apical incision cut $^1/_{16}$–$^1/_8$ of the distance to midrib, midvein, or junction of primary veins; and **retuse**, having an apical incision cut up to $^1/_{16}$ of the distance to midrib, midvein, or junction of primary veins.

Apical process (Figure 9.50) generally denotes an extension of a vein (typically the midvein); thus, most of the apical process is vascular tissue. A given apical process can be associated with virtually any type of apical shape.

Common apical processes are **apiculate**, with a flexible apical process, length : width ratio >3:1, usually slightly curled; **aristate**, with a stiff apical process, length : width ratio >3:1, usually prolonged and straight; **cirrhose**, with a

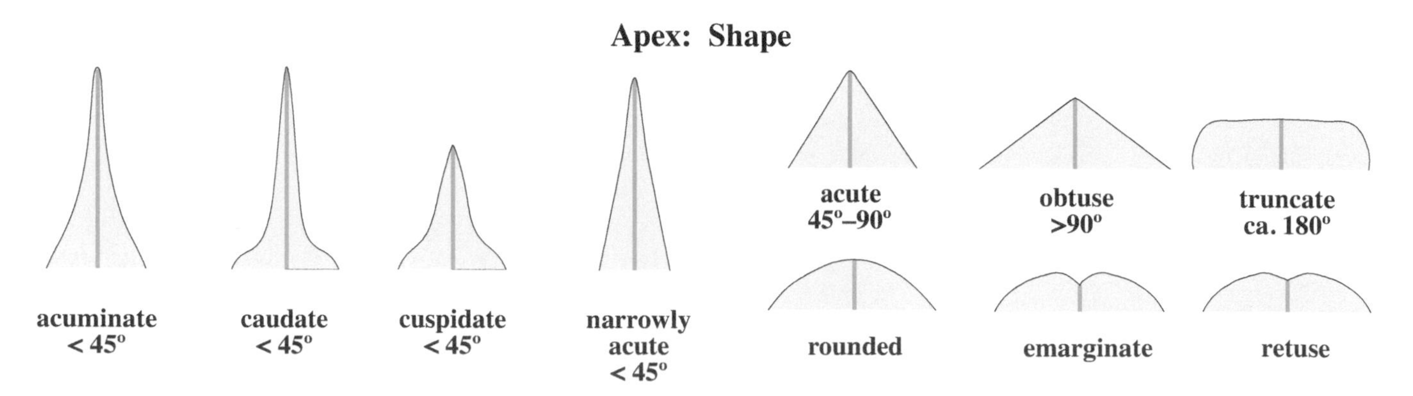

Figure 9.50 Apices. Note that bottoms of drawings are basal.

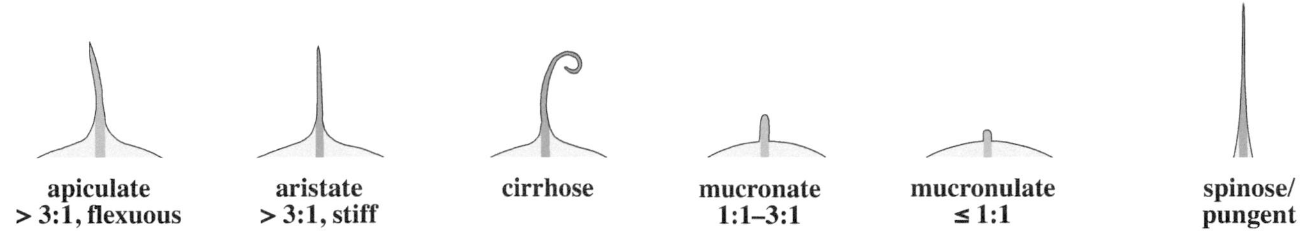

Figure 9.51 Apical processes.

flexible, greatly curled apical process; **mucronate**, with a stiff, straight apical process, the length:width ratio 1:1–3:1; **mucronulate**, with a stiff, straight apical process, length :width ratio ≤ 1:1; and **spinose** or **pungent**, with a sharp, stiff, spinelike apical process.

Shape Combinations (Figure 9.52)

The overall shape, base shape, and apex shape can be used in combination to describe a variety of two-dimensional forms. For example, Figure 9.51 shows five leaves, all with a more or less elliptic overall shape, but differing in the shape of the base and apex.

Margin (Figure 9.53)

Margin refers to the sides of an object, usually a leaf, bract, sepal, or petal. Many margin terms describe the presence and morphology of *teeth*, small sharp-pointed or rounded projections or lobes along the sides. Technically, teeth extend no more than $^1/_8$ of the distance to the midrib, midvein, or (in a palmately lobed leaf) junction of the primary veins; if further than $^1/_8$ of this distance, then the object is described as *lobed*, *cleft*, *parted*, or *divided* (see Division).

A margin without teeth is termed **entire**. (However, the plane may be divided; see later discussion.) A margin with teeth can be generally termed "toothed," but more specific terms are preferable. Margin terms describing sawlike teeth, i.e., sharp-pointed and ascending (the lower side longer than the upper) are **serrate**, teeth cut $^1/_{16}$–$^1/_8$ of the distance to midrib, midvein, or junction of primary veins; **serrulate**, diminutive of serrate, teeth cut to $^1/_{16}$ of the distance to midrib, midvein, or junction of primary veins; and **doubly serrate**, with large, serrate teeth having along the margin smaller, serrate teeth.

Margin terms describing sharklike teeth that point outward at right angles to the margin outline (the upper and lower sides about the same length) are **dentate**, with teeth cut $^1/_{16}$–$^1/_8$ of the distance to midrib, midvein, or junction of primary veins; and **denticulate**, diminutive of dentate, cut to $^1/_{16}$ of the distance to midrib, midvein, or junction of primary veins.

Margin terms describing rounded to obtuse teeth, that point outward at right angles or shallowly ascend, are **cre-**

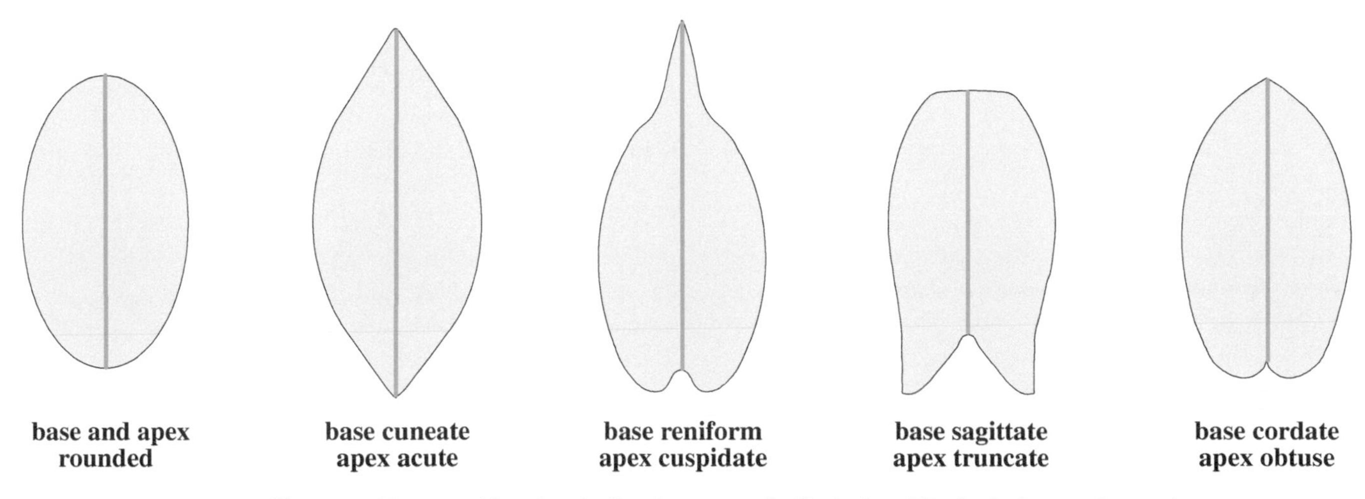

Figure 9.52 Shape combinations. Note that the five shapes are all elliptic, but differ in the base and apex shapes.

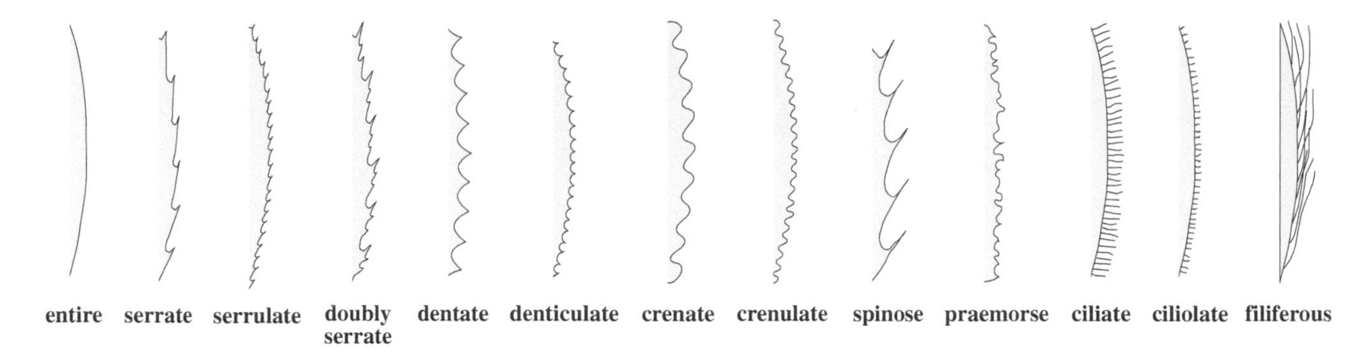

Figure 9.53 Margins.

nate, with teeth cut $^1/_{16}$–$^1/_8$ of the distance to midrib, midvein, or junction of primary veins; and **crenulate**, diminutive of crenate, teeth cut to $^1/_{16}$ of the distance to midrib, midvein, or junction of primary veins.

The relative size and density of teeth may also be described, with terms such as *coarsely*, to describe large and uneven teeth (e.g., "coarsely crenate"), *finely*, to describe relatively small, evenly spaced teeth (e.g., "finely denticulate"), or *sparsely*, to describe teeth that are few in number or spaced well apart (e.g., "sparsely serrate").

Margins with teeth bearing sharp, stiff, spinelike processes are termed **spinose**. **Praemorse** describes a margin having a jagged, chewed appearance, as in some palms.

Terms describing margins with trichomes (plant hairs; see later discussion) are **ciliate**, with trichomes protruding from margins, and **ciliolate**, with minute trichomes protruding from margins, minutely ciliate. The term **eciliate** describes a margin without trichomes, regardless of presence or absence of teeth. Finally, the term **filiferous** refers to margins bearing coarse, fiberlike structures (e.g., fibrovascular bundles, as in the leaf margins of some *Yucca* species).

(Terms that are often treated as features of margin, but treated here as "longitudinal posture," are **involute**, with margins rolled upward, and **revolute**, with margins rolled under; see later discussion.)

Division (Figure 9.54)

Division is a shape character referring to the presence and characteristics of **sinuses** (incisions), the sinuses defining **lobes** or **segments**. Division character states are sometimes treated as features of "margin."

Four division terms that precisely denote the degree of division are: **lobed**, sinuses extending $^1/_8$ to $^1/_4$ of the distance to the midrib, midvein, or vein junction; **cleft**, sinuses extending $^1/_4$ to $^1/_2$ of the distance to the midrib, midvein, or vein junction; **parted**, sinuses extending $^1/_2$ to $^3/_4$ of the distance to the midrib, midvein, or vein junction; and **divided**, sinuses extending $^3/_4$ to almost to the midrib, midvein, or vein junction.

Lobed, *cleft*, *parted*, and *divided* should be prefaced by terms that denote further the type of division: **pinnately** (e.g., "pinnately lobed" or "pinnately cleft") to specify a division along a central axis (typically the midvein), and **palmately** (e.g., "palmately divided") to specify a division relative to a point (typically the basal junction of major veins). (Note that the terms *lobed* and *divided* are sometimes used generally, for any extent of division; as used here, these terms refer to specific degrees of division.)

Some useful, general terms that indicate the general form, but not the extent, of division are **pinnatifid**, pinnately lobed to divided; **pinnatisect**, pinnately divided, almost into discrete leaflets but confluent at the midrib; **bipinnatifid**, bipinnately lobed to divided; **palmatifid**, palmately lobed to divided; and **palmatisect**, palmately divided, almost into discrete leaflets but confluent at the lobe bases. **Decompound** denotes deeply divided into numerous segments such that leaflets are not clearly defined. (Note that *decompound* can also be used for a multiply compound leaf; see Leaf Type.) **Pedate** refers to a palmately divided leaf in which the lateral lobes are further divided, as in some ferns.

Some division terms refer specifically to the shape of the sinuses and lobes. **Dissected** means divided into very fine, often indistinct segments. **Bifid** means 2-lobed to 2-divided, especially at the apex. **Incised** means the sinuses are sharp and deeply cut, usually jaggedly. **Sinuate**, in contrast, refers to sinuses being shallow and smooth, wavy in a horizontal plane (compare with *undulate*, under Longitudinal Posture). **Lacerate** refers to sinuses that are irregularly cut, the lobes appearing torn. **Laciniate** denotes lobes that are cut into narrow, ribbonlike segments. **Pectinate** means comblike, being pinnately divided with close, very narrow lobes.

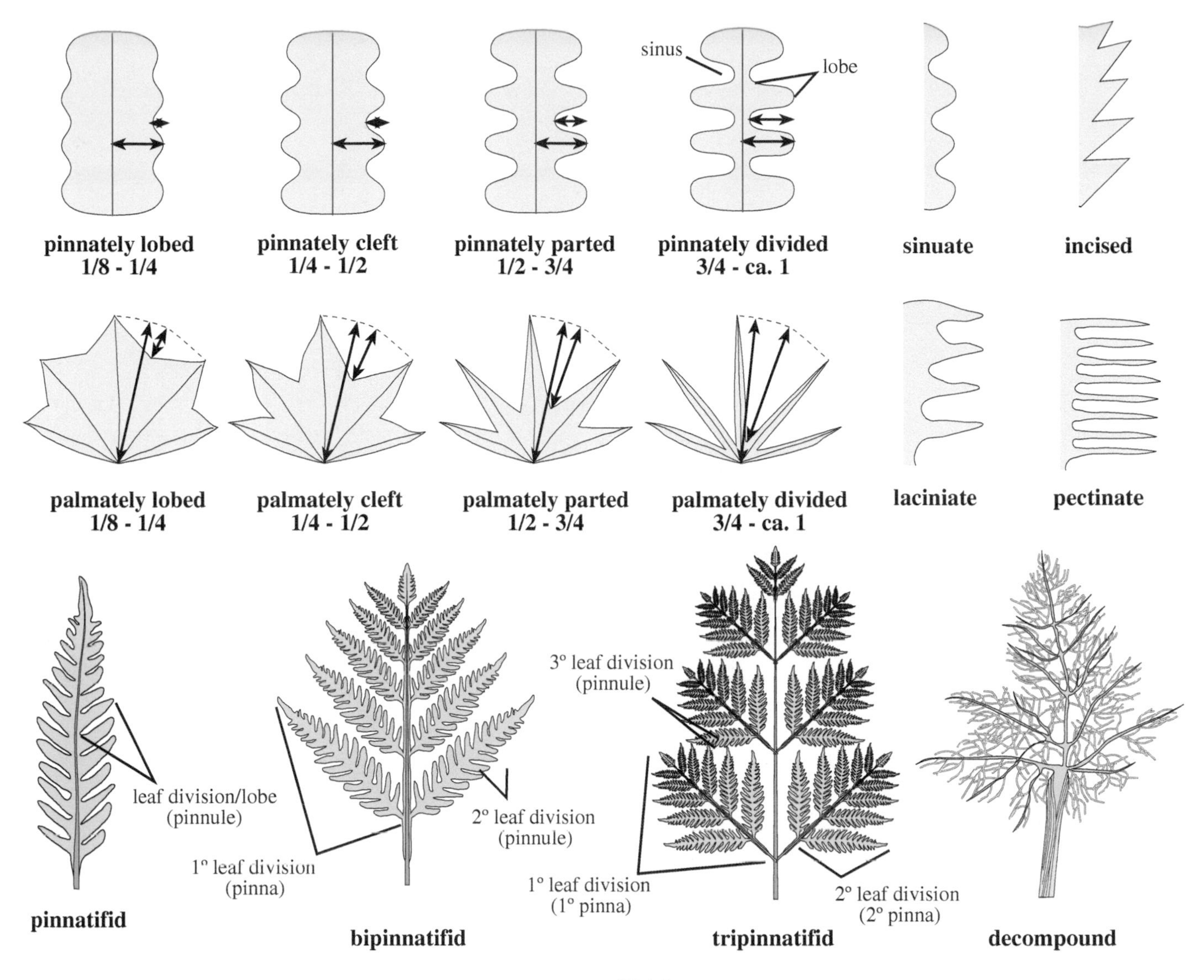

Figure 9.54 Division.

DISPOSITION

Disposition refers to the relative placement of objects, e.g., of plant organs or parts of plant organs. Disposition is logically broken down into position, arrangement, orientation, and posture.

Position (Figure 9.55)

Position is the placement of parts relative to other, *unlike* parts. Some general position terms, which may apply to leaves, bracts, and flower parts, have to do with development. **Adaxial** (also known as **ventral**) corresponds to the *upper* or *inner* surface of an organ. Adaxial literally means toward the axis; in early development of the primordia of leaves or floral parts, the surface that is initially facing toward or nearest the axis will typically become the upper surface. Confusion arises when the organ in question bends downward or twists later in development; in such cases, it is clearer to state that a particular surface is *developmentally adaxial*. Correspondingly, **abaxial** (also known as **dorsal**) corresponds to the lower or outer surface of an organ, i.e., the surface most distant from the axis early in development. (Note that *ventral* and *dorsal* are used in an opposite sense to that for animals; for this reason, these terms are best avoided in plant descriptions, although they are still frequently used to refer to certain inflorescence, floral, or fruit features.)

With respect to a horizontally oriented structure, **posterior** refers to the upper lobe or part; **anterior** refers to the lower lobe or part. *Posterior* and *anterior* are widely used for horizontally oriented floral parts and correspond to *adaxial* and *abaxial*, respectively. **Basal** or **radical** (not to be con-

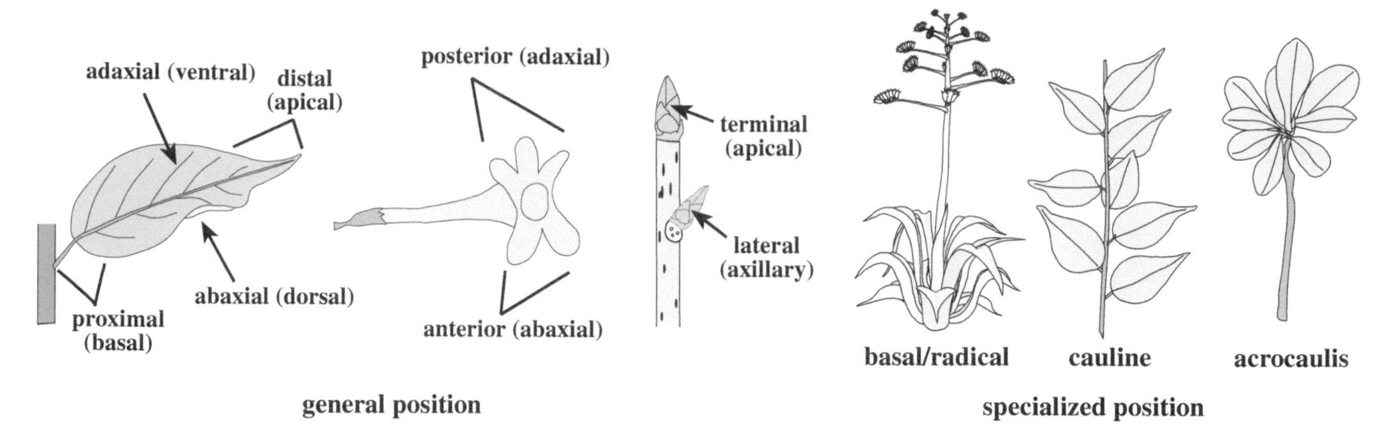

Figure 9.55 Position.

fused with the *radicle* of a seed embryo) indicates at or near the bottom or base of a structure. **Proximal** is similar to basal or radical and means near the point of origin or attachment, as in the point of attachment of a leaf. **Apical** or **terminal** means at or near the top, tip, or end of a structure. **Distal** is similar to *apical* or *terminal* and means away from the point of origin or attachment, e.g., the apex of a structure. Proximal and distal always refer to the point of attachment of an organ and are especially valuable for structures that loop around, in which the original "base" and "apex" are obscured. Some other general position terms are **lateral** or **axillary**, on the side of a structure or at the **axil** (the adaxial region of a node), as in a "lateral" or "axillary" bud; **central**, at or near the middle or middle plane of a structure; and **circumferential**, at or near the circumference of a rounded structure.

Some position terms are used primarily to describe the position of structures relative to the stem. In this sense, the general terms **radical** or **basal** mean positioned at the base of the stem; **cauline** means positioned along the length of the stem (as in cauline leaves or flowers); and **acrocaulis** means positioned at the apex of the stem.

Arrangement (Figure 9.56)

Arrangement is the placement of parts with respect to *similar, like* parts. Some arrangement terms, used primarily for leaves, bracts, or flower parts, describe the number of organs per node. **Alternate** refers to one leaf or other structure per node. Subcategories of alternate are **monistichous**, alternate with points of attachment in one, vertical row/rank, e.g., as in the Costaceae; **distichous**, alternate, with points of attachment in two vertical rows/ranks, e.g., as in the grasses (Poaceae); **tristichous**, alternate, with points of attachment in three rows/ranks, as in the sedges (Cyperaceae); and **spiral** (also termed **polystichous**), alternate, with points of attachment in more than three rows/ranks.

Opposite describes two leaves or other structures per node, i.e., on opposite sides of a stem or central axis. Two subcategories of opposite are **decussate**, opposite leaves or other structures at right angles to preceding pair; and **nondecussate**, opposite leaves or other structures not at right angles to preceding pair. Most leaves, if opposite, are decussate; in fact, nondecussate leaves may be superficially the result of stem twisting. Leaflets of a compound leaf are typically nondecussate.

The term **subopposite** refers to two leaves or other structures on opposite sides of stem or central axis but at different nodes slightly displaced relative to one another. **Whorled** or **verticillate** means having three or more leaves or other structures per node.

More arrangement terms denote more specialized conditions. **Equitant** refers to leaves with overlapping bases, usually sharply folded along the midrib. **Fasciculate** refers to leaves or other structures in a fascicle or short shoot, a cluster with short internodes. **Imbricate** is a general term for leaves or other structures overlapping. **Valvate** means the sides are enrolled so that the margins touch. **Rosulate** means in a **rosette**, an arrangement in which parts (usually leaves) radiate from a central point at ground level (e.g., the leaves of *Taraxacum officinale*, dandelion). **Secund** or **unilateral** refers generally to flowers, inflorescences, or other structures on one side of the axis, often due to twisting of stalks.

Orientation (Figure 9.57)

Orientation denotes the angle of a structure relative to a central (often vertical) axis. Precise orientation terms utilize ranges of angles in degrees, 0–15° or 15°–45°, from the upper axis, the horizontal axis, or the lower axis. These terms are

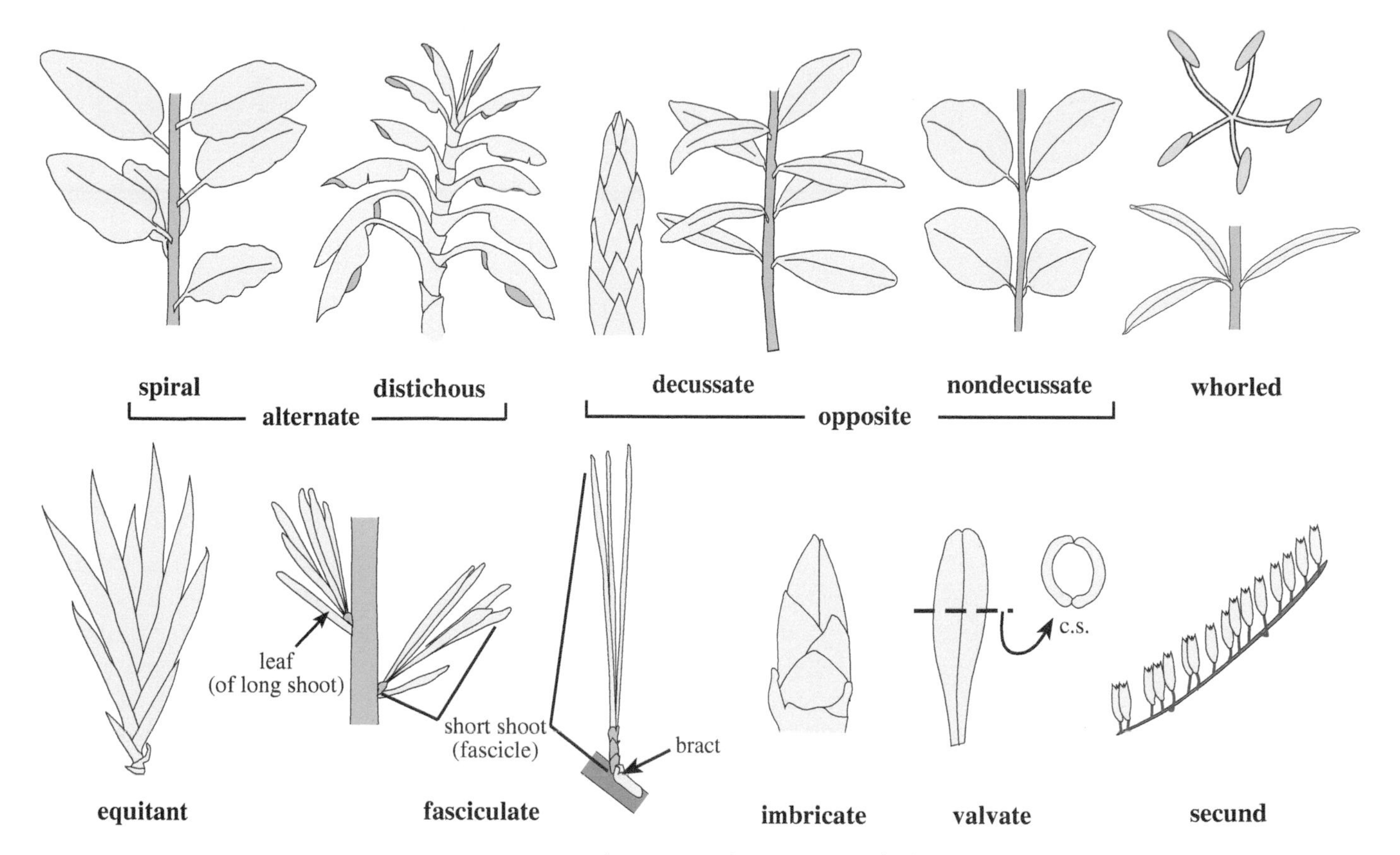

Figure 9.56 Arrangement (c.s. = cross section).

(from top to bottom): **appressed**, pressed closely to axis upward, with divergence angle of 0–15° from upper axis; **ascending**, directed upward, with divergence angle of 15°–45° from upper axis; **inclined**, directed upward, with divergence angle of 15°–45° from horizontal axis; **divergent** or **horizontal**, more or less horizontally spreading with divergence angle of ≤15° up or down from the horizontal axis (also termed *divaricate* or *patent*); **reclined** or **reclinate**, directed downward, with divergence angle of 15°–45° from horizontal axis; **descending**, directed downward, with divergence angle of 15°–45° from lower axis; and **depressed**, pressed closely to axis downward, with divergence angle of 0–15° from lower axis. **Declinate** is a geneal term meaning angled downward or forward.

Other orientation terms are more general. **Antrorse** means bent or directed upward, usually referring to small appendages; **retrorse** means bent or directed downward. **Connivent** means convergent apically without fusion, as in anthers that come together in a flower (e.g., *Solanum*). **Erect** is pointing upward (usually without reference to an axis). **Pendant** or **pendulous** means hanging downward loosely or freely. **Deflexed** means bent abruptly downward, and **reflexed** means bent or turned downward.

Posture (Figure 9.58)

Posture refers to the placement relative to a flat plane. It may be further classified as transverse, longitudinal, twisting/bending posture, or ptyxis/vernation.

Transverse posture is the placement of the *tip* (distal end) of an object with respect to a starting plane. Transverse posture terms are **recurved**, tip gradually curved outward or downward (abaxially); **cernuous**, tip *drooping* downward (abaxially); **squarrose**, sharply curved downward or outward (abaxially) near the apex, as phyllaries of some Asteraceae; **incurved**, tip gradually curved inward or upward (adaxially); **plane** or **straight**, flat, without vertical curves or bends; and **flexuous**, the central axis and tip curved up and down.

Longitudinal posture is the placement of the *margins* of an object with respect to a starting plane. (Note: see also Aestivation.) Common longitudinal posture terms are **conduplicate**, longitudinally folded at central axis, with adjacent adaxial sides facing one another; **revolute**, with margins or outer portion of sides rolled outward or downward over the abaxial surface; **involute**, with margins or outer portion of sides rolled inward or upward over adaxial surface; **cup-shaped**, concave-convex along entire surface (may be abaxially or adaxially concave); **plicate**, pleated, with a series of longitudinal

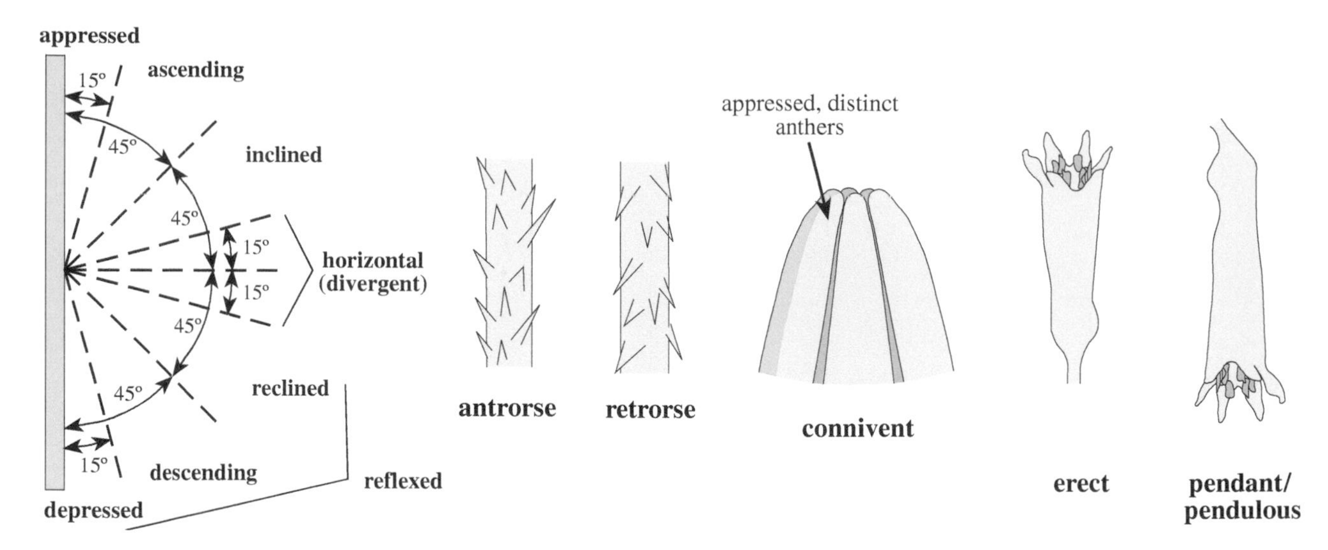

Figure 9.57 Orientation.

folds (subcategories of plicate used, e.g., for palm leaves are **induplicate**, plicate with adjacent adaxial sides facing one another, V-shaped in cross section; or **reduplicate**, plicate with adjacent abaxial sides facing one another, Λ-shaped in cross -section); and **undulate** or **repand**, the margins wavy in a vertical plane (compare *sinuate*, under Division).

Twisting/bending posture refers to the posture of a twisting or bending object relative to a starting plane. **Resupinate** means inverted or twisted 180° during development, as in leaves of Alstroemeriaceae or ovaries of most Orchidaceae flowers. [Note that with resupinate structures, a distinction should be made between positions such as abaxial or adaxial, being *early in development* (*developmentally*) versus *at maturity* (*functionally*), because such positions are reversed during resupination.] **Geniculate** is having a "zig-zag" posture, as in the inflorescence rachis of some grasses. **Twining** is twisted around a central axis, as in many vines. The stems of twining vines may be **dextrorse**, twining helically like a typical, right-handed screw, as in some Convolvulaceae; or **sinistrorse**, twining helically like a left-handed screw, as in some Caprifoliaceae.

Ptyxis, also termed **vernation**, refers to the posture of embryonic structures, such as cotyledons within a seed or immature leaves or leaf parts. Many of the same terms used for posture of mature organs can be used to designate ptyxis. Some specialized ptyxis terms include **circinate**, with the blade (including rachis and rachillae, if present) coiled from apex to base, as in young fern and cycad leaves (see Chapters 4, 5); and **supervolute**, with one half of a simple leaf coiled tightly around the midrib, the other half coiled (in the opposite direction) around the first half, as in members of the Zingiberales (see Chapter 7).

SURFACE

Numerous terms describe the surface of organs or plant parts. Surface features can be broken down into three characters: configuration, epidermal excrescence, and vestiture. Aspects of all three characters may be described as surface features. In addition, trichome type and bristle type may be described as surface features.

Configuration (Figure 9.59)

Configuration refers to the gross surface patterns of the epidermal cells other than that caused by venation (see Leaf Venation, earlier), or excrescences (next). Configuration terms include **canaliculate**, longitudinally grooved, usually in relation to petioles or midribs; **fenestrate**, having windowlike holes in the surface (e.g., *Monstera deliciosa*, Araceae); **lacunose**, having a surface with cavities, pits, or indentations; **mammillate**, having small, nipple-shaped projections; **punctate**, covered with minute, pitlike depressions; **rugose** or **bullate**, covered with coarse reticulate lines, usually with raised blisterlike areas between; **ruminate,** unevenly textured, coarsely wrinkled, looking as if chewed (also used for texture); **smooth** or **plane**, with a smooth configuration; **striate**, with fine longitudinal lines; **tessellate**, with small, defined areas, having a cobblestone or checkerboard-like sculpturing; **wrinkled**, with irregular, fine lines or deformations.

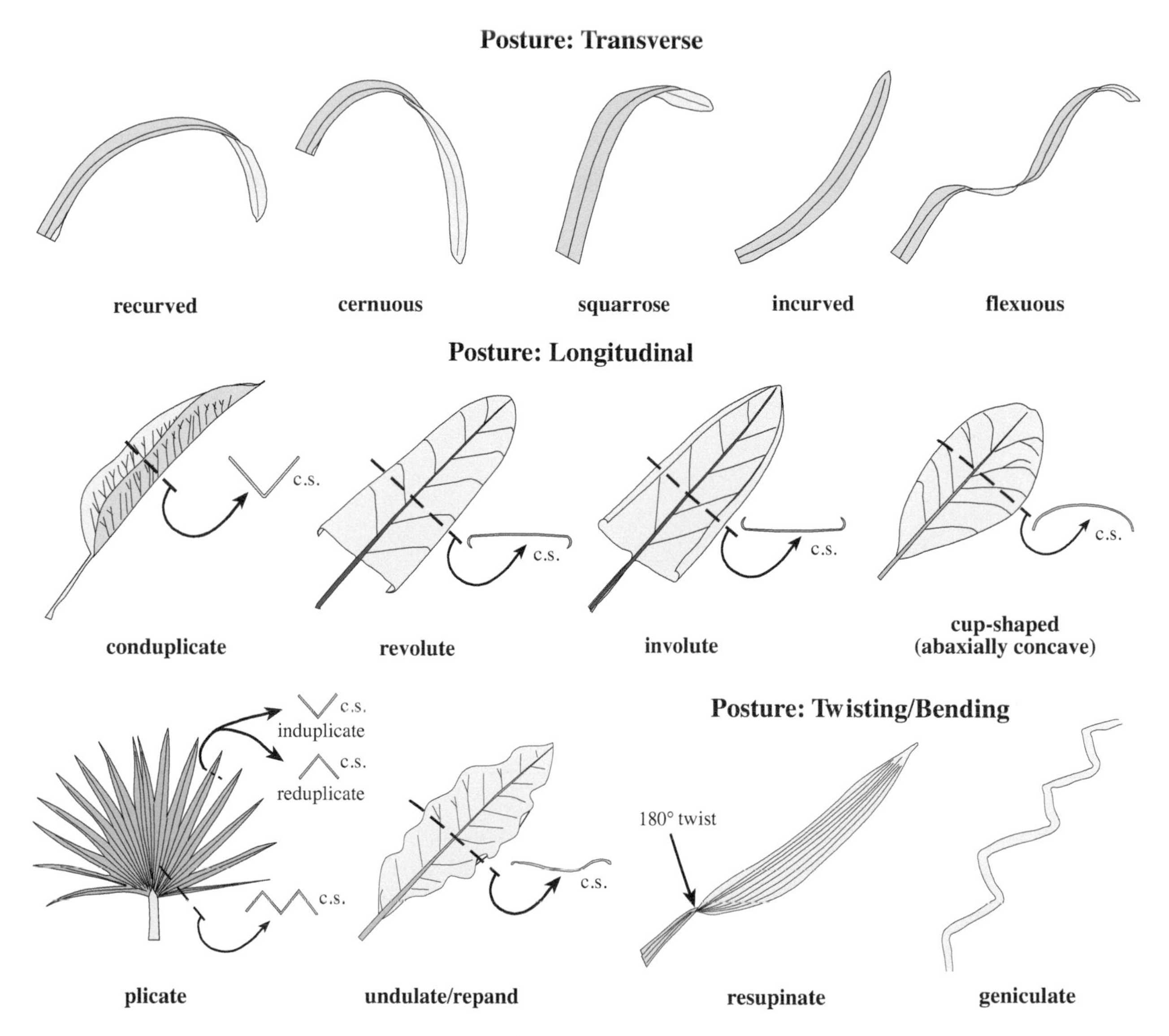

Figure 9.58 Posture (c.s. = cross-section).

Epidermal Excrescence (Figure 9.59)

Epidermal excrescence refers to surface patterns caused by secretions or structural outgrowths of the epidermis (other than trichomes or bristles). Terms that denote epidermal secretions are **glandular**, covered with minute, blackish to translucent glands; **glaucous**, covered with a smooth, usually whitish, waxy coating (that can be rubbed off with touch); **shining** (**nitid** or **laevigate**), appearing lustrous or polished; and **viscid** or **glutinous**, having a shiny, sticky surface. Terms that denote epidermal outgrowths are **aculeate** or **prickly**, with prickles, sharp nonspine, nonthorn appendages; **farinaceous** (*scurfy* or *granular*), finely mealy, covered with small granules; **muricate**, having coarse, radially elongate, rounded protuberances; **papillate**, **tuberculate**, or **verrucate**, having minute, rounded protuberances; and **scabrous**, having a rough surface, like that of sandpaper (also treated under Vestiture).

Vestiture (Figure 9.59)

Vestiture denotes trichome cover. Trichomes are surface hair-like structures that may function by protecting the plant from herbivory, reflecting visible and UV light, and inhibiting water transpiration. Vestiture terms encompass a combination of trichome type, length, strength, shape, density, and color.

Glabrous is the term meaning without trichomes at all. (A glabrous surface is often, but not necessarily, smooth or plane, which is used to denote a flat configuration.) **Subglabrous** means nearly glabrous (with just a few, scattered trichomes) and **glabrate** means the same or can mean becoming glabrous with age by loss of trichomes.

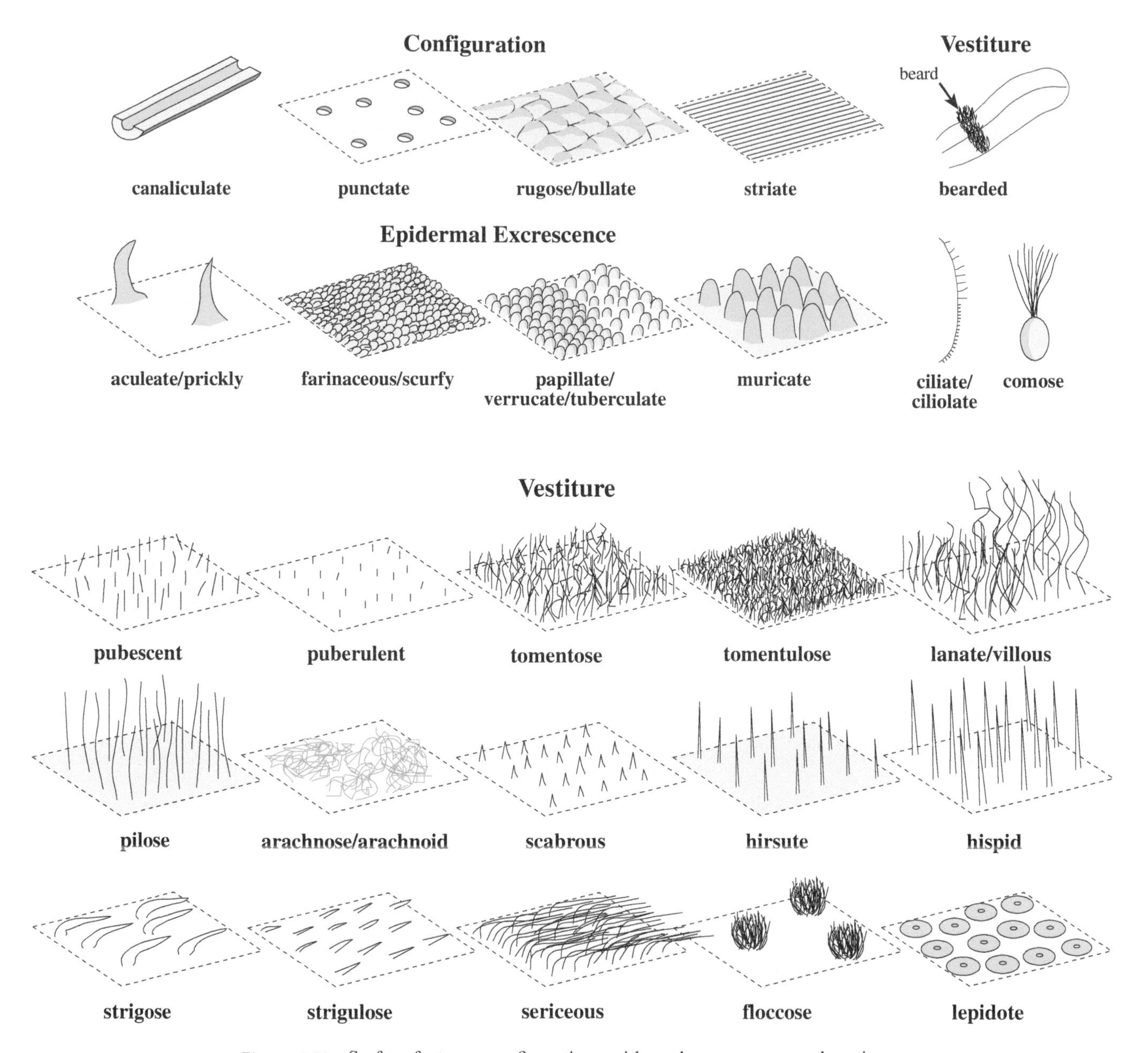

Figure 9.59 Surface features: configuration, epidermal excrescence, and vestiture.

Bearded means with a single tuft or patch of trichomes arising from the surface of an object, e.g., from a petal. **Comose** is similar, but refers to an apical tuft of trichomes, e.g., from a seed. **Penicillate** means tufted, like an artist's brush. Two terms that are also treated under "margin" are **ciliate**, with conspicuous marginal trichomes, and **ciliolate**, with tiny or small marginal trichomes.

Pubescent is a common vestiture type meaning with more or less straight, short, soft, somewhat scattered, slender trichomes. (Note that *pubescent* can be used as a general term, meaning simply "having trichomes.") **Puberulent** means minutely pubescent, i.e., pubescent but with very short or sparse (scattered) trichomes. **Canescent** or **incanous** means whitish-pubescent, covered with dense, fine grayish to white trichomes. **Tomentose** means covered with very dense, interwoven trichomes. **Tomentulose** is minutely tomentose, i.e., tomentose but with very short trichomes.

Villous or **lanate** means covered with long, soft, crooked trichomes; although *lanate* has shorter trichomes than *villous*, these terms intergrade and are probably best treated synonymously. **Pilose** means having soft, straight to slightly shaggy trichomes, generally at right angles to the surface. **Arachnose** or **arachnoid** means having trichomes forming a dense, cobwebby mass (but which resemble *villous* and can be confused with that type).

Scabrous means having rough trichomes, like that of sandpaper; **scaberulous** means minutely scabrous. (*Scabrous* is also treated under Epidermal Excrescence, because the scabrosity can be caused by either outgrowths or trichomes.) **Hirsute** means having long, rather stiff trichomes (but not quite skin-penetrating), whereas **hispid** means having very long, stiff trichomes, often capable of penetrating skin. The term **urent** means having hispid trichomes that are stinging, as in *Urtica*, stinging nettle.

Strigose is covered with dense, coarse, bent and mostly flat trichomes often with a bulbous base. **Strigulose** is minutely strigose, i.e., with the same morphology but a much smaller size. **Sericeous** describes long, appressed trichomes that have a silky appearance or sheen.

Floccose means having dense trichomes that are in several patches or tufts. **Lepidote** means covered with scales or scalelike structures (intergrading with an epidermal excrescence character).

Trichome and Bristle Type (Figure 9.60)

Trichome type refers to the specific, microscopic structure of individual trichomes and may come under the realm of plant anatomy (see Chapter 10) and constitute a systematically valuable character. Although trichome type and vestiture may be correlated, *vestiture* refers to the gross appearance of masses of trichome. For example, a tomentose vestiture could have any number of trichome types, e.g., dendritic, stellate, multiseriate tapering, or uniseriate tapering.

Trichome type may assess the number of cells per trichome. A **unicellular** trichome consists of a single cell and is usually quite small. A **multicellular** trichome contains two or more cells. Multicellular trichomes can be either **uniseriate**, having a single vertical row of cells, or **multiseriate**, having more than one vertical row of cells. The number of cell layers in a trichome can also be diagnostic.

Many trichomes are diagnosed based on their general shape and morphology. **Tapering** trichomes are those ending in a sharp apex. **Malpighian** or **dolabriform** (also termed "two-armed" or "T-shaped") trichomes are those with two arms arising from a common base. (Malpighian is named after the family Malpighiaceae, where this trichome type is common.) **Glandular** trichomes are secretory or excretory trichomes, usually having an apical glandular cell. Glandular trichomes can be **pilate-glandular**, with a glandular cell atop an elongate basal stalk, or **capitate-glandular**, with a glandular cell having a very short or no basal stalk. Branched trichomes include two types: **stellate**, which are star-shaped trichomes having several arms arising from a common base (either stalked or sessile); and **dendritic**, which are treelike trichomes with multiple lateral branches. **Peltate** trichomes are those with a disk-shaped apical portion atop a peltately attached stalk.

Trichomes may also be delimited based on their position and function. For example, trichomes found in the axils of typically sheathing leaves, which may function in secreting protective mucilage, are termed **intravaginal** (or **axillary**) **squamules** (found, e.g., in many Alismatales).

Bristles are similar to trichomes but are generally much stouter (although bristles and trichomes may intergrade). Some so-called bristles are actually modified leaves, such as

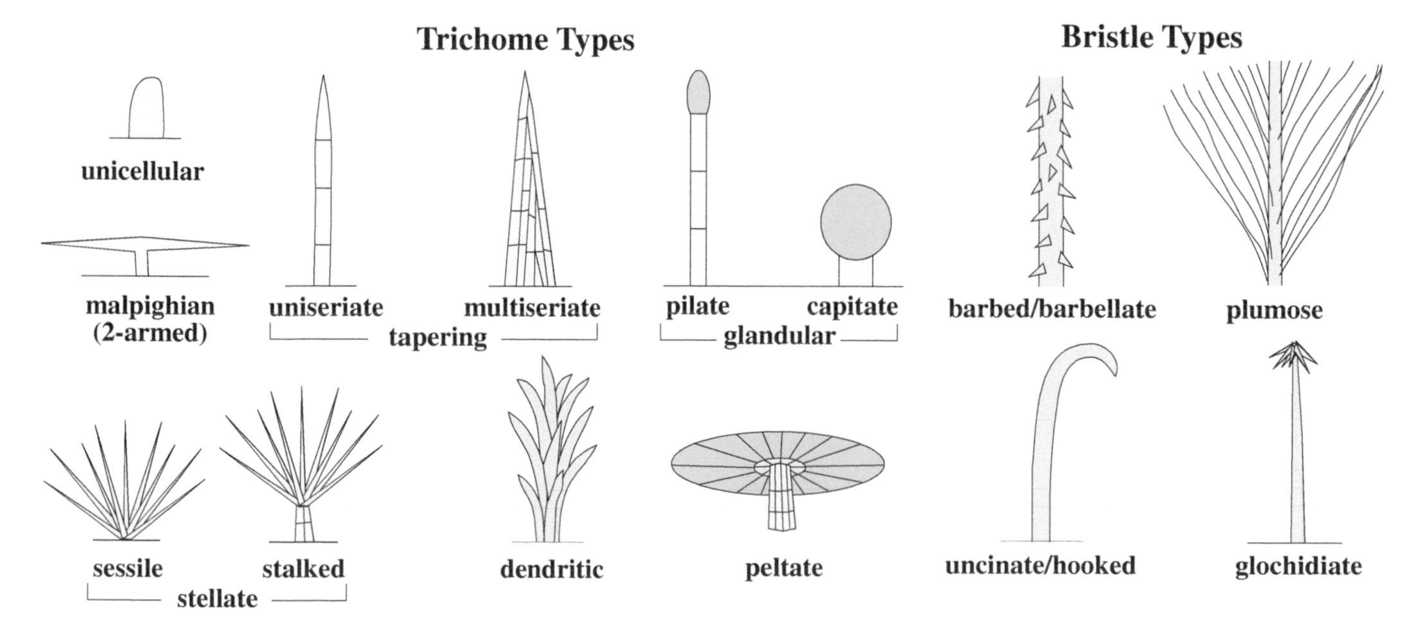

Figure 9.60 Trichome and bristle types.

the glochidia of cacti. Major bristle types include **barbed** or **barbellate**, with minute, lateral, sharp appendages (*barbs*, which may be antrorse or retrorse in orientation) arising along the entire bristle surface; **plumose**, featherlike, covered with fine, elongate, ciliate appendages; **uncinate** or **hooked**, with an apical hooklike structure; and **glochidiate**, with apical, clustered barblike structures.

SYMMETRY

Although symmetry is usually used with reference to flowers (see Flower Symmetry), it can be a general feature to describe any plant organ or part. Symmetry is defined by the presence and number of mirror-image planes of symmetry. **Zygomorphic** or **bilateral symmetry** [**monosymmetric**, **irregular**] is that in which there is only one plane of symmetry. **Biradial symmetry** [**disymmetric**] means having two (and only two) planes of symmetry. **Actinomorphic** or **radial symmetry** [**polysymmetric**, **regular**] is that with three or more planes of symmetry. **Asymmetric** describes a structure lacking any plane of symmetry. (Note that the distinction between biradial and radial symmetry is sometimes not recognized, both being termed *radial symmetry* or *actinomorphy*; however, that distinction is often useful and is recognized here.)

TEMPORAL PHENOMENA

Temporal phenomena deal with any consideration specifically time-based. These are logically broken down into duration, maturation, and periodicity.

Duration refers to the length of life of a plant or part of a plant. **Plant duration** describes the length of life of an entire plant: annual, biennial, and perennial. An **annual** is a plant living 1 year or less, typically living for one growing season within the year. Annual plants are herbs (although herbs can be either annuals, biennials, or perennials). Annuals can usually be detected in that they lack an underground rootstock and show no evidence of growth from a previous season (e.g., there are no thickened, woody stems, dormant buds, or old fruits). **Biennials** are plants living 2 years (or two seasons), usually flowering in the second year. Biennial plants typically form a basal rosette of leaves during the first year and "bolt" (grow an elongate inflorescence stalk) in the second year. Biennials may be hard to detect without actually observing plants over two seasons. A **perennial** is a plant living more than 2 years. Perennials include herbs with rootstocks, shrubs, lianas, and trees.

Other duration terms describe plant parts, e.g., of leaves (in which the term **leaf duration** is used). **Evergreen** means persistent two or more growing seasons, as in the leaves of most conifers. **Deciduous** means parts persistent for one growing season, then falling off, as the leaves of *Acer*, maples. (Note that *evergreen* and *deciduous* can refer to the plants themselves, as in "eastern deciduous forest.") **Cladoptosic** refers to dead foliage falling with the accompanying shoot, rather than as individual leaves, e.g., Cupressaceae such as *Taxodium*. **Caducous** or **fugacious** refers to dropping off very early (compared with what is typical) and usually applies to floral parts. **Deliquescent** means becoming slimey or mucilaginous, e.g., after death. **Marcescent** means ephemeral but with persistent remains, withering but persistent, such as corollas that remain attached during fruit formation. **Accrescent** refers to plant parts that persist and continue to grow beyond what is normal or typical, as with the calyx of *Physalis* (Solanaceae), which expands considerably and functions as an accessory part enclosing the fruit. Finally, **monocarpic** refers to a plant that flowers and fruits only once, then dies; the plant itself can be an annual or perennial, but the term is usually used only for perennials (because all annuals are, by definition, monocarpic).

Maturation refers to the *relative* time of development of plant parts. The term **anthesis** refers to the time of flowering, when flowers open with parts available for pollination. **Protandrous** (meaning "male first") refers to stamens or anthers developing before the carpels or stigma. **Protogynous** (meaning "female first") refers to the stigma or carpels maturing before the stamens or anthers. Both protandry and protogyny are general mechanisms to promote outcrossing within a species (see Chapter 13).

Periodicity refers to periodically repeating phenomena. Terms that refer to the time of day are **diurnal** (during the day), **nocturnal** (at night), **matutinal** (in the morning), and **vespertine** (in the evening). These terms are usually used with respect to when flowers of a given taxon open in a twenty-four hour period. Other terms correspond to seasons, such as **vernal** (appearing in spring), **aestival** (appearing in summer), or **autumnal** (appearing in fall).

Phenology refers to the general timing of reproduction, usually the specific date (month and day) of flowering or fruiting. Phenological data can also include stages of reproduction, such as timing of developmental stages of reproductive organs, such as flower buds, young fruits, mature fruits, etc.

REVIEW QUESTIONS

PLANT STRUCTURE: GENERAL, ROOTS, AND STEMS/SHOOTS

1. Name the major plant organs.
2. What are the continuously actively dividing cell regions of a plant called and where are they located?
3. What is meant by plant habit and what are the types of plant habit?
4. Name various types of plant habitat.
5. Name and define five different types of plant life forms.
6. What is the function of roots?
7. What are the root cap, root hair, adventitious root, and lateral root?
8. What is the difference between a taproot and a fibrous root system?
9. What is a shoot? Define node, internode.
10. What is a bud, where do buds typically develop, and what do they develop into?
11. What is the difference between a bulb, corm, and tuber? between a rhizome, caudex, and stolon (runner)?
12. What is the difference between a caudiciform stem and a pachycaul?
13. What is thorn and how does it differ from a spine or prickle?
14. Define: tiller, burl, pseudobulb, short shoot, tendril.
15. Name the difference between acaulescent and caulescent; between prostrate, repent, and decumbent. What is the corresponding character for all of these?
16. What is the difference between monopodial and sympodial? Plagiotropic and orthotropic? Hapaxanthic and monocarpic?
17. Name three features that various tree branching models may be based upon.
18. Draw a typical twig and label terminal bud, axillary bud, leaf scar, vascular bundle scars, lenticels.
19. What is the difference between an axillary, terminal, and pseudoterminal bud? A collateral and superposed bud?

PLANT STRUCTURE: LEAVES

20. What is the difference between a bract and a scale?
21. Name some specialized modifications of leaves associated with flowers or inflorescences.
22. From what is a phyllode derived?
23. What is a spine and what are the three major types?
24. What are three modifications of leaves found in carnivorous plants.
25. Name five leaf types.
26. What are the basic components of a simple leaf?
27. Draw a bipinnately compound leaf and label: leaflet, petiole, petiolule, rachis, rachilla, stipule, stipel.
28. What is the difference between imparipinnate and paripinnate? Trifoliolate and palmate? Geminate-pinnate and bipinnately compound? Unifoliolate and simple?
29. Name four different types of leaf attachment.
30. What is the difference between parallel and penni-parallel? Between pinnate-netted, palmate-netted, and ternate-netted?
31. Name four major types of specialized venation types.

PLANT STRUCTURE: FLOWERS AND PERIANTH

32. Draw a typical flower and label all the parts, including collective terms.
33. Name the two basic types of flower sex.
34. Name the three basic types of plant sex. What is the corresponding type of flower sex for each?
35. Draw a zygomorphic corolla and label anterior lobe(s) and posterior lobe(s).
36. What is the difference between radial and biradial symmetry?
37. What is the difference between protandrous and protogynous? Between centrifugal and centripetal? Between cleistogamous and chasmogamous?
38. What is a claw, corona, hypanthium, limb, lip, lobe, spur, throat, tube?
39. What are the two major types of perianth arrangement?
40. What is perianth cycly?

41. What is the difference between dichlamydeous and homochlamydeous?
42. What are the two types of calyx fusion? The two types of corolla fusion?
43. Define or draw the following perianth types: bilabiate, campanulate, rotate, salverform, urceolate.
44. Draw and label a petal with a claw and limb. What is the name of this perianth type?
45. Define convolute, imbricate, and valvate. What is the corresponding character?

PLANT STRUCTURE: ANDROECIUM

46. Name the two parts of a stamen; the two parts of an anther.
47. What is the difference between stamen arrangement and stamen position?
48. What is the difference between didymous, didynamous, and tetradynamous? What is the character?
49. What is the difference between antipetalous, antisepalous, and diplostemonous?
50. Do the above terms refer to stamen arrangement or to stamen position?
51. What is the difference between exserted and inserted? What is the character?
52. What is the term for fusion of stamens to the corolla?
53. What is the term for fusion of all the filaments together?
54. What is the term for fusion of the filaments into two groups?
55. What is a monothecal anther?
56. Name three types of anther attachment.
57. Name two types of anther dehiscence with regard to (a) the shape of the opening; (b) the direction of the opening.
58. What is a nectary and what are some types of nectaries?

PLANT STRUCTURE: GYNOECIUM

59. What is the difference between a gynoecium, carpel, and pistil?
60. What are the three parts of a pistil? What is a locule?
61. Name the two types of gynoecial fusion.
62. How is carpel number determined?
63. Name and draw the two basic types of ovary attachment and ovary position.
64. What does perianth/androecial position mean? Name and distinguish between four of these.
65. What is the difference between axile and parietal placentation? Between basal and apical?
66. Name the basic ovule parts, and name and define three ovule types and three ovule positions.
67. What is a gynobasic style?

PLANT STRUCTURE: INFLORESCENCES

68. What are two types of specialized bracts associated with inflorescences?
69. What is the difference between a pedicel and a peduncle?
70. Define compound receptacle.
71. What are three types of inflorescence position?
72. What is the difference between determinate and indeterminate inflorescence development?
73. What is a dichasium?
74. How does a monochasium differ and what are two major types?
75. What is a "ray" in an inflorescence?
76. What is the difference between a raceme and a spike? What is the inflorescence development of both?
77. What is the difference between a raceme and a panicle?
78. What is the difference between an umbel and a corymb? Between an umbel and compound umbel?
79. What is the difference between a thyrse and verticillaster?
80. Name a taxonomic group characterized by a compound umbel; cyathium; head; hypanthodium; spadix.

PLANT STRUCTURE: FRUITS AND SEEDS

81. What are the differences between simple, aggregate, and multiple fruits?
82. What features are used to define and classify fruit types?
83. What is a schizocarp? A mericarp? A valve?
84. What are the similarities and differences between an achene grain (caryopsis) and a nut?
85. What are the differences between loculididal, septicidal, and circumscissile capsules?
86. What are the similarities and differences between a follicle, legume, and silique?
87. What is the difference between a silique and a silicle? What family do they occur in?
88. What is the name given to a winged fruit?
89. How does a berry differ from a drupe or a hesperidium?
90. What is the placentation, ovary position, and texture of a pepo? In what family are they found?
91. A pome consists of much outer fleshy tissue derived from what? What is the ovary position? What is an example of a plant with pomes?
92. Name two types of aggregate fruits.
93. What types of fruits are burs, soroses, and syconia?
94. Name two types of seed based on endosperm type; seed germination type.
95. What is a diaspore? Name three types of diaspore dispersal.

GENERAL TERMINOLOGY: COLOR, NUMBER, TEXTURE, AND FUSION

96. What is the difference between color and color pattern? Name and define a color pattern character state.
97. What is the difference between cycly and merosity? Give an example of each.
98. What is the difference between coriaceous and indurate? Between scarious and succulent? What is the character for these?
99. What is the difference between connate and distinct? Between adnate and free? Between adherent and coherent? What is the character for these?

GENERAL TERMINOLOGY: SHAPE

100. Define the following terms for three-dimensional shapes: capitate, clavate, filiform, pilate, terete.
101. What is the difference between lanceolate, ovate, and lance-ovate? Between lanceolate and oblanceolate? (Draw.)
102. What is the difference between ovate and obovate? Between oblanceolate and spatulate? (Draw.)
103. What is the difference between elliptic and oblong? Between oblong and linear? (Draw.)
104. What is the difference between cordate and reniform? (Draw.)
105. What is the difference between hastate and sagittate? (Draw.)
106. What does peltate mean?
107. What does an attenuate base mean? An oblique base? A sagittate base? A cuneate base?
108. What is the difference between entire, crenate, serrate, and dentate? What character do these refer to?
109. What is the difference between crenate and crenulate? Serrate and serrulate? Dentate and denticulate?
110. Define ciliate, ciliolate, filiferous.
111. What is the difference between acuminate, narrowly acute, acute, and obtuse (apex)? What are the corresponding base terms?
112. What is the difference between rounded and truncate (apex and base)?
113. What is the difference between mucronate, aristate, and apiculate?
114. What does emarginate or retuse mean (apex)?
115. What is the difference between lobed, cleft, parted, and divided?
116. What do pinnatifid and bipinnatifid mean?
117. Draw the following: (a) simple, sessile, ovate, acute, crenate leaf; (b) simple, petiolate, oblanceolate, serrulate leaf; (c) pinnately compound, petiolate, stipulate and stipellate leaf with sessile, entire, narrowly elliptic, cuneate, acuminate leaflets; (d) trifoliolate (ternately compound), petiolate leaf with petiolulate obovate, narrowly cuneate, apically obtuse, mucronate leaflets; (e) simple, lanceolate, mucronate, sagittate, dentate leaf.

GENERAL TERMINOLOGY: DISPOSITION

118. What is the difference between position and arrangement?
119. What is the difference between abaxial (dorsal) and adaxial (ventral)?
120. What is the difference between proximal and distal?
121. What does radical mean?
122. Name the three basic (general) types of arrangement (e.g., for leaves).
123. What is the difference between alternate and distichous? Between opposite and decussate?
124. Describe the difference between equitant, imbricate, secund, valvate, and rosulate.
125. Define orientation and name three types (character states).
126. What is the difference between transverse posture and longitudinal posture? Give two examples of each.
127. Distinguish between conduplicate, revolute, sinuate, and undulate.

GENERAL TERMINOLOGY: SURFACE, SYMMETRY, TEMPORAL PHENOMENA

128. Surface refers to three features: configuration, epidermal excrescence, and vestiture. How do they differ?
129. Define rugose. For what character is this a character state?
130. What is the difference between glaucous, scabrous, and viscid? For what character are these character states?
131. What is the difference between hirsute, pubescent, and tomentose? For what character are these character states?
132. What do stellate, pilate, and uniseriate refer to? For what character are these character states?
133. What is the difference between actinomorphic (radial) and zygomorphic (bilateral)? For what character are these character states?
134. What is the difference between annual, biennial, and perennial? For what character are these character states?
135. What is the difference between caducous and accrescent? For what character are these character states?
136. What is the difference between protandrous and protogynous?

EXERCISES

1. Select a plant species and thoroughly describe its morphology using the comprehensive character list of Appendix 1. Fill in every applicable character with a character state, noting that several characters will not apply to your taxon. Try to examine several populations, individuals, or plant organs/parts and note the range of variation. For characters that are variable, either list the range of variation (e.g., "Leaves oblanceolate to narrowly elliptic, crenate to dentate . . .") or list the most common morphology and in brackets list the exceptions (e.g., "Leaves trifoliolate [rarely pinnate with 5 leaflets]" or "Leaves 4–7 [2.5–10] cm long...").
2. From the character listing of Appendix 1, write a detailed description, using the Plant Description Example of Appendix 1. Note to list only the plant organ or plant part, not the character. For example, the description format should be "**Leaves** are opposite, simple, and evergreen" and *not* "**Leaf** arrangement is opposite, leaf type is simple, leaf duration is evergreen." (Note that a word processing "merge" file is often useful for this.) Edit this description such that it reads smoothly and avoids repetition.
3. Make detailed drawings, using a hard (2H or 3H) pencil, of various parts of your species, such as leaves, inflorescence, whole flower, flower in longitudinal section, anther close-up, ovary close-up, ovary cross- or longitudinal section, fruit, and seed. Be sure to include a scale bar, in metric measurements, beside each drawing. Make copies or tracings of these drawings and trace the outlines with a fine, black ink rapidograph. Attempt to do some fine stippling in various regions to show venation, shading, and depth. (See Appendix 2.)
4. Compare your description and drawings with that of standard references, including floras and monographic treatments. Note that yours is probably much more detailed and comprehensive than that of most floras, but perhaps comparable in detail to a monograph.

REFERENCES FOR FURTHER STUDY

Bell, A. D. 2008. Plant Form: an Illustrated Guide to Flowering Plant Morphology. With line drawings by Alan Bryan. Timber Press, Portland, Oregon.

Brouk, B. 1975. Plants Consumed by Man. Academic Press, London.

Ellison, A. M. and L. Adamec. 2018. Carnivorous Plants: Physiology, Ecology, and Evolution. Oxford University Press, Oxford.

Endress, P. 1994. Diversity and Evolutionary Biology of Tropical Flowers. Cambridge University Press, Cambridge.

Hallé, F., R. A. A. Oldeman, and P. B. Tomlinson. 1978. Tropical Trees and Forests: An Architectural Analysis. Springer, Berlin.

Hallé, F.. 2004. Architecture de Plantes. JPC, Montpellier.

Harris, J. G. and M. W. Harris. 2001. Plant identification terminology. Spring Lake Publishing, Spring Lake, Utah.

Hickey, L. J. 1973. Classification of the architecture of dicotyledonous leaves. American Journal of Botany 60: 17–33.

Hickey, L. J., and J. A. Wolf. 1975. The bases of angiosperm phylogeny: vegetative morphology. Annals of the Missouri Botanical Garden 62: 538–589.

Hilger, H. H. 2014. Ontogeny, morphology, and systematic significance of glochidiate and winged fruits of Cynoglosseae and Eritrichieae (Boraginaceae). Plant Diversity and Evolution 131.

Lawrence, G. H. M. 1951. Taxonomy of Vascular Plants. Macmillan, New York.

Radford, A. E., W. C. Dickison, J. R. Massey, and C. R. Bell. 1974. Vascular Plant Systematics. Harper & Row, New York.

Raunkiaer, C. 1934. The Life Forms of Plants and Statistical Plant Geography. Clarendon Press, Oxford.

Rowley, Gordon D. 1987. Caudiciform and Pachycaul Succulents—Pachycauls, Bottle-, Barrel- and Elephant-Trees and Their Kin: A Collector's Miscellany. Strawberry Press, Mill Valley, CA.

Tucker, A. O., M. J. Maciarello, and S. S. Tucker. 1991. A survey of color charts for biological descriptions. Taxon 40: 201–214.

Systematics Association Committee for Descriptive Terminology. 1962. II. Terminology of Simple Symmetrical Plane Shapes (Chart I). Taxon 11(5): 145–156.

Weberling, F. 1989. Morphology of Flowers and Inflorescences. Cambridge University Press, Cambridge, United Kingdom.

WEBSITES

Stevens, P. F. (2001 onwards). Angiosperm Phylogeny Website. *http://www.mobot.org/MOBOT/research/APweb*
In addition to being a source of information on plant classification and phylogeny, this site also has an excellent glossary of plant terms.

Wikipedia contributors. August 11, 2017. Plant Morphology. *https://en.wikipedia.org/wiki/Category:Plant_morphology*

10

PLANT ANATOMY AND PHYSIOLOGY

Plant anatomy is the study of the tissue and cell structure of plant organs. The term *anatomy*, as applied to plants, generally deals with structures that are observed under a high-powered light microscope or electron microscope. (In zoology, the term *anatomy* refers to the study of internal organs; *histology* is the study of cells and tissues of animals.) **Plant physiology** is the study of metabolic processes in plants. A limited explanation of plant physiology is presented, dealing specifically with photosynthesis. Physiology and anatomy are tightly correlated, as cell and tissue structure has changed with respect to the evolution of novel functional mechanisms.

The following is a summary of basic plant anatomy and physiology, with a focus on the sporophytes of vascular plants. Plant anatomical and physiological features may provide valuable characters and character states in assessing homology and elucidating phylogenetic relationships among plants. Moreover, the evolution of anatomical and physiological characters is of great interest, as many of these features are of significant adaptive value and have been among the major selective pressures in plant evolution.

PLANT CELL STRUCTURE

In the 1600s the English "biologist" Robert Hooke first coined the term **cell** after observing that plant tissues, such as cork, are divided into little cavities separated by walls. Since then, the **cell theory** has been perfected, providing a unifying theme in biology, stating the following: First, all life is composed of one or more cells. Second, cells arise only from preexisting cells, occurring either through cell division (meiosis or mitosis) or cell fusion (e.g., fertilization of egg and sperm). Third, cells are the units of metabolic processes; thus, each cell contains the necessary chemical compounds and cellular components to carry on the biosynthetic pathways needed for basic physiological processes. Fourth, each cell contains a set of DNA, the hereditary material that is transferred from one cell to another that codes for the structural and functional features of the organism.

Plant cells are bounded by a **plasma membrane** that is composed of a phospholipid bilayer with embedded proteins. The membrane functions as the boundary of the cell, to contain the cellular components. It also functions in cell–cell recognition and in transport of compounds. Everything inside the plasma membrane is called the **protoplasm**. The protoplasm contains **organelles**, which carry on some vital metabolic function, and **ergastic substances**, which do not function in metabolism and have a variety of functions such as storage, waste secretion, and protection.

Cellular organelles include the following (Figure 10.1): (1) a **nucleus**, which is double membrane-bound and contains DNA, the hereditary material of the cell (Note: Everything

https://doi.org/10.1016/B978-0-12-812628-8.50010-9,

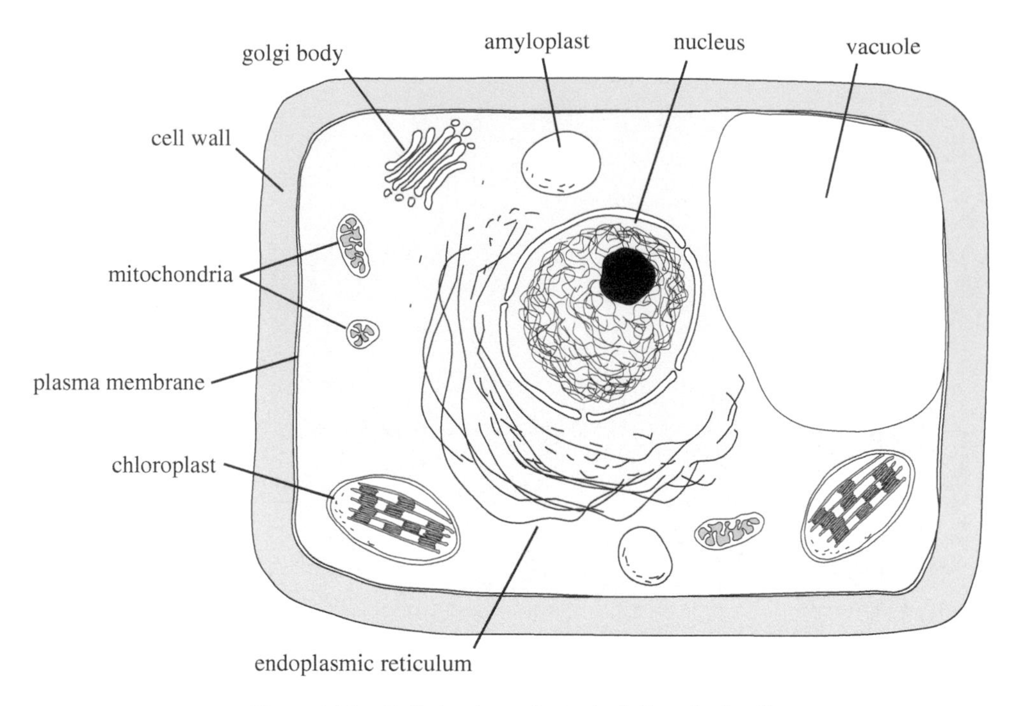

Figure 10.1 Cell structure of a typical, live plant cell.

inside the plasma membrane but not including the nucleus is called **cytoplasm**); (2) **mitochondria** (singular, *mitochondrion*), which are double membrane-bound, with invaginations called *cristae* that function in the electron transport reactions of respiration; (3) a **vacuole**, which is a large (often occupying *most* of the volume of plant cells), internal, membrane-bound sac that functions in storage of compounds such as pigments (e.g., anthocyanins or betalains), acids (e.g., malic acid involved in CAM photosynthesis), or ergastic substances (see later discussion); (4) **endoplasmic reticulum**, which is composed of interconnected phospholipid membranes and functions as the site of protein synthesis and material transport; (5) **golgi bodies**, which are composed of parallel stacks of flattened membranes and function in transport and modification of compounds; (6) **chloroplasts**, which are double membrane-bound with internal thylakoid membranes (composed of lamellae and grana in the green plants), functioning in the reactions of photosynthesis; and (7) **ribosomes**, which function as the site of protein synthesis.

Ergastic substances are cellular materials that are not actively metabolized, functioning, e.g., as storage reserves or wastes. Ergastic substances include (1) **chromoplasts** (Figure 10.2A), which are carotenoid-containing bodies that function to provide yellow, orange, or red pigmentation for a plant organ, as in petals or fruits; (2) **amyloplasts** or **starch grains** (Figure 10.2B,C), which are lamellate deposits of starch (alpha-1,4-glucopyranoside, a polysaccharide polymer of glucose units with alpha-1,4 chemical bonding) which functions as the high-energy storage compound in green plants; (3) **aleurone grains** (=**proteinoplasts**), which are granular protein deposits, functioning as storage compounds; (4) **tannins**, which are phenol derivatives that may function to deter herbivory and parasite growth; (5) **fats, oils, waxes**, which are types of triglyceride compounds that may function as high energy storage compounds or secretion products; and (6) **crystals**, which may be composed of silica or calcium oxalate in various forms, such as **druses** (spherical crystals with protruding spikes; Figure 10.2D), **raphides** (bundles of needle-like crystals; Figure 10.2E), **styloids** (single, elongate, angular crystals; Figure 10.2F), or **prismatic** (shorter, prism-shaped crystals; Figure 10.2G). Crystals may function as waste products, as calcium ion sinks (a means of removing excess calcium for certain cellular functions), or as an irritant to deter herbivory.

In land plants, a pectic-rich **middle lamella** layer is formed between the plasma membrane of adjacent cells (Figure 10.3). The middle lamella functions to bind adjacent cells together. During plant cell development, a **cell wall** is secreted outside the plasma membrane. The cell wall that is secreted soon after cell division and that is maintained during cell growth is called the **primary (1y) cell wall** (Figure 10.3). As discussed earlier, an apomorphy for the green plants is a cell wall composed of **cellulose**, a polysaccharide polymer of glucose sugar units chemically bonded in the beta-1,4 position (=beta-1,4-glucopyranoside). Recall that cellulose is constructed of

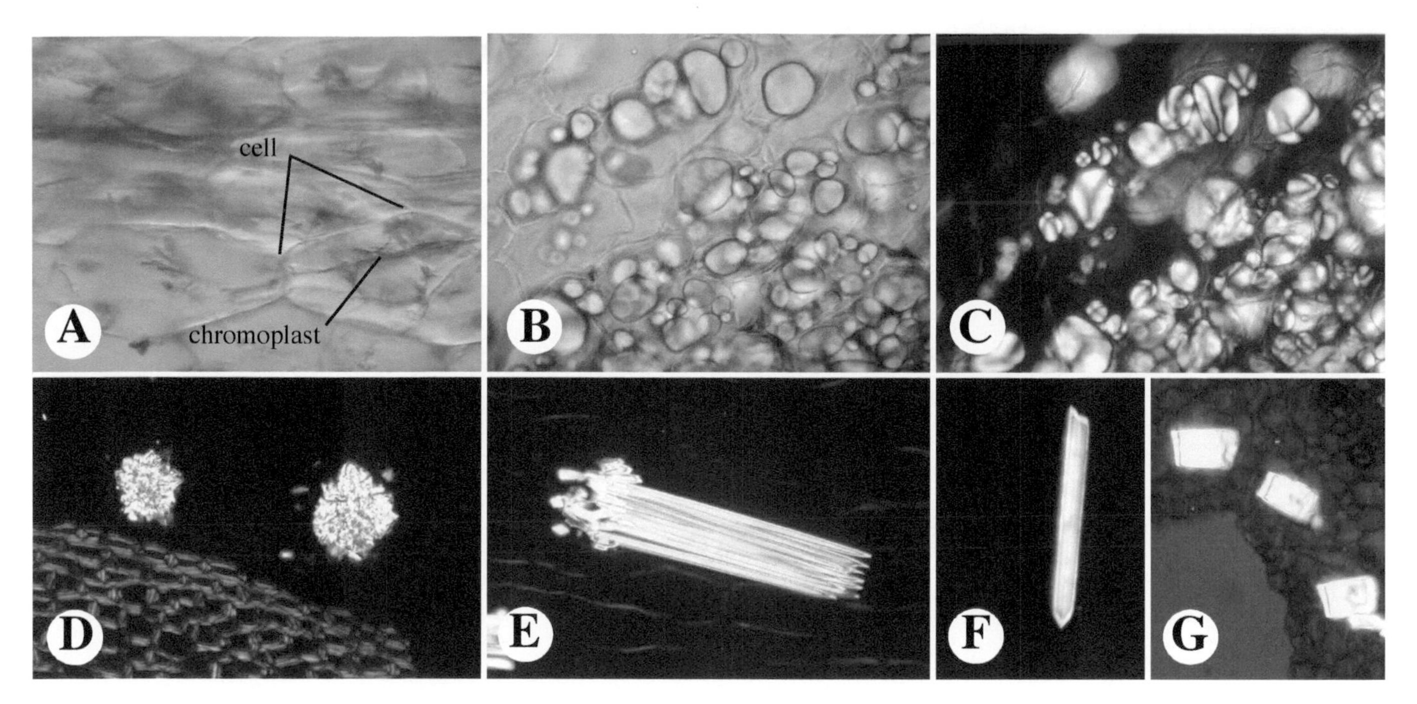

Figure 10.2 **A.** Chromoplasts, beta-carotene deposits in cultivated root of carrot (*Daucus carota*). **B,C.** Amyloplasts of potato (*Solanum tuberosum*). **B.** Brightfield microscopy. **C.** Polarization microscopy, showing typical "Maltese cross" optical pattern of starch. **D–G.** Calcium oxalate rystals, all viewed with polarization microscopy. **D.** Druses. **E.** Raphides. **F.** Styloids. **G.** Prismatic crystals.

microscopic fiberlike units (called **microfibrils**) that are further intertwined into larger fibril units, forming a meshwork outside the plasma membrane. Its function is to impart rigidity to the cells, acting as a cellular skeleton. Within the primary cell wall, ultramicroscopic pores may form, termed **plasmodesmata**. These tiny openings function to allow for a continuity of membranes between cells, allowing for intercell exchange of compounds. A group of plasmodesmata is called a **primary pit field** (Figure 10.3).

In certain plant cells (e.g., sclerenchyma and tracheary elements) an additional wall layer, called a **secondary (2j) wall**, is secreted externally, between the primary cell wall and plasma membrane (Figure 10.4). A secondary wall is generally formed after the plant cell has ceased growth. In vascular land plants the secondary wall is composed partly of cellulose but also contains **lignin**, a complex polymer of phenolic compounds that binds the cellulose microfibrils together. Lignin imparts significant strength and rigidity to the cell wall.

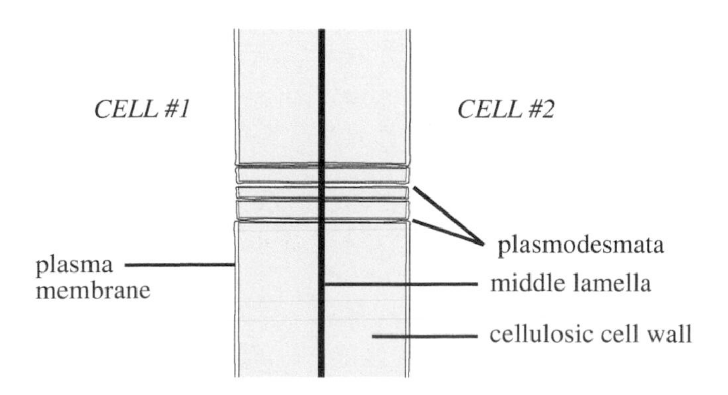

Figure 10.3 Cellulosic cell wall. Note plasmodesmata, small pores or cavities in the cellulosic structure of the primary cell wall.

In virtually all plant cells with lignified cell walls, there are holes in the secondary wall called **pits**. Pits of adjacent cells often occur opposite one another, as **pit pairs** (Figure 10.4). The actual chamber and opening of a pit may assume different morphological forms. Pits function in allowing communication between cells during their development and differentiation. They may also have specialized functions in fluid conducting cells (see later discussion).

PLANT GROWTH

Plant cell growth is defined here in the broad sense as the initiation, expansion, and specialization of cells. The haploid spore or diploid zygote of land plants initially undergoes more or less continuous, sequential mitotic cell divisions. Later, as gametophytes or sporophytes mature, active cell divisions become restricted to certain regions of the plant. This region of actively dividing cells is known as a **meristem**. In the vascular plants **apical meristems** are located at the

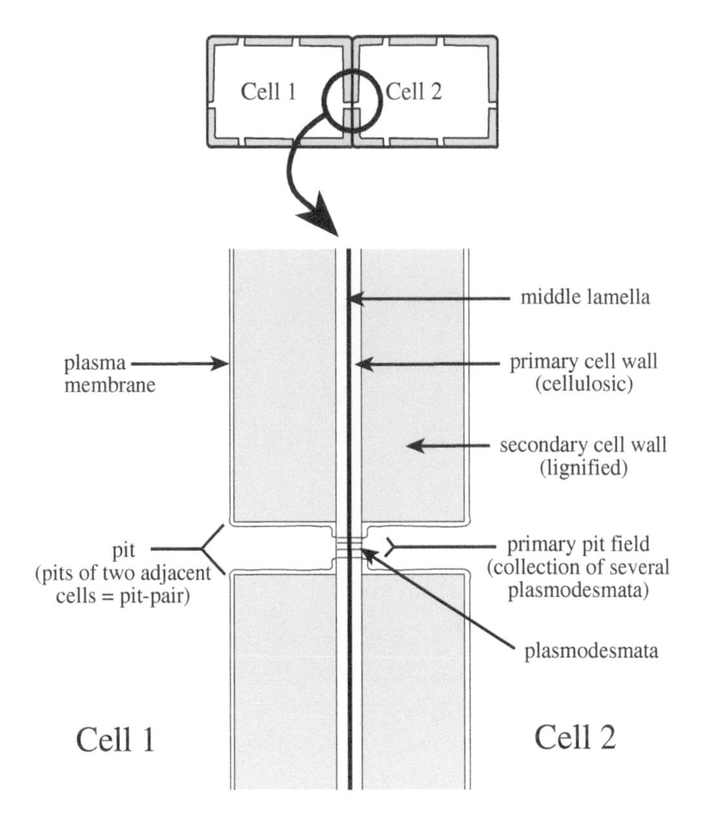

Figure 10.4 Lignified secondary cell wall of specialized cells of vascular plants. Note pit pair.

apices of roots and shoots (Figure 10.5), resulting in growth in height or length. Apical meristems may contain a single, enlarged apical initial cell (found in Selaginellaceae and monilophytes) or a group of actively dividing cells (known as *complex*, found in Lycopodiaceae, Isoetaceae, and seed plants). In woody seed plants both apical meristems and lateral meristems occur. **Lateral meristems** are cylindrical sheaths of cells (Figures 10.18, 10.19), which function in growth that increases width or girth (see later discussion).

In both apical and lateral meristems a single meristematic cell undergoes a mitotic cell division, giving rise to two cells. Each of these two "daughter cells" undergoes some initial expansion. The derivatives themselves may continue to divide several more times, but only those cells that remain near the meristem will do so indefinitely. The others eventually cease mitosis and undergo further differentiation.

Cell differentiation refers to the series of changes that a cell undergoes from the point of inception to maturity, involving the transformation of a meristematic cell into one that assumes a particular structure and function. Differentiation involves two processes: **cell expansion**, in which the cell grows in size (often by **elongation**, in which growth in the axial direction is greatest); and **maturation** or **specialization**, in which the cell acquires the structural and functional features at maturity. Cell specialization means simply that cells may differ from one another, becoming specialized for a particular structure and function within the whole plant. Cell differentiation results in the development of various **cell types** (discussed next).

PLANT TISSUES AND SPECIALIZED CELL TYPES

A **tissue** is a group of cells having a common function or structure. Plant tissues of the vascular plants are often categorized into three broad classes: **ground, vascular,** and **dermal**; see later discussion. In addition, tissues may be classified as simple or complex. A **simple tissue** consists of only one type

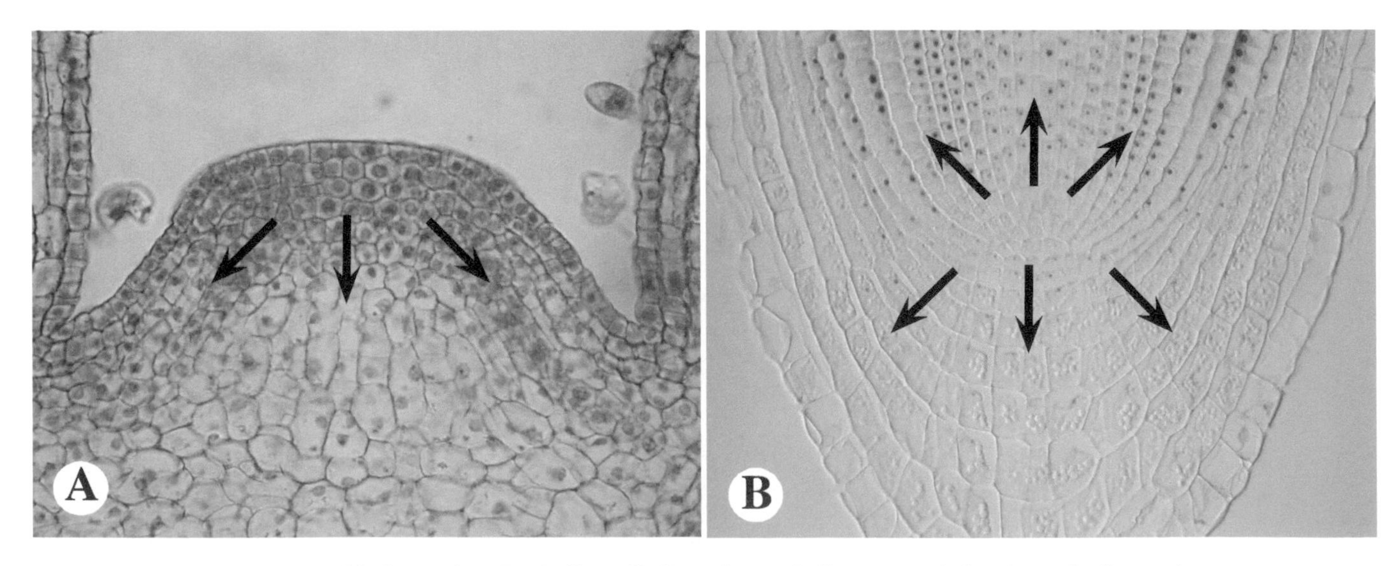

Figure 10.5 Meristematic cells. **A.** Shoot. **B.** Root. Arrows indicate general directions of cell growth.

of cell; thus, a particular term may refer either to the simple tissue or the cell type. A **complex tissue** contains more than one cell type.

Ground tissue is that occurring inside the epidermis but not part of the vascular tissue. Three cell types (which are simple tissues) make up the ground tissue: parenchyma, collenchyma, and sclerenchyma. **Parenchyma** (Gr. *para*, beside + *en-chein*, to pour; in reference to the analogy that parenchyma is "poured" beside other tissues to fill up space) are cells that most resemble the unspecialized, undifferentiated cells of actively dividing meristematic tissue (Figure 10.6A). Structurally, parenchyma cells are (1) isodiametric to elongate; (2) have a primary (1°) cell wall only (rarely with secondary wall); and (3) are living at maturity and potentially capable of cell division. Parenchyma cells function in metabolic activities (e.g., respiration, photosynthesis, transport, storage) and in wound healing and regeneration, being capable of transforming into a meristem to form new roots or shoots.

Collenchyma (Gr. *colla*, glue + *enchyma*, infusion; in ref. to thick, glistening cell walls) are cells that structurally (1) are elongated; (2) have only a primary cell wall that is unevenly thickened and rich in pectins (glistening white in the light microscope); and (3) are living at maturity (Figure 10.6B). Collenchyma cells function in mechanical support and are often found at the periphery of stems or leaves. They can be stretched during elongation growth of the organ.

Sclerenchyma (Gr. *scleros*, hard + *enchyma*, infusion, in ref. to hard, lignified cell walls) are cells that structurally (1) have thick, lignified secondary (2°) cell walls, which may have pits; and (2) are (usually) dead at maturity. There are two general types of sclerenchyma, which sometimes intergrade: (a) **fibers**, which are long, very narrow cells with sharply tapering end wall (Figure 10.7A–C); and (b) **sclereids**, which are isodiametric to irregular in shape and often branched (Figure 10.7D–F). Fibers function in mechanical support in various organs and tissues, sometimes making up the bulk of the tissue. Fibers often occur in groups (or bundles) and may be components of the xylem and/or phloem or may occur independently of the vascular tissue. Fibers may be *septate*, in which they have septa (cross-walls), or *non-septate*, in which septa are absent; the presence or absence of septa in fibers can be systematically significant. Sclereids, like fibers, may also function in structural support, but their role in some plant organs is unclear (e.g., possibly aiding in providing protection from herbivory).

Vascular tissues are made up of xylem and phloem, each of which are complex tissues (having more than one cell type). **Xylem** (Gr. *xylos*, wood) is a tissue composed of **tracheary elements** plus some parenchyma and sometimes sclerenchyma. Structurally, **tracheary elements** (1) are elongate to short; (2) have lignified, secondary (2°) cell walls, with **pits;** and (3) are dead at maturity, in which protoplasts degrade, leaving only cell walls (Figure 10.8). Tracheary elements are joined end-to-end, forming a tubelike continuum. They function to conduct water and dissolved essential mineral nutrients, generally from the roots to other parts of the plant.

There are two types of tracheary elements: **tracheids** and **vessel members**. These differ with regard to the junction between adjacent end-to-end cells, whether *imperforate* or *perforate*. **Tracheids** are imperforate, meaning that water

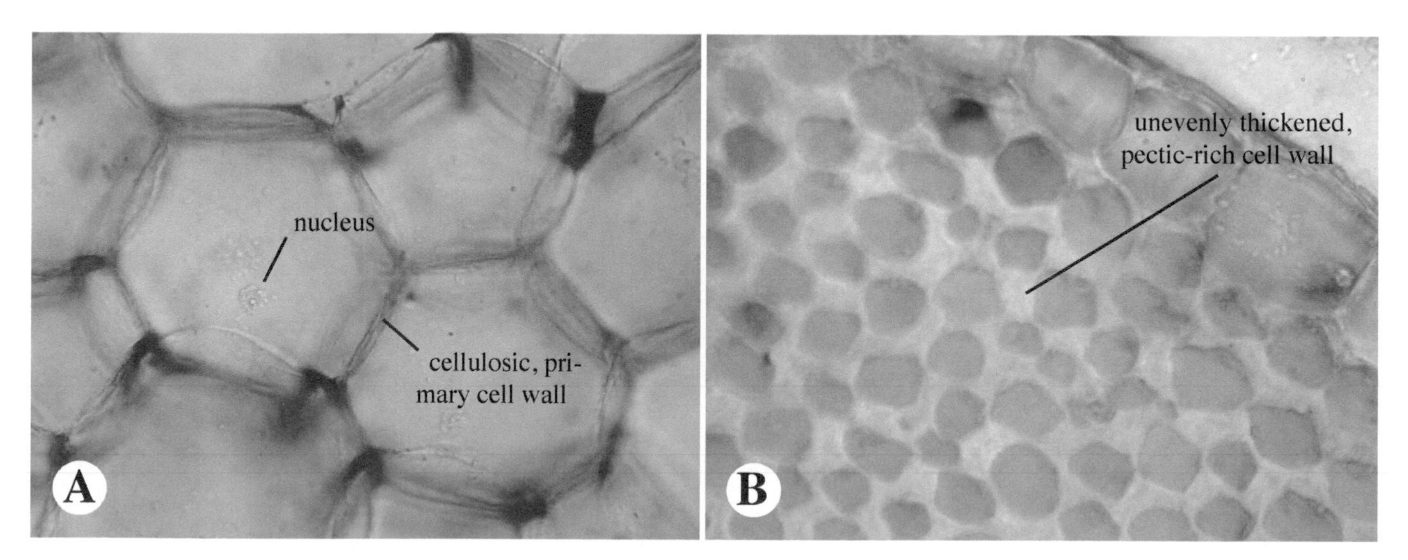

Figure 10.6 **A.** Parenchyma cells, live at maturity, with cellulosic primary cell wall. Note nucleus. **B.** Collenchyma cells (cross-sectional view) live at maturity, with unevenly thickened, pectic-rich primary cell walls.

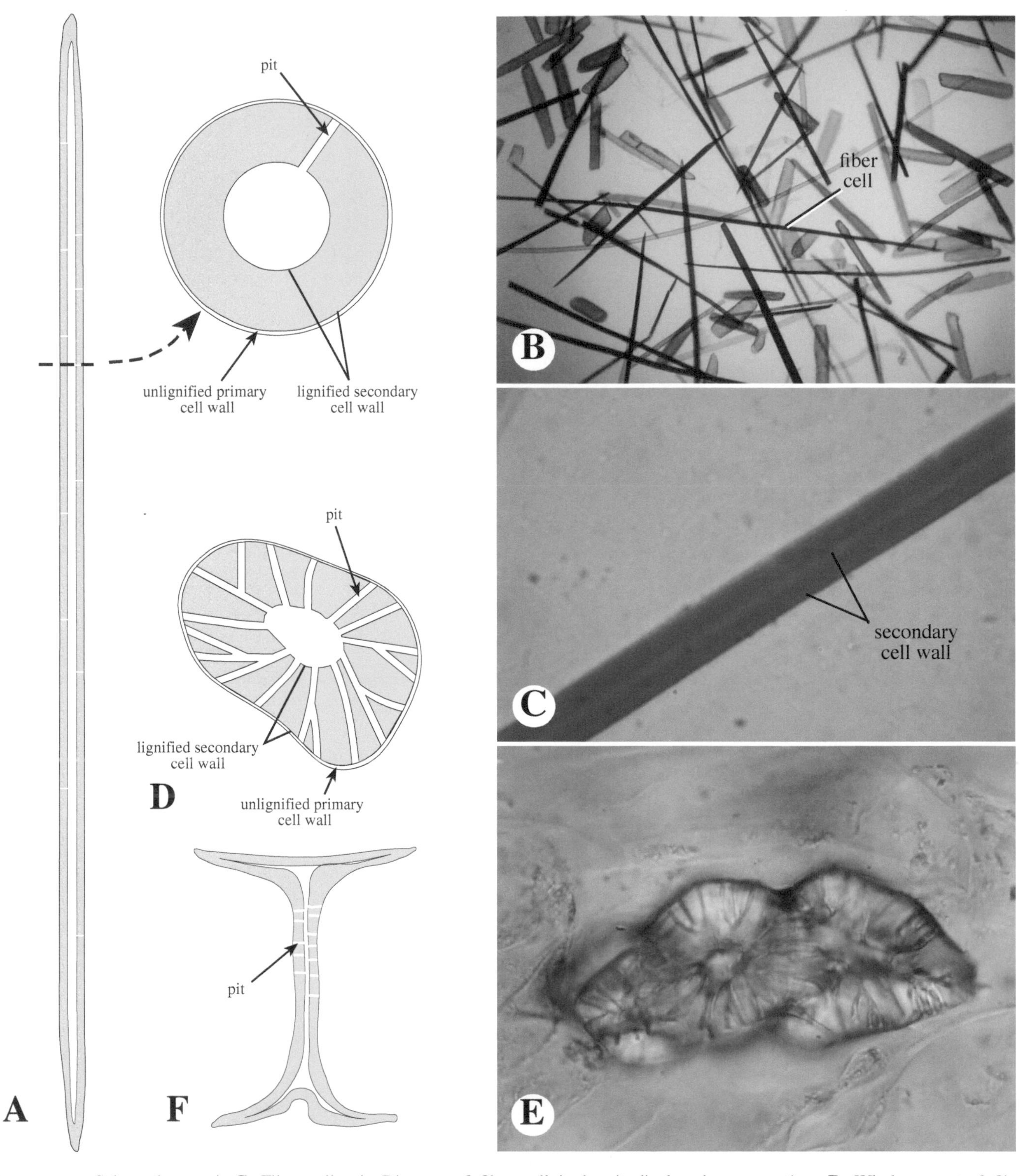

Figure 10.7 Sclerenchyma. **A–C.** Fiber cells. **A.** Diagram of fiber cell in longitudinal and cross section. **B.** Whole mount of fiber cells, from macerated tissue of *Yucca*. **C.** Close-up of fiber cell, side view. Note thick, lignified secondary cell wall. **D–F.** Sclereid cells. **D,E.** Sclereids of pear (*Pyrus*) fruit tissue. Note thick, lignified secondary cell wall with numerous, canal-like pits. **F.** Sclereid of *Ficus* leaf.

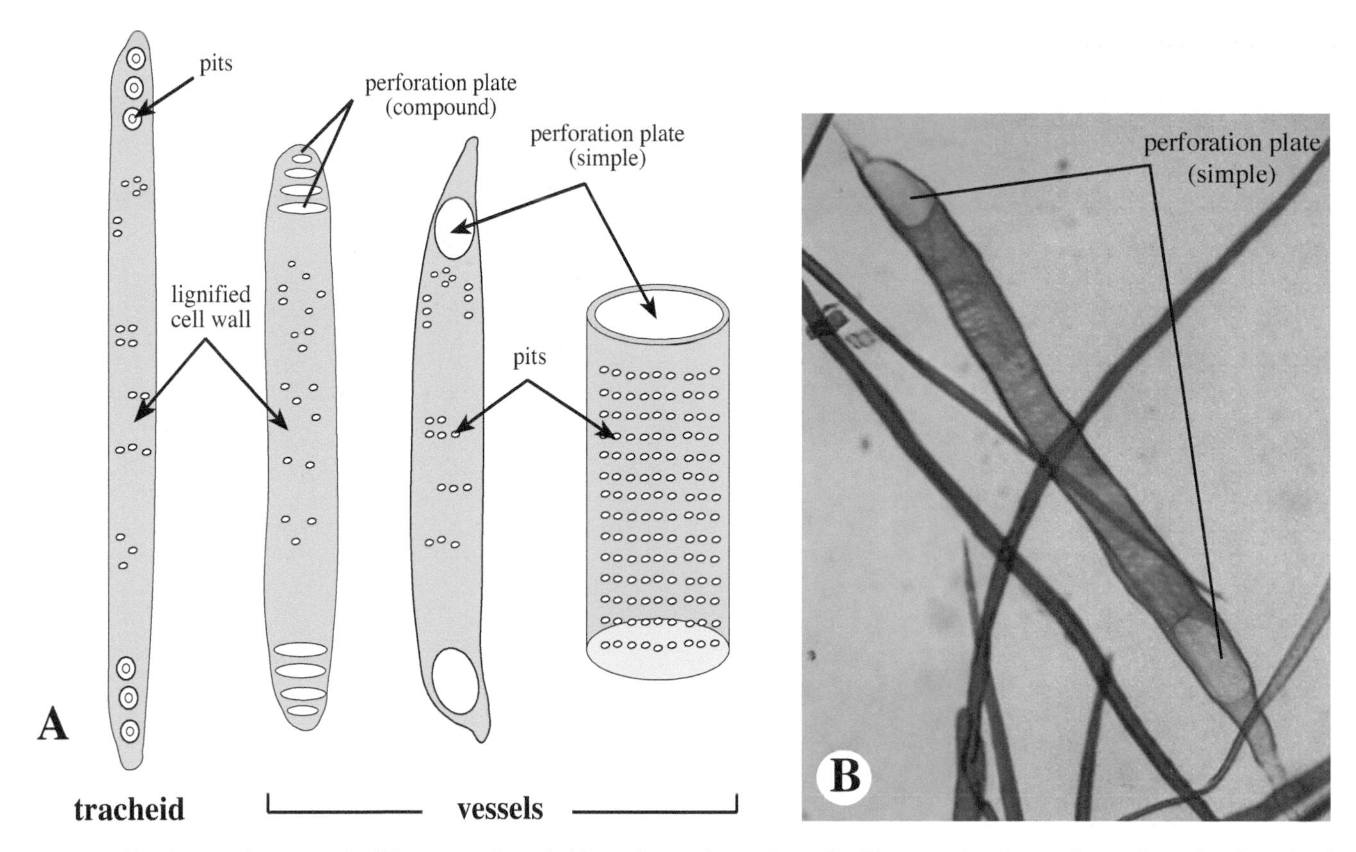

Figure 10.8 Tracheary elements. **A.** Diagrams of tracheids and vessel members. **B.** Photograph of vessel member, showing simple perforation plates.

and mineral nutrients flow between adjacent cells through pit pairs (holes in the lignified 2° cell wall), in which there are intermediate primary cell walls (Figure 10.8A). **Vessel members** are perforate, meaning that there are one or more continuous holes between adjacent cells through which water and minerals may pass. (The term **vessel** refers to several vessel members attached end-to-end, forming a continuous, conductive tube.) The contact area of two adjacent vessel members is called the **perforation plate**. The perforation plate may be **compound**, if composed of several pores (pit pairs with no primary cell walls) or **simple** if composed of a single opening (Figure 10.8A,B). Vessel members may differ considerably in length, width, angle of the end walls, and degree of perforation. As previously discussed, tracheids are the primitive type of tracheary element. Vessels are thought to have evolved from preexisting tracheids *independently* in several different groups, including a few species of *Equisetum*, a few leptosporangiate ferns, all Gnetales, and almost all angiosperms (although not always found in all plant organs).

Phloem (Gr. *phloem*, bark) is a tissue composed of specialized cells called **sieve elements** plus some parenchyma and often some sclerenchyma. Structurally, **sieve elements** (1) are elongated cells; (2) have only a primary wall (no lignified 2° wall); (3) are "semialive" at maturity, losing their nucleus and other organelles but retaining the endoplasmic reticulum, mitochondria, and plastids; and (4) have specialized pores, aggregated together into **sieve areas** (Figure 10.9). Each pore of the sieve area is a continuous hole that is lined with a substance called **callose**, a polysaccharide composed of beta-1,3-glucose. (Note the different linkage from cellulose, which is a polymer of beta-1,4-glucose.) Like tracheary elements, sieve elements are oriented end-to-end, forming a tubelike continuum. Sieve elements function to conduct dissolved sugars from sugar-rich to sugar-poor regions of the plant. Sugar-rich regions include the leaves or other photosynthetic regions, where sugars are synthesized during photosynthesis, or storage roots or stems, where sugars may be produced by the hydrolysis of starch.

There are two types of sieve elements: **sieve cells** and **sieve tube members**. **Sieve cells** have only sieve areas on both end and side walls (Figure 10.9A). **Sieve tube members** have both sieve areas and **sieve plates** (Figure 10.9A–C). Sieve plates consist of one or more sieve areas at the end wall

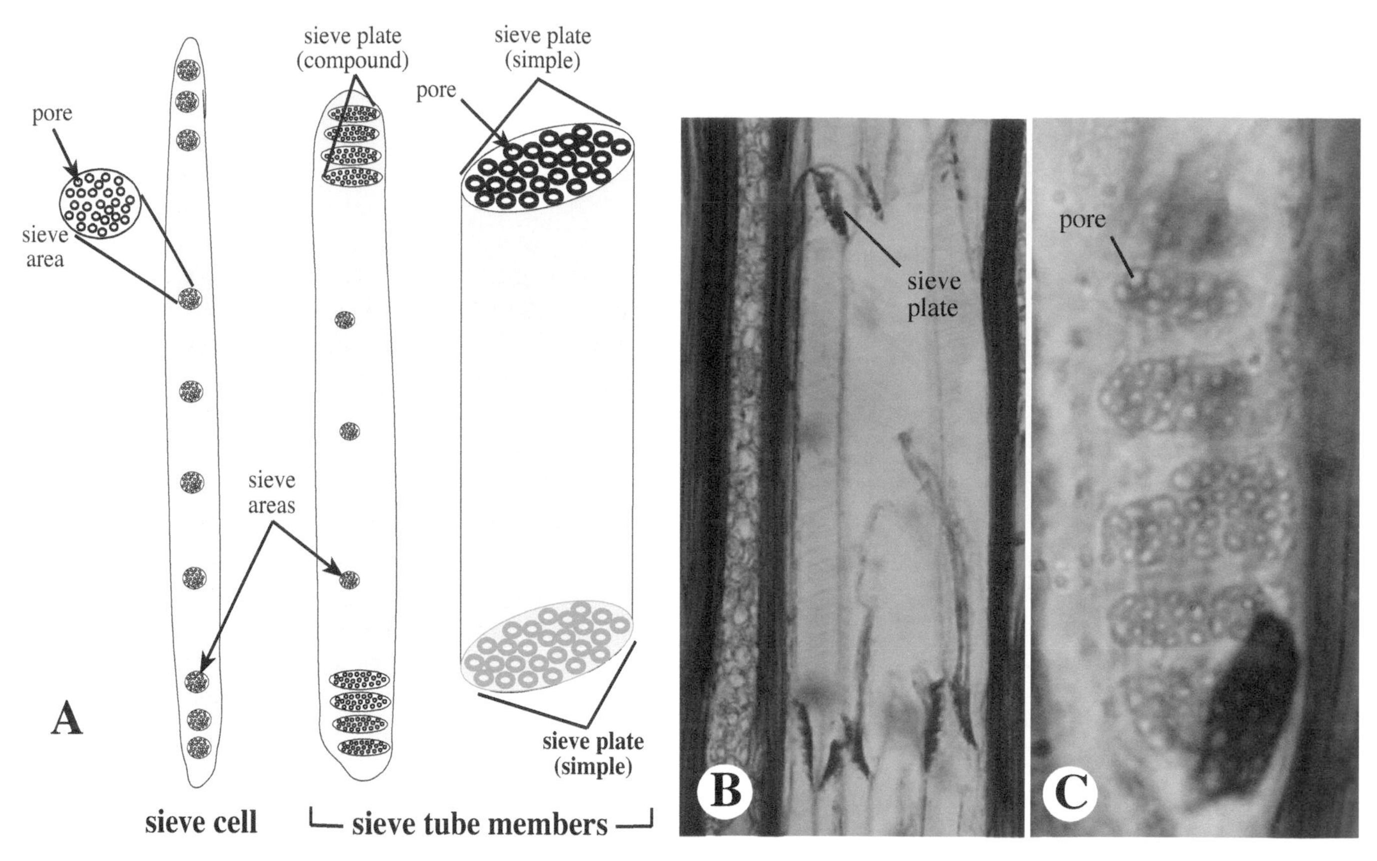

Figure 10.9 Sieve elements. **A.** Diagrams of sieve cells and sieve tube members. **B.** Sieve tube members, showing sieve plates at end walls. **C.** Close-up of compound sieve plate, showing callose-lined pores.

junction of two sieve tube members; the sieve pores of a sieve plate, however, are significantly larger than are those of sieve areas located on the side wall (Figure 10.9C). Both sieve cells and sieve tube members have parenchyma cells associated with them. Parenchyma cells associated with sieve cells are called **albuminous cells**; those associated with sieve tube members are called **companion cells**. The two differ in that companion cells are derived from the same parent cell as sieve tube members, whereas albuminous cells and sieve cells are usually derived from different parent cells. Both albuminous cells and companion cells function to load and unload sugars into the cavity of the sieve cells or sieve tube members. Sieve cells (and associated albuminous cells) are the primitive sugar-conducting cell and are found in all nonflowering vascular plants. Sieve tube members (and associated companion cells) are found only in angiosperms, the flowering plants.

Dermal tissue makes up the outer region of the plant and functions in mechanical protection of inner tissues and inhibition of water loss. Dermal tissue consists of the **epidermis** or, in woody plants, the **periderm** (see later discussion). The **epidermis** (Figure 10.10) makes up the outermost layer of all primary plant organs. Structurally, epidermal cells (1) are usually tabular (flattened, tilelike) in shape; (2) have a cutinized (infiltrated with **cutin**, a polymer of fatty acids) or suberized (infiltrated with **suberin**) outer cell wall; (3) secrete a layer of cutin (called a **cuticle**) *outside* the cell wall; and (4) are usually living at maturity. As previously discussed, the **cuticle** (Figure 10.10) was a major innovation in the evolution of land plants, providing the primary protection from desiccation.

Specialized types of epidermal cells include stomates and trichomes. **Trichomes** (plant hairs) are cellular appendages that grow from the epidermal cells. They come in an amazing variety of shapes, sizes, and densities (see Chapter 9 for vestiture and trichome types). Trichomes may function in providing protection from UV light or herbivory; trichomes of carnivorous plants even function in digestion. **Stomates** are epidermal cells specialized for gas exchange. (See Leaf Structure and Function.)

Secretory structures are cells that secrete compounds, either internally (and stored within the cell) or externally

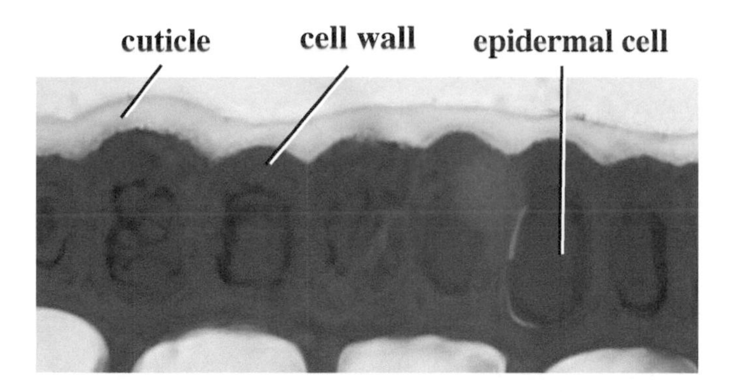

Figure 10.10 Epidermis, showing outer cuticle.

(outside epidermis or into a canal or duct). These include (1) **glandular** (Figure 10.11A) or **stinging** (urent) trichomes that secrete fluid to outside at tip of trichome; (2) **nectaries** (Figure 10.11B), specialized cells secreting sugar (or protein)-rich fluids to the outside that may be floral (associated with flowers as a reward for pollination) or extrafloral (often as a reward for protection); (3) **hydathodes**, which secrete excess transported water (usually due to root pressure) from leaf margins; (4) **resin/oil/mucilage** ducts or canals, which contain cells a lining of cells that secrete resin, oil, or mucilage; and (5) **laticifers** (Figure 10.11C), cells located in the periphery of some tissues that secrete and store latex. The last two may function both to deter herbivory and to seal and protect plant tissue upon wounding.

ROOT STRUCTURE AND FUNCTION

Roots are plant organs that generally function in anchorage and in absorption of water and minerals. Roots are found in the sporophytes of all land plants except for the nonvascular liverworts, hornworts, and mosses (in which the sporophytes are attached to the gametophytes), the psilotophytes (e.g., *Psilotum*), and a few other, specialized taxa. Land plants lacking roots generally have uniseriate (one-cell-thick), filamentous **rhizoids** that assume a similar function.

The first root to develop, in the embryo, is termed the **radicle**. If the radical continues to develop after embryo growth, it is known as the **primary root**. Additional roots may arise from internal tissue of either another root, the stem/shoot (often near buds), or (rarely) a leaf. Roots that arise from other roots are called **lateral roots**. Roots that arise from a nonroot organ (stem or leaf) are called **adventitious roots**. Various modifications of roots have evolved, such as storage roots, aerial roots, fibrous roots, tap roots, contractile roots, haustoria, prop roots, and pneumatophores (see Chapter 9).

Roots, like shoots, develop by the formation of new cells within the actively growing **apical meristem** of the root tip (Figures 10.5B, 10.12A), a region of continuous mitotic divisions. At a later age (and further up the root) these cell derivatives elongate significantly. This cell growth, which occurs by considerable expansion both horizontally and vertically, literally pushes the apical meristem tissue downward. Even later in age and further up the root, the fully grown cells differentiate into specialized cells.

Roots can be characterized by several anatomical features. First, the apical meristem is covered on the outside by a **root cap** (Figure 10.5B, 10.12A). The root cap functions both to protect the root apical meristem from mechanical damage as the root grows into the soil and to provide lubrication as the outer cells slough off. Second, the epidermal cells proximal to the root tip develop hair-like extensions called **root hairs** (Figure 10.12A; see Figure 4.10A); root hairs function to *greatly* increase the surface area available for water and mineral absorption. Third, roots have no exogenous (externally developing) organs; all **lateral roots arise endogenously** from the internal tissues of the root. Lateral roots grow from cell divisions of the **pericycle**, a cylindrical layer of parenchyma cells located just inside the endodermis, or from the **endodermis** itself (see later discussion).

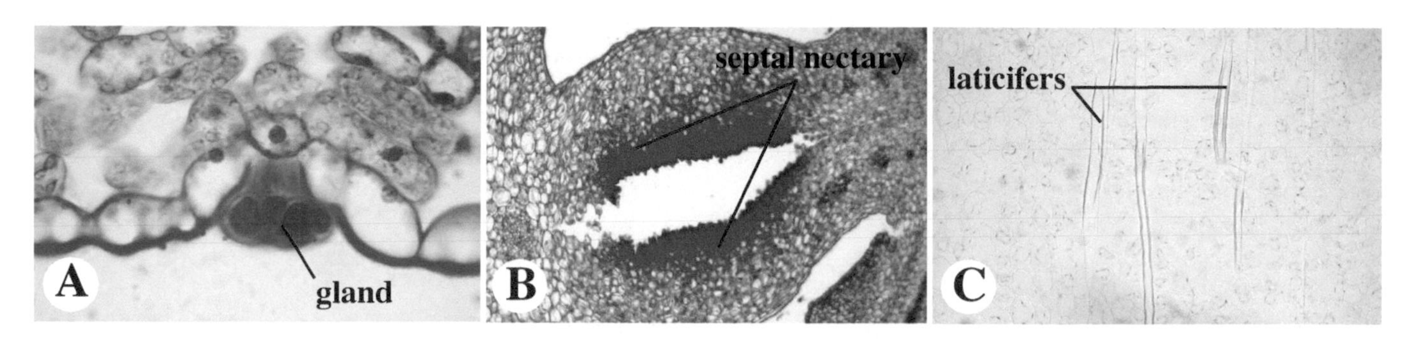

Figure 10.11 **A.** Secretory gland, on leaf surface. **B.** Septal nectary, the cells of which secrete sugar-rich nectar into cavity. **C.** Laticifers.

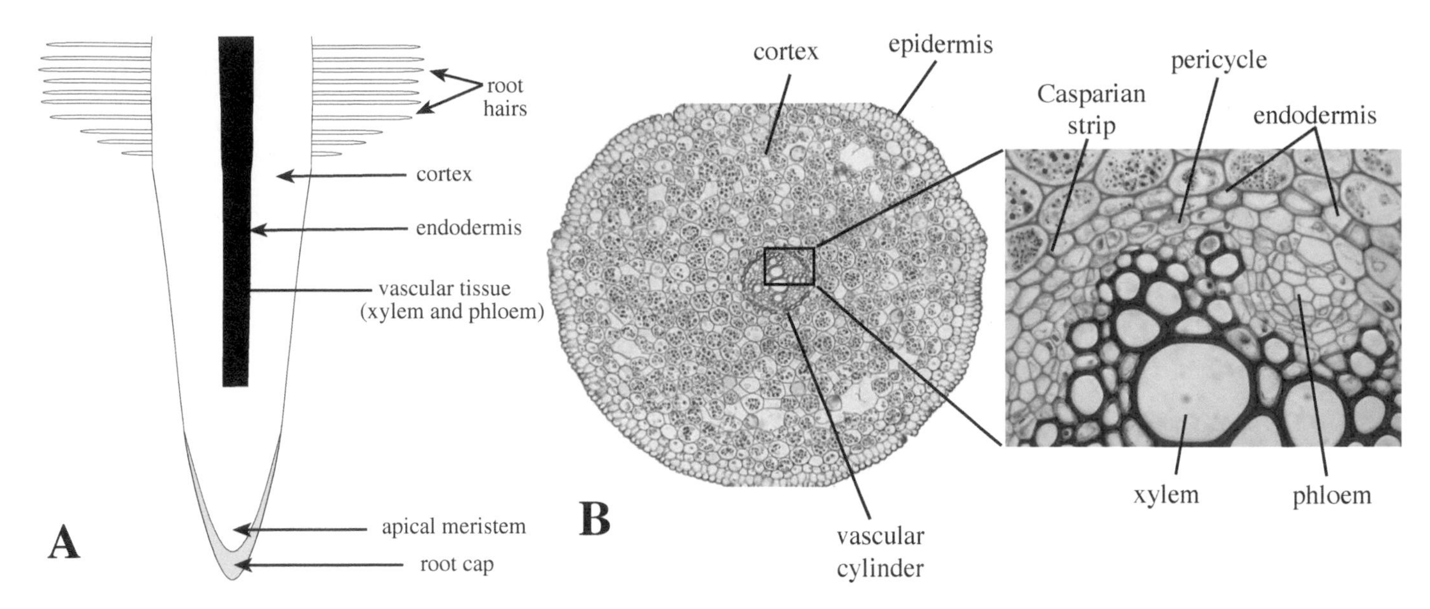

Figure 10.12 Root anatomy. **A.** Root longitudinal section. **B.** Root cross section, close-up at right.

Two other features of roots may or may not distinguish them from stems, as the stems of some land plants are very similar in these features to roots. All roots have a central **vascular cylinder** of xylem and phloem. Often, ridges of xylem alternate with cylinders of phloem (i.e., the **xylem and phloem are on alternate radii**). As with stems, the mostly parenchymatous region between the vasculature and epidermis is called the **cortex**; the center of the vascular cylinder, if vascular tissue is lacking there, is called a **pith**. In addition, the vascular tissue of all roots is surrounded by a special cylinder of cells known as the **endodermis**. In the general region of the root hairs, where most absorption takes place, each cell of the endodermis has a **Casparian strip**, which is a tangential band of suberin that infiltrates the cell wall (Figure 10.13). As discussed in Chapter 4, the Casparian strip functions as a water-impermeable binding material to the plasma membrane of the endodermal cells. This forces absorbed water and nutrients to flow through the endodermal plasma membrane, as opposed to within the intercellular spaces (between the cells or through the cell wall). The function of the Casparian strips is to allow selectivity as to what mineral nutrients are and are not absorbed by the plant; e.g., toxic minerals may be selectively excluded. (Note that further away from the root apical meristem, away from root hairs, the endodermal cells become completely suberized, preventing fluid transport altogether.)

Some root anatomical specializations are found in certain taxa. For example, the aerial roots of many Orchidaceae and Araceae lack root hairs and have a multilayered epidermis called a **velamen**. The velamen may function in protection, prevention of water loss, or water and mineral absorption.

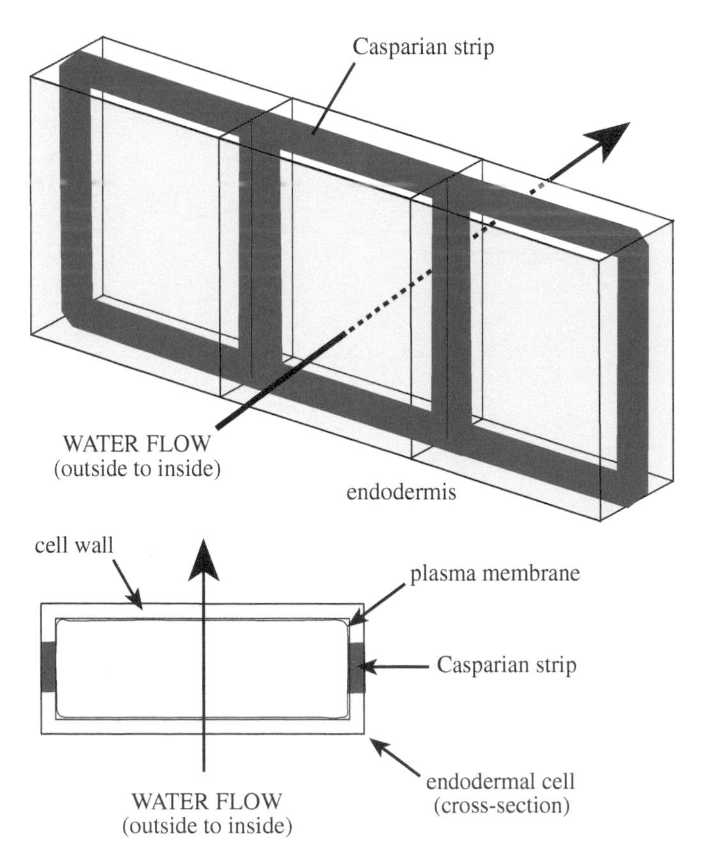

Figure 10.13 The Casparian strip, a specialized feature of cells of the endodermis.

SHOOT AND STEM STRUCTURE AND FUNCTION

A **shoot** is a stem plus its associated leaves. Sporophytic shoots apparently evolved twice, in the lycophytes and separately for the euphyllophytes, associated with lycophylls and euphylls, respectively (see later discussion). The leafy shoot-like structures of mosses and some liverworts are gametophytic and not homologous with shoots of vascular plants.

The first shoot of a vascular plant develops from the **epicotyl** of the embryo. The epicotyl elongates after embryo growth into an axis (the stem), which bears leaves from its outer surface. The tip of a shoot contains the actively dividing cells of the **apical meristem** (Figure 10.14). As in the root, these cells undergo continuous mitotic divisions. A bit down from the apical meristem, the cells undergo considerable expansion, literally pushing the cells of the apical meristem upward (or forward). Proximal to the shoot tip, the fully expanded cells differentiate into specialized cell types.

Slightly down from the apical meristem region, the outermost cell layers of a shoot begin to repeatedly divide. Further cell divisions and growth result in the development of a mass of tissue that forms an immature leaf, the **leaf primordium**. **Vascular strands** run between stem and leaf, providing a connection for fluid transport (Figures 10.14, 10.15). As the shoot matures, the leaves fully differentiate into an amazing variety of forms. In many taxa and a bit later in development, the tissue at or near the upper junction of leaf and stem (called the **axil**) may begin to divide and differentiate into a **bud primordium**. The bud primordium matures into a **bud**, defined as an immature shoot system. Buds have an architecture identical to that of the original "parent" shoot. Buds may develop into a vegetative, lateral branch or may terminate by developing into a flower or inflorescence. Note that buds may also develop at a later time from the stem surface; these are known as *adventitious* buds.

Stems generally function both as supportive organs (supporting and usually elevating leaves and reproductive organs) and as conductive organs (conducting both water/minerals and sugars through the vascular tissue between leaves, roots, and reproductive organs). Stems can be distinguished from roots in at least three ways. First, the apical meristem of stems is not covered by an outer protective layer (like the root cap; Figure 10.14). Second, the epidermal cells of the stem do not form structures resembling root hairs. However, the epidermal cells of stems and leaves may divide and differentiate into separate, one-to-many-celled **trichomes**, described earlier (see also Chapter 9). Third, **stems bear leaves exogenously**; no organs are born endogenously (except in cases of adventitious roots potentially arising from the internal parenchyma cells of stems).

Stems, particularly underground stems, may possess an endodermis similar to that of roots in structure and function. The aerial stems of many plants lack an endodermis. Numerous modifications of stems and shoots have evolved, such as bulbs, corms, caudices, rhizomes, stolons (=runners), cladodes, pachycauls, and thorns (see Chapter 9).

The primary vasculature of stems is organized into arrangements of xylem and phloem known as **steles** (Figure 10.16). In some groups of non-seed vascular plants, such as the lycophytes, the stem stelar type is a **protostele** (Figures 10.15A, 10.16A), in which there is a central cylinder of vascular tissue,

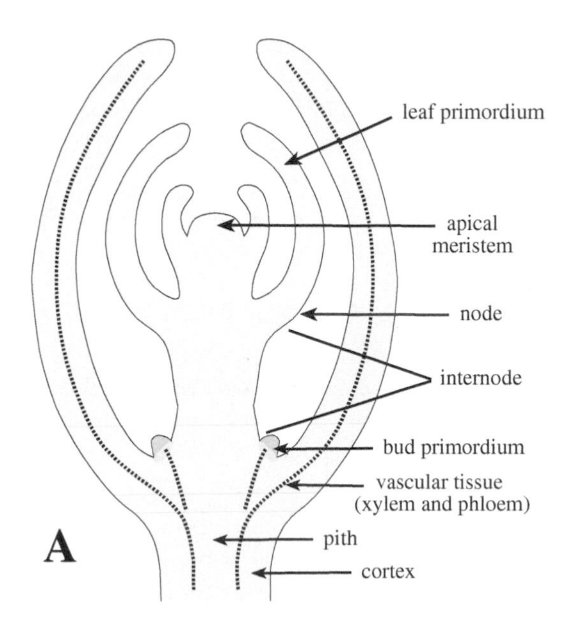

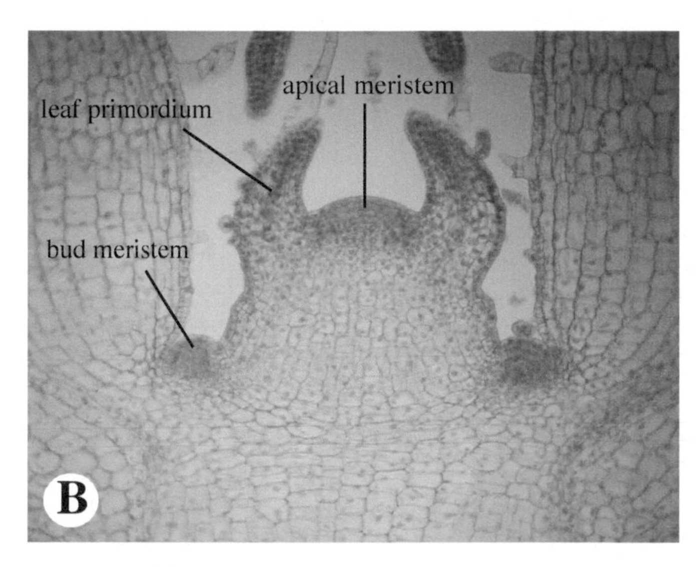

Figure 10.14 Shoot longitudinal section. **A.** Diagram. **B.** Photograph, *Plectranthus* [*Coleus*] shoot.

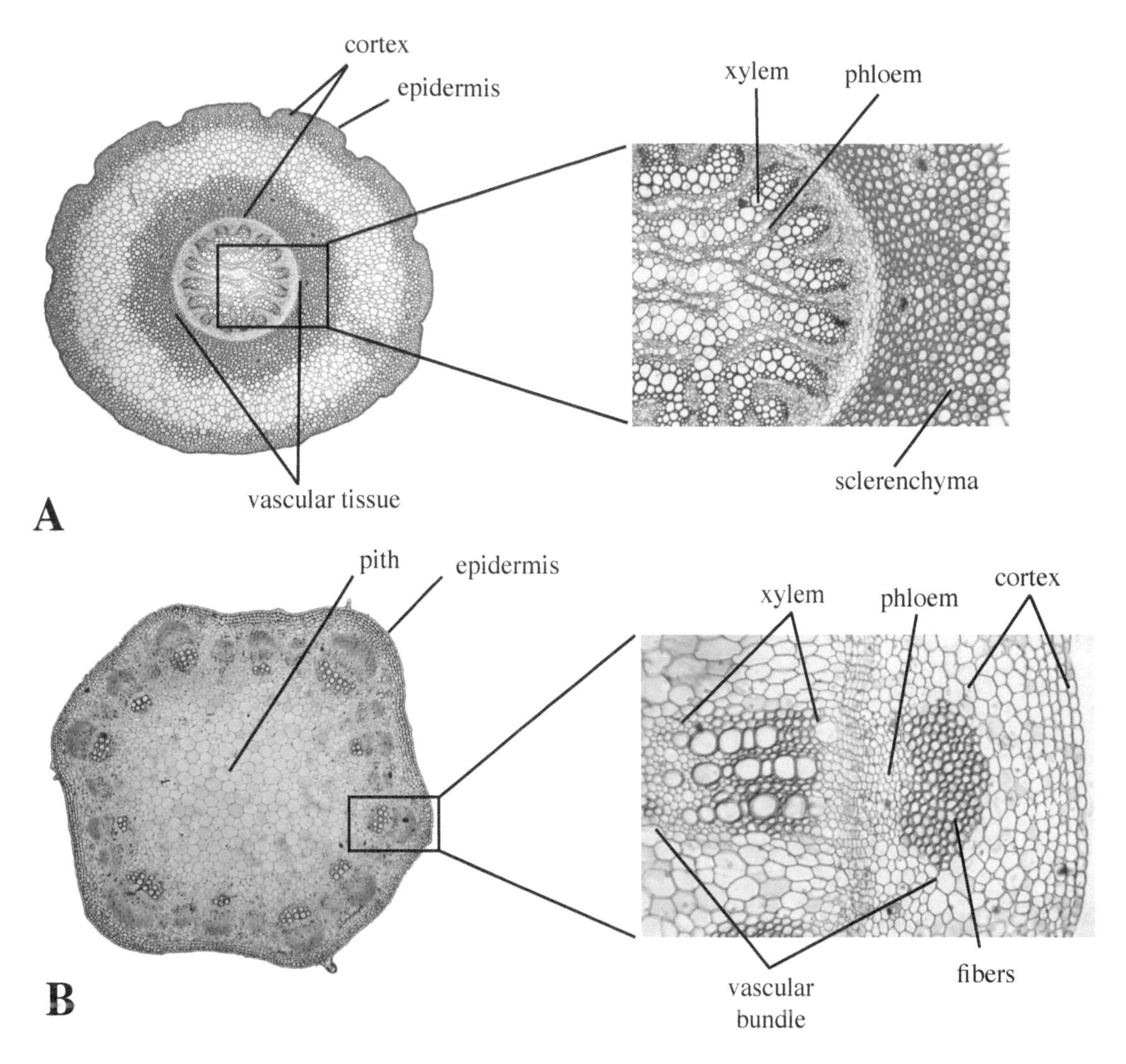

Figure 10.15 Stem anatomy. **A.** *Lycopodium* cross section of stem, a protostele, showing close-up of vascular tissue, with interdigitated xylem and phloem and outer fibers. **B.** *Helianthus* cross section of stem, a eustele, showing close-up of vascular bundle, with xylem, phloem, and associated fibers.

either xylem surrounded by phloem or the two tissues interdigitated (Figures 10.15A, 10.16A). Protosteles are the most ancestral type of stem vasculature, one that most resembles the vasculature of a root. The vasculature of monilophyte stems is typically a **siphonostele**, in which a ring of xylem is surrounded by a continuous layer of phloem, either on the outside only (an ectophloic siphonostele; Figure 10.16B) or on the outside and inside (an amphiphloic siphonostele; Figure 10.16C); if the latter is much dissected, it is known as a dictyostele (Figure 10.16D). The stems of seed plants contain discrete **vascular bundles** in which xylem and phloem are grouped together along a common radius, usually with xylem to the inside and phloem to the outside, a type known as a **collateral** vascular bundle. (In some angiosperms the stem vascular bundles have phloem to both the inside and outside of the xylem, a type known as **bicollateral**.) These collateral vascular bundles may be organized as a single ring, known as a **eustele** (Figures 10.15B, 10.16E). The eustele is an apomorphy for many seed plants, including all that are extant. For both siphonosteles and eusteles, the central region of tissue in the stem is called **pith**; the region between the vasculature and the outer epidermis is called the **cortex**. Stems of monocots (of the angiosperms) have a modification of the eustele called an **atactostele** (Figure 10.16F). The atactostele, which represents an apomorphy for the monocots, consists of numerous, collateral vascular bundles positioned throughout the stem tissue (appearing "scattered" but actually having a precise and complex disposition). In an atactostele, there is no pith; the region of tissue between vascular bundles is called **ground meristem**. Vascular bundles typically are associated with sclerenchyma fibers, which may surround the entire bundle or occur in outer patches called "**bundle caps**" (e.g., Figure 10.15B). Parenchyma, collenchyma, or scleren-

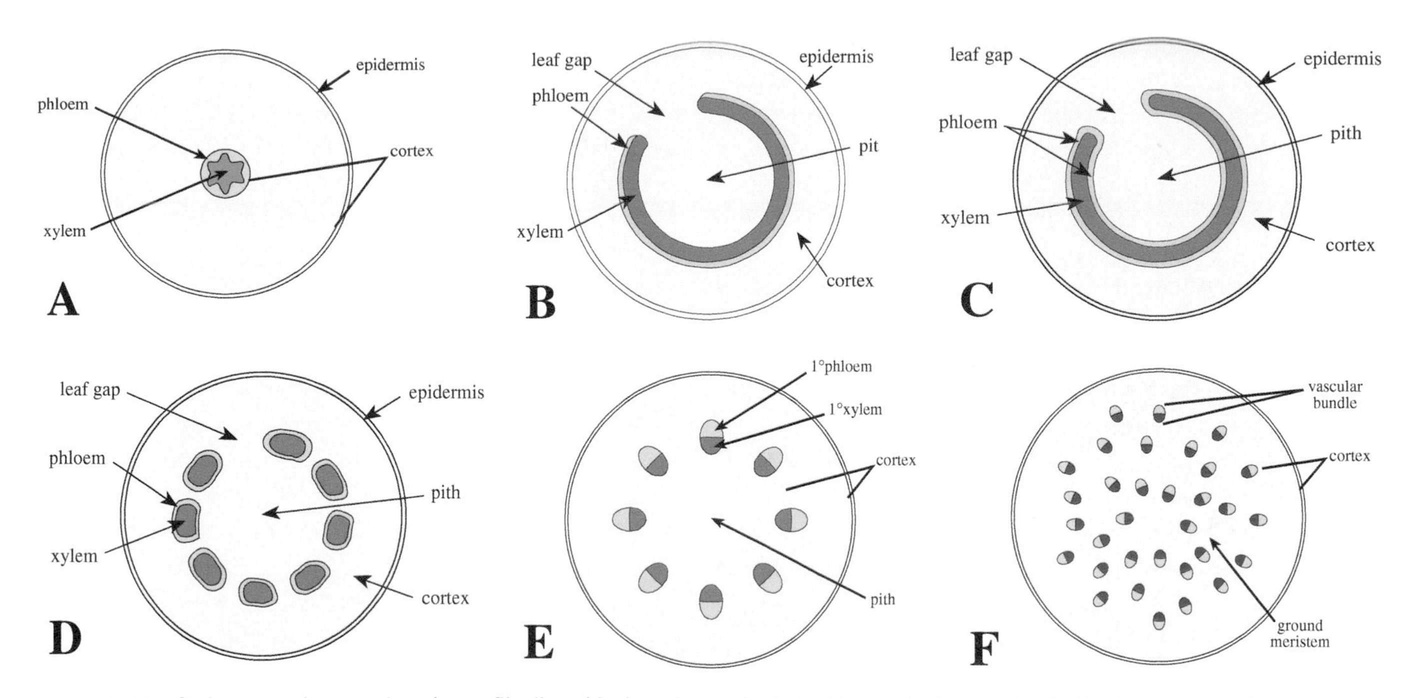

Figure 10.16 Stelar types in vascular plants. Shading: black=xylem; stippled=phloem. **A.** Protostele. **B–D.** Siphonostele. **B.** Ectophloic siphonostele. **C.** Amphiphloic siphonostele. **D.** Dictyostele. **E.** Eustele. **F.** Atactostele.

chyma cells make up the tissues of the pith, cortex, and ground meristem.

The vasculature of stems (and roots) can vary also with respect to the sequence of maturation of tracheary elements in the xylem. **Protoxylem** refers to the first xylem that matures in a group of vascular tissue; protoxylem cells are often smaller in diameter. **Metaxylem** is the xylem that develops later and usually consists of larger diameter cells. Three general types of protoxylem orientation are recognized: (1) **exarch**, in which the protoxylem is oriented toward the outside relative to metaxylem, as occurs in some protosteles (Figure 10.17A); (2) **endarch**, in which the protoxylem is oriented toward the center of the stele, relative to the metaxylem, as occurs in eusteles and atactosteles (Figure 10.17B); and (3) **mesarch**, in which the protoxylem is surrounded by metaxylem within the vascular tissue, as can occur in siphonosteles.

SECONDARY GROWTH

In vascular plants, the growth in height or length of a stem or root is brought about by the elongation and differentiation of cells derived from the apical meristem. This is termed

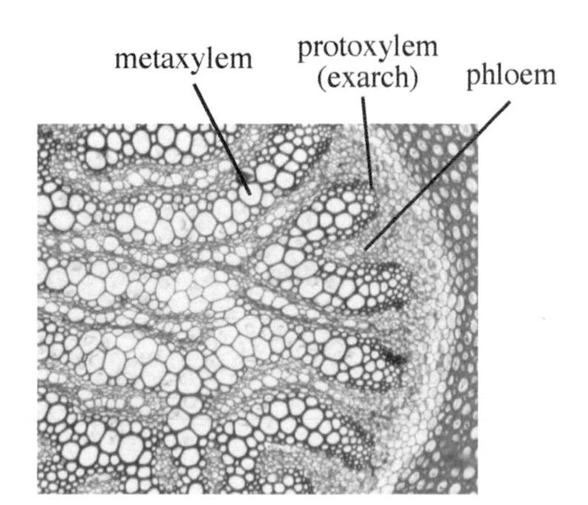

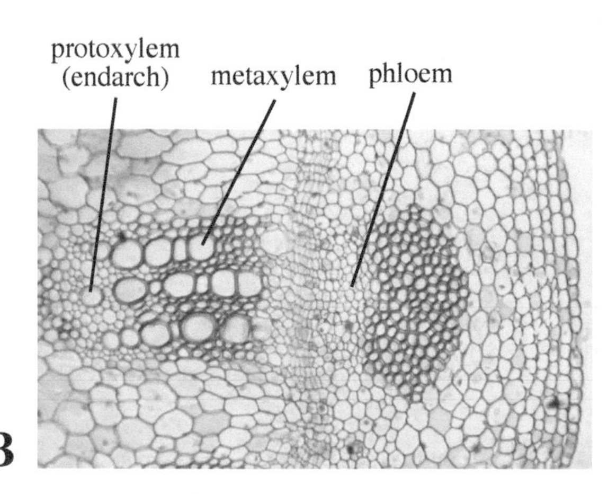

Figure 10.17 **A.** *Lycopodium* stem, showing exarch protoxylem development of protostele. **B.** *Helianthus* stem, showing endarch protoxylem development of eustele.

primary growth, and the tissues formed by primary growth are called **primary tissues** (e.g., as in primary xylem or primary phloem). However, in many seed plants, roots and stems may grow in girth or width by means of cells produced not from the apical meristems, but from **lateral meristems**. This process is termed **secondary growth**, and the tissues formed by secondary growth are called **secondary tissues**.

Two types of lateral meristems function in secondary growth: the **vascular cambium** and the **cork cambium**. These lateral meristems represent apomorphies for the woody plants, including all extant seed plants plus several fossil groups (although lateral meristems have been lost in some angiosperms, most notably the monocots and many annual eudicots). The **vascular cambium** is a cylindrical sheath of cells that typically forms by cell divisions of undifferentiated parenchyma cells. In eustelic stems the vascular cambium forms from parenchyma cells both between the primary xylem and phloem of vascular bundles and in the adjacent region between the bundles (Figures 10.18, 10.19). In woody roots the vascular cambium develops from parenchyma cells between xylem and phloem and from the adjacent pericycle. The cells of the vascular cambium divide more or less synchronously, and mostly in a tangential plane, the initial result being the formation of two layers of cells (Figure 10.18). One of these layers continues as the vascular cambium and divides indefinitely; the other layer eventually differentiates into either **secondary xylem** (=**wood**), if produced to the inside of the cambium, or **secondary phloem**, if produced to the outside of the cambium (Figures 10.18, 10.20A). As discussed in Chapter 5, because derivatives of the vascular cambium are produced in two directions, this growth is known as *bifacial*. (Note that a type of secondary growth occurs in some mono-

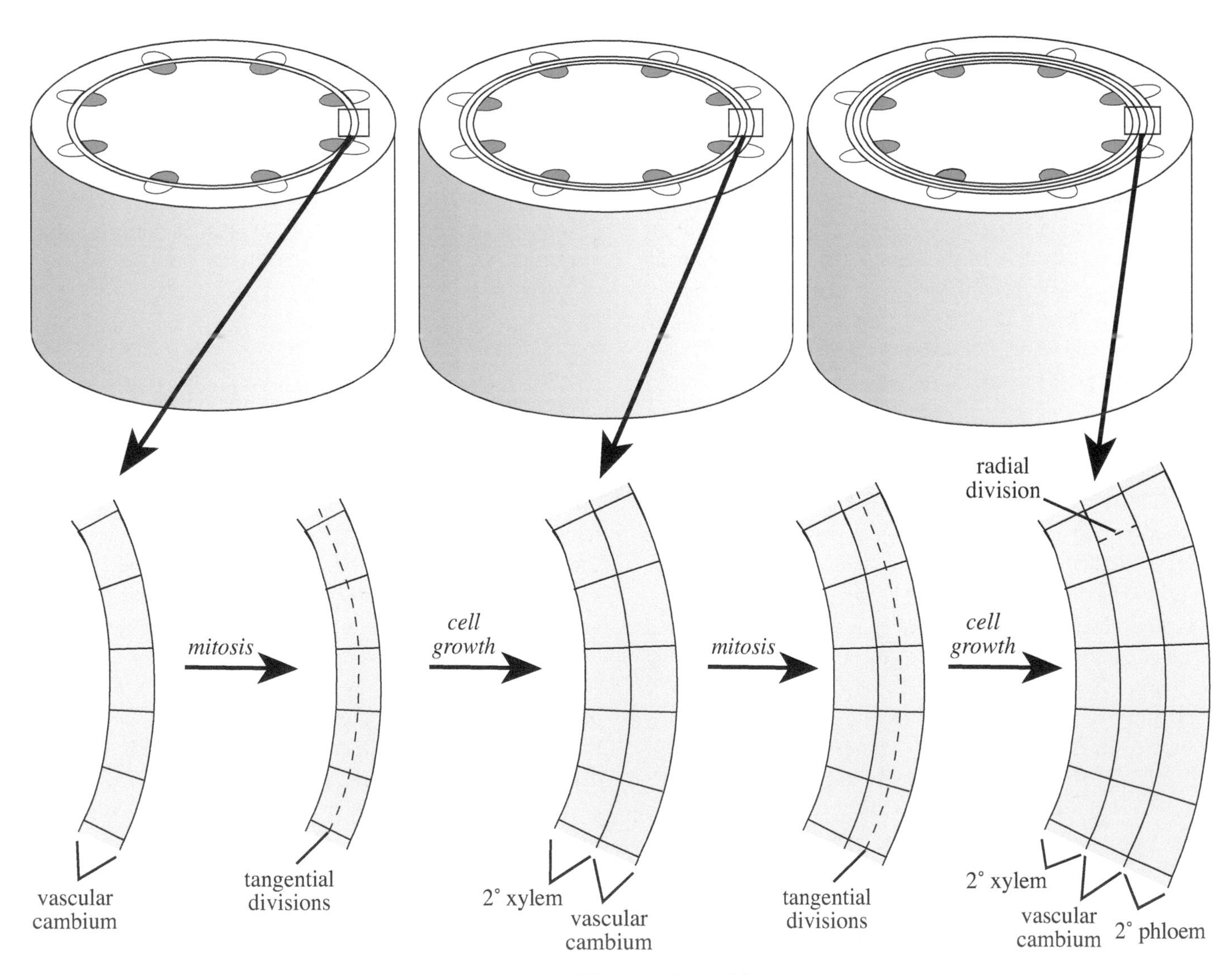

Figure 10.18 The vascular cambium.

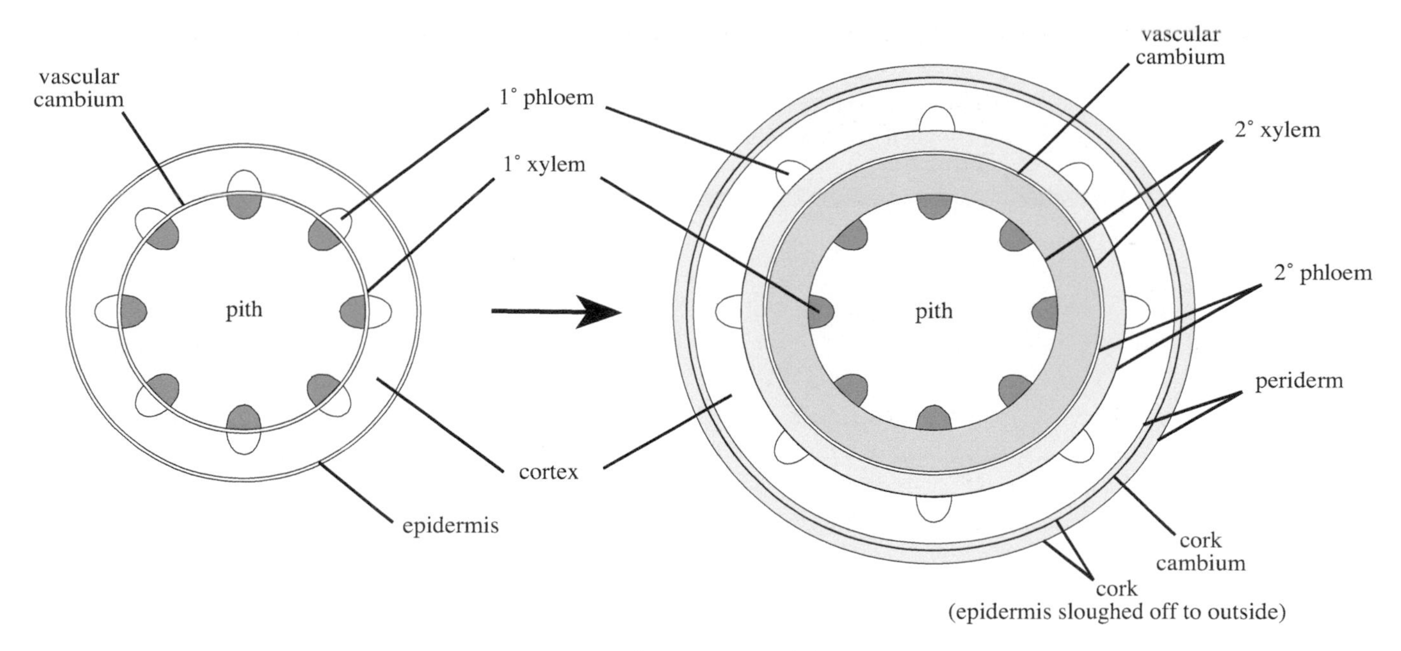

Figure 10.19 Secondary growth in stems.

cot groups but not by means of a continuous, bifacial, vascular cambium and is thought to have evolved independently.)

Generally, much more secondary xylem is produced than is secondary phloem. As the secondary tissue is formed, the inner cylinder of wood expands outward by the continuous deposition of concentric layers of secondary xylem cells. As this growth in girth continues, some cells of the vascular cambium and its derivatives undergo radial divisions (parallel to a radius), enabling the vascular cambium, secondary xylem layers, and secondary phloem layers to grow larger in circumference.

The **cork cambium** is similar to the vascular cambium, only it differentiates near the periphery of the stem or root axis. The cork cambium forms **cork** to the outside and **phelloderm** to the inside, the latter usually much thinner (Figure 10.20G). The cork cambium and all of its derivatives constitute the **periderm**. The outer cork cells contain a waxy polymer called **suberin** (chemically related to cutin), which is quite resistant to water loss. In the wood industry, the term **inner bark** refers to all the tissues between the vascular cambium and the periderm (including all of the secondary phloem). **Outer bark** is equivalent to the periderm.

The vascular cambium and cork cambium are of significant adaptive value. Secondary xylem (wood) functions in structural support, enabling the plant to grow tall and acquire massive systems of lateral branches. Thus, the vascular cambium was a precursor to the formation of intricately branched shrubs or trees with tall overstory canopies, a significant ecological adaptation. Cork produced by the cork cambium functions as a thick layer of dermal tissue cells that protects the delicate vascular cambium and secondary phloem from mechanical damage, predation, and desiccation.

Secondary xylem, or **wood**, consists mostly of longitudinally oriented tracheary elements, either tracheids (in cycads, *Ginkgo*, and conifers, excluding Gnetales) or vessels (Gnetales and almost all angiosperms; Figures 10.18–10.21). Other longitudinally oriented cells may include **fibers** and **axial parenchyma**. The vascular cambium also forms cells that are radially oriented (parallel with a stem or root radius). These radially oriented cells occur in bandlike strands called **rays** (Figure 10.19B–E); their function is lateral translocation of water, minerals, and sugars.

In most woody plants with regular, annual growth seasons (in temperate regions caused by seasonal cold, in tropical regions by seasonal drought), the vascular cambium and cork cambium actively divide only near the start of the growing season; further secondary growth is delayed until the next growing season. As a result of this periodic growth, there are differences in the structure of the secondary xylem from the first part of the growing season (**spring wood**) versus the latter part of the growth season (**summer wood**). For example, the tracheary elements of spring wood tend to be larger in diameter with thinner walls; those of summer wood tend to be smaller in diameter with thicker walls (Figure 10.20C,D). The overall result of this discrepancy in structure between spring and summer wood results in the formation of **annual rings** (Figures 10.20A,B, 10.21A,B). Each annual ring represents the accumulation of secondary xylem (or phloem) over a single growing season. Annual rings are evident because of

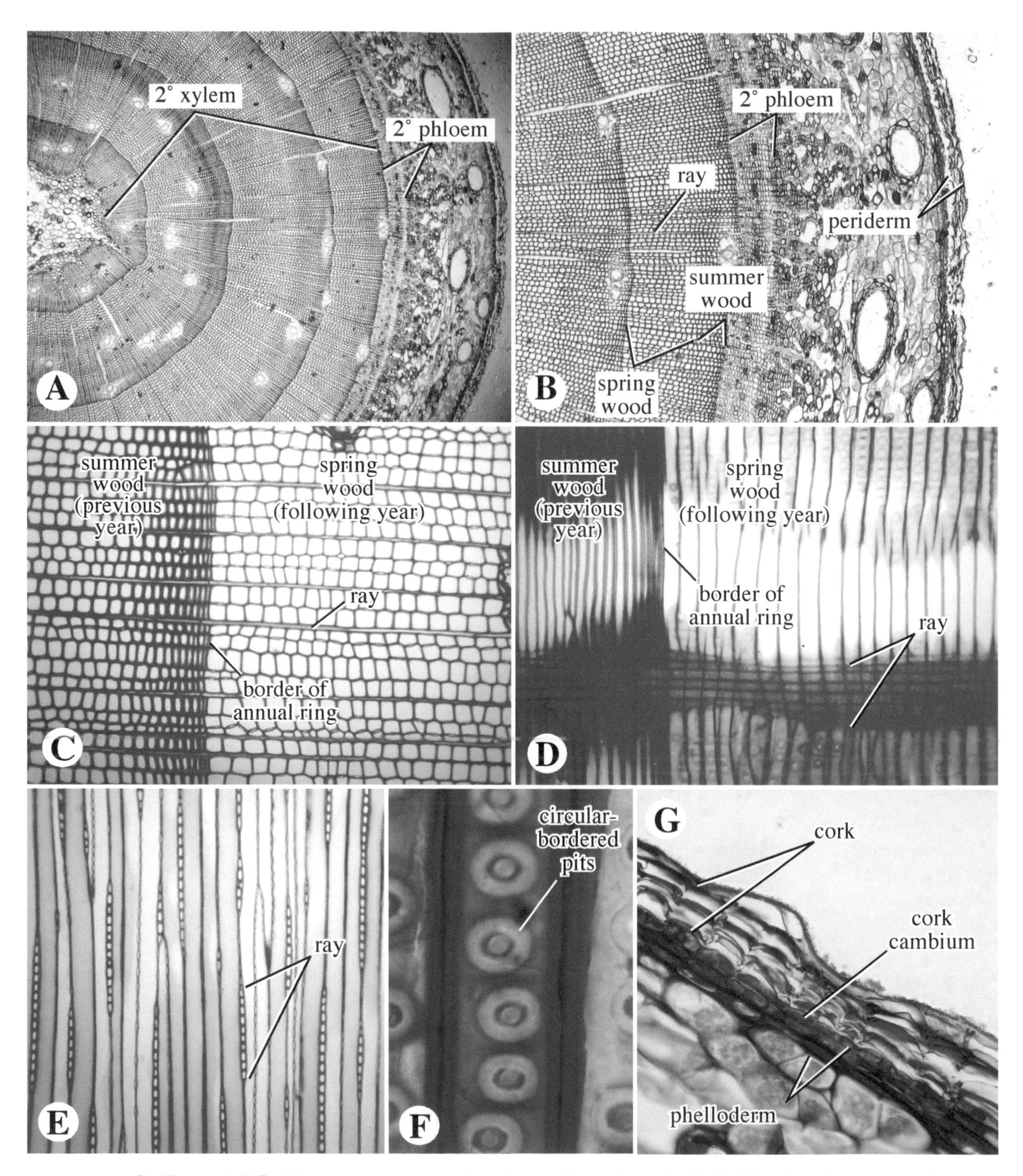

Figure 10.20 Conifer wood. **A,B.** *Pinus* sp. stem cross section, showing 3 years' growth. **C–E.** *Libocedrus decurrens* wood sections. **C.** Transverse or cross section, showing junction of summer wood (of previous year) and spring wood (of following year). **D.** Radial section, showing transverse ray and border of annual ring (summer wood to left; spring wood to right). **E.** Tangential section, showing vertical tracheids and rays. **F.** *Pinus*, circular bordered pits of tracheids (radial face). **G.** *Pinus* periderm, showing cork cambium, phelloderm, and cork.

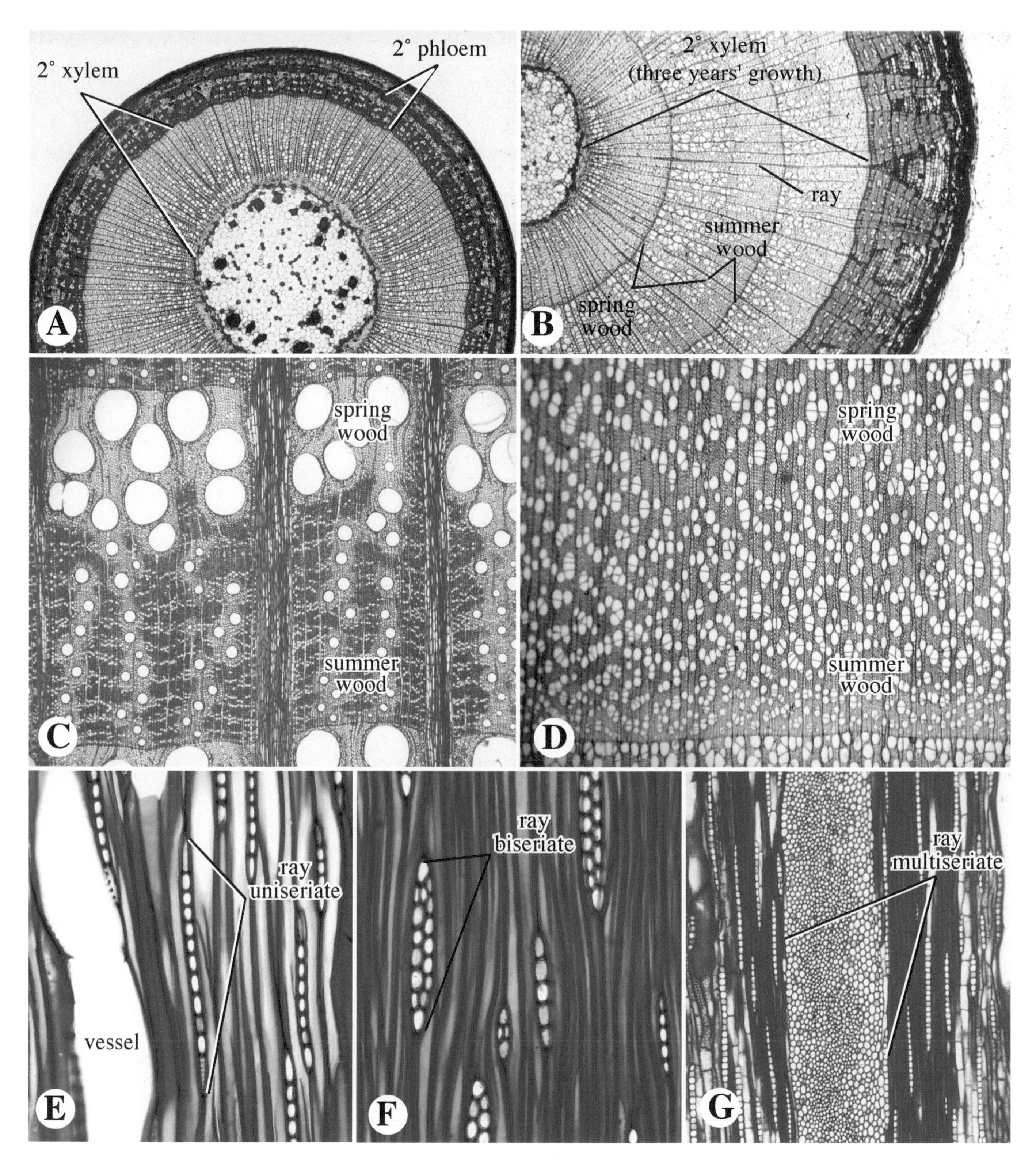

Figure 10.21 Eudicot wood. **A,B.** Woody stem cross section, *Tilia*. sp. **A.** One year's growth. **B.** Three years' growth. Note rays and rings with spring and summer wood. **C.** Ring-porous wood, *Quercus*, with vessels much larger in spring wood (above), smaller to absent in summer wood (below). **D.** Diffuse-porous wood, *Salix*, having vessels evenly distributed in annual ring. **E,F.** Ray types. **E.** Uniseriate rays. **F.** Biseriate rays (some uniseriate also present). **G.** Multiseriate rays (some uniseriate also present).

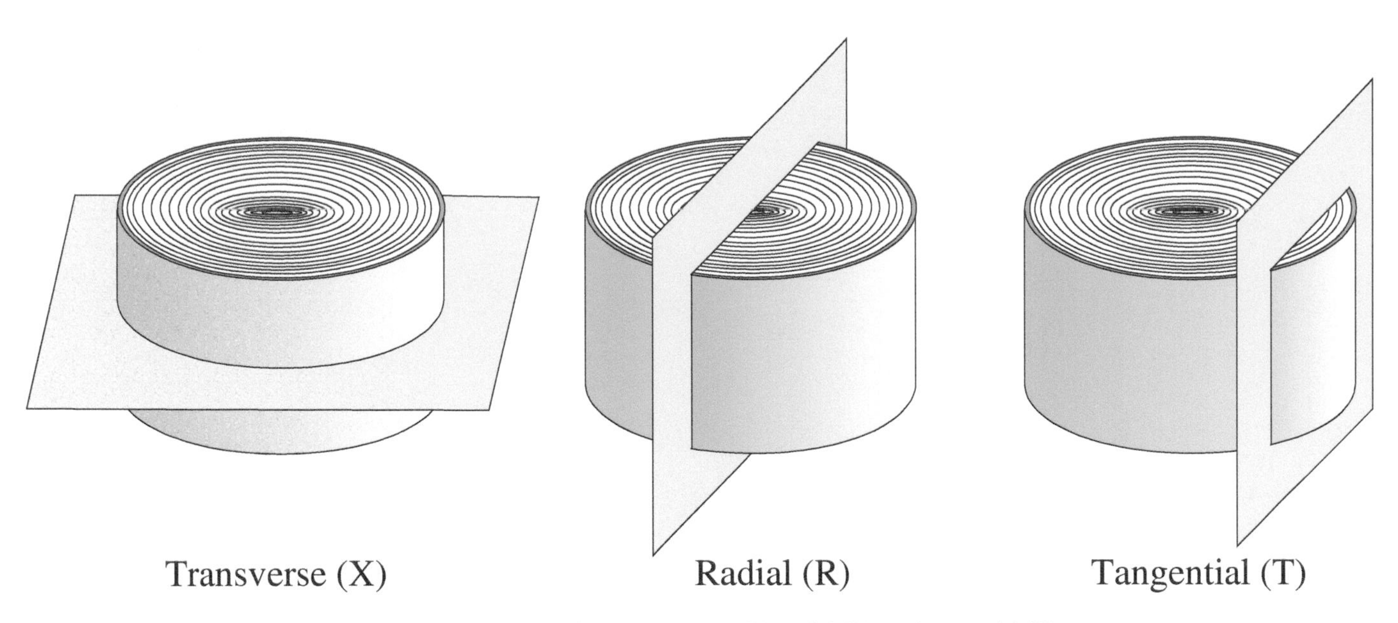

Figure 10.22 Wood sections: transverse (X), radial (R), and tangential (T).

the structural difference between the last cells of the summer wood and the first cells of the subsequent spring wood.

Wood may be cut in three major planes: transverse (cross-sectional), radial (longitudinal and ca. parallel to a stem radius), or tangential (longitudinal and perpendicular to a stem radius); these planes of section are often abbreviated **X, R, T**, respectively (Figure 10.22). These different cuts are used for different purposes in the wood industry and influence the *figure*, or general appearance of the wood. The three cuts are also used by wood anatomists to view the cells from three different directions, often necessary for precise description or identification of wood samples (see Figures 10.20, 10.21).

In the wood industry, the term **softwood** is used for a wood product derived from a conifer and **hardwood** is used for one derived from a nonmonocotyledonous angiosperm. Softwoods from conifers (such as pine) are indeed usually softer and easier to work with than hardwoods (such as oak), as the latter typically contain numerous wood fiber cells. However, there are exceptions; some so-called hardwoods, such as balsa, are quite soft.

Wood anatomy may be very complex. Its structure may provide several characters that may be of systematic importance; these characters include tracheary element type (whether having only tracheids, termed **nonporous**, or having vessels, termed **porous**), tracheary element anatomy (size, shape, and pit or perforation plate structure), distribution of vessels (if present), ray anatomy, presence of resin ducts, distribution of axial parenchyma, and presence/distribution of fibers or fiberlike cells. For example, in some angiosperms there may be differences in the formation of vessel elements associated with the annual rings. The vessels may form only in the spring wood, with summer wood either lacking or having relatively small vessels and usually containing mostly fibers; this type of growth is called **ring-porous** (the term *porous* referring to the presence of "**pores**," the vernacular term for vessels; Figure 10.21C). The alternative, in which vessels develop more or less uniformly throughout the growth season, is called **diffuse-porous** (Figure 10.21D). Another feature of systematic importance is ray anatomy. Rays can be **uniseriate** (with a single, vertical row of cells, as in Figure 10.21E), **biseriate** (with two vertical rows of cells, as in Figure 10.21F), or **multiseriate** (with many vertical rows of cells, as in Figure 10.21G). Wood anatomical characters may be useful in phylogenetic inference and are valuable for microscopic identification of the species.

Some aspects of wood anatomy are ecologically significant. In fact, wood (both extant and fossil) may be used to trace the history of climatic conditions in a given region. This field of study is called **dendrochronology**. When growth conditions are good (e.g., high rainfall), annual rings will be wide; when conditions are poor, they will be narrow. By correlating the width of annual rings with time, assessment of past conditions may be made, e.g., cycles of cold or drought and even sunspot cycles.

LEAF STRUCTURE AND FUNCTION

Leaves are the plant organs that function primarily in photosynthesis. However, leaves or leaflike homologs have been co-opted for innumerable other functions in plants.

As discussed earlier, leaflike structures occur on the gametophytes of mosses and leafy liverworts. However, sporophytic leaves evolved first in the vascular plants; thus, *leaf* is here equated with sporophytic leaf. True leaves evolved with the development of a continuous strand of vascular tissue running from the stem into the leaf. As discussed in Chapter 4, sporophytic leaves evolved twice. Lycophytes leaves, which have a single, generally unbranched vein, lack a leaf gap (see later discussion), and grow by means of an intercalary meristem are called **lycophylls** (essentially equivalent to "microphylls"; Figure 10.23). Lycophylls may have evolved by the innervation and planation of peg-like enations (see Chapter 4). The leaves of euphyllophytes (the monilophytes and lignophytes/seed plants) are called euphylls, and are thought to have evolved independently by the planation and webbing of lateral branch systems (Chapter 4). **Euphylls** (essentially synonymous with "megaphyll," a more traditional term) are characterized in having (1) multiple, branched vascular strands in the leaf blade; (2) a **leaf gap**, in which parenchymatous tissue replaces vascular tissue in the region just distal to the point of departure of the vasculature from stem to leaf (Figure 10.23); and (3) growth by means of either marginal or apical meristems. The evolution of euphylls allowed for a bigger, broader, more morphologically diverse leaf structure. This has undoubtedly been adaptive in several habitats, permitting, e.g., greater photosynthetic output.

Leaves have a characteristic development and structure (Figure 10.24). As previously discussed, leaves (both lycophylls and euphylls) arise as **leaf primordia** on shoots. Differential cell growth results in a flattened, dorsiventral structure with an upper (adaxial) and lower (abaxial) surface. Thus, leaves have both an **upper epidermis** and **lower epidermis** (Figure 10.24). The cuticle, which is an apomorphy of all land plants, is often thickened on leaf epidermal cells. As discussed in Chapter 3, the **stomate** was a major innovation in the evolution of land plants. Stomates consist of two chlorophyllous **guard cells**, between which is an opening, the **stomatal pore** or **stoma** (Figure 10.25A,B). The guard cells can alter turgor pressure by changes in ion gradients, which results in opening or closing the stomatal pore. In vascular plants, stomates occur mostly on the leaves, and are predominate on the lower (abaxial) surface. Stomates function to regulate gas exchange. An open stomate permits carbon dioxide to enter the leaf, and oxygen and water to exit. Stomates are the only epidermal cell to have chloroplasts (which function in regulation of the stoma). Stomates are often associated with **subsidiary cells**, specialized epidermal cells that are contiguous with the stomate and that may function in ion exchange and therefore stomate opening and closing (the stomate plus subsidiary cells called a *stomatal apparatus*). The number, size, and placement of subsidiary cells vary among taxa and can be useful anatomical systematic characters (Figure 10.25C–E).

The nonvascular cells located between upper and lower epidermal layers comprise the **mesophyll** (Figure 10.24). The mesophyll is composed primarily of chlorophyllous cells, the chloroplast-containing parenchyma cells that function as the site of photosynthesis. Typically (but not always), they are of two morphological types: (1) columnar **palisade meso-**

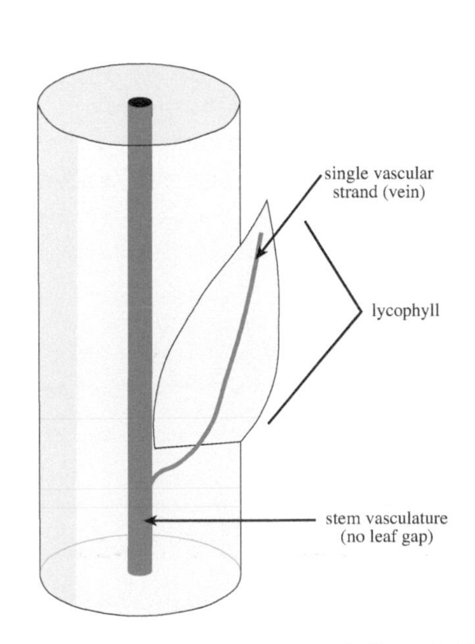

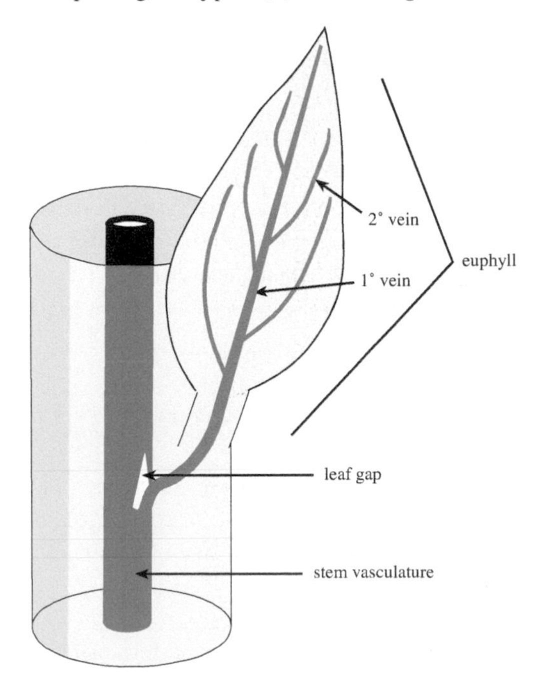

Figure 10.23 **A.** Lycophyll. **B.** Euphyll. Note leaf gap and branched venation.

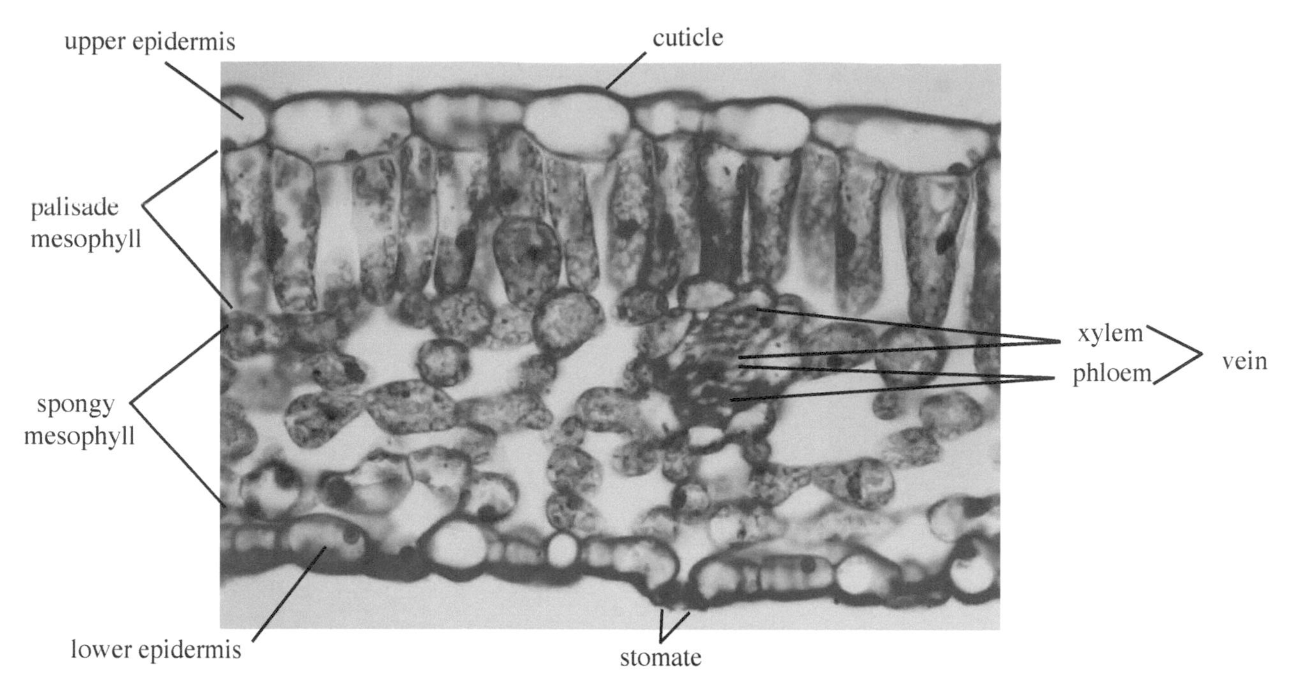

Figure 10.24 Cross section of a typical vascular plant leaf.

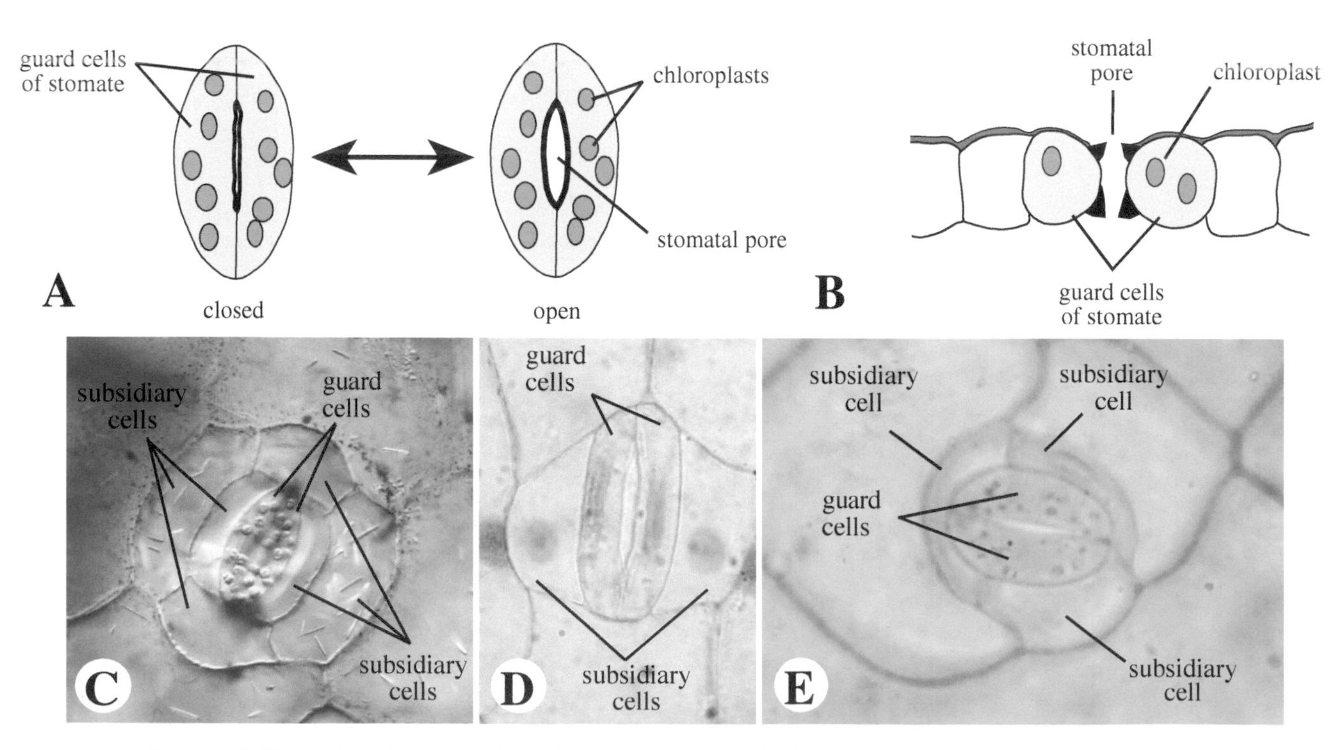

Figure 10.25 Stomata. **A.** Illustration of stomate in face view, closed and open. **B.** Stomate in cross-sectional view. **C–E.** Stomates of various taxa, showing the differences in subsidiary cells.

phyll cells, which occur in the upper (adaxial) region and have relatively small intercellular spaces; and (2) irregularly shaped **spongy mesophyll** cells, which occur in the lower (abaxial) region and have large intercellular spaces. The **veins** of a leaf have the anatomy of typical vascular bundles. In almost all veins the xylem is oriented to the adaxial side, phloem to the abaxial side, corresponding to their orientation in the stem. Veins may very often have a ring of cells surrounding the xylem and phloem called a **vascular bundle sheath**. This sheath may be composed of fiber cells, which function in structural support of the vein within the leaf tissue, or of parenchyma cells. The parenchymatous, chloroplast-containing bundle sheath cells of some plants function in C4 photosynthesis (discussed later).

PHOTOSYNTHESIS

The tremendous importance of plants is directly related to the photosynthetic process. Photosynthesis occurs by the fixation of carbon dioxide in the following net reaction: $nCO_2 + nH_2O \rightarrow (CH_2O)n + nO_2$. Interestingly, this net reaction actually occurs via two series of interdependent reactions: light reactions and dark reactions. In the **light reactions**, which occur within the thylakoid membranes and require photons of light, water (H_2O) is broken down into hydrogen ions (H^+), electrons (–), and molecular oxygen (also called dioxygen, O_2). This splitting of water molecules occurs via a complex series of enzymes and cofactors embedded within the thylakoid membranes of the chloroplast (Figure 10.26). The hydrogen ions resulting from the splitting of water become concentrated in the space within the thylakoids. These hydrogen ions are transported across the thylakoid membrane into the outer region called the stroma; that transport results in a net transfer of energy, used to synthesize a high-energy molecule of ATP (adenosine triphosphate; Figure 10.26). The electrons produced by the splitting of water are also transported across the thylakoid membrane to the stroma. There, the electrons react with hydrogen ions and a compound called $NADP^+$ (nicotinamide adenine dinucleotide phosphate) to produce a higher energy product, NADPH.

In the **dark reactions** (or Calvin cycle) atmospheric carbon dioxide (CO_2) makes its way into the stroma of the chloroplast, where it reacts with a five-carbon molecule to form two molecules, each containing three carbon atoms; hence, photosynthesis in these plants is called **C3 photosynthesis** and the plants are called **C3 plants** (Figure 10.26). This initial binding, or *fixation*, of CO_2 is catalyzed by a very important enzyme called **ribulose-bisphosphate carboxylase** (**RuBP-carboxylase**, which is thought to be the enzyme with the greatest worldwide biomass). The two three-carbon molecules then undergo a series of further reactions, each catalyzed by a separate enzyme, to ultimately produce a net molecule of glucose. The chemical reactions resulting in glucose production require the input of high-energy compounds, notably ATP and NADPH. As these compounds are converted into lower energy products in the dark reactions, they are regenerated in the light reactions. Thus, light and dark reactions are interdependent; each comes to a halt without the concerted action of the other (Figure 10.26).

In some species of vascular plants, the parenchymatous bundle sheath cells function in a different type of photosynthesis called **C4 photosynthesis** (Figure 10.27). In C4 plants carbon dioxide is initially fixed in the mesophyll cells by a different enzyme, **PEP carboxylase**. The initial molecule of carbon fixation is a four-carbon molecule, which, in the form of malic acid, is then transported to the bundle sheath cells. Chloroplasts of the bundle sheath cells are typically much larger than those of the mesophyll cells, this type of anatomy termed **Kranz anatomy** (Figure 10.28). In the bundle sheath cells, the carbon dioxide is released and fixed by the typical (and ancestral) enzyme, **ribulose-bisphosphate carboxylase** (RuBP-carboxylase). C4 photosynthesis actually requires more energy (one more ATP per CO_2 molecule) than C3 photosynthesis. However, C4 photosynthesis has apparently been selected for in plants growing under conditions of high light intensity or drought. Under water-stressed conditions, the stomata of plants generally remain closed to inhibit excess water loss, but this also inhibits the flow of CO_2 into the leaf. The enzyme PEP carboxylase has a much greater affinity for CO_2 molecules than does the enzyme RuBP carboxylase. Thus, under conditions of low CO_2 (occurring under drought conditions), the initial fixation of CO_2 is much more efficient in C4 plants than in C3 plants. By fixing, transporting, and releasing CO_2 into the bundle sheath cells, it becomes more concentrated than in the mesophyll cells and can more readily be catalyzed by RuBP carboxylase in the Calvin Cycle. C4 photosynthesis has evolved in a number of angiosperms, one common example being corn (*Zea mays*, Figure 10.28).

C4 photosynthesis is found only in angiosperms, in about 8000 members of 17 families (Table 10.1), constituting about 3% of all land plants. C4 photosynthesis is particular common in the Cyperaceae (sedges) and Poaceae (grasses), together making up about 79% of all C4 plants (Sage et al. 1999). There are three biochemical subtypes of C4 photosynthesis, based on the enzyme used to decarboxylate (remove CO_2 from) the C4 acid of the bundle sheath cells (Sage et al. 1999). Interestingly, C4 photosynthesis has been calculated to have evolved a minimum of 31 times (Kellogg 1999).

Another different mechanism of photosynthesis is **CAM: crassulacean acid metabolism** (Figure 10.29), named after the family Crassulaceae. CAM plants are often succulents.

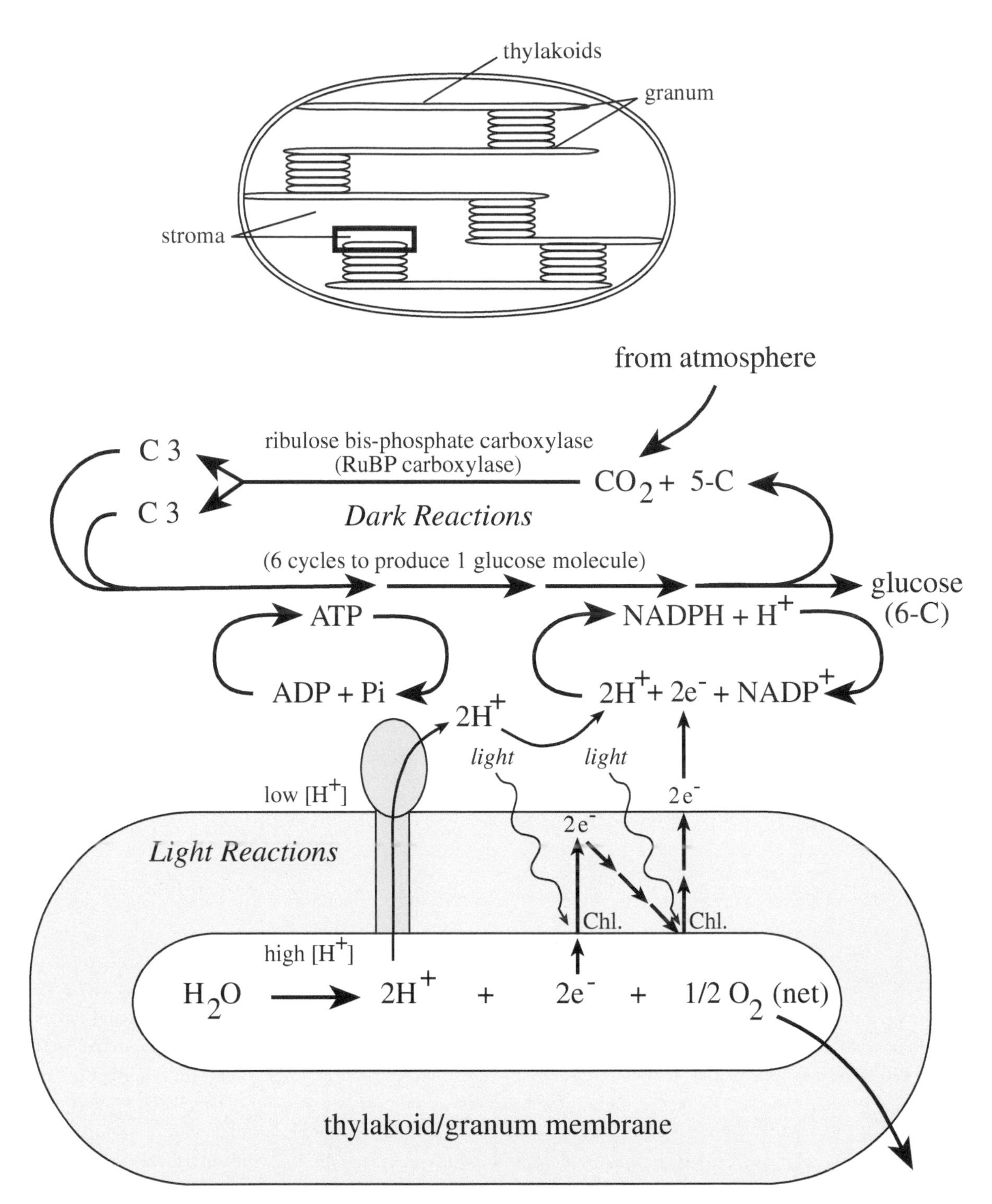

Figure 10.26 Photosynthesis (C3). ADP = adenosine diphosphate; Pi = phosphate. See text for other abbreviations.

As with C4 plants, CAM plants are generally adapted to xeric conditions. CAM photosynthesis may also be adaptive in minimizing water loss due to evapotranspiration (although some CAM plants are aquatic; see later discussion). In CAM plants the initial fixation of CO_2 occurs at night, when (unlike other plants) stomata are open. The CO_2 is initially fixed by the enzyme **PEP carboxylase** to form malic acid, which is temporarily stored within vacuoles of the mesophyll cells. (This is experimentally detected by a lowering of the pH.) During the day the stomata close and CO_2 is released from the vacuoles into the cytoplasm, where it is fixed in the chloroplasts by the Calvin cycle.

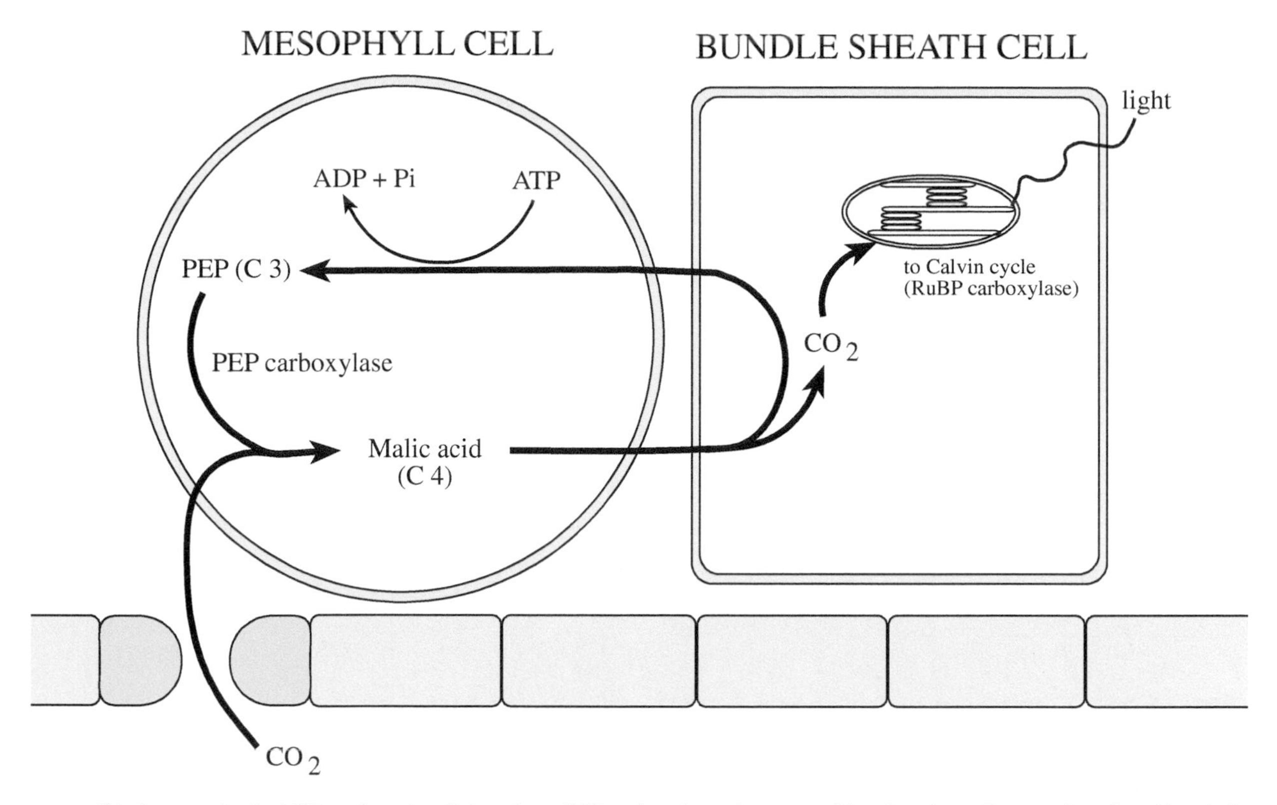

Figure 10.27 C4 photosynthesis. ADP = adenosine diphosphate; PEP = phosphoenolpyruvate; Pi = phosphate. See text for other abbreviations.

The number of CAM species is estimated at up to 20,000, in about 40 plant families (Table 10.2), constituting roughly 8% of all land plants. CAM photosynthesis is found in a number of land plant groups, including some lycophytes (Isoetes), leptosporangiate ferns, cycads, Gnetales, and many angiosperms, especially desert succulents (e.g., in Aizoaceae, Cactaceae, Portulacaceae, Crassulaceae, and Euphorbiaceae), some epiphytic plants (e.g., in Bromeliaceae and Orchidaceae), and even aquatic groups (e.g., in Crassulaceae, Isoetaceae, and Plantaginaceae). See Smith and Winter (1996) for information

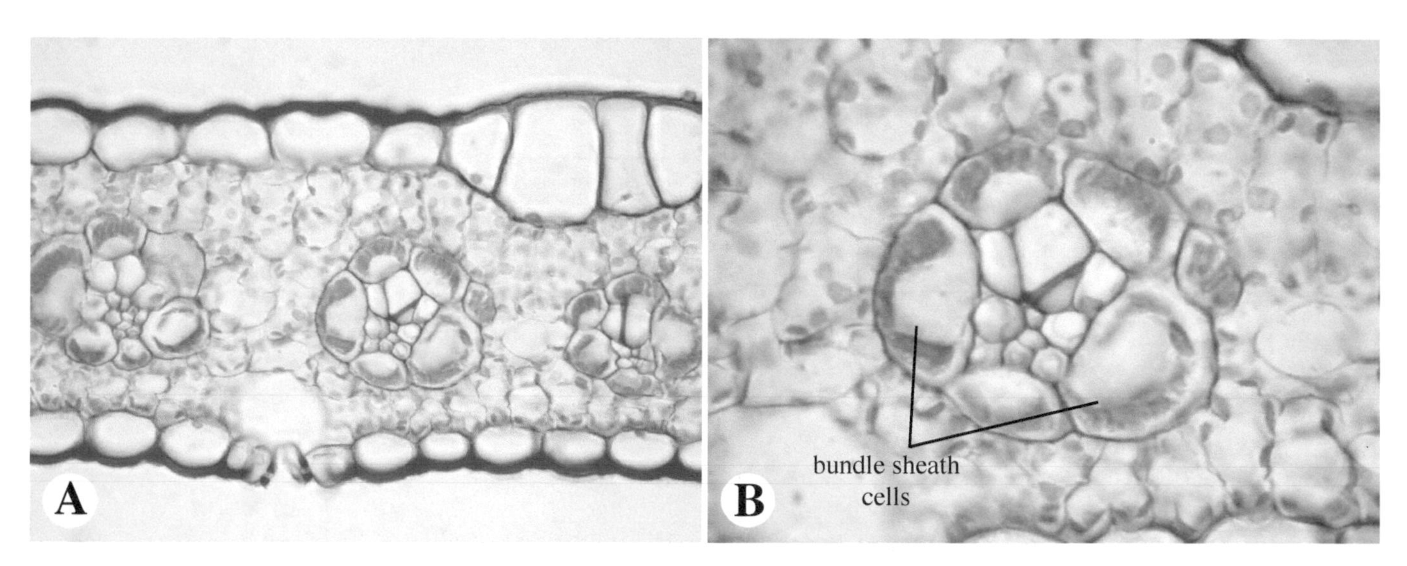

Figure 10.28 Kranz anatomy, illustrated by *Zea mays* leaf cross section. **A.** Low magnification. **B.** Close-up of vascular bundle with enlarged bundle sheath cells, surrounded by mesophyll cells.

Monocots	**Asterids**
Alismatales	Asterales
Hydrocharitaceae	Asteraceae
Poales	Boraginales
Cyperaceae	Boraginaceae
Poaceae	Lamiales
Eudicots	Acanthaceae
Caryophyllales	Scrophulariaceae
Aizoaceae	**Rosids**
Chenopodiaceae	Brassicales
Caryophyllaceae	Capparidaceae
Molluginaceae	Malpighiales
Nyctaginaceae	Euphorbiaceae
Polygonaceae	Zygophyllales
Portulacaceae	Zygophyllaceae

TABLE 10.1 Angiosperm families, listed within orders and higher groups, having at least one member with C4 photosynthesis. After Sage et al. (1999).

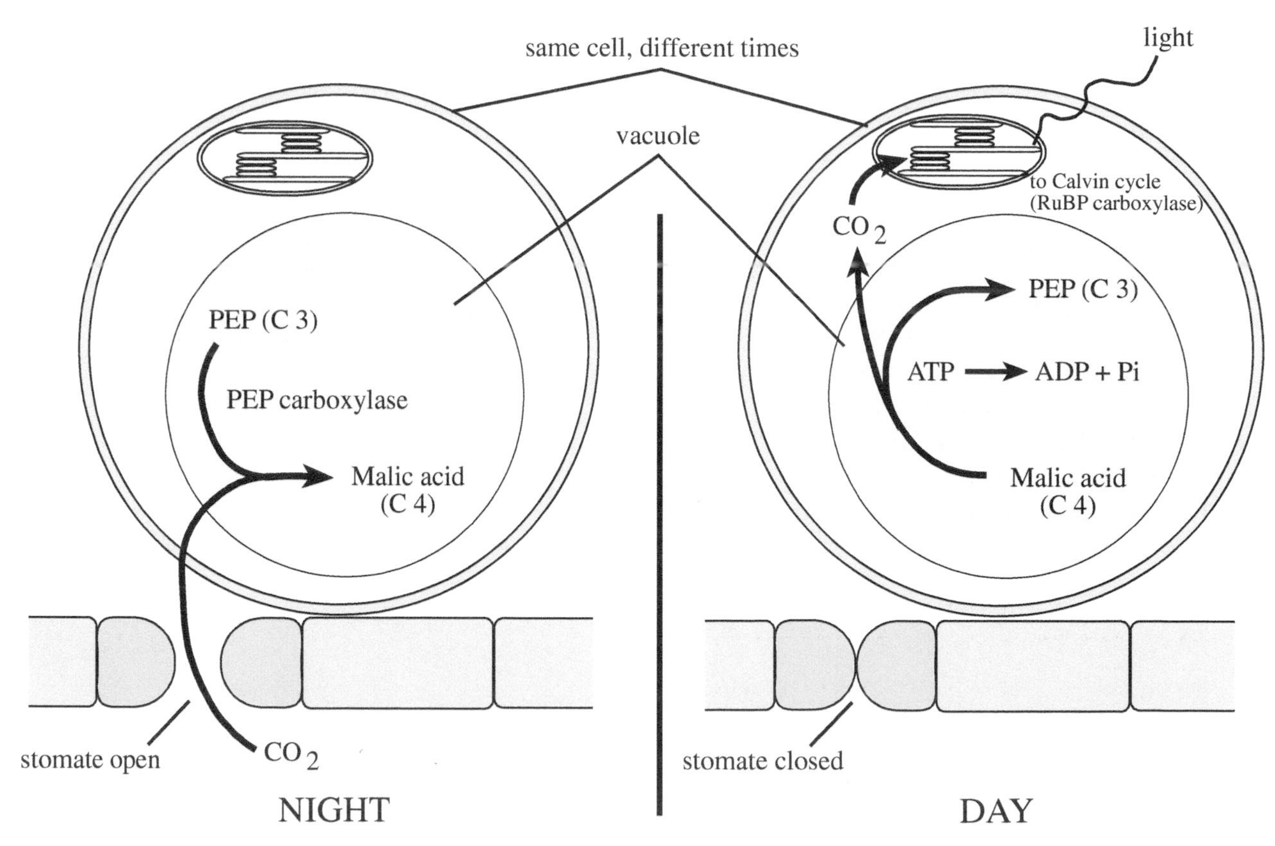

Figure 10.29 CAM photosynthesis. ADP = adenosine diphosphate; PEP = phosphoenolpyruvate; Pi = phosphate.

Lycopodiophyta
- Isoetaceae (*Isoetes*)

Polypodiopsida
- Polypodiaceae
- Pteridaceae

Cycadophyta
- Zamiaceae (*Dioon edule*)

Gnetales
- Welwitschiaceae (*Welwitschia mirabilis*)

Angiosperms
- Piperales
    - Piperaceae
- **Monocots**
    - Alismatales
        - Araceae
        - Alismataceae
        - Hydrocharitaceae
    - Asparagales
        - Agavaceae
        - Asphodelaceae
        - Orchidaceae
        - Ruscaceae
    - Commelinales
        - Commelinaceae
    - Poales
        - Bromeliaceae
        - Cyperaceae

Eudicots

Superosids
- Saxifragales
    - Crassulaceae
- Vitales
    - Vitaceae
- **Rosids**
    - Brassicales
        - Moringaceae
    - Celastrales
        - Celastraceae
    - Cucurbitales
        - Begoniaceae
        - Cucurbitaceae
    - Geraniales
        - Geraniaceae
    - Malpighiales
        - Clusiaceae
        - Euphorbiaceae
        - Passifloraceae
    - Oxalidales
        - Oxalidaceae
    - Sapindales
        - Sapindaceae
    - Zygophyllales
        - Zygophyllaceae

Superasterids
- Caryophyllales
    - Aizoaceae
    - Cactaceae
    - Didiereaceae
    - Portulacaceae
- **Asterids**
    - Apiales
        - Apiaceae
    - Asterales
        - Asteraceae
    - Ericales
        - Ebenaceae
    - Gentianales
        - Apocynaceae
        - Rubiaceae
    - Lamiales
        - Gesneriaceae
        - Lamiaceae
    - Solanales
        - Convolvulaceae

TABLE 10.2 Angiosperm families, listed within orders and higher groups, having at least one member with CAM photosynthesis. After Smith and Winter (1996).

on the taxonomic distribution of CAM species in the land plants.

CAM and C4 photosynthesis are very similar to one another. Both involve initial fixation of CO_2 utilizing the enzyme PEP carboxylase and final fixation of CO_2 with RuBP carboxylase. The essential difference between the two is that initial and final CO_2 fixation differ *spatially* in C4 plants (mesophyll versus bundle sheath cells) and *temporally* in CAM plants (night versus day).

ANATOMY AND SYSTEMATICS

Plant anatomy can provide valuable characteristics in phylogenetic analyses, but these are less frequently acquired today than in the past. However, anatomical features, whether used directly to generate a cladogram or merely traced on an existing cladogram, can give insight into major adaptational shifts. In that sense, they are quite important in understanding different selective pressures.

A summary of major anatomical apomorphies for the land plants is seen in Figure 10.30, taken from the cladograms of Chapters 3–6. As can be seen, many of the apomorphies discussed and presented in these chapters are anatomical. Anatomical and physiological traits are worthy of study at a lower taxonomic level as well, and are often correlated with adaptational strategies and ecological shifts.

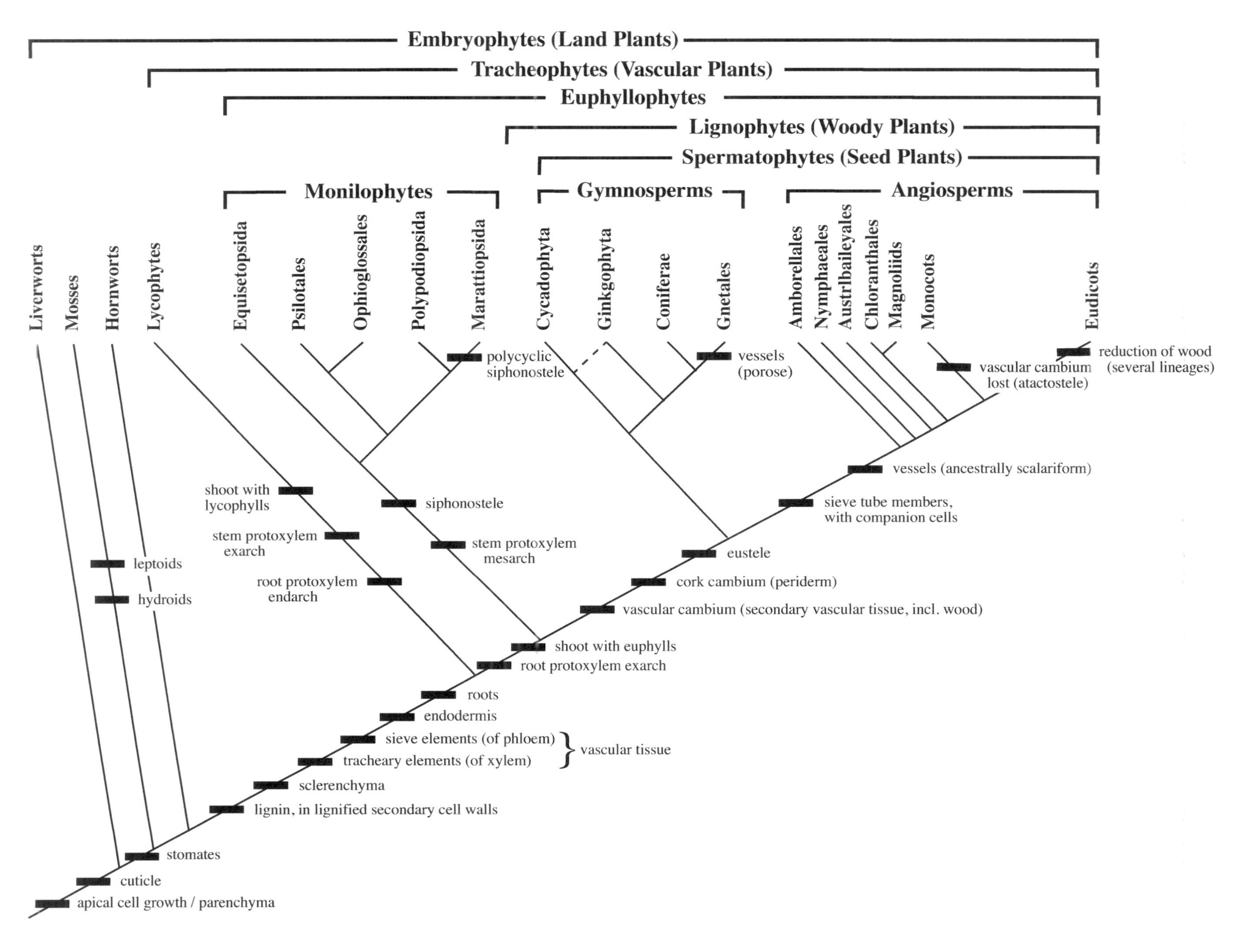

Figure 10.30 Summary cladogram of Land Plants, showing major anatomical apomorphies.

PLANT ANATOMY TECHNIQUE

Material dissection and preparation:

A wealth of information can be gained by careful dissection and observation of plants. Look first at the outer form of the plant, noting the basic plant organs (root, stem, leaves, buds, flowers, fruits) and specific aspects of these organs. Gently pull apart the plant organs to better see their morphology. For flowers and fruits, use both your hands and naked eye and dissecting needles under a dissecting scope to examine the components.

Careful anatomical studies usually involve time-consuming embedding and microtome sectioning. However, a simple technique of hand sectioning with a razor blade will allow you to see considerable detail of cell and tissue anatomy. Stout material, such as an herbaceous stem, can be held *upright* in the left hand between thumb and index finger (assuming you are right-handed). More flimsy material, such as a leaf, can be sandwiched between two small pieces (cut only slightly larger than the material) of Styrofoam; the end is moistened and both Styrofoam and plant material are sectioned together. In either case, rest the side of the razor blade on your index finger and position your thumb a bit lower (so that if you do slip, you won't cut yourself). There are tricks of the trade to successful sections:

1. As you cut, move the razor blade toward you, as well as across the material; thus, the cut is somewhat diagonal.
2. Make an initial cut to level off (discarding this piece) and then make several thin slices, keeping the sections on the razor blade until they get too crowded; then, transfer the sections to water in a Syracuse dish or Petri plate. Clean your razor blade and make a few more sections.
3. Select the thinnest sections, pull out with a brush, and place in a few drops of stain in another dish. After staining, rinse your sections very briefly in water and place in a drop of water or (for a semipermanent mount) 50% glycerol. Cover with a cover slip, avoiding air bubbles and adding more fluid to the side if necessary.

Most important is to make those sections THIN!! Although you will want at least one complete section, other sections may be partial, as long as they are thin. Clean your razor blade afterward and you may reuse it.

For tough, fibrous, or woody tissue place the material down on a plastic Petri plate and make downward slices with your razor blade. This same technique can be used with softer, small plant material if it is sandwiched between two layers of Parafilm and the material sectioned in a "dicing" motion.

The following are some "vital" stains (i.e., used with live material):

Stain	Compound for which stain is specific	Color
Alcian blue	Pectins	Blue
Aniline blue	Callose	Blue (UV-fluoresces Yellow)
IKI	Starch	Blue to black
Phloroglucinol/HCl	Lignin	Red (NOTE: Takes sev. mins. to react)
Sudan III or Sudan IV	Oil droplets	Reddish
Toluidine blue	Metachromatic (will stain a variety of cell walls different shades of blue/green):	
	Lignified tracheary elements	Dark blue
	Sclerenchyma	Blue to blue-green
	Parenchyma	Light blue
	Collenchyma	Reddish-purple
	Sieve tubes and companion cells	Greenish
	Callose/starch	Unstained

Drawings:

Making careful drawings not only gives you a record of what you observe, it also helps you become a careful observer. When "forced" to draw it, you often see more than you otherwise would. Make drawings with a #2 or #3 hard lead pencil. Draw the outlines of organs or tissues (e.g., of a root cross section) at low magnification to record the overall structure. Then draw a portion of the whole (e.g., a "pie slice" of the root section, showing some of the individual cells of a vascular bundle) to show details.

REVIEW QUESTIONS

PLANT CELL STRUCTURE AND PLANT GROWTH

1. What is plant anatomy and how does it differ from animal anatomy?
2. What is the cell theory and its four tenets?
3. Name and give the function of the major components and organelles of a typical plant cell.
4. Name the various types of ergastic substances.
5. What are chromoplasts and what is their function?
6. What is an amyloplast and what is its chemical composition?
7. Of what is an aleurone grain composed?
8. Name four types of crystals based on their shape. What two different substances make up plant crystals?
9. Characterize a primary cell wall in terms of development and structure.
10. How does cellulose differ from starch?
11. What are plasmodesmata and what is their function?
12. Where is the secondary cell wall formed in relation to the plasma membrane and the primary cell wall?
13. What are the name, properties, and function of the compound (other than cellulose) making up a secondary cell wall?
14. What is a pit and what is the function of pit pairs?
15. What are meristems and what are the two major types?
16. Explain the processes of cell differentiation.

PLANT TISSUES AND SPECIALIZED CELL TYPES

17. What is a tissue and what are the three general tissue types?
18. How are parenchyma and collenchyma similar and how different with respect to structure and function?
19. What are the characteristics and two general cell types of sclerenchyma?
20. What is the function of xylem and why is it a complex tissue?
21. What are the names and characteristics of the two types of water-conductive cells of xylem?
22. What is the function of phloem and why is it a complex tissue?
23. What are the names and characteristics of the two types of sugar-conductive cells of phloem?
24. What tissue occurs as the outermost cell layer of plant organs?
25. Describe the characteristics and function of the epidermis, stomata, trichomes, and secretory structures.

ROOT STRUCTURE AND FUNCTION

26. From what in an embryo does the first root arise?
27. Distinguish between a primary, lateral, and adventitious root.
28. Name three ways that roots can be distinguished from shoots/stems.
29. What is a Casparian strip and endodermis and what are their function?
30. What is the function of the pericycle?
31. What is a velamen?

SHOOT/STEM STRUCTURE AND FUNCTION

32. What is the definition of a shoot?
33. From what in an embryo does the first shoot arise?
34. What is a bud primordium and where are buds typically located?
35. What are three ways that stems differ from roots?
36. What is a stele? Name five stele types and distinguish between them.
37. Distinguish between protoxylem and metaxylem; between exarch, endarch, and mesarch.

SECONDARY GROWTH

38. What is secondary growth and from what general type of meristem does it arise?
39. Where does the vascular cambium arise?
40. What two products does the vascular cambium give rise to and in what direction?
41. What is the technical name for wood?
42. Where does the cork cambium form, and what two tissues does it give rise to?
43. Describe the adaptive significance of the lateral meristems.
44. What is a ray and what is its function?
45. What is an annual ring and what is the structural difference between spring wood and summer wood?
46. Define and draw the three major sections of wood.
47. What is the difference between a softwood and a hardwood?
48. Distinguish between nonporous, ring-porous, and diffuse-porous wood.
49. Distinguish between uniseriate, biseriate, and multiseriate rays.
50. What is dendrochronology and for what can it be used?

LEAF STRUCTURE AND FUNCTION

51. What is the difference between a lycophyll and a euphyll?
52. What are the structure and function of stomates and of subsidiary cells?
53. What is the name of the leaf cells located between upper and lower epidermal layers? What are the two types called?

PHOTOSYNTHESIS

54. Describe the basic pathway of C3 photosynthesis.
55. What enzyme functions to fix carbon dioxide in C3 photosynthesis?
56. How does C4 photosynthesis differ from C3?
57. What is Kranz anatomy?
58. What enzyme functions to fix carbon dioxide in C4 photosynthesis?
59. How does CAM photosynthesis differ from C3 and C4 and how does this function for plants living in dry conditions?

ANATOMY AND SYSTEMATICS

60. Draw a general cladogram of land plants, illustrating several anatomical apomorphies.

EXERCISES

1. Obtain live material of a plant species and prepare hand sections of the root, stem, and leaf, if feasible. Stain these with the appropriate stain (see PLANT ANATOMY TECHNIQUE), and describe all the cell and tissue types. Note the differences between the three organs.
2. Obtain live material of the leaves of a few monocot and eudicot species. Prepare epidermal peels of the leaves and note the differences between the stomata and subsidiary cells. Can you determine a correlation with taxonomic group?
3. Observe the trichomes of various plant organs (e.g., leaves, stem axes, or flower parts) by peeling the epidermal tissue bearing the trichomes or scraping them from the surface. Place this material on a microscope slide in a drop of water or (to preserve for some time) 50% glycerol. The material may be stained with, e.g., toluidine blue. Carefully draw the various trichome types. Is the trichome anatomy the same from organ to organ or does it vary? What might be the adaptive significance of trichomes?
4. Peruse journal articles in plant systematics, e.g., *American Journal of Botany*, *Systematic Botany*, or *International Journal of Plant Sciences*, and note those that describe plant anatomical features in relation to systematic studies. Identify all anatomical characters and character states used.

REFERENCES FOR FURTHER STUDY

Ayensu, E. S. 1972. Dioscoreales. In: Anatomy of the Monocotyledons. Volume 6. Clarendon Press. Oxford.

Carlquist S. 2001. Comparative wood anatomy, 2nd ed. Springer-Verlag, Berlin.

Cutler, D. F. 1969. Juncales. In: Anatomy of the Monocotyledons. Volume 4. Clarendon Press. Oxford.

Crang, R., S. Lyons-Sobaski, and R. Wise. 2019. Plant Anatomy: A Concept-Based Approach to the Structure of Seed Plants. Springer International Publishing AG, Cham, Switzerland.

Dengler, N. G., and T. Nelson. 1999. Leaf structure and development in C4 plants. In: R. F. Sage and R. K. Monson [eds.], C4 plant biology. Pp. 133–172. Academic Press, San Diego, California, USA.

Dickison, W. C. 2000. Integrative Plant Anatomy. Harcourt/Academic Press, New York.

Esau, K. 1965. Plant Anatomy, 2nd ed. John Wiley and Sons, New York.

Esau, K. 1977. Anatomy of Seed Plants, 2nd ed. John Wiley and Sons, New York.

Evert, R. F. and S. E. Eichhorn. 2006. Esau's Plant Anatomy: Meristems, Cells, and Tissues of the Plant Body: Their Structure, Function, and Development. 3rd Edition. John Wiley & Sons, Inc., Hoboken, New Jersey.

Fahn, A. 1982. Plant Anatomy. 3rd ed. Pergamon Press, Oxford.

Keating, R. C. 2003. Acoraceae and Araceae. In: Anatomy of the Monocotyledons. Volume 9. Clarendon Press. Oxford.

Keeley, J. E., and R. W. Rundel. 2003. Evolution of CAM and C4 carbon-concentrating mechanisms. International Journal of Plant Sciences 164: S55-S77.

Kellogg, E. A. 1999. Phylogenetic aspects of the evolution of C4 photosynthesis. In: R. F. Sage and R. K. Monson (eds.). C4 Plant Biology. Pp. 411-444. Academic Press, San Diego.

Mauseth, J. D. 1988. Plant Anatomy. Benjamin Cummings.

Metcalfe, C. R. 1960. Gramineae. In: Anatomy of the Monocotyledons. Volume 1. Clarendon Press. Oxford.

Metcalfe, C. R. 1963. Comparative Anatomy as a Modern Botanical Discipline: with Special Reference to Recent Advances in the Systematic Anatomy of Monocotyledons. Academic Press, London, New York.

Metcalfe, C. R. 1971. Cyperaceae. In: Anatomy of the Monocotyledons. Volume 5. Clarendon Press. Oxford.

Metcalfe, C. R. and L. Chalk. 1979. Systematic anatomy of leaf and stem, with a brief history of the subject. In: Anatomy of the Dicotyledons. 2nd edition. Volume 1. Clarendon Press, Oxford.

Metcalfe, C. R. and L. Chalk. 1983. Wood structure and conclusion of the general introduction. In: Anatomy of the Dicotyledons. 2nd edition. Volume 2. Clarendon Press, Oxford.

Metcalfe, C. R. and L. Chalk. 1987. Magnoliales, Illiciales, and Laurales (sensu Armen Takhtajan). In: Anatomy of the Dicotyledons. 2nd edition. Volume 3. Clarendon Press, Oxford.

Muhaidat, R., R. F. Sage, and N. G. Dengler. 2007. Diversity of Kranz anatomy and biochemistry in C4 eudicots. American Journal of Botany 94: 362–381. 2007.

Rudall, P. 1995. Iridaceae. In: Anatomy of the Monocotyledons. Volume 8. Clarendon Press. Oxford.

Rudall, P. J. 2007. Anatomy of Flowering Plants: An Introduction to Structure and Development. 3rd ed. Cambridge University Press, New York.

Sage, R. F., M. Li, and R. K. Monson. 1999. The taxonomic distribution of C4 photosynthesis. In: R. F. Sage and R. K. Monson (eds.). C4 Plant Biology. Pp. 551-584. Academic Press, San Diego.

Smith, J. A. C., and K. Winter. 1996. Taxonomic distribution of crassulacean acid metabolism. In: K. Winter, J. A. C. Smith, eds. Crassulacean Acid Metabolism: Biochemistry, Ecophysiology and Evolution. Pages 427–436. Springer, New York.

Tomlinson, P. B. 1961. Palmae. In: Anatomy of the Monocotyledons. Volume 2. Clarendon Press. Oxford.

Tomlinson, P. B. 1969. Commelinales-Zingiberales. In: Anatomy of the Monocotyledons. Volume 3. Clarendon Press. Oxford.

Tomlinson, P. B. 1982. Helobiae (Alismatidae). In: Anatomy of the Monocotyledons. Volume 7. Clarendon Press. Oxford.

11

PLANT EMBRYOLOGY

Plant embryology is the study of the development of sporangia, gametophytes, and embryos in the land plants, the embryophytes. Among the seed plants, the spermatophytes, embryological studies encompass the development of microsporangia (within anthers in the angiosperms), microspores, pollen grains, ovules, megaspores, female gametophytes, and seeds. Because most plant embryological data have been acquired and utilized for the flowering plants, this chapter focuses on processes and terminology for the angiosperms.

As characters used in phylogenetic studies, plant embryological data are generally most useful at higher taxonomic levels, as in the characterization of traditional plant families. However, the data may be useful at any taxonomic level.

ANTHER AND POLLEN DEVELOPMENT

ANTHER TYPE

In the angiosperms an important embryological character, one often treated as a standard morphological character, is the number of microsporangia per anther. Microsporangia are typically tubular in shape and occur in pairs, which coalesce during development by the breakdown of the cell layers between them. Each pair of microsporangia is termed a **theca**. The great majority of angiosperm species have anthers composed of two thecae, termed **bithecal** or **tetrasporangiate** (Figure 11.1A), which is the ancestral condition. However, some angiosperm taxa, such as the Malvaceae, Cannaceae, Marantaceae, and species of *Salvia* (Lamiaceae), have a derived anther type with only one theca, termed **monothecal** or **bisporangiate** (Figure 11.1B).

ANTHER WALL DEVELOPMENT

A cross section of an anther reveals a division between the internal microsporangium, the cells of which undergo meiosis, and an outer anther wall. The development of the anther wall has provided some useful embryological features. A mature anther wall consists of few to several layers of cells. The outermost cell layer (just inside the epidermis) is termed the **endothecium**, which typically consists of enlarged cells with secondary wall thickenings functioning in anther dehiscence. The secondary wall thickenings function by providing tensile force that pulls back the anther walls from the line or region of dehiscence. The innermost cell layer is termed the **tapetum**, which consists of metabolically active cells that function in the development of pollen grains. Additional

https://doi.org/10.1016/B978-0-12-812628-8.50011-0,

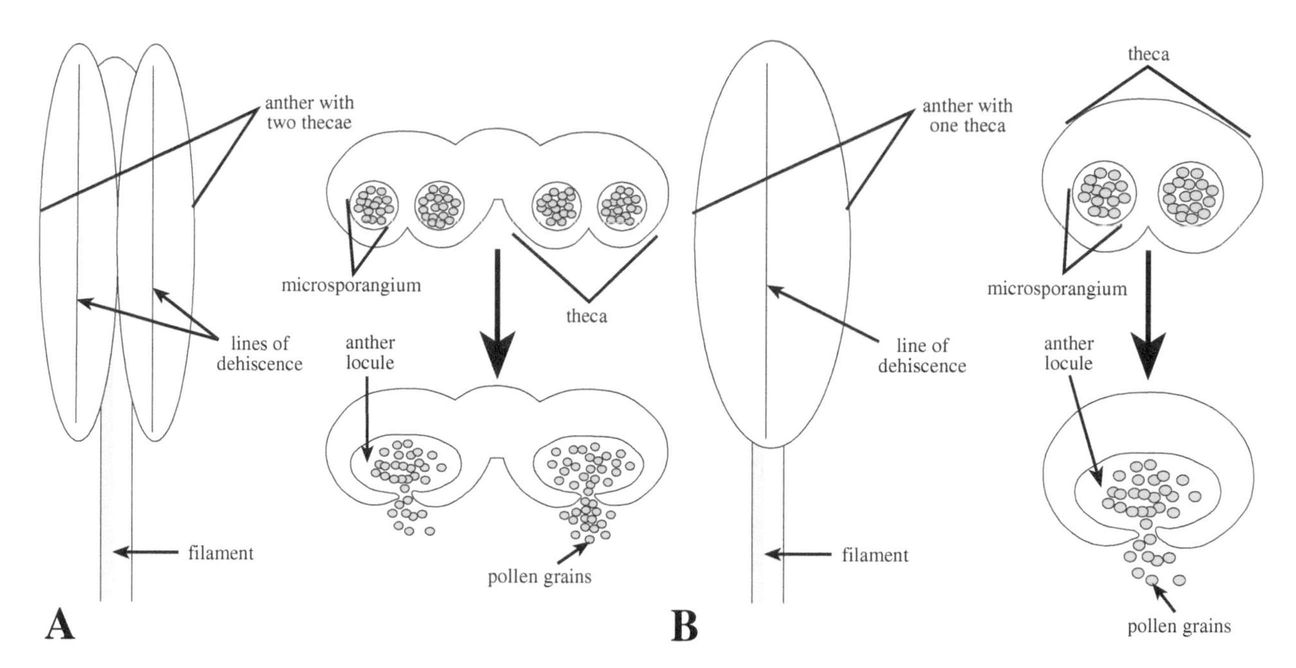

Figure 11.1 Anther types in the angiosperms. **A.** Dithecal. **B.** Monothecal.

wall layers, termed middle layers, may occur between the endothecium and tapetum. Both the total number of wall layers and their developmental origin define various anther wall types. Early in development an anther contains *two* layers of cells, an outer epidermis and an inner layer of **primary parietal cells**. Cells of the primary parietal layer divide tangentially (parallel to the outer surface) to give rise to two layers of cells, **secondary parietal cells**. Based on the derivation of cell lineages, four general types of anther wall development have been defined (Figure 11.2): (1) **basic**, in which both secondary parietal cell layers divide to yield two middle layers; (2) **dicotyledonous**, in which only the *outer* secondary parietal cell layer divides to yield the endothecium and a single middle layer; (3) **monocotyledonous**, in which

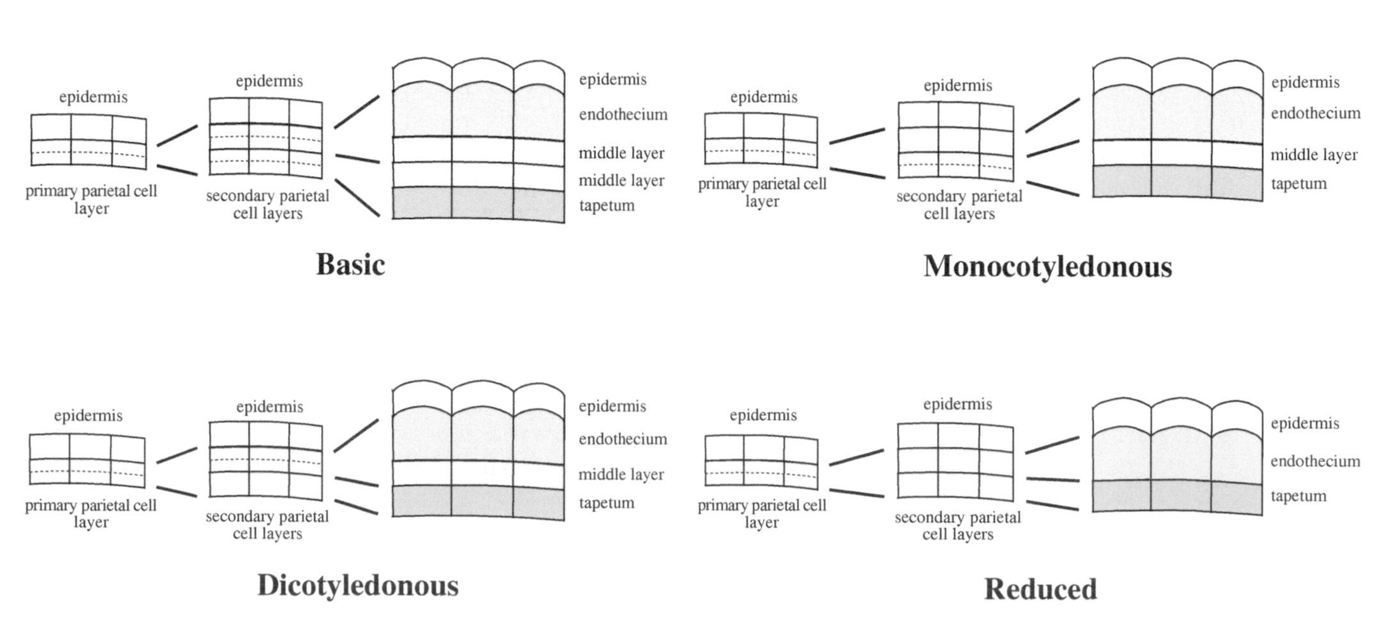

Figure 11.2 Anther wall development, outer epidermis at top.

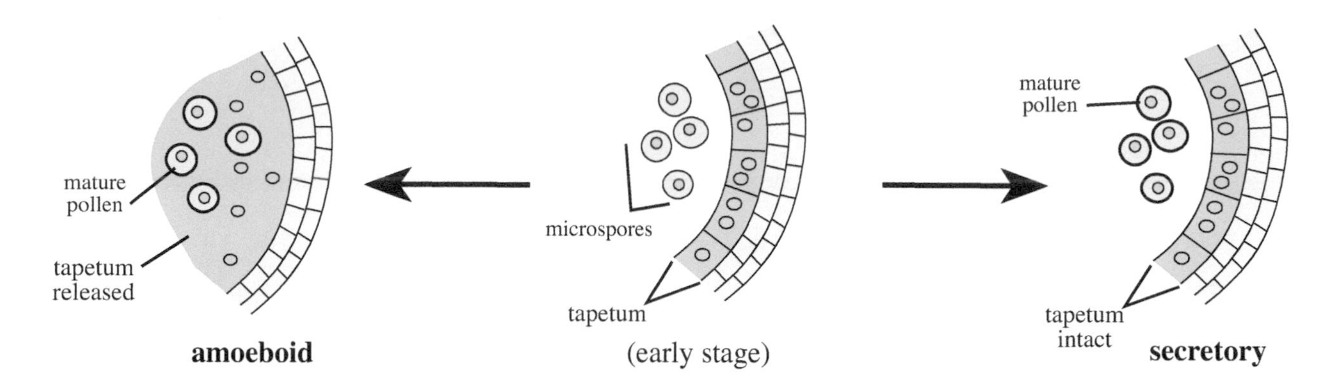

Figure 11.3 Tapetum types.

only the *inner* secondary parietal cell layer divides to yield the tapetum and a single middle layer; and (4) **reduced**, in which the secondary parietal cells do not divide further and develop directly into the endothecium and tapetum, respectively.

Another embryological character concerns the development of the tapetum, with two basic types defined (Figure 11.3). In some angiosperms the tapetum remains intact with no breakdown of cell walls. This tapetal type is called **secretory** (or glandular; Figure 11.4A,B) because of the implication that compounds are secreted into the locule of the anther that function in pollen development. In other angiosperm taxa the tapetal cell walls break down, with release of the cytoplasm of the tapetal cells into the locule. This latter tapetal type is called **amoeboid** (plasmodial or periplasmodial; Figure 11.4C,D) because the cytoplasmic contents surround developing pollen grains like an amoeba surrounds food. Subtypes of the secretory and amoeboid tapetal types have been proposed by some, based on fine developmental differences.

A final embryological character dealing with the anther wall is endothecial anatomy. Two basic types of endothecial cells have been defined based on the structure of the secondary wall thickenings. A **girdling** endothecium is one in which the secondary wall thickenings form rings with cross bridges between them (Figure 11.5). A **spiral** endothecium is one in which the secondary wall thickenings are spiral or helical in shape.

POLLEN DEVELOPMENT

Development of microspores from microsporocytes is termed **microsporogenesis**. There are two basic types of microsporogenesis as determined by the timing of cytokinesis, which is the formation of a plasma membrane and cell wall that divides one cell into two (Figure 11.6A). If cytokinesis occurs after meiosis I, then microsporogenesis is **successive** (Figure 11.6B). Successive microsporogenesis results in two cells after meiosis I and four cells after meiosis II. If cytokinesis does not occur until after meiosis II, then microsporogenesis is **simultaneous** (Figure 11.6C). Simultaneous microsporogenesis results in cell formation only after meiosis II.

Development of pollen grains (male gametophytes) from microspores is called **microgametogenesis**, technically beginning with the first mitotic division of the single microspore nucleus. One embryological character concerning microgametogenesis is the number of nuclei present in the pollen grain at the time of anthesis, or flower maturation (Figure 11.7). Most angiosperms have pollen grains that are **binucleate** (Figure 11.7), containing one tube cell/nucleus and one generative cell/nucleus. The generative cell divides to form two sperm cells only after pollen tube formation. In many angiosperm taxa, however, the pollen at anthesis is **trinucleate** (Figure 11.7), caused by division of the generative cell prior to pollen release.

OVULE DEVELOPMENT

The development of the ovule provides a number of significant embryological characters used in plant systematics studies.

OVULE PARTS

Ovules are immature seeds, consisting of a stalk, the **funiculus**, a **megasporangium** (also called the **nucellus**), from which develops the megasporocyte and female gametophyte, plus one or two surrounding **integuments**. The **micropyle** is the pore or canal within one or more integuments through which (in angiosperms) a pollen tube traverses prior to fertilization. (In nonangiospermous seed plants, the micropyle receives

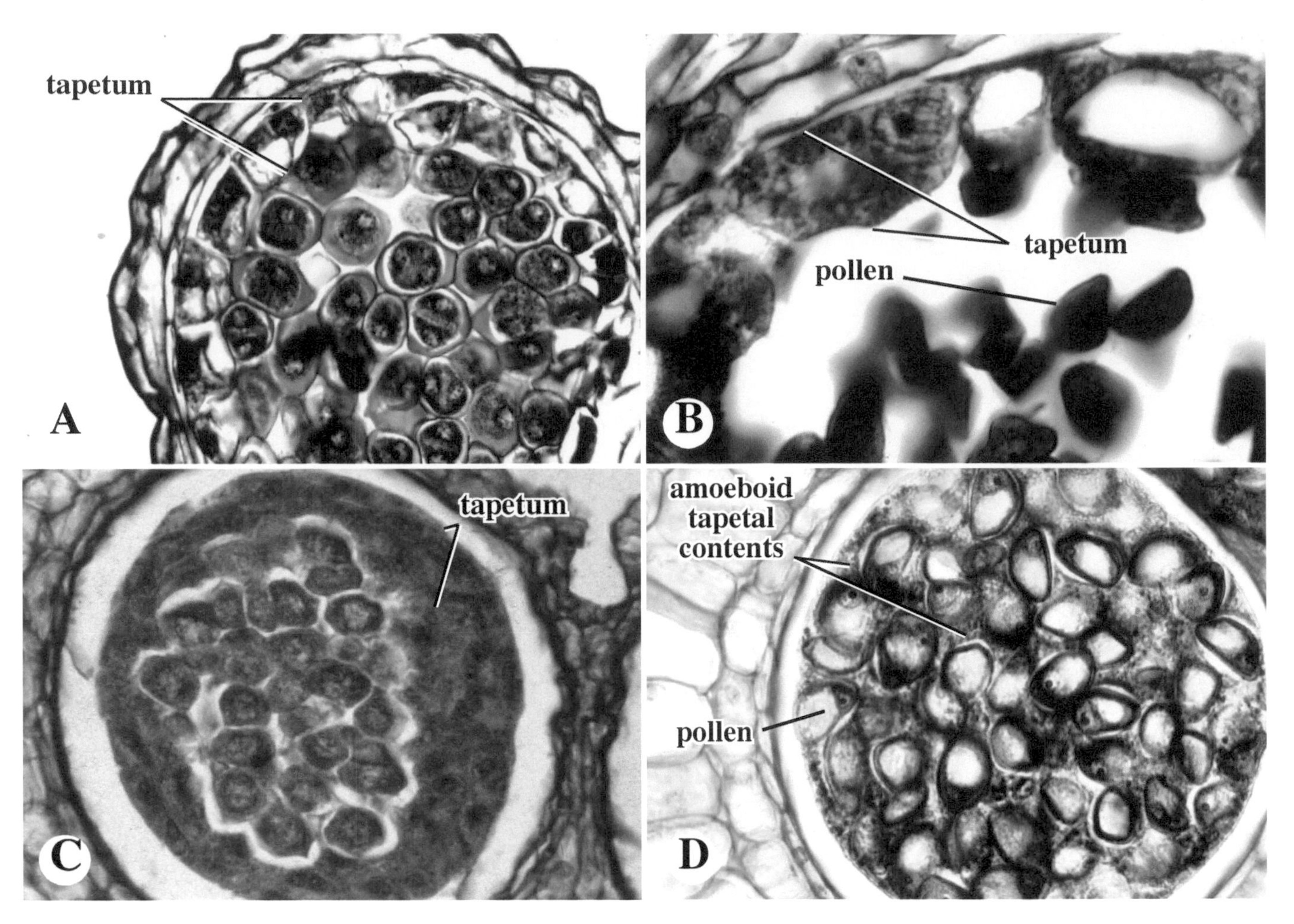

Figure 11.4 Anther cross sections, showing different tapetum types. **A,B.** Secretory tapetum, the cells of which remain intact during pollen development (*Lophiola aurea*). **C,D.** Amoeboid tapetum, in which the cells break down, releasing their cytoplasmic contents into the anther locule. Early stage at left, later stage at right for both types (*Lachnanthes caroliniana*).

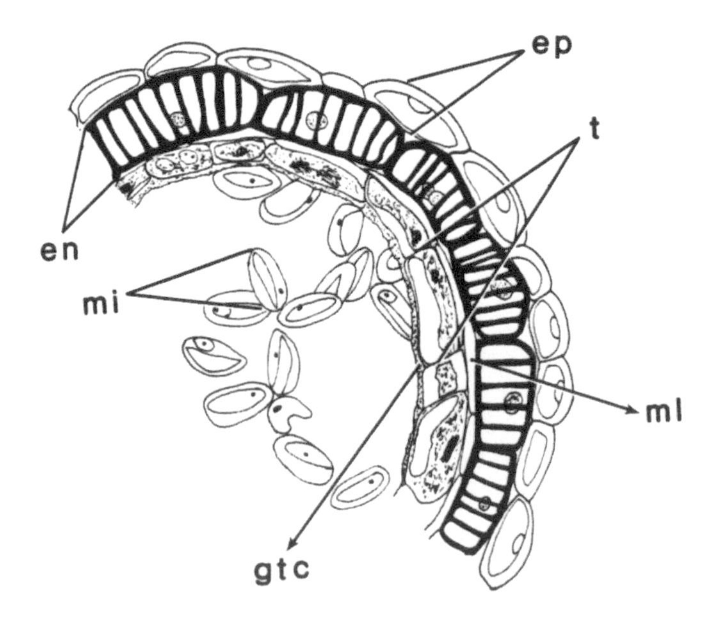

Figure 11.5 A girdling anther endothecium type. Symbols: **en** = endothecium; **ep** = epidermis; **gtc** = glandular tapetal cell; **mi** = microspore; **ml** = middle layer; **t** = tapetum.

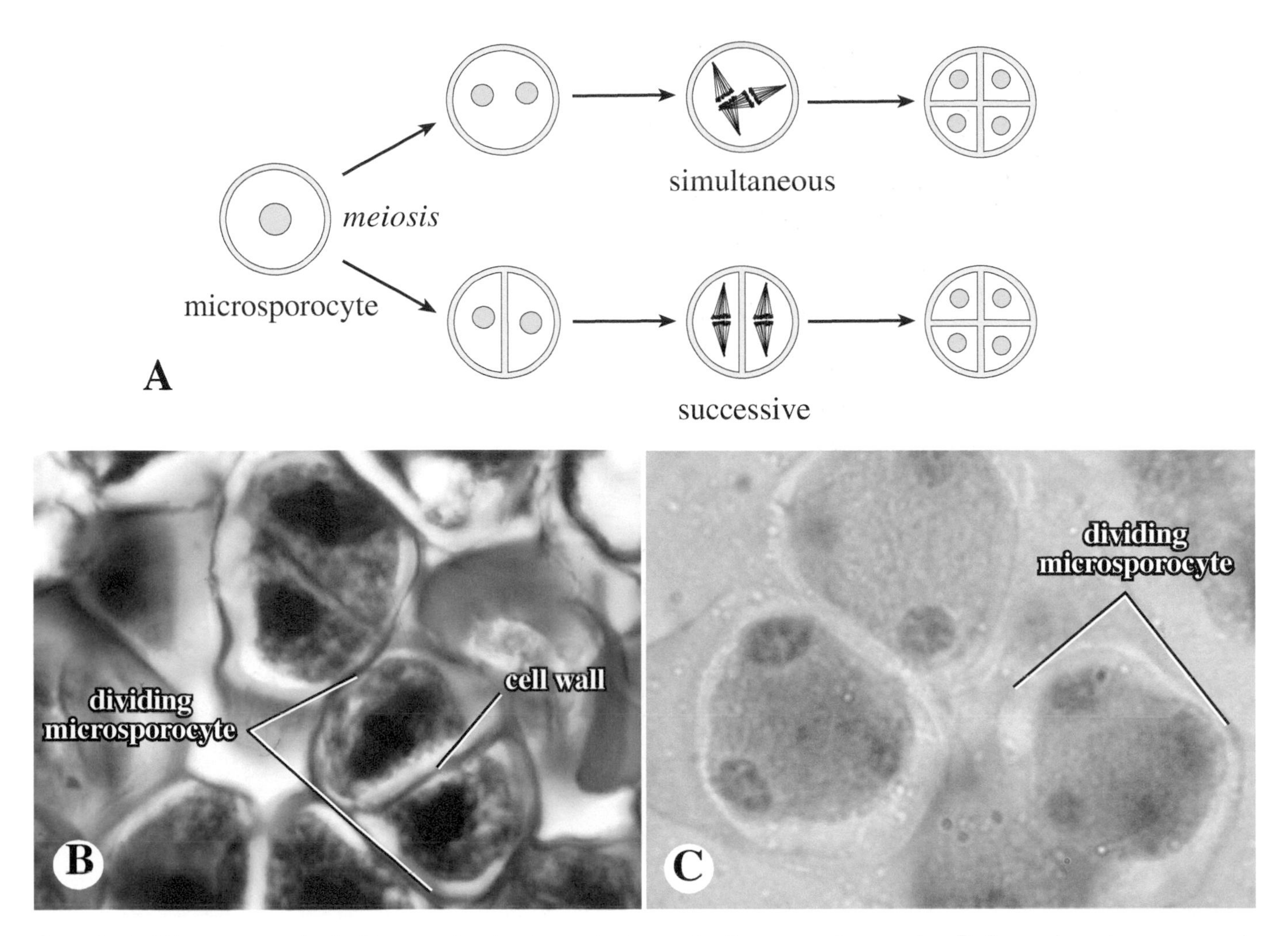

Figure 11.6 Microsporogenesis. **A.** Diagram showing two major types, simultaneous and successive. **B.** Successive microsporogenesis (*Lophiola aurea*). Microsporocyte at anaphase II of meiosis. Note that cytokinesis, resulting in cell wall formation, has occurred after meiosis I. **C.** Simultaneous microsporogenesis. Note lack of cell wall after anaphase II of meiosis.

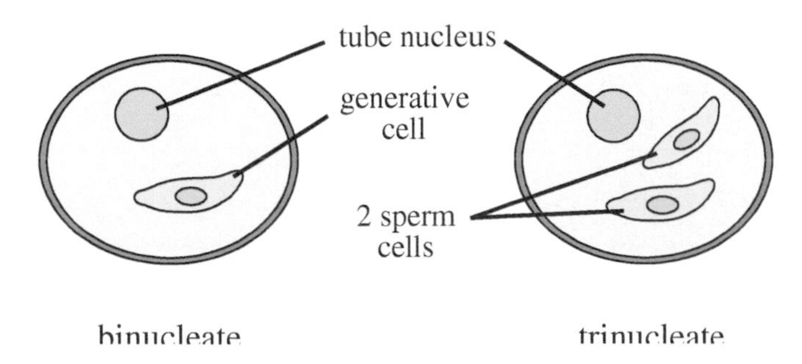

Figure 11.7 Microgametogenesis. Pollen nucleus number at anthesis.

pollen grains directly.) The structure of the outer versus inner integument can be used to define various micropylar types (below).

The region of the nucellus where the micropyle is located is called the **micropylar** region; that opposite the micropyle is called the **chalazal** region. A vascular strand typically traverses from the base of the funiculus to the nucellus. In most angiosperm ovules, the ovule curves during development (see Ovule Type), displacing the micropyle to a location near the funiculus base. In this type, the body of the funiculus appears fused to the body of the nucellus. This region where the funiculus is adnate (or decurrent) to the nucellus is called the **raphe**, which is sometimes visible in the mature seed as a ridge (see later discussion).

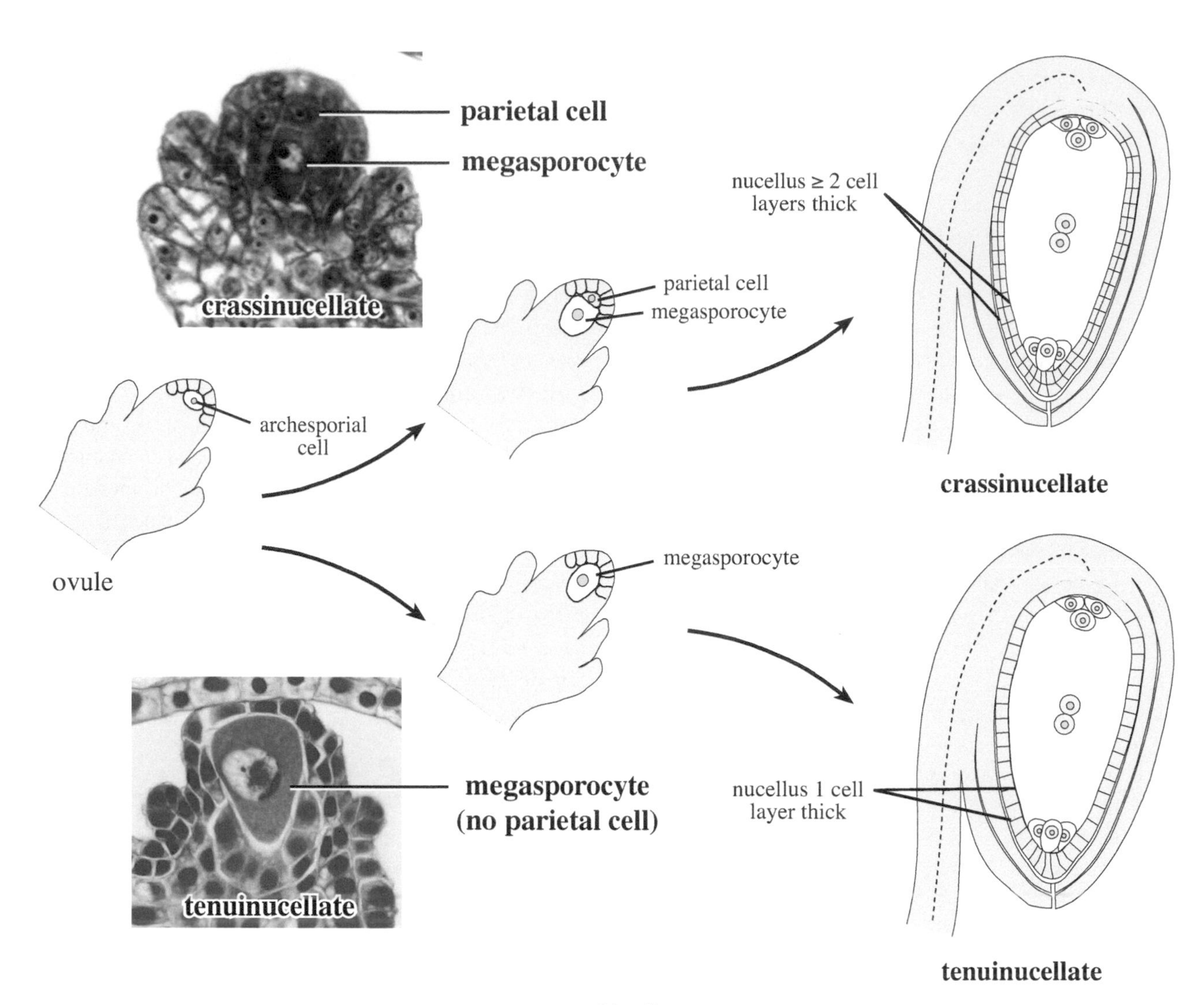

Figure 11.8 Nucellar type.

NUCELLUS TYPE

The type of nucellus, or megasporangium, is defined based on the number of cell layers comprising it (especially at the micropylar end) and the derivation of those cells. An immature ovule contains a single large cell known as an **archesporial cell**. In some taxa the archesporial cell undergoes a single periclinal division, resulting in the formation of an outer **parietal cell** and an inner **megasporocyte** (Figure 11.8). The parietal cell undergoes additional mitotic divisions, the products of which form an inner layer of nucellus cells; this type of nucellus is called **crassinucellate**, composed of two (sometimes more by additional divisions) layers of cells (Figure 11.8). On the contrary, if the archesporial cell does not divide and develops directly into the megasporocyte, the nucellus will generally be composed of a single layer of cells, the original outer layer; this type of nucellus is called **tenuinucellate** (Figure 11.8). However, in a few taxa, no parietal cell is formed, yet periclinal divisions occur in the single outer layer, forming an additional inner layer of nucellar cells; this type of nucellus is called **pseudocrassinucellate** because it appears at maturity to resemble the crassinucellate type in having two nucellar layers, but the inner layer has a different derivation. The fact that crassinucellate and pseudocrassinucellate ovules resemble one another at maturity emphasizes the need for early developmental studies to distinguish between them.

Some taxa may have a proliferation of cell divisions of the nucellus at the micropylar region of the ovule; this mass of cells is typically termed a **nucellar beak**.

MEGASPOROGENESIS

Megasporogenesis refers to the development of megaspores from the megasporocyte, the cell that undergoes meiosis. Meiosis of the megasporocyte nucleus results in the formation of four haploid megaspore nuclei. In most taxa, meiosis is followed by cytokinesis, resulting in four megaspore cells. This pattern is termed **monosporic** megasporogenesis; because of the four megaspores produced, only one of them contributes to the female gametophyte (Figures 11.9, 11.10A–D). In some angiosperm taxa, however, cytokinesis occurs after the first meiotic division, but not the second, resulting in two cells, each of which contain two haploid nuclei. This developmental pattern is termed **bisporic** megasporogenesis because one of the binucleate cells, containing two megaspore nuclei, contributes to the female gametophyte (Figure 11.9). Finally, in other taxa cytokinesis does not occur at all after meiosis, resulting in a single cell with four haploid nuclei. Because all four haploid megaspore nuclei contribute to the female gametophyte, this pattern is termed **tetrasporic** megasporogenesis (Figure 11.9).

MEGAGAMETOGENESIS

Megagametogenesis is development of the female gametophyte from the haploid product(s) of meiosis. The particular type of megagametogenesis is a function of mitotic divisions, the formation of new cells, and the fusion of existing nuclei or cells. This sequence of events defines what are termed **female gametophyte** (or **embryo sac**) **types**. The type of female gametophyte is dependent in part on the pattern of megasporogenesis, whether tetrasporic, bisporic, or monosporic. The most common and presumably ancestral type of female gametophyte in the angiosperms is one that develops from the chalazal haploid megaspore, the result of monosporic megasporogenesis. This haploid megaspore nucleus then divides mitotically to yield two nuclei, each of those two nuclei divide to yield four, and each of those four divide to yield eight. The eight nuclei arrange themselves into seven cells: three **antipodals** at the chalazal end, a large **central cell** having two **polar nuclei**, and one **egg** cell flanked by two **synergids** at the micropylar end. (The egg and two synergid cells are together termed the *egg apparatus*.) This sequence of nuclear and cell divisions gives rise to the ***Polygonum* type** of female gametophyte (named after the genus *Polygonum* where it was first described), the most common and the ancestral type among the angiosperms (Figures 11.9, 11.10E). However, numerous other types of female gametophytes occur in various taxa of angiosperms (Figure 11.9). For example, the *Fritillaria* type develops from a tetrasporic megasporogenesis in which three of the four megapores fuse to form a triploid nucleus (Figures 11.9, 11.10F,G). Two sequential mitotic divisions of the haploid and triploid nuclei ultimately result in an 8-nucleate female gametophyte in which the three antipodals and one of the polar nuclei are triploid (the other polar nucleus and the cells of the egg apparatus remaining haploid).

A recent theory of female gametophyte evolution in the angiosperms suggests that the ancestral condition was not the common monosporic, 8-nucleate, 7-celled *Polygonum* type, but was instead a monosporic, 4-nucleate, 4-celled condition found in virtually all Nymphaeales and Austrobaileyales (termed the *Nuphar/Schisandra* type; Figure 11.11A). (Note that this female gametophyte type is identical to the *Oenothera* type of Figure 11.9, which is presumed to be independently derived.) This 4-nucleate condition, with one polar nucleus in a central cell and 3 cells (the egg apparatus) at the micropylar end could represent an ancestral *module*. This module would subsequently have been doubled (a third sequence of mitotic divisions) to yield the *Polygonum* type (Figure 11.11B) or quadrupled to yield something like the 16-nucleate *Penaea* type (Figure 11.11C). In fact, the most basal angiosperm, *Amborella trichopoda* (Chapters 6, 7) has a modified type of female gametophyte, being 9-nucleate and 8-celled via an extra mitotic division in the egg apparatus, producing a third synergid cell; this type has been termed the *Amborella* type and it thought to have evolved independently of the *Polygonum* type common of the great majority of angiosperms. See Friedman and Williams (2004) and Friedman and Ryerson (2009) for more information on this idea.

INTEGUMENT TYPE (Figure 11.12)

The ovules of angiosperms have either one or two integuments. If two, the ancestral condition for the angiosperms, the ovule is called **bitegmic**. If one, the ovule is **unitegmic**. Unitegmic ovules have evolved in several different angiosperm groups, including the bulk of the Asteridae. Very rarely, ovules may lack any integument; this condition is termed **ategmic**.

MICROPYLE TYPE (Figure 11.13)

In a typical, bitegmic angiosperm ovule, the micropyle is typically formed or delimited by both integuments; this is termed an **amphistomal** micropyle type. If the micropyle is delimited by only the inner integument (the outer one being foreshortened), it is termed **endostomal**; if by only the outer integument (the inner one foreshortened), it is termed **exostomal**. In some angiosperms the micropyle is **zig-zag**, meaning that the micropylar pore of the outer integument is spatially displaced relative to the inner integument. If the ovule is unitegmic, the micropyle type may be called **unistomal** by default.

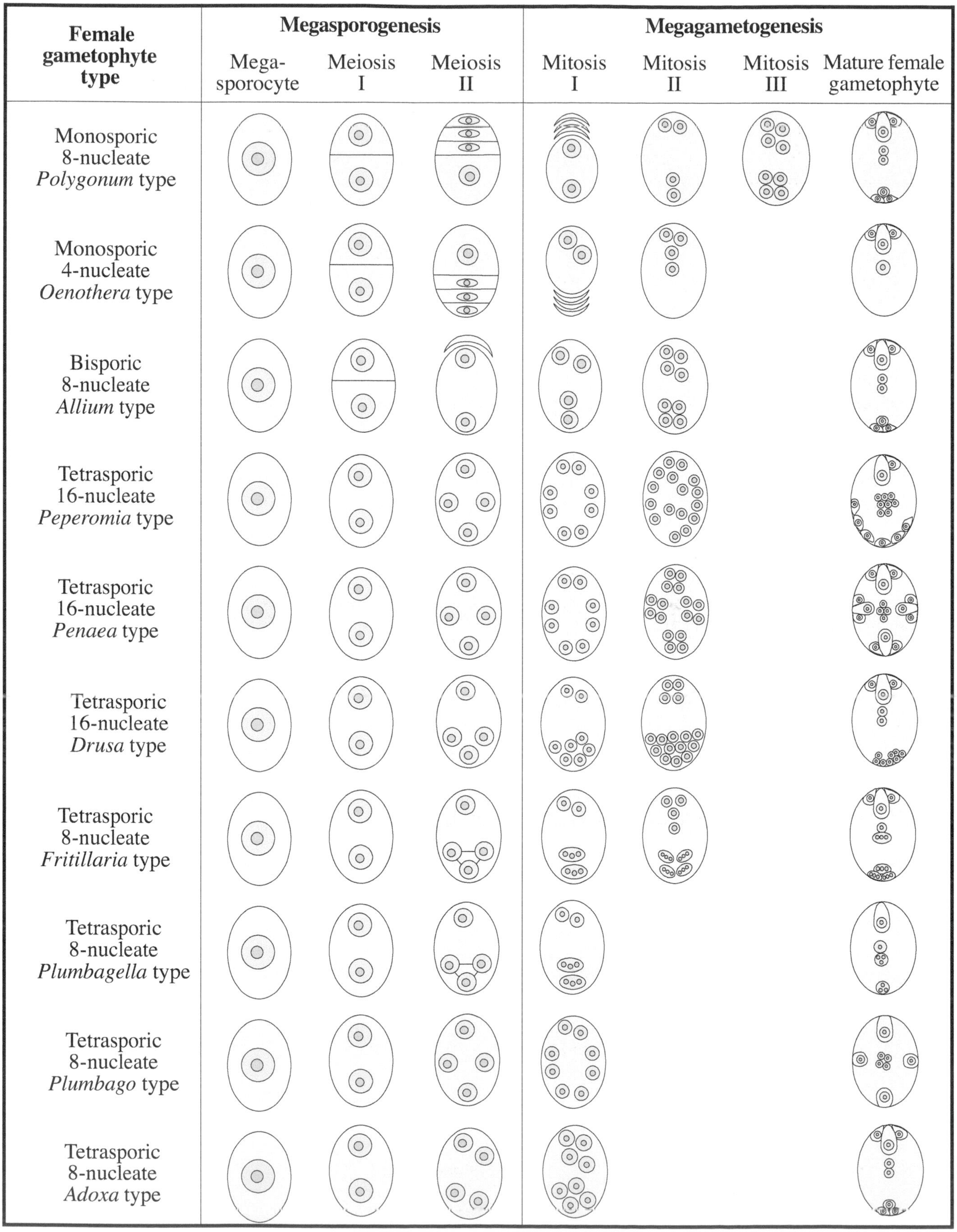

Female gametophyte type	Megasporogenesis			Megagametogenesis			
	Mega-sporocyte	Meiosis I	Meiosis II	Mitosis I	Mitosis II	Mitosis III	Mature female gametophyte
Monosporic 8-nucleate *Polygonum* type							
Monosporic 4-nucleate *Oenothera* type							
Bisporic 8-nucleate *Allium* type							
Tetrasporic 16-nucleate *Peperomia* type							
Tetrasporic 16-nucleate *Penaea* type							
Tetrasporic 16-nucleate *Drusa* type							
Tetrasporic 8-nucleate *Fritillaria* type							
Tetrasporic 8-nucleate *Plumbagella* type							
Tetrasporic 8-nucleate *Plumbago* type							
Tetrasporic 8-nucleate *Adoxa* type							

Figure 11.9 Female gametophyte types, based on type of megasporogenesis and sequence of divisions and cell fusions during megagametogenesis. Note: micropyle above in all illustrations. Terminology after Maheshwari (1950).

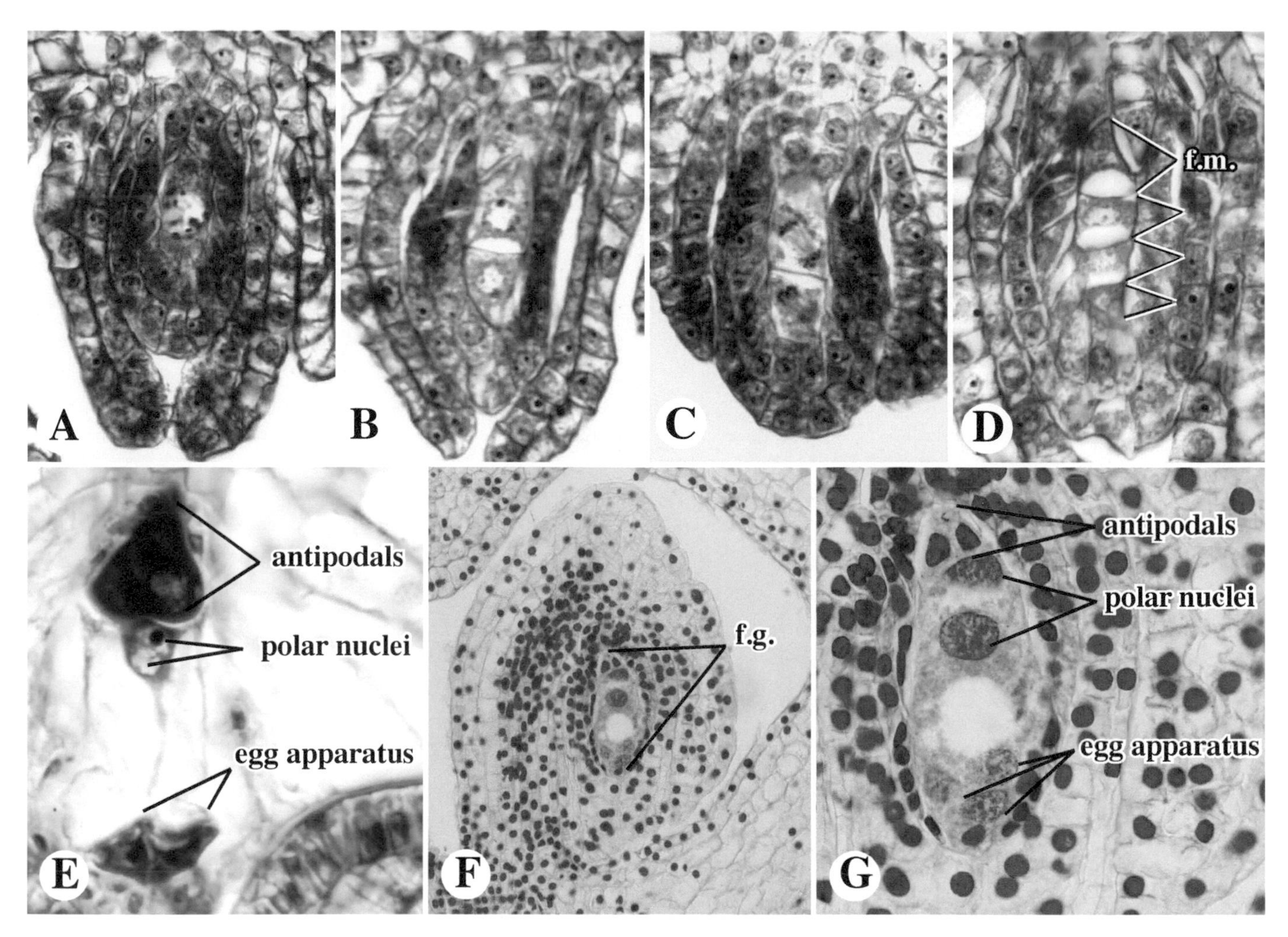

Figure 11.10 Examples of embryological development. (Note: micropyle below in all.) **A–D.** Monosporic megasporogenesis. **A.** Ovule with single megasporocyte. **B.** After first meiotic division. **C.** Second meiotic division. **D.** After second meiotic division, with four megaspores; f.m. = functional (proximal) megaspore. **E.** *Polygonum*-type female gametophyte (embryo sac), showing 3 antipodals, 2 polar nuclei, and 2 of 3 cells of egg apparatus. **F,G.** Megagametogenesis of *Lilium* sp., having tetrasporic, *Fritillaria*-type female gametophyte development. **F.** Whole (anatropous) ovule, with mature female gametophyte (f.g.). **G.** Mature female gametophyte close-up, showing 2 of 3 triploid antipodals, 2 polar nuclei (one haploid, the other triploid), and haploid cells of egg apparatus.

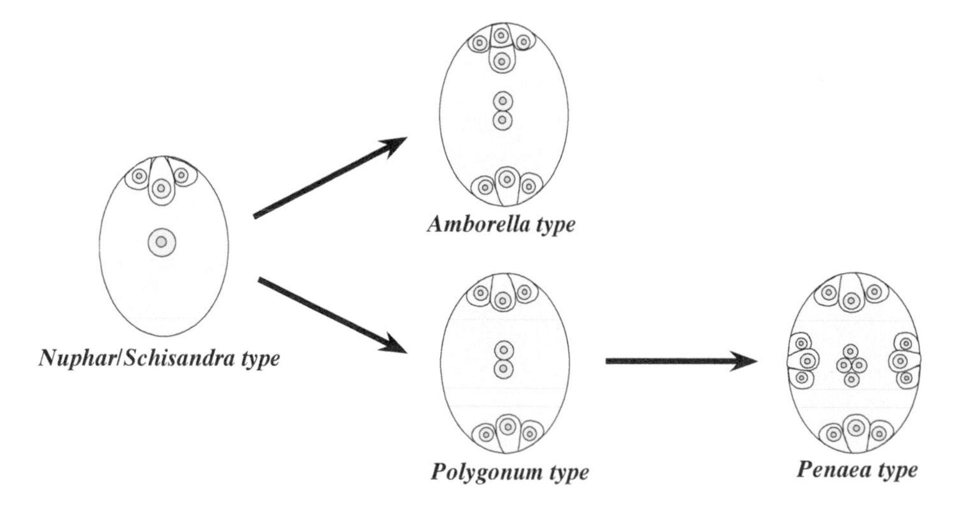

Figure 11.11 Modular hypothesis of female gametophyte evolution. The monosporic, 4-nucleate *Nuphar/Schisandra* type may represent the ancestral condition in the angiosperms, independently giving rise to the 9-nucleate *Amborella* type and 8-nucleate *Polygonum* type by duplication of the 4-nucleate module. A quadrupling of the module yields the 16-nucleate *Penaea* type. (After Friedman and Williams, 2004.)

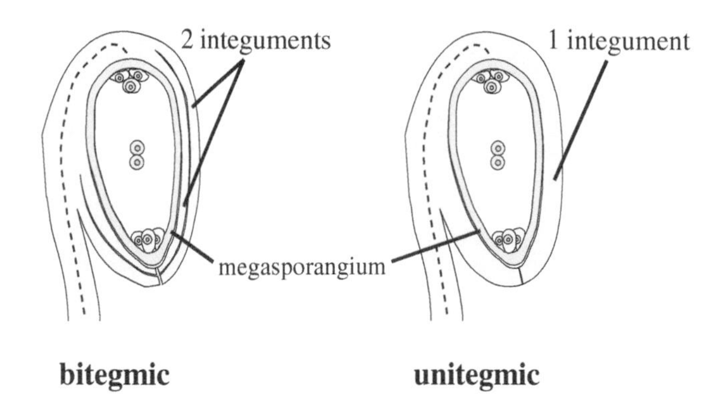

Figure 11.12 Ovule integument types.

OVULE TYPE (Figure 11.14)

Ovule types are defined primarily on the curvature of the funiculus and nucellus/female gametophyte. The following terms are useful, yet different ovule types can be difficult to define and may require quantitative analyses. An **anatropous** ovule is one in which curvature during development results in displacement of the micropyle to a position adjacent to the funiculus base; a vasculature strand traverses from the base of the funiculus to the nucellar region opposite the micropyle. The anatropous ovule type is the most common in the angiosperms and is presumed to be ancestral for the group. An **orthotropous [atropous]** ovule is one in which no curvature takes place during development; the micropyle is positioned opposite the funiculus base, and the vasculature traverses from the base of the funiculus to the chalazal nucellar region. Orthotropous ovules have evolved independently in various groups of angiosperms. Both anatropous and orthotropous ovules have a straight (unbent) nucellus. (An ovule somewhat intermediate in curvature between anatropous and orthotropous is sometimes termed **hemitropous** or **hemianatropous**.)

Four other ovule types that have been defined exhibit a curvature of the ovule during development such that the micropyle is displaced adjacent to the funiculus base, similar to an anatropous ovule. These four additional ovule types differ from an anatropous ovule in having a bent or curved nucellus, as viewed in mid-sagittal section (i.e., a section along the plane of symmetry). Traditionally, these four types were divided into only two: the **amphitropous** type, in which the nucellus is bent along both upper and lower sides, and the **campylotropous** type, in which the nucellus is bent only along the lower side. The amphitropous and campylotropous ovule types may often be cited in plant systematic literature. However, these may be subdivided into additional types ("ana-" and "ortho-") based on the the orientation of the vasculature. An **ana-amphitropous** ovule is one in which a vascular strand curves, traversing from the base of funiculus to the chalazal region of the nucellus; the nucellus is bent sharply in the middle along both the lower and upper sides, often with differentiated cells (called a "basal body") at the angle of the bend. An **ana-campylotropous** ovule is similar to the ana-amphitropous type in vasculature, differing in that the nucellus is bent only along the lower side, with no "basal body." An **ortho-amphitropous** ovule is one in which the vasculature is straight, leading from the funiculus base to the middle of the nucellus; the nucellus is bent sharply in the middle along both the lower and upper sides, often with a "basal body" present. An **ortho-campylotropous** ovule is similar to that of the ortho-amphitropous type, except that the nucellar body is bent only along the lower side, with no "basal body."

OVULE POSITION (Figure 11.15)

Ovule position refers to the placement of the micropyle and raphe relative to the distal end (apex), proximal end (base), or sides of the floral axis.

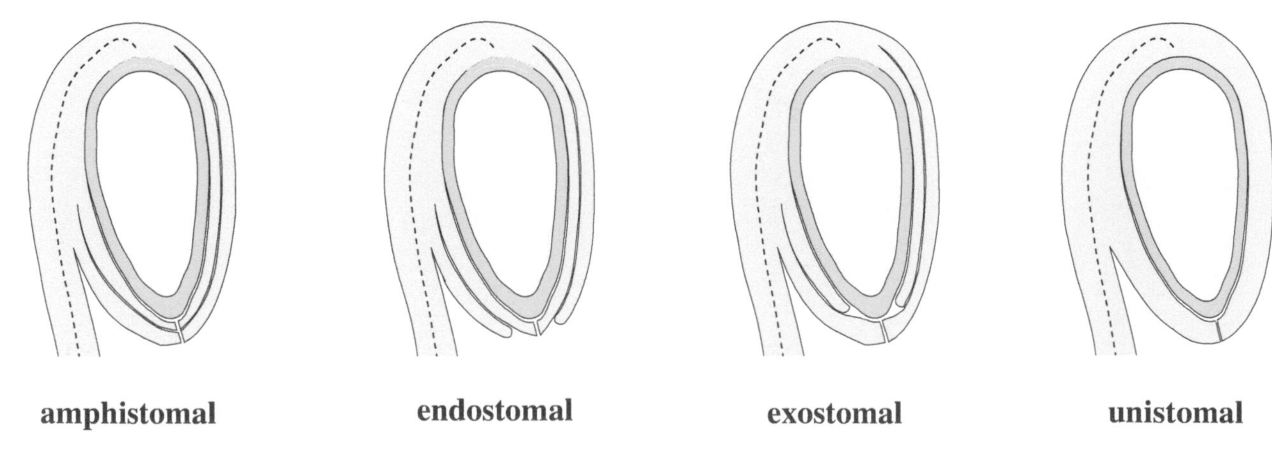

Figure 11.13 Ovule micropyle types.

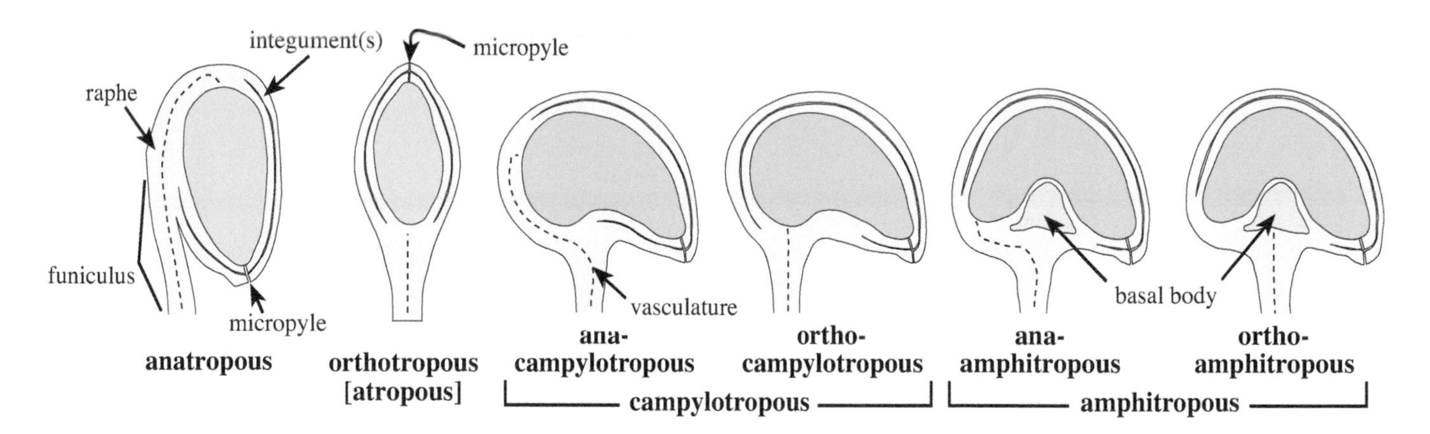

Figure 11.14 Ovule types.

An **epitropous** ovule is one in which the micropyle points distally. This type can be further divided into **epitropous-dorsal**, in which the raphe is dorsal (abaxial, pointing away from the central floral or ovary axis) or **epitropous-ventral**, in which the raphe is ventral (adaxial, pointing toward the central floral or ovary axis).

A **hypotropous** ovule is one in which the micropyle points proximally. This type can be further divided into **hypotropous-dorsal**, in which the raphe is dorsal (abaxial, pointing away from the central floral or ovary axis) or **hypotropous-ventral**, in which the raphe is ventral (adaxial, pointing toward the central floral or ovary axis).

A **pleurotropous** ovule is one in which the micropyle points to the side. This type can be further divided into **pleurotropous-dorsal**, in which the raphe is above or **pleurotropous-ventral**, in which the raphe is below.

A **heterotropous** ovule is one that varies in orientation.

OBTURATOR PRESENCE/ABSENCE

Rarely, a protuberance of tissue, typically arising from the funiculus or placenta, may develop at the base of the ovule. This mound of tissue, termed an **obturator**, may be typical of certain groups, e.g., the Euphorbiaceae.

SEED DEVELOPMENT

EMBRYOGENY

Embryogeny refers to the development of the embryo within the seed. The sequence of divisions of the zygote (the product of fertilization of egg and sperm) can define various embryogeny types, which have been named after the major taxonomic groups where they occur.

Typically, the first division of the zygote is transverse (perpendicular to the long axis of the female gametophyte and nucellus), initiating the formation of a very young embryo, often termed the **proembryo**. This transverse division delimits two cells, a basal cell at the micropylar end and an apical (terminal) cell at the chalazal end. The terminal cell will divide prolifically, generally forming all or most of the **embryo proper**, which will eventually grow into the new sporophyte. Mitotic divisions of the original basal cell may also contribute to the mature embryo and/or may develop into column of cells termed the **suspensor** (Figure 11.16A,B), a nonpersistent structure that functions in transport of nutrients to the mature embryo during its development.

Five embryogeny types have been defined based on the sequence of divisions of the basal and terminal cells and

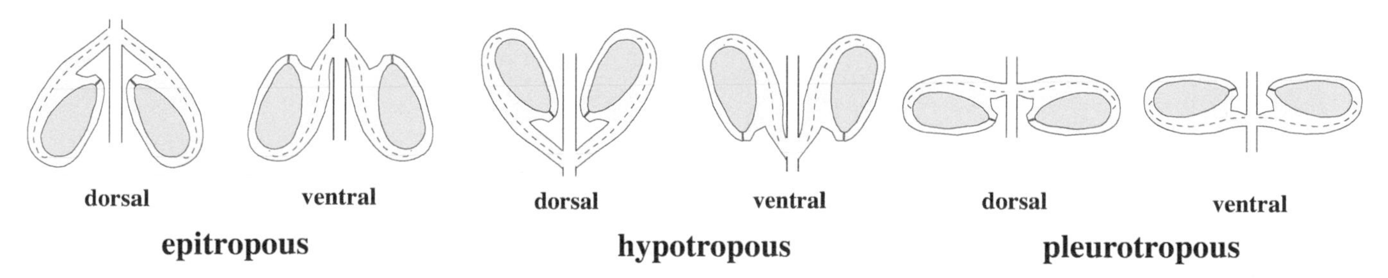

Figure 11.15 Ovule position, illustrated with anatropous ovules.

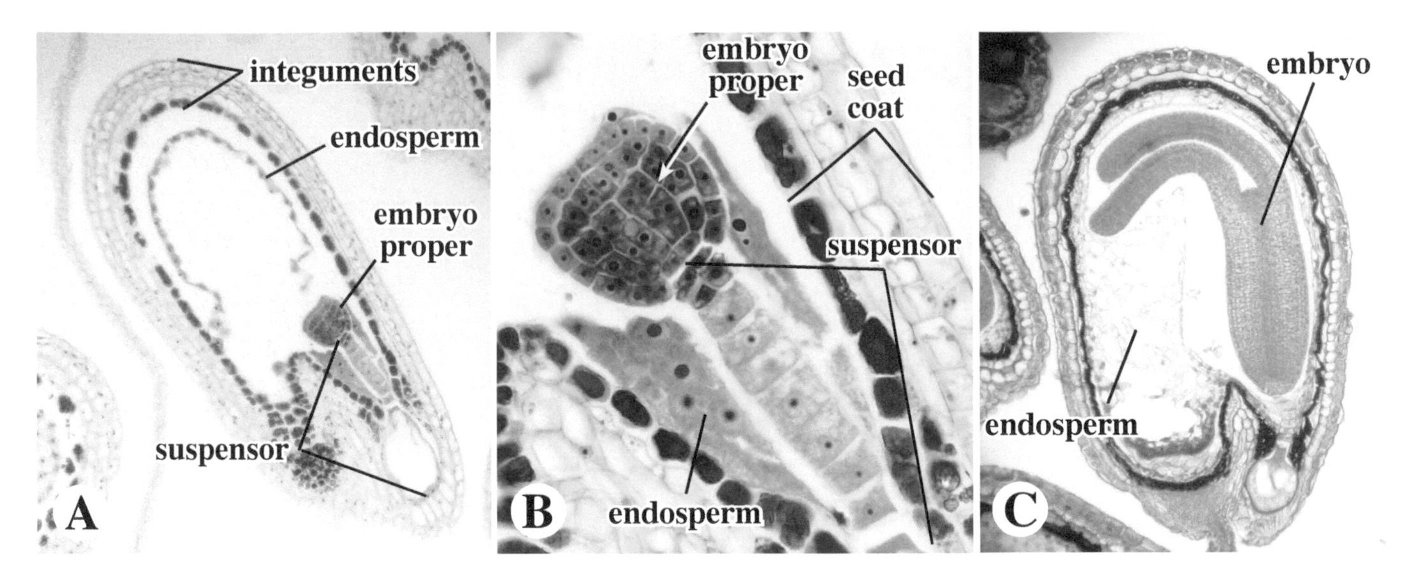

Figure 11.16 Endosperm and embryo development, *Capsella bursa-pastoris*. **A.** Very young seed, showing immature embryo (composed of suspensor and embryo proper), early endosperm, and seed coat. **B.** Close-up of young embryo, showing basal suspensor cells and terminal, actively dividing cells of globular embryo proper. **C.** More mature seed, showing embryo with two cotyledons.

which cell derivatives contribute to the mature embryo: (1) **asterad type**, in which the terminal cell divides longitudinally, with both basal and terminal cell derivatives contributing to the mature embryo; (2) **caryophyllid** type, in which the terminal cell divides transversely, with only terminal cell derivatives contributing to the mature embryo; (3) **chenopodiad** type, in which the terminal cell divides transversely, with both basal and terminal cell derivatives contributing to the mature embryo; (4) **crucifer** or **onagrad type**, in which the terminal cell divides longitudinally, with only terminal cell derivatives contributing to the mature embryo; (5) **solanad type**, in which the terminal cell divides transversely, the basal cells derivatives forming a suspensor but otherwise not contributing to mature embryo development.

Finally, a sixth embryogeny type, the **piperad type**, is defined if the zygote divides longitudinally (i.e., parallel to the axis of the female gametophyte and nucellus), thus not forming a basal and terminal cell.

EMBRYO TYPE

The mature embryo type is based on its form and size. The shape and size of the radicle and cotyledons are most important. Various embryo types have been defined primarily on **ptyxis**, the aestivation of the cotyledons (Chapter 9). A mature embryo may be either **achlorophyllous** (lacking chloroplasts) or **chlorophyllous** (green, having chloroplasts).

ENDOSPERM DEVELOPMENT

Development of the endosperm (Figure 11.16A–C) is described based on early mitosis and cytokinesis of the usually triploid, endosperm cell (the second product of double fertilization). A **cellular** endosperm is one in which the endosperm cell divides mitotically, regularly followed by cytokinesis. Thus, each endosperm nucleus is contained within a cell wall from the beginning. A **nuclear** endosperm is one in which the early mitotic divisions are not followed by cytokinesis. Thus, numerous nuclei are contained within a single cell, at least early in development; later, cell walls typically surround the nuclei. A **helobial** endosperm is one in which the first mitotic division is followed by cytokinesis, delimiting two cells. However, the nucleus of one cell continues a nuclear type of development; that of the other cell divides in a cellular fashion.

SEED STORAGE TISSUE ORIGIN

The most common, and ancestral, type of storage tissue in angiospermous seeds is endosperm. This typical seed type is called **endospermous** or **albuminous** (Figures 11.16, 11.17). In some taxa, however, double fertilization and endosperm development occur, but the endosperm soon stops growing; the mature seed is termed **exalbuminous**. This is typical, for example, of all orchids, which have very reduced seeds in general. In other exalbuminous taxa, the early endosperm tissue may be absorbed, with other tissues taking its place as a storage tissue. A **cotylespermous** seed storage tissue type is one in which the cotyledons enlarge and assume the function of storage tissue (Figure 11.17). Cotylespermous seeds are found, e.g., in many legumes, such as peas and beans. A **perispermous** type of seed storage tissue is one in which the chalazal nucellar cells enlarge and store energy-rich compounds.

SEED STORAGE TISSUE COMPOSITION

The storage tissue of a seed (usually endosperm) can be defined by the chemical composition of the energy-rich compounds within its cells. Storage tissue can contain primarily **starch** (in the form of starch grains or amyloplasts), **oil** (in the form of oil bodies), or **protein** (in the form of protein bodies).

SEED COAT ANATOMY

The integument(s) of the ovule matures into the **seed coat** of the seed. As discussed in Chapter 9, the seed coat of angiosperms consists of two, postgenitally fused layers, an outer **testa** derived from the outer integument (itself sometimes divided into layers, an inner **endotesta**, middle **mesotesta**, and outer **exotesta**) and an inner **tegmen** derived from the inner integument (which can be divided into similar layers, the **endotegmen**, **mesotegmen**, and **exotegmen**). A seed coat that is fleshy at maturity may be termed a **sarcotesta**. Features of the anatomy of the mature seed coat can be significant embryological characters. These include the number of cell layers in each seed coat layer (versus the number in the integuments) and specialized cell anatomy of the cells (including cell shape, cell wall thickness, and cell wall composition) of each seed coat layer.

In addition, in some taxa an extra, fleshy layer may form outside the seed coat. If the fleshy layer more or less envelops the seed coat, it is known as an **aril** (Figure 11.17). The aril generally functions as an attractant in animal dispersal. A **caruncle** or **strophiole** is a fleshy layer that does not surround the seed coat, but forms as a basal appendage, typically near the **hilum** (the scar of the funiculus). The caruncle functions like an aril, as a food reward in animal seed dispersal (Figure 11.17).

SEED MORPHOLOGY

Aspects of mature seed morphology include shape, size, color, and sculpturing. Also important in seed morphology are the shape, size, and color of the **hilum**, the scar of attachment of the funiculus on the seed coat, and of the **raphe**, a ridge on the seed coat formed from an adnate funiculus (Figure 11.17).

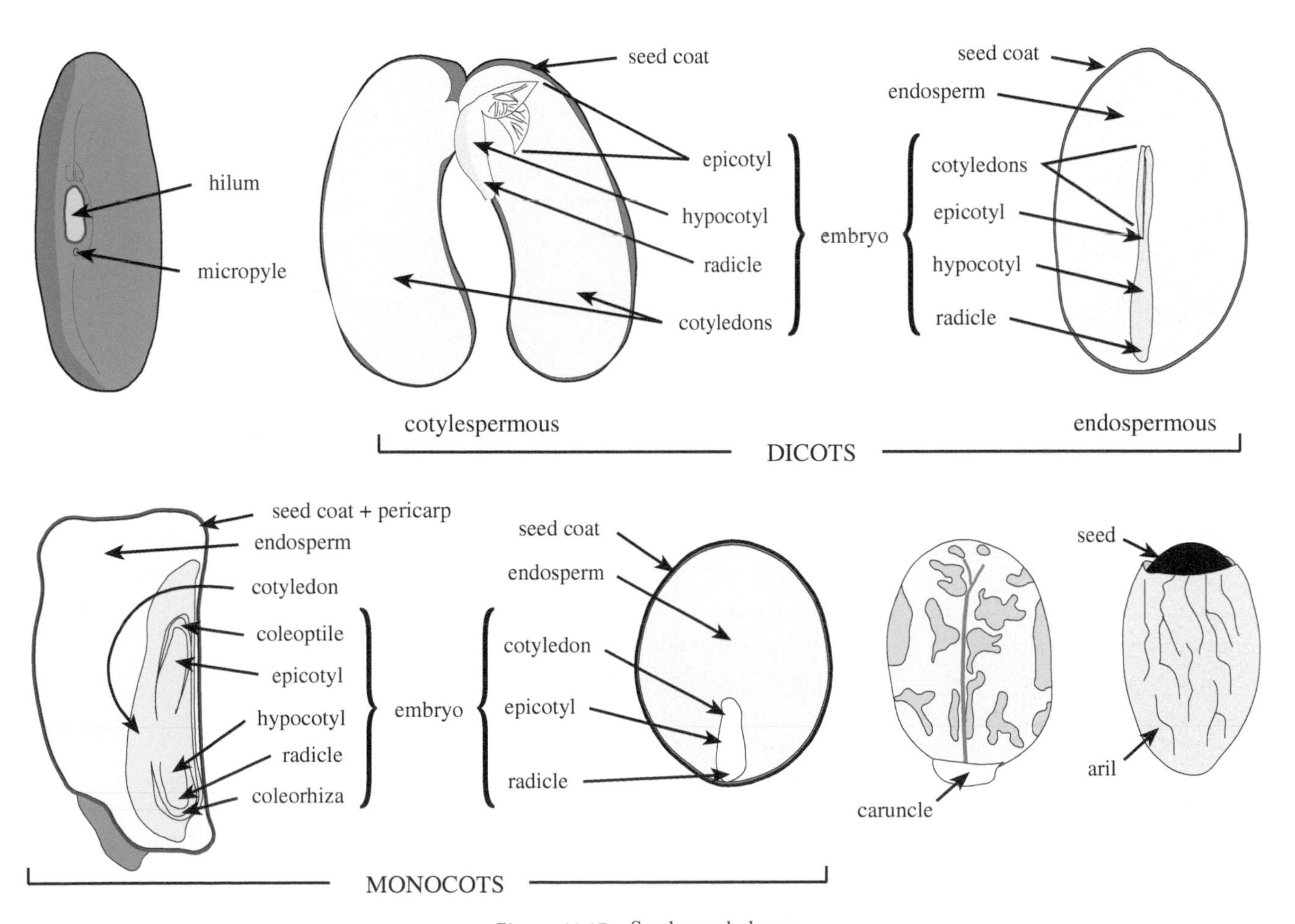

Figure 11.17 Seed morphology.

TABLE 11.1 Example of embryological features of the Caryophyllids, after Rodman, 1990. Centrospermae revisited, Part I. Taxon 39: 353–393. Character names are bolded in brackets.

Ovule anatropous, (ana-)campylotropous, or orthotropous **[Ovule type]**
Micropyle formed by both integuments or inner only **[Micropyle type]**
Nucellar beak (or cap) absent or present **[Nucellus type]**
Embryogeny: onagrad, asterad, solanad, caryophyllad, or chenopodiad type **[Embryogeny]**
Embryo sac development Polygonum-type or Plumbago-type **[Female gametophyte (embryo sac) type]**
Funicular obturator absent or present **[Obturator presence/absence]**
Endosperm present or absent in seed **[Seed storage tissue origin]**
Perisperm not persisting or persisting in seed **[Seed storage tissue origin]**
Embryo without chlorophyll or with chlorophyll **[Embryo type]**
Seeds arillate (funicular aril) or not **[Seed coat anatomy]**
Exotestal layer of seed not thickened or thickened **[Seed coat anatomy]**
Endotestal layer of seed thickened or not **[Seed coat anatomy]**
Exotegmic layer of seed tracheidal, fibrous or palisade, or unspecialized **[Seed coat anatomy]**
Endotegmic layer of seed not thickened or thickened **[Seed coat anatomy]**

EMBRYOLOGY AND SYSTEMATICS

As noted earlier, the collection of embryological features can be very valuable in delimiting or aiding in phylogenetic inference. An example of embryological features in a group of angiosperms is portrayed in Table 11.1.

EMBRYOLOGICAL TECHNIQUE

Material Dissection and Preparation:

As with anatomical studies, the study of plant embryology can involve very time-consuming embedding, microtome sectioning, staining, and slide preparation. Often, it is critical to obtain a range of developmental stages, in order to trace the changes that occur from inception to maturity. Female gametophyte development is particularly difficult to study, as some developmental changes occur rapidly and are hard to catch.

For all embryological studies, plant material must first be *fixed* in a chemical solution. The *fixative* preserves the material close to its original state and often clears it somewhat such that it can be better resolved under the microscope. Fix a number of flowers, from very young buds to developing fruits, by placing the material in a jar or vial of 70%–95% ethanol. The buds and flowers (or large ovaries or anthers) should generally be cut open to allow for better penetration of the fixative. Fix for a minimum of 10–15 minutes, although 1–2 days is better; store in 70% ethanol. [Note that, for more detailed studies, the material should be fixed in *FAA*, which is a mixture of formalin, acetic acid, alcohol. One recipe for FAA is: 66 ml of 95% ethanol, 21 ml water, 8 ml commercial (37%) formalin, and 5 ml glacial acetic acid. Formalin is dangerous to inhale, and glacial acetic acid is very caustic, so the solution should be mixed very carefully in a laboratory hood. Once mixed, it can be stored indefinitely, in a properly sealed container.]

Ovule Morphology:

Many features of the ovule can be observed with some relatively simple techniques. Place mature flowers that have been chemically fixed in a small dish (Petri or Syracuse dish) filled with 70% ethanol. Dissect the material with needles and forceps to remove and open up the ovary. Use fine needles to detach the ovules. During the dissection, observe the general ovule type (e.g., is the micropyle pointing toward the point of attachment of the funiculus [anatropous] or away from it [orthotropous]) and ovule position (the placement of the micropyle and funicular raphe relative to the flower axis). Place some ovules (using forceps or a pipette) on a microscope slide in a drop of water or 50% glycerol and cover with a cover slip.

For more detailed studies, the ovules can be cleared and observed using phase contrast or (preferably) differential interference contrast (DIC or "Nomarski") optics. One useful clearing fluid is Herr's solution. [1 part 85% lactic acid : 1 part chloral hydrate : 1 part phenol crystals : 1 part clove oil : ½ part Histoclear (less toxic) or xylene (all parts by weight; after Herr 1971; Rudall and Clark 1992)]

Microspore and Pollen Development:

One simple technique to observe the development of pollen grains is to dissect the internal contents of fixed anthers onto a microscope slide, stain, and cover with a cover slip. The material may be stained with toluidine blue or (to observe meiotic stages) acetocarmine. After staining, the material may be "squashed" by placing a cork on top of the cover slip and applying gentle, firm pressure. Squashing spreads the cells out into a thin layer, allowing for better observation of cell divisions and morphology.

REVIEW QUESTIONS

ANTHER AND POLLEN DEVELOPMENT

1. What are the two major anther types and how do they differ?
2. What criteria are used to define the four types of anther wall development?
3. What is the tapetum? What are the two types of tapetum development and how do they differ?
4. What are two types of anther endothecial anatomy?
5. What is microsporogenesis and what are the two major types?
6. What is microgametogenesis and what are the two major types?

OVULE DEVELOPMENT

7. Name the parts of a typical ovule.
8. What is meant by the chalazal region? A raphe?
9. Name and distinguish between the three types of nucellus. Which two resemble one another at maturity?
10. Name and distinguish between the three types of megasporogenesis.
11. What criteria are used to distinguish between the numerous female gametophyte development types?
12. Which female gametophyte type is most common and probably ancestral in the angiosperms?
13. What are the two major integument types?
14. What are the differences between endostomal, exostomal, amphistomal, unistomal, and zig-zag micropylar types?
15. What criteria are used to distinguish between ovule types? Which type is most common and ancestral in the angiosperms?
16. Define and give three examples of ovule position.

SEED DEVELOPMENT

17. What is embryogeny and on what criteria are different embryogeny types based?
18. What is ptyxis and what does it define?
19. Name the three basic types of endosperm development and describe how they differ.
20. Other than endosperm, what two other seed storage tissue origins occur in angiosperms?
21. Name four seed storage tissue origin types. Of what three major chemicals are seed storage tissue composed?
22. What are arils and caruncles, and what is their function?

EMBRYOLOGY AND SYSTEMATICS

23. Name some features of embryology that may be valuable in plant systematics.

EXERCISES

1. Obtain flowering material of a species and fix the material according to the procedures noted above (Embryological Technique). Dissect ovules from the ovaries and prepare a slide for light microscope observations. If possible, clear the ovules for phase contrast or differential interference contrast (DIC) microscopy. Note, from the dissected or cleared ovules, the (a) ovule position (epitropous, hypotropous, pleurotropous (and whether dorsal or ventral), or heterotropous; (b) integument type (bitegmic or unitegmic); (c) female gametophyte shape; (d) vasculature; and (e) specific ovule type (ana-amphitropous, ana-campylotropous, ortho-amphitropous, or ortho-campylotropous). Draw and record this information.
2. Obtain anthers of various stages and dissect them open to make slide preparations. Stain with acetocarmine. Observe stages of meiosis and pollen development.
3. Obtain mature seeds of various flowering plants. Observe outer components of the seed, including seed coat morphology, funicular scar, raphe, caruncle, or aril (if present). Dissect the seeds by cutting with a razor blade. Observe the embryo and the seed nutritive type (endospermous or albuminous, exalbuminous, cotylespermous, or perispermous). Stain the seed sections with IKI, which stains starch a dark purple or brown, to determine if the nutritive tissue is starchy at maturity.
4. Peruse journal articles in plant systematics, e.g., *American Journal of Botany*, *Systematic Botany*, or *International Journal of Plant Sciences*, and note those that describe plant embryological features in relation to systematic studies. Identify all embryological characters and character states described.

REFERENCES FOR FURTHER STUDY

Björnstad, I. N. 1970. Comparative embryology of Asparagoideae-Polygonateae, Liliaceae. Nytt Magasin for Botanikk 17(3–4): 169–207.

Davis, G. L. 1966. Systematic Embryology of the Angiosperms. John Wiley and Sons, New York.

Friedman, W. E. and J. H. Williams. 2004. Developmental evolution of the sexual process in ancient flowering plant lineages. The Plant Cell 16, S119–S132, Supplement.

Friedman, W. E. and K. C. Ryerson. 2009. Reconstructing the ancestral female gametophyte of angiosperms: insights from Amborella and other ancient lineages of flowering plants. American Journal of Botany 96: 129–143.

Herr, J. M., Jr. 1971. American Journal of Botany 58: 785–790.

Johri, B. M. (ed.). 1984. Embryology of angiosperms. Springer-Verlag, Berlin, New York.

Johri, B. M., K. B. Ambegaokar, and P. S. Srivastava. 1992. Comparative embryology of angiosperms. Springer-Verlag, Berlin, New York.

Maheshwari, P. 1950. An Introduction to the Embryology of Angiosperms. McGraw-Hill, New York.

Maheshwari, P. (ed.). 1963. Recent Advances in the Embryology of Angiosperms. International Society of Plant Morphologists, Delhi.

Rudall, P. J. and L. Clark. 1992. The megagametophyte in Labiatae. In: Harley, R. M. and T. Reynolds (eds), Advances in Labiatae Science, Royal Botanic Gardens, Kew, Pp. 65-84.

12

PALYNOLOGY

INTRODUCTION

Palynology (Gr. *palynos*, dust) is the study of spores and pollen grains. Spores and pollen grains have a number of morphological and ultrastructural features. These palynological features have provided a wealth of characters that have been important in inferring phylogenetic relationships of plants. In addition, the features of spores and pollen grains can often be used to identify a particular plant taxon. For this reason, palynological studies are used extensively to examine the fossil record, a field called paleo-palynology. The identity, density, and frequency of pollen grains at a particular stratigraphic level can give information as to the plant species present at that time and place. Paleopalynolgical studies are thus used to determine plant community structure and to gauge, by extrapolation over time, shifts in climate.

PALYNOLOGICAL TERMINOLOGY

The terminology applied to pollen morphology and ultrastructure varies from author to author. The following terminology follows the suggestions of Reitsma (1970), Walker and Doyle (1975), and Punt et al. (1994).

POLLEN NUCLEUS NUMBER

The number of nuclei at the time of pollen release can be phylogenetically informative. Two types occur in angiosperms. **Binucleate** grains (Figure 12.1) contain one tube cell and nucleus and one generative cell and nucleus; this is the most common and ancestral type in the angiosperms. **Trinucleate** grains contain one tube cell and nucleus and two sperm cells, the latter resulting from precocious division of the generative cell. Trinucleate grains are relatively rare in the angiosperms, being a diagnostic feature and possible apomorphy for some Caryophyllales. Pollen nuclear number is also listed as an embryological character (see Chapter 11).

POLLEN STORAGE PRODUCT

Pollen grains contain high-energy storage reserves. These are composed of either **starch** or **oil**. The distribution of storage product type can be phylogenetically informative in the angiosperms.

POLLEN UNIT

Pollen unit refers to the number of pollen grains united together at the time of release. Most commonly, the four microspores formed after microsporogenesis separate prior to pollen (or spore) release. Such single, unfused pollen grains are called **monads**, found in the great majority of angiosperms.

https://doi.org/10.1016/B978-0-12-812628-8.50012-2,

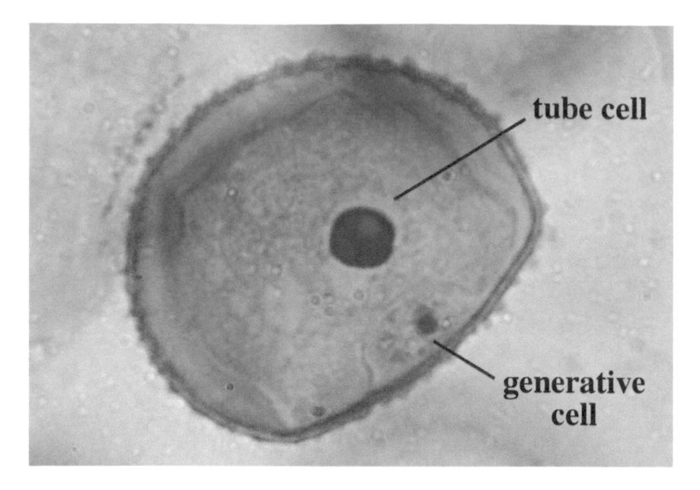

Figure 12.1 Binucleate pollen nuclear number. Note generative cell and tube cell nuclei.

Rarely, pollen grains will fuse in pairs, each pair known as a **dyad**. More commonly, the four haploid products of meiosis remain fused together, comprising a **tetrad**. Five types of tetrads are recognized, based on the arrangement of pollen grains: (1) **tetrahedral tetrads** (Figure 12.2A), in which the four grains form the points of a tetrahedron, e.g., as in most members of the Ericaceae; (2) **linear tetrads**, in which the four pollen grains are arranged in a straight line, e.g., as in *Typha* spp.; (3) rhomboidal tetrads, in which the four grains are in one plane, with two of the grains separated from one another by the close contact of the other two; (4) **tetragonal tetrads** (Figure 12.2B), in which the four grains are in one plane and are equally spaced apart; and (5) **decussate tetrads** (Figure 12.2C), in which the four grains are in two pairs arranged at right angles to one another.

Pollen grains that are connate in precise units of more than four are called **polyads** (Figure 12.2D). Polyads are common in the Mimosoid clade of the Fabaceae and generally consist of a multiple of eight fused grains. Fusion of pollen grains in large, often irregular numbers, but less than an entire theca,

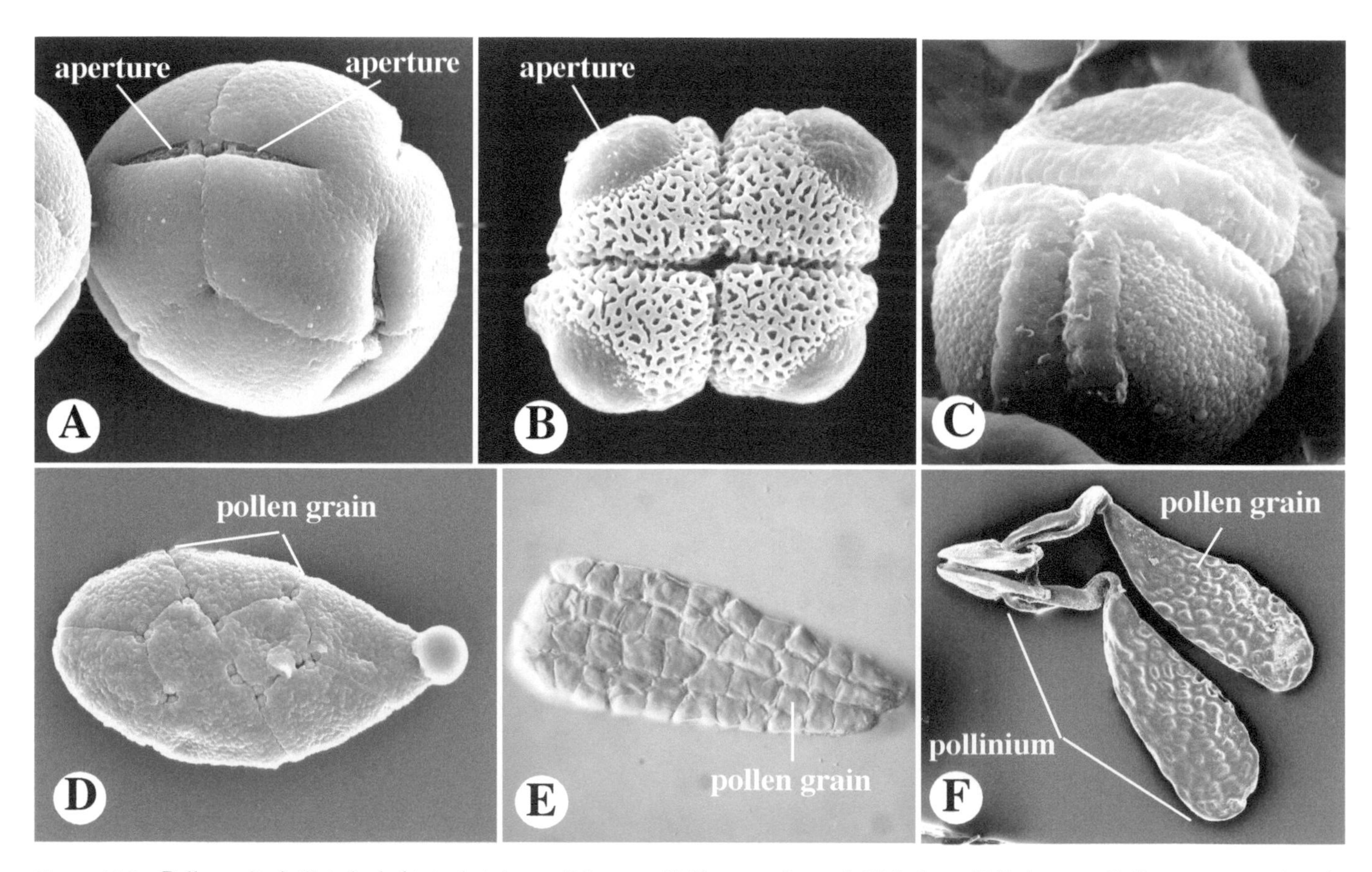

Figure 12.2 Pollen unit. **A.** Tetrahedral tetrad, *Arbutus*, Ericaceae. **B.** Tetragonal tetrad, *Philydrum*, Philydraceae. **C.** Decussate tetrad, early developmental stage, *Lachnanthes*, Haemodoraceae. **D.** Polyad of eight fused pollen gains, *Calliandra*, Mimosoideae, Fabaceae. **E.** Massula of *Piperia*, Orchidaceae. **F.** Pollinium of *Asclepias*, Apocynaceae.

are called **massulae** (singular **massula**) (Figure 12.2E). Finally, the fusion of all pollen grains of an entire theca is called a **pollinium** (plural **pollinia**), found in the families Apocynaceae (Figure 12.2F) and Orchidaceae (Chapter 7).

POLLEN POLARITY

Pollen polarity refers to the position of one or more apertures (see later discussion) relative to a spatial reference. This spatial reference defines a **polar axis** as the extended pollen grain diameter that passes through the center of the original pollen tetrad (Figure 12.3). The intersection of the polar axis with the grain surface near the center of the tetrad is the **proximal pole**, the surrounding area being the proximal face or proximal hemisphere; that away from the tetrad center is the **distal pole**, the surrounding area being the distal face or distal hemisphere. Just as with a globe, the intersection with the pollen surface of a plane at a right angle to the pole and passing through the center of the grain defines the pollen **equator**, the surrounding area being the equatorial region. Observing a pollen grain from the direction of either pole is known as a **polar view**; observing from the equatorial direction is an **equatorial view** (Figure 12.3).

The three general types of pollen polarity are (1) **isopolar**, in which the two polar hemispheres are the same but can be distinguished from the equatorial region; (2) **heteropolar**, in which the two polar hemispheres are different, because of differential displacement of one or more apertures; and (3) **apolar**, in which polar and equatorial regions cannot be distinguished after pollen grain separation from the tetrad. Note that pollen polarity is with reference to the microspore or pollen tetrad. Unless the mature pollen unit is a tetrad (above), pollen grain polarity can be directly determined only by observing the position of apertures during the early tetrad stage. Because this is rarely observed, polarity is generally inferred by comparison with taxa for which polarity has been directly observed.

POLLEN APERTURE

A pollen **aperture** (Figures 12.4, 12.5) is a specially delimited region of the pollen grain wall. (See Pollen Wall Structure.)

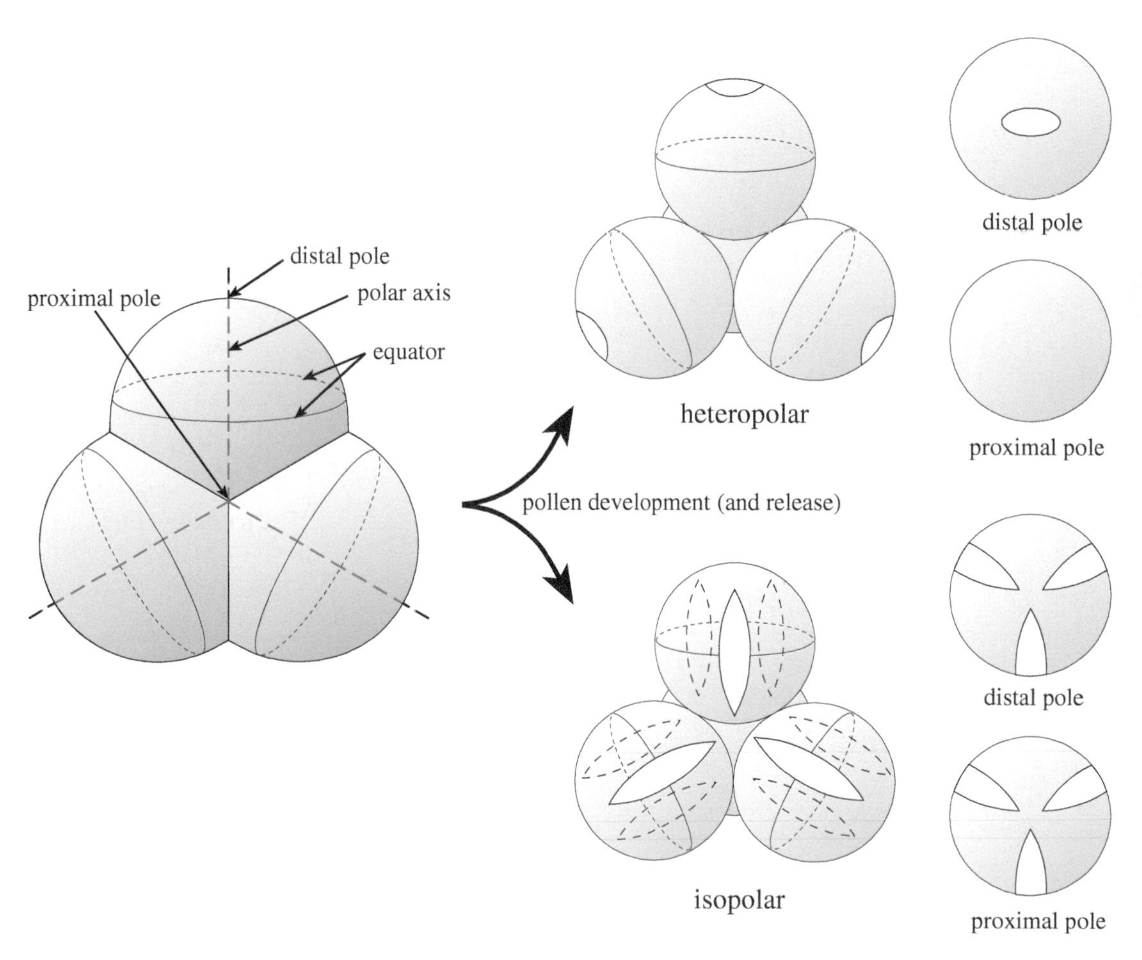

Figure 12.3 Pollen polarity.

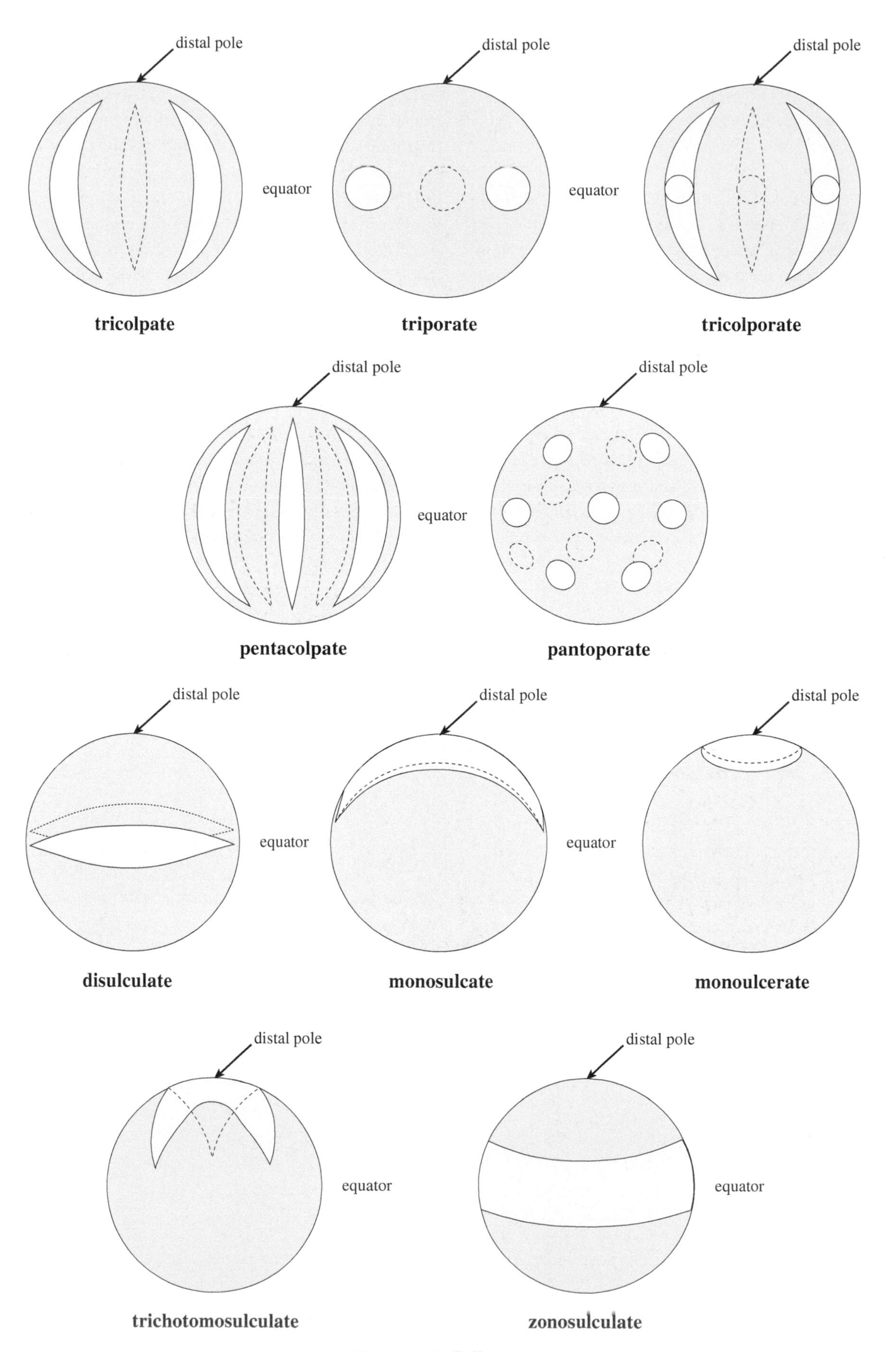

Figure 12.4 Pollen aperture.

Figure 12.5 Pollen aperture examples. **A.** Monosulcate. **B.** Tricolporate. **C.** Triporate. **D.** Diporate. **E.** Spiraperturate. **F.** Pentaporate. **G.** Ulcerate. **H.** Disulculate. **I.** Tricolporate (one aperture visible, with flanking "pseudo-apertures"). **J.** Pantoporate, with echinate sculpturing. **K.** Pantoporate. **L.** Zonosulculate.

The function of the aperture is primarily to serve as the site of formation of a pollen tube exiting from the pollen grain body. Apertures may also function to allow volume changes of the pollen grain with changes in water content, e.g., humidity. This feature is known as **harmomegathy**. Harmomegathy allows the pollen grain apertures to contract with water loss, effectively sealing the apertures via the surrounding desiccation resistant exine wall (see later discussion).

Pollen aperture type refers to the shape, number, position, and arrangement of the aperture(s) of a pollen grain, often with an implied reference to the polar axis. Rarely, pollen grains lack any recognizable aperture; these are termed **inaperturate**.

Two general types of apertures correspond to shape. A **colpus** (plural, colpi) is an elongate aperture with a length/width ratio of greater than 2:1 (Figure 12.4). Colpi can be elliptic, oblong, or fusiform in outline shape. A **porus** (plural, pori) is a circular to slightly elliptic aperture with a length/width ratio of less than 2:1 (Figures 12.4, 12.5C,D,F); if pori occur globally on the pollen grain surface, the aperture type is called **pantoporate** (Figures 12.4, 12.5J,K). An aperture that is shaped like a colpus but has a circular region in the center (corresponding to a different wall architecture) is termed **colporate** (Figures 12.4, 12.5B,I).

Pollen grains with apertures occurring in the equatorial region may be generally termed **zono-aperturate** (or stephanoaperturate), e.g., as in **zonocolpate** or **zonoporate**. However, the terms **colpus** and **porus** are often restricted to apertures occurring in a region of the pollen grain other than the poles (often centered at the equator), with the long axis oriented perpendicular to the equator. In contrast, an elongated aperture similar in shape to a colpus (length/width ratio >2:1) but either centered at a (usually distal) pole or, more rarely, parallel to the equator is called a **sulcus** (Figure 12.5A). For example, **disulculate** refers to a pollen grain with two elongated apertures on opposite sides of the grain and parallel to the equatorial plane (e.g., Pontederiaceae; Figure 12.5H). Comparably, a circular to slightly elliptic aperture similar in shape to a porus (length/width ratio <2:1) but occurring at the (usually distal) pole is called an **ulcus** (Figure 12.5G).

The number of apertures of any shape can be designated by appending the prefix mono-, di-, tri-, tetra-, penta-, hexa-, or poly- (more than six) to the terms **colpate** or **porate**. Thus, a tricolpate pollen grain is one with three, elongated apertures occuring in the equatorial region. A pentaporate pollen grain is one with five, approximately circular apertures occuring in the equatorial region.

Some aperture types are rather rare and specialized. **Syncolpate** refers to a pollen grain in which the colpi are joined, e.g., at the poles. **Trichotomosulcate** refers to an aperture type that is three-branched. Sulcate and ulcerate pollen grains typically have only a single aperture; these terms are usually equivalent to monosulcate and monoulcerate, respectively. **Spiraperturate** refers to one or more apertures that are spirally shaped (Figure 12.5E).

POLLEN SYMMETRY

Pollen symmetry is generally either radially symmetric, i.e., with two or more planes of symmetry, or bilaterally symmetric, with a single plane of symmetry. Symmetry is often incorporated or assumed as part of a shape term (see Pollen Shape).

POLLEN SIZE

Pollen size can vary tremendously across taxa. Size is typically measured in terms of *both* the polar diameter and the equatorial diameter (see Pollen Shape).

Typical pollen grains are ca. 25–50 μm in diameter, but pollen diameter can range from < 5 μm (approaching the size of some bacteria!) to > 200 μm.

POLLEN SHAPE

Pollen shape (Figure 12.6) may refer to the three-dimensional shape of a pollen grain; e.g., **boat-shaped**, **ellipsoid**, **fusiform**, or **globose/spheroidal**. Shape may also be assessed by the two-dimensional outline shape either in polar view or equatorial view, e.g., as viewed by focusing under a light microscope. The outline shape in polar view is known as **amb**. Amb can be nonangular, e.g., circular, elliptic, or angular, e.g., triangular, rhombic, rectangular, five-angled. For angular ambs, the shape of the sides may be described as straight, concave, or convex (Figure 12.6).

Another measure of pollen shape is the ratio of the polar diameter to the equatorial diameter, termed the **P/E ratio**. If the P/E ratio is approximately equal to 1, the grains are termed **spheroidal**. If P/E is greater than 1, the grains are termed **prolate** (i.e., elongated along the polar axis, like a cucumber); if less than 1, the grains are **oblate** (compressed along the polar axis, like a tangerine). The subcategories **prolate-spheroidal** and **oblate-spheroidal** are sometimes used for grains that are slightly prolate or oblate, respectively.

POLLEN SCULPTURING

Pollen sculpturing (Figure 12.7) refers to the *external* features of the pollen grain wall. Sculpturing features may be viewed with light microscopy, but much greater detail can be detected with scanning electron microscopy.

Specialized pollen sculpturing terms include:

baculate, having rod-shaped elements, each element termed a baculum, plural baculi;

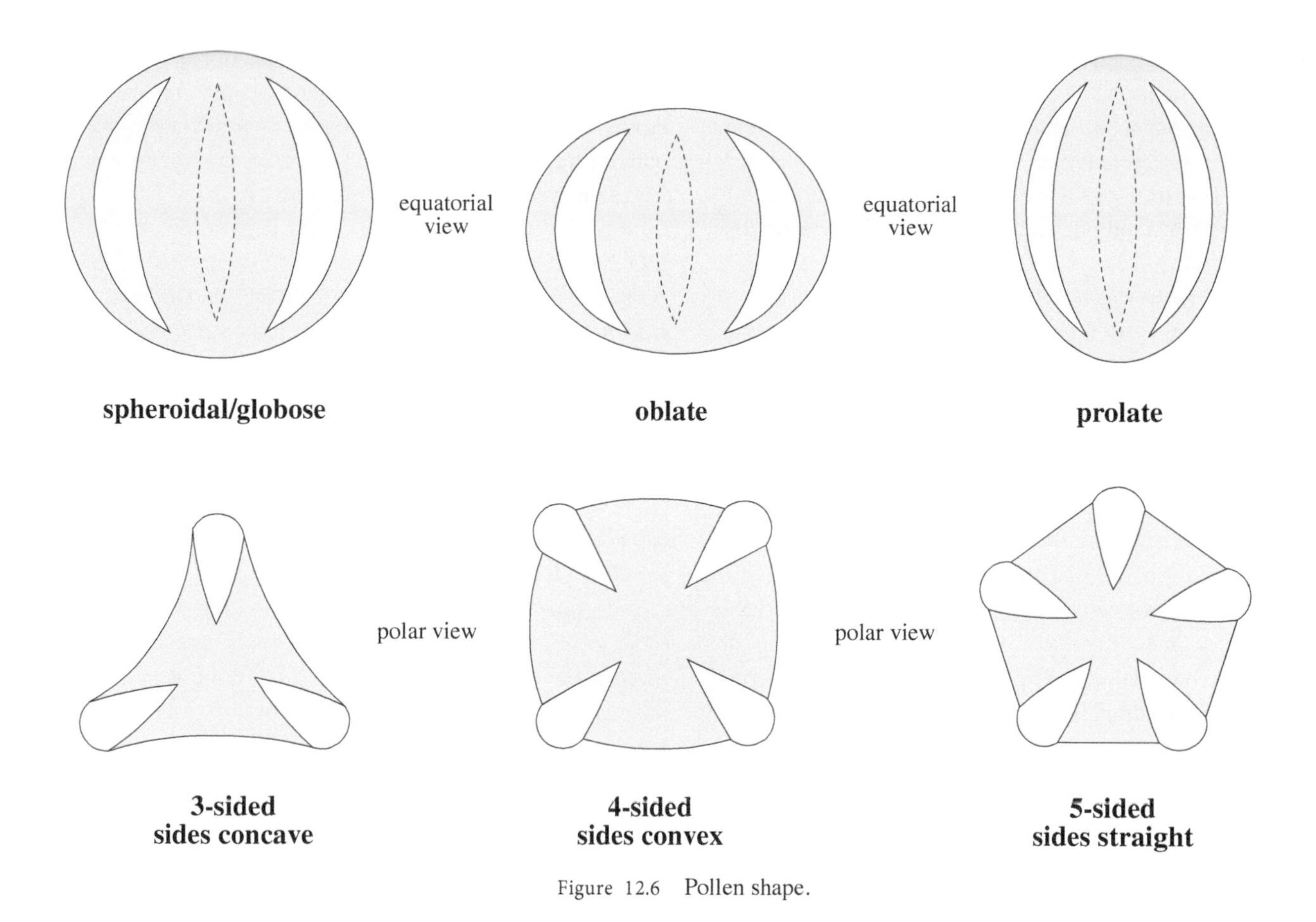

Figure 12.6 Pollen shape.

Figure 12.7 Pollen sculpturing. **A.** Echinate. **B.** Verrucate. **C.** Foveolate. **D.** Rugulose. **E.** Striate. **F,G.** Reticulate.

clavate, having club-shaped elements, each element called a clava, plural clavae;
echinate, having spinelike elements > 1 μm long, each element termed an echina, plural echinae;
fossulate, having longitudinal grooves.
foveolate, having a pitted surface caused by pores in the surface;
gemmate, having globose or ellipsoid elements, each element termed a gemma, plural gemmae;
psilate, having a smooth sculpturing;
reticulate, having a netlike sculpturing, each element termed a murus (plural muri) and the space between muri termed a lumen (plural lumina);
rugulate, having irregular to sinuous, tangentially oriented elements, often appearing brainlike;
spinulose (also termed **scabrate**), having spinelike elements < 1 μm long, each element termed a spinulum, plural spinuli;
striate, having thin, cylindrical, tangentially oriented elements;
verrucate, having short, wart-like elements, each element termed a verruca, plural verrucae.

POLLEN WALL STRUCTURE

The pollen grain wall functions primarily to provide structural support and protection of the cytoplasm from mechanical damage and desiccation. The wall may also function to facilitate pollination. For example, entomophilous (insect-pollinated) flowers tend to have elaborately sculptured pollen; these sculpturing elements may function to attach pollen grains to one another in masses and to appendages on the insect. Anemophilous (wind-pollinated) flowers tend to be smooth (psilate), functioning as a more efficient aerodynamic mechanism for wind transport.

The pollen grain wall may also function to store proteins involved in incompatibility reactions. Sporophytic incompatible taxa tend to have incompatibility proteins stored in cavities of the exine, derived from the sporophyte tapetum. Gametophyte incompatible taxa tend to have incompatibility proteins stored in the intine, derived from the microspore/pollen cytoplasm.

Pollen wall structure (Figure 12.8) refers to the *internal* form of the pollen grain wall. Early in development, microspores typically have a thick cell wall composed of callose, the same substance that lines the pores of sieve elements. During pollen development, however, the callose wall breaks down completely. Mature pollen walls almost always consist of two major layers: intine and exine. The **intine** is the innermost layer, which is composed primarily of cellulose and pectines, resembling the primary cell wall of a typical parenchyma cell. The **exine** is the hard, outermost, desiccation-resistant wall layer that provides the major structural support for the cytoplasm. Exine is impregnated with a substance called **sporopollenin**, a complex polymer of carotenoids, fatty acids, phenolics, phenylpropanoids, and carotenoids. Sporopollenin is very tough and resistant to mechanical damage and decay. The presence of sporopollenin accounts for the fact that pollen grains may often be preserved in the fossil record. The sporopollenin-impregnated exine is also resistant to **acetolysis**, which is a standard acid treatment used to dissolve all but the exine in order to better observe pollen wall structure with the light microscope.

The exine of many taxa may be divided into two layers, an inner **endexine** and an outer **ektexine**. These two layers differ chemically and have different staining properties as viewed with transmission electron microscopy. The endexine typically forms a more or less homogeneous, inner layer. The ektexine may exhibit a variety of structural forms. The most common type of ektexine in angiosperms is termed **tectate-columellate** (Figures 12.8, 12.9) and consists of an inner **foot layer**, a middle layer of radially elongated **columellae**, and an outer, rooflike layer called the **tectum**. In some taxa, the middle layer (given the generalized term *interstitium*) may

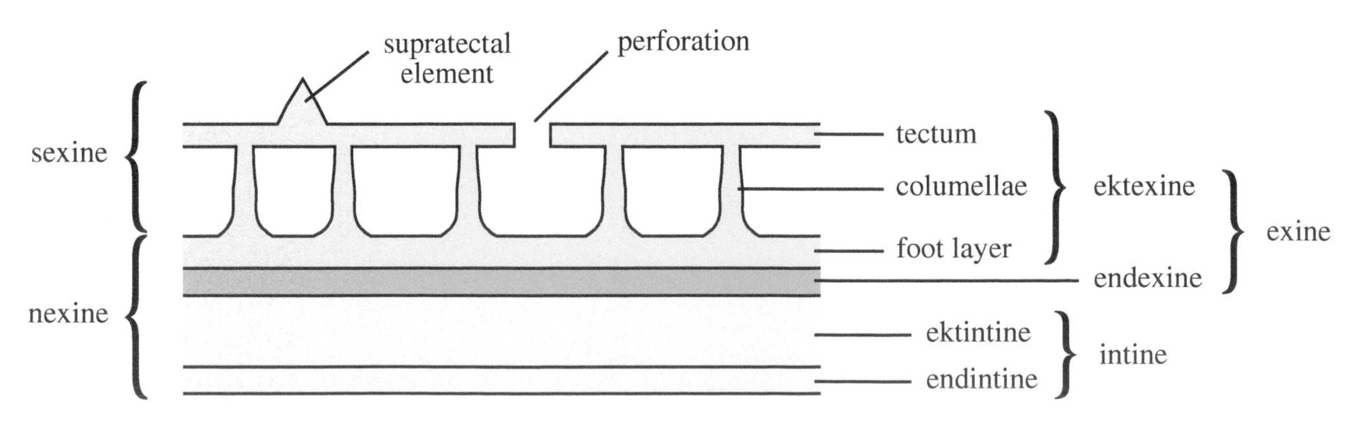

Figure 12.8 Pollen wall structure.

Figure 12.9 Example of tectate-columellate wall structure.

not be composed of columellae, but may instead consist of irregular or granular elements.

A tectate-columellate wall structure that lacks pores or perforations in the tectum is termed tectate-imperforate. In some taxa, the tectum contains tiny pores, a structure known as semi-tectate. A semi-tectate structure typically corresponds with a foveolate sculpturing type. A wall structure in which the tectum has large openings is called tectate-perforate. This may correspond, e.g., with a reticulate sculpturing.

Exinous elements on top of the tectum (described as supratectal) may account for sculpturing types such as baculate or echinate. However, in some taxa, a tectum may be absent; in these taxa protruding sculpturing elements such as baculae or echinae may be homologous to modified columellae. Only by viewing the wall structure internally may these differences be noted.

In addition to the tectate-columellate wall structure, various angiosperm taxa may have a wall that ancestrally lacks a tectum, termed an **atectate** wall structure. The exine wall of atectate taxa may be structurally solid, termed **homogeneous**, or **granular**, containing small, granular elements with intervening air spaces. Additional types of exine wall structure include **lamellar**, having stacked, tangentially oriented, planar structures, often constituting the inner wall layer; and **alveolar**, having numerous, spherical air pockets within the exine.

The layers of the exine can be precisely observed only using transmission electron microscopy. However, because many aspects of pollen wall structure may be observed using light microscopy, the terms **nexine** and **sexine** are sometimes applied to describe exine wall layers (Figure 12.8). **Nexine** refers to the inner layers, which may include both endexine and the foot layer of the ektexine. **Sexine** refers to the outer, protruding layers, which may include columellae, tectum, and supratectal sculpturing elements (if present).

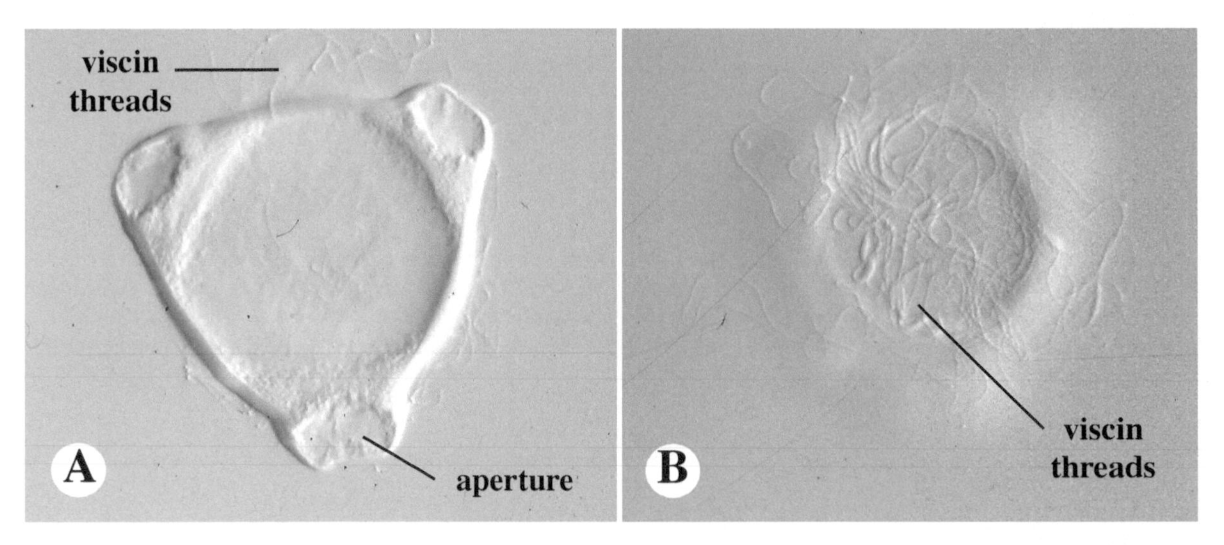

Figure 12.10 *Fuchsia* sp. (Onagraceae), having triangular, triporate pollen grains with viscin threads (arrow). **A.** Equatorial region in focus. **B.** Polar region in focus.

POLLENKIT AND VISCIN THREADS

Pollenkit is a yellowish or orange, carotenoid-like material adhering to the exine. It functions to stick pollen grains in masses, better effecting transfer of pollen by animal (esp. insect) pollinators.

Viscin threads (Figure 12.10) arc long strands of carbohydrate material that, like pollenkit, function in sticking pollen grains together.

PALYNOLOGY AND SYSTEMATICS

Palynological features have been very valuable in delimiting taxa or aiding in phylogenetic inference. An example of palynological features in a group of angiosperms is portrayed in Table 12.1.

TABLE 12.1 Example of palynological features of the Caryophyllales (Caryophyllidae), after Rodman. 1990. Centrospermae revisited, Part I. Taxon 39: 353–393. Character names are in brackets.

Pollen shed in binucleate or trinucleate condition **[Pollen nuclear number]**
Plastid-DNA not transmitted through pollen or transmitted
Aperturate pollen with furrows or not **[Pollen aperture]**
Aperturate pollen without pores or with pores **[Pollen aperture]**
Apertures simple or compound **[Pollen aperture]**
Number of apertures basically three, 4–7, or >7 **[Pollen aperture]**
Zonocolpate apertures (>3) absent or present **[Pollen aperture]**
Pollen surface not spinulose or spinulose **[Pollen sculpturing]**
Pollen surface punctate/perforate or not punctate/perforate **[Pollen sculpturing]**
Pollen surface not reticulate or reticulate **[Pollen sculpturing]**

PALYNOLOGICAL TECHNIQUE

Material preparation:

For studies of pollen morphology, it is best to obtain living material of anthers, just at the time they are dehiscing, and fix these in a chemical solution, such as alcohol or FAA (see Chapter 10). [For transmission electron microscope studies, other fixatives, such as glutaraldehyde, formalin, or osmium tetroxide, are used.] Collect plenty of material, and store in vials.

Light microscopic observations:

Pollen grains can be observed simply by making a "wet mount" on a microscope slide. A single anther can be removed from the fixative material, placed in a drop of water or 50% glycerol (the latter to prevent the material from drying out), and dissected with needles to extrude the pollen grains; the anther wall material should then be removed and a cover slip applied.

In addition, the pollen can be stained with either toluidine blue or basic fuchsin, in order to better visualize details of the apertures and wall sculpturing. Simply dissect the anthers in a drop of stain, remove the anther wall, and add a cover slip.

Another technique is to clear the pollen grains in a clearing solution and visualize them using phase contrast or differential interference contrast (DIC, also called "Nomarski") optics. A useful clearing solution is called "Hoyer's" clearing fluid. (Recipe: soak 30 g arabic gum lump in 50 g of water for 24 hours; add 200 g chloral hydrate (note: a controlled substance) until all the material dissolves; then add 20 g glycerine.) Dissect the anthers in a drop of Hoyer's as before, add a cover slip, and observe under phase contast or DIC optics. The pollen grains may need time to clear, but once they do, you can visualize many details of the wall and apertures.

The presence or absence of starch in pollen grains can be examined by staining the pollen with IKI stain; starch changes to a dark blue or black in the presence of this stain. In addition, pollen grains can be mounted in 50% glycerol and viewed with polarization optics; starch grains are birefringent and show a "Maltese cross" type pattern under polarized light.

REVIEW QUESTIONS

1. What is the study of spores and pollen called?
2. What are the two types of pollen nuclear number? Of pollen storage product?
3. What does pollen unit refer to?
4. What is the difference between a monad, tetrad, polyad, and pollinium?
5. What is pollen polarity?
6. What is the difference between an isopolar and heteropolar pollen grain?
7. What is the definition and function of a pollen aperture?
8. What is the difference between a colpus, porus, sulcus, and ulcerus?
9. What is a tricolporate pollen grain? A pentaporate pollen grain?
10. What is the size range of angiospermous pollen grains?
11. Name and define six terms that specify pollen sculpturing.
12. Name three functions of the pollen grain wall.
13. What are the two major layers of a pollen grain wall and how do they differ in chemical composition?
14. Name the two layers of exine.
15. Name and describe the most common type of exine wall structure.
16. What do nexine and sexine refer to?
17. What is the function of pollenkit or viscin threads?

EXERCISES

1. Using the simple procedures described earlier (Palynological Technique), examine pollen grains of various groups of angiosperms, including Magnoliids, Monocots, and several Eudicots, including a member of the Ericaceae (with permanent tetrads). Tabulate the differences in pollen unit, aperture type, aperture number, sculpturing type. Also, note the presence or absence of starch in the pollen grains.
2. Peruse journal articles in plant systematics, e.g., *American Journal of Botany*, *Systematic Botany*, or *International Journal of Plant Sciences*, or in specific palynological journals such as *Grana* or *Pollen et Spores* (see Appendix 3: Scientific Journals in Plant Systematics). Note those that describe palynological features in relation to systematic studies. Identify all pollen characters and character states described.

REFERENCES FOR FURTHER STUDY

Erdtman, G. 1966. Pollen Morphology and Plant Taxonomy. Angiosperms. Corrected reprint and new addendum. Hafner, New York.

Faegri, K., and J. Iversen. 1964. Textbook of Pollen Analysis. 1964. Blackwell Scientific, Oxford.

Hoen, P. Glossary of Pollen and Spore Terminology. Second and revised edition. *http://www.bio.uu.nl/~palaeo/glossary/glos-txt.htm* (An excellent Web page for pollen terminology, based on Punt et al. 1994).

Lewis, W. H., P. Vinay, and V. E. Zenger. 1983. Airborne and Allergic Pollen of North America. The Johns Hopkins University Press, Baltimore. [QK 658 L48 1983]

Moore, P. D., J. A. Webb, and M. E. Collinson. 1991. Pollen Analysis, 2nd ed. Blackwell Scientific, Oxford.

Nilsson, S., and J. Praglowski (eds). 1992. Erdtman's Handbook of Palynology. Munksgaard International, Copenhagen.

Punt, W., S. Blackmore, S. Nilsson, and A. le Thomas. 1994. Glossary of Pollen and Spore Terminology. LPP Foundation, Utrecht.

Radford, A. E., W. C. Dickison, J. R. Massey, and C. R. Bell. 1974. Vascular Plant Systematics. Harper & Row, New York.

Reitsma, T. 1970. Suggestions towards unification of descriptive terminology of angiosperm pollen grains. Reviews in Palaeobotany and Palynology 10: 39–60.

Walker, J. W., and J. A. Doyle. 1975. The Bases of Angiosperm Phylogeny: Palynology. Annals of the Missouri Botanical Garden 62: 664–723.

13

PLANT REPRODUCTIVE BIOLOGY

Plant reproductive biology is the study of the mechanisms and processes of sexual and asexual reproduction in plants. It may encompass the study of pollination mechanisms, gene flow, genetic variation, and propagule dispersal between and within populations. A knowledge of the reproductive mechanisms of plants can help assess the adaptive significance and homology of features studied in plant systematics. Assessing reproductive biology can also give an insight into the delimitation and classification of species and infraspecies.

The following is a very abbreviated summary of the concepts and terms used in reproductive biology as they may be significant in studies of plant systematics.

SEXUAL REPRODUCTION

In nonseed plants, sexual reproduction entails the release of motile sperm from a free-living gametophyte into the outside environment. The sperm swims in a film of water into the neck of an archegonium, fertilizing the egg to form a zygote and then embryo. Completion of this phase of the life cycle is dependent on survivorship of the gametophytes, on the effective development and operation of antheridia and archegonia, and on the proper external conditions. The sporophytes of nonseed plants generally release massive numbers of spores into the environment, which are transported by wind or, more rarely, by water. These spores may, upon encountering the proper environmental conditions, germinate and grow into a gametophyte, completing the cycle (see Chapters 3, 4).

In seed plants, separate male and female gametophytes are produced within male and female spores (endosporic microspores and megaspores). Sex involves the transfer of male gametophytes, the pollen grains, either to the micropyle of an ovule (in gymnosperms) or to the stigma of a pistil (in angiosperms). Sperm cells are ultimately released (into or just outside of the female gametophyte of the ovule), where one sperm cell fuses with the egg, initiating development of an embryo within the seed. Seeds are then transported by a variety of mechanisms to a new environment (see Chapters 5, 6, 9).

Thus, two major processes in sexual reproduction of seed plants are **pollination**, the transfer of pollen grains from microsporangia to the ovule or stigma, and **fertilization**, union of sperm and egg. Many of the structural modifications of seed plants function in this transfer of pollen and the subsequent development and propagation of seeds.

In gymnosperms—cycads, *Ginkgo*, conifers, including Gnetales—pollen grains are mostly transported by wind. Because transport by wind is indirect, it necessitates the production of relatively large numbers of pollen grains to overcome the very low probability that any given pollen grain will make it to the ovule. In contrast, the great majority of angiosperms are animal (mostly insect) pollinated, which appears to be the ancestral condition for the group (Chapter 6), although wind pollination has arisen secondarily in several groups of flowering plants (see later discussion).

https://doi.org/10.1016/B978-0-12-812628-8.50013-4,

FLOWERING PLANTS

Angiosperms have largely evolved very specialized floral structures that are adaptive in promoting animal pollination. Animal pollination is much more directed and precise, necessitating the synthesis of many fewer pollen grains to effect fertilization of the eggs within ovules.

The basic adaptive "strategy" of animal-pollinated flowering plants has been the evolution of an **attractant**. The attractant works to entice the animal to the flower, either by vision or by odor. A visual attractant is usually a showy perianth (corolla and/or calyx) that may be brightly colored or otherwise contrasting with the external environment, e.g., a white perianth at night. Other floral parts, such as stamens (e.g., *Hibiscus*), staminodes (e.g., members of the Aizoaceae, Cannaceae, or Zingiberaceae), corona (e.g., *Crinum*, *Narcissus*, *Passiflora*), or even the gynoecium, may replace or augment the perianth as a visual attractant. Individual flowers may actually be small, but the accumulation of flowers in an inflorescence may provide a significant visual attractant. Olfactory attractants include the volatile compounds emitted by flowers, usually from the surface of the perianth. Most odiferous flowers have a sweetish smell, but some smell like rotting flesh or feces (see below).

Many species of flowering plants have evolved structures or exudates that act as a **reward**, ensuring that the animal pollinator will consistently return to transport pollen. The most common floral reward is **nectar**, a fluid primarily rich in sugars, secreted from specialized regions or organs of the flower called **nectaries** (Chapter 9). Nectaries are specialized tissues or organs that may be located within the gynoecium (e.g., the "septal" nectaries of many monocots), on the perianth, or at the base of and often surrounding the gynoecium or androecium. Another pollination reward is pollen itself, which is a relatively rich source of protein. Some flowering plants produce waxes (e.g., *Krameria*) or oils as a reward. Finally, in some rare cases, insects may obtain specific chemical compounds that are used to attract a mate.

Although nectar usually functions as a food source and reward for the prospective pollinator, some nectaries are "extra-floral," produced outside the flower, and function as a reward not for pollination but for protection of the plant. This is most common in plant interactions with ants, known as **myrmecophily**. Plant structures such as extra-floral nectaries, **domatia** (specialized chambers that house ants or other arthropods), and **Beltian bodies** (nutritious structures that ants eat) all may serve as rewards for ants. The ants do not generally pollinate the plant, but protect the plant from herbivores or parasites, a mutualistic symbiotic association.

Although the general strategy of pollination in most flowering plants is to provide a reward (thus, presumably, increasing the fitness of both plant and animal), not all animal-pollinated flowers do this. Some flowers have evolved structures or mechanisms to **deceive** the animal to transport pollen, possibly with an adverse affect on the reproductive success of the animal. For example, in certain water lilies and orchids, the nectar may actually function to trap or even drown the insect to promote pollination. Other species of orchids actually mimic (visually and olfactorily) the female of an insect (usually a wasp), fooling the male to attempt to copulate with the flower, which, in the process, transports pollen. Flowers of some species emit compounds that mimic the smell of rotting flesh or feces, attracting flies (see **Fly pollination**, below). The flies enter sometimes complex floral structures, perceiving the tissue as real rotting flesh, and in the process effecting pollination.

POLLINATION MECHANISMS

Many, if not most, species of angiosperms have evolved specialized pollination mechanisms in which structural modifications are correlated with a specific agent of transferring pollen. Knowledge of the pollination agent can give an insight into the function, homology, and evolution of associated floral features. The following are a summary of these pollination mechanisms or "syndromes."

Insect pollination (or **entomophily**) is undoubtedly the most common type in angiosperms. **Bee pollination** (**melittophily** or **hymenopterophily**) is correlated with flowers that tend to be showy, colorful, and fragrant. The flowers often have specialized color patterns called nectar guides (Figure 13.1A), which function to attract and orient the bee to maximally effect pollination. In many bee pollinated flowers, nectar guides may be located on the anterior perianth part(s) (usually petals or corolla lobes), modified as landing platforms (Figure 13.1A), on which the bee lands to more efficiently gather nectar or pollen and more effectively cause pollination.

Beetle pollination (**cantharophily**), often thought to have been the ancestral type in the angiosperms, is correlated with open flowers, the sexual organs exposed, often with a fruity or foul odor. Some beetle pollinated flowers (and cycad cones) produce heat internally, presumed to function in more effectively dispersing odor-generating compounds.

Butterfly pollination (**psychophily**) is associated with showy, colorful, and fragrant flowers, usually with no nectar guides. The flowers tend to have long, nectar-filled tubes or spurs (Figure 13.1B), preventing all but an insect with a long proboscis from acquiring the nectar.

Moth pollination (**phalaenophily**) is correlated with night-blooming large, white, and fragrant flowers, with no nectar guides; as with butterfly pollination, the flowers often have long, nectar-filled tubes (Figure 13.1C) or spurs. One interesting example of moth pollination occurs in species of *Yucca*

Figure 13.1 Flower modifications associated with specific pollination mechanisms. **A.** *Penstemon eximius*, bee-pollinated. Note nectar guide and landing platform. **B.** *Nuttallanthus texanus*, toad-flax, butterfly pollinated, with corolla-spur. **C.** *Ipomoea alba*, moon flower, moth pollinated, white with long corolla tube. **D.** *Stapelia gigantea*, star flower, fly-pollinated. **E.** *Selenicereus* sp., night-blooming cereus, bat-pollinated. **F.** *Couroupita guianensis*, cannonball tree, bat pollinated. **G.** *Erythrina caffra*, coral bean tree, bird pollinated. **H.** *Anigozanthos manglesii*, kangaroo-paw, bird-pollinated. **I.** *Bromus* sp., brome grass, wind-pollinated.

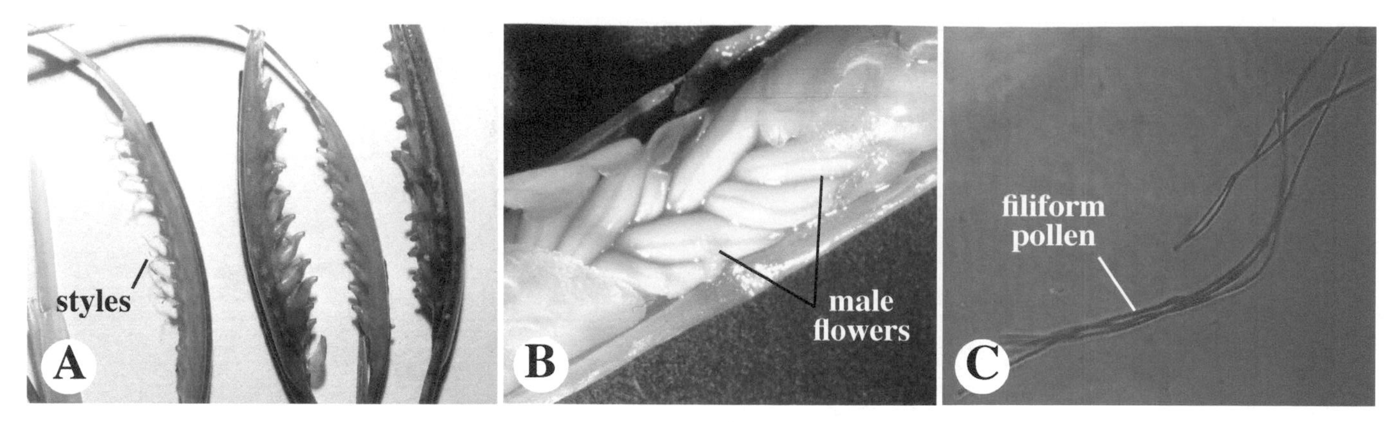

Figure 13.2 *Phyllospadix torreyi*, surf-grass, a water-pollinated angiosperm. **A.** Female plant. Note styles at left. **B.** Close-up of male flowers. **C.** Elongated, filiform pollen grains.

and relatives (Agavaceae), which are exclusively pollinated by yucca moths (*Parategeticula* and *Tegeticula* spp.). Yucca moths, in addtion to pollinating *Yucca* flowers, deposit their eggs only within the ovary of *Yucca* plant species. Thus, the *Yucca* plant and yucca moths are obligately dependent upon each other for procreation.

Fly pollination (**sapromyiophily**) is correlated with flowers that are often maroon or brown in color and emit a fetid odor that simulates the smell of rotting flesh, these visual and olfactory attactants effecting deceptive pollination. Examples of these are *Aristolochia, Arum,* and *Stapelia* spp. (Figure 13.1D). In some of these, flies may lay their eggs, which fail to develop because of the absence of a suitable food source.

Bat pollination (**cheiropterophily**) is correlated with flowers that open at night (have a nocturnal anthesis), and are large, white or colorful, with copious production of pollen or nectar (often secreted into a hypanthium or perianth tube), either or both of which may serve as a reward. When pollen is the reward, stamens tend to be numerous (Figure 13.1E,F).

Bird pollination (**ornithophily**) tends to be correlated with red, relatively large, and often tubular flowers that secrete copious nectar as a reward (Figure 13.1G,H). Sometimes the "tube" results from tightly wrapped but distinct perianth parts (e.g., the cactus *Cleistocactus*, pollinated by hummingbirds).

Wind pollination (or **anemophily**) is correlated with small, numerous, often unisexual flowers that tend to have a reduced, nonshowy, or absent perianth (Figure 13.1I). Pollen is produced in large quantities and pollen grains tend to have a smooth (*psilate*) wall sculpturing. Styles tend to be highly branched as a more efficient means of catching pollen grains in air currents. Anthers and styles may be erect or pendant. Wind pollination is found in several flowering plant groups, such as the Fagaceae (e.g., oaks), Betulaceae (e.g., birches), Salicaceae (poplars and willows), and many Poales (grasses and their close relatives). Some wind-pollinated taxa are quite specialized, such as *Alexgeorgea* (Restionaceae, a grass relative), in which the flowers are underground but in which the emergent styles and stamens undergo wind pollination. As discussed earlier, wind pollination is predominant in the nonflowering seed plants, the gymnosperms (see Chapter 5).

Water pollination (**hydrophily**) may occur in aquatic plants with flowers either at or under the water surface. For example, *Vallisneria* (Hydrocharitaceae) releases tiny male flowers that float to the surface, where they may float to the enlarged stigmas of a relatively large female flower. Some sea grasses, such as *Phyllospadix* (Figure 13.2) have very elongate, filiform pollen grains (Figure 13.2C), making them considerably more efficient in being captured by the styles and stigmas of female flowers in ocean currents (Figure 13.2A).

BREEDING SYSTEMS

Plants can be predominately outbreeding, inbreeding, or some mixture of the two. In many flowering plants specific mechanisms have evolved that promote one of these systems.

Outbreeding, also called **outcrossing**, **allogamy**, or **xenogamy**, is the transfer of gametes from one individual to another, genetically different individual. The general advantage of outbreeding is to promote an increase in phenotypic variability within a population. This generally enables plants to adapt to a wider range of environmental conditions and increases the likelihood for survival and evolutionary change. One disadvantage of outbreeding is that it requires a transfer of gametes between individuals. If individuals are far apart, or if pollinators are scarce, sexual reproduction may not occur at all in obligately outbreeding species.

The probability of outbreeding can be increased by a variety of mechanisms. In the flowering plants, **dioecy**, in which individual plants have either male (staminate) or female (pistillate) flowers, ensures that outbreeding will always occur. Many flowering plants exhibit a modified type of dioecy in

Figure 13.3 **A,B.** Protandry in *Encelia californica* (Asteraceae). **A.** Style is elongating through anther tube, pushing pollen outward. **B.** Style has finished elongating, and style branches have unfolded. **C.** Protogyny in *Suaeda* (Chenopodiaceae). Styles have already matured and are drying up at time of pollen release.

which some individuals have flowers of one sex but others have bisexual (perfect) flowers. These include **gynodioecy** (some individuals with pistillate flowers only, others with perfect flowers), **androdioecy** (some individuals with staminate flowers only, others with perfect flowers), and **trioecy** (some individuals with staminate flowers only, some with pistillate flowers only, and some with perfect flowers). These alternative mechanisms may promote outcrossing but also allow for some inbreeding (see later discussion), ensuring that at least some seed will be set. (As discussed earlier, most gymnosperms are dioecious; see Chapter 6.)

Another outcrossing mechanism arises from differences in *timing* of maturation of male and female floral parts, a feature known as **dichogamy**, of which there are two general types. **Protandry** (Figure 13.3A,B) is the precocious development of the androecium, as occurs, e.g., in many members of the Apiaceae, Asteraceae, and Campanulaceae. In protandrous species the pollen matures and is released prior to the maturation and receptivity of the gynoecium. **Protogyny** (Figure 13.3C) is the precocious development of the gynoecium, as occurs, e.g., in some Chenopodiaceae. Both protandry and protogyny promote outcrossing when flowers of different individuals mature at slightly different rates. Thus, the pollen from one flower will not normally pollinate that same flower, but can pollinate a different flower in which the gynoecium is receptive. In protandrous and protogynous species, outcrossing is ensured only if the flowers from a given individual mature at the same time. In reality, most of these species have flowers aggregated together into inflorescences, in which a range of developmental stages may be present. In any case, at least some outcrossing may occur, and pollination within a single flower is normally prevented. However, if the pollen is not removed from a flower, it may in some cases pollinate the gynoecium of that same flower. This provides a fail-safe mechanism for producing seeds even in times or environments where pollinators are lacking.

Outcrossing has also been enhanced by evolutionary changes in floral structure, particularly the *spatial separation* of anthers and stigmas, a phenomenon known as **hercogamy**, also spelled "herkogamy." (However, hercogamy may also function to prevent interference of male and female functions in the flower, by physically separating them.) One type of hercogamy is **heterostyly**, in which the relative lengths or heights of stigmas versus anthers vary among different flowers. (Most flowers are monomorphic or homostylous, whereby the height of stigmas and anthers is relatively constant.) In so-called **distylous** species, two floral morphologies occur: **pin** flowers, with a long style and short stamens, and **thrum** flowers, with a short style and long stamens (Figure 13.4A). In this syndrome, an insect visiting a pin flower is likely to have pollen deposited on its body in a location that would effect pollination of a thrum flower rather than another pin flower, and vice versa. This increases the probability of pollination between flowers rather than within flowers. If individuals tend to have one floral type or the other, outcrossing would be ensured. A rarer, more complex situation occurs in species that are **tristylous** (Figure 13.4B), with three heights of styles and stamens; the principle for cross-pollination is the same.

Another type of hercogamy is **enantiostyly** (or *enantiomorphy*), the curvature of the style to either the left or the right (as seen in face-view), defining "left-styled" and "right-

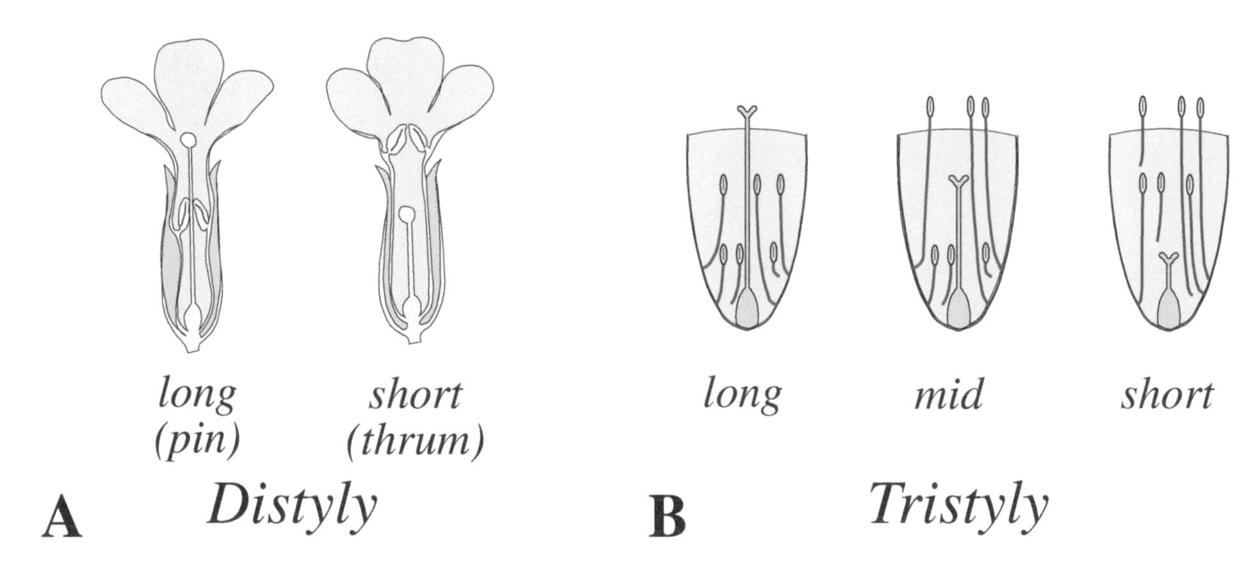

Figure 13.4 Flower heteromorphism. **A.** Distyly. (Redrawn from Weberling. 1989. Morphology of flowers and inflorescences. Cambridge University Press, Cambridge, New York.) **B.** Tristyly. (Redrawn from Kohn et al. 1996. Evolution 50: 1454–1469.)

styled" flowers (Figure 13.5). This style curvature usually corresponds with a curvature of at least one stamen to the side opposite the style. As with heterostyly, enantiostyly results in the preferential deposition of pollen on one side of, say, an insect pollinator's body. For example, an insect visiting a left-styled flower would tend to get pollen deposited on the right side of its body. Thus, the insect would more likely pollinate a right-styled flower as opposed to another left-styled flower. If an individual plant is constant as to the curvature of the style, enantiostyly will greatly promote the probability of outcrossing.

Yet another type of hercogamy involves movement of floral parts. One type of **movement hercogamy** is the rapid closure of the stigmas upon their being touched by a potential animal pollinator (e.g., *Diplacus*, Phyrmaceae). If the stigmas first receive pollen from a pollinator, their rapid closure can physically prevent pollen from the same flower being transferred to the stigmatic region, effectively preventing intra-floral self-pollination; however, this mechanism may actually function to clear the stigma from the path of a potential pollinator. Another type of movement hercogamy involves **trigger mechanisms** (e.g., *Kalmia*, *Stylidium*), whereby an insect pollinator triggers the sudden movement of one or more stamens, dusting the insect with pollen at the point of contact (Figure 13.6). The pollen is then at a location to be effectively transmitted to the stigma of another flower.

Finally, outcrossing can be promoted by genetically determined self-incompatibility mechanisms. **Self-incompatibility**

Figure 13.5 Example of enantiostyly in *Wachendorfia thyrsiflora* (Haemodoraceae). Note curvature of style with corresponding opposite positioning of stamen. **A.** Right-styled flower. **B.** Left-styled flower.

Figure 13.6 Example of a trigger mechanism in *Stylidium* (Stylidiaceae). **A.** Prior to being "triggered." **B.** Pencil has triggered single stamen to release pollen grains.

refers to the inability for fertilization to occur between gametes derived from an individual genotype. Because this is genetically determined, the incompatibility operates both within a single flower and between flowers of one individual. There are two basic types of self-incompatibility. Gametophytic self-incompatibility is controlled by the genetic composition of the male gametophyte. Sporophytic self-incompatibility is controlled by the genetic composition of the sporophyte, specifically the stigma and style of the pistil.

Inbreeding, also called **selfing**, is the union of gametes derived from a single individual. In flowering plants, inbreeding may occur either within a single flower, known as **autogamy** (**infrafloral selfing**) or between flowers derived from one individual, known as **geitonogamy**. (The genetic product of autogamy is identical with that of geitonogamy.) A major evolutionary advantage of inbreeding is enabling reproduction to occur when there are relatively few (or even one) individuals present in a population or at times when pollinators are rare, e.g., in ephemeral habitats. The disadvantage of inbreeding is that it reduces variation in a population and can even result in the accumulation of deleterious alleles, a phenomenon known as inbreeding depression.

Some plant species have both outcrossing and selfing flowers, a breeding system known as **allautogamy**. For example, species of *Viola* (Violets) and *Clarkia* have two types of flowers. **Chasmogamous** flowers are typical ones in which the perianth opens and exposes the sexual organs, with subsequent cross-pollination common. Other flowers, however, are **cleistogamous**, in which the perianth remains closed, such that pollen produced from within the flower pollinates only the stigma(s) of that flower. In still other species, both self- and cross-pollination may occur within the same flower. For example, *Myosurus*, mousetail, has numerous, small pistils born on a receptacle that elongates during flower maturation. When the flower first blooms, the receptacle is short, and the pistils that mature tend to be pollinated by the low, surrounding anthers. As the receptacle elongates, however, the pistils are positioned high above the anthers and are more likely to be pollinated by an insect visitor carrying pollen from another flower. Allautogamous breeding systems are adaptive in promoting some outcrossing, which may increase overall genetic variation, but also ensuring seed set regardless of availability of pollinators via inbreeding.

Additional reproductive strategies may be correlated with overall timing of sexual reproduction. **Iteroparous** plants are those that reproduce more than one time in the life of the plant, typically in regular cycles. Iteroparity is very common in angiosperms, ensuring regular seed set. **Semelparous** plants are those in which plant resources are utilized entirely for one episode of reproduction, followed by degeneration and death of the entire plant (i.e., the plants are **monocarpic**; Chapter 9). Semelparity occurs in all annual and biennial plants, but in very few perennials (occurring, e.g., in *Agave* and *Bambusa* spp.). Semelparity in perennial plants may be a strategy for deceiving or overwhelming potential seed predators, the former by not reproducing seasonally, the latter by producing seeds in such numbers that predators cannot consume them all.

Other temporal phenomena may be correlated with the breeding system of a plant species. For example, in annual or deciduous plants, the relative timing of leaf versus flower development may influence pollination and/or seed dispersal. Two general temporal patterns are **synanthous**, in which leaves and flowers develop at the same time, and **hysteranthous**, in which leaf and flower development do not coincide. Wind-pollinated plants are sometimes hysteranthous (e.g., in the willows, *Salix* spp.), with flowers maturing and releasing pollen before leaves form, thus, more effectively transmitting pollen in the canopy region of a community of trees.

SEED AND FRUIT DISPERSAL

The evolution of the numerous types of fruits and seeds (Chapter 9), which are used to delimit many taxa, is strongly correlated with their function as dispersal devices. Many mechanisms for dispersal of seeds and fruits have evolved in the angiosperms, including (1) **wind dispersal** (or **anemochory**), as in the samaras of *Ulmus*, elms, and *Acer*, maples, and the winged seeds of *Liquidambar*, sweetgum), including those wind dispersed by tumbling (e.g., the tumble weeds, such as *Salsola*); (2) **water dispersal** (or **hydrochory**), as occurs in the ocean-dispersed fruits of *Cocos nucifera*, coconut; (3) dispersal by **explosive dehiscence** of fruits, as in the explosively dehiscent capsules of *Ceanothus* or *Impatiens*; (4) **self-dispersal** (or **autochory**), as in *Arachis hypogaea*, peanut, which buries its own fruits; or (5) **animal dispersal** (or **zoochory**). Zoochory is divided into **epizoochory** (**ectozoochory**), in which propagules are carried on the outside of an animal (as in the burs of *Xanthium*, cocklebur, or the loments of *Desmodium*, sticktight), and **endozoochory**, in which seeds are eaten (the fruit pericarp or fleshy seed coat or aril being an award or attractant) but are passed through the gut of the animal unharmed. In endozoochorous plants, animal ingestion is often obligatory, the digestive fluids releasing dormancy of the seed.

ASEXUAL REPRODUCTION

Many species of land plants will regularly produce offspring without sex. The advantage of asexual reproduction is that numerous propagules can be generated relatively quickly and efficiently, without reliance on the transfer of gametes. However, the major evolutionary disadvantage is that no genetic variability results. One type of asexual reproduction is vegetative reproduction, the production of genetic clones from vegetative tissue. (Clones of an individual are known as **ramets**, whereas genetically different individuals are called **genets**.) Vegetative plant clones may form by the development of aerial plantlets (e.g., develop along the leaf margins of *Kalanchoe daigremontiana*, maternity plant). Cloning may also result from stolons, rhizomes, bulbels, cormels, etc., that may become dispersed or physically detached from the genetically identical parent plant. Some clones of creosote bush (*Larrea* sp.) are calculated to have persisted in nature for several thousands of years and may represent the oldest known clonal organisms on earth.

Agamospermy is the production of seeds without fertilization. In some species, agamospermy requires pollination to form seeds, though fertilization never occurs. Embryos of agamospermic seeds are genetically identical to the parent plant. The embryo may develop from a cell of an abnormal, diploid female gametophyte, such as a diploid egg, this being **parthenogenesis**. Alternatively, the embryo may arise from a cell of the surrounding tissue, such as from megasporangial (nucellar) or integument tissue, which is called **adventive embryony**. Agamospermy may be facultative, occurring in addition to normal sexual reproduction if flowers are unpollinated. Alternatively, some species are obligately agamospermous, an example being *Taraxacum officinale*, the common dandelion. Evidence for agamospermy includes (1) the occurrence of viable seed in absence of males or after experimental bagging or emasculation of flowers (see later discussion); (2) precocious embryo formation, i.e., prior to anthesis or pollen maturation; (3) adventive embryos, budding from nucellar or integument tissues; (4) multiple embryo and seedling formation from a single seed; (5) formation of seeds in the absence of males in nature, such as in dioecious species.

HYBRIDIZATION, INTROGRESSION, AND POLYPLOIDY

Hybridization is usually defined as sexual reproduction between different species, specifically termed *interspecific hybridization* (although the term can be used for sexual reproduction between different populations or infraspecific taxa within a species). Hybridization has been observed to be relatively common in plants, more so than in most groups of animals.

Two different species or populations of plants cannot interbreed if they are geographically isolated. However, interbreeding also may not occur if plants exhibit one or more genetically determined traits that prevent or inhibit gene exchange. These genetic, reproductive isolating features include (1) differences in habitat; e.g., one species adapted to a wet environment may not be capable of interbreeding with one that is adapted to a dry environment simply because the two species are rarely in close enough proximity to allow crossing; (2) differences in timing of reproduction; e.g., one plant flowering in the spring, the other in the late summer; (3) differences in floral morphology, such that the pollinators of one do not visit the flowers of the other; and (4) genetic incompatibility, such that the pollen of one species will not germinate on the stigma of the other (or traverse the style) or sperm of one is incompatible with the egg of the other.

Hybridization between two plant taxa may only occur if these taxa are genetically similar enough. Any hybrid progeny that are produced may be fertile (capable of sexual reproduction) or sterile, the latter often the result of

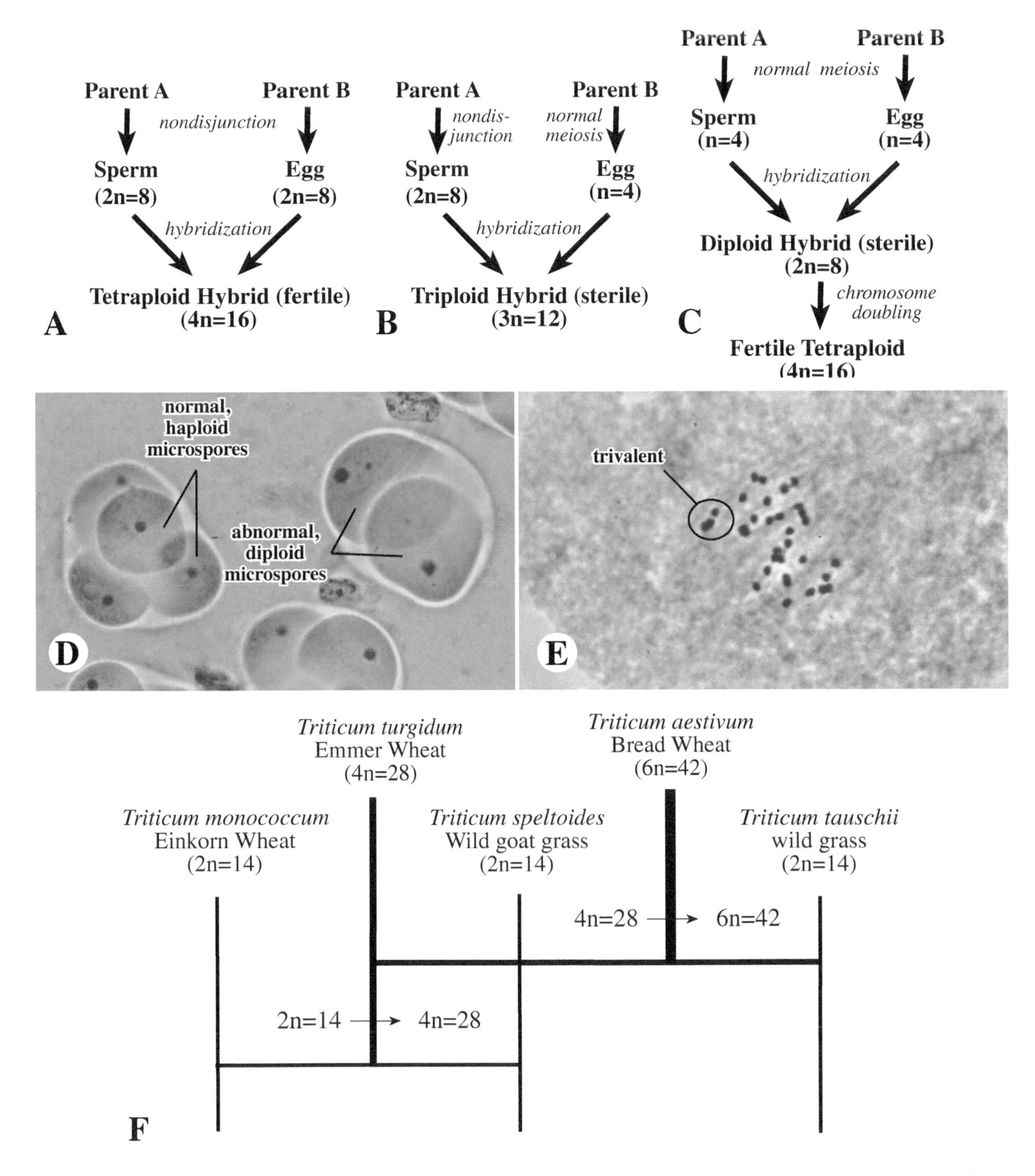

Figure 13.7 Polyploidy. **A,B.** Mechanisms by which tetraploidy and triploidy can arise in nature by meiotic nondisjunction, resulting in diploid gametes. **C.** Mechanism by which a tetraploid individual can arise by somatic chromosome doubling in a sterile hybrid. **D.** Pollen development in *Cylindropuntia* sp., showing normal tetrad of haploid microspores (left) and abnormal dyad of diploid microspores (right), the latter precursors to diploid pollen grains. **E.** Chromosome squash (*Cylindropuntia prolifera*, a sterile triploid), showing groupings of three homologous chromosomes (trivalents) during meiosis, indicative of triploidy. **F.** Evolution of wheat, *Triticum* spp., via polyploid events. Horizontal lines represent interbreeding between species, the hybrid subsequently undergoing polyploidy. (Photos at D and E courtesy of Jon Rebman.)

irregularities in meiosis, resulting in sterile or noncompatible gametes.

Introgression is hybridization between two species followed by backcrossing to one or both parents. The importance of introgression is that it can be a mechanism of promoting some gene flow between two different species, ultimately increasing the genetic variability or fitness of one or the other species.

Polyploidy is a large-scale mutation in which offspring have a multiple of some ancestral set of chromosomes. Polyploidy can occur either within a species (**autopolyploidy**) or between different species (**allopolyploidy**).

Polyploidy can occur in two general ways. One way that polyploidy can occur is by the production of gametes that have more than one set of chromosomes (Figure 13.7A,B). Diploid gametes can result from an irregularity during meiosis termed **nondisjunction**, in which homologous chromosomes do not segregate; if this occurs with all homologous chromosome pairs, then the daughter cells may be unreduced (i.e., diploid, not haploid). [An unreduced (diploid) pollen grain can sometimes be detected microscopically, whereby only two, larger microspores (see Chapter 12) are detected in the tetrad phase (Figure 13.7D).] If both parents (either of the same or different species) produce diploid gametes, then the offspring possesses four sets of chromosomes, which is a **tetraploid** (Figure 13.7A). Tetraploids are normally fertile, as they can produce viable, diploid gametes. If, however, one parent contributes a haploid gamete and the other a diploid gamete, the offspring will be **triploid** (Figure 13.7B). Triploids are generally sterile, as any gametes they produce will generally lack a full complement of chromosomes because of meiotic irregularities, forming groupings of three homologs (trivalents; e.g., Figure 13.7E), instead of the normal two (bivalents). Triploids might persist as a population, however, if they can continue to reproduce asexually.

A second way that polyploidy can occur is by the spontaneous doubling of chromosome number in an individual plant *after* normal sexual reproduction. For example, hybridization between two different species might produce living diploid offspring, but the offspring often cannot produce viable gametes because of the genetic dissimilarities between the two parents (Figure 13.7C). However, if this sterile offspring can persist, e.g., if it can also reproduce vegetatively, it might (rarely) undergo a rare **somatic** (i.e. in a nonreproductive cell) chromosome doubling during mitosis in a critical region, e.g., the apical meristem of a shoot, such that this entire shoot becomes tetraploid. The tetraploid is now potentially capable of producing viable, diploid gametes and, therefore, fertile offspring (Figure 13.7C). This type of polyploid event may be rare, but has been documented in species of *Spartina* (cordgrass), in which a new tetraploid species evolved from two, separate diploid parents.

Polyploidy is thought to be a major mechanism in plant evolution, as chromosome studies have demonstrated that most plants have undergone at least one polyploid event during some time of their evolutionary history (see References). For example, the evolution of both emmer wheat and bread wheat occurred via ancestral, allopolyploidy events, resulting in tetraploid (4n) and hexaploid (6n) individuals (Figure 13.7F).

TESTING FOR BREEDING MECHANISM

Experimental methods may be used to assess the type of breeding mechanism/system. Flower buds may be "bagged" or "caged," i.e., covered with a fine netting that excludes potential pollinators. Also, flower buds may be "emasculated," in which the anthers are removed prior to pollen release. An example of an experimental regime to test the breeding system is seen in Table 13.1. The determination of seed set for each experiment allows inference as to the breeding system.

In addition to these manipulative experiments, embryological observations may be made to determine, e.g., if pollen tubes are growing through the style of the flower. (Fluorescence microscopy may be used to detect pollen tubes; Figure 13.8.) Absence or inhibition of pollen tube growth indicates some type of genetic incompatibility. In addition, observations of female gametophyte or embryo development and/or chromosome counts of these tissues may detect the occurrence of agamospermy.

TABLE 13.1 Five experimental manipulations 1–5, with explanation of results. Normal seed set = "+" and greatly reduced to zero seed set = "–".

	SEED PRODUCTION	
1. Flowers left to develop normally, a control.	+ **Fertile**	– **Infertile**
2. Flowers caged, then self-pollinated by hand.	+ **Self-fertile**	– **Not self-fertile**
3. Flowers caged, then left alone.	+ **Self-pollinating**	– **Not self-pollinating**
4. Flowers emasculated and caged.	+ **Agamospermous**	– **Not agamospermous**
5. Flowers caged, emasculated, and outcrossed.	+ **Outcrossing**	– **Not outcrossing**

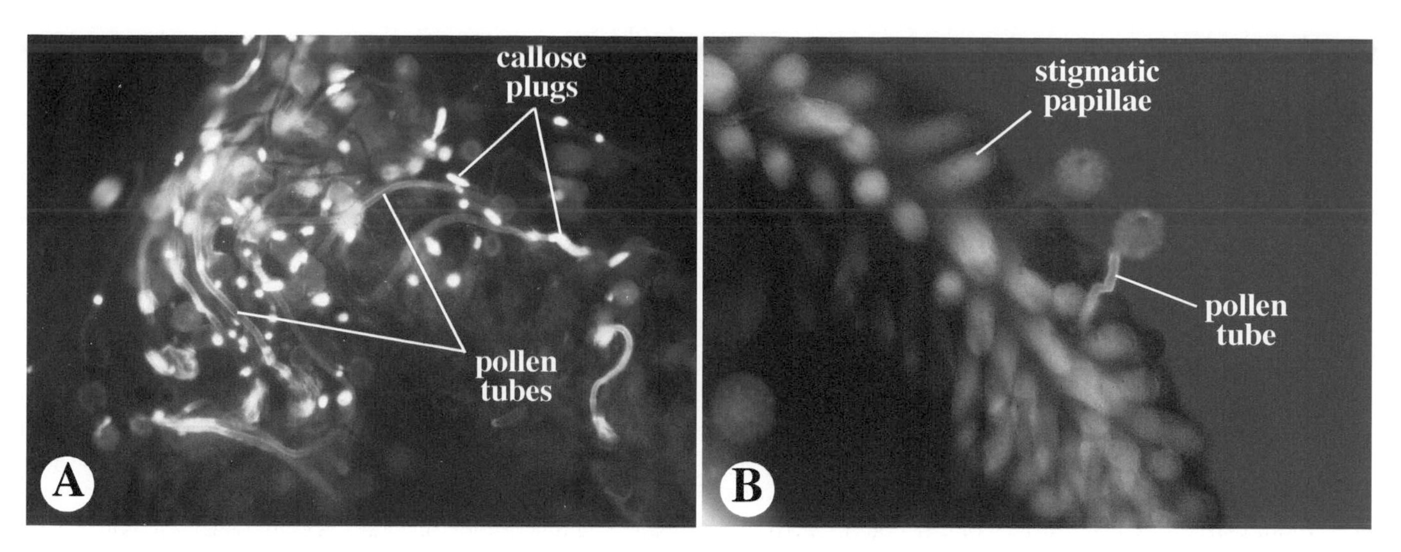

Figure 13.8 **A,B.** Pollen tube growth. Fluorescent stain, positive for callose, allows visualization of pollen tubes down length of style.

REVIEW QUESTIONS

1. What is plant reproductive biology?
2. What is pollination?
3. What two general features have evolved in flowers that function to effect animal pollination?
4. What products serve as a reward for animal pollinators?
5. What is a nectar guide? A landing platform?
6. For the following pollination mechanisms, name some floral syndromes (correlated structural modifications): (a) bee; (b) butterfly; (c) moth; (d) wind; (e) bird; (f) bat.
7. What are the two major or extreme types of breeding systems?
8. What are the advantages and disadvantages of outbreeding?
9. What is dichogamy? Name two specific types of dichogamy that can promote outcrossing.
10. What is hercogamy?
11. Define and explain: heterostyly, distyly (pin and thrum), tristyly, enantiostyly, movement hercogamy, trigger mechanisms. What is the overall function of these floral mechanisms?
12. What is self-incompatibility and what is its significance in plant reproductive biology?
13. What are the advantages and disadvantages of inbreeding?
14. What is the difference between allogamy, autogamy, geitonogamy, and allautogamy?
15. Name some types of inbreeding mechanisms.
16. What is agamospermy? How can it be detected?
17. What is hybridization? Introgression?
18. Define polyploidy.
19. Cite the ways that polyploidy can occur.
20. How can one test the breeding mechanism in plants? (cite specific ways to test)
21. If the following experiments are performed for plant species A–D, what can you say about the breeding mechanism based on the pattern of seed set?

	Seed Production			
	A	*B*	*C*	*D*
1. Flowers left to develop normally, a control.	+	+	+	+
2. Flowers caged, then self-pollinated by hand.	-	+	+	+
3. Flowers caged, then left alone.	-	-	+	+
4. Flowers emasculated and caged.	-	-	-	+
5. Flowers caged, emasculated, and outcrossed.	+	+	+	+

EXERCISES

1. Examine specimens of two species of plants plus any putative hybrids between them. (a) Study both vegetative and floral characters, from original observations or using a manual of the area, and note which diagnostic features distinguish the two species. (b) Decide upon which characters to measure in the specimens available. (c) Record 10–25 measurements of each of the parameters chosen. Compare these by preparing graphs in order to recognize discontinuities (or lack thereof) of the three taxa.
2. Locate a population of a Composite (Asteraceae) species that has both disk and ray flowers. Observe insect visitors (potential pollinators) in each of two subsets of plants (or inflorescences): one undisturbed and another with all ray flowers removed. Count the number and type of visitors over a time period (e.g., 10–30 minutes) and record.
3. If material is available, observe ultraviolet light-sensitive regions in the perianth by placing a flower into a jar saturated with ammonium vapors. Bees can detect these UV-reflective regions of the flower, enabling them to find flowers and orient to pollen or nectar more efficiently.
4. Fix the styles of a species of flowering plant in 70% alcohol. Remove the style and place in drops of aniline blue on a microscope slide, covered by a cover slip. If this style is small enough, it may be "squashed" by applying firm pressure on the cover slip (using, e.g., a cork). Observe under fluorescence microscopy. Pollen tubes regularly deposit callose, which differentially picks up the aniline blue stain. This method allows for detection of pollen tube growth and can be used to test whether self-incompatibility is occuring.
5. If time permits, select a plant species and perform the crossing and caging experiments described in the text. These techniques are used to test the potential and degree of self-pollination versus cross-pollination.
6. Peruse journal articles on plant systematics, e.g., *American Journal of Botany*, *Systematic Botany*, or *International Journal of Plant Sciences*, and note those that describe aspects of reproductive biology in relation to systematic studies. Identify the techniques used and the problems addressed.

REFERENCES FOR FURTHER STUDY

Bernhardt, P. 1989. Wily Violets and Underground Orchids. W. Morro & Co., New York.

Chittka, L., A. Shmida, N. Troje, and R. Menzel. 1994. Ultraviolet as a component of flower reflections, and the colour perception of Hymenoptera. Vision Research 34: 1489–1508.

Faegri, K., and L. van der Pijl. 1979. The Principles of Pollination Ecology, 3rd ed. Pergamon Press, Oxford.

Jiao, Y., N. J.Wickett, S. Ayyampalayam, A. S. Chanderbali, L. Landherr, P. E. Ralph, L. P. Tomsho, Y. Hu, H. Liang, P. S. Soltis, D. E. Soltis, S. W. Clifton, S. E. Schlarbaum, S. C. Schuster, H. Ma, J. Leebens-Mack, and C. W. dePamphilis. 2011. Ancestral polyploidy in seed plants and angiosperms. Nature 473:97–100.

Li, Z., A. E. Baniaga, E. B. Sessa, M. Scascitelli, S. W. Graham, L. H. Rieseberg, and M. S. Barker. 2015. Early genome duplications in conifers and other seed plants. Science Advances 1:e1501084.

Soltis, D. E., V. A. Albert, J. Leebens-Mack, C. D. Bell, A. H. Paterson, C. Zheng, D. Sankoff, C. W. dePamphilis, P. K. Wall, and P. S. Soltis. 2009. Polyploidy and Angiosperm diversification. American Journal of Botany 96:336–348.

Soltis, D. E., C. J. Visger, D. B. Marchant, and P. S. Soltis. 2009. Polyploidy: Pitfalls and paths to a paradigm. American Journal of Botany 103:1-21.

Stace, C. A. 1989. Plant Taxonomy and Biosystematics, 2nd ed. Edward Arnold. Distributed in the USA by Routledge, Chapman, and Hall, New York.

van der Pijl, L. 1982. Principles of Dispersal in Higher Plants. Springer-Verlag, Berlin.

Weber, M. G. and K. H. Keeler. 2012. The phylogenetic distribution of extrafloral nectaries in plants. Annals of Botany 111:1251-1261.

Wendel, J. F. 2015. The wondrous cycles of polyploidy in plants. American Journal of Botany 102:1753-1756.

Zhan, S. H., M. Drori, E. E. Goldberg, S. P. Otto, and I. Mayrose. 2016. Phylogenetic evidence for cladogenetic polyploidization in land plants. American Journal of Botany 103:1-7.

14

PLANT MOLECULAR SYSTEMATICS

Molecular systematics encompasses a series of approaches in which phylogenetic relationships are inferred using information from macromolecules of the organisms under study. Most molecular data acquired and analyzed today are direct DNA (or RNA) sequences, but indirect measures of sequence data, such as RFLPs and AFLPs, microsatellite data, and RAPDs, are still being used.

The use of data from other, generally smaller molecules, such as secondary compounds in plants, is usually relegated to the field of "chemosystematics" and will not be reviewed here. However, these chemical data are becoming increasingly important, as more sophisticated techniques are elucidating the developmental evolution of biochemical pathways and the adaptive significance of secondary compounds produced. (See Smith et al. 2019.)

A revolution in inferring the phylogenetic relationships of life is occurring with the use of molecular data. The following is a review of the types of data, methods of acquisition, and methods of analysis in plant molecular systematics.

ACQUISITION OF MOLECULAR DATA

Plant samples from which DNA is to be isolated may be acquired by various means. It is vital to always collect a proper voucher specimen, properly mounted and accessioned in an accredited herbarium, to serve as documentation for any molecular systematic study (see Chapters 17, 18). Live samples may be collected and immediately subjected to chemical processing. For many DNA methods, pieces of leaves (from which chloroplast, mitochondrial, and nuclear DNA can be isolated) are removed from the live plant and immediately dried, often in a container of silica gel. Alternatively, plant samples may be frozen (e.g., in dry ice or liquid nitrogen) or placed in concentrated extraction buffer. With any of these procedures, DNA is usually preserved intact. Usable DNA is often successfully isolated from dried herbarium sheets, attesting to the "toughness" of the molecule.

TYPES OF PLANT DNA SEQUENCE DATA

Perhaps the most important data for inferring phylogenetic relationships of life are DNA sequences. (Note that the beginning student may wish to review the basics of DNA structure, replication, and protein synthesis.) DNA sequence data refers to the linear arrangement of nucleotides (adenine = A, cytosine = C, guanine = G, or thymine = T; Figure 14.1) in a particular region of the DNA of a given taxon. Comparisons of homologous regions of DNA among the taxa under study yield the characters and character states that are used to infer relationships in phylogenetic analyses.

For plants, the three basic types of DNA sequence data stem from the three major sources of DNA: nuclear (**nDNA**),

https://doi.org/10.1016/B978-0-12-812628-8.50014-6,

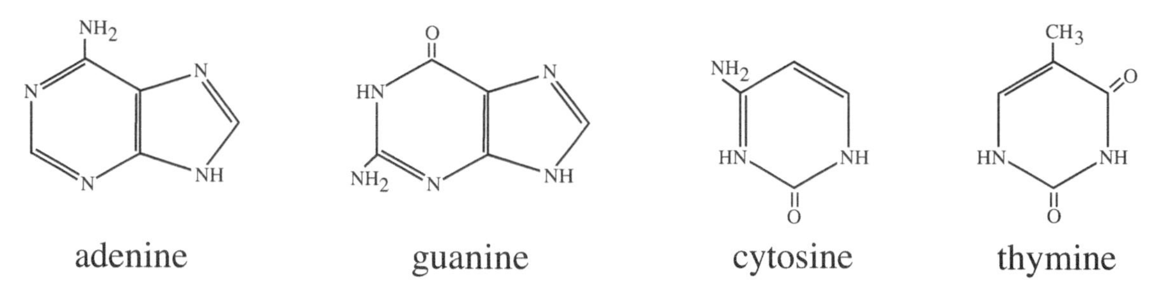

Figure 14.1 Molecular structure of the four DNA nucleotides. Adenine and guanine are chemically similar *purines*; cytosine and thymine are chemically similar *pyrimidines*.

chloroplast (**cpDNA**), and mitochondrial (**mtDNA**). Nuclear DNA is, of course, transmitted from parent(s) to offspring by nuclear division (meiosis or mitosis) via sexual or asexual (somatic) reproduction. Chloroplasts and mitochondria, however, replicate and divide independently of the nucleus and may be transmitted to offspring in a different fashion. For example, in angiosperms these organelles are usually (with some exceptions) sexually transmitted only maternally, being retained in the egg but excluded in sperm cells. (In conifers, interestingly, chloroplast DNA is transmitted paternally, not maternally.)

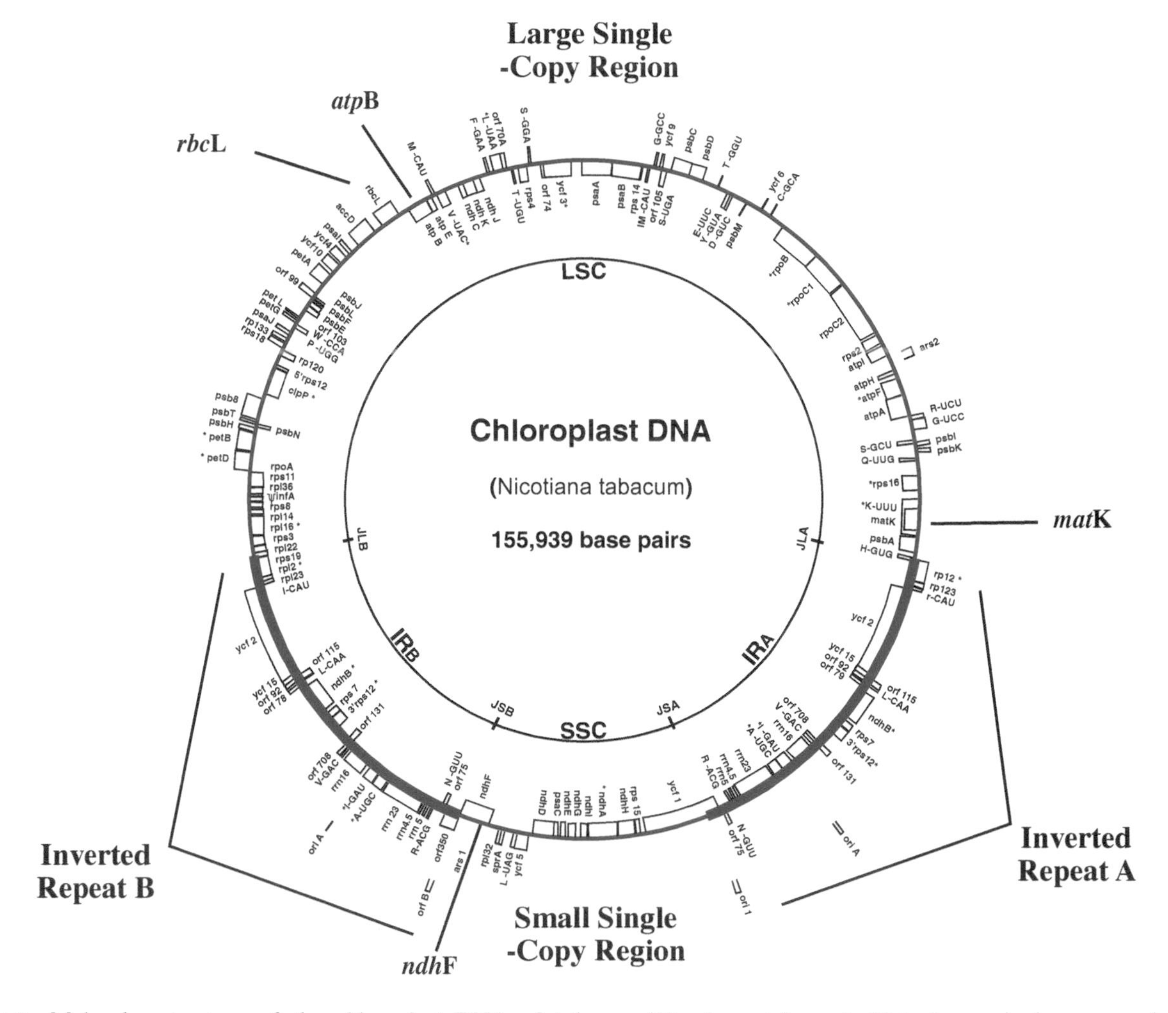

Figure 14.2 Molecular structure of the chloroplast DNA of tobacco (*Nicotiana tabacum*). Note large single-copy region (LSC), small single-copy region (SSC), and the two inverted repeats (IRA and IRB). Also note location of *atp*B, *rbc*L, *mat*K, and *ndh*F genes (see Table 14.1). (Redrawn from Wakasugi, T., M. Sugita, T. Tsudzuki, and M. Sugiura. 1998. Updated gene map of tobacco chloroplast DNA. Plant Molecular Biology Reporter 16: 231–241, by permission.)

TABLE 14.1 Some chloroplast genes that have been used in plant molecular systematics, after Soltis et al. 1998.

CHLOROPLAST GENES		
GENE	**LOCATION**	**FUNCTION**
*atp*B	Large single-copy region of chloroplast	Beta subunit of ATP synthethase, which functions in the synthesis of ATP via proton translocation
*rbc*L	Large single-copy region of chloroplast	Large subunit of ribulose-1,5-bisphosphate carboxylase/oxygenase (RUBISCO), which functions in the initial fixation of carbon dioxide in the dark reactions
*mat*K	Large single-copy region of chloroplast	Maturase, which functions in splicing type II introns from RNA transcripts
*ndh*F	Small single-copy region of chloroplast	Subunit of chloroplast NADH dehydrogenase, which functions in converting NADH to NAD + H^+, driving various reactions of respiration

The use of sequence data from the DNA of chloroplasts (known as the **plastome**) has been highly used in elucidating plant relationships, given the large number of copies per cell and relative ease of PCR amplification (see below). The basic structure of chloroplast DNA for a flowering plant, with coding genes indicated, is shown in Figure 14.2. Most plant DNA consists of a **large single-copy region**, a **small single -copy region**, and two flanking **inverted repeats**, the latter mirror images of one another (Figure 14.2). Like all organelle and prokaryotic DNA, chloroplast DNA is circular.

Some of the more commonly sequenced chloroplast DNA coding genes are listed in Table 14.1, although many more have been utilized. In addition to coding genes of chloroplast DNA, the sequences between genes, known as **intergenic spacers**, may be used in phylogenetic analyses. Intergenic spacer regions often show a higher degree of variability than the coding genes, making the former more useful for analyses at a lower taxonomic level, such as species or infraspecies. A list of some commonly used chloroplast intergenic spacers is seen in Table 14.2.

Like that of chloroplasts, plant mitochondria also have circular DNA that may be utilized in phylogenetic studies. However, mitochondrial DNA in plants is known to have a high degree of plasticity in terms of genomic rearrangements, insertion or transfer of DNA to the chloroplast or nucleus, disruption of intron/exon gene continuity, and evolutionary changes in gene expression (Knoop, 2004) and is generally less useful in phylogenetic studies of plants.

Until recently, nuclear DNA sequencing has been used to a lesser degree in plant systematics. One of the more useful types of nuclear DNA sequences has been the 18S-26S **nuclear ribosomal cistron**, a repeating region of the nuclear ribosomal DNA (**nrDNA**). The nrDNA cistron functions in ribosome synthesis and consists of multiple copies, as opposed to single copies found in most protein-coding genes. One component of this cistron is the **internal transcribed spacer** (**ITS**) region. The ITS region lies between the 18S and 26S nuclear ribosomal DNA (nrDNA); the ITS region is divided into two sub-regions, ITS1 and ITS2, separated by a the 5.8S nrDNA (Figure 14.3). ITS sequence data has been most valuable for inferring phylogenetic relationships at a lower level, e.g., between closely related species. However, it has also been used in elucidating higher level relationships (Baldwin et al. 1995).

A related DNA sequence region is the **external transcribed spacer** (**ETS**) region. The ETS region lies between 26S and 18S nrDNA, adjacent to the latter (Figure 14.3). (The entire region, including both the ETS and the non-transcribed spacer region (NTS) is known as the intergenic spacer region, or IGS; see Figure 14.3.) The ETS region contains even more sequence variation than ITS and is useful in analyses at lower taxonomic levels (Baldwin and Markos 1998).

TABLE 14.2 Some chloroplast intergenic spacer regions that have been used in plant molecular systematics, after Shaw et al. (2005, 2007).

CHLOROPLAST INTERGENIC SPACER REGIONS			
3'rps16-5'trnK	*petL-psbE*	*rpl32-trnL*	*trnL* intron
3'trnK-matK intron	*psal-accD*	*rpoB-trnC*	*trnL-trnF*
3'trnV-ndhC	*psbA-3'trnK*	*rps16* intron	*trnQ-5'rps16*
5'rpS12-rpL20	*psbB-psbH*	*rps4-trnT*	*trnS-rps4*
atpI-atpH	*psbD-trnT*	*trnC-ycf6*	*trnS-trnfM*
matK-5'trnK intron	*psbJ-petA*	*trnD-trnT*	*trnS-trnG*
ndhA intron	*psbM-trnD*	*trnG* intron	*trnT-trnL*
ndhF-rpl32	*rpl14-rps8-infA-rpl36*	*trnH-psbA*	*ycf6-psbM*
ndhJ-trnF	*rpl16* intron		

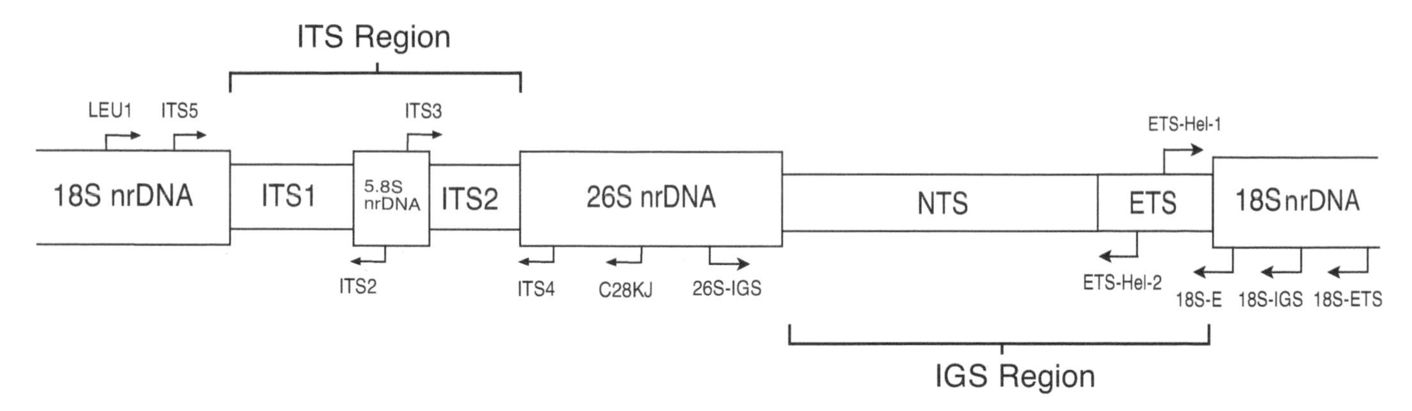

Figure 14.3 Nuclear ribosomal DNA cistron, showing (1) internal transcribed spacers (ITSs) of nuclear ribosomal DNA, illustrating the ITS region and flanking subunits, and showing the orientations and locations of primer sites; and (2) external transcribed spacer (ETS) of the intergenic spacer (IGS) region, also showing orientations and locations of primer sites. After Baldwin et al. (1995) and Baldwin and Markos (1998).

Cistron sequences, especially ITS and ETS, have been very valuable in a number of plant phylogenetic studies. However, a starting assumption made is that the thousands of copies of the cistron are all identical, which can be effected by a process called concerted evolution. This may not always be the case, a criticism of using these data. (See, e.g., Alvarez and Wendel 2003.)

DNA SEQUENCING

POLYMERASE CHAIN REACTION (PCR)

In some methodologies, a given region of DNA is first identified by a variety of methods. The gene region(s) of interest is then **amplified**, in which multiple copies of that DNA are generated. Until relatively recently, DNA sequences of interest were mostly amplified using the **polymerase chain reaction** (or **PCR**). The invention of this technology was crucial to modern DNA sequencing, as it permitted rapid and efficient DNA **amplification**, the replication of thousands of copies of DNA. PCR amplification is still used today in various methodologies, including some high-throughput sequencing library preparations (see below).

The polymerase chain reaction works as follows (see Figure 14.4). Prior research establishes the occurrence of relatively short regions of DNA that flank (occur at each end of) the gene or DNA sequence of interest and that are both unique (not occurring elsewhere in the genome) and conserved (i.e., invariant) in all taxa to be investigated. These short, conserved, flanking regions are used as a template for the synthesis of multiple, *complementary* copies, known as **primers**. Primers ideally are constructed such that they do not bind with one another.

In the polymerase chain reaction, a solution is prepared, made up of (1) the isolated and purified DNA of a sample; (2) multiple copies of primers; (3) free nucleotides; (4) DNA polymerase molecules (typically *Taq* polymerase, which can tolerate heat); and (5) buffer and salts. This solution is heated to a point at which the sample DNA **denatures**, whereby the double helix unwinds and the two complementary DNA chains separate from one another. Once the sample DNA denatures, the primers in solution may bind with the corresponding, complementary DNA of the sample (Figure 14.4). Following binding of the primer to the sample DNA, individual nucleotides in solution attach to the 3′ end of the primer, with the sample DNA acting as a template; DNA polymerase catalyzes this reaction. A second primer, at the opposite end of the DNA sequence of importance, is used for the complementary, denatured DNA strand. Thus, the two denatured strands of DNA are replicated. After replication, the solution is cooled to allow for annealing of the replicated DNA with the complementary DNA single strands. This is followed by heating to the point of DNA denaturation, and repeating the process. A typical PCR reaction can produce more than a million copies of DNA in a matter of hours, these amplified copies known commonly as **amplicons**.

SANGER SEQUENCING

After DNA is replicated, it was traditionally sequenced by a technique known as **Sanger sequencing**, although today this method has been largely replaced by high-throughput sequencing (see below). Sanger sequencing today uses an apparatus that reads fluorescent dyes with a laser detector. The production of dye-labeled DNA is very similar to DNA replication using the PCR. The replicated DNA is placed into solution with DNA polymerase, primers, free nucleotides, and a small concentration of synthesized compounds called

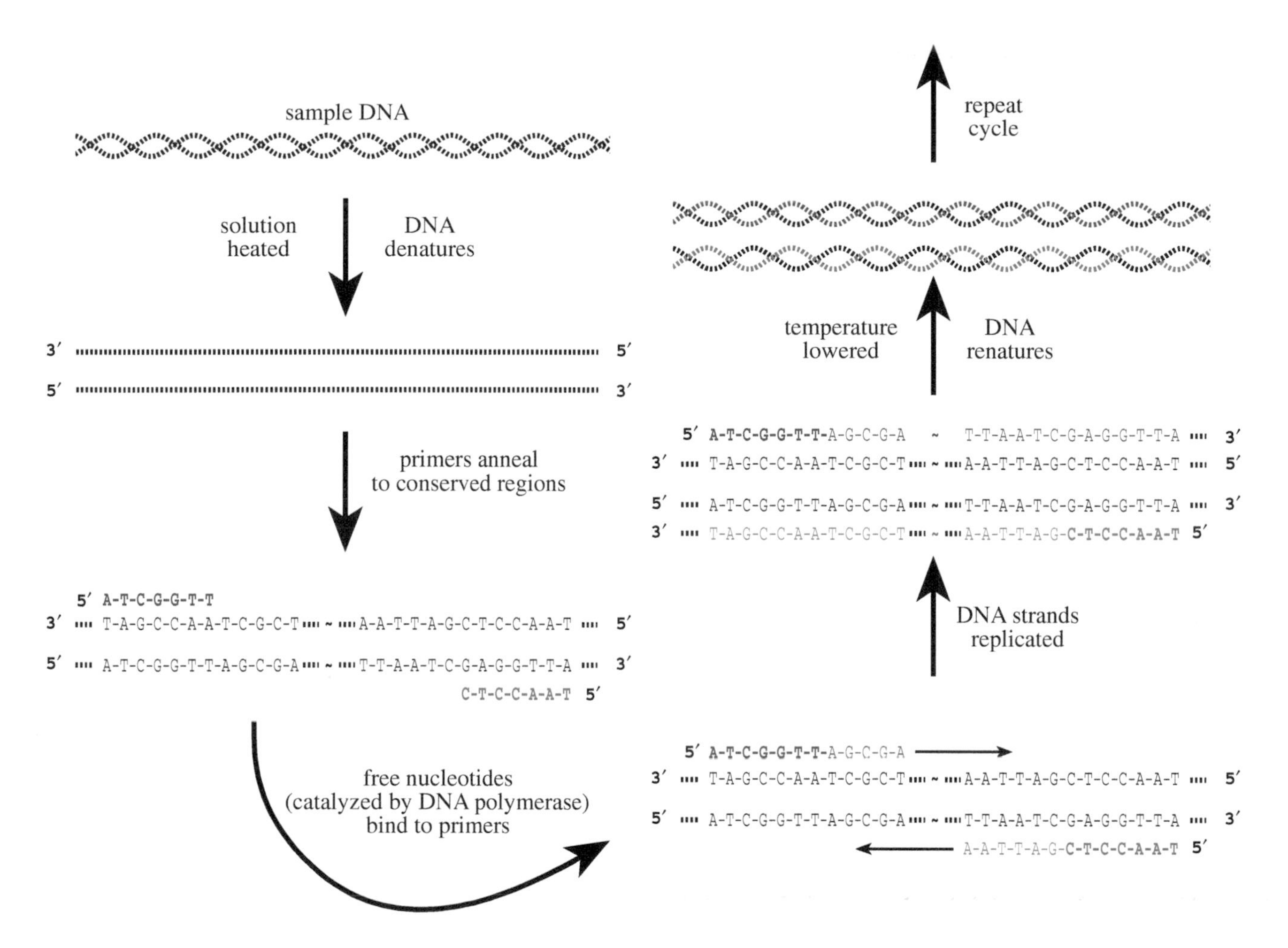

Figure 14.4 Polymerase chain reaction, using cycle sequencing to produce multiple copies of a stretch of DNA.

dideoxynucleotides (discussed later) that are each attached to a different type of fluorescent dye. As in the polymerase chain reaction, the sample DNA is heated and denatures (Figure 14.5). At this point, a primer attaches to a conserved region of one of the strands of DNA, and free nucleotides in solution join to the 3′ end of the primer, using the sample DNA as a template and catalyzed by DNA polymerase (Figure 14.5). Thus, a replicated copy of the DNA strand begins to form. However, at some point a dideoxynucleotide joins to the new strand instead of a nucleotide doing so. The dideoxynucleotides (dideoxyadenine, dideoxycytosine, dideoxyguanine, and dideoxythymine) resemble the four nucleotides, except that they lack a hydroxyl group. Once a dideoxynucleotide is joined to the chain, absence of the hydroxyl group prevents the DNA polymerase from joining it to anything else. Thus, with the addition of a dideoxynucleotide, synthesis of the new DNA strand terminates (Figure 14.5).

The ratio of dideoxynucleotides to nucleotides in the reaction mixture is carefully set and is such that the concentration of dideoxynucleotides is always much smaller than that of normal nucleotides. Thus, the dideoxynucleotides may terminate the new DNA strand at any point along the gene being replicated. For example, some of the new DNA strands will be the length of the primer plus one additional base (in this case the dideoxynucleotide); some will be the primer length plus two bases (a nucleotide plus the terminal dideoxynucleotide); some will be the primer length plus three bases (two nucleotides plus the terminal dideoxynucleotide); etc. There are many thousands, if not millions, of copies of the sample DNA. Thus, there will be an equivalent number of newly replicated DNA strands, of all different lengths.

The final step of DNA sequencing entails subjecting the DNA strands to **electrophoresis**, in which the DNA is loaded onto a flat gel plate or in a thin capillary tube subjected to an electric current. Because the phosphate components of nucleic acids give DNA a net negative charge, the molecules are attracted to the positive pole. The DNA strands migrate through the medium over time, the amount of migration inversely proportional to the molecular weight of the strand

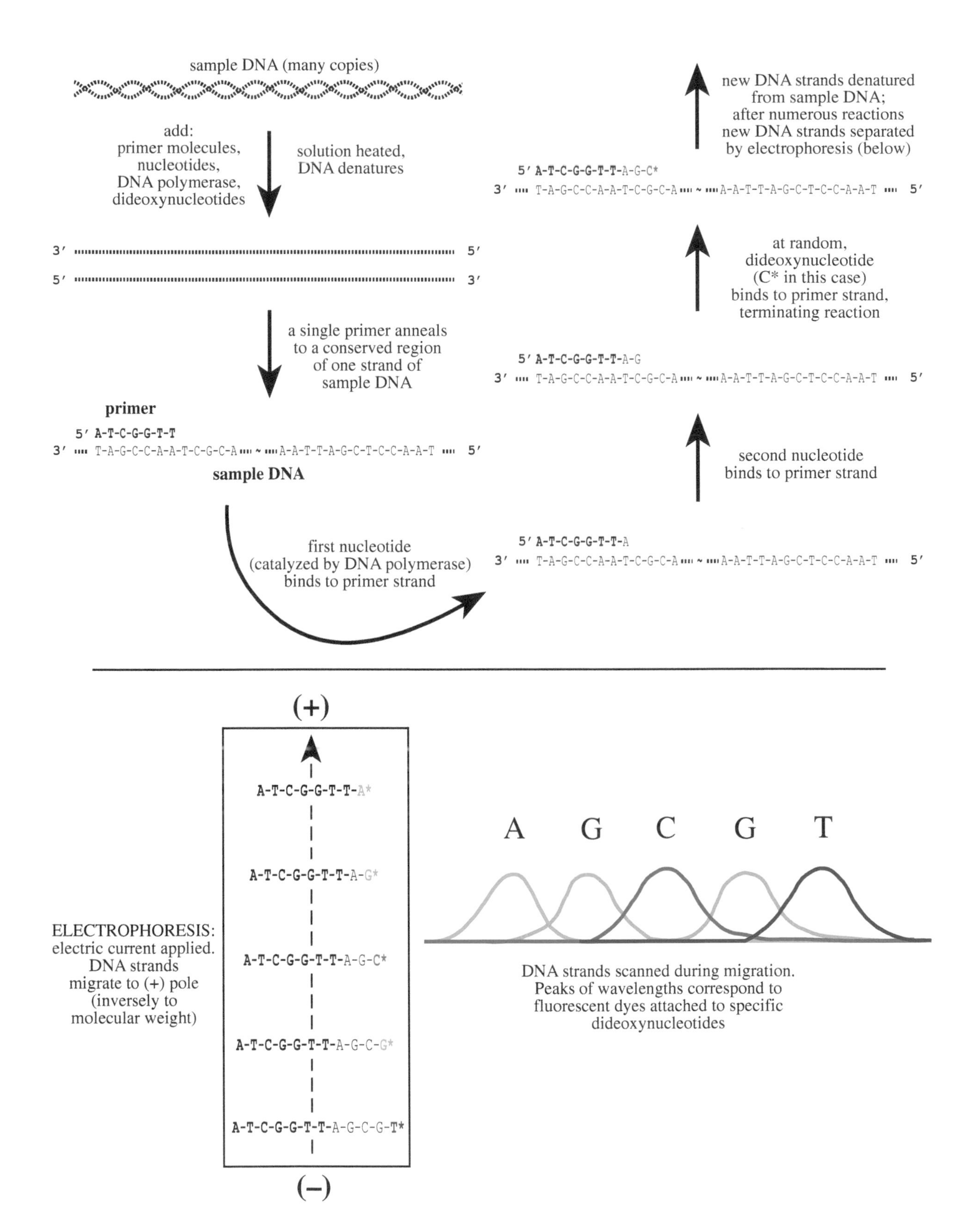

Figure 14.5 Sanger DNA sequencing reactions. A* = dideoxyadenine; C* = dideoxycytosine; G* = dideoxyguanine; T* = dideoxythymine.

(i.e., lighter strands migrate farther). Each strand is terminated with a dideoxynucleotide to which a fluorescent dye is attached; each of the four dideoxynucleotides has a different type of fluorescent dye, which (upon excitation) emits light of a different wavelength. Thus, as the multiple copies of DNA of one particular length migrate along the gel or capillary, the wavelength of emitted light is detected and recorded as a peak, which measures the light intensity. Because a given emitted wavelength ("color") is determined by one of the four dideoxynucleotides, the corresponding nucleotide can be inferred and its position identified by the timing of migration of the DNA strands. In this way, the sequence of nucleotides of the DNA strand can be inferred (Figure 14.5).

HIGH THROUGHPUT DNA SEQUENCING

There has been a revolution in DNA sequence acquisition by the relatively recent development of a class of techniques collectively termed **high-throughput sequencing** (commonly termed "**next generation sequencing**" or "**NGS**"). These have essentially replaced traditional Sanger sequencing, as they can produce considerably more sequence data at a much lower cost.

There are several different methodologies and technologies under the rubric of high-throughput sequencing. One of these, **sequencing by synthesis** (used by the company Illumina) is described here. This method, like others, is accomplished in a number of sequential steps, these collectively known as the **workflow**. The first step is extracting and purifying the DNA from a given sample, e.g., from dried plant tissue of a particular plant species of interest, just as one would do for Sanger sequencing. Next, a series of steps are done to "prepare" the DNA, collectively termed **library preparation**. First, DNA is typically broken or sheared into fragments (Figure 14.6A), either chemically or with sonication (high frequency sound vibrations). In some workflows the fragments are **size-selected**, such that they are filtered to a particular ideal size. Following or during shearing, the individual fragments are **ligated** (chemically bound) to primers, barcodes, and adapters. The **primer** acts as in Sanger sequencing, as the binding site for initiation of DNA sequencing. The **barcode** (or **index**) is a short segment of DNA that acts as an identifier each samples, if the samples need to be differentiated. In the simultaneous sequencing of large numbers of library preparations, a process known as **multiplexing**, the final sequence products can later be identified from barcode sequences. **Adapters** are short segments of DNA ligated to ends of the DNA fragments involved in the sequencing reaction.

The next step is **amplification**, producing thousands of copies of the DNA fragments. In one method, termed clonal bridge amplification, a glass slide with lanes is prepared, with millions of copies of two types of oligos attached to the surface of the lanes, these complementary to the two types of adapters used in prepping the fragments. The prepped, single-stranded fragments are added, and the adapters of these fragments bind ("hybridize") with an oligo on the slide. Also added are DNA polymerase and multiple copies of the four nucleotides, as in Sanger sequencing. A polymerase reaction results in the synthesis of a complementary strand of DNA, using the fragment as the template (Figure 14.6B). This double-stranded DNA is then also denatured, and the original single-stranded fragment detaches and is washed away. The remaining strand bends over such that the adapter at the free end hybridizes with a complementary adapter on the slide. This is then amplified, followed by denaturation, with the two complementary strands remaining attached to the slide. This process, repeated multiple times, results in amplified clusters of the original DNA fragments (Figure 14.6C). The reverse strands are removed and washed away, leaving only single-stranded DNA clusters of one complement of the fragments.

Sequencing is accomplished by adding a complementary primer, DNA polymerase, and fluorescently "tagged" nucleotides (Figure 14.6D). In each cycle, a fluorescently tagged nucleotide binds with the complementary nucleotide of the DNA to be sequenced. A light source induces the fluorescent tag to excite and emit light at one of four frequencies, each corresponding to one of the four nucleotides, as in Sanger sequencing. The light is emitted simultaneously from all the fragments of a cluster at an intensity sufficient to be recorded by a detector (Figure 14.6E). Unlike the dideoxynucleotides of Sanger sequencing, however, the fluorescent tags subsequently detach, and additional fluorescently tagged nucleotides bind to the fragment in the next cycle. The flashes of fluorescent light are recorded by a detector, and the sequences of light emitted correspond to the sequence of the DNA fragment. **Sequencing by synthesis** occurs simultaneously for numerous DNA clusters, resulting in millions of sequence "**reads**" (Figure 14.6F). Recall that the barcode that was originally ligated to the fragment is also sequenced, allowing each read to be associated with a given taxon.

The process outlined above may be repeated for the complement of the original DNA fragment. This method, termed **paired-end sequencing**, acquires sequence data in both directions of a DNA strand, resulting in longer reads and greater accuracy.

Thousands of reads for a given taxon are then analyzed using a computer program. Recall that each read corresponds to the sequence information of one of the fragments of the original DNA of interest. The sequence of the entire DNA is determined by algorithms that align the reads with one

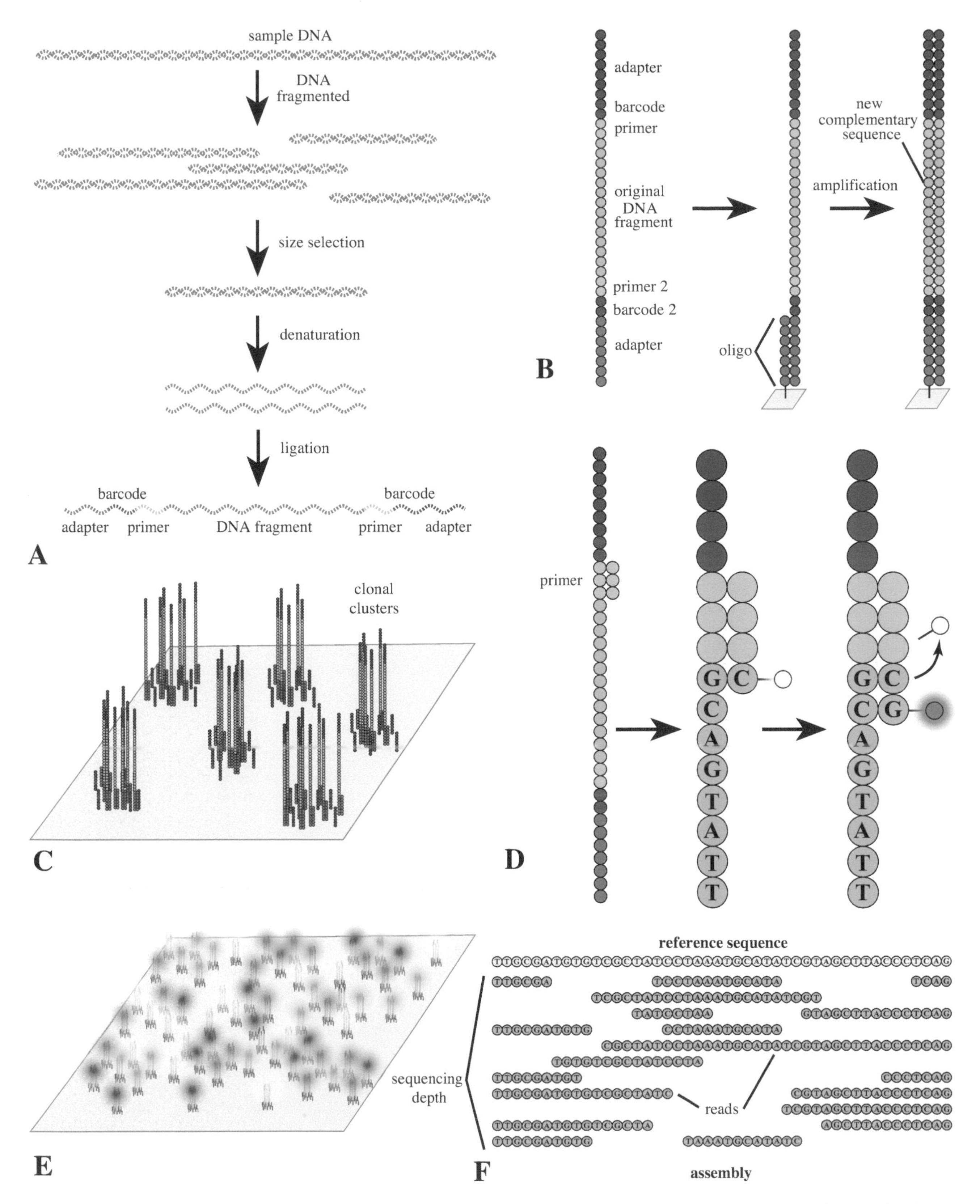

Figure 14.6 High-throughput sequencing, illustrated for sequencing by synthesis. **A.** DNA fragmentation, size selection (optional), denaturation, and ligation of adapters, primers, and barcodes. **B.** DNA amplification, in which ligated adapters hybridize with oligos on a glass slide. **C.** Synthesis of clonal clusters, derived from one original DNA molecular. **D.** Sequencing, in which nucleotides are bound to a fluorescent dye, which emits light of a frequency corresponding to one of the four nucleotides. **E.** Flow cell, showing simultaneous fluorescence from multiple clonal clusters. **F.** Assembly, in which the reads, each derived from sequence data of a given clonal cluster, are aligned with respect to a reference sequence. Based in part on information at *https://www.illumina.com/techniques/sequencing.html*.

another, given that the read sequences will overlap. This process is known as fragment **assembly**. The term **contig** (from "contiguous") refers to a set of overlapping DNA segments that together represent a consensus region of DNA, generally corresponding to overlapping sequence data reads. DNA assembly permits reconstruction of the entire DNA sequence. Its accuracy is enhanced by a greater **sequencing depth** (**depth of coverage**), which refers to the number of overlapping assemblies of DNA fragments (reads). The accuracy of the overall DNA sequence is also enhanced by assembling the reads with a known DNA sequence from a taxon closely related, and therefore likely similar, to those of the taxa being analyzed, the process known as **reference guided assembly**. If no reference DNA sequence is used (e.g., not available), the process is termed **de novo assembly**. Metrics of quality control are used, most commonly the Phred score, to filter out less reliable reads.

What is described above constitutes only one of several types of high-throughput sequencing, others involving different techniques and technology. The field of DNA sequencing is rapidly changing, with new advances continuously being made.

WHOLE GENOME METHODS

GENOME SKIMMING

If the total DNA from a sample of a taxon is extracted, some components of the DNA will have many more copies than others. Within a given cell there may be hundreds of chloroplasts and mitochondria, so the DNA extracted from these organelles will have a comparable number of copies. Most genes of the nucleus will have only a single copy, but some nuclear regions, e.g., the nrDNA cistron (discussed earlier) will also have hundreds of copies. Thus, these three components of the cellular DNA will generally be in much greater proportion relative to the single copy nuclear DNA and will thus outcompete the latter in the proportion of DNA fragments sequenced in high-throughput sequencing. It is possible, therefore, to acquire sequence data for the entire chloroplast, cistron, and mitochondrial DNA from fragmentation of the total DNA. In contrast, the nuclear DNA will not be in enough concentration to have the depth of coverage to be high-throughput sequencing.

Genome skimming refers to the acquisition of these three DNA components found in high concentration in most plant cells. The total genomic DNA is randomly fragmented and the fragments sequenced and assembled in a high-throughput method (termed "shotgun" sequencing). The advantage of genome skimming is that no additional preparations are needed other than extraction and fragmentation of the total cellular DNA. In addition, dried herbarium material may be well suited for genome skimming analyses, as the high copy DNA may still be in relatively high concentration despite the DNA degradation that occurs in dried specimens over the years. The disadvantage of genome skimming is that the data are limited to that of two organelles, which have a different means of inheritance than that of nuclear DNA, and to one high-copy region of the nucleus, the cistron. Single-copy nuclear genes, which may be essential for species tree construction (see Chapter 2), are generally not sequenced or only partially sequenced. However, partially sequenced low copy nuclear loci may be sufficient for designing PCR primers or probes for hybridization type techniques (see Targeted Sequencing).

WHOLE GENOME SEQUENCING

Whole genome sequencing (**WGS**), in which sequences of the entire DNA of an organism are obtained, is becoming much more common, even in "non-model" organisms and may be the norm in the future. Sequencing machines may now produce reads thousands (instead of a hundred or so) base pairs long, greatly enhancing assembly.

GENOME RESEQUENCING

Genome Resequencing is obtaining the whole genome of numerous, closely related individuals and comparing their sequence data to an existing reference genome, the latter obtained using data from several individuals. This allows precise genetic comparisons between individuals of a population, including single nucleotide polymorphisms (SNPs), indels, and even large scale mutations (see below).

REDUCED REPRESENTATION METHODS

The remaining techniques described below can be grouped as **reduced representation** methods, as they are designed to acquire data from only a subset of the total genomic DNA.

TARGETED SEQUENCING

Targeted sequencing refers to a class of techniques that obtain sequence data for specific low copy nuclear genes, which in a typical DNA sample would be in very low concentration compared to the high copy chloroplast, mitochondrial, or cistron genes. Some targeted methods seek to detect and sequence conserved (e.g., **Conserved Ortholog Set**, or **COS**) genes, which may be present in a number of representative taxa in a study group.

Various techniques have been developed for "targeting" these low copy nuclear genes. One technique uses **PCR amplification**. If a primer can be identified that is conserved for a range of single copy nuclear loci, then a PCR reaction can produce thousands of **amplicons** (see PCR) of these loci, which can then be subjected to high-throughput sequencing.

A second targeted technique is **hybridization enrichment sequencing** (often termed **HybSeq**). In this methodology, highly specific short (oligonucleotide) probes are used to hybridize with sample DNA, either in solution or on a solid support (array). The probes target selected regions of the DNA, with unwanted DNA removed. These hybridized DNA regions are then prepped for high-throughput sequencing.

TRANSCRIPTOMICS

The **transcriptome** refers to that part of the total genome that is transcribed from DNA to RNA. **Transcriptomics** (or **RNASeq**) is the acquisition of sequence data from the RNA produced in a given tissue, usually focusing on messenger RNA (mRNA). The mRNA is derived from expressed genes and is generally subsequently translated into proteins, presumed to be involved in the development and function of the tissue of interest.

Transcriptomics entails the extraction of RNA from a particular plant tissue of interest, e.g., roots, shoots, leaves, flowers, fruits, seeds, or their components. The total RNA consists mostly of ribosomal RNA (rRNA), so methods are used to "filter" this out and enrich the mRNA. Other methods may be used to capture the mRNA that is in relatively low concentration by effectively reducing the concentration of mRNA in high concentration. The mRNA may be converted to cDNA (complementary DNA, the complement of RNA) using *reverse transcriptase*. The cDNA may then be sequenced, e.g., using high-throughput methods. An advantage of transcriptomics is that the process does not require prior knowledge of the genetic makeup of individuals, as would be needed for PCR amplification or hybridization enrichment.

The sequence data from transcriptomes may be used in a standard phylogenetic analysis. These data often yield robust results (with high support values), as these sequence data represent a significant portion of the coding portion of the nuclear DNA and may be used, e.g., more effectively with coalescent methods (Chapter 2).

In addition, transcriptome data may be correlated with a given plant tissue. The genes producing the transcript products are assessed for their involvement in the development and/or function of that tissue or the organs they produce.

RESTRICTION SITE METHODS

A **restriction site** is a sequence of approximately 6–8 base pairs of DNA that binds to a given restriction enzyme. These **restriction enzymes**, of which there are many, have been isolated from bacteria. Their natural function is to inactivate invading viruses by cleaving the viral DNA. Restriction enzymes known as type II recognize restriction sites and cleave the DNA at particular locations within or near the restriction site. An example is the restriction enzyme *Eco*RI (named after *E. coli*, from which it was first isolated), which recognizes the DNA sequence seen in Figure 14.7 and cleaves the DNA at the sites indicated by the arrows in this figure. Four types of restriction site methods are discussed below.

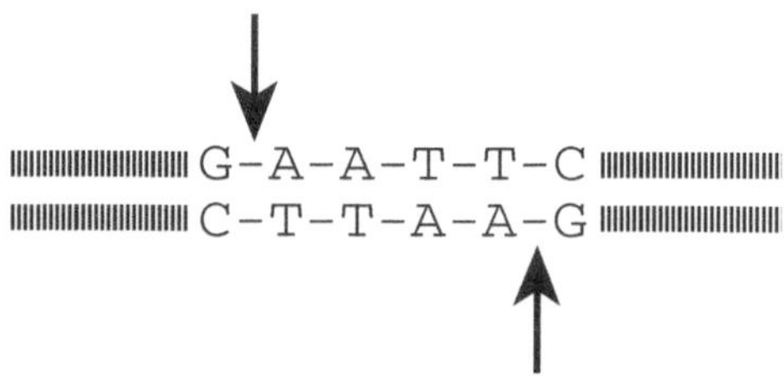

Figure 14.7 A DNA restriction site, cleaved (at arrows) by the restriction site enzyme *Eco*RI.

Restriction site Associated DNA Sequencing, or **RADSeq**, uses restriction enzymes to break up the DNA into fragments, each end having a component of the restriction site (Figure 14.8). (If two restriction enzymes are used, the technique is known as **ddRADSeq**, for "double-digest.") However, the fragments are then ligated with a adapters, barcodes, and primers, as described in high throughput sequencing. These may be further sheared and size-selected and then sequenced, either from one end (single end sequencing) or from both ends (pared end sequencing), the end product being multiple contigs that are subsequently assembled.

RADSeq yields only a limited picture of the genome, but with high coverage near the restriction sites. The advantage of RADSeq is that the sequence data, in the form of SNPs (single nucleotide polymorphisms; see below), are spread out across the entire genome. This gives much greater representation of the nuclear genome than, e.g., genome skimming. RADSeq has proven useful for both population-level studies and studies of groups with rapidly evolving lineages.

A modification of RADSeq, known as **Multiplexed Shotgun Genotyping**, or **MSG**, also uses restriction enzymes, but those that bring about more frequent cleavages to the DNA, resulting in numerous fragments that are generally not sheared or size-selected. Libraries of these fragments are prepared and sequenced using high-throughput methods. MSG generally results in more sequence data, particularly valuable for population and species level analyses.

NON-SEQUENCING METHODS

A class of techniques rely on determining lengths of homologous regions of DNA, as opposed to direct DNA sequences. These are less commonly used today.

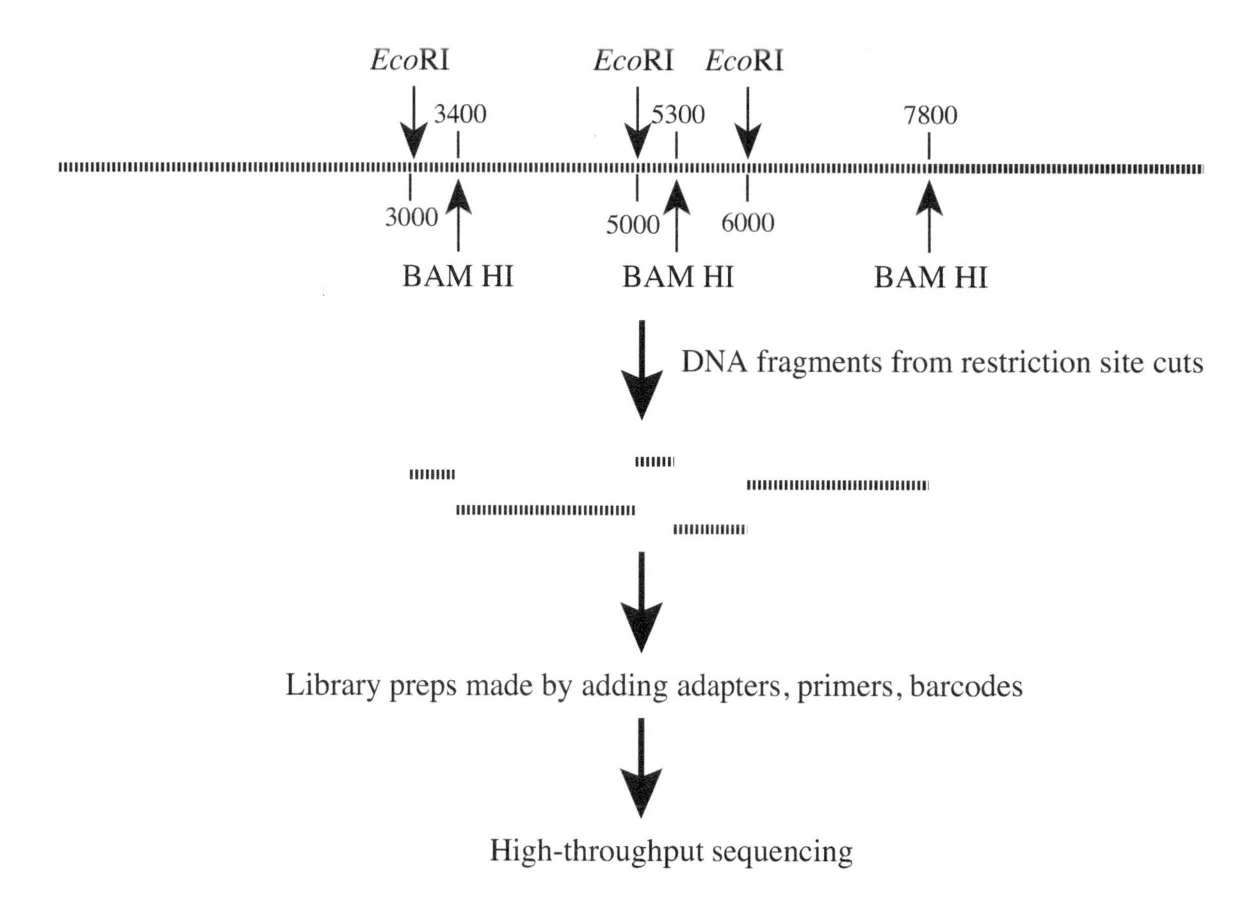

Figure 14.8 Illustration of ddRADSeq, in which two restriction enzymes cleave DNA into fragments (which may be size selected), followed by library preparation for high-throughput sequencing.

RFLPs AND AFLPs

Restriction fragment length polymorphism, or **RFLP**, refers to differences between taxa in restriction sites, and therefore the lengths of fragments of DNA following cleavage with restriction enzymes. For example, Figure 14.9A illustrates, for two hypothetical species, amplified DNA lengths of 10,000 base pairs that are subjected to ("digested with") the restriction enzyme *Eco*RI. Note, after a reaction with the *Eco*RI enzyme, that the DNA of species *A* is cleaved into three fragments, corresponding to two *Eco*RI restriction sites, whereas that of species *B* is cleaved into four fragments, corresponding to three *Eco*RI restriction sites. The relative locations of these restriction sites on the DNA can be mapped; one possibility is seen at Figure 14.9B. (Note that there are other possibilities for this map; precise mapping requires additional work.) More than one (typically two) restriction enzymes can be used.

Restriction site fragment data can be coded as characters and character states in a phylogenetic analysis, based on the presence of different sized fragments. Restriction site analysis contains far less data than complete DNA sequencing, accounting only for the presence or absence of sites 6–8 base pairs long. It has the advantage, however, of surveying considerably larger segments of DNA. However, with improved and less expensive sequencing techniques, like RADSeq (above), it is much less often used than in the past.

Similar to analysis of RFLPs is that of **AFLPs, Amplified Fragment Length Polymorphism**. This method also uses a restriction enzyme to cut DNA into numerous, smaller pieces, each of which (because of the action of the restriction enzymes) terminates in a characteristic nucleotide sequence. The numerous, cut DNA fragments are then modified by binding to each end a primer adapter. Primers are then constructed that bind to the primer adapters and amplify the DNA fragments using a polymerase chain reaction. Electrophoresis separates the **amplified** DNA **fragments** that exhibit **length polymorphism** (hence, **AFLP**), enabling the recognition of numerous genetic markers.

MICROSATELLITE DATA

Microsatellites, also called **Single Sequence Repeats** (**SSRs**), are regions of DNA that contain short repeats of nucleotides, an example being GTCGTCGTCGTCGTC, in which three base pairs repeat, the regions known as **tandem repeats**. If these vary within a population or species, they are termed **variable-number tandem repeats** (**VNTR**). (Other

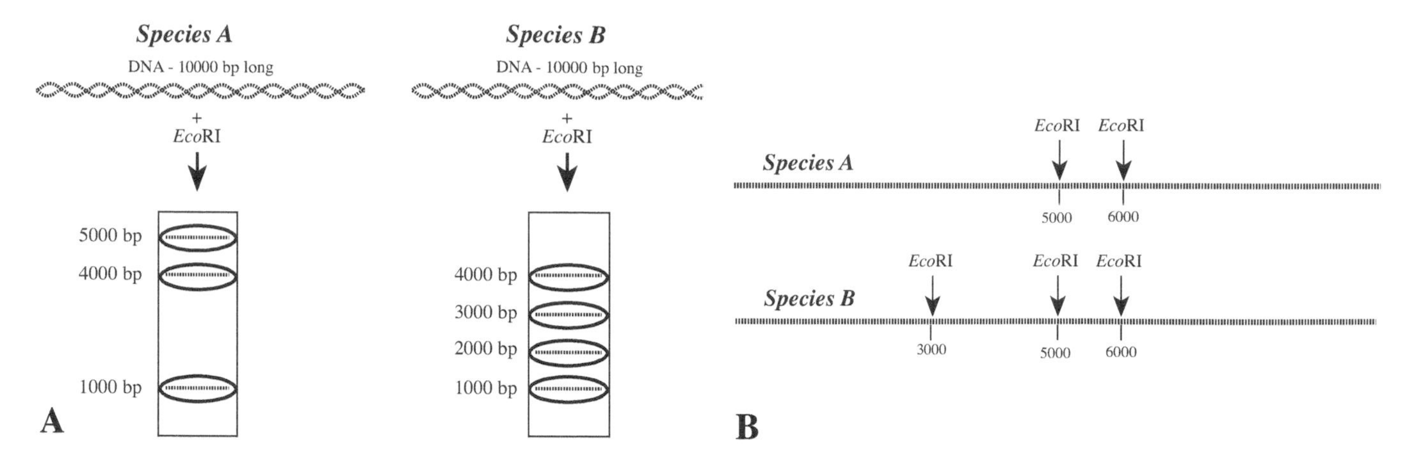

Figure 14.9 Example of restriction site analysis of species A and B, using restriction site enzyme *Eco*RI. **A.** Electrophoresis of fragments from original 10,000 base pair region, showing differences in fragment lengths in the two species. **B.** Restriction site maps of species A and B, showing the location of *Eco*RI restriction sites in the two species.

designations and acronyms are used, depending on the particular field of study.) These tandem repeats can be located all across the genome; at a given location (locus), the repeat will tend to be of a certain length. However, individuals within or between populations may vary in the number of tandem repeats at a given locus (or even show allelic variation) because of irregularities in crossing-over and replication. Thus, variable-number tandem repeats can be used as a genetic marker.

Microsatellites are identified by constructing primers that flank the tandem repeats and then using PCR technology. The primers are initially identified for a species by the time-consuming process of synthesizing genetic probes of a tandem repeat, screening DNA for binding to these probes, and sequencing these regions to design primers that flank the tandem repeats. Once the primers are identified, PCR can be used to quickly generate multiple copies of the tandem repeat DNA, the length of which, for a given individual at a given locus or allele, can be determined by gel electrophoresis. (See example in Figure 14.10.)

Microsatellite analysis can generate data quickly and efficiently (once the primers are identified for a given group) for a large number of individuals. It is most often used for population studies, e.g., to assess genetic variation or homozygosity. Its use in systematics is largely in examining relationships within a species, such as to assess infraspecific classifications, or between very closely related species.

RANDOM AMPLIFICATION OF POLYMORPHIC DNA (RAPD)

Another method of identifying genetic markers uses **randomly** synthesized primers to **amplify polymorphic DNA** (**RAPD** analysis). In this technique, the primer will anneal to complementary regions located in various locations of iso-

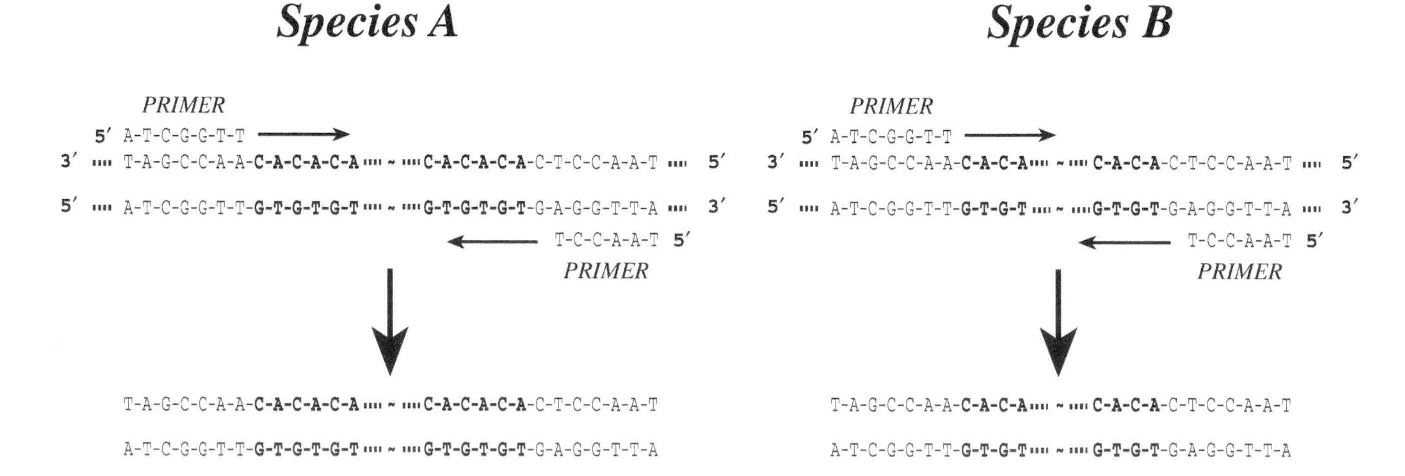

Figure 14.10 Microsatellite data. Primers were constructed to flank regions of tandem repeats. Note that tandem repeat region of species *A* is longer than that of species *B* and is thus a genetic difference between the two.

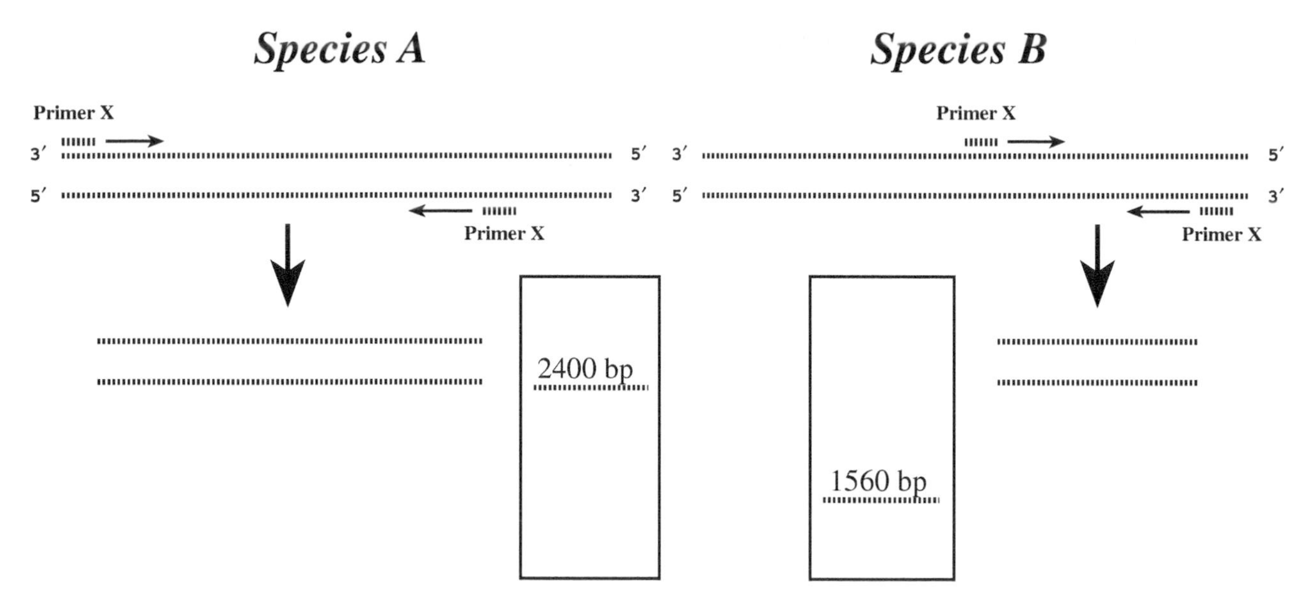

Figure 14.11 RAPDs data. In this example the same DNA regions for species *A* and *B* anneal to different randomly generated primers, resulting in amplified DNA of different lengths, a genetic difference between the two taxa.

lated DNA. If another complementary site is present on the opposing DNA strand at a distance that is not too distant (i.e., within the limits of PCR), then the reaction will amplify this region of DNA (Figure 14.11). Because many sections of DNA complementary to the primer may occur, the PCR reaction will result in DNA strands of many different lengths, which can be size-separated by electrophoresis. Because even closely related individuals may show some sequence variation that could determine potential primer sites, these different individuals will show different amplification products. The data are known as **RAPD**s. (See example in Figure 14.11.)

RAPDs, like microsatellites, may often be used for within -species genetic studies, but may also be successfully employed in phylogenetic studies to address relationships within a species or between closely related species. However, RAPD analysis has the major disadvantages in that results are difficult to replicate (being very sensitive to PCR conditions) and in that the homology of similar bands in different taxa may be unclear.

ANALYSIS OF MOLECULAR DATA

The scientific field encompassing the study and analysis of genetic data is known as **bioinformatics**, which has been defined as "conceptualizing biology in terms of macromolecules ... and then applying 'informatics' techniques (derived from disciplines such as applied maths, computer science, and statistics) to understand and organize the information associated with these molecules, on a large-scale" (Luscombe et al. 2001). A knowledge of bioinformatics theory and techniques is now part of standard training for graduate students in plant systematics. The following is a brief summary of how molecular data may be analyzed in a molecular phylogenetic study.

As discussed in Chapter 2, DNA or RNA sequence data is converted to characters and characters states to be used in phylogenetic analyses. The sequences of a given length of DNA or RNA of different taxa of the ingroup are **aligned**, in which homologous nucleotide positions (e.g., if coding DNA, corresponding to the same codon position of a given gene) are arranged in matching columns (Figure 14.12). For some sequences that are relatively conserved, alignment is straightforward, as taxa may have the same number of nucleotides per gene. Mutational variation among taxa may be in the form of **single nucleotide polymorphisms** (**SNPs**, e.g., as in Figure 14.2, characters 1–5). For other sequences or gene regions, taxa may have one or more additions or deletions, known as **indels** (e.g., as in Figure 14.12, character 6). In addition, major structural mutations, such as inversions or translocations, may be detected and coded. The occurrence of numerous mutations can make alignment of DNA sequences difficult. Sequences derived from RNA will be missing *introns*, non-coding segments of the coding DNA. In addition, multiple copies of a gene that are variable can make homology assessment difficult. If gene duplication occurred, distinguishing **orthologs** (homologous sequences from one of the duplicated sets) from **paralogs** (sequences from different duplicated sets, having different functions) may be prob-

	DNA Alignment	***Character Coding***					
	000000000000000000111111111111111111111111						
	888888889999999999000000000011111111112222						
	234567890123456789012345678901234567890123	1	2	3	4	5	6
Taxon 1	**G**CCTACCC**A**AAGCTC**T**TCCAAG**G**TG**A**CTCTCAA**GTTCAA**GCT	2	0	3	2	0	**4**
Taxon 2	**G**CCTACCC**A**AAGCTC**T**TCCAAG**C**TG**A**CTCTCAA**------**GCT	2	0	3	1	0	**5**
Taxon 3	**G**CCTACCC**T**AAGCTC**A**TCCAAG**G**TG**T**CTCTCAA**GTTCAA**GCT	2	3	0	2	3	**4**
Taxon 4	**G**CCTACCC**T**AAGCTC**T**TCCAAG**G**TG**T**CTCTCAA**GTTCAA**GCT	2	3	3	2	3	**4**
Taxon 5	**G**CCTACCC**A**AAGCTC**T**TCCAAG**C**TG**A**CTCTCAA**------**GCT	2	0	3	1	0	**5**
Taxon 6	**C**CCTACCC**A**AAGCTC**T**TCCAAG**C**TG**A**CTCTCAA**GTTCAA**GCT	1	0	3	1	0	**4**
Taxon 7	**C**CCTACCC**A**AAGCTC**T**TCCAAG**C**TG**A**CTCTCAA**GTTCAA**GCT	1	0	3	1	0	**4**
Taxon 8	**G**CCTACCC**T**AAGCTC**T**TCCAAG**C**TG**A**CTCTCAA**GTTCAA**GCT	2	3	3	1	0	**4**

Figure 14.12 Example of alignment of DNA sequences of 41 nucleotide sites (positions 82–123) for eight taxa; character x taxon matrix at right. Variable nucleotide sites are in **bold**. Note indel (insertion-deletion) of six bases at positions 115–120. Coding of nucleotides is as follows: A = 0; C = 1; G = 2; T = 3. Characters 1–5 are coded as single nucleotide polymorphisms. The indel is coded as a single binary character (character 6), as state 4 = nucleotides absent and state 5 = nucleotides present.

lematic. Computer algorithms can be used to automatically align sequences of the taxa being studied, but these have assumptions that must be carefully assessed.

As discussed in Chapter 2, in using DNA sequence data in a phylogenetic analysis, a character may be equivalent to the nucleotide position, and a character state of that character is the specific nucleotide at that position (there being four possible character states, corresponding to the four nucleotides; see Figure 14.12). A large number (often the great majority) of nucleotide positions are generally invariant among taxa, and some of the variable ones are often uninformative by being autapomorphic for a given taxon; thus, relatively few sites are informative and therefore useful in phylogenetic reconstruction (Figure 14.12).

An indel, inversion, or translocation can in itself be identified as an evolutionary novelty (apomorphy), used in grouping lineages together. For example, members of the Dalbergioid clade (Faboideae, Fabaceae; see Chapter 8) lack, by deletion, one of the inverted repeats found in the chloroplasts of most angiosperms (see Figure 14.2). Chromosomal mutations such as these may be coded separately from single base differences (e.g., as in Figure 14.12, character 6) and may be given a different weight in inferring relationships.

Parsimony, maximum likelihood, Bayesian, and coalescence are the most common methods used to infer phylogenetic relationships using DNA sequence data (Chapter 2). Robust hypotheses of relationship are generally those using a large taxon sampling and sequence data from multiple genes and/or sequence regions. As reviewed in Chapter 2, **evolutionary models** may be calculated using the sequence data in these methods of phylogenetic analysis. These models evaluate the structure of the sequence data. For example, for protein encoding genes, the codon position may be differentially weighted in these models, given that the third codon position is generally more labile because of redundancy in the code, and may not affect the resulting amino acid in a synthesized protein (a **synonymous mutation**). Another model parameter concerns transitions, changes between purines or between pyrimidines, versus transversions, changes from a purine to a pyrimidine or vice versa (Chapter 2).

DNA sequence data can also be used to evaluate the secondary structure of a molecule. Thus, nucleotide differences that result in major changes in the conformation of the product (whether ribosomal RNA or protein) may have a much greater physiological effect than those that do not and might receive a higher weight. Computer algorithms can evaluate this to some degree.

Finally, as alluded to earlier, analysis of sequence data can yield information on the functional features of genes , such as Gene Ontology (GO) categories which provide broad categories of functions in which genes may play. Evolution occurs from mutations and genetic recombination, their persistence and spread in populations directed by natural selection (Chapter 1). Elucidating how these molecular changes direct the development of the morphology and physiology will help us to understand the evolution of these functional adaptations, used in the characterization and classification of plants.

REVIEW QUESTIONS

1. How are samples used to acquire molecular data typically processed, and what is vital to obtain with these samples?
2. Name the three major types of plant DNA sequenced in plant phylogenetic systematics studies.
3. Describe the major DNA regions of the chloroplast.
4. What is the cistron, what is its function, and which two regions are most commonly sequenced in phylogentic studies?
5. Explain the polymerase chain reaction and its importance in molecular systematics.
6. What is a primer?
7. Explain the basic process of Sanger DNA sequencing. How do dideoxynucleotides function in Sanger sequencing?
8. What are the advantages of high-throughput (next generation) sequencing?
9. What is a barcode, and how does it relate to multiplexing?
10. Describe how sequencing by synthesis works.
11. What is meant by: (a) a read? (b) assembly? Sequencing depth?
12. What is the difference between reference guided and de novo sequencing?
13. What is meant by genome skimming, and what types of data are acquired?
14. What is meant by targeted sequencing? Name two methods of targeted sequencing.
15. What does transcriptomics refer to?
16. What are the advantages and potential uses of transcriptome data?
17. What is a restriction site? A restriction enzyme?
18. Describe how RADSeq works and the type of data derived from this technique.
19. How does Multiplexed Shotgun Genotyping differ from RADSeq, and what are its advantages?
20. What type of data is derived from RFLPs and AFLPs?
21. What are microsatellites (single sequence repeats)? What type of data is generated from their analysis?
22. What are RAPDs and what type of data is generated from this technique?
23. What does the field of bioinformatics include?
24. What is DNA alignment, and what are potential problems with this?
25. In general, what are the characters and character states for DNA sequence data?
26. What is a single nucleotide polymorphism (SNP)? An indel?
27. What are orthologs versus paralogs, and how might paralogs from gene duplication be problematic in genetic coding?
28. What is the importance of evaluating functional features of sequence data in plant systematics?

EXERCISES

1. If possible, get a demonstration of the various techniques of molecular systematics, e.g., DNA extraction and sequencing. Consider a special topics project in which you define a problem and use these techniques to acquire the data to answer the problem.
2. Access GenBank (*http://www.ncbi.nih.gov/Genbank*) and acquire molecular data on a particular group of choice. Consider analyzing these data using phylogenetic inference software (see Chapter 2).
3. Peruse journal articles in plant systematics, e.g., *American Journal of Botany, Annals of the Missouri Botanical Garden, Systematic Botany*, or *Taxon*, and note those that describe the use of molecular data in relation to systematic studies. Identify the techniques used, data acquired, and problems addressed.

REFERENCES FOR FURTHER STUDY

Alvarez, I. and J. F. Wendel. 2003. Ribosomal ITS sequences and plant phylogenetic inference. Molecular Phylogenetics and Evolution 29:417-434.

Andolfatto, P., D. Davison, D. Erezyilmaz, T. T. Hu, J. Mast, T. Sunayama-Morita, and D. L. Stern. 2011. Multiplexed shotgun genotyping for rapid and effi cient genetic mapping. Genome Research 21:610-617.

Baldwin, B. G., M. J. Sanderson, J. M. Porter, M. F. Wojciechowski, C. S. Campell, and M. J. Donoghue. 1995. The ITS region of nuclear ribosomal DNA: a valuable source of evidence on angiosperm phylogeny. Annals of the Missouri Botanical Garden 82:247–277.

Baldwin, B. G., and S. Markos. 1998. Phylogenetic utility of the External Transcribed Spacer (ETS) of 18S-26S rDNA: congruence of ETS and ITS Trees of *Calycadenia* (Compositae). Molecular Phylogenetics and Evolution 10:449-463.

Buggs, R. J. A., S. Renny-Byfield, M. Chester, I. E. Jordon-Thaden, L. F. Viccini, S. Chamala, A. R. Leitch, P. S. Schnable, W. B. Barbazuk, P. S. Soltis, and D. E. Soltis. 2012. Next-generation sequencing and genome evolution in allopolyploids. American Journal of Botany 99:372-382.

Cronn, R., B. J. Knaus, A. Liston, P. J. Maughan, M. Parks, J. V. Syring, and J. Udall. 2012. Targeted enrichment strategies for next-generation plant biology. American Journal of Botany 99:291-311.

Doyle, J. J. and J. L. Doyle. 1987. A rapid DNA isolation procedure for small quantities of fresh leaf tissue. Phytochemical Bulletin 19:11–15.

Fulton, T., R. V. D. Hoeven, N. Eannetta, and S. Tanksley. 2002. Identification, analysis and utilization of a conserved ortholog set (COS) markers for comparative genomics in higher plants. Plant Cell 14:1457-1467.

Grover, C. E., A. Salmon, and J. F. Wendel. 2012. Targeted sequence capture as a powerful tool for evolutionary analysis. American Journal of Botany 99:312-319.

Hillis, D. M., C. Moritz, and B. K. Mable. 1996. Molecular Systematics. 2nd ed. Sinauer, Sunderland, Massachusetts.

Huang, X., Q. Feng, Q. Qian, Q. Zhao, L. Wang, A. Wang, J. Guan, D. Fan, Q. Weng, T. Huang, G. Dong, T. Sang, and B. Han. 2009. High-throughput genotyping by whole-genome resequencing. Genome Research 19:1068-1076.

Knoop, V. 2004. The mitochondrial DNA of land plants: peculiarities in phylogenetic perspective. Current Genetics 46:123-139.

Lemmon, M. E. and A. R. Lemmon. 2013. High-throughput genomic data in systematics and phylogenetics. Annual Review of Ecology, Evolution and Systematics 44:99-121.

Luscombe, N. M., G. D., and G. M. 2001. What is bioinformatics? A proposed definition and overview of the field. Methods of Information in Medicine 40:346-358.

Schmieder, R. and R. Edwards. 2011. Quality control and preprocessing of metagenomic datasets. Bioinformatics 27:863-864.

Shaw, J., E. B. Lickey, J. T. Beck, S. B. Farmer, W. Liu, J. Miller, K. C. Siripun, C. T. Winder, E. E. Schilling, and R. L. Small. 2005. The Tortoise and the hare II: relative utility of 21 noncoding chloroplast DNA sequences for phylogenetic analysis. American Journal of Botany 92:142-166.

Shaw, J., E. B. Lickey, E. E. Schilling, and R. L. Small. 2007. Comparison of whole chloroplast genome sequences to choose noncoding regions for phylogenetic studies in angiosperms: the tortoise and the hare III. American Journal of Botany 94:275-288.

Small, R. L., J. A. Ryburn, R. C. Cronn, T. Seelanan, and J. F. Wendel. 1998. The tortoise and the hare: choosing between noncoding plastome and nuclear Adh sequences for phylogenetic reconstruction in a recently diverged plant group. American Journal of Botany 85:1301–1315.

Smith, S. D., R. Angelovici, K. Heyduk, H. A. Maeda, G. D. Moghe, J. C. Pires, J. R. Widhalm, and J. H. Wisecaver. 2019. The renaissance of comparative biochemistry. American Journal of Botany 106:1-11.

Soltis, P. S., D. E. Soltis, and J. J. Doyle (eds.). 1992. Molecular Systematics of Plants. Chapman and Hall, New York.

Soltis, D. E., P. S. Soltis, and J. J. Doyle (eds.). 1998. Molecular Systematics of Plants II: DNA Sequencing. Kluwer Academic, Boston.

Soltis, D. E., M. A. Gitzendanner, G. Stull, M. Chester, A. Chanderbali, S. Chamala, I. Jordon-Thaden, P. S. Soltis, P. S. Schnable, and W. B. Barbazuk. 2013. The potential of genomics in plant systematics. Taxon 62:886-898.

Straub, S. C. K., M. Parks, K. Weitemier, M. Fishbein, R. C. Cronn, and A. Liston. 2012. Navigating the tip of the genomic iceberg: Next-generation sequencing for plant systematics. American Journal of Botany 99:349–364.

Weitemier, K., S. C. K. Straub, R. C. Cronn, M. Fishbein, R. Schmickl, A. McDonnell, and A. Liston. 2014. Hyb-Seq: Combining target enrichment and genome skimming for plant phylogenomics. Applications in Plant Sciences 2:1400042.

MOLECULAR SYSTEMATICS COMPUTER PROGRAMS AND SERVICES

DArTSeq. Diversity Arrays Technology Sequencing. Bruce, Australia. *https://www.diversityarrays.com*

FastQC. A quality control tool for high throughput sequence data. *https://www.bioinformatics.babraham.ac.uk/projects/fastqc*

GenBank. *https://www.ncbi.nlm.nih.gov/genbank*

Geneious. Biomatters Ltd., Auckland, New Zealand. *http://www.geneious.com*

MAFFT. Multiple alignment program for amino acid or nucleotide sequences. *https://mafft.cbrc.jp/alignment/software*

PRINSEQ. PReprocessing and INformation of SEQuences. *http://edwards.sdsu.edu/cgi-bin/prinseq/prinseq.cgi*

TransDecoder (Find Coding Regions Within Transcripts). *https://github.com/TransDecoder/TransDecoder/wiki*

IV

RESOURCES IN PLANT SYSTEMATICS

CAROLI LINNÆI
S:Æ R:GIÆ M:TIS SVECIÆ ARCHIATRI; MEDIC. & BOTAN.
PROFESS. UPSAL; EQUITIS AUR. DE STELLA POLARI;
nec non ACAD. IMPER. MONSPEL. BEROL. TOLOS.
UPSAL. STOCKH. SOC. & PARIS. CORESP.

SPECIES
PLANTARUM,
EXHIBENTES
PLANTAS RITE COGNITAS,
AD
GENERA RELATAS,
CUM
DIFFERENTIIS SPECIFICIS,
NOMINIBUS TRIVIALIBUS,
SYNONYMIS SELECTIS,
LOCIS NATALIBUS,
SECUNDUM
SYSTEMA SEXUALE
DIGESTAS.

TOMUS I.

Cum Privilegio S. R. M:tis Sueciæ & S. R. M:tis Poloniæ ac Electoris Saxon.

HOLMIÆ,
IMPENSIS LAURENTII SALVII.
1753.

Welcome to the Consortium of California Herbaria Portal (CCH2)

CCH2 serves data from specimens housed in CCH member herbaria. The data included in this database represents all specimen records from partner institutions. The data served through this portal are currently growing due to the work of the **California Phenology Thematic Collections Network (CAP-TCN)**. This collaboration of 22 California universities, research stations, natural history collections, and botanical gardens aims to capture images, label data, and phenological (i.e., flowering time) data from nearly 1 million herbarium specimens by 2022. Data contained in the CCH2 portal will continue to grow even after this time through the activities of the CCH member institutions.

For more information about the California Phenology TCN, visit the project website:

https://www.capturingcaliforniasflowers.org

For more information about the California Consortium of Herbaria (CCH) see:

http://ucjeps.berkeley.edu/consortium/about.html

The California Phenology TCN is made possible by the National Science Foundation Award 1802312. Any opinions, findings, and conclusions or recommendations expressed in this material are those of the author(s) and do not necessarily reflect the views of the National Science Foundation.

Special thanks to the National Park Service who provided funds for the initial setup of the CCH2 website and database (November 2016)

Note also these other portals that will better serve the data needs of more-specialized users:

California vascular plants - CCH1: For California vascular plants linked to the statewide flora project (the Jepson eFlora: http://ucjeps.berkeley.edu/eflora/), please see the original the CCH1 portal (active since 2003).

Pteridophytes - fern portal: A new world-wide Pteridophyte portal is currently in development. Please check back later for a website URL

Macroalgae: For algae specimens, see the Macroalgal Herbarium Consortium Portal

Brytophytes: For bryophyte specimens, see the Consortium of North American Bryophyte Herbaria (CNABH)

Lichens: For lichen specimens, see the Consortium of North American Lichen Herbaria (CNALH)

Fungi: For fungi, see the Mycology Collections data Portal (MyCoPortal)

15

PLANT IDENTIFICATION

Identification is the process of *associating* an unknown entity with a known one (or recognizing that the unknown entity does not have a known counterpart). In other words, identification is a judgement that some perceived entity is similar enough to a known entity that it falls within the criteria of belonging to the same "class" as the known entity. The entity itself can be a physical object or a mental concept. Identification is a basic activity of humans and other animals, and perhaps of all life at some level of organization. The ability to recognize, e.g., edible from toxic or friend from foe has undoubtedly evolved via strong selective pressure.

Because two entities are never exactly the same, a critical consideration in identification is determining the characteristics or boundaries of the known. For example, plant identification entails studying a plant, plant specimen, or plant image and making a decision as to whether the plant "belongs to" a particular taxon, e.g., to a species. This identification rests on some type of prior description, both of the unknown plant and of the taxa that are possibilities. In considering the possible taxa, it is critical to evaluate the **diagnostic characterization** (= **diagnosis**) of each, which is a listing or assessment of the features of a taxon that distinguish it from all other taxa. If the characteristics of the unknown entity fall within the range of the diagnosis of a known one, then an identification is made.

MOLECULAR PLANT IDENTIFICATION

One more commonly used method of plant identification entails acquiring one or more DNA sequences of one or more candidate genes and comparing these with a repository of genes associated with plant taxa. Generally, short orthologous sequences, known as "barcodes," are used (see Kress et al. 2005). These may be obtained quickly at relatively low cost. They may be used not just to identify an individual to species, but also to assess stages of the lifecycle and sex (in the case of dioecious taxa). Barcode sequences are not always diagnostic for a given species within, say a genus, but progress has been made (Hollingsworth 2011), even to the point of using whole chloroplast genomes (e.g., Li et al. 2015). The use of DNA sequences from plant material for identification will only continue in importance, and has held especial importance in the forensic sciences.

MORPHOLOGICAL PLANT IDENTIFICATION

Plant identification using morphological features can utilize virtually all of our senses. One can learn to identify a plant by the smell of its leaf or flower, by the taste of its fruit, or by the feel of its stem's surface. In fact, experts in a given plant group often rely on smell, touch, and (occasionally) taste to identify or confirm the identity of an unknown plant in the field. Our sense of hearing would not normally be used in plant identification, although one can certainly learn to distinguish between trees by the sound of the wind rustling their leaves; in identification of plants by various animals, such as birds or bats, sound might be the major sense used.

However, for the great majority of us and for most of the time, vision is the primary means of plant identification. Our brains can almost instantaneously process how light is reflected from a plant or plant part into what we call a visual image. This visual pattern, at least for a fraction of a second, is unknown to us. Identification entails associating that unknown visual pattern with a known one, the latter either

PLANT SYSTEMATICS
https://doi.org/10.1016/B978-0-12-812628-8.50015-8,

stored in our brain or deduced using various tools of identification (see later discussion).

In identifying a plant (or any entity) there are two major conceptual ways that our brains process information: holistically and analytically (Gauthier and Tarr 2002; see Kirchoff et al. 2007). **Holistic** mental processing is that in which an object is viewed or emphasized as a whole. Most trained systematists familiar with a given group or floristic region identify plants holistically. They can, at a glance, process the visual pattern of light reflected from, say, leaves or flowers and almost instantaneously associate that pattern with their memory of a similar pattern for a given plant. The associated pattern is also typically, though not strictly necessarily, linked with a name. **Analytic** processing is that in which the parts of an object are emphasized. Analytic identification entails breaking down or subdividing the whole object into parts, typically using specialized descriptive terminology (see Chapter 9). For example, a leaf could be analyzed in terms of its arrangement, position, orientation, type, size, attachment, base, margin, apex, venation, texture, and surface features. Analytic processing might be used more often by someone with limited prior knowledge of a plant group, or for plants in which fine or obscure characters must be examined to distinguish between taxa, requiring, e.g., careful dissection and inference.

Practically speaking, the identification of a plant to one among hundreds of thousands of possibilities often requires an initial narrowing down of those possibilities. If a plant is native or naturalized, regional floras are typically used in the process of identification, although there is always the possibility that the plant in question is new to the region and not included in the flora. One should check the geographic range of the flora used to be sure that it encompasses that of the unknown plant. Cultivated plants can be particularly difficult to identify. This is true in part because the number of plants taken into cultivation is quite large and continues to expand every year. Because a cultivated plant can be native to any region of the world, one can rarely be certain of having a reference that will include the correct taxon. Cultivated plant species may also be difficult to identify because they may include a great number of cultivars, hybrids, or other breeds that are continually being introduced and may be quite different in appearance from an original native species.

There are several methods of identifying plants. These are described below.

TAXONOMIC KEYS

Perhaps the most useful method of identification is a taxonomic key. A **key** is an identification device that consists of sequentially choosing among a list of possibilities until the possibilities are narrowed down to one. Most keys are practical, narrowing down the identity of a taxon in the most efficient and effective means. The key may or may not split a larger group into smaller, natural (monophyletic) subgroups.

As reviewed in Chapter 1, the most common type of key, particularly in floras and monographic treatments, is a **dichotomous key**. This consists of a sequence of two contrasting statements, each statement known as a **lead**; the two leads together comprise a **couplet**. The leads of a couplet may be indented and/or numbered. Identification proceeds by choosing between the contrasting leads of a couplet. That lead which best fits the organism to be identified is selected; then all couplets hierarchically beneath that lead (either by indentation or numbering) are sequentially checked until an identification is obtained (see Figure 15.1). A well-written dichotomous key may have several types of evidence presented, with every character of the first lead matched, respectively, in the second lead (Figure 15.1).

Most keys are *artificial* or *practical*, meaning that the sequential groupings of the key do not intentionally reflect natural groups; their goal is to most easily and efficiently identify a given taxon, with no concern about classification into other groups. Rarely, a key may be *natural* or *phylogenetic*, in which diagnostic (or even apomorphic) features are used to delimit "natural" groups, which are usually formal taxa. An example of a natural key might be one to the tribes of the Asteraceae. More technical, but less obvious, characters are used in natural keys, so they are generally less useful in practical identification, but may denote the features used to separate taxonomic groups.

Some precautions should be taken in using a dichotomous key. Most important is to *read all parts of both leads* before making a decision as to which fits the plant best. Never read just the first lead; although it might seem to fit, the second lead may fit even better. If, after reading both leads of a couplet, you are not certain which is correct, both should be considered. The two (or more) possibilities attained can then be checked against descriptions, illustrations, or specimen comparisons.

Another type of identification device is the polyclave key. A **polyclave key** consists of a list of numerous character states, whereby the user selects all of states that match the specimen (e.g., Figure 15.2). Based on which of the many character states are a match, the correct taxon (or closest match) can be determined or narrowed down to a smaller subset of the possibilities. All polyclave keys in use today are implemented by a computer algorithm.

The great advantage of polyclave keys over dichotomous ones is that they permit the use of a limited subset of information to at least narrow down the possibilities. For example, if a dichotomous key lists only floral characters, its usefulness may be limited if your plant specimen lacks flowers. A polyclave key, however, will have a listing of not only floral

Papaveraceae POPPY FAMILY

1. Flower bilateral or biradial; petals 4, 1 or both outer ones spurred or pouched at base (Fumarioideae)
 2. Flower biradial; outer 2 petals alike, both pouched at base
  3. Plant without leafy stems .... **DICENTRA**
  3′ Plant with leafy stems .... **EHRENDORFERIA**
 2′ Flower bilateral; outer 2 petals not alike, upper petal spurred at base
  4. Fruit a capsule, generally linear to oblong, several to many-seeded .... **CORYDALIS**
  4′ Fruit a nut, round, 1-seeded .... **FUMARIA**
1′ Flower radial; petals ≥4, not spurred or pouched at base (Papaveroideae)
 5. Subshrub to small tree
  6. Petals 4(6), yellow, 2–3 cm; leaf entire or minutely toothed .... **DENDROMECON**
  6′ Petals 6, white, 4–10 cm; leaf deeply lobed .... **ROMNEYA**
 5′ Annual, biennial, perennial herb, occasionally woody at base
  7. Leaves entire to minutely toothed, cauline opposite; petals generally 6
   8. Plant generally glabrous; stamens 4–15
    9. Basal leaves sessile, linear, fleshy; petals persistent; capsule ovoid, 1.5–2.5 mm .... **CANBYA**
    9′ Basal leaves petioled, spoon-shaped, not fleshy; petals deciduous; capsule linear, to 50 mm .... **MECONELLA**
   8′ Plant long-hairy; stamens > 12
    10. Stigmas 3; fruit ovoid, opening by 3 valves, not breaking transversely into units .... **HESPEROMECON**
    10′ Stigmas ≥6; fruit ovoid to widely linear, breaking transversely into 1-seeded, indehiscent units .... **PLATYSTEMON**
  7′ Leaves toothed to dissected; petals 4, 6, or more
   11. Plant spiny; leaves generally cauline .... **ARGEMONE**
   11′ Plant unarmed; leaves basal, cauline, or both
    12. Hairs generally ±5–15 mm, wavy; leaves generally basal .... **ARCTOMECON**
    12′ Hairs 0 or to 3 mm, straight; leaves basal, cauline, or both
     13. Ovary, fruit <3x longer than wide; carpels > 2; fruit opening by pores .... **PAPAVER**
     13′ Ovary generally >3x (fruit >4x) longer than wide; carpels 2; fruit splitting into 2 valves
      14. Leaf dissected into ± linear segments; sepals 2, fused into conical cap; receptacle cup-shaped around ovary base; stigma lobes 4–8, linear .... **ESCHSCHOLZIA**
      14′ Leaf deeply pinnate-lobed; sepals 2, free; receptacle not cup-shaped, not around ovary base; stigma lobes 2, not linear .... **GLAUCIUM**

Figure 15.1 Example of indented and numbered dichotomous key, to the genera of the Papaveraceae in California. Note that character states of the first lead of a couplet are matched by corresponding character states in the second lead. Authored by Gary L. Hannan & Curtis Clark, in Jepson Flora Project (eds.) Jepson eFlora, *http://ucjeps.berkeley.edu/cgi-bin/get_IJM.pl?key=214*, reprinted by special permission.

characters, but also features of the roots, stems, leaves, fruits, and seeds. Thus, the polyclave key will often enable the user to identify the plant, even if one or more types of data are missing from the specimen. A second advantage of polyclave keys is that if the specimen cannot be absolutely identified, its identity may at least be narrowed down to a few alternatives, which can then be checked by other means. The only major disadvantage of polyclave keys is their availability; they have generally been written only for a limited number of taxonomic groups.

Most keys, whether dichotomous or polyclave, rely solely on written statements, which the user chooses between in narrowing down the identity. A very useful tool in identification is the use of visual images. Images, in the form of a photograph or drawing, can be used to clarify the meaning of contrasting characters or character states. Images of the taxa themselves (showing their distinguishing features) are particularly useful in checking the accuracy of an identification. In fact, images alone may be used in constructing a key. (See Kirchoff et al. 2008 for a discussion of a key that uses images instead of descriptive statements, emphasizing holistic mental processing.)

Although keys are probably the most practical and utilized method of identification, they should be regarded as guides, not foolproof methods. Any identification attained should be checked by other means, such as specimen comparison or expert determination (discussed later).

WRITTEN DESCRIPTION

A second means of identification is to compare features of the unknown plant with written descriptions of the possible known taxa. This is a good method of determining with certainty whether the range of variation of the unknown plant corresponds to that listed in the description of a known plant. However, because reading all of the written descriptions of a flora is impractical, this method relies on narrowing down the possibilities first. In addition, gleaning the diagnostic characteristics from a long list of features may be difficult.

1 Woody plants (excl. suffrutices)
2 Herbaceous plants (incl. suffrutices)
3 Aquatic plants, leaves floating or submerged
4 Chlorophyll absent (parasites or saprophytes)
5 Bulb present (monocots only)
6 Milky juice present
7 Spiny stems or leaves
8 Tendrils present
9 Cladodes or phyllodes (modified branches or petioles)
10 Hairs glandular
11 Hairs stellate (also 2-armed, branched and tufted)
12 Hairs stellate (not 2-armed, branched and tufted)
13 Hairs 2-armed or t-shaped, nonglandular
14 Hairs branched
15 Hairs tufted, nonglandular
16 Hairs peltate or scalelike
17 Hairs vesicular or bladderlike
18 Hairs stinging
19 Cystoliths present (dicots only)
20 Leaves opposite or verticillate
21 Leaves alternate (excl. distichous monocots)
22 Leaves distichous (monocots only)
23 Leaves equitant (e.g., *Iris*)
24 Leaves not compound
25 Leaves compound
26 Leaves pinnately compound (4 or more leaflets)
27 Leaves ternately compound (3 leaflets)
28 Leaves palmately compound (4 or more leaflets)
29 Venation pinnate or hardly visible in leaves or leaflets (incl. no. 30)
30 Venation invisible or leaves 1-nerved (monocots only)
31 Venation longitudinal in leaves or leaflets (incl. 3-nerved leaves)
32 Venation palmate in leaves or leaflets

Figure 15.2 Example of a polyclave key. Note that the key consists of a list of numerous character states. Identification proceeds by indicating which subset of character states describes the unknown plant. Not all characters need to be considered.

Thus, written descriptions are best used to verify an identity after one or a few possibilities are presented.

SPECIMEN COMPARISON

A third method of identification is to compare the plant in question to a live or preserved plant collection, usually an identified herbarium specimen. This is an excellent method of identification, as many features of a plant (e.g., subtle coloration and surface features) are often not adequately denoted in written descriptions or visible from photographs or illustrations. As with the foregoing methods, comparison to an herbarium specimen is practically limited to verifying an identity after a subset of possibilities is narrowed down. **Synoptic collections**, which house generally one specimen of each taxon for a given region (e.g., a county), are very useful in this regard. If a taxon can be narrowed to a smaller group, such as a family or genus, a quick search through a synoptic collection for that region may often allow for site identification of the unknown. One precaution about this method, however, is that it is dependent on the fact that the herbarium specimens are themselves correctly identified. Thus, a possible match should always be verified with a written description.

IMAGE COMPARISON

A fourth method by which an unknown plant may be identified is by visually comparing it to photographs or illustrations of known taxa. These are usually obtained from books, although Webpage images have now become a very useful resource. A practical problem with this method is that photographs and illustrations are usually available only for a small subset of possible taxa. In addition, it may be cumbersome to locate the matching photograph or illustration, necessitating an examination of all those available. However, visual comparison to an image can still be an excellent way to identify a plant, particularly if the possibilities can be narrowed down beforehand. The major precaution about this method is that two or more taxa may look very similar to one another as based on a photograph or illustration; the differences between them may reside on obscure morphological features that are not easily visible. Thus, any match of the unknown to a visual image should be confirmed with a technical description of the plant.

Interactive computer programs, especially those available as a phone app, such as *iNaturalist* (see web page citation), have begun to revolutionize the identification of plants (and animals). Both professionals and members of the public (citizen scientists) may photograph a plant in the field. (Note that the photographs should be down properly for identification, including shots of the whole plant, leaves, flowers and inflorescences from side- and face-view, and fruits if available.) The image and its geographic location are instantly available to other members. Experts may confirm or suggest alternative identifications, making the observations of greater scientific value. The data may be mapped or downloaded, and can tremendously expand our knowledge or the geographic distribution of a taxon. Observations on platforms such as *iNaturalist* have already led to the discovery of plant taxa that are new to a given region, previously thought extinct, or even new to science. Although field images cannot replace voucher herbarium specimens (see Chapter 18), and their permanency is still something to be worked out, these have become a significant resource in plant systematic research. In addition, and very importantly, these programs have increased the interest of members of the public in learning about our natural world.

EXPERT DETERMINATION

A fifth and final means of identification is simply to ask someone else, preferably an expert in the group in question. This method may be time-consuming, as it usually requires sending a specimen away for identification (as well as knowing *who* the experts of a given group are). However, expert identification is perhaps the best way to identify a specimen, as the expert will usually know the taxa of that group over a wide geographic range.

Of course, experts rely on all the aforementioned means of identification, especially specimen comparison. Experts of a plant group have likely examined hundreds of herbarium specimens and noted features that may not be in any key or even in any description. Moreover, an expert in the group is likely to be familiar with the literature on these plants, including recent treatments and new taxon discoveries. Their determination is often more accurate and current than any flora.

Expert determination is often essential for certain groups in which species or infraspecific identification is very difficult.

PRACTICAL IDENTIFICATION

The practical steps taken in identification of plant taxa often depend on the experience of the person making the determination. Obviously, the more you know, the easier it is to identify a plant. For example, most floras begin with a key to the plant families, which may be cumbersome because they must take into account the variation within the total flora. Thus, knowing the general characteristics of several families ahead of time helps, as you may proceed directly to the key of genera within that family. Similarly, if you have an idea as to the general group within a family to which the taxon belongs (e.g., a suspected genus), you may wish to check the keys, illustrations, descriptions, or specimens within that group first. However, when in doubt, it is best to start from the beginning to be certain of eliminating the close but incorrect choices.

The importance of correctly identifying a specimen cannot be overstated. Once a determination is made, it should be viewed as only tentative. Never assume you have reached "the correct answer" in using any one method; it is important to check your determination by all available means. Be your own devil's advocate; check and recheck yourself. Verify every identification against a written description and comparison to an herbarium specimen. Some groups may be particularly difficult to identify, being composed of a great number of taxa that differ from one another by obscure features. Proceed very carefully, and do not hesitate to send off specimens for expert determination if needed.

Finally, one should always be conscious of the possibility that the identification process points to a new taxon. If a thorough evaluation of available references indicates that the unknown plant in question does not match any known taxa listed in a flora, then the plant may be a new record (either native or naturalized) for the geographic range of that flora. In some cases, the unidentifiable taxon may be new to science, warranting the valid publication of a new taxon.

REVIEW QUESTIONS

1. What is identification?
2. Describe how identification is used in your everyday life.
3. What is a means of identification using molecular sequence data and what is its increasing importance?
4. What is a diagnostic characterization (= diagnosis) and how is this important in identification?
5. What is the difference between holistic and analytic mental processing?
6. What are the difficulties with identifying cultivated plant taxa?
7. Name five methods used to identify plant taxa, citing the advantages and disadvantages of each.
8. What is a dichotomous key?
9. What is a couplet? A lead?
10. What precautions should be made in using a dichotomous key?
11. What is the difference between a *natural* and an *artificial* key?

12. What is a polyclave key?
13. What are the advantages and disadvantages of a polyclave key?
14. How are images important in plant identification?
15. Name ways, other than a taxonomic key, to identify a plant specimen.
16. What is a synoptic collection and what is its advantage in plant identification?
17. State the practical steps made in identifying a plant specimen.

EXERCISES

1. Select an unknown cultivated plant and attempt to identify it using all of the methods discussed in the chapter. What difficulties did you encounter with any of these?
2. Select an unknown native plant and attempt to identify it using local floras or manuals or by using an herbarium collection, such as a synoptic collection.
3. Do a Web search for a polyclave key (see Reference below), either one on-line or one that may be downloaded. Test this on a given unknown. If possible, create a key for a set of 5–10 plants.
4. Download an app, such as *iNaturalist* (see below) and photograph one or more plants in the field. Attempt to provide an identification, but note what experts have determined to be the plant. Consider doing this all the plants of a given region.

REFERENCES FOR FURTHER STUDY

Gauthier, I. & Tarr, M.J. 2002. Unraveling mechanisms for expert object recognition: bridging brain activity and behavior. Journal of Experimental Psychology of Human Perception and Performance 28: 431–436.

Hollingsworth, P. M., S. W. Graham, and D. P. Little. 2011. Choosing and using a plant DNA barcode. PLOS ONE 6:e19254.

Kirchoff, B., D. Remington, L. Fu, and F. Sadri. 2008. A New Type of Image-Based Key. International Conference on BioMedical Engineering and Informatics.

Kress, W. J., K. J. Wurdack, E. A. Zimmer, L. A. Weigt, and D. H. Janzen. 2005. Use of DNA barcodes to identify flowering plants. Proceedings of the National Academy of Sciences of the United States of America 102:8369.

Li, X., Y. Yang, R. J. Henry, M. Rossetto, Y. Wang, and S. Chen. 2015. Plant DNA barcoding: from gene to genome. Biological Reviews 90:157-166.

Radford, A. E., W. C. Dickison, J. R. Massey, and C. R. Bell. 1974. Vascular Plant Systematics. Harper and Row, New York. 891 pp.

WEBSITES

Dallwitz, M. J. 1996 onwards. Programs for interactive identification and information retrieval.
http://delta-intkey.com/www/idprogs.htm
A comprehensive listing of computerized key programs, their platforms, cost (if any), and if web-based.

iNaturalist. Available from *http://www.inaturalist.org*

Morphbank. 2009. Florida State University, School of Computational Science, Tallahassee, FL 32306-4026 USA.
http://www.morphbank.net
A database of biological images, using open source software and having a Fair Use Web Site policy.

USDA, NRCS. 2009. The PLANTS Database. National Plant Data Center, Baton Rouge, LA 70874-4490 USA.
http://plants.usda.gov
Provides standardized information about land plants of the U.S. and its territories. It includes names, plant symbols, checklists, distributional data, species abstracts, characteristics, images, crop information, automated tools, onward Web links, and references.

16

PLANT NOMENCLATURE

Nomenclature is the assignment of names utilizing a formal system. The criteria for formally naming organisms "traditionally treated as plants" are based on the rules and recommendations of the **International Code of Nomenclature for algae, fungi, and plants** or **ICN** (Turland et al. 2018). Botanical names serve as symbols of a group of natural entities for the purpose of communication and data reference.

The ICN deals with the names of extant or extinct (fossil) organisms encompassed by the field of botany (see Chapter 1 for a definition of *botany*). These include not only the land plants, but also the "blue-green algae (Cyanobacteria); fungi, including chytrids, oomycetes, and slime moulds; photosynthetic protists and taxonomically related non-photosynthetic groups." As discussed in Chapter 1, it is now known that many of these groups are not closely related phylogenetically. Yet, the ICN deals with these taxa, as they were historically treated as plants.

Separate nomenclatural codes exist for traditional zoology (International Code of Zoological Nomenclature) and for prokaryotes (International Code of Nomenclature of Bacteria). One difficulty with this is that photosynthetic bacteria are named both under the ICN and under the Bacteria Code. Similarly, some of the so-called protists (itself a paraphyletic assemblage) are named both under the ICN and the Zoological Code. Thus, some organisms have two names, from two different nomenclatural codes. A draft of a future universal code covering all forms of life, termed the *BioCode* (Greuter et al. 1997), has been prepared, but has not progressed. A separate code using a phylogenetic definition, termed the *PhyloCode* (Cantino and de Queiroz 2007), has also been proposed, recently for the naming of "higher" land plant taxa (see Cantino et al. 2007).

The International Code of Nomenclature governs the rules both for the names assigned to taxa and for the name endings that denote taxon rank (see below). The ICN is utilized in two basic activities: (1) naming new taxa, which were previously unnamed and often not described; and (2) determining the correct name for previously named taxa, which may have been divided, united, transferred, or changed in rank (see later discussion). The rules of the ICN can be somewhat complex, often necessitating careful scrutiny, and a lawyerlike mentality. (Note that a supplementary code is utilized for cultivated plants, the "International Code of Nomenclature for Cultivated Plants." This code operates within the framework of the ICN and its provisions do not override those of the ICN.)

Any name governed by the ICN must be **validly published** (see later discussion). **Legitimate names** are those that are

https://doi.org/10.1016/B978-0-12-812628-8.50016-X,

validly published in accordance with the rules of the ICN. A validly published name that is not in accordance with the rules is an **illegitimate name**.

Changes to the International Code of Nomenclature are voted upon at the Nomenclatural Session of the International Botanical Congress, which assembles about every 6 years in some city around the world. As of this writing, the last Congress was held in Shenzhen, China, July 2017. The following summary is based on the ICN resulting from that Congress (Turland et al. 2018).

PRINCIPLES OF NOMENCLATURE

The **Principles** of the International Code of Nomenclature are stated verbatim below from the 2018 Shenzhen Code. Each of these will be covered in detail.

I. The nomenclature of algae, fungi, and plants is independent of zoological and prokaryotic nomenclature. This *Code* applies equally to names of taxonomic groups treated as algae, fungi, or plants whether or not these groups were originally so treated.
II. The application of names of taxonomic groups is determined by means of nomenclatural types.
III. The nomenclature of a taxonomic group is based upon priority of publication.
IV. Each taxonomic group with a particular circumscription, position, and rank can bear only one correct name, the earliest that is in accordance with the rules, except in specified cases.
V. Scientific names of taxonomic groups are treated as Latin regardless of their derivation.
VI. The rules of nomenclature are retroactive unless expressly limited.

The details of the International Code of Nomenclature are organized into a number of **Rules**, organized as Articles (which are binding), **Recommendations** (which are subsidiary and non-binding but to be preferred), and explanatory **Notes** and **Examples**.

Currently, the entire International Code of Nomenclature is available on a website (see Web Pages).

SCIENTIFIC NAMES

The *fundamental* principle of nomenclature is the fourth principle of the ICN, stating that every **taxon** (a taxonomic group of any rank), whether species, genus, family, etc., can bear only one correct name (see below for precise definition of *correct* name). This is only common sense. Confusion would reign if taxonomic entities could bear more than one name or if one name could refer to more than one entity. The names assigned to individual plant groups by the rules of the ICN are known as **scientific names**. Scientific names are, by convention, treated as Latin (see later discussion).

As reviewed in Chapter 1, the scientific names of species are **binomials** (or **binary combinations**), consisting of two parts, the genus name plus a specific epithet. The binomial convention was first consistently used by Carolus Linnaeus (also known as Carl Linné or Carl von Linné), a Swedish botanist, who is often referred to as the "father of taxonomy." Prior to the use of binomials, the designation of species was descriptive in nature; a name commonly utilized many more words than just two, and names often varied from one author to the next.

As an example of a binomial, the species commonly known as "sweetgum" has the scientific name *Liquidambar styraciflua*. Note that botanical names at the rank of genus and below are typically italicized or underlined; conventions vary above the rank of genus. The first name of the binomial, *Liquidambar* in this case, is the **genus name** and is always capitalized. The second name of the binomial, *styraciflua* in this example, is the **specific epithet**. The specific epithet may be capitalized if it is a commemorative (named after a person or place), but the ICN recommends that epithets not be capitalized. Recall from Chapter 1 that a species name is always the entire binomial. It is incorrect to say that the species name for sweetgum is *styraciflua*, as this is the specific epithet; the species name is *Liquidambar styraciflua*. The genus name may be abbreviated by its first letter, but only after it is first spelled out in its entirety and if it would not be confused with another genus name starting with that letter; thus, the above may be abbreviated as *L. styraciflua*.

In contrast to scientific names, many taxa also bear **common names** (also called vernacular names), which are generally used by people within a limited geographic region. Common names are not formally published and are not governed by the ICN. Scientific names are much preferable to common names for several reasons. First, only scientific names are universal, used the same world-wide; common names may vary from region to region, even within a country or within regions of a country. For example, species of the genus *Ipomoea* are known commonly as "morning glory" in the United States, but as "woodbine" in England. Differences in language will, of course, further increase the number of different common names. In addition, a single taxon may bear more than one common name, these often varying in different regions. For example, *Adenostoma fasciculatum* of the Rosaceae is known by at least two common names, "chamise" and "greasewood." Alternatively, a single common name may refer to more than one taxon. "Hemlock" may refer to two quite different plants, either a species of *Tsuga*, a coniferous tree of the Pinaceae, or *Conium maculatum*, an herb of the Apiaceae (the extract of which Socrates drank in execution). Second, common names tell nothing about rank and often nothing about classification, whereas scientific names generally indicate rank and yield at least some information about their classification. For example, "pygmy weed" tells nothing about rank; it could be variety, species, genus, or family. However, one immediately knows that *Crassula argentea* is at the rank of species and is a close relative to other species of *Crassula*. Third, many, if not most, organisms have no common name in any language; thus, scientific names alone must be used to refer to them. This is especially true for plants that are not showy, occur in remote areas, or belong to groups whose members are difficult to distinguish from one another.

There is a tendency in some works to arbitrarily convert all scientific species names into common names by translating from the Latin, even when these common names are *not* used by the native people. For example, *Carex aurea* might be designated "golden carex" or "golden sedge," even if these names are not in common usage. It is the author's opinion that this is less than ideal policy and that it is preferable simply to utilize scientific names and refer to common names only if they are, in fact, commonly used. However, the trend seems to be to assign standardized common names, at least for some regional floras.

RANKS

Recall from Chapter 1 that taxa are classified hierarchically by **rank**, in which a higher rank is inclusive of all lower ranks (Figure 16.1). [Note that there are "principal" ranks, "secondary" ranks, and additional ranks (if needed) that may be used by adding the prefix "sub"; see Figure 16.1.] Each scientific name of a particular rank ends in a certain suffix according to the rules and recommendations of the ICN (Figure 16.1). For example, Asteridae is a taxon at the rank of subclass, Asterales is at the rank of order, and Asteraceae is at the rank of family, etc. Note that taxa above the rank of genus are *not* underlined or italicized.

An exception to standard rank endings of taxa is the acceptance of nine traditional family names. These are Compositae (= Asteraceae), Cruciferae (= Brassicaceae), Gramineae (= Poaceae), Guttiferae (= Clusiaceae), Labiatae (= Lamiaceae), Leguminosae or Papilionaceae (= Fabaceae), Palmae (= Arecaceae), and Umbelliferae (= Apiaceae). In addition, within the Fabaceae (= Leguminosae), the subfamily name Papilionoideae is an acceptable alternative to the Faboideae. The trend today is to consistently apply the type principle (see later discussion) by using the standardized family names that end in "-aceae" and to use subfamily names that are based on these (e.g., to use "Faboideae" over "Papilionoideae"). However, plant taxonomists should know

TAXONOMIC RANKS OF LAND PLANTS	ENDING	EXAMPLE TAXON
Kingdom	(various)	**Plantae**
Phylum = Division	**-phyta**	**Magnoliophyta**
Subphylum = Subdivision	-phytina	Magnoliophytina
Class [cl.]	**-opsida**	**Asteropsida**
Subclass [subcl.]	-idae	Asteridae
Order [ord.]	**-ales**	**Asterales**
Suborder [subord.]	-ineae	Asterineae
Family [fam.]	**-aceae**	**Asteraceae**
Subfamily [subfam.]	-oideae	Asteroideae
Tribe [tr.]	-eae	Heliantheae
Subtribe [subtr.]	-inae	Helianthinae
Genus [gen.]	(various)	***Helianthus***
Subgenus [subg.]	(various)	*Helianthus* subg. *Helianthus*
Section [sect.]	(various)	*Helianthus* sect. *Helianthus*
Series [ser.]	(various)	*Helianthus* ser. *Helianthus*
Species [sp.]	(various)	***Helianthus annuus***
Subspecies [subsp.]	(various)	*Helianthus annuus* subsp. *annuus*
Variety [var.]	(various)	*Helianthus annuus* var. *annuus*
Form [f.]	(various)	*Helianthus annuus* f. *annuus*

Figure 16.1 Some of the taxonomic ranks and endings recognized by the International Code of Nomenclature. "Principal" ranks are in bold. "Secondary" ranks are underlined. "Sub" ranks may be used as needed, some of the possibilities indicated. Phylum, subphylum, class, and subclass may utilize different endings for Fungi or "Algae." Standard endings above the rank of genus are required or recommended and indicate the rank of that name. "Division" may be used interchangeably with "Phylum." Standard abbreviations are in brackets.

these classical names, as they are often used in older, as well as some current floras and other taxonomic works.

Position is the placement of a taxon as a member of another taxon of the next higher rank. For example, the position of the genus *Aster* is as a member of the family Asteraceae. Taxa may be the same in rank but differ in position. *Rosa* and *Aster* are both at the rank of genus but differ in position, the former in the Rosaceae, the latter in the Asteraceae.

As mentioned earlier, the prefix *sub-* can be used formally in a rank name in more categories are needed, such as *subgenus* or *subspecies*. The term *infraspecific* taxon can be used to denote a taxon below the rank of species, including, among others, subspecies and varieties.

A subspecies or variety name is a **ternary name** and consists of three parts, e.g., *Toxicodendron radicans* subsp. *diversilobum* or *Brickellia arguta* var. *odontolepis*. In these examples, the **subspecific epithet** is *diversilobum*; the **varietal epithet** is *odontolepis*. Note that, technically, the rank of subspecies is above that of variety (Figure 16.1). However, in practice, subspecies and variety are used interchangeably (Hamilton and Reichard 1992).

AUTHORSHIP

Scientific names are associated with one or more **authors**, the person(s) who validly published the name (see later discussion). For example, the family Rosaceae can be cited as "Rosaceae Jussieu" because de Jussieu validly published the family. In other examples, the name of the tribe Conostylideae with authorship is "Conostylideae Lindley"; that of the genus *Mohavea* is "*Mohavea* A.Gray"; that of the species *Mohavea confertiflora* is "*Mohavea confertiflora* (Bentham) Heller"; and that of the subspecies *Monardella linoides* subsp. *viminea* is "*Monardella linoides* A.Gray subsp. *viminea* (Greene) Abrams." Author names are often abbreviated, such as Haemodoraceae R. Br. (for Robert Brown) or *Liquidambar styraciflua* L. ("L." being the standardized abbreviation for Linnaeus). See Brummitt and Powell 1992, International Plant Names Index, and Index Fungorum for standardized author abbreviations.

Authorship should be cited in all scientific publications at least once, in order to clarify the name's origin (valid publication). In practice the author is not typically memorized or recited as part of a scientific name. The authors of higher taxa are sometimes omitted in print even in scientific publications, except in detailed monographic treatments in which the nomenclatural history of the taxa under study is described. In many floras and journal publications, only species and infraspecific taxa may be listed with full authorship.

LEARNING SCIENTIFIC NAMES

As argued earlier, it is important to learn the scientific names of plants, correctly spelled. The serious plant taxonomist will learn many hundreds of scientific names in his/her lifetime, still just a tiny fraction of the more than 250,000 described land plant species. Beginners may at first have difficulty learning scientific names. Some suggestions for mastering them are as follows.

First, learn to divide into syllables and accent scientific names (see **BOTANICAL NAMES**). It is often easier to recite and spell a scientific name if it is consciously broken down into syllables, each of which is separately pronounced.

Second, use mnemonic devices. Select one distinctive feature about the plant. Then find a common word that *sounds* somewhat similar to the scientific name. Link the distinctive plant feature with the similar sounding word in an active, vivid mental image, the weirder and more active the better. Thus, when you see the plant, you associate it with the mental image, which sounds like (and reminds you of) the scientific name. For example, visualizing "liquid amber" flowing from the distinctive, ball-shaped fruits of sweetgum may help you remember the genus name, *Liquidambar*.

Third, learn the etymology (meaning) of scientific names. Scientific names often are descriptive about the morphology of the plant. Once you know, for example, that the Latin word *alba* means "white" or that *leptophylla* means "narrow-leaved," you can better associate the name with the organism. Other scientific names may be named after a person or place of significance; learning the history of these commemorative names may be helpful in memorizing them.

Finally, there is no substitute for continual practice and review. Use a combination of both oral and (for correct spelling) written recitation, with the plant, plant specimen, photograph, or mental image in view.

NOMENCLATURAL TYPES

The second principle of the ICN states that scientific names must be associated with an "element," known as a **nomenclatural type** or simply **type**. A nomenclatural type is almost always a single specimen, e.g., a standard herbarium "sheet" for vascular plants, but it may also be an illustration. The type serves the purpose of acting as a reference for the name. If there is ever any doubt as to whether a name is correct or not, the type must be consulted.

There are different "types of types." A **holotype** is the one specimen or illustration upon which a name is based, originally used or designated at the time of valid publication. It serves as the definitive reference source for any questions of identity or nomenclature. It is recommended that a holotype be deposited in a public herbarium (see Chapter 18) or other public collection, so that it is available for study by systematists. Indication of the holotype is one of the criteria for the valid publication of a name (see later discussion). Holotypes constitute the most valuable of specimens and are kept under safe keeping, usually in a large herbarium. An **isotype** is a duplicate specimen of the holotype, collected at the same time by the same person from the same population. Isotypes are valuable in that they are reliable duplicates of the same taxon and may be distributed to numerous other herbaria to make it easier for taxonomists of various regions to obtain a specimen of the new taxon. A **lectotype** is a specimen that is selected from the original material to serve as the type when no holotype was designated at the time of publication, if the holotype is missing, or if the original type consisted of more than one specimen or taxon. Lectotypes must be selected from among isotypes, syntypes, or isosyntypes (below), if these are available. A **neotype** is a specimen derived from a nonoriginal collection that is selected to serve as the type as long as all of the material on which the name was originally based is missing. Other types of types include (1) **syntype**, which is any specimen that was cited in the original work when a holotype was not designated; alternatively, a syntype can be one of two or more specimens that were all designated as types; a duplicate of a syntype is an "isosyntype"; (2) **paratype**, a specimen cited in the valid publication (see later discussion), but that is not a holotype, isotype, or syntype; and (3) **epitype**, a specimen (or illustration) that is selected to serve as an "interpretive" type if the holotype, lectotype, or neotype is ambiguous with respect to the identification and diagnosis of the taxon.

Normally, we think of types as referring to genera, species, or infraspecific taxa. However, type specimens may serve as references for higher taxonomic ranks as well. For example, the type for the family Poaceae Barnh. is the genus *Poa* L., which has as its type *Poa annua* L.

VALID PUBLICATION

According to the ICN, in order for a scientific name to be formally recognized, it must be **validly published**. There are four general criteria for valid publication of a name. First, the name must be effectively published, which means that it must be published "by distribution of printed matter (through sale, exchange, or gift) to the general public or at least to scientific institutions with generally accessible libraries." As of January 1, 2012, the ICN allows effective publication to include "electronic material in Portable Document Format (PDF ... [with the possibility of this technology being superseded in the future] in an online publication with an International Standard Serial Number (ISSN) or an International Standard Book Number (ISBN)." This newer policy, although representative of the electronic age, has caused some problems in that peer review is not a necessity (see Specht et al. 2018). Second, the name must be published in the correct form, i.e., properly Latinized (see later discussion), with the rank indicated (e.g., as "sp. nov." or "gen. nov."; see **Abbreviations).** Such a name in correct form is known as an *admissible name*. Third, the name must be published with a Latin or English description or diagnosis or with a reference to such. This diagnosis may be brief, e.g., listing how the new taxon is different from a similar, related taxon. (In addition, a more detailed description in some vernacular language, or, with a new combination, a reference to a previous description is usually included but not required.) Fourth, a nomenclatural type must be indicated. For species and below the type is refers to a specimen, the location of which is also indicated (using the acronyms of *Index Herbariorum*; Thiers, continuously updated; Chapter 18). For genera and above, the type is a reference to a species; e.g., the type for the genus *Stebbinsoseris* is the species *S. heterocarpa* (Nutt.) Chamb. An example of a valid publication, illustrating these criteria, is seen in Figure 16.2. The term **protologue** is "everything associated with a name at its valid publication, e.g. description, diagnosis, illustrations, references, synonymy, geographical data, citation of specimens, discussion, and comments" (Turland et al. 2018).

A full citation of a scientific name may include the authorship and the journal, volume, page numbers, and date of publication. For example, a complete citation for the species cited in Figure 16.2 is "*Perityle vigilans* Spellenb. & A. Powell, Syst. Bot. 15: 252. 1990." Full citations are listed in the International Plant Names Index (see References for Further Study, Web Pages).

PRIORITY OF PUBLICATION

The third principle of the ICN is priority of publication, which generally states that of two or more competing possibilities for a correct name, the one published *first* is the correct one, with some exceptions. Priority of publication only applies to taxa at the rank of family and below, and priority does not apply outside a particular rank (e.g., with the change of a name in rank; see later discussion). For example, of two competing names (both legitimate and validly published)—*Mimulus* (published in 1753) and *Diplacus* (published in 1838)—*Mimulus* has priority and is the correct name if and when the two genera are combined into one. The principle of priority for vascular plants starts 1 May 1753 with the publication of *Species Plantarum* by Linnaeus; names published prior to that are not considered for priority. (Different groups covered by the ICN have various starting dates.)

Effective publication, in a journal commonly available to botanists

Systematic Botany (1990), 15(2): pp. 252-255

A New Species of *Perityle* (Asteraceae) from Southwestern Chihuahua, Mexico

RICHARD SPELLENBERG

Department of Biology, New Mexico State University,
Las Cruces, New Mexico 88003

A. MICHAEL POWELL

Department of Biology, Sul Ross State University,
Alpine Texas 79832

ABSTRACT. **Perityle vigilans** is described from the Sierra Madre Occidental of southwestern Chihuahua. It differs from other *Perityle* by the combination of its white ligules, sparsely setose-hispid achenes, finely grayish villosulous tomentum, and absence of pappus bristles. The new species is believed to be most closely related to *P. rosei* and *P. trichodonta*, which occur about 800 km to the south. A key is presented that distinguishes the 12 known taxa of *Perityle* occurring in the northern Sierra Madre Occidental, an area of high diversity in the genus.

In a series of three papers Powell (1969, 1973, 1974) revised the genus *Perityle*, recognizing 53 species. Except for one amphitropical disjunct, *P. emoryi* Torrey, the genus is restricted to southwestern North America. The treatment by Niles (1970) is in concurrence with regard to the cir-cumscription of species. In his series of papers Powell proposed that speciation in *Perityle* occurs primarily by geographic isolation, that the populations of species, many of which inhabit nearly barren rock cliffs, were derived from more widespread ancestral species whose ranges were divided by geologic uplift, igneous intrusion, and subsequent erosion. Powell explained that this isolation on island-like habitats of exposed rock has resulted in a high degree of endemism, and he noted (1974) that the diversity in the genus is highest in the Sierra Madre Occidental of northwestern Mexico. He postulated that this may be the center of origin for the genus. Since that revision, seven narrowly endemic species have been added to the genus as remote areas have been explored: *P. ajoensis* Todsen (1974), *P. batopilensis* Powell (1983), *P. carmenensis* Powell (1976), *P. fosteri* Powell (1983), *P. huecoensis* Powell (1983), *P. specuicola* Welsh and Neese (1983), and *P. vandevenderi* B. Turner (1989). A few varieties also have been described. This article adds an eighth spe-cies, this from the isolated mid-elevations of the west slope of the Sierra Madre Occidental.

Name in Latin — **Perityle vigilans** Spellenb. & A. Powell, sp. nov. (fig. 1).--TYPE: Mexico, Chihuahua, Municipio Maguarichi, on igneous rocks at Maguarichi and to 3 mi. NE along road, just below oak zone and in lower edge of zone, elev. 1700 m, 27°52'30"N, 107°59'30"W, 25 Apr 1985, Spellenberg, Soreng, Corral, Todsen 8104 (holotype: NMC; isotypes: ENCB, NY, MEXU, SRSC, TEX, UC, Escuela Superior Agricultura "Hermanos Escobar" [Cd. Juarez].

Rank indicated

Type specimen indicated

Latin (or English) description or diagnosis

Plantae perennes suffrutescentes subpulvinatae, caulibus 2.5-8.5 cm longis. Indumentum densum, griseum, minute villosulum. Folia opposita, periolata, laminae ovataeve rhombeae-ovatae, 2.5-5.5 mm longae, 1.7-4 mm latae, dentibus 0-3 brevibus obtusis in uterque marginibus. Capitula terminales solitaria radiata; corollae radii 6-8, ligulis 2.5-3.0 mm longis, 1.5-2.5 mm latis, in tubis extus et laminis subtus glandulis aureis sparsis obsitis; corollae disci ca. 35-40, ca. 2.0 mm longae. Achaenia anguste obdeltata nigra 2.2-2.5 mm longa modice pilosa-hispida. Pappus obsoletus constans ex corona dentibus tenuibus triangularibus minoribus quam 0.1 mm longis. Fig. 1.

Vernacular description

Plants suffruticose, more or less cushion-like, the stem tips ascending. Stems 2.5-8.5 cm long, densely and finely grayish-villosulous, the fine kinked hairs over-topping yellowish spherical glands. Leaves opposite, pubescent as the stem, the upper leaves slightly more densely so than the lower; petioles slender, expanding into the cuneate leaf base, 2-3 mm long; blades ovate to rhombic-ovate, broadly cuneate at the base, 2.5-5.5 mm long, 1.7-4.0 mm wide, with 0-3 low blunt teeth on each margin, when only 1 tooth the blade then subhastate. Capitulescence of a

Figure 16.2 Example of a new species description, illustrating the components of validly publication. Note the abbreviation "sp. nov." after the scientific name, meaning *species nova*, Latin for "new species." Rewritten from *Systematic Botany*, by permission.

CONSERVATION OF NAMES

One adverse effect of the principle of priority is that scientific names that are well known and frequently used may be replaced by some other name if the latter was discovered to have been published earlier. This lends a degree of instability to nomenclature. However, in such a case, a petition may be presented and voted upon at the International Botanical Congress to conserve one name over another that actually has priority. Such a procedure is outlined as three Amendments to the ICN: Nomina familiarum conservanda, Nomina generica conservanda et rejicienda, and Nomina specifica conservanda et rejicienda. The rationale for the conservation of names is to provide greater stability in nomenclature by permitting names that are well known and widely used to persist, even upon the discovery of an earlier, but more obscure, name.

NAME CHANGES

Occasionally, the name of a taxon will change. Name changes can occur for only two reasons: (1) because of the recognition that one name is illegitimate (contrary to the rules), and, thus, another name must take its place; or (2) because additional taxonomic study or research (for example, a cladistic analysis) has resulted in a change of the definition and delimitation of a taxon; this process is known as a **taxonomic revision**.

There are four basic types of nomenclatural activities that can result in a name change. First, a single taxon may be **divided** into two or more taxa, often called "segregate" taxa because they are segregated from one another relative to the original classification. This is done generally via the recognition of features that clearly distinguish two or more groups from one another. For example, the genus *Langloisia* has been split into two genera, *Langloisia* and *Loeseliastrum*, based on a number of morphological, anatomical, and palynological (pollen) features that distinguish them. Ideally, the segregate groups should be monophyletic, as based upon a rigorous cladistic analysis (see Chapter 2). Other examples of taxa being divided are:

1. The genus *Carduus* of the family Asteraceae is often split into two genera: *Carduus*, having barbellate pappus bristles, and *Cirsium*, having plumose pappus bristles.
2. The genus *Rhus* of the Anacardiaceae has been split into several segregate genera, such as *Malosma*, *Rhus*, and *Toxicodendron*, the last including poison-oak and poison-ivy.
3. The classical family Liliaceae has been split into numerous families, such as the Liliaceae s.s., Melanthiaceae, Philesiaceae, and Smilacaceae.
4. The large genus *Haplopappus* of the Asteraceae has been split into several genera, including *Anisocoma*, *Ericameria*, *Hazardia*, *Haplopappus*, and *Isocoma*.

Note that when a larger taxon is divided into two or more smaller taxa of the same rank, the terms **sensu lato** (abbreviated **s.l.**) and **sensu stricto** (abbreviated **s.str.** or **s.s.**) may be used to distinguish the more inclusive and less inclusive treatments, respectively. For example, *Haplopappus* s.l. contains many more species than *Haplopappus* s.s., the latter of which is what remains after *Haplopappus* s.l. is split into many segregate genera.

A second, major name change occurs when two or more separate taxa are **united** into one. One reason for uniting taxa is the recognition that features previously used to distinguish them are, upon more detailed study, unsupportive of their being different; i.e., there is no clear character state discontinuity. Another reason to unite taxa may be based on cladistic studies, in which of two or more separate taxa, one (or more) is demonstrated to be paraphyletic; thus, one way to eliminate a paraphyletic taxon is to unite it with other taxa such that the new inclusive group is now monophyletic (see Chapter 1). In cases of taxa being united, the final name used is that which was *published earliest*, according to the principle of priority. Examples of taxa being united are:

1. The species *Bebbia juncea* and *Bebbia aspera*, which were considered indistinct as species and were united into one, *B. juncea*.
2. The genera *Salvia* and *Rosmarinus*, which by some have been united into one genus, *Salvia*.
3. The families Apocynaceae and Asclepiadacccac, which have been united into one family, the Apocynaceac (which could be designated Apocynaceae s.l. to distinguish it from the earlier, less inclusive family).

Third, a taxon may be **transferred in position**, i.e., from one taxon to another of the *same rank*. Examples of this are:

1. The species *Rhus laurina* was transferred in position as a member of the genus *Malosma*, the new species name being *Malosma laurina*.
2. The species *Sedum variegata* was transferred to the genus *Dudleya*, the new species name being *Dudleya variegata*.

Note that a transfer in position may be an automatic result of uniting or dividing taxa of higher rank. For example, if the *genera Diplacus* and *Mimulus* are united into the genus *Mimulus*, then all of the *species* of *Diplacus* must be transferred in position.

Fourth, a taxon may be **changed in rank**. Examples include:

1. The species *Eruca sativa* was changed to the rank of subspecies (of the species *E. vesicaria*), the new combination being *Eruca vesicaria* subsp. *sativa*.

2. The variety *Cryptantha pterocarya* var. *cycloptera* was changed to the rank of species, the new name being *Cryptantha cycloptera*.

Note in the two rank change examples just given that the original names for the epithets are retained. A name that is changed in rank may be retained, but only if an earlier name for the same taxon had not already been published at that rank (and also, only if the *same* name had not already been used for another taxon; see **homonym**). The principle of priority does not apply outside the rank of a taxon, however; this means that if a name is changed in rank, the date of publication of the original name (before being changed in rank) cannot be considered in evaluating priority of publication with respect to the change.

In some cases a taxonomic study results in **remodeling** of a taxon, i.e., a change in diagnostic characteristics, those that distinguish the taxon from others. In this case, a name change is not warranted and the rules of the ICN need not apply.

A **basionym** is the "name-bringing or epithet-bringing synonym," i.e., the original (but now not accepted) name, part of which has been used in a new combination. As seen earlier, if a species or infraspecific name is transferred in position or rank, the specific or infraspecific epithet of the unaccepted basionym may be retained (unless violating another rule of the code, such as priority of publication, e.g., if the taxon had already been named, or if the name had already been used for another taxon at that rank). The name of the author(s) who originally named the basionym is also retained and placed in parentheses ahead of the author who made the change. Thus, botanical names may have two sets of authors: the author(s) — set in parentheses — who originally named the basionym, and the author(s) who made the name change. (This is true only for names at and below the rank of genus; for suprageneric names, these "parenthetical" authors are not cited.) From some earlier cited examples:

1. When *Sedum variegata* Wats. was transferred to the genus *Dudleya* by Moran, the new species name became *Dudleya variegata* (Wats.) Moran. The original epithet, *variegata*, is retained, and the author associated with that epithet, Watson in this case, is also retained, but is placed in parentheses preceding the new author. The basionym in this case is *Sedum variegata* Wats., the original name.
2. When *Dilatris caroliniana* Lam. was transferred to the genus *Lachnanthes* by Dandy, the new species name became *Lachnanthes caroliniana* (Lam.) Dandy. The basionym in this case is *Dilatris caroliniana* Lam.
3. When *Fumaria bulbosa* L. var. *solida* L. was elevated to the rank of species by Miller, the new name became *Fumaria solida* (L.) Miller. The basionym in this case is *Fumaria bulbosa* L. var. *solida* L. Subsequent to this change, *Fumaria solida* (L.) Miller was transferred in position by Clairv to the genus *Corydalis*, the new name becoming *Corydalis solida* (L.) Clairv [*not Corydalis solida* (Mill.) Clairv]. Note that it is the author of the varietal epithet of the basionym, *Fumaria bulbosa* L. var. *solida* L., which is retained in parentheses.

An **autonym** is an automatically created name for infrafamilial, infrageneric, and infraspecific taxa. Autonyms are automatically created whenever a family is divided into subfamilies, tribes, or subtribes; a genus is divided into subgenera or sections; or a species is divided into subspecies or varieties. Of the two or more subtaxa formed, the autonym is used for the one that corresponds to the original type specimen. Autonyms have no authors; only the higher taxa upon which they are based and the other subtaxa have formal authorship. For example, I.M. Johnston split *Cryptantha crassisepala* (Torr. & A.Gray) Greene into two varieties: *C. crassisepala* var. *elachantha* I.M.Johnst. and *C. crassisepala* var. *crassisepala*; note that the latter variety, containing the autonym, lacks authorship because its type is the same as that for the originally described species.

For infrafamilial taxa, the autonym has the same root name as the family but a different ending that corresponds to the infrafamilial rank. For example, the family Euphorbiaceae is usually divided into subfamilies, one of which, the Euphorbioideae, is the autonym; this subfamily, of course, contains the genus *Euphorbia*, the type for the family. For infrageneric taxa, the autonym is identical to the genus name and should be preceded by the name of the rank to avoid confusion. For example, *Ceanothus* (a genus) consists of two subgenera, subgenus *Ceanothus* and subgenus *Cerastes*; subgenus *Ceanothus* is the one that includes the type for the genus itself. For infraspecific taxa, autonyms are identical to the specific epithet. For example, *Eriogonum fasciculatum* is divided into several varieties, one of which, *Eriogonum fasciculatum* var. *fasciculatum*, is the autonym, based on the original type specimen for the species.

SYNONYMS

A **synonym** is an unaccepted name, *by a particular author or authors*, applying to the same taxon as the accepted name. Synonyms are rejected for either of two reasons: (1) because they are illegitimate, i.e., contrary to the rules of the ICN; or (2) because of taxonomic judgment, i.e., a particular author rejects the classification represented by the synonym.

A **homotypic** (or **nomenclatural**) **synonym** is an unaccepted name that is based on the same type as that of the accepted name. For example, *Krynitzkia decipiens* M.E.Jones is a homotypic synonym of *Cryptantha decipiens* (M.E.Jones) A.Heller because both names are based on the same type specimen. A **heterotypic** (or **taxonomic**) **synonym** is an unaccepted name based on a type different from that of the correct name. For example, *Amsinckia lemmonii* J.F.Macbr. (1916) is a heterotypic synonym of *Amsinckia douglasiana* A.DC. (1846) because they have different holotypes, the former determined to be the same species as the latter.

Synonyms, of whatever type, are typically indicated in brackets following the correct name, such as *Malosma laurina* (Nutt.) Abrams [*Rhus laurina* Nutt.].

A **correct name** is a validly published, legitimate name that is adopted *by a particular author or authors*. Recall that the fundamental principle of the ICN is that each taxon can have only *one* correct name. Thus, if there are two or more competing names for the same taxon, e.g., *Malosma laurina* (Nutt.) Abrams and *Rhus laurina* Nutt., only one of them can be correct. However, *which* name is correct may depend on the author(s) of a given reference book or journal. For example, according to one author, *Rhus laurina* Nutt. is the correct name and *Malosma laurina* (Nutt.) Abrams is the synonym. According to other authors, *Malosma laurina* (Nutt.) Abrams is the correct name and *Rhus laurina* Nutt. is the synonym.

A **homonym** is one of two (or more) *identical* names (not including authorship) that are based on different type specimens. The **later homonym**, based on publication date, is illegitimate (unless it is conserved; see earlier discussion). For example, *Cryptantha foliosa* Reiche and *Cryptantha foliosa* Greene are homonyms. *Cryptantha foliosa* Reiche is a later homonym and illegitimate because its publication date (Anales Univ. Chile 121: 827. 1907.) was after that of *Cryptantha foliosa* Greene (Pittonia 1: 113. 1887.). The two names in this case refer to different taxa.

A **tautonym** is a binomial in which the genus name and specific epithet are identical in spelling. Tautonyms are not permitted in botanical nomenclature. For example, the name *Helianthus helianthus* would be a tautonym and could not be validly published, whereas *Helianthus helianthoides* is not and would be permitted. (Note that zoological nomenclature does permit tautonyms, as in *Gorilla gorilla*.)

LATIN TERMS AND ABBREVIATIONS

Certain Latin terms and abbreviations are often used in scientific names. For example, the word **ex** essentially means "validly published by." For example, *Microseris elegans* Greene ex A.Gray means that Asa Gray validly published the name *Microseris elegans* that was originally proposed (but not validly published) by Greene. The "ex" plus the author(s) *preceding* it may be omitted, as in *Microseris elegans* A.Gray.

The word **in** means "in the publication of," referring to a name published within a larger work authored by the person(s) following the "in." For example, *Arabis sparsiflora* Nutt. in T. & G. means that Nuttall validly published the name *Arabis sparsiflora* in another work authored by Torrey & Gray. The "in" plus the author(s) *following* it may be omitted for brevity, as in *Arabis sparsiflora* Nutt. (The use of "in" is not recommended by the ICN.)

An × indicates a hybrid. For example, *Delphinium* × *inflexum* Davidson is a named (validly published) taxon representing a hybrid between *D. cardinale* Hook. and *D. parryi* A.Gray. Alternatively, a hybrid can be represented as *D. cardinale* Hook. × *D. parryi* A.Gray. Hybrids may also be indicated by placing the prefix *notho-* prior to the rank name, as in *Polypodium vulgare* nothosubsp. *mantoniae* (Rothm.) Schidlay, indicating that the named subspecies is of hybrid origin.

The abbreviation **sp. nov.** following a binomial (e.g., "*Eryngium pendletonense*, sp. nov.") refers to the Latin *species nova* and means that the species is new to science. Similarly, **gen. nov.** (*genus novum*) cites a new genus name. The abbreviation **comb. nov.** following a name refers to the Latin *combinatio nova* and means that the taxon has recently been transferred to a new position or rank. An example of a change in position is:

> ***Lithospermum album*** (G. L. Nesom) J. Cohen, **comb. nov.** Basionym: *Macromeria alba* G. L. Nesom, Madroño 36: 28. 1989. Type: Mexico. Tamaulipas, Mpio. Gomez Farias, 5–7 km, NW of Gomez Farias, just S of Agua del Indio, 30 May 1969, A. Richardson 1763 (holotype: TEX)."

Note that to be validly published, a comb. nov. only requires reference to the basionym and to the protologue.

Two abbreviations — "aff." and "cf." — are used to describe plant specimens whose identity is uncertain. The distinction between the two abbreviations is unclear, as different taxonomists have used them with slightly different meanings. The abbreviation **aff.** preceding a taxon name literally means "related to" (Latin *affinis*, "related, connected"), as in "*Calyptridium* aff. *monandrum*" or "aff. *Calyptridium monandrum*." This abbreviation implies some type of close relationship, presumably an evolutionary relationship, but also that the specimen differs from the cited taxon in some way, e.g., beyond the described range of variation for one or more characters; the cited specimen might, in fact, be indicative of a new taxon. The abbreviation "**cf.**" (Latin *confer*, "compare") preceding a taxon name, as in "*Calyptridium* cf. *monandrum*" or "cf. *Calyptridium monandrum*," indicates that the identity of a specimen is more questionable or uncertain (perhaps because references or comparative specimens are not available), and should be compared with specimens of the taxon indicated (i.e., the name following "cf.") for more detailed study.

Circumscription refers to the boundaries of a taxon, i.e., what is included within it and how it is delimited from other taxa. As mentioned earlier, **s.l.** (*sensu lato*) means "in the broad sense," referring to a broad, inclusive taxon circumscription, and **s.str.** or **s.s.** (*sensu stricto*) means "in the strict sense," referring to a narrow, exclusive circumscription.

A list of some standard terms, abbreviations, and symbols, including the preceding, are as follows:

1. **aff.** (*affine*) means "related to"
2. **auct. non** (*auctorum non*) refers to a misapplication of a name (a "misapplied" name), such that the type specimen of the name does not fall within the circumscription of the taxon being referred to by that name
3. **cf.** (*confer*) means "compare to"
4. **comb. nov.** (*combinatio nova*) means a new nomenclatural combination
5. **emend.** (*emendatio*) means a correction or amendment
6. **et** is Latin for "and"
7. **ex** is Latin for "from," meaning validly published by
8. **f.** is Latin *filius*, meaning "son of," e.g., L.f.
9. **gen. nov.** (*genus novum*) means a new genus
10. **in** is Latin for "in," meaning in the publication of
11. **ined.** (*ineditus*) means not validly published
12. **nom. cons.** (*nomen conservandum*) means a conserved name
13. **nom. nov.** (*nomen novum*) means a new name, e.g., proposed as a substitute for an older name (e.g., an illegitimate homonym, in which case the older name serves as the type for the new one)
14. **nom. nud.** (*nomen nudum*) means published without a description or diagnosis, making the name invalid
15. **non** is Latin for "not"
16. **n. v.** (*non visus*) is Latin for "not seen," typically meaning that authors did not see a specimen, such as a type.
17. **orth. cons.** (*orthographia conservanda*) means a conserved spelling
18. **s. l.** (*sensu lato*) means "in the broad sense"
19. **s. s.** or **s. str.** (*sensu stricto*) means "in the strict sense"
20. **sp. nov.** (*species novus*) means a new species
21. **stat. nov.** (*status novus*) means a change in rank, e.g., elevating a varietal name to specific status
22. **typ. cons.** (*typus conservandus*) means a conserved type specimen
23. **typ. des.** (*typus designatus*) means the designation of a type specimen
24. **vide** (*video*) means to cite a reference
25. × indicates a hybrid
26. **!** (symbol for *vidi*, "I have seen it") can mean (a) confirmation of a name, as on an annotation label agreeing with the name on the original herbarium label; or (b) indication that a given specimen has been seen by the author(s).
27. = a heterotypic synonym, based on a different type
28. ≡ a homotypic synonym, based on the same type

INDEPENDENCE OF BOTANICAL NOMENCLATURE

The first principle of the International Code of Nomenclature for algae, fungi, and plants is that it is independent of zoological and prokaryotic nomenclature, these governed by the International Code of Zoological Nomenclature and the International Code of Nomenclature of Prokaryotes, respectively. Because of these separate codes, there are some names of plants that are identical to those of animals. For example, the genus *Morus* refers both to a flowering plant, the mulberry, and to a bird, the gannett.

RETROACTIVITY OF THE ICN

The sixth principle of the *ICN* is that the rules of nomenclature are retroactive unless expressly limited (e.g., taking place after a designated starting date).

BOTANICAL NAMES

"Those who wish to remain ignorant of the Latin language, have no business with the study of botany" (J. Berkenhout, 1789, cited in Stern, 1992).

"Botanical Latin is best described as a modern Romance language of special technical application" (Stern 1992). The fifth principle of the ICN is that botanical names are treated as Latin, a language chosen because of its classical history (in the past being the language of scholars) and perhaps because it is no longer in active use (minimizing international rivalries). No matter what the language of the person who published a name, the name itself must consist of direct Latin words or be "Latinized," i.e., converted from the vernacular to the Latin. Thus, the Latin alphabet (which is almost identical to the English alphabet) and grammatical rules must be used.

GENDER

All Latin words have a gender: masculine, feminine, or neuter. Gender determination is usually only of concern for names at the rank of genus or below. The standardized gender endings are:

Masculine	Feminine	Neuter
-us	-a	-um
-er	-ra	-rum
-is	-is	-e
-r	-ris	-re

The first row of endings (*-us*, *-a*, and *-um*) are most commonly used. For example, the gender of the genus *Amaranthus* is masculine, *Crassula* is feminine, and *Polygonum* is neuter. Specific or infraspecific epithets are usually adjectives, the endings of which must agree in gender with that of the genus name, as in *Amaranthus albus*, *Crassula connata*, and *Eriogonum fasciculatum* subsp. *polifolium*. However, in rare cases the specific or infraspecific epithet is a noun (in apposition), in which case it retains its original gender. For example, in *Cypripedium calceolus*, *calceolus* is a noun and retains the masculine gender despite the neuter gender of the genus name.

One exception to the standardized gender endings is that many tree genera are typically treated as *feminine*, regardless of the ending. For example, genera *Quercus*, *Pinus*, and *Liquidambar* are feminine in gender, even though they have masculine endings. Thus, specific or infraspecific adjectival epithets of these genera names must be feminine (to agree in gender), as in *Quercus alba*, *Pinus ponderosa*, and *Liquidambar styraciflua*.

Note that a **name change** (divided, united, transferred in position, or changed in rank) can necessitate a *change in the gender ending of a specific epithet*. For example, for species *Haplopappus squarrosus*, the ending (*-us*) is masculine. When this species is transferred to the genus *Hazardia*, the new name becomes *Hazardia squarrosa*. Although the root of the specific epithet does not change, its ending may, in order to agree in gender with the new genus name.

NUMBER

Names of genera, infrageneric names (such as subgenera or sections), and species or infraspecific combinations are all treated as the singular case in Latin. However, all taxon names above the rank of genus are treated as Latin plural nouns. Thus, it is correct to say, e.g., "The Orchidaceae *are* a large family of monocotyledons" and "The Rosales *consist* of many species."

COMMEMORATIVES

Commemorative names are those named after a person or place. Generic commemoratives are generally treated as nouns, e.g., *Linnaea*. Specific or infraspecific commemorative names are usually treated as the genitive case (denoting possession) and must have genitive endings. For male commemoratives, the ending is (1) *-ii*, if the name ends in a consonant (as in *Isoetes orcuttii*); or (2) *-i*, if the terminal consonant is *-r* (as in *Erigeron breweri*) or *-y* (as in *Yucca baileyi*) or if the name ends in a vowel other than *a* (as in *Arctostaphylos pringlei*). The ending *-orum* or *-iorum* can be used for plural commemoratives, e.g., *Ambrosia johnstoniorum* (named for I.M. and M.C. Johnston). For commemorative names ending in *-a* (regardless of sex) an *-e* is added, as in *Baccharis vanessae* (female) or *Aphelandra trianae* (male). For female commemoratives, regardless of the last letter of the name, the ending is *-ae* or *-iae*, e.g., *Hippeastrum wilsoniae* or *Carex barbarae*.

In some cases a commemorative name is treated as an adjective, in which case the endings *-ianus*, *-iana*, or *-ianum* may be used. These endings agree in gender with the genus name, as in *Astragalus nuttallianus* (after Thomas Nuttall), *Prunus caroliniana* (after the Carolinas), or *Antirrhinum coulterianum* (after John M. Coulter). Another suffix ending used for place commemoratives is *-ensis*, as in *virginiensis* ("of Virginia") or *capensis* ("of the Cape").

Finally, if the personal name is already converted to Latin, e.g., Linnaeus or Xantus, then the commemoratives are given appropriate Latin genitive endings, e.g., *linnaei* (not *linnaeusii*) or *xanti* (not *xantii*).

PRONUNCIATION OF NAMES

Although scientific names are universally in the Latin language, their pronunciation often varies from region to region, especially between different countries. For example, European or Latin American pronunciations are often different from those of most United States botanists. There are no international agreements as to how scientific names should be pronounced, even no agreement as to whether classical or ecclesiastical Latin should be used. Very often, pronunciations are influenced by one's native language. One should be flexible and adaptive with regard to pronunciations, as the overriding goal is communication.

The rules cited in Figures 16.3 and 16.4 are recommended here. These generally use traditional English ("Anglicized Latin") for pronunciation of diphthongs, long and short vowels, and consonants and "reformed" academic pronunciation (based on classical Latin) for converting to syllables and for accenting. (See Stern, 1992; however, see also Weber, 1986.)

DIPHTHONGS

Diphthongs are two vowels that are combined together and treated as the equivalent of a single vowel. The Latin diphthongs and their "traditional" English pronunciations are:

Diphthong	English Pronunciation	Example
ae	long "e"	*Tropaeolum*
oe	long "e"	*Kallstroemia*
au	"aw"	*Daucus*
ei	long "i"	*Eichhornia*
eu	long "u"	*Teucrium*
ui	as in "quick"	*Equisetum*

Note that "ie" is not a Latin diphthong, but two separate vowels, each of which would be pronounced separately, as in the genus *Parietaria* (Pa-ri-e-ta-ri-a). Also note that "oi" is not a Latin diphthong. Technically, each vowel should be pronounced separately, as in *Langloisia* (Lan-glo-i-si-a).

Rule	Example	Syllables
A single consonant between two vowels or diphthongs goes with the second one	*Tridens*	Tri-dens
Exception:		
an "x" between two vowels or diphthongs goes with the preceding one	*exaltataus*	ex-al-ta-tus
Two adjacent consonants between vowels or diphthongs are split evenly	*guttatus*	gut-ta-tus
Exceptions: the combinations bl, cl, dl, gl, kl, pl, tl;		
br, cr, dr, gr, kr, pr, tr; and ch, ph, th	*scabra*	sca-bra
go together with the following vowel	*leptocladus*	lep-to-cla-dus
	Ephedra	E-phe-dra
	agrifolia	a-gri-fo-li-a
	brachypoda	bra-chy-po-da
	eremophila	e-re-mo-phi-la
	Notholaena	No-tho-lae-na
Of three or more consonants between two vowels or diphthongs, all but the first goes with the second vowel or diphthong	*absconditus*	ab-scon-di-tus

Figure 16.3 Rules for converting Latinized scientific names into syllables.

However, by convention "oi" is often pronounced like the English language diphthong, as in "oil." Thus, the genus *Langloisia* is often heard as Lan-gloi-si-a.

Occasionally, adjacent vowels will resemble a diphthong, but are actually separate vowels. In "ligatured" typesetting, the two letters of a diphthong are connected together, such as "æ," to distinguish the diphthong from two adjacent vowels. However, in cases where the diphthong is not specially indicated (most print these days), a diaeresis (¨) is permitted to indicate that the vowel combination is not a diphthong. For example, in the genera *Aloë* and *Kalanchoë*, there is no diphthong; the diaeresis shows that the "o" and "e" are separate vowels and are pronounced separately. (Sometimes these are ignored in practice; for example *Aloë* is usually pronounced as if the ë were absent, as in Ah´-loh, not Ah´-lo-e).

SYLLABLES

Latin words have as many syllables as there are vowels and diphthongs. *Every* syllable of a Latin word is pronounced. Thus, it is often valuable to convert scientific names to syllables in order to pronounce them properly and better memorize them. Some of the rules for this are enumerated in Figure 16.3. Special rules for the pronunciation of consonants and vowels are cited in Figure 16.4.

ACCENTING

Determining the accent of a scientific name may be difficult without actually looking up the word in a Latin dictionary or other reference. However, if these are not available, the following general rules may be used to determine which syllable is accented and whether the vowel of that syllable is long or short. (One format for denoting accent, used here, is ` for a "grave" accent denoting a long vowel, ´ for an "acute" accent denoting a short vowel.)

Determining whether a vowel is long or short generally requires consulting a Latin dictionary. The last syllable of a word, the "ultimate," is *never accented*, unless of course the word has only one syllable; e.g., "*max*" of *Glycine máx*. If a word has two syllables, the accent always goes with the next to the last syllable, termed the "penult"; e.g., *Àcer*. If a word has three or more syllables, the accent always goes either with the next to the last (penult) or the third from the last, termed the "antepenult." The next to the last (penult) is accented if (1) it ends in a consonant, in which case the vowel is short, as in *perennis* (pe-rén-nis); (2) it ends in a diphthong, which is treated as long, as in *amoenus* (a-moè-nus); or (3) it ends in a long vowel, e.g., *alsine* (al-sì-ne). If none of these conditions is met, then the accent goes with the third from the last syllable (antepenult), which can be short or long, e.g., *dracontium* (dra-cón-ti-um), or *densifolius* (den-si-fò-li-us).

COMMEMORATIVES

Although commemoratives are Latinized and divided into syllables and accented according to the rules of Latin, they may by convention be pronounced as the person or place would be pronounced in the native language. For example, the specific epithet of *Hesperoyucca whipplei* may be pronounced "wíp-pull-i" (as the person's name is pronounced plus the letter "i") as opposed to the Latinized pronunciation "wíp-plee-i." The general pronunciation rule is to simply pronounce the commemorative as it would be pronounced in the language of that person, and then add the ending. However, in practice the commemorative pronunciation is usually converted to the language of the speaker, as pronunciation in the original language of that person may be unknown or unwieldy. (Remember, the overriding goal is communication!)

Pronunciation Rule	Example	Syllabizing and Accenting	Sounds like
"C" or "g" is hard (pronounced like "k" or a hard "g," respectively)	*Cakile*	Ca-kì-le	Kah-kì-lee
	Garcinia	Gar-cí-ni-a	Gar-cíh-nee-ah
Exceptions: "c" or "g" is soft (pronounced like "s" or "j," respectively) when followed by the letters/diphthongs e, i, y, ae, or oe	*Cedrus*	Cè-drus	Seè-druhs
	cinerea	ci-nè-re-a	sigh-neè-ree-ah
	coccinea	coc-cí-ne-a	Kahk-síh-nee-ah
	cyaneus	cy-à-ne-us	sigh-à-nee-us
	caerulea	cae-rù-le-a	see-rù-lee-ah
	Geranium	Ge-rà-ni-um	Jeh-rà-nee-uhm
	Gibasis	Gi-bà-sis	Jih-bà-sis
	Gypsophila	Gyp-só-phi-la	Jip-só-fi-lah
When a word or root begins with cn, ct, gn, mn, pn, ps, pt, or tm, the first letter is silent; only the second letter is pronounced.	*Cneoridium*	Cne-o-rí-di-um	Nee-oh-rí-di-um
	Ctenium	Ctè-ni-um	Teè-nih-um
	Gnetales	Gne-tà-les	Nee-tày-lees
	Mniodes	Mni-ò-des	Ni-ò-des
	Pneumatopteris	Pneu-ma-to-pté-ris	Noo-ma-to-té-ris
	Psilotum	Psi-lò-tum	Sigh-lò-tum
	Pteridium	Pte-rí-di-um	Teh-rí-di-um
	Tmesipteris	Tme-sí-pte-ris	Meh-sí-te-ris
"Ch" is hard, pronounced like "k"	*Chilopsis*	Chi-lóp-sis	Ki-lóp-sis
"X" at the beginning of word or root is pronounced like a "z"	*Xylococcus*	Xy-lo-cóc-cus	Zy-lo-cóc-cus
	Xanthium	Xán-thi-um	Zán-thi-um
An "x" within a word is pronounced like "ks"	*Zanthoxylum*	Zan-thóx-y-lum	Zan-thóks-i-lum
A final "e" or "es" is long	*Anemone*	A-né-mo-ne	A-né-mo-nee
	Rosales	Ro-sà-les	Ro-sày-lees
A final "a" is short	*Nicotiana*	Ni-co-ti-à-na	Ni-co-ti-à-nah
A "y" is pronounced like a short "i"	*argophyllus*	ar-go-phy´l-lus	ar-go-fi´l-lus
For "uu," both "u"s are pronounced, the first long, the second short	*Carduus*	Cár-du-us	cár-doo-us
An "i" at the end of a syllable is short	*crassifolius*	cras-si-fò-li-us	cras-si-fòh-li-us
An "e" is long if it is derived from the Greek diphthong "ei"	*Achillea*	A-chil-lè-a	a-kil-leè-a

Figure 16.4 Rules for pronunciation of Latinized scientific names, ` representing a grave accent denoting a long vowel, ´ representing an acute accent denoting a short vowel.

REVIEW QUESTIONS

NOMENCLATURE

1. What is nomenclature?
2. What is the name (and abbreviation) of the work that provides the rules and recommendations for plant nomenclature?
3. What groups of organisms are covered by this reference? What organisms are not?
4. What are the two basic activities governed by nomenclature (and the ICN)?
5. What are legitimate and illegitimate names?
6. How are changes to the ICN made?
7. Name the six principles of botanical nomenclature. Which of these is considered the fundamental principle?
8. What is the difference between the rules and the recommendations of the ICN?
9. What is meant by a scientific name? Give three examples.
10. Which scientific names (i.e., at which rank) are always binomials (binary combinations)? Give an example of a binomial.
11. Who first consistently used the binomial and is called the "father of taxonomy"?
12. What is the correct form of a binomial?
13. For *Quercus dumosa* Nuttall, what is (a) *Quercus*; (b) *dumosa*; (c) *Quercus dumosa*; (d) Nuttall?
14. What are common names?
15. Name the reasons that scientific names are advantageous over common names.
16. What is the difference between rank and position?
17. Name the standardized or recommended endings for scientific names at the ranks of phylum, class, subclass, superorder, order, family, subfamily, tribe.
18. What is the rank of the following: (a) Conostyloideae; (b) Flacourtiaceae; (c) Haemodoreae; (d) Hamamelidae; (e) *Linnaea borealis* var. *longiflora*; (f) Liliopsida; (g) Magnoliophyta; (h) Rosales; (i) *Tribonanthes*; (j) *Tribonanthes variegata*; (k) *Phlebocarya ciliata* subsp. *pilosissima*?
19. What is the additional, validly published name for the Apiaceae; Arecaceae; Asteraceae; Brassicaceae; Fabaceae; Faboideae; Clusiaceae; Lamiaceae; Poaceae?
20. What is a ternary name? What are two infraspecific ranks and which is "higher"?
21. What does the author of a scientific name refer to?
22. Name four suggestions for memorizing scientific names.
23. What is meant by a nomenclatural "type"?
24. What is the difference between a holotype, isotype, lectotype, and neotype?
25. What is the nomenclatural type of a family name?
26. What is meant by "priority of publication"?
27. When and with what publication does priority of publication officially begin?
28. What is an adverse consequence of priority of publication?
29. What is conservation of names and how is this accomplished?
30. What are the two basic reasons for changing a scientific name?
31. Give the four major ways that names are changed and give an example of each.
32. What is remodeling? Does it require a name change?
33. What is a basionym?
34. What does it mean if an author's name is in parentheses, e.g., *Machaeranthera juncea* (Greene) Hartman?
35. You decide that the taxon *Xiphidium coeruleum* Aublet should be transferred in position to the genus *Schiekia*. What is the *required* new name (including authorship)? What if the name *Schiekia coeruleum* had already been validly published?
36. You decide that the taxon *Quercus albiniana* (C. Jones) G. Smith subsp. *tomentosa* H. Carlisle should be elevated to the rank of species. What is the new name called (including authorship)?
37. What is an autonym? Give an example of an autonym at the rank of subfamily, subgenus, or subspecies.
38. What are the main criteria of a validly published name?
39. What is a protologue?
40. What is a synonym?
41. What are the two major reasons that a name may be rejected?

42. How can a name be legitimate yet not be correct?
43. What can you infer from: *Malacothrix incana* (Nutt.) T. & G. [*Malacomeris i.* Nutt.]?
44. What can you infer from: *Porophyllum gracile* Benth. [*P. caesium* Greene; *P. vaseyi* Greene]?
45. What can you infer from: *Gilia diegensis* (Munz) A. & V. Grant [*G. inconspicua* (Sm.) Sweet var. *diegensis* Munz]?
46. What is a homonym? Is a later homonym legitimate or illegitimate?
47. What is a tautonym? Are tautonyms acceptable in (a) botanical nomenclature; (b) zoological nomenclature?
48. What is meant by "in" in authorship designations? How may such a designation be simplified?
49. What is meant by "ex" in authorship designations? How may such a designation be simplified?
50. What is the meaning of an × in a scientific name, as in *Quercus* ×*morehus*?
51. What is the meaning of auct. non, emend., ined., nom. nov., nom. nud., s.l., s.s., and vide (!)?
52. How does the fact that a plant and a bird have the same scientific name *not* violate the principles of the ICN?

BOTANICAL NAMES

53. In what language are scientific names treated?
54. Name the three Latin genders and give the standardized genus endings.
55. What is one prominent exception to these gender endings?
56. Names at which taxonomic ranks are always Latin plurals?
57. What is a commemorative name?
58. What endings may commemorative names have?
59. Are there universal rules for the pronunciation of scientific names?
60. What are the Latin diphthongs and how are they pronounced?
61. How is the combination "oi" properly pronounced in Latin?
62. What is the rule determining the number of syllables in a scientific name?
63. Name some of the specific rules for converting scientific names to syllables (refer to Figure 16.3).
64. Name some of the specific rules for pronouncing scientific names (refer to Figure 16.4).
65. Name the basic rules for accenting scientific names.
66. Convert to syllables and pronounce the following names: *Cleistes*, *Eucalyptus*, *microcarpa*, *Oenothera*, *Pyrus*.
67. A commemorative (named after a person or place) may be pronounced in what two basic ways?

EXERCISES

1. Using a manual or flora of local, native plants, record 12 scientific names plus the listed synonymy for these names. Trace the nomenclatural history of these taxa names as best you can from the data given, especially noting author names in parentheses.
2. Look up these 12 scientific names using the International Plant Names Index (*http://www.ipni.org*). Record the date and journal/book of publication of these names. Also record the synonymy indicated. Does this added information elucidate the nomenclatural history of the taxa?
3. Divide into syllables, accent, and pronounce these 12 scientific names, using any available references.

REFERENCES FOR FURTHER STUDY

Bailey, L. H. 1963. How Plants Get Their Names. Dover Publications, Inc., New York.

Borror, Donald J. 1960. Dictionary of Word Roots and Combining Forms. Mayfield, Palo Alto, California.

Brummitt, R. K., and C. E. Powell (eds.). 1992. Authors of Plant Names: A List of Authors of Scientific Names of Plants, with Recommended Standard Forms of Their Names, Including Abbreviations. Royal Botanic Gardens, Kew, London.

Cantino, P. D., J. A. Doyle, S. W. Graham, W. S. Judd, R. G. Olmstead, D. E. Soltis, P. S. Soltis, and M. J. Donoghue. 2007. Towards a phylogenetic nomenclature of Tracheophyta. Taxon 56: 822-846.

Gledhill, D. 1989. The Names of Plants. 2nd ed. Cambridge University Press, Cambridge, UK.

Hamilton, C. W., and S. H. Reichard. 1992. Current practice in the use of subspecies, variety, and forma in the classification of wild plants. Taxon 41: 485-498.

Jaeger, E. C. 1959. A Source Book of Biological Names and Terms. 3rd ed. Charles C. Thomas, Springfield, Illinois.
McDade, L. A. 1995. Species concepts and problems in practice: insight from botanical monographs. Systematic Botany 20:606-622.
McVaugh, R., R. Ross, and F. A. Stafleu. 1964. An annotated glossary of botanical nomenclature. Regnum Vegetabile 56: 1–31.
Radford, A. E., W. C. Dickison, J. R. Massey, and C. R. Bell. 1974. Vascular Plant Systematics. Harper and Row, New York.
Specht, C., M. Fishbein, J. Doyle, L. Struwe, H. Ballard, S. Krosnick, D. Keil, G. Yatskievych, A. Mast, C. Barrett, T. Stoughton, S. Ickert-Bond, I. Jordon-Thaden, J. Smith, C. Martine, and B. Thiers. 2018. Editorial. Systematic Botany 43:1-3.
Stern, William T. 1992. Botanical Latin: History, Grammar, Syntax, Terminology and Vocabulary. David & Charles, Brunel House, Newton Abbot, Devon, UK.
Stern, William T. 2002. Stearn's Dictionary of Plant Names for Gardeners: A Handbook on the Origin and Meaning of the Botanical Names of Some Cultivated Plants. Timber Press, Portland.
Turland, N. J., Wiersema, J. H., Barrie, F. R., Greuter, W., Hawksworth, D. L., Herendeen, P. S., Knapp, S., Kusber, W.-H., Li, D.-Z., Marhold, K., May, T. W., McNeill, J., Monro, A. M., Prado, J., Price, M. J. & Smith, G. F. (eds.). 2018. International Code of Nomenclature for algae, fungi, and plants (Shenzhen Code) adopted by the Nineteenth International Botanical Congress Shenzhen, China, July 2017. Koeltz Botanical Books, Glashütten.
Weber, William. 1986. Pronunciation of scientific names. Madroño 33(3): 234–235.

WEBSITES

Cantino, P. D. and K. de Queiroz. 2007. International Code of Phylogenetic Nomenclature. Version 4b. *http://www.ohio.edu/phylocode*
Greuter, W., D. L. Hawksworth, J. McNeill, A. Minelli, B. J. Tindall, P. Trehane, and P. Tubbs. 1997. Draft BioCode (1997): the prospective international rules for the scientific names of organisms. *http://www.bgbm.org/iapt/biocode/biocode1997.html*
Griffith, C. 1996 onwards. Dictionary of Botanical Epithets. *http://www.winternet.com/~chuckg/dictionary.html*
Index Fungorum. A list of the names of fungi (including yeasts, lichens, chromistan fungi, protozoan fungi and fossil forms) at species level and below. *http://www.indexfungorum.org*
International Code of Nomenclature for algae, fungi, and plants. 2018. *https://www.iapt-taxon.org/nomen/main.php*
IPNI. 2019. International Plant Names Index. The Royal Botanic Gardens, Kew, Harvard University Herbaria & Libraries and Australian National Botanic Gardens. Published on the Internet. *http://www.ipni.org*
Plants of the World Online. 2019. *http://plantsoftheworldonline.org*
Taxacom Discussion List. 1992 onwards. A listserv of discussions and queries of taxonomic issues, for plants and animals. Useful for querying specific information, listing images for identification, and searching for topics. *http://mailman.nhm.ku.edu/mailman/listinfo/taxacom*
The Plant List. 2019. *http://www.theplantlist.org*
Thiers, B. [continuously updated]. Index Herbariorum: A global directory of public herbaria and associated staff. New York Botanical Garden's Virtual Herbarium. *http://sycamore.nybg.org/science/ih*
WCSP. 2019. World Checklist of Selected Plant Families. Facilitated by the Royal Botanic Gardens, Kew. Published on the Internet. *http://apps.kew.org/wcsp*

17

PLANT COLLECTING AND DOCUMENTATION

INTRODUCTION

Plant collections are essential components of systematic research. Collections generally consist of samples of plants that are preserved by drying or by means of liquid preservation. They may also include live plants or propagules taken from the wild and grown in an artificial environment.

Collections of plants serve several purposes. One is to provide resource material in systematic research. Although systematists should attempt to study plants in the wild, in practice almost all research in plant systematics is done using preserved or living plant collections.

Another function of plant collections is to serve as reference material for named taxa. Such a reference plant collection is known as a **voucher specimen**. (Voucher specimens are almost always dried herbarium specimens; see later discussion.) Voucher specimens are required by the International Code of Botanical Nomenclature to serve as types in the valid publication of new taxa names (Chapter 16). Thus, every botanical name at and below the rank of family is associated with the **type specimen** (generally the holotype; see Chapter 16), which is almost always a voucher specimen selected from the original plant collection.

In addition, voucher specimens may serve as a reference in verifying the identity of a plant taxon. If there is ever any doubt as to a taxon's identity, the voucher can be studied to check a prior identification. Reference voucher specimens are essential to obtain and cite in any systematic study. This is true whether the actual data are acquired from study of morphology, chemistry, anatomy, ultrastructure, reproduction, or molecular biology. Reference vouchers are also essential in field surveys involving the species composition of a given region. Thus, studies of floristic diversity, ecological mechanisms, or environmental assessment (e.g., environmental impact reports) must include plant collections and voucher specimens as a component of the study. Otherwise, the scientific validity of the conclusions may be in doubt. (See Funk et al. 2005.)

Finally, the information recorded in the field as part of a plant collection is very important and can be utilized for a number of purposes. Many larger herbaria and some smaller ones have now initiated projects to input data from the labels of herbarium collections into a computerized database system. The database system allows information on plant morphology, ecology, phenology, and geography to be summarized and categorized in order to gain insight on a number of biological questions (see Documentation of Plant Collections). Thus, as these database systems are implemented, plant collections and their associated data are becoming increasingly valuable for fully documenting biodiversity in studies of systematics, ecology, and conservation biology.

METHODS OF COLLECTING PLANTS

Documentation of plant taxa necessitates not only thoroughly recording data in the field about the plant and its habitat but also procuring a physical specimen. This specimen is obtained

PLANT SYSTEMATICS
https://doi.org/10.1016/B978-0-12-812628-8.50017-1,

by (1) collecting the plant; (2) pressing and drying the plant; and (3) preparing a mounted herbarium specimen by gluing the plants and a label (listing the field data) to a sheet of standard herbarium paper. The specimen is deposited and maintained in an herbarium in order to be accessible for future study, e.g., to verify its identity or prepare a taxonomic revision.

FIELD COLLECTING

Locating specific plants may be by chance or can involve prior checking of specific collection records (e.g., herbarium sheet label information) or pertinent maps to locate the likely location of a plant in a specialized habitat. The collector should obtain prior permission or the proper permit for collecting on a tract of land.

Once a plant of interest is located in the field, the conscientious botanist must evaluate whether or not the species *should* be collected. The first guideline is to become aware of and be able to recognize any possible "sensitive" species, i.e., those that are rare, threatened, or endangered. These are typically protected by law and may not be collected legally without special permits. Second, regardless of the legal status of a plant species, any collecting should not endanger the local population. A good rule of thumb is the so-called "1 to 20" rule: for every one plant sample you collect, there should be at least 20 more present in the surrounding population. (For herbs, the "1 to 20" rule applies to individual plants; for shrubs and trees, it applies to shoots removed.)

In collecting an herb, at least one whole plant must be completely dug up to show roots and/or rootstocks. (The exception might be a plant that is extremely rare or endangered.) This is often necessary to determine whether the plant is an annual, biennial, or perennial and to determine the type of root (e.g., fibrous or tap) or underground stem (e.g., corm, bulb, or rhizome). With shrubs, trees, or vines, only one or more branches need be clipped off, using hand clippers to minimize damage to the plant. An attempt should be made to collect plants at flowering and/or fruiting stage and to collect enough individual specimens (population size permitting) to represent the range of individual variation.

It is strongly urged that plants be pressed immediately upon collecting in the field. Portable **field plant presses** can be obtained from herbarium supply companies. A simple, inexpensive field plant press can be made by placing several single, folded sheets of used newspaper (preferably ca. 11.5" × 16.5" when closed), between two adjacent 12" × 18" cardboards, all secured with two small elastic cords or straps. Plants are pressed by placing the specimen inside one of the single sheets of folded newspapers, all of which are temporarily stacked atop one another and sandwiched between the two cardboards. Each newspaper should be labeled with the collection number, referencing that recorded in the field notebook. Plants initially pressed in the field are then later transferred to a standard plant press prior to drying. (See later discussion for details of preparing pressed plant specimens.)

If collected plants are not immediately pressed, they should be stored to prevent wilting. Identifying string tags may be attached to the plant with the collector's name and collection number. Plants then may be stored in a plastic bag. Alternatively, plants may be wrapped in newspaper (open at the top end), wetted, and stored in a large plastic bag; evaporation from the newspaper keeps the plants cool and moist. Ideally, unpressed plants should be kept in an ice chest or refrigerator.

PREPARATION OF PLANT SPECIMENS

The standard method of preserving plants for future study and reference is by the preparation of a specimen that is deposited in an herbarium. An **herbarium specimen** (see Chapter 18) consists of a pressed and dried plant sample that is permanently glued and strapped to a sheet of paper (of standard weight and type, measuring 11.5" × 16.5" in most U.S. herbaria) along with a documentation label (see later discussion). Herbarium specimens or "sheets" will last for hundreds of years if properly maintained. They are still the most efficient and economical means of preserving a sample of plant diversity.

To prepare an herbarium specimen, material from the field plant press or bag is transferred to a standard plant press to be pressed flat and air-dried. A **plant press** consists of several 12" × 18" pieces of standard cardboard that are placed between two outer 12" × 18" frames or 1/4" plywood pieces all secured by two straps (Figure 17.1). Optionally, two 12" × 18" felts may be placed between adjacent cardboards to help absorb moisture, but good results can be obtained without felt. Plants are pressed by placing the specimen inside a single page of folded paper (again, used newspaper, preferably close to the size of herbarium paper, ca. 11.5" × 16.5"), which is then placed between two adjacent cardboards (or felts and cardboards) in the plant press.

The plants to be pressed should be positioned on the newspaper in a way that best represents the plant in the wild and maximizes information content, according to the following guidelines. Open the single sheet of newspaper and carefully place plant organs in a position that allows full view of morphology. Press herbs to show roots and underground stems, which should first be rinsed to remove dirt. Place whole, small herbs on the newspaper with several plants on a single sheet, enough to fill up the space (Figure 17.2B). Taller herbs may be bent into a "V," "N," or "M" shape (Figure 17.2A) in

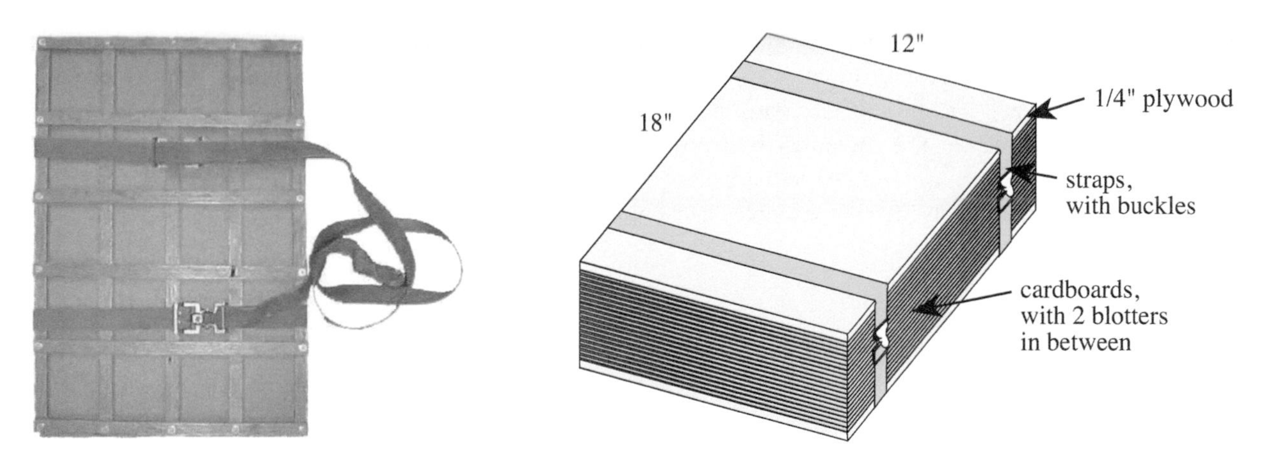

Figure 17.1 A standard herbarium plant press.

order to fit the entire plant on one sheet. If necessary, cut a tall herb into two or more pieces, preparing a separate newspaper for each. Slice large rhizomes, corms, or bulbs longitudinally and place one cut side face down and the other face *up* to show internal structure. For larger or highly branched specimens (Figure 17.2C–E), clip back the shoots or leaves (leaving the shoot or leaf base) in order to minimize overlapping of parts. Orient at least one leaf up and one leaf down, so that both leaf surfaces will be in full view upon drying. To dry succulent plants properly, cut their leaves or stems longitudinally and, if large, scoop out the fleshy tissue, placing the cut side face down. Cacti and other succulents may be soaked in 95% alcohol for 1–2 days before drying. Arrange flowers or flower parts carefully; section larger flowers to allow viewing of internal organs. Place extra flowers or inflorescences to one side in order to provide extra material for morphological study. Fruits may be sectioned to illustrate internal wall layers or placentation and to facilitate drying. Use wax paper on both sides of fleshy, aquatic, or delicate plant samples in order to prevent adhering to the newspaper. Place folded sections of newspaper on top of leaves or flowers in order to press them flat when the adjacent stems are thick. For all pressed plants, keep the space at the lower right corner (ca. 3" × 5" area) free, as this is where the herbarium label will be glued on the herbarium specimen. After final positioning of the plant sample, carefully fold the newspaper over the plant and place between two cardboards in the press.

After all plants have been placed in the plant press, the straps are tightened and the press is positioned on its long edge (with buckles on the opposite side) inside a plant drier. The plant drier consists of a ventilated box or cabinet having at its base either heating elements or light bulbs plus a fan to provide air circulation. Because modern techniques permit removal and amplification of DNA from herbarium material, it is important that plant specimens be dried at not too high a temperature, to prevent DNA degradation. Heated and circulated air rises through the cardboards and newspapers, drying most plants in 2 or 3 days. After this time, the plant specimens should be removed and checked individually; if any specimen feels cool to the touch, water is still evaporating from its tissues, necessitating a longer drying time.

DOCUMENTATION OF PLANT COLLECTIONS

It is critical that certain data be recorded at the time of collecting a plant. Such data will be typed onto an herbarium label and may be entered into computerized database systems. The following is an explanation of the data categories to be recorded at the time of collecting: Figure 17.3 illustrates an example data page for this documentation.

Field Site Data

[List a locality number to cross-reference to other collections.]

Date of collection: List day — month (*spell out* to avoid confusion) — year

Time (optional): Sometimes important for noting the time of flowering.

Country/state/province/county/city: List as needed.

Specific locality information: List complete locality data for possible relocation of habitat in the future, including measured or estimated distance on roads or trails.

Latitude and longitude: Important to list for biogeographic data systems. Use GPS device or put dot on topographic map to reference plant collection numbers.

Source/accuracy of lat./long.: List how lat./long. is determined, e.g., by USGS 7.5' quad or GPS device. List (in seconds) accuracy of determination.

Figure 17.2 Examples of plants collected and pressed. **A.** Herb, stem bent twice to fit on newspaper. **B.** Herb, in which whole plant is collected, including rootstock. **C.** Small shrub, whole plant collected, including roots. **D.** Tree, in which a branch (in fruit, in this example) is collected. **E.** Vine, in flower; rootstock not collected.

LOCALITY #: 2-A

FIELD SITE DATA

Date of collection: 24 April 1994	Time (optional): 10 AM
Country/State/Province/County/City: CA / San Diego Co.	
Specific Locality information: Near hiking trail, just east of Oak Canyon, ca. 1.5 miles north of trailhead at Hwy 83 and Ventura Rd., Pickwood State Reserve	
Latitude: 32 ° 50 ' 28 " (N) S	Longitude: 117 ° 02 ' 59 " (W) E
Source/Accuracy of Lat./Long.: USGS 7.5' topo. quad (La Mesa); +/-1"	
Township & Range:	
Elevation (ft or m): 1,100 feet	
Landmark information: Ca. 4.7 miles northwest of Wilson Peak	

PLANT DATA

Coll. No.: 702	Photo. doc.: Roll #2, slides 13-14	
Collector (primary): Cynthia D. Jones		
Associated collector(s): John J. Smith		
Taxon: *Porophyllum gracile* Benth.		
Ann./Bien./(Per.) Habit, Height, Branching: Subshrub, 30-40 cm tall, with several branches from base, densely branched above		
(Fl.)/(Fr.) colors, other notes: Involucre purple. Corolla white to greenish-yellow. Pappus bristles white to purplish. Leaves strongly pungent. NOTE: Fls. visited by Checkerspot butterflies. Material preserved in Carnoy's fixative for chromosomal studies		
Physical Habitat, Substrate: Mountain slope. Rocky, sandy loam soil.		
Slope, Aspect, Exposure: Slope ca. 30 degrees, south-facing, exposed.		
Community/Vegetation type: Open *Eriogonum fasciculatum* - mixed (*Artemisia californica*, *Malosma laurina*) scrub.		
I.d. by: Cynthia D. Jones	Date: 4/1994	I.d. source: Jepson Manual, 1993
Accession Number: SDSU 12837		

Figure 17.3 Plant collecting documentation sheet.

Township and range: May be listed instead of lat./long., but less preferable.

Elevation (ft or m): List in units appropriate for source of determination.

Landmark information: Describe nearest major landmark (preferably one listed on standard topographic map) and list distance and direction from landmark.

Plant Data

Collection number: A unique number associated with the primary collector. Standard format is for a given person to begin with "1" for the first plant collected, "2" for the second, etc. Another format is to transform the date into a collection number, e.g., "10VI94A," in which the month is in Roman numerals, "A" represents the first plant collected that day, "B" the second plant, etc. Note: Duplicate specimens of a taxon collected at the same site and time receive the same collection number. If one plant specimen is divided into two (or more) parts, the labels for the pressed sheets are listed as "1 of 2," "2 of 2," etc.

Photograph documentation: For keeping track of photos or other images.

Collector (primary): The *one* person associated with a plant collection.

Associated collector(s): Other people present or aiding in collecting. These names are not directly associated with the collection number.

Taxon: Scientific name of species (a binomial of genus + specific epithet), including authorship. If applicable, also list the subspecies (ssp.) or variety (var.) name, including authorship. In final form, the scientific name is always underlined or italicized.

Ann./bien./per., habit, height, branching: Circle or list duration (annual, biennial, or perennial), habit (herb, shrub, subshrub, vine, or tree), height from ground level (in metric, not essential if entire plant is collected), and any distinguishing features of the branching pattern that are not apparent from the specimen itself.

Fl./Fr., colors, other notes: Circle or list phenology, whether plants are in flower and/or fruit. Precisely describe the colors of unusual vegetative parts and of all flower parts (e.g., of calyx, corolla, anthers). If precision needed for colors, use a chart (e.g., Royal Horticultural Society Color Charts; see Tucker, et al. 1991. Taxon 40: 201–214.) Describe features that are obscure or might be lost from specimen upon drying.

Other field notes may include references to additional research studies or additional field observations, such as observed visitors/pollinators.

Population size/distribution: A few notes about the size and distribution of the population are useful, such as "very rare," "population very large (>1000 individuals per hectare)," or "plants locally common."

Physical habitat/substrate: Physical habitat refers to abiotic features, such as "dry creek bed," "granite outcrop," or "flood plain." For substrate, list color and basic soil type (e.g., clay, clay-loam, loam, sandy-loam, sand, gravel, boulder, or rock). More detailed information can include soil series and/or rock type.

Slope/aspect/exposure: List angle of slope, from none (flat) to 90° (cliff face). Aspect is general compass direction toward which slope is facing. Exposure is either exposed, partly shaded, or shaded.

Community/vegetation type: Both immediate and surrounding plant communities/vegetation types may be listed for a single plant collection. Community/vegetation type may be general (e.g., "chaparral" or "woodland") or precise. A precise designation of community type (modified from Radford, A. E., et al. 1981. Natural Heritage: Classification, Inventory, and Information. University of North Carolina Press, Chapel Hill) is as follows:

1. Determine the **boundaries** of the community, based on overall similarity of species composition. This may not be clear cut, as one community may intergrade with another or show much variation.
2. Identify **layers** present in the community: *canopy* (tall tree and lianas, if present) / *subcanopy* (smaller tree layer under canopy) / *shrub* or *subshrub* / *herb*. A vine, epiphyte, moss, or lichen layer also may be defined if a major component of the community.
3. For each layer of vegetation, assess the **total cover**, measured as the degree to which the total area of the community is covered by members of a given layer. Designations of cover are (1) *closed* (50–100% cover); (2) *open* (25–50% cover); or (3) *sparse* (<25% cover).
4. For each of the common species of a given layer (e.g., the shrub layer), assess **relative cover**, measured as the degree to which each species of the layer contributes to the total cover *of that layer alone*. (Other ecological measures, such as importance value, may be used instead, but relative cover is perhaps easiest to "eyeball" in the field.) Assess relative cover as (a) *dominant* = > 50% relative cover; (b) *codominant* = 25–50% relative cover.
5. Summarize the **community type** by listing layers separated by a " / " in sequence from tall to short layers, e.g., tree/shrub/herb/moss. Note: Dominant or co-dominant vines and epiphytes are listed at the end separated by, respectively, a double slash ("//") or triple slash ("///").

List as follows: (a) total cover; (b) dominant species, if *one* species is dominant (50–100% cover); or (c) codominant species, if two or more species are codominant (25–50% cover), each species present separated by a hyphen.

Note: You may use "mixed trees," "mixed shrubs," or "mixed herbs" as a layer designation where collectively the group of "mixed" species is dominant or codominant, but each individual species is <25% relative cover. This designation may be followed by a listing of the more common species (those with at least 10% relative cover) in parentheses.

6. Follow the community type with a designation of **vegetation type**. This is based on habit, habitat, and cover of species present. Examples include *forest* (closed trees), *woodland* (open trees), *savanna* (sparse trees with intervening grassland), *chaparral* (closed, evergreen, sclerophyllous shrubs), *scrub* (open to sparse shrubs), *grassland* or *meadow*, *strand* (sparse, low shrubs/herbs), *marsh* (aquatic shrubs and/or herbs in slow-moving water), *swamp* (closed to sparse aquatic trees), *pond*, *vernal pool*.

Example: Open *Malosma laurina – Artemisia californica* / closed *Erodium botrys* scrub community

Meaning: The total cover of the shrub layer is open (25–50% total site area). *Malosma laurina* and *Artemisia californica* are a codominant in the shrub layer (25–50% relative cover). *Erodium botrys* is a dominant in the herb layer, which is closed (>50% relative cover). The vegetation type is a scrub (open shrubs).

I.D. by/date/source: List the person who identified the taxon, even if it is the same as the primary collector. Also list the date, usually just the month and year, and the source or reference of determining the taxon identity. The source will generally be a flora of the region, but could include monographic treatments or expert determination.

Accession number: After the plant collection is processed into an herbarium sheet and deposited in an herbarium, list the herbarium acronym and accession number for a complete record of the collection. Accession numbers are usually cited in publications to document a collection (see later discussion).

LIQUID-PRESERVED COLLECTIONS

It is often valuable to preserve samples of a plant collection in a liquid preservative. Liquid preservation maintains the shape, size, and internal structure of plant tissues. This is particularly valuable to do for delicate floral parts, whose form is easily distorted or even destroyed from standard herbarium specimen drying techniques. Liquid preservation is also essential for anatomical, developmental, or ultrastructural studies, in which the internal structure of cells and tissues must be maintained.

The most commonly used, general liquid preservative (known as a "fixative") is FAA, one recipe being 10 parts 70% ethanol:1 part commercial (37%) formalin : 1 part glacial acetic acid (all by volume). (Note: FAA is toxic; avoid getting on skin or breathing the fumes!) Plant samples are simply placed into a glass or plastic vial or jar filled with FAA. Although FAA penetrates most plant tissues rapidly, some plant samples should be cut open to allow the fixative to fully infiltrate into the tissues. At least some closed flower buds or ovaries, leaves, and stems should generally be sectioned with a razor blade prior to fixation.

For cytological studies, e.g., chromosome counts, flower buds or root tips may be fixed in Carnoy's fixative (3 parts 100% ethanol:1 part glacial acetic acid). For detailed ultrastructural studies, e.g., using electron microscopy, other fixatives may be needed, such as glutaraldehyde or osmium tetroxide. These compounds are dangerously toxic and should only be handled in a laboratory hood. Because they penetrate less rapidly than FAA, the material must be cut into much smaller pieces, generally 1 mm or less.

Plant material may be fixed in 70–100% ethanol and used for general morphological studies and sometimes DNA analysis. This is not commonly done for the latter, as material dried in silica gel is better preserved (see later discussion).

Any liquid preserved material should have a corresponding herbarium voucher specimen to serve as a reference for identification. The vial or jar should be labeled both on the outside and on a strip of paper (using a pencil) placed into the fixative. Label information should include the species name and collector and collection number; other data are optional and can be obtained from the field collection notebook or voucher.

LIVING COLLECTIONS

A very valuable type of plant collection is a live specimen removed from the wild. This may be either a whole plant, a vegetative propagule, or a seed. Living plant collections are typically grown in a greenhouse or botanic garden, where they can be accessible to a researcher. Growing them and keeping them alive requires some horticultural experience and may involve trial and error under different regimes of potting or soil mixture, moisture, and photoperiod. As with liquid-preserved collections, they should be properly labeled with permanent metal or plastic tags, with collection information corresponding to a voucher specimen deposited in an herbarium.

A living plant collection has the great advantage of permitting long-term observations, e.g., through an entire reproductive stage, or experimental manipulations, such as breeding studies. It also permits removing fresh samples of material for study over an extended period of time (rather than from a single field expedition). However, one precaution about studying live plant collections is that their morphology may be altered in cultivation from that in the wild. In addition, pollinators normally present in the wild will not normally be present in an artificial environment, perhaps preventing normal seed set.

COLLECTIONS FOR MOLECULAR STUDIES

A standard method for collecting material for studies of DNA is to cut pieces of leaves or other plant tissue and immerse these in a container (vial or plastic bag) of silica gel. A paper label, indicating the taxon and the name and number of the collector (corresponding to an herbarium voucher collection), is placed in the container. The silica gel rapidly dehydrates the material, preserving the DNA for future extraction, purification, and amplification. Extracted plant material is usually frozen at −80°C to prevent degradation of the DNA. Plant material to be used for DNA analysis may also be fixed in 70–100% ethanol, but this may not preserve the DNA as well.

For allozyme analysis, fresh material must be used, as enzymes degrade very rapidly. Extra plant material is placed in a plastic bag (again with a slip of paper or label indicating the voucher information) and kept in a cooler until it is transported to the lab.

IMAGES AS PLANT SPECIMENS

A novel idea is to use photographic images of plants as specimens themselves. However, this necessitates strict standards of image acquisition, such that the images contain maximum information on many aspects of the plant and ideally at different stages of the life cycle. For example, for a woody tree, "specimens" might include images of the whole plant, bark of a mature and immature tree, twig during growth, twig in winter, adaxial and abaxial leaf surfaces showing surfaces, adaxial and abaxial leaf surfaces showing venation and margins, inflorescence, lateral and front views of a flower, lateral view of a fruit. Such images would be in a high resolution digital format and accompanied by detailed information on the location (latitude and longitude), date and time, size scale, community structure, abiotic environmental factors, etc. Although images will never replace physical specimens, their standardization may augment and serve many of the functions of physical specimens. (See Baskauf and Kirchoff 2008.)

REVIEW QUESTIONS

1. What are the different types of plant collections?
2. List the several uses of plant collections.
3. What is a voucher specimen?
4. What is the purpose of a voucher specimen?
5. Review the preparations needed for collecting plants in the field.
6. What are the general rules for assessing whether a plant *should* be collected?
7. List the guidelines for properly collecting plants in the field.
8. How should plant collections be stored prior to processing?
9. What is a standard herbarium specimen?
10. What are the components of a plant press?
11. Review the guidelines for properly pressing plants, including special requirements for processing (a) herbs; (b) tall herbs; (c) shrubs or trees; (d) highly branched specimens; (e) rootstocks (such as rhizomes, bulbs, or corms); (f) succulent plants; (g) flowers and fruits.
12. How are herbarium specimens dried?
13. List all the data that should be recorded in the field at the time of collecting.
14. Review in detail how specific plant community types can be assessed.
15. Why is it important to list the person who determined the identity of the plant and the determination source?
16. Review the guidelines for preparing liquid-preserved collections.
17. What is the most common type of liquid preservative?
18. What liquid preservatives must be used in ultrastructural studies?
19. Review the guidelines for obtaining living plant collections.
20. What are the advantages and disadvantages of living plant collections?
21. How is material for molecular studies normally collected?

EXERCISES

1. Collect six plants, including at least two herbs, one shrub, and one tree. Record all pertinent information in the field, using Figure 17.3 as a guide.
2. For at least one of the above, collect liquid-preserved material for both anatomical and cytological studies.
3. For at least one of the above, collect material for DNA sequence studies.

REFERENCES FOR FURTHER STUDY

Baskauf, S. J. and B. K. Kirchoff. 2008. Digital plant images as specimens: toward standards for photographing living plants. Vulpia 7: 16–30.
Funk, V. A., P. C. Hoch, L. A. Prather, and W. L. Wagner. 2005. The importance of vouchers in science. Taxon 54: 127-129.
Radford, A. E., W. C. Dickison, J. R. Massey, and C. R. Bell. 1974. Vascular Plant Systematics. Harper & Row, New York.
Radford, A. E., et al. 1981. Natural Heritage: Classification, Inventory, and Information. University of North Carolina Press, Chapel Hill.
Tucker, A. O., M. J. Maciarello, and S. S. Tucker. 1991. A survey of color charts for biological descriptions. Taxon 40: 201–214.

18

HERBARIA AND DATA INFORMATION SYSTEMS

Herbaria are repositories of preserved plant collections; these are usually in the form of pressed and dried plant specimens mounted on a sheet of paper. The purpose of herbaria is both to physically contain the plant collections and to act as centers for research. The plant collections themselves function as vouchers for identification and as sources of material for systematic work. Herbaria also may house numerous geographic and taxonomic references, particularly floras or manuals that may aid in plant identification. In addition to housing plant collections, many herbaria today have initiated computerized data information systems to record and access the collection information of the plant specimens, as well as to access information from other collections worldwide (see **Data Information Systems**).

As of this writing, there are over 3,000 herbaria in the world, housing a total of over 387 million specimens (Thiers, accessed 2019). Information about the world's herbaria is contained in *Index Herbariorum* (Thiers, continuously updated; see also listing of online computer access in References for Further Study), which lists the names, addresses, curators, and number and types of specimens. Each herbarium listed in *Index Herbariorum* is assigned an acronym. It is this acronym that is cited in publications in order to specify where voucher specimens are deposited. Herbaria are typically associated with universities or colleges, botanic gardens, museums, or other research institutions. The 15 largest herbaria, their acronyms, and the number of specimens they contain are listed in Figure 18.1.

HERBARIUM SPECIMENS

An **herbarium specimen** consists of a pressed and dried plant sample that is permanently glued and/or strapped to a sheet of paper along with a documentation label. The herbarium paper is of high quality, heavyweight, and acid-free to inhibit yellowing. In most American herbaria, standard herbarium paper measures 11.5" wide × 16.5" tall; in other countries the dimensions may be slightly different. An herbarium label (see below) is glued to the lower right corner of the herbarium specimen. An example of an herbarium specimen is seen in Figure 18.2. Herbarium specimens (also called *herbarium sheets*) will last for hundreds of years if properly maintained. They are still the most efficient and economical means of preserving a record of plant diversity.

https://doi.org/10.1016/B978-0-12-812628-8.50018-3,

Herbarium	Year Founded	Acronym	Number of Specimens
Muséum National d'Histoire Naturelle, Paris, France	1635,1904	**P (8M), PC (2M)**	10,000,000
New York Botanical Garden, Bronx, New York, U.S.A.	1891	**NY**	7,800,000
Royal Botanic Gardens, Kew, England, U.K.	1841	**K**	7,000,000
Naturalis, Leiden, Netherlands	1829,1816,1896	**L,U,WAG**	6,900,000
Missouri botanical Garden, Saint Louis, Missouri, U.S.A.	1859	**MO**	6,600,000
Conservatoire et Jardin botaniques, Genève, Switzerland	1824	**G**	6,000,000
Komarov Botanical Institute of RAS, St. Petersburg, Russia	1823	**LE**	6,000,000
Naturhistorisches Museum Wien, Vienna, Austria	1807	**W**	5,500,000
The Natural History Museum, London, England, U.K.	1753	**BM**	5,200,000
Smithsonian Institution, Washington, D.C., U.S.A.	1848	**US**	5,100,000
Harvard University, Cambridge, Massachusetts, U.S.A.	1864	**A, AMES, ECON, FH, GH, NEBC**	5,005,000
Herbarium Universitatis Florentinae, Florence, Italy	1842	**FI**	5,000,000
Swedish Museum of Natural History, Stockholm, Sweden	1739	**S**	4,570,000
Université Claude Bernard, Villeurbanne, France	1924	**LY**	4,400,000
Botanic Garden Meise, Meise, Belgium	1870	**BR**	4,000,000
Botanischer Garten und Botanisches Museum Berlin- Dahlem	1815	**B**	3,800,000
Université de Montpellier, Montpellier, France	1809	**MPU**	3,500,000
Friedrich-Schiller-Universität Jena, Jena, Germany	1896	**JE**	3,500,000
University of Helsinki, Helsinki, Finland	1751	**H**	3,350,501
Botanische Staatssammlung München, München, Germany	1813	**M**	3,200,000
Museum of Evolution, Uppsala, Sweden	1785	**UPS**	3,100,000
Royal Botanic Garden Edinburgh, Edinburgh, Scotland, U.K.	1839	**E**	3,000,000
University of Copenhagen, Copenhagen, Denmark	1759	**C**	2,900,000
Field Museum of Natural History, Chicago, Illinois, U.S.A.	1893	**F**	2,700,000
Institute of Botany, Chinese Academy of Sciences, Beijing, China	1928	**PE**	2,650,000

Figure 18.1 Names, years founded, acronyms, and number of specimens of the 25 largest herbaria (or herbaria complexes) in the world. After Thiers (continuously updated), http://sweetgum.nybg.org/science/ih. (See **Websites**.)

HERBARIUM LABELS

An **herbarium label** is affixed to each specimen, usually at the lower right-hand corner. Herbarium labels are typically computer generated using a laser or ink jet printer. Label sizes vary, but are generally about 4–5" (10–12 cm) wide and 2–3" (5–7 cm) tall, using high-quality, thick-weight (20- or 24-lb), acid-free bond paper. Virtually all of the information recorded at the time of collecting should be placed on the herbarium label. An example of a typical label format, containing all information from the collecting event, is seen in Figure 18.3. A convenient formatting is to list (following the taxon name) all characteristics about the plant itself in the first paragraph, including duration/habit/height/branching pattern and phenology, colors, and other features. The second paragraph contains information about the habitat and locality of the plant, including physical habitat/substrate, slope/aspect/exposure, community/vegetation type, specific locality information, landmark information, latitude and longitude, source/accuracy of lat./long., and elevation. A third paragraph may include other field notes and photograph/image documentation. At the bottom of the label, the collector, collection number, and date of collection are listed. (The abbreviation **s.n.,** Latin for *sine numero*, without a number, sometimes follows a collector's name to indicate that the collector did not designate a personal collecting number.) The last item on the herbarium label may list by whom and when the identity was determined (even if by the same person who collected the material) and what the source of that identification was. Information on taxon determination is important to include on the label, as it cannot be assumed that the person who collected a plant identified it. In addition, the source or means of identification (whether a flora, monograph, or expert determination) may constitute valuable information in verification of identities.

If the plant specimen is so large that it must be divided between two or more herbarium sheets, a separate label must be prepared for each of these parts. All labels referring to the same plant have the same collection number (but different accession numbers if on different herbarium sheets; see later discussion). The two herbarium sheets may be differentiated by the designation, e.g., "1 of 2," "2 of 2."

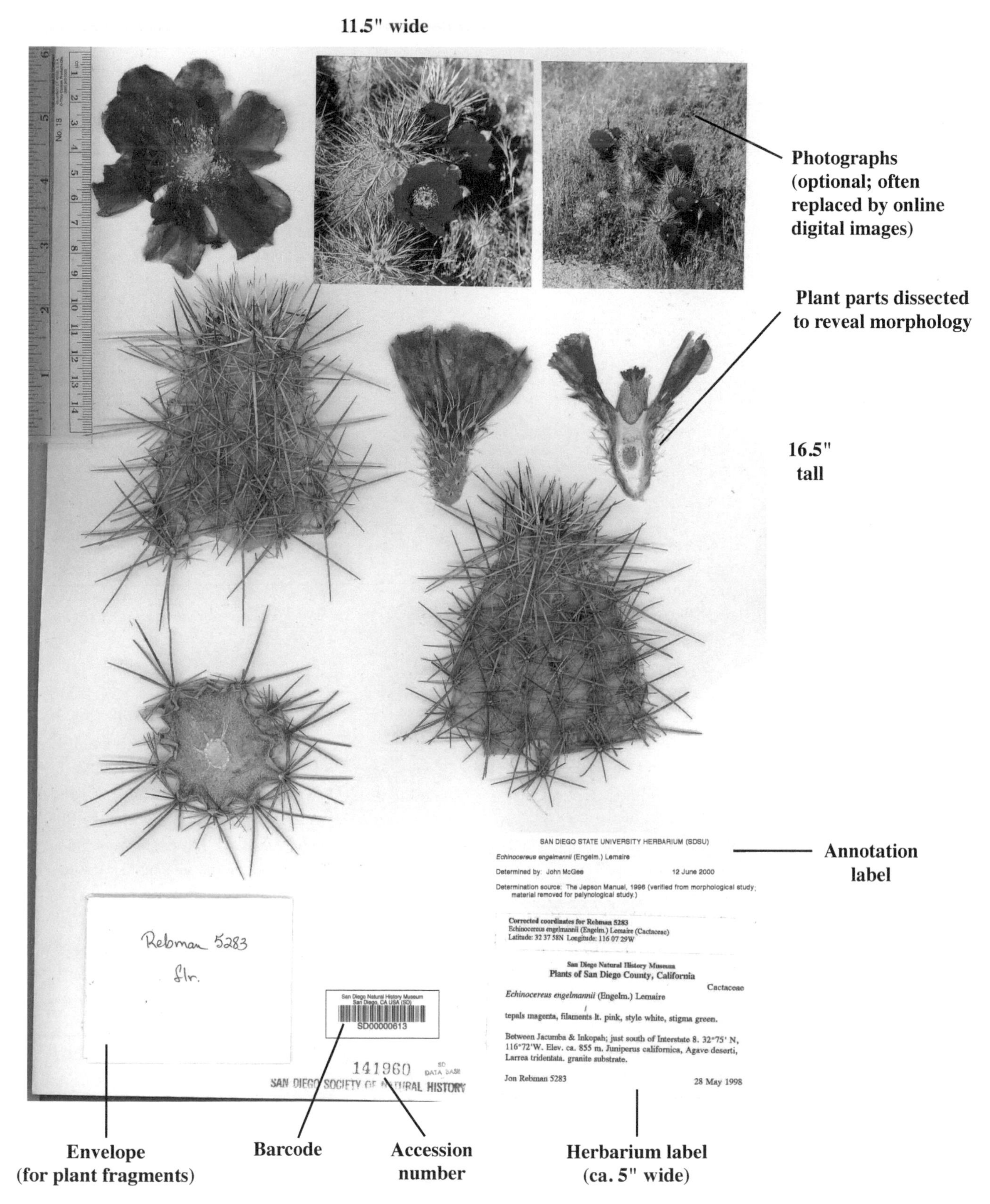

Figure 18.2 Example of a typical herbarium sheet. (Image modified, courtesy of Dr. Jon P. Rebman.)

ca. 5" wide

SAN DIEGO STATE UNIVERSITY HERBARIUM
USA
CALIFORNIA
San Diego Co.

Porophyllum gracile Benth.

Perennial subshrub, 30–40 cm tall, with several branches from base, densely branched above. In flower and fruit. Involucre purple. Corolla white to greenish yellow. Pappus bristles white to purplish. Leaves strongly pungent. Note: Flowers visited by checkerspot butterflies. Material preserved in Carnoy's fixative for chromosomal studies.

Near hiking trail, just east of Oak Canyon, ca. 1.5 miles north of trailhead at Hwy 83 and Ventura Rd., Pickwood State Reserve. Mountain slope. Rocky, sandy loam soil. Slope ca. 30 degrees, south facing, exposed. Open *Eriogonum fasciculatum*–mixed (*Artemisia californica, Malosma laurina*) scrub. 32.841111 N 117.049722 W (USGS 7.5' La Mesa quad, ±1" accuracy.) Elevation 335 meters (1100 feet).Ca 4.7 miles northwest of Wilson Peak.

Cynthia D. Jones 702 24 April 1994
with John J. Smith

Figure 18.3 Example of a typical herbarium label (this one hypothetical).

MOUNTING HERBARIUM SPECIMENS

Plant specimens are affixed to herbarium paper with glue and/or straps. The glue used may be standard white glue or a solution of methyl cellulose, available from chemical supply and some herbarium supply companies. White glue is best diluted slightly, ca. 9 parts glue to 1 part tap water, stirred well. Methyl cellulose is prepared by adding about 70 grams of methyl cellulose powder to a liter of warm tap water and stirring briskly until well mixed; more water or powder may be added to achieve a thick, viscous solution. The advantage of methyl cellulose is that, with minimal moistening, it will soften or dissolve, allowing for relatively easy removal of dried plant material from the herbarium specimen (see later discussion). Glues containing organic solvents are not recommended, as they are toxic and require special ventilation.

The following is one useful method to glue a dried plant specimen and label to a sheet of herbarium paper. Have the following supplies on hand: herbarium paper, cardboard (12" × 18"), a flat sheet (ca. 12" × 18"), paintbrush (2–4" wide), glue, two pairs of forceps, spatula, and weights (standard bathroom tiles, measuring 4", 6", and 8" square work well). First, place a sheet of herbarium paper on top of a cardboard. Position the herbarium label (without gluing yet) on the lower right corner of the herbarium paper, leaving ca. 1/8–1/4" space between the label and the margins of the paper. Place the pressed plant specimen (also without gluing yet) on the paper in order to test the final positioning. Make sure the specimen does not overlap the label or go beyond the edges of the herbarium paper; if overlap occurs, the plant must be cut. Also, try to leave some room above and to the left of label (or top of sheet) for placing an accession number or barcode and possible annotation labels (see later discussion). Extra pieces of the plant specimen (e.g., individual flowers, fruits, or inflorescence) may be placed on the sheet as well. Smaller pieces are best placed in a separate small envelope that may be glued to the final specimen, such that it may be opened to remove the material for study. (Envelopes may be constructed by cutting heavyweight, 100% bond typing/printing paper (e.g., 8.5" × 11") into two pieces; each 8.5" × 5.5" piece is then folded to make a 4.25" × 5.5" rectangle, which is then folded to overlap ca. 1/4" along the three cut margins.)

Next, using a paintbrush, coat a large (at least 12" × 18") sheet (e.g., of glass, Plexiglas, or metal) thoroughly with a layer of glue. Transfer the plant specimen from the paper to the glue-covered sheet, gently press down, carefully remove (using forceps for delicate material to prevent damage), and place back onto the herbarium paper, positioning the plant as originally placed. You may use a scalpel or squirt jar to transfer glue directly to plant surfaces that require greater adhesion. Continue this until all plant components are glued to the sheet. Finally, in a smaller region of the sheet, paint a very thin layer of glue (preferably white glue, diluted as specified above) on the sheet, place and press down the herbarium label onto this region, and transfer back to the herbarium specimen with forceps, being careful to correctly position it about 1/4" from the edges. Flatten and smooth the herbarium label by placing a paper towel or scrap paper (to absorb excess glue) over the label and pressing down firmly. Finally, place weights (e.g. different-sized ceramic tiles or lead weights) over various locations of the plant material. Leave the specimen overnight to dry thoroughly. Specimens, with underlying cardboard, may be stacked to conserve space.

After the glue has dried, remove the weights and check the specimen. Reapply glue to individual spots as needed. Place narrow (ca. 1/8" wide) strips of strapping tape (available from herbarium supply companies) over stout stems to better secure them to the sheet. Some herbaria use little to no glue, relying on heavy use of strips of strapping tape to secure the specimen. Although this may not secure some plant specimens as well, it has that advantage of making removal of plant material from a mounted specimen (e.g., for detailed study) much easier.

HERBARIUM OPERATIONS

CURATORS

The person in charge of the day-to-day running of an herbarium is known as a **curator**. The duties of the curator (and assistant curators or collections managers, if any) are to (1) manage the existing collection, including the mechanics of proper storage and regular treatment to control insect pests; (2) mount, label, and accession new additions to the herbarium collection; (3) distribute requested loans from scientific institutions and receive loans from other herbaria; and (4) act as a resource person for the identification of regional plants or plants of special collections. In addition, curators today are often involved in transferring herbarium collection data to a computerized data information system for interactive access to that information. (See Rabeler et al. 2019 on herbarium practices and ethics.)

ACCESSIONING

Accessioning refers to the designation of a unique number, known as an **accession (or catalog) number**, to each specimen of a collection. (In the case of a mixed specimen, each different plant on the herbarium sheet may get a separate number.) The accession number is written or imprinted onto the herbarium sheet, often along with the international acronym of the herbarium (e.g., **UC1118485**, where **UC** is the herbarium acronym, here referring to University of California at Berkeley). With the advent of computerized data information systems, accession numbers are typically designated with a barcode label that may be scanned, this barcode number often different from the original accession number.

The purpose of the accession number is to provide a permanent reference for each specimen of the plant collection. Accession numbers (plus the collector and collection number and sometimes the date of collection) are often cited in journal publications and may be valuable in tracking down the exact specimen for verifying its identity or for further study.

STORAGE AND CLASSIFICATION OF SPECIMENS

Standard procedure is to store herbarium specimens by genus in a **genus folder**. There are different constructions of genus folders, but all consist of stout, heavyweight paper folded along at least one crease. The genus folder is labeled, typically on the lower right corner of the outer cover. The specimens within a genus folder are usually arranged alphabetically by species. If a particular genus has numerous collections, two to many genus folders may be used to house them.

Genus folders are sometimes **color-coded** to represent different geographic regions for the plant collections. For example, different colors may represent various counties, states, regional areas (e.g., Southwestern states), countries, blocks of countries, or continents. Thus, specimens of the same species may occur in two or more genus folders if these were collected from different geographic regions. Color-coded genus folders are typically stacked one on top of the other according to a standard order.

Plant specimens are usually stored in **herbarium cabinets** (Figure 18.4). Herbarium cabinets are usually made of metal and have sealed doors to inhibit insect migration or to prevent possible diffusion of pesticides. A standard, full-sized herbarium cabinet is typically 7 feet tall and 2.5 feet wide, with approximately 26 shelved compartments arranged in two columns, having a capacity of approximately 1,000 herbarium specimens (depending on the bulkiness of the plants). In many of the larger herbaria, standard herbarium cabinets have been replaced by **compactors**, which allow for a greater number of specimens to be stored. Compactor systems consist of rows of attached cabinets (or shelves), each row mounted on floor tracks. Entire rows of cabinets can be moved as a unit to abut against an adjacent row. Thus, compactors generally allow for only one (temporary) aisle space, maximizing the storage space available.

Genus folders are usually classified alphabetically within a given plant family. In many herbaria, families are arranged according to someone's formal classification system, a common one still in use being the Dalle Torre and Harms, which is based on the antiquated Englerian system. In other herbaria, families are simply classified alphabetically, with the exception that the major plant groups (e.g., lycophytes, equisetophytes, leptosporangiate ferns, conifers, Gnetales, monocots, or eudicots) are usually stored separately.

Figure 18.4 Example of standard herbarium cabinets. Note genus folders (color-coded here) on cart.

USING HERBARIA

In general, use of an herbarium requires prior approval and/or an appointment made through the herbarium curator. When using the herbarium, please be considerate of dissecting microscopes, tools, and references in the collection. Clean up after yourself; brush the table clean of debris (into a trash can) as needed.

REMOVING AND HANDLING SPECIMENS

You will, of course, need to remove herbarium specimens from the collection for observation, for example, to check the identity of your own plant or to study a given taxonomic group. Herbarium specimens may be rather fragile and should be handled very carefully, as follows.

Note taxa lists or maps to locate the family and genus of interest. Remove the entire genus folder from the cabinet. You may wish to *slightly* pull out the genus folders above or below the desired folder to mark the location. (However, always recheck the labels when filing!) *Close the herbarium cabinet door* immediately in order to inhibit insect infestation. Carefully transfer the genus folder to a table (with plenty of space) for observation and open it.

Always hold an individual specimen with both hands to prevent it from inadvertently bending. Never place anything (e.g., books) on top of a specimen. *Never turn a sheet upside down*, as this may result in the plant material breaking or falling off the sheet. Remove each specimen, one at a time from the top, and stack (in reverse order) to the side. Avoid sliding stacked specimens against one another, as this can result in damage to the plants. To find a specific collection, you may very gently shuffle through the labels at the lower right-hand corner of the specimens. Then, move aside the group of sheets on top of the desired specimen to expose it. When finished, replace the removed specimens in the genus folder, generally classified in alphabetical order by species or infraspecific name.

SYNOPTIC COLLECTIONS

Synoptic collections are those that contain generally one specimen (of all available specimens) of each taxon for a given region. Synoptic collections are very useful for quickly perusing the possible taxa in a region, such as a state, county, park, reserve, or some other political boundary. The disadvantage of synoptic collections is that they are generally limited to one specimen per taxon. Thus, it is imperative to always check an identity with the main collection, to note the entire range of variation of the taxon.

ANNOTATION LABELS

An **annotation label** is a label that verifies or changes the identity of a specimen or that documents the removal of plant material from the specimen (Figure 18.5). Annotation labels are permanently glued to the plant specimen, typically just above the standard herbarium label (see Figure 18.2). Annotation labels are typically placed on herbarium specimens by experts in a particular group, often as part of a

A

SAN DIEGO STATE UNIVERSITY HERBARIUM (**SDSU**)
Cryptantha leiocarpa (Fischer & C. A. Meyer) Greene
Determined by: Stella C. Danner 7 April 2016
Determination source: Jepson Flora Project (eds.) 2019. Jepson eFlora, http://ucjeps.berkeley.edu/eflor

B

FIELD MUSEUM OF NATURAL HISTORY (**F**)
Eriogonum tomentosum Michaux
Determined by: C.A.T. Mistoffelees 1 January 2001
Determination source: Monograph (in preparation)

C

ROYAL BOTANIC GARDEN EDINBURGH (**E**)
Fragment removed for study
Name: Date:
Institution:

Figure 18.5 Examples of annotation labels citing identity (**A,B**) or documenting removal of plant material (**C**).

research project. The labels vary in format, but generally measure about 4" wide and 1" high, using paper like that of herbarium labels.

Annotation labels that verify identity (Figure 18.5A,B) generally include (1) the name of the herbarium; (2) the species, subspecies, or variety name, including full authorship; an exclamation point "!" (symbol for the Latin *vidi*, "I have seen it") indicates a confirmation and is sometimes written instead of the full taxon name, if the annotated name is the *same* as that on the herbarium label; (3) the name of the person who made the correct name determination (often listed after "Det."); (4) the date of the determination; and (5) the determination source. As with herbarium labels, the determination source is often omitted from annotation labels, but is nonetheless valuable to include. The determination source refers to which, if any, references were used in the determination (usually a specific flora) and how the correct identity was ascertained.

One type of annotation is an update of the nomenclature of a species, without verifying the identity by morphological examination. Such annotations are not ideal, but may be necessary in large collections in order to cite the specimens in terms of the nomenclature of a recent flora or monograph. In such a case, the original identity is assumed to have been correct and now a synonym of the new name. Another type of annotation would list the reference and indicate that the person making the determination did examine the specimen critically, examining its morphology. An example of this type of annotation label is seen in Figure 18.5A. A third type of determination source might cite an original monographic treatment, published or unpublished. An example of this type of annotation label is seen in Figure 18.5B.

REMOVING PLANT MATERIAL

Annotation labels may also document the removal of dried plant material from an herbarium specimen. Removal of material may be needed to verify the identity of the specimen or to study some detail of the plant, e.g., anatomy or palynology. Even DNA may be successfully extracted from dried plant fragments.

You must *always get permission* from the herbarium curator before removing any material. Once permission is obtained, first see if the material you need is contained in an envelope attached to the sheet. If not, you will need to remove a piece from the plant specimen that is affixed to the sheet. Be very careful and conscientious doing this, trying to minimize the damage done to the specimen. Generally, material may simply be clipped, cut, or pulled off with forceps. In some cases the material that you need will be directly glued to the sheet. This may be difficult to remove, requiring the use of a razor blade to gently cut under the specimen (but above the paper). Plant material that is glued with methyl cellulose is easily removed by adding a few drops of water to soften the adhesive. Dried material can be reconstituted in boiling water and/or a detergent solution (such as Aerosol OT). It may then be observed and dissected in water or fixed in a liquid preservative for long-term storage.

Annotation labels should be used to document the removal of plant material. The person, institute, date, and reason for removing material should be indicated. Additional information indicates the type or purpose of the study, e.g., for anatomical, morphological, palynological (pollen), embryological, ultrastructural, or molecular analysis (Figure 18.5C).

REFILING HERBARIUM SPECIMENS

At some herbaria the staff do all specimen refiling; others allow (and expect) the user to refile anything removed. Generally, if a genus folder was removed for a short period of time and *not* placed in proximity to plant debris, it may be refiled immediately into the collection. However, if possible insect contamination is suspected, the genus folder and its contents should be treated for insect control (see later discussion).

If you are refiling a folder yourself, you should be extremely careful to file it in the correct location, by both taxonomic category and geographic region (if color-coded).

HERBARIUM LOANS

Those doing research on a plant group do not generally need to visit herbaria to examine the specimens. Typically, herbarium specimens may be sent out (via standard mail) on loan. Loans are typically granted only to members of universities, museums, or other research institutions. A request for a loan requires a letter to the curator, justifying the research needs for examining the specimens. The period of a loan is often 6 months, but this varies at different herbaria and may be extended upon request and approval.

INSECT CONTROL

An essential component of maintenance of herbarium collections is insect control. If herbarium specimens are kept dry and free of insects, they may be preserved in good condition for hundreds, if not thousands, of years. However, if insects infect a specimen, it may quickly be reduced to rubble.

There are two general ways to control insects: by chemicals and by freezing. Chemical control generally involves placing a volatile chemical insecticide within each sealed cabinet. Moth balls or moth chips have been used in the past, but these have been shown to be extremely dangerous to people. Other types of insecticides include insecticide "strips," which can be placed directly into a cabinet and which may last for 3–6 months.

Another method of insect control is freezing. Genus folders are periodically placed in a freezer for 3 to 7 days at at least −20°C before being refiled back into the herbarium cabinet. The advantage of freezing is that it eliminates toxic fumes, which could cause health problems to those working in the herbarium. The disadvantage of freezing is that it is more labor-intensive and potentially may result in greater damage to the specimens because of the regular removal and refiling required.

FLORAS AND MONOGRAPHS

Herbaria are particularly essential in two important activities in plant systematics: floristics and monographic treatments. **Floristics** is the documentation of all plant species in a given geographic region. Floristics may also entail documentation of plant communities and abiotic factors as well. Floristic studies may be published in taxonomic journals or may result in the publication of a **flora** or plant **manual** of a given region, such as *Flora of North America*. Floristic studies are vital in the documentation of plant biodiversity.

A **monograph** is a detailed taxonomic study of all species and infraspecific taxa of a given taxonomic group, generally a genus or family. Unlike floristic studies, the goal of which is to document taxa for a given area, monographic treatments focus on a particular taxonomic group, over its entire geographic range. For example, see the Systematic Botany Monographic series (American Society of Plant Taxonomy, *http://www.sysbot.org*).

Taxonomic Number
Group
Family
Genus
Specific Epithet
Species Author
Infraspecific Rank
Infraspecific Name
Infraspecific Author
Country
State/Province
County
Municipality
Locality
Latitude
Longitude
Elevation
Date of Collection
Collector
Collector Number
Associate Collectors
Habitat
Substrate
Plant Description
Phenology
Accession Number
Determination
Determiner
Determination Date

Figure 18.6 Example of data fields of a computerized plant inventory data information system.

HERBARIUM DATABASES AND THEIR IMPORTANCE

A **database** (or **data information**) system refers to the organization, inputting, and accessing of information. The accumulated information (known as the database) from herbarium collections may subsequently be "mined" to address a number of questions in plant taxonomy and systematics. It is important that students today be trained in the basics of accessing and manipulating information available from these systems.

Digitization is a now popular term for the electronic recording of data, including both word descriptors and images. All data information systems utilize computer hardware and software to store and access the information. Word descriptors are classified in discrete **fields**. For example, for an herbarium specimen, typical fields might be species name; authorship; flower color; phenology; soil type; topography; community type; specific locality; latitude; longitude; collector; collection number; determiner; accession number; annotation information; etc. (see Figure 18.6). Basically all of the discrete items recorded at the time of collection or as part of accessioning or identifying the plant may be entered into the database.

One problem with data information systems resides not with the system itself but with the collections. Many herbarium specimens lack much of the critical information needed for a particular question. For example, information about plant characteristics, phenology, ecology, or latititude/longitude is often not recorded on labels. In fact, on many older "historical" specimens, locality information may be very scanty, corresponding to a very broad or ill-defined region. Thus, depending on the quality of the collection, the amount of useful information obtained from herbarium specimens may be quite limited. This drawback only emphasizes the need to continue to collect specimens of a given region and rigorously record associated data (see Whitfield 2012).

Database systems may aid in the day-to-day organization of herbarium operations. For example, inputting accession numbers, which are now commonly scanned with bar codes, may automatically keep track of both outgoing loans and incoming loan returns. In addition, it is now not unusual to electronically connect directly to the databases of a host herbarium, e.g., to annotate specimens, valuable in biogeographic or taxonomic/systematic studies.

Finally, a major emphasis today is to obtain high resolution, digitized images of plant herbarium specimens. These images are quite useful (see below). Although images cannot always replace physical specimens (e.g., for minute, diagnostic features requiring dissection), images do allow for instant access to a great number of morphological features.

Digitized herbarium specimens have a great number of (overlapping) uses. With greater development of and emphasis on database systems, this information in plant collections has begun to be fully utilized. (See Soltis 2017 and Soltis et al. 2018 for summaries.)

First, digitized specimen data are used in basic systematic and taxonomic research. Herbarium images may be examined by experts to check identifications and annotate them remotely, essential to maintaining accuracy in our determinations. Researchers of a particular group can use specimen images to recognize that a taxon is new to science (Bowdler 2010), or to re-discover a taxon previously thought to be extinct (Simpson et al. 2013). Specimen images also allow for physical measurements, which may provide evidence for description of a new taxon from quantitative comparisons (e.g., Jiménez-Mejías et al. 2017). And, of course, specimen digitization ameliorates the use of material from physical specimens for molecular phylogenetic or population genetic studies (e.g., Hart et al. 2016).

Second, these data are used in biogeographic studies to map the distribution of species. There has been an emphasis to **georeference** (determine latitude and longitude, along with an estimated error radius) plant specimen collections for this purpose. More detailed distribution maps can lead to the recognition of range outliers, which could be the result of a misidentification (making checking the identity of the outlier a priority), a previously unrecognized range expansion, or possibly even the recognition of a new taxon. (See Lavoie 2013.)

Third, digitized data can be a useful tool in evaluating the biodiversity of a region and the conservation needs of given taxa. A list or map of all plant species within a region gives a quantitative measure of species diversity. More sophisticated techniques can use distribution data, along with molecular phylogenetic studies, to evaluate regions of endemism (e.g., neo-endemism versus paleoendemism, as in Kraft et al. 2010), providing data for decisions on conservation of various regions. Also, collection data can be used to determine the rarity and distribution of a single taxon, in evaluating the need to list it, e.g., as threatened or endangered.

Fourth, digitized specimen data may be used to map and evaluate specific habitat features of taxa from their distribution data. These data can be used in ecological niche analysis, which may evaluate, e.g., expansion of contraction of plant colonization as a function of climatic shifts (Villaverde et al. 2017).

Fifth, digitized data, especially specimen images, are a tremendous source of information for assessing changes in phenology over time. However, phenological data (reproductive timing, such as time of flowering or fruiting) are often not recorded at the time of collection. Specimen images may be used to fill in these phenological data in a quite precise way (Miller-Rushing et al. 2006; Yost et al. 2018; Brenskelle et al. 2019; Ellwood et al. 2019a,b; Lorieul et al. 2019; Pearson 2019a,b). Robust phenological data can be used to evaluate long-term climatic changes, e.g., as the result of global warming. This has been done for some regions (e.g., Houle 2007), with more studies to come.

Sixth, perhaps one of the most valuable used of digitized biological collections in general is in education and public awareness (Cook et al. 2016). **Citizen science** has emerged as a means of engaging the public in scientific research. For example, with proper training, "citizen scientists" can aid in data entry, georeferencing, and phenological scoring of plant specimens, which can be largely done online. Students, from grade school to graduate school, can engage in digitization projects, encompassing all of the use of specimens cited above. This involvement of students and non-scientist members of the public has the beneficial side product of promoting an increased awareness of plant biology and evolution. Such involvement may even promote the importance of biological collections in understanding human culture (Ickert-Bond 2017).

Herbarium collections are tremendous resources for science and education, and for understanding the natural world. But, we in the field must continually engage (voting) members of the general community about the importance of biological collections, and persevere in the fight for resources to maintain them (Kemp 2015).

REVIEW QUESTIONS

1. What is an herbarium?
2. What is the function or purpose of herbaria?
3. What is the name and most recent version of the reference book that lists the names, acronyms, and details of herbaria worldwide?
4. What is an herbarium specimen?
5. What are the characteristics of an herbarium specimen, including the standard size (in the United States)?
6. Describe a standard format and list the information that is contained in an herbarium label.
7. If a plant specimen is divided among two or more herbarium sheets, how is the herbarium label written?
8. Describe the procedure for mounting plants on herbarium paper.
9. What are two type of glues used for the above? How do they differ?
10. What is an herbarium curator and what are his/her duties?
11. What is an accession number and what is its function?
12. What is a genus folder?
13. Why are many genus folders color-coded?
14. How are herbarium specimens typically stored?
15. Review the procedures for handling herbarium specimens.
16. What is an annotation label?
17. What are the different types or designations on annotation labels?
18. Review the procedures for removing material from an herbarium specimen.
19. How may removed material be reconstituted for observations?
20. What is an herbarium loan and what is its purpose?
21. How may insects be controlled in herbaria?
22. What is the difference between a flora (or floristic study) and a monograph, and what are examples of each?
23. What is a database (data information) system?
24. What types of data are "digitized" in data information systems?
25. How might data information systems be valuable in: (a) taxonomic and systematic research; (b) biogeographic studies; (c) biodiversity studies and conservation biology; (d) ecological studies; (e) phenological studies?
26. What is citizen science, and how is this valuable with respect to digitization of herbarium plant collections?

EXERCISES

1. Obtain a list of plant species from your instructor, including a lycopod, fern, conifer, monocot, and eudicot. Become familiar with the system of classification of your herbarium (or a herbarium that you visit). Locate and remove the genus folder for a species. Carefully transfer the genus folder to an open-space table, remove one specimen, and study it. (Be sure to handle the herbarium sheets correctly.) Note how the specimen is attached to the herbarium paper. Also note the label, accession number (possibly on a bar code), and annotation labels (if present). Write down the collector, collection number, date of collection, herbarium acronym (see *Index Herbariorum*), and accession number (e.g., *Smith 762*, 23 Oct. 2013; NY 1120387). Do this for each of the species on the list. When finished with each specimen, replace it in correct order (usually alphabetical by species) within the genus folder. If permitted, refile the genus folder in its correct location in the herbarium cabinet, both by taxonomy and (if used) color coding by region. [Caution: Be sure to check yourself carefully! If misfiled, specimens could be lost for some time.]
2. Check Index Herbariorum (Thiers, continuously updated; see below) for several large herbaria (e.g., K, MO, NY, P, UC, US) and determine (a) the curator's name, address, e-mail; (b) the number of specimens currently accessioned; (c) the general types of collections in the herbarium; (d) number of type specimens in the herbarium.
3. Obtain and study an example of a flora and one of a monograph. Copy an example entry from each, indicating how they differ.
4. Access at least two plant specimen databases (e.g., the Field Museum herbarium) by doing a website search or using *Index Herbariorum*. Look up one species and tabulate the data obtained by downloading it to a spreadsheet. How do the two sources differ with respect to database fields?
5. Do a search on a particular plant database (e.g., SEINet) for two species of the same genus. Generate a map for all collection records of each of the species. Do you see any overlap in range of the two species? Is there any correlation of either or both of the species with region, elevation, or habitat?

REFERENCES FOR FURTHER STUDY

Allkin, R. and F. A., Bisby (eds.). 1984. Databases in Systematics. Academic Press, London. Systematics Association special volume, no. 26. Proceedings of an International Symposium held in Southampton, UK.

Bowdler, N. 2010. Thousands of plant species 'undiscovered in cupboards'. BBC News Science & Environment. http://www.bbc.co.uk/news/science-environment-11913076. Accessed on 7 December, 2010.

Brenskelle, L., B. J. Stucky, J. Deck, R. Walls, and R. P. Guralnick. 2019. Integrating herbarium specimen observations into global phenology data systems. Applications in Plant Sciences 7:e01231.

Cook, J. A., S. V. Edwards, E. A. Lacey, R. P. Guralnick, P. S. Soltis, D. E. Soltis, C. K. Welch, K. C. Bell, K. E. Galbreath, C. Himes, J. M. Allen, T. A. Heath, A. C. Carnaval, K. L. Cooper, M. Liu, J. Hanken, and S. Ickert-Bond. 2014. Natural history collections as emerging resources for innovative education. Bioscience 64:725-734.

Ellwood, E. R., K. D. Pearson, and G. Nelson. 2019. Emerging frontiers in phenological research. Applications in Plant Sciences 7:e01234.

Ellwood, E. R., R. B. Primack, C. G. Willis, and J. HilleRisLambers. 2019. Phenology models using herbarium specimens are only slightly improved by using finer-scale stages of reproduction. Applications in Plant Sciences 7:e01225.

Forman, L., and D. M., Bridson (eds.). 1992. The Herbarium Handbook, rev. ed. Royal Botanic Gardens, Kew, UK.

Funk, V. A. 2006. Floras: A model for biodiversity studies or a thing of the past? Taxon 55: 581-588.

Hart , M. L., L. L. Forrest, J. A. Nicholls, and C. A. Kidner. 2016. Retrieval of hundreds of nuclear loci from herbarium specimens. Taxon 65:1081-1092.

Houle, G. 2007. Spring-flowering herbaceous plant species of the deciduous forests of eastern Canada and 20th century climate warming. Canadian Journal of Forest Research 37:505-551.

Ickert-Bond, S. M. 2017. Arctic Museum collections: Documenting and understanding changes in biological and cultural diversity through time and space. Arctic Science 3:i–ii.

Jiménez-Mejías, P., J. I. Cohen, and R. F. C. Naczi. 2017. The study of online digitized specimens revalidates Andersonglossum boreale as a species different from A. virginianum (Boraginaceae). Phytotaxa 295:22-34.

Kemp, C. 2015. The endangered dead: The billions of specimens in natural-history museums are becoming more useful for tracking Earth's shrinking biodiversity, but the collections also face grave threats. Nature 518:292-294.

Kraft, N. J. B., B. G. Baldwin, and D. D. Ackerly. 2010. Range size, taxon age and hotspots of neoendemism in the California flora. Diversity and Distributions 16:403-413.

Kress, W.J. and Funk, V.A. 2005. Herbarium, floras, and checklists. In: Krupnick, G. A. and W. J. Kress (eds.). Plant Conservation: A Natural History Approach. Pp. 209-217. University of Chicago Press, Chicago.

Lavoie, C. 2013. Biological collections in an ever changing world: Herbaria as tools for biogeographical and environmental studies. Perspectives in Plant Ecology, Evolution and Systematics 15:68-76.

Lorieul, T., K. D. Pearson, E. R. Ellwood, H. Goëau, J.-F. Molino, P. W. Sweeney, J. M. Yost, J. Sachs, E. Mata-Montero, G. Nelson, P. S. Soltis, P. Bonnet, and A. Joly. 2019. Toward a large-scale and deep phenological stage annotation of herbarium specimens: Case studies from temperate, tropical, and equatorial floras. Applications in Plant Sciences 7:e01233.

Miller-Rushing, A. J., R. B. Primack, D. Primack, and S. Mukunda. 2006. Photographs and herbarium specimens as tools to document phenological changes in response to global warming. American Journal of Botany 93:1667-1674.

Pearson, K. D. 2019a. A new method and insights for estimating phenological events from herbarium specimens. Applications in Plant Sciences 7:e01224.

Pearson, K. D. 2019b. Spring- and fall-flowering species show diverging phenological responses to climate in the Southeast USA. International Journal of Biometeorology 63:481-492.

Rabeler, R. K., H. T. Svoboda, B. Thiers, A. Prather, J. A. Macklin, L. P. Lagomarsino, L. C. Majure, and C. J. Ferguson. 2019. Herbarium practices and ethics, III. Systematic Botany 44:7-13.

Radford, A. E., W. C. Dickison, J. R. Massey, and C. R. Bell. 1974. Vascular Plant Systematics. Harper & Row, New York.

Simpson, M. G., J. P. Rebman, K. E. Hasenstab-Lehman, C. M. Guilliams, and P. O. McConnell. 2013. Cryptantha wigginsii (Boraginaceae): A presumed extinct species rediscovered. Madroño 60:24-34.

Soltis , D. E. and P. S. Soltis. 2016. Mobilizing and integrating big data in studies of spatial and phylogenetic patterns of biodiversity. Plant Diversity 38:264-270.

Soltis, P. 2017. Digitization of herbaria enables novel research. American Journal of Botany 104:1281-1284.

Soltis, P. S., G. Nelson, and S. A. James. 2018. Green digitization: Online botanical collections data answering real-world questions. Applications in Plant Sciences 6:e1028.

Takano, A., Y. Horiuchi, Y. Fujimoto, K. Aoki, H. Mitsuhashi, and A. Takahashi. 2019. Simple but long-lasting: A specimen imaging method applicable for small- and medium-sized herbaria. PhytoKeys 188:1-14.

Villaverde, T., P. González-Moreno, F. Rodríguez-Sánchez, and M. Escudero. 2017b. Niche shifts after long-distance dispersal events in bipolar sedges (Carex, Cyperaceae). American Journal of Botany 104:1765-1774.

Whitfield, J. 2012. Rare specimens: A handful of plant collectors has shaped the field of botany. Now they are disappearing, and there are no clear successors. Nature 484:436-438.

Yost, J. M., P. W. Sweeney, E. Gilbert, G. Nelson, R. Guralnick, A. S. Gallinat, E. R. Ellwood, N. Rossington, C. G. Willis, S. D. Blum, R. L. Walls, E. M. Haston, M. W. Denslow, C. M. Zohner, A. B. Morris, B. J. Stucky, J. R. Carter, D. G. Baxter, K. Bolmgren, E. G. Denny, E. Dean, K. D. Pearson, C. C. Davis, B. D. Mishler, P. S. Soltis, and S. J. Mazer. 2018. Digitization protocol for scoring reproductive phenology from herbarium specimens of seed plants. Applications in Plant Sciences 6:e1022.

WEBSITES

ALA. 2019. Atlas of Living Australia. *https://www.ala.org.au*

GBIF. 2017. Global Biodiversity Information Facility. *http://www.gbif.org*

Global Plants. 2019. *https://plants.jstor.org*

iDigBio. 2019. Integrated Digitized Biocollections. *https://www.idigbio.org*

Thiers, B. [continuously updated]. Index Herbariorum: A global directory of public herbaria and associated staff. New York Botanical Garden's Virtual Herbarium. *http://sycamore.nybg.org/science/ih*

V

SPECIES CONCEPTS AND CONSERVATION BIOLOGY

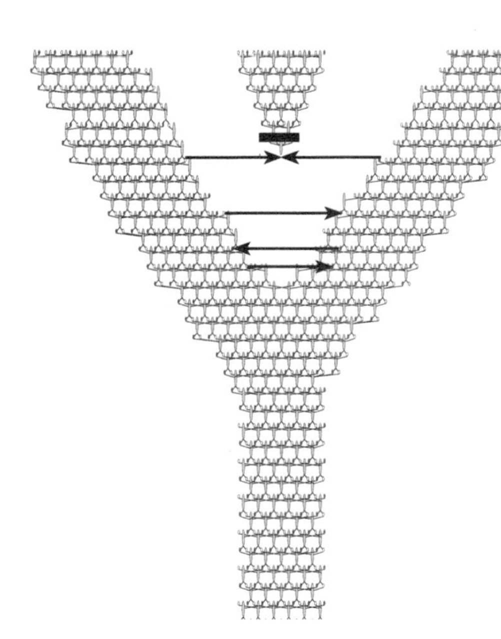

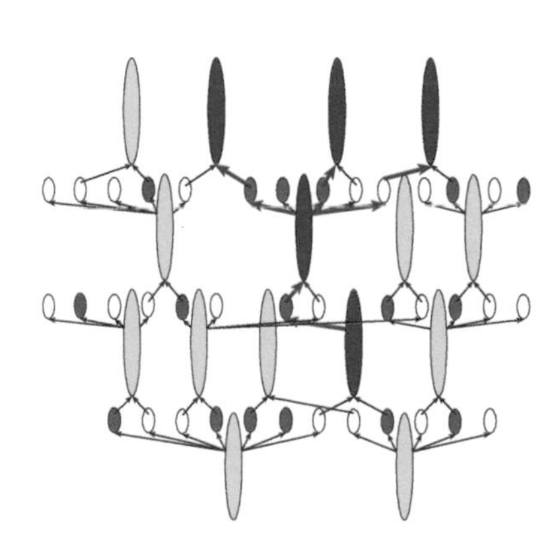

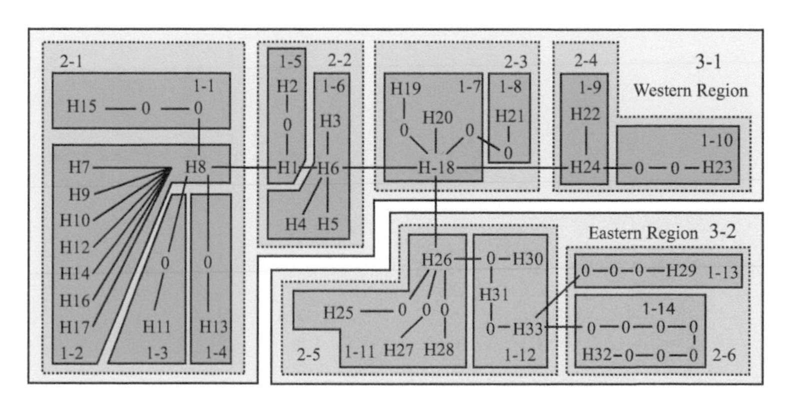

19

SPECIES AND CONSERVATION IN PLANT SYSTEMATICS

INTRODUCTION

Over the years, biologists have proposed different defi One of the most intractable definitions in biology is that of the species. Species may be defined in different ways and using different criteria, and the species "concept" used may depend on the particular group studied or on a particular goal. However, it is important to precisely define the basis or concept of a species and infraspecies in any systematic or taxonomic work (see Hamilton and Reichard 1992; McDade 1995; and Baum 2009).

What is a species? This question has generated and continues to foster considerable discussion and debate among biologists, from systematists concerned with the units and classification of life to geneticists concerned with the evolutionary processes of population divergence. The idea of a "species" (which in Latin literally means a "kind" or "type," from the Latin *specere*, "to look at, see") is an ancient human construct and obviously related to the attempt to categorize the vastness of life into manageable units. The ability to recognize, describe, name, and classify these units has had and continues to have practical, or even survival, value.

The International Code of Nomenclature for algae, fungi, and plants (Turland et al. 2018) defines a species as a taxon of the "basic" rank. But, are species "real" units of nature, or arbitrary human concepts? Is there an intrinsic difference between species and groups of populations? Several so-called species concepts have been formulated over the years, with different or overlapping criteria used to define them (see later discussion). The suggestion made here is not to decide on some given species concept and force it upon natural systems, but, foremost, to look at what happens in nature and to use these examples as a guide in assessing how to usefully define the species unit. It has also been pointed out that different species concepts have different emphases, some concerned with understanding the processes of evolution, others with defining and grouping the end products of evolution. This utilitarian approach might necessitate using different species concepts or definitions for different natural systems or for differing degrees of knowledge about what happens in nature.

Systematics plays a vital role in conservation biology in precisely determining the limits of species and infraspecies; only if these taxonomic entities are clearly defined can they be evaluated for rarity and the threat of extinction. In addition, floristic studies and phylogenetic analyses may have an impact on which species or biogeographic regions are most worthy of protection, given limited resources.

https://doi.org/10.1016/B978-0-12-812628-8.50019-5,

REPRODUCTION, LINEAGES, AND CLADES

SEXUAL AND ASEXUAL REPRODUCTION

In order to better understand species concepts, a review of plant reproduction is valuable. Plants, as well as many other eukaryotes, can reproduce both sexually and asexually. This entails **gene flow**, considered narrowly here as the transfer of the genetic material from parent to offspring.

Sexual reproduction in all land plants involves two "phases" of a haplodiplontic life cycle (Chapter 3). Haploid gametophytes produce egg and sperm, which fuse and develop into a diploid sporophyte. Cells of the sporangia produced by the diploid sporophyte undergo meiosis, generating (by recombination and independent assortment) genetically different haploid spores, which develop into gametophytes.

In non-vascular land plants (liverworts, mosses, and hornworts), the gametophyte is generally long-lived and can produce multiple "generations" of sporophytes, which are relatively short-lived (see Chapter 3). Each gametophyte can produce egg and/or sperm cells, some of which (from the same or different gametophytes) may fuse, producing a diploid zygote that differentiates into a new sporophyte (represented diagramatically in Figure 19.1A). In the vascular plants the reverse is true; the sporophyte is generally long-lived and the gametophyte of short duration (Figure 19.1B; see Chapter 4). The generally long-lived sporophyte of vascular plants can produce multiple generations of short-lived gametophytes; one or two gametophytes from the same generation may give rise to another long-lived sporophyte (Figure 19.1B).

There are some differences in the life cycles of the vascular plants. Within most *non-seed* vascular plants (most lycophytes and monilophytes), the sporophyte is generally long-lived and homosporous; only one type of gametophyte is formed, producing both egg and sperm (Figure 19.1C). In contrast, the Isoetales of the lycophytes, the Salviniales of the monilophytes, and all seed plants are heterosporous, producing separate male and female gametophytes, each of which produce *only* sperm or eggs, respectively (Figure 19.1D). The great majority of seed plants, including most angiosperms (but excepting the cycads, *Ginkgo*, Gnetales, and some conifers) have sporophytic individuals that can give rise to *both* male and female gametophytes (i.e., their plant sex is either hermaphroditic or monoecious). A twist to the general seed plant life cycle is found in many angiosperms with an annual plant duration (Chapter 9). In this case, gametophytes are short-lived (Figure 19.1E), but the new sporophytes are also relatively short-lived, mostly persisting less than a year (although in some cases remaining dormant over many years in the soil bank). Thus, a land plant lineage is a little complicated, with gene flow occurring via both gametophytes and sporophytes (Figure 19.1F). Variation in the lifetime of the gametophyte, the lifetime of the sporophyte, spore output, egg and sperm output, gametophyte sex, sporophyte sex, and general viability of stages are all contributing factors of gene flow.

Asexual reproduction often occurs in land plants. Asexual reproduction is simply the transfer of an exact copy of DNA (bearing the lack of a somatic mutation) from a parent to an offspring, the latter viewed as a separate, independent organism. The gametophytes of liverworts, mosses, and hornworts can reproduce asexually by breaking apart and forming separate individuals or by means of specialized asexual propagules such as gemmae (Chapter 3; see Figure 19.1A). In the vascular plants, sporophytes can reproduce asexually (Figure 19.1B) by means of rootstocks (e.g., bulbs, corms, rhizomes) that can become detached and separated from the parent plant or stolons that end in a pre-formed plantlet. Asexual reproduction can also occur by formation and release of aerial bulbs, termed bulbils (e.g., *Allium* spp.) or aerial plantlets (e.g., *Agave* or *Kalanchoe* spp.). In addition, sometimes seeds develop asexually by apomixis (see Chapter 13). Many land plants can and do reproduce both sexually and asexually (Figure 19.1A,B). The latter is considered a more fail-safe means of propagating to the next generation, not requiring, e.g., the sometimes precarious and uncertain processes of gamete transfer (including pollination) and fertilization. However, there are a few plants that reproduce exclusively by asexual means (Figure 19.1G), whether by vegetative propagules (e.g., the aerial sporophytic bulbils of *Poa bulbosa*), by seed (e.g., the apomictic, sporophytic, clonal seeds of dandelion, *Taraxacum officinale*), or by gemmae (e.g., the persistent gametophytes of some ferns, e.g., *Vittaria*).

LINEAGES AND CLADES

The transfer of the genetic material from parent to offspring over time (generation to generation) constitutes descent. A **lineage** can be defined as a sequence of ancestral-descendent populations, in which the members are linked or connected by gene flow (Figure 19.2A). **Tokogenetic** relationships are those between individual organisms with regard to this gene flow in ancestral-descendent lineages (Figure 19.1, 19.2). In sexual systems, tokogenetic relationships are "reticulate," in which gene transmission to any offspring (diploid sporophytes in the case of land plants) always comes from two sources, involving multiple lines of gene flow in a lineage (Figure 19.1F). In asexual systems, tokogenetic relationships are "divergent," in which gene flow comes from one source; a single line of gene transmission occurs between two individual organisms (Figure 19.1G). **Evolutionary divergence** is the splitting of one lineage into two, producing two new lineage segments from one (Figure 19.2B). **Phylogenetic** relationships refer to

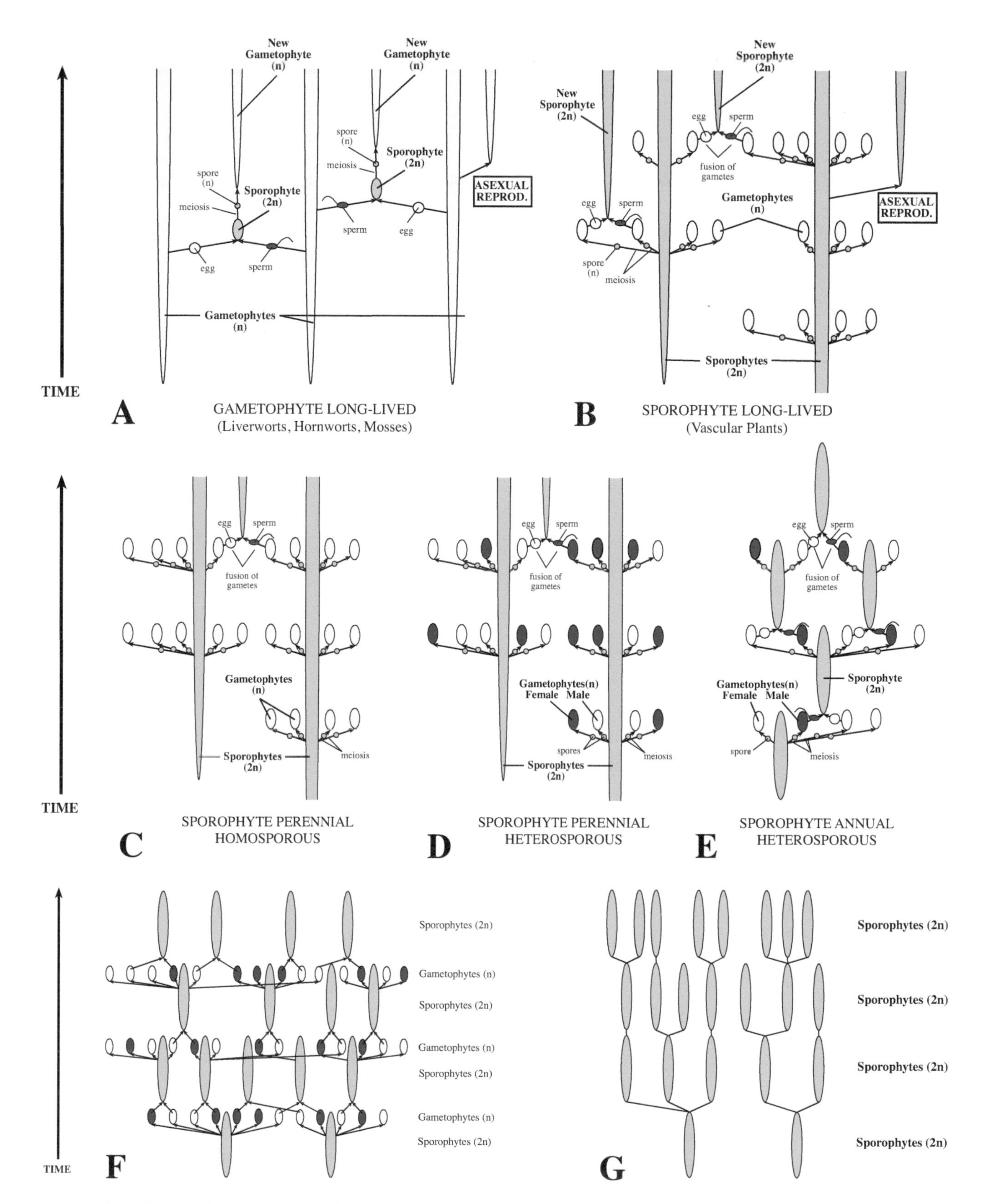

Figure 19.1 Gene flow in the land plants. **A.** Non-vascular land plants, with long-lived gametophytes and short-lived sporophytes. **B.** Vascular land plants, with long-lived sporophytes and short-lived gametophytes. **C–E.** Gene flow in the vascular plants. **C.** Non-seed vascular plants, homosporous. **D.** Seed plants, heterosporous. **E.** Annual flowering plants, with short-lived sporophytes. **F.** Simplified diagram of gene flow in a heterosporous land plant with short-lived (annual) sporophytes (spores and gametes omitted). **G.** Simplified diagram of gene flow in a land plant with short-lived (annual) sporophytes, reproducing exclusively asexually.

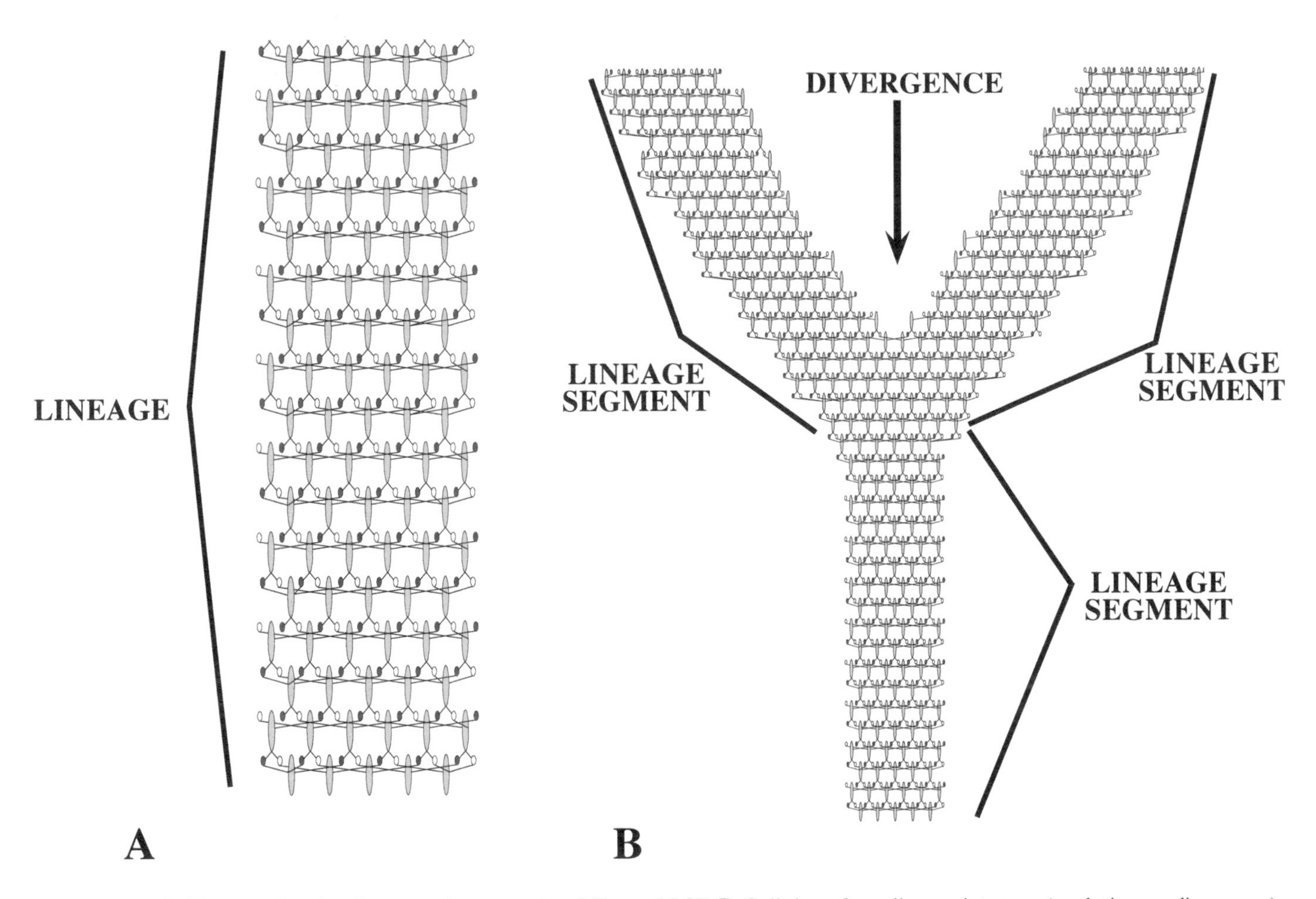

Figure 19.2 **A.** Diagram showing lineage, using example of Figure 19.2F. **B.** Splitting of one lineage into two (evolutionary divergence), showing lineage segments.

those between groups of individuals occurring at a given point in time.

Lineages can be envisioned as a continuous transfer of the pattern of DNA from parent to offspring, ultimately from the origin of life to all extant (or extinct) species (example in Figure 19.3A). The term **lineage segment** can be used for a portion of a lineage, e.g., from one divergence point (or "node" of a cladogram) to another (Figures 19.2B, 19.3B). A **clade** is equivalent to a monophyletic group, consisting of the lineages arising from any given common ancestor (Figure 19.3C; see Chapter 2).

The tokogenetic history of an allele or gene copy, from a particular gene, is termed a **gene genealogy** or **gene lineage**. Figure 19.3D,E illustrates a gene lineage showing the transfer of a unique mutation from parent to offspring in sexual and asexual systems, respectively. Individual gene trees, as derived from phylogenetic methods (such as parsimony or maximum likelihood) can differ from the best estimate of the cladogram depicting taxa relationships, or **species tree**.

Lineage sorting refers to the process by which, following evolutionary divergence, several gene lineages inherited from an ancestor converge (are reduced to) to a single gene lineage within a given lineage segment. Lineage sorting can occur by the random extinction of all but one of these ancestral gene lineages, or by strong selection that favors a single gene lineage. Complete lineage sorting results in terminal taxa that are monophyletic, with all individuals of a taxon at a given point in time having a common ancestor more recent in time that the common ancestor with any other taxa. For example, Figure 19.4A shows a clade in which lineage sorting has completed at time t_2. **Incomplete lineage sorting** occurs when a lineage segment at a given point in time contains more than one gene lineage. In this same example (Figure 19.4A) each lineage segment contains more than one gene lineage at time t_1. At this time, none of the lineages have sorted completely and the individuals of each lineage segment at time t_1 do not comprise a monophyletic group. Incomplete lineage sorting is more likely to occur soon after evolutionary divergence and is influenced by population size and life cycle type.

If ancestral polymorphisms of a gene occur in a clade,

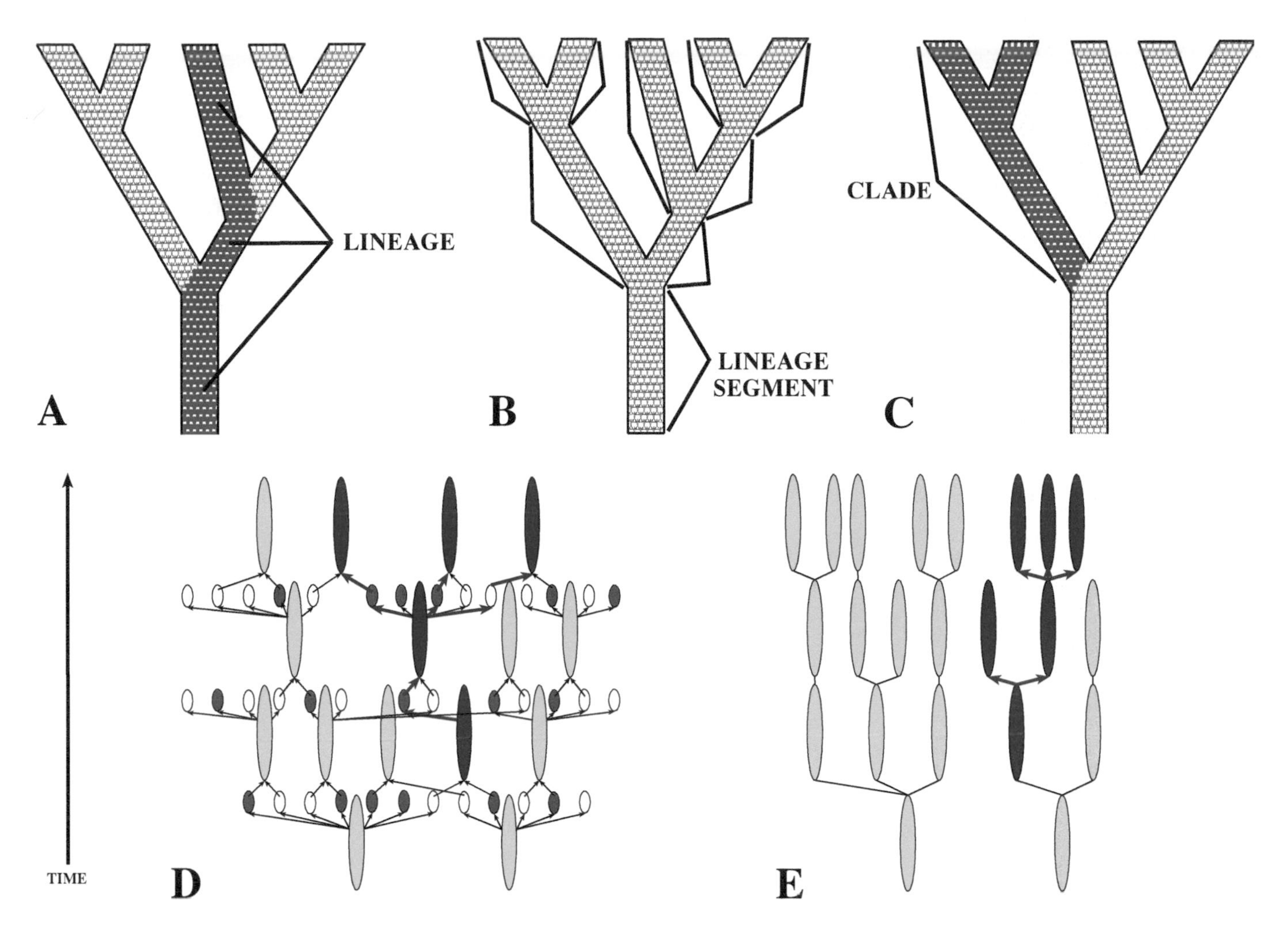

Figure 19.3 **A.** Diagram showing a continuous lineage within a hypothetical phylogenetic tree. **B.** Designation of lineage segments. **C.** One of several clades (monophyletic groups) of same phylogenetic tree. **D–E.** Gene flow of derived mutation (dark blue) in sporophytes of a sexual (**D**) and asexual (**E**) system.

then gene lineages may not correspond to species lineages. For example, Figure 19.4B shows the same cladogram (of Figure 19.4A), in which a mutation in a gene lineage is transferred to taxa *B* and *C*. This mutation might be detected as an apomorphy, linking taxa *B* and *C* in a clade (inset). However, in the separate example of Figure 19.4C, the mutation is transferred to gene lineages that terminate in taxa *X* and *Y*. This apomorphy of the gene lineages would result in the reconstruction of a species tree linking taxa *X* and *Y*, which would differ from the true species tree, in which *Y* and *Z* are closest relatives.

INTROGRESSION, HYBRIDIZATION, AND POLYPLOIDY

After a lineage splits (e.g., due to some barrier between the new lineage segments), there may still occur limited gene flow from one lineage segment to the next, known as **introgression**. Introgression can occur by (e.g., in seed plants) by transfer of pollen, containing sperm, from one lineage to the ovule, containing an egg, of another (Figure 19.5A). Introgression may be particularly common soon after evolutionary divergence, as the new lineages may not have changed much genetically and the ability to sexually reproduce is an ancestral feature, often retained in both lineage segments. Over time, however, introgression between lineages may decrease in intensity, possibly due to the evolution of barriers to gene exchange (Figure 19.5A). However, there are numerous examples of plants in which occasional introgression can be detected, yet the species maintain their integrity (or cohesion; see below).

In addition to introgression, different plant species may occasionally undergo **hybridization**, sexual reproduction between two forms (e.g., separate populations, species, or infraspecies) that results in a lineage independent of either parent (Figure 19.5B). In other cases, hybrid populations are fertile and undergo subsequent gene flow with either or both parent lineages, an extension of introgression. Often, hybrids

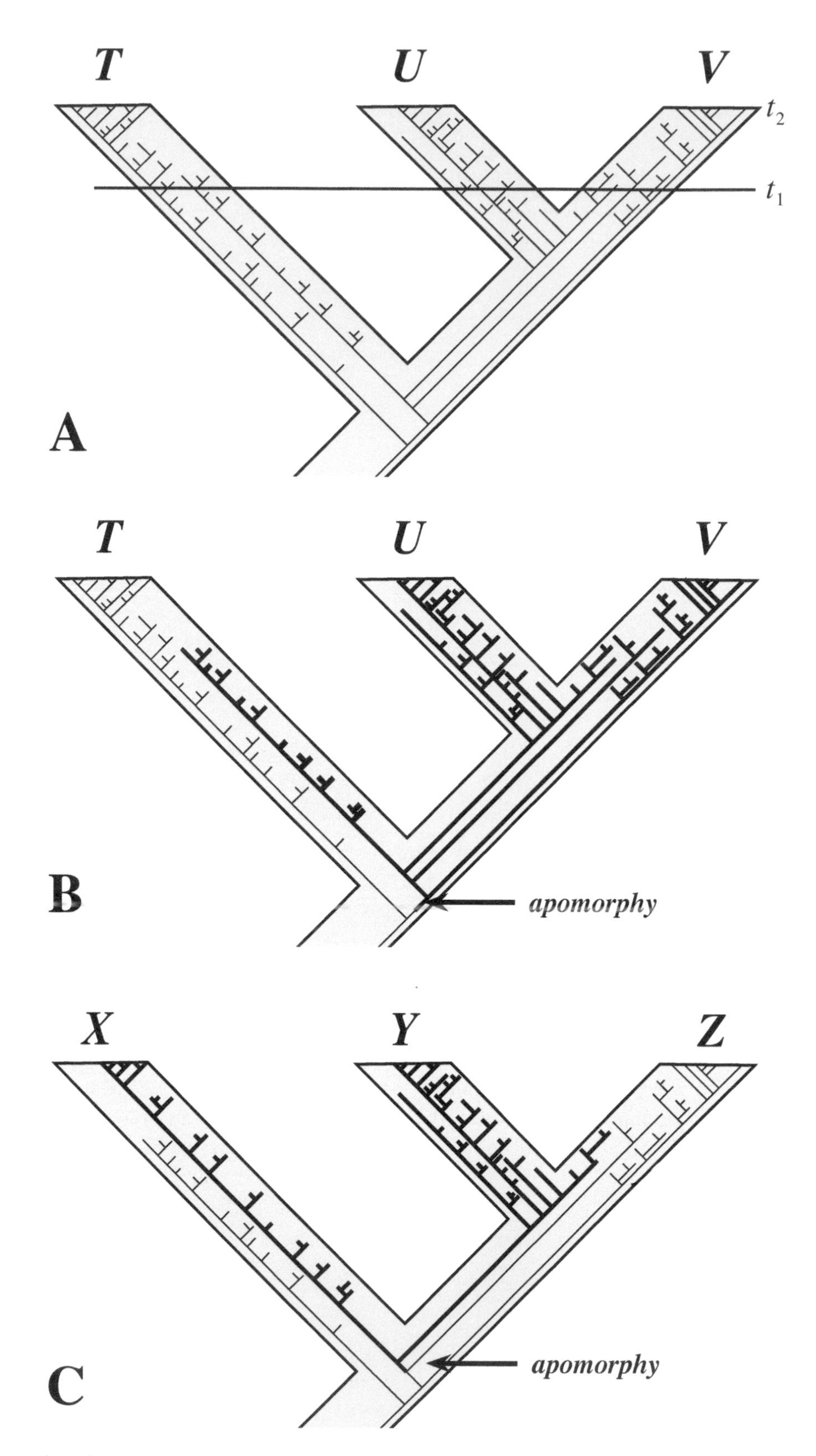

Figure 19.4 Lineage sorting. **A.** Note that gene lineages are not completely sorted at time t_1, but at time t_2, complete lineage sorting has occurred. **B.** Unique mutation (dark lines) is transmitted to all three lineage segments (*T–V*), but extinction of gene lineage in "*T*" make this mutation detectable as an apomorphy for taxa *U* and *V*. **C.** Novel mutation (dark line), transmitted to lineages *X* and *Y*, but not *Z*, representing incomplete lineage sorting. This feature, interpreted as an apomorphy, would give misleading results in a phylogenetic analysis.

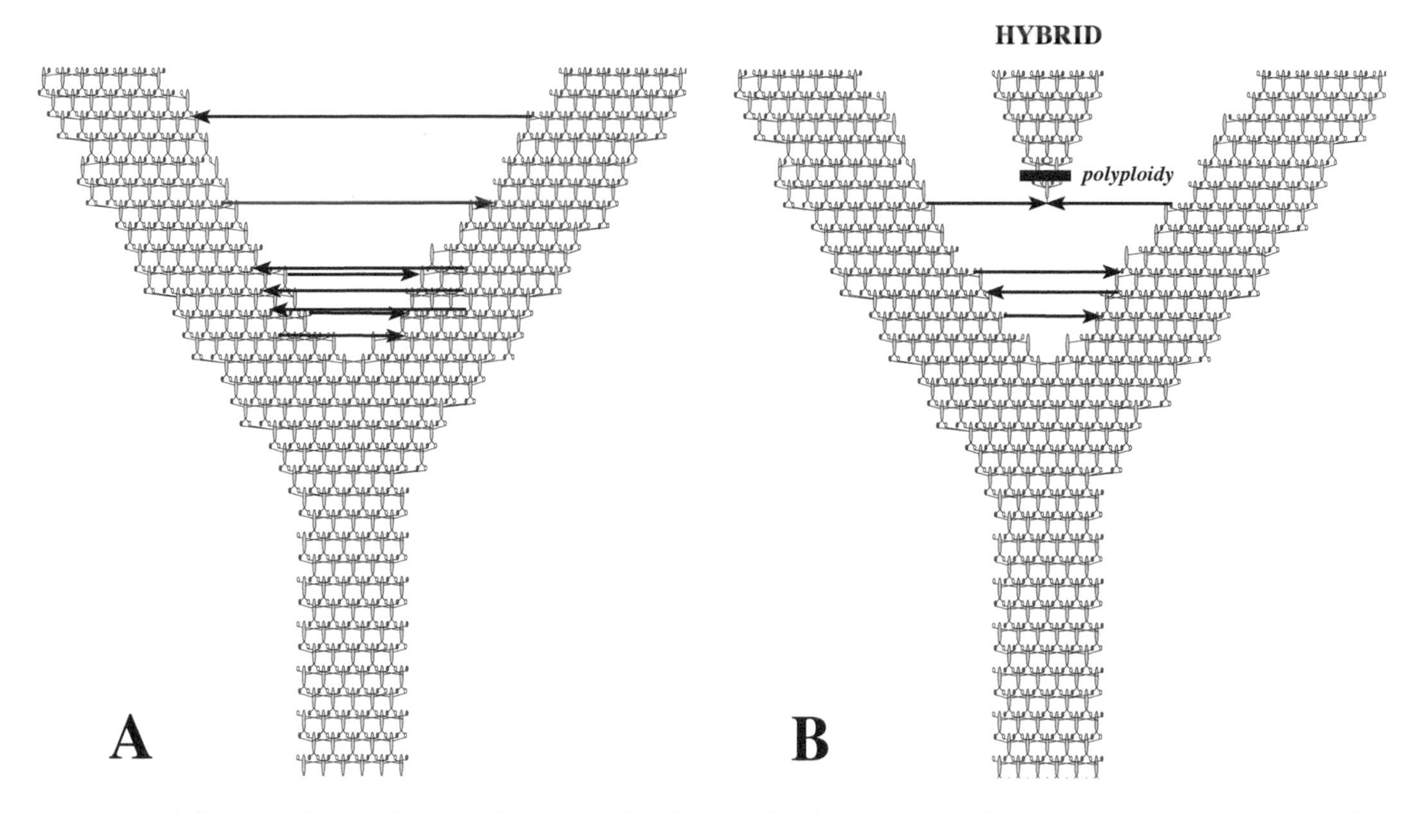

Figure 19.5 **A.** Divergent lineage, illustrating limited gene flow (introgression) between the two lineage segments, which decreases in frequency over time. **B.** Illustration of the evolution of a hybrid lineage from two parents of separate lineage segments. In this example, the hybrid lineage does not show introgression with the parental lineages due to the evolution of polyploidy, which prevents genetic exchange with the hybrids, maintaining a discrete hybrid lineage.

are effectively sterile because of genetic incompatibilities, e.g., the prevention of normal meiosis from taking place. In these cases, the evolution of **polyploidy** (Chapter 13), an increase in chromosome number by a multiple of some ancestral set (e.g., a doubling of the chromosome number) will allow normal meiosis to take place such that the hybrids are now fertile and can form a sexual lineage. Such polyploids, derived from an interspecific hybrid, would now be reproductively isolated from either parent species.

PHYLOGEOGRAPHY

Phylogeography is a field of study that attempts to tease apart relationships among individual genotypes within a species or within a group of very closely related species and correlate those relationships with their spatial distribution. The genealogical relationships inferred can be used to trace the biogeographic history of infraspecific populations as well as to address evolutionary questions such as gene flow, fragmentation, range expansion, and colonization. One way to address phylogeography is **nested clade analysis**, a methodology that reconstructs the genealogical relationships of individuals using **haplotypes**, unique alleles of a chromosome or organelle (mitochondrion or chloroplast). Sequence data from haplotypes of individual samples are connected in a **network** using a statistical algorithm. Nested clade analysis can identify clusters of genetically similar individuals and is useful for grouping them into subpopulations or subtaxa. This fine-scale phylogenetic pattern can then be correlated with geography, in what is called **nested clade phylogeographic analysis** (Figure 19.6). Hypotheses about the mechanism of evolution can be inferred from the nested clade pattern (Templeton et al. 1988; Templeton 1989, 1998, 2001, 2004). However, the statistical basis for these evolutionary inferences has been criticized (Knowles and Maddison 2002; Knowles 2008).

SPECIES CONCEPTS

Over the years, biologists have proposed different definitions or "concepts" of species, which are not necessarily mutually exclusive. As discussed earlier, the approach adopted here is to consider what may be present in nature and adapt a species definition to that natural system. A given system may

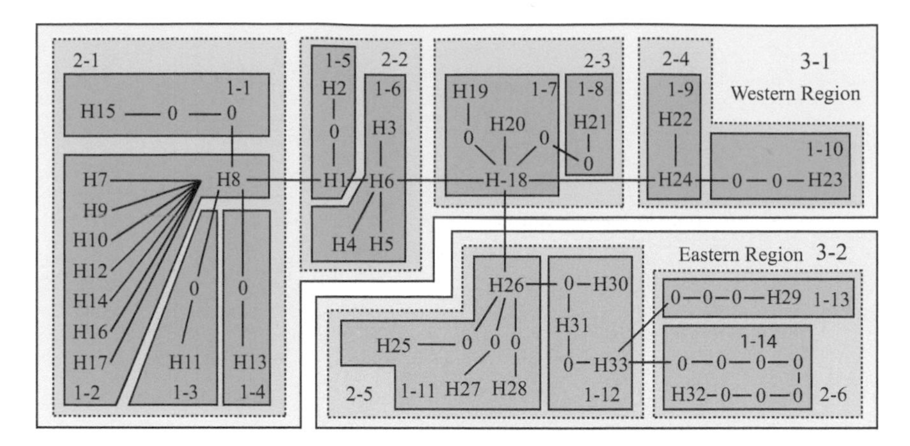

Figure 19.6 Example of nested clade analysis of *Hygrophila pogonocalyx* (Acanthaceae). Plant samples are indicated by "H" plus a number; a "0" represents a hypothetical ancestral or extinct population. Note nested subpopulations and correspondence of major groups with western and eastern regions of the study area. Redrawn, by permission, from Huang, J.-C., W.-K. Wang, C.-I. Peng, and T.-Y. Chiang. 2005. Phylogeography and conservation genetics of *Hygrophila pogonocalyx* (Acanthaceae) based on atpB–rbcL noncoding spacer cpDNA. Journal of Plant Research 118:1–11.

fit one or more species concepts but not another. Thus, it may be useful to designate one system by one species type and a second system by another species type.

Species concepts can be classified into two general groups. Several species concepts emphasize **processes** of evolution that maintain the species as a unit, as well as those that can result in evolutionary divergence and speciation. Other species concepts emphasize the **pattern** or end result of evolution in defining a species. Thus, different species concepts may have different aims. A review of these is insightful to understanding both what a species is and how they may be studied.

CONCEPTS EMPHASIZING PROCESS

Species concepts that emphasize evolutionary processes include the biological, recognition, evolutionary, and cohesion concepts. **Biological species** (also know as **isolation species**) are groups of populations that **interbreed** (actually or potentially) and that are **reproductively isolated** from other such groups in nature (Mayr 1963). Note that this definition specifies "in nature," because members of a species that normally do not interbreed in a natural system can be induced to do so in captivity. Also note that interbreeding is specified as "actually or potentially," meaning that if certain populations are not at present interbreeding because of geographical isolation, they would do so if brought together in the same region. An emphasis of the biological species concept is how different species have become reproductively isolated. A number of different types of **barriers to gene exchange** (often referred to as "**reproductive isolating mechanisms**") have been described (Table 19.1); these could be viewed as derived suites of characters in one or more lineages that inhibit or prevent interbreeding. Barriers to gene exchange are of interest to evolutionary biologists studying the process of speciation and the genetic mechanisms that help maintain species.

One practical problem with a biological/isolation species concept is that it requires a detailed knowledge of reproductive biology to assess; this is known for a relatively small fraction of plants. Second, biological/isolation species do not encompass the numerous examples of discrete evolutionary units in the natural world (described as "species") that reproduce asexually or that sexually reproduce but are highly self-mating. (In fact, there is a genetic grade between species that reproduce asexually, those that are highly self-mating, and those that are highly out-crossing.) Third, the biological/isolating species concept is faulty as an explanation of the process of species formation (i.e., speciation) in that these barriers to gene exchange may not necessarily have had anything to do with the formation of new species (speciation), but may just be by-products of lineage evolution.

Another issue with the biological/isolation species concept is that any type of gene flow between different species that results in viable progeny would collapse those populations into a single species. And yet, it is quite common for different plant species to occasionally hybridize or exchange genes. (A term for an entire group of populations or "species" that exchange at least some genetic material is a "syngameon.") Plant species that occasionally interbreed typically maintain their integrity as units that are discrete in morphology, ecology, and physiology and that (aside from this occasional gene flow) are maintained as separate lineages.

Recognition species refer to sexually reproducing systems that are maintained by genetically based features that promote reproduction (Paterson 1985). In this concept, the emphasis is on mechanisms or processes that promote gene flow in a

Pre-mating, i.e., prevent mating from occurring
(1) ecological or habitat isolation, in which different species have a genetic basis for occupying different habitats (e.g., dry versus wet soils) and therefore do not normally come into contact to successfully breed
(2) temporal isolation, in which different species breed at different times, e.g., spring versus fall or day versus night blooming flowering plants
(3) behavioral/ethological isolation, in which different species have different behaviors that inhibit interbreeding, such as (for flowering plants) different pollination mechanisms, preventing the transfer of pollen between different flower types
Post-mating, i.e., allowing mating (equivalent to pollination in seed plants), but not zygote formation
(1) mechanical or physiological isolation, such as self-incompatibility mechanisms, in which pollen that lands on the stigma of another species will not germinate or will not form a functional pollen tube
(2) gametic isolation, in which sperm and egg come into contact, but these gametes are incompatible the sperm cannot successfully fertilize the egg
Post-zygotic, i.e., occurring after a zygote forms
(1) hybrid inviability, in which hybrids between species cannot develop properly to adulthood
(2) hybrid sterility, in which hybrids develop to adulthood but are sterile or have reduced fertility
(3) hybrid breakdown, in which hybrids have are fertile but their F2 generations have reduced viability or fertility

TABLE 19.1 Barriers to gene exchange (reproductive isolating "mechanisms").

sexual system, including things such as fertility, behavior, and gamete propagation and recognition. Barriers to gene flow between species (emphasized in the biological/isolating species concept) may simply be by-products of the maintenance of features that promote sexual reproduction within species.

Evolutionary species refer basically to a single, continuous lineage of ancestral-descendent populations (e.g., Figure 19.3A,B) that retains its "identity" from other such lineages and has its own "evolutionary tendencies and historical fate" (after Wiley 1978).

Cohesion species are defined as the largest or most inclusive group of individuals that maintains genetic and phenotypic "cohesion" (Templeton 1989). Members of a cohesion species comprise an evolutionary lineage and maintain some type of similarity to one another over time via a number of evolutionary processes, the latter constituting "cohesion mechanisms." It is the study of these cohesion mechanisms that is the focus of this species concept, in trying to understand both how species are maintained and how they diverge (speciate).

Cohesion mechanisms include not only gene flow itself, but also genetic drift and natural selection. The last two may help to maintain the *fundamental niche* of the species, i.e., their genetically determined environmental tolerances. (See Templeton 1989 for details.)

The cohesion species concept has an advantage in that cohesion species can be either asexually or sexually reproducing; the latter can range from those that are rampantly out-crossing to those that are exclusively selfing. In addition, cohesion species can allow limited exchange of genetic material with other species, either from one lineage to another or in formation of hybrids, as long as the integrity of the cohesion species is maintained. Thus, cohesions species can be maintained within a so-called "syngameon" and may be more representative of real, natural systems.

CONCEPTS EMPHASIZING PRODUCT

Species concepts emphasizing the product of evolution include taxonomic, phylogenetic, and genealogical species. Perhaps the most basic concept of species is the **taxonomic** or **morphologic species**, which could be defined as the smallest group (or class) of individuals that are similar to one another in one or more features and different from other such groups (see Cronquist 1988). This phenetic notion uses overall similarity as the criterion for grouping individuals together into a common species and some measure of dissimilarity for separating different species. As long as the taxonomic species are diagnosable from other such species by one or more features, with no (or with statistically insignificant) intergradation in these features, they can be considered discrete entities, worthy of recognition as species.

This taxonomic species concept is associated with a concept that came about prior to the acceptance of evolutionary theory and detailed knowledge of genetics and sexual reproduction. It is derived from the historical *typological concept*, in which species were deemed to have "essential" properties, making them unique. The taxonomic species concept is neutral with regard to history and processes, neutral as to how species got here and what keeps them apart. Taxonomic species can be viewed as first approximations, in the absence of other information, to be tested further with more refined techniques. And yet it must be admitted that the great majority of named species are defined and categorized in this way, with no knowledge of phylogenetic history, gene flow, or detailed reproductive mechanisms. Thus, this taxonomic species concept can be viewed as a practical, utilitarian, but limited means of recognizing and categorizing biodiversity.

Phylogenetic species use some notion of phylogenetic relationships to recognize species units. One definition of a phylogenetic species is character-based, defined as the small-

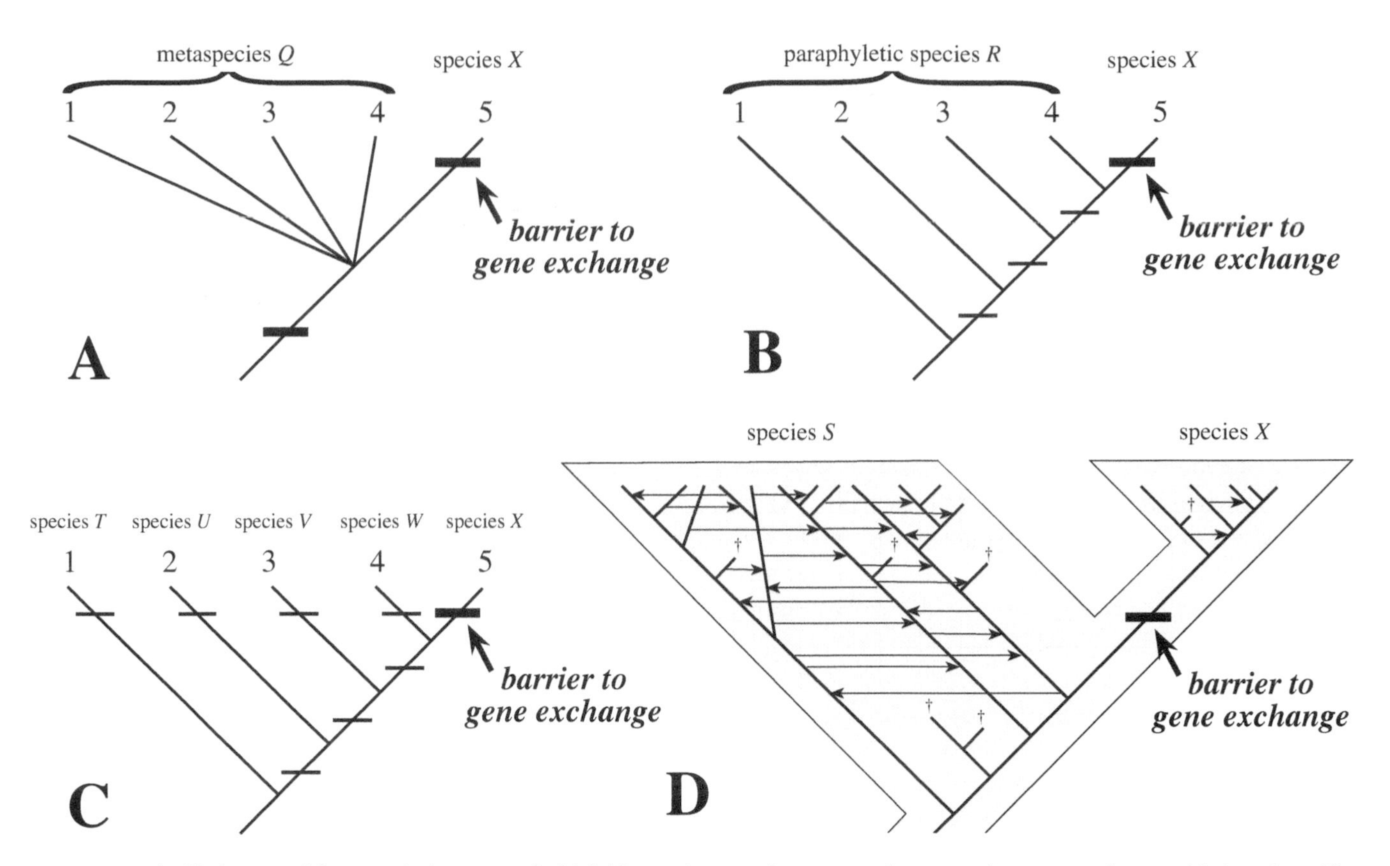

Figure 19.7 **A.** Cladogram of five populations, one of which (5) acquires a major apomorphy preventing gene exchange with the others. Note that the other four populations are unresolved, designated "metaspecies *Y*." **B.** Additional data reveals that populations 1–4 form a cladistic grade, designated here as a paraphyletic species *Y*. **C.** Discovery of apomorphies for populations 1–4 warrants their classification as separate species (*T–W*). **D.** Over time, populations 1–4 undergo extensive introgression, warranting their designation as a single lineage segment and species *Y*. Modified from Donoghue (1985) and Olmstead (1995).

est group of populations or lineages that are "diagnosable by a unique combination of character states in comparable individuals ..." (Nixon and Wheeler 1990). This concept does not require a strict knowledge of pattern of phylogenetic relationships to implement. In another sense, a phylogenetic species is equivalent to a the smallest monophyletic group, a common ancestor and all descendents of that common ancestor, recognizable by one or more shared derived features or apomorphies (de Queiroz and Donoghue 1988); this type of phylogenetic species has also been termed an "autapomorphic species" (termed "apomorphic species" here) because of the necessity of recognizing an apomorphy for the species lineage alone. One advantage of an apomorphic phylogenetic species is that it is by definition a monophyletic group and thus a direct reflection of the pattern of evolutionary history. However, problems with the latter definition is that it does require a phylogenetic study, ideally with a relatively large sample size, such as members of numerous populations. From this analysis, the species would need to be identified as a monophyletic group with some degree of robustness (e.g., a standard bootstrap or Bayesian posterior probability value). Perhaps a more serious concern about the phylogenetic species concept in practice is that there are often populations in natural systems—all belonging to what might normally be considered the same species (by other criteria, such as overall similarity, gene flow, or "cohesion" mechanisms)—but one or more of which might be identified as separate monophyletic groups, identifiable by one or more apomorphies. In reality, these monophyletic populations may likely be temporarily isolated assemblages that have diverged slightly but that are not reproductively isolated and that, if brought together, would exchange genes. Thus, although having species as monophyletic units is desirable, phylogenetic species, defined as the smallest monophyletic group, may be impractical and unworkable as a useful species concept.

Associated with phylogenetic species are metaspecies and paraphyletic species. A **metaspecies** is defined as two or more lineages segments (e.g., from samples of separate populations) that can be resolved as neither monophyletic or paraphyletic (Donoghue 1988). For example, Figure 19.7A shows a cladogram in which population 5 is resolved as monophyletic (qualifying in this case as both biological and apomorphic phylogenetic species *X*), whereas populations 1–4 are unresolved; the latter could be designated as metaspecies *Q*. Metaspecies are phylogenetically unresolved either because of lack of data or because the lineage segments of the populations diverged very rapidly, leaving behind no differentiating apomorphies (or possibly because the separate lineage segments actually diverged contemporaneously). In the example of Figure 19.7, with a finer-scaled study, e.g., the use of numerous, rapidly evolving molecular markers, it might be possible to resolve these populations further. Figure 19.7B shows populations 1–4 appearing as a paraphyletic grade, each lineage segment lacking an apomorphy. These lineage segments might be designated as a **paraphyletic species** (*R* in Figure 19.7B). Paraphyletic species (similar to the "plesiospecies" of Olmstead 1995) may represent temporarily separate lineage segments that have diverged relative to one another but that still may be capable of gene introgression. This may represent a not uncommon phenomenon in nature, in which a broadly ranging set of populations, which are morphologically similar to one another and at least potentially interbreeding, gives rise to isolated lineages that diverge. One or more of these divergent lineages from the "parent" paraphyletic species may acquire a unique apomorphy, e.g., becoming adapted to a specialized, restrictive habitat, or that otherwise prevents gene exchange with the parent lineages (Figure 19.7B). Another possibility, however, is that these divergent population lineage segments may represent incipient species, which in time acquire unique apomorphies, each of which may qualify as an apomorphic phylogenetic species (*T–W* in Figure 19.7C). Alternatively, these segments could undergo future, rampant gene exchange, effectively merging into a single lineage segment (species *S* of Figure 19.7D).

Genealogical species are those in which all members of the group are more closely related to one another than to any organisms outside the group (Baum and Shaw 1995). Relationship is assessed in terms of tokogenetic genealogies, gene flow between parent and offspring. The example of Figure 19.8 shows the tokogenetic relationships between organisms. At the terminus (time t_1) of the lineages, taxon B would be an genealogical species, having the property of "exclusivity" because all individuals share a common ancestor that is more recent in time than any shared by individuals of taxon *A* at time t_1. However, this is not true for the terminal members of taxon *A*, whose common ancestor is equally closely related to taxon *B*. Thus, taxon *A* would not qualify as a genealogical species. Practical ways of accessing genealogical species have been proposed (see Baum 2007).

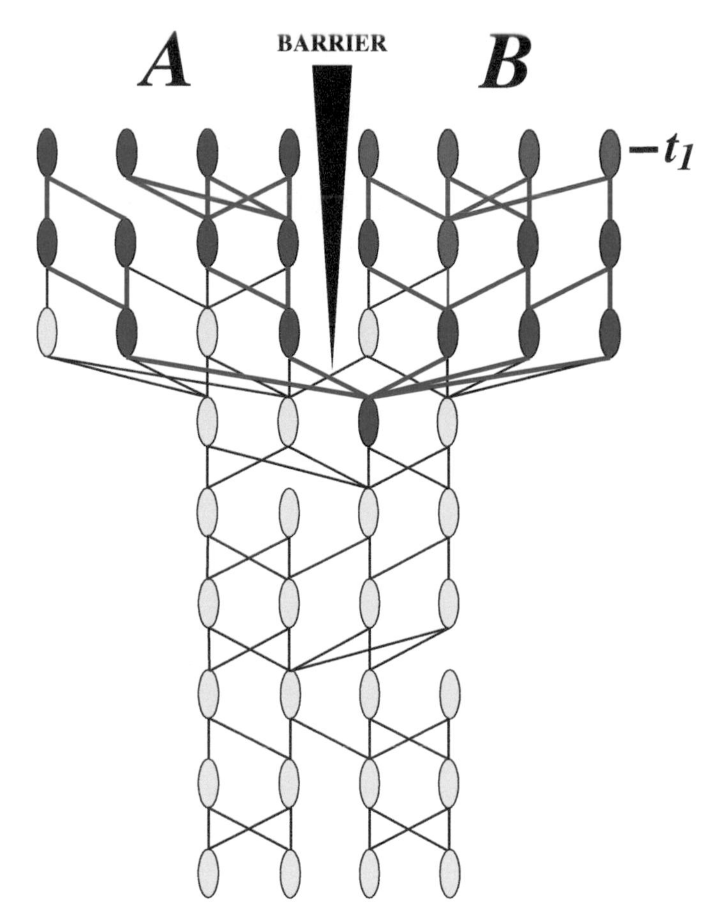

Figure 19.8 Genealogical Species. Note that individuals at the terminus of *B* qualify as a genealogical species because their most recent common ancestor (red in generation below) is more recent in time than any common ancestor shared with any other organisms. Taxon *A* does not qualify as a genealogical species. (Modified from Baum 2009.)

OTHER SPECIES TYPES

Two more general species types that may overlap with any of the previous ones are asexual and cryptic species. **Asexual species** are those reproducing without any sexual reproduction, yet are recognizable morphological units. For example, apomictic populations that develop seed without fertilization, wherein the embryo is a clone of the original parent, have been documented in plants. Asexual species have a divergent tokogenetic pattern (Figure 19.1G). They may be maintained over time by cohesion mechanisms.

Cryptic species are those that are not *morphologically* distinguishable from another species. The rationale for distinguishing cryptic taxa from one another is based on the fact that they are genetically different from other populations. Molecular phylogenetic studies may demonstrate that a population that is essentially indistinguishable morphologically from, e.g., those of a more widespread species is actually phylogenetically

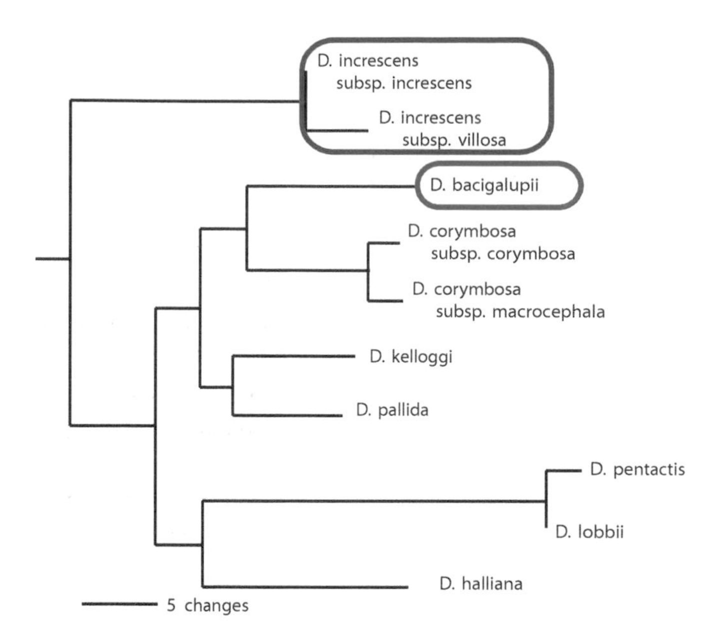

Figure 19.9 Cladogram of a complex of *Deinandra* species. *D. bacigalupii* represents an originally cryptic species, represented by a small, disjunct population, originally thought to belong to *D. increscens* (being virtually indistinguishable from the latter morphologically). However, this molecular phylogenetic analysis demonstrated that *D. bacigalupii* is well separated from *D. increscens* phylogenetically and geographically. (Redrawn from Baldwin 2000.)

distinct, warranting recognition of the former as a separate species (see example in Figure 19.9; from Baldwin 2000).

INFRASPECIES

For a number of plant species, **infraspecies**, including subspecies, varieties, or rarely forms (formae), are recognized. The great majority of species with infraspecies list either subspecies or varieties (not both), as in the trinomials *Camissonia claviformis* subsp. *peirsonii* and *Quercus agrifolia* var. *oxyadenia*. Technically, subspecies are of higher rank than varieties and should show greater differences from one another than varieties. In practice, these two infraspecies ranks are used interchangeably, the distinction between them arbitrary to most taxonomists (Hamilton and Reichard 1992); however, see Clausen (1941) and Stebbins (1950) for an alternative viewpoint.

Subspecies or varieties are often characterized as separate groups of populations within a species that show slight and often intergrading morphological differences as well as some geographic, ecological, and/or phylogenetic distinctions. Infraspecies could be the result of **incipient speciation**, the early and incomplete divergence of one lineage into two (or more), such that the terminal entities of those lineages are not fully separated from one another (e.g., exhibit incomplete lineage sorting). Thus, the lineages may have no recognizable apomorphies and/or may intergrade with one another, with possible active gene exchange through introgression or hybridization still occurring. By this incomplete divergence hypothesis, infraspecies might be **sympatric** (occurring in the same geographic range) or **allopatric** (occurring in difference ranges).

Infraspecies could also potentially represent **secondary contact**, in which two (or more) lineages that had diverged and were likely geographically separated (allopatric) in the past, have come to occupy the same or overlapping ranges. This hypothesis seems to necessitate that the infraspecies are now sympatric. In this case, gene exchange may have become re-established or the two lineages are still so similar that teasing them apart is difficult.

It is possible that some infraspecies may represent the end result of **fragmentation**, in which an originally broadly distributed species with more or less continuous gene exchange and clinal intergradation became split into two or more populations. The populations or sets of populations resulting from this fragmentation may not have diverged but are the remains of an ancestrally continuous population, their differences the result of the extirpation of intervening populations.

If infraspecies represent either incipient speciation, secondary contact, or fragmentation, they might be viewed as incompletely distinct evolutionary lineages. Thus, the characterization and naming of infraspecies in these examples may be based on unique evolutionary histories. However, infraspecies could also represent a recurring phenomenon, one that repeats itself over and over. For example, when a species ranges from a mesic habitat to a xeric ones, or from low elevation to high elevation, changes in morphology could readily be selected that distinguish the extremes in the two habitats.

In summary, an author who uses infraspecies should as clearly as possible state the criteria, rationale, and evidence for this classification. Further, more detailed studies (such as phylogenetic analyses or nested clade analyses using a "fast" molecular marker) may help to elucidate the validity of infraspecies designations as evolutionary lineages (albeit imperfect and perhaps intergrading) versus simple genetic variation or spurious local adaptation. Infraspecies may still be of significance in representing the total biodiversity in nature.

SPECIES PROGRESSIONS OR LIFE HISTORIES

An interesting concept is that lineages undergo a transformation over time, such that they correspond with different species types at different times (Harrison 1998; see Figure 19.10). For example, immediately following evolutionary

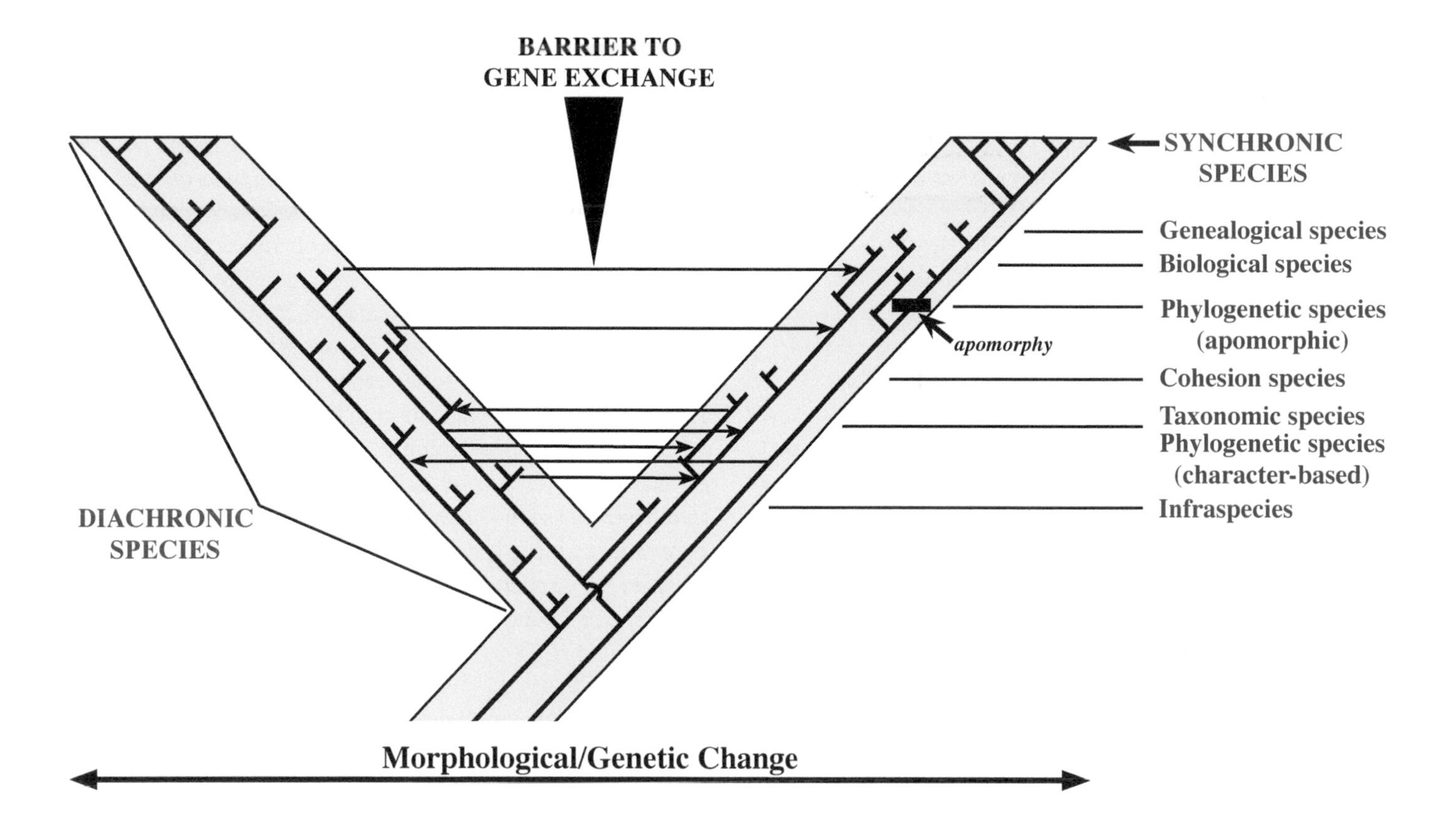

Figure 19.10 Illustration of species progressions of life history. Following the initial evolutionary divergence, conemporaneous members of a segment acquire different characteristics, corresponding to different species concepts. (Modified from Harrison 1998.)

divergence, members of the two lineages may intergrade morphologically, show incomplete lineage sorting, and have considerable introgression, perhaps warranting the use of an infraspecific rank (subspecies or variety) as most useful. At a later time after this initial divergence, species may be morphologically distinct in one or more features, fitting the definition of a **taxonomic species**. At some point, these lineage segments might be diagnosable by a suite of character states, meeting the criteria of **character-based phylogenetic species**. However, it may take some time for lineage sorting to occur. Initially, one or both lineage segments may be paraphyletic or polyphyletic (Chapter 2; see Neigel and Avise 1986). Only in time will both lineage segments acquire unique, shared derived character states, meeting the criteria of **apomorphic phylogenetic species**. In sexually reproducing plants, unless a major genetic change occurs at the time of divergence, gene flow (a primitive character) may occur or be capable of occurring between both lineage segments. If and when barriers to gene exchange (so-called isolating mechanisms) evolve, the two lineage segments exhibit criteria for **isolation (biological) species**. As features for maintaining fertility evolve (such as selection for gamete recognition, mating systems, etc.), lineage segments qualify as **recognition species**. These lineages may acquire distinctive ecological niche specializations, maintained through gene flow, genetic drift, and selection, qualifying them as **cohesion species**. If all members of the lineage at a given point in time are genetically more closely related to one another than to any individuals of another species, they become **genealogical species**. These progressions can occur in different sequences. The criteria for species that are present in a given species will depend on when in its evolutionary track that lineage is studied following divergence, influenced by the particular evolutionary processes of and following divergence, impacted by population size, generation time, and genetic features of that system.

In addition, the temporal extent of species can be evaluated. Species can be defined or conceptualized as either synchronic or diachronic. **Synchronic species** refer to a group of *contemporaneous* organisms, living at the same general point in time. **Diachronic species** refer to a group of organisms that span versus a lineage of ancestral-descendent populations, extending through a period of time (Figure 19.10). (See Baum 2009 for a summary of these and an argument for species as ranked taxa.)

PRACTICAL SPECIES CONCEPTS

Virtually all systematic studies begin with a selection of ingroup taxa (OTUs) that have been previously identified and names based on a **taxonomic/morphological species** concepts. Most species and infraspecies have been recognized from taxonomic studies in which an author determined that one group of organisms was consistently (and, ideally heritably, although this is rarely tested) different from all other species in at least one morphological feature, without intergradation between different species with regard to the feature(s). Using pre-existing taxonomic species is generally not a problem for higher level analyses, as long as, for example, the boundaries of more inclusive taxa are questioned. For example, in a phylogenetic study of a genus or family, it should not be assumed that these higher taxa are monophyletic; a thorough sampling of species within these higher units should be sampled, and other taxa outside the original study ingroup should be included, especially those that have been problematic in past classifications. Thus, most higher level, floristic, and even monographic studies will typically utilize a taxonomic/morphologic species concept.

More detailed studies of a species (or **species complex**, i.e., a group of very similar and presumably closely related species and/or infraspecies) may reveal more complicated and interesting facets of evolution. For example, a molecular phylogenetic analysis of numerous individuals and populations of a species could reveal unexpected results. Phylogenetic studies may show that what was considered to be a single species actually consists of more than one discrete clade (see example in Figure 19.9). A closer examination of these population clusters may reveal that, indeed, the disparate lineages have subtle, morphological or ecological differences.

If one or more of the lineages has a unique combination of features it may fit the definition of a **character-based phylogenetic species**; if these features are identified as shared derived character states, then a monophyletic clade can be recognized, fitting the definition of an **apomorphy-based phylogenetic species**.

Studies of the reproductive biology of this new, phylogenetic species may reveal that it is reproductively isolated from other members of the original "parent" species, e.g., through the evolution of a genetically based reproductive incompatibility (Chapter 13). Thus, the new, phylogenetic species would then also correspond to a **biological/isolation species** or **recognition species**. If it can be demonstrated to maintain an ecological niche, separate from members of the "parent" species, it might fit the definition of a **cohesion species**. Finally, detailed genetic studies, using numerous samples from numerous populations, may demonstrate that the new clade is genealogically unique, fitting the definition of a **genealogical species**.

The remaining populations of the original "parent" species would be assessed independently for species status by the criteria cited above. It may also be concluded to be a reciprocally monophyletic clade (fitting the apomorphy-based phylogenetic species) and can be evaluated for fitting the definition of a biological/isolation, recognition, cohesion, or genealogical species. However, a very real possibility in nature is that what remains of the original species is not demonstrably a monophyletic clade. This may be common in situations in which a "species" occupies an extensive range, within which it has given rise to isolated populations (e.g., adapated to a restricted soil type) within the range. If the wide-ranging remnant populations are still diagnosably distinct from the new, isolated species, it might fit the criterion of a character-based phylogenetic species. If it is not, what remains could be termed either a **metaspecies** (if studies show that it is neither monophyletic nor paraphyletic) or a **paraphyletic species** (if demonstrably paraphyletic). The latter case may warrant further, detailed studies to evaluate the certainty and efficacy of splitting this metaspecies or paraphyletic species into further units. However, if so, these units should be clearly diagnosable, preferably morphologically so.

In summary, detailed studies of the relationships of organisms, populations, lineages, or clades may give insight into both the processes and patterns of biological evolution. The natural entities recognized may then be evaluated for their properties, and fit into one or several species definitions. Which "species" these entities correspond to may depend both on the amount of knowledge we possess about them and on the evolutionary process and pattern itself, such as the time since divergence, population size, and aspects of their life cycle.

SYSTEMATICS AND CONSERVATION BIOLOGY

Conservation biology is that branch of biology dealing with the preservation of biodiversity. **Biodiversity** refers to the totality of life within a given region and can be assessed or calculated in different ways. **Species richness** refers to a simple count of the number of species (generally of a certain group, e. g., vascular plants) within a given geographical region. For example, a grassland with a total of 63 plant species has a greater species richness than a woodland with a total of 48. Comparisons of species richness are generally made as a function of the area of the geographic region being considered. **Species evenness** takes into account not only the number of species but also how evenly they are distributed within the region. Evenness of distribution can be measured in terms of the number of individuals or various ecological parameters such as cover, density, frequency, or importance value of species within the region. For example, two regions or habitats may have the same number of species, but that region having the species relatively evenly distributed will have a greater species evenness than one having a few dominants with the other species being sparse.

Because organisms are dependent on their biotic and abiotic environment for survival, conservation biology by necessity deals with the preservation of both organisms and their habitat. Conservation biology is an integrative discipline and encompasses the fields of systematics, ecology, geography, geology, and geochemistry.

CONSERVATION BIOLOGY AND THE GOALS OF SYSTEMATICS

How does plant systematics relate to conservation biology? Perhaps the most important ways that systematics impacts conservation biology is in taxon diagnosis, floristic surveys, evaluation of taxon rarity, and use of phylogenetic information in evaluating conservation decisions.

A primary way that plant systematists impact conservation biology is **taxon diagnosis**. Systematists and taxonomists are the ones who ultimately name and categorize species and infraspecies. This is a critical factor in evaluating whether a given taxon is endangered or threatened. For example, floristic studies may reveal the existence of a rare, isolated, morphologically or genetically distinct population within the range of an otherwise widespread, common species. Systematic studies must be done to evaluate this rare population and decide if it should be classified as a distinct species or infraspecies. (Note that the reverse can happen; what was thought to be a rare taxon can be ascertained, through detailed studies, to be natural variation within a common taxon, not warranting recognition as a distinct taxon.) Detailed molecular studies of species complexes, combined with studies of ecology and reproductive biology, may reveal additional evolutionary units (e.g., genetically distinctive populations) worthy of conservation, as in the example of cryptic species (Figure 19.9). These may warrant naming as a new species or infraspecies (Chapter 16), or recognition as a genetically unique population, one worthy of environmental listing.

A second way that systematics impacts conservation is through **floristic surveys**, determining what plant species and infraspecies occur in a given region (see Chapter 18). Studies of broad regions, such as a state flora, are starting points for assessing biodiversity and geographic range of plant taxa, both native and exotic/naturalized (non-native, but established in natural habitats). These floristic studies yield valuable information on the species and infraspecies **range**, which is the collective geographic distribution of a taxon. Range information on exotic taxa can be particularly important, as exotics often negatively impact the biodiversity of native taxa. A **disjunct** population is one that occurs well outside the range of other members of that taxon. Disjunct populations are often particularly important to study, as they may reveal genetically or morphologically distinctive organisms, in some cases warranting recognition as a distinct taxon. Floristic surveys are the basis for assessing if species or infraspecies are **endemic** to a given region, i.e., found naturally only in a particular, often limited, geographic location and in no other place in the world. Regions of high endemism are termed biodiversity "hotspots" (Meyers 1988, 1990; Meyers et al. 2000).

More detailed surveys of smaller regions, e.g., in a sensitive habitat, yield better information about the number of individuals, number of populations, and geographic and ecological range of plant taxa. However, it cannot be overstated that the thoroughness and validity of floristic surveys depend on the experience of the person(s) conducting it. Floristic surveys should not be just a list, but must include preparing a **voucher specimen** (deposited in an accredited herbarium) of every taxon observed (Chapter 17); only with a voucher specimen can a permanent record exist and identities be verified by trained and experienced plant taxonomists (or even evaluated from DNA data obtained from that specimen). In addition, a good floristic survey depends on conducting the survey during different seasons (e.g., spring, summer, and fall for some plants) and over more than one year (e.g., during wet and dry seasons), to be sure that a thorough inventory of the plants in the region has been attempted.

A third way that systematics influences conservation is in estimating **taxon rarity**. Evaluating the rarity of taxa usually begins with floristic surveys, including the records and observations of professional botanists and information recorded in regional floras. Today, however, with the increased databasing of field collections (Chapter 18), the total number of plant collections of a given species can be instantly assembled and mapped. Follow-up studies of these records should be done, determining if taxa or populations that were historically cited are still present. An **extinct** species or taxon is one no longer known to exist in the wild. An **extirpated** species or taxon is one that is extinct within a particular portion of its range but present within other parts of its range. For rare, historic populations of a taxon, making careful estimates of the number of individuals persisting is important. Finally, ecological or reproductive studies can further evaluated the significance of plant species to a given ecosystem. A **keystone species** is one that, by its absence from an ecosystem, results in the disappearance (directly or indirectly) of several other species, causing an "extinction cascade."

These data on presence and frequency in nature, along

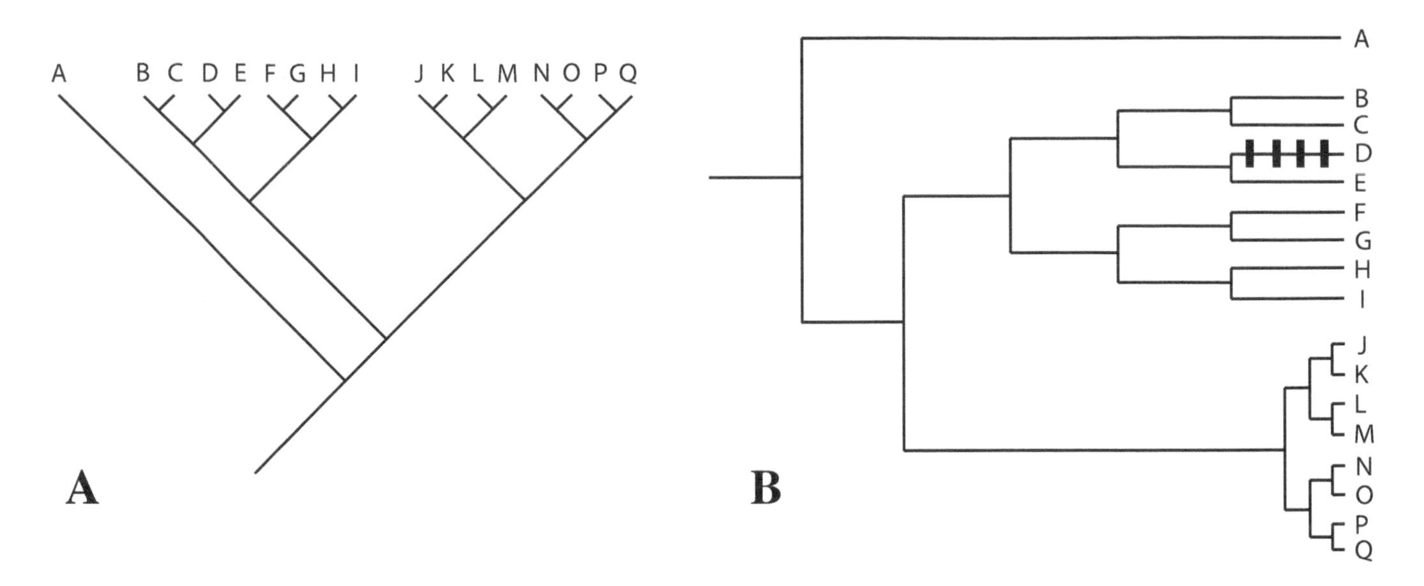

Figure 19.11 **A.** Cladogram showing early diverging (basal) lineage segment terminating in species *A*, representing a potential paleoendemic, and species *B–Q*, which diverged in a cladistically equivalent manner. **B.** Phylogram, representative of absolute time, of same group. Note that the divergences of lineages terminating in taxa *B–I* occurred much earlier in time than the rapid diversification of those terminating in taxa *J–Q*, the latter representing neoendemics. Also note taxon *D* with major, adaptively significant apomorphy.

with an evaluation of the threat to existing populations, are used by systematists and ecologists to make formal assessments of taxon rarity and propose legal designations in order to help protect these taxa. As defined by the 1973 U. S. Federal Endangered Species Act, an **endangered taxon** (which can be a species, subspecies, or distinct population segment) is one that is in danger of becoming extinct in the near future within all or a significant part of its geographical range due to one or more causal factors. A **threatened taxon** is one thought likely to become endangered in the wild in the near future within all or part of its geographical range if the same causal factors continue to operate. According to the Endangered Species Act, the actions of agencies of the federal government must not "jeopardize the continued existence of endangered and threatened species or result in the destruction or modification of habitat of such species which is determined to be critical."

A fourth way that systematics can aid in conservation biology is by using information derived from **phylogenetic analyses** to inform our decisions as to which *taxa* should be preferentially conserved, given limited resources. For example, in the cladogram of Figure 19.11A, taxon *A* has a uniquely basal cladistic position, being sister to all the other taxa of this group. In contrast, the individual taxa of group *B–Q* are equal in cladistic relationships. Thus, it could be argued that taxon *A* should be preferentially conserved over taxa *B–Q* in having this unique phylogenetic history, possibly representing the descendent of an ancient relic lineage (e.g., like *Ginkgo biloba*). If the study of this group was refined in having a more absolute time scale, rates of evolutionary divergence might be inferred. For example, the phylogram of Figure 19.11B shows the contrast between taxa *B–I* and taxa *J–Q*; the latter group has a much more recent origin and shows much more rapid rates of speciation. Finally, one might also give credence to taxa having unique apomorphies, such as taxon *D* of Figure 19.11B; these apomorphies might include major genetic novelties (e.g., a novel chromosome number in the group) or some major adaptive feature important in the ecosystem (e.g., a novel symbiotic relationship), warranting preferential conservation.

A correlation of phylogenetic relationships with biogeographic distribution can be used to make decisions about *geographic regions* to preferentially conserve. For example, a tabulation of geographically restricted (endemic) species that represent the products of either ancient lineages (termed **paleoendemics**) or relatively recently evolved lineages (termed **neoendemics**) might identify geographic regions where these endemics are concentrated. The former could be given conservation priority as a region of unique, ancient relictual species. The latter might be given priority as a region of active speciation. Although choices of conservation are difficult to make, and need to consider a great number of factors, phylogenetic information can at least provide data that can add to a decision that is well informed. (See Moritz 2002; Mishler et al. 2014; González-Orozco et al. 2016; Thornhill et al. 2016, 2017; and Kling et al. 2019.)

CONSERVATION BIOLOGY AND STEWARDSHIP

Species are being lost at an alarming rate, particularly in tropical rain forests, but also in other ecosystems. Thus, it is urgent that systematists focus on documenting life on our planet. Floristic surveys and other research studies by plant systematists will inform us of the plant component of a given region, habitat, or ecosystem; that information can be added to our total knowledge of biodiversity in making informed decisions in ways discussed earlier. Unfortunately, botanical studies of some regions may only enable us to discover what life used to be present in a given region, prior to its permanent destruction. However, preservation of propagules (in, e.g., seed banks) and repositories of DNA will enable some future study of these plants and at least the possibility of their reintroduction into habitats in the future.

Why save biodiversity? One reason is aesthetic. We may gain much joy from viewing or studying nature. It is worth preserving and nurturing the natural world for our direct enjoyment and those of future generations. Philosophically, all living organisms on the earth should be allowed to exist. It is anthropocentric chauvinism to assume that humans are all important and other creatures unimportant. A second reason is economic importance. Many species of plants could have untold economic importance to humans. Undiscovered or unutilized plant species may have future important uses as sources of food, fiber, oil, rubber, wood, etc. Many of these have had or will have medicinal uses. In fact, it is estimated that 25% of all medicines have been derived (directly or indirectly) from plant compounds such as alkaloids or glycosides, and yet relatively few have been studied for their medicinal properties. There are undoubtedly many more species of plants that have the potential for medicinal uses. A third reason is maintenance of genetic diversity. Native plants that are close relatives to important cultivated agricultural plant species may be important in contributing genes, either directly or via interbreeding or genetic engineering, for valued traits such as disease resistance, increased vigor and yield, or food quality.

If we are interested in preserving the diversity of life on the planet, it is up to each of us, whether academics or students, to make it a priority in our lives. We can act to train more systematists to describe, inventory, and map the biotic diversity that still exists. Documentation of both native plants and introduced and potentially competitive plants are both important. We can act to help direct more money into these training efforts, as well as to museums, botanic gardens, zoos, and herbaria, which maintain collections of species. We can act to prioritize our own research programs—to preferentially study plant groups containing endangered species, local endemics in vulnerable environments, paleoendemic "relicts," or neoendemics representing recent speciation. Understanding the phylogeny of members of a complex may give insight into their adaptation to a particular environment (both abiotic and biotic factors); this understanding may aid in the future management of the endangered taxon. Understanding the reproductive biology of plants can aid in their preservation by considering plant pollinators as a key component of the ecosystem, knowledge that may also be important economically. As academics, we can remove ourselves from the "ivory tower" at least occasionally and work toward the environmental listing (as a rare, threatened, or endangered taxon) of plants with which we have expertise. And, perhaps most important, we can strive to educate others regarding nature, conservation, and biodiversity and to cultivate (especially in our children) the aesthetics of enjoying and revering the natural world.

Note: Recommended reading on the importance of biodiversity to "human well-being" is the Millennium Ecosystem Assessment Report (2005) on Biodiversity, which addresses these questions: **1.** *Biodiversity: What is it, where is it, and why is it important?* **2.** *Why is biodiversity loss a concern?* **3.** *What are the current trends and drivers of biodiversity loss?* **4.** *What is the future for biodiversity and ecosystem services under plausible scenarios?* **5.** *What response options can conserve biodiversity and promote human well-being?* **6.** *What are the prospects for reducing the rate of loss of biodiversity by 2010 or beyond and what are the implications for the Convention on Biological Diversity?*

REVIEW QUESTIONS

SPECIES

1. How can gene flow be defined?
2. Draw a diagram of the haplodiplontic life cycle of land plants, with regard to gene flow over multiple generations.
3. How do mosses, ferns, conifers, and an annual flowering plant differ with respect to this gene flow?
4. How are plants capable of undergoing asexual reproduction?
5. What is a lineage?
6. What is the difference between tokogenetic and phylogenetic relationships?
7. What is the difference between a lineage segment and a clade?
8. What is a gene genealogy/lineage versus a species tree?
9. What is lineage sorting? How can lineage sorting be incomplete, and what is a result of this?
10. Define introgression, hybridization, and polyploidy and their significance in plant evolution.
11. What is meant by nested clade analysis and for what can this technique be used?
12. Define what is meant by the biological (isolating) species concept.
13. Name some barriers to genetic exchange that are pre-mating, post-mating, and post-zygotic.
14. How is the emphasis of the recognition species concept different from that of the biological (isolating) species concept?
15. What are cohesion species and how are they maintained?
16. How are evolutionary species defined?
17. What is a taxonomic/morphological species? How are these often the starting point for species assessment?
18. What is a phylogenetic species? In what two ways can these be defined?
19. Explain the concept of a metaspecies. How does this differ from a paraphyletic species?
20. Why might paraphyletic species be not uncommon in nature?
21. What is a genealogical species? An asexual species?
22. Define cryptic species. How might they be made less "cryptic"?
23. What are infraspecies and what are the two most common ranks used in their classification?
24. What three things, in terms of evolutionary processes, might infraspecies represent?
25. Describe in detail how species might undergo an evolutionary progression over time, indicating the transition and overlap among different species types or concepts.
26. What is the distinction between synchronic and diachronic species?
27. Review how species can be evaluated practically in plant systematics studies.

SYSTEMATICS AND CONSERVATION

28. What is the concern of conservation biology?
29. What is the difference between species richness and species evenness?
30. In what four ways might plant systematics impact or relate to conservation biology?
31. How is taxon diagnosis fundamental to conservation biology?
32. How are floristic studies important in the conservation of plants?
33. What is a disjunct? An endemic?
34. Why are voucher specimens essential to any rigorous floristic survey?
35. What is the differences between extinct and extirpated taxa?
36. What is a keystone species?
37. How are endangered and threatened species actually legally defined?
38. Give two examples as to how phylogenetic studies can be valuable in making conservation decisions.
39. What are paleoendemics? Neoendemics?
40. Describe in an essay how plant systematists may act to save biodiversity, and if you think we have a moral obligation to do so.

EXERCISES

1. Find five descriptions of a new species in a systematic journal (see Appendix 3) and ascertain what, if any, concept or rationale for the definition of the species is stated in the article.
2. If available, lay out several herbarium specimens of a plant species having two or more infraspecies. Evaluate the infraspecies with regard to morphological distinctiveness and intergradation, geographical range, and differing habitat requirements. Assess whether you think an infraspecies rank is appropriate for these different populations and what it might mean biologically.
3. Select a rare, threatened, or endangered plant species native to your area. Determine from taxonomic treatments, how this species differs from close relatives and evaluate its classification as a separate species. In addition, obtain information on this species size (in terms of number of individuals or populations), range, and distribution (the last from distribution maps, if possible). Evaluate why the species may be sensitive, and what has or can be done to help preserve it.

REFERENCES FOR FURTHER STUDY

Avise J. C. and R. M. Ball 1990. Principles of genealogical concordance in species concepts and biological taxonomy. Oxford Surveys in Evolutionary Biology 7: 45-67.

Baldwin, B. G. 2000. Roles for modern plant systematics in discovery and conservation of fine-scale biodiversity. Madroño 47: 219-229.

Baum D.A. 2009. Species as ranked taxa. Systematic Biology 58: 74-86.

Baum D. A., Shaw K. L. 1995. Genealogical perspectives on the species problem. In: P. C. Hoch and A. G. Stephenson (eds.), Experimental and molecular approaches to plant biosystematics. Pp. 289–303. Missouri Botanical Garden Press, St. Louis.

Clausen, R. T. 1941. On the use of the terms "subspecies" and "varieties". Rhodora 43:157-167.

Cronquist, A. 1988. The Evolution and Classification of Flowering Plants. Ed. 2. New York Botanic Garden, New York.

de Queiroz K., and M. J. Donoghue. 1988. Phylogenetic systematics and the species problem. Cladistics. 4:317–338.

de Queiroz K. 1998. The general lineage concept of species, species criteria, and the process of speciation: a conceptual unification and terminological recommendations. In: Howard D.J., Berlocher S.H. (eds.), Endless forms: species and speciation. Pp. 57-75. New York: Oxford University Press.

Donoghue, M. J. 1985. A critique of the biological species concept and recommendations for a phylogenetic alternative. The Bryologist 88: 172-181.

Erwin, T. L. 1991. An evolutionary basis for conservation strategies. Science 253: 750-752.

Faith, D. P. 1992. Conservation evaluation and phylogenetic diversity. Biological Conservation 61:1-10.

González-Orozco, C. E., Laura J. Pollock, Andrew H. Thornhill, Brent D. Mishler, N. Knerr, Shawn W. Laffan, Joseph T. Miller, Dan F. Rosauer, Daniel P. Faith, David A. Nipperess, H. Kujala, S. Linke, N. Butt, C. Külheim, Michael D. Crisp, and B. Gruber. 2016. Phylogenetic approaches reveal biodiversity threats under climate change. Nature Climate Change 6:1110.

Hamilton, C. W., and S. H. Reichard. 1992. Current practice in the use of subspecies, variety, and forma in the classification of wild plants. Taxon 41: 485-498.

Harrison, R. G. 1998. Linking evolutionary pattern and process: the relevance of species concepts for the study of speciation. In: D. J. Howard and S. H. Berlocher (eds.), Endless forms: species and speciation. Pp. 19-31. Oxford University Press, New York.

Heywood, V. H., R. K. Brummitt, A. Culham, and O. Seberg. 2007. Flowering Plant Families of the World. Royal Botanic Gardens, Kew, Richmond, Surrey.

Kling, M., M., B. Mishler, D., A. Thornhill, H., B. Baldwin, G., and D. Ackerly, D. 2019. Facets of phylodiversity: evolutionary diversification, divergence and survival as conservation targets. Philosophical Transactions of the Royal Society B: Biological Sciences 374:20170397.

Knowles, L. L. 2008. Why does a method that fails continue to be used? Evolution 62: 2713-2717.

Knowles, L. L. and W. P. Maddison. 2002. Statistical phylogeography. Molecular Ecology 11: 2623-2635.

Mayr, E. 1963. Animal Species and Evolution. Belknap Press, Cambridge, Mass.

McDade, L. A. 1995. Species concepts and problems in practice: Insight from botanical monographs. Systematic Botany 20: 606-622.

Millennium Ecosystem Assessment. 2005. Ecosystems and Human Well-being: Biodiversity Synthesis. World Resources Institute, Washington, DC.

Mishler, B. D., N. Knerr, C. E. González-Orozco, A. H. Thornhill, S. W. Laffan, and J. T. Miller. 2014. Phylogenetic measures of biodiversity and neo- and paleo-endemism in Australian Acacia. Nature Communications 5:4473.

Moritz, C. 2002. Strategies to protect biological diversity and the evolutionary processes that sustain it. Systematic Biology 51:238-254.

Myers, N. 1988. Threatened Biotas: "Hot Spots" in Tropical Forests. The Environmentalist 8: 187-208.

Myers, N. 1990. The Biodiversity Challenge: Expanded Hot-spots Analysis. The Environmentalist 10: 243-256.

Myers, N., R. A. Mittermeier, C. G. Mittermeier, G. A. B. da Fonseca, and J. Kent. 2000. Biodiversity hotspots for conservation priorities. Nature 403:853-858.

Neigel, J. E. and J. C. Avise. 1986. Phylogenetic relationships of mitochondrial DNA under various demographic models of speciation. In: S. Karlin and E. Nevo (eds.), Evolutionary Processes and Theory. Pp. 515-534. Academic Press, New York.

Nixon K. C., Wheeler Q. D. 1990. An amplification of the phylogenetic species concept. Cladistics 6: 211–223.

Olmstead, R. G. 1995. Species concepts and plesiomorphic species. Systematic Botany 20: 623-630.

Paterson, H. E. H. 1985. The recognition concept of species. In: E. S. Vrba (ed.), Species and Speciation. Pp. 21-29. Transvaal Museum, Pretoria.

Soulé, Michael E. 1990. The real work of systematics. Annals of the Missouri Botanical Garden 77: 4-12.

Stebbins, G. L. 1950. Variation and evolution in plants. Columbia University Press,

Systematics Agenda 2000:Charting the Biosphere. 1994. Technical Report. Society of Systematic Biologists, American Society of Plant Taxonomists, Willi Hennig Society, Association of Systematics Collections. New York. [An excellent introduction to the goals and rationale of systematic studies, described as "a global initiative to discover, describe and classify the world's species." Available through SA2000, Herbarium, New York Botanical Garden, Bronx, New York 10458, USA]

Templeton, A. R. 1989. The meaning of species and speciation: a genetic perspective. In: D. Otte and J.A. Endler (eds.), Speciation and its Consequences. Pp. 3-27. Sinauer, Sunderland, Massachusetts.

Templeton, A. R. 1998. Nested clade analyses of phylogeographic data: testing hypotheses about gene flow and population history. Molecular Ecology 7: 413-418.

Templeton, A. R. 2001. Using phylogeographic analyses of gene trees to test species status and processes. Molecular Ecology 10: 779-791.

Templeton, A. R. 2004. Statistical phylogeography: methods of evaluating and minimizing inference errors. Molecular Ecology 13: 789-809.

Thornhill, A. H., B. D. Mishler, N. J. Knerr, C. E. González-Orozco, C. M. Costion, D. M. Crayn, S. W. Laffan, and J. T. Miller. 2016. Continental-scale spatial phylogenetics of Australian angiosperms provides insights into ecology, evolution and conservation. Journal of Biogeography 43:2085-2098.

Thornhill, A. H., B. G. Baldwin, W. A. Freyman, S. Nosratinia, M. M. Kling, N. Morueta-Holme, T. P. Madsen, D. D. Ackerly, and B. D. Mishler. 2017. Spatial phylogenetics of the native California flora. BMC Biology 15:96.

Turland, N. J., Wiersema, J. H., Barrie, F. R., Greuter, W., Hawksworth, D. L., Herendeen, P. S., Knapp, S., Kusber, W.-H., Li, D.-Z., Marhold, K., May, T. W., McNeill, J., Monro, A. M., Prado, J., Price, M. J. & Smith, G. F. (eds.). 2018. International Code of Nomenclature for algae, fungi, and plants (Shenzhen Code) adopted by the Nineteenth International Botanical Congress Shenzhen, China, July 2017. Koeltz Botanical Books, Glashütten.

Wiley, E. O. 1978. The evolutionary species concept reconsidered. Systematic Zoology 27: 17-26.

Williams, P. H., R. I. Vane-Wright, and C. J. Humphries. 1993. Measuring biodiversity for choosing conservation areas. In: J. LaSalle and I. D. Gauld (eds.), Hymenoptera and Biodiversity. Oxford University Press, New York.

Wilson, Edward O. and F. M. Peters (editors). 1988. Biodiversity. National Academy Press, Washington, D. C.

WEBSITES

International Union for the Conservation of Nature and Natural Resources (IUCN) Red List of Threatened Species.
http://www.iucnredlist.org

Millenium Ecosystem Assessment.
http://www.millenniumassessment.org/en/index.aspx

APPENDIX 1

PLANT DESCRIPTION

WRITING A PLANT DESCRIPTION

The following list of characters can serve as the basis for a detailed plant description. The basic form of the description is to list the plant organ (noted in **bold** in the character list below), followed by a listing of all character states that apply for that plant organ, with each character state separated by commas. Note that, for any particular species, not all characters will apply; these are simply omitted. Also note that some characters are listed with multiple character names, e.g., "**Sepal/Calyx lobes/Outer tepals**." This is designed as a guide, with the intention that only one of these three will be used, depending on whether the outer whorl of the perianth consists of distinct sepals (**Sepal** used), of fused sepals (**Calyx lobes** used), or of tepals (**Outer tepals** used).

There are different styles in writing a detailed plant description. Some use a telegraphic style, e.g., "Leaves simple, sessile, whorled, ovate, entire, glabrous." This style is common in floras, where space for text may be at a premium. Other descriptions use complete sentences, e.g., "Leaves are simple, sessile, whorled, ovate, entire, and glabrous." The use of "the" at the beginning of a sentence is optional, as in "The leaves are simple, sessile, whorled, ovate, entire, and glabrous."

Some general suggestions are as follows:

1. Be sure to *only* list the plant organs (and list only once), followed by the character states that apply to that plant organ. The major plant organs are sometimes placed in **bold** text to highlight them. *Do not* list the specific character names, unless a clarification is needed. Examples:
   Do write: "**Flowers** are bisexual, actinomorphic, pedicellate, 1.5–2.2 cm long (including pedicel) . . ." ["Flowers" refers to the plant organ; all other terms are character states.]
   Do not write: "**Flower sex** is bisexual, **symmetry** is actinomorphic, **attachment** is pedicellate, **length** is 1.5–2.2 cm ..." ["Sex," "symmetry," "attachment," and "length" are characters and should not be listed.]
   However, *do* write: "**Leaf blades** are elliptic, serrate, rounded at base, obtuse at apex." ["Rounded" and "obtuse" could refer to either of the characters base or apex, so these characters should be listed for clarification.]
2. Description of the major organs may be written in the singular or plural form, but the latter should be used only if more than one such organ occurs in an individual. If only one organ occurs per individual, the singular should be used.
   Do write: "**Leaves** are trifoliolate, alternate, . . ." or "The **leaf** is trifoliolate, alternate, . . ." if there are multiple leaves.
   Do write: "The **inflorescence** is a solitary raceme,. . ." if there is a single raceme per individual.
3. Always use metric for plant or plant organ heights, lengths, and widths. Always abbreviate these: "mm" for millimeters, "cm" for centimeters, "dm" for decimeters, "m" for meters. Use mm and cm for smaller structures, dm or m for larger. Use the appropriate unit of measure to avoid values less than 1, if possible. (E.g., write "2–5 mm" instead of "0.2–0.5 cm.") Always place a "0" before a decimal point, as in "0.5 mm." Be clear about what you are describing. Examples:
   Do write: "**Flowers** are 0.5–1.3 mm long (excluding pedicel), 2–3 mm wide when fully opened."
   Do not write: "**Flowers** are 0.5–1.3 mm."
4. For characters that are variable, either list the range of variation (e.g., "**Leaves** oblanceolate to narrowly elliptic, crenate to dentate . . .") or list the most common morphology and in brackets list the exceptions (e.g., "**Leaves** trifoliolate [rarely pinnate with 5 leaflets]" or "**Leaves** 4–7 [2.5–10] cm long . . .").

COMPLETE MORPHOLOGICAL CHARACTER LIST

[Available as download from Website; Note: Not all characters apply to a given taxon; add characters for specialized structures.]

Species/Infraspecies Name (with authorship) [Common Name]: ______

Family: ______

Native locality: ______

Plant Habitat: ______

Plant Duration: ______

Plant Sex if not hermaphroditic: ______

Plant Habit: ______

Plant Height: ______

Root Type: ______

Root Origin (e.g., primary, adventitious): ______

Underground Stem Type if specialized: ______

Underground Stem Branching Pattern: ______

Underground Stem Size: ______

Aerial Stem Habit: ______

Aerial Stem Branching Pattern: ______

Bark Type: ______

Bark Lenticels presence/shape: ______

Twig Surface/Shape: ______

Twig Lenticel presence/shape: ______

Twig Shape/Cross-Sectional Outline: ______

Pith Type: ______

Pith Cross-Sectional Outline: ______

Fruit Scar presence/shape: ______

Leaf Scar Size/Shape: ______

Vascular Bundle Scar Number/Pattern: ______

Stipule Scar presence: ______

Stipule Scar Position/Shape if present: ______

Terminal Bud Scale Scars presence/absence: ______

Bud Type: ______

Bud Orientation: ______

Bud Shape/Size: ______

Bud Position: ______

Bud Scale Arrangement: ______

Bud Scale Surface/Texture: ______

Thorns if present: ______

Spines if present: ______

Prickles if present: ______

Spur Shoot if present: ______

Leaves/Leaf Number if unusual: ______

Leaf Type: ______

Leaf Length/Width: ______

Leaf Attachment: ______

Leaf stipule presence: ______

Leaf Duration: ______

Leaf Position if not cauline: ______

Leaf Arrangement: ______

Leaf Orientation if discrete: ______

Leaf Posture if discrete: ______

Petiole Shape: ______

Petiole Color: ______

Petiole Length: ______

Petiole Orientation: ______

Stipule Shape: ______

Stipule Surface adaxial: ______

Stipule Surface abaxial: ______

Stipule Length: ______

IF LEAVES SIMPLE:

Leaf Blade Color if unusual: ______

Leaf Blade Shape: ______

Leaf Blade Length: ______

Leaf Blade Width: ______

Leaf Blade Base: ______

Leaf Blade Margin: ______

Leaf Blade Apex: ______

Leaf Blade Apical Process: ______

Leaf Blade Division: ______

Leaf Blade Venation: ______

Leaf Blade Surface adaxial: ______

Leaf Blade Surface abaxial: ______

Leaf Blade Texture: ______

IF LEAVES COMPOUND:

Leaf Outline Shape: ______

Rachillae Number if decompound: ______

Leaflets Number if not very large: ______

Leaflet Arrangement: ______

Leaflet Blade Shape: ______

Leaflet Blade Attachment: ______

Leaflet Blade Color if unusual: ______

Leaflet Blade Length: ______

Leaflet Blade Width: ______

Leaflet Blade Base: ______

Leaflet Blade Margin: ______

Leaflet Blade Apex: ______

Leaflet Blade Apical Process: ______

Leaflet Blade Division: ______

Leaflet Blade Venation: ______

Leaflet Blade Surface adaxial: ______

Leaflet Blade Surface abaxial: ______

Leaflet Blade Texture: ______

Petiolule Shape: ______

Petiolule Color: ______
Petiolule Length: ______
Stipel presence: ______
Stipel Shape: ______
Stipel Surface adaxial: ______
Stipel Surface abaxial: ______
Stipel Length: ______

Inflorescence Position: ______
Inflorescence Bract presence: ______
Inflorescence Type: ______
Inflorescence Length: ______
Inflorescence Width: ______
Inflorescence Branch Orientation: ______
Inflorescence Sex: ______
Inflorescence Axis Surface: ______
Flower Sex: ______
Flower Bract presence: ______
Flower Length minus pedicel: ______
Flower Width minus pedicel: ______
Flower Arrangement: ______
Flower Orientation: ______
Flower Posture: ______
Flower Symmetry overall: ______
Flower Attachment: ______
Pedicel if present Length: ______
Pedicel if present Shape if unusual: ______
Bracts/Bractlets No (note inflorescence vs. flower): ______
Bracts Position: ______
Bracts Length: ______
Bracts Color if unusual: ______
Bracts Attachment: ______
Bracts Shape: ______
Bracts Base: ______
Bracts Margin: ______
Bracts Apex: ______
Bracts Apical Process: ______
Bracts Division: ______
Bracts Venation if unusual: ______
Bracts Texture if unusual: ______
Bracts Surface adaxial: ______
Bracts Surface abaxial: ______
Receptacle Size if unusual: ______
Receptacle Shape if evident: ______
Hypanthium presence: ______
Hypanthium Shape: ______
Hypanthium Length: ______
Hypanthium Width: ______
Perianth Cycly: ______
Perianth Arrangement if not whorled: ______
Perianth Type (if homochlamydeous): ______
Calyx/Outer Tepals Aestivation: ______
Calyx/Outer Tepals Fusion: ______
Calyx/Outer Tepals Symmetry: ______
Calyx/Outer Tepals Length: ______
Calyx/Outer Tepals Color if not green: ______
Calyx/Outer Tepals Surface adaxial: ______
Calyx/Outer Tepals Surface abaxial: ______
Calyx/Outer Tepals Venation if unusual: ______
Calyx/Outer Tepals Texture if unusual: ______
Sepal/Calyx Lobes/Outer Tepals Merosity: ______
Sepal/Calyx Lobes/O.T. Length: ______
Sepal/Calyx Lobes/O.T. Shape: ______
Sepal/Calyx Lobes/O.T. Base: ______
Sepal/Calyx Lobes/O.T. Margin: ______
Sepal/Calyx Lobes/O.T. Apex: ______
Sepal/Calyx Lobes/O.T. Apical Process: ______
Corolla Type (if dichlamydeous): ______
Corolla/Inner Tepals Aestivation: ______
Corolla/Inner Tepals Fusion: ______
Corolla/Inner Tepals Cycly if not uniseriate: ______
Corolla/Inner Tepals Color: ______
Corolla/Inner Tepals Symmetry: ______
Corolla/Inner Tepals Length: ______
Corolla/Inner Tepals Surface: ______
Corolla/Inner Tepals Venation if unusual: ______
Corolla/Inner Tepals Texture if unusual: ______
Petal/Corolla Lobes/Inner Tepals Merosity: ______
Petal/Corolla Lobe/I.T. Shape: ______
Petal/Corolla Lobe/I.T. Base: ______
Petal/Corolla Lobe/I.T. Margin: ______
Petal/Corolla Lobe/I.T. Apex: ______
Petal/Corolla Lobe/I.T. Length: ______
Petal/Corolla Lobe/I.T. Orientation: ______
Petal/Corolla Lobe/I.T. Posture: ______
Stamens (Androecium) Cycly: ______
Stamens (Androecium) Merosity: ______
Stamen Type: ______
Stamen Attachment: ______
Stamen Arrangement: ______
Stamen Position: ______
Stamen Insertion if applicable: ______
Stamen Fusion: ______
Staminodes if present No: ______
Staminodes if present Pos: ______
Staminodes if present Size: ______
Staminodes if present Shape: ______
Filament Shape/Color: ______
Filament Length: ______
Anthers Attachment: ______

Anther Type: ______
Anther Dehiscence Type: ______
Anther Dehiscence Direction: ______
Anther Color: ______
Anther Length: ______
Anther Shape: ______
Anther Thecae Arrangement: ______
Connective Morphology if unusual: ______
Pollen color: ______
Gynoecium Fusion: ______
Perianth Androecial Position: ______
Ovary Position: ______
Ovary Attachment if not sessile: ______
Ovary Color: ______
Ovary Length: ______
Ovary Shape: ______
Ovary Surface: ______
Styles Number per pistil: ______
Style Position: ______
Style Shape/Color: ______
Style Disposition/Length: ______
Stigmas Number: ______
Stigmas Position: ______
Stigmas Shape: ______
Stigmas Surface: ______
Nectaries presence/absence: ______
Nectary Type/Position: ______
Carpels Number: ______
Median Carpel Position relative to stem axis: ______
Locules Number: ______
Placentation: ______
Placenta Shape/Position if unusual: ______
Ovules Number per carpel: ______
Ovule Type: ______
Ovule Position: ______
Fruit Type: ______
Fruit Color: ______
Fruit Shape: ______
Fruit Length/Width: ______
Fruit Surface: ______

Seed Color: ______
Seed Shape: ______
Seed Length: ______
Seed Surface: ______
Funiculus Length: ______
Funiculus Shape: ______
Aril presence: ______
Aril Size: ______
Aril Shape: ______
Aril Position: ______
Seeds Nutritive Tissue: ______
Embryo Type Size/Shape/Position: ______
Cotyledon Position: ______
Radicle Position: ______
Seedling Type: ______

FLORAL FORMULA:
P ______ A ______ G
or K ______ C ______ A ______ G

Note: List number of parts after each symbol:
P = # perianth parts or tepals (outer + inner whorls)
or **K** = # sepals or calyx lobes **C** = # petals or corolla lobes
A = # stamens of androecium (outer + inner whorls)
G = # carpels of gynoecium (add ovary position)
( ) = fusion of parts [ ] = rare numbers of parts
Optional:
$\mathbf{K}_z$ = zygomorphic calyx; $\mathbf{C}_z$ = zygomorphic corolla; etc.

E. g., **K** (5) $\mathbf{C}_z$ (5) **A** 5 [4] **G** (2), inferior
= calyx synsepalous with 5 lobes
corolla zygomorphic, sympetalous with 5 lobes
stamens 5, rarely 4, distinct, in one whorl
gynoecium syncarpous, carpels 2, ovary inferior

E.g., **P** 3+3 **A** 3+3 **G** 3, superior
= perianth apotepalous with 3 outer and 3 inner tepals
stamens 6, distinct, in two whorls: 3 outer + 3 inner
gynoecium apocarpous, carpels (pistils) 3, ovaries superior

PLANT DESCRIPTION EXAMPLE

Tecoma capensis (Thunb.) Lindl. [*Tecomaria capensis* (Thunb.) Spach], Cape-Honeysuckle (native to S. Africa). Bignoniaceae.

Plant a shrub, up to ca. 5 m tall. **Root** a woody taproot with numerous lateral roots. **Stems** (aerial) highly and sympodially branched by abortion of terminal inflorescence meristems, branches basally inclined. **Bark** brown, smooth to minutely furrowed, lenticels orbicular to vertically elliptic with raised borders, ca. 1–2 mm wide. **Twigs** terete, minutely puberulent. **Pith** solid, circular in outline. **Fruit scars** (of infructescence) raised, circular, typically at junction of two, lateral branches. **Leaf scars** slightly raised below, orbicular with truncate apex. **Vascular bundle scar** U-shaped. **Buds** in leaf axils small (ca. 2 mm long), with outer two scales in a plane tangential to stem axis, scales valvate, lance-ovate and strongly cup-shaped, densely pubescent; terminal buds naked, elongate, to 5 mm long. **Leaves** 10–12 cm long, imparipinnate, petiolate, exstipulate, evergreen, cauline, opposite-decussate, divergent to inclined, and planar to recurved. **Petiole** green, terete to canaliculate, 1–3 cm long. **Leaf outline** elliptic. **Leaflets** 9 [11], opposite. Lateral leaflets elliptic to widely elliptic, subsessile, 15–17 mm long, 10–14 mm wide, base attenuate to obtuse, sometimes oblique, margin usually proximally entire and serrate to crenate distally, apex acuminate (caudate), tip minutely mucronulate, apical leaflet widely elliptic, usually petiolulate (**petiolule** green, narrowly winged, 3–13 mm long) 24–30 mm long, 15–20 mm wide, cuneate, entire at base and distally serrate to crenate, acute to accuminate, mucronulate. All leaflets pinnate-netted, midvein and secondary veins sunken above and raised below, mostly glabrous but with arachnose trichomes near abaxial vein junctions, mesophyllous. **Inflorescence** a terminal thyrse with several bracteate units of simple dichasia or of solitary flowers, the latter often with two abortive, lateral flower buds or with two sub-basal bracts (indicative of a vestigial dichasium). **Flowers** perfect, ca. 50 mm long, ca. 25 mm wide, opposite, appressed, recurved, zygomorphic, pedicellate. **Pedicel** ca. 7 mm long, terete. **Bract** 1 subtending each unit of inflorescence, 1–5 mm long, lanceolate, mucronulate. **Bractlets** 2, sub-basal, subtending lateral flowers if simple dichasium present. **Perianth** biseriate, dichlamydeous. **Calyx** synsepalous, actinomorphic, ca. 5 mm long. **Calyx lobes** acute, mucronulate, ca. 1 mm long. **Corolla** sympetalous, orange, zygomorphic, salverform-bilabiate with enlarged throat, ca. 45 mm long, recurved, inner surface pubescent. **Corolla lobes** 5 (2 posterior, 2 lateral, and 1 anterior), oblong to elliptic, apices rounded to emarginate, 7–12 mm long, 5–7 mm wide, inclined to divergent and recurved relative to floral axis. **Stamens** 4 fertile, uniseriate, filamentous, epipetalous, didynamous, alternipetalous, exserted, apostemonous. **Staminodium** 1, medio-posterior, reduced, up to ca. 10 mm long. Filaments (of fertile stamens) terete, yellow-orange, 35–40 mm long. **Anthers** versatile, basifixed, longitudinally and introrsely dehiscent (downwardly dehiscent at maturity), ca. 3 mm long, thecae divergent. **Pollen** orange. **Gynoecium** syncarpous. **Perianth/Androecial position** hypogynous. **Ovary** superior, green, 4–5 mm long, narrowly obloid, glabrous. **Carpels** 2. **Locules** 2. **Placentation** parietal-axile. **Ovules** many. **Styles** 1, terminal, apically recurved, purple-brown. **Stigmas** 2, ovate, divergent to appressed. **Nectary** dark maroon, doughnut-shaped, surrounding ovary base. **Fruit** a brown loculicidal capsule (with persistent replum), narrowly oblong, up to ca. 1 cm wide and 6 cm long. **Seeds** flat, with surrounding, yellowish, translucent wing, ca. 15 mm long and 8 mm wide (including wing), seed body roughly orbicular, ca. 6 mm in diameter.

FLORAL FORMULA: **K** (5) **C** (5) **A** $2+2+1_{\text{staminode}}$ **G** (2), superior.

APPENDIX 2

BOTANICAL ILLUSTRATIONS

Botanical illustration is the preparation and presentation of line drawings or paintings of plants and plant parts. Although photography is perhaps the primary medium of image documentation (and should be encouraged in field and laboratory classes), illustration is an important component of plant systematics studies and is generally required in publications to document features that are described. Illustrations are also important didactic tools in that they promote careful and complete observation of features; going through the process of drawing often helps an investigator to see more and in greater detail.

Illustrations in plant systematic research are almost always line drawings. Line drawings typically begin with a pencil drawing, which alone may be sufficient for personal observations and data collection. For publication-quality illustrations, pencil drawings must be retraced in black ink and are typically stippled for a shaded, three-dimensional appearance (see Figure A2.1). Alternatively, line drawings with shading may be computer generated using graphics software (below).

Basic supplies needed for pencil line drawings include a 2H or 3H pencil, high-quality eraser, and drawing paper (ranging from generic white to artist's drawing paper). A clipboard or artist's drawing board is handy to secure the paper. Drawings may be made free-hand. More precise drawings can be made using a camera-lucida/drawing tube device. Such an optical device allows one to see a double image of the object to be drawn plus the hand and pencil, allowing quick and easy tracing of object features. Pencil drawings should consist of outlines of whole organs/parts and of individual components. Lines should be crisp and precise. Minimize shading; shade only when objects are darker and then only lightly with pencil.

Drawings should be labeled clearly, including (a) name of taxon and documentation of material (e.g., reference to a voucher specimen), and (b) names of structures, indicating all pertinent terms, with lines or arrows leading from the structure to the term label. A metric magnification scale bar should accompany each drawing. A scale bar (e.g., "5 mm") is much preferred over a simple magnification listing (e.g., "25×") because the scale bar remains to scale with any subsequent enlargement or reduction of the drawing. The size of drawings should be planned for a final size reduction (after inking and stippling) of approximately 50–75%, which yields a better final product.

WHAT TO DRAW

A complete illustration of plant morphology may include drawings of the following (Figure A2.1): (a) the whole plant at low magnification, showing the plant habit, branching pattern and overall form; (b) one or more leaves, showing leaf attachment to the stem; (c) a flower in front, oblique, and/or side view; (d) a flower in median longitudinal (sagittal) section; (e) androecium, especially stamen/anther close-up; (f) gynoecium; (g) ovary longitudinal and/or cross section; and (h) close-ups of other floral parts of significance.

In addition to drawing real views of plant parts, diagrams may be drawn to illustrate the relative position of parts. **Floral diagrams** (Figures A2.1H, A2.2) show the relative position, aestivation, and fusion of perianth parts, stamens, and pistil(s)/carpels. For showing the relative position of floral parts, a diagram of the floral axis is typically indicated at the top of the drawing, corresponding to the posterior side of the flower. The floral diagram begins at the center of the flower. On a sheet of paper, draw the pistil(s) as appearing in cross -sectional view, carefully denoting ovary wall, septa, ovules, and placentation. Next, stamens are drawn surrounding the gynoecium. Stamens are drawn as anther cross sections (internal contents such as microsporangia usually not denoted),

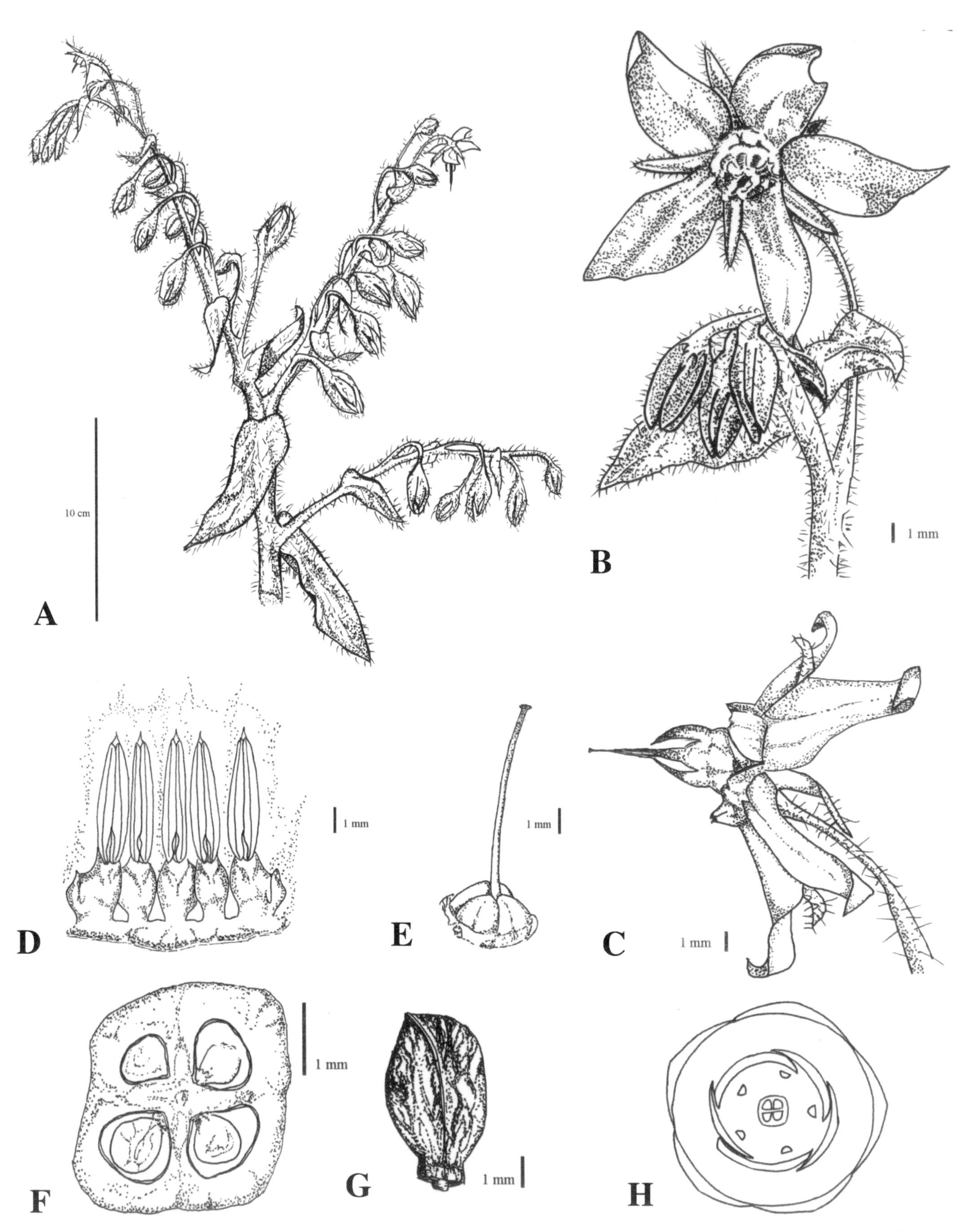

Figure A2.1 Example of illustrations, showing stippling and scale bars. *Borago officinalis*. **A.** Whole inflorescence and leaves. **B.** Flower, face view. **C.** Flower, side view. **D.** Androecium, spread flat, adaxial view. **E.** Gynoecium. **F.** Ovary cross section, base of ovary. **G.** Fruit. **H.** Floral diagram. (Contributed by Dinna Estrella, student in Plant Systematics class.)

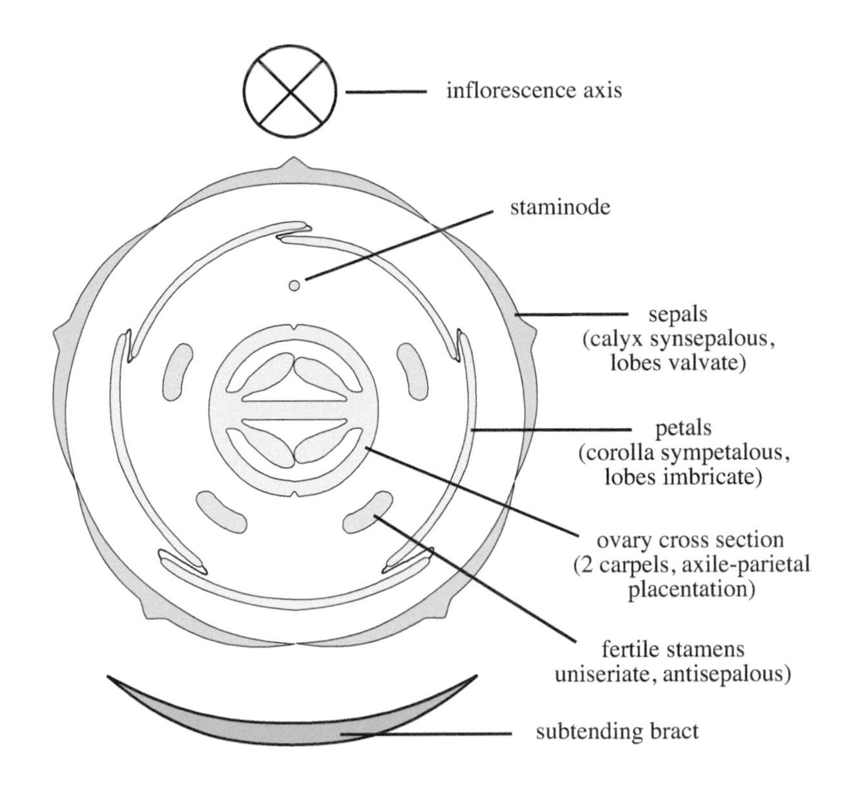

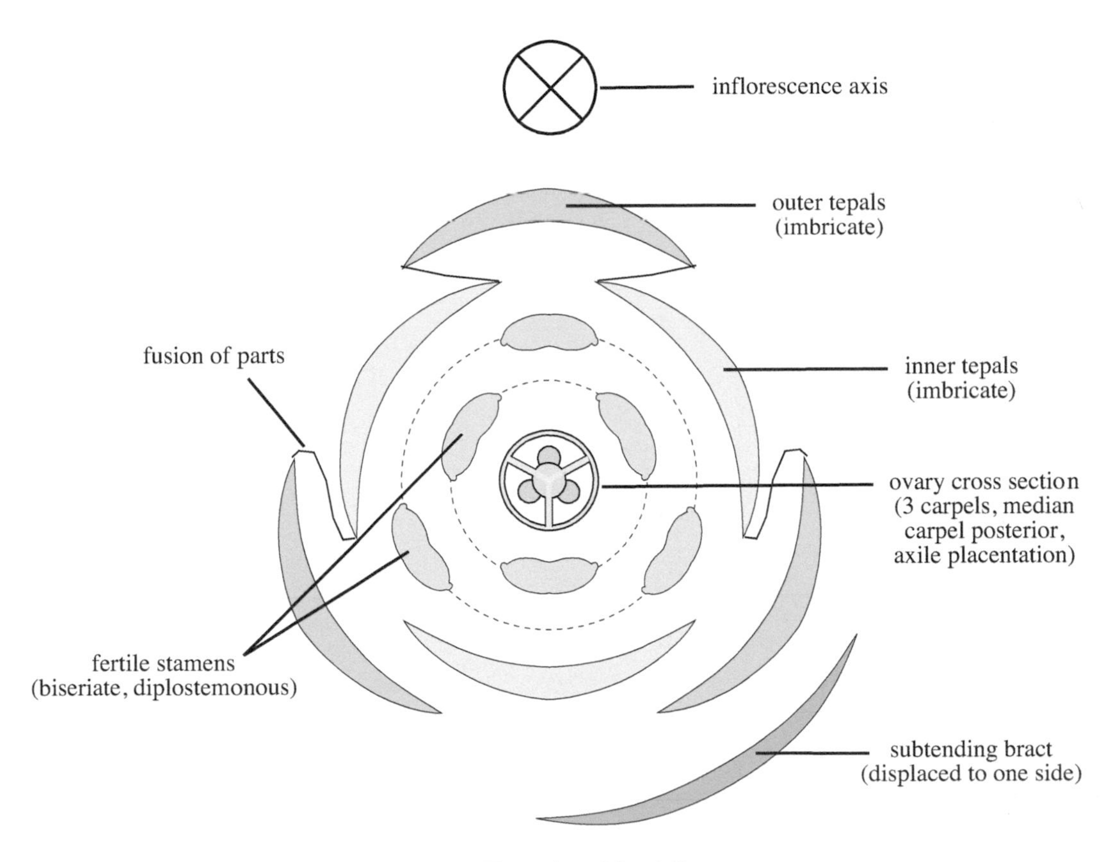

Figure A2.2 Examples of floral diagrams.

with the direction of dehiscence indicated. Petals (or corolla lobes) and sepals (or calyx lobes) are drawn surrounding the gynoecium and androecium. These perianth parts are drawn as in cross section, with careful attention to relative position and aestivation. Bracts are drawn similar to perianth parts, at their position of attachment. Any connation or adnation of parts is drawn either as organs contacting one another or as lines drawn between fused structures. If the flower is more or less erect, a circle is drawn that indicates the relative position of the axis to which the flower is attached. If the flower is horizontal in orientation, parts may be drawn as if the opening of the flower is facing the observer. If flowers are unisexual, male and female flowers should be drawn separately, of course.

REFERENCES FOR FURTHER STUDY

Cook, C. D. K. 1998. A quick method for making accurate botanical illustrations. Taxon 47: 317–380.

Holmgren, N. H., and B. Angell. 1986. Botanical Illustration: Preparation for Publication. The New York Botanical Garden, Bronx, New York.

Simpson, N. 2010. Botanical symbols: a new symbol set for new images. Botanical Journal of the Linnean Society 162:117-129.

APPENDIX 3

SCIENTIFIC JOURNALS IN PLANT SYSTEMATICS

JOURNALS

JOURNAL NAME	WEB SITE (at time of publication)
Aliso	https://scholarship.claremont.edu/aliso
American Fern Journal	https://www.amerfernsoc.org/american-fern-journal
American Journal of Botany	https://bsapubs.onlinelibrary.wiley.com/journal/15372197
Annals of the Missouri Botanic Garden	https://annals.mobot.org/index.php/annals/index
Australian Journal of Botany	https://www.publish.csiro.au/bt
Australian Systematic Botany	https://www.publish.csiro.au/sb
Biochemical Systematics and Ecology	https://www.journals.elsevier.com/biochemical-systematics-and-ecology
Blumea	https://www.ingentaconnect.com/content/nhn/blumea
Botanical Journal of the Linnaean Society	https://academic.oup.com/botlinnean
Botanical Review	https://link.springer.com/journal/12229
Botany	https://www.nrcresearchpress.com/journal/cjb
Brittonia	https://link.springer.com/journal/12228
Cladistics	https://onlinelibrary.wiley.com/journal/10960031
Edinburgh Journal of Botany	https://www.cambridge.org/core/journals/edinburgh-journal-of-botany
Fieldiana (Botany)	https://bioone.org/journals/fieldiana-botany
Fremontia	https://www.cnps.org/publications/fremontia
Grana	https://www.tandfonline.com/loi/sgra20
International Journal of Plant Sciences	https://www.journals.uchicago.edu/toc/ijps/current
Journal of Biogeography	https://onlinelibrary.wiley.com/journal/13652699
Journal of the Botanical Research Institute of Texas [formerly SIDA]	https://www.brit.org/journal-botanical-research-institute-texas
Journal of the North Carolina Academy of Science	https://www.jncas.org
Journal of Plant Research	https://link.springer.com/journal/10265
Journal of Systematics and Evolution	https://onlinelibrary.wiley.com/journal/17596831
Kew Bulletin	https://link.springer.com/journal/12225
Madroño	https://calbotsoc.org/madrono

Molecular Biology and Evolution	https://academic.oup.com/mbe
Molecular Phylogenetics and Evolution	https://www.journals.elsevier.com/molecular-phylogenetics-and-evolution
Muelleria	https://www.rbg.vic.gov.au/science/publications/muelleria/muelleria-online
New Phytologist	https://nph.onlinelibrary.wiley.com/journal/14698137
New Zealand Journal of Botany	https://www.tandfonline.com/loi/tnzb20
Nordic Journal of Botany	https://onlinelibrary.wiley.com/journal/17561051
Novon	https://bioone.org/journals/novon-a-journal-for-botanical-nomenclature
Pesquisas. Botanica	http://www.anchietano.unisinos.br/publicacoes/botanica/botanica.htm
Phytokeys	https://phytokeys.pensoft.net
Phytologia	https://www.phytologia.org
Phytoneuron	http://www.phytoneuron.net
Phytotaxa	https://www.mapress.com/j/pt
Plant Ecology and Evolution	https://www.plecevo.eu/index.php/plecevo
Plant Systematics and Evolution	https://link.springer.com/journal/606
Pollen et Spores	https://www.worldcat.org/title/pollen-et-spores/oclc/825130
Proceedings of the National Academy of Sciences of the United States of America (PNAS)	https://www.pnas.org
Schumannia	https://www.dkg.eu/cs/index.pl?navid=Schumannia_1025&sid=de
Selbyana	https://selby.org/botany/botany-resources/sbg-press
Smithsonian Contributions to Botany	https://opensi.si.edu/index.php/smithsonian/catalog/series/SCB
South African Journal of Botany	https://www.journals.elsevier.com/south-african-journal-of-botany
Systematic Biology	https://academic.oup.com/sysbio
Systematic Botany	http://www.sysbot.org
Taxon	https://onlinelibrary.wiley.com/journal/19968175
Tropical Plant Biology	https://link.springer.com/journal/12042
Willdenowia	https://www.bgbm.org/en/willdenowia

EXERCISES

1. Examine several of the botanical journals listed above. For each of the ones selected by your instructor: (a) record the general types of papers for which it specializes by reading the journal's mission and by perusing the index or website (above); (b) cite one specific article (of interest to you) *that deals with plant systematics* (*not* other fields such as physiology or ecology). For this, copy the citation format used in that particular journal.
2. Of the plant systematics articles that you recorded above, select the one of greatest interest to you and make a photocopy or download to an electronic device. Read the abstract of this article thoroughly. Then look over all the figures and tables. Next, skim through the introduction, materials and methods, and discussion/conclusion. Be prepared to give a 5-minute oral presentation (with projected tables, figures, or illustrations) on this article.
3. If a pdf is not freely available, email the author(s) and request a pdf be mailed to you. Many older articles have reprints available on JSTOR (https://www.jstor.org).
4. Search the International Plant Names Index (*http://www.ipni.org*). Find and record the full citation (including author, journal, date, etc.) of *any one* of the species that were studied in the article you chose for discussion.
5. Optional: Do a literature search on a particular item in plant systematics. It may be of taxon or topic. (Note: Consult your library as to how to access literature databases; also try https://scholar.google.com for general scientific literature.)

APPENDIX 4

STATISTICS AND MORPHOMETRICS IN PLANT SYSTEMATICS

Statistics is a branch of mathematics dealing with collecting, organizing, analyzing, and interpreting numerical data. **Morphometrics** is the study of shape or form, and generally uses statistical methods. Statistical and morphometric analyses often go hand in hand and are used in plant systematics to quantify structural features. The quantified features can be used to derive a new character or to refine or define two or more character states of an existing character. These quantitative characters may be used in two major ways: (1) to discriminate between two or more taxonomic entities; e.g., to assess the distinctiveness of taxa, such as a species, infraspecies, or hybrids; and (2) to serve as data in phylogenetic analyses.

The following is a very brief overview of these statistic and morphometric techniques and how they might be used to address taxonomic and systematic problems. See References for Further Study for detailed information.

QUALITATIVE AND QUANTITATIVE CHARACTERS

Characters may be divided into two general types: qualitative and quantitative. **Qualitative** (or **categorical**) characters are those in which the states are not directly measured but are based on defined classes of attributes. Examples of qualitative characters are: (1) petal color, with states red and yellow; (2) ovary position, with states superior and inferior; and (3) leaf shape, with states ovate and elliptic. Qualitative characters are assessed by observation and comparison with self-evident or pre-defined terms. Nothing is measured; features are observed and compared to the definitions of our terminology.

Quantitative characters are those in which the states are measured and based on numbers. There are two general types of quantitative characters: continuous and discrete. **Continuous** quantitative characters are those in which individual measurements are not necessarily integers and may potentially form a continuum. Examples would include: (1) inflorescence length, with states ranging from 6.2–10.4 cm versus 14.6–28.1 cm; or (2) leaf area, with states 4–8 cm^2 and 11–22 cm^2. **Discrete** quantitative characters are those in which measurements of are always integers, including (1) number of parts (also called **meristic** characters), such as stamen number (e.g., 5, 10, or 15 per flower); and (2) presence/absence data (e.g., presence/absence of a corolla).

The difference between qualitative and quantitative characters is arbitrary. Any qualitative character can be quantified (e.g., "petal color" can be quantified in terms of ranges of visible light wavelength), and quantitative characters can be made qualitative by defining and naming classes of attributes (e.g., a leaf area of 4–8 cm^2 can be arbitrarily termed "small," and one of 11–22 cm^2 can be termed "large"). Qualitative characters are in reality a product of our terminology and may actually represent continuous variation or be arbitrarily divided into discrete states made to conform to these pre-existing terms. When variation is carefully examined, clear breaks in the character may not be present. In these cases, it may be necessary to standardize or more precisely define qualitative characters and character states (see Stevens 1991). Perhaps the most important criterion in defining either type of character is whether (and by what criteria) the states are non-overlapping versus overlapping. Statistical methods are used to evaluate this.

STATISTICS IN TAXONOMY

One use of statistical methods in taxonomy is ascertaining if and how two or more taxa are different from one another, e.g., to evaluate if two or more infraspecies or morphologically similar species are the same or different with respect to the features that have been used to distinguish them. The past differentiation of these taxa may have been based on observations of specimens, but with no supporting data presented. A careful statistical study can corroborate or refute these past classifications, or provide evidence for new groupings.

For example, suppose two very similar plant species have been distinguished solely by fruit length: species *Q* with fruits 0.4–1 mm long, species *R* with fruits 1.2–1.8 mm long. Are these species really distinct from one another? Or do the features used to differentiate them form a continuum with no clear breaks? The following summarizes some steps in acquiring data, calculating simple statistics, graphing the data, and evaluating differences by statistical tests.

DATA

One consideration in statistical methods is the **sample**, the subset of the total that is actually measured (given that the entire "population" cannot be feasibly measured). Ideally, the sample should be large enough in size (represented by **n**), and random enough in distribution to be representative of the statistical population. However, taxonomic studies are often restricted to herbarium specimens, which, unfortunately, may be limited in number and not necessarily representative of the range of variation of the taxon. The ideal situation, in a rigorous study, is to make extensive, new collections of the taxa of interest, over their entire geographic and habitat range.

A second consideration in such a study is **comparability**, an assessment that the features measured are homologous and heritable, in order that a study can assess evolutionary change. Plants and plant parts can show environmental plasticity. Things like temperature and sunlight exposure, soil moisture and mineral content, and interactions with other organisms can cause considerable variation in the size, number, and shape of plant parts. In addition, one should always measure features that are comparable sexually (e.g., males compared with males and females compared with females in a dioecious species), positionally (e.g., leaves, flowers, or fruits comparable in position on the plant, e.g., the periphery versus center of the inflorescence, or base versus apex of the plant), and developmentally (having comparable stages of maturation).

A third consideration is **precision**. Features to be measured should, of course, be defined carefully, so there is no ambiguity as to what is actually quantified. In addition, the device used to make the measurements must be precise. Use of a digital camera (mounted, as needed, on a microscope), scale, and computer-interfaced software (see References for Further Study) may give the best and most consistent type of data, one that has a permanent, easily accessed record.

SIMPLE STATISTICS

Recall the taxonomic problem cited above: Is species *Q*, traditionally defined as having fruits 0.4–1 mm long, different in this feature from species *R*, defined with fruits 1.2–1.8 mm long? One method of tackling this question is simply to measure the fruits of numerous individuals of the two putative species, and determine if they: (1) sort into two groups, one with shorter fruits and one with longer, corresponding to the two groups; (2) do not sort into groups at all; or (3) sort into groups that are novel with respect to the previous taxonomy.

Because this problem is to assess if two previously defined and name species are indeed different, the appropriate approach is to examine features of separate individuals of each of the two species, typically from plant herbarium specimens (each of which is assumed to represent a separate individual plant). The factors mentioned earlier, a large and unbiased sample size, comparability, and precision of definition and measurements should be taken into account.

Given that numerous fruits from each sample (individual plant specimen) are measured, several standard statistics are typically calculated: (1) an average of the values, typically the **mean** (the numeric average; Table A4.1, illustrated in Figure A4.1), used if the features follow a normal (Gaussian) distribution; if the data are skewed, a better average might be the **median**, the middle value of the list of values; (2) the **range** of values, from the minimum to maximum measurements (e.g., as in Figure A4.1); (3) **quartiles**, the three data points (1st, 2nd, and 3rd quartiles) that divide the data into four equal parts (example in Figure A4.4); the second quartile is the median.

Also often calculated are simple statistical parameters that measure how much the data varies relative to the mean. For example, of two samples of n=5, one with fruit lengths of 0.5, 0.6, 0.7, 0.8, and 0.9, and another with fruit lengths of 0.65, 0.675, 0.7, 0.725, and 0.75, both have the same mean (0.7 mm), but the former has a greater variability relative to the mean. The **standard deviation** of a sample is a measure of the differences of individual values from the mean of all values, calculated as the square root of the "sum of squares" (SS_x) divided by the sample size minus one (Table A4.1). With a normal distribution of the data, ±1 standard deviation from the mean encompasses approximately 68% of the data; ±2 standard deviations of the mean account for approximately 95% of the data. **Standard error** is a measure of how close the sample mean is to the mean of the entire population. Standard error is calculated as the standard deviation divided by the square root of the sample size (Table A4.1). A range of ±2 standard errors will include about 95% of all sample means. As with standard deviation, calculations of standard error are accurate only if the data have a normal distribution (conforming to *parametric* statistics).

Sample size (total number of observations): n

Mean: $\bar{x} = \dfrac{\sum x}{n}$

Standard deviation: $s_x = \sqrt{\dfrac{SS_x}{n-1}} = \sqrt{\dfrac{\sum(x-\bar{x})^2}{n-1}}$

Standard error: $se_x = \dfrac{s_x}{\sqrt{n}}$

TABLE A4.1 Some simple statistics.

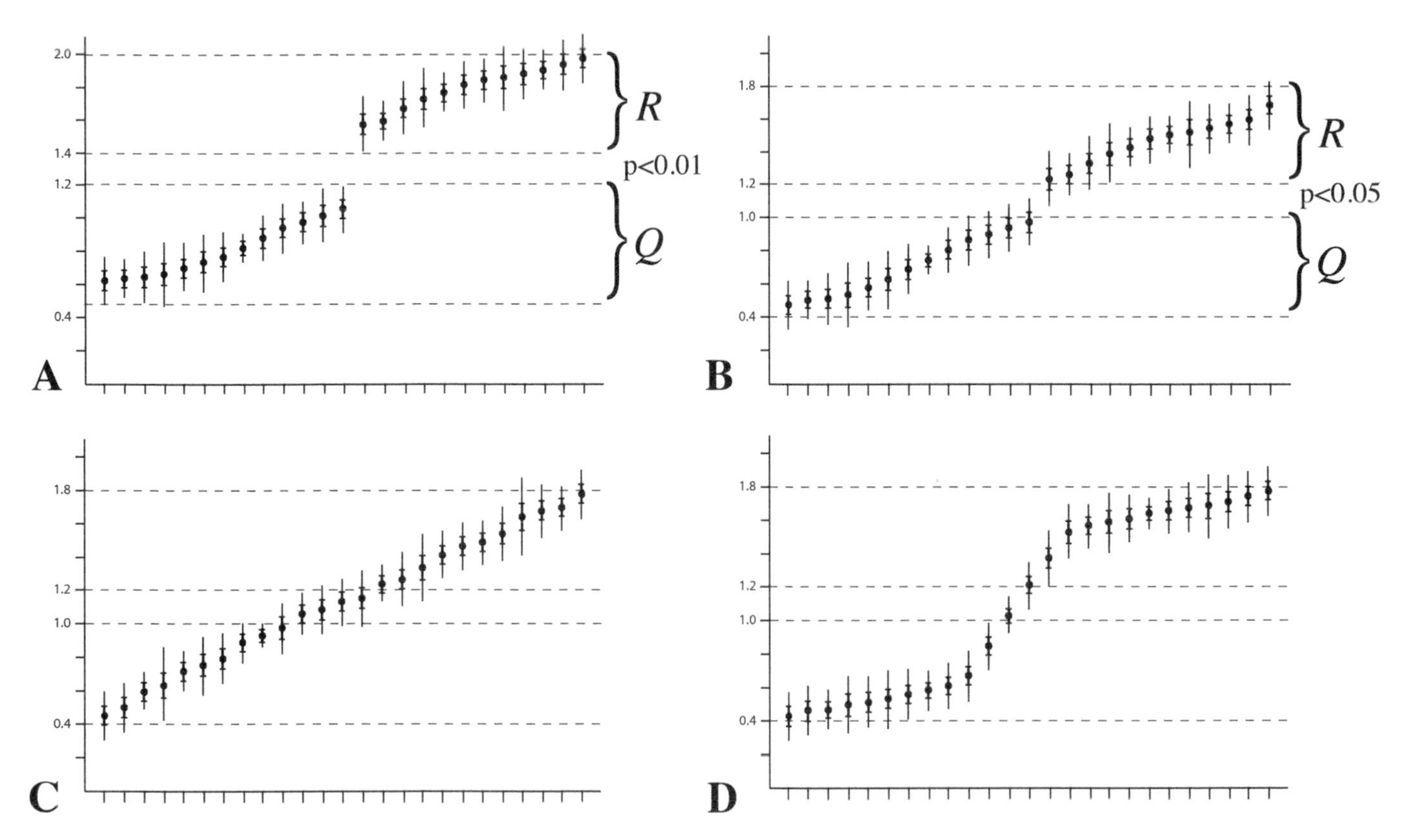

Figure A4.1 Univariate graphs and associated statistics of fruit length (mm), plotted by increasing mean for individual specimens of species *Q* (fruits defined as 0.4–1.0 mm long) and species *R* (fruits defined as 1.2–1.8 mm long). Mean is indicated by dots, ranges by vertical line, and ±1 standard deviation by bars on each side of the mean. P values indicate results of t-test, comparing species *Q* and *R*. **A.** Plot showing non-overlapping fruit lengths between species, although the boundaries of the features are different than originally defined: *Q* now defined by fruits 0.5–1.2 mm long, *R* by fruits 1.4–2.0 mm long. Note small p value. **B.** Plot showing mostly clear break in fruit length between species (by original definitions), but with some samples overlapping in range. Note larger p value. **C.** Plot showing continuous grade of fruit length, with no clear breaks. **D.** Plot showing generally smaller and larger-fruited groups, but with some individuals intermediate in fruit size.

UNIVARIATE AND BIVARIATE PLOTS

In the example cited above (assessing whether fruit length differentiates between species *Q* and *R*), a graph or plot of this feature can be informative (Figure A4.1). Note in this example that individual specimens are sorted on the X-axis by increasing sample means, with range and ±1 standard deviation of values shown. Because only a single feature (fruit length in this case) is being graphed, this is called a **univariate plot**.

This simple plot may reveal any number of trends. One scenario is that the two species show a clear break in fruit length (Figure A4.1A), corroborating their distinctiveness and recognition as separate entities; however, note that the ranges of fruit length characterizing the two taxa are different than originally proposed. A second scenario is similar to the first, only in this case the ranges of the two groups overlap for some specimens (Figure A4.1B). Can we be sure that the two species are distinct? A third scenario is that the samples show a continuously overlapping grade in fruit length (Figure A4.1C). In this case, one might conclude that the fruit length character originally used to distinguish the two species was arbitrarily determined and does not reflect real taxonomic entities. Thus, the separation of these two species comes into question; perhaps they are best merged into one species. Yet another scenario is that two entities are generally recognizable in the original definition of fruit length differences, but with some intermediate individuals (Figure A4.1D). The two species in this case could be genetically different but intergrading, e.g., among samples intermediate to their geographic ranges. This might be interpreted as evidence that the two "species" have incompletely diverged and are best treated as infraspecies of one species. Or, the intermediates could represent occasional hybrids between the two parent species, where they come into contact. (Other types of data, e.g., detailed molecular studies or correlation with biogeography or habitat, would be needed to tease apart these possibilities.) Tests of statistical differences (see later discussion) can be done to assess the delimitation of taxa in the various scenarios of Figure A4.1.

Another graph sometimes used to assess character differences among taxonomic groups is a **bivariate plot**, in which the relationship between two variables is graphed (Figure

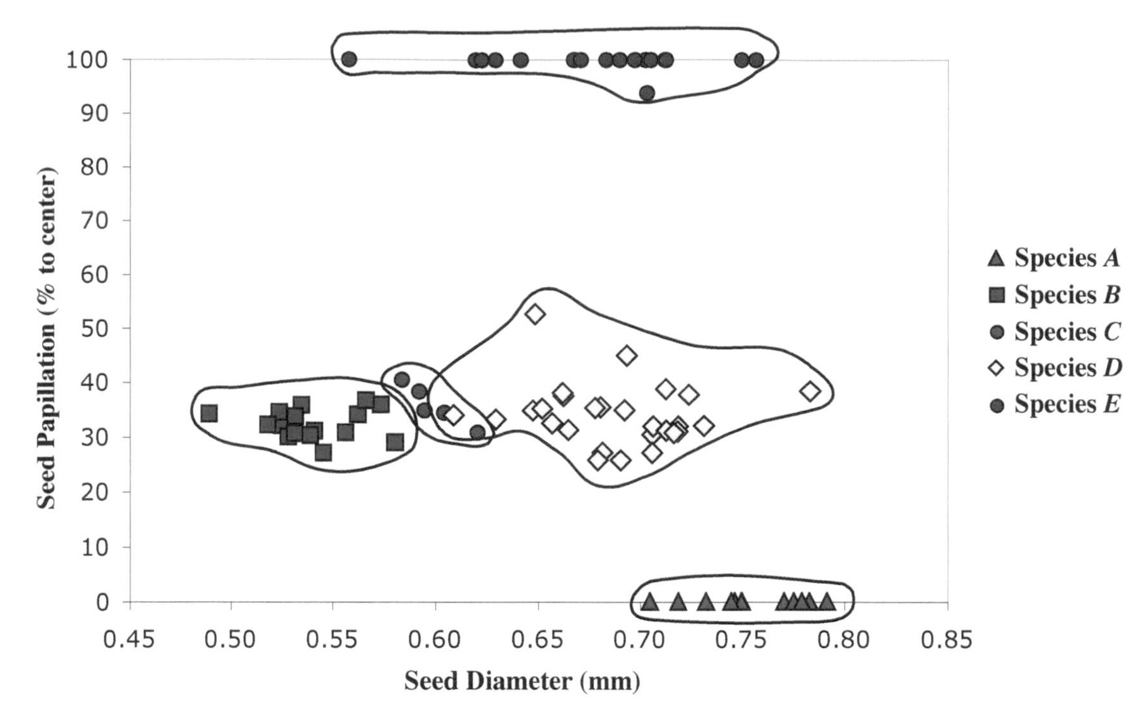

Figure A4.2 Bivariate scatter plot of two morphological features, in this case seed papillation and seed diameter, for species *A–E*. Note clustering of points into five species groups, with some overlap between species *C* and *D*.

A4.2). In this example, species *A–E* are assumed to be discrete entities, and the mean values for each individual within the taxa are plotted. The advantage of a bivariate plot is that boundaries between taxa can often be better established by using two characters. Note that in this example, seed diameter alone would show considerable overlap among taxa, but a combination of seed diameter and seed papillation separates the taxa into more differentiated groups (Figure A4.2).

One type of bivariate plot traces the development (or ontogeny) of a feature. Developmental data require plotting one variable, Y, against time, either as real time or as time estimated by some other criterion (e. g., the size of a particular organ, which increases with time). Such a plot of a variable as a function of real or relative time for a given taxon yields a so-called ontogenetic trajectory (Figure A4.3). A comparison of the ontogenetic trajectories between different taxa may be used both to refine character definition and to assess the evolutionary change of a given feature.

STATISTICAL TESTS

Graphing measurements from a study can give insight as to the differentiation of taxa. However, statistical tests are often needed to determine the degree of differences among groups, given the variation in the samples. In these tests, the "null hypothesis" is that there are no differences between the groups being compared. If any differences between populations are due to chance or random variation, then the null hypothesis is not rejected. However, if the differences are likely *not* due to chance, then differences between groups then the null hypothesis is rejected. Typically, the null hypothesis is rejected if the probability of differences between groups being due to chance alone is less than 5% ($p < 0.05$), though sometimes values of less than 1% ($p < 0.01$) are considered. Although these p values are widely accepted, they are ultimately arbitrary. Note: The term "statistically significant" has been rejected by a number of leaders in the field; see, e.g., Hurlbert et al. 2019.

The commonly used tests for significance are valid only with data that have a normal (Gaussian) distribution, and in the purview of *parametric* statistics (as mentioned earlier with standard deviation and standard error calculations). Most morphological data used in taxonomic studies will apply to parametric statistics, if the sample size is large enough and there are no biases in selecting the individual plant or character. [Note that there are statistical tests to determine if the data are normally distributed. If they are not, the data can be **transformed**, e.g., by converting all values to their logarithms; these transformed data may often have a normal distribution and be subject to parametric statistics.]

Linear regression analysis is a statistical procedure for fitting a straight line onto a bivariate plot of two variables, X and Y and testing the significance of relationship between them. Linear regression analysis assumes that the average relationship of Y to X can be represented as a straight line. This is a big assumption, and non-linear methods may need to be considered. Linear regression is not often used in systematic or taxonomic work, but might possibly be used to evaluate slopes

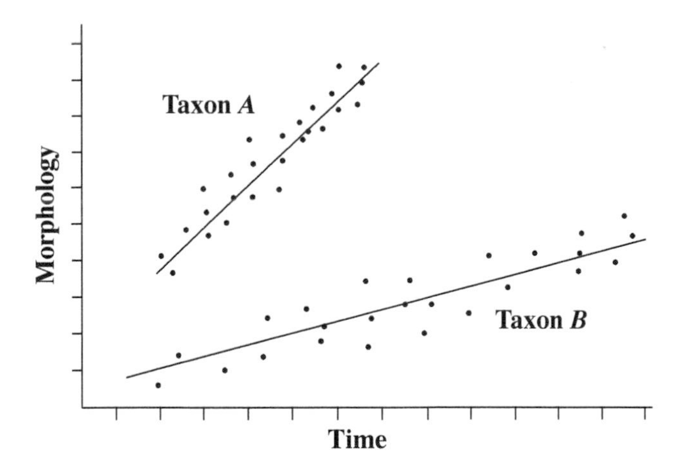

Figure A4.3 Ontogenetic bivariate plot of a morphological feature over time of two taxa (*A* & *B*), with linear regression line superposed.

of curves in tracing a developmental pathway (Figure A4.3) or a bivariate plot of two morphological features. The character states in these cases could be the slopes of the curves.

One test of significance between sample means is the **t-test** (or **student's t-test**). One form of this, the "unpaired t-test," may be used to statistically compare two samples in order to evaluate the probability of their being the same (the null hypothesis). The relationship of this statistic to probability is a function of the number of *degrees of freedom*, which is related to sample size. For example, in the scenarios of Figure A4.1, a t-test of the two groups corresponding to species *Q* and *R* of Figure A4.1A, in which all samples of the two groups are combined, yields a probability of <0.01, meaning that the two species are statistically different for fruit length. Similarly, a t-test for the combined species samples of Figure A4.1B, showing a slight gap in the data, yields a higher probability but one <0.05, a general cut-off that the two groups show differences.

Analysis of variables (ANOVA) is used to evaluate differences among more than two groups in a single test of significance. ANOVA is somewhat complex (see references listed), but basically compares variability between samples with variability within samples. A statistic, known as the F-value (= between-sample variance/within-sample variance), is calculated, along with the p significance value. Additional tests, known as **post hoc tests**, are used to determine which samples are different from others. For example, Figure A4.4 represents a graph of a feature (fruit length) derived for six taxa. An ANOVA, followed by a *Tukey post hoc test* showed that taxon *T* and taxon *X* are different from all other taxa in the analysis at the $p < 0.01$ level, but no other combinations show pair-wise differences.

MULTIVARIATE STATISTICS

Multivariate statistical methods are able to evaluate the analysis of two or more (often numerous) variables simultaneously. These separate variables might include morphological characters such as leaf length, leaf width, calyx length, corolla tube length, corolla limb width, stamen length, and ovary width.

Principal components analysis (PCA) is a multivariate method that transforms the data variables into other variables called principal components. The first principal component projects the greatest amount of variance in the data, the second principal component the second greatest amount of the remaining variance, and so forth. One use of these new variables is a "scatter plot," which permits visualization of the relationships of the totality of data variables as clusters, which might otherwise not be apparent. For example, Figure A4.5A shows the plot of the first and second "factors" of a PCA, calculated from 12 different morphological variables. Note that the five species are (to various degrees) clustered and separable from the others, although the outgroup taxon overlaps (Figure A4.5A).

Discriminate function analysis is a similar multivariate technique that determines the variables that discriminate among (in this case) taxonomic groups.

MAPPING OF MORPHOLOGICAL DATA

Data obtained from statistical taxonomic studies may be mapped to illustrate correlations between morphological features and geographic distribution. Various graphics and symbols are often used to represent one or more characters (Figure A4.5B).

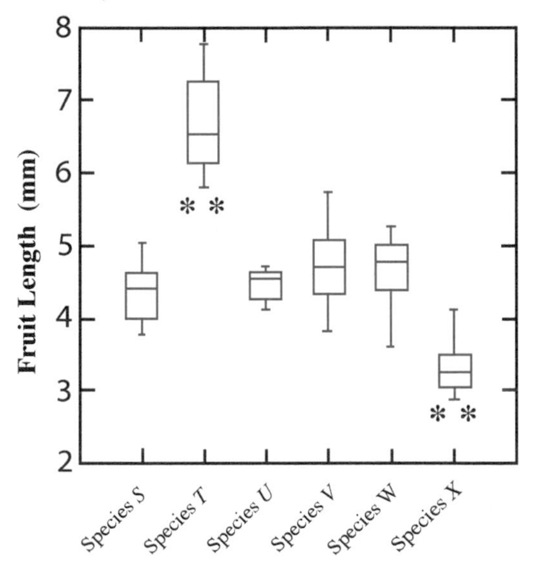

Figure A4.4 Box plots of fruit length for six hypothetical taxa, showing results of ANOVA and Tukey post hoc test. The "box" shows limits of the first and third quartiles (horizontal line = median), with uppermost and lowermost data points indicated by "tails." Note difference of species *T* and species *X* from all other species in this character at $p < 0.01$ (**).

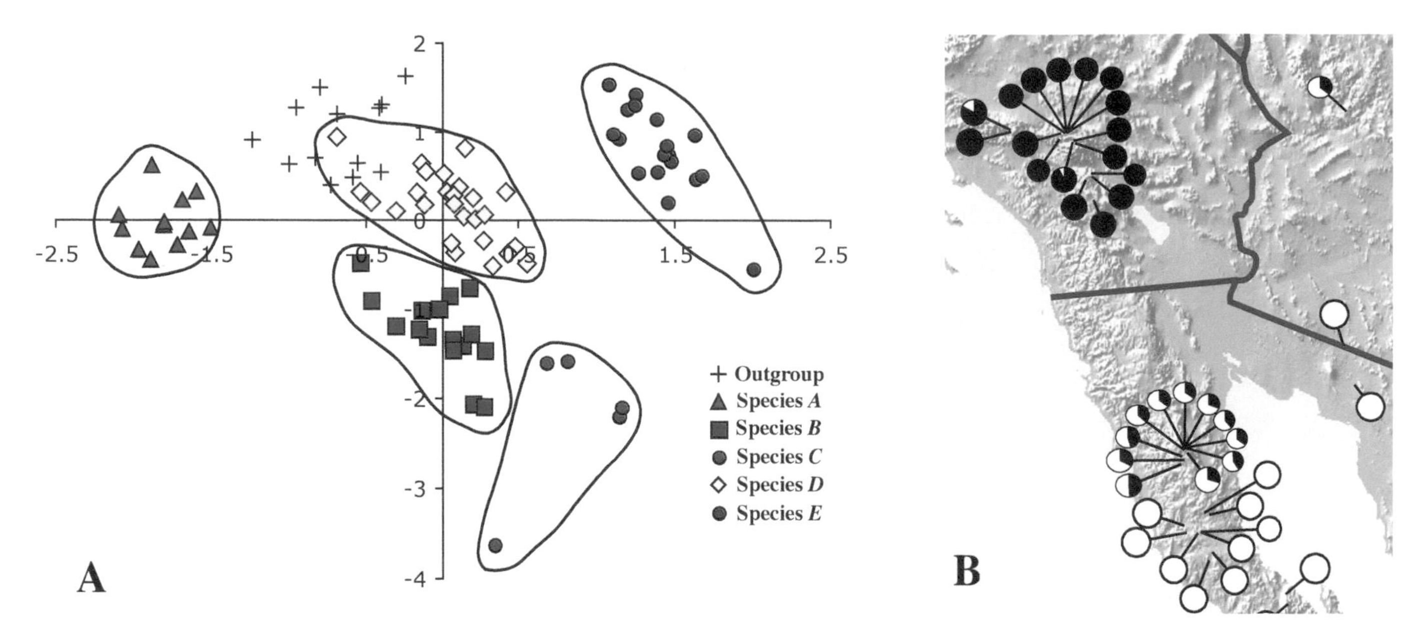

Figure A4.5 **A.** Graph of results of first and second principal components from analysis of six taxa. Note clear separation of most taxa, but overlap of species *D* with outgroup taxon. **B.** Map, showing distribution and two morphological features of taxa. Diameter of circle is representative of seed size, shading with quantification of seed sculpturing.

MORPHOMETRICS

Morphometrics is the study of shape and form and may utilize the statistical methods mentioned earlier. Morphometric techniques can be used in systematics and taxonomy to characterize differences in morphology between two or more taxa, typically for a complicated structure that may be broken down into numerous, separate features. It can be used to assess if two or more taxa are different, derive new characters and character states, or to trace and compare the development of a given structure through time.

Geometric morphometrics is a class of techniques that utilized **landmarks**, points on the object that generally correspsond to homologous features (Bookstein 1991, Zelditch et al. 2004). For example, a shape comparison of three species differing in corolla and androecium position could be conducted by taking measurements from longitudinally-sectioned corollas, spread flat; from these structures, homologous points are defined, such as the corolla base, corolla lobe apices and sinuses, point of stamen insertion, filament apex, stamen apex, etc. (Figure A4.6A–C). (In addition, sometimes "pseudolandmarks" may be defined, these being relative and not directly attributed to a homologous point but formed, e.g., by the intersection of or arbitrary distances from true landmark points and lines between them.)

One morphometric technique involves connecting every homologous point with every other homologous point and measuring the real distance between them (Figure A4.6A–C), this pattern of lines sometimes called a "truss" network (Strauss and Bookstein 1982). These data are typically log transformed. The log-transformed measures are then used as parameters in a principal components analysis. The resulting pattern of principal components can be used to assess differences between complicated structures among taxa, e.g., those distinguished by fine differences of morphology (Figure A4.6D). (See Dickinson et al. 1987 and Shipunov and Bateman 2005 for examples.)

However, there are many other morphometric techniques (some quite mathematically sophisticated), including elliptic Fourier functions, which analysis shape outlines (Kuhl and Giradina 1982, Premoli 1996, McLellan and Endler 1998, Olsson et al. 2000, Jensen et al. 2002) and relative warp scores (Bookstein 1989, Rohlf 1993, Jensen et al. 2002). An explanation is beyond the scope of this introduction, but see references sited for more information.

QUANTITATIVE CHARACTERS IN PHYLOGENETIC ANALYSES

The statistical methods reviewed earlier can be used to more precisely define a morphological character and its character states, often used to help differentiate between two or more taxa, e.g., in defining the circumscription of infraspecies, close species, or hybrids. Another use of statistical methods is to clearly derive characters and character states that may be either used directly in phylogenetic analyses or traced (optimized) on a tree to evaluate evolutionary changes in these features.

Although some feel that character states used in phylogenetic analyses should always be discrete and non-overlapping, there are many cases of characters that show no clear breaks between taxa, and yet obviously have some information reflect-

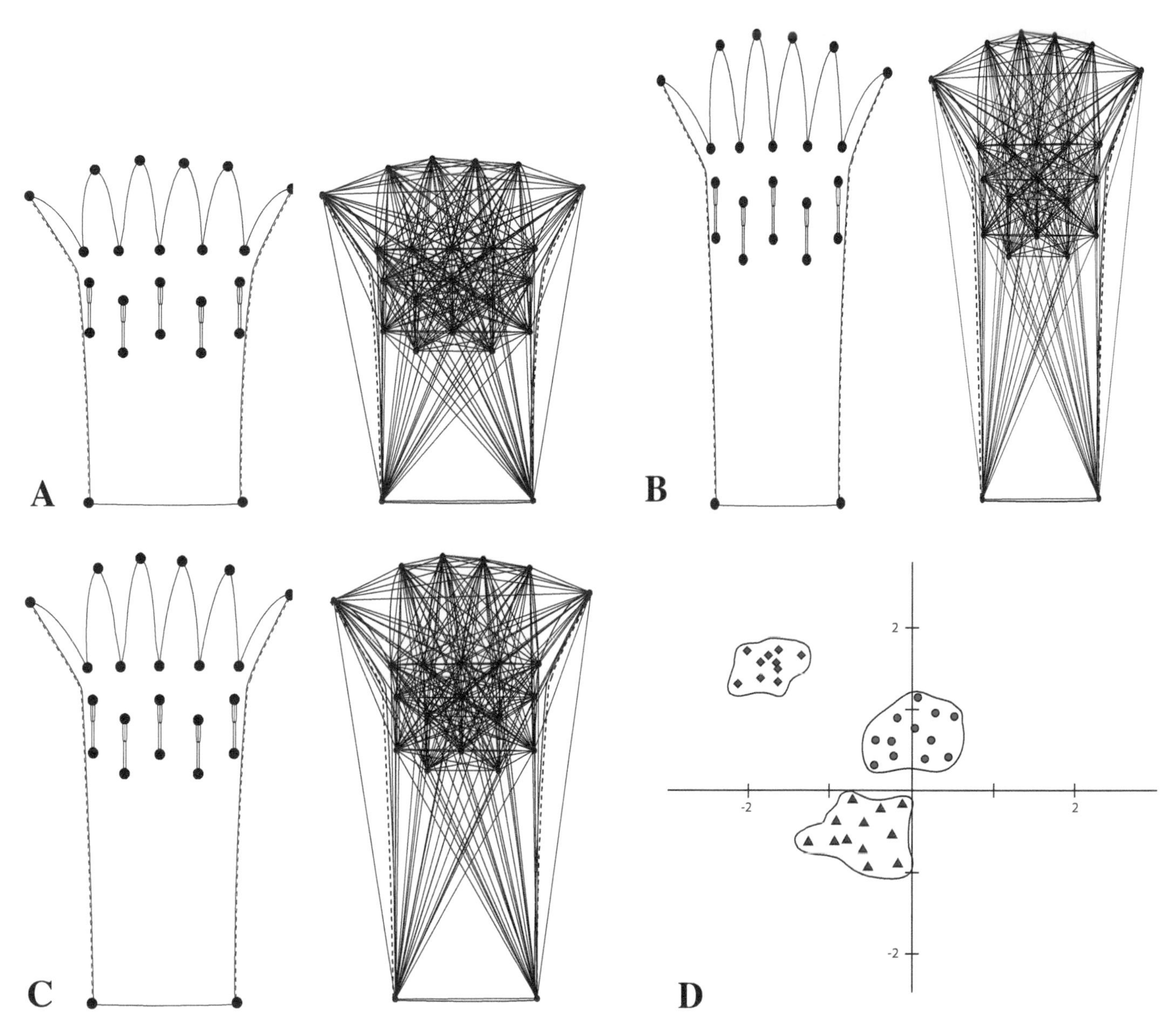

Figure A4.6 **A–C.** Example of morphometric comparison of three taxa. Illustrated are corollas which are longitudinally cut and flattened, with adnate stamens. Homologous points are shown at left, and "truss" network of lines connecting dots shown at right. **D.** Principle components analysis plot of first and second axes, showing multivariate analysis using log-transformed distances between homologous points.

ing evolutionary history. For example Figure A4.7A shows a plot of a feature (sepal width) in ten taxa. Although no taxa are clearly different from all other taxa (as would be assessed, e.g., with a p value using ANOVA), there is obviously information that might be valuable in elucidating relationships or at least trends in this feature.

One method of dealing with continuous data like this is to refine the morphological analysis to smaller clades. Morphological data of the entire study group (ingroups and outgroups) may overlap, but within a smaller subset (e.g., the outgroup alone or a well-supported clade within the ingroup) a character may show discrete character states.

Another method of dealing with this type of continuous data is dividing more or less continuous variation into discrete states, a process known as **gap-coding**, **gap-weighting**, or **homogeneous-subset coding** (Mickevich and Johnson 1976, Almeida and Bisby 1984, Archie 1985, Baum 1988, Goldman 1988, Chappill 1989). In these methodologies, character states that overlap are placed into different subsets by some criterion. The *pooled within-group* standard deviation (s_p, calculated from the total data of all taxa studied) multiplied by some constant (c, which is often =1) is often used as a measure to obtain discrete states from a continuum. One method of deriving these subsets is to compare the difference of means

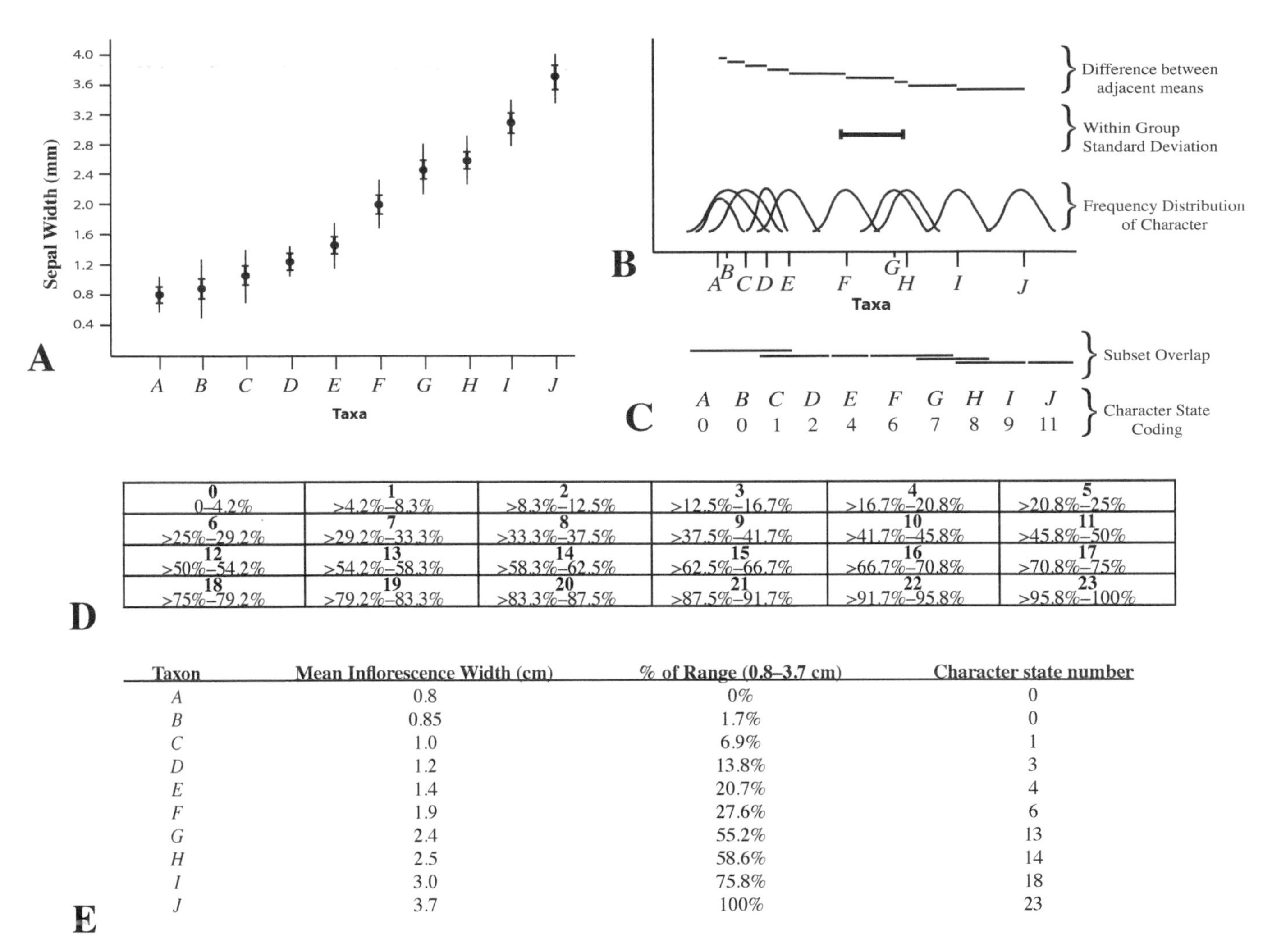

0	1	2	3	4	5
0–4.2%	>4.2%–8.3%	>8.3%–12.5%	>12.5%–16.7%	>16.7%–20.8%	>20.8%–25%
6	7	8	9	10	11
>25%–29.2%	>29.2%–33.3%	>33.3%–37.5%	>37.5%–41.7%	>41.7%–45.8%	>45.8%–50%
12	13	14	15	16	17
>50%–54.2%	>54.2%–58.3%	>58.3%–62.5%	>62.5%–66.7%	>66.7%–70.8%	>70.8%–75%
18	19	20	21	22	23
>75%–79.2%	>79.2%–83.3%	>83.3%–87.5%	>87.5%–91.7%	>91.7%–95.8%	>95.8%–100%

Taxon	Mean Inflorescence Width (cm)	% of Range (0.8–3.7 cm)	Character state number
A	0.8	0%	0
B	0.85	1.7%	0
C	1.0	6.9%	1
D	1.2	13.8%	3
E	1.4	20.7%	4
F	1.9	27.6%	6
G	2.4	55.2%	13
H	2.5	58.6%	14
I	3.0	75.8%	18
J	3.7	100%	23

figure A4.7 **A.** Graph of 10 taxa, showing continuous grade of feature, with no clear breaks into character states. **B.** Frequency coding (after Weins, 1995). Note that the 24 frequency "bins" correspond to ranges of the data, each assigned to a character state between 0 and 23. Character weight is scaled to 1/23 (= min. no. state transformations) = 0.0434.

between adjacent taxa (Figure A4.7B). If this difference is less than the cs_p value, they are included within the same subset; if adjacent means are greater than cs_p, they are placed in different subsets, which may overlap (Figure A4.7C). Finally, these subsets form the basis for character coding. For example, in homogeneous-subset coding (Archie 1985), taxa in the lowest subset receive a character state of 0. For successive taxa, the state value is increased by 1 if the taxon is part of a different subset and increases by another 1 if it is no longer part of the previous subset (Figure A4.7C). In this fashion, a continuously overlapping character (as in Figure A4.7A) can be subdivided into character states (typically treated as "ordered"; see Chapter 2) that reflect differences between subgroups (Figure A4.7C). Other methods for dealing with character coding have been proposed (e.g., Strait et al. 1996).

Another method for coding overlapping characters uses **gap-weighting**. In one gap-weighing technique (Thiele 1993), character states are placed into subsets according to the formula: $x_s = n\,(x - \min) / (\max - \min)$, where x is the mean value of the original character state, x_s is the gap-weighted character state, "max" and "min" are the maximum and minimum state values for all taxa, and n is the maximum allowable number of character states (dependent on the computer algorithm). From the same example of overlapping character states cited earlier (Figure A4.7A), the mean of each taxon is placed into a subset, often called a "frequency bin" (Figure A4.7D), based on its linear distance from the minimum to maximum values. The character states obtained are, again, typically treated as an ordered character, which may be proportionally scaled relative to other characters. Thus, in this example, a greater distance, in terms of state changes, occurs between adjacent taxa *I* and *J* (5 steps) than between adjacent taxa *A* and *B* (zero steps). (Frequency bins can also be used to code polymorphic characters; see Wiens 1995.)

REVIEW QUESTIONS

1. What is the difference between a qualitative and quantitative character? Give two examples of each.
2. What features of data and data acquisition need to be assessed?
3. Define mean, median, range, quartiles.
4. What does standard deviation measure? Standard error?
5. What are univariate plots and how might these be used in systematics and taxonomy?
6. What is a bivariate plot and what two types are used in systematics? What is its advantage over a univariate plot?
7. What is linear regression analysis and how might it be used?
8. Explain what is meant by a test of significance in terms of the null hypothesis and p values.
9. What is the basic function of a t-test?
10. When and how are analysis of variables (ANOVA) used? What is a post hoc test?
11. What are the two major types of multivariate statistics and how are they used in systematics?
12. How might mapping of statistical data be done, and how is it useful?
13. What is morphometrics?
14. Explain one way that morphometric techniques might be used in systematic studies.
15. How might statistics be used in coding quantitative characters?
16. Name one method of coding overlapping features into character states that may be used in phylogenetic analyses.

EXERCISES

1. Select about ten herbarium specimens (or samples of live plants) of 2–3 morphologically similar species or infraspecies. For each specimen, measure two features (e.g., leaf length and width), from a minimum of 10 organs. Carefully define the features to be measured, taking into account homology of structures, variation due to position, and environmental plasticity. For each sample calculate (e.g., in a spreadsheet) mean, range, and standard deviation. Prepare a bivariate plot of the two features and assess the discreteness of taxa from the results. If statistical software is available, run a t-test (for two taxa) or ANOVA (for three taxa) and assess if the differences between taxa are statistically different.
2. For ten herbarium specimens each of two taxa, measure up to ten features. Pool these features by taxon and conduct a principal components analysis. Graph the first two components and evaluate if the clusters of points are discrete.
3. Select three or more closely related taxa and identify a complex organ or plant part (e.g., one used primarily to distinguish between the taxa) to subject to a morphometric analysis. Identify homologous points on the structure in question. If possible, quantify at least 10 objects per taxon for these points and conduct any number of analyses, e.g., a principal components analysis from a "truss" network. How might the data be valuable in systematic studies?

REFERENCES FOR FURTHER STUDY

RESEARCH ARTICLES AND CHAPTERS

Almeida, M.T. and F.A. Bisby. 1984. A simple method for establishing taxonomic characters from measurement data. Taxon 33: 405-409.
Archie, J. W. 1985. Methods for coding variable morphological features for numerical taxonomic analysis. Systematic Zoology 34: 326-345.
Baum, B. R. 1988. A simple procedure for establishing discrete characters from measurement data, applicable to cladistics. Taxon 37: 63-70.
Bookstein, F. L. 1989. Principal warps: thin-plate splines and the decomposition of deformations. IEEE Trans. Pattern Analysis and Machine Intelligence 11: 567-585.
Chappill, J. A. 1989. Quantitative characters in phylogenetic analysis. Cladistics 6: 217-234.
Dickinson, T. A., W. H. Parker and R. E. Strauss. 1987. Another approach to leaf shape comparisons. Taxon 36: 1-20.
Goldman, N. 1988. Methods for discrete coding of morphological characters for numerical analysis. Cladistics 4: 59-71.
Hurlbert, S. H., R. A. Levine, and J. Utts. 2019. Coup de grâce for a tough old bull: "Statistically significant" expires. The American Statistician 73:352-357.
James, F. C. and C.E. McCulloch. 1990. Multivariate Analysis in Ecology and Systematics: Panacea or Pandora's box? Annual Review of Ecology and Systematics 21: 129-166.
Jensen, R. J., K. M. Ciofani and L. C Miramontes. 2002. Lines, outlines, and landmarks: morphometric analyses of leaves of Acer rubrum, Acer saccharinum (Aceraceae) and their hybrid. Taxon 51: 475-492.
Kuhl, E. P. and C.R. Giardina. 1982. Elliptic Fourier features of a closed contour. Computer Graphics and Image Processing 18: 236-258.
McLellan, T. and J. A. Endler. 1998. The relative success of some methods for measuring and describing the shape of complex objects. Systematic Biology 47: 264-281.
Pimentel, R.A. and R. Riggins. 1987. The nature of cladistic data. Cladistics 3: 201-209.

Olsson, A., H. Nyborn, and H. C. Prentice. 2000. Relationships between Nordic dogroses (*Rosa* L. sect. *Caninae*, Rosaceae) assessed by RAPDs and elliptic Fourier analysis of leaflet shape. Systematic Botany 25: 511-521.

Premoli, A. C. 1996. Leaf architecture of South American *Nothofagus* (Nothofagaceae) using traditional and new methods in morphometrics. Botanical Journal of the Linnean Society 121: 25-40.

Rohlf, F. J. 1993. Relative-warp analysis and an example of its application to mosquito wings. In: L. F. Marcus, E. Bello, and A. Garcia-Valdecasas eds. Contribution to Morphometrics, v8. Pp. 131-159. Museo Nacional de Ciencias Naturales, Madrid, Spain.

Rohlf, F. J. and D. Slice. 1990. Extensions of the Procrustes method for the optimal superimposition of landmarks. Systematic Zoology 39: 40-59.

Roth, V. L. 1993. On three-dimensional morphometrics, and on the identification of landmark points. Pages 42-61 in Contributions to morphometrics, Volume 8 (L. F. Marcus, E. Bello, and A. Garcia-Valdecasas eds.). Museo Nacional de Ciencias Naturales, Madrid.

Shipunov, A. B. and R. M. Bateman. 2005. Geometric morphometrics as a tool for understanding *Dactylorhiza* (Orchidaceae) diversity in European Russia. Biological Journal of the Linnean Society 85: 1-12.

Stevens, P. F. 1991. Character states, morphological variation, and phylogenetic analysis: a review. Systematic Botany 16: 553-583.

Strait, D. S., M. A. Moniz, and P. T. Strait. 1996. Finite mixture coding: A new approach to coding continuous characters. Systematic Biology 45:67-78.

Strauss, R. E. and F. L. Bookstein. 1982. The Truss: body form reconstructions in morphometrics. Systematic Zoology 3: 113-35.

Thiele, K. 1993. The Holy Grail of the perfect character: The cladistic treatment of morphometric data. Cladistics 9: 275-304.

Weins, J.J. 1995. Polymorphic characters in phylogenetic systematics. Systematic Biology 44: 482-500.

SOME STATISTICS AND MORPHOMETRIC BOOKS

Bookstein, F. L. 1991. Morphometric tools for landmark data: geometry and biology. Cambridge University Press, Cambridge, U. K.

Gould, S. J. 1977. Ontogeny and phylogeny. Belknap Press of Harvard University Press, Cambridge, Mass.

Humphries, C. J. (ed.) 1988. Ontogeny and Systematics. Columbia University Press, New York.

Ott, R. L. and M. T. Longnecker. 2008. An Introduction to Statistical Methods and Data Analysis. 6th ed. Duxbury Press, North Scituate, Massachusetts.

Quinn, G. P. and M. J. Keogh. 2002. Experimental Design and Data Analysis for Biologists. Cambridge University Press, Cambridge.

Rosner, B. 2005. Fundamentals of Biostatistics. 5th ed. Duxbury, Pacific Grove, California.

Tabachnick, B. G. and L. S. Fidell. 2006. Using Multivariate Statistics. 5th ed. Harper Collins, New York.

Zar, J. H. 2009. Biostatistical Analysis. 5th ed. Prentice Hall, Upper Saddle River, New Jersey.

Zelditch, M. L., D. L. Swiderski, H. D. Sheets, and W. L. Fink. 2004. Geometric morphometrics for biologists: a primer. Elsevier Academic Press, San Diego, California.

Wiens, J. J. (ed.). 2000. Phylogenetic analysis of morphological data. Smithsonian Institution Press, Washington, D.C.

COMPUTER SOFTWARE

Abramoff, M. D., P. J. Magelhaes, and S. J. Ram. 2004. Image Processing with ImageJ. Biophotonics International 11: 36-42.

Rasband, W. S. 1997-2007. ImageJ, U. S. National Institutes of Health, Bethesda, Maryland, USA
http://rsb.info.nih.gov/ij
Free, multiplatform software for morphometric analyses.

Rohlf, F. J. 2009 onwards. Morphometrics at SUNY, Stony Brook, NY.
http://life.bio.sunysb.edu/morph
Extensive software downloads and information on hardware, date, courses, and investigators in morphometrics.

Sheets, H.D. 2002. IMP (Integrated Morphometrics Package) Suite. Dept. of Physics, Canisius College, Buffalo, NY 14208.
http://www3.canisius.edu/~sheets/morphsoft.html

SYSTAT software. 2009 onwards.
http://www.systat.com
Very widely used commercial software in the biological sciences.

GLOSSARY OF TERMS

The following is a glossary of terms used in all chapters of the book. For the terms from Chapter 9 (Plant Morphology), the character to which a character state belongs is noted in parentheses following the definition. Symbols used are: Abbr = abbreviation; Adj = adjective; Cf = compare; Pl = plural; Syn = synonym.

! Abbreviation for the confirmation of a name in an annotation label.

abaxial Surface most distant or away from the axis, the lower or outer surface of organ. Syn: *dorsal*. (position)

ABC model A model of floral development, in which gene products of the so-called A, B, and C classes combine to produce the four major floral organs: sepals, petals, stamens, and carpels.

acaulescent Lacking an above-ground stem other than the inflorescence axis. (stem habit)

accession number A number assigned to each specimen placed into a permanent herbarium collection.

accessioning The assignment of a number to all new specimens placed into a permanent herbarium collection.

accessory bud Bud(s) lateral to or above axillary buds. (bud type)

accessory part A portion of the mature fruit that is not directly derived from the ovary or ovaries, may include bract(s), stem axes, receptacle, hypanthium, or perianth. (fruit part)

accrescent Parts persistent and continuing to grow beyond what is normal or typical, e.g., calyx of *Physalis*, Solanaceae. (duration)

acetolysis A standard acid treatment used to dissolve all but the exine of pollen grains in order to better observe pollen wall structure with the light microscope.

achene A one-seeded, dry, indehiscent fruit with seed attached to pericarp at one point only, e.g., unit fruits of sunflowers and other Asteraceae. (fruit type)

achenecetum An aggregate fruit of achenes, e.g., *Fragaria* (strawberry), in which the achenes are on the surface of accessory tissue, an enlarged, fleshy receptacle. (fruit type)

achlamydeous Lacking a perianth. (perianth merosity)

achlorophyllous Lacking chlorophyll/chloroplasts.

acicular Needlelike, often round in cross-section, with margins straight and parallel, length:width ratio >12:1. (shape)

acrocaulis Positioned at the apex of the stem. (position)

acrocidal capsule A capsule dehiscing by means of apical slits. (fruit type)

acrodromous With two or more primary veins or strongly developed secondary veins running in convergent arches toward the leaf apex but not recurved at the base. (leaf venation)

actinodromous With three or more primary veins diverging from one point, inclusive of ternate or palmate venation. (leaf venation)

actinomorphic Radially symmetrical, with 3 or more planes of symmetry. (symmetry) Syn: *polysymmetric, radial*.

aculeate With prickles, sharp nonspine, nonthorn appendages. Syn: *prickly*. (epidermal excrescence)

acuminate Apical margins abruptly incurved (concave), the apical intersection angle <45°. (apex)

acute Apical margins approximately straight, the intersection angle 45°–90°. (apex)

adaptation A structure or feature that performs a particular function and which results in increased survival or reproduction.

adapter A short segment of DNA ligated to ends of the DNA fragments involved in a high-throughput sequencing reaction.

adaxial Surface toward or nearest the axis, the upper or inner surface of organ. Syn: *ventral*. (position)

adherent With unlike parts joined, but only superficially and easily separable. (fusion)

adnate With unlike parts integrally fused, not easily separable. (fusion)

advanced *Derived*.

adventitious roots A root arising from an organ other than a root, usually from a stem. (root type)

adventive embryony Development of an embryo from a cell of the surrounding tissue, such as megasporangial or integument tissue.

aerial roots Adventitious roots that absorb moisture and minerals from the air or runoff, common in some epiphytic plants, e.g., of Araceae and Orchidaceae. (root type)

aestival Appearing in summer. (periodicity)

aestivation Referring to position, arrangement, and overlapping of floral perianth parts. (perianth aestivation)

affinis Implying, within a taxon name, some type of close relationship, presumably an evolutionary relationship, but also that the specimen differs from the cited taxon in some way, e.g., beyond the described range of variation for one or more characters. Abbr: *aff.*

AFLP *Amplified fragment length polymorphism*.

agamospermy The production of seeds without fertilization.

ageotropic Growing upward; used for a plant part that normally grows downward; e.g., roots of some cycads, *Avicennia*.

aggregate fruit A fruit derived from two or more pistils (ovaries) of one flower. (fruit type)

alate Winged. (shape: plane)

albuminous *Endospermous.* (seed endosperm type)

albuminous cell A parenchyma cell associated with a sieve cell, derived from a different parent cell than is the sieve cell.

alete Spores lacking any evidence of a laesura.

aleurone grain Granular protein deposits in plant cells, functioning as storage compounds; a type of ergastic substance. Syn: *proteinoplast.*

alignment The process of matching homologous nucleotide positions of two or more sequences of DNA in order to code the data for phylogenetic or other types of analysis.

allautogamy Having both outcrossing and selfing flowers. Adj: *allautogamous.*

allogamy *Outbreeding.* Adj: *allogamous.*

allopatric Occurring in different geographic ranges.

allopolyploidy Polyploidy occurring between different species.

allozyme One of two or more different molecular forms of an enzyme, corresponding to different alleles of a common gene.

alternate One leaf or other structure per node. (arrangement)

alternation of generations *Haplodiplontic life cycle.*

alternipetalous Stamens with point of attachment between the petals or corolla lobes. (stamen position)

alternisepalous Stamens with point of attachment between the sepals or calyx lobes. (stamen position)

alveolar An exine wall structure having numerous, spherical air pockets within the exine.

amb The outline shape of a pollen grain in polar view.

ament *Catkin.* (inflorescence type)

amoeboid An anther tapetum type in which the tapetal cell walls break down, with release of the cytoplasm of the tapetal cells into the locule. Syn: *plasmodial*; *periplasmodial.*

amphiphloic siphonostele A siphonostele in which a ring of xylem is surrounded by an outer and an inner layer of phloem.

amphistomal Referring to the micropyle of a bitegmic ovule formed or delimited by both integuments.

amphitropous A general ovule type in which curvature of the ovule during development displaces the micropyle adjacent to the funiculus base, with the nucellus bent along both upper and lower sides.

amplexicaul Sessile and clasping most of stem circumference. (leaf attachment)

amplification The replication of numerous copies of DNA.

amplification fragment length polymorphism The use of amplified DNA fragments that exhibit length polymorphism, enabling the recognition of numerous genetic markers. Abbr: *AFLP.*

amplification The replication of numerous copies of DNA.

amplicon One of the numerous copies of amplified DNA.

ana-amphitropous An ovule type in which a vascular strand curves, traversing from the base of funiculus to the chalazal region of the nucellus, with the nucellus bent sharply in the middle along both the lower and upper sides, often with differentiated cells (basal body) at the angle of the bend.

ana-campylotropous An ovule type in which a vascular strand curves, traversing from the base of funiculus to the chalazal region of the nucellus, with the nucellus bent only along the lower side, with no basal body.

analysis of variables A statistical test to evaluate differences among two or more groups in a single test of significance. Abbr: *ANOVA.*

anastomosing *Netted, reticulate.*

anatropous A type of ovule in which curvature during development results in displacement of micropyle to a position adjacent to the funiculus base; most common and apomorphic for the angiosperms.

ancestral Referring to a preexisting condition or character state. Syn: *plesiomorphic*; *primitive.*

ancestral state reconstruction A method of assessing character evolution by calculating the maximum probability of a state at each node, using a selected model of evolution.

androdioecious/androdioecy Having male flowers on some individuals and perfect flowers on other individuals. (plant sex)

androecium The male organ(s) of a flower; collectively all stamens of a flower. (flower part)

androgynophore A stalklike structure that bears the gynoecium and androecium; e.g., *Passiflora.* (flower part)

andromonoecious Having both staminate and perfect flowers on the same individual. (plant sex)

androphore A stalk-like structure that bears stamens. (flower part)

anemochory Dispersal of propagules by wind. Adj: *anemochorous.*

anemophily Pollination by wind. Adj: *anemophilous.*

anisomerous Having a different number of members in different whorls. (merosity, perianth merosity)

annotated phylogenetic classification A classification in which monophyletic groups are ordered by a sequential listing of derivative lineages from the base to the apex of the cladogram.

annotation label A label affixed to an herbarium specimen that verifies or changes the identity of a specimen or that documents the removal of plant material from the specimen.

annual Plant living 1 year or less. (duration)

annual ring The accumulation of secondary xylem (or phloem) over a single growing season, being evident because of the structural difference between the last cells of the summer wood and the first cells of the subsequent spring wood.

annulus A single row of specialized cells, having differentially thickened cell walls, on the outer rim of a *leptosporangium*, functioning in its dehiscence.

antepelatous *Antipetalous.* (stamen position)

anterior Referring to the lower lobe or part, especially in a horizontally oriented structure. (position)

antesepalous *Antisepalous.* (stamen position)

anther The pollen-bearing part of a filamentous stamen. (stamen part)

anther sac *Microsporangium.* (anther part)

antheridial wall The outer, sterile layer of cells of the antheridium. Syn: *jacket layer*; *sterile jacket layer.*

antheridiophore A specialized, stalked, generally peltate structure that grows from the gametophyte of some liverworts and bears antheridia.

antheridium The male gametangium of the gametophyte of land plants, producing sperm cells and surrounded by an outer layer of sterile cells, the antheridial wall or jacket layer.

antherode The sterile anther of some staminodes. (stamen type)

anthesis Time of flowering; the opening of flower with parts available for pollination. (maturation, flower maturation)

anthocarp A fruit in which one or more flower parts functions as accessory tissues, e.g., *Pontederia* (Pontederiaceae) or *Boerhavia* (Nyctaginaceae), in which an accrescent perianth surrounds and fuses to the achene. (fruit type)

antipetalous Stamens with point of attachment in line with (opposite) the petals or corolla lobes, e.g., Primulaceae, Rhamnaceae. Syn: *antepetalous, obhaplostemonous.* (stamen position)

antipodal cells In a typical angiosperm female gametophyte, the three haploid cells that are positioned opposite the micropyle, i.e., in the chalazal region.

antisepalous Stamens with point of attachment in line with (opposite) the sepals or calyx lobes. Syn: *antesepalous, haplostemonous*. (stamen position)

antrorse Bent or directed upward, usually referring to small appendages. (orientation)

aperture A specially delimited region of the pollen grain wall.

apetalous Having no petals or corolla. (perianth merosity)

apical (a) At or near the top, tip, or end of a structure. (position) (b) Style arising at the apex of the ovary. (style position) (c) With the placenta at the top of the ovary. Syn: *pendulous*. (placentation)

apical bud *Terminal bud*. (bud type)

apical meristem A region of actively dividing cells in the land plants, located at the apex of the thallus, shoot, or root.

apical-axile With two or more placentae at the top of a septate ovary, e.g., Apiaceae. (placentation)

apiculate With a flexible, apical process, usually slightly curled, length:width ratio >3:1. (apical process)

apocarpous With carpels distinct, the pistil or ovary simple. (gynoecial fusion)

apolar Pollen polarity in which polar and equatorial regions cannot be distinguished after pollen grain separation from the microspore tetrad.

apomorphic *Derived*.

apomorphy A derived condition or character state, representing an evolutionary novelty. Adj: *apomorphic*.

apomorphy-based clade A clade in which all members of the group share a given, unique evolutionary event.

apopetalous With distinct petals. Syn: *choripetalous, polypetalous*. (perianth fusion)

aposepalous With distinct sepals. Syn: *chorisepalous, polysepalous*. (perianth fusion)

apostemonous With distinct stamens. (stamen fusion)

apotepalous With distinct tepals. Syn: *choritepalous, polytepalous*. (perianth fusion)

appressed Pressed closely to an axis oriented upward, with a divergence angle of 0°–15° from upper axis. (orientation)

aquatic Growing in water. (plant habitat)

arachnoid/arachnose With trichomes forming a cobwebby mass. (vestiture)

arborescent Treelike in appearance and size. (stem habit)

archegonial chamber A cavity between the megasporangium and female gametophyte of gymnosperm seeds, into which sperm cells are released by the male gametophyte.

archegonial wall The outer, sterile layer of cells of the archegonium. Syn: *jacket layer*; *sterile jacket layer*.

archegoniophore A specialized, stalked, generally peltate structure that grows from the gametophyte of some liverworts and bears archegonia.

archegonium The female gametangium of the gametophyte, containing a basal egg cell and surrounded by an outer layer of sterile cells, the archegonial wall, which differentiates into a basal venter and proximal neck.

archesporial cell A single, large cell of an immature ovule that either directly becomes the megasporocyte or that divides once to form a parietal cell and a megasporocyte.

areole A modified, reduced, nonelongating shoot apical meristem bearing spines, e.g., Cactaceae. (stem/shoot type)

aril A fleshy outgrowth of the funiculus, raphe, or integuments (but separate from the integuments), generally functioning in animal seed dispersal, e.g., Sapindaceae. Adj: *arillate*. (seed part)

aristate With a stiff apical process, usually prolonged and straight, length:width ratio >3:1. (apical process)

arrangement Placement with respect to similar parts. (disposition)

arylphenalenones A class of chemical compounds common in the Haemodoraceae and also found in closely related families.

ascending Directed upward, with a divergence angle of 15°–45° from upper axis. (orientation)

ascidiate Referring to a carpel that is not leaflike, but develops from a ring of tissue that grows upward, sometimes assuming a somewhat peltate form.

asepalous Having no sepals or calyx. (perianth merosity)

asexual reproduction The transfer of an exact copy of DNA (barring a somatic mutation) from a parent to offspring.

asexual species Species reproducing without sex, yet are recognizable morphological or genetic units.

assembly A computer algorithm that generates a final DNA sequence from the overlapping of read sequences in high-throughput methods.

asterad type A type of embryo development in which the terminal cell divides longitudinally, with both basal and terminal cell derivatives contributing to the mature embryo.

asymmetric Lacking a plane of symmetry. (symmetry)

atactostele A stele consisting of numerous, collateral vascular bundles positioned throughout the stem tissue.

atectate A pollen grain that ancestrally lacks a tectum.

ategmic An ovule lacking an integument.

attenuate Basal margins abruptly incurved (concave), the basal intersection angle <45°. (base)

attractant An aspect of floral morphology that functions to entice an animal pollinator to the flower, either by vision or by odor.

auctorum non The misapplication of a name, such that the type specimen of the name does not fall within the circumscription of the taxon being referred to by that name. Abbr: *auct. non*.

auriculate With two rounded, basal lobes, margins above lobes concave. (base)

autapomorphy An apomorphy that occurs for a single lineage or taxon.

author(s) The name of the person(s) who validly published a scientific name.

autochory *Self-dispersal*. Adj: *autochorous*.

autogamy Inbreeding occurring within a single flower. Adj: *autogamous*. Syn: *infrafloral selfing*.

autonym An automatically created name for infrafamilial, infrageneric, and infraspecific taxa.

autopolyploidy Polyploidy occurring within a species.

autumnal Appearing in fall. (periodicity)

awn (a) Bristlelike, apical appendage, e.g., on the glumes or lemmas of grass spikelets. (b) A unit of a pappus type in the Asteraceae that is narrow, elongated, straight, and stiff. (inflorescence part)

axial parenchyma Longitudinally oriented parenchyma cells that occur in some secondary xylem (wood) tissues.

axil The region at the upper (adaxial) junction of leaf and stem. (position)

axile With the placentae positioned along the column in a septate, compound ovary. (placentation)

axillary (a) On the side of a structure or at the nodes of an axis. (position). (b) With the inflorescence positioned in the axil of the nearest vegetative leaf. Syn: *lateral*. (inflorescence position)

axillary bud Bud in axils of leaves or leaf scars. Syn: *lateral bud*. (bud type)

baccate Fleshy or juicy, often with reference to a fruit. Syn: *succulent*, *carnose*. (texture)

baculate A pollen sculpturing with rod-shaped elements, each element termed a baculum.

baculum A rod-shaped element, as in the wall sculpturing of some pollen grains. Pl: *baculi*. Adj: *baculate*.

ballochory Seed self-dispersal by fruit explosive dehiscence. (fruit and seed dispersal)

barbed/barbellate With minute, lateral, sharp appendages (barbs) arising along the surface or margin of a bristle, the barbs typically antrorse or retrorse in orientation. (bristle type)

barcode (a) A short segment of DNA that acts as an identifier for each sample in high-throughput sequencing. (b) A DNA marker used to differentiate and identify taxa.

bark Tissues external to the vascular cambium in stem (and roots) of woody plants, consisting of secondary phloem (inner bark) and derivatives of the cork cambium (outer bark or periderm). (stem/shoot parts)

barochory Fruit or seed dispersal by gravity. (fruit/seed dispersal)

basal (a) At or near the bottom or base of a structure. Syn: *radical*. (position) (b) With the placenta at the base of the ovary, e.g., Asteraceae, Poaceae. (placentation) (c) With three or more primary veins diverging from one point at the base of the blade, a subcategory of actinodromous and palinactinodromous. (leaf venation)

basic A type of anther wall development in which both secondary parietal cell layers divide to yield two middle layers.

basicidal capsule A capsule dehiscing by means of basal slits, as in *Aristolochia* spp. (fruit type)

basifixed Anther attached at its base to the apex of the filament. (anther attachment)

basionym The original, but now not accepted, name, part of which has been used in a new combination.

Bayesian analysis/inference A method of phylogenetic inference based upon the posterior probability of a phylogenetic tree.

beak An extended, usually accrescent, basal stylar region, typically functioning in fruit dispersal, e.g., *Taraxacum*, dandelion. (style structural type)

beard A tuft, line, or zone of trichomes on a perianth or perianth part. Adj: *bearded*. (perianth part, vestiture)

Beltian body A nutritious structure produced by a plant that ants eat, serving as a reward for protecting the plant.

berry A fleshy fruit with a succulent pericarp, e.g., *Vitis*, grape. (fruit type)

bicollateral bundle A vascular bundle with phloem to both the inside and outside of the xylem.

bicolor unit An inflorescence unit, possibly an apomorphy of the Malvaceae s.l., consisting of a modified, three-bracted cyme, the bracts modified into an epicalyx in members of the group.

biennial A plant living 2 years, typically forming a basal rosette of leaves during the first year and flowering with an elongated inflorescence stalk in the second year. (duration)

bifacial Secondary growth in which layers of cells are produced both to the inside and outside of a continuously generated cambium.

bifid Two-lobed to two-divided, especially at the apex. (division)

bigeminate A compound leaf with two rachillae, each bearing two leaflets. (leaf type)

bilabiate Two-lipped, with two, generally upper and lower segments, e.g., many Lamiaceae. (perianth type)

bilateral *Zygomorphic, irregular, monosymmetric*. (symmetry)

bimerous Referring to a whorl with two members. (merosity, perianth merosity)

binary character A character with only two character states.

binary combination *Binomial*.

binomial Format for the scientific name of species, composed of two names, the genus name and the specific epithet, italicized or underlined. Syn: *binary combination*.

binucleate Having two nuclei, referring to some angiosperm pollen grains at the time of release.

biodiversity The totality of life within a given region.

bioinformatics The scientific field encompassing the study and analysis of genetic data.

biological species Groups of populations that interbreed (actually or potentially) and that are reproductively isolated from other such groups in nature. Syn: *isolation species*.

bipinnate/bipinnately compound A compound leaf with two orders of axes, the second axes (rachillae) bearing leaflets. (leaf type)

bipinnatifid Bipinnately lobed to divided. (division)

biradial Having two planes of symmetry. (symmetry) Adj. *disymmetric*.

biseriate (a) With two whorls of parts. (cycly) (b) Perianth parts in two distinct whorls. (perianth cycly) (c) Having two whorls or cycles of stamens (stamen cycly) Syn (a–c): *dicyclic*. (d) Rays in wood that are made up of two vertical rows of cells.

bisexual Flowers having both carpel(s) and stamen(s). Syn: *perfect*. (flower sex)

bisporangiate Anther with two microsporangia and typically one theca. Cf: *monothecal*. (anther type)

bisporic Megasporogenesis in which cytokinesis occurs after the first meiotic division, but not the second, resulting in two cells, each of which contain two haploid nuclei, with one of the binucleate cells contributing to the female gametophyte.

bitegmic An ovule with two integuments, apomorphic for the angiosperms.

biternate/biternately compound A compound leaf with three axes, each of which is ternately compound. (leaf type)

bivariate plot A graph of two variables, for several individuals, samples, or taxa.

blade The flat, expanded portion of leaf. Syn: *lamina*. (leaf part)

bootstrap/bootstrapping A method of evaluating cladogram robustness that reanalyzes the data of the original character × taxon matrix by selecting (resampling) characters at random, such that a given character can be selected more than once.

bostryx *Helicoid cyme*. (inflorescence type)

botany The traditional study of photosynthetic organisms (including the green plants, red algae, brown plants, dinoflagellates, and euglenoids, but excepting the photosynthetic bacteria), the true fungi, and groups that used to be treated as fungi, such as the Oomycota and slime molds; inclusive of the plant sciences.

bract A modified, generally reduced leaf, generally found associated with reproductive organs, e.g., subtending the ovuliferous scale of conifers or the flowers or inflorescence axes of flowering plants. Adj: *bracteate*. (leaf structural type)

bractlet/bracteole A smaller or secondary bract, often borne on the side of a pedicel. Syn: *prophyllum*. (leaf structural type, flower part)

Bremer support *Decay index*.

bristle An external hairlike plant structure, but stouter than a trichome.

brochidodromous Pinnate venation in which secondary veins do not terminate at the margin, forming prominent upward loops near the margin, joining other, more distal, secondary veins. (leaf venation)

bud An immature shoot system, often surrounded by protective scale leaves, developing into a lateral branch, a flower, or an inflorescence; may be gametophytic or sporophytic. (plant part, stem/shoot parts, twig part)

bud primordium An immature bud of the shoot, typically located in the leaf axil. (stem/shoot parts)

bulb A short, erect, underground stem surrounded by fleshy leaves, e.g., *Allium* spp., onions. (stem/shoot type)

bulbel A proliferative bulb arising from existing bulbs at or below ground level. (stem/shoot type)

bulbil A proliferative bulb arising from shoots above ground, typically within an inflorescence. (stem/shoot type)

bullate *Rugose*. (configuration)

bundle cap An outer patch of sclerenchyma fibers associated with a vascular bundle.

bur A multiple fruit of achenes or grains surrounded by a prickly involucre, e.g., *Cenchrus* (Poaceae), *Xanthium* (Asteraceae). (fruit type)

burl A swollen, woody underground stem from which arises persistent, woody, aerial branches, e.g., fire regenerative stems in some *Arctostaphylos* spp., Manzanita. Syn: *lignotuber*. (stem/shoot type)

buttress roots Enlarged, horizontally spreading and often vertically thickened roots at the base of trees that aid in mechanical support. (root type)

C4 photosynthesis An alternate photosynthetic pathway of some land plants in which carbon dioxide is initially fixed in the mesophyll cells by the enzyme PEP carboxylase, producing a four-carbon molecule, which is transported to bundle sheath cells, where the carbon dioxide is released and fixed by ribulose-bisphosphate carboxylase in the typical dark reactions.

caducous Dropping off very early compared with what is typical, usually applied to floral parts. Syn: *fugacious*. (duration)

callose A polysaccharide, composed of beta-1,3-glucose units, which lines the pores of sieve areas and sieve plates of sieve elements and is commonly deposited within pollen tubes.

calyculus The outer often reduced bracts of the head of the Asteraceae. Alternatively, a synonym for *epicalyx*.

calyptra An apical region of archegonial tissue that is torn from and lifted up by the elongating sporophyte during the latter's development and that may function to protect the young sporophyte apex.

calyptrate Having calyx and corolla fused into a cap that falls off as a unit, e.g., *Eucalyptus*. (perianth type)

calyx The outermost series or whorl of modified leaves in the perianth, the units of which are sepals. (flower part)

calyx lobes The segments of a calyx that is synsepalous (with connate sepals).

campanulate Bell-shaped, i.e., with a basally rounded, flaring tube about as broad as long and flaring lobes, e.g., *Campanula*; may also be used for bell-shaped apopetalous corolla or apotepalous perianth. (perianth type)

camptodromous Pinnate venation in which secondary veins do not terminate at the margin. (leaf venation)

campylodromous With several primary veins running in prominent, recurved arches at the base, curving upward to converge at the leaf apex. (leaf venation)

campylotropous A general ovule type in which curvature of the ovule during development displaces the micropyle adjacent to the funiculus base, with the nucellus bent only along the lower side.

canaliculate Longitudinally grooved, usually in relation to petioles or midribs. (configuration)

canescent Covered with dense, fine grayish-white trichomes; whitish-pubescent. (vestiture)

capillary bristle A unit of a pappus type in the Asteraceae that is generally straight, very thin, and threadlike, often barbellate.

capitate Head-shaped; spherical with a short basal stalk. (shape)

capitulum *Head*. (inflorescence type)

capsule (a) The spore-producing component of the sporophytes of liverworts, hornworts, and mosses. (plant part). (b) A dry, dehiscent fruit derived from a compound ovary. (fruit type)

carinate Keeled, having a sharp, median fold projected on the abaxial side; sharply conduplicate. (perianth type)

carnose *Baccate*; *succulent*. (texture)

carpel The unit of the gynoecium of angiosperm flowers, arising as a ring of tissue (ascidiate) or as a modified, conduplicate female megasporophyll of a flower, (plicate) at maturity enclosing one or more ovules. (flower part, gynoecium part)

carpophore The stalklike structure that is attached to the mericarps in a schizocarp of mericarps, e.g., Apiaceae. (fruit part)

cartilaginous With the texture of cartilage; hard and tough but flexible, usually whitish. (texture)

caruncle A fleshy outgrowth at the base of the seed, functioning in animal seed dispersal. Syn: *elaiosome, strophiole*. (seed part)

caryophyllid type A type of embryo development in which the terminal cell divides transversely, with only terminal cell derivatives contributing to the mature embryo.

caryopsis *Grain*. (fruit type)

Casparian strip A band or ring of mostly suberin that infiltrates the cell wall of endodermal cells, functioning to force water and mineral solutes to pass through the plasma membrane of these cells.

cataphyll A scale-like, often non-green, meristem-protecting leaf (e.g., in cycads or palms); a rudimentary scale leaf found in usually hypogeous (cryptocotylar) seedlings. (leaf structural type)

categorical characters *Qualitative characters*.

catkin A unisexual, typically male spike or elongated axis that falls as a unit after flowering or fruiting, e.g., *Quercus*. Syn: *ament*. (inflorescence type)

caudate Abruptly acuminate into a long, narrowly triangular (tail-like) apical region. (apex)

caudex A short, thick, vertical or branched perennial stem, underground or at/near ground level. (stem/shoot type)

caudiciform stem A low (at or above ground level), swollen, perennial storage stem from which arise annual or otherwise nonpersistent photosynthetic shoots, e.g., some *Dioscorea* spp., *Calibanus*. (stem/shoot type)

caulescent Having an above ground, vegetative stem. (stem habit)

cauliflorus Inflorescence growing directly from a woody trunk. (inflorescence position)

cauline Positioned along the length of the stem. (position)

cell The structural unit of all life.

cell differentiation The series of changes that a cell undergoes from the point of inception to maturity, involving the transformation of a meristematic cell into one that assumes a particular structure and function.

cell expansion Growth of a cell in size, often involving axial elongation.

cell theory The postulate that all life is composed of one or more cells, that cells arise only from preexisting cells, and that cells are the units of metabolic processes.

cell wall A layer of the plant cell that is secreted outside the plasma membrane.

cellular endosperm An endosperm in which the endosperm cell divides mitotically, regularly followed by cytokinesis, each endosperm nucleus contained within a cell wall.

cellulose A polymer of glucose sugar units (= polysaccharide) in which the glucose molecules are chemically bonded in the beta-1,4 position (= β-1,4-glucopyranoside); a major component of the cell wall of green plants.

cellulose A polysaccharide of glucose units (β-1,4-glucopyranoside), a primary component of plant cell walls.

central At or near the middle or middle plane of a structure. (position)

central cell In a typical angiosperm female gametophyte, the single, relatively large cell in the central region that contains the two polar nuclei.

centric leaf A leaf that is cylindrical in shape, e.g., *Fenestraria* of the Aizoaceae. (leaf structural type)

centrifugal Developing from the center region toward the outside or periphery; can be applied to parts of perianth, calyx, corolla, androecium, or gynoecium. (flower maturation)

centripetal Developing from the outside or periphery toward the center region; can be applied to parts of perianth, calyx, corolla, androecium, or gynoecium. (flower maturation)

cernuous With tip drooping downward, abaxially. (transverse posture)

cespitose/caespitose Referring to a generally short, bunched, much-branched plant forming a cushion. (stem habit)

chaff One of the bracts subtending flowers in some Asteraceae, e.g., tribe Heliantheae. Syn: *palea*. (leaf structural type)

chalazal Describing the region of an ovule that is opposite the micropyle.

chalazal Referring to the proximal region of the ovule, opposite the micropyle.

chamaephyte An overwintering perennial with buds at or just below ground level. (plant life form)

character A feature or attribute of a taxon

character correlation The condition in which one character is interdependent upon and influenced by another character.

character evolution The sequence of evolutionary changes occurring for a given character.

character optimization The representation of characters in a cladogram in a most parsimonious way, such that the minimal number of character state changes occur.

character state One of two or more forms of a character.

character step matrix A numerical tabulation of the number of changes occurring between all pairwise combinations of character states for a given character.

character i taxon matrix A numerical tabulation of the characters and corresponding character states for each taxon in a phylogenetic analysis.

chartaceous Opaque and of the texture of writing paper. (texture)

chasmogamy Referring to typical flowers in which the perianth opens and exposes the sexual organs, with subsequent cross-pollination common. Adj: *chasmogamous*.

chasmophyte Referring to a saxicolous plant growing specifically in the cracks of rocks or boulders.

cheiropterophily Pollination by bats. Adj: *cheiropterophilous*.

chenopodiad type A type of embryo development in which the terminal cell divides transversely, with both basal and terminal cell derivatives contributing to the mature embryo.

Chlorobionta A monophyletic group of eukaryotes, consisting of the "green algae" and the land plants, united having chloroplasts with chlorophyll a and b, starch, and thylakoids stacked as grana [green plants].

chlorophyll a The primary pigment in the light reactions of photosynthesis, found in the chloroplasts of all photosynthetic eukaryotes and some bacteria.

chlorophyll b An accessory pigment in the light reactions of photosynthesis, an apomorphy of the green plant chloroplast.

chlorophyllous Having chloroplasts at maturity; green.

chlorophyllous cell 1. Any general, chloroplast-containing cell. 2. One of the chloroplast-containing cells in the specialized leaves of *Sphagnum* moss the border and surround a hyaline cell.

chloroplast A double membrane-bound organelle with internal thylakoid membranes (lamellae and grana in the green plants), functioning in the reactions of photosynthesis.

chloroplast DNA The DNA restricted to the chloroplast. Abbr: *cpDNA*.

choripetalous *Apopetalous, polypetalous*. (perianth fusion)

chorisepalous *Aposepalous, polysepalous*. (perianth fusion)

choritepalous *Apotepalous, polytepalous*. (perianth fusion)

chromoplast Carotenoid-containing bodies that function to provide yellow, orange, or red pigmentation for a plant organ, as in petals or fruits; a type of ergastic substance.

ciliate With conspicuous marginal trichomes. (margin, vestiture)

ciliolate With minute trichomes protruding from margins; minutely ciliate. (margin, vestiture)

cincinnus *Scorpioid cyme*. (inflorescence type)

circinate Coiled from apex to base, as with the leaf blade in young fern and cycads or the inflorescence unit of most Boraginaceae. (posture: ptyxis/aestivation)

circinate vernation Descriptive of a leaf (including the blade and rachis/rachillae, if present) that is coiled from apex to base when immature, as in young fern and cycad leaves (posture: ptyxis/aestivation).

circumferential At or near the circumference; surrounding a rounded structure. (position)

circumscissile capsule A capsule having a transverse line of dehiscence, e.g., *Plantago*. Syn: *pyxis/pyxide*. (fruit type)

circumscription The boundaries of a taxon, what is included within it and how it is delimited from other taxa.

cirrhose With a flexible, greatly curled apical process. (apical process)

cistron A repeating DNA sequence of the nuclear ribosomal DNA (nrDNA), containing ETS and ITS regions, their sequence data traditionally valuable in plant phylogenetic systematics.

clade A monophyletic group, consisting of a common ancestor and all lineages arising from that common ancestor. Cf: *lineage*.

cladistics A methodology for inferring the pattern of evolutionary history of a group of organisms, utilizing grouping of taxa by apomorphies [phylogenetic systematics].

cladistics *Phylogenetic systematics*.

cladode A flattened photosynthetic stem, functioning as and resembling a leaf. Syn: *cladophyll, phylloclade*. (stem/shoot type)

cladodromous Pinnate venation in which secondary veins do not terminate at the margin and branch near the margin. (leaf venation)

cladogram A branching diagram that conceptually represents the best estimate of phylogeny. Syn: *phylogenetic tree*.

cladophyll *Cladode*. (stem/shoot type)

cladogram robustness A measure of the confidence for which a cladogram actually denotes phylogenetic relationships, e.g., by bootstrapping.

cladoptosic Referring to dead foliage falling with an accompanying shoot, rather than as individual leaves, e.g., Cupressaceae such as *Taxodium*. (duration)

clambering Sprawling across objects, without specialized climbing structures. Syn: *scandent*. (stem habit)

classification The arrangement of taxa (or other entities) into some type of order or grouping.

clathrate Fern scales in which the cell walls of adjacent cells ("anticlinal" walls) are thick. Cf: non-clathrate.

clava A club-shaped element, as in the sculpturing of some pollen grains. Pl: *clavae*.

clavate (a) Club-shaped; terete with a gradually tapering thickened and rounded end. (shape) (b) A pollen sculpturing with club-shaped elements, each element termed a clava.

claw An attenuate base of a sepal or petal. (perianth part)

cleft Sinuses extending (pinnately or palmately) one quarter to one half the distance to the midrib, midvein, or vein junction. (division)

cleistogamy Referring to flowers in which the perianth remains closed, such that pollen produced from within the flower pollinates only the stigma(s) of that flower. Adj: *cleistogamous*.

climbing Growing upward by means of tendrils, petioles, or adventitious roots. (stem habit)

crown clade A clade in which both or all branches from the common ancestor contain extant members.

coalescent methods/analysis Phylogenetic inference inferring the most likely species tree from separate gene trees.

coherent With like parts joined, but only superficially and easily separable. (fusion)

cohesion species The largest or most inclusive group of individuals that maintains genetic and phenotypic cohesion via gene flow, genetic drift, and natural selection, maintaining the fundamental niche through genetically determined environmental tolerances.

coleoptile A protective sheath surrounding the epicotyl, e.g., in some members of the Poaceae. (seed/embryo part)

coleorhiza A protective sheath surrounding the radicle, e.g., in some members of the Poaceae. (seed/embryo part)

collateral A vascular bundle with an internal strand of xylem and an external strand of phloem.

collateral bud Bud(s) lateral to the axillary bud. (bud type)

collateral bundle A vascular bundle with xylem to the inside and phloem to the outside.

collenchyma A cell type that is live at maturity, has unevenly thickened, pectic-rich, primary cell walls, and functions in structural support, often found at the periphery of stems or leaves.

colleter A structure on the inner surface of connate stipules that secretes mucilage, aiding in protection of young, developing shoots, e.g., Rubiaceae. (leaf parts)

color pattern The distribution of colors on an object. (color)

colporate A pollen grain aperture type that is shaped like a colpus but has a circular region in the center.

colpus A pollen grain aperture that is elongated with a length:width ratio of greater than 2:1. Pl: *colpi*. Adj: *colpate*.

columella (a) A central column of sterile (non-spore-producing) tissue within the sporophyte capsule of hornworts. (b) One of the middle, radially elongated elements of a tectate-columellate pollen exine wall. Pl: *columellae*.

column (a) The central axis to which septae and/or placentae are attached in axile or free-central placentation. (gynoecium part) (b) *gynandrium*, *gynostegium*, *gynostemium*. (flower part)

combinatio nova Indication that a taxon has recently been transferred to a new position or rank. Abbr: *comb. nov.*

commemorative A name that is after a person or place.

common name A vernacular name, used by people within a limited geographic region, not formally published and governed by no rules.

comose With an apical tuft of trichomes. (vestiture)

companion cell A parenchyma cell associated with a sieve tube member, derived from the same parent cell as is the sieve tube member and functioning to load and unload sugars into the cavity of the sieve tube member.

complementary DNA DNA synthesized from mRNA using reverse transcriptase. Abbr: *cDNA*.

complete Having all four major whorls or floral parts: sepals, petals, stamens, carpels. (flower cycly)

complex tissue A tissue that contains more than one cell type.

compound cone The cone of a conifer, consisting of an axis bearing bracts, each of which subtends a modified branch systems, the ovuliferous scale.

compound corymb A branched corymb, consisting of two or more orders of inflorescence axes bearing flat-topped or convex, pedicellate flowers. (inflorescence type)

compound cyme A branched determinate inflorescence, similar to a compound dichasium but lacking a consistent dichasial branching pattern, often by reduction of internodal axes. (inflorescence type)

compound dichasium A many-flowered, determinate inflorescence of repeatedly branching simple dichasia units. (inflorescence type)

compound leaf A leaf divided into two or more discrete leaflets. (leaf type)

compound ovary/pistil An ovary/pistil composed of a two or more carpels, the gynoecium syncarpous. (ovary/pistil type)

compound perforation plate A perforation plate composed of several openings.

compound receptacle A mass of tissue at the apex of a peduncle that bears more than one flower. Syn: *torus*. (inflorescence part)

compound sieve plate A sieve plate that is made up of two or more aggregations of pores.

compound umbel An umbel with the peduncle bearing rays attached at one point and unit simple umbels attached at the tip of the rays, e.g., many Apiaceae. (inflorescence type)

conduplicate Longitudinally folded at the central axis, with adjacent adaxial sides facing one another. (longitudinal posture)

cone A modified, determinate, reproductive shoot system of many nonflowering vascular plants, consisting of a stem axis bearing either sporophylls (in simple cones) or ovuliferous scales subtended by bracts (in compound cones of conifers). Syn: *strobilus*. (plant part)

confer Indication that the identity of a specimen is questionable or uncertain and should be compared with specimens of the taxon indicated for more detailed study. Abbr: *cf.*

configuration Referring to gross surface patterns other than venation or epidermal excrescences. (surface)

connate With like parts integrally fused, not easily separable. (fusion)

connate-perfoliate Two opposite leaves fused basally, such that the blade base of each leaf completely surrounds the stem. (leaf attachment)

connective The tissue or filament extension between the thecae of an anther. (anther part)

connivent Convergent apically without fusion. (orientation)

consensus tree A cladogram derived by combining features in common between two or more cladograms.

conservation biology That branch of biology dealing with the preservation of biodiversity.

conservation of names A principle of the International Code of Nomenclature for algae, fungi, and plants, stating that scientific names that are well known and frequently used may be retained over other, earlier, but less well-known, names.

conserved ortholog set Conserved genes present in a number of representative taxa in a study group. Abbr: *COS.*

consistency index (CI) A measure of the relative amount of homoplasy in a cladogram, equal to the ratio of the minimum possible number of character state changes to the actual number of changes that occur.

contiguous With parts touching but not fused. (fusion)

continuous quantitative characters Characters in which individual measurements are not necessarily integers and may potentially form a continuum. Cf: *discrete quantitative characters.*

contorted *Convolute.* (perianth aestivation)

convergence Homoplasy occurring by the independent evolution of a similar feature in two or more lineages. Syn: *parallelism.*

convolute Perianth parts of a single whorl overlapping at one margin, being overlapped at the other, e.g., corolla of Malvaceae. Syn: *contorted.* (perianth aestivation)

cordate With two rounded, basal lobes intersecting at sharp angle, the margins above lobes smoothly rounded. (base)

cordate/cordiform Shaped like an upside-down Valentine heart; approximately ovate with a cordate base. (shape)

coriaceous Thick and leathery, but somewhat flexible. (texture)

cork The outermost layer of the periderm, generated by the cork cambium.

cork cambium A sheath or hollow cylinder of cells that develops near the periphery of the stem or root, undergoing tangential divisions to form phelloderm to the inside and cork to the outside.

corm An enlarged, solid underground storage stem or stem base, with outer, protective scales. (stem/shoot type)

cormel A proliferative corm arising from existing corms. (stem/shoot type)

corolla The innermost series or whorl of modified leaves in the perianth, the units of which are petals. (flower part)

corolla lobe A segment of a sympetalous corolla (with connate petals).

corona A crownlike outgrowth between stamens and corolla; may originate from petals or stamens. (perianth part, perianth type)

coronate With a tubular or flaring perianth or staminal outgrowth, e.g. *Narscissus*, *Asclepias*. (perianth type)

correct name A validly published, legitimate name that is accepted by a particular author or authors.

cortex The outer, mostly parenchymatous tissue inner to the epidermis and external to the vasculature. (root part, stem/shoot parts)

corymb An indeterminate inflorescence consisting of a single axis with lateral axes and/or pedicels bearing flat-topped or convex flowers. (inflorescence type)

cosmopolitan Having a worldwide distribution.

COS *Conserved ortholog set.*

costa (a) *Midrib*. (b) The nonvascularized conductive tissue found in the gametophytic leaves of some mosses. (leaf part)

costapalmate Essentially palmately lobed to compound but with an elongated, rachislike extension of the petiole, as in some palms. (leaf type)

cotyledon A first (seed) leaf of the embryo, often functioning in storage of food reserves. (seed part)

cotylespermous With the food reserve in the cotyledon, part of the embryo. (seed endosperm type)

couplet The pair of contrasting leads in a dichotomous key.

craspedodromous Pinnate venation in which secondary veins terminate at the leaf margin. (leaf venation)

crassinucellate An ovule in which the nucellus develops two or more layers of cells, the inner ones from divisions of a parietal cell.

crassulacean acid metabolism An alternate photosynthetic pathway in some xeric, generally succulent plants and functioning to conserve water, in which initial fixation of carbon dioxide occurs at night (when stomata are open) by the enzyme PEP carboxylase to form malic acid, which is stored within vacuoles of the mesophyll cells; during the day the stomata close and CO_2 is released from the vacuoles into the cytoplasm, where it is fixed in the chloroplasts. Abbr: *CAM.*

crenate With rounded to obtuse, shallowly ascending teeth, cut $^1/_{16}$ to $^1/_8$ the distance to the midrib, midvein, or junction of primary veins. (margin)

crenulate Diminutive of crenate, teeth cut $\leq {}^1/_{16}$ the distance to the midrib, midvein, or junction of primary veins. (margin)

crown clade A clade in which both or all branches from the common ancestor contain extant members.

crownshaft The collection of overlapping, sheathing leaf bases at the apex of a palm trunk.

crozier A leaf that is coiled during its development, characteristic of the leptosporangiate ferns (Polypodiales) and Marattiales. Syn: *fiddlehead.*

cruciate With four, distinct petals in a cross form, e.g., many Brassicaceae. (perianth type)

crucifer type A type of embryo development in which the terminal cell divides longitudinally, with only terminal cell derivatives contributing to the mature embryo. Syn: *onagrad type.*

crumpled Having a wrinkled or crinkled appearance, particularly in bud. (perianth aestivation)

cryptantherous *Inserted.* (stamen insertion)

cryptic species Species that are not morphologically distinguishable from another species.

cryptocoylar *Hypogeous.* (seed germination type)

crystal A deposit of silica or calcium oxalate in plant cells that may function as waste products, as calcium ion sinks, or as an irritant to deter herbivory; a type of ergastic substance.

cucullate Hooded; with an abaxially concave posterior lip. (perianth type)

culm The flowering and fruiting stem(s) of grasses and sedges. (stem/shoot type)

cuneate With basal margins approximately straight, intersection angle 45°–90°. (base)

cup-shaped Concave–convex along entire surface; may be abaxially or adaxially concave. (longitudinal posture)

cupule A structure that encloses a cluster of unitegmic ovules/seeds, with a small opening near the proximal end, through which pollen grains entered; characteristic of some Pteridosperms, e.g., *Caytonia*.

curator The person in charge of the day-to-day running of an herbarium.

cuspidate Abruptly acuminate into a triangular, stiff or sharp apex. (apex)

cuticle A protective layer, containing cutin, that is secreted to the outside of epidermal cells and functions to inhibit water loss; found in all land plants.

cutin A polymer of fatty acids, functioning as a sealant in the cuticle layer of land plants, inhibiting water loss.

cyathium An inflorescence bearing small, unisexual flowers and subtended by an involucre (frequently with petaloid glands), the entire inflorescence resembling a single flower, e.g., *Euphorbia*. (inflorescence type)

cycly Number of cycles or whorls of parts. (number)

cymbiform Boat-shaped, e.g., glumes of many grasses. (shape)

cyme General term for a determinate inflorescence. (inflorescence type)

cymule A small, simple dichasium. (inflorescence type)

cypsela An achene derived from an inferior ovary, e.g., Asteraceae. Syn: *achene,* in general sense.

cystolith A mass of calcium carbonate attached to a stalk from the cell wall, occurring within specialized cells termed *lithocysts*.

cytoplasm Everything inside the plasma membrane but not including the nucleus.

dark reactions A series of biochemical reactions of photosynthesis in plants, occurring in the stroma of the chloroplast, during which atmospheric carbon dioxide reacts to produce a molecule of glucose, requiring the input of the high-energy compounds ATP and $NADPH_2$.

data information system/database system Referring to the (computerized) organization, inputting, and accessing of information.

ddRADSeq Double digest restriction-site associated DNA sequencing, using two restriction enzymes.

decay index A method of evaluating cladogram robustness by calculating how many extra steps are needed (beyond the number in the most parsimonious cladograms) before the original clade is no longer supported; the greater this value, the greater the confidence in a given clade. Syn: *Bremer support*.

deciduous Parts persistent for one growing season, then falling off, e.g., leaves of deciduous plants. (duration)

declinate Angled downward or forward. (orientation)

decompound A general term for a leaf having leaflets in two or more orders: bi-, tri-, and so on pinnately, palmately, or ternately, compound; also used for a highly divided leaf. (leaf type, division)

decumbent Basally prostrate but ascending apically. (stem habit)

decurrent Appearing to extend down the stem from the point of attachment, as if fused to the stem, e.g., many Cupressaceae. (leaf attachment)

decussate Opposite leaves or other structures at right angles to the preceding pair. (arrangement)

decussate tetrad A tetrad in which the four grains are in two pairs arranged at right angles to one another.

deflexed Bent abruptly downward. (orientation)

deltate Three-sided, length:width ratio ca. 1:1. (shape)

dendritic Trichomes treelike, with multiple lateral branches. (trichome type)

dendrochronology The scientific study of wood anatomy to infer details about past events.

de novo assembly Assembly of DNA sequences without using an already acquired reference DNA sequence for alignment.

dentate With sharp, coarse teeth that point outward at right angles to margin outline, cut $^1/_{16}$ to $^1/_8$ distance to midrib, midvein, or junction of primary veins. (margin)

denticulate Diminutive of dentate, teeth cut to $^1/_{16}$ or less the distance to the midrib, midvein, or junction of primary veins. (margin)

depth of coverage *Sequencing depth*.

depressed Pressed closely to axis downward, with divergence angle of 0°–15° from lower axis. (orientation)

derived Referring to a new condition or character state. Syn: *apomorphic, advanced*.

dermal tissue The outer region of plant organs, composed of the **epidermis**.

descending Directed downward, with divergence angle of 15°–45° from lower axis. (orientation)

descent The sequence of ancestral-descendant populations through time.

description The assignment of features or attributes to a taxon or other entity.

determinate (a) A shoot that terminates growth after a certain period, the apical meristem aborting or converting into a flower, inflorescence, or other specialized structure (stem branching pattern). (b) An inflorescence in which the terminal flower matures first, maturating from apex to base. (inflorescence development)

dextrorse Twining helically like a typical, right-handed screw, e.g., some Convolvulaceae. (twisting/bending posture)

diachronic species A species corresponding to a group of organisms that span a lineage of ancestral-descendent populations, extending through a period of time.

diadelphous With two groups of stamens, each connate by filaments, e.g., many Faboideae (Fabaceae). (stamen fusion)

dichasium A determinate inflorescence that develops along two axes, forming one or more pairs of opposite, lateral axes. (inflorescence type)

dichlamydeous Perianth composed of a distinct outer calyx and inner corolla, regardless of total number of whorls. (perianth cycly)

dichogamy A type of outcrossing mechanism that is the result of differences in timing of maturation of male and female floral parts. Adj: *dichogamous*.

dichopodial Roots in which the apical meristem branches into two roots, as in the lycophytes.

dichotomous With veins successively branching distally into two veins of equal size and orientation, e.g., *Ginkgo biloba*. (leaf venation)

dichotomous key A key utilizing series of two contrasting statements, each statement a lead, the pair of leads a couplet.

diclesium *Anthocarp*.

dicotyledonous A type of anther wall development in which only the outer secondary parietal cell layer divides to yield the endothecium and a single middle layer.

discriminate function analysis A multivariate statistical method that transforms numerous data variables into other variables in order to determine the variables that discriminate among taxa.

dictyostele A dissected amphiphloic siphonostele.

dicyclic *Biseriate*. (cycly, perianth cycly, stamen cycly)

didymous With stamens in two equal pairs. (stamen arrangement)

didynamous With stamens in two unequal pairs, e.g., many Bignoniaceae, Lamiaceae, Scophulariaceae. (stamen arrangement)

diffuse-porous Wood in which vessels develop more or less uniformly throughout the growth season.

dimorphic Having two discrete forms, as leaves of some species.

dioecious/dioecy Having unisexual flowers, staminate and pistillate on separate individual plants. (plant sex)

diphthong A two-vowel combination in Latin that is treated as the equivalent of a single vowel.

diplohaplontic life cycle *Haplodiplontic life cycle*.

diplostemonous Stamens in two whorls, the outer opposite the sepals, the inner opposite petals. (stamen position)

discoid (a) Discus-shaped. (shape). (b) Stigma(s) disk-shaped. (stigma/stigmatic region type)

discrete quantitative characters Characters in which measurements of are always integers. Syn: *meristic characters*.

disjunct A population that occurs well outside the geographic range of other members of that taxon.

disk A discoid or doughnut-shaped structure arising from the receptacle at the outside and surrounding the stamens. (*extrastaminal disk*), at the base of the stamens (*staminal disk*), or at the inside of the stamens and/or base of the ovary (*intrastaminal disk*); may be nectar-bearing (*nectariferous disk*). (flower part)

disk flower Having an actinomorphic, tubular corolla with flaring lobes, e.g., some Asteraceae. (perianth type)

dispersal The movement of an organism or propagule from one region to another, such as the transport of a seed or fruit (by wind, water, or bird) from a continent to an island.

disposition Relative placement of objects or parts of objects, inclusive of position, arrangement, and orientation.

dissected Divided into very fine, often indistinct segments. (division)

distal Away from the point of origin or attachment. (position)

distal pole The intersection of the pollen grain polar axis with the grain surface that is away from the center of the microspore tetrad.

distichous Alternate, with points of attachment in two vertical rows/ranks, e.g., the grasses, Poaceae. (arrangement)

distinct With like parts unfused and separate. (fusion)

distyly Hercogamy in which two floral morphologies occur: pin flowers, with a long style and short stamens, and thrum flowers, with a short style and long stamens. Adj: *distylous*.

disulculate A pollen grain with two elongated apertures on opposite sides of the grain, parallel to the equatorial plane.

disymmetric *Biradial*. (symmetry)

dithecal Anther with two thecae and typically four microsporangia. Cf: *tetrasporangiate*. (anther type)

diurnal During the day, typically with respect to when flowers open. (periodicity)

divaricate *Divergent*; *horizontal*; *patent*. (orientation)

divergence/diversification The formation of two (or more), separate lineages from one common ancestor. Syn: *evolutionary divergence*.

divided Sinuses extending (pinnately or palmately) $^3/_4$ to almost to midrib, midvein, or vein junction. (division)

dolabriform *Malpighian*. (trichome type)

domatium (*plural*, **domatia**) A specialized chamber formed on a plant that houses ants or other arthropods, like mites.

dorsal *Abaxial*. (position)

dorsal vein The central vein of a carpel, corresponding to the midvein of a leaf. Syn: *median vein*.

dorsifixed Anther attached dorsally and medially to apex of the filament. (anther attachment)

dorsiventral Having a flattened shape, with an upper (adaxial) and lower (abaxial) surface, characteristic, e.g., of leaves.

doubly serrate With large, serrate teeth that have along the margin smaller, serrate teeth. (margin)

downward Anther dehiscing toward the ground in a horizontally oriented flower. (anther dehiscence direction)

drepanium A monochasium in which the axes develop on only one side of each sequential axis, typically appearing coiled at least early in development; sometimes equated with helicoid cymes. (inflorescence type)

drupe A fleshy fruit with a hard, stony endocarp, e.g., *Prunus*: peach, plum, cherry. (fruit type)

drupecetum An aggregate fruit of drupes, e.g., *Rubus*: raspberry, blackberry. (fruit type)

druse A spherical crystal with protruding spikes, composed of calcium oxalate; a type of ergastic substance.

duration The length of life of a plant or plant part. (temporal phenomena)

dyad A fusion product of two pollen grains.

ebracteate Lacking bracts. (flower attachment)

echina A spinelike sculpturing element >1 μm long, as in some pollen grain walls. Pl: *echinae*. Adj: *echinate*.

eciliate Without trichomes on the margins, regardless of presence or absence of teeth. (margin)

ectophloic siphonostele A siphonostele in which a ring of xylem is surrounded by an outer layer of phloem only.

ectozoochory *Epizoochory*. (fruit and seed dispersal)

egg A nonmotile, evolutionarily enlarged gamete, the end product of oogamy.

ektexine An outer layer of the pollen grain exine wall.

elaiosome *Caruncle, strophiole*. (seed part)

elater (a) One of the hygroscopic appendages arising from the spores of *Equisetum*, functioning in spore dispersal. (b) A nonsporogenous, elongated, hygroscopic cell with spiral wall thickenings that develops within the sporangia of some liverworts and that functions in spore dispersal.

electrophoresis Separation of DNA strands/fragments by loading the DNA onto a flat gel plate or in a thin capillary tube subjected to an electric current, the amount of migration inversely proportional to the molecular weight of the strand.

eligulate Lacking a ligule; usually used for taxa whose close relatives are ligulate. Cf: *ligulate*.

elliptic Margins curved, widest at the midpoint, the length:width ratio 2:1 to 3:2. (shape)

emarginate Having an apical incision, cut $^1/_{16}$ to $^1/_8$ the distance to midrib, midvein, or junction of primary veins. (apex)

embryo An immature, diploid sporophyte developing from the zygote of land plants. (seed part; plant part)

embryo proper That portion of the proembryo that will eventually grow into the new sporophyte.

embryo sac Term for the female gametophyte of angiosperms.

embryogeny The development of the embryo within the seed.

Embryophyta/embryophytes A monophyletic group of eukaryotes united by an outer cuticle, specialized gametangia—antheridia and archegonia—and an intercalated diploid phase in the life cycle, including the embryo [land plants].

emendatio A correction or amendment of a name. Abbr: *emend.*

emergent With roots or stems anchored to substrate under water, the aerial shoots above water. (plant habitat)

emersed Occurring under water. (plant habitat)

enantiostyly A type of hercogamy in which the style of different flowers curves either to the left or the right. Adj: *enantiostylous*.

enation A small appendage arising from the stem, resembling a rudimentary leaf but lacking vascular tissue.

endarch An orientation of xylem maturation in which the protoxylem is oriented toward the center of the stele relative to the metaxylem, as occurs in eusteles and atactosteles.

endexine An inner layer of the pollen grain exine wall.

endocarp The innermost wall layer of the pericarp, if the latter is divided into layers. (fruit part)

endodermis A hollow cylinder of cells in roots and some stems that surrounds the vasculature and functions to selectively control passage of solutes from the outside, via Casparian strips. Adj: *endodermal*. (root part, stem part)

endogenous Arising from the internal tissues, as in the growth of secondary roots from within a primary root.

endoplasmic reticulum A cellular organelle consisting of interconnected phospholipid membranes that may function in material transport and as the site of protein synthesis.

endosperm The triploid tissue that develops from mitotic divisions of the endosperm cell (the product of double fertilization), ultimately enveloping or abutting the embryo and functioning as the nutritive tissue of angiosperm seeds.

endospermous With endosperm as the food reserve in mature seeds. Syn: *albuminous*. (seed endosperm type)

endospory The development of a gametophyte within the original spore wall. Adj: *endosporic*.

endostomal Referring to the micropyle of a bitegmic ovule delimited by only the inner integument, the outer one being foreshortened.

endosymbiosis The intracellular cohabitation of one cell within another cell; the process that gave rise to mitochondria and chloroplasts by engulfment of a prokaryote by a eukaryotic cell.

endothecium The outermost cell layer of an anther, typically of enlarged cells with secondary wall thickenings functioning in anther dehiscence.

endozoochory Dispersal by animals in which propagules are eaten but are passed through the gut of the animal unharmed. Adj: *endozoochorous*. Syn: *endozoic*. (fruit and seed dispersal)

ensiform Sword-shaped, with length : width ratio greater than 12:1, e.g., leaves of *Iris* spp. (shape)

entire Without teeth on margins; locally smooth. (Note, however, that surface may be divided.) (margin)

entomophily Pollination by insects. Adj: *entomophilous*.

epicalyx A series of bracts immediately subtending the calyx, e.g., *Hibiscus*, other Malvaceae. Syn: *calyculus*. (leaf structural type)

epicotyl (a) The first shoot of a vascular plant that develops from the embryo. (b) The first shoot of a seed plant, derived from the embryo of the seed. (stem/shoot parts, seed part)

epidermal excrescence Referring to surface patterns from structural outgrowths or secretions of the epidermis. (surface)

epidermis The outermost cell layer of all land plant organs, functioning to provide mechanical protection of inner tissue and to inhibit water loss.

epigeous With cotyledon(s) elevated above the ground during germination. Syn: *phanerocotylar*. (seed germination type)

epigynous With sepals, petals, and stamens attached at the apex of the ovary, the ovary inferior. (perianth/androecial position)

epihypogynous With sepals, petals, and stamens attached at middle of the ovary, the ovary half-inferior. (perianth/androecial position)

epihypoperigynous Hypanthium present, attached at middle of ovary, with sepals, petals, and stamens attached to hypanthium rim, the ovary half-inferior. (perianth/androecial position)

epiperigynous Hypanthium present, attached at apex of ovary, with sepals, petals, and stamens attached to hypanthium rim, the ovary inferior. (perianth/androecial position)

epipetalous With stamens adnate to (inserted upon) petals or corolla. Syn: *petalostemonous*. (stamen fusion)

epiphyte A plant that grows upon another plant. (plant life form) Adj: *epiphytic*.

episepalous With stamens adnate to (inserted upon) sepals or calyx. (stamen fusion)

epitepalous With stamens adnate to tepals or the perianth as a whole. (stamen fusion)

epitropous An ovule position in which the micropyle points distally.

epitropous-dorsal An epitropous ovule position in which the raphe is dorsal (abaxial), pointing away from the central floral or ovary axis.

epitropous-ventral An epitropous ovule position in which the raphe is ventral (adaxial), pointing toward the central floral or ovary axis.

epitype A specimen (or illustration) that is selected to serve as an "interpretive" type if the holotype, lectotype, or neotype is ambiguous with respect to the identification.

epizoochory Dispersal by animals in which a fruit or seed becomes attached to and is carried away by an animal. Adj: *epizoochorous*. Syn: *ectozoochory*. (fruit and seed dispersal)

equator The intersection with the pollen surface of a plane at a right angle to the polar axis and passing through the center of the grain.

equatorial view Observing a pollen grain from the equatorial region.

equitant Leaves with overlapping bases, usually sharply folded along midrib. (arrangement)

erect Pointing upward. (orientation)

eremocarp One of the fruit units developing from ovary lobes that are separate from their inception, as in the Boraginaceae and Lamiaceae. Cf: *nutlet, schizocarp of nutlets*. (fruit type)

ergastic substance A cellular substance that does not function in metabolism, generally functioning in storage, waste secretion, and protection.

et Latin for "and," used in scientific name combinations.

ETS *External transcribed spacer*.

eucamptodromous Pinnate venation in which secondary veins do not terminate at the margin, curving upward near the margin but not directly joining adjacent secondaries. (leaf venation)

euphyll The sporophytic leaf of the euphyll group, growing by means of either marginal or apical meristems, having multiple, branched veins, and having an associated leaf gap. Adj: *euphyllous*. Syn: *megaphyll*. (leaf structural type)

eusporangium/eusporangiate sporangium A relatively large sporanigum that is derived from several epidermal cells and having a sporangial wall composed of more than one cell layer.

eustele A primary stem vasculature that consists of a single ring of discrete collateral or bicollateral vascular bundles.

even-pinnate *Paripinnate / paripinnately compound.*

evergreen Persistent two or more growing seasons, e.g., leaves of most conifers. (duration)

evolution Descent with modification; the transfer of genetic material from parent(s) to offspring over time, with a corresponding change in that genetic material.

evolutionary species A single, continuous lineage of ancestral-descendent populations, retaining its identity from other such lineages and has its own.

evolutionary divergence *Divergence.*

ex Latin for "from;" in nomenclature meaning validly published by.

exalbuminous Lacking endosperm as the food reserve in mature seeds. Syn: *nonendospermous*. (seed endosperm type)

exarch An orientation of xylem maturation in which the protoxylem is oriented toward the organ periphery relative to metaxylem, as occurs in some protosteles.

exarillate Lacking an aril on the seed. Cf: *aril.*

exfoliating Bark cracking and splitting off in large sheets. (bark type)

exindusiate Referring to a sorus that lacks an indusium.

exine The hard, outermost, desiccation-resistant layer of a pollen grain wall, providing structural support and inhibiting desiccation.

exocarp The outermost wall layer of the pericarp, if the latter is divided into layers. (fruit part)

exospory The formation of a gametophyte external to the original spore wall. Adj: *exosporic.*

exostomal Referring to the micropyle of a bitegmic ovule delimited by only the outer integument, the inner one being foreshortened.

explosive dehiscence Referring to a dehiscent fruit that opens with force, in the process ejecting the seeds some distance away. (fruit type)

exserted With stamens protruding beyond the perianth. Syn: *phanerantherous*. (stamen insertion)

exstipellate Without stipels. (leaf part)

exstipulate Without stipules. (leaf part)

external transcribed spacer A region of the cistron nrDNA occurring between 26S and 18S nrDNA, adjacent to the latter. Abbr: *ETS.*

extrafloral nectary A nectary produced outside the flower that functions as a reward, generally for ants, which in turn protect the plant from herbivores or parasites.

extrastaminal disk A discoid or doughnut-shaped structure arising from the receptacle at the outside and surrounding the stamens; may be nectar-bearing (nectariferous disk). (flower part)

extrorse Dehiscing outward, away from the flower center. (anther dehiscence direction)

facultative hemiparasite A chlorophyllous, normally parasitic plant not requiring a host to survive and reproduce. (plant life form)

falcate/falciform Lanceolate to linear and curved to one side; scimitar-shaped. (shape)

false indusium An extension of the blade margin that overlaps the sorus of a leptosporangiate fern.

false septum The membranous, intervening tissue of a cross-wall that persists after fruit dehiscence in silicles and siliques of Brassicaceae. (fruit part)

farinaceous Finely mealy, covered with small granules. Adj: *granular, scurfy*. (epidermal excrescence)

fascicle (a) A shoot with very short internodes on which flowers or leaves are borne. Syn: *short shoot*; *spur*; *spur shoot*. (stem/shoot type). (b) A raceme-like or panicle-like inflorescence with pedicellate flowers in which internodes between flowers are very short, with pedicel bases appearing congested. (inflorescence type) Adj: *fasciculate.*

fat A type of triglyceride compound that may function as high-energy storage compounds or secretion products in plants; a type of ergastic substance.

female (a) Individual with female reproductive organs only. (plant sex) (b) *Pistillate*. (flower sex)

female gametophyte A gametophyte that bears only archegonia, housing the egg cell. Syn: *megagametophyte*, *embryo sac.*

female sporophyll A sporophyll that bears one or more megasporangia or seeds. Syn: *megasporophyll.*

fenestrate Having windowlike holes in the surface, e.g., *Monstera deliciosa*, Araceae. (configuration)

fertile segment The sporangia containing component of the leaf of an ophioglossoid fern.

fiber A sclerenchyma cell that is long and very narrow, with sharply tapering end walls, functioning in mechanical support and often occurring in bundles.

fibrous roots Roots that are adventitious and typically fine and numerous. (root type)

fiddlehead *Crozier.*

filament A stamen stalk, generally terete in shape. (stamen part)

filamentous (a) With a more or less terete stamen stalk, as opposed to a laminar body. (stamen type) (b) Filament present, as opposed to absent and anther sessile. (stamen attachment)

filiferous Bearing coarse, fiberlike structures. (margin)

filiform Long, thin, and typically flexuous, threadlike, filamentous. (shape)

fissured Bark split or cracked into vertical or horizontal, usually coarse grooves. (bark type)

fistulose/fistular Cylindrical and hollow within. (shape)

flabellate With three or more primary veins diverging from one point and several, equal, fine veins branching toward the leaf apex, a subcategory of actinodromous. (leaf venation)

flexuous Central axis and tip alternately curved up and down. (transverse posture)

floating Occurring at the water surface. (plant habitat)

floccose With dense trichomes in several patches or tufts. (vestiture)

flora A listing of the plant taxa of a given region, usually accompanied by keys and descriptions. Syn: *manual.*

floral diagram A diagrammatic, cross-sectional view of a flower bud, showing the relative relationship of perianth, androecial, and gynoecial components, and illustrating things such as stamen position, placentation, and perianth, calyx, or corolla aestivation.

floral formula A symbolic representation of floral morphology, including cycly (number or whorls or series), merosity (number of parts per whorl), fusion of parts, and ovary position.

floral receptacle *Receptacle*. (flower part)

floral tube *Hypanthium*. (flower part)

floret A unit of a grass (Poaceae) spikelet, consisting of a short, lateral axis bearing two bracts (lemma and palea) that subtend a terminal, reduced flower. (inflorescence type)

floristics The documentation of all plant species in a given geographic region.

flower The reproductive organ of flowering plants; a modified, determinate shoot bearing sporophylls (stamens and/or carpels) with or without outer modified leaves, the perianth. (plant part, inflorescence part)

flower bract A modified, generally reduced leaf subtending a flower. (flower part)

flower bud A bud that develops into a flower. (bud type)

follicetum An aggregate fruit of follicles, e.g., *Magnolia*. (fruit type)

follicle A dry, dehiscent fruit derived from one carpel that splits along one suture, e.g., *Asclepias*, milkweed. (fruit type)

foot-layer The inner layer of a tectate-columellate pollen exine wall.

form genus A genus that corresponds to a particular organ of a fossil plant.

fossulate A pollen sculpturing with longitudinal grooves.

foveolate A pollen sculpturing with a pitted surface caused by pores in the surface.

fragmentation Whereby an originally broadly distributed species with more or less continuous gene exchange and clinal intergradation becomes split into two or more populations.

free With unlike parts unfused, separate. (fusion)

free-central With the placenta along the column in a compound ovary lacking septa, e.g., Caryophyllaceae. (placentation)

free venation *Open venation*.

***Fritillaria* type** A type of tetrasporic female gametophyte in which three of the four megaspores fuse to form a triploid nucleus, followed by two sequential mitotic divisions of the haploid and triploid nuclei, resulting in an 8-nucleate female gametophyte in which the three antipodals and one of the polar nuclei are triploid, the other polar nucleus and the cells of the egg apparatus remaining haploid.

frond Specialized term for a fern leaf.

fruit The mature ovary of flowering plants, consisting of the pericarp (mature ovary wall), seeds, and (if present) accessory parts. (plant part)

frutescent Having the habit of a shrub, with numerous, woody, aerial trunks. (stem habit)

fugacious *Caducous*. (duration)

fumatory A substance that is smoked by humans, usually for its pleasing or euphoric effects, e.g., tobacco, *Nicotiana tabacum*.

funiculus A stalk that attaches the ovule to the placenta. (gynoecium part)

fusiform Spindle-shaped; narrowly ellipsoid with two attenuate ends. (shape)

galeate Hooded; with an abaxially concave posterior lip. (perianth type)

gamete A specialized, haploid cell that fuses with another gamete (in sexual reproduction) to form a diploid zygote.

gametophyte The haploid phase in the life cycle of all land plants.

gamopetalous *Sympetalous*. (perianth fusion)

gamosepalous *Synsepalous*. (perianth fusion)

gamotepalous *Syntepalous*. (perianth fusion)

gap-coding A method of dividing more or less continuous variation into discrete states.

geitonogamy Inbreeding occurring between different flowers derived from one individual. Adj: *geitonogamous*.

geminate A compound leaf with two leaflets arising from a petiole and no rachillae. (leaf type)

geminate-pinnate A compound leaf with two rachillae, each bearing a pinnate arrangement of leaflets. (leaf type)

gemma (a) An asexual propagule produced within the gemmae cups of some thalloid liverworts. (b) One of the globose or ellipsoid elements of a gemmate pollen grain. Pl: *gemmae*.

gemma cup A cup-shaped organ on the upper surface of the gametophytes of some thalloid liverworts, containing gemmae propagules.

gemmate A pollen sculpturing with globose or ellipsoid elements, each element termed a gemma.

gender The designation of masculine, feminine, or neuter in Latin names.

gene flow The transfer of the genetic material from parent to offspring.

gene genealogy/gene lineage The tokogenetic history of an allele or gene copy, from a particular gene. Cf: *tokogenetic*.

genealogical species Species in which all individuals of the group are more closely related to one another than to any organisms outside the group.

generative cell One of the two initial, haploid cells in the male gametophyte of angiosperms that mitotically divides to form two sperm cells.

genetic drift Random genetic modification of a population or species, not the result of natural selection.

genet A genetically different individual of a population. Cf: *ramet*.

geniculate Having a zig-zag posture, e.g., the inflorescence rachis of some grasses. (twisting/bending posture)

genome resequencing Obtaining the whole genome of numerous, closely related individuals and comparing their sequence data to a single reference genome permitting genetic comparisons between individuals of a population.

genome skimming A whole genome method of DNA squencing in which sequences of high concentration in the cell are obtained.

genus name The first component of a binomial, always capitalized.

genus novum Meaning that a taxon name, at the rank of genus, is new to science. Abbr: *gen. nov.*

geocarpy Dispersal in which the plant pushes fruits into the ground; e.g., *Arachis hypogaea*, peanut. (fruit and seed dispersal)

geophyte A perennial herb, typically with a bulb, corm, rhizome, or tuber underground stem. (plant habit, plant life form)

girdling A type of anther endothecium in which the secondary wall thickenings form rings with cross bridges between them.

glabrate Nearly glabrous or becoming glabrous with age. (vestiture)

glabrous Without trichomes. (vestiture)

glandular (a) Covered with minute, blackish to translucent glands (epidermal excrescence). (b) Trichomes secretory or excretory, usually having an apical glandular cell. (trichome type)

glandular (tapetum type) *Secretory*.

glaucous Covered with a smooth, usually whitish, waxy coating, which can be rubbed off with touch. (epidermal excrescence)

globose (a) Spherical in shape (shape). (b) Stigma(s) spherical in shape. (stigma/stigmatic region type)

glochidiate With apical, clustered barblike structures. (bristle type)

glochidium A very small leaf spine with numerous, retrorse barbs along its length, produced in the areoles of opuntioid cacti. Pl: *glochidia*; *glochids*. (leaf structural type)

glomerule An inflorescence of sessile or subsessile flowers in which internodes between flowers are very short, with flowers appearing congested. (inflorescence type)

glucosinolate A secondary chemical compound found in many Brassicales that functions to deter herbivory and parasitism and also serves as a flavoring agent in the commercially important members of the Brassicaceae.

glume One of usually two bracts occurring at the base of a grass spikelet. (leaf structural type)

glutinous *Viscid*. (epidermal excrescence)

golgi body A cellular organelle comprised of parallel stacks of flattened membranes, functioning in transport and modification of compounds.

gradate Development in which the sporangia of a fern sorus mature in succession from the base (periphery) toward the apex (acropetalous) or from the apex toward the base (basipetalous). Syn: *sequential*.

grain A one-seeded, dry, indehiscent fruit with the seed coat adnate to pericarp wall, e.g., Poaceae, grasses. Syn: *caryopsis*. (fruit type)

granular *Farinaceous*.

granum A pancakelike aggregation of thylakoid membranes within the chloroplast of green plants. Pl: *grana*.

grass spikelet The inflorescence unit of the Poaceae, grass family, consisting of an axis (rachilla) bearing distichous parts: two basal bracts (glumes, sometimes modified or absent) and one or more florets, each floret consisting of a minute lateral axis with two additional bracts (lemma and palea) plus the flower.

green plants A monophyletic group of eukaryotes, consisting of the "green algae" and the land plants, united having chloroplasts with chlorophyll a and b, starch, and thylakoids stacked as grana [Chlorobionta].

ground meristem The nonvascular, usually parenchymatous tissue between and among the vascular bundles of an atactostele. (stem/shoot parts)

ground tissue Tissue that is inside the epidermis and not part of the vascular tissue, composed of parenchyma, sclerenchyma, and collenchyma cells.

guard cell One of the two cells that together make up a *stomate*.

gynandrium A fusion product of androecium and gynoecium, e.g., Aristolochiaceae, Orchidaceae. Syn: *column*, *gynostegium*, *gynostemium*. (flower part)

gynobasic With style arising at the base and center of a lobed ovary, e.g., Boraginaceae, Lamiaceae. (style position)

gynodioecious/gynodioecy Having female flowers on some individuals and perfect flowers on other individuals. (plant sex)

gynoecium The female organ(s) of a flower, collectively all carpels of a flower. (flower part)

gynomonoecious/gynomonoecy Having both pistillate and perfect flowers on the same individual. (plant sex)

gynophore A stalk of the pistil, usually absent. Syn: *stipe*. (gynoecium part)

gynostegium *Column*, *gynandrium*, *gynostemium*. (flower part)

gynostemium *Column*, *gynandrium*, *gynostegium*. (flower part)

half-inferior With sepals, petals, stamens, and/or hypanthium attached at the middle of the ovary. (ovary position)

halophyte A salt-adapted plant. (plant life form)

hapaxanthic A determinate shoot that completely transforms into a flower or inflorescence. (tree branching model)

haplodiplontic life cycle A life cycle having both haploid and diploid phases, occurring in all land plants. Syn: *alternation of generations*; *diplobiontic life cycle*.

haplomorphic Appearing radially symmetric but not having strict mirror image halves because the parts are numerous and/or spirally inserted. (flower symmetry)

haplontic life cycle A type of sexual life cycle in which the mature, adult phase is haploid, which produces gametes (egg and sperm) that fuse to form a diploid zygote, the latter undergoing meiosis to produce haploid spores, which develop into new haploid adults. Syn: *haplobiontic life cycle*.

haplostemonous Stamens uniseriate, equal in number to the petals, and opposite the sepals. Syn: *antesepalous, antisepalous*. (stamen cycly, number)

haplotype A unique allele of a chromosome or organelle (mitochondrial or chloroplast DNA).

hardwood Wood derived from a nonmonocotyledonous angiosperm, generally (but not always) harder than a softwood because of a greater concentration of fiber cells.

harmomegathy Volume changes of the pollen grain with changes in water content, e.g., humidity, functioning to inhibit desiccation.

hastate With two basal lobes, more or less pointed and oriented outwardly approximately 90° relative to central axis. (base)

hastula An appendage or projection at the junction of petiole and blade, as in some palms. (leaf part)

haustoria Parasitic roots that penetrate the tissues of a host plant. (root type)

head A determinate or indeterminate, crowded group of sessile or subsessile flowers on a compound receptacle, often subtended by an involucre, e.g., Asteraceae. Syn: *capitulum*. (inflorescence type)

helicoid cyme A monochasium in which the branches develop on only one side of each sequential axis, appearing coiled at least early in development; may intergrade with scorpioid cyme. Syn: *bostryx*. (inflorescence type)

helobial An endosperm in which the first mitotic division is followed by cytokinesis, delimiting two cells, with the nucleus of one of the cells dividing without cytokinesis, that of the other cell dividing with cytokinesis.

hemiparasite A chlorophyllous, parasitic plant. (plant life form)

hemispheric Half-sphere-shaped. (shape)

hemitropous/hemianatropous An ovule somewhat intermediate in curvature between anatropous and orthotropous types.

herb A plant with annual above-ground shoots, including a flower or inflorescence, the plant itself being annual, biennial, or perennial. (plant habit)

herbaceous Having a soft or slightly succulent texture. (texture)

herbarium specimen A pressed and dried plant sample that is permanently glued and/or strapped to a sheet of paper, along with a documentation label.

hercogamy/herkogamy The spatial separation of anthers and stigmas, generally enhancing outbreeding. Adj: *hercogamous/herkogamous*.

hermaphroditic A plant with bisexual flowers. (plant sex)

hesperidium A septate berry with a thick-skinned, leathery pericarp wall and fleshy modified trichomes (juice sacs) arising from the inner walls, e.g., *Citris* (orange, lemon, grapefruit, etc.). (fruit type)

heteroblasty The condition in which the juvenile leaves are distinctly different in size or shape from the adult leaves, e.g. many Araceae. Adj: *heteroblastic*. (leaf type)

heterochrony An evolutionary change in the rate or timing of development.

heteropolar Pollen polarity in which the two polar hemispheres are different because of displacement of one or more apertures.

heterospory The formation of two types of haploid spores, microspores and megaspores, within two types of sporangia. Adj: *heterosporic*.

heterostyly Hercogamy in which the relative lengths or heights of stigmas versus anthers vary among different flowers. Adj: *heterostylous*.

heterotropous An ovule that varies in orientation.

heterotypic synonym An unaccepted name based on a type different from that of the correct name. Syn: *taxonomic synonym*.

high-throughput sequencing A class of relatively new sequencing methods that have largely replaced Sanger sequencing. Syn: *Next Generation Sequencing (NGS)*.

hilum Funicular scar on the seed coat. (seed part)

hirsute With long, rather stiff trichomes. (vestiture)

hispid With very long, stiff trichomes, often capable of penetrating skin. (vestiture)

holoparasite A parasitic plant lacking chloroplasts/photosynthesis. (plant life form)

holotype The one specimen or illustration upon which a name is based, originally used or designated at the time of publication.

homochlamydeous Perianth composed of similar parts, each part termed a tepal. (perianth cycly)

homology Similarity that is the result of common ancestry. Adj: *homologous*.

homolog/homologue A specific feature that is homologous to another, cited feature. Cf: *homology*.

homonym One of two (or more) identical names that are based on different type specimens.

homoplasy Similarity that is not due to homology or common ancestry, but the result of independent evolutionary change.

homospory The formation of one type of haploid spore, by one type of sporangium. Adj: *homosporic*.

homotypic synonym An unaccepted name that is based on the same type as that of the accepted name. Syn: *nomenclatural synonym*.

hood A hoodlike appendage arising from the gynostegium of some Asclepiadoids of the Apocynaceae.

hooked With apical hooklike structure. Syn: *uncinate*. (bristle type)

horizontal More or less horizontally spreading with divergence angle of ≤15° up or down from horizontal axis. Syn: *divaricate*; *divergent*; *patent*. (orientation)

horn A hornlike appendage, often associated with a hood, arising from the gynostegium of some Asclepiadoids of the Apocynaceae.

hyaline cell One of the nonchlorophyllous cells in the specialized leaves of *Sphagnum* moss, having characteristic pores and helical thickenings and functioning in water absorption and retention.

hybridization Sexual reproduction between different species (interspecific hybridization) or between different populations, infraspecific taxa, or forms within a species.

hybridization enrichment sequencing A technique in which specific, short (oligonucleotide) probes are hybridized with sample DNA, targeting selected regions of the DNA, which are subsequently prepped for high-throughput sequencing. Abbr: HybSeq.

HybSeq *Hybridization enrichment sequencing*.

hydathode A group of specialized cells that secrete excess, transported water (usually due to root pressure) from leaf margins.

hydrochory Dispersal of propagules by water. Adj: *hydrochorous*.

hydroid A specialized cell that functions in water conduction in some mosses.

hydrophily Pollination by water. Adj: *hydrophilous*.

hygroscopic Absorbing moisture from the air, often resulting in movement.

hymenopterophily *Melittophily*.

hypanthium A cuplike or tubular structure around or atop the ovary, bearing along its margin the sepals, petals, and stamens. Syn: *floral tube*. (flower part)

hypanthodium An inflorescence bearing numerous flowers on the inside of a convex or involuted compound receptacle, e.g., *Ficus*. (inflorescence type)

hyphodromous Pinnate venation with only the primary midrib vein present or evident, the secondary veins absent, very reduced, or hidden within the leaf mesophyll. (leaf venation)

hypocotyl A region of the embryo between the root and epicotyl; may function in seedling development and as an anatomical transition between root and shoot. (seed part)

hypocrateriform A corolla with a tube having abruptly spreading lobes, encompassing both rotate and salverform. (corolla type)

hypogeous With cotyledon(s) remaining in the ground during germination. Syn: *cryptocoylar*. (seed germination type)

hypogynous With sepals, petals, and stamens attached at the base of a superior ovary. (perianth/androecial position)

hypotropous An ovule in which the micropyle points proximally.

hypotropous-dorsal A hypotropous ovule in which the raphe is dorsal (abaxial), pointing away from the central floral or ovary axis.

hypotropous-ventral A hypotropous ovule in which the raphe is ventral (adaxial), pointing toward the central floral or ovary axis.

hysteranthy Timing in which leaf and flower development do not coincide. Adj: *hysteranthous*.

identification The process of associating an unknown taxon or other entity with a known one.

illegitimate name A name that is validly published name but not in accordance with the rules of the International Code of Nomenclature for algae, fungi, and plants.

imbricate (a) Leaves or other structures overlapping. (arrangement). (b) With overlapping perianth parts. (perianth aestivation)

imbricate-alternate Outer whorl of perianth parts (sepals or outer tepals) alternating with, along different radii, the inner whorl of perianth parts (petals or inner tepals), a typical perianth aestivation. (perianth aestivation)

imparipinnate/imparipinnately compound A pinnately compound leaf with a terminal leaflet, typically odd-pinnate. (leaf type) Syn: *odd-pinnate*.

imperfect (a) *Unisexual*. (flower sex) (b) With lateral primary veins covering less than two thirds of the leaf blade area, a subcategory of actinodromous and of acrodromous. (leaf venation)

in Latin for "in," for "in the publication of," referring to a name published within a larger work authored by the person(s) following the "in."

inaperturate A pollen grain that lacks any recognizable aperture.

inbreeding The union of gametes derived from a single individual. Syn: *selfing*.

incanous Covered with dense, fine, grayish-white trichomes; whitish-pubescent. (vestiture)

incipient speciation Early and incomplete divergence of one lineage into two (or more), such that the terminal entities of those lineages are not fully separated from one another (e.g., exhibit incomplete lineage sorting).

incised With margins sharply and deeply cut, usually jaggedly. (margin, division)

inclined Directed upward, with a divergence angle of 15°–45° from horizontal axis. (orientation)

incompatibility reaction The inhibition of pollen germination or pollen tube growth between genetically similar individuals, mediated by incompatibility genes and functioning to promote outcrossing.

incomplete Lacking one or more of the four major whorls or floral parts: sepals, petals, stamens, carpels. (flower cycly)

incomplete lineage sorting Condition whereby a lineage segment at a given point in time contains more than one gene lineage.

incurved Tip gradually curved inward or upward (adaxially). (transverse posture)

indehiscent legume A secondarily modified legume does not split open, e.g., *Arachis hypogaea*, peanut. (fruit type)

indel A homologous region of a gene that may represent an insertion or a deletion of nucleotides.

indented phylogenetic classification A classifcation in which monophyletic groups are ordered in a sequential, hierarchical method.

indeterminate (a) A shoot that has the potential for unlimited growth, the apical meristem continuing to grow. (stem/shoot type, stem branching pattern). (b) An inflorescence in which the basal flower matures first; maturation from base to apex. (inflorescence development)

induplicate Plicate, with adjacent adaxial sides facing one another, being V-shaped in cross section. (longitudinal posture)

indurate Hardened and inflexible. (texture)

indusium A flap of tissue that covers a *sorus*, found in some leptosporangiate ferns. Adj: *indusiate*.

ineditus Not validly published. Abbr: *ined.*

inferior With sepals, petals, stamens, and/or hypanthium attached at the apex of the ovary. (ovary position)

inflorescence An aggregate of one or more flowers, the boundaries of which generally occur with the presence of vegetative leaves below; may be composed of unit inflorescences. (plant part)

inflorescence bract A modified, generally reduced leaf subtending an inflorescence axis. (inflorescence part)

inflorescence bud A bud that develops into an inflorescence. (bud type)

infrafloral selfing *Autogamy.*

infrafoliar Descriptive of a palm inflorescence that is positioned below the leaves of the crownshaft. (inflorescence position)

infrapetiolar bud An axillary bud surrounded by a petiole base, e.g., *Platanus*, sycamore. (bud type)

infraspecies Subspecies, varieties, or rarely forms (formae) of a species, generally showing slight and often intergrading morphological differences as well as some geographic, ecological, and/or phylogenetic distinctions.

infructescence The complete inflorescence at the stage of fruiting. (fruit part)

infundibular Funnel-shaped; with a narrow base and greatly expanded apex, e.g., *Ipomoea*. (perianth type)

ingroup The study group as a whole in a phylogenetic analysis.

inner bark *Secondary phloem.*

inserted With stamens included within the perianth. Syn: *cryptantherous*. (stamen insertion)

insignificant Condition in which the differences between two or more groups are due to chance alone; i.e., the probability of the differences being due to chance is calculated as greater than some standard value (generally 5%, or $p > 0.05$).

integument A sheath or flap of tissue that surrounds the megasporangium (nucellus) of an ovule and develops into the seed coat of the seed.

intercalary meristem An indeterminate (having potentially continuous growth), basal or sub-basal region of actively dividing cells.

interfoliar Descriptive of a palm inflorescence that is positioned among the leaves of the crownshaft. (inflorescence position)

intergenic spacers Non-coding regions of DNA occurring between coding genes in chloroplast DNA.

intermingled With no consistent developmental pattern of sporangia in a fern sorus.

internal transcribed spacer A region of the cistron nrDNA occurring between the 18S and 26S, divided into two sub-regions, ITS1 and ITS2, separated by a the 5.8S nrDNA. Abbr: *ITS*.

International Code of Nomenclature for algae, fungi, and plants The standardized system of rules for naming plants, "algae," fungi, and organisms traditionally treated as fungi, governing specific names assigned to taxa and the endings that denote taxon rank, and utilized for naming new taxa and determining the correct name for previously named taxa. (Formerly the International Code of Botanical Nomenclature.) Abbr: *ICN*.

internode (a) The region between two adjacent nodes of a shoot. (stem/shoot parts, twig part) (b) A cladogram lineage that spaces between two nodes (points of divergence). Syn: *stem*.

intine The innermost layer of a pollen grain wall, composed primarily of cellulose and pectines.

intrastaminal disk A discoid or doughnut-shaped structure arising from the receptacle at the inside of the stamens and/or base of the ovary; may be nectar-bearing. ("nectariferous disk") (flower part)

intravaginal (axillary) squamules Trichomes found in the axils of sheathing leaves, possibly functioning in secreting a protective mucilage, e.g., many Alismatales. (trichome type)

introgression Hybridization between two species followed by backcrossing to one or both parents.

introrse Dehiscing inward, toward the flower center. (anther dehiscence direction)

inversion A mutation resulting in the 180° flipping of a segment of DNA.

inverted repeats Two stretches of DNA, each an inverted copy of the other, occurring in the chloroplast of most plant cells, flanked by the large and small single copy regions.

involucel A group or cluster of bracts subtending a unit of an inflorescence. (inflorescence part)

involucral bract *Phyllary*. (leaf structural type)

involucre A group or cluster of bracts subtending an inflorescence. Adj: *involucrate*. (inflorescence part)

involute (a) Margins or outer portion of sides rolled inward or upward over adaxial surface (longitudinal posture, margin). (b) Valvate with each perianth part induplicate, folded longitudinally inward along central axis. (perianth aestivation)

iridoid A secondary chemical compound characteristic of many Asterids.

irregular *Zygomorphic, bilateral, monosymmetric*. (symmetry)

isolation species *Biological species*.

isomerous Having the same number of parts in different whorls. (merosity, perianth merosity)

isomorphic Appearing identical, e.g., the gametes of some green plants.

isopolar Pollen polarity in which the two polar hemispheres are the same but can be distinguished from the equatorial region.

isotype A duplicate specimen of the holotype, collected at the same time by the same person from the same population.

iteropary Referring to plants that reproduce more than one time in the life of the plant, typically in regular cycles. Adj: *iteroparous*.

ITS *Internal transcribed spacer*.

jacket layer *Antheridial wall; archegonial wall*.

jacknife/jackniﬁng A method of evaluating cladogram robustness that reanalyzes the data of the original character i taxon matrix by selecting (resampling) characters at random, such that a given character can be selected only once, the resultant resampled data matrix being smaller than the original.

jaculator Funiculi of the seeds that are modified into rigid, often hook-shaped structures that function to disperse the seeds by a catapulting mechanism, characteristic of the Acanthaceae. Syn: *retinaculum*.

key/taxonomic key An identification device, consisting of contrasting statements used to narrow down the identity of a taxon.

Kranz anatomy A leaf anatomy in which chloroplasts of the bundle sheath cells are typically much larger than those of the mesophyll cells, correlated with C4 photosynthesis.

labellum A modified, typically expanded, median petal, tepal, or perianth lobe, such as in the Orchidaceae. (perianth part)

lacerate With sinuses irregularly cut, lobes appearing torn. (division)

laciniate Cut into narrow, ribbonlike segments. (division)

lacunose Having a surface with cavities, pits, or indentations. (configuration)

laevigate Lustrous, polished. (epidermal excrescence)

lagenostome A rim or ring of tissue at the apex of the megasporangium, which functioned to funnel pollen grains to a pollination chamber in ancestral seeds.

lamellar An exine wall structure having stacked, tangentially oriented, planar structures, often constituting the inner wall layer.

lamina *Blade*. (leaf part)

laminar (a) With a dorsiventrally flattened, leaflike structure bearing the thecae. (stamen type) (b) With ovules arising from the surface of the septae. (placentation)

lanate *Villous*. (vestiture)

lanceolate Margins curved, widest near base, length:width ratio between 6:1 and 3:1. (shape)

lance-ovate Margins curved, widest near base, length:width ratio between 3:1 and 2:1. (shape)

landmarks Defined points of an object that generally correspond to homologous features, used in morphometric analyses.

land plants A monophyletic group of eukaryotes united by an outer cuticle, specialized gametangia—antheridia and archegonia—and an intercalated diploid phase in the life cycle, including the embryo [embryophytes/Embryophyta].

later homonym A homonym that is illegitimate (unless conserved), being published at a date later than the earliest published homonym.

lateral (a) *Axillary* (position, inflorescence position). (b) Style arising at the side of an ovary. (style position)

lateral bud *Axillary bud* (bud type)

lateral meristem A cylindrical sheath of cells, functioning in secondary growth, that increases width or girth of stems or roots in woody plants; includes the vascular cambium and cork cambium.

lateral root A root that arises from another root, derived endogenously from the pericycle. (root type)

lateral vein *Ventral vein*.

laticifer Cells located in the periphery of some tissues that secrete and store latex, functioning to deter herbivory and to seal and protect plant tissue upon wounding.

latrorse Dehiscing laterally relative to the flower center. (anther dehiscence direction)

layer One of the ecological criteria of plant communities based on height and plant habit, including the canopy, subcanopy, shrub or subshrub layer, and herb layer.

lead One of the two contrasting statements in a dichotomous key.

leaf A generally dorsi-ventrally flattened organ, usually functioning in photosynthesis and transpiration, either gametophytic (in mosses and some liverworts) or sporophytic (in vascular plants), often variously modified. (plant part)

leaf gap A region of nonvascular parenchyma tissue interrupting the vasculature of the stem at a node, associated with euphylls.

leaf primordium An immature leaf of the shoot. (stem/shoot parts)

leaf scar A mark indicating the former place of attachment of a leaf. (twig part)

leaf spine A sharp-pointed leaf, e.g., cactus spines or glochidia. Cf: *prickle*; *thorn*. (leaf structural type)

leaflet A distinct and separate segment of a leaf. (leaf part)

leaflet spine A sharp-pointed leaflet, e.g., some palms, such as *Phoenix*. (leaf structural type)

laesura The differentially thickened wall region corresponding to the tetrad attachment scar on each of the four immature spores following meiosis. Pl: *laesurae*.

large single copy region A relatively long stretch of DNA occurring as one copy in the chloroplast of plant cells.

lectotype A specimen that is selected from the original material to serve as the type when no holotype was designated at the time of publication, if the holotype is missing, or if the original type consisted of more than one specimen or taxon.

legitimate name A name that is validly published in accordance with the rules of the International Code of Nomenclature for algae, fungi, and plants.

legume A dry, dehiscent fruit derived from one carpel that splits along two sutures, e.g., Fabaceae. (fruit type)

lemma The outer and lower bract at the base of the grass floret. (leaf structural type)

lenticel A pore in the bark, generally functioning in gas exchange. (twig part)

lenticular Lens-shaped; disk-shaped with two convex sides. (shape)

lepidote Covered with scales or scalelike structures. (vestiture)

leptoid A specialized cell that functions in sugar conduction in some mosses.

leptosporangium The sporangia of the leptosporangiate ferns (Polypodiales), characterized by developing from a single cell and having a single layer of cells making up the sporangium wall. Pl: *leptosporangia*.

liana/liane A woody, perennial vine, in tropical forests often a component of the canopy layer. (plant habit)

library preparation The steps, involving ligation, of preparing DNA for a particular technique of molecular data acquisition.

ligate To chemically attach a molecular component to another molecule, e.g., to a region of DNA.

light reactions A series of biochemical reactions of photosynthesis in plants, occurring in the thylakoid membranes of the chloroplast and requiring light as an energy source, during which water is broken down into hydrogen ions, electrons, and molecular oxygen, and producing high-energy ATP and $NADPH_2$, which are utilized in the dark reactions.

lignin A complex polymer of phenolic compounds that impregnates the *secondary cell wall* of some cells (including tracheary elements and sclerenchyma), functioning to impart strength and rigidity to the wall.

lignotuber *Burl.* (stem/shoot type)

ligulate (a) Strap- or tongue-shaped; flattened and somewhat oblong in shape, e.g., ray flowers of some Asteraceae. (perianth type, shape) (b) Having a ligule arising from top of leaf sheath, at junction with blade. Cf: *eligulate*.

ligule (a) A small appendage on the upper (adaxial) side of the leaf, near the leaf base, found in the Selaginellaceae and Isoetaceae of the lycophytes. (b) An outgrowth or projection from the top of a leaf sheath at its junction with the blade, as in the Poaceae. (leaf part)

limb The expanded portion of the corolla or calyx above the tube, throat, or claw. (perianth part)

lineage A sequence of ancestral-descendent populations, in which the members are linked or connected by gene flow. Cf: *clade*.

lineage segment A portion of a lineage, e.g., from one divergence point (or "node" of a cladogram) to another.

lineage sorting The process by which, following evolutionary divergence, several gene lineages inherited from an ancestor converge (are reduced to) to a single gene lineage.

linear (a) With margins straight, parallel, length:width ratio between 12:1 and 6:1. (shape) (b) Stigmas or stigmatic tissue long and narrow in shape. (stigma/stigmatic region type)

linear regression analysis A statistical procedure for fitting a straight line onto a bivariate plot of two variables and testing the significance of relationship between them.

linear tetrad A tetrad in which the four pollen grains are arranged in a straight line, e.g., *Typha*.

lip Either of two variously shaped parts into which a calyx or corolla is divided, usually into upper (posterior) and/or lower (anterior) lips, each lip often composed of one or more lobes, e.g., Lamiaceae, Orchidaceae. Cf: *labellum*. (perianth part)

lithocyst A specialized cell that contains a cystolith. Cf: *cystolith*.

lithophyte Referring to a saxicolous plant growing specifically on the surface or within a rock or boulder.

lobe (a) A segment of a synsepalous calyx or sympetalous corolla. (perianth part) (b) A segment of a divided leaf. (leaf part)

lobed (a) Sinuses extending (pinnately or palmately) one eighth to one fourth the distance to midrib, midvein, or vein junction. (b) A general term meaning having lobes. (division)

locule An ovary cavity, bounded by ovary walls and septa. (gynoecium part) (b) A compartment of the anther, usually the result of two microsporangia fusing within a theca. (anther part)

loculicidal capsule A capsule in which longitudinal lines of dehiscence are radially aligned with the locules. (fruit type)

lodicule One of the (2–3) modified perianth parts of a grass (Poaceae) flower, which collectively upon swelling function to open the floret by separating the lemma from palea. (perianth type)

loment A secondarily modified legume that splits transversely into 1-seeded segments. (fruit type)

long-branch attraction A condition in which taxa with relatively long branches (having numerous character state changes) tend to come out as close relatives of one another in a phylogenetic analysis because of random effects.

longitudinal dehiscence Dehiscing along a suture parallel to the long axis of the thecae. (anther dehiscence type)

longitudinal posture Placement of margins with respect to a horizontal plane. (disposition)

lower epidermis The abaxial epidermis of a leaf.

lumen The space between muri in a reticulate pollen grain. Pl: *lumina*.

lycophyll The sporophytic leaf of the lycophytes, characterized by an intercalary meristem, having a single vein, and lacking a gap in the vasculature of the stem. Adj: *lycophyllous*. Syn: *microphyll*. (leaf structural type)

lyrate Pinnatifid, but with a large terminal lobe and smaller basal and lateral lobes. (shape)

maculate Spotted; with small spots on a more or less uniform background. (color pattern)

majority consensus tree A consensus tree in which only those clades that are retained 50% or more of the time are retained (i.e., not collapsed to a polytomy).

male (a) Individual with male reproductive organs only. (plant sex) (b) *Staminate*. (flower sex)

male gametophyte A gametophyte that bears only antheridia. Syn: *microgametophyte*.

male sporophyll A sporophyll that bears one or more microsporangia. Syn: *microsporophyll*.

malpighian Trichomes with two arms arising from a common base, e.g., Malpighiaceae. Syn: *dolabriform*. (trichome type)

mammillate Having small, nipple-shaped projections. (configuration)

manual *Flora*.

marcescent Ephemeral, but with persistent remains; withering persistent. (duration)

marginal (a) With the placenta along the margin of a unicarpellate (simple) ovary, e.g., Fabaceae. (placentation) (b) With three or more primary veins diverging from one point and reaching the blade margin, a subcategory of actinodromous. (leaf venation)

massula A group of fused pollen grains in large, often irregular numbers, but less than an entire theca. Pl: *massulae*.

masticatory A substance that is chewed by humans, usually for its pleasing or euphoric effects, e.g., peyote, *Lophophora williamsii*.

maturation (a) Acquisition of the mature structural and functional features of a cell following cell expansion. Syn: *specialization*. (b) Relative time of development of parts. (temporal phenomena)

matutinal In the morning, typically with respect to when flowers open. (periodicity)

maximum likelihood A method of phylogenetic inference that considers the probability, based on some selected model of evolution, that each tree explains the data.

mealy Covered with small, fine granules. (epidermal excrescence)

median The middle value of a list of values. Cf: *mean*.

median vein *Dorsal vein*.

mean The numeric average. Cf: *median*.

megagametogenesis The process of development of the female gametophyte from a megaspore.

megagametophyte *Female gametophyte*.

megaphyll *Euphyll*.

megasporangium A female sporangium, within which megasporocytes undergo meiosis to produce haploid megaspores. Pl: *megasporangia*. Syn (in seed plants only): *nucellus*. (plant part)

megaspore A female spore, produced via meiosis in the megasporangium and giving rise to a female gametophyte.

megaspore mother cell *Megasporocyte*.

megasporocyte A cell within the megasporangium that undergoes meiosis, forming four megaspores. Syn: *megaspore mother cell*.

megasporogenesis The process of development of megaspores from the megasporocyte.

megasporophyll *Female sporophyll.*

melittophily Pollination by bees. Adj: *melittophilous*. Syn: *hymenopterophily*.

membranous Thin and somewhat translucent; membranelike. (texture)

mericarp A portion of a fruit that separates from the ovary as a distinct unit that completely encloses the seed(s). (fruit part)

meristem A region of actively dividing cells.

meristic characters *Discrete quantitative characters.*

merosity Number of parts per whorl or cycle. (number)

mesarch An orientation of xylem maturation in which the protoxylem is surrounded by metaxylem within the vascular tissue, as can occur in siphonosteles.

mesocarp A middle wall layer of the pericarp, if the latter is divided into layers. (fruit part)

mesophyll The region of a sporophytic leaf between the outer epidermal layers and exclusive of the vasculature, containing the chlorophyllous cells.

mesophytic Having an intermediate texture, between coriaceous and membranous. (texture)

metaspecies Two or more lineage segments that can be resolved as neither monophyletic nor paraphyletic.

metaxylem The xylem of a group of tracheary elements that matures later, consisting of larger-diameter cells.

microfibril Microscopic fiberlike units of intertwined cellulose molecules, forming a meshwork within the cell wall.

microgametogenesis The process of development of pollen grains from haploid microspores via mitosis and differentiation.

microgametophyte *Male gametophyte.*

microhair A very small trichome, as in the three-celled, glandular microhairs of the Commelinaceae.

microphyll Essentially equivalent to a lycophyll.

micropylar Describing the region of an ovule that is near the micropyle.

micropylar Referring to the distal region of the ovule, near the micropyle.

micropyle A small pore at the distal end of the integument of a seed, functioning as the site of entry of pollen grains, or in angiosperms of pollen tubes.

microsatellites Regions of DNA that contains tandem repeats, short repeats of nucleotides. Syn: *Single sequence repeats.*

microsporangium A male sporangium, within which microsporocytes undergo meiosis to produce haploid microspores. Pl: *microsporangia*. (plant part)

microspore A male spore, produced via meiosis in the microsporangium and giving rise to a male gametophyte.

microsporogenesis The process of development of haploid microspores from diploid microsporocytes via meiosis.

microsporophyll A sporophyll bearing one or more microsporangia; a male sporophyll.

middle lamella A pectic-rich layer formed between the plasma membrane of adjacent cells in land plants, functioning to bind adjacent cells together.

middle layers Anther wall layers that may occur between the endothecium and tapetum.

midrib The central, main vein of the blade of a simple leaf or of a compound leaf in some palms. Syn: *costa*. (leaf part)

midvein The central, main vein of the blade of a leaflet. (leaf part)

mitochondrial DNA The DNA restricted to the mitochondrion. Abbr: *mDNA*.

mitochondrion A double membrane-bound, cellular organelle with invaginations called cristae that function in the electron transport reactions of respiration. Pl: *mitochondria*.

mixed bud A bud that produces both flowers and leaves. (bud type)

mixed craspedodromous Pinnate venation in which some secondary veins terminate at the margin, but with many terminating away from the margin. (leaf venation)

mixed development A combination of gradate and simultaneous development of sporangia in a fern sorus.

molecular clock An average rate of change of nucleotides of one or more gene sequences, used to date divergence times.

monad A single, unfused pollen grain. (pollen unit)

monadelphous With one group of stamens connate by their filaments, e.g., Malvaceae. (stamen fusion)

monaxial Referring to a tree that is unbranched with a single (vegetative) apical meristem. (tree branching model)

monistichous Alternate, with points of attachment in one vertical row/rank, e.g., Costaceae. (arrangement)

monocarpic A perennial or annual plant, flowering and fruiting once, then dying. (tree branching model, temporal phenomena, duration)

monochasium A determinate inflorescence that develops along one axis only. (inflorescence type)

monocotyledonous A type of anther wall development in which only the inner secondary parietal cell layer divides to yield the tapetum and a single middle layer.

monocyclic *Uniseriate*. (cycly, perianth cycly)

monoecious Having unisexual flowers, both staminate and pistillate on the same individual plant. (plant sex)

monograph A detailed taxonomic study of all species and infraspecific taxa of a given taxonomic group.

monolete Spores, with an unbranched, linear laesura that is linear and unbranched.

monomorphic Being similar to one another, as in leaves.

monomorphic character A character that is invariable in character state values within an OTU.

monophyletic/monophyly Referring to a group that consists of a common ancestor plus all (and only all) descendants of that ancestor.

monopodial A branching pattern in which a given axis is derived from a single apical meristem. (stem branching pattern)

monosporic Megasporogenesis in which meiosis of the megasporocyte nucleus results in the formation of four haploid megaspore nuclei, followed by cytokinesis, resulting in four megaspore cells, only one of which contributes to the female gametophyte.

monosulcate A pollen grain with a single, sulcate aperture.

monosymmetric *Zygomorphic, bilateral, irregular*. (symmetry)

monothecal Anther with one theca and typically two microsporangia. Cf: *bisporangiate*. (anther type)

monoulcerate A pollen grain with a single, ulcerate aperture.

morphocline *Transformation series.*

morphologic species *Taxonomic species.*

morphometrics The study of shape or form, generally useing statistical methods.

movement hercogamy A type of hercogamy involving movement of floral parts, e.g., the rapid closure of the stigmas upon their being touched by a potential animal pollinator.

MSG *Multiplexed shotgun genotyping.*

mucilage ducts/canals Specialized cells that secrete mucilage.

mucronate With a stiff, straight apical process, length:width ratio 1:1 to 3:1. (apical process)

mucronulate With a stiff, straight apical process, length:width ratio ≤ 1:1. (apical process)

multicellular Trichomes having two or more cells. (trichome type)

multiple fruit A fruit derived from two or more flowers. (fruit type)

multiplexed shotgun genotyping A molecular technique using restriction enzymes that result in more frequent cleavages to the DNA, resulting in numerous fragments that are prepared and sequenced using high-throughput methods. Abbr: *MSG*.

multiplexing The simultaneous sequencing of large numbers of DNA library preparations.

multiseriate (a) Perianth composed of three or more distinct whorls. (perianth cycly) (b) Trichomes having more than one vertical row of cells. (trichome type) (c) Rays in wood that are made up of many vertical rows of cells.

multistate character A character with three or more character states.

muricate Having coarse, radially elongated, rounded protuberances. (epidermal excrescence)

murus The structural element of a reticulate pollen grain. Pl: *muri*.

mycorrhiza A symbiotic association between a fungus and roots, functioning to increase absorptive surface area and mineral uptake. Pl: *mycorrhizae*. (root part)

mycorrhizae A symbiotic association between the root of a vascular plant and a fungus.

mycotroph A plant obtaining nutrition from mycorrhizal fungi in the soil, as in some Ericaceae. (plant life form) Adj: *mycotrophic*.

myrmecophily A symbiotic association between a plant and ants, in which the plant provides a reward for the ant, e.g., extrafloral nectaries, domatia, or Beltian bodies, and the ant in turn protects the plant from herbivores or parasites.

naked bud A bud lacking surrounding protective scales, e.g. *Viburnum*, Caprifoliaceae. (bud type)

narrowly acute Apical margins approximately straight, the intersection angle <45°. (apex)

narrowly cuneate Basal margins approximately straight, the intersection angle <45°. (base)

narrowly elliptic Margins curved, widest near midpoint, length :width ratio between 6:1 and 3:1. (shape)

narrowly oblong Margins straight, parallel, length:width ratio between 6:1 and 3:1. (shape)

narrowly triangular Three-sided, length:width ratio between 6:1 and 3:1. (shape)

natural selection The directed and nonrandom genetic modification of a population or species, in which genetic changes that result in an increase in survivorship and/or reproduction are contributed to the next generation more.

neck The distal, narrow extension of the sterile jacket cells of the archegonium, through which a sperm cell must travel to fertilize the egg.

neck canal cells Cells located within the neck of the archegonium that break down and are secreted from the pore of the neck at maturity.

nectariferous disk A nectary consisting of a disklike or doughnut-shaped mass of tissue surrounding the ovary base or top; may be inner to (intrastaminal), beneath (staminal), or outer to (extrastaminal) the androecium. (flower part, nectary type)

nectary A group of specialized cells that secrete sugar- (or protein-) rich fluids to the outside, as a reward for pollination or protection. (flower part)

neotony A type of paedomorphosis caused by a decrease in the rate of development of a feature.

neotropical Distributed in tropical areas of the New World (tropical Mexico, Caribbean, Central/South America).

neotype A specimen derived from a nonoriginal collection that is selected to serve as the type as long as all of the material on which the name was originally based is missing.

nerve *Vein*. (leaf part)

nested clade analysis A methodology that reconstructs the genealogical relationships of individuals using haplotypes.

nested clade phylogeographic analysis Nested clade analysis correlated with geographic distribution of individuals.

netted With ultimate veinlets forming a reticulum or netlike pattern. Syn: *anastomosing*, *reticulate*. (leaf venation)

network *Unrooted tree*.

nexine The inner layers of the exine, including both endexine and the foot-layer of the ektexine.

Next Generation Sequencing *High-throughput sequencing*. Abbr: *NGS*.

nitid Appearing lustrous, polished. Syn: *shining*. (epidermal excrescence)

nocturnal Occurring during the night, typically with respect to when flowers open. (periodicity)

node (a) The point of attachment of a leaf to a stem. (stem/shoot parts) (b) The region of stem at which leaf, leaves, or branches arise. (twig part) (c) The point of divergence of one clade into two; the point in time and space of the most common ancestor of the two divergent clades.

node-based A type of phylogenetic classification in which a node (common ancestor) of the cladogram (and all descendants of that common ancestor) serves as the basis for grouping.

nomen conservandum A conserved name. Abbr: *nom. cons.*

nomen novum Meaning a new name. Abbr: *nom. nov.*

nomen nudum Meaning a name published without a description or diagnosis, making the name invalid. Abbr: *nom. nud.*

nomenclatural synonym *Homotypic synonym*.

nomenclatural type A specimen or illustration that acts as a reference for a scientific name, upon which the name is based. Syn: *type*; *type specimen*.

nomenclature The formal naming of taxa according to some standardized system; for plants, "algae," fungi, and organisms traditionally treated as fungi, governed by the International Code of Nomenclature for algae, fungi, and plants.

non Latin for "not."

non-clathrate Fern scales in which the cell walls of adjacent cells ("anticlinal" walls) are thin. Cf: clathrate.

nondecussate Opposite leaves or other structures (e.g., leaflets) not at right angles to the preceding pair; may be superficially the result of stem twisting. (arrangement)

nondisjunction An irregularity during meiosis in which homologous chromosomes do not segregate, which may result in the production of gametes that are unreduced, i.e., have two sets of chromosomes.

nonendospermous *Exalbuminous*. (seed endosperm type)

nonporous Referring to wood having only tracheids.

non visus Latin for "not seen," typically meaning that authors did not see a specimen, such as a type. Abbr: *n. v.*

nucellar beak A proliferation of cell divisions of the nucellus at the micropylar region of the ovule.

nucellus Term for the megasporangium of a seed. Adj: *nucellar*.

nuclear DNA The DNA restricted to the nucleus of a cell. Abbr: *nDNA*.

nuclear ribosomal cistron A repeating region of the nuclear ribosomal DNA, functioning in ribosome synthesis and consists of multiple copies, including the ITS and ETS regions.

nuclear ribosomal DNA A region of the nuclear DNA involved in ribosome production, containing multiple copies of the cistron. Abbr: *nrDNA*.

nuclear endosperm An endosperm in which the early mitotic divisions are not followed by cytokinesis.

nucleus A double membrane-bound, cellular organelle that contains DNA.

nude Descriptive of the ovary of a female flower lacking a perianth; thus, ovary position not evident, e.g., Betulaceae.

number Whether a Latin name is singular or plural.

nut A one-seeded, dry indehiscent fruit with a hard pericarp, usually derived from a 1-loculed ovary. (fruit type)

nutlet A small nut, usually refering to the schizocarp of nutlets in fruits of the Boraginaceae and Lamiaceae. Cf: *eremocarp*. (fruit type)

nutritive tissue Tissue that surrounds or abuts the embryo of a seed and that consists of female gametophyte (in nonangiosperms) or endosperm (in angiosperms).

nyctinasty Movement (closing) of the leaflets of a compound leaf as a response to darkness, in photoperiodism. (leaf behavior)

obdiplostemonous Stamens in two whorls, the outer opposite petals, inner opposite the sepals, e.g., Simaroubaceae. (stamen position)

obhaplostemonous Stamens uniseriate, equal in number to the petals, and opposite the petals. Syn: *antepetalous, antipetalous* (stamen cycly, number)

oblanceolate Margins curved, widest near the apex, length:width ratio 6:1 to 3:1. (shape)

oblance-ovate Margins curved, widest near the apex, length:width ratio 3:1 to 2:1. (shape)

oblate A pollen grain in which the P/E ratio is less than 1.

obligate hemiparasite A chlorophyllous, parasitic plant requiring a host to survive and reproduce. (plant life form)

oblique With an asymmetrical apex or base. (apex, base)

oblong Margins straight, parallel, length : width ratio 2:1 to 3:2. (shape)

obovate Margins curved, widest near apex, length:width ratio 2:1 to 3:2. (shape)

obturator A protuberance of tissue, typically arising from the funiculus or placenta at the base of the ovule, e.g., Euphorbiaceae.

obtuse Apical or basal margins approximately straight, intersection angle >90°. (apex, base).

ocrea A specialized, scarious, sheathlike structure arising above the node in some members of the family Polygonaceae, interpreted as modified stipules. (leaf part)

odd-pinnate *Imparipinnate/imparipinnately compound*.

oil A type of triglyceride compound that may function as high-energy storage compounds or secretion products; a type of ergastic substance.

oil bodies Oil-containing structures found within certain cells of most liverworts.

oil ducts/canals Specialized cells that secrete oil.

oligomerous Having a fewer than typical number of parts. (merosity)

onagrad type *Crucifer type*.

ontogenetic sequence The discrete stages of the developmental sequence of a given feature.

ontogenetic trajectory A plot of developmental change as a function of time.

ontogeny The developmental sequence of a given feature.

oogamy A type of sexual reproduction in which one gamete, the egg, becomes larger and nonflagellate and the other gamete, the sperm cell, remains relatively small and flagellate; found in all land plants and independently evolved in many other eukaryotes.

open venation Fern venation in which the veins arising from the midvein or base of a pinnule do not join back together. Syn: *free venation*.

operational taxonomic unit (OTU) One of the individual, unit taxa of a phylogenetic analysis.

operculate Having calyx and corolla fused into a cap that falls off as a unit, e.g. *Eucalyptus*. (perianth type)

operculum An apical lid, as in the capsule of most mosses that falls off during spore release.

opposite With two leaves or other structures per node, on opposite sides of stem or central axis. (arrangement)

orbicular Circular in outline; margins curved, length:width ratio approximately 1:1. Adj: *circular*. (shape)

ordered Referring to a transformation series in which the character states occur in a predetermined sequence.

organelles Structural, membrane-bound units of the cell that provide some vital metabolic function.

orientation Referring to the angle relative to a central, usually vertical, axis. (disposition)

ornithophily Pollination by birds. Adj: *ornithophilous*.

ortho-amphitropous An ovule type in which the vasculature is straight, leading from the funiculus base to the middle of the nucellus, with the nucellus bent sharply in the middle along both the lower and upper sides, often with a basal body present.

ortho-campylotropous An ovule type in which the vasculature is straight, leading from the funiculus base to the middle of the nucellus, with the nucellus bent only along the lower side, with no basal body.

orthographia conservanda Meaning a conserved spelling in a scientific name. Abbr: *orth. cons*.

orthologs Functionally similar DNA sequences that are homologous among compared taxa.

orthotropic Shoots that are erect and essentially radially symmetric, the branching three-dimensional. (tree branching model)

orthotropous A type of ovule in which no curvature takes place during development, the micropyle being positioned opposite the funiculus base.

outbreeding The transfer of gametes from one individual to another, genetically different individual. Syn: *outcrossing*; *allogamy*; *xenogamy*.

outcrossing *Outbreeding*.

outer bark *Periderm*.

outgroup A taxon or group of taxa that is not a member of the ingroup.

ovary The part of the pistil containing the ovules. (gynoecium part)

ovate Margins curved, widest near base, length:width ratio 2:1 to 3:2. (shape)

ovule An immature seed, prior to fertilization; a megasporangium enveloped by one or more integuments. (plant part, gynoecium part)

ovuliferous scale The modified lateral branch system of a conifer cone that bears one or more seeds.

P/E ratio The ratio of the polar diameter to the equatorial diameter.

pachycaul An erect, woody, trunklike stem that is swollen basally, the swollen region functioning in storage of food reserves or water. (stem/shoot type)

padlike nectary Developing as a discrete pad of tissue extending partway around the base of the flower. (nectary type)

paedomorphosis A type of heterochrony in which the mature or adult stage of the derived ontogenetic sequence resembles a juvenile ontogenetic stage of the ancestral condition.

paired end sequencing A type of high-throughput sequencing in which sequences are obtained from both directions, 3′ and 5′.

palea The inner/upper bract at the base of the grass floret, or, in some Asteraceae, the bracts subtending flowers, for the latter Syn: *chaff*. Pl: *paleae*. (leaf structural type; flower/inflorescence part)

paleotropical Distributed in tropical areas of the Old World. (tropical Africa, Asia, Australasia).

palinactinodromous With three or more primary veins diverging from one point, the primary veins having additional branching above their main point of divergence. (leaf venation)

palisade mesophyll The usually upper (adaxial), columnar cells of the mesophyll of some leaves.

palmate/palmately compound A compound leaf with four or more leaflets arising from a common point, usually at the end of the petiole. (leaf type)

palmately netted Netted, with four or more primary veins arising from a common basal point. (leaf venation)

palmately veined With four or more primary veins arising from a common basal point. (leaf venation)

palmate-netted Palmately veined, with four or more primary veins arising from a common basal point, the ultimate veins forming a fine reticulum. (leaf venation)

palmate-parallel With several primary veins arising from one point, the adjacent secondary veins that are parallel to these having transverse, interconnecting veins, e.g., "fan" palms. (leaf venation).

palmate-ternate Ternate, with the three leaflets joined at a common point. (leaf type)

palmatifid Palmately lobed to divided. (division)

palmatisect Palmately divided, almost into discrete leaflets but confluent at the lobe bases. (division)

palynology The study of spores and pollen grains.

pandurate Violin-shaped, obovate with the side margins concave. (shape)

panicle An indeterminate inflorescence, consisting of several branched axes bearing pedicellate flowers. (inflorescence type)

pantropical Distributed worldwide in tropical areas.

pantoporate A pollen grain aperture type in which pori occur globally on the pollen grain surface.

papilionaceous A flower with one large posterior petal (banner or standard), two inner, lateral petals (wings), and two usually apically connate lower petals (keel); floral structure of the Faboideae-Fabaceae. (perianth type)

papillate Having minute, rounded protuberances. Syn: *tuberculate*; *verrucate*. (epidermal excrescence)

pappus The calyx of the Asteraceae, modified as awns, scales, or capillary bristles.

paracladium A unit inflorescence that resembles the secondary inflorescence; e.g., an umbel of umbels or compound umbel. (secondary inflorescence)

parallel With primary or secondary veins essentially parallel to one another, generally converging at the apex, the ultimate veinlets transverse. (leaf venation)

parallelism *Convergence*.

parallelodromous *Parallel*. (leaf venation)

paralogs Functionally dissimilar DNA sequences, the result of gene duplication.

paraphyletic/paraphyly Referring to a group that includes a common ancestor plus some, but not all, descendants of that common ancestor.

paraphyletic species A species consisting of lineage segments that form a paraphyletic grade, each lineage segment lacking an apomorphy.

paraphyses Hair-like structures arising from the receptacle of a fern sorus.

paratype A specimen cited in a publication but that is neither a holotype, isotype, or syntype.

parenchyma Cells that, at maturity, are alive and potentially capable of cell division, are isodiametric to elongated in shape, and have a primary cell wall only (rarely with secondary wall), forming a solid mass of tissue and functioning in metabolic activities and in wound healing and regeneration.

parietal With the placentae on the inner wall or on intruding partitions of a unilocular, compound ovary, e.g., Violaceae. (placentation)

parietal cell The outer cell formed if the archesporial cell of an ovule divides to form an inner megasporocyte.

parietal-axile With the placentae at the junction of the septum and pericarp of a 2- or more loculate ovary, e.g., Brassicaceae. (placentation)

parietal-septate With placentae on the inner ovary walls but within septate locules, as in some Aizoaceae. (placentation)

paripinnate/paripinnately compound A pinnately compound leaf without a terminal leaflet, typically even-pinnate. (leaf type) Syn: *even-pinnate*.

parsimony analysis *Principle of parsimony*.

parted With sinuses extending (pinnately or palmately) one half to three quarters of the distance to the midrib, midvein, or vein junction. (division)

parthenogenesis Development of an embryo from a cell of an abnormal, diploid female gametophyte, such as a diploid egg.

partial inflorescence *Unit inflorescence*. (inflorescence part)

patent *Horizontal*; *divaricate; divergent*. (orientation)

PCR *Polymerase chain reaction*.

peat Fossilized and partially decomposed *Sphagnum* (peat moss).

pectinate Comblike; pinnately divided with close, narrow lobes. (division)

pedate Refers to a palmately divided leaf in which the lateral lobes are further divided, as in some ferns. (division)

pedicel A flower stalk. Adj: *pedicellate*. (flower part; inflorescence part)

peduncle The stalk of an entire inflorescence. Adj: *pedunculate*. (inflorescence part)

pellucid Having translucent spots or patches. (color pattern)

peltate (a) Having a stalk attached away from the margin of a flattened structure, e.g., a petiole attached on the underside of blade. (base) (b) Trichomes with stalk attached on the underside of a disk-shaped apical portion. (trichome type)

pendant/pendulous Hanging downward loosely or freely (orientation)

pendulous *Apical*. (placentation)

penicillate Tufted, like an artist's brush. (vestiture)

penni-parallel With secondary veins arising along the length of a single primary vein region, the former essentially parallel to one another and interconnected by tranverse veins. Syn: *pinnate-parallel*. (leaf venation)

pentamerous Having a whorl with five members. (merosity, perianth merosity)

pepo A nonseptate, fleshy fruit with parietal placentation and a leathery exocarp derived from an inferior ovary, e.g. Cucurbitaceae. (fruit type)

peramorphosis A type of heterochrony in which ontogeny passes through and goes beyond the stages or trajectory of the ancestral condition.

perennial A plant living more than 2 years. (duration)

perfect (a) *Bisexual*. (flower sex) (b) With lateral primary veins covering at least two thirds of the leaf blade area, a subcategory of actinodromous and of acrodromous. (leaf venation)

perfoliate A leaf blade that is sessile with the base completely surrounding the stem. (leaf attachment)

perforation plate The contact area of two adjacent vessel members, may be compound, if composed of several pores or simple if composed of a single opening.

perforation plate The region of one or more perforations at the end wall of a vessel member, where one cell makes contact with another; may be simple or compound.

perianth The outermost, nonreproductive group of modified leaves of a flower, composed of the combined calyx and corolla or of tepals. Syn: *perigonium*. (flower part)

pericarp The fruit wall, derived from the mature ovary wall, sometimes divisible into layers: endocarp, mesocarp, and exocarp. Syn: *rind*. (fruit part)

pericycle A cylindrical sheath of parenchyma cells just inside the endodermis, functioning as the site of resumed meristematic growth, forming a secondary root or (in woody plants) contributing to the vascular cambium.

periderm The cork cambium and its derivatives: phelloderm and cork tissue. Syn: *outer bark*.

perigonal nectary Nectaries on the perianth, usually at the base of sepals, petals, or tepals. (nectary type)

perigonium *Perianth*. (flower part)

perigynous Hypanthium present, attached at base of ovary, with sepals, petals, and stamens attached to hypanthium rim, the ovary superior. (perianth/androecial position)

perine/perine layer A thick, outer layer in the spores of mosses and ferns.

periodicity Referring to periodically repeating phenomena. (temporal phenomena)

periplasmodial *Amoeboid*.

perispermous Having a seed storage tissue in which the chalazal nucellar cells enlarge and store energy-rich compounds.

peristome teeth Hygroscopic, teethlike structures that occur in a whorl along the margin of the opening of a moss capsule and that function in spore release.

personate Two-lipped, with the upper arched and the lower protruding into corolla throat, e.g., *Antirrhinum*. (perianth type)

petal An individual member or segment of the corolla, typically (nongreen) colored and functioning as an attractant for pollination. (flower part, perianth part)

petalostemonous *Epipetalous*. (stamen fusion)

petiolar spine A sharp-pointed leaf petiole, e.g., *Fouquieria* spp. Cf: *thorn*; *prickle*. (leaf structural type)

petiole A leaf stalk. Adj: *petiolate*. Syn: *stipe*. (leaf part)

petiolule A leaflet stalk. Adj: *petiolulate*. (leaf part)

phalaenophily Pollination by moths. Adj: *phalaenophilous*.

phanerantherous *Exserted*. (stamen insertion)

phanerocotylar *Epigeous*. (seed germination type)

phelloderm The inner layers of cells produced by the cork cambium.

phenetic A classification system in which taxa are grouped by some measure of overall similarity.

phenogram A branching diagram representing a phenetic classification.

phenology The timing of reproduction, generally referring to the date of flowering or fruiting, or stages thereof.

phloem A tissue composed of specialized sieve elements plus some parenchyma and often some sclerenchyma, functioning in conduction of sugars.

photosynthesis The series of biochemical reactions in which light energy is used to produce high-energy compounds, in land plants involving reactions of carbon dioxide and water to produce polysaccharides.

phreatophyte A plant with a long taproot, in contact with ground water, e.g., mesquite, *Prosopis*. (plant life form)

phyllary One of the involucral bracts subtending a head, as in the Asteraceae. Syn: *involucral bract*. (leaf structural type)

phylloclade *Cladode*. (stem/shoot type)

phyllode A leaf consisting of a flattened bladelike petiole. (leaf structural type)

phylogenetic Referring to a classification that is based on evolutionary history, or pattern of descent; referring to relationships between groups of individuals at a given point in time.

phylogenetic species A species recognized as the smallest group of populations or lineages diagnosable by a unique combination of character states; a monophyletic group recognized and defined by one or more apomorphies.

phylogenetic systematics A methodology for inferring the pattern of evolutionary history of a group of organisms by grouping taxa based on apomorphies. Syn: *cladistics*.

phylogenetic tree *Cladogram*.

phylogeny The evolutionary history or pattern of descent of a group of organisms.

phylogeography A field of study that assesses relationships among genotypes of a species or closely related species and correlates those relationships with their spatial distribution.

phylogram A cladogram that has an estimated absolute time scale, such that nodes and branch lengths are calibrated and correspond more closely to real elapsed time.

pilate With a long terete stalk terminating in a globose or ellipsoid apical thickening. (shape)

pilate-glandular Having a glandular cell atop an elongated basal stalk. (trichome type)

pilose With soft, straight to slightly shaggy trichomes at right angles to the surface. (vestiture)

pin Flowers with a long style and short stamens, found in distylous flowers.

pinna The first discrete leaflets or blade divisions of a fern leaf. Pl: *pinnae*. (leaf part)

pinnate/pinnately compound A compound leaf with leaflets arranged oppositely or alternately along a central axis, the rachis. (leaf type)

pinnately veined With secondary veins arising along length of a single primary vein, the latter a midrib or leaflet midvein. (leaf venation)

pinnate-netted Pinnately veined, with secondary veins arising along length of a single primary vein, the latter a midrib or midvein, the ultimate veins forming a fine reticulum. (leaf venation)

pinnate-parallel *Penni-parallel*. (leaf venation)

pinnate-ternate Ternately compound, with the terminal leaflet arising from the tip of a rachis, evolutionarily derived from a pinnately compound leaf, e.g., some Fabaceae. (leaf type)

pinnatifid Pinnately lobed to divided (division)

pinnatisect Pinnately divided, almost into discrete leaflets but confluent at the midrib. (division)

pinnule The ultimate divisions or leaflets of a leaf, usually applied to ferns. (leaf part)

piperad type A type of embryo development in which the zygote divides longitudinally (i.e., parallel to axis of the female gametophyte and nucellus), thus not forming a basal and terminal cell.

pistil That part of the gynoecium composed of an ovary, one or more styles (which may be absent), and one or more stigmas. (flower part, gynoecium part)

pistillate Flowers unisexual, with carpel(s) only, lacking fertile stamen(s). Syn: *female*. (flower sex)

pit A hole in a secondary cell wall that functions in cell-to-cell communication during development and that may function in water conduction in some tracheary elements.

pit pairs Adjacent holes in the lignified secondary cell walls of two adjacent cells.

pitcher leaf A leaf shaped like a container that bears an internal fluid and that functions in the capture and digestion of small animals, e.g., leaves of *Darlingtonia*, *Nepenthes*, *Sarracenia*, pitcher plants. Syn: pitfall. (leaf structural type)

pitfall leaf *Pitcher*. (leaf structural type)

pith The central, mostly parenchymatous tissue, internal to vasculature of siphonosteles and eusteles and within the vascular cylinder of some roots. (root parts, stem/shoot parts)

pit-pair Pits of adjacent cells occurring opposite one another, functioning in allowing communication between cells during their development and differentiation.

placenta The ovule-bearing tissue of the ovary. (gynoecium part)

placental vein *Ventral vein*.

placentation Referring to the position of the ovules within the ovary. (gynoecium, carpel, and pistil)

plagiotropic More or less horizontal with dorsiventral symmetry, the branching two-dimensional and leaves generally in one plane (either distichous or secund). (tree branching model)

plagiotropy by apposition Plagiotropy in which extension growth of the branch is taken over by an axillary meristem, but with the original branch terminal meristem continuing growth, usually as a short shoot. (tree branching model)

plagiotropy by substitution Plagiotropy in which the original branch terminal meristem aborts or converts into a terminal inflorescence or flower, extension growth of the branch being taken over by an axillary meristem. (tree branching model)

planation Evolutionary flattening, into a dorsiventral, planar posture.

plane (a) Flat, without vertical curves or bends. (transverse posture). (b) With a smooth configuration. Syn: *smooth*. (configuration)

plant A group of organisms, defined either by characteristics (possessing photosynthesis, cell walls, spores, and a more or less sedentary behavior), or by phylogenetic relationships, equivalent in this text to the land plants, Embryophyta.

plant anatomy The study of tissue and cell structure of plant organs.

plant habit General form of plant, including aspects of stem duration, branching pattern, development, and texture. (plant organs)

plant physiology The study of metabolic processes in plants.

plant press A device used to press and then dry plant specimens, such that they can be effectively used in an herbarium specimen.

plant sciences The study of plants, which are here equivalent to the land plants.

plasma membrane A phospholipid bilayer with embedded proteins that envelops all cells, functioning as the cell boundary, in cell–cell recognition, and in transport of compounds.

plasmodesmata Minute pores in the primary cell wall through which membranes traverse between cells, allowing for interchange of compounds between cells; an apomorphy for the Charophytes of the green plants.

plasmodial *Amoeboid*.

plated Bark split or cracked, with flat plates between fissures. (bark type)

pleonanthic An indeterminate shoot that bears lateral flowers but that continues vegetative growth. (tree branching model)

plesiomorphic *Ancestral*.

pleurotropous An ovule position in which the micropyle points to the side.

pleurotropous-dorsal A pleurotropous ovule in which the raphe is above.

pleurotropous-ventral A pleurotropous ovule in which the raphe is below.

plicate With a series of longitudinal folds; pleated. (longitudinal posture)

plumose (a) Covered with fine, elongated, ciliate appendages; featherlike, e.g., pappus of some Asteraceae. (bristle type) (b) Stigmas with feathery, trichomelike extensions, often found in wind-pollinated taxa such as Cyperaceae, Poaceae. (stigma/stigmatic region type)

plurilocular Referring to an ovary with two or more locules. (locule number)

pneumatophores Roots that grow upwardly from soil to air, functioning in obtaining additional oxygen for the plant. (root type)

polar axis An extended pollen grain diameter that passes through the center of the original pollen tetrad.

polar nuclei In a typical angiosperm female gametophyte, the two haploid nuclei of the central cell that ultimately fuse with a sperm cell (via double fertilization) to form a triploid endosperm cell.

polar view Observing a pollen grain from the direction of either pole.

polarity The designation of relative ancestry to the character states of a transformation series/morphocline.

pollen grain An immature, endosporic male gametophyte of seed plants. (anther part)

pollen sac A microsporangium, usually one-half of a theca in an angiosperm anther. (anther part)

pollen tube An exosporic process that grows from a pollen grain, functioning as a haustorial organ or to deliver sperm cells to the egg.

pollenkit A yellowish or orange, carotenoid-like material adhering to the exine, functioning to stick pollen grains in masses.

pollinarium In an orchid flower, the pollinia plus a sticky stalk (derived from either the anther or stigma), the unit of transport during pollination.

pollination The transfer of pollen grains from microsporangia either directly to the ovule (in gymnosperms) or to the stigma (in angiosperms).

pollination chamber A cavity formed by the breakdown of cells at the distal end of the megasporangium (nucellus) in gymnosperm seeds.

pollination droplet A droplet of liquid secreted by the young ovule through the micropyle, functioning to transport pollen grains by resorption.

pollinium Anther in which all pollen grains of both thecae (Orchidaceae) or of adjacent thecae (Asclepiadaceae) are fused together as a single mass. Pl: *pollinia*. (anther type)

polyad A group of pollen grains that are fused in precise units of more than four, e.g., Mimosoid clade, Fabaceae. (pollen unit)

polyaxial Referring to a tree that is branched, with more than one apical meristem. (tree branching model)

polyclave key A key in which all of the known character states that match a specimen are selected in order to narrow down the identity to a smaller subset of the possibilities.

polygamodioecious A plant with staminate and perfect flowers on some individuals, pistillate and perfect flowers on other individuals. (plant sex)

polygamomonoecious A plant with staminate, pistillate, and perfect flowers on the same individual. Syn: trimonoecious (plant sex)

polygamous A plant with both bisexual and unisexual flowers. (plant sex)

***Polygonum* type** A type of monosporic female gametophyte in which the megaspore nucleus undergoes three, sequential mitotic divisions, producing eight nuclei and seven cells, the most common and ancestral type in the angiosperms.

polymerase chain reaction A procedure for the rapid amplification of DNA using primers, free nucleotides, and DNA polymerase in solution and heating the solution to effect denaturation and replication of the DNA. Abbr: *PCR*.

polymerous Having a larger than typical number of parts. (merosity)

polymorphic character A character that has variable character state values within an OTU.

polypetalous *Apopetalous*. (perianth fusion)

polyphyletic group A group that consists of two or more, separate monophyletic or paraphyletic groups, each with a separate common ancestor; a group in which the common ancestor of all members is not itself a member of the group.

polyploidy A mutation in which offspring have an increase in chromosome number by a multiple of some ancestral set.

polysepalous *Aposepalous*. (perianth fusion)

polystemonous Referring to an androecium with many stamens. (stamen number)

polystichous *Spiral*. (arrangement)

polysymmetric *Actinomorphic, radial, regular*. (symmetry)

polytomy Three or more lineages arising from a single common ancestor in a cladogram, representing conflicting data or the lack of resolution.

pome A fleshy fruit with a cartilaginous endocarp derived from an inferior ovary, with the bulk of the fleshy tissue derived from the outer, adnate hypanthial tissue, e.g., *Malus*, apple. (fruit type)

population A group of individuals of the same species that is usually geographically delimited and that typically have a significant amount of gene exchange.

pore (a) A specialized, permanent opening in the upper epidermis of the thallus of some liverworts, functioning in gas exchange. (b) A single hole in the primary cell wall of sieve elements that is lined with callose and through which solutes flow in sugar conduction. (c) Vernacular term for a wood vessel, used in the wood industry.

poricidal Dehiscing by a pore at one end of the thecae, e.g., Ericaceae. (anther dehiscence type)

poricidal capsule A capsule in which the dehiscence occurs by means of pores, e.g., *Papaver*, poppy. (fruit type)

porose Referring to vessel cells with more or less circular, porelike perforations.

porous Referring to wood that contains vessel cells.

porus A pollen grain aperture that is circular to slightly elliptic with a length:width ratio of less than 2:1. Pl: *pori*. Adj: *porate*.

position (a) Placement relative to other, unlike parts. (disposition) (b) The placement of a taxon as a member of another taxon of the next higher rank.

posterior Referring to the upper lobe or part, especially in a horizontally oriented structure. (position)

posture Placement relative to a flat plane. (disposition)

praemorse Having a jagged, chewed appearance, e.g., some palms. (margin)

prickle A nonspine, nonthorn, sharp-pointed outgrowth from the surface of any organ. Adj: *prickly*. (plant, twig part)

primary cell wall The first, mostly cellulosic cell wall layer that is secreted external to the plasma membrane during cell growth.

primary endosymbiosis Endosymbiosis involving the engulfment of an ancestral bacterium by a eukaryotic cell.

primary growth Growth in height or length of a stem or root, brought about by the elongation and differentiation of cells derived from the apical meristem.

primary inflorescence *Unit inflorescence*. (inflorescence part)

primary parietal cells The inner layer of cells in an early stage of anther wall development.

primary pit field A group of numerous plasmodesmata in the primary cell wall, spatially associated with secondary cell wall pit pairs.

primary root The root of the sporophyte that develops from the radical of the embryo. (root type)

primary tissue A tissue formed by primary growth, e.g., as in primary xylem or primary phloem.

primary vein The major vein or veins of a leaf with respect to size. (leaf venation)

primer A complementary copy of a short, conserved, flanking region of a region of DNA of interest, used to amplify and sequence the DNA.

primitive *Ancestral*.

principal components analysis A multivariate statistical method that transforms numerous data variables into other variables called principal components, in order to study relationships among taxa. Abbr: *PCA*.

principle of parsimony The principle stating that the cladogram exhibiting the fewest number of evolutionary steps is accepted as the best estimate of phylogeny; a corollary of the general principle of Ockham's Razor. Syn: *parsimony analysis*.

priority of publication A principle of the International Code of Nomenclature for algae, fungi, and plants stating that of two or more competing possibilities for a name, the one published first is the correct one, with some exceptions.

prismatic A short, prism-shaped crystal; a type of ergastic substance.

procumbent *Prostrate*. (stem habit)

proembryo A very young embryo.

prolate A pollen grain in which the P/E ratio is greater than 1.

prolepsis Growth of an axillary bud into a shoot only after a period of rest. Adj: *Proleptic*. (tree branching model)

prop root Above-ground, adventitious roots that function in supporting the stem. (root type).

prophyll/prophyllum *Bracteole/bractlet*. (leaf structural type, flower part)

prostrate Trailing or lying flat, not rooting at the nodes. Syn: *procumbent*; *reclining*. (stem habit)

protandry With stamens or anthers developing before carpels or stigma. Adj: *protandrous*. (maturation, flower maturation)

protologue Everything associated with a name at its valid publication, i.e., description or diagnosis, illustrations, references, synonymy, geographical data, citation of specimens, discussion, and comments.

proteinoplast *Aleurone grain*.

protogyny With carpels or stigma maturing before stamens or anthers. Adj: *protogynous*. (maturation, flower maturation)

protonema An initial, filamentous form of a gametophyte (e.g., in mosses), prior to its differentiation into parenchymatous tissue.

protoplasm Everything inside the plasma membrane of a cell.

protostele A stele with a central, solid cylinder of vascular tissue.

protoxylem Referring to the first tracheary elements that develop within a patch of xylem, being typically smaller and with thinner cell walls than the later formed metaxylem.

proximal Near the point of origin or attachment. (position)

proximal pole The intersection of the pollen grain polar axis with the grain surface that is near the center of the microspore tetrad.

pseudanthium A unit that appears as and may function like a single flower, but that typically consists of two or more flowers fused or grouped together. (inflorescence type)

pseudobulb A short, erect, aerial storage stem of certain epiphytic orchids. (stem/shoot type)

pseudocrassinucellate An ovule type in which an second inner layer of nucellar cells by periclinal divisions in the single outer layer, a parietal cell not forming.

pseudodrupe A nut surrounded by a fleshy, indehiscent involucre, resembling a true drupe, e.g., *Juglans*, walnut. (fruit type)

pseudoelaters Groups of elongated, cohering, nonsporogenous, generally hygroscopic cells that develop within the sporangia of hornworts and function in spore dispersal.

pseudomonopodial A branching pattern that initiates as dichotomous, but in which one of the two axes becomes dominant and overtops the other, the other axis appearing lateral. (stem branching pattern)

pseudopetiole Term sometimes used for the petiole-like structure arising between a leaf sheath and blade, found in several monocots, such as bananas, bamboos. (leaf part).

pseudoterminal bud A bud appearing to be apical but that is actually lateral near the apex, assuming a terminal position with the death or nondevelopment of the true terminal bud. (bud type)

pseudoumbel An inflorescence appearing like a simple umbel, but actually composed of condensed, monochasial cymes, as in the Alliaceae and Amaryllidaceae. (inflorescence)

psilate A pollen grain having a smooth sculpturing.

psychophily Pollination by butterflies. Adj: *psychophilous*.

ptyxis The posture of embryonic structures, e.g., cotyledons within a seed or immature leaves or leaf parts in a bud. Syn: *vernation*.

puberulent Minutely pubescent. (vestiture)

pubescent (a) With straight, short, soft, somewhat scattered, slender trichomes. (b) A general term, meaning having trichomes. (vestiture)

pulvinus The swollen base of a petiole or petiolule, e.g., in some Fabaceae. (leaf part)

punctate Covered with minute, pitlike depressions. (configuration)

pungent *Spinose*. (apical process)

pyrene A fleshy fruit in which each of two or more seeds is enclosed by a usually bony-textured endocarp, or the seed covered by a hard endocarp unit itself, regardless of the number. (fruit type)

pyxide/pyxis *Circumscissile capsule*. (fruit type)

qualitative characters Characters in which the states are not directly measured but are based on defined classes of attributes. Syn: *categorical characters*.

quantitative characters Characters in which the states are measured and based on numbers. Cf: *continuous quantitative characters, discrete quantitative characters*.

quincuncial Perianth parts of a single pentamerous whorl having two members overlapping, two being overlapped, and one overlapping only at one margin. (perianth aestivation)

raceme An indeterminate inflorescence consisting of a single axis bearing pedicellate flowers. (inflorescence type)

rachilla A lateral or secondary axis of a bipinnate leaf. (leaf part) (b) The central axis of a grass or sedge spikelet. (inflorescence part) Pl: *rachillae*.

rachis (a) The main axis of a pinnately compound leaf. (leaf part) (b) A major, central axis within an inflorescence. (inflorescence part)

radial (a) *Actinomorphic, polysymmetric, regular*. (symmetry) (b) Referring to a longitudinal section of wood, parallel to a stem radius.

radical *Basal*. (position)

radicle The first root of a seed embryo. (root type, seed part)

RADSeq *Restriction-site associated DNA sequencing*.

ramet A clonal unit of a genet that is, at least potentially, independent from other ramets. Cf: *genet*.

random amplified polymorphic DNA The use of using randomly generated primers for the amplification of DNA to identify polymorphic DNA regions of different individuals or taxa. Abbr: *RAPD*.

range The collective geographic distribution of a taxon; the extent of measurements, from the minimum to the maximum values.

rank One of the hierarchical taxonomic categories, in which a higher rank is inclusive of all lower ranks.

RAPD *Random amplified polymorphic DNA*.

raphe Ridge on seed coat formed from adnate funiculus. (seed part)

raphide A needlelike crystal of calcium oxalate, typically occurring in bundles; a type of ergastic substance.

ray (a) A secondary axis in a compound umbel. (inflorescence part) (b) A corolla with a short tube and a single, elongated, straplike apical extension, e.g., some Asteraceae. Syn: *ligulate*. (perianth type) (c) Radially oriented cells that occur in bandlike strands in the secondary xylem (wood), functioning in lateral translocation of water, minerals, and sugars.

read One of the thousands of individual sequences obtained with high-throughput sequencing, these used in assembling a final DNA sequence region.

recency of common ancestry A measure of phylogenetic relationship, stating that two taxa are more closely related to one another if they share a common ancestor that is more recent in time than the common ancestor they share with any other taxon.

receptacle The tissue or region of a flower to which the other floral parts are attached. Syn: *torus*. (flower part)

reclinate *Reclined*. (orientation)

reclined Directed downward, with divergence angle of 15°–45° from horizontal axis. Syn: *reclinate*. (orientation)

reclining *Prostrate*. (stem habit)

recognition species Sexually reproducing systems that are maintained by genetically based features that promote reproduction.

recurved With tip gradually curved outward or downward, abaxially. (transverse posture)

reduced A type of anther wall development in which the secondary parietal cells do not divide further and develop directly into the endothecium and tapetum, respectively.

reduplicate Plicate, with adjacent abaxial sides facing one another, Λ-shaped in cross-section. (longitudinal posture)

reference guided assembly Assembly of DNA sequences using an already acquired reference DNA sequence for alignment.

reflexed Bent or turned downward. (orientation)

regular *Actinomorphic, polysymmetric, radial*.

reiteration The growth of shoots not conforming to the parameters of the model, e.g., due to environmental stress, such as mechanical or animal damage, obscuring its normal expression. (branching model)

relative cover A measure of the degree to which each species of a community layer contributes to the total cover of that layer alone.

remodel/remodeling To change the diagnostic characteristics of a taxon.

reniform (a) Kidney-shaped; wider than long with a rounded apex and reniform base. (shape) (b) With two rounded, basal lobes, smoothly concave at intersection of lobes. (base)

repand Margins wavy in a vertical plane. (longitudinal posture)

repent Creeping or lying flat and rooting at the nodes. (stem habit)

replum A peripheral rim of a cross-wall that persists after fruit dehiscence in silicles and siliques of Brassicaceae. (fruit part)

rescaled consistency index (RC) A measure of the relative amount of homoplasy in a cladogram, equal to the product of the consistency index and retention index.

resin ducts/canals Specialized cells that secrete resin.

restriction enzyme An enzyme, typically derived from bacteria, that are used to cleave DNA samples at particular locations at the restriction site.

restriction fragment length polymorphism Differences between taxa in restriction sites, and therefore the lengths of fragments of DNA after cleavage with restriction enzymes. Abbr: *RFLP*.

restriction site A sequence of approximately six to eight base pairs of DNA that binds to a given restriction enzyme.

Restriction-site associated DNA sequencing A technique in which restriction enzymes are used to break up the DNA into fragments, the fragments then sequenced in high throughput sequencing. Abbr: *RADSeq*.

resupinate Inverted or twisted 180°, as leaves of Alstroemeriaceae or ovaries of most Orchidaceae flowers. (twisting/bending posture)

retention index (RI) A measure of the relative amount of homoplasy in a cladogram, equal to the ratio $(g-s)/(g-m)$, where g is the maximum possible number of state changes that could occur on any conceivable tree, m is the minimum possible number of character state changes, and s is the actual number of state changes that occur.

reticulate (a) Venation in which the veins appear to join back together, forming a net-like "reticulum." (leaf venation) Syn: *anastomosing, netted*. (b) A pollen sculpturing with a netlike sculpturing., each element termed a murus and the space between termed a lumen.

reticulation Hybridization of two previously divergent lineages, forming a new lineage.

reticulodromous (a) Pinnate venation in which secondary veins do not terminate at the margin and branch repeatedly, forming a very dense, netlike structure. (leaf venation) (b) With three or more primary veins diverging from one point but not reaching the blade margin, a subcategory of actinodromous. (leaf venation)

retinaculum *Jaculator*. Pl: *retinacula*.

retrorse Bent or directed downward, usually referring to small appendages. (orientation)

retuse Having an apical incision, cut up to 1/16 the distance to the midrib, midvein, or junction of primary veins. (apex)

reversal Homoplasy by the loss of a derived feature and the reestablishment of an ancestral feature.

revolute Margins or outer portion of sides rolled outward or downward over the abaxial surface. (longitudinal posture, margin)

RFLP *Restriction fragment length polymorphism*.

reward A floral structure, such as pollen, or exudate, such as nectar, that functions to ensure that an animal pollinator will consistently return to transport pollen.

rheophyte A plant found along (often swiftly flowing) streams and river banks. (plant habitat)

rhipidium A monochasium in which the branches develop on alternating sides of each sequential axis, typically having a geniculate (zig-zag) appearance; sometimes equated with scorpioid cymes. (inflorescence type)

rhizoid One of several uniseriate (one-cell-thick), filamentous processes that function in anchorage and water/mineral absorption, arising from the gametophytes of free living gametophytes and from the underground, sporophytic stems of some ferns (e.g., the psilotophytes).

rhizome A horizontal, underground stem, generally with short internodes and scalelike leaves, e.g., *Zingiber officinale*. (b) A horizontal stem of a fern that grows at ground level. (stem/shoot type)

rhombic Parallelogram-shaped, widest near the middle, the length:width ratio 2:1 to 3:2. (shape)

rhomboidal tetrad A tetrad in which the four grains are in one plane, with two of the grains separated from one another by the close contact of the other two.

ribosome A cellular organelle that functions in protein synthesis.

ribulose-bisphosphate carboxylase The enzyme that catalyzes the initial binding (fixation) of carbon dioxided in the dark reactions of photosynthesis. Abbr: *RuBP-carboxylase*.

rind *Pericarp*.

ring-porous Wood in which the vessels form only in the spring wood, with summer wood either lacking or having relatively small vessels and usually containing mostly fibers.

robustness The confidence for which a tree or particular clade actually denotes true phylogenetic relationships.

root A cylindrical organ of virtually all vascular plants, consisting of an apical meristem that gives rise to a protective root cap, a central endodermis-bounded vascular system, absorptive epidermal root hairs, and endogenously developed lateral roots; usually functioning in anchorage and absorption of water and minerals; initially derived from the radicle of the embryo and typically growing downward. (plant part)

root apical meristem Region of continuously dividing cells from which all cells of the root are derived. (root part)

root cap The outer cell layer at the root tip, functioning in protection and lubrication. (root part)

root hair One of several hairlike extensions from an epidermal cell of a root, functioning to greatly increase the surface area available for water and mineral absorption. (root part)

root tuber A swollen taproot containing concentrations of high-energy compounds such as starch. (root type)

root cap An outer layer of cells at the tip of a root, functioning to protect the root apical meristem and to provide lubrication for the growing root.

rootstock A general term for an underground stem or shoot, these generally giving rise to aerial shoots either by direct conversion of the terminal apical meristem or via lateral buds. (stem/shoot type)

rosette An arrangement in which parts, usually leaves, radiate from a central point at ground level, e.g., the leaves of *Taraxacum officinale*, dandelion. Adj: *rosulate*. (arrangement)

rotate With a short tube and wide limbs oriented at right angles to the tube. (perianth type)

rounded With apical or basal margins convex, forming a single, smooth arc. (apex, base)

rugose Covered with coarse reticulate lines, usually with raised, blisterlike areas between. Syn: *bullate*. (configuration)

rugulate A pollen sculpturing having irregular to sinuous, tangentially oriented elements, often appearing brainlike.

ruminate Unevenly textured, coarsely wrinkled, looking as if chewed, e.g., the endosperm of the Annonaceae. (surface, texture)

runner *Stolon*. (stem/shoot type)

saccate Having a pouchlike evagination. (perianth type)

sagittate With two basal lobes, more or less pointed and oriented downward, away from apex. (base)

salverform Trumpet-shaped; with a long, slender tube and flaring limbs at right angles to tube. (perianth type)

samara A winged, dry fruit, e.g., *Acer*, maple; *Ulmus*, elm. (fruit type)

Sanger sequencing A traditional DNA sequencing method, using DNA polymerase, primers, free nucleotides, and typically fluorescent-labeled dideoxynucleotides.

sapromyiophily Pollination by flies. Adj: *sapromyiophilous*.

saprophyte A plant living off dead organic matter. (plant life form)

sarcotesta A seed coat that is fleshy at maturity. (seed part)

sarmentose *Stoloniferous*. (stem habit)

saxicolous Occurring on rock. Syn. *epipetric*. (plant habitat)

scaberulous Minutely scabrous. (epidermal excrescence, vestiture)

scabrate *Spinulose*.

scabrous Having a rough surface or trichomes, resembling sandpaper. (epidermal excrescence, vestiture)

scale (a) A small, nongreen leaf of a bud or underground rootstock. (leaf structural type) (b) A bract of a sedge spikelet. (inflorescence part)

scaling Assigning a weight in a phylogenetic analysis to correlated characters that is the inverse of the total number of the correlated characters. Adj: *scaled*.

scandent *Clambering*.

scape A "naked" (leafless) peduncle, generally arising from a basal rosette of vegetative leaves. Adj: *scapose*. (stem/shoot type, inflorescence part)

scarious Thin and appearing dry, usually whitish or brownish. (texture)

schizocarp Derived from a two or more loculed compound ovary in which the locules separate at inception or at maturity. (fruit type)

schizocarp of follicles A schizocarp in which the carpels of a pistil split at fruit maturity, each carpel developing into a unit follicle, e.g. *Asclepias*. (fruit type)

schizocarp of mericarps A schizocarp in which (generally two) carpels of a single, unlobed ovary split during fruit maturation, the carpels developing into unit mericarps and attached to one another via a stalklike carpophore, e.g., Apiaceae. (fruit type)

schizocarp of nutlets A schizocarp in which a single ovary becomes lobed during development, each lobe developing at maturity into a nutlet, e.g., Boraginaceae, Lamiaceae. Cf: *eremocarp*. (fruit type)

scientific name A formal, universally accepted name, the rules and regulations of which (for plants, "algae," fungi, and organisms traditionally treated as such) are provided by the International Code of Nomenclature for algae, fungi, and plants.

sclereid A sclerenchyma cell that is isodiametric to irregular in shape and often branched, functioning in structural support or possibly aiding in providing protection from herbivory.

sclerenchyma A tissue composed of nonconductive cells that are dead at maturity, that have a thick, lignified, generally pitted secondary cell wall, and that function in structural support and/or to deter herbivory; composed of fibers and sclereids.

scorpioid cyme A monochasium in which the branches develop on alternating sides of each sequential axis, typically appearing geniculate (zig-zag); may intergrade with helicoid cyme. Syn: *cincinnus*. (inflorescence type)

scurfy *Farinaceous*. (epidermal excrescence)

secondary cell wall An additional wall layer, composed of cellulose and lignin, secreted between the plasma membrane and the primary cell wall following cell elongation, found in some cell types (including tracheary elements and fibers).

secondary contact Whereby two (or more) lineages that had diverged and were likely geographically separated (allopatric) in the past, have come to occupy the same or overlapping ranges.

secondary endosymbiosis Endosymbiosis involving the engulfment of an ancestral eukaryotic cell by another eukaryotic cell, a possible mechanism of chloroplast exchange between eukaryotes.

secondary growth Growth in girth or width by means of cells produced from lateral meristems.

secondary parietal cells Two layers of cells arising by tangential divisions of the primary parietal cells of the anther wall.

secondary phloem Sugar-conducting tissue produced by the vascular cambium to the outside of a woody stem or root. Syn: *inner bark*.

secondary root A root derived endogenously from the pericycle of an existing root. (root type)

secondary tissue A tissue formed by secondary growth, via lateral meristems.

secondary vein A lateral vein that branches from and is smaller than a primary vein. (leaf venation)

secondary xylem Water- and mineral-conducting tissue produced by the vascular cambium to the inside of a woody stem or root. Syn: *wood*.

secretory An anther tapetum type in which the tapetum remains intact with no breakdown of cell walls. Syn: *glandular*.

secretory structure A collection of cells that secrete compounds, either internally or externally.

secund Flowers, inflorescences, or other structures on one side of axis, often due to twisting of stalks. Syn: *unilateral*. (arrangement)

sedge spikelet The inflorescence unit of the Cyperaceae, sedge family, consisting of a central axis (the rachilla), bearing spiral or distichous bracts (also called scales or glumes), each subtending a single flower.

seed An embryo surrounded by nutritive tissue and enveloped by a seed coat; the propagule of the seed plants.

seed coat The outer protective covering of seed, developing from one or two integuments. Syn: *testa*. (seed part)

seismonasty *Thigmonasty*.

self-dispersal Process by which a flowering plant buries its own fruits, e.g., *Arachis hypogaea*, peanut. Syn: *autochory*.

self-incompatibility Outcrossing occurring by the genetic inability for fertilization to occur between gametes derived from a single genotype.

selfing *Inbreeding*.

semelpary Referring to plants that have one episode of reproduction, followed by degeneration and death of the entire plant. Adj: *semelparous*.

semicraspedodromous Pinnate venation in which the secondary veins branch near the margin, one terminating at the margin, the other looping upward to join the next secondary vein. (leaf venation)

sensu lato Meaning in the broad sense; designated for a taxon name that is used inclusively, for a broad, inclusive taxon circumscription to include other, previously recognized taxa. Abbr: *s.l.*

sensu stricto Meaning in the strict sense; designated for a taxon name used exclusively, to exclude other, previously recognized taxa. Abbr: *s.s.*; *s.str.*

sepal An individual member or segment of the calyx, typically green, leaflike, and functioning to protect the young flower. (flower part, perianth part)

septal nectary A nectary embedded within the ovary septae, secreting nectar via a pore at the ovary base or top. (nectary type)

septicidal capsule A capsule in which longitudinal lines of dehiscence are radially aligned with the septa. (fruit type)

septifragal capsule A capsule in which the valves break off from the septa, as in *Ipomoea*, morning glory. Syn: *valvular capsule*. (fruit type)

septum A partition or cross-wall of the ovary. (gynoecium part, fruit part)

sequencing depth The number of overlapping assemblies of DNA fragments (reads) in high-throughput sequencing. Syn: *depth of coverage*.

sequencing by synthesis A method of high-throughput sequencing in which DNA fragments are ligated with adapters, primers, and (possibly) barcodes, amplified into clusters, and sequenced in cycles with fluorescent-labeled nucleotides, the light released used to determine the sequence.

sequential *Gradate*.

sericeous With long, appressed, silky trichomes. (vestiture)

series *Whorl*. (flower part)

serotinous Opening not at maturity, but generally with the stimulus of fire, e.g., cones of some conifers.

serrate Saw-toothed; teeth sharp and ascending, the lower side longer, cut $^1/_{16}$ to $^1/_8$ the distance to midrib, midvein, or junction of primary veins. (margin)

serrulate Diminutive of serrate, teeth cut to $^1/_{16}$ the distance to midrib, midvein, or junction of primary veins. (margin)

sessile (a) Without a petiole or, for leaflets, without a petiolule. (leaf attachment) (b) Lacking a pedicel. (flower attachment) (c) With filament absent, the anther attached directly. (stamen attachment) (d) Ovary lacking a basal stalk. (ovary attachment)

sessile-glandular Glandular cell with a very short or no basal stalk. (trichome type)

sexine The outer, protruding layers of the exine, including columellae, tectum, and supratectal sculpturing elements.

sexual reproduction Reproduction involving meiosis with recombination and fusion of egg and sperm, in land plants entailing two "phases" of a haplodiplontic life cycle.

sheath A flattened leaf base or petiole that partially or fully clasps the stem, e.g., Poaceae and many Apiaceae. (leaf part)

sheathing Attached by a curved or tubular structure that partially or totally encloses the stem. (leaf attachment, base)

shining *Nitid*. (epidermal excrescence)

shoot A stem plus associated, derivative leaves, initially formed by an apical meristem that gives rise to the stem and external (exogenous) leaf primordia; may be gametophytic or sporophytic. (plant part)

short shoot *Fascicle*; *spur*; *spur shoot*. (stem/shoot type)

shreddy Bark coarsely fibrous, often fissured. (bark type)

shrub A perennial, woody plant having several main stems arising at ground level. (plant habit)

sieve area An aggregation of callose-lined pores in *sieve elements*.

sieve cell A type of sieve element that has only sieve areas on both end and side walls, found in nonangiospermous vascular plants.

sieve element A generally elongated cell that is semialive at maturity, has a nonlignified, primary cell wall with specialized, callose-lined pores aggregated into sieve areas and/or sieve plates, and that functions in conduction of sugars.

sieve plate One or more sieve areas at the end wall junction of two sieve tube members, the pores of which are significantly larger than are those of sieve areas located on the side wall; characteristic of angiosperms.

sieve tube member A type of sieve element that has both sieve areas and sieve plates, the latter at the end wall junction of two sieve tube members and having larger pores; an apomorphy of angiosperms.

sieve tube plastid A membrane-bound plastid found in sieve tube members, the contents of which (whether starch or protein) can vary between taxa and be systematically informative.

significant Condition in which the differences between two or more groups are not due to chance alone; i.e., the probability of the differences being due to chance is calculated as less than some standard value (generally 5%, or $p<0.05$).

silicle A dry, dehiscent, 2-carpeled fruit that dehisces along two sutures, has a persistent partition (replum), and is as broad or broader than it is long, e.g., Brassicaceae. Cf: *silique*. (fruit type)

silique A dry, dehiscent, 2-carpeled fruit that dehisces along two sutures, has a persistent partition (replum), and is longer than it is broad, e.g., Brassicaceae. Cf: *silicle*. (fruit type)

simple cone A cone consisting of an axis bearing sporophylls.

simple corymb An unbranched corymb, consisting of a central axis bearing pedicellate flowers, the collection of flowers being flat-topped or convex. (inflorescence type)

simple craspedodromous Pinnate venation in which all secondary veins terminate at the margin. (leaf venation)

simple dichasium A three-flowered determinate inflorescence with a single terminal flower and two, opposite lateral flowers, the pedicels of all of equal length. (inflorescence type)

simple fruit A fruit derived from one pistil (ovary) of one flower. (fruit type)

simple leaf A leaf not divided into leaflets, bearing a single blade. (leaf type)

simple ovary/pistil An ovary/pistil composed of a single carpel, the gynoecium apocarpous. (ovary/pistil type)

simple perforation plate A perforation plate composed of a single opening.

simple sieve plate A sieve plate that is composed of one pore region.

simple tissue A tissue that consists of only one type of cell.

simple umbel An umbel with the peduncle bearing pedicellate flowers attached at one point, e.g., Alliaceae. (inflorescence type)

simultaneous (a) A type of microsporogenesis in which cytokinesis does not occur until after meiosis II. (b) Development of a fern sorus in which sprorangia mature at the same time. Syn: *simple*.

single nucleotide polymorphism Variation among samples/taxa in a nucleotide at a given position of a gene.

single sequence repeats *Microsatellites*.

sinistrorse Twining helically like a left-handed screw, e.g. some Caprifoliaceae. (twisting/bending posture)

sinuate Sinuses shallow and smooth, wavy in a horizontal plane. (division)

sinus The indentation of a divided planar structure. (division)

siphonogamy Gymnospermous pollen tubes that are haustorial, but deliver non-motile sperm cells to the archegonium or egg. Cf: *zooidogamy*.

siphonostele A type of stem vasculature in which a ring of xylem is surrounded by an outer layer of phloem (ectophloic siphonostele) or by an outer and inner layer of phloem (amphiphloic siphonostele; if dissected, called a dictyostele).

sister group/sister taxa Two taxa that are each other's closest relative, representing two descendant lineages from a common ancestor.

small single copy region A relatively short stretch of DNA occuring as one copy in the chloroplast of plant cells.

smooth (a) *Plane*. (configuration) (b) A nonfibrous bark without fissures, plates, or exfoliating sheets. (bark type)

snap-trap leaf A leaf that mechanically moves after being triggered, capturing and digesting small animals, e.g., *Aldrovanda*; *Dionaea*, venus fly trap. (leaf structural type)

SNP *Single nucleotide polymorphism*.

softwood Wood derived from a conifer, generally (but not always) softer in texture than a hardwood.

solanad type A type of embryo development in which the terminal cell divides transversely, the basal-cell derivatives forming a suspensor but otherwise not contributing to mature embryo development.

solitary A one-flowered inflorescence. (inflorescence type)

somatic Vegetative; referring to a cell not directly involved in sexual reproduction.

sorosis A multiple fruit in which the unit fruits are fleshy berries and are laterally fused along a central axis, e.g., *Ananas*, pineapple. (fruit type)

sorus A discrete cluster or aggregation of leptosporangia. Pl: *sori*.

spadix A spike with a thickened or fleshy central axis, typically with congested flowers, e.g., Araceae. (inflorescence type)

spathaceous Having a spathe. (leaf structural type, inflorescence part)

spathe An enlarged, sometimes colored bract subtending and usually enclosing an inflorescence, e.g., in the Araceae. (leaf structural type, inflorescence part)

spatulate A shape that is oblong, obovate, or oblanceolate with a long attenuate base. (shape)

speciation The formation of new species from preexisting species.

species A group of individuals that are related to one another by certain criteria and distinct from other such groups of individuals. Cf: *biological (isolation) species, cohesion species, evolutionary species, genealogical species, metaspecies, paraphyletic species, phylogenetic species, recognition species, taxonomic (morphologic) species*.

species complex A group of very similar and presumably closely related species and/or infraspecies.

species diversification rate A measure of how rapidly new species form in a given clade.

species evenness A measure of the number and evenness of distribution of species occurring within a given geographical region.

species nova Meaning that a taxon name, at the rank of species, is new to science. Abbr: *sp. nov.*

species richness A count of the number of species occurring within a given geographical region.

specific epithet The second name of a binomial; may be capitalized or not.

sperm A haploid, motile or nonmotile, male gamete that functions to fuse with an egg cell, producing a diploid zygote.

spheroidal Approximately spherical in shape, e.g., referring to pollen grains.

spike An indeterminate inflorescence consisting of a single axis bearing sessile flowers. (inflorescence type)

spikelet A small spike; the basic inflorescence unit in the Cyperaceae (sedges) and Poaceae (grasses). (inflorescence type)

spine A sharp-pointed leaf (e.g., cactus spines or glochidia) or leaf part, including petiole (e.g., *Fouquieria* spp.), midrib, secondary vein, leaflet (e.g., *Phoenix* spp.), or stipule (e.g., *Euphorbia* spp.). Cf: *thorn*; *prickle*. (leaf structural type)

spinose (a) Margins with teeth bearing sharp, stiff, spinelike processes. (margin) (b) With a sharp, stiff, spinelike apical process. Syn: *pungent*. (apical process)

spinulose A pollen sculpturing having spinelike elements <1 μm long, each element termed a spinulum. Syn: *scabrate*.

spinulum One of the spinelike elements of a spinulose pollen grain. Pl: *spinuli*.

spiral (a) Alternate, with points of attachment in more than three rows/ranks. Syn: *polystichous*. (arrangement) (b) Perianth parts arranged in a spiral, one per node, not in distinct whorls. (perianth arrangement) (c) Stamens arranged in a spiral, one per node; not whorled. (stamen arrangement) (c) A type of endothecium in which the secondary wall thickenings are spiral or helical in shape.

spiraperturate A pollen grain having one or more apertures that are spirally shaped.

spongy mesophyll The usually lower (abaxial), irregular cells of the mesophyll that generally have large intercellular spaces, found in some leaves.

sporangial wall One or more outer layers of sterile, non-spore-producing cells of the sporangium.

sporangiophore A unit of the strobilus of the Equisetales, consisting of a peltate axis bearing pendant (ancestrally recurved), longitudinally dehiscent sporangia.

sporangium The spore-producing organ of the sporophyte. Pl: *sporangia*. (plant part)

spore A haploid cell that, in the land plants, originates from meiotic divisions of sporocytes within a sporangium, ultimately growing into a gametophyte.

sporocarp The generally spherical reproductive structure of aquatic ferns, functioning in allowing the sporangia inside to remain dormant and resist desiccation for a long time.

sporocyte A cell of the sporangium that undergoes meiosis, producing (generally four) haploid spores.

sporogenous tissue Internal cells of the sporangium that mature into sporocytes, the latter forming spores by meiotic divisions.

sporophyll A specialized leaf that bears one or more sporangia.

sporophyte The diploid (2n) phase in the haplodiplontic life cycle of all land plants.

sporophytic leaf The leaf of a vascular plant sporophyte.

sporopollenin A polymer of oxidative carotenoids or carotenoid esters that impregnates the exine of pollen grain walls.

spring wood The secondary xylem from the first part of the growing season, the cells of which tend to be larger in diameter with thinner walls.

spur A tubular, rounded or pointed projection from the calyx or corolla, functioning to contain nectar. (perianth part)

spur/spur shoot *Fascicle*. (stem/shoot type)

squamate Having or producing scales. (leaf structure type)

squarrose Sharply curved downward or outward (abaxially) near the apex, as the phyllaries of some Asteraceae. (transverse posture)

stamen The unit of the androecium; a microsporophyll (typically modified as a filament) bearing generally two pollen-bearing thecae, fused into an anther. (flower part)

stamen insertion (a) Referring to whether or not stamens protrude beyond the perianth. (b) Denoting the "point of insertion," the point of adnation of an epipetalous stamen to the corolla. (androecium)

staminal disk A discoid or doughnut-shaped structure arising from the receptacle at the base of the stamens; may be nectar-bearing. Syn: *nectariferous disk*. (flower part)

staminate Flowers unisexual, with stamen(s) only, lacking fertile carpel(s). Syn: *male*. (flower sex)

staminode/staminodium A sterile stamen, producing no functional pollen; may be modified as a nectary or petaloid structure; may or may not contain an antherode, a sterile anther. (stamen type)

standard deviation A measure of the differences of individual values of a sample from the mean of all values, calculated as the square root of the "sum of squares" divided by the sample size minus one.

standard error A measure of how close the sample mean is to the mean of the entire population, calculated as the standard deviation divided by the square root of the sample size.

starch A polysaccharide of glucose units (alpha-1,4-glucopyranoside) in green plants, functioning as the high-energy storage compound.

starch grain Lamellate deposits of starch in green plant cells, functioning as the high-energy storage compound; a type of ergastic substance. Syn: *amyloplast*.

statistics A branch of mathematics dealing with collecting, organizing, analyzing, and interpreting numerical data.

status novus Indication of a change in rank of a scientific name. Abbr: *stat. nov.*

stele The spatial distribution of the primary vasculature of a stem, organized into arrangements of xylem and phloem.

stellate Star-shaped trichomes having several arms arising from a common base (stalked or sessile), e.g., many Malvaceae. (trichome type)

stem A generally cylindrical organ that bears leaves, typically functioning in support and elevation of leaves and reproductive structures and in conduction of water, minerals, and sugars; in vascular plants initially derived from the epicotyl of the embryo and generally growing upward. (plant part)

stem (cladogram) *Internode*.

stem-based clade A clade that includes the "stem" (internode) region just above a common ancestor plus all descendants of that stem.

stem habit Stem feature based on structure, position, growth, and orientation of above-ground stems. (stems and shoots)

stem-based A type of phylogenetic classification in which the stem (internode) region just above a common ancestor plus all descendants of that stem are the basis for grouping.

sterile jacket layer *Antheridial wall; archegonial wall*.

sterile segment The photosynthetic blade or lamina component of the leaf of an ophioglossoid fern.

stigma The pollen-receptive portion of a pistil; may be a discrete structure or a region of a style or style branch. (gynoecium part)

stinging trichome A sharp-pointed trichome that secretes an irritating fluid upon penetrating the tissue of an animal.

stipe (a) A leaf stalk, often used for ferns. Syn: *petiole*. (leaf part) (b) *Gynophore*. Adj: *stipitate*. (gynoecium part) (c) The stalk of the sporophyte of mosses.

stipel One of a pair of leaflike structures, which may be modified as spines or glands, at either side of the base of a petiolule, e.g., some Fabaceae. Adj: *stipellate*. (leaf part)

stipular spine A sharp-pointed stipule, e.g., *Euphorbia* spp. Cf: *thorn*; *prickle*. (leaf structural type)

stipule One of a pair of leaflike structures, which may be modified as spines or glands, at either side of the base of a petiole. Adj: *stipulate*. (leaf part)

stipule scar A mark indicating the former place of attachment of a stipule. (twig part)

stolon An indeterminate, elongated, slightly underground or above-ground propagative stem, with long internodes, rooting at the tip forming new plants. Syn: *runner*. (stem/shoot type)

stoloniferous With elongated propagative shoots (stolons) rooting at the tip producing new plants. Syn: *sarmentose*. (stem habit)

stoma The opening between two guard cells of a stomate.

stomate Specialized epidermal cells, consisting of two guard cells, which, by changes in turgor pressure, can increase or decrease the size of the opening (stoma) between them. Pl: *stomata*; *stomates*.

storage root A swollen taproot containing concentrations of high-energy compounds such as starch. (root type)

straight Flat, without vertical curves or bends. (transverse posture)

strap-shaped Flat, not needlelike but with length:width ratio greater than 12:1. (shape)

striate (a) With fine longitudinal lines. (configuration) (b) A pollen sculpturing having thin, cylindrical, tangentially oriented elements.

strict consensus tree A consensus tree that collapses any differences in branching pattern between two or more cladograms to a polytomy.

strigose Covered with dense, coarse, bent, and mostly flat trichomes, often with a bulbous base. (vestiture)

strigulose Minutely strigose. (vestiture)

strobilus *Cone*. (plant part)

strophiole *Caruncle, elaiosome*. (seed part)

student's t-test *T-test*.

style A stalklike portion of the pistil between the stigma and ovary; may be absent. (gynoecium part)

styloid A single, elongated, angular crystal of calcium oxalate; a type of ergastic substance.

stylopodium A swollen region at apex of the ovary, below the styles/stigmas, e.g., Apiaceae.

subapical With style arising at one side near the apex of the ovary. (style position)

subbasifixed Anther attached just above its base to the filament. (anther attachment)

subcosmopolitan Having a broad, but not worldwide, distribution.

suberin A waxy, water-resistant substance, found in stomata, Casparian strips of the endodermis, and the cell walls of cork cells.

subglabrous Nearly glabrous, with just a few, scattered trichomes. (vestiture)

subopposite With two leaves or other structures on opposite sides of stem or central axis but at different nodes, slightly displaced relative to one another. (arrangement)

subsessile Having a very short, rudimentary petiole (leaf attachment), petiolule (leaflet attachment), pedicel (flower attachment), or filament (stamen attachment).

subshrub A short shrub that is woody only at the base and that seasonally bears new, nonwoody, annual shoots above. (plant habit)

subsidiary cell An epidermal cell adjacent to a stomate and somewhat different from other epidermal cells in shape and size, functioning in regulating stomatal opening.

subspecific epithet The third component of a ternary name at the rank of subspecies.

subulate Awl-shaped; approximately narrowly oblong to narrowly triangular. (shape)

successive A type of microsporogenesis in which cytokinesis occurs after meiosis I.

succulent (a) Fleshy or juicy. Syn: *baccate*, *carnose*. (texture) (b) A plant with fleshy stems or leaves. (plant life form)

suffrutescent Woody basally, herbaceous apically; the habit of a subshrub. (stem habit)

sulcus An elongated aperture similar in shape to a colpus (length :width ratio >2:1) occurring at the (usually distal) pole. Adj: *sulcate*.

summer wood The secondary xylem from the latter part of the growing season, the cells of which tend to be smaller in diameter with thicker walls.

superior With sepals, petals, stamens, and/or hypanthium attached at the base of the ovary. (ovary position)

superposed bud Bud(s) occurring above the axillary bud. (bud type)

supervolute With one half of a simple leaf coiled tightly around the midrib, the other half coiled (in the opposite direction) around the first half, as in members of the Zingiberales. (posture: ptyxis/ aestivation)

suprabasal With three or more primary veins diverging from one point above the base of the blade, a subcategory of actinodromous and palinactinodromous. (leaf venation)

suprafoliar Descriptive of a palm inflorescence that is positioned above the leaves of the crownshaft.

surficial Growing upon the ground.

suspensor A nonpersistent column of cells of the embryo, functioning in transport of nutrients to the mature, persistent embryo during its development.

syconium A multiple fruit in which the unit fruits are small achenes covering the surface of a fleshy, inverted compound receptacle, e.g., *Ficus*, fig. (fruit type)

syllepsis Growth of an axillary bud into a shoot without a period of rest. Adj: *Sylleptic*. (tree branching model)

symmetry Referring to the presence and number of mirror-image planes of symmetry.

sympatric Occurring in the same geographic range.

sympetalous With connate petals. Syn: *gamopetalous*. (perianth fusion)

symplesiomorphy A shared, ancestral features among taxa, which may be used as a basis for grouping in a phenetic classification.

sympodial A branching pattern in which a given axis is made up of several units, each of which is derived from a separate apical meristem, the units themselves determinate or indeterminate. (stem branching pattern)

synangium A fusion product of two or more sporangia, e.g., in the psilotophytes and some marattioid ferns.

synonomous mutation A mutation of a coding gene that results in no change in the amino acid of a synthesized protein.

synanthy Timing in which leaves and flowers develop at the same time. Adj: *synanthous*.

synapomorphy An apomorphy that unites two or more taxa or lineages.

syncarp An aggregate fruit, typically of berries, in which the unit fruits fuse together, e.g., *Annona*. (fruit type)

syncarpous With carpels (at least at the base) connate, the pistil or ovary being compound. (gynoecial fusion)

synchronic species A species corresponding to a group of contemporaneous organisms, living at the same general point in time.

syncolpate A pollen grain in which the colpi are joined, e.g., at the poles.

synergid cells In a typical angiosperm female gametophyte, the two haploid cells that flank the egg and that may function in pollen tube entry.

syngenesious With anthers connate and filaments distinct, e.g., Asteraceae. (stamen fusion)

synonym An unaccepted name, by a particular author or authors, applying to the same taxon as the accepted name.

synoptic collection A collection of plant specimens that contain one specimen (of all available specimens) of each taxon for a given region.

synsepalous With connate sepals. Syn: *gamosepalous*. (perianth fusion)

syntepalous With connate tepals. Syn: *gamotepalous*. (perianth fusion)

syntype A specimen that was cited in the original work when a holotype was not designated, or one of two or more specimens that were all designated as types.

systematics A science that includes and encompasses traditional taxonomy and that has, as its primary goal, the reconstruction of phylogeny. Cf: *taxonomy*.

***t*-test** A statistical test of two samples used to evaluate the probability of their being the same. Syn: *Student's t-test*.

tandem repeat A sequence of DNA that repeats multiple times within a given region.

tangential Referring to a longitudinal section of wood, perpendicular to a stem radius.

tannin Phenol derivatives in plant cells that may function to deter herbivory and parasite growth; a type of ergastic substance.

tapering Trichomes ending in a sharp apex. (trichome type)

tapetum The innermost cell layer of the anther wall, consisting of metabolically active cells that function in the development of pollen grains.

taproot A persistent, well-developed primary root. (root type)

targeted sequencing A class of techniques that obtain sequence data for specific low copy nuclear genes.

tautonym A binomial in which the genus name and specific epithet are identical in spelling, illegitimate in the International Code of Nomenclature for algae, fungi, and plants.

taxon (*plural*, **taxa**) A group of organisms, ideally monophyletic and traditionally treated at a particular rank.

taxonomic key *Key*.

taxonomic revision A change of the definition and delimitation of a taxon.

taxonomic species The smallest group (or class) of individuals that are similar to one another in one or more features and different from other such groups. Syn: *morphologic species*.

taxonomy A field of science (and major component of systematics) that encompasses description, identification, nomenclature, and classification. Cf: *systematics*.

tectate-columellate A common exine structure that consists of an inner foot layer, a middle layer of radially elongated columellae, and an outer, rooflike layer called the tectum.

tectum The outer, rooflike layer of a tectate-columellate pollen exine wall.

telomes Ancestral branches, e.g., those surrounding the megasporangium of ancestral seeds.

temperate Distributed in regions between the tropics and Arctic/Antarctic, having a relatively moderate climate.

tendril (a) A coiled and twining leaf part, usually a modified rachis or leaflet. (leaf structural type) (b) A long, slender, coiling branch, adapted for climbing. (stem/shoot type)

tentacular leaf A leaf bearing numerous, sticky, glandular hairs or bristles that function in capturing and digesting small animals, e.g., *Drosera*, sundew. (leaf structural type)

tenuinucellate An ovule in which the nucellus is composed of a single layer of cells, with no formation of a parietal cell.

tepal A component of the perianth in which the parts intergrade or in which the perianth is undifferentiated into distinctive sepals and petals. (flower part, perianth part)

terete Cylindrical. (shape)

terminal (a) At or near the top, tip, or end of a structure. (position) (b) Entire inflorescence positioned as a terminal shoot relative to the nearest vegetative leaves. (inflorescence position) (c) Style arising at the apex of the ovary. (style position)

terminal bud Bud at the apex or end of a stem. Syn: *apical bud*. (bud type)

terminal bud scale scar Ringlike marks indicating the former place of attachment of the terminal bud scales. (twig part)

ternary name A scientific name composed of three names, such as a subspecies or variety name.

ternate/ternately compound A compound leaf with three leaflets. Syn: *trifoliolate*. (leaf type)

ternately veined With three primary veins arising from a common basal point. (leaf venation)

ternate-netted/ternately netted Ternately veined, with three primary veins arising from a common basal point, the ultimate veins forming a fine reticulum. (leaf venation)

terrestrial Growing on land. (plant habitat)

tertiary vein A lateral vein that branches from and is smaller than a secondary vein. (leaf venation)

testa *Seed coat*. (seed part)

tetrad A fusion product of four pollen grains, developing from the four products of microsporogenesis.

tetradynamous With stamens in two groups of four long and two short, e.g., Brassicaceae. (stamen arrangement)

tetragonal tetrad A tetrad in which the four grains are in one plane and are equally spaced apart.

tetrahedral tetrad A tetrad in which the four grains form the points of a tetrahedron, e.g., members of the Ericaceae.

tetramerous A whorl with four members. (merosity, perianth merosity)

tetraploid A polyploid having four sets of chromosomes.

tetrasporangiate Anther with four microsporangia and typically two thecae. Cf: *dithecal*. (anther type)

tetrasporic Megasporogenesis in which cytokinesis does not occur after meiosis, resulting in a single cell with four haploid nuclei, all of which contribute to the female gametophyte.

texture Internal structural consistency, sometimes incorporating color.

thalloid Referring to or having a thallus.

thallus The flattened (dorsi-ventral) gametophyte of some liverworts and all hornworts. (plant part)

theca One half of typical anther containing two microsporangia. (anther part)

therophyte An annual plant. (plant life form)

thigmonasty Movement (closing) of the leaflets of a compound leaf as a response to touch, vibration, or heat (e.g., as in *Mimosa pudica*, a sensitive plant). Syn: *seismonasty*. (leaf behavior)

thorn A sharp-pointed stem or shoot (cf: prickle, spine). (stem/shoot type)

throat An open, expanded region of a perianth, usually of a sympetalous corolla. (perianth part)

thrum Flowers with a short style and long stamens, found in distylous flowers.

thylakoid A membrane of chloroplasts and photosynthetic bacteria that contains compounds involved in the light reactions of photosynthesis.

thyrse A secondary inflorescence with an indeterminate central axis bearing opposite, lateral, pedicellate cymes, e.g., *Echium* (Boraginaceae), *Penstemon* (Plantaginaceae). (inflorescence type)

tiller A grass shoot produced from the base of the stem. (stem/shoot type)

tissue A group of cells having a common function or structure; may be simple or complex.

tokogenetic Relationships between individual organisms with regard to gene flow in ancestral-descendent lineages.

tomentose Covered with dense, interwoven trichomes. (vestiture)

tomentulose Minutely tomentose. (vestiture)

topology The structure of a branching diagram such as a cladogram, including how lineages are connected together.

torus *Receptacle* (flower part) or *compound receptacle*. (inflorescence part)

total clade A clade that includes a crown clade plus all other taxa that share a recent common ancestor with the crown clade but not with other crown clades.

total cover A measure of the degree to which the total area of the community is covered by members of a given layer.

tracheary element A generally elongated cell that is dead at maturity, has a lignified, secondary cell wall, and is positioned end-to-end with other tracheary elements, forming a continuous tube that functions in water and mineral conduction; the major component of xylem tissue.

tracheid The ancestral type of tracheary element that is imperforate, in which water and mineral nutrients flow between adjacent cells through the primary cell walls at pit-pairs.

transcriptome That part of the total genome that is transcribed from DNA to RNA.

transcriptomics The acquisition of sequence data from the RNA produced in a given tissue, usually mRNA derived from expressed genes.

transfer in position To reclassify a taxon without a change in rank; to place within a different taxon of the same rank.

transformation series The hypothesized sequence of evolutionary change, from one character state to another, in terms of direction and probability. Syn: *morphocline*.

transverse (a) Dehiscing at right angles to the long axis of the theca. (anther dehiscence type) (b) Referring to a cross-section of wood.

transverse posture Placement of tip with respect to the horizontal plane of object. (disposition)

tree A generally tall, perennial, woody plant having one main stem (the trunk) arising at ground level. (plant habit)

triangular Three-sided, length:width ratio 2:1 to 3:2. (shape)

trichome An external, hairlike plant structure.

trichotomosulcate A pollen grain aperture type that is three -branched.

tricyclic *Triseriate*. (cycly)

trifoliolate *Tenate/ternately compound*. (leaf type)

trigger mechanism A type of movement hercogamy in which an insect pollinator triggers the sudden movement of one or more stamens, dusting the insect with pollen at the point of contact.

trilete mark A three-lined differentially thickened wall (laesura) on the spore of some land plants, being the remnant (scar) of attachment of the adjacent three spores of a tetrad.

trilete Spores with a 3-branched laesura.

trimerous A whorl with three members. (merosity, perianth merosity)

trimonoecious *Polygamomonoecious*. (plant sex)

trinucleate Having three nuclei, referring to some angiosperm pollen grains at the time of release.

trioecious/trioecy A plant with pistillate, staminate, and perfect flowers on different individuals. (plant sex)

tripinnate/tripinnately compound A compound leaf with three orders of axes, the third (rachillae) bearing leaflets. (leaf type)

triploid A polyploid having three sets of chromosomes.

triseriate With three whorls of parts. Syn: *tricyclic*. (cycly)

tristichous Alternate, with points of attachment in three vertical rows/ranks, e.g., the sedges, Cyperaceae. (arrangement)

tristyly Hercogamy in which there are three heights of styles and stamens. Adj: *tristylous*.

trisulculate A pollen grain with three elongated apertures on opposite sides of the grain, parallel to the equatorial plane.

trullate Parallelogram-shaped, widest near base and the length:width ratio 2:1 to 3:2. (shape)

truncate Apical or basal margin cut straight across, the angle approximately 180°. (apex, base) .

tryma A nut surrounded by an involucre that dehisces at maturity, e.g., *Carya*, pecan. (fruit type)

T-shaped *Malpighian*. (trichome type)

tube A cylindrically shaped perianth or region of the perianth, usually of a sympetalous corolla. (perianth part)

tube cell One of the two initial, haploid cells in the male gametophyte of angiosperms that remains near the tip of the growing pollen tube and may function in its development.

tuber A thick, underground storage stem, usually not upright, typically bearing outer buds and lacking surrounding storage leaves or protective scales, e.g., *Solanum tuberosum*, potato. (stem/shoot type)

tuberculate *Papillate*. (epidermal excrescence)

tubular Cylindrical. (perianth type)

turbinate Top- or turban-shaped. (shape)

twining Twisted around a central axis. (twisting/bending posture)

two-armed *Malpighian*. (trichome type)

type/type specimen *Nomenclatural type*.

typus conservandus Meaning a conserved type specimen. Abbr: *typ. cons*.

typus designatus Referring to the designation of a type specimen. Abbr: *typ. des*.

ulcerus A circular to slightly elliptic aperture similar in shape to a porus (length:width ratio <2:1) occurring at the (usually distal) pole. Adj: *ulcerate*.

umbel A determinate or indeterminate, flat-topped or convex inflorescence with pedicels attached at one point. (inflorescence type)

uncinate *Hooked*. (bristle type)

undulate Margins wavy in a vertical plane. (longitudinal posture)

unguiculate Clawed, e.g., many Brassicaceae, Caryophyllaceae. (perianth type)

unicarpellous With gynoecium composed of one carpel, the pistil or ovary being simple. (gynoecial fusion)

unicellular Referring to a trichome (or other structure) consisting of a single cell. (trichome type)

unifacial growth Growth of a secondary cambium in which secondary xylem (wood) is produced to the inside but with no secondary phloem produced to the outside, as in fossil lycophytes and equisetophytes.

unifacial leaf A leaf that is isobilateral and flattened side-to-side, having left and right sides, except at the base, where it is often sheathing, e.g., members of the Iridaceae. (leaf structural type).

unifoliolate A leaf bearing a single leaflet with petiolule distinct from the petiole, interpreted as the derived reduction of an ancestral compound leaf, e.g., *Cercis*, redbud. (leaf type)

unilateral *Secund*. (arrangement)

unilabiate One-lipped; with one, generally lower, segment, e.g., many Goodeniaceae. (perianth type)

unilocular An ovary with a single locule. (locule number)

uninervous Having a central midrib with no lateral veins, e.g. lycopods, psilotophytes, and equisetophytes, many conifers. (leaf venation)

uniseriate (a) With a single whorl of parts. (cycly, stamen cycly, perianth cycly) Syn: *monocyclic*. (b) Trichomes having a single vertical row of cells. (trichome type) (c) Rays in wood that are composed of a single, vertical row of cells.

unisexual Flowers having only carpel(s) or only stamen(s). Syn: *imperfect*. (flower sex)

unistomal Referring to the micropyle of a unitegmic ovule.

unit inflorescence A subunit of the entire secondary inflorescence. Syn: *partial inflorescence*; *primary inflorescence*. (inflorescence part)

unitegmic An ovule with a single integument, found in all gymnosperms and derived in some angiosperms.

univariate plot A graph of a single variable, usually per individual, sample, or taxon.

unordered Referring to a character transformation series in which each character state is allowed to evolve into every other character state with equal probability, generally in a single evolutionary step.

unrooted tree A branching diagram in which polarity is not indicated, representing the relative character state changes between taxa. Syn: *network*.

upper epidermis The adaxial epidermis of a leaf.

upward Dehiscing toward the sky in a horizontally oriented flower. (anther dehiscence direction)

urceolate Urn-shaped; expanded at the base and constricted at the apex, e.g., many Ericaceae. (perianth type)

urent With hispid and stinging trichomes. (vestiture)

utricle A small, bladdery or inflated, one-seeded, dry fruit; an achene in which the pericarp is significantly larger than the mature seed, e.g., *Atriplex*, salt bush. (fruit type)

vacuole A large, internal, membrane-bound sac of plant cells that functions in storage of compounds such as pigments, acids, or ergastic substances.

valid name A name that is validly published.

validly published Referring to the criteria needed for a scientific name to be formally recognized, including effective publication, publication in correct form, publication with a Latin description or diagnosis or with a reference to such, and indication of a nomenclatural type.

valvate (a) Sides enrolled, so that margins touch. (arrangement) (b) Having a whorl of perianth parts meeting at the margins and not overlapping. (perianth aestivation)

valve A portion of the pericarp of a dehiscent fruit that splits off but does not enclose the seed(s). (fruit part)

valvular Anther dehiscing through a pore covered by a flap of tissue, e.g., Lauraceae. (anther dehiscence type)

valvular capsule *Septifragal capsule*. (fruit type)

variable-number tandem repeats Tandem repeats that vary within a population or species. Abbr: *VNTR*.

variegated With two or more colors occurring in various irregular patterns, generally used for leaves. (color pattern)

varietal epithet The third component of a ternary name at the rank of variety.

vascular bundle A strand of associated xylem and phloem tissue. Syn: *vascular strand*.

vascular bundle scar A mark within the leaf scar indicating the former position of a vascular bundle that extended from stem to leaf. (twig part)

vascular bundle sheath A ring of cells, composed of fiber or parenchyma cells, surrounding the xylem and phloem of a vascular bundle, functioning in C4 photosynthesis.

vascular cambium A cylindrical sheath of cells that undergoes primarily tangential divisions, producing secondary xylem (wood) to the inside and secondary phloem (inner bark) to the outside.

vascular cylinder A central region of vascular tissue (xylem and phloem) in a root (root part) or in some stems (stem part).

vascular cylinder A central region of xylem and phloem in the root.

vascular strand *Vascular bundle*. (stem/shoot parts)

vascular tissue A tissue made up of xylem and phloem, functioning mainly in conduction of water, minerals, and sugars.

vegetation type An assessment of the habit, habitat, and cover of plant species present, e.g., forest, woodland, savanna, chaparral, scrub, grassland, meadow, strand marsh, swamp, pond, and vernal pool.

vegetative bud A bud that develops into a vegetative shoot, bearing leaves. (bud type)

vein The vascular bundle of a leaf or leaf homologue such as a sepal, petal, stamen, or carpel, containing the conductive tissues. Syn: *nerve*. (plant part, leaf part)

velamen A specialized, multilayered epidermis of some roots, functioning in protection, prevention of water loss, or water and mineral absorption, e.g., Araceae, Orchidaceae.

venation The pattern of veins and vein branching of a leaf or leaf homologue. (leaf venation)

venter The swollen, basal portion of an archegonium, containing the egg.

ventral *Adaxial*. (position)

ventral canal cell A second cell within the swollen base of the archegonium, located just distal to the egg.

ventral vein One of the two veins of a carpel near the carpel margins. Syn: *lateral vein*, *placental vein*.

vernal Appearing in spring. (periodicity)

verruca One of the short, wartlike elements of a verrucate pollen grain. Pl: *verrucae*.

verrucate (a) *Papillate*. (epidermal excrescence) (b) A pollen sculpturing having short, wartlike elements, each element termed a verruca.

versatile With anther freely pivoting at the point of attachment with the filament. (anther attachment)

verticillaster A secondary inflorescence with an indeterminate central axis bearing opposite, lateral, sessile cymes, the flowers appearing congested, e.g., some Lamiaceae. (inflorescence type)

verticillate Having three or more leaves or other structures per node. (arrangement)

very widely obovate Margins curved, widest near the apex, length:width ratio ca. 1:1. (shape)

very widely ovate Margins curved, widest near the base, length:width ratio ca. 1:1. (shape)

vernation *Ptyxis*. (posture)

vespertine In the evening, typically with respect to when flowers open. (periodicity)

vessel Term for several vessel members attached end-to-end, forming a continuous, conductive tube. Syn: *pore*.

vessel member A type of tracheary element that is perforate, in which continuous holes or perforations in the cell walls occur through which water and mineral nutrients flow between cells.

vestiture Trichome cover, a combination of trichome type, length, strength, shape, density, and color. (surface)

vicariance The splitting of one ancestral population into two (or more) populations, such as by continental drift or the formation of a geographic barrier.

video To cite a reference. Abbr: *vid*.

villous Covered with very long, soft, crooked trichomes. Syn: *lanate*. (vestiture)

vine A plant with elongated, weak stems, supported by means of scrambling, twining, tendrils, or roots; may be annual or perennial, herbaceous or woody. (plant habit)

viscid Having a shiny, sticky surface. Syn: *glutinous*. (epidermal excrescence)

viscin thread One of the long strands of carbohydrate material that function in sticking pollen grains together.

vivipary Referring to a seed that germinates into a seedling before being shed from the parent plant, e.g., *Rhizophora*, red mangrove. (seed germination type)

VNTR *Variable-number tandem repeats*.

voucher specimen An herbarium specimen in a plant collection serving as reference material for a named taxon or as part of a research project.

wax A type of triglyceride compound that may function as high-energy storage compounds or secretion products; a type of ergastic substance.

webbing Evolutionary development of thin tissue between the axes of the branches, thought to have occurred in euphylls.

weight The specific assignment of taxonomic importance when weighting a character in a phylogenetic analysis.

weighting/character weighting The assignment of greater or lesser taxonomic importance to certain characters over other characters in determining phylogenetic relationships.

whole genome sequencing Sequencing of the entire DNA of an organism.

whorl A cyclic group of floral parts, e.g., of sepals, petals, stamens, or carpels. Syn: *series*. (flower part)

whorled (a) With three or more leaves or other structures per node. (leaf arrangement) (b) Perianth parts in distinct whorls or series, with parts arising from the same nodal region. (perianth arrangement) (c) With stamens in one or more whorls or series. (stamen arrangement)

widely elliptic Margins curved, widest near the midpoint, length:width ratio ca. 6:5. (shape)

widely obovate Margins curved, widest near the apex, length:width ratio ca. 6:5. (shape)

widely ovate Margins curved, widest near the base, length:width ratio ca. 6:5. (shape)

widely triangular Three-sided, length : width ratio ca. 6:5. (shape)

wood *Secondary xylem*.

woody Having a hard, woodlike texture. (texture)

workflow The sequence of steps used in techniques of molecular data acquisition.

wrinkled With irregular, fine lines or deformations. (configuration)

× Abbreviation, used within a scientific name, that indicates a hybrid.

xenogamy *Outbreeding*. Adj: *xenogamous*.

xerophyte A plant adapted to live in a dry, generally hot environment. (plant habitat)

xylem A tissue composed of tracheary elements plus some parenchyma and sometimes sclerenchyma, functioning in conduction of water and mineral nutrients.

zig-zag Referring to the micropyle of a bitegmic ovule in which the micropylar pore of the outer integument is spatially displaced relative to the inner integument.

zonoaperturate Descriptive of pollen grains with apertures occurring in the equatorial region. Syn: *stephanoaperturate*.

zonocolpate A pollen grain with colpi occurring in the equatorial region.

zonoporate A pollen grain with pori occurring in the equatorial region.

zoochory Dispersal of propagules by animals.

zooidogamy Gymnospermous pollen tubes that are haustorial and deliver motile sperm cells to a fertilization chamber, where the sperm swims to the archegonium or egg. Cf: *siphonogamy*.

zygomorphic Bilaterally symmetrical, with one plane of symmetry. Syn: *bilateral, irregular, monosymmetric*. (symmetry)

zygote A diploid cell that results from the fusion of two haploid cells (egg and sperm) and that ultimately matures into a new sporophyte in the land plants.

INDEX

M